Handbook on
Theoretical and Algorithmic Aspects of Sensor, Ad Hoc Wireless, and Peer-to-Peer Networks

Handbook on
Theoretical and Algorithmic Aspects of Sensor, Ad Hoc Wireless, and Peer-to-Peer Networks

Edited by Jie Wu

Auerbach Publications
Taylor & Francis Group

Boca Raton   London   New York   Singapore

Published in 2006 by
Auerbach Publications
Taylor & Francis Group
6000 Broken Sound Parkway NW, Suite 300
Boca Raton, FL 33487-2742

International Standard Book Number-10: 0-8493-2832-2 (Hardcover)
International Standard Book Number-13: 978-0-8493-2832-9 (Hardcover)
Library of Congress Card Number 2004065010

Library of Congress Cataloging-in-Publication Data

Handbook on theoretical and algorithmic aspect of sensor, ad hoc wireless, and peer-to-peer networks /
    edited by Jie Wu.
        p. cm.
    Includes bibliographical references and index.
    ISBN 0-8493-2832-2 (alk. paper)
    1. Wireless LANs. I. Wu, Jie, 1961-

TK5105.78.H37 2005
004.6'8--dc22   2004065010

Visit the Taylor & Francis Web site at
http://www.taylorandfrancis.com

and the Auerbach Publications Web site at
http://www.auerbach-publications.com

Preface

Overview

Theoretical and algorithmic approaches in Sensor networks, Ad hoc wireless networks, and Peer-to-peer networks (together called SAP networks) have played a central role in the development of emerging network paradigms. These three networks are characterized by their ad hoc nature without infrastructure or centralized administration. Unlike infrastructured networks, such as cellular networks, where nodes interact through a centralized base station, nodes in a SAP network interact in a peer-to-peer fashion. As a result of the mobility (including joining/leaving the network) of their nodes, SAP networks are characterized by dynamically changing topologies. The applications of SAP networks range from civilian (file-sharing) to disaster recovery (search-and-rescue), to military (battlefield).

The main goal of this book is to fill the need for comprehensive reference material on the recent development on theoretical and algorithmic aspects of three related fields. Topics covered include: theoretical and algorithmic methods/tools for optimization, computational geometry, graph theory, and combinatorics; protocol security and privacy; scalability design; distributed and localized solutions; database and data management; operating systems and middleware support; power control systems and energy efficient design; applications; and performance and simulations.

This book brings together different research disciplines to initiate a comprehensive technical discussion on theoretical and algorithmic approaches to three related fields: sensor networks, ad hoc wireless networks, and peer-to-peer networks. The objective is to identify several common theoretical and algorithmic approaches that can address issues related to SAP networks. The central topic is defined by the following two questions: What are the central technical issues in SAP networks? What are the possible solutions/tools available to address these issues?

This book is expected to serve as a reference book for developers in the telecommunication industry or as a textbook for a graduate course in computer science and engineering. It is organized in the following three groups as 47 chapters.

- Ad-Hoc Wireless Networks (19 chapters)
- Sensor Networks (16 chapters)
- Peer-to-Peer Networks (12 chapters)

Although many books have emerged recently in this area, none of them address all three fields in terms of common issues. This book has the following features and benefits:

- Coverage of three related fields, ad hoc wireless, sensor, and peer-to-peer networks, allows the reader to easily cross-reference similar results in three fields.
- International groups of authors present balanced coverage of research results.
- Systematic treatment of theoretical and algorithmic aspects allows the reader easy access to some important results.

- Applications and uses of these networks offer good motivation for research in these fields.
- Authoritative materials on a broad range of topics provide a comprehensive treatment of various important topics by some of the leading researchers in the field.

Common Theoretical and Algorithmic Issues

The following preliminary set of common theoretical and algorithmic issues is identified for SAP networks.

Location Management (in sensor networks, ad hoc wireless networks, and peer-to-peer networks): This issue addresses the problem of "Where is X." This problem can be analyzed from two aspects: *update* and *page*. The updating process notifies the location servers of the current locations of nodes. In search of a node, the paging process queries the servers to identify the exact/possible locations of the mobile station before the actual search. This avoids the potentially high costs of doing a global search. Updating and paging costs are tradeoffs. More frequent updates can improve the accuracy of the information in location servers, thus reducing the paging costs. On the contrary, less frequent updates can save updating costs, but may incur higher paging costs, especially for highly mobile stations. Many analytical tools such as queueing analysis and Markov chain analysis are used in this area. Graph theoretical models are used in peer-to-peer networks based on building an overlay network.

Security and Privacy (in sensor networks, ad hoc wireless networks, and peer-to-peer networks): Security is the possibility of a system withstanding an attack. There are two types of security mechanisms: preventive and detective. The majority of the preventive mechanisms have cryptography as building components. The goal of system security is to have controlled access to resources. The key requirements for SAP networks are confidentiality, authentication, integrity, non-repudiation and availability. SAP networks are more prone to attack because of their dynamic and/or infrastructureless nature. The attacks on networks can be categorized into interruption, interception, modification, and fabrication. In addition to various "attacks," a number of "trust" issues also occur in SAP networks. Cryptographic algorithms are widely used in this area.

Topology Design and Control (in sensor networks, ad hoc wireless networks, and peer-to-peer networks): Topology design deals with the way to control the network topology to achieve several desirable properties in SAP networks, including small diameter and small average node distance in peer-to-peer networks, and a certain level of node connectivity in sensor and ad hoc wireless networks. In general, each node has a similar number of neighbors, and the average nodal degree should be small. Regular and uniform structures are usually preferred. In many cases, topology control is tied to energy-efficient design. Traditional graph theory is usually used to deal with topology control.

Scalable Design (in sensor networks, ad hoc wireless networks, and peer-to-peer networks): Scalable design deals with how to increase the number of nodes without degrading system or protocol performance. The most common approach for supporting scalability is the clustering approach used both in sensor networks and ad hoc wireless networks. Basically, the network is partitioned into a set of clusterheads, with one clusterhead in each cluster. Clusterheads do not have direct connectivity to each other, but each clusterhead directly connects to all of its members. In sensor networks, the clustering approach is used to reduce the number of forward nodes (which contact the base station directly), and hence, to reduce overall energy consumption. The traditional scalability analysis is normally used.

Energy-Aware Design (in sensor networks and ad hoc wireless networks): Energy-aware design has been applied to various levels of protocol stacks. Most works have been done at the network layer. Several different protocols have been proposed to manage energy consumption by adjusting transmission ranges. In the source-independent approach, all nodes can be a source and are able to reach all other nodes by assigning appropriate ranges. The problem of minimizing the total transmission power consumption (based on an assigned model) is NP-complete for both 2-D and 3-D space. Various heuristic solutions exists for this problem. At the MAC layer, power saving techniques for ad hoc and sensor networks can be divided into two categories: *sleeping* and *power controlling*. The sleeping methods put wireless nodes into periodic sleep states in order to reduce power consumption in the *idle listening* mode. Both graph theory and optimization methods are widely used in this area.

Routing and Broadcasting (in sensor networks and ad hoc wireless networks): This issue deals with trade-offs between proactive and reactive routing, flat and hierarchical routing, location-assist and non-location-assist routing and source-dependent and source-independent broadcasting. These trade-offs focus on cost and efficiency and are dependent on various parameters, such as network topology, host mobility, and network and traffic density. Various graph theoretical models (such as dominating set) and computational geometric models (such as Yao graph, RNG (relative neighborhood graph), and Gabriel graph) have been used. Graph theory, distributed algorithms, and computational geometry are widely used in this area.

Acknowledgments

I wish to thank all the authors for their contributions to the quality of the book. The support from NSF for an international workshop, held at Fort Lauderdale, Florida in early 2004, is greatly appreciated. Many chapters come from the extension of presentations at that workshop.

Special thanks to Rich O'Hanley, the managing editor, for his guidance and support throughout the process. It has been a pleasure to work with Andrea Demby and Claire Miller, who collected and edited all chapters. I am grateful to them for their continuous support and professionalism. Thanks to my students, Eyra Bethancourt and Max Haider, for their assistance.

Finally, I thank my children, NiNi and YaoYao, and my wife, Ruiguang Zhang, for making this all worthwhile and for their patience during my numerous hours working both at home and at the office.

Contributors

Mehran Abolhasan
Telecommunication and IT
    Research Institute (TITR)
University of Wollongong
Wollongong, NSW,
    Australia

Dharma P. Agrawal
OBR Research Center for
    Distributed and Mobile
    Computing
ECECS Department
University of Cincinnati
Cincinnati, Ohio

Anish Arora
Department of Computer
    Science and Engineering
The Ohio State University
Columbus, Ohio

James Aspnes
Department of Computer
    Science
Yale University
New Haven,
    Connecticut

Rimon Barr
Computer Science and
    Electrical
    Engineering
Cornell University
Ithaca, New York

Ratnabali Biswas
OBR Research Center for
    Distributed and Mobile
    Computing
ECECS Department
University of Cincinnati
Cincinnati, Ohio

Douglas M. Blough
School of Electrical and
    Computer Engineering
Georgia Institute of Technology
Atlanta, Georgia

Andrija M. Bosnjakovic
Faculty of Electrical
    Engineering
University of Belgrade
Belgrade, Yugoslavia

Virgil Bourassa
Panthesis, Inc.
Bellevue, Washington

Aharon S. Brodie
Wayne State University
Detroit, Michigan

Gruia Calinescu
Department of Computer
    Science
Illinois Institute of Technology
Chicago, Illinois

Edgar H. Callaway, Jr.
Florida Communication
    Research Laboratory
Motorola Labs
Plantation, Florida

Guohong Cao
Department of Computer
    Science and Engineering
Pennsylvania State University
University Park, Pennsylvania

Ionu Cârdei
Department of Computer
    Science and Engineering
Florida Atlantic University
Boca Raton, Florida

Krishnendu Chakrabarty
Department of Electrical and
    Computer Engineering
Duke University
Durham, North Carolina

Chih-Yung Chang
Department of Computer
    Science and Information
    Engineering
Tamkang University
Taipei, Taiwan

Sriram Chellappan
Department of Computer
    Science and Engineering
Ohio State University
Columbus, Ohio

Po-Yu Chen
Institute of Communications
    Engineering
National Tsing Hua University
Hsin-Chu, Taiwan

Wen-Tsuen Chen
Department of Computer
  Science
National Tsing Hua University
Hsin-Chu, Taiwan

Xiao Chen
Department of Computer
  Science
Texas State University
San Marcas, Texas

Yuh-Shyan Chen
Department of Computer
  Science and Information
  Engineering
National Chung Cheng
  University
Chia-Yi, Taiwan

Liang Cheng
Laboratory of Networking
  Group (LONGLAB)
Department of Computer
  Science and Engineering
Lehigh University
Bethlehem, Pennsylvania

Young-ri Choi
Department of Computer
  Sciences
The University of Texas at
  Austin
Austin, Texas

Marco Conti
Institute for Informatics and
  Telematics (IIT)
National Research Council
  (CNR)
Pisa, Italy

Jon Crowcroft
Computer Laboratory
University of Cambridge
Cambridge, UK

Arindam Kumar Das
Department of Electrical
  Engineering
University of Washington
Seattle, Washington

Saumitra M. Das
School of Electrical and
  Computer Engineering
Purdue University
West Lafayette, Indiana

Haitao Dong
Department of Computer
  Science and Technology
Tsinghua University
Beijing, China

Sameh El-Ansary
Swedish Institute of
  Computer Science
  (SICS)
Sweden

Mohamed Eltoweissy
Department of Computer
  Science
Virginia Tech
Falls Church, Virginia

Jakob Eriksson
Department of Computer
  Science and Engineering
University of California,
  Riverside
Riverside, California

Patrick Th. Eugster
Sun Microsystems, Inc.
Client Solutions Volketscuil
  Switzerland and School of
  Information and
  Communication Sciences
Swiss Federal Institute of
  Technology
Lausanne, Switzerland

Michalis Faloutsos
Department of Computer
  Science and Engineering
University of California,
  Riverside
Riverside, California

Yuguang Fang
Department of Electrical and
  Computer Engineering
University of Florida
Gainesville, Florida

Ophir Frieder
Department of Computer
  Science
Illinois Institute of Technology
Chicago, Illinois

Luca M. Gambardella
Istituto Dalle Molle di Studi
  sull'Intelligenza Artificiale
  (IDSIA)
Manno-Lugano
Switzerland

Mohamed G. Gouda
Department of Computer
  Sciences
The University of Texas at
  Austin
Austin, Texas

Aditya Gupta
OBR Research Center for
  Distributed and Mobile
  Computing
ECECS Department
University of Cincinnati
Cincinnati, Ohio

Sandeep K.S. Gupta
Department of Computer
  Science and Engineering
Arizona State University
Tempe, Arizona

Zygmunt J. Haas
Department of Electrical and
    Computer Engineering
Cornell University
Ithaca, New York

Joseph Y. Halpern
Department of Computer
    Science
Cornell University
Ithaca, New York

Seif Haridi
Royal Institute of Technology
    (IMIT/KTH)
Sweden

Fred B. Holt
Panthesis, Inc.
Bellevue, Washington

Jennifer C. Hou
Department of Computer
    Science
University of Illinois at
    Urbana-Champaign
Urbana, Illinois

Hung-Chang Hsiao
Computer and
    Communications Research
    Center
National Tsing-Hua University
Hsin-Chu, Taiwan

Jinfeng Hu
Department of Computer
    Science and Technology
Tsinghua University
Beijing, China

Y. Charlie Hu
School of Electrical and
    Computer Engineering
Purdue University
West Lafayette, Indiana

Yiming Hu
Department of Electrical
    and Computer
    Engineering and
    Computer Science
University of Cincinnati
Cincinnati, Ohio

Chi-Fu Huang
Department of Computer
    Science and Information
    Engineering
National Chiao Tung
    University
Hsin-Chu, Taiwan

Zhuochuan Huang
Department of Computer
    and Information
    Sciences
University of Delaware
Newark, Delaware

François Ingelrest
IRCICA/LIFL
University of Lille
INRIA futurs
France

Neha Jain
OBR Research Center for
    Distributed and Mobile
    Computing
ECECS Department
University of Cincinnati
Cincinnati, Ohio

Xiaohua Jia
Department of Computer
    Science
City University of Hong Kong
Hong Kong

Kennie Jones
Department of Computer
    Science
Old Dominion University
Norfolk, Virginia

Dongsoo S. Kim
Electrical and Computer
    Engineering
Indiana University, Purdue
    University
Indianapolis, Indiana

Chung-Ta King
Department of Computer
    Science
National Tsing-Hua University
Hsin-Chu, Taiwan

Manish Kochhal
Department of Electrical
    and Computer
    Engineering
Wayne State University
Detroit, Michigan

Odysseas Koufopavlou
Electrical and Computer
    Engineering Department
University of Patras
Greece

Srikanth Krishnamurthy
Department of Computer
    Science and Engineering
University of California,
    Riverside
Riverside, California

Tom La Porta
Pennsylvania State
    University
University Park,
    Pennsylvania

Mauro Leoncini
Dipartimento dilngegneria dell'
    in Formazione
UniversitÂ di Modena e Reggio
    Emilia
Modena, Italy

Dongsheng Li
School of Computer
National University
  of Defense Technology
Changsha, China

Li (Erran) Li
Center for Networking
  Research
Bell Labs, Lucent
Holmdel, New Jersey

Xiang-Yang Li
Department of Computer
  Science
Illinois Institute of Technology
Chicago, Illinois

Xiuqi Li
Department of
  Computer Science
  and Engineering
Florida Atlantic University
Boca Raton, Florida

Hai Liu
Department of Computer
  Science
City University of
  Hong Kong
Hong Kong

Xuezheng Liu
Department of Computer
  Science and Technology
Tsinghua University
Beijing, China

Yunhao Liu
Department of
  Computer Science
  and Engineering
Michigan State
  University
East Lansing, Michigan

Xicheng Lu
School of Computer
National University of
  Defense Technology
Changsha, China

B. S. Manoj
Department of
  Computer Science
  and Engineering
Indian Institute of
  Technology
Chennai, India

Gaia Maselli
Institute for Informatics and
  Telematic (IIT) National
  Research Council (CNR)
Pisa, Italy

Jelena Mišić
University of Manitoba
Winnipeg, Manitoba
Canada

Vojislav B. Mišić
University of Manitoba
Winnipeg, Manitoba
Canada

Nikolay A. Moldovyan
Specialized Center of
  Program Systems
  (SPECTR)
St. Petersburg, Russia

Roberto Montemanni
Istituto Dalle Molle di Studi
  sull'Intelligenza Artificiale
  (IDSIA)
Manno-Lugano, Switzerland

Thomas Moscibroda
Department of Computer
  Science
Swiss Federal Institute of
  Technology
Zurich, Switzerland

Anindo Mukherjee
OBR Research Center for
  Distributed and Mobile
  Computing
ECECS Department
University of Cincinnati
Cincinnati, Ohio

C. Siva Ram Murthy
Department of
  Computer Science
  and Engineering
Indian Institute of
  Technology
Chennai, India

Lionel M. Ni
Department of Computer
  Science
Hong Kong University of
  Science and Technology
Hong Kong

Stephan Olariu
Department of Computer
  Science
Old Dominion University
Norfolk, Virginia

Shashi Phoha
Pennsylvania State
  University
University Park,
  Pennsylvania

Jovan Popovic
Faculty of Electrical
  Engineering
University of Belgrade
Belgrade, Yugoslavia

Himabindu Pucha
School of Electrical and
  Computer Engineering
Purdue University
West Lafayette, Indiana

Cauligi S. Raghavendra
Departments of Electrical
  Engineering-Systems and
  Computer Science
University of Southern
  California
Los Angeles, California

Giovanni Resta
Instituto di Informaticae
  Telematica
Area della Ricerca del CNR
  Pisa, Italy

Paolo Santi
Instituto di Informaticae
  Telematica
Area della Ricerca del CNR
  Pisa, Italy

Loren Schwiebert
Department of Computer
  Science
Wayne State University
Detroit, Michigan

Sandhya Sekhar
OBR Research Center for
  Distributed and Mobile
  Computing
ECECS Department
University of Cincinnati
Cincinnati, Ohio

Gauri Shah
IBM Almaden Research
  Center
San Jose, California

Chien-Chung Shen
Department of Computer
  and Information Sciences
University of Delaware
Newark, Delaware

Haiying Shen
Department of Electrical and
  Computer Engineering
Wayne State University
Detroit, Michigan

Jian Shen
Department of Mathematics
Texas State University
San Marcas, Texas

Jang-Ping Sheu
Department of Computer
  Science and Information
  Engineering
National Central University
Chung-Li, Taiwan

Shuming Shi
Department of Computer
  Science and Technology
Tsinghua University
Beijing, China

Weisong Shi
Department of Computer
  Science
Wayne State University
Detroit, Michigan

Zhenghan Shi
Department of Computer
  Science
Clemson University
Clemson, South Carolina

David Simplot-Ryl
IRCICA/LIFL
University of Lille
INRIA futurs
France

Nicolas Sklavos
VLSI Design Laboratory
University of Patras
Patras, Greece

Pradip K Srimani
Department of Computer
  Science
Clemson University
Clemson, South Carolina

Ivan Stojmenovic
Computer Science
SITE
University of Ottawa
Ottawa, Ontario
Canada

Caimu Tang
Department of Computer
  Science
University of Southern
  California
Los Angeles, California

Yu-Chee Tseng
Department of Computer
  Science and Information
  Engineering
National Chiao Tung University
Hsin-Chu, Taiwan

Giovanni Turi
Institute for Informatics and
  Telematics (IIT) National
  Research Council (CNR)
Pisa, Italy

Robbert van Renesse
Department of Computer
  Science
Cornell University
Ithaca, New York

Ashraf Wadaa
Department of Computer
  Science
Old Dominion University
Norfolk, Virginia

Peng-Jun Wan
Department of Computer
  Science
Illinois Institute of
  Technology
Chicago, Illinois

Wheizhao Wang
Department of Computer
  Science
Illinois Institute of Technology
Chicago, Illinois

Guiling Wang
Department of Computer
    Science and Engineering
Pennsylvania State
    University
University Park,
    Pennsylvania

Xun Wang
Department of Computer
    Science and Engineering
Ohio State University
Columbus, Ohio

Roger Wattenhofer
Department of Computer
    Science
Swiss Federal Institute of
    Technology
Zurich, Switzerland

Larry Wilson
Department of Computer
    Science
Old Dominion University
Norfolk, Virginia

Jie Wu
Department of Computer
    Science and Engineering
Florida Atlantic University
Boca Raton, Florida

Tadeusz Wysocki
Telecommunication and IT
    Research Institute (TITR)
University of Wollongong
Wollongong, New South Wales
Australia

Li Xiao
Department of Computer
    Science and Engineering
Michigan State University
East Lansing, Michigan

Cheng-Zhong Xu
Department of Electrical and
    Computer Engineering
Wayne State University
Detroit, Michigan

Chuanfu Xu
School of Computer
National University of Defense
    Technology
Changsha, China

Dong Xuan
Department of Computer
    Science and Engineering
Ohio State University
Columbus, Ohio

Qing Ye
Laboratory of Networking
    Group (LONGLAB)
Department of Computer
    Science and Engineering
Lehigh University
Bethlehem, Pennsylvania

Hongqiang Zhai
Department of Electrical
    and Computer
    Engineering
University of Florida
Gainesville, Florida

Honghai Zhang
Department of Computer
    Science
University of Illinois at
    Urbana-Champaign
Urbana, Illinois

Wensheng Zhang
Pennsylvania State
    University
University Park,
    Pennsylvania

Weimin Zheng
Department of Computer
    Science and Technology
Tsinghua University
Beijing, China

Yingwu Zhu
Department of Electrical
    and Computer
    Engineering and
    Computer Science
University of Cincinnati
Cincinnati, Ohio

Yi Zou
Department of Electrical
    and Computer
    Engineering
Duke University
Durham, North Carolina

Contents

Section II   Sensor Networks   313

Section III   Peer-to-Peer Networks   587

I

Ad Hoc Wireless Networks

The maturity of wireless transmissions and the popularity of portable computing devices have made the dream of "communication anytime and anywhere" possible. An ad hoc wireless network is a good choice for fulfilling this dream. An ad hoc wireless network consists of a set of mobile hosts operating without the aid of an established infrastructure of centralized administration. Communication is done through wireless links among mobile hosts using their antennas. Due to concerns such as radio power limits and channel utilization, a mobile host may not be able to communicate directly with other hosts in a single-hop fashion. In this case, a multihop scenario occurs, in which the packets sent by the source host must be relayed by several intermediate hosts before reaching the destination host.

Although military tactical communication is still considered the primary application for ad hoc wireless networks, commercial interest in this type of network continues to grow. Applications such as law enforcement operation, commercial and educational use, and sensor networks are just a few possible commercial examples. There are several technical challenges related to the ad hoc wireless network. In ad hoc wireless networks, the topology is highly dynamic, and frequent changes in the topology may be difficult to predict. With the use of wireless links, the network suffers from higher loss rates, and can experience more delays and jitter. In addition, physical security is limited due to wireless transmission. Finally, as ad hoc wireless network nodes rely on batteries, energy saving is an important system design criterion.

Most of the existing works on ad hoc wireless networks focus on issues related to the network layer, such as routing and broadcasting. Routing protocols in ad hoc wireless networks are either *proactive* or *reactive*, although a combination of proactive and reactive is also possible. In proactive routing, routes to all destinations are computed *a priori* and are maintained in the background via a periodic update process. Route information is maintained either as routing tables or as global link state information. In reactive routing, a route to a specific destination is computed "on demand," that is, only when needed. To efficiently use resources in controlling large dynamic networks, hierarchical routing, including cluster based and dominating set based, is normally used.

Many other technical issues are discussed through the use of protocol stacks where at least four layers are used:

1. Physical layer: responsible for frequency selection, carrier frequency generation, signal detection, modulation, and data encryption.
2. Data link layer: responsible for the multiplexing of data streams, data frame detection, medium access, and error control.
3. Network layer: responsible for forwarding the data to appropriate destinations.
4. Application layer: responsible for supporting various applications.

The 19 chapters in this section cover a wide range of topics across multiple layers: MAC (part of the data link layer), network, and applications. One chapter is devoted to the cross-layer architecture for ad hoc wireless networks. Several chapters deal with various techniques for efficient and scalable routing, including multicasting and geocasting, in ad hoc wireless networks. One chapter discusses routing in a selfish wireless network. Three chapters present some recent results on topology control while three other chapters are dedicated to energy-efficient design under several different system settings. The security and reliability issues are covered in two separate chapters. MAC protocols are given in one dedicated chapter. Of the three chapters about applications, one discusses ad hoc relaying in cellular networks, one uses ad hoc wireless networks for robust data communication, and one is devoted to the application in Bluetooth. This section ends with a chapter on scalable simulation for ad hoc wireless networks.

1

A Modular Cross-Layer Architecture for Ad Hoc Networks

Marco Conti

Jon Crowcroft

Gaia Maselli

Giovanni Turi

The success of the cleanly layered Internet Architecture has promoted its adoption for wireless and mobile networks, including ad hoc networks. This has also fostered skepticism toward alternative approaches. However, a strict-layered design is not flexible enough to cope with the dynamics of mobile networks and can prevent many classes of performance optimizations. To what extent, then, must developers modify the pure layered approach by introducing closer cooperation among protocols belonging to different layers?

Although the debate on cross-layer versus legacy-layer architecture has been around for a while, we propose a novel solution based on loosely coupled cross-layering, which constitutes a trade-off between the two extremes. Our solution allows for performance optimizations, but at the same time maintains flexibility.

This innovative architecture not only makes room for techniques to design new ad hoc protocols, for which we present specific examples, but also opens up the possibility of research into the usage of cross-layering for the Internet more generally.

1.1   Introduction

The Internet transparently connects millions of heterogeneous devices, supporting a huge variety of communications. From a networking standpoint, its popularity is due to a core design that has made it extensible, and robust against evolving usage as well as failures. Now, mobile devices and wireless communications prompt the vision of networking without a network (ad hoc networking). This brings new challenging issues where the need for flexibility confronts ad hoc constraints. A careful architectural design for the ad hoc protocol stack is necessary to incorporate this emerging technology.

The Internet architecture layers protocol and network responsibilities, breaking down the networking system into modular components, and allowing for transparent improvements of single modules. In a *strict-layered* system, protocols are independent of each other and interact through well-defined (and static) interfaces: each layer implementation depends on the interfaces available from the lower layer, and those exported to the upper layer. Strict-layering provides flexibility to a system's architecture: extensions introduced into single levels do not affect the rest of the system. The separation of concerns brings the added benefits of minimizing development costs by re-using existing code. This design approach relies on "horizontal" communication between peer protocol layers on the sender and receiver devices (the dashed arrows in Figure 1.1). The result is a trend to spend bandwidth (an abundant resource in the Internet) instead of processing power and storage.

Several aspects of the Internet architecture have led to the adoption of this strict-layer approach also for mobile ad hoc networks. Some of these aspects include (1) the "IP-centric" view of ad hoc networks; and (2) the flexibility offered by independent layers, which allows for reuse of existing software. The choice of the layered approach is supported by the fact that ad hoc networks are considered mobile extensions of the Internet, and hence the protocol stack must be suitable. However, this design principle clashes with the following facts:

1. Issues such as energy management, security, and cooperation characterize the whole stack and cannot be solved inside a single layer.
2. Ad hoc networks and the Internet have conflicting constraints; and while the former are dynamic, the latter is relatively static.

Some guidelines to approach these problems point to an enhancement of "vertical" communication in a protocol stack (see Figure 1.1),[1,2] as a way to reduce peer (horizontal) communication, and hence conserve bandwidth. Vertical communication, especially between nonadjacent layers, facilitates local data retrieval, otherwise carried out through network communication. The practice of accessing not only the next lower layer, but also other layers of the protocol stack, leads to *cross-layering* to allow performance improvements. The main downside of strict-layering is that it hinders extensibility: a new, higher-level component can only build on what is provided by the next lower layer.[3] Hence, if one layer needs to access functionality or information provided by a nonadjacent layer, then an intermediate extension should be devised. Cross-layering allows nonadjacent protocols to directly interact, making overall optimizations possible and achieving extensibility at the eventual expense of flexibility.

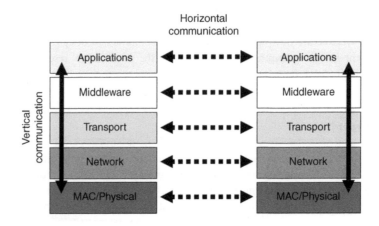

FIGURE 1.1 The Internet emphasizes horizontal communication between peer protocol layers to save router resources, while ad hoc networking promotes vertical interaction to conserve bandwidth.

In the literature there is much work showing the potential of cross-layering for isolated performance improvements in ad hoc networks. However, the focus of that work is on specific problems, as it looks at the joint design of two to three layers only. For example, cross-layer interactions between the routing and the middleware layers allow the two levels to share information with each other through system profiles, in order to achieve high quality in accessing data.[4] An analogous example is given by Schollmeier et al.,[5] where a direct interaction between the network and the middleware layers, termed Mobile Peer Control Protocol, is used to push a reactive routing to maintain existing routes to members of a peer-to-peer overlay network. Yuen et al.[6] propose an interaction between the MAC and routing layers, where information like signal-to-noise ratio, link capacity, and MAC packet delay is communicated to the routing protocol for the selection of optimal routes. Another example is the joint balancing of optimal congestion control at the transport layer with power control at the physical layer.[7] This work observes how congestion control is solved in the Internet at the transport layer, assuming that link capacities are fixed quantities. In ad hoc networks, this is not a good assumption, as transmission power, and hence throughput, can be dynamically adapted on each link. Last, but not least, Kozat et al.[8] propose cross-layer interaction between physical, MAC, and routing layers to perform joint power control and link scheduling as an optimized objective.

Although these solutions are clear examples of optimization introduced by cross-layering, the drawback on the resulting systems is that they contain tightly coupled, and therefore mutually dependent, components. Additionally, while an individual suggestion for cross-layer design, in isolation, may appear appealing, combining them all together could result in interference among the various optimizations.[9] From an architectural point of view, this approach leads to an "unbridled" stack design, difficult to maintain efficiently, because every modification must be propagated across all protocols. To give an example of interfering optimizations, let us consider an adaptation loop between a rate-adaptive MAC and minimal hop routing protocol (most ad hoc routing protocols are minimum hop). A rate-adaptive MAC would be able to analyze the quality of channels, suggesting higher layers on the outgoing links, which provide the higher data rates in correspondence with shorter distances. This conflicts with typical decisions of a minimum hop routing protocol, which chooses a longer link (for which the signal strength and data rate are typically lower) to reach the destination while using as few hops as possible.

We claim that cross-layering can be achieved, maintaining the layer separation principle, with the introduction of a vertical module, called *Network Status** (NeSt), which controls all cross-layer interactions (see Section 1.2). The NeSt aims at generalizing and abstracting vertical communications, getting rid of the tight coupling from an architectural standpoint. The key aspect is that protocols are still implemented in isolation inside each layer, offering the advantages of:

- Allowing for full compatibility with standards, as NeSt does not modify each layer's core functions
- Providing a robust upgrade environment, which allows the addition or removal of protocols belonging to different layers from the stack, without modifying operations at other layers
- Maintaining the benefits of a modular architecture (layer separation is achieved by standardizing access to the NeSt)

In addition to the advantages of a full cross-layer design, which still satisfies the layer separation principle, the NeSt provides full context awareness at all layers. Information regarding the network topology, energy level, local position, etc. is made available by the NeSt to all layers, to achieve optimizations, and offers performance gains from an overhead point of view. Although this awareness is restricted to the node's local view, protocols can be designed so as to adapt the system to highly variable network conditions (the typical ad hoc characteristic).

---

*This term indicates the collection of network information that a node gathers at all layers. It should not be confused with a concept of globally shared network context.

This innovative architecture opens research opportunities for techniques to design and evaluate new ad hoc protocols (see Section 1.3), but also remains compliant with the usage of legacy implementations, introducing new challenging issues concerning the usage of cross-layering for the Internet more generally (see Section 1.4).

1.2   Toward Loosely Coupled Cross-Layering

One of the main problems caused by direct cross-layer interactions (as already discussed in Section 1.1) is the resulting *tight coupling* of interested entities. To solve this problem, the NeSt stands vertically beside the network stack (as shown in Figure 1.2), handling eventual cross-layer interactions among protocols. That is, the NeSt plays the role of intermediary, providing standard models to design protocol interactions. While the new component uniformly manages vertical exchange of information between protocols, usual network functions still take place layer-by-layer through standardized interfaces, which remain unaltered. This introduced level of indirection maintains the *loosely coupled* characteristic of Internet protocols, preserving the flexible nature of a layered architecture.

The idea is to have the NeSt exporting an interface toward protocols, so as to allow sharing of information and reaction to particular events. In this way, cross-layer interactions do not directly take place between the interested protocols, but are implemented using the abstractions exported by the NeSt. This approach allows protocol designers to handle new cross-layer interactions apart, without modifying the interfaces between adjacent layers. The work described by Conti et al.[10,11] introduces this idea in the context of pure ad hoc networking. This work extends the definition of the NeSt interaction models, presenting the exported interface. This is to evolve toward a general-purpose component, eventually suitable for cross-layering in a future Internet architecture.

1.2.1   Overview of NeSt Functionalities

The NeSt supports cross-layering implementation with two models of interaction between protocols: *synchronous* and *asynchronous*. Protocols interact synchronously when they share private data (i.e., internal status collected during their normal functioning). A request for private data takes place on-demand, with a protocol querying the NeSt to retrieve data produced at other layers, and waiting for the result. Asynchronous interactions characterize the occurrence of specified conditions to which protocols may be willing to react. As such conditions are occasional (i.e., not deliberate), protocols are required to subscribe for their occurrences. In other words, protocols subscribe for events they are interested in, and then return to their work. The NeSt, in turn, is responsible for delivering eventual occurrences to the right subscribers. Specifically, we consider two types of events: *internal* and *external*. Internal events are directly generated inside the protocols. Picking just one example, the routing protocol notifies the rest of the stack about a "broken route" event, whenever it discovers the failure of a preexisting route. On the other side, external events are discovered inside the NeSt on the basis of instructions provided by subscriber protocols.

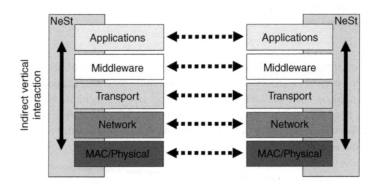

FIGURE 1.2   An architectural trade-off for loosely coupled vertical protocol interactions.

An example of an external event is a condition on the host energy level. A protocol can subscribe for a "battery-low" event, specifying an energy threshold to the NeSt, which in turn will notify the protocol when the battery power falls below the given value.

As the NeSt represents a level of indirection in the treatment of cross-layer interactions, an agreement for common-data and events representation inside the vertical component is a fundamental requirement. Protocols must agree on a common representation of shared information, in order to guarantee loose coupling. To this end, the NeSt works with *abstractions* of data and events, intended as a set of data structures that comprehensively reflect the relevant (from a cross-layering standpoint) information and special conditions used throughout the stack. A straightforward example is the topology information collected by a routing protocol. To abstract from implementation details of particular routing protocols, topology data can be represented as a graph inside the NeSt. Therefore, the NeSt becomes the provider of shared data, which appear independent of its origin and hence usable by each protocol.

How is protocol internal data exported into NeSt abstractions? The NeSt accomplishes this task using *callback* functions, which are defined and installed by protocols themselves. A callback is a procedure that is registered to a library (the NeSt interface) at one point in time, and later on invoked (by the NeSt). Each callback contains the instructions to encode private data into an associated NeSt abstraction. In this way, the protocol designer provides a tool for transparently accessing protocol internal data.

1.2.2   The NeSt Interface

To give a technical view of the vertical functionalities, we assume that the language used by the NeSt to interface the protocol stack allows for declaration of functions, procedures, and common data structures. We adopt the following notation to describe the NeSt interface:

$$functionName : (input) \rightarrow output$$

Each protocol starts its interaction with the NeSt by *registering* to the vertical component. This operation assigns to each protocol a unique identifier (PID), as shown by step *a* in Figure 1.3. The registration is expected to happen once for all at protocol bootstrap time by calling

$$register : () \rightarrow PID$$

As described in the previous section, the NeSt does not generate shared data, but acts as an intermediary between protocols. More precisely, a protocol *seizes* the NeSt abstractions related to its internal functionalities and data structures. The example of the network topology suggests the routing protocol to acquire ownership of an abstract graph containing the collected routing information. This operation requires a protocol to identify itself, providing the PID, and to specify the abstraction's identifier (AID) together with the associated callback function (see step *b* in Figure 1.3). When invoked, the callback function fills out the

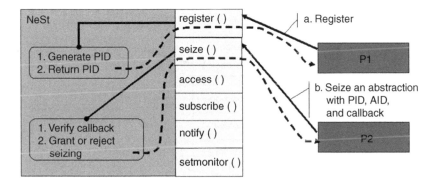

FIGURE 1.3   NeSt functionalities: register and seize.

abstraction, encoding protocol internal representation in NeSt format. Note that the callback invocation takes place asynchronously with the seizing operation, every time a fresh copy of the associated data is needed inside the NeSt. The entire process begins by calling

$$seize : (PID, AID, readCallBack()) \rightarrow result$$

The result of a call to $seize()$ indicates the outcome of the ownership request.

Once an abstraction has been seized, the NeSt is able to satisfy queries of interested entities. A protocol *accesses* an abstraction by calling

$$access : (PID, AID, filter()) \rightarrow result$$

This function shows that the caller must identify itself with a valid PID, providing also the abstraction identifier and a *filter* function. The latter parameter is a container of instructions for analyzing and selecting only information relevant to the caller's needs. The NeSt executes this call by spawning an internal computation that performs the following steps (see Figure 1.4):

1. Invoke the callback installed by the abstraction's owner (if any).
2. Filter the returned data locally (i.e., in the context of the NeSt).
3. Deliver the filtering result to the caller.

The remaining functions of the NeSt interface cope with asynchronous interactions. In the case of internal events, the role of the NeSt is to collect subscriptions, wait for notifications, and vertically dispatch occurrences to the appropriate subscribers, as shown in Figure 1.5. A protocol *subscribes* for an event by identifying itself and providing the event's identifier (EID), calling the function

$$subscribe : (PID, EID) \rightarrow result$$

To *notify* the occurrence of an event, a protocol must specify in addition to the event identifier (EID), information regarding the occurrence. This happens by calling the function

$$notify : (PID, EID, info) \rightarrow result$$

After the notification of an event, the NeSt checks it against subscriptions, and dispatches the occurrence to each subscriber.

In the case of external events, protocols subscribe by instructing the NeSt on how to detect the event. The rules to detect an external event are represented by a monitor function that periodically checks the status of a NeSt abstraction. When the monitor detects the specified condition, the NeSt dispatches the information to the subscriber protocol. As shown in Figure 1.6, a protocol delegates the *monitoring* of an

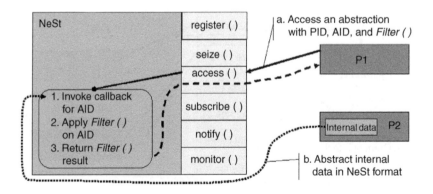

FIGURE 1.4   NeSt functionalities: access an abstraction.

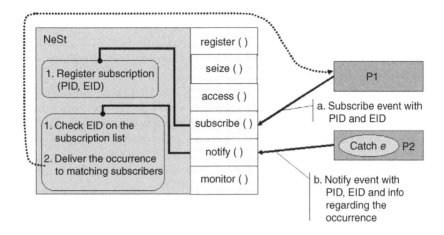

FIGURE 1.5 NeSt functionalities: management of internal events.

external event by passing to the NeSt a monitor function and the identifier of the target abstraction. This happens by calling

$$set\ monitor : (PID, AID, monitor()) \rightarrow result$$

The NeSt serves this call by spawning a *persistent* computation (see Figure 1.6) that executes the following steps:

1. Verify the monitor (e.g., type checking).
2. While (true):

    a. Refresh the abstraction invoking the associated callback.
    b. Apply the monitor to the resulting content.
    c. If the monitor detects the special condition, then notify the requesting protocol.

The result of a call to *set monitor()* only returns the outcome of the monitor's installation, while the notification of external events takes place asynchronously.

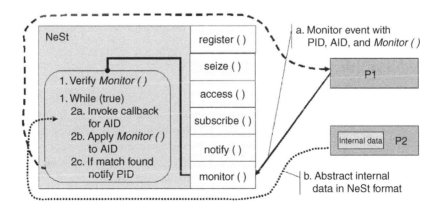

FIGURE 1.6 NeSt functionalities: management of external events.

1.2.3   Design and Implementation Remarks

It is difficult to find comparisons to the proposed architecture as, to the best of our knowledge, there are no similar approaches in the organization of a protocol stack. However, there are important observations and remarks to be given.

First of all, the NeSt is a component dedicated to enabling optimization. If on the one hand it helps maintain the layering principle allowing loosely coupled interactions, on the other hand it must guarantee the appropriate level of real-timing. That is, when subjected to a heavy load of cross-layering, the NeSt should be responsive, avoiding making protocol efforts fruitless. For example, in the case of synchronous interactions where call-backs are employed, the NeSt should not degrade the performance of both the requestor and provider protocols. For these reasons, it advisable to pre-fetch and cache exported data (when possible), serving a series of accesses to the same abstraction with fewer callback executions. However, this approach also requires the presence of cache invalidation mechanisms, which protocols can use to stale pre-fetched or cached abstractions.

As presented here, the NeSt should come with an *a priori* set of abstractions for data and events, to which protocols adapt in order to cross-interact. A more mature and desirable approach would reverse the adaptation process, having the vertical component adapt to whatever the protocols provide. For example, this adaptation issue could be solved through the use of *reflection*, a characteristic of some modern programming languages[12] that enables introspection of software components, allowing for dynamic changes in behavior. A NeSt reflective API would allow each protocol to define its contribution to cross-layer interactions, providing an initial registration of profiles describing the data and the events it is able to share. The resulting data and event sharing would be more *content-based* than the presented *subject-based* mechanism. With this approach, the sole agreement between the two parties would regard the representation of protocol profiles. A solution could be the usage of a language that provides rules to define both profiles data and metadata, as for example the eXtensible Markup Language (XML). This solution would restrict the agreement on the set of tags (i.e., the *grammar*) to use in building profiles. Note that such use of higher-level programming languages would interest only initial negotiation phases between the NeSt and the protocols, without affecting the runtime performance.

One might argue that the NeSt exhibits some conceptual similarities with a management information base (MIB). An MIB is a collection of network-management information that can be accessed, for example, through the Simple Network Management Protocol (SNMP). SNMP facilitates the exchange of information between network devices and enables network administrators to manage performances, find and solve problems, and plan for network growth. Some NeSt functionalities could be realized through a local MIB (storing protocols information), to which other protocols can access in order to read and write data. However, the NeSt and MIBs target different goals. MIBs are designed for network statistics and *remote* management purposes, while the NeSt aims at overall *local* performance improvements. Furthermore, the MIB's nature makes it unsuitable for the real-time tasks typical of NeSt optimizations, which require only local accesses and fine-grained time scales (e.g., in the order of single packets sent/received).

1.3   The Need for Global Evaluation

The NeSt architecture, as described in previous sections, is a *full* cross-layer approach where protocols become adaptive to both application and underlying network conditions. Such an approach brings the stack as a whole to the best operating trade-off. This has been highlighted by Goldsmith and Wicker,[13] where the authors point at global system requirements, like energy saving and mobility management, as design guidelines for a joint optimization. Our approach opens up different perspectives in the evaluation of network protocols. We claim that in a full cross-layer framework, the performance of a protocol should not only be evaluated by looking at its particular functionalities, but also by studying its contribution in cross-layer activities. Therefore, a stack designed to exploit joint optimizations might outperform a "team" of individually optimized protocols.

To give an example, let us consider ad hoc routing, which is responsible for finding a route toward a destination in order to forward packets. With reference to the classifications reported by Royer and Toh[14] and Chlamtac et al.,[15] the main classes of routing protocols are proactive and reactive. While reactive protocols establish routes only toward destinations that are in use, proactive approaches compute all the possible routes, even if they are not (and eventually will never be) in use. Typically, reactive approaches represent the best option: they minimize flooding, computing and maintaining only indispensable routes (even if they incur an initial delay for any new session to a new destination). But what happens when we consider the cross-layer contribution that a routing protocol might introduce in a NeSt framework?

To answer this question, we provide an example of cross-layer interaction between a routing protocol and middleware platform for building overlay networks, where the former contributes exporting the locally collected knowledge of the network topology. Building an overlay network mainly consists of discovering service peers, and establishing and maintaining routes toward them, as they will constitute the backbone of a distributed service. The overlay network is normally constituted by a subset of the network nodes, and a connection between two peers exists when a route in the underlay (or physical) network can be established. The task of building and maintaining an overlay is carried out at the middleware layer, with a cost that is proportional to the dynamics of the physical network. Overlay platforms for the fixed Internet assume no knowledge of the physical topology, and each peer collects information about the overlay structure in a distributed manner. This is possible because the fixed network offers enough stability, in terms of topology, and bandwidth to exchange messages. Of course, similar conditions do not apply for ad hoc environments, where bandwidth is a precious (and scarce) resource and the topology is dynamic. In ad hoc networks, cross-layering can be exploited, offering the information exported by the network routing to the middleware layer. The key idea is that most of the overlay management can be simplified (and eventually avoided) on the basis of already available topology information.[16] In this case, the more information available, the easier the overlay management; and for this reason, a proactive routing approach becomes more appealing. To support this claim, let us look at what is described by Schollmeier et al.[5] This article describes a cross-layer interaction between a middleware that builds an overlay for peer-to-peer computing and a dynamic source routing (DSR) at the network layer. In this work, the DSR algorithm is forced to maintain valid routes toward the overlay peers, even if these routes are not in use. That is, a reactive routing is forced to behave proactively, with the additional overhead of reactive control packets. The same cross-layer approach with a proactive protocol would probably represent the best joint optimization.

Another example of joint optimization is the extension of routing to support service discovery. A service discovery protocol works at the middleware layer to find out what kind of services are available in the network. As the dynamics of the ad hoc environment determines frequent changes in both available services and hosting devices, service discovery is of fundamental support. The IETF proposes the Service Location Protocol (SLP)[17] to realize service discovery in both Internet and ad hoc networks. Recently, they also underlined the similarity of the messages exchanged in SLP, with those used in a reactive routing such as the Ad hoc On-demand Distance Vector (AODV) protocol.[18] This proposal discusses an extension of AODV to allow service request/reply messages in conjunction with route request/reply. In this proposal, there is a background cross-layer interaction that allows SLP to interface directly with AODV, asking for service-related messages, providing local service data, and receiving service information coming from other nodes. The proposed joint optimization would work even better in the case of a proactive routing protocol such as DSDV or OLSR (see Chlamtac et al.[15] for details). In case of proactive routing, the service information regarding the local services offered on each node could be *piggybacked* on routing control packets and proactively spread around the network, together with local connectivity information. The service discovery communication could be significantly reduced, at the expense of broadcasting routing control packets a few bytes longer. This optimization would result in a proactive service discovery, where a component such as the NeSt supports the exchange of service information from the service discovery protocol, at the middleware, with the routing protocol, and vice versa.

1.4   Discussion and Conclusions

Typically, cross-layering is emphasized as a way to work around the TCP/IP implementation limits, as it introduces direct interaction between protocols to enable smarter adaptation or better performance, at the cost of a "spaghetti-like" code.[9] We believe that cross-layering is possible while keeping the layer separation principle, and that the Internet community is incrementally moving toward cross-layering. The simplest step in this direction is represented by layer triggers, which are predefined signals that notify events between protocols. An example is given by the Explicit Congestion Notification (ECN) mechanism, which notifies the TCP layer about congestion detected by intermediate routers (IP layer). In this way, the source can be informed of congestion quickly and unambiguously, without resorting to inferring mechanisms based on retransmit timer or repeated duplicate ACKs. Another example is given by L2 triggers,[19] added between the data link and IP layers to efficiently detect changes in the wireless links' status. A further step toward cross-layering is presented in by Waldvogel and Rinaldi.[20] This work proposes a way to build topology-aware overlay networks, where logical neighbors are also close in the physical network, according to metrics coming from different levels. These include metrics typically used in routing protocols, such as the physical distance and the bandwidth achieved by a TCP stream.

An open question is to understand if a NeSt-like approach can support cross-layering in the future Internet architecture. Considering the above examples, the NeSt could easily handle the described interactions in the following way:

- The signaling of the ECN bit might correspond to an internal event generated by the IP protocol, previously subscribed by the TCP. Analogously, the MAC layer could generate link-related events to notify the IP layer.
- The metrics used in the construction of the topology-aware overlay network could be associated to NeSt abstractions, seized by the routing and transport protocols, and accessible through the NeSt API.

The NeSt could also be employed to realize optimizations, proposed for the Internet architecture, which are not cross-layered but sidestep the standard layer interfacing. For example, Application Level Framing (ALF)[21] aims at minimizing retransmissions due to data loss, enabling the application level to break the data (to transmit) into suitable aggregates, and the lower level to preserve these frame boundaries when processing the data. Thus, the data segmentation functionality is moved from the transport layer to the application layer. Although the vertical data pipeline inside the protocol stack is not altered, the ALF approach needs a customized implementation, which complicates the maintainability of the overall stack and disagrees with standard interfacing. In the NeSt architecture, the same goal could be achieved keeping the data segmentation functionality at the transport layer and allowing the application layer to instruct, through information sharing, the transport protocol on the way to break data.

Finally, cross-layering provides an effective step toward the mobile Internet. The NeSt architecture is a building block for context-aware computing and networking, a novel paradigm in which a system shows the ability to discover and take advantage of contextual information. Context can be defined as the set of environmental states and settings that determines a system's behavior, and in which system events of interest for the user occurs.[22] While current distributed Internet applications and middleware platforms tend to provide a transparent representation of the underlying execution environment,[23] context awareness fits well in the area of mobile computing, where applications and software components have to cope with dynamic environments, determining changes in context. In mobile networks, applications and middleware platforms need to be aware of context details, leaking out information such as device location, network bandwidth, or surrounding environment, to and from adjacent layers. The NeSt approach goes toward full awareness of networking context: if applications and network protocols are mutually aware of their operating conditions, then they can adjust their behavior to achieve functional trade-offs and deliver the best end-to-end performance.[13]

Awareness is an important requisite for extending the mobile Internet toward 4th Generation wireless technology. The computing world is experiencing a seamless integration of mobile ad hoc networks with other wireless networks and the fixed Internet infrastructure. The global system presents different characteristics, depending on both physical constraints (e.g., bandwidth, energy, processing power) and usage patterns. The key requirement for the operation of this *heterogeneous* Internet is the protocol's ability to globally adapt to application requirements and underlying network conditions. The need for adaptive networking becomes a challenging issue, the solution of which requires context-awareness.

Acknowledgments

This work was partially funded by the Information Society Technologies programme of the European Commission, Future and Emerging Technologies, under the IST-2001-38113 MOBILEMAN project, and by the Italian Ministry for Education and Scientific Research in the framework of the FIRB-VICOM project.

References

1. J.P. Macker and M.S. Corson. Mobile ad hoc networking and the IETF. *ACM Mobile Computing and Communications Review,* 2(3):7–9, 1998.
2. M.S. Corson, J.P. Macker, and G.H. Cirincione. Internet-based mobile ad hoc networking. *IEEE Internet Computing,* 3(4):63–70, 1999.
3. C. Szyperski. *Component Software,* pp. 140–141. Addison-Wesley, 1998.
4. K. Chen, S.H. Shah, and K. Nahrstedt. Cross-layer design for data accessibility in mobile ad hoc networks. *Wireless Personal Communications,* 21(1):49–76, 2002.
5. R. Schollmeier, I. Gruber, and F. Niethammer. Protocol for peer-to-peer networking in mobile environments. In *Proc. 12th IEEE Int. Conf. Computer Communications and Networks,* Dallas, TX, 2003.
6. W.H. Yuen, H. Lee, and T.D. Andersen. A simple and effective cross layer networking system for mobile ad hoc networks. In *Proc. IEEE PIMRC 2002,* Lisbon, Portugal, 2002.
7. M. Chiang. To Layer or not to layer: balancing transport and physical layers in wireless multihop networks. In *Proc. IEEE INFOCOM 2004,* Hong Kong, China, 2004.
8. U.C. Kozat, I. Koutsopoulus, and L. Tassiulas. A framework for cross-layer design of energy-efficient communication with QoS provisioning in multi-hop wireless netwroks. In *Proc. IEEE INFOCOM 2004,* Hong Kong, China, 2004.
9. V. Kawadia and P.R. Kumar. A Cautionary Perspective on Cross Layer Design. In *IEEE Wireless Communications,* 12(2):3–11, 2005.
10. M. Conti, S. Giordano, G. Maselli, and G. Turi. MobileMAN: mobile metropolitan ad hoc networks. In *Proc. 8th IFIP-TC6 Int. Conf. on Personal Wireless Communications,* pp. 169–174, Venice, Italy, 2003.
11. M. Conti, G. Maselli, G. Turi, and S. Giordano. Cross-layering in mobile ad hoc network design. *IEEE Computer, Special Issue on Ad Hoc Networks,* 37(2):48–51, 2004.
12. Sun Microsystems. The JAVA Reflection API. http://java.sun.com/j2se/1.4.2/docs/guide/reflection/index.html, 2002.
13. A.J. Goldsmith and S.B. Wicker. Design challenges for energy-constrained ad hoc wireless networks. *IEEE Wireless Communication,* 9(4):8–27, 2002.
14. E.M. Royer and C.-K. Toh. A review of current routing protocols for ad hoc mobile wireless networks. *IEEE Wireless Communications,* 6(2):46–55, 1999.
15. I. Chlamtac, M. Conti, and J.J.-N. Liu. Mobile ad hoc networking: imperatives and challenges. *Ad Hoc Networks Journal,* 1(1):13–64, 2003.
16. M. Conti, E. Gregori, and G. Turi. Towards scalable P2P computing for mobile ad hoc networks. In *Proc. First Int. Workshop on Mobile Peer-to-Peer Computing (MP2P'04), in conjunction with IEEE PerCom 2004,* Orlando, FL, 2004.
17. E. Guttman, C.E. Perkins, J. Veizades, and M. Day. Service Location Protocol, Version 2. IETF RFC 2608, June 1999.
18. R. Koodli and C.E. Perkins. Service Discovery in On-Demand Ad Hoc Networks. Internet Draft, October 2002.

19. S. Corson. A Triggered Interface. http://www.flarion.com/products/drafts/draft-corson-triggered-00.txt, May 2002.

20. M. Waldvogel and R. Rinaldi. Efficient topology-aware overlay network. *ACM Computer Commun. Rev.*, 33(1):101–106, 2003.

21. D.D. Clark and D.L. Tennenhouse. Architectural considerations for a new generation of protocols. In *Proc. ACM Symp. Communications Architectures and Protocols*, pp. 200–208. ACM Press, 1990.

22. G. Chen and D. Kotz. A Survey of Context-Aware Mobile Computing Research. Technical Report TR2000-381, Dept. of Computer Science, Dartmouth College, 2000.

23. C. Mascolo, L. Capra, and W. Emmerich. Middleware for mobile computing (a survey). In E. Gregori, G. Anastasi, and S. Basagni, Editors, *Neworking 2002 Tutorial Papers*, LNCS 2497, pp. 20–58, 2002.

2

Routing Scalability in MANETs

Jakob Eriksson,

Srikanth Krishnamurthy

Michalis Faloutsos

With today's rapidly improving link-layer technology, and the widespread adoption of wireless networking, the creation of large-scale ad hoc networks could be construed as all but inevitable. However, for routing in such a network to be feasible, there is a pressing need for a scalable ad hoc routing protocol. Applications for large-scale ad hoc networking include consumer-owned networks, tactical military networks, natural disaster recovery services, and vehicular networks.

Ad hoc routing protocols used experimentally today, such as DSDV, OLSR, AODV, and DSR, only scale reasonably well to dozens or sometimes hundreds of nodes. To support networks one or several orders of magnitude larger, there is a need for routing protocols designed specifically to scale to large networks. Under certain limiting assumptions, geographical location information can be used to help the routing layer scale to support very large networks. However, this chapter focuses on the more generally viable approach of multilevel clustering, which to some extent is what has made made the Internet scale as well as it does.

We study various aspects of routing protocol scalability. First, we take a look at the analytical results thus far, with regard to ad hoc network scalability. These results assess the theoretical limits for ad hoc network scalability in terms of the capacity achievable per node in the network. To set the stage for the scalable routing techniques, and to introduce the reader to the issues that impact scalability, we briefly discuss several techniques used for ad hoc routing. These include flat proactive routing, pure reactive routing, geographic routing, and zone-based hybrid protocols. We then take a more detailed look at routing based on clustering, in its single-level and multilevel forms. Finally, we spend the last third of the chapter describing a recent promising scalable routing technique based on multilevel clustering, called Dynamic Address Routing.

2.1 Defining Scalability

The scalability of a network protocol can potentially be defined in many different ways, and at several different levels. In this chapter, we use the following high-level definition of scalability.

> **Scalability** *is the ability of a routing protocol to perform efficiently as one or more inherent parameters of the network grow to be large in value.*

Typical parameters that are studied for ad hoc networks are the number of **nodes (N)** and the average rate of **mobility(M)** in m/s under various mobility models. Other parameters that have an impact on scalability include **node density (D)**, number of **links (L)**, the **frequency of connection establishment (F)**, and the average number of **concurrent connections (C)**. Measuring performance can also be done in several ways. Typical metrics used to evaluate routing protocols are overall message or byte overhead, amount of per-node state to be maintained, latency, and total network throughput. In this chapter, we primarily discuss the overhead aspect. However, we also discuss the other metrics briefly in the sections that follow.

In the remainder of this chapter, we use notation commonly employed in asymptotic analysis to describe various scalability characteristics. In particular, we use the $\Omega(X)$ to denote a lower asymptotic bound, $O(X)$ to denote an upper asymptotic bound, and $\Theta(X)$ to denote a simultaneous upper *and* lower asymptotic bound. By asymptotic bound, we refer to the scaling behavior of the protocol with respect to a given variable. For example, if a protocol is said to have an overhead of $O(N)$, this means that there exists a constant c such that the amount of overhead incurred in a network of N nodes is at most cN, where N can take on any finite value. Except where explicitly stated, node identifiers are taken to be 48-bit MAC addresses. It is reasonable to assume that a 48-bit identifier space will not be exhausted within the foreseeable future ($2^{48} = 281, 474, 976, 710, 656$ or about 281 trillion unique identifiers).

2.2 Analytical Results on Ad Hoc Network Scalability

The analytical study of scalability relationships in ad hoc networks can provide us with valuable insights into the proper design of ad hoc routing protocols and possibly related mechanisms at other layers. So far, the study of scalability in ad hoc networks has been mostly limited to simulation. However, a few significant analytical results have emerged fairly recently, and we introduce them in this section.

2.2.1 Link Layer

Even without considering the effects of routing overhead on the performance of ad hoc networks, there are several concerns regarding the scalability of current wireless networking link-layer technology.

It is easily seen that the popular 802.11 link layer, when deployed with omnidirectional antennas, does not scale with respect to node density, D. Clearly, as D grows, each node will receive only a proportional share of the channel capacity. The upper limit on the average link layer capacity made available to each node decreases as $1/D$. A well-known solution to this problem is to reduce the transmission range of each node, thereby reducing D. The effect achieved is called *spatial reuse*, where several transmissions can take place on the same frequency band simultaneously, due to the limited spatial overlap of the transmitters involved.

However, as a direct effect of reducing the transmission range, packets in some cases must be forwarded over an increased number of wireless links to reach their respective destinations. Increasing the number of hops is likely to lead to longer end-to-end delays, lower packet delivery ratios, and in some cases, increased traffic congestion.

A fundamental result in multihop ad hoc networking was shown by Gupta and Kumar.[11] A simplified argument for their result follows. In a network of nodes with omnidirectional antennas, and with a constant node density, we can expect the average path length to be $\Theta(\sqrt{N})$, where N is the number of nodes in the network. Therefore, for every packet a node generates, it will see, on average, $\Theta(\sqrt{N})$ packets originated by other nodes. Thus, with a channel capacity of C, the capacity available for a node's own packets will be:

$$O\left(\frac{C}{\sqrt{N}}\right) \tag{2.1}$$

where C is the total channel capacity, that is, the maximum throughput achievable by a single link when there are no other links competing for the channel. The unfortunate conclusion is that under certain reasonable assumptions, purely omnidirectional ad hoc networks cannot grow beyond certain fairly restrictive limits. However, we would like to point out that all hope is not lost. As link layer technologies evolve, the channel capacity C will continue to increase. And for every increase in channel capacity, the feasible network size grows by the square of this increase, as per Equation 2.1. Since the publication of the article by Gupta and Kumar,[11] channel capacity has grown by approximately 100 times. Our conclusion is that whatever the feasible network size was at the time of publication (1999), the upper limit today is 10,000 times higher. Clearly, this shows that link layer capacity by itself is not the limiting factor in multihop ad hoc networks. Note that this highly theoretical result does not take into account any routing layer overhead, the scalability of which is the topic of this chapter.

In addition, there is the prospect of using directional antennas[16] and adaptive beamforming antennas.[26] These could be employed to have nodes dynamically direct a narrow transmission beam toward the neighbor it wishes to communicate with, thereby greatly improving both transmission range and spatial reuse.*

Grossglauser and Tse[10] published a somewhat controversial result. The authors show that if nodes are mobile, then each node could potentially achieve a throughput that *does not* decrease with the size of the network. By relaying each packet only once, to a random one-hop neighbor, a source can achieve a stationary uniform distribution of its packets throughout the network. Subsequently, as the destination moves around, each of its neighbors will always have packets to deliver to it. As each packet only traverses two hops, the throughput of the node can be expected to remain the same, regardless of the size of the network.

This result relies on strong assumptions with regard to the mobility patterns of the nodes, and even given those assumptions, the expected delay is of the order of node mobility, in the sense that nodes have to move considerable distances before a packet can be delivered. In our opinion, although this result holds in theory, it is unclear as yet if it will have much practical relevance.

2.2.2 Hierarchical Routing

Hierarchical routing protocols, such as those based on multilevel clustering, consist of a number of different components, such as clustering, routing, and location management. Here, clustering is the process by which nearby nodes form groups, called *clusters*. For the purpose of routing, clusters can be treated as a single destination, thereby reducing the amount of routing state that must be maintained at each node. Location management is any technique by which a source can determine the current address or *location* of an intended destination node, given its identifier.

When studying the scalability of such protocols, the scaling properties of each of these components must be considered. Sucec and Marsic[29,30] have studied the theoretical scalability aspects of multilevel hierarchical

---

*To see how both range and spatial reuse can be improved simultaneously, consider the extreme case of directional transmission, a point-to-point laser link.

routing in ad hoc networks. In the general scheme they analyze, nodes are organized in clusters, which are then grouped in higher-level clusters. The number of levels is logarithmic in the network size. The location management technique they analyze is a distributed location server, where each node stores the current address of $\Theta(\log N)$ other nodes, where N is the number of nodes in the network. A similar location management scheme is discussed in Section 2.9. Specifically, their analysis focuses on the number of routing-related control datagrams that a node needs to transmit per unit of time, on average, given a wide variety of parameters.

Their main result is that routing overhead is polylogarithmic in the size of the network. More specifically, the channel capacity required for routing control messages sent by each node, on average, is:

$$\Omega(\log^3 N)$$

Interestingly, they show that the dominating factor in the overhead calculation is not routing updates, but the retransmission of location information due to changes in the clustering hierarchy, called *location management handoff*. Other potentially valuable results of the same article include the overhead incurred by cluster formation and maintenance, which is computed to be

$$O(\log N)$$

packet transmissions per node per second, and the overhead for location management handoff, which is shown to be

$$\Theta(\log^2 N)$$

packet transmissions per node per second, where the size of every control packet is $\Theta(\log N)$. Note that this study is targeted at a particular group of clustering schemes (Max-Min D-hop clustering[1]). These are based on finding the node with the maximum node identifier in a D-hop neighborhood, and assigning that node to be the cluster head. Other types of clustering, such as that described in Section 2.9, could potentially have different scaling behaviors. It is also geared toward a scalable hierarchical location management scheme similar to that used by Eriksson et al.[7,8] and described in Section 2.9. Again, other types of location management will exhibit different scaling behavior. Nevertheless, these results offer valuable insights into the scalability of hierarchical, multilevel clustering ad hoc routing protocols. To our knowledge,[30] is the first paper with comprehensive theoretical results on the overhead of multilevel hierarchical routing protocols.

2.3   Flat Proactive Routing

Flat proactive routing scales very well with respect to the frequency of connection establishment (F) and the number of concurrent connections (C). However, the number of control packet transmissions per node is $\Theta(N)$.

In proactive routing, the routing protocol periodically disseminates routing information throughout the network. With flat proactive routing, every node keeps routing information for every other node; there is no abstraction for nodes far away. This strategy generally leads to close to optimal paths, but this is achieved at the cost of lacking scalability. The flat proactive routing protocols proposed so far can be roughly divided into two subcategories: (1) link-state (LS) and (2) distributed Bellman-Ford (DBF) algorithms.

In LS algorithms such as Fisheye State Routing,[22] Global State Routing,[4] and Optimized Link-State Routing,[6] each node has complete, although not always accurate, knowledge of the state of every link in the network. Using this information, it can calculate the entire path to the destination on its own accord. This has many advantages. In particular, recovery from link failure is typically very quick in LS protocols. With large N or D, the number of links in the network, and thus the routing table size, may be prohibitive. Fisheye State Routing (FSR)[22] tries to reduce the overall overhead by limiting the rate of link-state updates far away from the source of the update. The idea in FSR is that link changes far away generally have a small effect on local routing decisions.

In DBF algorithms, such as Destination Sequenced Distance Vector routing[24] and Wireless Routing Protocol,[19] each node has much less information about the network. For every destination, a node maintains a routing table consisting of the distance to the destination, and the next hop neighbor on the shortest route toward the destination. Typically, after a link failure, there is an interval of time where faulty routes may exist, until the protocol has settled on a new route.

Common for all of these protocols is that the necessary amount of state kept at each node scales at least linearly with N. In a mobile network, this state must be updated frequently, resulting in protocol overhead on the order of $O(N)$.[30] For this reason, flat proactive routing protocols are only feasible for small networks.

2.4 Pure Reactive Routing

Reactive routing is scalable with respect to most parameters, as long as the frequency of connection establishment (F), and the average number of concurrent connections (C), remain low. Control packet transmissions per node grow as $O(F + C)$, which is $\Omega(1)$, but $O(N^2)$.

In an effort to address the problem of maintaining state for all nodes in the network, reactive protocols such as Ad hoc On-demand Distance Vector routing,[25] Dynamic Source Routing,[13] Associativity Based Routing,[31] and Labeled Distance Routing[9] defer the expenditure of routing overhead until the time of connection establishment. With this technique, nodes keep completely quiet as long as there is no data to transmit. If a connection is to be established, the source node S needs to flood the network with a route request, as shown in Figure 2.1. When the intended destination D receives the route request, it responds to the source with a route response, using one of the routes discovered during the route request phase. In networks where a large majority of nodes have nothing to send, and where connections involve more than just a few packets, this strategy pays off in terms of reducing the overall routing overhead.

Reactive routing protocols have seen much popularity in ad hoc networks research. This is due to several good reasons, including the battery savings achieved by not transmitting anything during idle periods. Other important reasons are the good performance and the straightforward design principles of AODV and DSR, the two most well-known reactive ad hoc routing protocols.

However, by deferring the routing overhead, these protocols lose many potential aggregation benefits made possible by proactively distributing routing information. In contrast with flat proactive routing, every connection establishment sets off a reactive route request with an asymptotic cost of $O(N)$, as a nonnegligible constant fraction of all nodes will rebroadcast the request packet. In addition, in a mobile network, established connections will fail regularly due to link breakages caused by node motion, thereby initiating additional route requests. This gives an overhead complexity of $\Theta(F + C)$ for the number of route requests per second. Putting the two terms together, the expected number of control packet transmissions is $O(N(F + C))$. With N nodes to share the burden, the average per-node cost is $O(F + C)$. Note that if a

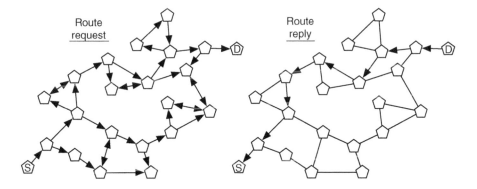

FIGURE 2.1 Reactive routing. A route request is flooded throughout the network. Once the request reaches the intended destination, a route reply is sent back along a discovered path.

constant fraction of the nodes can be expected to start or maintain a connection every second, this reverts back to the $O(N)$ per-node cost of flat proactive routing. In the worst case, where a constant fraction of the nodes can be expected to set up k connections, and k grows linearly with N, the overhead incurred will be $O(N^2)$.

A performance optimization used aggressively in DSR[13] is route caching, where intermediate nodes are allowed to send a route response, if they have recently observed a route to the desired destination. This can result in greatly improved performance but there is also a high risk of *route poisoning*, where intermediate nodes unwittingly return routes that are no longer accurate. In general, reactive routing has been shown through simulation to scale better than flat proactive routing in most considered scenarios.

As we see below, proactive routing has an advantage that a purely reactive protocol lacks: the ability to cluster nodes and aggregate routes. As we see in the following sections, clustering and address aggregation have the potential to drastically reduce the protocol overhead of a proactive routing protocol, as network size increases. The relationship between the overhead of reactive and proactive routing under different traffic scenarios is discussed in more detail by Eriksson et al.[8]

2.5   Geographical Routing

The control overhead of geographical routing is typically $O(1)$, not counting location management. However, geographical routing relies heavily on two assumptions: (1) that each node knows its position, and (2) that the geographical distance between nodes corresponds well to the distance between these nodes in the network topology. In many situations, these assumptions are unacceptable.

Geographical routing protocols make use of the geographical location of a node to make routing decisions. Such location information would generally be acquired either from GPS satellites, or from location interpolation given the positions of neighboring nodes.

In addition to knowing its own geographical location, a node also needs to know the locations of its neighbors, as well as the location of its intended destination. Dream (Distance Routing Effect Algorithm for Mobility)[2] and Grid Location Service[17] are mechanisms for finding out the location of any given node in the network. In Dream, nodes periodically flood their location information throughout the network. However, as the flood travels away from the source, the speed with which updates are propagated decreases, thereby drastically reducing the overall overhead of the protocol. This is similar to the technique used in Fisheye routing[22] to reduce the cost of disseminating link state updates. In GLS, the location for a given node is stored at an *anchor node*. An anchor node is defined as the node positioned closest to a geographic location that is determined by hashing the node identifier. Every node is responsible for keeping its anchor node up-to-date on its current location. This method of distributing responsibility for storing location information is highly scalable and efficient, given that some characteristics of the network are known, such as the extent of the network in geographical terms.

Geographic routing protocols include Greedy Perimeter State-less Routing (GPSR)[14] and Location Aided Routing (LAR).[15] GPSR greedily routes packets to the one-hop neighbor that is closest to the destination. Should an obstacle appear between source and destination, GPSR uses a planarized version of the network graph and follows the "right hand rule" to route around the obstacle. The technique is illustrated in Figure 2.2, where a packet destined for node D is originated at S. When the large obstacle in the middle of the network is encountered, the *right-hand rule* is triggered, routing the packet around it. The use of the *right-hand rule* for routing around obstacles can result in paths of length $O(N)$, making greedy routing a risky proposition unless the characteristics of the topology are known in advance. LAR uses the geographic location of the destination to guide a reactive route lookup. By limiting the route request flood to neighbors in the approximate direction of the destination, the cost of route setup is reduced.

Any geographical routing protocol relies on the assumption that the geographical distance between two nodes corresponds well with their distance in the network topology. In scenarios where this is not the

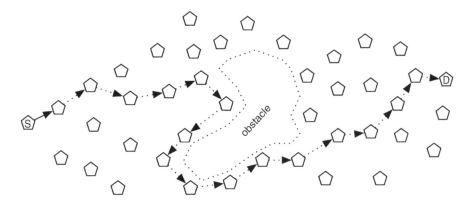

FIGURE 2.2 Geographical routing. The next hop is selected on greedily, until there is no neighbor that is closer to the destination than the current node. When this happens, routing switches to the *right-hand rule* until the obstacle has been successfully routed around.

case, such as sparse or heterogeneous networks, or networks with directional or wired links, geographical routing is unlikely to achieve acceptable performance.

2.6 Zone-Based Routing

Zone-based routing combines the merits of flat proactive and pure reactive routing. However, while these hybrid protocols are more efficient than the component protocols they are made up out of, the asymptotic scalability of zone-based routing is the same as that of other flat routing protocols.

In the Zone Routing Protocol (ZRP)[12] and Sharp Hybrid Adaptive Routing Protocol (SHARP),[28] the merits of proactive and reactive routing are combined to form two hybrid proactive–reactive protocols. Both protocols follow a similar architecture. Around every node, a *zone* of d hops is maintained in which proactive routing is performed. For all destinations outside the zone, reactive route requests are used to establish a route. As soon as a route request reaches a node in the zone of the intended destination, this node replies with a route response.

In ZRP, the size of the zones can be varied, depending on the mobility and traffic characteristics of the network.[20] Thanks to the proactive routing information available within the zone, the damaging effect of the flood is limited, as route requests can be efficiently routed to the edges of the zone, using a technique called *bordercasting*. Several other techniques are introduced to minimize the duplication of effort that could otherwise happen due to zone overlap.

While ZRP introduces its own routing components, such as *bordercasting*, SHARP is a straightforward combination of proactive and reactive routing (Figure 2.3). In SHARP, every node individually adapts the size of its zone, that is, the distance (in hops) up to which its proactive routing updates should be forwarded. Reactive routing is done according to whatever reactive protocol is used, with the modification that intermediate nodes that have proactive routing information for the desired destination node are allowed to reply to the route request. SHARP trades off the constant overhead of proactive routing against the high incremental cost of reactive routing by adaptively tuning the zone size of a node to correspond to the popularity or usage profile of the node. In addition to improving performance for popular nodes, the same trade-off is used to achieve desired packet delivery ratio and delay characteristics.

However, if properly done, route caching in reactive routing protocols can likely achieve a constant-term savings in terms of protocol overhead. Moreover, neither route caching nor the hybrid approaches taken in ZRP and SHARP can efficiently handle the case where there are frequent connection establishments, unless traffic is concentrated to a small number of nodes. For SHARP to achieve a successful trade-off

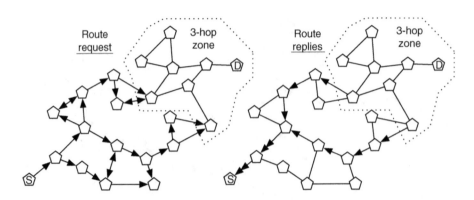

FIGURE 2.3 Hybrid proactive–reactive routing with SHARP. Route requests are flooded until they reach a node within the destination's proactive zone.

between flat proactive and reactive routing, it is necessary for a few nodes to receive the majority of the network traffic.

Compared to flat proactive routing, or pure reactive routing, this middle ground between reactive and proactive routing can be expected to achieve lower overhead and delay under many traffic scenarios. However, although hybrid methods can be expected to reduce routing overhead, they only do so by a constant factor.

2.7   Single-Level Clustering

Single-level clustering improves scalability with respect to the network size (N) if the size of each cluster can be set to $\sqrt{(N)}$. In this case, the control packet overhead is $\Theta(\sqrt{N})$. With constant size clusters, overhead remains at $\Theta(N)$. Certain node mobility patterns (M) can have a larger detrimental effect on the performance of clustering protocols than they have on flat protocols.

Several protocols propose to use clustering to improve routing protocol scalability. *Clustering* is a process by which neighboring nodes form connected subsets, with one node elected as the cluster head. Depending on the clustering technique used, clusters can be of radius of one or more hops from the cluster head. The cluster heads may have responsibilities in addition to that of a regular node, such as inter-cluster routing and intra-cluster coordination.

In hierarchical protocols, such as LANMAR[21] and CGSR,[5] routes are aggregated by cluster. Inside a cluster, every node has complete routing information for every other node in the cluster. Externally, however, only a route to the cluster as a whole is published. Packets are first routed toward the cluster head of the destination. Once the cluster head, or simply any node within the destination cluster, has been reached, the packet is routed directly toward its final destination within the cluster. Through this technique, a smaller amount of routing state is necessary on each node, and intra-cluster changes in the topology do not affect external routes. Note that all routing schemes that use clustering for routing will incur a cost in terms of increased path length. However, this cost is usually negligible compared to the savings achieved by reducing the amount of routing overhead incurred.

In contrast to the hybrid schemes mentioned earlier, which rely on flooding, hierarchical schemes also need to keep track of which cluster a node belongs to. This is sometimes referred to as *location management*. Depending on the assumptions used, location management can be a crucial factor in the performance of a clustering-based routing protocol.

In LANMAR (Figure 2.4), it is assumed that most nodes will remain in the same cluster throughout their lifetimes, and group membership is determined at network initialization. The authors use a *group mobility* model, which applies mostly to military scenarios. LANMAR builds on ideas from Landmark Routing[32]

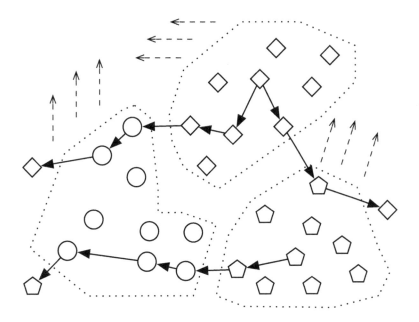

FIGURE 2.4 Mobile groups and stray nodes in LANMAR. Nodes move together in groups, while stray nodes are handled with separate distance vector entries.

and Fisheye State Routing (FSR).[22] Nodes within a cluster exchange link-state information using FSR. In addition to this link-state information, each node keeps a distance vector table for a specific node in each cluster. This node is referred to as the Landmark. Any stray nodes, that is, nodes that are not directly connected to their home cluster, are handled as special cases: a separate distance vector routing entry is kept by every node on the shortest path between the Landmark node and the stray node. Assuming that only a constant number of nodes stray from their home clusters, the asymptotic control packet overhead is $\Theta(\sqrt{N})$.

In CGSR,[5] one-hop clustering is performed, and is mainly used for transmission scheduling. A technique is also proposed in which each node globally advertises its cluster membership, and routing entries are kept only for cluster heads. Because the cluster radius is limited to a single hop, a cluster will contain only a constant number of nodes, leading to, at best, a constant improvement in the overhead incurred.

Both of these protocols rely on nodes staying within their original clusters throughout their lifetimes, or overhead will grow quickly. More flexible and scalable location management methods have been developed, and these are discussed in the upcoming sections.

As with the hybrid scheme above, these single-level clustering protocols only reduce overhead to at best $O(\sqrt{N})$, depending on the cluster size. In the next section, we discuss how to extend the idea of cluster-based routing to reduce the size of the routing tables from $O(\sqrt{N})$ to $O(\log N)$.

2.8 Multilevel Clustering

Multilevel clustering protocols scale well with network size (N), frequency of connection establishment (F), and the number of concurrent connections (C). The number of control packet transmissions per node is $\Omega(\log^2 N)$.

For large networks, the address size in bits is $\Omega(\log^2 N)$, which in practice could easily grow beyond the limit of feasibility.

The ability to achieve true routing scalability with respect to network size (N), under most common scenarios, has so far only been demonstrated through the use of multilevel clustering. In these protocols,

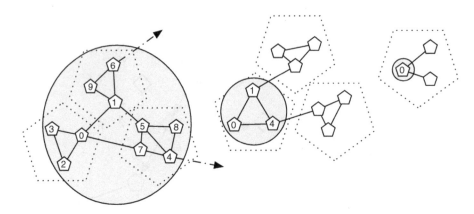

FIGURE 2.5   An example multilevel cluster hierarchy. At the left, individual nodes with their respective node IDs, partial view. In the middle, level-1 clusters forming level-2 clusters; and to the right, a level-2 cluster view of the entire network.

physical nodes cluster first into level-1 clusters. Then, up to d level-1 clusters are further clustered into level-2 clusters, etc. (Figure 2.5). In general, with a *clustering degree* of d, the size of the routing table will be on the order of $O(d \log_d N)$.

In addition to reducing the size of the routing table, multilevel clustering will also make the network appear much less dynamic, as link-state changes within a given cluster generally are not propagated to nodes that are not part of the cluster. This will reduce the overall control packet overhead under node mobility.

Examples of multilevel clustering are Hierarchical State Routing (HSR)[23] and MMWN.[27] These are link-state protocols, and they use the clustering abstraction to define *virtual links* between clusters. Instead of keeping track of all links in the network, a node now only needs to maintain entries for the virtual links going to or from a cluster in which the node is a member, a much smaller number.

Initially, one-hop physical-level clusters are formed, as in the previous section, by electing a cluster head and having nodes in the k-hop neighborhood of that node join the cluster head to form a cluster. To build the next higher clustering level, the cluster heads of neighboring clusters elect a higher-level cluster head from among themselves. Once the cluster hierarchy is formed, each node creates its own hierarchical identifier (HID) by concatenating the identifiers of all the clusterheads from the root of the hierarchy to the node in question. In theory, the size of a node identifier is $\Theta(\log N)$ bits, which results in an asymptotic HID size of $\Theta(\log^2 N)$. However, in practice, node identifiers are typically 48-bit MAC addresses.*

Data packet headers contain their destination HID. Routing is performed one level at a time: first, a packet is routed directly to the lowest-level cluster head in the HID which exists in the current node's routing table. Once this cluster head has been reached, the routing proceeds to the next lower level, as indicated in the HID. Eventually, the intended destination is reached, through the recursive application of this procedure.

Another example is Landmark routing, which is similar to HSR and MMWN but uses a Distributed Bellman-Ford-like routing scheme and does not concentrate traffic to cluster heads to the same extent. First described by Tsuchiya,[32] Landmark routing establishes a set of self-elected Landmark nodes in multiple levels. The main difference in Landmark routing is how the hierarchy is formed. Here, each Landmark periodically broadcasts an advertisement, announcing its presence. Depending on the level k

---

*In addition to the obvious practical reasons for using MAC addresses, it is worth noting that if a node identifier is to be constant throughout the lifetime of a node, it must be unique not only in the current network, but in every network it could conceivably be part of. The only way to feasibly assign identifiers to ensure this is to assign every node a globally unique ID, which is the sole purpose of the current 48-bit MAC addresses used by network interface cards.

of the Landmark, the advertisement will travel r_k hops. A new node initially assigns itself level 0 and sends an advertisement. If it can hear the advertisement of a level-1 Landmark, it can remain at level 0 and select that Landmark as its parent. If it does not, it cooperates with its level-0 neighbors to elect a new level-1 node. This process is repeated until a small subset of the nodes in the network are level-d Landmarks, where r_d is larger than the diameter of the network. At this point, the Landmark hierarchy is complete.

An interesting difference between Landmark routing other multilevel clustering schemes is that several Landmarks can cover a single node, giving the node several valid addresses. Landmark routing was the basis for LANMAR mentioned in the previous section, and was later extended in L+ routing[3] and Safari routing.[18]

Safari routing[18] is similar in many respects to Landmark routing. Landmark nodes, here called *drums*, self-elect and form a multilevel Landmark hierarchy. One major difference is that the Safari hierarchy does not extend all the way to the physical (node) level. Instead, it extends down to the level of a *fundamental cell*, consisting of approximately 10 to 100 nodes. Inside a fundamental cell, routing is done by Dynamic Source Routing (DSR).[13] Note that this is the opposite of Zone-based routing, where the local scope is handled by proactive routing, and distant nodes are served through reactive route requests. In Safari, the local scope is handled by DSR, and proactive routing is used for computing routes to more distant nodes, to avoid the high cost of long-range reactive route requests. If the size of the fundamental cell is kept constant, the size of the routing table in Safari is $O(\log N)$.

Routing based on multilevel clustering, just like single-level clustering and geographical routing, needs a mechanism through which a node can acquire the current location (HID, or Landmark address) of its intended destination. This has been addressed in a variety of ways, including assumptions of group mobility[21] (which makes the problem go away by assuming that nodes stay with their original clusters), flooding,[5] Mobile IP-style home agents,[23] and distributed location servers.[3,7,8,18,27,30] The distributed location server is the most versatile and scalable of these options. Here, the responsibility for storing the current location of a given node is distributed across the network. MMWN uses a combination of a hierarchical organization of location servers together with *paging*, essentially a restricted flood, to find the current location of a node. A different method is used by Chen and Morris,[3] Mohammed et al.,[18] and Susec and Marsic,[30] similar to the *anchor node* idea in GLS.[17] To find the anchor node of a node i, a function $hash(ID_i)$ is computed. For every level, the cluster with the identifier most similar to $hash(ID_i)$ is selected as the cluster to which the anchor node should belong, until eventually a level-0 cluster (a single node) has been reached. This is the *anchor node* that is responsible for storing the current location of node i. A similar $hash(ID_i)$ is computed, and the node with the *routing address* most similar to $hash(ID_i)$ is the anchor node for node i.[7,8] This is discussed in more detail in Section 2.9. In several of these location management schemes, multiple anchor nodes are selected, such that there are many anchor nodes close to the node and fewer anchor nodes far away from it. This improves the scalability of the distributed location server, as local changes and requests only have an effect on local network resources.

Multilevel clustering protocols that depend on hierarchical identifiers (including Landmark addresses) are highly sensitive to changes in the clustering hierarchy: whenever a cluster head gets disconnected or otherwise leaves the cluster, a new cluster head must be elected, and all the nodes within the affected cluster need to update their hierarchical identifiers. In addition, the election of a new cluster head will change the *anchor node* relationships, causing a necessity for a location-handoff mechanism. This has been identified[30] as the dominating component of the total routing overhead of multilevel hierarchical routing protocols. This takes the total number of control packet transmissions per node in routing protocols based on multilevel clustering to $\Omega(\log^2 N)$, where every packet is of length $\Omega(\log N)$ bits.

2.9 Dynamic Address Routing

Dynamic address routing is similar to routing based on multilevel clustering but the address size is reduced to $\Omega(\log N)$ from the $\Omega(\log^2 N)$ address size required with the previous multilevel clustering protocols. Dynamic address routing is also less sensitive to node movement than previous multilevel clustering approaches, because its routing addresses are not built up from individual node identifiers.

TABLE 2.1   Routing Table and Address Sizes for a 1024-Node Network Varying d.

d	Routing Table Size	Routing Address Size	Hierarchical ID Size
2	10	10	480
4	15	10	240
16	45	12	144
64	126	12	96
1024	1023	10	48

Note: Address and ID sizes in bits. Changing routing address size is due to rounding to the nearest $\log_2 d$ bit word.

Dynamic address routing, described by Eriksson et al.,[7,8] takes the idea of multilevel clustering one step further. Whereas previous multilevel clustering schemes use cluster identifiers to form addresses (HIDs), dynamic address routing dynamically assigns a considerably shorter *index* to each cluster. The *routing address* of a node is formed by concatenating the *indices* of the cluster that the node belongs to at every level. In more detail, with a clustering degree of d, each of the $1 \ldots d$ clusters belonging to the same higher-level cluster gets an index in the range $0 \ldots d - 1$. The more lengthy cluster identifiers are used only to ensure that there is a single, unique cluster per index.

As shown in Table 2.1, the difference in size between hierarchical identifiers and routing addresses is dramatic. With a low clustering degree d, the size of the hierarchical identifiers can be quite daunting. In contrast, the routing addresses used in dynamic address routing are roughly constant with respect to d.

Moreover, the size of the routing table in a multilevel clustering hierarchy is equal to

$$(d - 1) \log_d N$$

There are $\log_d$ levels and $d - 1$ routing entries per level. As shown in Table 2.1, selecting $d = 2$ minimizes the size of the routing table. Clearly, $d = 2$ is not feasible with previous multilevel clustering protocols, as the size of the hierarchical identifier is unacceptably large. Instead, these protocols are forced to use higher clustering degrees. Regardless of the choice of d, the address size in a regular multilevel clustering protocol is likely to exceed that of a dynamic address routing protocol by at least an order of magnitude, together with a marked increase in routing table size. In the remainder of this section, we assume $d = 2$ for dynamic address routing because this choice of d minimizes the size of the routing table.

Conveniently, with $d = 2$ the *index* can be represented using a single bit. Figure 2.6 shows an example address allocation for a six-node network. Because $\log_2 6 < 3$, 3 bits of address is sufficient for this small network. The most significant bit of the address selects the top-level cluster.

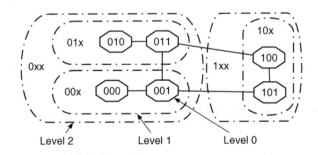

FIGURE 2.6   A network topology and three-level clustering. Nodes have 3-bit routing addresses, with each bit selecting one out of two possible clusters at a given level in the hierarchy.

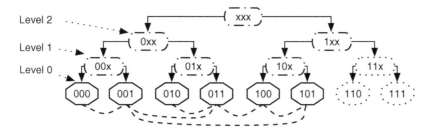

FIGURE 2.7 The address tree of a 3-bit binary address space. Leaves represent actual addresses, whereas inner nodes represent groups of addresses with a common prefix. Dashed lines show physical connectivity between nodes, corresponding to Figure 2.6.

Because $d = 2$, we can also think of the cluster hierarchy as a binary tree, as shown in Figure 2.7. The root of the tree represents the entire network. The leaves of the tree represent nodes and the internal nodes of the tree represent clusters. Each node (leaf) has one routing entry for every level of the tree. This routing entry indicates the path to the other subtree (cluster) at any given level. For example, the node with address [000] would have routing entries for subtrees [001], [01x], and [1xx]. If it wanted to route a packet to the node with address [100], it would look up the routing entry for the subtree [1xx]. After an additional routing step, the packet reaches the node with address [101]. This node has routing entries for subtrees [100], [11x], and [0xx], and is able to forward the packet to its final destination.

One definition of a cluster in the routing context is that the nodes in a cluster form a connected subgraph in the network topology. Because address prefixes uniquely identify clusters in dynamic address routing, nodes with a common address prefix need to have the same property, which is called the *prefix subgraph* constraint. Ensuring that this constraint is satisfied is the primary objective of dynamic address allocation. Next, we describe how this is handled in DART, the Dynamic Address Routing Protocol described by Eriksson et al.[8]

2.9.1 Address Allocation

Dynamic address allocation has many things in common with clustering in multilevel hierarchical networks. However, because dynamic address routing does not rely on concatenating unique identifiers to form its *routing address*, a major concern is to ensure the uniqueness of the addresses allocated.

When a node joins an existing network, it uses the periodic routing updates of its neighbors to identify and select an unoccupied and legitimate address. In more detail, every null entry in a neighbor's routing update indicates an empty subtree. This subtree represents a block of free and valid routing addresses. By definition, the prefix constraint is satisfied if the two subtrees under a given parent are connected, and any empty subtree in a neighbor's routing update by definition shares a parent with the neighbor's subtree at the same level.

Let us see an example of address allocation. Figure 2.8 illustrates the address allocation procedure for a 3-bit address space. Node A starts out alone with address [000]. When node B joins the network, it observes that A has a null routing entry corresponding to the subtree [1xx] and picks the address [100]. Similarly, when C joins the network by connecting to B, C picks the address [110]. Finally, when D joins via A, A's [1xx] routing entry is now occupied. However, the entry corresponding to sibling [01x] is still empty, and so D takes the address [010].

To handle cluster merging and splitting, each cluster, or subtree in this case, is loosely associated with the lowest of all the identifiers of the nodes that belong to that subtree. This is called the *subtree identifier*. With node mobility, subtree identifiers may need to be updated, but these updates are piggybacked on the periodic routing updates at little extra cost. When the node with the lowest identifier within any subtree leaves the subtree, the identifier of that subtree must be recomputed. However, this is generally a nondisruptive process because the route updates from the new lowest identifier node in the subtree will

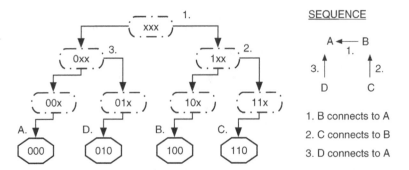

FIGURE 2.8   Address tree for a small network topology. The numbers 1, 2 and 3 show the order in which nodes were added to the network.

propagate and eventually reach all the concerned nodes without forcing any address changes in the process. Note that because of this, the routing address of a node does not depend directly on the identifiers of a set of cluster heads. Therefore, if the node with the lowest identifier gets disconnected, we can expect to see a smaller effect on the cluster hierarchy.

Due to node mobility, clusters will sometimes be partitioned into two or more parts. When this happens, the prefix subgraph constraint does not hold, and the clustering is thus invalid. The solution is to have one of the two partitions acquire new addresses as soon as the partitioning event is detected. The remaining problem is to detect such an event. As described above, subtree identifiers are assigned to be the minimum identifier in the cluster. If the cluster partitions, one of the two partitions will quickly compute a new identifier, as routing updates propagate through the cluster. However, a mere change of the cluster identifier is not enough to accurately diagnose a partitioning event. It could simply be that the node with the lowest identifier went out of range or ran out of battery power. Instead, all route advertisements are made to contain the identifier of the destination subtree. The idea is that in the event that a node receives two routing updates for the same subtree, but with different identifiers, only the update with the lower identifier prevails and gets forwarded further. In addition, when a node perceives a route to its own address subtree, but with a lower identifier, it must acquire a new address. This solution also solves the problem of network merging: if two networks merge, this event will be detected as one or more cluster partitionings, causing some or all of the nodes in one of the two networks to immediately acquire new, valid addresses.

2.9.2   Distributed Location Server

As in several other types of routing protocols, dynamic address routing protocols need a distributed location server. The main problem in designing any distributed location server is to find an effective method to select the *anchor node* of any given identifier. The solution proposed for dynamic address routing protocols is similar to that used in multilevel clustering protocols. However, the methods do not depend on any node identifier, except that of the destination node. A global and *a priori* known function $hash(ID_i)$, which takes a node identifier ID_i and returns a bit string with the same length as the *routing address*, is defined. Second, the $hash(ID_i)$ for the desired destination node i is calculated. If there exists a node that occupies this address, then that node is the *anchor node*. If there is no node with that address, then the node with the most similar address* is the anchor node.

---

*The metric used here for similarity between addresses can be described as the integer value of the XOR result of the two addresses.

For example, using Figure 2.7 for reference, consider a node with identifier ID_1 that has a current routing address of [010]. This node will periodically send an updated entry to the lookup table, namely $\langle ID_1, 010 \rangle$. To figure out where to send the entry, the node uses the hash function to calculate an address: $hash(ID_1)$. If the return value of the hash function is [100], the packet will simply be routed to the node with that address. However, if the returned bit string was instead [111], the packet could not be routed to the node with address [111] because there is no such node. In such a situation, the packet gets routed to the node with the most similar address to [111], which in this case would be [101].

To improve the scalability of the distributed location server, each lookup entry is stored in several locations, at increasing distances from the destination node. By starting with a small, local lookup and gradually going to further away locations, nodes can avoid sending lookup requests across long distances to find a node that is nearby. Similarly, when a node makes a small address change, it need only contact nearby location servers with the location update, as the records at distant location servers will still be sufficiently accurate to guide the packets to the correct neighborhood, where more recent information is readily available.

2.9.2.1 Coping with Temporary Route Failures

On occasion, due to link or node failure, a node will not have a completely accurate routing table. This could potentially lead to lookup packets, both updates and requests, terminating at the wrong node. The end result of this is that requests cannot be promptly served. In an effort to reduce the effect of such intermittent errors, a node can periodically check the lookup entries it stores to see if a route to a more suitable host has been found. If this should be the case, the entry is forwarded in the direction of this more suitable host.

Requests are handled in a similar manner: if the request cannot be responded to with an address, it is kept in a buffer awaiting either the arrival of the requested information, or the appearance of a route to a node that more closely matches the *hash* of the identifier the request was in regard to. This way, even if a request packet arrives at the anchor node before the update has the anchor, the request will be buffered and served as soon as the update information is available.

Dynamic address routing has size $O(\log N)$ routing tables and an $O(\log N)$ address size. The $\Omega(\log^2 N)$ result for location management handoff shown by Susec and Morsic has not yet been shown for dynamic address routing. However, there are considerable structural similarities between the distributed location server described in that article and the one described for dynamic address routing. Moreover, dynamic address routing has a decreased reliance on node identifiers for clustering and addressing. These observations lead us to conjecture that the lower bound on per-node channel utilization for control packets is, at most, $\Omega(\log^3 N)$.

2.10 Conclusion

This chapter discussed a variety of aspects of ad hoc routing scalability. We deliberated the various routing protocols that have been proposed over the past decade in an effort to understand how these scale with respect to various parameters. To achieve true scalability, it is the belief of the authors that multilevel clustering is the only viable option. While geographic routing is an attractive alternative for certain niche applications, multilevel clustering applies well to all scenarios, except for those with extremely high mobility.

From a scalability perspective, dynamic address routing represents the current state-of-the-art in scalable ad hoc routing. The use of dynamic address routing, a variation on multilevel clustering, results in addresses of length $\Omega(\log N)$. This is considerably shorter than the hierarchical identifiers used in previous multilevel clustering protocols, which are of size $\Omega(\log^2 N)$. Dynamic address routing achieves a similar average routing table size of $\Theta(\log N)$ and offers a reduced dependence on node identifiers for ensuring the stability of clustering and location management.

In summary, scalable ad hoc routing remains a focal point of interest in terms of making the deployment of large-scale ad hoc networks a reality.

References

1. A.D. Amis, R. Prakash, D. Huynh, and T. Vuong. Max-min d-cluster formation in wireless ad hoc networks. In *INFOCOM (1)*, pp. 32–41, 2000.
2. S. Basagni, I. Chlamtac, V.R. Syrotiuk, and B.A. Woodward. A distance routing effect algorithm for mobility (DREAM). In *ACM/IEEE MOBICOM*, 1998.
3. B. Chen and R. Morris. L+: scalable Landmark routing and address lookup for multi-hop wireless networks, Tech Report, Massachusetts Institute of Technology, 2002.
4. T. Chen and M. Gerla. Global state routing: a new routing scheme for ad-hoc wireless networks. In *Proc. of IEEE ICC*, 1998.
5. C. Chiang, H. Wu, W. Liu, and M. Gerla. Routing in clustered multihop, mobile wireless networks. In *The IEEE Singapore Int. Conf. on Networks*, 1997.
6. T. Clausen and P. Jaquet. Rfc 3626: Optimized link state routing, 2003.
7. J. Eriksson, M. Faloutsos, and S. Krishnamurthy. Peernet: pushing peer-2-peer down the stack. In *IPTPS:* International Workshop on Peer-to-Peer Systems, 2003.
8. J. Eriksson, M. Faloutsos, and S. Krishnamurthy. Scalable ad hoc routing: the case for dynamic addressing. In *IEEE INFOCOM*, 2004.
9. J.J. Garcia-Luna-Aceves, M. Mosko, and C.E. Perkins. A new approach to on-demand loop-free routing in ad hoc networks. In *Proc. 22nd Annu. Symp. on Principles of Distributed Computing*, pp. 53–62. ACM Press, 2003.
10. M. Grossglauser and D.N.C. Tse. Mobility increases the capacity of ad-hoc wireless networks. In *INFOCOM*, pp. 1360–1369, 2001.
11. P. Gupta and P. Kumar. Capacity of wireless networks, Tech Report, University of Illinois, Urbana-Champaign, 1999.
12. Z. Haas. A new routing protocol for the reconfigurable wireless networks, IEEE International Conference on Universal Personal Communications (ICUPC), 1997.
13. D.B. Johnson and D.A. Maltz. Dynamic source routing in ad hoc wireless networks. In *Mobile Computing*, Vol. 353. Kluwer Academic Publishers, 1996.
14. Brad Karp and H. T. Kung. GPSR: greedy perimeter stateless routing for wireless networks. In *Mobile Computing and Networking*, pages 243–254, 2000.
15. Y.-B. Ko and N.H. Vaidya. Location-aided routing (LAR) in mobile ad hoc networks. In *ACM/IEEE MOBICOM*, 1998.
16. Y.-B. Ko, V. Shankarkumar, and N.H. Vaidya. Medium access control protocols using directional antennas in ad hoc networks. In *INFOCOM (1)*, pp. 13–21, 2000.
17. J. Li, J. Jannotti, D. De Couto, D. Karger, and R. Morris. A scalable location service for geographic ad-hoc routing. In *Proc. 6th ACM Int. Conf. on Mobile Computing and Networking (MOBICOM '00)*, pp. 120–130, August 2000.
18. A.K. Mohammed, R.H. Johnson Reidi, David B. Johnson, P. Druschel, and R. Baraniuk. Analysis of safari: an architecture for scalable ad hoc networking and services. Technical Report TREE 0304, Rice University, 2004.
19. S. Murthy and J.J. Garcia-Luna-Aceves. An efficient routing protocol for wireless networks. *Mob. Netw. Appl.*, 1(2):183–197, 1996.
20. M.R. Pearlman and Z.J. Haas. Determining the optimal configuration for the zone routing protocol. *IEEE J. Selected Areas in Communication*, 17(8), August 1999.
21. G. Pei, M. Gerla, and X. Hong. LANMAR: landmark routing for large-scale wireless ad hoc networks with group mobility. In *ACM MobiHOC'00*, 2000.
22. G. Pei, M. Gerla, and T.-W. Chen. Fisheye state routing: a routing scheme for ad hoc wireless networks. In *ICC (1)*, pp. 70–74, 2000.
23. G. Pei, M. Gerla, X. Hong, and C.-C. Chiang. A wireless hierarchical routing protocol with group mobility. In *WCNC:* IEEE Wireless Communications and Networking Conference, 1999.

24. C. Perkins and P. Bhagwat. Highly dynamic destination-sequenced distance-vector routing (DSDV) for mobile computers. In *ACM SIGCOMM'94*, 1994.

25. C.E. Perkins and E.M. Royer. Ad-hoc on-demand distance vector routing. In *Proc. Second IEEE Workshop on Mobile Computer Systems and Applications*, pp. 90. IEEE Computer Society, 1999.

26. R. Ramanathan. On the performance of ad hoc networks with beamforming antennas. In *Proc. 2nd ACM Int. Symp. on Mobile Ad Hoc Networking and Computing*, pp. 95–105. ACM Press, 2001.

27. R. Ramanathan and M. Steenstrup. Hierarchically-organized, multihop mobile wireless networks for quality-of-service support. *Mobile Networks and Applications*, 3(1):101–119, 1998.

28. V. Ramasubramanian, Z.J. Haas, and E.G.ün SIRER. SHARP: a hybrid adaptive routing protocol for mobile ad hoc networks. In *Proc. 4th ACM Int. Symp. on Mobile Ad Hoc Networking and Computing*, pp. 303–314. ACM Press, 2003.

29. J. Sucec and I. Marsic. Clustering overhead for hierarchical routing in mobile ad hoc networks. In *IEEE INFOCOM 2002*, New York, June 23–27, 2002.

30. J. Sucec and I. Marsic. Hierarchical routing overhead in mobile ad hoc networks. *IEEE Trans. on Mobile Computing*, 3, Jan. 2004.

31. C.-K. Toh. Associativity-based routing for ad hoc mobile networks. *Wireless Personal Commun. J.*, 4(2):103–139, March 1997.

32. P.F. Tsuchiya. The Landmark hierarchy: a new hierarchy for routing in very large networks. In *SIGCOMM*. ACM, Press, 1988.

3

Uniformly Distributed Algorithm for Virtual Backbone Routing in Ad Hoc Wireless Networks

Dongsoo S. Kim

Ad hoc wireless networks consist of a group of mobile wireless devices. The transmission of a mobile host is received by all hosts within its transmission range due to the broadcast nature of wireless communication and omnidirectional antenna of the mobile hosts. If two wireless hosts are out of their transmission ranges, other mobile hosts residing between them can forward their messages and construct connected networks. Because of the mobility of wireless hosts, each host must be equipped with the capability of an autonomous system, or a router without any statically established infrastructure or centralized administration. In wireless networks, each host can move arbitrarily and can be turned on or off without notifying other hosts. The mobility and autonomy introduces a dynamic topology of the networks not only because end-hosts are transient, but also because intermediate hosts of a communication path are transient.

The ad hoc wireless networks can be modeled using a graph $G = (V, E)$, a vertex $v \in V$ represents a host, and an edge (u, v) indicates that two hosts u and v are within their transmission range. For simplicity, we assume that the communication is bidirectional in that a host u is reachable from a host v iff v is reachable from u, although two or more hosts within a transmission range may not be reachable to each other in the wireless communication due to the hidden-station problem.

The routing problem in ad hoc wireless networks is to find a set of hosts that perform message forwarding, while hosts not included in the set have direct links to at least one host in the set. To construct connected networks, the nodes in the set form a virtual backbone network. A critical issue of such networks is to design efficient routing schemes that can readily adapt to the dynamic topology of the networks. Numerous routing algorithms have been proposed[1,5,7–11,13] recently for ad hoc wireless networks. There are several good surveys on the ad hoc mobile wireless routing protocols.[4,11,12]

3.1   Characteristics of Ad Hoc Networks

Mobility is one of the major characteristics of most ad hoc wireless environments. There are several exceptions where mobility does not need to be considered. In sensor networks, for example, hosts equipped with communication and sensor devices are interconnected to collect data such as temperature or vibration and to report the data to a monitoring center. Most applications of sensor networks require each host to be deployed at a fixed location during its operation. In many other circumstances, however, wireless hosts can move freely in a deployment area, and the connectivity of the hosts is dynamically varying with time. When constructing a virtual backbone, it is necessary to take into account mobility. A global positioning system (GPS) can be used to handle the mobility problem,[15] but its usage must be limited as secondary information because the location information such as latitude, longitude, and height does not directly relate to the connectivity. For example, consider ad hoc networks in transportation trains. Hosts in this application move fast with respect to global position, but their relative locations are practically fixed during travel. In other perspectives, noise in wireless environments and interference by obstacles can make a disconnection between wireless hosts although they are within the normal transmission range. Figure 3.1 shows the relation between motion(M) and connection (C). The shaded area indicates the motion of mobile hosts related to the connection of hosts, in which a global location method can be used to approximate the connectivity of wireless hosts. However, the train example belongs to the category of $M \setminus C$ and the problems of noises and obstacles are in $C \setminus M$, in which GPS-based virtual backbone finding methods are not efficient.

In wireless networks, each host can move arbitrarily (*mobile host's movement*) and can be turned on (*mobile host's switch on*) or off (*mobile host's switch off*) without notifying other hosts.[17] The mobility and autonomy introduce a dynamic topology to the networks not only because end-hosts are transient and mobile, but also because intermediate hosts of a routing path are transient and have the same mobile characteristic. Even in a sensor network where each host tends to stay in a location, moving obstacles such as animals and vehicles introduce the dynamic topology of networks. Every host in ad hoc networks or sensor networks is autonomous, in that their functionalities are identical in the aspect of communication networks (space uniformity). Although it is possible to designate a host as a request initiator or a data collector in some applications, they are generally operating in a distributed environment, and a backbone construction method in ad hoc networks has more constraints than general grid computing. The goal of grid computing is to minimize the computation time to search a final solution utilizing a cluster of computing powers. The physical topology is not altered during its computation and a central node can orchestrate the remaining nodes as needed by an algorithm for finding the solution. In the meantime, a physical topology can be continuously transformed in ad hoc networks due to movement, switch-on, and switch-off. The method of finding a virtual backbone must be able to gradually adapt to the network transformation (time uniformity). This progressive computation plays a critical role in searching virtual backbones in networks. A network state can alter to another before an algorithm finds a backbone unless the transient period is ignorable, meaning the problem set of finding a virtual backbone is time-variant so that its solution must fit into the variation. As a fact of the time-variant, it is desired that the network supports some partial solutions during the transient period. A searching method preferably minimizes the impacts of the transient period. A non-optimal partial solution is acceptable but can provide the network connectivities among mobile hosts before obtaining a final solution.

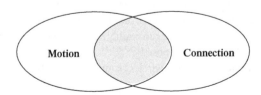

FIGURE 3.1   Set relation of motion and connection.

3.2 Searching Virtual Backbone

For a graph $G = (V, E)$, a subset of the vertex, $V_D \subseteq V$, is a dominating set if each vertex in $V - V_D$ has at least one neighbor in V_D. A virtual backbone network is constructed by connecting the dominating set. To construct the virtual backbone as simple and small as possible, we can use a *minimum connected dominating set* (MCDS). Because finding such an MCDS is an intractable problem,[6] we develop an algorithm for approximating the MCDS that is suitable for use in independent mobile hosts.

For simplicity, we assume that each host is assigned a globally unique identifier and maintains local information, including its dominator, state, a set of neighbors, etc. A host locally broadcasts hello messages in a fixed interval or when its local information is modified so that its neighbors need to update their neighbor data. The local broadcast can be achieved within $O(\Delta)$ under the assumption of a perfect underlying Medium Access Control protocol, where Δ is the maximum degree of the graph.[2] Each host determines its state by comparing its local information to the neighbor information collected through the hello message using a state machine. The machine consists of six states: four permanent states (dominator, essential dominator, dominatee, and absolute dominatee) and two transient states (candidate and dominatee candidate). Each host is initially in the candidate state and moves to one of the permanent states by considering neighbor information.

Many approximation algorithms[3,8,14,16,17] try to find a dominating node set (V_D) in a unit-disk graph $G = (V, E)$ and then add more nodes in V_D using a global leader or a state machine to make the nodes connected. As we see, the resulting CDS is a spanning tree; these approaches are to identify internal nodes, or backbone nodes, first and then to decide leaf nodes, dominated nodes. Our approach, however, uses an opposite direction that recognizes dominated nodes and searches the best dominator for each dominatee by comparing neighbor information. This approach came from a simple observation of geometric graphs, realizing that some nodes do not need to be dominators by directly examining their neighbors, called absolute dominatees. Figure 3.2 illustrates examples of the absolute dominatees. Node 1 has only one neighbor 2, meaning that the node is a leaf in a resulting spanning tree. In other words, edge $(1, 2)$ forms a complete subgraph K_2 and node 1 has no other edge than the subgraph. For the same reason, nodes 3, 4, 5, and 6 are forming a maximum complete subgraph K_4 and nodes 3, 4, and 5 have no additional edge other than K_4. With the maximum complete subgraph, we can designate nodes 3, 4, and 5 as absolute dominatees. Unfortunately, finding a maximum complete subgraph is impractical because it is identical to the NP-complete k-clique problem. For searching an absolute dominatee, however, we can relieve the constraint of the complete subgraph; that is, a node becomes an absolute dominatee if it has a neighbor whose neighbors cover all neighbors. Let N_a denote the neighbors of node a, including the node itself. Node a becomes an absolute dominatee if it has a neighbor b such as $N_a \subset N_b$. For example, node 8 becomes an absolute dominatee because the set of its neighbors $\{7, 8, 9\}$ is included by the neighbor set of node 10, or $\{2, 6, 7, 8, 9, 10\}$. We generated many random connected unit-disk graphs and found out that 40 percent of nodes were absolute dominatees, which validates our approach of quickly identifying dominatees with limited information.

After a node becomes an absolute dominatee, it searches its dominator among its neighbors. A node becomes an essential dominator if its absolute dominatee neighbor considers it as a dominator.

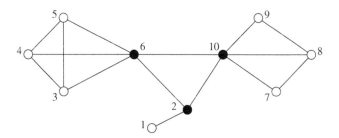

FIGURE 3.2 White nodes are absolute dominatees and black nodes are essential dominators.

For example, nodes 2, 6, and 10 are essential dominators because their neighbors are pointing them as dominators. The absolute dominatees and essential dominators construct trees that will merge together to build a single tree in a connected graph. There are some cases in which an absolute dominatee and an essential dominator cannot be found at all (for example, cyclic graphs or complete graphs), in which a node is a candidate, or the initial state, as all its neighbors are candidates. In this case, the group of nodes start a contention procedure to select a local leader. A simple contention can be achieved using unique identifiers.

Level information and join messages play a critical role in merging subtrees. Each node is initially in level 0, indicating that the node itself is a tree. A node in candidate state looks for its dominator among its neighbors in the dominator state. If the candidate node has a neighbor belonging to a different tree, we can say the node is on a border between two or more subtrees. Each tree has a precedence indicated by its root ID. The precedence of the border node is compared with the precedence of its nearby tree and the node locally broadcasts a join message if it has a higher precedence than the nearby tree. The join message contains the root ID of the nearby tree (old root ID) and the sender's root ID (new root ID). Upon receiving a join message whose old root ID is equal to its own root ID, a node joins to the new tree and searches the best dominator with the new root ID. If the node was not in one of dominatee states, it regenerates and broadcasts a new join message to its neighborhood. For examining the best dominator, the level, the number of children, and the state are taken into consideration.

Figure 3.3 illustrates the algorithm for constructing a virtual backbone. All hosts are initially candidates. After exchanging hello messages at the first round, nodes 0, 1, 6, and 8 identify themselves as absolute dominatees because node 5 (for node 0 and 1), node 2 (for node 6), and node 7 (for node 8) cover the set of neighbors of the corresponding dominatee nodes. Three subtrees are merged together by sending join messages in the next steps. In step 4, node 1 changes its dominator from node 5 to node 2 because nodes 2 and 5 are in the same tree and the level of node 2 in the tree is smaller than the level of node 5.

Figure 3.4 summarizes the state transition of the algorithm. Note that a node state does not move from dominators to dominatees directly, or vice versa. A hello message received by a node does not reflect the network topology at the time when the node calculates its state; rather, the message contains the topology at the previous time interval in the distributed computing environment. The anachronistic information makes it difficult to calculate the dominating set, especially when a node can move, or switch on or off. If the direct transition between dominators and dominatee is allowed, its neighbor can make a premature decision based on timely incorrect information. In the worst case, this miscalculation results in an endless state transition among a set of nodes. To avoid an unstable transition, the candidate states are employed so that a node informs the trends of its state transition to the neighbors.

As explained previously, the topological changes of ad hoc networks are described as three different types. We discuss the recalculation of a connected dominating set caused by topological changes. When a mobile host u switches on, it is in candidate state and all neighbors will not recognize its existence and their existing connectivity will not be affected by the switch-on of node u so that the only possible transition is to be in a dominatee state. The node stays in the dominatee state unless some of its neighbors consider node u as their dominator. When a mobile host is turned off without any notification, its neighbors will not receive a hello message from the node. Each entry in the table for maintaining neighbor information is associated with a timer. If a node does not receive a hello message from a neighbor before the timer expires, it deletes the entry of the neighbor from the table. Eliminating a dominatee node u will not affect the dominatee neighbors; u's dominator might transit to a dominatee if it has no children but node u. If the node turned off is a dominator, its children will look for other nodes as their dominators by deleting the entry for the node after its timer expires. A mobile host movement is viewed as a sequence of connections and disconnections. The event of disconnection can be considered a mobile host switch-off because the event of "out of transmission range" cannot be practically distinguished from the disappearance of the node if we have no geometrical information. The differences between a new connection and switch-on are when the counterpart of a new connection is not in a candidate state. However, when the counterpart v of a node u for a new connection is in a dominatee state, the connection will not affect the computation of u's state because u already has its dominator.

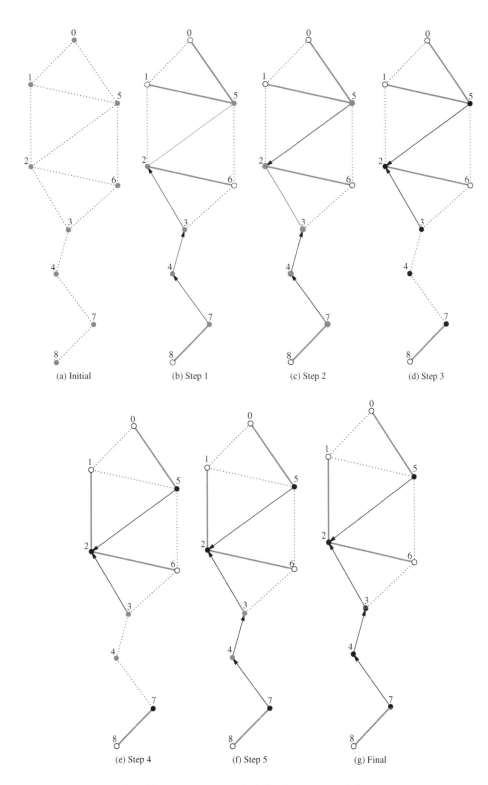

FIGURE 3.3 A sample graph and its state transition. Shaded nodes are in candidate states, hollow nodes are dominatees, and solid nodes are dominators. A dotted line indicates that the edge has no use for building a dominating set. A gray line is an edge between a dominatee and a dominator. A solid line is an edge between two dominators, and their dominating relation is indicated by an arrow.

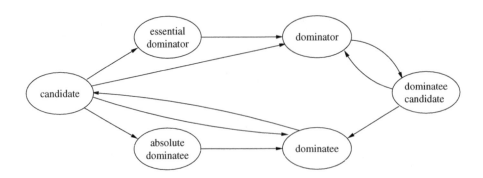

FIGURE 3.4   State transition diagram.

If v is in a dominator state, u will consider v as its dominator when v is beneficial over its previous dominator.

3.3   Conclusion

This chapter proposed a distributed algorithm for constructing a connected dominating set in wireless ad hoc networks. It is distinguished from other approximation algorithms in the aspect of using a bottom-up approach realizing absolute dominates. The proposed algorithm does not require any geometric information but uses only data that can be obtained from messages exchanged among neighbors. In addition, the algorithm is uniform in terms of space and time so that it is adequate for being deployed in ad hoc networks with no central contol host and no interruption.

References

1. K.M. Alzoubi, P.-J. Wan, and O. Fieder. New distributed algorithm for connected dominating set in wireless ad hoc networks. In *Proc. of HICSS*, 2002.
2. B. Awerbuch. Optimal distributed algorithm for minimum weight spanning tree, counting, leader election and related problems. In *Proc. 19th ACM Symp. on Theory of Computing*, pp. 230–240, 1987.
3. M. Cardei, X. Cheng, and D.Z. Du. Connected domination in multihop ad hoc wireless networks. In *Proc. 6th Int. Conf. on Computer Science and Informatics (CS&I 2002)*, North Carolina, March 2002.
4. Y. Chen and A. Liestman. Approximating minimum size weakly-connected dominating sets for clustering mobile ad hoc Networks. In *Proc. of the Symp. on Mobile Ad Hoc Networking and Computing*, 2002.
5. X. Cheng and D.Z. Du. Virtual backbone-based routing in multihop ad hoc wireless networks. In *preparation*, 2002.
6. B.N. Clark, C.J. Colbourn, and D.S. Johnson. Unit disk graph. *Discrete Mathematics*, 86:165–177, 1990.
7. M.S. Corson and A. Ephremides. A distributed routing algorithm for mobile wireless networks. *ACM J. Wireless Networks*, 1(1):61–81, 1995.
8. B. Das and V. Bharghaven. Routing in ad-hoc networks using minimum connected dominating sets. In *Proc. Int. Conf. on Communication (ICC97)*, pp. 376–380, 1997.
9. D.B.Johnson. Routing in ad hoc networks of mobile hosts. In *Proc. of Workshop on Mobile Computing Systems and Applications*, pp. 158–163, Dec. 1994.
10. S. Guha and S. Khuller. Approximation algorithms for connected dominating sets. *Algorithmica*, 20(4):374–387, 1998.
11. P. Krishna, M. Chatterjee, N.H. Vaidya, and D.K. Pradhan. A cluster-based approach for routing in ad-hoc networks. In *Proc. Second USENIX Symposium on Mobile and Location-Independent Computing*, pp. 1–10, 1995.

12. E.M. Royer and C.-K. Toh. A review of current routing protocols for ad-hoc mobile wireless networks. *IEEE Personal Communications,* pp. 46–55, 1999.

13. R. Sivakumar, B. Das, and V. Bharghavan. An improved spine-based infrastructure for routing in ad hoc networks. In *Proc. Int. Symp. on Computers and Communications (ISCC'98),* 1998.

14. I. Stojmenovic, M. Seddigh, and J. Zunic. Dominating sets and neighbor elimination-based broadcasting algorithm in wireless networks. In *Proc. IEEE Hawaii International, Conference on System Science,* January 2001.

15. I. Stojmenovic, M. Seddigh, and J. Zunic. Dominating sets and neighbor elimination-based broadcasting algorithms in wireless networks. *IEEE Trans. on Parallel and Distributed Systems,* 12(12), December 2001.

16. P.-J. Wan, K.M. Alzoubi, and O. Frieder. Distributed construction of connected dominating set in wireless ad hoc networks. In *Proc. on the Joint Conf. of the IEEE Computer and Communication Societies (INFOCOM),* pp. 1597–1604, March 2002.

17. J. Wu and H. Li. On calculating connected dominating set for efficient routing in ad hoc wireless networks. In *Proc. 3rd Int. Workshop on Discrete Algorithms and Methods for Mobile Computing and Communication,* pp. 7–14, Seattle, WA, 1999.

4

Maximum Necessary Hop Count for Packet Routing in MANETs

Xiao Chen

Jian Shen

This chapter investigates a fundamental characteristic of a mobile ad hoc network (MANET): the maximum necessary number of hops needed to deliver a packet from a source to a destination. In this chapter, without loss of generality, we assume that the area is a circle with a radius of r, $r > 1$, and the transmission range of each mobile station is 1. We prove that the maximum necessary number of hops needed to deliver a packet from a source to a destination is $\frac{4\pi}{\sqrt{3}}(r + \frac{1}{\sqrt{3}})^2 - 1 = \frac{4\pi r^2}{\sqrt{3}} + O(r) \approx 7.255 r^2 + O(r)$. We show that this result is very close to optimum with only a difference of $O(r)$.

4.1   Introduction

Recent advances in technology have provided portable computers with wireless interfaces that allow networked communication among mobile users. The resulting environment no longer requires users to maintain a fixed and universally known position in the network and enables almost unrestricted mobility.

A *mobile ad hoc network (MANET)* is formed by a cluster of mobile stations randomly located within a certain area without the infrastructure of base stations. The applications of MANETs appear in places where predeployment of network infrastructure is difficult or unavailable (e.g., fleets in oceans, armies in march, natural disasters, battlefield, festival field grounds, and historic sites).

In a MANET, stations communicate with each other by sending and receiving packets. The delivery of a packet from a source station to a destination station is called *routing*. Particularly in a MANET, two stations can communicate directly with each other if and only if they are within each other's *wireless transmission range*. Otherwise, the communication between them must rely on other stations. For example, in the

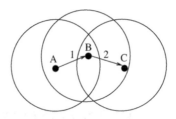

FIGURE 4.1   Example ad hoc wireless network.

network shown in Figure 4.1, stations A and B are within each other's transmission range (indicated by the circles around A and B, respectively). If A wants to send a packet m to B, A can send it directly in one hop. A and C are not within each other's transmission range. If A wants to send a packet to C, it must first forward the packet to B and then use B to route the packet to C. Therefore, it takes two hops to deliver a packet from A to C.

Many routing algorithms have been designed[1,3,5,6,8] but less work has been done to investigate the fundamental properties of MANETs in a mathematical way. This chapter addresses the issue of finding the maximum necessary number of hops needed to deliver a packet from a source to a destination in a MANET within a certain area. An initial attempt by us to solve the problem was made.[2] In this chapter we reconsider the problem and provide a much better maximum necessary hop count. Without loss of generality, we assume that all the mobile stations are located within an area of a circle with a radius r, $r > 1$, and the transmission range of all the nodes is 1, assuming they use the fixed transmission power.

This chapter is organized as follows. Section 4.2 provides the notations. Section 4.3 puts forward the problem. Section 4.4 introduces the circle packing problem, and Section 4.5 provides our solution to the problem. Section 4.6 shows the sharpness of our solution, and Section 4.7 is the conclusion.

4.2   Notations

We can use a simple graph $G = G(V, E)$ to represent a MANET, where the vertex set V is the collection of mobile stations within the wireless network. An edge between two stations u and v denoted by $u \leftrightarrow v$ means that both of them are within each other's transmission range. We assume that this graph G is finite and connected.

See an example of a wireless network in Figure 4.2. There are five stations A, B, C, D, and E in a MANET. The circle around each one represents its transmission range. Two vertices are connected if and only if they are within each other's transmission range. The resultant graph is shown in Figure 4.3.

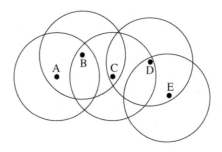

FIGURE 4.2   Example ad hoc wireless network.

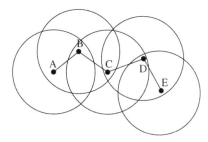

FIGURE 4.3 The graph representing an ad hoc wireless network.

4.3 The Problem

This chapter addresses the issue of finding the maximum necessary number of hops needed to deliver a packet from a source to a destination in a MANET within a certain area. Without loss of generality, we assume that all the mobile stations are within an area of a circle with a radius r, $r > 1$. Two stations u and v can communicate with each other if and only if their geographic distance is less than or equal to 1. Based on the above description, a simple graph G can be drawn to represent the MANET within this circle. The maximum necessary number of hops to deliver a packet from a source to a destination is actually the diameter of the graph G.

4.4 Circle Packing Problem

Before presenting our result, we introduce the circle packing problem that leads to our solution.

The *circle/sphere packing* problem is to consider how to effectively pack non-overlapping small circles/spheres of the same size into a large circle/sphere as many as possible so that the *density of a packing*, which is the ratio of the region/space occupied by the circles/spheres to the whole region/space, is as large as possible.

The history of the circle/sphere packing problem goes back to the early 1600s when astronomer-mathematician Johannes Kepler asserted that no sphere packing could be better than face-centered cubic (FFC) packing. FFC packing is the natural one that arises from packing spheres in a pyramid, as shown in Figure 4.4.

The Kepler Conjecture. *The density of any sphere packing in three-dimensional space is at most $\pi/\sqrt{18}$, which is the density of the FCC packing.*

Although the Kepler conjecture looks natural, it is very difficult to prove. Recently, a 250-page proof of the Kepler conjecture was claimed.[4]

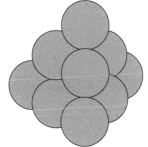

FIGURE 4.4 A pyramid in face-centered cubic sphere packing.

Compared with the difficulties of the Kepler conjecture, the problem of circle packing in the plane has been solved with ease. The result is stated in the following lemma. It was first proved by Axel Thue in 1890. To make this chapter self-contained, we include a proof provided by Surendran[7] with the following slight change: while the proof in Ref. 7 uses unit circles to pack the plane, we use circles of radius 1/2 to pack the plane in order to be consistent with the proof of Theorem 4.1 in Section 4.5.

Lemma 4.1 *The optimum packing of circles in the plane has density $\pi/\sqrt{12}$, which is that of the hexagonal packing shown in Figure 4.5.*

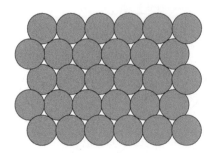

FIGURE 4.5   The hexagonal circle packing of the plane.

Proof   We can suppose without loss of generality that we use circles of radius $1/2$ to pack the plane. Start with an arbitrary packing of circles of radius $1/2$ in the plane. Around each circle draw a concentric circle of radius $1/\sqrt{3}$. (The reason for choosing $1/\sqrt{3}$ as the radius of the concentric circle is that when three $1/2$ radius circles are packed next to each other just as shown in the left part of Figure 4.6, the three centers can form an equilateral triangle, then the three cooresponding concentric circles can intersect at the middle of the equilateral triangle.) Where a pair of concentric circles overlap, join their points of intersection and use this as the base of two isosceles triangles. The centers of each circle form the third vertex of each triangle. In this way, the plane can be partitioned into three regions, as depicted in Figure 4.6.

1. The isosceles triangles: if the top angle of the triangle is θ radians, then its area is $(1/\sqrt{3})^2 \sin\theta/2 = \sin\theta/6$ and the area of the sector is $(1/2)^2\theta/2 = \theta/8$, so that the packing density is

$$\frac{\theta/8}{\sin\theta/6} = \frac{3\theta}{4\sin\theta},$$

   where θ ranges between 0 and $\pi/3$ radians. The maximum value of the density is $\pi/\sqrt{12}$, attained at $\theta = \pi/3$ radians.
2. Regions of the larger circles not in a triangle: here the regions are sectors of a pair of concentric circles of radius $1/2$ and $1/\sqrt{3}$, so the density is

$$\left(\frac{1/2}{1/\sqrt{3}}\right)^2 = \frac{3}{4} < \frac{\pi}{\sqrt{12}}$$

3. Regions not in any circle: here the density is zero.

Because the density of each region of space is at most $\pi/\sqrt{12}$, the density of this circle packing is at most $\pi/\sqrt{12}$. Furthermore, a packing can only be optimum when it causes space to be divided into equilateral triangles. This only happens for the hexagonal packing of Figure 4.5.   □

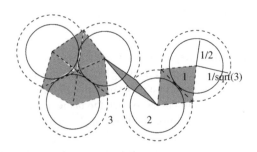

FIGURE 4.6   The plane partition induced by a sample circle packing.

Remarks   The above lemma concerns circle packing into an unlimited space. Now consider that some circles C_i of radius $1/2$ are packed into a limited region such as a large circle with a radius t. From now on, we use the notation $C(x)$ to represent a circle with a radius x. In the proof of Lemma 4.1, we draw a concentric circle $C(1/\sqrt{3})$ around each C_i. Because all circles C_i are within a circle $C(t)$, all those circles $C(1/\sqrt{3})$ must be within the circle $C(t + 1/\sqrt{3} - 1/2)$, which is concentric with $C(t)$. Thus, the proof of Lemma 4.1 can still be applied to the limited region $C(t + 1/\sqrt{3} - 1/2)$; that is,

$$\frac{\sum_i (\text{area of } C_i)}{\text{area of } C(t + 1/\sqrt{3} - 1/2)} \leq \frac{\pi}{\sqrt{12}}$$

Furthermore, Lemma 4.1 shows that $\pi/\sqrt{12}$ is the best upper bound for the ratio of the area of all C_i's to the area of $C(t + 1/\sqrt{3} - 1/2)$, in the case that $C(t)$ is sufficiently larger than each of the small circles C_i.

4.5   Our Solution

Based on the above lemma, the following is our result of the maximum necessary hop count to deliver a packet from a source to a destination in MANET.

Theorem 4.1   *Assume that all the mobile stations are within an area of a circle with a radius r, $r > 1$. The transmission range of each mobile station is 1. Denote the graph generated by connecting all pairs of vertices within each other's transmission range as G; that is, two vertices are connected if and only if their geographic distance is less than or equal to 1. Then an upper bound for the diameter of G is $\frac{4\pi}{\sqrt{3}}(r + \frac{1}{\sqrt{3}})^2 - 1$; in other words, it takes maximum $\frac{4\pi}{\sqrt{3}}(r + \frac{1}{\sqrt{3}})^2 - 1 = \frac{4\pi r^2}{\sqrt{3}} + O(r)$ necessary hops to deliver a packet from a source to a destination.*

Proof   Let D be the diameter of the graph G. Choose two vertices u, v such that the distance in G between u and v, $d_G(u, v)$, is D; that is, $d_G(u, v) = D$. Then there exist distinct vertices u_i, $1 \leq i \leq D+1$, such that

$$u = u_1 \leftrightarrow u_2 \leftrightarrow u_3 \leftrightarrow \cdots \leftrightarrow u_D \leftrightarrow u_{D+1} = v$$

is a shortest path of length D between u and v (see Figure 4.7).

We define a set $I = \{u_i : i \text{ is odd}\}$. Then the size of I, denoted by $|I|$, is $\lceil (D+1)/2 \rceil$. We can prove that I is an *independent set* of vertices. An independent set of vertices is defined as a set of vertices in which there is no edge between any pair of vertices in the set. $\qquad \square$

Claim 4.1   *I is an independent set of vertices in graph G.*

Proof of Claim 4.1   Suppose that I is not an independent set; that is, there exists at least one edge between some pair of vertices in I. Without loss of generality, we assume that there are two vertices u_{2j+1}, u_{2k+1} in I such that $j < k$ and $u_{2j+1} \leftrightarrow u_{2k+1}$. By the definition of I, the two vertices are on the shortest path from u to v. Then the shortest path can be represented as

$$u = u_1 \leftrightarrow \cdots \leftrightarrow u_{2j} \leftrightarrow u_{2j+1} \leftrightarrow u_{2k+1}$$

$$\leftrightarrow u_{2k+2} \leftrightarrow \cdots \leftrightarrow u_{D+1} = v$$

FIGURE 4.7   A shortest path between u and v.

The length of this path, as represented, is $D - [(2k + 1) - (2j + 1)] + 1 = D + 2j - 2k + 1$. It is less than D. This contradicts to $d_G(u, v) = D$. Therefore, there is no edge between any pair of vertices in I; in other words, I is an independent set in G.

By Claim 4.1 and the definition of G, we have the following claim.

Claim 4.2   The geographic distance between any pair of vertices in I is larger than 1.

Then, for each vertex $u_i \in I$, we define a circle S_i with a center at u_i and with a radius of $1/2$. By Claim 4.2, two circles S_j, S_k are disjoint if $j \neq k$. Because all the vertices in I are covered by the circle $C(r)$, all the disjoint circles S_i (i is odd) can be covered by a larger circle named $C(r + 1/2)$, which has the same center as the circle $C(r)$ and has a radius of $r + 1/2$ (see Figure 4.8 for an example). Now we can relate this diameter problem to the circle packing problem, that is, how to effectively pack these non-overlapping circles S_i (i is odd) as many as possible into the larger circle $C(r + 1/2)$. By the remarks following the proof of Lemma 4.1 in Section 4.4 (note that the circle $C(t + 1/\sqrt{3} - 1/2)$ in the remark should be converted to $C(r + 1/\sqrt{3})$ since $t = r + 1/2$.), we have

$$\frac{\sum_i (\text{Area of } S_i)}{\text{Area of } C(r + 1/\sqrt{3})} \leq \frac{\pi}{\sqrt{12}}$$

where $C(r + 1/\sqrt{3})$ is the circle that has the same center as the circle $C(r)$ and has a radius of $r + 1/\sqrt{3}$. Thus,

$$\frac{|I|\pi \left(\frac{1}{2}\right)^2}{\pi \left(r + \frac{1}{\sqrt{3}}\right)^2} \leq \frac{\pi}{\sqrt{12}}$$

Solving for $|I|$,

$$|I| \leq \frac{2\pi}{\sqrt{3}} \left(r + \frac{1}{\sqrt{3}}\right)^2$$

Since $|I| = \lceil (D + 1)/2 \rceil \geq (D + 1)/2$, we have

$$D \leq 2|I| - 1 \leq \frac{4\pi}{\sqrt{3}} \left(r + \frac{1}{\sqrt{3}}\right)^2 - 1$$

So, it takes maximum $\frac{4\pi}{\sqrt{3}}(r + \frac{1}{\sqrt{3}})^2 - 1 = \frac{4\pi r^2}{\sqrt{3}} + O(r)$ necessary hops to deliver a packet from a source to a destination.

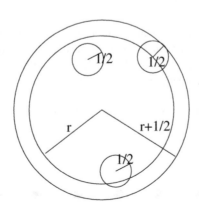

FIGURE 4.8   All the S_i's are covered by a circle of radius $r + 1/2$.

4.6 Sharpness of the Maximum Necessary Hop Count

In this section, we show the sharpness of the maximum necessary hop count. The idea is like this: if we can actually construct a graph G with a diameter of $4\pi r^2/\sqrt{3} + O(r)$ in a circle $C(r)$, then our maximum necessary hop count $\frac{4\pi}{\sqrt{3}}(r + \frac{1}{\sqrt{3}})^2 - 1$ is very close to optimum, with only a difference of $O(r)$. The construction of such a G is based on a parallelogram packing into the circle $C(r)$. Before the construction, we present some assumptions and properties of the packing unit "parallelogram."

Let ϵ $(0 < \epsilon < 1)$ be a positive real number. We draw a parallelogram $ABDE$. Then we choose points C and G on the side of BD, and choose a point F on the side of AE such that $|AC| = |AG| = 1 + \epsilon$, $|AF| = 1$, $|CG| = 1 - \epsilon$, and $|BG| = |CD| = |EF| = \epsilon$, where $|\ |$ represents the length of a line segment. (See Figure 4.9.) Then $|AB|$ and $|DE|$ can be uniquely determined in terms of ϵ because $ABDE$ is a parallelogram. Next we give some properties of the parallelogram.

Property 4.0 $|AF| = |BC| = 1$, $|EF| = |CD| = \epsilon$, and $|AE| = |BD| = 1 + \epsilon$.

This property is obvious from the above assumptions.

Property 4.1 Any two vertices chosen from sets $\{A, E, F\}$ and $\{B, C, D\}$, respectively, are at least $1 + \epsilon$ distance apart.

Proof of Property 4.1 First, since $|AC| = |AG|$ and $|CD| = |GB|$, elementary geometry shows that the triangles ACD and AGB are identical. Thus, $|AD| = |AB|$, angle $\angle ADC$ is an acute angle, and angle $\angle ACD$ is an obtuse angle. Then, $\angle ACD > \angle ADC$ implies $|AD| > |AC|$, so

$$|AB| = |AD| > |AC| = 1 + \epsilon$$

Second, because $|AF| = |GD| = 1$, we know that $AGDF$ is a parallelogram. This implies $|FD| = |AG|$ and $\angle FDB = \angle AGB = \angle ACD > \pi/2$ radians. Thus,

$$|FB| > |FC| > |FD| = |AG| = 1 + \epsilon$$

Third, because $\angle EDB > \angle FDB > \pi/2$ radians, we have

$$|EB| > |EC| > |ED| = |AB| > |AC| = 1 + \epsilon.$$

Therefore, any two vertices chosen from sets $\{A, E, F\}$ and $\{B, C, D\}$, respectively, are at least $1 + \epsilon$ distance apart.

Property 4.2 The area of the parallelogram $ABDE$ is $\frac{1+\epsilon}{2}\sqrt{3 + 10\epsilon + 3\epsilon^2}$.

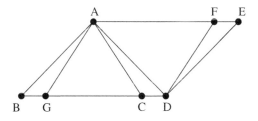

FIGURE 4.9 A parallelogram ABDE.

Proof of Property 4.2   By the Pythagoras theorem, the height of the parallelogram $ABDE$ is $\sqrt{|AC|^2 - (|CG|/2)^2}$. Thus, the area of the parallelogram is

$$|BD| \cdot \sqrt{|AC|^2 - \left(\frac{|CG|}{2}\right)^2} = (1+\epsilon) \cdot \sqrt{(1+\epsilon)^2 - \left(\frac{1-\epsilon}{2}\right)^2}$$

$$= \frac{1+\epsilon}{2}\sqrt{3 + 10\epsilon + 3\epsilon^2}$$

Property 4.3 $|BE| = \sqrt{3 + 7\epsilon + 3\epsilon^2}$.

Proof of Property 4.3   Draw a line segment EH perpendicular to the line BD, as shown in Figure 4.10. Then $|EH|$ is the height of the parallelogram, and so

$$|EH| = \sqrt{|AC|^2 - \left(\frac{|CG|}{2}\right)^2} = \frac{1}{2}\sqrt{3 + 10\epsilon + 3\epsilon^2}$$

Applying the Pythagoras theorem to the right triangle BHE,

$$|BE| = \sqrt{|EH|^2 + |BH|^2} = \sqrt{|EH|^2 + \left(|AE| + \frac{|BD|}{2}\right)^2}$$

$$= \sqrt{\frac{3 + 10\epsilon + 3\epsilon^2}{4} + \left(\frac{3(1+\epsilon)}{2}\right)^2} = \sqrt{3 + 7\epsilon + 3\epsilon^2}$$

Now let P denote the parallelogram $ABDE$ with six points A, B, C, D, E, F on its boundary. This is our packing unit parallelogram. We use parallel packing to pack as many non-overlapping P's as possible into the circle $C(r)$, as shown in Figure 4.11. Let

$$l = \sqrt{3 + 7\epsilon + 3\epsilon^2}$$

By Property 4.3, any two points in P are at most l distance apart. Then the circle $C(r-l)$ concentric with $C(r)$ is entirely covered by P's because otherwise we could have packed one more P without reaching the boundary of $C(r)$. Thus, by Property 4.2, at least n parallelograms P can be packed into $C(r)$, where

$$n \geq \frac{\text{Area of } C(r-l)}{\text{Area of } P} = \frac{\pi(r-l)^2}{(1+\epsilon)/2\sqrt{3 + 10\epsilon + 3\epsilon^2}} = \frac{2\pi(r - \sqrt{3 + 7\epsilon + 3\epsilon^2})^2}{(1+\epsilon)\sqrt{3 + 10\epsilon + 3\epsilon^2}}$$

Let V be the set of all points A, B, C, D, E, F in P's; that is, $V = \cup_P \{A, B, C, D, E, F\}$ and let $|V|$ denote the number of points in V. We have the following property:

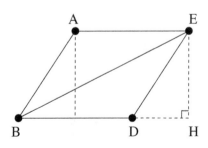

FIGURE 4.10   Computation of $|BE|$.

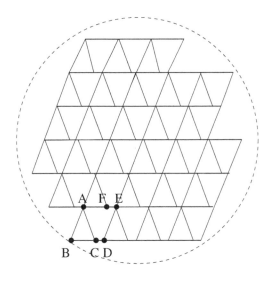

FIGURE 4.11 Parallel packing of parallelograms into a circle.

Property 4.4 $|V| \geq 2n$.

Proof of Property 4.4   We can assume that the parallelograms are packed one by one into $C(r)$, from the lower layer to the higher layer, and in each layer from the left to the right. Because each time adding a parallelogram P increases $|V|$ by at least 2, Property 4.4 holds by induction.

Having the above assumptions and properties of parallelogram, now we construct a graph G within $C(r)$ as follows: let V be the vertex set of G and two vertices in V are connected if and only if they are at most 1 distance apart. By Property 0, vertices in each layer are connected to form a path; and by Property 4.1, any two vertices in different layers are not connected. So G is a union of disjoint paths. We can add some new vertices to G and let them be the intermediate vertices connecting paths of two adjacent layers, as shown in Figure 4.12. Because at least 1 new vertex is added to G, the final graph is a path with at least

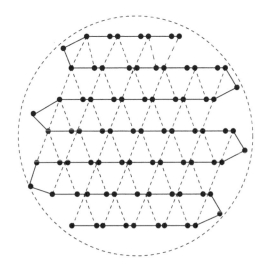

FIGURE 4.12 The construction of a path.

$|V| + 1$ vertices. So it has a diameter of at least

$$|V| \geq 2n \geq \frac{4\pi (r - \sqrt{3 + 7\epsilon + 3\epsilon^2})^2}{(1 + \epsilon)\sqrt{3 + 10\epsilon + 3\epsilon^2}}$$

Because the above argument holds for all the real numbers ϵ with $0 < \epsilon < 1$, we can make ϵ approaching 0. Because

$$\lim_{\epsilon \to 0} \frac{4\pi (r - \sqrt{3 + 7\epsilon + 3\epsilon^2})^2}{(1 + \epsilon)\sqrt{3 + 10\epsilon + 3\epsilon^2}} = \frac{4\pi}{\sqrt{3}}(r - \sqrt{3})^2 = \frac{4\pi r^2}{\sqrt{3}} + O(r)$$

a graph with a diameter of $4\pi r^2/\sqrt{3} + O(r)$ always exists (by choosing ϵ very close to 0). Therefore, our upper bound for the diameter (maximum necessary hop count) in Theorem 4.1,

$$\frac{4\pi}{\sqrt{3}} \left(r + \frac{1}{\sqrt{3}} \right)^2 - 1 = \frac{4\pi r^2}{\sqrt{3}} + O(r)$$

is very close to optimum with only a difference of $O(r)$.

4.7   Conclusion

The maximum necessary number of hops needed to deliver a packet from a source to a destination has been found for a MANET within a circle of radius r, $r > 1$, assuming the transmission range of each mobile station is 1. Our proofs show that our result is very close to optimum, with only a difference of $O(r)$.

References

1. V. Bharghavan, A. Demers, S. Shenker, and L. Zhang, MACAW: A Medium Access Protocol for Wireless LANs, *Proc. of SIGCOMM'94*, 1994.
2. X. Chen, J. Shen, and X.D. Jia, An Upper Bound for a Mobile Ad-Hoc Wireless Network, *Proc. of the Int. Conf. on Parallel and Distributed Processing Techniques and Applications*, June 2001, pp. 1617–1620.
3. C.L. Fullmer and J.J. Garcia-Luna-Aceves, Floor Acquisition Multiple Access (FAMA) for Packet-Radio Networks, *Proc. of SIGCOMM'95*, 1995.
4. Web page of Prof. T. Hales, http://www.math.pitt.edu/~thales/
5. C.R. Lin and M. Gerla, MACA/PR: An Asynchronous Multimedia Multihop Wireless Network, *Proc. of INFOCOM'97*, 1997.
6. C.R. Lin and M. Gerla, Real-Time Support in Multihop Wireless Network, *ACM/Baltzer Wireless Networks*, 5(2), 1999.
7. D. Surendran, The Conquest of the Kepler Conjecture, *Math Horizons*, 8, 8–12, (2001).
8. J. Wu and H. Li, A dominating-Set-Based Routing Scheme in Ad Hoc Wireless Networks, special issue on Wireless Networks in the *Telecommunication Systems Journal*, 3, 63–84, 2001.

5

Efficient Strategy-Proof Multicast in Selfish Wireless Networks

Xiang-Yang Li

Weizhao Wang

In this chapter, we study how to perform routing when each wireless node is selfish; that is, a wireless node will always maximize its own benefit. Traditionally, it is assumed by the majority of the routing protocols for wireless networks that each wireless node will forward the packets for other nodes if it is asked to do so. However, this assumption may not be true in practice, especially when the wireless devices are owned by individual users. A node will deviate from a routing protocol if it will gain more benefit by doing so. In this chapter, we assume that each wireless node will incur a cost when it forwards a unit of data for some other nodes. A node will forward the data only if it gets a payment to compensate its cost. Its profit (or called utility) will then be the payment minus its cost if it did forward the data. For a multicast with a source node and a set of receiver nodes, we assume that they will pay the relay nodes to carry the traffic from the source to receivers. We assume that the cost of each agent is private and each agent can manipulate its reported cost to maximize its utility. A payment scheme is *strategyproof* if every agent maximizes its utility when it reports its cost truthfully. In this chapter, we propose several strategyproof mechanisms for multicast in selfish wireless networks when each node has a cost of forwarding a unit data based on various structures. We prove that each of our payment schemes is optimum for the corresponding structure used.

5.1   Introduction

Recent years saw a great amount of research into wireless ad hoc networks on various important problems such as routing, quality of service, security, power management, and traffic and mobility modeling. However, there are still many challenges left. In wireless ad hoc networks, each host contributes its local resources to forward the data for other nodes to serve the common good, and may benefit from resources contributed by other hosts in return. Based on such a fundamental design philosophy, wireless networks provide appealing features of enhanced system robustness, high service availability, and scalability. However, the critical observation that users are generally selfish and noncooperative may severely undermine the expected wireless structure. For example, for a routing algorithm based on the least cost path (LCP), the individual wireless node may declare an arbitrarily high cost for forwarding a data packet to other nodes because wireless nodes are energy-constrainted and it is often not in the interest of a node to always relay the messages for other nodes. The root cause of the problem is, obviously, that there exist no incentives for users to be altruistic. Following the common belief in neoclassic economics, it is more reasonable to assume that all wireless terminals are *rational*: they try to maximize their benefits instead of conforming to the existing protocols. Thus, we need to design some mechanisms to ensure that these *rational* wireless terminals will conform to our protocols without any deviation.

How to achieve cooperation among wireless terminals in network was previously addressed.[2–4,11,13,16,17] The key idea behind these approaches is that terminals providing a service should be remunerated, while terminals receiving a service should be charged. Both of these methods belong to the so-called *credit-based method*. Some of these algorithms need some special hardware that is not very practical in the real world. In recent years, *incentive-based methods* have been proposed to solve the noncooperative problem. The most well-known and widely used incentive-based methods is the so-called family of VCG mechanisms by Vickrey,[18] Clarke,[5] and Groves.[9] Nisan and Ronen[14] provided a mechanism belonging to the VCG family to assure the cooperation for the unicast problem in a general network where each communication link is assumed to be selfish and rational.

While unicast in wireless network has been studied extensively in the literature and deployed in practice for years, several important issues about multicast over wireless networks have not been explored fully. In practice, multicast is a more efficient way to support group communication than unicast or broadcast, as it can transmit packets to destinations using fewer network resources, which is critical in wireless networks. Typical wireless multicast applications include group-oriented mobile commerce, military command and control, distance education, and intelligent transportation systems. For a multicast routing, usually a tree with the minimum cost that spans the sources and receivers is used because it requires less network resources than other structures. Finding such a minimum cost tree is known to be NP-hard. Thus, some multicast trees with good practical performances have been proposed in the literature. Unlike unicast problem, as we will show later, if we simply apply a VCG mechanism to those commonly used multicast tree structures, we cannot guarantee that all wireless devices will follow our prescribed protocols. In this chapter, we discuss how to design truthful non-VCG mechanisms for those multicast structures in *selfish* wireless networks.

The rest of the chapter is organized as follows. In Section 5.2, we review some definitions and *priori* art on truthful mechanism design for multicast. In Section 5.3, we present the first strategyproof mechanism for the Steiner tree problem (or multicast). The output of our mechanism (a tree) has a cost within a constant factor of the optimum, and the payment is minimum among any truthful mechanism having this output. In Section 5.4, we show our experimental study of our proposed mechanisms. We conclude the chapter in Section 5.5 by pointing out some possible future work.

5.2   Preliminaries and Priori Art

5.2.1   Wireless Ad Hoc Networks

Wireless *ad hoc* networks are emerging as a flexible and powerful wireless architecture that does not rely on a fixed networking infrastructure. Wireless ad hoc networks have received significant attention over the past few years due to their potential applications in various situations such as battlefield, emergency relief and environmental monitoring, etc. In a wireless ad hoc network, each mobile node has a transmission range and energy cost. A node v can receive the signal from another node u iff node v is within node u's transmission range. We assume that when node u received a message and then forwarded the message to another node, it would consume node u some energy, which will be categorized as the cost of node u forwarding the data for other nodes. If the receiving node is not within the sender's transmission range, then it must choose some intermediate nodes to relay the message. So unlike wired networks, all nodes in the wireless ad hoc network should be able to act as a router. On the other hand, the wireless node usually uses omnidirectional antennas, which means that it can use a broadcasting-like manner to distribute the message to all nodes within its transmission range.

Usually, there are two different categories of wireless ad hoc nodes: *fixed* transmission range and *adjustable* transmission range. For *fixed* transmission range nodes, their transmission range has been fixed and cannot be adjusted afterward. So there is a directed arc from u to v if node v is within the transmission range of node u. Here the transmission cost depends on node u, regardless of the distance between two nodes. Thus, the wireless ad hoc network can be considered a *node weighted graph*, where the weight of each node is its cost to forward a unit data. If all nodes' transmission range is the same, by properly scaling, we can assume all nodes have transmission range 1. Thus, wireless topology can be modeled by a *unit disk graph* (*UDG*).

The second type of wireless network is that each wireless node can adjust its transmission range: they can adjust their transmission power to the amount needed to reach the next relay node. The power needed to send a packet from node u to v consists of three parts. First, the source node u needs to consume some power to prepare the packet. Second, node u needs to consume some power to send the message to v. The power required to support the transmission between u and v not only depends on u, but also depends on the geometry distance of u and v. In the literature, it is often assumed that the power needed to support a link uv is $d_u \cdot |uv|^\beta$, where $2 \le \beta \le 5$ depends on the transmission environment, $|uv|$ is the Euclidean distance between u and v, and d_u is a positive number depending on node u only. Finally, when v receives the packet, it needs to consume some power to receive, store, and then process that packet. Thus, the weight of an edge uv is the power consumed for transmitting packet from u to v plus some possible energy consumed by u and v to process the signal. The wireless network under this model can be considered a *link weighted graph*: all wireless devices are the vertices of the graph, and the weight of each link uv is the total energy cost of communication using link uv.

5.2.2   Algorithm Mechanism Design

In designing efficient, centralized (with input from individual agents) or distributed algorithms and network protocols, the computational agents are typically assumed to be either *correct/obedient* or *faulty* (also called adversarial). Here, agents are said to be *correct/obedient* if they follow the protocol correctly; agents are said to be *faulty* if (1) they stop working, or (2) they drop messages, or (3) they act arbitrarily, which is also called *Byzantine failure* (i.e., they may deviate from the protocol in arbitrary ways that harm other users, even if the deviant behavior does not bring them any obvious tangible benefits).

In contrast, as mentioned before, economists design market mechanisms in which it is assumed that agents are *rational*. The rational agents respond to well-defined incentives and will deviate from the protocol only if it improves its gain. A rational agent is neither correct/obedient nor adversarial.

A standard economic model for analyzing scenarios in which the agents act according to their own self-interests is as follows. There are n agents. Each agent i, for $i \in \{1, \cdots, n\}$, has some private information t^i, called its *type*. The type t^i could be its cost to forward a packet in a network environment, or it could be its willing payment for a good in an auction environment. Then the set of n agents define a type vector $t = (t^1, t^2, \cdots, t^n)$, which is called the *profile*. There is an output specification that maps each type vector t to a set of allowed outputs. Agent i's preferences over the possible outputs are given by a valuation function v^i that assigns a real number $v^i(t^i, o)$ to each possible output o. Here, notice that the valuation of an agent does not depend on other agents' types. Everything in the scenario is public knowledge except the type t^i, which is a private information to agent i.

Definition 5.1 A Mechanism $M = (A, \mathcal{O}, p)$ defines three functions: a set of strategies A for all agents, an output function $\mathcal{O}$, and a payment function $p = (p^1, \ldots, p^n)$:

1. *For each agent i, it has a set of strategies A^i. Agent i can only choose a strategy $a \in A^i$.*
2. *For each strategy vector $a = (a^1, \cdots, a^n)$, i.e., the agent i plays a strategy $a^i \in A^i$, the mechanism computes an output $o = \mathcal{O}(a^1, \cdots, a^n)$ and a payment $p^i = p^i(a)$. Here, the payment p^i is the money given to each participating agent i. If $p^i < 0$, it means that the agent has to pay $-p^i$ to participate in the action.*

For an agent i, given the output o and the payment p^i, its utility is $u^i(o, t^i) = v^i(t^i, o) + p^i$. If a strategy a^i by an agent i is *dominant*, the agent i maximizes its utility *regardless* of whatever other agents do. Considering all different strategies, there will be too many candidate mechanisms; but with the Revelation Principle, we only need to focus our attention on these *direct revelation mechanisms*. A mechanism is a *direct revelation mechanism* if the types are the strategy space A^i. In this chapter, we only consider the direct revelation mechanisms. In practice, a mechanism should satisfy the following properties:

1. **Incentive Compatibility (IC).** The payment function should satisfy the *incentive compatibility*; that is, for each agent i,

$$v^i(t^i, o(a^{-i}, t^i)) + p^i(a^{-i}, t^i) \geq v^i(t^i, o(a^{-i}, a^i)) + p^i(a^{-i}, a^i).$$

In other words, revealing the type t^i is the *dominating strategy*. If the payment were computed by a strategyproof mechanism, he would have no incentive to lie about its type because his overall utility would be not greater than it would have been if he had told the truth.

2. **Individual Rationality (IR).** It is also called *Voluntary Participation*. For each agent i and any a^{-i}, it should have non-negative utilities. That is, if agent i reveals its true type t^i, then its utility should be non-negative.

3. **Polynomial Time Computability (PC).** All computation is done in polynomial time. Notice that after every agent declares its type, the mechanism must compute an output o and a payment vector to all agents. For example, for the family of VCG mechanisms, the output that maximizes the summation of the valuations of all nodes must be found. When the optimal output cannot be found exactly, the individual agent may have incentives to misreport its type initially.

A mechanism is *strategyproof* or *truthful* if it satisfies both IR and IC properties. In the remainder of this chapter, we focus attention on these *truthful* mechanisms only.

Arguably the most important positive result in mechanism design is what is usually called the generalized Vickrey-Clarke-Groves (VCG) mechanism by Vickrey,[18] Clarke,[5] and Groves.[9] The VCG mechanism applies to mechanism design maximization problems where the objective function is simply the sum of all agents' valuations and the set of possible outputs is assumed to be finite.

A maximization mechanism design problem is called *utilitarian* if its objective function satisfies that $g(o,t) = \sum_i v^i(t^i, o)$. A direct revelation mechanism $m = (o(t), p(t))$ belongs to the VCG family if (1) the output $o(t)$ computed based on the type vector t maximizes the objective function $g(o,t) = \sum_i v^i(t^i, o)$, and (2) the payment to agent i is

$$p^i(t) = \sum_{j \neq i} v^j(t^j, o(t)) + h^i(t^{-i}).$$

Here, $h^i()$ is an arbitrary function of t^{-i} and a different agent could have different function $h^i()$ as long as it is defined on t^{-i}. It is proved by Groves[9] that a VCG mechanism is truthful. Green and Laffont[8] proved that, under mild assumptions, VCG mechanisms are the only truthful implementations for utilitarian problems.

An output function of a VCG mechanism is required to maximize the objective function. This makes the mechanism computationally intractable in many cases. Notice that replacing the optimal algorithm with a non-optimal approximation usually leads to untruthful mechanisms. In this chapter, we study how to perform truthful routing for multicast, which is known to be NP-hard and thus VCG mechanisms cannot be applied.

5.2.3   Priori Arts

There are generally two ways to implement the truthful computing: (1) credit-based method and (2) incentive-based method.

The first category uses various non-monetary approaches, including auditing, systemwide optimal point analysis, and some hardwares. Credit-based methods have been studied for several years, and most of them are based on simulation and are heuristic.

Nodes that agree to relay traffic but do not are termed "misbehaving." Marti et al.[13] used *Watchdog* and *Pathrater* to identify misbehaving users and avoid routing through these nodes. *Watchdog* runs on every node, keeping track of how the other nodes behave; *Pathrater* uses this information to calculate the route with the highest reliability. Notice that this method ignores the reason why a node refused to relay the transit traffics for other nodes. A node will be wrongfully labeled as misbehaving when its battery power cannot support many relay requests and thus refuses to relay. It also does not provide any incentives to encourage nodes to relay the message for other nodes.

Buttyan and Hubaux[3] focused on the problem of how to stimulate selfish nodes to forward the packets for other nodes. Their approach is based on a so-called *nuglet counter* in each node. A node's counter is decreased when sending its own packet, and is increased when forwarding other nodes' packets. All counters should always remain positive. In order to protect the proposed mechanism against misuse, they presented a scheme based on a trusted and tamper-resistant hardware module in each node that generates cryptographically protected security headers for packets and maintains the nuglet counters of the nodes. They also studied the behavior of the proposed mechanism analytically and by means of simulations, and showed that it indeed stimulates the nodes for packet forwarding.

They still use a nuglet counter to store the nuglets and they use a fine that decreases the nuglet counter to prevent the node from not relaying the packet. They use the *packet purse model* to discourage the user from sending useless traffic and overloading the network. The basic idea presented in Ref. 4 is similar to Ref. 3 but different in the implementation.

Srinivasan et al.[16] proposed two acceptance algorithms. These algorithms are used by the network nodes to decide whether to relay traffic on a per-session basis. The goal is to balance* the energy consumed by a node in relaying traffics for others with energy consumed by other nodes to relay its traffic and to find an optimal trade-off between energy consumption and session blocking probability. By taking decisions

---

*It is impossible to strictly balance the number of packets a node has relayed for other nodes and the number of packets of this node relayed by other nodes because, in a wireless ad hoc network, the majority of packet transmissions are relayed packets. For example, consider a path of h hops. $h - 1$ nodes on the path relay the packets for others. If the average path length of all routes is h, then $1 - 1/h$ fractions of the transmissions are transit traffic.

on a per session basis, the per packet processing overhead of previous schemes is eliminated. In Ref. 17, a distributed and scalable acceptance algorithm called GTFT is proposed. They proved that GTFT results in Nash equilibrium and the system converges to the rational and optimal operating point. Notice that they assumed that each path is h hops long and the h relay nodes are chosen with equal probability from the remaining $n - 1$ nodes, which may be unrealistic.

Salem et al.[15] presented a charging and rewarding scheme for packet forwarding in multihop cellular networks. In their network model, there is a base station to forward the packets. They use symmetric cryptography to cope with the lying. To count several possible attacks, it precharges some nodes and then refunds them only if a proper acknowledgment is received. Their basic payment scheme is still based on nuglets.

Jakobsson et al.[11] described an architecture for fostering collaboration between selfish nodes of multihop cellular networks. Based on this architecture, they provided mechanisms based on per-packet charge to encourage honest behavior and to discourage dishonest behavior. In their approach, all packet originators attach a payment token to each packet, and all intermediaries on the packet's path to the base station verify whether this token corresponds to a special token called *winning ticket*. Winning tickets are reported to nearby base stations at regular intervals. The base stations, therefore, receive both reward claims (which are forwarded to some accounting center) and packets with payment tokens. After verifying the validity of the payment tokens, base stations send the packets to their desired destinations, over the backbone network. The base stations also send the payment tokens to an accounting center. Their method also involves some traditional security methods, including auditing, node abuse detection and encryption, etc.

The *incentive-based methods* borrow some ideas from the micro-economic and game-theoretic world, which involve monetary transfer. The key result of this category is that all nodes will not deviate from their normal activities because they will benefit most when they reveal their true cost, even knowing all other nodes' true costs. We can thus achieve the optimal system performance. This idea has been introduced by Nisan and Ronen[14] and is known as the algorithm mechanism design.

Nisan and Ronen[14] provided a polynomial-time strategyproof mechanism for optimal unicast route selection in a centralized computational model. In their formulation, the network is modeled as an abstract graph $G = (V, E)$. Each edge e of the graph is an agent and has a private type t_e, which represents the cost of sending a message along this edge. The mechanism-design goal is to find a Least Cost Path (LCP) $LCP(x, y)$ between two designated nodes x and y. The valuation of an agent e is $-t_e$ if the edge e is part of the path $LCP(x, y)$ and 0 otherwise. Nisan and Ronen used the VCG mechanism for payment. The payment to an agent e is $D_{G-\{e\}}(x, y) - D_G(x, y)$, where $D_{G-\{e\}}(x, y)$ is the cost of the LCP through G when edge e is not presented and $D_G(x, y)$ is the cost of the least cost path $LCP(x, y)$ through G. Clearly, there must be two link disjoint paths connecting x and y to prevent the monopoly. The result can be easily extended to deal with wireless unicast problems for an arbitrary pair of terminals.

Feigenbaum et al.[6] then addressed truthful low-cost routing in a different network model. They assume that each node k incurs a transit cost c_k for each transit packet it carries. For any two nodes i and j of the network, $T_{i,j}$ is the intensity of the traffic (number of packets) originating from i and destined for node j. Their strategyproof mechanism again is essentially the VCG mechanism. They gave a distributed method such that each node i can compute a payment $p_{ij}^k > 0$ to node k for carrying the transit traffic from node i to node j if node k is on the LCP $LCP(i, j)$. Anderegg and Eidenbenz[1] recently proposed a similar routing protocol for wireless ad hoc networks based on the VCG mechanism again. They assumed that each link has a cost and each node is a selfish agent.

For multicast flow, Feigenbaum et al.[7] assumed that there is *fixed* multicast infrastructure, given any set of receivers $Q \subset V$, that connects the source node to the receivers. Additionally, for each user $q_i \in Q$, they assumed a *fixed* path from the source to it, determined by the multicast routing infrastructure. Then for every subset R of receivers, the delivery tree $T(R)$ is merely the union of the fixed paths from the source to the receivers R. They also assumed that there is a link cost associated with each communication link in the network and the link cost is *known* to everyone. For each receiver q_i, there is a valuation w_i that this user values the reception of the data from the source. This information w_i is only known to q_i. User q_i will report a number w_i', which is the amount of money he is willing to pay to receive the data. The source node then selects a subset $R \subset Q$ of receivers to maximize the difference $\sum_{i \in R} w_i' - C(R)$, where $C(R)$ is the cost of the multicast tree $T(R)$ to send data to all nodes in R. The approach of fixing the multicast tree is relatively simple to implement but could not model the greedy nature of all network terminals in the network.

5.2.4 Problem Statement and Network Model

In this chapter, we consider a wireless ad hoc network composed of n selfish nodes $V = \{v_1, v_2, \cdots, v_n\}$. Every node v_i has a fixed transmission range r_i, and nodes u and v can communicate with each other if and only if $|v_i v_j| \leq \min\{r_i, r_j\}$. When node v_i sends a packet to one of its neighbors, say v_j, all v_i's neighbors can receive this packet. Thus, every node will broadcast its packet. We assume each node v_i has a private cost c_i to broadcast a unit data (the unit data could be 1 byte or 1 Megabyte). In this chapter, we model this wireless ad hoc network as a node weighted graph $G = (V, E, c)$, where V is the set of wireless nodes and $e = v_i v_j \in E$ if and only if v_i and v_j can communicate with each other directly. Here, $c = \{c_1, c_2, \cdots, c_n\}$ is the cost profile of all nodes. Notice here that the graph is undirected.

Based on the node weighted graph, we now define the multicast problem as follows. Given a set of receivers $Q = \{q_0, q_1, q_2, \cdots, q_{r-1}\} \subset V$, when selecting node $q_i \in Q$ as the source, the multicast problem is to find a tree $T \subset G$ spanning all receiving terminals Q. For simplicity, we assume that q_0 is the source of the multicast. Each node v_i is required to declare a cost d_i of relaying the message. Based on the declared cost profile $d = \{d_1, d_2, \cdots, d_n\}$, the source node constructs the multicast tree and decides the payment for each node. It is well known[10,12] that it is NP-hard to find the minimum-cost multicast tree when given an arbitrary node weighted graph G, and it is at least as hard to approximate as the set cover problem. Klein and Ravi[12] showed that it can be approximated within $O(\ln r)$, where r is the number of receivers. The utility of an agent is its payment received, minus its cost if it is selected in the multicast tree. Instead of reinventing the wheel, we will still use the previously proposed structures for multicast as the output of our mechanism. Given a multicast tree, we will study the design of strategyproof payment schemes based on these trees.

Given a graph G, we use $\omega(G)$ to denote the total cost of all nodes in this network. If we change the cost of any agent i (link e_i or node v_i) to c_i', we denote the new network as $G' = (V, E, c|^i c_i')$, or simply $c|^i c_i'$. If we remove one agent i from the network, we denote it as $c|^i \infty$. Denote $G \backslash e_i$ as the network without link e_i, and denote $G \backslash v_i$ as the network without node v_i and all its incident links. For simplicity of notation, we use the cost vector c to denote the network $G = (V, E, c)$ if no confusion is caused.

5.3 Strategyproof Multicast

In this section, we discuss in detail how to conduct truthful multicast when the network is modeled by a node weighed communication graph. We specifically study the following three structures: (1) least cost path star (LCPS), (2) virtual minimum spanning tree (VMST), (3) and node weighted Steiner tree (NST). In practice, for various applications, receivers and senders in the same multicast group usually belong to the same organization or company, so their behavior can be expected to be cooperative instead of uncooperative. Thus, we assume that every receiver will relay the packet for other receivers for free.

5.3.1 Strategyproof Mechanism Based on LCPS

5.3.1.1 Least Cost Path Star (LCPS)

Given a network modeled by the graph G, a source node s, and a set of r receivers Q, the least cost path star is the union of all r shortest paths from the source to each of the receivers in Q. In practice, this is one of the most widely used methods of constructing the multicast tree because it takes advantage of the unicast routing information collected by the distance-vector algorithm or link-state algorithm. Notice that, although here we only discuss the use of least cost path star for the node weighted case, all results presented in this subsection can be extended to the link weighted scenario without any difficulty, when each link will incur a cost when transmitting data.

5.3.1.2   Constructing LCPS

For each receiver $q_i \neq s$, we compute the shortest path (least cost path), denoted by $LCP(s, q_i, d)$, from the source s to q_i under the reported cost profile d. The union of all least cost paths from the source to receivers is called *least cost path star*, denoted by $LCPS(d)$. Clearly, we can construct LCPS in time $O(n \log n + m)$. The remaining issue is how to design a truthful payment scheme while using LCPS as output.

5.3.1.3   VCG Mechanism on LCPS Is Not Strategyproof

Intuitively, we would like to use the VCG payment scheme in conjunction with the LCPS tree structure as follows. The payment $p_k(d)$ to every node v_k is

$$p_k(d) = \omega(LCPS(d \mid^k \infty)) - \omega(LCPS(d)) + d_k.$$

We show by example that the above payment scheme is not strategyproof. Figure 5.1 illustrates such an example where node V_2 will have a negative utility when it reveals its true cost.

Notice that $\omega(LCPS(c)) = 2M + \epsilon$ and $\omega(LCPS(c \mid^1 \infty)) = M + \epsilon$. If v_1 reveals its true cost, its payment is $p_1(c) = \omega(LCPS(c \mid^1 \infty)) - \omega(LCPS(c)) + M = M + \epsilon - (2M + \epsilon) + M = 0$. Thus, its utility is $p^1(c) - C_1 = 0 - M < 0$, which violates the IR property.

5.3.1.4   Strategyproof Mechanism on LCPS

Now we describe our strategyproof mechanism that does not rely on VCG payment. For each receiver $q_i \neq s$, we compute the least cost path from the source s to q_i, and compute a payment $p_k^i(d)$ to every node v_k on the $LCP(s, q_i, d)$ using the scheme for unicast

$$p_k^i(d) = d_k + |LCP(s, q_i, d \mid^k \infty)| - |LCP(s, q_i, d)|.$$

Here, $|LCP(s, q_i, d)|$ denotes the total cost of the least cost path $LCP(s, q_i, d)$. The payment $p_k^i(d) = 0$ if node v_k is not on $LCP(s, q_i, d)$. The total payment to a link v_k is then

$$p_k(d) = \max_{q_i \in Q} p_k^i(d). \tag{5.1}$$

Theorem 5.1   *Payment (5.1) based on LCPS is truthful and it is minimum among all truthful payments based on LCPS.*

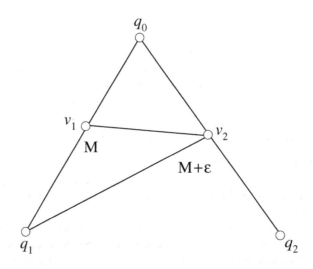

FIGURE 5.1   The cost of terminals are $v_1 = M$ and $v_2 = M + \epsilon$.

Proof   Clearly, when node v_k reports its cost truthfully, it has non-negative utility; that is, the payment scheme satisfies the IR property. In addition, because the payment scheme for unicast is truthful, v_k cannot lie its cost to increase its payment $p_k^i(c)$ based on $\text{LCP}(s, q_i, d)$. Thus, it cannot increase $\max_{q_i \in Q} p_k^i(d)$ by lying its cost. In other words, our payment scheme is truthful.

We then show that the above payment scheme pays the minimum among all strategyproof mechanisms using LCPS as the output. Before showing the optimality of our payment scheme, we give some definitions first. Consider all paths from source node s to a receiver q_i; they can be divided into two categories: with node v_k or without node v_k. The path with the minimum length among these paths with edge v_k is denoted as $\text{LCP}_{v_k}(s, q_i, d)$; and the path with the minimum length among these paths without edge v_k is denoted $\text{LCP}_{-v_k}(s, q_i, d)$.

Assume that there is another payment scheme $\bar{p}$ that pays less for a node v_k in a network G under a cost profile d. Let $\delta = p_k(d) - \bar{p}_k(d)$; then $\delta > 0$. Without loss of generality, assume that $p_k(d) = p_k^i(d)$. Thus, node v_k is on $\text{LCP}(s, q_i, d)$ and the definition of $p_k^i(d)$ implies that

$$|\text{LCP}_{-v_k}(s, q_i, d)| - |\text{LCP}(s, q_i, d)| = p_k(d) - d_k.$$

Then consider another cost profile $d' = d \mid^k (p_k(d) - \frac{\delta}{2})$, where the true cost of node v_k is $p_k(d) - \frac{\delta}{2}$. Under profile d', since $|\text{LCP}_{-v_k}(s, q_i, d')| = |\text{LCP}_{-e_k}(s, q_i, d)|$, we have

$$|\text{LCP}_{v_k}(s, q_i, d')| = |\text{LCP}_{v_k}(s, q_i, d \mid^k 0)| + p_k(d) - \frac{\delta}{2}$$
$$= |\text{LCP}_{v_k}(s, q_i, d)| + p_k(d) - \frac{\delta}{2} - d_k$$
$$= |\text{LCP}(s, q_i, d)| + p_k(d) - \frac{\delta}{2} - d_k$$
$$= |\text{LCP}_{-v_k}(s, q_i, d)| - \frac{\delta}{2}$$
$$< |\text{LCP}_{-v_k}(s, q_i, d)| = |\text{LCP}_{-v_k}(s, q_i, d')|.$$

Thus, $v_k \in LCPS(d')$. From the following Lemma 5.1, we know that the payment to node v_k is the same for cost profile d and d'. Thus, the utility of node v_k under the profile d' by the payment scheme $\bar{p}$ becomes $\bar{p}_k(d') - c_k = \bar{p}_k(d) - c_k = \bar{p}_k(d) - (p_k(d) - \frac{\delta}{2}) = -\frac{\delta}{2} < 0$. In other words, under the profile d', when the node e_k reports its true cost, it gets a negative utility under payment scheme $\bar{p}$. Thus, $\bar{p}$ is not strategyproof. This finishes our proof.   □

Lemma 5.1   *If a mechanism with output T and the payment function $\bar{p}$ is truthful, then for every node v_k in network, if $v_k \in T$, then payment function $\bar{p}_k(d)$ should be independent of d_k.*

Proof   We prove it by contradiction. Suppose that there exists a truthful payment scheme such that $\bar{p}_k(d)$ depends on d_k. There must exist two valid declared costs x_1 and x_2 for node v_k such that $x_1 \neq x_2$ and $\bar{p}_k(d \mid^k x_1) \neq \bar{p}_k(d \mid^k x_2)$. Without loss of generality we assume that $\bar{p}_k(d \mid^k x_1) > \bar{p}_k(d \mid^k x_2)$. Now consider the situation when node v_k has an actual cost $c_k = x_2$. Obviously, node v_k can lie its cost as x_1 to increase his utility, which violates the incentive compatibility (IC) property. This finishes the proof.   □

Notice that the payment based on $p_k(d) = \min_{q_i \in Q} p_k^i(d)$ is not truthful because a node can lie its cost upward so it can discard some low payment from some receiver. In addition, the payment $p_k(d) = \sum_{q_i \in Q} p_k^i(d)$ is *not* truthful either.

5.3.2   Strategyproof Mechanism Based on VMST

5.3.2.1   Constructing VMST

We first describe our method to construct the virtual minimum spanning tree.

Algorithm 5.1

Virtual MST Algorithm.

1. First calculate the pairwise least cost path $\text{LCP}(q_i, q_j, d)$ between any two terminals $q_i, q_j \in Q$ when the cost vector is d.
2. Construct a virtual complete link weighted network $K(d)$ using Q (including the source node here) as its terminals, where the link $q_i q_j$ corresponds to the least cost path $\text{LCP}(q_i, q_j, d)$, and its weight $w(q_i q_j)$ is the cost of the path $\text{LCP}(q_i, q_j, d)$, i.e., $w(q_i q_j) = |\text{LCP}(q_i, q_j, d)|$.
3. Build the minimum spanning tree (MST) on $K(d)$. The resulting MST is denoted as $VMST(d)$.
4. For every virtual link $q_i q_j$ in $VMST(d)$, we find the corresponding least cost path $\text{LCP}(q_i, q_j, d)$ in the original network. Combining all these paths can generate a subgraph of G, say $VMSTO(d)$.
5. All nodes on $VMSTO(d)$ will relay the packets.

Notice that a terminal v_k is on $VMSTO(d)$ iff v_k is on some virtual links in the $VMST(d)$, so we can focus our attention on these terminals in $VMST(d)$. It is not very difficult to show that the cost of VMST could be very large compared to the optimal. But when all nodes have the same transmission ranges in the original wireless ad hoc network, which can be modeled as UDG, the following theorem shows that the virtual minimum spanning tree can approximate the cost of the optimal tree within a constant factor.

Theorem 5.2 *$VMST(G)$ is a 5-approximation of the optimal solution in terms of the total cost if the wireless ad hoc network is modeled by a unit disk graph.*

Proof Assume that the optimal solution is a tree called T_{opt}. Let $V(T_{opt})$ be the set of nodes used in the tree T_{opt}. Clearly, $\omega(T_{opt}) = \sum_{v_i \in V(T_{opt})} c_i$. Similarly, for any spanning tree T of $K(G, Q)$, we define $\omega(T) = \sum_{e \in T} w(e)$. Following we will prove $5 \cdot \omega(T_{opt}) \geq \omega(VMST(G))$.

First, for all nodes in T_{opt}, when disregarding the node weight, there is a spanning tree T'_{opt} on $V(T_{opt})$ with node degree at most 5 because the wireless network is modeled by a unit disk graph. This is due to the well-known fact that there is a Euclidean minimum spanning tree with the maximum node degree at most 5 for any set of two-dimensional points. Note here that we only need to know the existence of T'_{opt}; we do not need to construct such a spanning tree explicitly. Obviously, $\omega(T_{opt}) = \omega(T'_{opt})$. Thus, tree T'_{opt} is also an optimal solution with maximal node degree of at most 5.

For spanning tree T'_{opt}, we root it at an arbitrary node and duplicate every link in T'_{opt} (the resulting structure is called DT'_{opt}). Clearly, every node in DT'_{opt} has even degree now. Thus, we can find a Euler circuit, denoted by $EC(DT'_{opt})$, that visits every vertex of DT'_{opt} and uses every edge of DT'_{opt} exactly once, which is equivalent to saying that every edge in $T'_{opt}(G)$ is used exactly twice. Consequently, we know that every node v_k in $V(T_{opt})$ is used exactly $deg_{T'_{opt}}(v_k)$ times. Here, $deg_G(v)$ denotes the degree of a node v in a graph G. Thus, the total weight of the Euler circuit is at most 5 times the weight $\omega(T'_{opt})$; that is,

$$\omega(EC(DT'_{opt})) \leq 5 \cdot \omega(T'_{opt}).$$

Notice that here if a node v_k appears multiple times in $EC(DT'_{opt})$, its weight is also counted multiple times in $\omega(EC(DT'_{opt}))$.

If we walk along $EC(DT'_{opt})$, we visit all receivers, and the length of any subpath between receivers q_i and q_j is no smaller than $|\text{LCP}(q_i, q_j, G)|$. Thus, the cost of $EC(DT'_{opt})$ is at least $\omega(VMST(G))$ because $VMST(G)$ is the minimum spanning tree spanning all receivers and the cost of the edge $q_i q_j$ in $VMST(G)$ corresponds to the path with the least cost $|\text{LCP}(q_i, q_j, G)|$. In other words,

$$\omega(EC(DT'_{opt})) \geq \omega(VMST(G)).$$

Consequently, we have

$$\omega(VMST(G)) \leq \omega(EC(DT'_{opt})) \leq 5 \cdot \omega(T'_{opt}).$$

This finishes the proof.   □

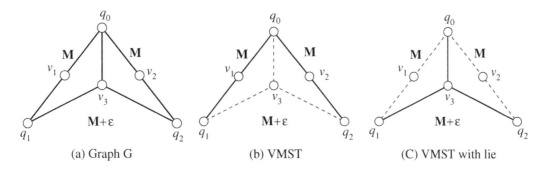

FIGURE 5.2   The cost of terminals are $c_4 = c_5 = M$ and $c_3 = M + \epsilon$.

5.3.2.2   VCG Mechanism on VMST Is Not Strategyproof

In this subsection, we show that a simple application of VCG mechanism on VMST is not strategyproof. Figure 5.2 illustrates such an example where terminal v_3 can lie its cost to improve its utility when the output is VMST. The payment to terminal v_3 is 0 and its utility is also 0 if it reports its cost truthfully. The total payment to terminal v_3 when v_3 reported a cost $d_3 = M - \epsilon$ is $\omega(VMST(c|^3\infty)) - \omega(VMST(c|^3d_3)) + d_3 = 2M - (M - \epsilon) + M - \epsilon = 2M$ and the utility of terminal v_3 becomes $u_3(c|^3d_3) = 2M - (M + \epsilon) = M - \epsilon$, which is larger than $u_3(c) = 0$. Thus, the VCG mechanism based on VMST is not strategyproof.

5.3.2.3   Strategyproof Mechanism on VMST

Before discussing the strategyproof mechanism based on VMST, we give some related definitions first. Given a spanning tree T and a pair of terminals p and q on T, clearly there is a unique path connecting them on T. We denote such path as $\Pi_T(p,q)$, and the edge with the maximum length on this path as $LE(p,q,T)$. For simplicity, we use $LE(p,q,d)$ to denote $LE(p,q,VMST(d))$ and use $LE(p,q \mid^k d'_k)$ to denote $LE(p,q,VMST(d \mid^k d'_k))$.

Following is our truthful payment scheme when the output is the multicast tree $VMST(d)$.

Algorithm 5.2

Truthful payment scheme based on VMST.

1. For every terminal $v_k \in V \setminus Q$ in G, first calculate $VMST(d)$ and $VMST(d \mid^k \infty)$ according to the terminals' declared costs vector d.
2. For any edge $e = q_i q_j \in VMST(d)$ and any terminal $v_k \in \text{LCP}(q_i, q_j, d)$, we define the payment to terminal v_k based on the virtual link $q_i q_j$ as follows:

$$p_k^{ij}(d) = |LE(q_i, q_j, d \mid^k \infty)| - |\text{LCP}(q_i, q_j, d)| + d_k.$$

Otherwise, $p_{ij}^k(d)$ is 0. The final payment to terminal v_k based on $VMST(d)$ is

$$p_k(d) = \max_{q_i q_j \in VMST(d)} p_k^{ij}(d). \tag{5.2}$$

Theorem 5.3   *Our payment scheme (5.2) is strategyproof and minimum among all truthful payment schemes based on VMST structure.*

Instead of proving Theorem 5.3, we prove Theorem 5.4, Theorem 5.5, and Theorem 5.6 in the remainder of this subsection.

Before the proof of Theorem 5.3, we give some related notations and observations. Considering the graph $K(d)$ and a node partition $\{Q_i, Q_j\}$ of Q, if an edge's two end-nodes belong to different node sets of the partition, we call it a *bridge*. All bridges $q_s q_t$ over node partition Q_i, Q_j in the graph $K(d)$ satisfying $v_k \notin LCP(q_s, q_t, d)$ form a bridge set $B^{-v_k}(Q_i, Q_j, d)$. Among them, the bridge with the minimum length is denoted as $MB^{-v_k}(Q_i, Q_j, d)$ when the nodes' declared cost vector is d. Similarly, all bridges $q_s q_t$ over node partition Q_i, Q_j in the graph $K(d)$ satisfying $v_k \in LCP(q_s, q_t, d)$ form a bridge set $B^{v_k}(Q_i, Q_j, d)$. The bridge in $B^{v_k}(Q_i, Q_j, d)$ with the minimum length is denoted as $BM^{v_k}(Q_i, Q_j, d)$. Obviously, we have

$$BM(Q_i, Q_j, d) = \min\{BM^{v_k}(Q_i, Q_j, d), BM^{-v_k}(Q_i, Q_j, d)\}.$$

We then state our main theorems for the payment scheme discussed above.

Theorem 5.4   *Our payment scheme satisfies IR.*

Proof   First of all, if terminal v_k is not chosen as a relay terminal, then its payment $p_k(d \mid^k c_k)$ is clearly 0 and its valuation is also 0. Thus, its utility $u_k(d \mid^k c_k)$ is 0.

When terminal v_k is chosen as a relay terminal when it reveals its true cost c_k, we have $|LE(q_i, q_j, d \mid^k \infty)| \geq |LCP(q_i, q_j, d \mid^k c_k)|$. This is due to the following observation: for any cycle C in a graph G, assume e_c is the longest edge in the cycle; then $e_c \notin MST(G)$. The lemma immediately follows from

$$p_{ij}^k(d \mid^k c_k) = |LE(q_i, q_j, d \mid^k \infty)| - |LCP(q_i, q_j, d \mid^k c_k)| + c_k > c_k.$$

This finishes the proof.   □

From the definition of the incentive compatibility (IC), we assume that the d_{-k} is fixed throughout the proof. For our convenience, we will use $G(d_k)$ to represent the graph $G(d \mid^k d_k)$. We first prove a series of lemmas that will be used to prove that our payment scheme satisfies IC.

Lemma 5.2   *If $v_k \in q_i q_j \in VMST(d)$, then $p_k^{ij}(d)$ does not depend on d_k.*

Proof   Remember that the payment based on a link $q_i q_j$ is $p_k^{ij}(d) = |LE(q_i, q_j, d \mid^k \infty)| - |LCP(q_i, q_j, d)| + d_k$. The first part $LE(q_i, q_j, d \mid^k \infty)$ is the longest edge of the unique path from q_i to q_j on tree $VMST(d \mid^k \infty)$. Clearly, it is independent of d_k. Now consider the second part $LCP(q_i, q_j d) - d_k$. From the assumption, we know that $v_k \in LCP(q_i, q_j, d)$, so the path $LCP(q_i, q_j, d)$ remains the same regardless of v_k's declared cost d_k. Thus, the summation of all terminals' cost on $LCP(q_i, q_j, d)$ except terminal v_k equals

$$|LCP(q_i, q_j, d \mid^k 0)| = |LCP(q_i, q_j, d)| - d_k.$$

In other words, the second part is also independent of d_k. Now we can write the payment to a terminal v_k based on an edge $q_i q_j$ as follows:

$$p_k^{ij}(d) = |LE(q_i, q_j, d \mid^k \infty)| - |LCP(q_i, q_j, d \mid^k 0)|,$$

Here, terminal $v_k \in LCP(q_i, q_j, d)$ and $q_i q_j \in VMST(d)$.   □

If a terminal v_k lies its cost c_k upward, we denote the lied cost as $\overline{c_k}$. Similarly, if terminal v_k lies its cost c_k downward, we denote the lied cost as $\underline{c_k}$. Let $E_k(d_k)$ be the set of edges $q_i q_j$ such that $v_k \in LCP(q_i, q_j, d)$ and $q_i q_j \in VMST(d)$ when terminal v_k declares a cost d_k. From lemma 5.2, the non-zero payment to v_k is defined based on $E_k(d_k)$. The following lemma reveals the relationship between d_k and $E_k(d_k)$. The proof of the lemma is omitted due to its simplicity.

Lemma 5.3 $E_k(d_k) \leq E_k(d'_k)$.

We now state the proof that the payment scheme (5.2) satisfies IC.

Theorem 5.5   *Our payment scheme satisfies the incentive compatibility (IC).*

Proof   For terminal v_k, if it lies its cost from c_k to $\overline{c_k}$, then $E_k(\overline{c_k}) \subseteq E_k(c_k)$, which implies that payment

$$p_k(d \mid^k \overline{c_k}) = \max_{q_i q_j \in E_k(\overline{c_k})} p_k^{ij}(d \mid^k \overline{c_k})$$

$$\leq \max_{q_i q_j \in E_k(c_k)} p_k^{ij}(d \mid^k c_k) = p^k(d \mid^k c_k).$$

Thus, terminal v_k will not lie its cost upward, so we focus our attention on the case when terminal v_k lies its cost downward. □

From lemma 5.3, we know that $E_k(c_k) \subseteq E_k(\underline{c_k})$. Thus, we only need to consider the payment based on edges in $E_k(\underline{c_k}) - E_k(c_k)$. For edge $e = q_i q_j \in E_k(\underline{c_k}) - E_k(c_k)$, let $q_I^k q_J^k = LE(q_i, q_j, d \mid^k \infty)$ in the spanning tree $VMST(d \mid^k \infty)$. If we remove the edge $q_I^k q_J^k$, we have a vertex partition $\{Q_I^k, Q_J^k\}$, where $q_i \in Q_I^k$ and $q_j \in Q_J^k$. In the graph $K(d)$, we consider the bridge $BM(Q_I^k, Q_J^k, d)$ whose weight is minimum when the terminal cost vector is d. There are two cases to consider about $BM(Q_I^k, Q_J^k, d)$: (1) $v_k \notin BM(Q_I^k, Q_J^k, d \mid^k \underline{c_k})$ or (2) $v_k \in BM(Q_I^k, Q_J^k, d \mid^k \underline{c_k})$. We discuss them individually.

Case 5.1 $v_k \notin BM(Q_I^k, Q_J^k, d \mid^k \underline{c_k})$. In this case, edge $q_I^k q_J^k$ is the minimum bridge over Q_I^k and Q_J^k. In other words, we have $|LE(q_i, q_j, \mid^k \infty)| \leq |LCP(q_i, q_j, d \mid^k \underline{c_k})|$. Consequently

$$p_k^{ij}(d \mid^k \underline{c_k}) = |LE(q_i, q_j, d \mid^k \infty)| - |LCP(q_i, q_j, d \mid^k \underline{c_k})| + \underline{c_k}$$

$$= |LE(q_i, q_j, d \mid^k \infty)| - |LCP(q_i, q_j, d \mid^k \underline{c_k})| + c_k$$

$$\leq c_k,$$

which implies that v_k will not get benefit from lying its cost downward.

Case 5.2 $v_k \in BM(Q_I^k, Q_J^k, d \mid^k \underline{c_k})$. From the assumption that $q_i q_j \notin VMST(G(d \mid^k \underline{c_k}))$, we know edge $q_i q_j$ cannot be $BM(Q_I^k, Q_J^k, d \mid^k \underline{c_k})$. Thus, there exists an edge $q_s q_t \neq q_i q_j$ such that $v_k \in LCP(q_s, q_t, d \mid^k \underline{c_k})$ and $q_s q_t = BM(Q_I^k, Q_J^k, d \mid^k \underline{c_k})$. This guarantees that $q_s q_t \in VMST(d \mid^k \underline{c_k})$.

Obviously, $q_s q_t$ cannot appear in the same set of Q_I^k or Q_J^k. Thus, $q_I^k q_J^k$ is on the path from q_s to q_t in graph $VMST(d \mid^k \infty)$, which implies that $|LCP(q_I^k, q_J^k, d \mid^k \infty)| = |LE(q_i, q_j, d \mid^k \infty)| \leq |LE(q_s, q_t, d \mid^k \infty)|$. Using lemma 5.3, we have $LCP(q_s, q_t, d \mid^k \underline{c_k}) \in VMST(d \mid^k \underline{c_k}))$. Thus,

$$p_k^{ij}(d \mid^k \underline{c_k}) = |LE(q_i, q_j, d \mid^k \infty)| - |LCP(q_i, q_j, d \mid^k \underline{c_k})| + \underline{c_k}$$

$$= |LE(q_i, q_j, d \mid^k \infty)| - |LCP(q_i, q_j, d \mid^k \underline{c_k})| + c_k$$

$$\leq |LE(q_s, q_t, d \mid^k \infty)| - |LCP(q_i, q_j, d \mid^k \underline{c_k})| + c_k$$

$$\leq |LE(q_s, q_t, d \mid^k \infty)| - |LCP(q_s, q_t, d \mid^k \underline{c_k})| + c_k$$

$$= p_k^{st}(d \mid^k \underline{c_k}).$$

This inequality concludes that even if v_k lies its cost downward to introduce some new edges in $E_k(\underline{c_k})$, the payment based on these newly introduced edges is not larger than the payment on some edges already contained in $E_k(c_k)$. In summary, node v_i does not have the incentive to lie its cost upward or downward, which proves the IC property.

Before proving Theorem 5.6, we prove the following lemma regarding all truthful payment schemes based on VMST.

Lemma 5.4   *If $v_k \in VMST(d \mid^k c_k)$, then as long as $d_k < p_k(d \mid^k c_k)$ and d^{-k} fixed, $v_k \in VMST(d)$.*

Proof   Again, we prove it by contradiction. Assume that $v_k \notin VMST(d)$. Obviously, $VMST(d) = VMST(d \mid^k \infty)$. Assume that $p_k(d \mid^k c_k) = p_k^{ij}(d \mid^k c_k)$; that is, its payment is computed based on edge $q_i q_j$ in $VMST(d \mid^k c_k)$. Let $q_I q_J$ be the $LE(q_i, q_j, d \mid^k \infty)$ and $\{Q_I, Q_J\}$ be the vertex partition introduced by removing edge $q_I q_J$ from the tree $VMST(d \mid^k \infty)$, where $q_i \in Q_I$ and $q_j \in Q_J$. The payment to terminal v_k in $VMST(d \mid^k c_k)$ is $p_k(d \mid^k c_k) = |LCP(q_I, q_J, d \mid^k \infty)| - c_{ij}^{v_k}$, where $c_{ij}^{v_k} = |LCP(q_i, q_j, d \mid^k 0)|$. When v_k declares its cost as d_k, the length of the path $LCP(q_i, q_j, d)$ becomes $c_{ij}^{v_k} + d_k = |LCP(q_I, q_J, d \mid^k \infty)| - p_k(d \mid^k c_k) + d_k < |LCP(q_I, q_J, d \mid^k \infty)|$.

Now consider the spanning tree $VMST(d)$. We have assumed that $v_k \notin VMST(d)$, that is, $VMST(d) = VMST(d \mid^k \infty)$. Thus, among the bridge edges over Q_i, Q_j, edge $q_I q_J$ has the least cost when the graph is $G \backslash v_k$ or $G(d \mid^k d_k)$. However, this is a contradiction to what we just proved: $|LCP(q_i, q_j, d \mid^k d_k)| < |LCP(q_I, q_J, d \mid^k \infty)|$. This finishes the proof. □

We are now ready to show that our payment scheme is optimal among all truthful mechanisms using VMST.

Theorem 5.6   *Our payment scheme is the minimum among all truthful payment schemes based on the VMST structure.*

Proof   We prove it by contradiction. Assume that there is another truthful payment scheme, say $\mathcal{A}$, based on VMST, whose payment is smaller than our payment for a terminal v_k under a cost profile d. Assume that the payment calculated by $\mathcal{A}$ for terminal v_k is $\tilde{p}_k(d) = p_k(d) - \delta$, where $p_k(d)$ is the payment calculated by our algorithm and $\delta > 0$.

Now consider another profile $d \mid^k d_k'$, where the terminal v_k has the true cost $c_k = d_k' = p^k(d) - \frac{\delta}{2}$. From lemma 5.4, we know that v_k is still in $VMST(d \mid^k d_k')$. Using lemma 5.1, we know that the payment for terminal v_k using algorithm $\mathcal{A}$ is $p_k(c) - \delta$, which is independent of terminal v_k's declared cost. Notice that $d_k = p_k(d) - \frac{\delta}{2} > p_k(d) - \delta$. Thus, terminal v_k has negative utility under the payment scheme $\mathcal{A}$ when node v_k reveals it true cost under cost profile $d \mid^k d_k'$, which violates the incentive compatibility (IC). This finishes the proof. □

By summarizing Theorem 5.4, Theorem 5.5, and Theorem 5.6, we obtain Theorem 5.3.

5.3.3   Strategyproof Mechanism Based on Spider

For a general node weighted network, in the worst case, the cost of the structure LCPS and VMST could be $\theta(n)$ times the cost of the optimal tree. It is known[10,12] that it is NP-hard to find the minimum cost multicast tree when given an arbitrary node weighted graph G, and it is at least as hard to approximate as the set cover problem. Klein and Ravi[12] showed that it can be approximated within $O(\ln r)$, where r is the number of receivers, which is within a small constant factor of the best achievable approximation ratio among all polynomial time computable trees if $N \neq NP$.

5.3.3.1   Constructing the Spider

Here we review the method used by Klein and Ravi[12] to find a node weighted Steiner tree (NST). Klein and Ravi used a special structure called a *spider* to approximate the optimal solution. A spider is defined as a tree having at most one node of degree more than two. Such a node (if it exists) is called the center of the spider. Each path from the center to a leaf is called a *leg*. The *cost* of a spider S is defined as the sum of the cost of all nodes in spider S, denoted as $\omega(S)$. The number of terminals or legs of the spider is denoted

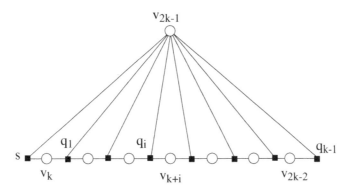

FIGURE 5.3 Terminals V_i, $1 \leq i \leq k$ are receivers; the cost of terminal v_{2k-1} is 1. The cost of each terminal $v_i, k \leq i \leq 2k_2$ is $\frac{2}{2k-i} - \epsilon$, where ϵ is a sufficiently small positive number.

by $t(S)$, and the ratio of a spider S is defined as

$$\rho(S) = \frac{\omega(S)}{t(S)}.$$

Contraction of a spider S is the operation of contracting all vertices of S to form one virtual terminal and connecting this virtual terminal to each vertex v when uv is a link before the contraction and $u \in S$. The new virtual terminal has weight zero.

Algorithm 5.3

Construct NST.

Repeat the following steps until no receivers are left and there is only one virtual terminal remaining.

1. Find the spider S with the minimum $\rho(S)$ that connects some receivers and virtual terminals.*
2. Contract the spider S by treating all nodes in it as one virtual terminal. We call this one *round*.

All nodes belong to the final unique virtual terminal to form the NST.
The following theorem is proved in Klein and Ravi.[12]

Theorem 5.7[12] *Given k receivers, the tree constructed above has cost at most $2 \ln k$ times of the optimal.*

5.3.3.2 VCG Mechanism on NST Is Not Strategyproof

Again, we may want to pay terminals based on the VCG scheme; that is, the payment to a terminal $v_k \in NST(d)$ is

$$p_k(d) = \omega(NST(d \mid^k \infty)) - \omega(NST(d)) + d_k.$$

We show by example that the payment scheme does *not* satisfy the IR property: it is possible that some terminal has negative utility under this payment scheme. Figure 5.3 illustrates such an example. It is not difficult to show that, in the first round, terminal v_k is selected to connect terminals s and q_1 with cost ratio $\frac{1}{k} - \frac{\epsilon}{2}$ (while all other spiders have cost ratios of at least $\frac{1}{k}$). Then, terminals s, v_k, and q_1 form a virtual

---

*For simplicity of the proof, we assume that there does not have to be two spiders with the same ratio. Dropping the assumption will not change our results.

terminal. At the beginning of round r, we have a virtual terminal, denoted by V_r formed by terminals v_{k+i-1}, $1 \le i \le r-1$, and receivers q_i, $1 \le i \le r$; all other receivers q_i, $r < i \le k$ are the remaining terminals. It is easy to show that we can select terminal q_{k+r-1} at round r to connect V_r and q_{r+1} with cost $\frac{1}{k+1-r} - \frac{\epsilon}{2}$. Thus, the total cost of the tree $NST(G)$ is $\sum_{i=1}^{k-1} (\frac{2}{k+1-i} - \epsilon) = 2H(k) - 2 - (k-1)\epsilon$.

When terminal v_k is not used, it is easy to see that the final tree $NST(G \backslash u_1)$ will only use the terminal v_{2k-1} to connect all receivers with cost $\frac{1}{k}$ when $\frac{1}{k-1} - \frac{\epsilon}{2} > \frac{1}{k}$. Notice that this condition can be trivially satisfied by letting $\epsilon = \frac{1}{k^2}$. Thus, the utility of terminal v_k is $p_1(d) - c(v_k) = \omega(NST(G \backslash v_k)) - \omega(NST(G)) = -2H(k) + 3 + (k-1)\epsilon$, which is negative when $k \ge 8$ and $\epsilon = 1/k^2$.

5.3.3.3 Strategyproof Mechanism on Spiders

Notice that the construction of NST is by rounds. Following, we show that if terminal v_k is selected as part of the spider with the minimum ratio under a cost profile d in a round i, then v_k is selected before or in round i under a cost profile $d' = d \mid^k d_k'$ for $d_k' < d_k$. We prove this by contradiction, which assumes that the terminal v_k will not appear before round $i+1$. Notice that the graph remains the same for round i after the profile changes, so spider $S_i(d)$ under the cost profile d is still a valid spider under the cost profile d'. Its ratio becomes $\omega_i^k(d) - d_k + d_k' < \omega_i^k(d)$, while all other spider ratios remain the same if they do not contain v_k. Thus, the spider $S_i^k(d)$ has the minimum ratio among all spiders under cost profile d', which is a contradiction. So, for terminal v_k, there exists a real value $B_k^i(d_{-k})$ such that the terminal v_k is selected before or in round i iff $d_k < B_k^i(d_{-k})$. If there are r rounds, we have an increasing sequence

$$B_k^1(d_{-k}) \le B_k^2(d_{-k}) \le \cdots \le B_k^r(d_{-k}) = B_k(d_{-k}).$$

Obviously, the terminal v_k is selected in the final multicast tree iff $d_k < B_k(d_{-k})$. Following is our payment scheme based on NST.

Definition 5.2   For a node v_k, if v_k is selected in NST, then it gets payment

$$p_k(d) = B_k(d_{-k}). \tag{5.3}$$

Otherwise, it gets payment 0.

Regarding this payment, we have the following theorem.

Theorem 5.8   *Our payment scheme (5.3) is truthful, and is minimum among all truthful payment schemes for multicast trees based on spider.*

Proof   To prove that it is truthful, we prove that it satisfies IR and IC, respectively. Notice that v_k is selected iff $d_k < B_k^i(d_{-k})$, and we have $u_k(d) = B_k(d_{-k}) - d_k > 0$, which implies IR. Now we prove that our payment scheme (5.3) satisfies IC by cases. Notice that when v_k is selected, its payment does not depend on d_k, so we only need to discuss the following two cases:

Case 5.1   When v_k declares c_k, it is not selected. What happens if it lies its cost upward as d_k to make it not selected? From the IR property, v_k gets positive utility when it reveals its true cost, while it gets utility 0 when it lies its cost as d_k. So, it better for v_k not to lie.

Case 5.2   When v_k declares a cost c_k, it is not selected. What happens if it lies its cost downward as d_k to make it selected? When v_k reveals c_k, it has utility 0; after lying, it has utility $B_k(d_{-k}) - c_k$. From the assumption that v_k is not selected under cost profile $d \mid^k c_k$, we have $B_k(d_{-k}) \le c_k$. Thus, v_k will get non-positive utility if it lies, which ensures v_k revealing its true cost c_k.

So overall, v_k will always choose to reveal its actual cost to maximize its utility (IC property).

Following we prove that our payment is minimal. We prove it by contradiction. Suppose that there exists such a payment scheme $\tilde{P}$ such that for a terminal v_k under a cost profile d, the payment to $\tilde{P}_i(d)$ is smaller than our payment. Notice that in order to satisfy the IR, the terminal must be selected, so we assume $\tilde{P}_i(d) = B_k(d_{-k}) - \delta$, where δ is a positive real number. Now consider the profile $d' = d \mid^k (B_k(d_{-k}) - \frac{\delta}{2})$ with v_k's actual cost being $c_k = B_k(d_{-k}) - \frac{\delta}{2}$. Obviously, v_k is selected; from lemma 5.1, the payment to v_k is $B_k(d_{-k}) - \delta$. Thus, the utility of v_k becomes $u_k(d') = B_k(d_{-k}) - B_k(d_{-k}) - \delta + \frac{\delta}{2} = -\frac{\delta}{2} < 0$, which violates the IR property. This finishes our proof. □

We then study how to compute such payment to a selected node v_k. With Theorem 5.8, we only need to focus attention on how to find the value $B_k^i(d_{-k})$. Before we present our algorithm to find $B_k^i(d_{-k})$, we first review in detail how to find the minimum ratio spider. To find the spider with the minimum ratio, we find the spider centered at every vertex v_j with the minimum ratio over all vertices $v_j \in V$ and choose the minimum among them. The algorithm is as follows.

Algorithm 5.4

Find the minimum ratio spider.

Do the following process for all $v_j \in V$:

1. Calculate the shortest path tree rooted at v_j that spans all terminals. We call each shortest path a *branch*. The weight of the branch is defined as the length of the shortest path. Here, the weight of the shortest path does not include the weight of the center node v_j of the spider.
2. Sort the branches according to their weights.
3. For every pair of branches, if they have terminals in common, then remove the branch with a larger weight. Assume that the remaining branches are

$$L(v_j) = \{L_1(v_j), L_2(v_j), \cdots, L_r(v_j),\}$$

   sorted in ascending order of their weights.
4. Find the minimum ratio spider with center v_j by linear scanning: the spider is formed by the first t branches such that $\frac{c_j + \Sigma_{k=1}^t L_k}{t} \leq \frac{c_j + \Sigma_{k=1}^h L_k}{h}$ for any $h \neq t$.
   Assume the spider with the minimum ratio centered at terminal v_j is $S(v_j)$ and its ratio is $\rho(v_j)$.
5. The spider with minimum ratio for this graph is then $S = \min_{v_j \in V} S(v_j)$.

In Algorithm 5.5, $\omega(L_i(v_j))$ is defined as the sum of the terminals' cost on this branch, and $\Omega_i(L(v_j)) = \Sigma_{s=1}^i \omega(L_s(v_j)) + c_j$. If we remove node v_k, the minimum ratio spider with center v_j is denoted as $S^{-v_k}(v_j)$ and its ratio is denoted as $\rho^{-v_k}(v_j)$. Assume that $L_1^{-v_k}(v_j), L_2^{-v_k}(v_j), \cdots, L_p^{-v_k}(v_j)$ are those branches in ascending order before linear scan.

From now on, we fix d_{-k} and the graph G to study the relationship between the minimum ratio $\rho(v_j)$ of the spider centered at v_j and the cost d_k of a node v_k.

Observation 5.1   *The number of the legs of the minimum ratio spider decreases over d_k.*

If the minimum ratio spider with the terminal v_k has t legs, then its ratio will be a line with slope of $\frac{1}{t}$. So, the ratio-cost function is formed by several line segments. From the Observation 5.1, these line segments have decreasing slopes and thus they have at most r segments, where r is the number of receivers. So, given a real value y, we can find the corresponding cost of v_k in time $O(\log r)$ such that the minimum cost ratio spider $S(v_j)$ centered at node v_j has a ratio y. We the present our algorithm to find these line segments as follows.

Algorithm 5.5

Find the ratio-cost function $y = \mathcal{R}_{v_j}(x)$ over the cost x of v_k.

There are two cases here: $j = k$ or $j \neq k$.

Case 5.1 $j = k$, we apply the following procedures:

Apply steps 1, 2, and 3 of Algorithm 5.4 to get $L(v_k)$.
Set the number of legs to $t = 1$, lower bound $lb = 0$, and upper bound $ub = 0$.
While $t < r$ do the following:
$\{ub = (t + 1) * \omega(L_{t+1}(v_k)) - \Omega_{t+1}(L(v_k))$
$y = \frac{\Omega_t(L(v_k)) + x}{t}$ for $x \in [lb, ub)$
Set $lb = ub$ and $t = t + 1$ $\}$
Let $y = \frac{\Omega_r(L(v_k)) + x}{r}$ for $x \in [lb, \infty)$.

Case 5.2 $j \neq k$, we apply the following procedures:

1. Remove terminal v_k; apply Algorithm 5.4 to find $S^{-v_k}(v_j)$.
2. Find the shortest path with terminal v_k from v_j to every receiver; sort these paths according to their length in descending order, say sequence

$$L^{v_k}(v_j) = \left\{ L_1^{v_k}(v_j), L_2^{v_k}(v_j), \cdots, L_r^{v_k}(v_j) \right\}.$$

Here, r is the number of terminals and $\omega(L_i^{v_k}(v_j))$ is the sum of terminals on path $L_i^{v_k}(v_j)$ *excluding* terminal v_k.
3. Hereafter, t is the index for branches in $L^{v_k}(v_j)$ and l is the index for paths in $L^{-v_k}(v_j)$.
4. For $L_t^{v_k}(v_j)$ $(1 \leq t \leq r)$, there may exist one or more branches in $L^{-v_k}(v_j)$ such that they have common terminals with $L_t^{v_k}(v_j)$. If there is more than one such branch, choose the branch with the minimum cost, say $L_l^{-v_k}(v_j)$. We define upper bound $upper_t$ for $L_t^{v_k}(v_j)$ equal to $\omega(L_l^{-v_k}(v_j)) - \omega(L_t^{v_k}(v_j))$. If there does not exist such a branch, we set $upper_t = \infty$.
5. Initialize lower bound $lb = 0$ and upper bound $ub = 0$. Apply the following algorithm:
For $t = 1$ to r do $\{$
      While $lb < upper_t$ do
            Set $l = 1$.
            Obtain a new sequence $LT^{-v_k}(v_j)$ from $L^{-v_k}(v_j)$ by removing all branches that have common nodes with $L_t^{v_k}(v_j)$. Let rt be the number of branches in sequence $LT^{-v_k}(v_j)$.
            While $l \leq rt$ do
                  While $\omega(L_t^{v_k}(v_j)) + lb > l\omega(LT_l^{-v_k}(v_j)) - \Omega_{l-1}(LT^{-v_k}(v_j)) - c_j$ and $l \leq rt$
                        $l = l + 1$
                  If $l \leq rt$ then
                        Set $ub = \omega(LT_l^{-v_k}(v_j)) - \Omega_{l-1}(LT^{-v_k}(v_j)) - \omega(L_t^{v_k}(v_j) - c_j$
                        If $ub \geq upper_t$ break;
                        Set $y = \frac{\Omega_{l-1}(LT^{-v_k}(v_j)) + \omega(LT_t^{v_k}(v_j)) + x}{l}$ for $x \in [lb, ub)$
                  Set $lb = ub$.
                  Set $l = l + 1$.
            Set $y = \frac{\Omega_{l-1}(LT^{-v_k}(v_j)) + \omega(L_t^{v_k}(v_j)) + x}{l}$ for $x \in [lb, upper_t)$.
            Set $lb = upper_t$.
$\}$

Given a real value x, the corresponding cost for terminal v_k is denoted as $\mathcal{R}_{v_j}^{-1}(x)$. Finally, we give the algorithm to find value $B_k(d_{-k})$.

Algorithm 5.6

Algorithm to find $B_k(d_{-k})$.

1. Remove the terminal v_k and find the multicast tree using the spider structure.
2. For every round i in the first step, we have a graph called G_i and a selected spider with ratio $\rho_i^{-v_k}$. Adding the node v_k and all its incident edges to G_i, we get a graph G_i'.

3. Find the function $y = \mathcal{R}_{v_j}^{-1}(x)$ for every terminal v_j in the graph G'_i using Algorithm 5.5.
4. Calculate $B_k^r(d_{-k}) = \max_{v_j \in V(G'_i)} \{\mathcal{R}_{v_j}^{-1}(\rho_i^{-v_k})\}$.
5. $B_k(d_{-k}) = \max_{1 \le i \le r} B_k^i(d_{-k})$.

The correctness of our algorithms is omitted due to space limitations. Notice that for practical implementations, we do not actually have to compute the functions. We are more interested in given some value y, what is the corresponding cost d_k such that the minimum ratio spider centered at the node v_k has a ratio y.

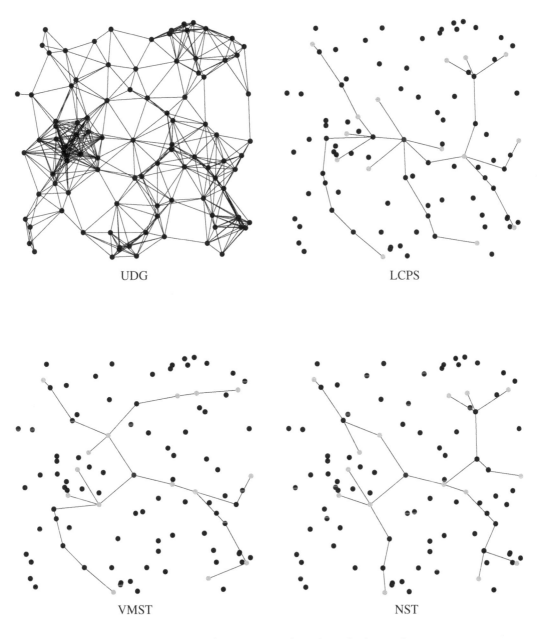

FIGURE 5.4 Multicast structures for node weighted network.

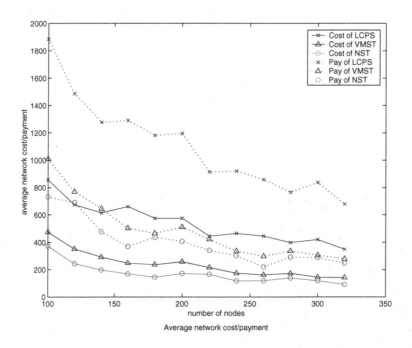

Average network cost/payment

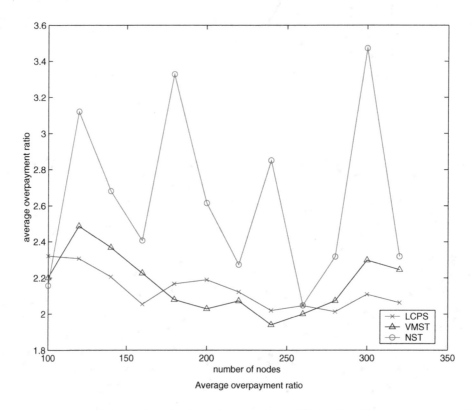

Average overpayment ratio

FIGURE 5.5 Results when the number of nodes in the networks are different (from 100 to 320). Here, we fix the transmission range at 300 ft.

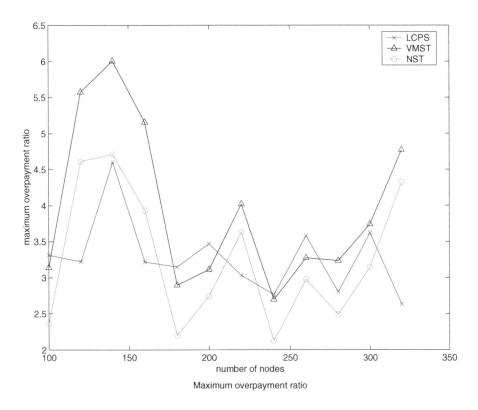

Maximum overpayment ratio

FIGURE 5.5 *(Continued)*

5.4 Experimental Studies

Remember that the payment of our structure is often larger than the structure's actual cost. For a structure H, let $c(H)$ be its cost and $p_s(H)$ be the payment of a scheme s based on this structure. We define the overpayment ratio of the payment scheme s based on structure H as

$$OR_s(H) = \frac{p_s(H)}{c(H)}. \tag{5.4}$$

When it is clear from the context, we often simplify the notation as $OR(H)$.

Actually, there are some other definitions about overpayment ratio in the literature. Some propose to compare the payment $p(H)$ with the cost of the new structure obtained from the graph $G - H$, that is, removing H from the original graph G. Here, we only focus attention on the overpayment ratio defined as (5.4).

We conducted extensive simulations to study the overpayment ratio of various schemes proposed in this chapter. In our experiments, we compare the different schemes proposed according to three different metrics: (1) actual cost, (2) total payment, and (3) overpayment ratio. Figure 5.4 shows the different multicast structures when the original graph is a unit disk graph. Here, the grey nodes are receivers.

In the first experiment, we randomly generate n nodes uniformly in a 2000 ft × 2000 ft region. The transmission range of each node is set as 300 ft. The weight of a node i is $c_i * 3^\kappa$, where c_i is randomly selected from a power level between 1 and 10. We vary the number of terminals in this region from 100 to 320, and fix the number of senders to 1 and the number of receivers to 15. For a specific number of terminals, we generate 500 different networks, and compare the average cost, maximum cost, average payment and maximum payment, average overpayment ratio, and maximum payment ratio.

In wireless ad hoc networks with nodes having a fixed transmission range, as shown in Figure 5.5, all structures' cost and payment decrease as the number of terminals increases. Notice that for all structures,

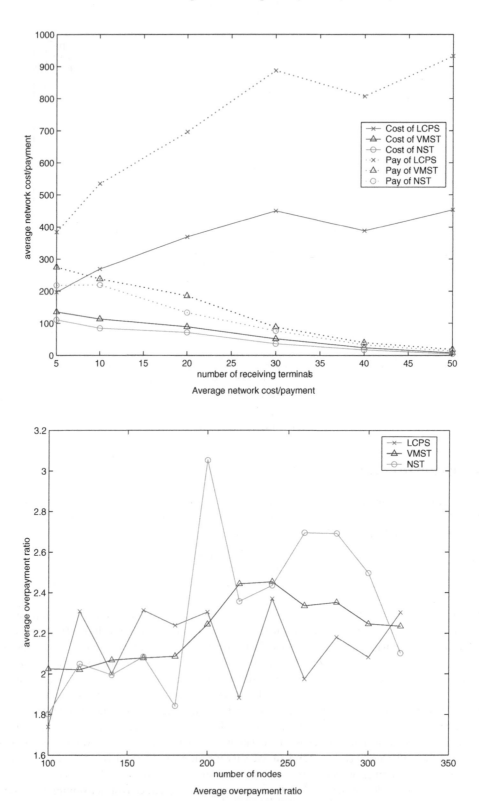

FIGURE 5.6 Results when the number of nodes in the networks are different (from 100 to 320). Here, we randomly set the transmission range from 100 to 500 ft.

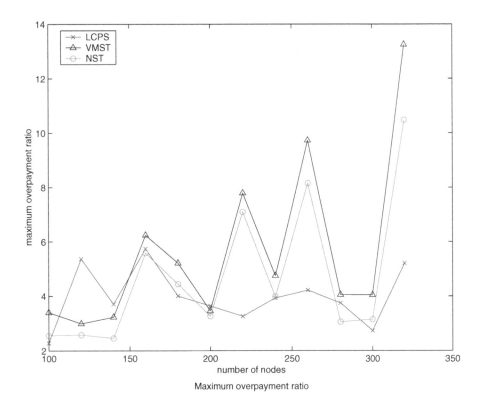

Maximum overpayment ratio

FIGURE 5.6 (*Continued*)

we assume all receivers (senders) will relay the message for free. The cost and payment of VMST and NST are much lower than the cost and payment of LCPS. As expected, due to the low cost of the VMST and NST structures, the maximum overpayment ratio of these two structures are very unsteady and much higher than the maximum overpayment ratio of LCPS. We still prefer the structures VMST and NST because they incur smaller costs and, more importantly, smaller payments.

In our second experiment, we vary the transmission range of each wireless node from 100 to 500 m. We assume that the cost c_i of a terminal v_i is $c_1 + c_2 r_i^{\kappa}$, where c_1 takes values from 300 to 500, c_2 takes values from 10 to 50, and r_i is v_i's transmission range. The ranges of c_1 and c_2 we used here reflect the actual power cost in one second of a node to send data at 2 Mbps rate.

Similar to the fixed transmission experiment, we vary the number of terminals in the region from 100 to 320, and fix the number of senders to 1 and the number of receivers to 15. For a specific number of terminals, we generate 500 different networks and compare the average cost, maximum cost, average payment and maximum payment, average overpayment ratio, and maximum payment ratio.

Figure 5.6 shows similar results for both the link weighted network and the node weighted network in fixed transmission range experiments.

5.4.1 Vary the Number of Receivers and Random Transmission Range

For a structure H, if there are r receivers, we define the cost density as

$$CD(H) = \frac{c(H)}{r}$$

and the payment density as

$$PD(H) = \frac{p(H)}{r}.$$

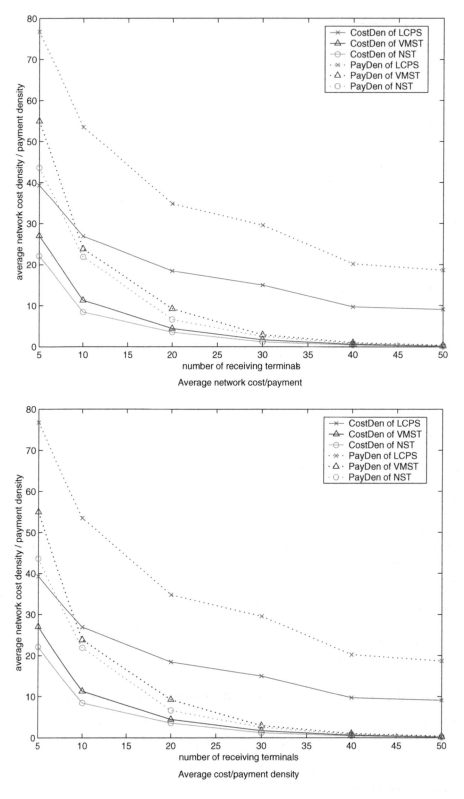

FIGURE 5.7 Results when the number of receivers in the networks are different (from 10 to 50). We randomly set the transmission range from 100 to 500 ft.

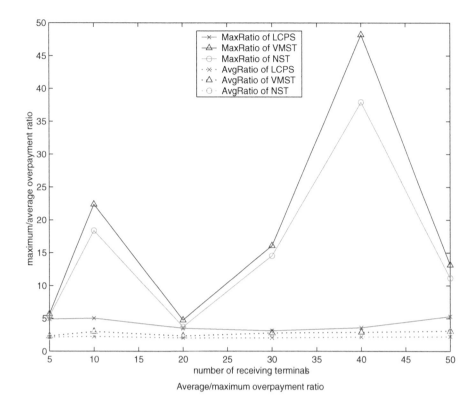

Average/maximum overpayment ratio

FIGURE 5.7 (*Continued*)

Now we study the relationship between cost, payment, overpayment ratio, cost density, payment density, and the number of the terminals. We use the same power cost model as in the previous experiment and the number of nodes in the region is set to 300. We fix the number of senders to 1 and vary the number of receivers from 5, 10, 20, · · ·, 50.

Figure 5.7 shows that when the number of the receivers increases, under most circumstances, the overall payment and cost increase while the average cost and payment for every terminal decrease. One exception is for a node weighted network. Notice that in a node weighted network, we set all terminals' cost to 0. It is natural to expect that when the number of receivers is larger than some threshold, the total cost and payment will decrease because we assume that receivers will relay for free. This experiment shows that more terminals in a multicast group can incur a lower cost and payment per terminal, which is an attractive property of multicast.

5.5 Conclusion

In this chapter, we studied how to conduct efficient multicast in *selfish* wireless networks by assuming that each wireless node will incur a cost when it has to transit some data and the cost is privately known to the wireless terminal node. For each of the widely used structures for multicast, we designed a strategyproof multicast mechanism such that each agent maximizes its profit when it truthfully reports its cost. The structures studied in this chapter are: least cost path star (LCPS), virtual minimum spanning tree (VMST), and the node weighted Steiner tree (NST). We showed that the VMST approximates the optimal multicast tree for the homogeneous network when all wireless nodes have the same transmission range.

Extensive simulations were conducted to study the practical performances of the proposed protocols, especially the total payment of these protocols compared with the actual cost of agents. Although, theoretically, the overpayment ratio of the protocols designed here could be large in the worst scenario, we found by experiments that the overpayment ratios of all our strategyproof mechanisms are small when the costs of agents are drawn from a random distribution. Recently, Wang et al.[19] studies more structures for multicast, and a general framework for designing truthful multicast protocols, and how to share the payment among receivers.[20]

References

1. Anderegg, L. and Eidenbenz, S. Ad hoc VCG: a truthful and cost-efficient routing protocol for mobile ad hoc networks with selfish agents. In *Proceedings of the 9th Annual International Conference on Mobile Computing and Networking* (2003), ACM Press, pp. 245–259.
2. Blazevic, L., Buttyan, L., Capkun, S., Giordano, S., Hubaux, J.P., and Boudec, J.Y. L. Self-organization in mobile ad-hoc networks: the approach of terminodes. *IEEE Communications Magazine, 39*, 6, June 2001.
3. Buttyan, L. and Hubaux, J. Stimulating cooperation in self-organizing mobile ad hoc networks. *ACM/Kluwer Mobile Networks and Applications, 5*, 8, October 2003.
4. Buttyan, L. and Hubaux, J.-P. Enforcing service availability in mobile ad-hoc WANs. In *Proceedings of the 1st ACM International Symposium on Mobile ad hoc Networking & Computing* (2000), pp. 87–96.
5. Clarke, E.H. Multipart pricing of public goods. *Public Choice* (1971), 17–33.
6. Feigenbaum, J., Papadimitriou, C., Sami, R., and Shenker, S. A BGP-based mechanism for lowest-cost routing. In *Proceedings of the 2002 ACM Symposium on Principles of Distributed Computing*, 2002, pp. 173–182.
7. Feigenbaum, J., Papadimitriou, C.H., and Shenker, S. Sharing the cost of multicast transmissions. *Journal of Computer and System Sciences, 63*, 21–41, 2001.
8. Green, J. and Laffont, J.J. Characterization of satisfactory mechanisms for the revelation of preferences for public goods. *Econometrica*, 1977, 427–438.
9. Groves, T. Incentives in teams. *Econometrica*, 1973, 617–631.
10. Guha, S. and Khuller, S. Improved methods for approximating node weighted Steiner trees and connected dominating sets. In *Foundations of Software Technology and Theoretical Computer Science*, 1998, pp. 54–65.
11. Jakobsson, M., Hubaux, J.-P., and Buttyan, L. A micro-payment scheme encouraging collaboration in multi-hop cellular networks. In *Proceedings of Financial Cryptography*, 2003.
12. Klein, P. and Ravi, R. A nearly best-possible approximation algorithm for node-weighted Steiner trees. Tech. Rep. CS-92-54, 1992.
13. Marti, S., Giuli, T. J., Lai, K., and Baker, M. Mitigating routing misbehavior in mobile ad hoc networks. In *Proc. of MobiCom*, 2000.
14. Nisan, N. and Ronen, A. Algorithmic mechanism design. In *Proc. 31st Annual Symposium on Theory of Computing (STOC99)*, 1999, pp. 129–140.
15. Salem, N.B., Buttyan, L., Hubaux, J.-P., and Kakobsson, M. A charging and rewarding scheme for packet forwarding in multi-hop cellular networks. In *ACM MobiHoc*, 2003.
16. Srinivasan, V., Nuggehalli, P., Chiasserini, C.F., and Rao, R.R. Energy efficiency of ad hoc wireless networks with selfish users. In *European Wireless Conference 2002 (EW2002)*, 2002.
17. Srinivasan, V., Nuggehalli, P., Chiasserini, C.F., and Rao, R.R. Cooperation in wireless ad hoc wireless networks. In *IEEE Infocom*, 2003.
18. Vickrey, W. Counterspeculation, auctions and competitive sealed tenders. *Journal of Finance* (1961), 8–37.
19. Wang, W., Li, X.Y., and Wang, Y. Truthful multicast in selfish wireless networks. In *Proc. of the 10th Annual International Conference on Mobile Computing and Networking*, 2004.
20. Wang, W., Li, X.Y., Sun, Z., and Wang, Y., Design multicast protocols for non-cooperative networks. In *Proc. of the 24th Annual Joint Conference of the IEEE Communication Society (INFOCOM)*, 2005.

6

Geocasting in Ad Hoc and Sensor Networks

Ivan Stojmenovic

6.1 Introduction

Ad hoc networks consist of laptops, PDAs, and other devices that can communicate in multihop fashion. Such networks emerge in conference, rescue, and battlefield environments, as well as for wireless Internet access. Ad hoc networks are often considered as mobile networks. However, this chapter is concerned with static nodes, nodes that can change activity status, or can arbitrarily fail. We are also interested in large-size networks and the design of corresponding scalable solutions. Sensor networks are a special case of ad hoc networks, with nodes that are generally densely deployed (hundreds or thousands of such nodes can be placed, mostly at random, either very close or inside the phenomenon to be studied), are small in size, static, with lower battery power, and smaller processing power. Recent technological advances have enabled the development of low-cost, low-power, and multifunctional sensor devices. These nodes are autonomous devices with integrated sensing, processing, and communication capabilities. Sensor networks consist of a large number of sensor nodes that collaborate together using wireless communication and asymmetric many-to-one data. Indeed, sensor nodes usually send their data to a specific node called the sink node or monitoring station, which collects the requested information. All nodes cannot communicate directly with the monitoring station because such communication may be over long distances that will drain the power quickly. Hence, sensors operate in a self-organized and decentralized manner, and message communication takes place via multihop spreading. To enable this, the network must maintain the best connectivity for as long as possible. The sensor's battery is not replaceable, and sensors can operate in hostile or remote environments. Therefore, energy consumption is considered the most important resource, and the network must be self-configured and self-organized. The best energy conservation method is to put as many sensors to sleep as possible. The network must be connected to remain functional, so that the monitoring station can receive messages sent by any of the active sensors. An intelligent strategy for selecting and updating a set of active sensors that are connected is needed to extend network lifetime. This problem is known as the *connected area coverage problem*, which aims to dynamically activate and deactivate sensors while maintaining full coverage of the monitoring area. Efficient solutions to the connected area coverage problem are discussed in References 1 through 3. When this coverage step is performed first, the large sensor network becomes reasonably sparse but remains connected.

If all active sensors are dedicated to monitoring the same event, then the monitoring center can spread the task and establish a reverse broadcast tree using any intelligent flooding protocol (see Ref. 4). If the network is reasonably sparse, even blind flooding (where each node receiving a message will retransmit it exactly once) is a viable option. Otherwise—that is, when the region to be monitored for a particular event contains only a small portion of active sensors—flooding the whole network may become an inefficient way of spreading the task. This chapter reviews existing solutions to the geocasting problem. In a multifunctional, multi-event sensor environment, the monitoring center can separately handle several geocasting regions and corresponding events. One particular application of geocasting is in tracking mobile objects. The monitoring center can collect reports from sensors in the vicinity of the object and send periodic signals to the sensor's adjusting geocasting region, following the object's movement.

We assume that each sensor is aware of its geographic position with respect to its neighbors and monitoring center. The problem of finding a reasonably accurate sensor location (when sensors are not directly equipped with a GPS receiver, which becomes technologically feasible) has been intensively studied recently.[5] We consider only the localized approach in this chapter. In a localized routing or geocasting algorithm, each node makes a decision to which neighbor(s) to forward the message based solely on the location of itself, its neighboring nodes, and destination. In the case of geocasting, D is a node approximately in the center of the region and includes a geocast region description. We also assume that the sensor network is static, and that the monitoring center is aware of the geocast region to be covered. For simplicity, we assume here that the geocast region is a circle. However, other shapes, such as convex polygons, can similarly be considered.

A number of localized geocasting protocols proposed in literature[6–14] do not guarantee delivery to all the nodes inside the geocasting region. Note that sensors can actually cover the geocast region but may not be connected inside it, because of possible obstacles in the region or differences between communication and sensing radii. Among localized geocasting algorithms recently claimed in literature, we show that the only one that is able to guarantee delivery is a flooding-based scheme (possibly dominating set based; see Ref. 15 and the survey in Ref. 4), which is inefficient when the geocast region is small compared to the whole area covered with active sensors. Further, a "forgotten" scheme[16] also guarantees delivery but has considerable message overhead. The chapter assumes that the medium access layer is ideal; that is, each message sent from a node to its neighbor is received properly by that neighbor. The guaranteed delivery property is conditional upon the availability of such an ideal MAC layer. A geocasting protocol therefore has a guaranteed delivery property if each node, located inside the geocasting region and connected to the source node, will receive the packet if the ideal MAC layer is applied. We describe in this chapter three nontrivial solutions for the geocasting task with guaranteed delivery. One is obtained from Ref. 16 by adding a preprocessing step, making the algorithm message efficient afterwards. Another is based on multicasting to entrance zones, followed by intelligent flooding.[17] We show that a recently proposed geocasting scheme[18] does not guarantee delivery, despite the claim, and then modify it to provide this property. All three algorithms are proven to guarantee delivery. Their expected performance is discussed.

This chapter is organized as follows. Section 6.22 presents localized location-based routing and geocasting algorithms. A geocasting algorithm with guaranteed delivery, based on a traversal of faces intersecting a geocast region boundary, is described in Section 6.33. Another algorithm,[16] based on a traversing face tree, is described in Section 6.4 (a variant of it, with preprocessing, and reducing message complexity, is proposed here). Section 6.5 describes a geocasting protocol that guarantees delivery[17] and is efficient compared to possible alternatives. Finally, conclusions and references complete this chapter.

6.2   Position-Based Localized Routing and Geocasting Algorithms

Finn[19] proposed the *greedy* routing algorithm for ad hoc networks. When node S wants to send a message to destination D, it uses the location information for D and for all its one-hop neighbors to determine the neighbor A that is closest to D among all neighbors of S. Figure 6.1 is an example where A is the closest to D among all neighbors of S. The message is forwarded to A, and the same procedure is repeated until D, if possible, is eventually reached. The greedy route in Figure 6.1 would be $SAFVD$ (note that D is geocast region center, not a real node). In the *MFR* (most forward progress within radius) method,[20] node S forwards the message to its neighbor that has the greatest progress toward the destination. In the directional-based method,[21] the message is forwarded to the neighbor whose direction (angle) is closest to the direction of destination. *Greedy* and *MFR* based routing and geocasting methods are loop free, while directional-based schemes are not.[22]

These basic methods were applied to describe geocasting protocols. In directional (*DIR*) geocasting methods,[8,11] the node S (the source or intermediate node) transmits message m to all neighbors located between the two tangents from S to the region that could contain the destination. The area containing the region and two tangents (marked by dotted lines in Figure 1) is referred as the *request zone*. For example, in Figure 6.1, where the geocast region is a circle centered at D, the request zone for S contains nodes A, B, and C. Continuing the process, nodes E, F, U, T, Q, R, and V in Figure 6.1 are those to receive the message from S. In Ko and Vaidya,[11] the request zone is fixed from the source, and a node that is not in the request zone does not forward a route request to its neighbors. If the source has no neighbors within the request zone, the zone is expanded to include some. In Basagni et al.,[8] the request zone is defined at each intermediate node.

An and Papavassiliou[6] describe a geocasting protocol that is basically a directional request zone method with blind flooding inside the request zone being replaced by a more intelligent method. Broadcasting

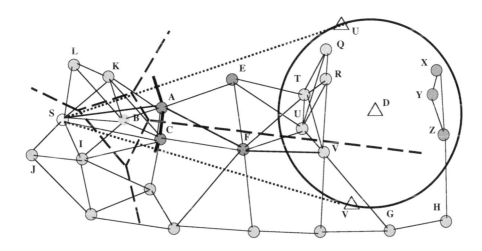

FIGURE 6.1   The request zone, Voronoi region, convex hull region, and nodes X, Y, and Z that are not reached.

(sending a message from one node to all the nodes inside a region) can be reliably performed (so that each node receives a copy of the packet) if only nodes from a connected dominating set retransmit the message.[15] A dominating set consists of nodes that "cover" the set; that is, each node not in the set must be a neighbor of at least one node from the set. The connectivity of such sets enables propagation of information to all the nodes. An and Papavassiliou[7] used a mobility-clustering method to construct one such dominating set. The mobility-clustering method is a well-known clustering method by Lin and Gerla,[23] where ID is replaced by mobility-based metrics, so that more stable nodes are more likely to become clusterheads. In addition, such restricted flooding (or directed guided routing[6]) in a cluster structure creates a mesh, established by links between clusterheads (connected via some gateway nodes), which can be used for faster and fault-tolerant geocast of subsequent traffic.

Directed diffusion[24] is a highly cited scheme for data gathering. The data sink identifies a set of attributes and propagates an interest message throughout the network. The interest message flooded the network (apparently blind flooding was used). Each receiving node records the interests and establishes a so-called gradient state indicating the next hop direction for other nodes to report the data of interest. When an interest arrives at a data producer, the data is forwarded to the sink along established gradients. Note that the algorithm is similar to the well-known *AODV* routing scheme,[25] considered as a possible routing standard. Flooding the interest, with attribute-based addressing, corresponds to the route discovery with IP or ID addressing. Therefore, various *AODV* optimizations that exist in the literature are applicable in the context of directed diffusion. The algorithm is also not scalable, despite claims that it is. Yu et al.[26] then considered the geocasting variant of the problem. They described the *GEAR* (geographic and energy aware routing) algorithm, which uses energy-aware neighbor selection to route the packet toward the target region, and recursive geographic forwarding, or restricted blind flooding algorithm, to disseminate the packet inside the destination region. Recursive forwarding applies *GEAR* to send messages to four sub-regions in the geocast region, which repeats until the region has a single node inside it. Both options are apparently less efficient than intelligent flooding schemes described later in this chapter. The *GEAR* algorithm selects a forwarding neighbor (among those that are closer to the destination) that minimizes a linear combination of their distance to the destination and the energy they already spent. This is almost equivalent to the cost-aware localized scheme by Stojmenovic and Lin[27] originally proposed in 1998. Yu et al.[26] also claimed that *GEAR* can avoid holes by applying a learning A^* algorithm based approach, without presenting details of it. To avoid holes, one can use, for example, a depth-first search (*DFS*) approach.[28] This approach requires memorizing past traffic at nodes. If, for example, *GFG*[16] is applied first, memorization is avoided (algorithm is more adaptive to topology changes), and some optimizations (described in Ref. 26) can follow later on recursively. Note also that Heidemann et al.[29] have further elaborated on the use of *GEAR* for various forms of data dissemination, but without giving any description.

Stojmenovic et al.[14] have described a general geocasting algorithm in which a message is forwarded to exactly those neighbors that may be best choices for a possible position of destination (using the appropriate criterion). If direction is used as a criterion, the corresponding *R-DIR* method may add one neighbor on each side of the tangent, if it is a better choice for the tangent than the closest direction toward the neighbor inside the tangent. Two other proposed criteria—geographic distance[19,27] and progress[20]—were used in Stojmenovic et al.[14] to describe the *VD-greedy* and *CH-MFR* methods, respectively. To avoid loops, only neighbors that are closer to the center of the considered geocasting region than the current node are considered.[14] Some of these neighbors are designated as forwarding nodes, and some are not. In the *VD-greedy* method, forwarding neighbors are determined by intersecting the Voronoi diagram of neighbors with the region (e.g., circle) of possible positions of destination, while the convex hull of neighboring nodes is analogously used in the *CH-MFR* method. Memoryless *VD-greedy* and *CH-MFR* algorithms are loop-free and have smaller flooding rates (total message count), with similar success rates, compared to the directional method.

We now describe the *VD-greedy* geocasting scheme[14] in more detail because it is a basic ingredient in the scheme that guarantees delivery,[17] as described in the next section. It is based on the Voronoi diagram concept, defined as follows. A Voronoi diagram of *n* distinct points in the plane is a partition of the plane into *n* Voronoi regions, one associated with each point. The Voronoi region associated with node *A* consists of all the points in the plane that are closer to A than to any other node. It can be shown that each region is a convex polygon (possibly unbounded) determined by bisectors of A and other nodes (more precisely, each region is the intersection of all such bisectors). There are several optimal $O(n \log n)$ algorithms[30] for constructing Voronoi diagrams. Each of these algorithms has sophisticated details. However, because the number of neighbors n is normally small, a node equipped with a processor can construct a Voronoi diagram with any of these methods in real-time. If a simple method is preferred, node S can make a decision in $O(n^2)$ time by constructing a bisector for each pair of its neighbors.

The *VD-greedy* geocasting method[14] is based on determining those neighbors of the current node that may be closest to a possible location of the destination. In Figure 6.1, nodes I, K, A, B, and C are closer to D than S, while L and J are not. The Voronoi diagram of neighbors I, K, A, B, and C of S is marked by dashed lines. Consider one of the bisectors, for example, for nodes B and C in Figure 6.1. Node S may determine that the circle is completely on one side of the bisector, and therefore node C is closer than node B for any possible location of destination within the circle. Node B is therefore deleted from the forwarding list of neighbors. Forwarding nodes for S are therefore A and C.

We also describe the corresponding selection of neighbors for the *CH-MFR* scheme,[14] which is based on the *MFR* routing algorithm.[20] Construct the convex hull $CH(S)$ of all eligible neighbors of a given node S, and tangents from S to the circle of possible location of destination (see Figure 6.1), which touch the circle at points U and V. In Figure 6.1, the convex hull is $CH(S) = KACI$. Then find the neighbors U' and V' that will be selected by S if D is located at U and V, respectively. The message is forwarded from S to all neighbors that are located on $CH(S)$ between U' and V' (including these two points). The forwarding neighbors for the *CH-MFR* algorithm in Figure 6.1 are nodes $U' = A$, and $V' = C$. The portion of $CH(S)$, "cut" by orthogonals from U' and V' to the two tangents, consists of three bold lines in Figure 6.1 that meet at A and C. The convex hull of n nodes can be constructed in $O(n \log n)$ time using Graham's scan algorithm.[31,32] Because the performance of *CH-MFR* is quite close to the performance of *VD-greedy*, which is conceptually simpler, we will not mention *CH-MFR*-based algorithms in the sequel, although they can be used as an alternative. Stojmenovic et al.[14] proposed the user of the dominating set concept[4] to reduce the flooding ratio significantly (in all proposed schemes), with a marginal impact on the success rate and hop count.

Liao et al.[12] suggest that to reduce flooding overhead one should divide the request zone into squares with diagonals approximately equal to the transmission radius (only one node in each square needs to retransmit the message). This is a useful approach for dense networks. However, its application as described requires that *a priori* network subdivision be communicated to all nodes. Further improvement along these lines can be obtained by applying the dominating set concept (see Section 6.4), which will further minimize the number of retransmissions and avoid the use of any division information.

Schwingenschlogl and Kosch[13] considered the geocasting protocol in the context of vehicular networks and suggested that one should define geocasting zones in accordance with the highway context and high-speed movement.

Camp and Liu[9] propose to create a mesh as a subset of the network topology that provides multiple paths between multicast senders and receivers. This mesh is used to establish paths between a source node and a geocast region, and it floods join request packets to mobile nodes within the geocast region via a forwarding zone instead of the entire ad hoc network. To establish multiple paths, they use blind flooding within the whole network, or within the cone or corridor shaped request zones between the source and geocast region.

All described geocasting algorithms fail to guarantee delivery because they either restrict the forwarding region between the source and geocasting region (connected via node outside the two regions), or they do not consider that nodes inside the geocasting region can be disconnected (but could be connected via nodes outside it). This is illustrated in Figure 6.2. Described methods are based on greedy forwarding restricted to neighbors inside a region, such as between tangents from the current node or source to the geocasting region, a rectangle containing the source and geocasting region, or all nodes closer to the geocasting region than the current node. In Figure 6.2, these regions are drawn by dashed, dot-dashed, and dotted lines, respectively. If white nodes are not in the network, the source is disconnected from the geocasting region,

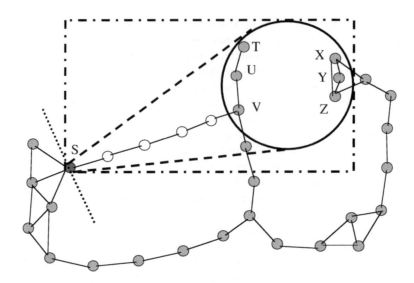

FIGURE 6.2 Example where most existing geocasting schemes do not guarantee delivery.

but could be connected via nodes that do not belong to indicated regions, as shown in Figure 6.2. Next, connecting the source to some nodes (e.g., T, U, V in Figure 6.2) does not mean that all nodes inside the geocasting region will be reached. For example, nodes $X, Y,$ and Z in Figure 6.2 are disconnected from T, U, V inside the geocasting region. However, they are connected via nodes outside it.

A routing algorithm that guarantees delivery by finding a simple path between source and destination (without any flooding effect) is described by Bose et al.[16] The *GFG* (greedy-face-greedy) algorithm[16] applies the greedy method until the current node A has no neighbor closer to the destination than itself (such a node is called a *concave* node[16]), or the message is delivered. Concave node A switches to the recovery mode by applying face routing,[21] and improved by Bose et al.[16] Face routing uses only the edges of a planar graph. Gabriel graph (*GG*) was used by Bose et al.,[16] constructed as follows. An edge AE belongs to *GG* if the disk with diameter AE contains no other nodes from the set. This can be tested by verifying whether the angles to this edge from all neighbors are acute, and the test does not require any message exchange between neighbors. In Figure 6.2, nodes belonging to *GG* are marked with bold lines. Face routing guarantees delivery in connected planar graphs, but is followed only until a node closer to the destination than the last concave node is encountered. Such a node switches back to the greedy mode. This mode alteration may repeat a few times, but the message is guaranteed delivery, and the *GFG* algorithm was shown to be competitive with respect to the shortest path, especially with some improvements given by Datta et al.[33] The improvements described by Datta et al.[33] include restricting face routing to nodes in a connected dominating set and applying a shortcut procedure, which is explained in a later section.

We illustrate the *GFG* routing example[16] in Figure 6.3 for the route from S to Y. Greedy routing *SAFV* is applied until concave node V is reached. Node V then switches to the recovery mode and applies face routing. Face routing follows faces along an imaginary line from source to destination, changing faces at intersections of the imaginary line with the faces of *GG*. Face routing from V to Y follows an open face, as marked with scribble line (route *VGHZ*). The return from recovery mode to greedy mode is possible at node Z, which is closer to the destination than the previous concave node V. Z then delivers to Y, and the imaginary line VY was never crossed in this example. To better understand face routing, we can show it through an example that is not applied in our scheme, one that uses *GFG* (that is, relies on greedy routing as much as possible). Consider face routing from S to M in Figure 6.3. From S, it starts in any direction (say in one that makes an initially smaller angle with respect to the imaginary line SM). It then traverses all faces that cross SM, switching faces at intersections. The face route in this case is drawn with a scribble line, and is *SBICPNFOM*. Note that this is after crossing face change-based face routing, before the crossing face change variant can be used instead, with similar performance.

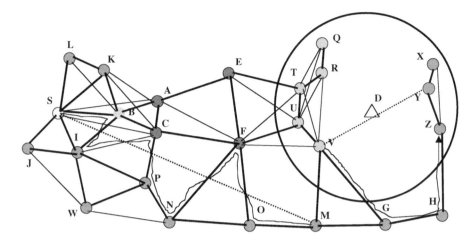

FIGURE 6.3   Gabriel graph: face routing and *GFG*.

A simple geocasting algorithm was proposed in a technical report version by Stojmenovic et al.[14] Source node *S* applies the *GFG* algorithm[16] to route toward center *D* of the region, until a node inside the region is encountered. In Figure 6.3, the route is equal to the greedy route *SAFV*. The first node inside the region then applies a flooding scheme, restricted to nodes inside the region. This surprisingly simple algorithm has a smaller flooding rate and increased delivery rate compared to all known methods. However, it also fails to guarantee delivery (see Figures 6.1, 6.2, and 6.3). Nevertheless, it is used as a basic ingredient in the scheme[17] that does guarantee delivery, which is described in Section 6.4.

6.3   Geocasting Based on Traversing Faces that Intersect the Boundary

Bose et al.[16] observed that a geocasting algorithm will guarantee delivery if all faces of a planar graph that are inside or intersect a geocasting region are traversed. The algorithm is based on a depth-first search of the face tree, and is described in the next section. We first describe a simpler algorithm, where only faces that intersect the region boundary are traversed.

Saeda and Helmy[18] observed that it is sufficient to traverse only faces that intersect the boundary of a given geocasting region, and proposed the following algorithm. The source node first uses the GFG algorithm[16] to forward the packet toward the region. Each node that is inside the region will retransmit the packet when receiving it for the first time ("regional flooding"). "A node is a region border node if it has neighbors outside of the region. By sending perimeter packets to neighbors outside the region (notice that perimeter packets are sent only to neighbors in the planar graph, and not to all physical neighbors), the faces intersecting the region are traversed. The node outside the region receiving the perimeter mode packet forwards the packet using the right-hand rule to its neighbor and so on. The packet goes around the face until it enters the region again. The first node inside the region to receive the perimeter packet floods it inside the region or ignores it if that packet was already received and flooded before."[18]

We first show that this algorithm by Saeda and Helmy[18] does not guarantee delivery, despite the claim that it does, and then we describe an algorithm, based on the same initial observation, that does guarantee delivery. Consider the example in Figure 6.4, where *S* is the source node. The message will enter the region on edge *MN*, and node *M*, according to Saeda and Helmy,[18] will not trigger face traversal (which would have reached the group of nodes containing node *I*, inside the region). Node *N* will initiate regional flooding (marked by dotted lines in Figure 6.4), which reaches nodes *W*, *P*, *D*, and *H*, having perimeter neighbors (*A*, *A*, *E*, *F*, *G*) outside the region. Nodes *A* (twice), *E*, *F*, and *G* are then instructed to perform right-hand face traversals, indicated by dashed lines inside corresponding faces in Figure 6.4 (dashed lines show full traversals for illustration of the intention, but traversals stop when entering the

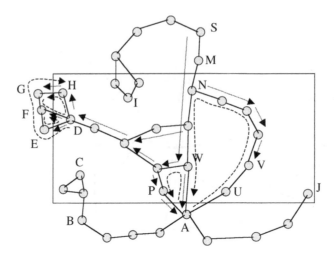

FIGURE 6.4   Face traversal based geocasting with right-hand rule.

region). According to the described scheme,[18] node U, receiving the packet from outside, does not instruct node A to perform a perimeter search, which misses the opportunity to "discover" nodes J and C that are inside the geocasting region (see Figure 6.4).

A geocasting algorithm that guarantees delivery can be described as follows:

Algorithm Geocast _traversal_intersecting_faces

> Source node S sends the message toward the geocasting region, using the *GFG* algorithm.[16]
>
> Each node, inside the region, retransmits the message when receiving it for the first time and ignores it when receiving it again.
>
> Each internal border node (node inside region having neighbor(s) on planar graph outside the region) will instruct (together with retransmission) all its perimeter neighbors outside the region to perform right-hand based face traversals.
>
> Each external border node (node outside region having neighbor(s) on planar graph inside the region) will initiate right-hand based face traversal(s) with respect to all edges leading to internal perimeter neighbors, after receiving the first copy of the message, and will ignore further received copies unless a packet is received from an external neighbor following a different "external" face (in which case it forwards it along that face as requested). Each traversal is performed until another node that is inside the region is found.

The main difference between this algorithm and the one described by Saeda and Helmy[18] is that external border nodes perform right-hand based face traversals with respect to all corresponding neighboring internal border nodes no matter how the message arrived there; it is activated only from an internal border neighbor, for one face at a time, and is not activated if an external border node transmits before the corresponding internal border neighbor. The example in Figure 8 in Ref. 18 shows such a node along the top border line of region; if one of the nodes in the top-right corner was inside the region, it would have been missed by the algorithm, but could be reachable by right-hand traversal starting at the indicated node.

We now illustrate in Figure 6.5 our algorithm on the same example, starting from source node S. Node M starts right-hand based face traversal (finding first neighbor in the clockwise direction with respect to edge MN). Note that right-hand based face traversal results in counterclockwise face traversal of closed faces, and clockwise traversal of the single open face. The face traversal from M reached edge KI in Figure 6.5. Node I floods the packet regionally, while node K initiates two face traversals, with respect to edges KI and KL. Face traversal with respect to KI ends at L, while face traversal with respect to KL follows the outer boundary until "seeing" edge MN again (which then ignores it). Regional flooding reaches nodes W and D. W "alerts" A to perform face traversals with respect to AP, AW, and AU. Neighbors P (by listening to all traffic from A) and U (as part of face traversal) receive packets from A, and can retransmit as part

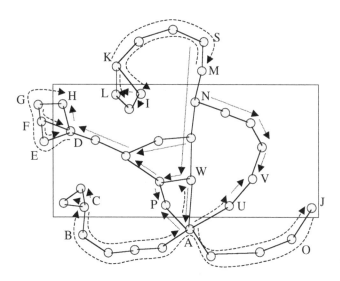

FIGURE 6.5 Face traversal based geocasting with guaranteed delivery.

of regional flooding. One face traversal reaches node J. Face traversal from O (neighbor of J) bypasses A and reaches nodes B and C. C floods to its neighborhood while B starts face traversal that reaches A again. Face traversal around node D does not go beyond that area.

We now *prove* that the algorithm indeed guarantees delivery to all nodes inside the geocasting region connected to the source node. The proof, in fact, is quite elegant, and is expressed in the following theorem.

Theorem 6.1   *The described geocasting algorithm, based on traversing faces that intersect the geocast region boundary, guarantees delivery to all nodes inside the geocast region, connected to the source.*

Proof   We can argue that every face, intersecting the geocasting region and connected to the source, was fully traversed by a combination of regional flooding and outer face traversals. Consider, for example, the outer boundary in Figure 6.5 (the proof is same for any face). Its traversal started at MN and reached I (by lower dashed line in Figure 6.5). With internal flooding, it reached LK from I. Then, from K it reached SMN (upper dashed line). By regional flooding it can reach UA. Then face traversals are used to follow AJ "piece," followed by $JABC$ "piece". Face traversal from C then reaches A again, and with regional flooding it reaches D, then face traversal $DEFGH$. Finally, flooding from H can reach back to MN, and the whole face is traversed. We could make the proof more formal, but believe that this informal exemplar explanation suffices. The main argument is that right-hand traversal of any face is composed of pieces containing regional flooding for consecutive face nodes inside the region, and pieces outside the region that are triggered when a packet arrives there. Regional flooding, piecewise face traversal, and connectivity assure reaching all possible nodes.

In addition to guaranteeing delivery, the proposed scheme is also close to a message optimal scheme, because each node inside the region retransmits the packet only once. Depending on face structure, some nodes outside the region can send several packets in the process. Let $N = n + k$ be the total number of nodes in the network, where n and k are the number of nodes outside and inside the geocasting regions, respectively. The number of messages sent inside the region by the described algorithm is then k. This number can be reduced if an intelligent flooding scheme is applied (see Ref. 4). Note, however, that nodes that have neighbors outside the geocasting region need to retransmit in order to "alert" their perimeter neighbors, although their neighbor inside the region may be "covered" by other transmissions. We will now find the upper limit for the number of messages outside the geocasting region. In the worst case, each edge outside the geocasting region can be traversed, at most, twice (once for each incident face). Because the graph is planar, the number of edges e is limited by $2e \leq 3n'$, where n' is the number of nodes on considered faces. In our case, we consider faces that intersect the geocasting region, and only a portion of these faces outside the geocasting region are of interest (we can treat the region border as being part of "shortened" faces, connecting two border nodes). Thus, n' is the number of nodes that are outside the geocasting region, belonging to a face that intersects the region. The inequality $2e \leq 3n'$ follows from the limit obtained when additional edges are added to triangulate all faces, and observing that then each edges belongs to one or two faces, but each face, being a triangle, contains three vertices. Therefore, the total number of messages is limited to $3n' + k < 3N$. A message from the source to a node inside the region, following the *GFG* scheme,[16] should be added as well. This worst-case limit is encouraging and appears smaller than in two other methods that guarantee delivery, as described here.

Note that Saeda and Helmy[18] proposed also another "practical" scheme that admittedly does not guarantee delivery. In that scheme, all nodes (with or without neighbors outside region) receiving a packet, in which at least one out of four quadrants is without a neighbor, will initiate counterclockwise face traversal along the empty quadrant. Each node can forward several face traversal packets in this algorithm, and every face needs to be fully traversed. This apparently causes significant message overhead inside the region, in addition to the lack of delivery guarantee.   □

6.4   Geocasting Based on Depth-First Search Traversal of Face Tree

Bose et al. (journal version of Ref. 16) proposed a geocasting algorithm that guarantees delivery to all nodes connected to the source, in which the packet follows a path from source node (thus, a single copy of the packet is in the network at any time). To improve latency, parallel paths (and multiple copies of the packet) can be explored at any branches of the face tree being used. The algorithm[16] (whose complete description is available in Ref. 34) does not require any memory to be left at the nodes, and needs only to carry some small amount of information with the packet (if entry edges are predetermined for a given source, then the message only needs to contain sender and source information).

6.4.1   Depth-First Search Based Traversal of Face Tree

The algorithm[16] first applies *GFG* to route toward a node inside the geocasting region. That node then selects a nearby point S inside the face to act as an artificial source. The face tree from S is constructed in the following way. Given a node S and a face f of a planar graph, the *entry edge entry* (f, S) is the edge from f that is closest to S. To break the ties, several keys for comparison of edges are used. The primary key is the distance from the edge to S, where the distance is decided by a point C from the edge that is closest to S. If the distances are same, a secondary key is used in the counterclockwise direction of vector SC. In case of further ties (which can occur only when two edges share common closer endpoint C), consider the size of the angle $\angle SCD$, where D is the other endpoint of the edge. If that still does not resolve, then consider the vector CD, which then must be different. Morin[34] proved that all entry edges are on the boundaries of two faces. In the face tree, the parent of a face f is the face $p(f)$ that contains its entry edge $e(f)$ on its boundary. Obviously then, $p(f)$ itself has other entry faces closer to S, which confirms that a tree of faces is indeed constructed. A face tree is dynamically constructed during geocasting operation. The geocasting algorithm follows a depth-first search based traversal of the face tree. For each node in the face tree, it actually traverses the corresponding face. When an entry edge is encountered, the traversal enters a new face. When traversal (which can recursively go to deeper levels) is completed, the traversal returns to the face. Traversal of each face begins from one end of its entry edge, and finishes at the other end of it. Figure 6.6 illustrates the algorithm. The face tree from S is drawn with directed edges intersecting entry edges (which are drawn with dashed lines). The path taken by the geocasting algorithm is shown with a dotted scribble line, starting from point S, where near real node U acts on behalf of S. The algorithm visits all edges along the path.

The algorithm[16,34] can be described as follows. In this scheme, opposite (e, f) is the other face containing the same edge e as the face f currently being traversed. The edge next (e, f) is the next edge being traversed by the right-hand rule from the current edge e on face f.

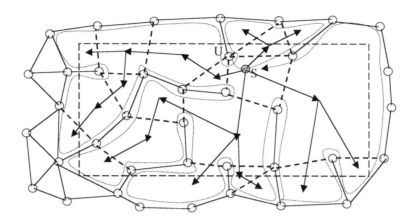

FIGURE 6.6 Face tree traversal based geocasting with guaranteed delivery.

Algorithm Geocast_Face_Tree_Traversal

$f \leftarrow$ face containing S
$e\_start \leftarrow e \leftarrow$ an edge of f
repeat
if e intersects geocast region then
if $e = $ entry(f, S) then $f \leftarrow$ opposite (e, f)
　$\{^*$ return to parent of $f^*\}$
else if $e = $ entry (opposite $(e, f), S)$
then $f \leftarrow$ opposite (e, f) $\{^*$ visit child of $f^*\}$
　$e \leftarrow$ next (e, f)
until e = e_start

In some applications (see Section 6.4.3), entry edges can be determined as the preprocessing step. For example, if the geocasting source is a fixed base station in sensor networks, then entry edges can be determined by flooding the network from the base station, and traversing all faces of the planar graph to determine and conveniently label entry edges. Afterward, geocasting regions can be dynamically determined (e.g, to follow a moving object) and geocasting can then proceed as described.

If entry edges are identified prior to geocasting communication, then the algorithm requires considerably fewer steps. Let n' and k be defined as in the previous scheme. We can consider face traversal as traversal of $F = 1$ face. The Euler formula for planar graphs is $F = E - N + 2$, where F, E, and $N = n' + k$ are numbers of faces, edges, and vertices, respectively. Thus, $E = N - 1$. The number of messages is at most $2E < 2(N - 1) < 2(n' + k)$. Therefore, the number of messages is limited to $2(n' + k)$. Communication steps from source to the geocasting region need to be added as well. Therefore, the scheme has reasonable communication overhead under the given assumptions. Compared to the previously described scheme, it has less communication overhead when $k < n'$. It does require preprocessing (or a significant additional number of messages at runtime) and offers the additional benefit of providing a single path in the network, which provides for time division that is suitable for applications when sensor networks alternate in reporting to the monitoring center directly (see details in Section 6.7).

6.4.2　Testing Entry Edges

For a dynamically selected source of geocasting message, such preprocessing is not possible, and we will now describe the scheme used by Bose et al.[16] and Morin[34] to decide, for each edge visited in the algorithm, whether or not it is an entry edge for the neighboring face to the face being currently traversed (if so, the traversals enters that new face). After the search arrives at a node A with the intention to continue along

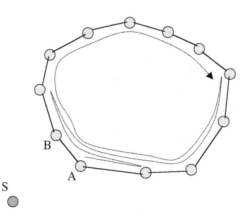

FIGURE 6.7   Testing whether or not *AB* is an entry edge for given face.

edge *AB*, the other face containing *AB* is tested by comparing the distance of *AB* with distances of other edges on that face. A full-face traversal will do, but would perform purely on average. Instead, starting from $p = 1$, and from node *A*, *p* edges are tested in given direction (e.g., initially clockwise). The search then returns to A, and, if *AB* is not "defeated," *p* doubles ($p \leftarrow 2p$), and the direction of face traversal changes. When *p* reaches its maximum (decided by the number of edges on the face, if known, or by recognizing on a path in a given direction that the first node has been already seen, as in Figure 6.7), and *AB* is still not defeated, *AB* is declared the entry edge, and the other face is entered. The order of comparison is illustrated in Figure 6.7 by dotted lines. The number of messages sent in the scheme overall is $O(N + k \log k)$,[16] where the latter term is due to entry edge tests. The proof that this geocasting algorithm guarantees delivery to all nodes connected to the source is given by Bose[16] and Morin.[34]

6.4.3   Application of Face Tree Traversal Based Geocasting for Data Gathering

Lindsey, Raghavendra, and Sivalingam[35] proposed a different framework for energy-efficient data gathering algorithms in sensor networks. They first organized sensors into a chain, by a centralized algorithm. Thus, sensors are initialized as $c_0, c_1, \ldots, c_{n-1}$. Data gathering is performed in rounds. In round *k*, first find $i = k \bmod n$. Each round consists of *n* iterations. In each iteration, only one sensor is sending a message, containing data gathered by that sensor. Iterations are performed as follows: c_0 sends to c_1, c_1 to $c_2, \ldots, c_{i-2}$ to c_{i-1}. Then c_{n-1} sends to c_{n-2}, c_{n-2} to $c_{n-3}, \ldots, c_i$ to c_{i-1}. Finally, c_i sends gathered data to the monitoring center (MC). Therefore, all sensors are assumed to know global information and the distance to all other sensors. The distance to MC is assumed to be larger than distances between sensors. There are several problems with the proposed solution. It is a centralized solution, and we believe in localized ones. Chains can be difficult to construct in multihop networks; for single-hop networks, an initialization algorithm needs to run, or the MC needs to assign IDs to individual sensors. Once constructed, when sensors change activity status (from active to passive) or stop functioning, the order scheme needs to run again. The scheme is also not sensitive to the power level of the sensors. Several localized solutions are proposed by Stojmenovic.[36] One is a localized algorithm that first constructs *LMST*[37] (or another sparse structure such as *RNG*). Instead of creating a chain, a token is circulated in the network. The node currently having the token will send it to one of its *LMST* neighbors. This can be done in different ways. Nodes can forward with equal probability to send to one of their neighbors (not returning to the neighbor it came from). Because the average degree of *LMST* is about 2.04,[37] there is, on average, one such neighbor to forward tokens. The forwarding probability may depend on node densities. Neighbors with more *LMST* neighbors should have a smaller probability of getting tokens (because they may get tokens from more neighbors in the process). Next, neighbors with more energy left may have a higher probability of getting tokens. Finally, in case of monitoring an event that can be geographically located, sensors nearby need to preserve more energy, and thus they may decide to postpone reporting to the MC. Thus, instead of reporting every *n*th time to the MC, the frequency can also be decided probabilistically, depending on the energy level of the node. *LMST* can be converted to *MST*,[38] or a spanning tree can be constructed by broadcasting a message from the token holding the node. The constructed tree can be used for data gathering from other sensors, before the token holding node sends a report to the MC.

Another solution proposed by Stojmenovic[36] is to apply a geocasting algorithm[16,34] that follows a single path from the source while visiting all nodes in the region. The algorithm guarantees to see all nodes and, on average, does so twice during a single geocasting process, which can be repeated periodically. If the base station is fixed, preprocessing can be done to decide entry edges and reduce communication time, as described in Section 6.3.2. The advantage over the solution described above is to guarantee the participation of each node on a fairly regular basis.

6.5   Multicasting and Geocasting with Guaranteed Delivery

We now describe an algorithm based on the following observation. If a node V inside a geocast region is connected to the source S, then the first node U on a route from S to V is no more than the transmission radius distance R from the border of the geocasting region. The set of points at distance $\leq R$ from the border of the geocasting region is called the *entrance ring*. The entrance ring is subdivided into *entrance zones*. The diameter of each entrance zone must be $\leq R$, and each such division can be used. The geocasting algorithm based on multicasting and entrance zones can be described as follows.
Algorithm Geocast_Entrance_Zones_Multicast

Determine entrance zones and their centers.

Multicast from source S toward the centers of each entrance zone, until a node inside the zone is reached (these nodes will be called multicast recipients), or a loop in recovery mode of routing scheme is identified.

Flood from each multicast recipient. This can be done by blind flooding restricted to nodes inside the region, or by some intelligent flooding scheme[4] that reduces the number of retransmissions. Each node memorizes received packets and ignores repeated copies of the same packet.

The next three sub-sections elaborate on these steps and prove that guaranteed delivery holds. We also illustrate the algorithm and discuss its message complexity.

6.5.1   Determining Entrance Zones

The entrance zones should be determined with the following two criteria in mind:

It is not possible to send a message directly from a node outside the geocasting region to a node inside it; this means that the width of all the zones together, measured as the minimum distance between a node outside the geocasting region to a node inside it that does not belong to any entrance ring, must be at least R, the transmission radius of the network.

The diameter of each entrance zone must be at most R; this means that if a node inside a zone receives the multicast packet, then all other nodes in the same zone will receive it after retransmission from that multicast recipient.

The exact construction of entrance zones to satisfy these criteria depends on the shape of the geocasting region. If the geocasting region is a rectangle, then the entrance zones may be composed of two layers of squares of edge length $R/2$. One dimension (not affecting overall width R) can be increased until the diameter becomes R. We illustrate here the case of a circular geocasting region. Let G be the radius of the circle. The entrance ring may be composed of two layers of zones, each of width $R/2$. The length along the circle for each zone may be extended until the diameter becomes R. Instead of doing an exact calculation, we suggest here a simpler determination of angle α from the center of the geocasting region toward each zone (see Figure 6.8). If we select α such that $\alpha = R/(2G)$, then the zone diameter (the distance between two diagonal points) trivially is less than R.

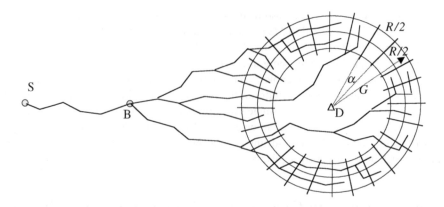

FIGURE 6.8   Covering entrance zones from a remote monitoring center.

6.5.2   Position-Based Multicasting

In a multicasting task, the sender node wishes to send the same packet to several other nodes in the network. Routing and broadcasting are all special cases of multicasting. Mauve et al.[39] proposed two similar multicasting schemes, with some optimizations. In the *optimal paths* method, each node receiving a multicasting message for a group of nodes will forward it to each neighbor that is closest to one of the group members. More precisely, each group member is assigned to a neighbor that is closest to it (provided that neighbor is closer to it than the current node). In the *aggregate paths* method, for each neighbor A, the number of destinations for which A is the closest node is determined. Then a covering algorithm is applied. Basically, a neighbor is chosen that covers the maximum number of destinations, these destinations (and other nodes for which the selected node makes some progress) are eliminated from the list, then another neighbor is chosen that covers a maximal number of remaining destinations, etc. The forwarding list of the multicast group similarly is changed as in the previous algorithm.[39] In both schemes, if no neighbor is closer to one or more destinations, then the recovery mode in the *GFG* algorithm[16] is applied. The virtual destination used for the recovery mode is calculated as the position representing the average of the positions of the affected destination nodes. When a node receives a multicast packet in recovery mode, it checks for each destination, if it is closer to that destination than the node where the packet entered recovery mode. For all destinations where this is the case, greedy multicast forwarding can be resumed as described in the corresponding scheme. For all other destinations, recovery mode is continued, with updated average of positions of affected nodes (those not recovered yet).

Note that the optimal paths method (without recovery scheme) corresponds to the *VD-greedy* scheme.[14] They both use hop count as the metric. Both optimal and aggregate paths methods can be modified by considering metrics different from hop count, such as power, cost, delay, or some other. Greedy routing can be replaced by power and/or cost-aware routing (see Section 6.6), and forwarding neighbors will be judged based on the metric in question, combined with their coverage ability, for their selection.

6.5.3   Entrance Zone Multicasting Based Geocasting
with Guaranteed Delivery

The algorithm consists of multicasting toward "centers" of all entrance zones, and flooding from the first nodes encountered in each non-empty zone. The zone center is any node inside it (for example, its center of mass or intersection of zone diagonals). In Figure 6.8, multicasting used in our scheme is illustrated. The source S initiates multicasting, which begins branching at B. This figure applies only greedy forwarding for simplicity. Several entrance zones in Figure 6.9 are empty, and the algorithm will make one loop in the *GFG* algorithm to confirm this (this loop is not drawn).

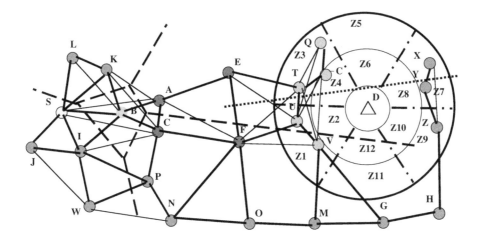

FIGURE 6.9 Optimal path-based multicasting to entrance zones.

A multicasting scheme can be followed in one of the ways described above. We will follow the optimal paths method on an example in Figure 6.9, where 12 entrance zones are denoted by $Z1$ *through* $Z12$ (we have used the position of zone names for zone centers). The monitoring center S considers a Voronoi diagram of its neighbors to make decisions. It assigns consecutive zones $Z2$ *through* $Z10$ to node A, and consecutive zones $Z1$, $Z11$, and $Z12$ to node C. Each intermediate node receives an interval of these zones that it is assigned to cover. Continuing the process, A forwards the message to E, assigning it zones $Z3$ *through* $Z6$, and to F, assigning it zones $Z2$, $Z7$ *through* $Z10$. Node C forwards in greedy mode to F, assigning it all its zones. Node E will forward the message to node T, covering $Z3$ with it, and assigning $Z4$ *through* $Z6$ to it to cover as well. To cover $Z4$, T forwards the message to C which is inside it, and assigns it $Z6$. Node C can conclude that $Z6$ is empty if it is at distance $<R$ from any point of it, or can conclude it by following the network perimeter in recovery mode with the message returning to it. To cover $Z5$, T (assuming it is at a distance greater than R from some points in $Z5$) sends packet to Q in greedy mode, then Q begins recovery mode, and follows a route along the perimeter $QCTUVGHZYXYZHGMONPWIJSLKBAETCQ$ at which point it discovers the loop, meaning that the zone $Z5$ is empty. Note that some optimizations can be made here. For example, Q and X collectively can conclude that $Z3$ is empty, or Q can change the virtual destination location $Z3$ to a center of subzone not in its reach. Node F is responsible for consecutive zones $Z1$–$Z2$, $Z7$–$Z12$, and "transfers" all these zones to V. We note here that optimization via merging assigned zones can be made when few neighbors assign tasks independently, that is, a node can wait for possible new assignment before starting its own forwarding and assignment; this is the case with node F here. V covers $Z1$, and may conclude that $Z2$ is empty since it is near its borders and thus at a distance less than R from any point of it. To reach $Z7$–$Z12$, node V converts to recovery mode and follows $VGHZ$ with face routing, then completes with ZY or ZX in greedy mode. Some of the zones are covered, and some may remain unresolved. To cover such zones (if any left), V initiates a face route until the face route makes a loop and returns to V.

Upon entering any zone, the protocol converts to an intelligent flooding inside the geocast region. In all existing intelligent flooding methods (see a review by Stojmenovic and Wu[4]), nodes may receive multiple copies of the same message, but forward it once only (or not forward at all), after a timeout which depends on the protocol selected. The intelligent flooding for geocasting inside the region, and existing flooding methods, differ in only one sense. Instead of having just one source for flooding, a geocast application can have several such sources, one per entrance zone. This difference requires adjusting timeouts to somewhat larger values than in regular flooding tasks, or memorizing past traffic somewhat longer, because some messages may be delayed by longer forced routes while in recovery mode, before arriving at an entrance zone. Also, the distances from a given node to entrance zones may be considerably different, adding to the differences in message arrival times.

The monitoring center S can be outside or inside the geocasting region. Although our description implicitly assumes that S is outside the region, the same algorithm works correctly if S is also inside it. We will now prove that this geocasting algorithm guarantees delivery.

Theorem 6.2   *The described geocasting algorithm, based on multicasting to entrance zones and flooding from multicast recipients, guarantees delivery to all nodes inside the geocast region, connected to the source.*

Proof   The proof that multicasting entrance zone based geocasting guarantees delivery is based on two key arguments. First of all, multicasting itself guarantees delivery, based on the guaranteeing delivery property of *GFG* (proven in Ref. 16), which is applied toward every destination. Guaranteed delivery of multicasting is also claimed by Mauve et al.[39] Next, we argue that any node inside the geocasting region, connected to the source, must be connected to at least one of the mentioned multicasting recipients. Suppose that a node X is inside the geocasting region. Then it is inside an entrance zone, or outside all entrance zones. If it is inside an entrance node, it is at distance $< R$ from a multicasting recipient, and therefore receives a retransmitted message from that recipient. If it is outside all entrance zones and connected to the source S, then the path from S to X needs to cross the entrance zone's ring somewhere. Because the width of that ring is R, it cannot "jump" over it and cross directly from outside the geocasting region to inside the geocasting region ("escaping" the entrance ring). Therefore, the path contains at least one node in one entrance ring. That node is connected to a multicast recipient, and flooding initiated from that multicast recipient will reach X. Therefore, all nodes connected to the source will receive the geocasting packet, and the algorithm then guarantees delivery.

It appears that (in dense networks) the proposed protocol[17] has small communication overhead with respect to listed methods that do not guarantee delivery. To convince the reader that this is the case, consider the case of a larger network in Figure 6.8. It demonstrates covering entrance zones from a remote monitoring center. The branching begins only when the message arrives close enough to the geocast region. The *VD-greedy* algorithm would approximately follow all branches from there until regions that are "visible" from B are encountered. The differences are only messages beyond these front entrance zones, toward entrance zones on the other side with respect to tangents from B to the geocast region. A more efficient protocol would merely continue the path from B without branching. Therefore, the comparative communication overhead depends on the relative distance from the monitoring center to the geocast region. It also depends on the existing coverage of the geocast region by active sensors. Obviously, several empty regions may cause long routes along the network perimeter to recognize them.

Entrance zone multicast based geocasting is expected to be competitive with face traversal based schemes, on average. However, in the worst case, it can exhibit excessive overhead due to potential face routing along the network perimeter for each empty region. Consecutive empty entrance zones, fortunately, do not necessarily require separate face routings, because the multicasting method merges them into a single destination. The worst case appears to be the scenario with every other entrance zone being empty, each of them thus requiring separate face routing to be confirmed. Note that the worst-case behavior of this algorithm can be reduced if the *GFG* procedure[16] is replaced by its enhancement,[40] which is based on applying *GFG* to a series of ellipses with doubling sizes. However, that procedure appears to somewhat worsen average case performance.

In conclusion, there are four geocasting algorithms that guarantee delivery. Intelligent flooding delivers to all nodes in the network (solves broadcasting task), and is best when the geocasting region is nearly covering the whole network. It is also competitive when the geocasting region covers a large portion of the network. The three methods presented here are designed for cases where the geocasting region is relatively small. Among the three proposed schemes, it is expected that (on average) traversing faces that intersect the boundary will perform best when there are many empty entrance zones; otherwise, a multicasting-based solution should be best. Depth-first search based face tree traversal requires preprocessing for reasonable performance, and has applications for sensor time division when reporting directly to the monitoring center. More reliable conclusions can be made only after their performance evaluation.   □

6.6   Conclusion

In large or dense ad hoc and sensor networks, it is not necessary to use all available nodes for performing data communication tasks. In sensor networks, for example, some nodes are sensing areas while some other nodes are there to support routing, a basic data communication protocol for data gathering. Some or all sensors can simultaneously perform sensing and forwarding traffic tasks. There are several reasons to reduce the number of nodes needed for monitoring or routing. Face routing, for example, has better performance on a connected dominating set than on a full set,[33] because there are fewer nodes, and consequently longer edges, to traverse in the considered planar graph. Intelligent flooding also is based on a connected dominating set, where nodes not belonging to it do not need to retransmit the message (see Ref. 4 for a survey on dominating set based broadcasting). To save energy, sensors may decide between active and sleeping nodes, with the goal of providing area coverage for monitoring reasons. Geocasting can be restricted to nodes in a connected dominating set, or nodes in an area coverage set. This is applicable to all methods described here.

There exist other relevant aspects, such as data gathering and aggregation, in the process of issuing requests and collecting data, with sensor ad hoc networks as the primary applications in mind. Protocol efficiency is the primary goal, with efficiency defined by some metrics or design characteristics (such as the localized behavior of protocols).

We assume that all active sensors within the monitored region need to be "alarmed," and assume the application of localized algorithms. Geocasting then needs to guarantee delivery, and this article describes all such existing methods. The next step is their analysis by means of performance evaluation. The proposed schemes have their variants, and allow for optimization with a variety of criteria and a variety of options for their implementation. A comparison of geocasting methods depends on the relative size of the geocast region compared to the size of the area containing all sensors, or, more precisely, on the ratio of the numbers of active sensors inside the geocast region. When compared to intelligent flooding, the smaller the size of geocast region, the more advantages the described methods provide. Their mutual comparison depends on parameters such as density and the existence of "holes" in the network. Performance evaluations may also lead to further improvements in each presented method, or their adjustments to particular scenarios or evaluation criteria. The performance evaluation is left for future work.

The described geocasting protocols can also be used for a few other related applications within sensor networks. One or more monitoring stations may simultaneously geocast to one or more regions. In the case of a monitoring region consisting of several disconnected sub-regions, the same protocol can still be followed. The protocol can also be used for geomulticast applications, such as reporting from one sensor to several monitoring stations.

Ad hoc and sensor network have recently attracted exponentially increasing interest, including the creation of new conferences, new journals, and publishing a number of books. We envision that this trend will continue in the short term and that network layer problems, discussed in this chapter, will continue to be intensively studied. We hope that the research efforts will lead toward real applications of ad hoc networks, especially sensor networks.

Acknowledgment

This research is partially supported by an NSERC Discovery grant.

References

1. J. Carle and D. Simplot-Ryl, Energy efficient area monitoring by sensor networks, *IEEE Computer,* February 2004, 40–47.
2. X. Wang, G. Xing, Y. Zhang, C. Lu, R. Pless, and C.D. Gill, Integrated coverage and connectivity configuration in wireless sensor networks, *Proc. ACM SenSys,* Los Angeles, November 2003.

3. H. Zhang and J.C. Hou, Maintaining Sensing Coverage and Connectivity in Large Sensor Networks, UIUCDCS-R-2003–2351, June 2003.

4. I. Stojmenovic I. and J. Wu, Broadcasting and activity scheduling in ad hoc networks. In *Ad Hoc Networking*, S. Basagni, M. Conti, S. Giordano, and I. Stojmenovic (Eds.), IEEE Press, 205–209, 2004.

5. D. Niculescu, Positioning in ad hoc sensor networks, *IEEE Networks*, 18(4), 24–29, July 2004.

6. B. An and S. Papavassiliou, Geomulticasting: architectures and protocols for mobile ad hoc networks, *J. Parallel and Distributed Computing*, 63, 182–195, 2003.

7. B. An and S. Papavassiliou, A mobility-based clustering approach to support mobility management and multicast routing in mobile ad hoc wireless networks, *Int. J. Network Management*, 11(6), 387–395, 2001.

8. S. Basagni, I. Chlamtac, and V.R. Syrotiuk, Geographic messaging in wireless ad hoc networks, *Proc. 49th IEEE Int. Vehicular Tech. Conf. VTC'99*, Houston, TX, May 1999, 3, 1957–1961.

9. T. Camp and Y. Liu, An adaptive mesh-based protocol for geocast routing, *J. Parallel Distributed Computing*, 63, 196213, 2003.

10. Q. Huang, C. Lu, and G.C. Roman, Mobicast: Just-In-Time Multicast for Sensor Networks under Spatiotemporal Constraints, TR WUCS-02-42, Washington University, St. Louis MOSA, Dec. 2002.

11. Y.B. Ko and N. Vaidya, Flooding-based geocasting protocols for mobile ad hoc networks, *Proc. WMCSA*, 1999, New Orleans; *Mobile Networks and Applications*, 7, 471–480, 2002.

12. W.H. Liao, Y.C. Tseng, K.L. Lo, and J.P. Sheu, Geogrid: a geocasting protocol for mobile ad hoc networks based on grid, *J. Internet Technol.*, 1, 23–32, 2000.

13. C. Schwingenschlogl and T. Kosch, Geocast enhancements of AODV for vehicular networks, *ACM Mobile Computing and Commun. Rev.*, 6(3), 96–97, 2002.

14. I. Stojmenovic, A.P. Ruhil, and D.K. Lobiyal, Voronoi diagram and convex hull based geocasting and routing in wireless networks, *Proc. IEEE Int. Symp. on Computers and Communications ISCC*, Kemer-Antalya, Turkey, June 30–July 3, 2003, 51–56; *SITE*, Univ. of Ottawa, TR-99-11, Dec. 1999.

15. I. Stojmenovic, M. Seddigh, and J. Zunic, Dominating sets and neighbor elimination based broadcasting algorithms in wireless networks, *IEEE Trans. Parallel and Distributed Syst.*, 13(1), 2002, 14–25.

16. P. Bose, P. Morin, I. Stojmenovic, and J. Urrutia, Routing with guaranteed delivery in ad hoc wireless networks, *ACM DIAL M*, Aug. 1999, 48–55; *Wireless Networks*, 7(6), 609–616, 2001.

17. I. Stojmenovic, Geocasting with guaranteed delivery in ad hoc and sensor networks, *Int. Workshop on Theoretical and Algorithmic Aspects of Sensor, Ad Hoc Wireless and Peer-to-Peer Networks*, Fort Lauderdale, FL, February 20–21, 2004.

18. K. Saeda and A. Helmy, Efficient geocasting with perfect delivery in wireless networks, *Proc. WCNC*, March 2004.

19. G.G. Finn, Routing and Addressing Problems in Large Metropolitan-Scale Internetworks, ISI Research Report ISU/RR-87-180, 1987.

20. H. Takagi and L. Kleinrock, Optimal transmission ranges for randomly distributed packet radio terminals, *IEEE Trans. Communications*, 32(3), 246–257, 1984.

21. E. Kranakis, H. Singh, and J. Urrutia, Compass routing on geometric networks, *Proc. 11th Canadian Conf. on Comp. Geom.*, Vancouver, August 1999.

22. I. Stojmenovic and Xu Lin, Loop-free hybrid single-path/flooding routing algorithms with guaranteed delivery for wireless networks, *IEEE Trans. Parallel Distr. Syst.*, 12(10), 1023–1032, 2001.

23. C.R. Lin and M. Gerla, Adaptive clustering for mobile wireless networks, *IEEE J. Selected Areas in Communications*, 15(7), 1265–1275, 1997.

24. C. Intanagonwiawat, R. Govindan, and D. Estrin, Directed diffusion: a scalable and robust communication paradigm for sensor networks, *Proc. MOBICOM*, 56–67, Boston, Aug. 2000.

25. C. Perkins and E.M. Royer, Ad hoc on-demand distance vector (AODV) routing, *Proc. IEEE Work. on Mobile Comp. Syst. and Applications*, Feb. 1999, 90–100.

26. Y. Yu, R. Govindan, and D. Estrin, Geographic and Energy Aware Routing: A Recursive Data Dissemination Protocol for Wireless Sensor Networks, TR-01-0023, Computer Science, University of California, Los Angeles, August 2001.

27. I. Stojmenovic and Xu Lin, Power aware localized routing in wireless networks, *IEEE Trans. Parallel and Distr. Syst.,* 12(11), 1122–1133, 2001.

28. I. Stojmenovic, M. Russell, and B. Vukojevic, Depth first search and location based localized routing and QoS routing in wireless networks, *Computers and Informatics,* 21(2), 149–165, 2002.

29. J. Heidemann, F. Silva, and D. Estrin, Matching data algorithms to application requirements, *Proc. ACM SenSys,* Los Angeles, November 2003.

30. A. Okabe, B. Boots, and K. Sugihara, *Spatial Tessellations: Concepts and Applications of Voronoi Diagrams,* Wiley, 1992.

31. J. O'Rourke, *Computational geometry in C,* Cambridge University Press, 1994.

32. D. Gries and I. Stojmenovic, A note on Graham's convex hull algorithm, *Information Processing Letters,* 25(5), 323–327, 1987.

33. S. Datta, I. Stojmenovic, and J. Wu, Internal nodes and shortcut based routing with guaranteed delivery in wireless networks, *Cluster Computing,* 5(2), 2002, 169–178.

34. P. Morin, Online Routing in Geometric Graphs, Ph.D. thesis, School of Computer Science, Carleton University, January 2001.

35. S. Lindsey, C. Raghavendra, and K. Sivalingam, Data gathering algorithms in sensor networks using energy metrics, *IEEE Trans. Parallel Distrib. Systems,* 13(9), 924–935, 2002.

36. I. Stojmenovic, Energy efficient data gathering algorithms in sensor networks, in preparation.

37. N. Li, J.C. Hou, and L. Sha, Design and analysis of an MST-based topology control algorithm, *Proc. IEEE INFOCOM, 2003,* San Francisco, CA, 2003.

38. F.J. Ovalle-Martinez, I. Stojmenovic, F. Garcia-Nocetti, and J. Solano-Gonzalez, Finding minimum transmission radii and constructing minimal spanning trees in ad hoc and sensor networks, *Journal of Parallel and Distributed Computing,* 65(2), 132–141, February 2005.

39. M. Mauve, H. Fusler, J. Widmer, and T. Lang, Position-Based Multicast Routing for Mobile Ad Hoc Networks, TR-03-004, Dept. Comp. Sci., University of Mannheim, 2003; ACM Mobihoc, 2003.

40. F. Kuhn, R. Wattenhoffer, and A. Zollinger, Worst-case optimal and average-case efficient geometric ad hoc routing, *Proc. ACM MobiHoc,* 2003.

Additional References

1. S. Giordano and I. Stojmenovic, Position based routing algorithms for ad hoc networks: a taxonomy, in *Ad Hoc Wireless Networking,* X. Cheng, X. Huang, and D.Z. Du (Eds.), Kluwer, 2003.

2. J. Wu and H. Li, A dominating set based routing scheme in ad hoc wireless networks, *Proc. DIAL M,* Seattle, Aug. 1999, 7–14; *Telecommun. Syst.,* 18(1–2), 13–36, 2001.

7

Topology Control for Ad Hoc Networks: Present Solutions and Open Issues

Chien-Chung Shen

Zhuochuan Huang

Ad hoc networks are constrained by the wireless communication capability and the limited battery energy. To address these issues, topology control aims to reduce interference, reduce energy consumption, and increase network capacity, while maintaining network connectivity. The primary method of topology control is to adjust the transmission power of the nodes. This chapter formulates the topology control problem and presents a comprehensive review of existing solutions. In addition, the chapter provides a qualitative comparison among existing solutions in terms of objectives, assumptions, approaches, and performance evaluation. Lastly, we describe open issues that need further investigation.

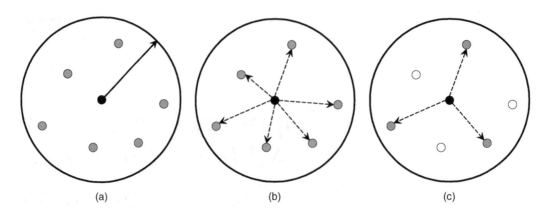

FIGURE 7.1   Physical topology versus logical topology: (a) node coverage with transmission power, (b) physical topology, and (c) logical topology.

7.1   Introduction

Ad hoc networks consist of mobile nodes that autonomously establish connectivity via multihop wireless communications without relying on any existing, preconfigured network infrastructure or centralized control. The topology of ad hoc networks is autonomously formed based on nodes' physical locations and transmission ranges. Ad hoc networks are constrained by the battery energy and the wireless interference, and network topology has a huge impact on the capacity and energy consumption of the networks. For instance, a dense topology may induce high interference, which, in turn, reduces the network capacity due to limited spatial reuse and causes unnecessarily high energy consumption. In contrast, a sparse topology is vulnerable to network partitioning due to node or link failures. **Topology control** for ad hoc networks aims to achieve network-wide or session-specific objectives such as reduced interference, reduced energy consumption, and increased network capacity, while maintaining network connectivity. The primary method of topology control is to adjust the transmission powers of the nodes, while maintaining connectivity. In this respect, topology control is sometimes termed *power control*. Reducing transmission power may encourage spatial reuse, mitigate wireless interference, improve network throughput, and reduce energy consumption.

In addition to adjusting transmission power, a node may further control topology by selecting a subset of those nodes reachable with its transmission power to form a *neighbor set*. As depicted in Figure 7.1, with respect to the node in the center of the circle, all other nodes located within the node's transmission range are included in the *physical* topology. In contrast, a *logical* topology may only include a subset of those nodes. Neighbor selection achieves certain desirable properties regarding the topology. For instance, a network-wide broadcast tree can be constructed such that no duplicate packet is received. As another example, *link symmetry*, a critical property for the proper operation of many MAC and ad hoc routing protocols, can be achieved. Because power adjustment alone usually results in asymmetric (unidirectional) links, neighbor selection is often used in conjunction with power adjustment. In this chapter, we term "power adjustment" as *physical topology control* (PTC), and neighbor selection jointly with power adjustment as *logical topology control* (LTC). In general, we formulate the topology control problem as follows.

Assumptions: Without topology control, all nodes in the network are transmitting using a common maximum power via an omnidirectional antenna. The network is connected, or, in some cases, k-connected.*

Objectives: The major objective of topology control is to reduce transmission power, and two optimization objectives are often used: (1) minimize the maximum power used by any node in the

---

*A network is k-connected if and only if there does not exist a set of $k - 1$ vertices whose removal partitions the network into two or more connected components.

network (MINMAX objective), and (2) minimize the total power used by all of the nodes in the network (MINTOTAL objective). Other objectives are to improve performance (throughput and capacity) by alleviating interference and contention, and to achieve fault tolerance such that the network is robust to node failure.

Constraints: Because topology control primarily reduces transmission power, one mandatory constraint is to maintain, strictly or with high fidelity, *connectivity** for the physical topology corresponding to the power assignment. Another constraint is to maintain *link symmetry*, which is important when the interoperability between topology control and MAC or routing is considered.

Approaches: PTC accomplishes topology control via power adjustment, and LTC accomplishes topology control via both power adjustment and neighbor selection.

Problem Formulation: Topology control is hereby formulated as *the problem of achieving one or more objectives under the assumptions, subject to connectivity or other constraints, by employing the approach of PTC or LTC.*

The remainder of the chapter is organized as follows. In the next section, we describe other power-related networking issues in the ad hoc networking context, and highlight the differences from topology control. In Section 7.3, we review existing topology control solutions and highlight their specific features. Detailed comparisons among these solutions are presented in Section 7.4. Section 7.5 concludes the paper with open research issues.

7.2 Related Topics

Power management schemes[1–3] aim to conserve energy by turning off nodes' *idle* network interfaces. The decisions on turning off idle network interfaces are made based on the prediction of nodes' impending activity. The main objective of power management is to conserve nodal energy and prolong network lifetime, while sustaining network forwarding capacity. In contrast to topology control, which adjusts nodes' transmission power, power management withdraws nodes from participating in forwarding activity.

Power-aware routing[4,5] focuses on choosing optimal routes based on some power-based criteria such as minimum (end-to-end) total power. When a node has more than one power level to choose for forwarding a packet, many possible connected routes can be examined. The issue is particularly important in a heterogeneous environment where different nodes may have different (full) transmission powers. In conventional ad hoc networks when one common transmission power level is used by all nodes, routing protocols based on the metric of hop-count implicitly achieve some power-based criteria such as minimum total power. When the ability of power control is enabled, nodes are generally assigned different power levels. Therefore, power-aware routing is necessary if power is the major concern in routing. Similar to power-aware routing, certain topology control algorithms[6,7] aim to maintain a minimum total power path between each pair of nodes.

7.3 Review of Existing Solutions

We categorize existing topology control solutions into either PTC or LTC approaches. PTC accomplishes topology control by only adjusting transmission power, while LTC selects neighbors in addition to adjusting power. The major objective of PTC is to reduce interference and conserve energy. Though also providing these benefits, LTC focuses on reducing the number of (logical) neighbors in order to reduce the routing overhead. Within each category, we further group solutions with common (or similar) characteristics together to present an organized taxonomy.

---

*Connectivity is regarded as an objective in some works and as a constraint in others.

7.3.1 Definitions

To aid the description, certain useful terms are first defined. A *topology* is modeled as a graph $G = (V, \vec{E})$, where V represents the set of nodes and $\vec{E}$ represents the set of (directed) edges. $Cover(v_i, v_j)$ means that node v_j is within node v_i's transmission range with a given power assignment for v_i. Graph $G_p = (V, \vec{E}_p)$ is called a *physical topology* where $\forall v_i, v_j \in V, (v_i, v_j) \in \vec{E}_p \Leftrightarrow Cover(v_i, v_j)$. Graph $G_l = (V, \vec{E}_l)$ is called a *logical topology* where $\forall v_i, v_j \in V, (v_i, v_j) \in \vec{E}_l \Rightarrow Cover(v_i, v_j)$. Note that there exists a unique, one-to-one correspondence between a physical topology and a power assignment of nodes. We define the *physical neighbor set* of node u as $N_p(u) = \{v | \forall (u, v) \in \vec{E}_p\}$, and $|N_p(u)|$ is node u's *physical node degree*. A logical topology is one of many subsets of the (unique) physical topology. The *logical neighbor set* of node u, denoted as $N_l(u)$, is a *purposely selected* subset of $N_p(u)$, and $|N_l(u)|$ denotes its *logical node degree*. It is clear that $|N_l(u)| \leq |N_p(u)|$. A link between node v_i and node v_j is *symmetric* if both $(v_i, v_j) \in \vec{E}$ and $(v_j, v_i) \in \vec{E}$; otherwise, it is *unidirectional* or *asymmetric*. A sub-graph of G_p, denoted by $G^- = (V, \vec{E}^-)$, is called its *symmetric sub-topology* when $\vec{E}^-$ only keeps the symmetric links in $\vec{E}_p$. A super-graph of G_p, denoted by $G^+ = (V, \vec{E}^+)$, is called its *symmetric super-topology* when $\vec{E}^+$ keeps all the symmetric links in $\vec{E}_p$, and makes all the asymmetric links in $\vec{E}_p$ to be symmetric (by potentially increasing transmission power). Note that since G_p is unique for a given power assignment of nodes, G^+ and G^- are also unique.

7.3.2 Logical Topology Control

Topology control was originally proposed for Packet Radio Networks (PRN) by Hu[8] as a solution to the problem of maintaining a network with good reliability and throughput, "in which a node can carefully select a set of neighbors to establish logical data links and dynamically adjust transmitting power (i.e., transmission radius) for different links." Its objectives are to maximize reliability* and throughput, given some traffic load and routing algorithms. Assuming the availability of location information at each node, Hu[8] presents a centralized topology control algorithm, termed *Novel Topology Control* (NTC), as well as a distributed implementation based on Delaunay triangulation. The algorithm and implementation take advantage of the derived property from the planarity of triangulation that the average degree of the graph is less than six. The algorithm utilizes heuristics to bound the degree of each node and to construct a regular and uniform graph structure. Performance evaluation exhibits higher throughput performance for different topologies of the network when the node degree is less than six.

Some LTC protocols emphasize reducing the logical node degree and transmission radius, that is, the longest distance from a node to its farthest neighbor.[7,9–13] They share the objective of reducing transmission power while maintaining connectivity and the link symmetry of the network.

Wattenhofer et al.[7] propose a distributed topology control algorithm that takes advantage of the *angle-of-arrival* (AOA) information. Its basic idea is for each node, v, to search for a minimum power $P_{v,\alpha}$ subject to a *local condition* that node v has a neighbor in every angle (cone) of degree α centered at v, or if not possible, still uses the original power. (Note that original power is termed full power or maximum power in other works. We use them interchangeably in accordance with the term used in each work.) It is proved that $\alpha \leq 2\pi/3$ is sufficient for the local condition to guarantee the connectivity of the symmetric super-topology of the physical topology corresponding to such a power assignment.** In addition to reducing power, the algorithm also includes a neighbor selection process, without adjusting power, to reduce the (logical) node degree and the total power for data forwarding. Furthermore, when $\alpha \leq \pi/2$, given that certain inequality in the neighbor selection process is satisfied, the node degree of any node is no greater

---

*The definition of reliability is equivalent to connectivity in this chapter.

**As a commonly agreed upon assumption, the original network, where no topology control is deployed and every node is using the common maximum power, is connected.

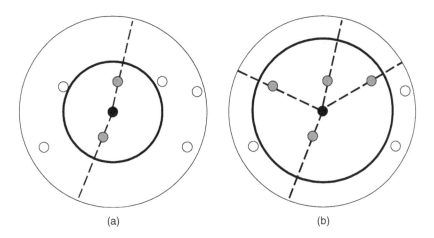

FIGURE 7.2 CBTC: Assume that α is $5\pi/6$. In (a), the black node has two neighbors and the radials toward them divide the circular angle into two, at least one of which is greater than α. Therefore, the condition is not satisfied. In (b), the black node has four neighbors and all four corresponding angles are no greater than α. Therefore, the local condition is satisfied.

than 6. The article evaluates the extension of network lifetime and the average node degree to demonstrate the effectiveness of its topology control schemes.

Li et al.[9] provide a more formal and well analyzed presentation of Wattenhofer et al.,[7] which is termed *Cone-Based Topology Control* (CBTC). In CBTC, the process of "searching for the minimum power subject to a local condition" is termed the *basic* algorithm. One way to reduce the transmission power in the case when the local condition is not satisfied is formalized as the *shrink-back* optimization. Figure 7.2 illustrates* the basic algorithm. In Li et al.,[9] the threshold of $2\pi/3$ for α in Wattenhofer[7] is relaxed to $5\pi/6$, and, furthermore, the condition of $\alpha \leq 5\pi/6$ is proved to be necessary, in addition to being sufficient. Li et al.[9] prove an interesting result that when $\alpha \leq 2\pi/3$, the corresponding symmetric sub-topology is connected. CBTC's operation of generating a symmetric topology from the physical topology is termed *asymmetric edge removal* optimization. A *pair-wise edge removal* optimization is also presented to further reduce transmission power; however, it requires additional distance information. Simulation results demonstrate the effectiveness of optimizing node degree and transmission radius. In addition, the paper proposes an event-driven strategy, based on neighbor discovery, to accommodate mobility. However, no simulation results were presented.

Bahramgiri et al.[10] describe an extension of CBTC's basic algorithm and shrink-back optimization in two aspects. First, it proves the result that $\alpha \leq 2\pi/(3k)$ is the necessary and sufficient condition that the resulting super-topology is k-connected, given that the original topology without topology control is k-connected. Such a result achieves fault tolerance for the network. Second, it proposes the CBTC-3D algorithm aiming to generalize CBTC into three-dimensional space. Experiment results show the behavior of the algorithm with varying k on node degree and transmission radius.

Huang et al.[11] introduce the usage of directional antennas to topology control, where using proper antenna pattern shares similar benefits of reducing power in that spatial reuse may be improved. On the other hand, because directional antennas could have higher gain in some direction(s), power may be further reduced while maintaining similar neighbor connectivity. This work, termed CBTC-DA, applies

---

*As a convention, the black node is the node in consideration and the outermost circle represents its full power range. The thick circle represents the corresponding power range when the black node chooses a lower power and the gray nodes are some or all of the neighboring nodes that are covered by this lower power.

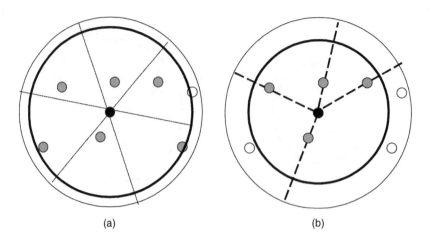

(a)   (b)

FIGURE 7.3 CBTC-DA: In (*a*), the black node finds the least power such that some node in each of the six sectors is covered. This power is no less than the one chosen by CBTC, as in (*b*), because the local condition in CBTC-DA leads to that of CBTC.

directional antenna to CBTC in two ways. First, instead of checking the existence of empty angles of degree $2\pi/3$, the power range is divided into six sectors of $\pi/3$ each, and the existence of empty sectors is checked, as illustrated in Figure 7.3. This condition is stronger, while much simpler to evaluate, than the original condition in CBTC. This approach significantly reduces the accuracy requirement of AOA. Second, in the sectors that are empty, the antenna gain is reduced such that only the closest neighbor is within the transmission range. Following CBTC, a simulation study on node degree and transmission radius shows comparable performance. In addition, network performance is evaluated over topology-controlled networks. The results show that the improvement of topology control on network performance is strongly correlated with the types of traffic, which suggests that good performance in topology metrics, such as node degree and transmission radius, does not necessarily imply good networking performance, such as throughput and delay.

Blough et al.[12] propose the *k*-NEIGH protocol, based on the principle of maintaining the number of neighbors of every node to be equal to or slightly below a specific value *k*, as illustrated in Figure 7.4. Instead of relying on location or angle information, it uses distance information from a node to its neighbors.

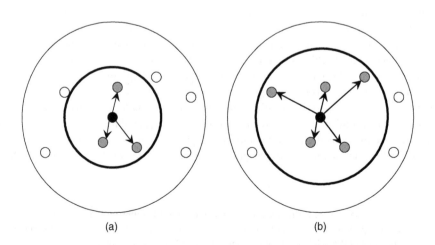

(a)   (b)

FIGURE 7.4 K-Neigh: (*a*) and (*b*) show the power chosen for *k* of 3 and 5, respectively.

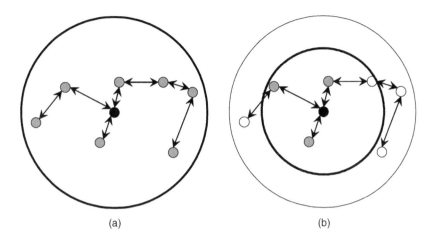

(a)   (b)

FIGURE 7.5 LMST: (*a*) shows the minimal spanning tree (MST) formed by the black node and all of its full-power neighbors. In (*b*), the gray nodes are the adjacent nodes of the black node in the MST. Therefore, the black node chooses the least power to cover all the gray nodes.

Basically, every node uses a common maximum power* for neighbor discovery, and searches for its k closest neighbors. It is proved that for some k of $O(log(n))$, where n is the number of nodes in the network, the probability that a physical topology G_k as well as its symmetric sub-topology G_k^-, that is, the subgraph of G_k that removes all asymmetric edges in G_k, is connected is 1 as n approaches infinity. An empirical distribution of the minimum k to maintain connectivity is obtained via simulation. The article finds out that for an n of 100, k's distribution centers around six; a k of 9 provides connectivity for a wide range of network sizes. The article also presents better performance in terms of *energy cost*, which is proportional to the total power of all nodes, and physical/logical node degree than CBTC.

Li et al.[13] present a minimum spanning tree (MST) based decentralized topology control algorithm termed *Local Minimum Spanning Tree* (LMST). The basic idea is that every node maintains an MST within its (maximum power) neighbors and chooses the minimum power such that all of its adjacent nodes in the MST are covered, as illustrated in Figure 7.5. It is proved that the physical topology G_0 corresponding to such power assignment has a degree bound of 6 for every node. It is also proved that the symmetric sub-topology of G_0, G_0^- is connected. A simulation study shows that LMST exhibits better performance on node degree, transmission radius, and average link length, which is a measurement of power level necessary for point-to-point communication, than CBTC.

CLTC and TMPO are two cluster-based LTC protocols.[14,15]

Shen et al.[14] present a *Cluster-Based Topology Control* (CLTC) framework for ad hoc networks that achieves the topology constraint of k-connectivity while reducing transmission power. The usage of clustering is motivated by the fact that centralized approaches, although able to achieve strong connectivity (k-connectivity for $k \geq 2$), suffers from scalability problems. In contrast, distributed approaches, although scalable, lack strong connectivity guarantees.** As a result, a hybrid approach is preferred. The basic idea of CLTC is to have nodes to autonomously form clusters, and each cluster head then executes a centralized topology control algorithm (such as those presented in Refs. 16 and 17) within its cluster to minimize the transmission power subject to the k-connectivity constraint. Clusters are then connected in such a way that each cluster has k disjoint symmetric links to some other cluster. It is proved that the resulting

---

*This power is empirically chosen based on the longest edge of the Euclidean minimum spanning tree, which is known to correspond to the critical transmission range for maintaining connectivity.

**The only distributed algorithm that achieves k-connectivity.[10]

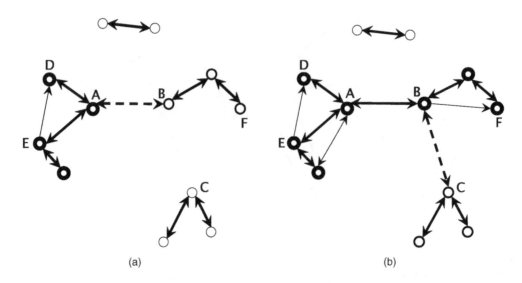

FIGURE 7.6 CONNECT: The algorithm basically executes the Kruskal MST algorithm while adjusting the nodes' powers. In (*a*), the *white* nodes and *black* nodes are in two trees, with (currently) the shortest distance between node A and node B. Therefore, the two trees are merged following the MST algorithm and A and B each can increase and keep its power to make sure it can cover the other. As a result, now B can cover node F now, as shown in (*b*).The process will continue, as shown in (*b*), until the MST algorithm terminates.

topology is *k*-connected and symmetric. Simulation studies also show significant improvement on node degree for the cases of *k* equal to 1 and 2.

Bao and Garcia-Luna-Aceves[15] present a hierarchical topology control approach, termed *Topology Management by Priority Ordering* (TMPO), via a novel clustering solution to the minimal dominating set (MDS) problem. TMPO connects members of the MDS to form a backbone infrastructure using links between the elected cluster heads, *doorways*, and *gateways*. Simulation results show that TMPO results in better network lifetime than using other MDS algorithms with mobility.

7.3.3 Physical Topology Control

In contrast to LTC, PTC accomplishes topology control by only adjusting transmission power. Ramanathan et al.,[16,17] propose centralized PTC algorithms.*

Ramanathan and Rosales-Hain[16] present two centralized greedy algorithms, *CONNECT* and *BICONN-AUGMENT*, to achieve the *maximum power minimization* (MINMAX) objective subject to the connectivity and bi-connectivity constraints for static networks. The algorithms basically apply the Kruskal's MST algorithm.** Figure 7.6 illustrates how CONNECT works. In addition, two distributed heuristic protocols, *Local Information No Topology* (LINT) and *Local Information Link-state Topology* (LILT), are proposed to adaptively adjust transmission power in response to the topological changes in mobile networks. LINT basically adjusts the transmission power of each node such that its (physical) node degree is within some predefined range; the power will be increased/decreased when the node degree is lower/higher than the low/high threshold. While LINT only relies on local neighbor information, LILT also exploits the global

---

*They are not classified as LTC since the neighbor-set is not *explicitly* specified.

**After running the MST algorithm, each node can also execute a *perNodeMinimalize* function to potentially further reduce its power as long as connectivity is maintained. However, such a post-processing phase can be ignored in practice.

topology information provided by link-state-based routing protocols to better adjust power in response to the change in topology connectivity. Simulation results show that CONNECT and BICONN-AUGMENT have higher throughput and lower maximum transmission power than the cases with no topology control. They also show that LINT and LILT have slightly better throughput than and delay comparable to that with no topology control.

Lloyd et al.[17] describe a general framework for topology control algorithms. They also propose a centralized approximation algorithm to achieve the challenging MINTOTAL objective subject to the bi-connectivity constraint, which has been proved to be NP-hard. Simulation results show the behavior of the algorithms on node degree and maximum/total power.

Li et al.[18] provide a PTC protocol that improves fault tolerance for the topology. They investigate the theoretical condition of a common transmission radius for all nodes such that the topology is k-connected with high probability, which is validated via simulation. They also present a localized method based on the *Yao structure*, as illustrated in Figure 7.7, to maintain k-connectivity for the network topology.

MobileGrid[19] and Poster[20] are two PTC protocols that reduce power not only to maintain connectivity, but also to improve network performance by taking into consideration certain network parameters.

Liu and Li[19] argue that the network performance of the ad hoc network may be affected by factors other than the transmission power of the node, such as node distribution. Via extensive performance evaluation, they observe that the *contention index* (CI), defined as the number of nodes within a transmission range, is closely related to network performance metrics such as network capacity and average path length. Inspired by this result, a distributed topology control algorithm, termed *MobileGrid*, is presented to improve network performance by adjusting transmission power based on the estimated contention index. Simulation results evidence the relation between CI and network capacity, power efficiency, and average path length.

Park and Sivakumar[20] challenge the idea that a *minimally connected topology* provides optimal performance in typical ad hoc networks. Via simulation, they discovered that the optimal transmission range varies with the traffic load. With this insight, they propose a set of topology control algorithms, termed *Adaptive Topology Control* (ATC), that are adaptive to traffic load, which is estimated by measuring local contention.

Narayanaswamy et al.[21] present a distributed power control protocol, termed *COMPOW* (COMmon POWer), in which a minimal common power is shared by all the nodes such that network connectivity is maintained. They argue that, by doing so, the network capacity is maximized, the (battery) lifetime of the network is extended, and the contention in the MAC layer is reduced. Furthermore, COMPOW is the only topology control protocol that produces a physical topology containing only symmetric links. To achieve

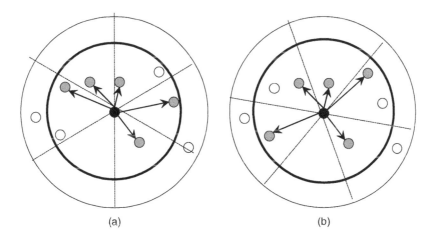

(a)   (b)

FIGURE 7.7   YAO(k): (*a*) shows that the black node selects a closest full-power neighbor (in gray), if any, in each of the $k = 6$ sectors and chooses the least power to cover all selected nodes. (*b*) shows that a different orientation of the sectors could lead to a different chosen power.

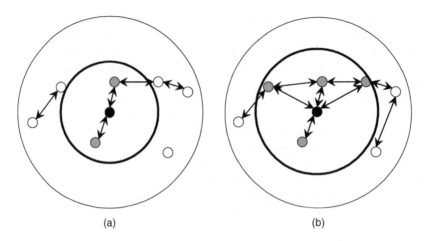

(a)   (b)

FIGURE 7.8   ABTC: In (*a*), when every full-power neighbor of the black node uses the black node's chosen power to originate/forward ant packets, the local condition is not satisfied. In (*b*), the condition is satisfied with a higher power assignment at the black node.

common power, COMPOW executes a routing daemon for each discrete power level in parallel, where each routing daemon maintains its own routing table by exchanging control messages at the corresponding power level. A unique feature of COMPOW is that, as in the case with no topology control, the physical topology is the same as its symmetric sub-topology and symmetric super-topology.

Huang and Shen[22] present a distributed topology control protocol, termed *Ant-Based Topology Control* (ABTC), by adapting the biological metaphor of swarm intelligence. The objective is to minimize the maximum power or the total power subject to connectivity constraint. The basic idea is to minimize the power for each node such that the local condition "the node can receive ant packets originated/forwarded with that power from all of its *full-power* neighbors" is satisfied, as illustrated in Figure 7.8. It is proved that such power assignments maintain network connectivity while reducing the maximum or the total power. Swarm intelligence is used as a heuristic search mechanism where ant packets originated and forwarded with different power levels are used to search for the minimal power such that the condition is satisfied locally at each node. This heuristic search mechanism lets the protocol converge quickly to a common maximum power or reduced total power, and allows the protocol to adapt well to mobility. In contrast to other distributed topology control protocols, the operations of ABTC do not require any location, AOA, or routing information. Simulation results demonstrate that ABTC effectively reduces the power while maintaining higher levels of connectivity with mobility.

7.4   Comparisons

We compare the reviewed topology control solutions in the following four aspects: *objective*, *constraint*, *approach*, and *performance evaluation*.

7.4.1   Objectives and Constraints

Table 7.1* summarizes the comparison in terms of objectives and constraints. It is observed that energy reduction is the major concern in most solutions, and network performance is only considered in a few

---

*The term "strict" used in Table 7.1 denotes that the satisfaction of the connectivity constraints can be proved for stationary networks.

TABLE 7.1 Comparison on Objectives and Constraints

		Objectives			Constraints	
	Protocol/Author	*Energy Conservation*	*Network Performance*	*Fault Tolerance*	*Connectivity*	*Symmetry*
LTC	NTC[8]		√		Strict	√
	CBTC(a)[7]	√			Strict	√
	CBTC(b)[9]	√			Strict	√
	CBTC(k,3D)[10]	√		√	Strict	√
	CBTC(DA)[11]	√	√		Strict	√
	K-Neigh[12]	√			Probabilistic	√
	LMST[13]	√			Strict	√
	CLTC[14]	√		√	Strict	√
	TMPO[15]	√			Strict	√
PTC	Ramanathan et al.[16]	√	√		*	*
	Lloyd et al.[17]	√			Strict	√
	Li et al.[18]			√	Probabilistic	
	MobileGrid[19]	√	√		Loose	
	ATC[20]		√		Loose	
	COMPOW[21]	√	√		Strict	√
	ABTC[22]	√			Strict	

recent efforts. In addition, several efforts address the challenging objective of fault tolerance, although its impact on network performance has not been evaluated.

Regarding constraints, Blough et al.[12] and Li et al.[18] present two efforts that adopt a probabilistic approach to guaranteeing connectivity or k-connectivity. Liu and Li,[19] Park and Sivakumar,[20] and LINT/LILT[16] adopt heuristic approaches that do not explicitly guarantee connectivity. Other efforts, including CONNECT and BICONN-AUGMENT,[16] strictly maintain connectivity, or k-connectivity, for (static) ad hoc networks. All LTC protocols satisfy the link symmetry constraint. PTC protocols of COMPOW, CONNECT, BICONN-AUGMENT, and the algorithm in Ref. 17 satisfy the link symmetry constraint. The centralized algorithms in Refs. 16 and 17 produce a physical topology whose symmetric sub-topology is connected.

7.4.2 Approaches

Table 7.2 summarizes the comparison of approaches. Different assumptions are made by the reviewed protocols. COMPOW[21] and ABTC[22] are the only efforts that use *finite discrete power levels*. It is also the only assumption (among all being investigated) that ABTC makes. CBTC and its derivatives assume the existence of *AOA* information. k-NEIGH[12] is the only work that relies on *distance*. A combination of AOA and distance is equivalent to relative location. Many efforts assume the availability of (absolute) *location*, usually via GPS. LILT[16] and COMPOW are the only efforts that rely on *routing* information. A number of efforts assume that the network is *two-dimensional*. Note that LMST[13] does not assume a two-dimensional network to work correctly, although its degree bound of six is based on this assumption. k-NEIGH and Ref. 18 base their analyses on the assumption of *uniform node distribution*, while COMPOW needs such assumption to achieve optimal performance.

Almost all these works provide a distributed protocol that utilizes local information in order to be practical for ad hoc networks. However, theoretical bounds derived from global topology information is important. NTC,[8] k-NEIGH, and LMST provide the bound of 6 for node degree. Most efforts claim to be adaptive to mobility, although some claim to work only with limited mobility, especially those that involve phases of operations, such as CBTC and its derivatives. LINT and LILT[16] adapt well to mobility while CONNECT and BICONN-AUGMENT do not. MobileGrid[19] and ATC [20] are the only efforts that adapt to traffic load.

TABLE 7.2   Comparison on Approaches

	Protocol/Author	Disk Pwr	AOA	Distance	Location	Routing	2D	Uniform	Local	Bound Degree	Mobil. Adapt	Traff. Adapt
LTC	NTC[8]			✓					✓	✓		
	CBTC(a)[7]	✓					✓		✓		Low	
	CBTC(b)[9]	✓					✓		✓		Low	
	CBTC(k,3D)[10]	✓							✓		Low	
	CBTC(DA)[11]	✓					✓		✓		Low	
	K-Neigh[12]		✓				✓	✓	✓	✓		
	LMST[13]				✓		*		✓	✓	Low	
	CLTC[14]				✓				✓			
	TMPO[15]								✓		Typical	
PTC	Ramanathan et al.[16]				✓	*			*		*	
	Lloyd et al.[17]				✓							
	Li et al.[18]						✓	✓	✓			
	MobileGrid[19]								✓		Typical	✓
	ATC[20]								✓		Typical	✓
	COMPOW[21]	✓			✓	✓		✓	✓		Low	
	ABTC[22]	✓							✓		Typical	

7.4.3   Performance Evaluation

Table 7.3 summarizes the comparison on performance evaluation.* It is observed that although most protocols are designed for (mobile) ad hoc networks, only a few of them clearly evaluate the impact of mobility. *Energy consumption, network lifetime, power or range (radius),* and *average link length* are four energy-related metrics. Although energy conservation is the main objective, a majority of the work does not evaluate these energy metrics directly. Power is widely used as an alternative, where power is usually proportional to $(distance)^\alpha$, where $2 \leq \alpha \leq 4$. Most work using power as a metric calculates a (weighted) sum of power. A simple and frequently used formula is the *total power*, where all weights are equal to 1, and the average link length is a more end-to-end oriented metric. Network performance metrics, such as *throughput* and *delay*, are largely ignored in most work. Almost all LTC work evaluates logical node degree, and Lloyd et al.[17] is the only PTC work that does that. Blough et al.[12] is the only work that contrasts physical and logical node degree. Because most approaches guarantee connectivity without mobility, the *level of connectivity* is only evaluated for two reasons: (1) when the connectivity guarantee is probabilistic[12,18]; and (2) to study the impact of mobility[22] or node/link failure[8] on connectivity.

7.5   Conclusion and Open Issues

We have introduced and formulated the topology control problem as well as categorized the solutions into physical topology control and logical topology control. We have reviewed several recent topology control solutions in each of these two categories, and compared them in terms of objective, constraint, approach, and performance evaluation. Next, we enumerate open issues that need further investigation.

7.5.1   Node Degree Versus Level of Contention

Most LTC protocols claim to reduce MAC layer contention by creating a topology with a low or bounded *logical* node degree. However, in fact, the level of contention is determined by the physical node degree. Blough et al.[12] is, to our knowledge, the only work that elucidates such nuance and presents simulation

---

*The column *network performance* in Table 7.1 is included here to contrast with *metrics*.

TABLE 7.3 Comparison of Performance Evaluation

	Protocol/Author	Performance Evaluation									
		Metrics									
		Network Performance	Mobility	Energy Consmpt.	Lifetime	Power or Range	Link Length	Throughput	Delay	Node Degree	Degree of Conn.
LTC	NTC[8]	✓						✓		Logical	✓
	CBTC(a)[7]			✓	✓	✓				Logical	
	CBTC(b)[9]					✓				Logical	
	CBTC(k,3D)[10]					✓				Logical	
	CBTC(DA)[11]	✓				✓		✓		Logical	
	K-Neigh[12]					✓				Both	✓
	LMST[13]					✓				Logical	
	CLTC[14]					✓	✓			Logical	
	TMPO[15]		✓	✓		✓					
PTC	Ramanathan et al.[16]	✓	✓			✓		✓		Logical	
	Lloyd et al.[17]					✓			✓		
	Li et al.[18]					✓	✓	✓			
	MobileGrid[19]	✓	✓			✓					
	ATC[20]	✓									
	COMPOW[21]	✓				✓					
	ABTC[22]		✓								✓

studies on both logical and physical node degree. In CBTC, for example, the physical node degree is around twice as much as the logical node degree. In addition, as the node density increases, the difference between them becomes significant. Therefore, we suggest using physical node degree, rather than the logical node degree, as an indicator of the level of contention. Therefore, logical node degree may not be a proper indicator of the level of contention. In addition, in a wireless network, node degree may not be as dominant a factor as in wired network because other factors (such as SINR) also play an important role. Further study is necessary to investigate the impact of topology control on the level of contention.

7.5.2   Estimation of Energy Consumption with Transmission Power

Energy conservation is a major objective of topology control. However, it is difficult to properly evaluate the effectiveness of energy conservation without taking network traffic, which is affected by many factors including source rate, MAC, routing, queuing, etc., into consideration. There have been few efforts to evaluate energy saving with network traffic. Most work, however, uses a static weighted sum of the transmission power of each node as an estimation of the energy consumption without network traffic. The former gives realistic results; however, the results are network traffic dependent. The latter makes implicit assumptions on network traffic that may not be realistic. For example, total power and its variants are used in several works to estimate energy consumption. Total power makes a good estimation of energy consumption when every node transmits with the same rate. A better estimation could be the total power over the shortest route, which approximates point-to-point energy consumption. In contrast, Bao and Garcia-Luna-Aceves[15] apply a flexible approach that gives different weights to different nodes, based on their expected loads, to calculate a weighted total energy. Weights could be obtained from the statistics of different traffic types. Therefore, efforts are necessary to provide more sophisticated and representative metrics for the energy consumption of the network with the consideration of network traffic.

7.5.3   Counteractive Impacts

Most topology control protocols strive to minimize the transmission power to maintain a topology with lower node degree, which may reduce MAC contention and expand capacity. However, lower node degree or shorter transmission range may also counteract the goal of energy conservation and network performance enhancement in many ways. Four examples are given as follows.

1. When transmission power is reduced, the physical node degree decreases. As a result, the level of route diversity is reduced, which leads to a bottleneck node that is prone to congestion and faster drainage of energy, as well as more frequent route discovery with mobility.
2. When transmission power is reduced, the transmission range is shorter and thus routes have higher hop counts, which leads to longer end-to-end delay to forward a same packet from the source to the destination.
3. Power control is correlated with several physical layer characteristics, such as modulation scheme, SINR, and BER, which could have significant impact on the link reliability and data rate, and in turn higher layer network performance. For example, when the transmission power is reduced, SINR decreases and BER increases. Consequently, the effective data rate of wireless links decreases and more retransmission may be needed.
4. The above described counteractive impacts, lower level of route diversity, longer routes, and higher BER all degrade the network performance. In addition, these counteractive impacts may potentially cause more redundant traffic to be injected into the network, which wastes energy. Other work[19,20] considers similar issues in terms of network performance and provides initial studies. More efforts are necessary to address these cross-feature trade-offs by exploring cross-layer design.

7.5.4   Interoperability

As an optimization in network design, it is desirable that topology control is capable of interoperating with other networking components. For example, topology control aims at conserving energy and improving network performance, which is closely correlated with power-aware routing. Therefore, it may be preferable to design an integrated solution. COMPOW[21] is one early effort that addresses this issue. As another example, directional antennas and power control are similar in that they both change the power coverage. When combined with directional antenna, power control can significantly expand the network capacity as well as allow more flexibility in controlling topology, such as achieving *partition bridging*, which may not be possible via power control alone. Huang and Shen[11,23] provide initial attempts on this issue, but more research is needed to incorporate directional antenna with topology control so as to maximize benefit. Finally, because several PTC protocols cannot maintain link symmetry, they will need extra functionality to interoperate with existing MAC and ad hoc routing protocols.

Acknowledgment

This work is supported in part by the National Science Foundation under grant ANI-0240398.

References

1. S. Singh and C.S. Raghavendra, PAMAS: Power Aware Multi-Access Protocol with Signalling for Ad Hoc Networks, *SIGCOMM Computer Communication Review,* Vol. 28, No. 3, July 1998.
2. R. Zheng and R. Kravets, On-Demand Power Management for Ad Hoc Networks, in *IEEE INFOCOM,* San Francisco, California, Mar. 30–Apr. 03 2003.
3. C. Srisathapornphat and C.-C. Shen, Ant-Based Energy Conservation for Ad Hoc Networks, in *12th International Conference on Computer Communications and Networks (ICCCN),* Dallas, TX, Oct. 20–22, 2003.
4. S. Singh, M. Woo, and C.S. Raghavendra, Power-Aware Routing in Mobile Ad Hoc Networks, in *ACM MobiCom,* Dallas, TX, 1998, pp. 181–190.
5. J. Gomez, A.T. Campbell, M. Naghshineh, and C. Bisdikian, Power-Aware Routing in Wireless Packet Networks, in *Sixth IEEE International Workshop on Mobile Multimedia Communications,* Nov. 1999.
6. V. Rodoplu and T.H. Meng, Minimum Energy Mobile Wireless Networks, *IEEE Transactions Selected Areas in Communications,* Vol. 17, No. 8, Aug. 1999.
7. R. Wattenhofer, L. Li, P. Bahl, and Y. Wang, Distributed Topology Control for Power Efficient Operation in Multihop Wireless Ad Hoc Networks, in *IEEE INFOCOM,* Anchorage, AK, Apr. 22–26, 2001.
8. L. Hu, Topology Control for Multihop Packet Radio Networks, *IEEE Transactions on Communications,* Vol. 41, No. 10, Oct. 1993.
9. L. Li, J.Y. Halpern, P. Bahl, Y. Wang, and R. Wattenhofer, Analysis of a Cone-Based Distributed Topology Control Algorithms for Wireless Multi-Hop Networks, in *ACM Symposium on Principles of Distributed Computing (PODC),* Aug. 26–29, 2001.
10. M. Bahramgiri, M. Hajiaghayi, and V.S. Mirrokni, Fault-Tolerant and 3-Dimensional Distributed Topology Control Algorithms in Wireless Multi-Hop Networks, in *IEEE ICCCN,* Miami, FL, Oct. 14–16, 2002.
11. Z. Huang, C.-C. Shen, C. Srisathapornphat, and C. Jaikaeo, Topology Control for Ad Hoc Networks with Directional Antennas, in *IEEE ICCCN,* Miami, FL, Oct. 14–16, 2002.
12. D.M. Blough, M. Leoncini, G. Resta, and P. Santi, The K-Neigh Protocol for Symmetric Topology Control in Ad Hoc Networks, in *ACM MobiHoc,* Annapolis, June 1–3, 2003.
13. N. Li, J.C. Hou, and L. Sha, Design and Analysis of an MST-Based Topology Control Algorithm, in *IEEE INFOCOM,* San Francisco, CA, Mar. 30–Apr. 03, 2003.
14. C.-C. Shen, C. Srisathapornphat, R. Liu, Z. Huang, C. Jaikaeo, and E.L. Lloyd, CLTC: A Cluster-Based Topology Control Framework for Ad Hoc Networks, *IEEE Transactions on Mobile Computing,* Vol. 3, No. 1, Jan. 2004.
15. L. Bao and J.J. Garcia-Luna-Aceves, Topology Management in Ad Hoc Networks, in *ACM MobiHoc,* Annapolis, MD, June 1–3, 2003.
16. R. Ramanathan and R. Rosales-Hain, Topology Control of Multihop Wireless Networks using Transmit Power Adjustment, in *IEEE INFOCOM,* Tel-Aviv, Israel, Mar. 26–30, 2000.

17. E.L. Lloyd, R. Liu, M.V. Marathe, R. Ramanathan, and S.S. Ravi, Algorithmic Aspects of Topology Control Problems for Ad Hoc Networks, in *ACM MobiHoc*, Lausanne, Switzerland, June 9–11, 2002.

18. X. Li, P. Wan, Y. Wang, and C. Yi, Fault Tolerant Deployment and Topology Control in Wireless Networks, in *ACM MobiHoc*, Annapolis, MD, June 1–3, 2003.

19. L. Liu and B. Li, MobileGrid: Capacity-Aware Topology Control in Mobile Ad Hoc Networks, in *IEEE ICCCN*, Miami, FL, Oct. 14–16, 2002.

20. S. Park and R. Sivakumar, Poster: Adaptive Topology Control for Wireless Ad-Hoc Networks, in *ACM MobiHoc* (also appears in *SIGCOMM Computer Communication Review*), Annapolis, MD, June 1–3, 2003.

21. S. Narayanaswamy, V. Kawadia, R.S. Sreenivas, and P.R. Kumar, Power Control in Ad-Hoc Networks: Theory, Architecture, Algorithm and Implementation of the COMPOW Protocol, in *Proceedings of European Wireless*, Feb. 2002.

22. Z. Huang and C.-C. Shen, Poster: Distributed Topology Control Mechanism for Mobile Ad Hoc Networks with Swarm Intelligence, in *ACM MobiHoc*, Annapolis, MD, June 1–3, 2003, (also appears in *ACM Mobile Computing and Communications Review*, Vol. 7, No. 3, July 2003, pp. 21–22).

23. Z. Huang and C.-C. Shen, Topology Control with Directional Power Intensity for Ad Hoc Networks, in *IEEE Wireless Communications and Networking Conference (WCNC)*, Atlanta, GA, Mar. 21–25, 2004.

8

Minimum-Energy Topology Control Algorithms in Ad Hoc Networks*

Joseph Y. Halpern

Li (Erran) Li

8.1   Introduction

Wireless ad hoc networks can be deployed in many settings such as environmental monitoring, disaster relief, and battlefield situations. In these settings, wireless devices such as sensors are often powered by an on-board battery. Many of these networks are expected to function for an extended period of time. To accomplish this without a renewable energy source, energy conservation is the key.

We consider how to adjust a node's transmission power to minimize its energy consumption and improve network performance in terms of network lifetime and throughput. We refer to this problem as the *topology-control* problem. Our focus here is on maximizing the time that the network is able to function, that is, the *network lifetime*. We discuss below how network lifetime can be increased,

*Based on "Minimum-Energy Mobile Wireless Networks Revisited" and "A Minimum-Eenergy Path-Preserving Topology-Control Algorithm," by Joseph Y. Halpern and Li (Erran) Li, which appear in the *IEEE International Conference on Communications*, 2001, pages 278–283, and *IEEE Transaction on Wireless Communications*, May, 2004, pages 910-921, respectively, 2004 IEEE. The work of Joseph Y. Halpern is supported in part by NSF under grant CTC-0208535, by ONR under grants N00014-00-1-03-41 and N00014-01-10-511, by the DoD Multidisciplinary University Research Initiative (MURI) program administered by the ONR under grant N00014-01-0795, and by AFOSR under grant F49620-02-1-0101.

the subtleties of defining it precisely, and the difficulties of achieving optimal network performance in practice.

In an ad hoc network, network lifetime can be increased by energy reduction in the hardware, the software (operating systems and applications), and the communication protocols. To reduce the energy consumption of hardware, low-power CPUs such as the Intel embedded StrongARM 1100 processor and low-power displays have been developed. To reduce the energy consumption of software, low-energy software can be developed through various techniques, including reducing the number of operations through code optimization and the use of multiple fidelity algorithms.[1] The synergy of hardware and application software can also be exploited by operating systems to reduce energy consumption. For example, CPU energy consumption can be reduced through dynamic voltage scaling if the computation workload decreases.[3] In addition, a disk can be spun down to reduce its idle-time energy consumption.

We focus on the design of energy-efficient communication protocols. A radio consumes energy at all times when sending, when receiving, and when idle. (Studies have shown that power consumption during the idle state cannot be ignored.[4]) This suggests two complementary approaches to reducing radio energy consumption: (1) minimizing energy consumption due to idle time or due to passively listening to transmissions not addressed to a node itself, and (2) minimizing energy consumption due to communication. Protocols that minimize idle-time energy consumption have been proposed.[5-7] We restrict our attention to minimizing energy consumption due to communication.

Ideally one would like to design a general-purpose communication protocol that maximizes network lifetime. However, the notion of network lifetime is application dependent. There are a number of reasonable notions. For example, for event-monitoring applications, one wants to maximize the time network monitoring centers are able to receive information about events happening in the field. For data-gathering applications, one may want to maximize the time until a certain percentage of nodes cannot deliver data to the data-gathering centers. For mission-critical applications, one might want to maximize the time until the first message cannot be delivered. Because of the application-dependent nature of the definition of network lifetime, it seems doubtful that there will be a general solution that is appropriate for all settings.

Of course, network lifetime is only one of several network performance metrics of interest. Other metrics, such as throughput and latency, are also important. Optimizing one metric can adversely impact another metric. For example, to maximize network lifetime, energy-efficient routes tend to be chosen. An energy-efficient route has more hops in general than the corresponding shortest route. This may lead to longer latency.

The problem of optimizing network performance is perhaps best viewed in terms of decision theory. Suppose we assume that, for each possible outcome of the algorithm, we can associate a utility. This utility would trade off the various features of an outcome, such as latency, throughput, and lifetime. If there is a probability distribution on outcomes, an optimal protocol is one that maximizes the expected utility. Because utilities and probability distributions on outcomes are difficult to obtain in practice, instead of trying to achieve application-specific optimal solutions, we focus on general heuristics for reducing communication-energy consumption.

To reduce energy consumption, it is typically better to relay messages through intermediate nodes rather than sending a message directly to the intended recipient. This is because radio-signal attenuation is inversely proportional to the nth power of the distance a signal propagates,[8] where n is between 2 and 6. Thus, relaying through intermediate nodes can reduce total power consumption. In addition, if a node sends a message directly to a distant receiver, it must use greater power and is more likely to interfere with the transmissions of other nodes.

While reducing broadcast power reduces power consumption and minimizes interference, we do not want to lose routes in the process. Suppose that each node u broadcasts with power $p(u)$. The resulting communication graph has an edge from u to v iff u can reach v when broadcasting with power $p(u)$. Because we do not want to lose routes, a minimal requirement on the choice of $p(u)$ is that if there is a route between a pair of nodes in the communication graph that results if each node broadcasts with maximum power, then there is a route in the communication graph that results if each node u broadcasts with power $p(u)$. But choosing $p(u)$ to satisfy this minimal constraint may not be the best choice in terms of reducing power consumption.

Let $p(u, v)$ denote the minimal power needed to send a point-to-point message between u and v. A *minimum-energy path* between a pair u and v of nodes is the path that requires the least amount of energy to send a message between u and v, provided that power $p(u', v')$ is used to transmit messages between neighboring nodes u' and v' on the path. To minimize power consumption for unicast (i.e., point-to-point) messages, it is typically best if each node broadcasts with enough power so that the minimal-energy path for any given node pair still exists in the resulting communication graph. A protocol for determining the broadcast power with this property is said to have the *minimum-energy property*. If a protocol has the minimum-energy property, then a suitable routing protocol can be used to find the minimum-energy path between any node pair. In this chapter, we present topology-control algorithms based on finding minimum-energy paths.

The rest of the chapter is organized as follows. Section 8.2 gives the network model. Section 8.3 identifies a condition necessary and sufficient for achieving the minimum-energy property. This characterization is used in Section 8.4 to construct the SMECN protocol. We prove that it has the minimum-energy property and that it constructs a network smaller than that constructed by Rodoplu and Meng[9] if the broadcast region is circular. Our SMECN requires location information, which is usually obtained from a GPS unit. In Section 8.5, we show how SMECN can be used to deal with topology changes as well. In Section 8.6, we give the results of simulations showing the energy savings obtained by using the network constructed by SMECN. We summarize in Section 8.7.

8.2   The Model

We assume that a set V of nodes is deployed in a two-dimensional region, where no two nodes are in the same physical location. Each node has a GPS receiver on board, so it knows its own location. It does not necessarily know the location of other nodes. Moreover, the location of nodes will, in general, change over time.

A transmission between nodes u and v takes power $p(u, v) = td(u, v)^n$ for some appropriate constant t, where $n \geq 2$ is the path-loss exponent of outdoor radio-propagation models,[8] and $d(u, v)$ is the distance between u and v. A reception at the receiver takes power c. This power expenditure at the receiver is referred to as the receiver power. Computational power consumption is ignored.

Suppose there is some maximum power p_{max} at which the nodes can transmit. Thus, there is a graph $G_R = (V, E_R)$ where $(u, v) \in E_R$ if u can reach v when using power p_{max}. Clearly, if $(u, v) \in E_R$, then $td(u, v)^n \leq p_{max}$. However, we do not assume that a node u can transmit to all nodes v such that $td(u, v)^n \leq p_{max}$. For one thing, there may be obstacles between u and v that prevent transmission. Even without obstacles, if a unit transmits using a directional transmit antenna, then only nodes in the region covered by the antenna (typically a cone-like region) will receive the message. Rodoplu and Meng[9] implicitly assume that every node can transmit to every other node. Here we take a first step in exploring what happens if this is not the case. However, we do assume that the graph G_R is connected, so that there is a potential communication path between every pair of nodes in V.

Because the power required to transmit between a pair of nodes increases as the nth power of the distance between them, for some $n \geq 2$, it may require less power to relay information than to transmit directly between two nodes. As usual, a *path* $r = (u_0, \ldots, u_k)$ in a graph $G = (V, E)$ is defined to be an ordered list of nodes such that $(u_i, u_{i+1}) \in E$. The *length* of $r = (u_0, \ldots, u_k)$, denoted $|r|$, is k. The total power consumption of a path $r = (u_0, u_1, \cdots, u_k)$ in G_R is the sum of the transmission and receiver power consumed; that is,

$$C(r) = \sum_{i=0}^{k-1} (p(u_i, u_{i+1}) + c).$$

A path $r = (u_0, \ldots, u_k)$ is a *minimum-energy path* from u_0 to u_k if $C(r) \leq C(r')$ for all paths r' in G_R from u_0 to u_k. For simplicity, we assume that $c > 0$. (Our results hold even without this assumption, but it makes the proofs a little easier.) A subgraph $G = (V, E)$ of G_R has the *minimum-energy property* if, for all $u, v \in V$, there is a path r in G that is a minimum-energy path in G_R from u to v.

8.3   A Characterization of Minimum-Energy Communication Networks

Our goal is to find a minimal subgraph G of G_R that has the minimum-energy property. Note that a graph G with the minimum-energy property must be connected because, by definition, it contains a path between every pair of nodes.

The intention is to have the nodes communicate using the links in G. To do this, it must be possible for each of the nodes in the network to construct G (or, at least, the relevant portion of G from their point of view) in a distributed way. In this section, we provide a condition that is necessary and sufficient for a subgraph of G_R to be minimal with respect to the minimum-energy property. In the next section, we use this characterization to provide an efficient distributed algorithm for constructing a graph G with the minimum-energy property that, while not necessarily minimal, still has relatively few edges.

Clearly, if a subgraph $G = (V, E)$ of G_R has the minimum-energy property, an edge $(u, v) \in E$ is *redundant* if there is a path r from u to v in G such that $|r| > 1$ and $C(r) \leq C(u, v)$. Let $G_{\min} = (V, E_{\min})$ be the subgraph of G_R such that $(u, v) \in E_{\min}$ iff there is no path r from u to v in G_R such that $|r| > 1$ and $C(r) \leq C(u, v)$. As the next result shows, $G_{\min}$ is the smallest subgraph of G_R with the minimum-energy property.

Theorem 8.1   *A subgraph G of G_R has the minimum-energy property iff it contains $G_{\min}$ as a subgraph. Thus, $G_{\min}$ is the smallest subgraph of G_R with the minimum-energy property.*

Proof   We first show that $G_{\min}$ has the minimum-energy property. Suppose, by way of contradiction, that there are nodes $u, v \in V$ and a path r in G_R from u to v such that $C(r) < C(r')$ for any path r' from u to v in $G_{\min}$. Suppose that $r = (u_0, \ldots, u_k)$, where $u = u_0$ and $v = u_k$. Without loss of generality, we can assume that r is the longest minimal-energy path from u to v. Note that r has no repeated nodes because any cycle can be removed to give a path that requires strictly less power. Thus, the length of a minimum-length path is bounded by $|V|$. Since $G_{\min}$ has no redundant edges, for all $i = 0, \ldots, k - 1$, it follows that $(u_i, u_{i+1}) \in E_{\min}$. For otherwise, there is a path r_i in G_R from u_i to u_{i+1} such that $|r_i| > 1$ and $C(r_i) \leq C(u_i, u_{i+1})$. But then it is immediate that there is a path r^* in G_R such that $C(r^*) \leq C(r)$ and r^* is longer than r, contradicting the choice of r.

To see that $G_{\min}$ is a subgraph of every subgraph of G_R with the minimum-energy property, suppose that there is some subgraph G of G_R with the minimum-energy property that does not contain the edge $(u, v) \in E_{\min}$. Thus, there is a minimum-energy path r from u to v in G. It must be the case that $C(r) \leq C(u, v)$. Since (u, v) is not an edge in G, we must have $|r| > 1$. But then $(u, v) \notin E_{\min}$, a contradiction.   □

This result shows that to find a subgraph of G with the minimum-energy property, it suffices to ensure that it contains $G_{\min}$ as a subgraph.

8.4   A Power-Efficient Protocol for Finding a Minimum-Energy Communication Network

Checking if an edge (u, v) is in $E_{\min}$ may require checking nodes that are located far from u. This may require a great deal of communication, possibly to distant nodes, and thus require a great deal of power. Because power efficiency is an important consideration in practice, we consider here an algorithm for constructing a communication network that contains G_{min} and can be constructed in a power-efficient manner rather than trying to construct G_{min} itself.

Say that an edge $(u, v) \in E_R$ is *k-redundant* if there is a path r in G_R such that $|r| = k$ and $C(r) \leq C(u, v)$. Notice that $(u, v) \in E_{min}$ iff it is not k-redundant for all $k > 1$. Let E_2 consist of all and only edges in E_R that are not 2-redundant. In our algorithm, we construct a graph $G = (V, E)$ where $E \supseteq E_2$; in fact, under appropriate assumptions, $E = E_2$. Clearly, $E_2 \supseteq E_{min}$, so G has the minimum-energy property.

There is a trivial algorithm for constructing E_2. Each node u starts the process by broadcasting a "Hello" at maximum power p_{max}, stating its own position. If a node v receives this message, it responds to u with a *Ack* message stating its location. Let $M(u)$ be the set of nodes that respond to u and let $N_2(u)$ denote u's neighbors in E_2. Clearly, $N_2(u) \subseteq M(u)$. Moreover, it is easy to check that $N_2(u)$ consists of all those nodes $v \in M(u)$ other than u such that there is no $w \in M(u)$ such that $C(u, w, v) \leq C(u, v)$. Since u has the location of all nodes in $M(u)$, $N_2(u)$ is easy to compute.

The problem with this algorithm is in the first step, which involves a broadcast using maximum power. While this expenditure of power may be necessary if there are relatively few nodes, so that power close to p_{max} will be required to transmit to some of u's neighbors in E_2, it is unnecessary in denser networks. In this case, it may require much less than p_{max} to find u's neighbors in E_2. We now present a more power-efficient algorithm for finding these neighbors than the one proposed by Rodoplu and Meng[9] (We refer to their protocol as MECN for *Minimum Energy Communication Network*.) Let $F(u, p)$ be the region that u can reach if it broadcasts with power p. For this algorithm, we assume that u knows $F(u, p)$. If there are no obstacles and the antenna is omnidirectional, then $F(u, p)$ is just a circle of radius d_p such that $t d_p^n = p$. We are implicitly assuming that even if there are obstacles or the antenna is not omnidirectional, a node u knows the terrain and the antenna characteristics well enough to compute $F(u, p)$.

Before presenting the algorithm, it is useful to define a few terms.

Definition 8.1 Given a node v, let $Loc(v)$ denote the physical location of v. The *relay region* of the transmit-relay node pair (u, v) is the physical region $R_{u \to v}$ such that relaying through v to any point in $R_{u \to v}$ takes less power than direct transmission. Formally,

$$R_{u \to v} = \{(x, y) : C(u, v, (x, y)) \leq C(u, (x, y))\},$$

where we abuse notation and take $C(u, (x, y))$ to be the cost of transmitting a message from u to a virtual node whose location is (x, y). That is, if there were a node v' such that $Loc(v') = (x, y)$, then $C(u, (x, y)) = C(u, v')$; similarly, $C(u, v, (x, y)) = C(u, v, v')$. Note that, if a node v is in the relay region $R_{u \to w}$, then the edge (u, v) is 2-redundant. Moreover, since $c > 0$, $R_{u \to u} = \emptyset$.

Given a region F, let

$$N_F = \{v \in V : Loc(v) \in F\};$$

if F contains u, let

$$R_F(u) = \bigcap_{w \in N_F} (F(u, p_{max}) - R_{u \to w}). \tag{8.1}$$

Intuitively, N_F consists of the nodes in region F, while $R_F(u)$ consists of those points that can be reached by u transmitting at maximum power other than those for which routing through some node in N_F would be more energy efficient than direct communication.

The following proposition gives a useful characterization of $N_2(u)$.

Proposition 8.1 Suppose that F is a region containing the node u. If $F \supseteq R_F(u)$, then $N_{R_F(u)} \supseteq N_2(u)$. Moreover, if F is a circular region with center u and $F \supseteq R_F(u)$, then $N_{R_F(u)} = N_2(u)$.

Proof   Suppose that $F \supseteq R_F(u)$ and that $N_{R_F(u)} \supseteq N_2(u)$. Suppose that $v \in N_2(u)$. Then clearly, $Loc(v) \notin \cup_{w \in V} R_{u \to w}$ and $Loc(v) \in F(u, p_{max})$. Thus, $Loc(v) \in R_F(u)$, so $v \in N_{R_F(u)}$. Since v was chosen arbitrarily, it follows that $N_2(u) \subseteq N_{R_F}(u)$.

Now suppose that F is a circular region with center u and $F \supseteq R_F(u)$. We now show that $N_{R_F(u)} \subseteq N_2(u)$. Suppose that $v \in N_{R_F(u)}$. If $v \notin N_2(u)$, then there exists some w such that $C(u, w, v) \leq C(u, v)$. Since transmission costs increase with distance, it must be the case that $d(u, w) \leq d(u, v)$. Since $v \in N_{R_F(u)} \subseteq N_F$ and F is a circular region with center u, it follows that $w \in N_F$. Since $C(u, w, v) \leq C(u, v)$, it follows that $Loc(v) \in R_{u \to w}$. Thus, $v \notin R_F(u)$, contradicting our original assumption. Thus, $v \in N_2(u)$.   □

The algorithm for node u constructs a set F such that $F \supseteq R_F(u)$, and tries to do so in a power-efficient fashion. By Proposition 8.1, the fact that $F \supseteq R_F(u)$ ensures that $N_{R_F(u)} \supseteq N_2(u)$. Thus, the nodes in $N_{R_F(u)}$ other than u itself are taken to be u's neighbors. By Theorem 8.1, the resulting graph has the minimum-energy property.

Essentially, the algorithm for node u starts by broadcasting a "Hello" message with some initial power p_0, getting Acks from all nodes in $F(u, p_0)$, and checking if $F(u, p_0) \supseteq R_{F(u, p_0)}(u)$. If not, it transmits with more power. It continues increasing the power p until $F(u, p) \supseteq R_{F(u, p)}(u)$. It is easy to see that $F(u, p_{max}) \supseteq R_{F(u, p_{max})}(u)$, so that as long as the power increases to p_{max} eventually, then this process is guaranteed to terminate. We do not investigate here how to increase the initial power p_0, nor do we investigate how to increase the power at each step. We simply assume some function *Increase* such that $Increase^k(p_0) = p_{max}$ for sufficiently large k. An obvious choice is to take $Increase(p) = 2p$. If the initial choice of p_0 is less than the power actually needed, then it is easy to see that this guarantees that u's estimate of the transmission power needed to reach a node v will be within a factor of 2 of the minimum transmission power actually needed to reach u.*

Thus, the protocol run by node u is simply

$$p = p_0;$$
$$\text{while } F(u, p) \not\supseteq R_{F(u,p)}(u) \text{ do } Increase(p);$$
$$N(u) = N_{R_{F(u,p)}}$$

A more careful implementation of this algorithm is given in Table 8.1. Note that we also compute the minimum power $p(u)$ required to reach all the nodes in $N(u)$. In the algorithm, A is the set of all the nodes that u has found so far in the search and M consists of the new nodes found in the current iteration. In the computation of η in the second-last line of the algorithm, we take $\cap_{v \in M}(F(u, p_{max}) - R_{u \to v})$ to be $F(u, p_{max})$ if $M = \emptyset$. For future reference, we note that it is easy to show that, after each iteration of the while loop, we have that $\eta = \cap_{v \in A}(F(u, p_{max}) - R_{u \to v})$.

Define the graph $G = (V, E)$ by taking $(u, v) \in E$ iff $v \in N(u)$, as constructed by the algorithm in Table 8.1. It is immediate from the earlier discussion that $E \supseteq E_2$. Thus, the following theorem holds.

Theorem 8.2   *G has the minimum-energy property.*

We next show that SMECN dominates MECN. MECN is described in Table 8.2. For easier comparison, we have made some inessential changes to MECN to make the notation and presentation more like that of SMECN. The main difference between SMECN and MECN is the computation of the region η. As we observed, in SMECN, $\eta = \cap_{v \in A}(F(u, p_{max}) - R_{u \to v})$ at the end of every iteration of the loop. On the other

---

*Note that, in practice, a node may control a number of directional transmit antennae. Our algorithm implicitly assumes that they all transmit at the same power. This was done for ease of exposition. It would be easy to modify the algorithm to allow each antenna to transmit using different power. All that is required is that after sufficiently many iterations, all antennae transmit at maximum power.

TABLE 8.1 Algorithm SMECN Running at Node u

$p = p_0$;
$A = \emptyset$;
$NonNbrs = \emptyset$;
$\eta = F(u, p_{max})$;
while $F(u, p) \not\supseteq \eta$ do
    $p = Increase(p)$;
    Broadcast "Hello" message with power p and gather $Acks$;
    $M = \{v \mid Loc(v) \in F(u, p), v \notin A, v \neq u\}$;
    $A = A \bigcup M$;
    for ëach $v \in M$ do
        for ëach $w \in A$ do
            if $Loc(v) \in R_{u \to w}$ then $NonNbrs = NonNbrs \bigcup \{v\}$;
            else if $Loc(w) \in R_{u \to v}$ then $NonNbrs = NonNbrs \bigcup \{w\}$;
    $\eta = \eta \cap \bigcap_{v \in M}(F(u, p_{max}) - R_{u \to v})$;
$N(u) = A - NonNbrs$;
$p(u) = \min\{p : F(u, p) \supseteq \eta\}$

---

hand, in MECN, $\eta = \bigcap_{v \in A - NonNbrs}(F(u, p_{max}) - R_{u \to v})$. Moreover, in SMECN, a node is never removed from *NonNbrs* once it is in the set, while in MECN, it is possible for a node to be removed from *NonNbrs* by the procedure *Flip*. Roughly speaking, if a node $v \in R_{u \to w}$, then, in the next iteration, if $w \in R_{u \to t}$ for a newly discovered node t, but $v \notin R_{u \to t}$, node v will be removed from *NonNbrs* by *Flip(v)*. In Rodoplu and Meng,[9] it is shown that MECN is correct (i.e., it computes a graph with the minimum-energy property) and terminates (and, in particular, the procedure *Flip* terminates). Here we show that, at least for circular search regions, SMECN does better than MECN.

Theorem 8.3 *If the search regions considered by the algorithm SMECN are circular, then the communication graph constructed by SMECN is a subgraph of the communication graph constructed by MECN.*

Proof For each variable x that appears in SMECN, let x_S^k denote the value of x after the kth iteration of the loop; similarly, for each variable in MECN, let x_M^k denote the value of x after the kth iteration of the loop. It is almost immediate that SMECN maintains the following invariant: $v \in NonNbrs_S^k$ iff

TABLE 8.2 Algorithm MECN Running at Node u

$p = p_0$;
$A = \emptyset$;
$NonNbrs = \emptyset$;
$\eta = F(u, p_{max})$;
while $F(u, p) \not\supseteq \eta$ do
    $p = Increase(p)$;
    Broadcast "Hello" message with power p and gather $Acks$;
    $M = \{v \mid Loc(v) \in F(u, p), v \notin A, v \neq u\}$;
    $A = A \bigcup M$;
    $NonNbrs = NonNbrs \bigcup M$;
    for each $v \in M$ do $Flip(v)$;
    $\eta = \bigcap_{v \in (A - NonNbrs)}(F(u, p_{max}) - R_{u \to v})$;
$N(u) = A - NonNbrs$;
$p(u) = \min\{p : F(u, p) \supseteq \eta\}$

Procedure $Flip(v)$
    if $v \notin NonNbrs$ then $NonNbrs = NonNbrs \bigcup \{v\}$;
        for each $w \in A$ such that $Loc(w) \in R_{u \to v}$ do $Flip(w)$;
    else if $Loc(v) \notin \cup_{w \in A - NonNbrs} R_{u \to w}$ then $NonNbrs = NonNbrs - \{v\}$;
        for each $w \in A$ such that $Loc(w) \in R_{u \to v}$ do $Flip(w)$;

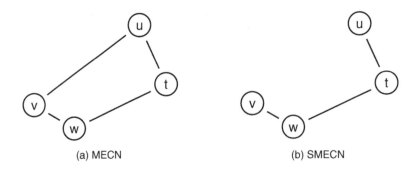

(a) MECN   (b) SMECN

FIGURE 8.1   A network where SMECN dominates MECN.

$v \in A_S^k$ and $Loc(v) \in \cup_{w \in A_S^k} R_{u \to w}$. Similarly, it is not hard to show that MECN maintains the following invariant: $v \in NonNbrs_M^k$ iff $v \in A_S^k$ and $Loc(v) \in \cup_{w \in A_M^k - NonNbrs_M^k} R_{u \to w}$. (Indeed, the whole point of the *Flip* procedure is to maintain this invariant.) Since it is easy to check that $A_S^k = A_M^k$, it is immediate that $NonNbrs_S^k \supseteq NonNbrs_M^k$. Suppose that SMECN terminates after k_S iterations of the loop and MECN terminates after k_M iterations of the loop. Hence, $\eta_S^k \subseteq \eta_M^k$ for all $k \leq \min(k_S, k_M)$. Since both algorithms use the condition $F(u, p) \supseteq \eta$ to determine termination, it follows that SMECN terminates no later than MECN; that is, $k_S \leq k_M$.

Because the search region used by SMECN is assumed to be circular, by Proposition 8.1, $A_S^{k_S} - NonNbrs_S^{k_S} = N_2(u)$. Moreover, even if we continue to iterate the loop of SMECN (ignoring the termination condition), then $F(u, p)$ keeps increasing while η keeps decreasing. Thus, by Proposition 8.1 again, we continue to have $A_S^k - NonNbrs_S^k = N_2(u)$ even if $k \geq k_S$. That means that if we were to continue with the loop after SMECN terminates, none of the new nodes discovered would be neighbors of u. Because the previous argument still applies to show that $NonNbrs_S^{k_M} \supseteq NonNbrs_M^{k_M}$, it follows that $N_2(u) = A_S^{k_M} - NonNbrs_S^{k_M} \subseteq A_M^{k_M} - NonNbrs_M^{k_M}$. That is, the communication graph constructed by SMECN has a subset of the edges of the communication graph constructed by MECN.   □

In the proof of Theorem 8.3, we implicitly assumed that both SMECN and MECN use the same value of initial value p_0 of p and the same function *Increase*. In fact, this assumption is not necessary, since the neighbors of u in the graph computed by SMECN are given by $N_2(u)$ independent of the choice of p_0 and *Increase*, as long as $F(u, p_0) \not\supseteq F(u, p_{max})$ and $Increase^k(p_0) \geq p_{max}$ for k sufficiently large. Similarly, the proof of Theorem 8.3 shows that the set of neighbors of u computed by MECN is a superset of $N_2(u)$, as long as *Increase* and p_0 satisfy these assumptions.

Theorem 8.3 shows that the neighbor set computed by MECN is a superset of $N_2(u)$. As the following example shows, it may be a strict superset (so that the communication graph computed by SMECN is a strict subgraph of that computed by MECN).

Example 8.1

Consider a network with four nodes t, u, v, w, where $Loc(v) \in R_{u \to w}, Loc(w) \in R_{u \to t}$, and $Loc(v) \notin R_{u \to t}$. As shown in Figure 8.1, it is not hard to choose power functions and locations for the nodes that have this property. It follows that $N_2(u) = \{t\}$. (It is easy to check that $Loc(t) \notin R_{u \to v} \cup R_{u \to w}$.) On the other hand, suppose that *Increase* is such that t, v, and w are added to A in the same step. Then all of them are added to *NonNbrs* in MECN.

8.5   Reconfiguration

In a multihop wireless network, nodes can be mobile. Even if nodes do not move, nodes may die if they run out of energy. In addition, new nodes may be added to the network. We assume that each node uses a Neighbor Discovery Protocol (NDP), a periodic message that provides all its neighbors with its current position (according to the GPS) to detect changes in the topology of the network. A node u sends out the message with just enough power to reach all the nodes that it currently considers its neighbors (i.e., the nodes in $N_2(u)$). Once a node detects a change, it may need to update its set of neighbors. This is done by a *reconfiguration protocol*. Rodoplu and Meng[9] do not provide an explicit reconfiguration protocol. Rather, they deal with changes in network topology by running MECN periodically at every node. While this will work, it is inefficient. If a node does not detect any changes, then there is no obvious need to run MECN. We now present a reconfiguration protocol where, in a precise sense, we run SMECN only when necessary (in the sense that it is run only when not running it may result in a network that does not satisfy the minimum-energy property).

There are three types of events that trigger the reconfiguration protocol: *leave events*, *join events*, and *move events*:

- A *leave$_u$(v)* event happens when a node v that was in u's neighborhood is detected to no longer be in the neighborhood (because its beaconing message is not received). This may happen because v is faulty or dies, or because it has in fact moved away.
- A *join$_u$(v)* event happens when a node v is detected to be within u's neighborhood by the NDP.
- A *move$_u$(v, L)* event happens when u detects that v has moved from the previous location to the current location L. (Node v's location L is relative to u's location, so the event could be due to u's own movement.)

It is straightforward to see how to update the neighbor set if u detects a single change. Suppose p^* is u's current power setting (that is, the final power setting used in the last invocation of SMECN by u); let $F^* = F(u, p^*)$ be the last region searched by u. Let A^* consist of all the nodes in F^* (that is, the set of all nodes discovered by the algorithm).

- If a single *leave$_u$(v)* or a *move$_u$(v, L)* is detected, let $A' = A^* - \{v\}$ if *leave$_u$(v)* is detected, and let $A' = A^*$ if *move(v, L)* is detected. Let $R'_F = \bigcap_{w \in A'}(F(u, p_{max}) - R_{u \to w})$, where the new location for v is used in the computation if $v \in A'$. (Note that R'_F is defined essentially in the same way as $R_F(u)$ in Equation 8.1.) If $F^* \supseteq R'_F$, then take u's updated neighbor set to be $N_{R'_F}$; otherwise, run SMECN taking $p_0 = p^*$.
- If a single *join$_u$(v)* is detected, recompute the neighbor set as follows. Let $A' = A^* \cup \{v\}$. Let $R'_F = \bigcap_{w \in A'}(F(u, p_{max}) - R_{u \to w})$. Take u's updated neighbor set to be $N_{R'_F}$. Then let $p' = \min\{p : F(u, p) \supseteq \bigcap_{w \in A'}(F(u, p_{max}) - R_{u \to w})\}$.

The following proposition is almost immediate from our earlier results.

Proposition 8.2 Suppose that a graph G has the minimum-energy property. If the nodes in G observe a sequence of single changes and update their edge sets as above, the resulting graph $G^*(V, E^*)$ still has the minimum-energy property for the new topology. Moreover, if $F(u, p)$ is a circular region for all p, then $E^* = E_2$.

In general, there may be more than one change event that is detected at a given time by a node u. (For example, if u moves, then there will in general several leave and move events detected by u.) If more than one change event is detected by u, we consider the events observed in some order. If we can perform all the updates without rerunning SMECN, we do so; otherwise, we rerun SMECN starting from p^*. By rerunning SMECN, we can deal with all the changes simultaneously.

Up to now we have assumed that no topology changes are detected while SMECN itself is being run. If changes are in fact detected while SMECN is run, then it is straightforward to incorporate the update into SMECN. For example, if u detects a *join$_u$(v)* event, then v is added to the set A in the algorithm; while if u detects a *leave$_u$(v)* event, u is dropped from A and η is recomputed. We leave the details to the reader.

As we mentioned earlier, there is no reconfiguration protocol given by Rodoplu and Meng.[9] However, it is easy to modify the reconfiguration algorithm protocol given above for SMECN so that it works for

MECN. If a $leave_u(v)$ or $move_u(v, L)$ is detected, then the same approach works (except that MECN rather than SMECN is called with $p_0 = p^*$). Similarly, if a $join_u(v)$ is detected, we update the neighbor set using the approach of MECN rather than SMECN.

Note that we have assumed a perfect MAC layer in our reconfiguration discussion. Our reconfiguration works fine even with a MAC layer that drops packets. The reason is as follows. If the Ack message of some nodes get dropped, then the final power setting p_a^* using an imperfect MAC layer will be bigger than the corresponding p^* using a perfect MAC layer. Because NDP beaconing with p^* reaches all nodes in $N_2(u)$, beaconing with a bigger power p_a^* will still reach all nodes in $N_2(u)$. Eventually, all the nodes in $N_2(u)$ whose *Acks* are lost will be detected by u through NDP beacons. Thus, the neighbor set computed using an imperfect MAC layer converges to a superset of $N_2(u)$. If the final search region is circular, then the neighbor set converges to the set $N_2(u)$.

8.6   Simulation Results and Evaluation

How can using the subnetwork computed by (S)MECN help performance? Clearly, sending messages on minimum-energy paths is more efficient than sending messages on arbitrary paths, but the algorithms are all local; that is, they do not actually find the minimum-energy path, they just construct a subnetwork in which it is guaranteed to exist.

There are actually two ways that the subnetwork constructed by (S)MECN helps. First, when sending periodic beaconing messages, it suffices for u to use power $p(u)$, the final power computed by (S)MECN. Second, the routing algorithm is restricted to using the edges $\cup_{u \in V} N(u)$. While this does not guarantee that a minimum-energy path is used, it makes it more likely that the path used is one that requires less energy consumption.

To measure the effect of focusing on energy efficiency, we compared the use of MECN and SMECN in a simulated application setting.

Both SMECN and MECN were implemented in ns-2,[10] using the wireless extension developed at Carnegie Mellon.[11] We generated 20 random networks, each with 100 nodes. The nodes were placed uniformly at random in a rectangular region of 1500 by 1500 meters. (There has been a great deal of work on realistic placement.[12,13] However, this work has the Internet in mind. Since the nodes in a multihop network are often best viewed as being deployed in a somewhat random fashion and move randomly, we believe that the uniform random placement assumption is reasonable in many large multihop wireless networks.)

We assume that the path-loss exponent for outdoor radio propagation models is 4. The carrier frequency is 914 MHz and transmission raw bandwidth is 2 MHz. We further assume that each node has an omnidirectional antenna with 0 dB gain and is placed 1.5 meters above the node. The receive threshold is −94 dBW, the carrier sense threshold is −108 dBW, and the capture threshold is 10 dB. These parameters simulate the 914 MHz Lucent WaveLAN DSSS radio interface. Given these parameters, the t parameter in the equation $p(u, v) = td(u, v)^n$ in Section 8.2 is −101 dBW. In WaveLAN radio, it has been measured that radio receiver power can be quite significant[4] and accounts for 75 percent of the fixed transmission power. However, techniques for reducing the power consumption of radio electronics are fast improving. A radio typically consists of transmitter electronics, receiver electronics, and a transmit amplifier. Low-power circuit designs and signal processing reduce the power expended in the transmitter and receiver electronics. As a result, the receiver power of future radios is likely quite small. However, the power needed by the transmit amplifier is constrained by the rapid radio attenuation in space. Therefore, transmission power is expected to dominate receiver power in the future. Because radio-receiver power varies from radio to radio and has an impact on the computation of the minimal-energy path, we vary the receiver power c to study its effect on MECN and SMECN.

Each node in our simulation has an initial energy of 1 Joule. We would like to evaluate the effect of using SMECN on network performance. To do this, we need to simulate the network's application traffic. We used the following application scenario. All nodes periodically send UDP traffic to a sink node situated at the boundary of the network. The sink node is viewed as the master data-collection site. The application traffic is assumed to be CBR (constant bit rate); application packets are all 128 bytes. The sending rate is 0.25 packets per second. This application scenario has also been used before.[14] Although this application scenario does not seem appropriate for telephone networks and the Internet (cf. Refs. 15, 16), it does seem reasonable for ad hoc networks, for example, in environment-monitoring sensor applications. In this setting, sensors periodically transmit data to a data-collection site, where the data is analyzed.

To find routes along which to send messages, we use AODV.[17] However, as mentioned above, we restrict AODV to finding routes that use only edges in $\cup_{u \in V} N(u)$. There are other routing protocols such as LAR,[18] GSPR,[19] and DREAM,[20] that take advantage of GPS hardware. We used AODV because it is readily available in our simulator and it is well studied. Because we would like to optimize with respect to the minimum-energy path metric, we modify the ns-2 AODV implementation to use the minimum-energy path metric instead of the current shortest-path metric. Although different routing protocols may result in different network performance, we do not believe that using a different routing protocol would significantly affect the relative merits of SMECN and MECN we present here.

To simulate the effect of power control, we made changes to the physical layer of the ns-2 simulation code. Specifically, when simulating SMECN (resp., MECN), a node u broadcasts to its neighbors using the final transmission power $p(u)$ of its neighbor-discovery process with SMECN (resp., MECN). Similarly, a node u sends a point-to-point message to a neighbor v using the minimum power required to reach v, as determined during the neighbor-discovery process. A node's energy reserve is then subtracted by the appropriate amount for each transmission or reception.

We assumed that each node in our simulation had an initial energy of 1 Joule and then ran the simulation for 1600 simulation seconds, using both SMECN and MECN. Each data point represents an average of 20 randomly generated networks. For the sake of fairness, identical traffic scenarios are used for both MECN and SMECN. We did not actually simulate the execution of SMECN and MECN. Rather, we assumed the neighbor set $N(u)$ and power $p(u)$ computed by (S)MECN each time it is run were given by an oracle. (Of course, it is easy to compute the neighbor set and power in the simulation because we have a global picture of the network.) Thus, in our simulation, we did not take into account one of the benefits of SMECN over MECN, that it stops earlier in the neighbor-search process. Because a node's available energy is decreased after each packet reception or transmission, nodes in the simulation die over time. After a node dies, the network must be reconfigured. In Ref. 9, this is done by running MECN periodically. In our simulation, the NDP triggers the reconfiguration protocol. (When running MECN, we use the same reconfiguration protocol as the one we use for SMECN, with the appropriate modifications, as discussed in Section 8.5.) The NDP beacon for MECN and SMECN is sent with a period of 1 second and uses the power $p(u)$ computed by the neighbor-discovery process of SMECN (resp., MECN).

For simplicity, we simulated only a static network (that is, we assumed that nodes did not move), although some of the effects of mobility — that is, the triggering of the reconfiguration protocol — can already be observed with node deaths.

In this setting, we are interested in network lifetime, as measured by two metrics: (1) the number of nodes that are still alive over time and (2) the number of nodes that are still connected to the sink. As we argued in the introduction, these are reasonable metrics. Of course, if we have more knowledge of the application, the definition of network lifetime can be made even more application-specific. For example, in a sensor network, it may be more appropriate to define network lifetime as the time that the sensors completely cover the deployment region.

We first report the experimental results when the receiver power is 0. Before describing the performance, we consider some features of the subnetworks computed by MECN and SMECN. Because the search regions will be circular with an omnidirectional antenna, Theorem 8.3 assures us that the network used by SMECN will be a subnetwork of that used by MECN, although it does not say how much smaller the subnetwork will be. The initial network in a typical execution of the MECN and SMECN is shown in Figure 8.2. The average number of neighbors of MECN and SMECN in the 20 networks are initially 3.21 and 2.71, respectively. Thus, each node running MECN has roughly 19 percent more links than the same node running SMECN. This makes it likely that the final power setting computed will be higher for MECN than for SMECN. In fact, our experiments show that it is roughly 38 percent higher, so more power will

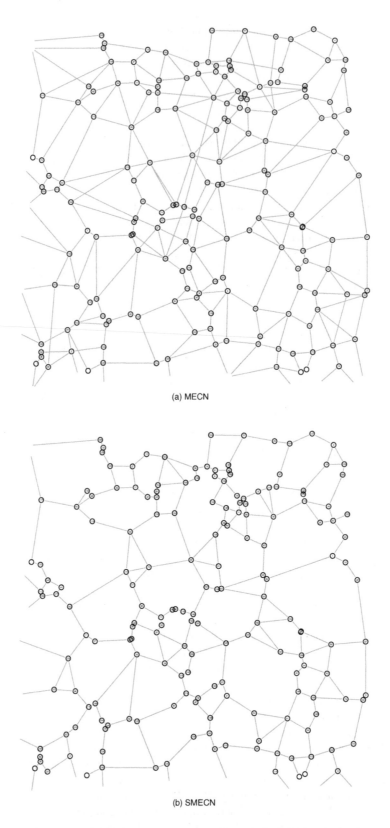

(a) MECN

(b) SMECN

FIGURE 8.2 Initial network computed by MECN and SMECN with $c = 0$ mW.

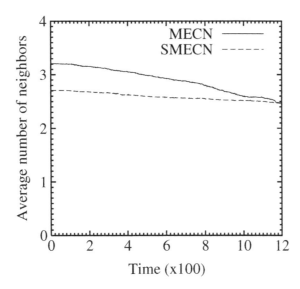

FIGURE 8.3   Average number of neighbors over time with $c = 0$ mW.

be used by nodes running MECN when sending messages. Moreover, AODV is unlikely to find routes that are as energy efficient with MECN.

As nodes die (due to running out of power), the network topology changes due to reconfiguration. Nevertheless, as shown in Figure 8.3, the average number of neighbors stays roughly the same over time, thanks to the reconfiguration protocol.

Turning to the network-lifetime metrics discussed above, as shown in Figure 8.4, SMECN performs consistently better than MECN for both. The number of nodes still alive and the number of nodes still connected to the sink decrease much more slowly in SMECN than in MECN. For example, in Figure 8.4(b), at time 800, 66.5 percent of the nodes have disconnected from the sink for MECN while only 36.4 percent of the nodes have disconnected from the sink for SMECN.

Finally, we collected data on average energy consumption per node at the end of the simulation, on the total number of packets delivered, and on end-to-end delay. MECN uses 21 percent more energy per node than SMECN. SMECN delivers more than 110 percent more packets than MECN by the end of the simulation; MECN's delivered packets have an average end-to-end delay that is 2 percent lower than SMECN. Overall, it is clear that the performance of SMECN is significantly better than MECN if the receiver power is negligible.

We now vary the receiver power c to study its impact on MECN and SMECN. As we discussed earlier in this section, the receiver power of a radio is expected to be small in the future. Hence, we set c to a small value (20 mW). A typical network topology maintained by MECN and SMECN is shown in Figure 8.5. Comparing Figures 8.2 and 8.5, it is clear that there tend to be more direct links with $c = 20$ mW than with $c = 0$ mW. The average number of neighbors and broadcast power using MECN and SMECN are quite similar with $c = 20$ mW. As a result, it is not surprising that the performance of the two algorithms is quite similar in this case. This is further substantiated by experimental results using the average number of neighbors metric (shown in Figure 8.6) and the two network-lifetime metrics (shown in Figure 8.7).

8.7   Summary

In this chapter, we presented a protocol SMECN that computes a network with the minimum-energy property. In the case of a circular search space, SMECN computes the set E_2 consisting of all edges that are not 2-redundant. Our protocol is localized in the sense that each node needs to know only

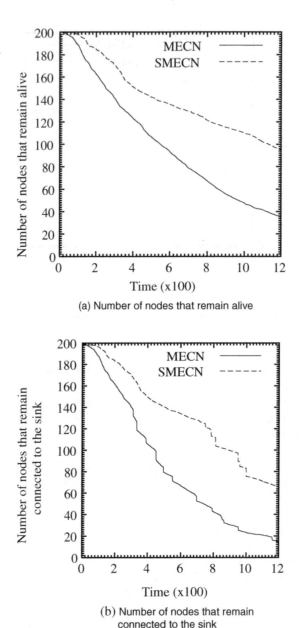

(a) Number of nodes that remain alive

(b) Number of nodes that remain
connected to the sink

FIGURE 8.4   Network lifetime for two different metrics with $c = 0$ mW.

about its local neighborhood (that is, those nodes that are a small number of hops away). In addition, we presented an energy-efficient reconfiguration protocol that maintains the minimum-energy path property despite changes in the network topology. The localized nature of our protocol makes it easy to deal with reconfiguration. We have shown by simulation that SMECN performs significantly better than MECN, while being computationally simpler.

There are a number of other localized topology-control algorithms.[21–24] CBTC [21] was the first algorithm that simultaneously achieved a variety of useful properties, such as symmetry (only symmetric links are used), sparseness (bounded degree), and good routes; CBTC achieves this without requiring each node to know its location; in particular, unlike SMECN, a GPS unit (or other means for knowing the location)

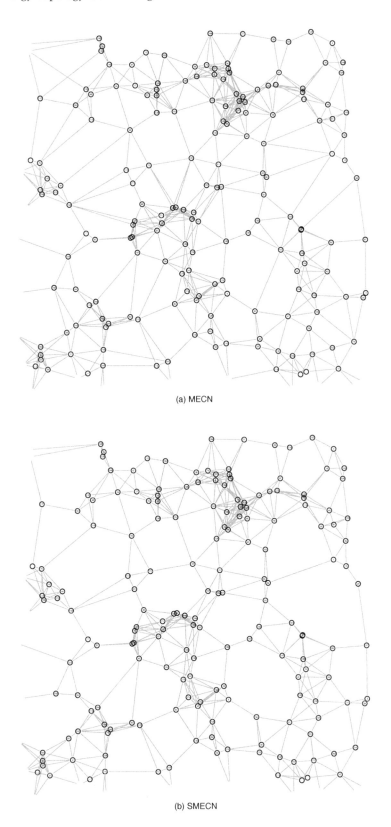

(a) MECN

(b) SMECN

FIGURE 8.5　Initial network computed by MECN and SMECN with $c = 20$ mW.

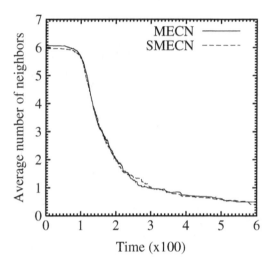

FIGURE 8.6 Average number of neighbors over time with $c = 20$ mW.

is not required. However, the network CBTC computes does not have the minimum-energy property. Consequently the total energy consumed in the network constructed by CBTC is likely to be greater than that in the network constructed by SMECN (see Ref. 21 for a detailed comparison of CBTC and SMECN). The subgraph G' of G_R constructed by Wang and Li's recent algorithm[22] has bounded degree and is a *k-spanner*, for a relatively small k, so that for every pair of nodes u and v, there is a path connecting them in G' whose length is no more than k times that of the shortest path from u to v in G_R. However, the network computed in Ref. 22 does not have the minimum-energy property and thus is unlikely to be as energy efficient as SMECN. The topology-control algorithm analyzed by Jia, Rajaraman, and Scheideler[23] constructs a graph with constant degree and constant energy-stretch (the minimum-energy path for any given pair of nodes in the subnetwork is within a constant factor of the minimum energy path in the original

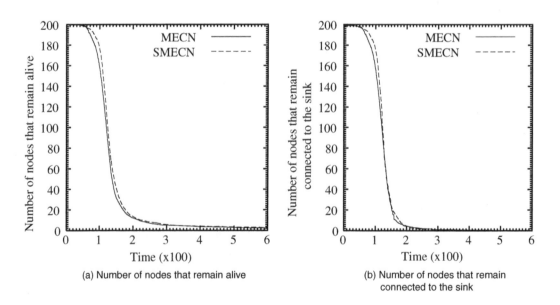

(a) Number of nodes that remain alive

(b) Number of nodes that remain connected to the sink

FIGURE 8.7 Network lifetime for two different metrics with $c = 20$ mW.

network). XTC[24] is similar in spirit to SMECN, but it does not assume any specific radio-propagation model. When running SMECN, each node u must know enough about the radio-propagation model to compute the sets $R_{u \to v}$; this knowledge is not required in XTC. Thus, XTC can be used in settings where the radio-propagation model is unknown. On the other hand, XTC must use maximum power in the neighbor-discovery process, so it is likely to be less energy efficient than SMECN in settings where the radio-propagation model is known.

With all these alternatives, it is clear that more work must be done to understand what the most appropriate algorithm is as a function of the demands of a specific application. We have focused here only on energy minimization, but there are clearly other relevant metrics as well, which further complicates the decision.

References

1. M. Satyanarayanan and D. Narayanan. Multifidelity algorithms for interactive mobile applications. In *Proc. 3th Int. Workshop on Discrete Algorithms and Methods for Mobile Computing and Communications (Dial M for Mobility)*, pp. 1–6, 1999.

2. J. Pouwelse, K. Langendoen, and H. Sips. Dynamic voltage scaling on a low-power microprocessor. In *Proc. Seventh Annu. ACM/IEEE Int. Conf. on Mobile Computing and Networking (MobiCom)*, pp. 251–259, 2001.

3. K. Flautner, S. Reinhardt, and T. Mudge. Automatic performance-setting for dynamic voltage scaling. In *Proc. Seventh Annu. ACM/IEEE Int. Conf. on Mobile Computing and Networking (MobiCom)*, pp. 260–271, 2001.

4. M. Stemm and R.H. Katz. Measuring and reducing energy consumption of network interfaces in hand-held devices. *IEICE Transactions on Fundamentals of Electronics, Communications, and Computer Science, Special Issue on Mobile Computing*, E80-B(8):1125–1131, 1997.

5. S. Singh and C.S. Raghavendra. PAMAS — power aware multi-access protocol with signaling for ad hoc networks. *ACM Computer Communications Review*, pp. 5–26, July 1998.

6. Y. Xu, J. Heidemann, and D. Estrin. Geography-informed energy conservation for ad-hoc routing. In *Proc. Seventh Annu. ACM/IEEE Int. Conf. on Mobile Computing and Networking (MobiCom)*, pp. 70–84, 2001.

7. B. Chen, K. Jamieson, H. Balakrishnan, and R. Morris. Span: an energy-efficient coordination algorithm for topology maintenance in ad hoc wireless networks. In *Proc. Seventh Annu. ACM/IEEE Int. Conf. on Mobile Computing and Networking (MobiCom)*, pp. 85–96, 2001.

8. T.S. Rappaport. *Wireless Communications: Principles and Practice.* Prentice Hall, 1996.

9. V. Rodoplu and T.H. Meng. Minimum energy mobile wireless networks. *IEEE J. Selected Areas in Communications*, 17(8):1333–1344, 1999.

10. VINT Project. The UCB/LBNL/VINT network simulator-ns (Version 2). http://www.isi.edu/nsnam/ns.

11. CMU Monarch Group. Wireless and mobility extensions to ns-2. http://www.monarch.cs.cmu.edu/cmu-ns.html, October 1999.

12. E.W. Zegura, K. Calvert, and S. Bhattacharjee. How to model an Internetwork. In *Proc. IEEE Infocom*, Vol. 2, pp. 594–602, 1996.

13. K. Calvert, M. Doar, and E.W. Zegura. Modeling Internet topology. *IEEE Communications Magazine*, 35(6):160–163, June 1997.

14. W.R. Heinzelman, A. Chandrakasan, and H. Balakrishnan. Energy-efficient communication protocol for wireless micro-sensor networks. In *Proc. IEEE Hawaii Int. Conf. on System Sciences*, pp. 4–7, January 2000.

15. V. Paxson and S. Floyd. Wide-area traffic: the failure of Poisson modeling. *IEEE/ACM Transactions on Networking*, 3(3):226–244, 1995.

16. V. Paxson and S. Floyd. Why we don't know how to simulate the Internet. *Proc. 1997 Winter Simulation Conference*, pp. 1037–1044, 1997.

17. C.E. Perkins and E.M. Royer. Ad-hoc on-demand distance vector routing. In *Proc. 2nd IEEE Workshop on Mobile Computing Systems and Applications*, pp. 90–100, 1999.
18. Y.B. Ko and N.H. Vaidya. Location-aided routing (LAR) in mobile ad hoc networks. In *Proc. Fourth Annu. ACM/IEEE Int. Conf. on Mobile Computing and Networking (MobiCom)*, pp. 66–75, 1998.
19. B. Karp and H.T. Kung. Greedy perimeter stateless routing (GPSR) for wireless networks. In *Proc. Sixth Annu. ACM/IEEE Int. Conf. on Mobile Computing and Networking (MobiCom)*, pp. 243–254, 2000.
20. S. Basagni, I. Chlamtac, V.R. Syrotiuk, and B.A. Woodward. A distance routing effect algorithm for mobility (DREAM). In *Proc. Fourth Annu. ACM/IEEE Int. Conf. on Mobile Computing and Networking (MobiCom)*, pp. 76–84, 1998.
21. L. Li, J.Y. Halpern, P. Bahl, Y.M. Wang, and R. Wattenhofer. Analysis of distributed topology control algorithms for wireless multi-hop networks. In *Proc. ACM Symp. on Principle of Distributed Computing (PODC)*, pp. 264–273, 2001.
22. Y. Wang and X.Y. Li. Localized construction of bounded degree and planar spanner for wireless ad hoc networks. In *Proc. ACM DIALM-POMC Joint Workshop on Foundations of Mobile Computing*, pp. 59–68, 2003.
23. L. Jia, R. Rajaraman, and C. Scheideler. On local algorithms for topology control and routing in ad hoc networks. In *Proc. SPAA*, pp. 220–229, 2003.
24. R. Wattenhofer and A. Zollinger. XTC: A Practical Topology Control Algorithm for Ad-Hoc Networks. Technical Report 407, Computer Science Department, ETH Zurich, 2003.

9

Models and Algorithms for the MPSCP: An Overview

Roberto Montemanni

Luca M. Gambardella

Arindam Kumar Das

9.1 Introduction

Ad hoc wireless networks have received significant attention in recent years due to their potential applications in battlefields, emergency disasters relief, and other application scenarios (see, for example, Blough et al.,[2] Chu and Nikolaidis,[4] Clementi et al.,[5,7] Kirousis et al.,[10] Lloyd et al.,[11] Ramanathan and Rosales-Hain,[17] Singh et al.,[19] Wan et al.,[20] and Wieselthier et al.[22]). Unlike wired networks of cellular networks, no wired backbone infrastructure is installed in ad hoc wireless networks. A communication session is achieved either through single-hop transmission if the recipient is within the transmission range of the source node, or by relaying through intermediate nodes.

We consider wireless networks where individual nodes are equipped with omnidirectional antennae. Typically, these nodes are also equipped with limited-capacity batteries and have a restricted communication radius. Topology control is one of the most fundamental and critical issues in multi-hop wireless networks that directly affects network performance. In wireless networks, topology control essentially involves choosing the right set of transmitter power to maintain adequate network connectivity. Incorrectly designed topologies can lead to higher end-to-end delays and reduced throughput in error-prone channels. In energy-constrained networks where replacement or periodic maintenance of node batteries is not feasible, the issue is all the more critical because it directly impacts the network lifetime.

In a seminal paper on topology control using transmission power control in wireless networks, Ramanathan and Rosales-Hain[17] approached the problem from an optimization viewpoint and showed that a network topology that minimizes the maximum transmitter power allocated to any node can be constructed in polynomial time. This is a critical criterion in battlefield applications because using higher transmitter power increases the probability of detection by enemy radar. This chapter attempts to solve the minimum power topology problem in wireless networks. Minimizing the total transmitter power has the effect of limiting the total interference in the network. It has been shown by Clementi et al.[6] that this problem is NP-complete. Related work in the area of minimum power topology construction includes Wattenhofer et al.,[21] Huang et al.,[9] and Borbash and Jennings,[3] all of which propose distributed algorithms. Specifically, Wattenhofer et al.[21] propose a cone-based distributed algorithm that relies only on angle-of-arrival estimates to establish a power-efficient connected topology. Huang et al.[9] describe a distributed protocol that is designed for sectored antenna systems. The work in Borbash and Jennings[3] explores the use of relative neighborhood graphs (RNGs) for topology control and suggests an algorithm for distributed computation of the RNG.

For a given set of nodes, the *min-power symmetric connectivity problem* (*MPSCP*), sometimes also referred to as the *minimum power topology problem*, is to assign transmission powers to the nodes of the network, which are equipped with omnidirectional antennae, in such a way that all the nodes are connected by bidirectional links and the total power consumption over the network is minimized. Having bidirectional links simplifies one-hop transmission protocols by allowing acknowledgment messages to be sent back for every packet (see Althaus et al.[1]). It is assumed that no power expenditure is involved in reception/processing activities, that a complete knowledge of pairwise distances between nodes is available, and that there is no mobility.

Unlike in wired networks, where a transmission from i to m generally reaches only node m, in wireless networks with omnidirectional antennae it is possible to reach several nodes with a single transmission (this is the so-called *wireless multicast advantage*; see Wieselthier et al.[22]). In the example of Figure 9.1, nodes j and k receive the signal originatating from node i and directed to node m because j and k are closer to i than m; that is, they are within the transmission range of a communication from i to m. This property is used to minimize the total transmission power required to connect all the nodes of the network.

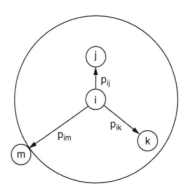

FIGURE 9.1 Communication model.

In Section 9.2 the *MPSCP* is formally described. Section 9.3 is devoted to an overview of the mathematical models and exact algorithms presented thus far in the literature. In Section 9.4 a preprocessing rule, useful to reduce problem dimensions, is described. In Section 9.5 a comparison of the performance of the available exact algorithms is proposed, while Section 9.6 is devoted to conclusions.

9.2 Problem Description

To represent the problem in mathematical terms, a model for signal propagation must be selected. We adopt the model presented in Rappaport[18] and used in most of the articles appearing in the literature (see, for example, Wieselthier et al.,[22] Montemanni et al.,[14] and Althaus et al.[1]). According to this model, signal power falls as $\frac{1}{d^\kappa}$, where d is the distance from the transmitter to the receiver and κ is a environment-dependent coefficient, typically between 2 and 4. Under this model, and adopting the usual convention (see, for example, Althaus et al.[1]) that every node has the same transmission efficiency and the same detection sensitivity threshold, the power requirement for supporting a link from node i to node j, separated by a distance d_{ij}, is then given by

$$p_{ij} = (d_{ij})^\kappa \tag{9.1}$$

Using the model described above, power requirements are symmetric; that is, $p_{ij} = p_{ji}$.

Constraints on maximum transmission powers of nodes can be treated by artificially modifying the power requirements. If, for example, node i cannot reach node j even when it is transmitting at its maximum power (i.e., $d_{ij}^\kappa >$ maximum power of node i), then p_{ij} can be redefined as $+\infty$.

MPSCP can be formally described as follows. Given the set V of the nodes of the network, a *range assignment* is a function $r : V \rightarrow \mathcal{R}^+$. A *bidirectional link* between nodes i and j is said to be established under the range assignment r if $r(i) \geq p_{ij}$ and $r(j) \geq p_{ij}$. Now let $B(r)$ denote the set of all bidirectional links established under the range assignment r. *MPSCP* is the problem of finding a range assignment r minimizing $\Sigma_{i \in V} r(i)$, subject to the constraint that the graph $(V, B(r))$ is connected.

As suggested by Althaus et al.,[1] a graph theoretical description of *MPSCP* can be given as follows. Let $G = (V, E, p)$ be an edge-weighted complete graph, where V is the set of vertices corresponding to the set of nodes of the network and E is the set of edges containing all the possible pairs $\{i, j\}$, with $i, j \in V$, $i \neq j$. A cost p_{ij} is associated with each edge $\{i, j\}$. It corresponds to the power requirement defined by Equation 9.1.

For a node i and a spanning tree T of G, let $\{i, i_T\}$ be the maximum cost edge incident to i in T; that is, $\{i, i_T\} \in T$ and $p_{ii_T} \geq p_{ij} \ \forall \{i, j\} \in T$. The *power cost* of a spanning tree T is then $c(T) = \Sigma_{i \in V} p_{ii_T}$. Because a spanning tree is contained in any connected graph, *MPSCP* can be described as the problem of finding the spanning tree T with minimum power cost $c(T)$.

9.3 Mathematical Models and Exact Algorithms

In this section we present four mathematical models for the *MPSCP* that have recently appeared in the literature, all based on mixed-integer programming. For each mathematical formulation discussed, some reinforcing inequalities and an exact algorithm, strongly based on the formulation, are also presented.

9.3.1 Althaus et al.[1]

Althaus et al.[1] have presented a mathematical formulation, with some reinforcing inequalities, and an exact algorithm. They are summarized in this section.

9.3.1.1 Mathematical Formulation

A weighted, directed, complete graph $G' = (V, A, p)$ is derived from G by defining $A = \{(i, j) | i, j \in V\}$; that is, for each edge in E, there are the respective two arcs in A and a dummy arc (i, i) with $p_{ii} = 0$ is inserted for each $i \in V$. p_{ij} is defined by Equation 9.1 when $i \neq j$.

In formulation AL, variables x define the spanning tree T on which the connectivity structure is based. $x_{ij} = 1$ if edge $\{i, j\}$ belongs to the spanning tree T, 0 otherwise. w variables represent the transmission range of nodes. $w_{ij} = 1$ if $i_T = j$ (see Section 9.2), $w_{ij} = 0$ otherwise.

$$(AL) \operatorname{Min} \sum_{(i,j)\in A} p_{ij} w_{ij} \tag{9.2}$$

$$\text{s.t.} \sum_{j\in V,(i,j)\in A} w_{ij} = 1 \qquad \forall i \in V \tag{9.3}$$

$$x_{ij} \le \sum_{\substack{(i,k)\in A, \\ p_{ik}\ge p_{ij}}} w_{ik} \qquad \forall \{i, j\} \in E \tag{9.4}$$

$$x_{ij} \le \sum_{\substack{(j,k)\in A, \\ p_{jk}\ge p_{ij}}} w_{jk} \qquad \forall \{i, j\} \in E \tag{9.5}$$

$$\sum_{\{i,j\}\in E} x_{ij} = |V| - 1 \tag{9.6}$$

$$\sum_{\substack{i,j\in S, \\ \{i,j\}\in E}} x_{ij} \le |S| - 1 \qquad \forall S \subset V \tag{9.7}$$

$$x_{ij} \in \{0, 1\} \qquad \forall \{i, j\} \in E \tag{9.8}$$

$$w_{ij} \in \{0, 1\} \qquad \forall (i, j) \in A \tag{9.9}$$

Constraints 9.3 enforce that exactly one range variable for every node $i \in V$ is selected; that is, the range of each node is properly defined. Constraints 9.4 and 9.5 enforce that an edge $\{i, j\}$ is included in the tree only if the range of each endpoint is at least the cost of the edge. Constraints 9.6 and 9.7 enforce that the tree variables indeed form a spanning tree. Constraints 9.8 and 9.9 define the domains of variables.

The bottleneck of formulation AL is represented by constraints 9.7, which are in exponential number and make difficult to handle the formulation in the case of real problems. An idea to overcome this problem is implemented within the exact algorithm described below.

9.3.1.2 Valid Inequalities

Althaus et al.[1] also discuss a set of valid inequalities for formulation AL, which is used within the exact algorithm presented by the authors. These inequalities are summarized in this section.

Definition 9.1 Given $W \subset V$, $\forall i \in W$ we define $i^W \in V \setminus W$ such that $p_{ii^W} \le p_{ij} \, \forall j \in V \setminus W$.

Theorem 9.1 (Crossing inequalities) *The set of inequalities*

$$\sum_{i\in W} \sum_{\substack{(i,j)\in A, \\ p_{ij}\ge p_{ii^W}}} w_{ij} \ge 1 \quad \forall W \subset V \tag{9.10}$$

is valid for formulation AL.

Proof Because T must be a spanning tree, at least one of its edges must cross the cut W. Let $\{i, j\}$ be such an edge, with $i \in W$. Then $p_{ij} \ge p_{ii^W}$ and the range of i is at least p_{ij}. Inequality 9.10 must then be satisfied. $\qquad\square$

9.3.1.3 Exact Algorithm

Althaus et al.[1] proposed a branch and cut algorithm based on formulation AL. Formulation AL_{LR}^R is considered by the algorithm. It is obtained from AL by substituting constraints 9.8 and 9.9 with their linear relaxation, formally

$$0 \le x_{ij} \le 1 \quad \forall \{i, j\} \in E \tag{9.11}$$

$$0 \le w_{ij} \le 1 \quad \forall (i, j) \in A \tag{9.12}$$

and by adding constraints 9.10.

Formally, the branch and cut algorithm works by solving formulation AL_{LR}^R. If the solution is integral, the optimal solution has been found, otherwise a variable with a fractional value is picked up and the problem is split into two subproblems by setting the variable to 0 and 1 in the subproblems. The subproblems are solved recursively and disregard a subproblem if the lower bound provided by AL_{LR}^R is worse than the best known solution.

Because there are an exponential number of inequalities of type 9.7, AL_{LR}^R cannot be solved directly at each node of the branching tree. Instead, the algorithm starts with a small subset of these inequalities and algorithmically tests whether the solution violates an inequality that is not in the current problem. If so, the inequality is added to it, otherwise the solution of AL_{LR}^R has been retrieved. The separation algorithm described by Padberg and Wolsey[15] has been used for these constraints.

A similar approach applies also for inequalities 9.10, which are again in exponential number. Because there was no known separation algorithm for them, the following heuristic was used. Capacity q_{kl} is defined for each edge $\{k, l\}$:

$$q_{kl} = \sum_{\substack{(k,r) \in A, \\ p_{kr} \ge p_{kl}}} w_{kr} \tag{9.13}$$

An arbitrary node i is chosen and for every node $j \in V \setminus \{i\}$, the minimal directed cut from i to j and from j to i (with capacities defined by Equation 9.13) is computed and it is tested whether or not the corresponding inequality is violated.

9.3.2 Das et al.[8]

Das et al.[8] have presented a mathematical formulation, with some reinforcing inequalities, and an exact algorithm. They are summarized in this section.

9.3.2.1 Mathematical Formulation

The mixed integer programming formulation DAS described in this section is based on a network flow model (see Magnanti and Wolsey[12]). A node s is elected as the source of the flow, and one unit of flow is sent from s to every other node. The meaning of variables in the formulation is as follows. Variable w_i contains the transmission power of node i. Variable t_{ij} represents the flow on arc (i, j), while u_{ij} is an indicator variable and assumes value 1 if the flow on arc (i, j) is greater than 0 (i.e., $t_{ij} > 0$), 0 otherwise.

$$(DAS) \text{ Min} \sum_{i \in N} w_i \tag{9.14}$$

$$\text{s.t.} \sum_{(i,j) \in A} t_{ij} - \sum_{(k,i) \in A} t_{ki} = \begin{cases} |V| - 1 & \text{if } i = s \\ -1 & \text{otherwise} \end{cases} \quad \forall i \in V \tag{9.15}$$

$$(|V| - 1)u_{ij} \ge t_{ij} \quad \forall (i, j) \in A \tag{9.16}$$

$$w_i \ge p_{ij} u_{ij} \quad \forall (i, j) \in A \tag{9.17}$$

$$u_{ij} = u_{ji} \quad \forall \{i, j\} \in E \tag{9.18}$$

$$u_{ij} \in \{0, 1\} \quad \forall (i, j) \in A \tag{9.19}$$

$$t_{ij} \ge 0 \quad \forall (i, j) \in A \tag{9.20}$$

$$w_i \ge 0 \quad \forall i \in V \tag{9.21}$$

Equations 9.15 define the flow problem on t variables. Constraints 9.16 are the activators for u variables; that is, they connect u and t variables. Inequalities 9.17 connect u variables to w variables, while Equations 9.18 force u variables corresponding to the arcs of a same edge to assume the same value (u variables must be symmetric because they regulate transmission powers in equalities 9.17). Constraints 9.19, 9.20, and 9.21 define the variables' domains.

The main drawback of formulation *DAS* is represented by constraints 9.16 and 9.17, which tend to push indicator variables u to assume fractional values, making the mixed-integer program very difficult to solve.

9.3.2.2   Valid Inequalities

Das et al.[8] also proposed some valid inequalities. They all rely on the symmetric nature of indicator variables u.

Theorem 9.2 (Connectivity inequalities 1)   *The set of inequalities*

$$\sum_{\substack{(i,j)\in A, \\ i\neq j}} u_{ij} \geq 1 \ \forall i \in V \tag{9.22}$$

is valid for formulation DAS.

Proof   To be connected to the rest of the network, each node i must be able to communicate with at least one other node; that is inequality 9.22 must be satisfied.   □

Theorem 9.3 (Connectivity inequalities 2)   *The set of inequalities*

$$\sum_{\substack{(i,j)\in A, \\ i\neq j}} u_{ji} \geq 1 \ \forall i \in V \tag{9.23}$$

is valid for formulation DAS.

Proof   To be connected to the rest of the network, each node i must receive the signal of at least one other node; that is, inequality 9.23 must be satisfied.   □

Theorem 9.4 (Connectivity inequality 3)   *The inequality*

$$\sum_{\substack{(i,j)\in A, \\ i\neq j}} u_{ij} \geq 2(|V| - 1) \tag{9.24}$$

is valid for formulation DAS.

Proof   To have a topology connected by bidirectional links, there must be at least $2(|V| - 1)$ active indicator variables (i.e., the number of edges of a spanning tree times 2), as stated by constraint 9.24.   □

9.3.2.3   Exact Algorithm

Formulation DAS^R is defined to be formulation *DAS* reinforced with the inequalities 9.22, 9.23, and 9.24. The exact algorithm described in Das et al.[8] works by directly solving formulation DAS^R. As observed by Das et al., experimental results suggest that solving DAS^R instead of *DAS* produces shorter computation times.

9.3.3   Montemanni and Gambardella[13] (a)

Montemanni and Gambardella[13] presented a mathematical formulation, with some reinforcing inequalities, and an exact algorithm. They are summarized in this section.

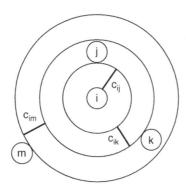

FIGURE 9.2 Incremental mechanism for costs.

9.3.3.1 Mathematical Formulation

To describe this mathematical formulation, the following definition is required.

Definition 9.2 Given $(i, j) \in A$, we define the *ancestor* of (i, j) as

$$a_j^i = \begin{cases} i & \text{if } p_{ij} = \min_{k \in V}\{p_{ik}\} \\ \arg\max_{k \in V}\{p_{ik} | p_{ik} < p_{ij}\} & \text{otherwise} \end{cases} \tag{9.25}$$

According to this definition, (i, a_j^i) is the arc that originated in node i with the highest cost such that $p_{ia_j^i} < p_{ij}$.* In case an *ancestor* does not exist for arc (i, j), vertex i is returned, that is, the dummy arc (i, i) is addressed.

In the example of Figure 9.1, arc (i, k) is the ancestor of arc (i, m); (i, j) is the ancestor of (i, k); and the dummy arc (i, i) is returned as the ancestor of (i, j).

The formulation is based on an incremental mechanism over the variables representing transmission powers. The costs associated with these variables in the objective function 9.27 will be given by the following formula:

$$c_{ij} = p_{ij} - p_{ia_j^i} \quad \forall(i, j) \in A \tag{9.26}$$

where c_{ij} is equal to the power required to establish a transmission from nodes i to node j (p_{ij}) minus the power required by nodes i to reach node a_j^i ($p_{ia_j^i}$). In Figure 9.2, the costs arising from the example of Figure 9.1 are depicted.

It is important to observe that the incremental mechanism is the most important element of the formulation. It will allow us to define very strong reinforcing inequalities, which are the basis for the good performance of the exact algorithm based on the formulation (see Section 9.5).

The mixed-integer programming formulation MGa described in this section is based on a network flow model (see Magnanti and Wolsey[12]). A node s is elected as the source of the flow, and one unit of flow is sent from s to every other node. Variable x_{ij} represents the flow on arc (i, j). Variable y_{ij} is 1 when

---

*For the sake of simplicity, we have considered the (usual) case where $\forall i \in V \, \nexists k, l \in V$ s.t. $p_{ik} = p_{il}$. In case this is not true, the following formula, which breaks ties, has to be used in place of 9.25:

$$a_j^i = \begin{cases} i & \text{if } p_{ij} = \min_{k \in V}\{p_{ik}\} \\ \arg\max_{k \in V}\left\{p_{ik} \middle| \begin{pmatrix} (p_{ik} < p_{ij} \wedge (\nexists l \in V \text{ s.t. } p_{ik} = p_{il} \wedge l > k)) \\ \vee (p_{ij} = p_{ij} \wedge (\nexists l \in V \text{ s.t. } p_{ik} = p_{il} \wedge j > l > k)) \end{pmatrix}\right\} & \text{otherwise} \end{cases}$$

node i has a transmission power that allows it to reach node j, $y_{ij} = 0$ otherwise.

$$(MGa) \quad \text{Min} \sum_{(i,j)\in A} c_{ij} y_{ij} \tag{9.27}$$

$$\text{s.t. } y_{ij} \leq y_{ia_j^i} \qquad \forall (i,j) \in A, a_j^i \neq i \tag{9.28}$$

$$x_{ij} \leq (|V| - 1) y_{ij} \qquad \forall (i,j) \in A \tag{9.29}$$

$$x_{ij} \leq (|V| - 1) y_{ji} \qquad \forall (i,j) \in A \tag{9.30}$$

$$\sum_{(i,j)\in A} x_{ij} - \sum_{(k,i)\in A} x_{ki} = \begin{cases} |V| - 1 & \text{if } i = s \\ -1 & \text{otherwise} \end{cases} \qquad \forall i \in V \tag{9.31}$$

$$x_{ij} \in \mathcal{R} \qquad \forall (i,j) \in A \tag{9.32}$$

$$y_{ij} \in \{0, 1\} \qquad \forall (i,j) \in A \tag{9.33}$$

Constraints 9.28 realize the incremental mechanism by forcing the variables associated with arc (i, a_j^i) to assume value 1 when the variable associated with arc (i, j) has value 1; that is, the arcs that originated in the same node are activated in increasing order of p. Inequalities 9.29 and 9.30 connect the flow variables x to y variables. Equations 9.31 define the flow problem, while 9.32 and 9.33 are domain definition constraints. We refer the interested reader to Magnanti and Wolsey[12] for a more detailed description of the spanning tree formulation behind the formulation presented above.

In formulation MGa, the bottleneck is represented by constraints 9.29 and 9.30, which tend to push y variables to be fractional. Fortunately, the incremental mechanism on which the mixed-integer program is based allows us to define strong reinforcing inequalities that help overcome this problem (see below).

9.3.3.2 Valid Inequalities

The valid inequalities presented in this section were proposed by Montemanni and Gambardella[13] to reinforce mathematical formulation MGa.

In the remainder of this section we refer to the subgraph of G' defined by the y variables with value 1 as G_y. Formally, $G_y = (V, A_y)$, where $A_y = \{(i, j) \in A | y_{ij} = 1 \text{ in the solution of } MPSC\}$.

Theorem 9.5 (Connectivity inequalities)   *The set of inequalities*

$$y_{ij} = 1 \; \forall (i,j) \in A \text{ s.t. } a_j^i = i \tag{9.34}$$

is valid for formulation MGa.

Proof   To have the graph G_y connected, each node must be able to communicate with at least one other node. Then its transmission power must be sufficient to reach at least the node which is closest to it; that is, $y_{ia_j^i} = 1$. □

Theorem 9.6 (Bidirectional inequalities 1)   *The set of inequalities*

$$y_{a_j^i i} \geq y_{ia_j^i} - y_{ij} \; \forall (i,j) \in A \text{ s.t. } a_j^i \neq i \tag{9.35}$$

is valid for formulation MGa.

Proof   If $y_{ij} = 1$, then $y_{ia_j^i} = 1$ because of inequalities 9.28 and consequently in this case the constraint does not give any new contribution.

If $y_{ij} = 0$ and $y_{ia_j^i} = 0$, then again the constraint does not give any new contribution.

If $y_{ij} = 0$ and $y_{ia_j^i} = 1$, then the transmission power of node i is set to reach node a_j^i and nothing more. The only reason for node i to reach node a_j^i and nothing more is the existence of a bidirectional link on edge $\{i, a_j^i\}$ in G_y. Consequently, $y_{a_j^i i}$ must be equal to 1, as stated by the constraint. □

Theorem 9.7 (Bidirectional inequalities 2)   *The set of inequalities*

$$y_{ji} \geq y_{ij} \; \forall (i,j) \in A \text{ s.t. } \not\exists (i,k) \in A, a_k^i = j \tag{9.36}$$

is valid for formulation MGa.

Proof   If $y_{ij} = 0$, the constraint does not give any new contribution.

If $y_{ij} = 1$, then the transmission power of node i is set in such a way to reach node j, which is the farthest node from i in G. The only reason for node i to reach node j is the existence of a bidirectional link on edge $\{i, j\}$ in G_y. Consequently, y_{ji} must be equal to 1, as stated by the constraint.   □

Theorem 9.8 (Tree inequality)   *The inequality*

$$\sum_{(i,j)\in A} y_{ij} \geq 2(|V| - 1) \tag{9.37}$$

is valid for formulation MGa.

Proof   To be strongly connected, the directed graph G_y must have at least $2(|V| - 1)$ arcs, as stated by constraint 9.37.   □

Definition 9.3 $G_a = (V, A_a)$ is the subgraph of the complete graph G' such that $A_a = \{(i, j)|\, a_j^i = i\}$.

Notice that $|A_a| = |V|$ by definition.

Definition 9.4 $\mathcal{R}_i = \{j \in V|\, j$ can be reached from i in $G_a\}$.

Theorem 9.9 (Reachability inequalities 1)   *The set of inequalities*

$$\sum_{\substack{(k,l)\in A \\ s.t. k \in \mathcal{R}_i, l \in V\setminus\mathcal{R}_i}} y_{kl} \geq 1 \; \forall i \in V \tag{9.38}$$

is valid for formulation MGa.

Proof   Because graph G_y must be strongly connected, it must be possible to reach every node j starting from each node i. This implies that at least one arc must exist between the nodes which is possible to reach from i in G_a (i.e., $\mathcal{R}_i$) and the other nodes of the graph (i.e., $V\setminus\mathcal{R}_i$).   □

Definition 9.5 $\mathcal{Q}_i = \{j \in V|\, i$ can be reached from j in $G_a\}$.

Theorem 9.10 (Reachability inequalities 2)   *The set of inequalities*

$$\sum_{\substack{(k,l)\in A \\ s.t. \; k \in \mathcal{Q}_i, l \in V\setminus\mathcal{Q}_i}} y_{kl} \geq 1 \; \forall i \in V \tag{9.39}$$

is valid for formulation MGa.

Proof   Because graph G_y must be strongly connected, it must be possible to reach every node i from every other node j of the graph. This means that at least one arc must exist between the nodes that cannot reach i in G_a (i.e., $V\setminus\mathcal{Q}_i$) and the other nodes of the graph (i.e., $\mathcal{Q}_i$).   □

9.3.3.3 Exact Algorithm

Formulation MGa^R is defined as the formulation MGa reinforced with the inequalities 9.34 through 9.40. The exact algorithm described by Montemanni and Gambardella[13] works by directly solving formulation MGa^R. The authors showed that solving MGa^R instead of MGa produces computation times that are shorter up to a factor of 1920 for some problems.

9.3.4 Montemanni and Gambardella[13] (b)

Montemanni and Gambardella[13] also presented a second mathematical formulation and a second exact algorithm. They are summarized in this section, together with a new valid inequality.

9.3.4.1 Mathematical Formulation

The mathematical model described in this section is based on the same incremental mechanism discussed in Section 9.3.3. In formulation MGb, a spanning tree is defined by z variables. Variable z_{ij} is 1 if edge $\{i, j\}$ is on the spanning tree, $z_{ij} = 0$ otherwise. Variable y_{ij} is 1 when node i has a transmission power, which allows it to reach node j, $y_{ij} = 0$ otherwise.

$$(MGb) \text{ Min} \quad \sum_{(i,j)\in A} c_{ij} y_{ij} \tag{9.40}$$

$$\text{s.t.} \quad y_{ij} \leq y_{ia_j^i} \qquad \forall (i, j) \in A, a_j^i \neq i \tag{9.41}$$

$$z_{ij} \leq y_{ij} \qquad \forall \{i, j\} \in E \tag{9.42}$$

$$z_{ij} \leq y_{ji} \qquad \forall \{i, j\} \in E \tag{9.43}$$

$$\sum_{\substack{(i,j)\in E, \\ i\in S, j\in V\setminus S}} z_{ij} \geq 1 \quad \forall S \subset V \tag{9.44}$$

$$z_{ij} \in \{0, 1\} \qquad \forall \{i, j\} \in E \tag{9.45}$$

$$y_{ij} \in \{0, 1\} \qquad \forall (i, j) \in A \tag{9.46}$$

Constraints 9.41 realize the incremental mechanism by forcing the variables associated with arc (i, a_j^i) to assume value 1 when the variable associated with arc (i, j) has value 1; that is, the arcs originated in the same node are activated in increasing order of p. Inequalities 9.42 and 9.43 connect the spanning tree variables z to y variables. Equations 9.44 state that all the vertices must be mutually connected in the subgraph induced by z variables, while 9.45 and 9.46 are domain definition constraints.

The main problem of formulation MGb is related to the exponential number of constraints (9.44). In the exact algorithm described below, a technique to overcome this bottleneck is implemented.

9.3.4.2 Valid Inequalities

Because all the inequalities presented in Section 9.3.3 use y variables only, and because the role of y variables is the same in both formulations MGa and MGb (i.e., y variables implement the incremental mechanism described in Section 9.3.3.1), the results presented in Section 9.3.3.2 for formulation MGa are valid also for formulation MGb.

In addition, the following inequality is also considered. It will be used within the exact algorithm described in the next section.

Theorem 9.11 (z tree inequalities)
Equation 9.47 is valid for formulation MGb.

$$\sum_{\{i,j\}\in E} z_{ij} \geq |V| - 1 \tag{9.47}$$

Proof   Inequality 9.47 forces the number of active z variables to be at least $|V| - 1$. This condition is necessary in order to have a spanning tree. $\square$

9.3.4.3 Exact Algorithm

The integer program MGb^R is defined as MGb without constraints 9.44 but with the inequalities 9.34, through 9.39 and 9.47. Notice that constraint 9.47 forces the active z variables to be at least $|V| - 1$ already during the very first iterations of the method we are describing in this section. This will contribute to speed up the algorithm.

The idea at the basis of the method is that it is very difficult to deal directly with constraints 9.44 of formulation MGb_R in case of large problems. For this reason, some techniques that leave some of them out must be considered. In this section we present an iterative approach that in the beginning does not consider any constraint 9.44, and adds them step by step in case they are violated. Formulation MGb is solved and the values of the z variables in the solution are examined. If the edges corresponding to variables with value 1 form a spanning tree, then the problem has been solved to optimality; otherwise, constraints 9.48, described below, are added to the integer program and the process is repeated.

At the end of each iteration, if edges corresponding to z variables with value 1 in the last solution generate a set $\mathcal{CC}$ of connected components, with $|\mathcal{CC}| > 1$, then the following inequalities are added to the formulation:

$$\sum_{\substack{\{i,j\}\in E, \\ i\in C, j\in V\setminus C}} z_{ij} \geq 1 \quad \forall C \in \mathcal{CC} \tag{9.48}$$

Inequalities 9.48 force z variables to connect the (otherwise disjoint) connected components of $\mathcal{CC}$.

9.4 Preprocessing Procedure

The results described in this section are used to delete some arcs of graph G' and consequently to speed up the exact algorithms previously presented. They were originally presented by Montemanni and Gambardella.[13]

We suppose a heuristic solution for the problem, heu, is available, and its cost is $cost(heu)$. All the variables that, if active, would induce a cost higher than $cost(heu)$ can be deleted from the problem.

Theorem 9.12 *If the following inequality holds:*

$$p_{ij} + p_{ji} + \sum_{\substack{k\in V\setminus\{i,j\}, \\ a_j^k=k}} p_{kl} > cost(heu) \tag{9.49}$$

then arc (i, j) can be deleted from A.

Proof Using the same intuition at the basis of the proofs of Theorems 9.6 and 9.7, we have that if p_{ij} is the power of node i in a solution, this means that the power of node j must be greater than or equal to p_{ji} (i.e., arc (j, i) must be in the solution), because otherwise there would be no reason for node i to reach node j. The left-hand side of inequality 9.49 represents then a lower bound for the total power required to maintain the network connected in case node i transmits to a power that allows it to reach node j and nothing farther. For this reason, if inequality 9.49 holds, arc (i, j) can be deleted from A. □

It is important to notice that once arc (i, j) is deleted from A, the value of the ancestor of node k (see Section 9.3.3.1), with $a_k^i = j$, must be updated to a_j^i.

9.5 Computational Results

Computational tests have been carried out on two different families of problems, randomly generated as described by Althaus et al.[1] and Das et al.,[8] respectively. In Althaus et al.,[1] $\kappa = 4$ and a problem with

TABLE 9.1   Average Computation Times (sec) on the Problems Described by Althaus et al.[1]

Algorithms	|V|						
	10	15	20	25	30	35	40
AL	2.144	18.176	71.040	188.480	643.200	2278.400	15120.000
MGa	0.192	0.736	8.576	33.152	221.408	1246.304	9886.080
Preprocessing + MGa	0.078	0.289	0.715	4.924	28.908	87.357	583.541
Preprocessing + MGb	0.052	0.196	0.601	2.181	13.481	28.172	79.544

$|V|$ nodes is obtained by choosing $|V|$ points uniformly at random from a grid of size 10000×10000. For the problems described in Das et al.,[8] the procedure is the same but the grid has dimension 5×5. In addition, for these last problems, a maximum transmission power, depending on the number of nodes of the network, is fixed. The following pairs (*number of nodes, maximum transmission power*) have been adopted: $(15, 3.00)$, $(20, 3.00)$, $(30, 2.50)$, $(40, 1.50)$, $(50, 0.75)$. ILOG CPLEX* 6.0 has been used to solve integer programs.

In the remainder of this section we refer to the algorithm presented by Althaus et al.[1] (see Section 9.3.1) as *AL*, to the one described in Das et al.[8] (see Section 9.3.2) as *DAS*, and to those proposed by Montemanni and Gambardella[13] (see Sections 9.3.3 and 9.3.4) as *MGa* and *MGb*, respectively.

In Table 9.1 we present the average computation times required (on a SUNW Ultra-30 machine) by some of the exact algorithms on the problems described by Althaus et al.[1] for different values of V. Fifty instances are considered for each value of $|V|$.

Table 9.1 shows that the *MGa* and *MGb* outperform *AL*. *MGb* also performs clearly better than *MGa*. In Table 9.1, the benefit derived from the use of the preprocessing technique described in Section 9.4 is highlighted. To apply this preprocessing procedure, a heuristic solution to the problem must be available. For this purpose we use one of the simplest algorithms available, which works by calculating the *Minimum Spanning Tree* (see Prim[16]) on the weighted graph with costs defined by Equation 9.1, and by assigning the power of each transmitter i to p_{ii_T}, as described near the end of Section 9.2. The computational times of the algorithm *MGa* are improved up to 17 times (for $|V| = 40$) when this technique is used (on average, 79 percent of the arcs were deleted for $|V| = 40$; see Montemanni and Gambardella[13]).

In Table 9.2 we present the average computational times required (on a Pentium 4 1.5-GHz machine) by some of the exact algorithms on the problems described by Das et al.[8] for different values of V. In brackets we also report the average standard deviation on solving times. Twenty-five instances are considered for each value of $|V|$. Some entries are marked with '—'; this means that the corresponding algorithms failed to solve some of the corresponding instances in less than 3600 seconds.

Table 9.2 suggests again that *MGa* and *MGb* obtain the best performance. For these problems, the algorithms highlight that all the algorithms are not extremely robust (see large standard deviation on solution times); that is, there are very different performances on instances of the same family. This could

TABLE 9.2   Average Computation Times (sec) on the Problems Described by Das et al.[8]

Algorithms	|V|				
	15	20	30	40	50
DAS	0.014 (0.018)	7.511 (36.697)	—	—	—
MGa	**0.008** (0.006)	**0.027** (0.013)	1.518 (4.401)	24.723 (111.378)	**12.233** (18.025)
MGb	0.019 (0.010)	0.058 (0.038)	**0.795** (1.093)	**9.906** (20.312)	47.756 (136.234)

---

*http://www.cplex.com.

depend on the small grid adopted, which tends to flatten power requirements, and this causes many almost equivalent solutions. On the other hand, average computational times are much shorter than those reported in Table 9.1, and this mainly depends on the maximum transmission power constraints, that substantially contribute to reduce the number of variables of the problems.

9.6 Conclusion

We have presented an overview of the mathematical formulations presented so far for the min-power symmetric connectivity problem in wireless networks. Some exact algorithms, strongly based on these formulations and on some reinforcing inequalities developed for them, have been discussed, together giving a preprocessing rule.

Computational results have been presented, aiming to compare the performance of the different exact approaches discussed in this chapter.

Acknowledgments

One of the authors (Montemanni) was partially supported by the Future & Emerging Technologies unit of the European Commission through Project "BISON: Biology-Inspired techniques for Self Organization in dynamic Networks"(IST-2001-38923) and by the Swiss National Science Foundation through project "Approximation Algorithms for Machine Scheduling Through Theory and Experiments" (200021–100539).

References

1. E. Althaus, G. Călinescu, I.I. Măndoiu, S. Prasad, N. Tchervenski, and A. Zelikovsky. Power efficient range assignment in ad-hoc wireless networks. In *Proceedings of the IEEE Wireless Communications and Networking Conference*, pages 1889–1894, 2003.
2. D. Blough, M. Leoncini, G. Resta, and P. Santi. On the symmetric range assignment problem in wireless ad hoc networks. In *Proceedings of the 2nd IFIP International Conference on Theoretical Computer Science (TCS 2002)*, pages 71–82, 2002.
3. S.A. Borbash and E.H. Jennings. Distributed topology control algorithm for multihop wireless networks. In *Proceedings of IJCNN*, 2002.
4. T. Chu and I. Nikolaidis. Energy efficient broadcast in modile ad hoc networks. In *Proceedings of AD-HOC NetwOrks and Wireless*, 2002.
5. A. Clementi, P. Crescenzi, P. Penna, G. Rossi, and P. Vocca. On the complexity of computing minimum energy consumption broadcast subgraphs. In *Symposium on Theoretical Aspects of Computer Science*, pages 121–131, 2001.
6. A. Clementi, P. Penna, and R. Silvestri. Hardness results for the power range assignment problem in packet radio networks. *Lectures Notes on Computer Science*, 1671:195–208, 1999.
7. A. Clementi, P. Penna, and R. Silvestri. On the power assignment problem in radio networks. Technical Report TR00-054, Electronic Colloquium on Computational Complexity, 2000.
8. A.K. Das, R.J. Marks, M. El-Sharkawi, P. Arabshani, and A. Gray. Optimization methods for minimum power bidirectional topology construction in wireless networks with sectored antennas. Submitted for publication. *Proceedings of the JEEE Wireless Comunications and Networking Conference*, 2005.
9. Z. Huang, C.-C. Shen, C. Srisathapornphat, and C. Jaikaeo. Topology control for ad hoc networks with directional antennas. In *Proceedings of the Eleventh International Conference on Computer Communications and Networks*, 2002.
10. L. Kirousis, E. Kranakis, D. Krizanc, and A. Pelc. Power consumption in packet radio networks. *Theoretical Computer Science*, 243:289–305, 2000.
11. E. Lloyd, R. Liu, M. Marathe, R. Ramanathan, and S. Ravi. Algorithmic aspects of topology control problems for ad hoc networks. In *Proceedings of the ACS MobiHoc*, pages 123–134, 2002.

12. T.L. Magnanti and L. Wolsey. Optimal trees. In *Network Models, Handbook in Operations Research and Management Science* (M.O. Ball et al., Eds.), Vol. 7, pages 503–615. North-Holland, 1995.
13. R. Montemanni and L.M. Gambardella. Exact algorithms for the minimum power symmetric connectivity problem in wireless networks. *Computers and Operations Research*, to appear.
14. R. Montemanni, L.M. Gambardella, and A.K. Das. The minimum power broadcast tree problem in wireless networks: a simulated annealing approach. Submitted for publication.
15. M. Padberg and L. Wolsey. Trees and cuts. *Annals of Discrete Mathematics*, 17:511–517, 1983.
16. R.C. Prim. Shortest connection networks and some generalizations. *Bell System Technical Journal*, 36:1389–1401, 1957.
17. R. Ramanathan and R. Rosales-Hain. Topology control of multihop wireless networks using transmit power adjustment. In *Proceedings of the IEEE Infocom*, pages 404–413, 2000.
18. T. Rappaport. *Wireless Communications: Principles and Practices*. Prentice Hall, 1996.
19. S. Singh, C. Raghavendra, and J. Stepanek. Power-aware broadcasting in mobile ad hoc networks. In *Proceedings of the IEEE International Symposium on Personal, Indoor and Mobile Radio Communications (PIMRC'99)*, 1999.
20. P.-J. Wan, G. Călinescu, X.-Y. Li, and O. Frieder. Minimum energy broadcast routing in static ad hoc wireless networks. In *Proceedings of the IEEE Infocom*, pages 1162–1171, 2001.
21. R. Wattenhofer, L. Li, P. Bahl, and Y.M. Wang. Distributed topology control for power efficient operation in multihop wireless ad hoc networks. In *Proceedings of the Infocom*, 2001.
22. J. Wieselthier, G. Nguyen, and A. Ephremides. On the construction of energy-efficient broadcast and multicast trees in wireless networks. In *Proceedings of the IEEE INFOCOM 2000 Conference*, pages 585–594, 2000.

10

A Survey of Algorithms for Power Assignment in Wireless Ad Hoc Networks

Gruia Calinescu

Ophir Frieder

Peng-Jun Wan

Abstract

Power is one of the most critical resources in wireless ad hoc networks. One way of conserving power is to assign transmission power levels to the wireless nodes. Recently, much progress has been made on algorithmic and probabilistic studies of various power assignment problems. These problems come in many flavors, depending on the power requirement function and the connectivity constraint, and minimizing the total power consumption is NP-hard for most versions. We present without proofs the best known approximation algorithms for minimizing the total power consumption in the network, and sketch useful heuristics with practical value.

10.1 Introduction

One of the major concerns in ad hoc wireless networks is reducing node power consumption. In fact, nodes are usually powered by batteries of limited capacity. Once the nodes are deployed, it is very difficult or even impossible to recharge or replace their batteries in many application scenarios. Hence, reducing power consumption is often the only way to extend network lifetime. For the purpose of energy conservation, each node can (possibly dynamically) adjust its transmitting power, based on the distance to the receiving node and the background noise.

In the most general model, a weighted directed graph $H = (V, E)$ with power requirements $c : E \rightarrow R^+$ is given by the positioning of the n wireless nodes, where $c(u, v)$ represents the power requirement for the node u to establish a unidirectional link to node v. But, reflecting the broadcast nature of ad hoc wireless networks, once a node u transmits with power $p(u)$, all nodes v with $c(u, v) \leq p(u)$ receive the signal. A function $p : V \rightarrow R^+$ is called a power assignment, and it induces a directed graph, always denoted by $G = (V, F)$, with links uv whenever $p(u) \geq c(u, v)$. The *symmetric restriction* of a directed graph $G = (V, F)$ is the undirected graph, always denoted by $\bar{G}$, with vertex set V and having an edge uv if and only if G has both uv and vu. An example is depicted in Figure 10.1.

Typically, we require that the graph induced by the power assignment respects certain connectivity constraints; assigning the power level to ensure that G or $\bar{G}$ have certain graph properties is called topology control. Many types of connectivity constraints have been studied in the literature and we will later address several. Here we mention just two:

1. Strong Connectivity: G must be strongly connected, thus guaranteeing that a packet can be relayed from any source to any destination with the current power assignment.
2. Symmetric Connectivity: $\bar{G}$ must be connected. This connectivity requirement is more stringent than strong connectivity, and is motivated by the advantage of having one-link acknowledgment packets.

Given a connectivity constraint, the objective of Power Assignment is minimizing the total power, given by $\Sigma_{u \in V} p(u)$. In addition to reducing energy consumption, having reduced transmission power creates less interference.

The following special cases of the power requirement have been considered in the literature, presented starting with the most general.

1. Arbitrary symmetric: an undirected graph $H = (V, E)$ with power requirements $c : E \rightarrow R^+$.
2. Euclidean: the nodes of the graph are embedded in the two-dimensional plane and $c(uv) = d(u, v)^\ell$, where d is the Euclidean distance.
3. Line: same power requirements as Euclidean, but the nodes lie on a line.

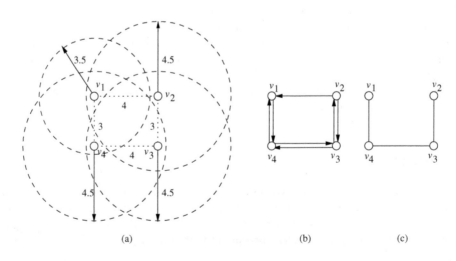

(a)   (b)   (c)

FIGURE 10.1 The network topology: (a) the nodes and their transmission power, (b) the directed graph G induced by the power assignment, and (c) the undirected graph $\bar{G}$ induced by the power assignment.

The Euclidean case is motivated by the fact that in the most common power-attenuation model, the signal power falls as $d^{-\ell}$, where d is the Euclidean distance from the transmitter antenna and ℓ is the pass-loss exponent of the wireless environment, a real constant typically between 2 and 5. The line case is a special case of the Euclidean case motivated by ad hoc networks with the nodes following a highway or other "linear" pattern.

The arbitrary symmetric cases handle the cases when the nodes are in three-dimensional space or obstacles completely block the communication between certain pairs of nodes, or if the signal attenuation is not uniform, or if there is a maximum transmission power, or if there is a discrete number of possible transmission power levels.

In addition, wireless nodes can have nonuniform power thresholds for signal detection, and we call *sensitivity* $s(v)$ the threshold of node v. Also, wireless nodes can have nonuniform transmission efficiency, and we use $e(v)$ to denote the efficiency of node v. Efficiency can also be used to simulate the fact that certain nodes have at some moment higher battery reserves.[2] With these definitions, node u reaches node v with power level $p(u)$ if and only if

$$p(u) \geq \frac{c(u,v)}{e(u)s(v)}.$$

Adding sensitivity or efficiency to any symmetric power requirement case creates asymmetric power requirements, while asymmetric power requirements can be adjusted to handle sensitivity and efficiency by simply redefining $c_{new}(u,v) = \frac{c(u,v)}{e(u)s(v)}$.

In this survey, we concentrate on centralized algorithms for static power assignment in static networks. We do point out when certain algorithms are suitable for distributed implementation or for mobility of the nodes, but space limitations do not allow a more general treatment. Also, we do not consider adjusting the transmission power for each packet transmitted, an approach suggested in the literature for further reducing power consumption.

Minimizing total power under most connectivity constraints generates NP-hard problems. Thus, efficient algorithms computing the exact optimum are unlikely to exist. Moreover, to our knowledge, Althaus et al.[1] is the only work that has computed optimum solutions, working with what we believe is the easiest of these NP-hard problems: Symmetric Connectivity in the Euclidean case. Customized state-of-the-art integer linear programming software solves randomly generated instances with up to 40 nodes in (at most) one hour. Our experience is that guaranteeing optimum solutions for 100 nodes or more is very unlikely.

This motivates the design of efficient heuristics and approximation algorithms. An approximation algorithm is a polynomial-time algorithm whose output is guaranteed to be (at most) α the optimum value, for some value α (which could depend on n, the number of vertices, but does not depend otherwise on the input). This value α is called the *approximation ratio*; and the smaller it is, the better the approximation algorithm. We call heuristic a proposed algorithm without a proven approximation ratio; such algorithms could have good practical performance.

At the end of the survey, we present a table with the best-known approximation ratios for the variety of existing problems. Due to the tight limitation on the number of references, only the best-known ratios will be quoted.

Before we proceed, we mention that in order to simplify the mathematical proofs, many articles use an alternative description of the optimization problem, described below: given (directed) spanning subgraph Q, define $p_Q(v) = \max_{uv \in Q} c(uv)$ and $p(Q) = \Sigma_{v \in V} P_Q(v)$; we call $p(Q)$ the *power* of Q. Since assigning $p(v) \geq p_Q(v)$ is necessary to produce the (directed) spanning subgraph Q, and $p(v) > p_Q(v)$ is just wasting power, the Power Assignment problem is equivalent to finding the (directed) graph Q satisfying the connectivity constraint with minimum $p(Q)$.

As an example, Min-Power Symmetric Connectivity asks for the spanning tree T with $p(T)$ minimum. The broadcast nature of the wireless communication explains the difference from classical graph optimization problems, such as the polynomial-time solvable minimum spanning tree. However, methods inspired by classical Steiner trees were useful in devising approximation algorithms.

10.2 Strong Connectivity

The study of the min-power Power Assignment for strong connectivity problems was initiated by Chen and Huang.[5] Assuming symmetric power requirements, they prove that a minimum (cost) spanning tree (*MST*) of the input graph H has power at most twice the optimum, and therefore the MST algorithm has approximation ratio at most 2. The fact that the power of the output is at most twice the optimum is simple enough to present in this survey, and we use the reversed model as follows.

Consider G, the digraph induced by an optimal power assignment. Let s be an arbitrary node and B be an inward branch rooted at s (a tree with edges oriented toward the root) contained in G. As every node $v \neq s$ with parent v' has $p_B(v) = c(vv')$, we have $c(B) = \Sigma_{v \in V \setminus \{s\}} c(vv') = \Sigma_{v \in V \setminus \{s\}} p_B(v) = p(B)$. On the other side,

$$p(MST) = \sum_{v \in V} p_{MST}(v) = \sum_{v \in V} \max_{vu \in MST} c(vu) \leq \sum_{v \in V} \sum_{vu \in MST} c(vu) = 2c(MST),$$

since every edge of MST is counted twice in $\Sigma_{v \in V} \Sigma_{vu \in MST} c(vu)$. Therefore, $p(MST) \leq 2c(MST) \leq 2c(B) = 2p(B) \leq 2p(G)$, where $c(MST) \leq c(B)$ follows from the fact that MST is a minimum spanning tree.

The following example shows that the ratio of 2 for the MST algorithm is tight. Consider $2n$ points located on a single line such that the distance between consecutive points alternates between 1 and $\epsilon < 1$ (see Figure 10.2) and let $\ell = 2$. Then the minimum spanning tree MST connects consecutive neighbors and has power $p(MST) = 2n$. On the other hand, the tree T with edges connecting each other node (see Figure 10.2(b)) has power equal $p(T) = n(1 + \epsilon)^2 + (n - 1)\epsilon^2 + 1$. When $n \to \infty$ and $\epsilon \to 0$, we obtain $p(MST)/p(T) \to 2$.

In the line case, Kirousis et al.[8] present a dynamic programming algorithm for Min-Power Strong Connectivity, and show the NP-hardness of the three-dimensional Euclidean case.

Min-Power Strong Connectivity with symmetric power requirements is APX-hard. This means that there is an $\epsilon > 0$ such that the existence of an approximation algorithm with ratio $1+\epsilon$ implies that $P = NP$. Clementi et al.[6] showed that Min-Power Strong Connectivity in the Euclidean case is NP-hard. With asymmetric power requirements (or even with symmetric power requirements modified by nonuniform efficiency), a standard reduction shows that Strong Connectivity is as hard as Set Cover, which implies that there is no polynomial-time algorithm with approximation ratio $(1 - \epsilon) \ln n$ for any $\epsilon > 0$ unless $P = NP$.

Calinescu et al.[2] present a greedy approximation algorithm with ratio $2 \ln n + 3$ for Strong Connectivity with asymmetric power requirements. This algorithm picks an arbitrary vertex s, and uses an approximation algorithm for Broadcast from s, which we discuss later, and Edmunds' algorithm for minimum cost incoming branch rooted at s.

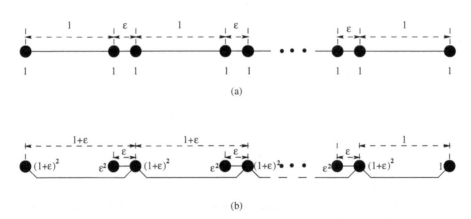

FIGURE 10.2 Tight example for the performance ratio of the MST algorithm ($\ell = 2$): (a) the MST-based power assignment needs total power $2n$; (b) optimum power assignment has total power $n(1 + \epsilon)^2 + (n - 1)\epsilon^2 + 1 \to n + 1$.

Improving the approximation ratio under 2 in the symmetric power requirements case appears to be a very difficult problem. For the Euclidean power requirements case, we have a candidate for an algorithm with approximation ratio under 2: run the two algorithms below and output the better solution. The two algorithms are the best approximation algorithm from the next section, and Christofidies' algorithm for the Traveling Salesman problem, followed by orienting all the edges of the Hamiltonian cycle to obtain a directed circuit.

10.3 Symmetric Connectivity

The connectivity of $\bar{G}$, the symmetric restriction of G, implies the connectivity of G. The reverse is not true, and in general it is harder for a power assignment to ensure symmetric connectivity. In fact, the power for the Min-Power Strong Connectivity can be half the power for Min-Power Symmetric Connectivity, as illustrated by the following example, in the Euclidean case with $\ell = 2$. The terminal set (see Figure 10.3) consists of n groups of $n + 1$ points each, located on the sides of a regular $2n$-gon. Each group has two terminals in distance 1 of each other (represented as thick circles in Figure 10.3) and $n - 1$ equally spaced points (dashes in Figure 10.3) on the line segment between them. It is easy to see that the minimum power assignment ensuring strong connectivity assigns power of 1 to the one thick terminal in each group and power of $\epsilon^2 = (1/n)^2$ to all other points in the group. The total power then equals $n + 1$. For symmetric connectivity it is necessary to assign power of 1 to all but two thick points, and power of ϵ^2 to the remaining points, which results in total power of $2n - 1 - 1/n + 2/n^2$.

Note that the *MST* algorithm of the previous section produces a symmetric output and therefore *MST* has an approximation ratio 2 for Symmetric Connectivity in the symmetric power requirements case, and this is tight, as shown in the example in Figure 10.2.

Better approximation algorithms were presented by Althaus et al.,[1] with the best achieving a ratio of $5/3 + \epsilon$ for any $\epsilon > 0$. This algorithm is based on an existing $1 + \epsilon$ approximation algorithm for the following "3-hypertree" problem (a particular case of matroid parity): given a weighted hypergraph $Q = (V, J)$ with the hyperedges of J being subsets of V of size 2 or 3, find a minimum weight set of edges K such that the hypergraph (V, K) is connected. A hypergraph is connected if one can reach any vertex from any other vertex by a path in which any two consecutive vertices are included in a hyperedge.

The "3-hypertree" algorithms are not practical. On random instances, both uniformly and with skewed distribution, the following heuristic, adapted from one of the approximation algorithms in Ref. 1 (but

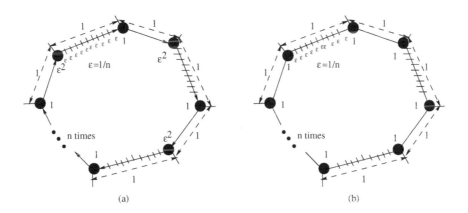

(a)   (b)

FIGURE 10.3 Total power for Min-Power Strong Connectivity can be almost as small as half the total power for Min-Power Symmetric Connectivity ($\ell = 2$) for the same input. (a) Minimum power assignment ensuring strong connectivity has total power $n + n^2 \epsilon^2 = n + n^2 \frac{1}{n^2} = n + 1$. (b) Minimum power assignment ensuring symmetric connectivity has the total power $(2n - 2) + (n^2 - n + 2)\epsilon^2 = 2n - 1 - \frac{1}{n} + \frac{2}{n^2}$.

without having a proof of approximation ratio better than 2), produced the best results:

1. Maintain a spanning tree T, initially the minimum spanning tree.
2. Find three vertices uvw and two edges e_1 and e_2 of T such that the $T' = T \backslash \{e_1, e_2\} \cup \{uv, uw\}$ has $p(T')$ minimum.
3. If $p(T') < p(T)$, replace T with T' and go to Step 2.

A simpler heuristic that only introduces one new edge uv in T with u and v at bounded distance in T, removes one edge e from T, as long as there is some improvement in the power, and also produces reasonable improvement over the minimum spanning tree and is suitable for distributed implementation.

With asymmetric power requirements (or even with symmetric power requirements modified by nonuniform efficiency), Min-Power Symmetric Connectivity is again as hard as set cover,[2] and the algorithm described below has approximation ratio at most $2 \ln n + 2$.

The algorithm starts iteration i with a graph G_i, seen as a set of edges with vertex set V. Unless G_i is connected, a star S (details below) is computed such that it achieves the biggest reduction in the number of components divided by the power of the star. The algorithm then adds the star (seen as a set of edges) to G_i to obtain G_{i+1}.

A *star* is a tree consisting of one *center* and several *leaves* adjacent to the center. Note that the power of the star is the maximum power requirement of the arcs from the center to the leaves plus the sum of power requirements of the arcs from the leaves to the center. With respect to G_i, let $d(S)$ be the number of different components of G_i to which the vertices of the star belong.

See Figure 10.4 of a star and its power. Our algorithm appears in Figure 10.5.

Next we describe how to find the star S minimizing $p(S)/(d(S) - 1)$. We search all vertices v and all power levels $p(v) \in \{c(vu) \mid u \in V\}$. For every connected component C_j of G_i, we find the vertex u_j (in case such a vertex exists; note that u_j might be v) such that $c(vu_j) \leq p(v)$ and $c(u_j v)$ is minimum. We sort the satisfied components C_j for which u_j exists in nondecreasing order of $c(u_j v)$. Then, for every $d \geq 2$, we try the star with center v and leaves $u_1, u_2, \ldots, u_d$. Thus, we search a total of at most n^3 stars and pick the one minimizing $p(S)/(d(S) - 1)$.

We conclude the section with a constant-ratio approximation algorithm for Min-Power Symmetric Connectivity in the Euclidean with efficiency model.[2] Define $w(u, v) = d(u, v)^\ell / e(u) + d(u, v)^\ell / e(v)$

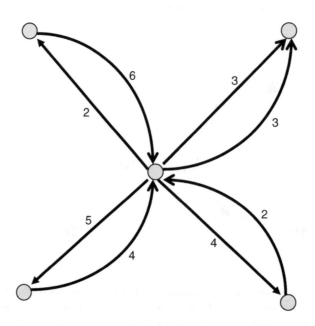

FIGURE 10.4   A star with four leaves, of power $\max\{2, 3, 4, 5\} + 6 + 3 + 2 + 4 = 20$.

Input: A complete directed graph $H = (V, E)$ with power requirements $c : E \rightarrow R^+$
Output: An undirected connected spanning graph G (seen as a set of edges, with $V(G) = V$)

---

0. Initialize $G = \emptyset$
1. While G has at least two connected components
1.1. Find the star S which minimizes $p(S)/(d(S) - 1)$ with respect to G
1.2 Set $G \leftarrow G \cup S$
2. For all vertices v, assign $p(v) = \max_{vu \in G} c(vu)$

FIGURE 10.5   The Greedy Algorithm for Symmetric Connectivity with asymmetric power requirements.

and construct a minimum spanning tree T of the node set V with respect to weight w. Assign power to vertices according to T; that is, set $p(v) = \max_{vu \in E(T)} d(v, u)^\ell / e(v)$.

10.4   Biconnectivity

In this section we discuss two connectivity constraints:

1. Biconnectivity: for any two vertices u and v, G must contain two paths from u to v that do not share nodes except for u and v. The biconnectivity requirement is motivated by reliability issues.
2. Symmetric Biconnectivity: $\bar{G}$ must be two-connected (that is, removing any vertex from G' results in a connected graph).

All the algorithms presented in this section produce outputs that satisfy the more stringent symmetric biconnectivity constraint. However, their approximation ratio is computed with respect to the optimum to Min-Power Biconnectivity.

Ramanathan and Rosales-Hain[10] proposed the first heuristic, which we describe in a slightly improved version. It uses the idea of Kruskal's algorithm for minimum spanning trees, and assumes symmetric power requirements:

- Sort the edges of H in non-decreasing cost order.
- Use binary search to find minimum i such that Q, the spanning subgraph of H with edge set $\{e_1, e_2, \ldots, e_i\}$, is biconnected.
- Starting with e_i and going downward, remove from Q any edge e for which $E(Q) \backslash \{e\}$ induces a biconnected spanning subgraph.
- Assign power to vertices according to Q; that is, set $p(v) = \max_{vu \in E(Q)} c(vu)$.
- For every vertex, reduce its power as long as the induced graph $\bar{G}$ is biconnected.

In the symmetric power requirements case, we have examples showing that the Ramanathan and Rosales-Hain heuristic has an approximation ratio of at least $n/2$. Its approximation ratio in the Euclidean power requirements case is not known, and our experiments show it is quite good on uniform random instances in the unit square.

Lloyd et al.[9] proposed using the approximation algorithm of Khuller and Vishkin,[7] which we refer to as **Algorithm KV** and which was designed for Minimum-Weight Biconnected Spanning Subgraph, and proved an approximation ratio of $2(2 - 2/n)(2 + 1/n)$ for Min-Power Symmetric Biconnectivity. Calinescu and Wan[3] showed that the power of the output of **Algorithm KV** is within 4 of the power of the (possibly nonsymmetric) best solution of the Min-Power Biconnectivity problem.

The **Algorithm KV** is complicated. In the Euclidean case, Calinescu and Wan[3] proposed **MST-Augmentation**, another algorithm with a constant approximation ratio, much faster, simpler, and better suited for distributed implementation. This $O(n \log n)$ algorithm first constructs a minimum spanning tree T over V. Then at any non-leaf node v of T, a local Euclidean minimum spanning tree T_v over all the neighbors of v in T is constructed. The output is Q, the union of T and T_v's for all non-leaf nodes v of the

T; power is assigned to vertices according to Q. Another advantage of this algorithm is its independence of the path-loss exponent.

10.5   *k*-Edge-Connectivity

In this section we discuss two connectivity constraints:

1. k-Edge-Connectivity: for any two vertices u and v, G must contain k edge-disjoint paths from u to v.
2. Symmetric k-Edge-Connectivity: $\bar{G}$ must be k-edge-connected (that is, removing any $k - 1$ edges from $\bar{G}$ results in a connected graph).

The results of this section are similar to those in the previous section. All the algorithms we present produce outputs that satisfy the more stringent symmetric k-edge-connectivity constraint. However, their approximation ratio is computed with respect to the optimum to k-Edge-Connectivity.

The Ramanathan and Rosales-Hain heuristic can be applied (with approximation ratio in the Euclidean case unknown) to k-Edge-Connectivity by checking for k-edge-connectivity instead of biconnectivity when removing the edges.

Lloyd et al.[9] proposed using the approximation algorithm of Khuller and Raghavachari,[7] which was designed for Minimum-Weight k-Edge-Connected Spanning Subgraph, and proved an approximation ratio of $8(1 - 1/n)$ for Min-Power Symmetric k-Edge-Connectivity. Calinescu and Wan[3] showed that the power of the output of the Khuller and Raghavachari algorithm is within $2k$ of the power of the (possibly non-symmetric) best solution of the Min-Power k-Edge-Connectivity problem.

MST-Augmentation can also be used for 2-Edge-Connectivity, with a constant approximation ratio in the Euclidean case.

10.6   Symmetric Unicast

The Min-Power Unicast problem requires that G contains a directed path from s to t, where s and t are given nodes. It is easily solvable by Dijkstra's shortest paths algorithm in the graph H.

In Min-Power Symmetric Unicast, we are given s and t, and $\bar{G}$ must contain a path from s to t. The following example in the Euclidean case shows that a straightforward application of Dijkstra's algorithm does not work; that is, a minimum cost (with cost function c given by the power requirements) $s - t$ path does not always have minimum power. Consider a network consisting of three nodes, $s = (0, 3)$, $t = (4, 0)$, and $x = (0, 0)$ (see Figure 10.6), and assume $\ell = 2$. Then, the two $s - t$ paths, namely, (s, t) and (s, v, t), have the same cost of 25 but different powers; the power; of (s, t) is $25 + 25 = 50$ while the power of (s, v, t) is $9 + 16 + 16 = 41$.

Althaus et al.[1] present a solution of MIN-POWER SYMMETRIC UNICAST that first modifies the given graph $H = (V, E, c)$ and then applies Dijkstra's algorithm to the resultant directed graph H'. We now describe

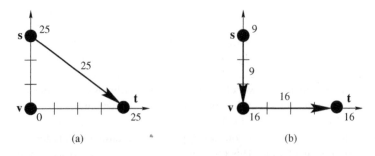

FIGURE 10.6   An example of two paths with the same cost and different powers: (a) the path (s, t) assigns powers 25 to s and to t; (b) the path (s, v, t) assigns powers 9 to s and 16 to v and t.

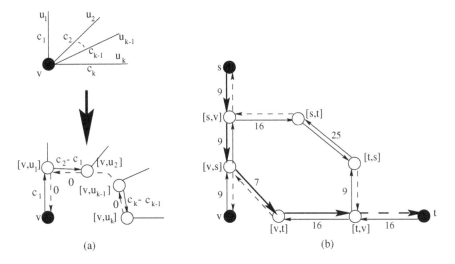

FIGURE 10.7 (a) A vertex v adjacent to k vertices $u_1,\ldots,u_k$ via edges of cost $c_1, c_2,\ldots, c_k$ and a gadget replacing v with a bidirectional path. The solid edges of the path $(v, [v, u_2]), ([v, u_2]), [v, u_3], \ldots, ([v, u_{k-1}], [v, u_k]$ have cost c_1, $c_2 - c_1, \ldots, c_k - c_{k-1}$, respectively. The dashed edges have zero cost. (b) The graph H' for the example in Figure 10.6. As the example in Figure 10.6 has symmetric power requirements, the pair of opposite directed edges of the type $([u, v], [v, u])$ and $([v, u], [u, v])$ have the same cost and the figure only has one number next to such a pair. Thick edges belong to the shortest path corresponding to the path (s, v, t) in H.

the construction of the directed graph $H' = (V', E', c')$, and note that it does not assume that the cost function is symmetric.

For any $u \in V$, we sort all adjacent vertices $\{v_1,\ldots,v_k\}$ in ascending order of costs of edges connecting them to u, i.e., $c(u, v_i) \leq c(u, v_{i+1})$. The vertex v is replaced by a *gadget* (see Figure 10.7a) as follows:

1. Each edge (u, v) is replaced by two vertices: $[u, v]$ and $[v, u]$.
2. For each u, we connect all vertices $[u, v_i]$'s by two directed paths:
   $P_1 = (u, [u, v_1], \ldots, [u, v_{k-1}], [u, v_{k-1}])$ and
   $P_2 = ([u, v_{k-1}], [u, v_{k-1}], \ldots, [u, v_1], u)$.
3. The costs of the arcs on path P_1 are $c(u, v_1), c(u, v_2) - c(u, v_1), \ldots, c(u, v_k) - c(u, v_{k-1})$, respectively; and the cost of all arcs on the path P_2 is zero.

Finally, each edge (u, v) of H is replaced in H' by one arc $([u, v], [v, u])$ of cost $c(v, u)$.

Figure 10.7b shows the graph H' for the example in Figure 10.6. It is easy to see that a shortest s–t path in H' corresponds to a minimum power s–t undirected path.

10.7 Broadcast and Multicast

In this section we discuss two related connectivity constraints:

1. Broadcast: G must contain a directed path from a given node called the root to every other node.
2. Multicast: G must contain a directed path from the root to a given set of nodes called terminals.

Min-Power Broadcast was first studied by Wieselthier et al.[12] They proposed three heuristics but did not prove approximation ratios. The first heuristic, SPT, uses a shortest-path tree from the root. The second heuristic, MST, uses a minimum spanning tree. In both cases, the resulting undirected graph is oriented away from the root, and power is assigned accordingly. The third heuristic, called BIP (broadcasting incremental power), is a Prim-like heuristic that starts with an outgoing branch consisting of the root, and iteratively adds an arc connecting the set of nodes currently reached to an outside node with minimum increase in power.

For the Euclidean case, Wan et al.[11] studied the approximation ratios of the above three heuristics. An instance was constructed to show that the approximation ratio of SPT is as large as $\frac{n}{2} - o(1)$. On the other hand, both MST and BIP have constant approximation ratios. The following geometric constant plays an important role in their analysis. Let

$$\sigma = \sup_{o \in P \subset \mathcal{D}, |P| < \infty} \sum_{e \in mst(P)} \|e\|^2,$$

where $\mathcal{D}$ is the closed disk of radius one centered at the origin $\mathbf{o}$ and $mst(P)$ is a Euclidean minimum spanning tree of P. It is proved that $6 \le \sigma \le 12$. In addition, any unidirectional broadcast routing has power at least $c(MST)/\sigma$. This immediately implies that the approximation ratio of MST is at most σ. It is also proved[11] that $p(BIP) \le c(MST)$ and therefore BIP also has an approximation ratio at most σ. Wan et al.[11] also constructed two instances that lead to a lower bound of 6 on the approximation ratio of MST and a lower bound of 13/4 on the approximation ratio of BIP, respectively. We conjecture that $\sigma = 6$ and therefore MST has an approximation ratio of 6.

For the Euclidean case, it follows immediately from the analysis in Ref. 11 that an α-approximate Steiner tree gives an approximation ratio $\alpha \cdot \sigma$ for Min-Power Multicast. With current best-known $\alpha = 1 + \frac{\ln 3}{2} + \epsilon$, we obtain an 18.59 approximation algorithm. Faster algorithms have been proposed by Wan (unpublished manuscript) for Multicast, based on faster Minimum Steiner Tree algorithms.

Now we move from the Euclidean case to symmetric power requirements. A standard reduction from Set Cover shows that no approximation ratio better than $O(\log n)$ is possible. It turns out that the best approximation algorithm[2] also works for Min-Power Broadcast with asymmetric power requirements. The algorithm is an extension of the algorithm for Min-Power Symmetric Connectivity presented in Figure 10.5, except that it uses a more complicated structure called *spider* and it aims to pick a spider with minimum ratio of its weight to the number of strongly connected components with no incoming edge it "hits."

For Min-Power Multicast in the symmetric power requirements case, the construction devised for Broadcast by Caragiannis and Kaklamanis[4] together with approximation algorithms for Node-Weighted Steiner Tree can be used to obtain an $O(\log n)$ approximation ratio. This result cannot be easily extended to Min-Power Multicast in the asymmetric power requirements case, as in fact this most general power assignment problem is equivalent to Directed Steiner Trees,[2] a problem seemingly harder than Node-Weighted Steiner Tree.

10.8   Summary of Approximability Results

We summarize the known results on Power Assignment in Table 10.1. Each cell of the table describes the best-known approximation ratio and lower bounds for each combination given by the connectivity constraint type and the power requirement case. NPH means NP-hard, and APXH means APX-hard

TABLE 10.1   Upper bounds (UB) and lower bounds (LB) of the Power Assignment complexity. Marked by * are the folklore results, while references preceded by ** indicate the result is implicit in the respective articles.

	Complexity of the Min-Power Assignment problems					
	Asymmetric Power Reqs.		Symmetric Power Reqs.		Euclidean ($\alpha \ge 2$)	
Connectivity Constraint	UB	LB	UB	LB	UB	LB
Strong Connectivity	$3 + 2\ln(n-1)^2$	SCH[2]	$2^{5,8}$	APXH*	2	NPH[6]
Broadcast	$2 + 2\ln(n-1)^2$	SCH	$2 + 2\ln(n-1)$	SCH[11]	12^{11}	NPH*
Multicast	DST*	DSTH[2]	$O(\ln n)^{**4}$	SCH[**11]	18.59^{**11}	NPH*
Symmetric Conn.	$2 + 2ln(n-1)^2$	SCH[2]	$\frac{5}{3} + \epsilon^1$	APXH*	$\frac{5}{3} + \epsilon$	NPH*
By Connectivity		NPH*	4^3	NPH*	4	
Symm. Biconn.		APXH*	4^3	APXH*	4	NPH[3]
k-Edge-Conn.		NPH*	$2k^3$	NPH*	$2k$	
Symm. k-Edge-Conn.		APXH*	$2k^3$	APXH*	$2k$	NPH[3]

(there is an $\epsilon > 0$ such that the existence of an approximation algorithm with ratio $1 + \epsilon$ implies that $P = NP$). SCH means the problem is as hard as Set Cover, which implies that there is no polynomial-time algorithm with approximation ratio $(1 - \epsilon) \ln n$ for any $\epsilon > 0$, unless $P = NP$. DST means that the problem reduces (approximation-preserving) to Directed Steiner Tree and DSTH means Directed Steiner Tree reduces (approximation-preserving) to the problem given by the cell. The best-known approximation ratio for Directed Steiner Tree is $O(n^\epsilon)$ for any $\epsilon > 0$ and finding a poly-logarithmic approximation ratio remains a major open problem in the field of approximation algorithms.

We omit the line case, where all problems can be solved in polynomial time. In line with the efficiency case, all the problems have dynamic programming polynomial-time algorithms,[2] while with sensitivity, surprisingly, even for the line nothing better than general asymmetric power requirements is known. We also omit Unicast, where shortest paths algorithms directly solve the problem, and Symmetric Unicast, where shortest paths algorithms in a specially constructed graph solve the problem (see Section 10.6).

In the Euclidean case with efficiency, constant-ratio algorithms exist for Min-Power Symmetric Connectivity,[2] Min-Power Symmetric Biconnectivity, and Min-Power Symmetric 2-Edge-Connectivity; while for the "directed" connectivity types, nothing better than the results for asymmetric power requirements is known.

References

1. E. Althaus, G. Calinescu, I. Mandoiu, S. Prasad, N. Tchervenski, and A. Zelikovsky, Power Efficient Range Assignment in Ad-hoc Wireless Networks, submitted for journal publication, Preliminary results in *Proc. IFIP-TCS 2002*, 119–130, and in *Proc. IEEE Wireless Communications and Networking Conference*, 2003.

2. G. Calinescu, S. Kapoor, A. Olshevsky, and A. Zelikovsky, Network Lifetime and Power Assignment in Ad-Hoc Wireless Networks, to appear in *Proc. 11th European Symp. on Algorithms*, 2003.

3. G. Calinescu and P.-J. Wan, Range Assignment for High Connectivitity in Wireless Ad Hoc Networks, submitted for publication, 2003.

4. P.K.I. Caragiannis and C. Kaklamanis, New Results for Energy-Efficient Broadcasting in Wireless Networks, in *ISAAC'2002*, 2002.

5. W.T. Chen and N.F. Huang, The Strongly Connecting Problem on Multihop Packet Radio Networks, *IEEE Transactions on Communications*, 37(3), 293–295, 1989.

6. A.E.F. Clementi, P. Penna, and R. Silvestri. On the Power Assignment Problem in Radio Networks. *Electronic Colloquium on Computational Complexity (ECCC)*, (054), 2000.

7. S. Khuller, Approximation Algorithms for Finding Highly Connected Subgraphs, in *Approximation Algorithms for NP-Hard Problems*, edited by D.S. Hochbaum, 1996, pp. 236–265.

8. L.M. Kirousis, E. Kranakis, D. Krizanc, and A. Pelc, Power Consumption in Packet Radio Networks. *Theoretical Computer Science*, 243, 289–305, 2000.

9. E. Lloyd, R. Liu, M. Marathe, R. Ramanathan, and S.S. Ravi, Algorithmic Aspects of Topology Control Problems for Ad hoc Networks, *Proc. 3rd ACM International Symposium on Mobile Ad Hoc Networking and Computing (MobiHoc)*, Lausanne, Switzerland, June 2002.

10. R. Ramanathan and R. Rosales-Hain, Topology Control of Multihop Wireless Networks Using Transmit Power Adjustment, *Proc. IEEE INFOCOM 2000*, pp. 404–413.

11. P.-J. Wan, G. Calinescu, X.-Y. Li, and O. Frieder, Minimum Energy Broadcast Routing in Static Ad Hoc Wireless Networks, *Wireless Networks*, 8(6), 607–617, 2002.

12. J.E. Wieselthier, G.D. Nguyen, and A. Ephremides, On the Construction of Energy-Efficient Broadcast and Multicast Trees in Wireless Networks, *Proc. IEEE INFOCOM 2000*, pp. 585–594, 2000.

11

Energy Conservation for Broadcast and Multicast Routings in Wireless Ad Hoc Networks

Jang-Ping Sheu

Yuh-Shyan Chen

Chih-Yung Chang

11.1   Introduction

Wireless ad hoc networks have received significant attention in recent years due to their potential applications on the battlefield, in disaster relief operations, festival field grounds, and historic sites. A wireless ad hoc network consists of mobile hosts dynamically forming a temporary network without the use of an existing network infrastructure. In such a network, each mobile host serves as a router. One important issue in ad hoc network routing is energy consumption. In MANETs, mobile hosts are powered by batteries and unable to recharge or replace batteries during a mission. Therefore, the limited battery lifetime imposes a constraint on network performance. To maximize the network lifetime, the traffic should be routed in such a way that energy consumption is minimized.

Broadcast and multicast are important operations for mobile hosts to construct a routing path in a MANET. Broadcast is a communication function in which a node, called the source, sends messages to all the other nodes in the network. Broadcast is an important function in applications of ad hoc networks, such as in cooperative operations, group discussions, and route discovery. Broadcast routing is usually constructing a broadcast tree, which is rooted from the source and contains all the nodes in the network. In addition to broadcasting, multicasting is also an important function in applications, including distributed games, replicated file systems, and teleconferencing. Multicast in a MANET is defined by delivering multicast packets from a single source node to all member nodes in a multi-hop communication manner. The energy cost of all the nodes that transmit the broadcast or multicast message in MANET should be minimized.

To overcome the problems of transmission collision, message storm, and battery exhaustion, several energy conservation schemes for broadcast and multicast routings are proposed in literature.[1–23] This chapter consists of two parts. The first part introduces novel energy conservation schemes for broadcast routing in MANETs, and the second part investigates existing energy conservation schemes for multicast routing in MANETs.

In the first part, existing energy-efficient broadcast protocols can be classified into tree-based and probability-based approaches. The tree-based broadcast protocol is to construct the *minimum-energy broadcast tree,*[1–9] which is a broadcast tree with minimum energy consumption. To establish the minimum-energy broadcast tree, centralized algorithms[1–3] and distributed algorithms[8,9] are investigated in wireless ad hoc networks. For centralized algorithms, we review centralized BIP[1] and EWMA[3] protocols. For distributed algorithms, we describe the DISP-BIP[8] and RBOP[9] protocols. In addition, integer-programming techniques can be used to establish the minimum-energy broadcast tree.[4] The *approximation ratio* of existing minimum-energy broadcast protocols is calculated in Wan et al.,[5] Clementi et al.,[6] and Li et al.[7] By considering the probability-based approach, the energy conservation for broadcast routing can be achieved by alleviating the "broadcast storm problem" with a high-performance probabilistic scheme.[10–12] A power-balance broadcast approach is then investigated in Sheu et al.[13] to extend the network lifetime using the probabilistic scheme to determine whether or not the host needs to rebroadcast.

In the second part, some existing power-efficient multicast protocols designed for MANETs are investigated. According to the topology constructed in the protocols, existing power-efficient multicast protocols can be classified into tree-based and cluster-based protocols. In tree-based multicast protocols, an energy-efficient broadcast tree is constructed first. By considering the power consumption of nodes in the tree, these protocols propose tree refining or pruning rules to construct a power-efficient multicast tree. According to the number of source nodes in the tree, the tree-based multicast protocols are further partitioned into two subsets: the single-source and multi-source multicast protocols. In the subset of single-source multicast protocols, power-efficient multicast protocols MIP,[2] S-REMiT,[14] and RBIP[19] are reviewed. The MIP and S-REMiT protocols apply refining and pruning rules on existing broadcast trees to construct a power-efficient multicast tree. Some applications require that the multicast be reliable. The RBIP protocol considers the reliable multicast and takes into consideration the retransmission cost in energy consumption. Another multicast protocol, G-REMiT,[15] is also reviewed in the tree-based multicast category. Different from the protocols mentioned above, the G-REMiT protocol[15] is mainly designed for multi-source energy-efficient multicast trees. In addition to the tree-based multicast protocols, Subsection 11.3.2 reviews the Cluster-Based Multicast Protocol (CBMP),[16] which applies the existing ODMRP[17] on cluster topology to achieve the purpose of energy-efficient multicast communication.

The remainder of this chapter is organized as follows. Section 11.2 reviews energy-efficient broadcast protocols in MANETs. Section 11.3 introduces energy-efficient multicast protocols in MANETs. Section 11.4 concludes this chapter and gives some possible future works.

11.2   Energy-Efficient Broadcast Protocols in MANETs

This section describes existing valuable energy-efficient broadcasting protocols in MANETs. These energy-efficient broadcast protocols are categorized according to the aspects of *tree-based* and *probability-based* approaches. The detailed operations of these energy-efficient broadcast protocols are described as follows.

11.2.1 Tree-Based Approach

The *minimum-energy broadcast tree* is formally defined in Cagalj et al.[3] as follows. Given the source node *r*, a set consisting of pairs of relaying nodes and their respective transmission levels is constructed such that all nodes in the network receive a message sent by *r*, and the total energy expenditure for this task is minimized. The objective of energy-efficient broadcasting protocols herein is to construct the *minimum-energy broadcast tree*. In the following, Section 2.1.1 describes centralized algorithms to establish the minimum-energy broadcast tree; Section 11.2.1.2 expresses distributed algorithms of constructing the minimum-energy broadcast tree.

11.2.1.1 Centralized Algorithms

To build a spanning tree with minimum energy consumption, one way is to construct a *minimum spanning tree* (MST).[1–3] A centralized algorithm, called a centralized BIP (*broadcast incremental power*), is developed in Wilson and Watkins[1] to construct a *minimum-energy broadcast tree* in MANETs. An improved centralized algorithm, called EWMA (embedded wireless multicast advantage), is proposed in Cagalj et al.[3] to construct a minimum-energy broadcast tree with less power consumption.

11.2.1.1.1 *Centralized BIP (Broadcast Incremental Power) Algorithm*

Section 11.2.1.3 investigates the establishment of minimum-energy broadcast tree using the integer programming technique. Finally, Section 11.2.1.4 calculates the approximation ratio of existing minimum-energy broadcast protocols. A centralized algorithm, called a BIP (*broadcast incremental power*) algorithm, is developed to build an energy-efficient broadcast tree in a MANET.[1] The BIP algorithm exploits the broadcast nature of the wireless communication environment and addresses the need for energy-efficient operation. The main objective of the BIP algorithm is to construct a minimum-energy broadcast tree. The BIP algorithm is based on Prim's algorithm,[2] which is an algorithm used to search for minimum spanning trees (MSTs). The wireless communication model is defined as follows. First, omni-directional antennas are used, such that every transmission by a node can be received by all nodes that lie within its communication range. Second, the connectivity of the network depends on the transmission power; each node can choose its power level, not to exceed some maximum value P_{max}. BIP assumed that the received signal power varies as $r^{-\alpha}$, where *r* is the range and α is a parameter that typically takes on a value between 2 and 4. Without loss of generality, P_{ij} = power needed for the link between nodes *i* and $j = r^{\alpha}$, where *r* is the distance between nodes *i* and *j*.

The BIP algorithm and the following protocols adopt the use of omni-directional antennas; thus, all nodes within the communication range of a transmitting node can receive its transmission. Consider the example shown in Figure 11.1, in which a subset of the multicast tree involves node *i*, which is transmitting data to its neighbors, node *j* and node *k*. The power required to reach node *j* is P_{ij} and the power required to reach node *k* is P_{ik}. A single transmission at power $P_{i,(j,k)} = \max\{P_{ij}, P_{ik}\}$ is sufficient to reach both node *j* and node *k*, based on the assumption of omni-directional antennas. The ability to exploit this property of wireless communication, which is called the *wireless multicast advantage*, makes multicasting an excellent setting in which to study the potential benefits of energy-efficient protocols.

One can explain the basic operation of BIP by offering a simple example of the construction of the broadcast tree, rooted at a source node.

1. Figure 11.2 shows a wireless network with ten nodes, in which node 10 is the source node. A propagation constant of $\alpha = 2$ is assumed. At first, the tree only consists of the source node. Then BIP begins by determining which node should be selected so that the source node can reach with minimum incremental power. The source node's nearest neighbor, which is node 9, should be added to the tree. The notation $10 \rightarrow 9$ means adding the transmission from node 10 to node 9.

2. BIP then determines which "new" node can be added to the tree at *minimum additional* cost. There are two alternatives. Either node 10 can increase its power to reach a second node, or node 9 can transmit to its nearest neighbor that is not already in the tree. In this example, node 10 increases its power level to reach node 6. Note that the cost associated with the addition of node 6 to the tree is the incremental cost associated with increasing node 10's power level sufficient to reach node 6.

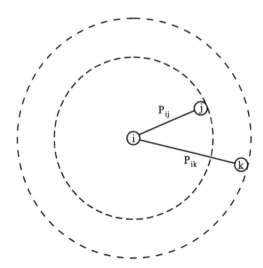

FIGURE 11.1   The "wireless broadcast advantage": $P_{i,(j,k)} = \max\{P_{ij}, P_{ik}\}$.

The cost of a transmission between nodes 10 and 9 is $r_{10,9}^{\alpha}$, and the cost of a transmission between nodes 10 and 6 is $r_{10,6}^{\alpha}$. The incremental cost associated with adding node 6 to the tree is $r_{10,6}^{\alpha} - r_{10,9}^{\alpha}$. BIP exploits the broadcast advantage because when node 10 has sufficient power to reach node 6, then node 10 also can reach node 9.

3. There are now three nodes in the tree, namely nodes 6, 9, and 10. For each of these nodes, BIP determines the incremental cost to reach a new node; that is, $6 \to 7$, as shown in Figure 11.2.

4. This procedure is repeatedly performed until all nodes are included in the tree. The order in which the nodes were added is: $6 \to 8, 6 \to 5, 9 \to 1, 9 \to 3, 9 \to 4, 9 \to 2$.

11.2.1.1.2   *EWMA (Embedded Wireless Multicast Advantage)*

The EWMA protocol[3] consists of two steps:

1. A minimum spanning tree (MST) for a broadcasting tree is initially established as shown in Figure 11.3, where node 10 is the *source* node and nodes 9, 1, 6, and 8 are *forwarding* nodes. The power consumptions of nodes 10, 9, 1, 6, and 8 are 2, 8, 4, 5, and 4, respectively. The total energy consumption of the MST is 23.

2. EWMA calculates the necessary power for every node from the constructed MST in Step 1. A node is said to be an *exclude node* if it is a transmitting node in MST but is not a transmitting node in the final EWMA broadcasting tree. The key idea of EWMA is to search for exclude nodes by increasing

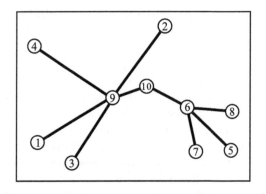

FIGURE 11.2   Broadcast tree using BIP.

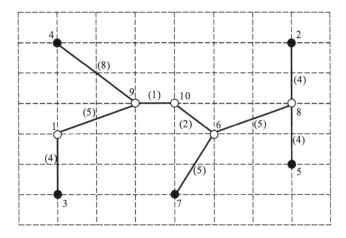

FIGURE 11.3 An MST broadcasting tree.

less power consumption for the exclude node to cover more forwarding nodes. For example, the resultant broadcast tree produced by EWMA is shown in Figure 11.4. After increasing the power consumption of node 10 (from 2 to 13), the original forwarding nodes 9, 6, and 8 in the MST can be excluded in the EWMA broadcast tree. Therefore, only nodes 10 and 1 are used in the EWMA broadcast tree. The total energy consumption of the EWMA broadcast tree is $13 + 4 = 17$. This result is illustrated in Figure 11.4.

11.2.1.1.3 *Integer Programming Technique*

It is interesting that three different integer programming models are used for an optimal solution of the minimum power broadcast problem.[4] The main idea is to use the *power matrix P*, where the (i, j)-th element of the power matrix P defines the power required for node i to transmit to node j. For example, as shown in Figure 11.5, the power matrix P is

$$\begin{bmatrix} 0 & 8.4645 & 12.5538 & 13.6351 \\ 8.4645 & 0 & 0.5470 & 3.8732 \\ 12.5538 & 0.5470 & 0 & 5.7910 \\ 13.6351 & 3.8732 & 5.7910 & 0 \end{bmatrix}.$$

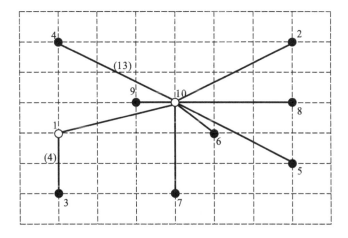

FIGURE 11.4 The EWMA broadcast tree.

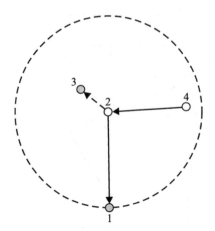

FIGURE 11.5 Example of a MANET and node 4 is the source node.

In addition, a *reward matrix R* is defined by

$$R_{mn}(p) = \begin{cases} 1, \text{ if } P_{mp} \leq P_{mn} \\ 0, \text{ otherwise.} \end{cases}$$

The example shown in Figure 11.5 explains the meaning of the reward matrix. A binary encoding is produced for all the nodes covered (or not covered) by all possible transmissions in the network. For example, the transmission $2 \rightarrow 1$ results in nodes 1, 3, and 4 being covered; therefore, $R_{21} = [1011]$ is encoded in the $(2, 1)$ cell of the reward matrix. Therefore, the reward matrix is:

$$R = \begin{bmatrix} [0 \ 0 \ 0 \ 0] & [0 \ 1 \ 0 \ 0] & [0 \ 1 \ 1 \ 0] & [0 \ 1 \ 1 \ 1] \\ [1 \ 0 \ 1 \ 1] & [0 \ 0 \ 0 \ 0] & [0 \ 0 \ 1 \ 0] & [0 \ 0 \ 1 \ 1] \\ [1 \ 1 \ 0 \ 1] & [0 \ 1 \ 0 \ 0] & [0 \ 0 \ 0 \ 0] & [0 \ 1 \ 0 \ 1] \\ [1 \ 1 \ 1 \ 0] & [0 \ 1 \ 0 \ 0] & [0 \ 1 \ 1 \ 0] & [0 \ 0 \ 0 \ 0] \end{bmatrix}.$$

To utilize the information of the calculated *power matrix P* and *reward matrix R*, the minimum power broadcast tree is constructed using integer programming formulations.[4]

11.2.1.1.4 *Calculating Approximation Ratios on Static Ad Hoc Networks*

A wireless ad hoc network is called a *static* ad hoc wireless network[5–7] if the nodes in the ad hoc network are assumed to be a point set randomly distributed in a two-dimensional plane and there is no mobility. The minimum-energy broadcast routing in static ad hoc wireless networks was first considered in Wan et al.[5] By exploring geometric structures of Euclidean MSTs, it is proven[5] that the *approximation ratios* of MST and centralized BIP are between 6 and 12, and between $\frac{13}{3}$ and 12, respectively, where the approximation ratio means that the results obtained by their executions are how close to the optimal value. Furthermore, the approximation ratio of the MST-based heuristic for the energy-efficient broadcast problem in static ad hoc networks is investigated in Clementi et al.[6] The main result of Clementi et al.'s[6] work shows that the approximation ratio is about 6.4. In addition, energy-efficient broadcasting routing is developed in static ad hoc wireless networks.[7] This work proposed three heuristic algorithms — (1) shortest path tree heuristic, (2) greedy heuristic, and (3) node weighted Steiner tree-based heuristic — which are centralized algorithms. The approximation ratio of the node weighted Steiner tree-based heuristic is proven to be $(1 + 2\ln(n - 1))$.[7]

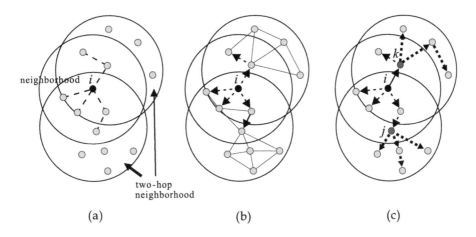

FIGURE 11.6   (a) Local BIP tree for node i, (b) Dist-BIP-A tree, and (c) Dist-BIP-G tree.

11.2.1.2   Distributed Algorithms

A distributed version of the BIP algorithm, called DIST-BIP, is then proposed in Wieselthier et al.[8] A localized minimum-energy broadcasting protocol is developed in Cartigny et al.[9] such that each node only requires the local information.

11.2.1.2.1   *DIST-BIP (Distributed Broadcast Incremental Power)*

Two distributed BIP algorithms[8] are proposed. One is Dist-BIP-A (distributed-BIP-All), and the other is Dist-BIP-G (distributed-BIP-gateway). In the Dist-BIP-A algorithm, each node constructs its local BIP tree using the *centralized-BIP* algorithm[1] within the one-hop transmission range. After constructing local BIP trees for every node, each node hears and broadcasts messages from or to its neighbors to connect many local BIP trees to form a global BIP tree. For example, node i constructs a local BIP tree as shown in Figure 11.6a. A Dist-BIP-A tree is established as shown in Figure 11.6b by connecting many local BIP trees, which are constructed by all neighboring nodes. The gateway nodes are jointed to hear and broadcast messages in the Dist-BIP-G protocol to form a Dist-BIP-G tree. An example of the Dist-BIP-G tree is illustrated in Figure 11.6c. Nodes i, j, and k are gateway nodes. The Dist-BIP-G tree is established by connecting local BIP trees, which are constructed by gateway nodes i, j, and k. In general, the message overhead of constructing a Dist-BIP-G tree is less than that of constructing a Dist-BIP-A tree. But the Dist-BIP-A tree is near the centralized BIP tree.

11.2.1.2.2   *RBOP (RNG Broadcast Oriented Protocol)*

A localized minimum-energy broadcasting protocol, called the RNG Broadcast Oriented Protocol (RBOP), which utilizes the relative neighborhood graph (RNG), is developed in Cartigny et al.[9] The protocol only requires the local information to design the minimum-energy broadcasting protocol. Unlike most existing minimum-energy broadcasting protocols that use the global network information, RBOP only maintains the local information, thus saving the communication overhead for obtaining global information.

To substitute minimum spanning tree (MST) in the protocol by utilizing the *relative neighborhood graph* (RNG), the wireless network is represented by a graph $G = (V, E)$, where V is the set of nodes and $E \subseteq V^2$ denotes the edge set that represents the available communications. Note that (u, v) belongs to E means that u can send message to v, and *RNG* is a sub-graph of G. An edge (u, v) belongs to the RNG if no node w exists in the intersection area for nodes u and v, as illustrated in Figure 11.7. This topology control scheme is called the RNG Topology Control Protocol (RTCP), which is used to build the relative neighborhood graph (RNG).

The main idea of the RBOP is that when a node u receives a message from neighbor nodes, the node selects an edge (u, v) in RNG as far as possible to broadcast the message within radius $d(u, v)$. For example, as shown in Figure 11.8, node S broadcasts a message to A, B, and C with radius $d(S, A)$,

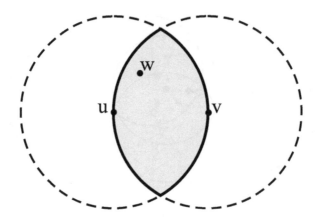

FIGURE 11.7   The edge (u, v) does not belong to RNG because of the existence of node w.

because $d(S, A) > d(S, C) > d(S, B)$, where (S, A), (S, C), and (S, B) are edges belonging to RNG. Then node C broadcasts with radius $d(C, D)$. Finally, node A broadcasts with radius $d(A, G)$. This method can reduce the total number of broadcast messages and efficiently transmit the broadcast messages. In the simulation results reported in Cartigny et al.,[9] the centralized BIP protocol can save about 50 percent energy compared to the RBOP. However, the communication overhead of centralized BIP is higher than that of the RBOP.

11.2.2   Probability-Based Approach

A probability-based approach also can be applied to determine whether or not a node should transmit the received packet during broadcasting. Some protocols[10–12] apply the probability-based approach to resolve the broadcast storm problem, hence saving the power consumption for redundant transmission. A power-balance protocol proposed in Sheu et al.[13] also adopts a probability-based approach to balance the power consumption on each node, thus improving the network lifetime. This subsection introduces the probability-based protocols that help improve the network lifetime.

In a MANET, flooding is a basic requirement and is frequently used to broadcast a message over the MANET. However, blind flooding will cause the broadcast storm problem,[10] resulting redundant message rebroadcasts, contentions, and collision. Alleviating the retransmission, contention, and collision situations will not only improve the success rate for receiving packets, but also reduce the power consumption.

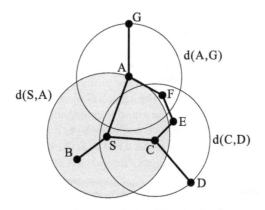

FIGURE 11.8   Example of the RNG Broadcast Oriented Protocol (RBOP).

To resolve the broadcast storm problem and achieve the goal of energy conservation, the *probabilistic*, *counter-based*, *location-based*, *polygon-based*, and *cluster-based* schemes were first investigated in Tseng et al.[10]

Sheu et al.[13] proposed a power-balance broadcast algorithm to extend the network lifetime. The power-balance broadcast algorithm uses the residual battery energy to determine whether or not the host needs to rebroadcast messages. The host with more residual energy will have higher probability to rebroadcast messages than the host with less residual energy. Therefore, the host with less residual energy will reduce the rebroadcast probability and reserve more energy for extending its lifetime. The proposed algorithm consists of two steps. First, each node i has an initial rebroadcast probability P_i according to its remaining energy. Second, the algorithm uses the average remaining energy of the neighbors of host i, the number of neighbors of host i, and the number of broadcast messages received by host i to refine the rebroadcast probability.

11.3 Energy-Efficient Multicast Protocol in MANETs

Energy-efficient multicasting has also been intensively discussed in wireless ad hoc networks. Multicasting is another important routing operation to transmit the message from one mobile host to a number of mobile hosts. Many applications require disseminating information to a group of mobile hosts in a MANET. These applications include distributed games, replicated file systems, teleconferencing, etc. A single-source multicasting in MANET is defined by delivering multicast packets from a single-source node to all member nodes in a multi-hop communication manner. A multi-source multicast is the one that each member can be the source of message sender of the other members. Although multicasting can be achieved by the multiple point-to-point routes, constructing a multicast topology for delivering the multicast packets always provides a better performance. A number of articles[20,21] have recently investigated multicast protocols in a MANET, by only considering how to reduce the tree level or the number of forwarding nodes. It is very important to take into consideration the factors of energy reservation and network lifetime to investigate the energy-efficient multicast protocol, because the wireless device in a MANET is mainly limited and constrained by the life of the battery. According to the topology constructed in the previous protocols, existing energy-efficient multicast protocols can be classified into *tree*-based and *cluster*-based protocols. This section reviews the existing power-efficient multicast protocols for MANETs.

11.3.1 Tree-Based Energy-Efficient Multicast Protocol

According to the number of source nodes in networks, existing tree-based energy-efficient multicast protocols are classified into two categories: the single-source and multi-source multicast protocols. Some articles construct the power-efficient multicast tree by pruning the broadcast tree, which is established by existing power-efficient broadcasting protocols such as MST,[2,3] BIP,[1,8] and BLiMST.[2] By taking into consideration the power consumption of the nodes in a broadcast tree, these protocols propose tree refining and pruning rules to construct a power-efficient multicast tree. Section 11.3.1.1 first reviews existing single-source multicast protocols, and Section 11.3.1.2 then reviews multi-source multicast protocols.

11.3.1.1 Single-Source Multicast Protocol

11.3.1.1.1 *MIP (Multicast Incremental Power) Algorithm*

Operations of the MIP algorithm can be partitioned into three phases. In the first phase, a power-efficient broadcast tree is constructed by a centralized BIP (*broadcast incremental power*) algorithm, as described in Section 11.2.1.1.1. By considering the characteristics of wireless transmission, the second phase applies sweep operations to the constructed broadcast tree to eliminate any unnecessary transmission. Nodes in the broadcast tree are examined in ascending ID order and leaf nodes are ignored because they do not transmit. The non-leaf node with the lowest ID will be the first candidate for restructuring. If the candidate's transmission range can reach a neighbor's node k and its downstream neighbor node j, then the link between node j and k can be eliminated. To obtain the multicast tree, the broadcast tree is pruned by eliminating all transmissions that are not needed to reach the members of the multicast group. More specifically, nodes with no downstream destinations will not transmit, and some nodes will be able to reduce their transmitted power. A similar technique can also be applied to broadcast trees produced by alternative algorithms, such as BLiMST (broadcast link-based MST), resulting in the algorithm of another energy-efficient multicast protocol MLiMST (multicast link-based MST).[2]

11.3.1.1.2 S-REMiT (Distributed Energy-Efficient Multicast) Protocol

Different from the MIP protocol,[2] S-REMiT tries to minimize the total energy cost for multicasting in a distributed manner.[14] The S-REMiT algorithm is divided into two phases. In the first phase, S-REMiT uses a minimum-weight spanning tree (MST) as the initial solution. In the second phase, S-REMiT tries to improve the energy efficiency of the multicast tree by switching some tree nodes from their respective parent nodes to new corresponding parent nodes. In the first phase, the algorithm starts with each individual node as a fragment. Each fragment finds its adjacent edge with minimum weight and attempts to combine with the fragment at the end of the edge. Finally, an MST that combines all the fragments will be constructed in a distributed manner.

The second phase of S-REMiT is organized in rounds in order to reduce the energy consumption of the constructed MST. In each round, the depth-first search (DFS) algorithm is used to pass the S-REMiT token, which gives permission to a node to refine the tree topology, thus improving the energy consumption of the tree. For each node i on the multicast tree T rooted by source s, S-REMiT uses $E_i(T, s)$ to evaluate the energy metric cost of each node i, where

$$E_i(T,s) = \begin{cases} E^T + Kd_i^\alpha & \text{if } i \text{ is the source node;} \\ E^T + Kd_i^\alpha + E^R & \text{if } i \text{ neither the source nor a leaf node;} \\ E^R & \text{if } i \text{ is a leaf node in } T; \end{cases}$$

where E^T denotes a constant that accounts for real-world overheads of electronics and digital processing, E^R denotes the energy cost at the receiver side, K is a constant that depends on the properties of the antenna, and α denotes a constant that depends on the propagation losses in the medium. Let $TEC(T, s)$ denote the total energy cost of nodes in the multicast tree T. In a round, assume that node i in the MST obtains the S-REMiT token. The S-REMiT protocol is described as follows.

S-REMiT multicast protocol:

Step 1: Node i selects a neighboring node x in MST that link $\overline{ix}$ has a highest energy cost tree. Node i then selects a new parent candidate j with the highest positive gain $g_i^{x,j} := ((E_x(T,s) + E_j(T,s)) - (E_x(T',s) + E_j(T',s)))$, which does not result in tree disconnection if node i replaces link $\overline{ix}$ with link $\overline{ij}$. If there is no such node j available, then it sets a token with $flag = false$.

Step 2: Node i replaces link $\overline{ix}$ with link $\overline{ij}$ and notifies nodes j, x, and its neighbors about the replacement.

Step 3: Node i passes the token to next hop node according to the DFS algorithm.

Step 4: If node s gets back the token with $flag = false$, which means there are no energy *gains* in this DFS round, s will request all of the tree nodes to prune the redundant transmissions that are not needed to reach the members of the multicast group from the tree.

Figure 11.9 provides an example of S-REMiT. The execution of Phase I will construct an MST T rooted by node 5, as shown in Figure 11.9. Assume node 1 obtains the S-REMiT token; it selects node 2 from tree neighbors because link $\overline{12}$ is the highest energy cost tree link of node 1. Then, node 1 will try to replace link $\overline{12}$ with some other link to reduce the total energy consumption of the tree. To achieve this goal, node 1 considers those communicative neighbors as candidates to refine the multicast tree. Node 1 selects node 4 from candidates and then evaluates the gain $g_1^{2,4} := (E_1(T,5) + E_2(T,5) + E_4(T,5)) - (E_1(T',5) + E_2(T',5) + E_4(T',5))$, where T' denotes the tree after replacing link $\overline{12}$ by link $\overline{14}$. In case that gain is positive, node 1 will replace link $\overline{12}$ by $\overline{14}$, and then notify its communicative neighbors about this change. Hereafter, node 1 passes the S-REMiT token to node 2 to refine the multicast tree.

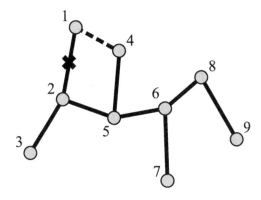

FIGURE 11.9 An example of executing the S-REMiT protocol.

11.3.1.1.3 *Reliable Energy-Efficient Multicast Protocol (RBIP)*

The BIP, BLU, and BLiMST heuristic algorithms for computing energy-efficient trees for unreliable wireless broadcasting and multicasting were presented in Wieselthier et al.[2] In wireless environments, individual links often have high error rates. This might result in reliable delivery potentially requiring one or more retransmissions because the number of retransmissions needed clearly depends on the error rates of the associated links. Banerjee et al.[19] present appropriate modifications to these algorithms (BIP, BLU, and BLiMST) to compute energy-efficient data delivery trees that take into account the costs for necessary retransmissions. Unlike most energy-efficient multicast protocols, this protocol selecting neighbors in the multicast tree is based not only on the link distance, but also on the error rates associated with the link.

Let $p_{i,j}$ denote the packet error probability of link (i, j). The expected number of transmissions to reliably transmit a single packet across this link is $1/(1 - p_{i,j})$. The expected energy requirements to reliably transmit a packet across the link (i, j) is given by $E_{i,j}$ (reliable) $= E_{i,j}/(1 - p_{i,j})$. The computation of a minimum-cost multicast tree will follow three steps as described below:

Step 1: Similar to Prim's algorithm, RBIP greedily adds links to an existing tree such that the incremental cost is minimized. However, because RBIP works on reliable transmission costs, these costs are a function of both the link distance and link error rates. The RBIP algorithm iteratively adds the minimum cost link from the set of eligible links to an existing tree. Hereafter, an energy-efficient broadcast tree has been formed.

Step 2: RBIP prunes those nodes from the tree that do not lead to any multicast group member. This processing is performed in a single post-order traversal.

Step 3: Finally, the sweep operations are performed on the remaining tree in post-order. A node x is transferred from being a child of its parent y to being a child of its grandparent z if doing so reduces overall energy requirements for reliable packet transmission costs.

The chapter also proposes two other reliable multicast protocols (RBLU and RBLiMST), which are the extensions of the protocols BLU and BLiMST, by considering $E_{i,j}$ (reliable) as the link cost in constructing the broadcast tree. Then Step 3 of RBIP can be applied to RBLU and RBLiMST to construct a reliable energy-efficient multicast tree.

11.3.1.2 Multi-Source Energy-Efficient Multicast Protocol

The multi-source multicasting problem is investigated in Wieselthier et al.[2] A multicast protocol G-REMiT is proposed[15] to reduce the energy cost of the constructed tree. G-REMiT consists of two phases. Similar to the S-REMiT protocol, G-REMiT constructs an MST in phase I and then refines the MST in phase II to reduce the energy cost of the constructed multicast tree.

G-REMiT employs an equation to evaluate the weight of each node. The energy consumption of each node in a multicast tree highly depends on the highest energy cost link and the second highest energy cost

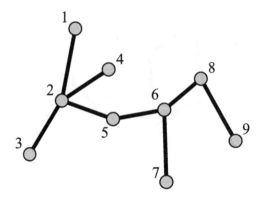

FIGURE 11.10   An example of evaluating the gain.

link. Take the multicast tree shown in Figure 11.10 as an example. Let the first and second highest energy cost links of node 2 be links $\overline{12}$ and $\overline{24}$, respectively. In the case that node 1 is a source node, node 2 will receive the multicast packet from node 1 and then transmit to its neighboring nodes 3, 4, and 5. The power consumption thus depends on the link $\overline{24}$, which is the second highest energy cost link. However, in the case that the source node is some other node rather than 1, node 2 will relay the message to neighboring nodes, including node 1. The power consumption of node 2 thus depends on the energy cost of link $\overline{12}$, which is the highest energy cost link. Thus, the energy cost of each node in MST could be evaluated by the following equation:

$$E_i = w_i[1](d_i[2])^\alpha + (|G| - w_i[1])(d_i[1])^\alpha + |G|E_{elec}$$

where $w_i[1]$ is the number of group nodes that depend on node i using the second furthest transmitted power to forward the multicast packets and G is the set of multicast group nodes; $d_i[j]$ is the distance of the j-th furthest neighboring node of node i; and E_{else} is a constant that accounts for real-world overheads of electronics and digital processing.

In phase I, a link-based minimum weight spanning tree is constructed as the initial tree. Phase II of G-REMiT improves the initial tree by exchanging some existing branches in the initial tree for new branches so that the total energy cost of the tree is lower. The difference in total energy cost of the trees before and after the branch exchange is called *gain*.

The second phase of S-REMiT is organized in rounds. In a round, assume node i in the MST obtains the G-REMiT token. One of the furthest connected neighbors in MST, say x, will be selected by node i. Another node j will be selected from candidate nodes that are communicative neighbors but not tree neighbors of i in the tree. Node i will replace link $\overline{ix}$ by link $\overline{ij}$ if this change improves the gain of power consumption of the tree.

Assume that node i obtains the G-REMiT token. Each node evaluates its energy cost E_i according to parameters that would include its largest link distance and the power consumption of data transmitting and receiving. The following algorithm details the second phase of the G-REMiT multicast protocol *G-REMiT multicast protocol*:

Step 1: Node i selects a farthest connected neighbor node x in the tree. If there is no such node x available, go to Step 6.

Step 2: Node i selects a new candidate node j that is located in its communicative range, to estimate the saving energy cost, called *gain*, after the link changes from $\overline{ix}$ to $\overline{ij}$. The gain $g_i^{x,j} := (E_i + E_x + E_j) - (E_i' + E_x' + E_j')$, where E_i, E_x, E_j respectively, denote the energy cost at nodes i, x, and j in the original tree; and E_i', E_x', E_j' respectively, denote the energy cost at nodes i, x, and j after link change.

Step 3: Node i sends *Path_Exploring(path_gain)* message along $path_{j,i}$. Every node on the $path_{j,i}$ may change the *path_gain* value if its longest link is on $path_{j,i}$, and forwards hop-by-hop along $path_{j,i}$.

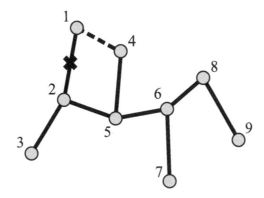

FIGURE 11.11   An example of executing the G-REMiT protocol.

When node i gets back *Path_Exploring*, it checks if *path_gain* is positive. Node i will go back to the first step to select another node x if *path_gain* is negative.

Step 4: Node i changes link $\overline{ix}$ to link $\overline{ij}$.

Step 5: Node i sends path-updating information along $path_{x,i}$ to update the local information of each node. Node i will locally broadcast to nodes located in its communicative range about the link change.

Step 6: Node i passes the token to the next node according to the DFS algorithm.

Figure 11.11 gives an example of G-REMiT. The execution of phase I will construct an MST as shown in Figure 11.11. Assume node 1 obtains the G-REMiT token; it selects node 2 from tree neighbors because node 2 has a largest energy cost. Then, node 1 replaces link $\overline{12}$ with some other link to reduce the total energy consumption of the tree. Node 1 considers those communicative neighbors as candidates to refine the multicast tree. Node 1 selects node 4 from candidates and then evaluates the gain $g_1^{2,4} :=$ $(E_1 + E_2 + E_4) - (E_1' + E_2' + E_4')$, and checks if the path gain of $Path_{41}$ is positive. In the case that both gains are positive, node 1 will replace link $\overline{12}$ by $\overline{14}$, and notify its communicative neighbors about this change. Node 1 then passes the G-REMiT token to node 2 to refine the multicast tree.

This subsection proposes a distributed multicast protocol that dynamically refines the tree topology to reduce the energy consumption of the tree node and extend the network lifetime. However, operations designed for preventing the constructed tree from disconnection also creates a lot of control overheads.

11.3.2   Cluster-Based Power-Efficient Multicast Protocol

Numerous mechanisms have been proposed for reducing packet retransmission. Cluster management has been widely discussed to alleviate the packet flooding phenomenon. A network can be partitioned into several clusters, each consisting of a header, gateway (optional), and members. The information from two clusters can be directly exchanged by their headers if their distance is smaller than the communicative range, or it can be relayed by gateway, which is a common member shared by more than one cluster. Cluster headers and gateways can be treated as the nodes of the backbone of the network and are responsible for relaying broadcast (or multicast) packets to all nodes (or all multicast members), thus preventing large amounts of packet retransmission and saving power consumption.

Tang et al.[16] applied the existing ODMRP[17] to cluster topology to achieve energy-efficient multicast communication. First, a clustering protocol was proposed for constructing a cluster where all nodes are capable of communicating with each other within that cluster. After executing the clustering algorithm, the network is partitioned into a set of disjoint clusters with a cluster head in each cluster. The cluster heads can be thought of as supernodes and they form a supernode network topology. The adaptation of ODMRP is proposed for the supernode topology. For balancing the energy consumption, nodes in the cluster take turns becoming cluster headers using some round-robin schedule. The work in Tang et al.[16]

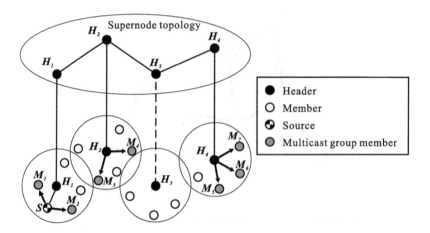

FIGURE 11.12   Adaptation of ODMRP.

takes advantages of balancing energy consumption from cluster management and the good multicast features of the existing multicast protocol to develop a power-efficient multicast protocol.

Based on the constructed supernode topology, the work in Tang et al.[16] proposes an adaptation scheme using the existing ODMRP to achieve the goal of energy conservation multicasting. Packets flow from the sender to its cluster header, then along the supernode topology, and finally get disseminated within the clusters. The following gives an example to illustrate the adaptation scheme. In Figure 11.12, a multicast source node S intends to send a multicast packet to receivers M_1, M_2, M_3, M_4, M_5, M_6, and M_7. Node S first broadcasts the message to all nodes within the same cluster. On receiving the multicast packets, header H_1 then forwards the packets to headers H_2 and H_4 along the supernode topology. The multicast packets thus can be received by all receivers from their headers.

The multicast data transmission highly relies on the supernode topology. Nodes in a cluster may take turns playing the header role, balancing the power consumption of nodes in the same cluster. However, supernode election in the clustering process does not take into consideration the energy cost among headers. This may introduce a large energy cost for transmitting multicast packets on supernode topology.

11.4   Conclusions and Future Works

Mobile ad hoc networks comprise mobile nodes that are power constrained because they operate with restricted battery power. Energy consumption is one of the most important issues in ad hoc networks. Selection of nodes to be active and control of the emitted transmission power are the most important issues in designing an energy-efficient protocol in MANETs. Broadcast and multicast routings are important operations in the network layer. Developing energy-efficient broadcast and multicast routing protocols reduces the power consumption of nodes and hence improves network lifetime.

This chapter reviewed existing, important energy-efficient broadcast and multicast protocols. Table 11.1 summarizes all reviewed energy-efficient broadcast protocols in this chapter. According to their different mechanisms, the broadcast routing protocols are categorized into two families: tree-based and probability-based approaches. The tree-based broadcast routing protocols[1–9] construct a *minimum-energy* broadcast tree by greedily selecting some nodes from networks and controlling their power level to maintain a broadcast tree with minimal energy consumption. By applying the probability-based approach, another family of protocols[10–13] was developed to reduce power consumption, alleviate the broadcast storm situation, or balance the power consumption. In addition to the study of broadcast routing protocols, this chapter also investigated some important energy-efficient multicast protocols. Table 11.2 summarizes all reviewed energy-efficient multicast protocols. According to the constructed topology, existing power-efficient multicast protocols are classified into tree-based and cluster-based protocols. The tree-based multicast

TABLE 11.1　Summary of Energy-Efficient Broadcast Protocols

Property Protocol [Ref.]	Tree-Based Approach				Probabilistic Approach	
	Centralized Algorithm	Distributed Algorithm	Integer Programming	Static Network	Broadcast Storm	Power-Balance
Centralized BIP [1]	•					
EWMA [3]	•					
IP [4]	•					
Minimum-Energy Broadcast in Static MANET [5]	•			•		
MST-Based Heuristic in Static MANET [6]	•			•		
Weighted Steiner Tree-Based [7]	•			•		
DIST-BIP [8]		•				
RBOP [9]		•				
Alleviating "Broadcast Storm Problem" [10–12]					•	
Power-Balance Broadcast Protocol [11]						•

protocols[2,4,15,19] consider the power consumption issue and obtain an energy-efficient multicast tree by applying refining and pruning rules to the existing energy-efficient broadcast tree. Another approach, which uses cluster topology to achieve the goal of energy-efficient multicasting, was also investigated in this chapter.

Numerous protocols address the broadcast and multicast problems with the goal of reduced power consumption, but most existing approaches were developed under the assumption of low mobility. Therefore, some future works should include the following:

1. A possible future work is how to design energy-efficient broadcast/multicast tree maintenance mechanisms with a mobility-tolerant capability. Because an ad hoc network is characterized by a highly dynamic topology, the impact of mobility should be incorporated into the protocol design, especially for some applications of wireless sensor networks (e.g., the object-tracking problem). Improved performance can be obtained by jointly considering the node failure, node move, and node join situations.

2. A major challenge in protocol design in MANETs is how to develop reliable broadcast and multicast routing protocols to simultaneously address the energy consumption cost and the number of packet retransmissions.

3. An interesting topic for future research is how to investigate the energy-efficient broadcast and multicast routing protocols by fully adopting the location information. Several algorithms are known to provide a node's location information in ad hoc and sensor networks. Location information is likely to be useful in calculating the node mobility and the power level required to maintain the constructed energy-efficient topology.

TABLE 11.2　Summary of Energy-Efficient Multicast Protocols

Property Protocol [Ref.]	Topology	Pruning or Refining Rules	Source in Tree	Characteristics
MIP [2]	Tree	Yes	Single	Power-efficient
S-REMiT [14]	Tree	Yes	Single	Power-efficient
RBIP [19]	Tree	No	Single	Reliable and power-efficient
G-REMiT [15]	Tree	Yes	Multiple	Group communication and power-efficient
CBMP [16]	Cluster/tree	Yes	Multiple	Clustering and power-efficient

4. In addition, the use of directional antennas may benefit from the elimination of unnecessary interferences and the less power consumption by focusing the transmitting power in a specific direction. Involving directional antennas and location information in the design of broadcast and multicast routing protocols expectably provides advantages of increasing the network lifetime.

Consequently, how to utilize the location information, along with consideration of mobility, unreliable transmission, and the use of directional antenna, will possibly be the next challenge in the design of energy-efficient broadcast and multicast protocols.

References

1. R.J. Wilson and J.J. Watkins, *Graphs: An Introductory Approach: A First Course in Discrete Mathematics,* John Wiley & Sons, New York, 1990.
2. J. Wieselthier, G. Nguyen, and A. Ephremides, On the construction of energy-efficient broadcast and multicast trees in wireless networks, in *Proc. IEEE Infocom'2000,* Tel Aviv, Israel, 2000, 585–594.
3. M. Cagalj, J.-P. Hubaux, and C. Enz, Minimum-energy broadcast in all-wireless networks: np-completeness and distribution issues, in *Proc. ACM MobiCom 2002,* Atlanta, GASept. 2002, 172–182.
4. A.K. Das, R.J. Marks, M. El-Sharkawi, P. Arabshahi, and A. Gray, Minimum power broadcast trees for wireless networks: integer programming formulations, in *Proc. IEEE Infocom'2003,* 2, 1001–1010, 2003.
5. P.-J. Wan, G. Calinescu, X.-Y. Li, and O. Frieder, Minimum-energy broadcast routing in static ad hoc wireless networks, in *Proc. IEEE Infocom'2001,* 2, 1162–1171, April 2001.
6. A.E.F. Clementi, G. Huiban, G. Rossi, Y.C. Verhoeven, and P. Penna, On the approximation ratio of the MST-based heuristic for the energy-efficient broadcast problem in static ad-hoc radio networks, in *Proc. PDPS'03: Parallel and Distributed Processing Symposium,* pp. April 22–26, 2003, pp. 222–229.
7. D. Li, X. Jia, and H. Liu, Energy efficient broadcast routing is static ad hoc wireless networks, *IEEE Trans. on Mobile Comput.,* 3(1), 144–151, Apr.–June 2004.
8. J. Wieselthier, G. Nguyen, and A. Ephremides, Distributed algorithms for energy-efficient broadcasting in ad hoc networks, in *Proc. MILCOM 2002,* 2, 7–10, October 2002.
9. J. Cartigny, D. Simplot, and I. Stojmenovic, Localized minimum-energy broadcasting in ad-hoc networks, in *Proc. IEEE Infocom'2003,* 3, 2210–2217, 2003.
10. Y.-C. Tseng, S.-Y. Ni, Y.-S. Chen, and J.-P. Sheu, The broadcast storm problem in a mobile ad hoc network, *ACM Wireless Networks,* 8(2), 153–167, March 2002.
11. Y. Sasson, D. Cavin, and A. Schiper, Probabilistic broadcast for flooding in wireless mobile ad hoc networks, in *Proc. IEEE WCNC'2003: Wireless Communications and Networking,* 2, 1124–1130, March 2003.
12. M.-T. Sun, W. Feng, and T.-H. Lai, Location aided broadcast in wireless ad hoc networks, in *Proc. IEEE GLOBECOM'2001: Global Telecommunications Conference,* 5, 2842–2846, Nov. 2001.
13. J.-P. Sheu, Y.-C. Chang, and H.-P. Tsai, Power-balance broadcast in wireless mobile ad hoc networks, in *Proc. EW2004: European Wireless Conference,* Spain, Feb. 2004.
14. B. Wang and Sandeep K.S. Gupta, S-REMiT: a distributed algorithm for source-based energy efficient of multicasting in wireless ad hoc networks, in *Proc. IEEE 2003 Global Communications Conference (GLOBECOM),* San Francisco, CA, 6, 3519–3524, December 2003.
15. B. Wang and Sandeep K.S. Gupta, G-REMiT: an algorithm for building energy efficient of multicast trees in wireless ad hoc networks, in *Proc. Second IEEE Int. Symp on Network Computing and Applications,* Cambridge, MA, April 2003, 265–272.
16. C. Tang, C.S. Raghavendra, and V. Prasanna, Energy efficient adaptation of multicast protocols in power controlled wireless ad hoc networks, in *2002 International Symposium on Parallel Architectures, Algorithms and Networks,* May 2002.

17. S. Lee, W. Su, and M. Gerla, *On demand multicast routing protocol in multihop mobile wireless networks*, ACM/Baltzer Mobile Networks and Applications, 2000.

18. H. Lim and C. Kim, Multicast tree construction and flooding in wireless ad hoc networks, in *Proc. Third ACM Int. Workshop on Modeling, Analysis and Simulation of Wireless and Mobile Systems (MSWiM)*, August 2000.

19. S. Banerjee, A. Misra, J. Yeo, and A. Agrawala, Energy-efficient broadcast and multicast trees for reliable wireless communication, in *IEEE Wireless Communications and Networking Conf. (WCNC)*, New Orleans, LA, March 2003.

20. C.-C. Chiang, M. Gerla, and L. Zhang, Adaptive shared tree multicast in mobile wireless networks, in *Proc. IEEE Globecom'98*, Sydney, Australia, November 1998, 1817–1822.

21. S.J. Lee, W. Su, J. Hsu, M. Gerla, and R. Bagrodia, A performance comparison study of ad hoc wireless multicast protocols, in *Proc. IEEE INFOCOM*, Tel Aviv, Israel, March 2000, 565–574.

22. J.E. Wieselthier, G.D. Nguyen, and A. Ephremides, On the construction of energy-efficient broadcast and multicast tree in wireless networks, in *Proc. IEEE INFOCOM*, Tel Aviv, Israel, March 2000, 585–594.

23. R. Gallager, P.A. Humblet, and P.M. Spira, A distributed algorithm for minimum weight spanning trees, in *ACM Trans. Programming Lang. & Systems*, v5(1), 66–77, January 1983.

12

Linear Programming Approaches to Optimization Problems of Energy Efficiency in Wireless Ad Hoc Networks

Hai Liu

Xiaohua Jia

12.1 Introduction

Linear programming/integer linear programming (LP/ILP) is a powerful and remarkably versatile tool that is widely applied in business activities, industry manufacturing, military activities, information techniques, etc. The techniques of linear programming, for the most part, have been developed over the past four to five decades. There are three basic steps in the linear programming model of formulations: (1) determination of the decision variables, (2) formulation of objective, and (3) formulation of the constraints.[12] Network optimization problems are a class of important applications of linear programming. Typically, the min-cut max-flow problem, the shortest path problem, and the minimum cost-flow problem can be formulated as linear programming problems.[11,12] Furthermore, in addition to the traditional *Simplex* and *Branch and*

Bound methods, there are some powerful software packages that can be used to compute LP/ILP, such as *Mathematic, Matlab, LPSolver*, etc.

Mobile Ad Hoc Networks (MANETs) are wireless networks consisting of entirely mobile nodes that communicate without using base stations. Nodes in these networks act as routers, as well as communication end-points. Rapid changes of connectivity, network partitioning, higher error rates, collision interference, and bandwidth and power constraints together pose new challenges for this type of network. In recent years, sensor networks have also received significant interest from the research community. Sensor networks comprise a new family of wireless networks and are different from traditional networks such as cellular networks or MANETs. A sensor network is composed of a large number of small sensor nodes, and energy efficiency is a more important issue in this kind of network.

This chapter introduces some applications of linear programming (integer linear programming) on the optimization problems in wireless networks. We focus on the most representative applications of LP/ILP in wireless networks. The remainder of the chapter is organized as follow. In Section 12.2 we discuss energy efficiency routing problems. Section 12.3 focuses on broadcast/multicast routing problems. Data extraction and gathering problems in sensor networks are discussed in Section 12.4. QoS topology control problems are first introduced in Section 12.5; and we discuss both traffic splittable and traffic non-splittable cases. We make conclusions in Section 12.6.

12.2   Energy Efficiency Routing

Energy efficiency routing is one of the most important issues in MANETs and sensor networks, because the nodes of wireless networks are usually battery-powered small devices, such as a laptop PC, PDA, etc. There are three basic metrics that are commonly included in energy-efficient routing mechanisms. They are (1) minimum energy routing,[19,20] (2) max-min routing,[3,20] and (3) minimum cost routing.[21−23]

The lifetime of MANETs or sensor networks often depends on the node with minimum residual energy in the network. Minimum-energy metric routing may not maximize network lifetime. This is because the nodes' residual energy (remaining battery capacity) is not taken into account. That is, some nodes on the minimum-energy routes will suffer early failure due to their heavy forwarding load. The max-min metric explicitly avoids this problem by selecting the route that maximizes the minimum residual energy of any node on the route. Routes selected using max-min metrics may be longer or have greater total energy consumption than the minimum-energy route. This increase in per-packet energy consumption could result in reducing the network lifetime in the long term. Minimum-cost routing minimizes the total *cost* of routing from source toward destination. It is often used to find a best trade-off among several parameters.

Energy-efficient routing in ad hoc disaster recovery networks was discussed by Zussman and Segall.[3] A network is modeled by a directed graph $G = (N, L)$, where N denotes the collection of nodes and L denotes the collection of directional links. A node could be a *badge* (sensor), a *receiver* (relay node and the collection of relay nodes is denoted by R), or the central unit (referred to as the destination and denoted by d). All nodes transmit at a constant power level. It is assumed that each node has an initial energy E_i, and the energy required by node i to transmit an information unit is denoted by e_i. It is further assumed that the ratio between the rate at which information is generated at sensor node i and the maximal possible flow on a link connecting sensors is denoted by r_i ($0 \leq r_i < 1$). The problem is to determine the data flow on each link such that all data generated by sensors can be forwarded to the destination and the network lifetime is maximized, where the network lifetime under a given flow is the time until the first battery drains out.

Let F_{ij} denote the average flow on link (i, j) ($F_{ij} \geq 0, \forall (i, j) \in L$), and let f_{ij} denote the ratio between F_{ij} and the maximal possible flow on a link connecting sensors ($0 \leq f_{ij} < 1$). The energy-efficient routing problem can be formulated as follows.

Objective: Maximize the network lifetime:

$$\max T = \max \left[\min_{i \in N} \frac{E_i}{e_i \sum_{j \in Z(i)} f_{ij}} \right] \tag{12.1}$$

Subject to: $f_{ij} \geq 0, \quad \forall (i,j) \in L$ $\tag{12.2}$

$$\sum_{k \in Z(i)} f_{ki} + r_i = \sum_{j \in Z(i)} f_{ij}, \quad \forall i \in N - \{R, d\} \tag{12.3}$$

$$\sum_{k \in Z(i)} f_{ki} = \sum_{j \in Z(i)} f_{ij}, \quad \forall i \in R \tag{12.4}$$

$$\sum_{k \in Z(i)} f_{ki} + \sum_{j \in Z(i)} f_{ij} \leq 1, \quad \forall i \in N - \{R, d\} \tag{12.5}$$

where $Z(i)$ denotes the collection of neighboring nodes of node i. Constraints 12.2, 12.3, and 12.4 are the usual flow conservation constraints. Constraint (12.5) means that the total flow through a node cannot exceed the maximal sensor node capacity, that is, the maximal data rate of a sensor.

Note that there is a nonlinear function (min) in (12.1); the above formulation is not a linear programming formulation. To deal with it, Zussman and Segall[3] used $\bar{f}_{ij}$, which denotes the amount of information transmitted from node i to node j until time T ($\bar{f}_{ij} = f_{ij} \times T$), to replace f_{ij}, where T is the network lifetime. Then, the problem was formulated as follows.

Objective: Maximize the network lifetime:

$$\max T \tag{12.6}$$

Subject to : $\bar{f}_{ij} \geq 0, \quad \forall (i,j) \in L$ $\tag{12.7}$

$$\sum_{k \in Z(i)} \bar{f}_{ki} + r_i.T = \sum_{j \in Z(i)} \bar{f}_{ij}, \quad \forall i \in N - \{R, d\} \tag{12.8}$$

$$\sum_{k \in Z(i)} \bar{f}_{ki} = \sum_{j \in Z(i)} \bar{f}_{ij}, \quad \forall i \in R \tag{12.9}$$

$$e_i \sum_{j \in Z(i)} \bar{f}_{ij} = E_i, \quad \forall i \in N - \{R, d\} \tag{12.10}$$

$$\sum_{k \in Z(i)} \bar{f}_{ki} + \sum_{j \in Z(i)} \bar{f}_{ij} \leq T, \quad \forall i \in N - \{R, d\} \tag{12.11}$$

Now the problem is formulated as a linear programming problem, and can be solved by traditional algorithms (e.g., Simplex). These algorithms, however, cannot be easily modified to allow distributed implementation. Based on this new formulation, constraint 12.11 was first removed to make the problem become a concurrent max-flow problem, which is a multi-commodity flow problem.[7-9] Then, an iterative algorithm was proposed to search suitable solutions such that the error of the removed constraint is less than a given tolerance, and an upper bound on the network lifetime was analyzed.

Similar to Zussman and Segall,[3] Sankar and Liu[1] were also able to maximize the network lifetime. Different from the many-to-one traffic model in Ref. 3, it is assumed that there is a set of source-destination pairs in the network and each pair has a throughput requirement. The definition of problem is as follows.

The network is modeled as a graph, with N nodes representing wireless devices and M edges representing the wireless links between nodes. Associated with each node i is a quantity E_i, representing the initial energy reserve of the device. Each edge ij has a cost e_{ij}, which is the energy required to transmit one packet of data across the corresponding link. Assume that there are K source-destination pairs in the network, and the number of packets per second that must be routed between connection $c(c = 1, 2, \ldots, K)$ is Q_c. The problem is to route packets such that the network lifetime is maximized while the throughput requirements are satisfied. Let $\hat{f}_{ij}^c$ denote the total number of packets for each connection c transmitted from each node i to node j over the lifetime of the network T. This problem can be formulated as follows.

Objective: Maximize the network lifetime:

$$\max T \tag{12.12}$$

$$\text{Subject to: } \hat{f}_{ij}^c \geq 0, \quad \forall i, j, c \tag{12.13}$$

$$\sum_{j,c} e_{ij} \hat{f}_{ij}^c \leq E_i, \quad \forall i \tag{12.14}$$

$$\sum_{j} \hat{f}_{ij}^c - \sum_{k} \hat{f}_{ki}^c = \begin{cases} Q_c T & i \text{ a source for } c \\ 0 & \text{otherwise} \end{cases} \quad \forall i, c \tag{12.15}$$

Constraint 12.14 is due to the finite power supplies at the nodes, and constraint 12.15 represents the throughput requirements at the sources and the conservation constraints at other nodes.

From the above formulations, we can see that routing problems in wireless networks can be generally interpreted as flow problems. The requirement that incoming packets arrive at the same rate as outgoing packets leave is equivalent to the flow conservation constraint that at any intermediate node, the amount of incoming flow equals the amount of outgoing flow. The sources and destinations of the routing problem become sources and sinks in the flow problem. Thus, the maximum lifetime routing problem can also be treated as the multi-commodity problem. Sankar and Liu[1] further proposed a distributed approximation algorithm, which is beyond the scope of this chapter.

Different from the work of Sankar and Liu[1] and Zussman and Segall,[3] where all nodes have uniform fixed power level, the energy-conserving routing problem addressed by Chang and Tassiulas[6] assumed that each node can adjust its power level within a certain range that determines the set of possible one hop away neighbors.

A wireless ad hoc network is modeled as a directed graph $G(N, A)$, where N is the set of all nodes and A is the set of all directed links (i, j) where $i, j \in N$. Let S_i be the set of all nodes that can be reached by node i with a certain power level in its dynamic range. Link (i, j) exists if and only if $j \in S_i$. Let each node i have the initial battery energy E_i, and let $Q_i^{(c)}$ be the rate at which information is generated at node i belonging to commodity $c \in C$, where C is the set of all commodities. Assume that the energy required for node i to transmit an information unit to its neighboring node j is e_{ij}, and $q_{ij}^{(c)}$ is the amount of information of commodity c that is transmitted from node i to node j during the network lifetime T. The goal is to find the flow that maximizes the system lifetime. (Note that the power level of a node depends on whether there exist flows on the neighboring links of this node.) The problem can be formulated as follows.

Objective: Maximize the network lifetime:

$$\max T \tag{12.16}$$

$$\text{Subject to: } q_{ij}^{(c)} \geq 0, \quad \forall i \in N, \forall j \in S_i, \forall c \in C, \tag{12.17}$$

$$\sum_{j \in S_i} e_{ij} \sum_{c \in C} q_{ij}^{(c)} \leq E_i, \quad \forall i \in N, \tag{12.18}$$

$$\sum_{j:i \in S_j} q_{ji}^{(c)} + T Q_i^{(c)} = \sum_{k \in S_i} q_{ik}^{(c)}, \quad \forall i \in N - D^{(c)}, \forall c \in C, \tag{12.19}$$

where $D^{(c)}$ is the set of destination nodes for each commodity c. Constraint 12.18 denotes that the energy consumed cannot exceed the initial battery energy, and constraint 12.19 means that for each non-destination node, the amount of received packets and its own generating packets should equal the amount of packets sent out. The linear programming formulation given above can be viewed as a variation of the conventional maximum flow problem with node capacities.[18]

To solve the problem, Chang and Tassiulas[6] proposed a local algorithm called the flow redirection algorithm, which is amenable to distributed implementation. The main idea of the algorithm is to balance the energy consumption among the nodes in proportion to the nodes' energy reserves. There are three steps. First, each node i for commodity c will determine two paths to the destination by finding the shortest cost path (as augmentation path). Second, compare the residual of nodes on the found path; if the lifetime of node i is the minimum over all nodes in the subnetwork consisting of node i and all its downstream nodes, the lifetime of node i should be increased by redirecting a flow at node i in the direction where the required transmission energy per information unit is smaller. Otherwise, the process is recursively run in the subnetwork consisting of node i and all its downstream nodes. Finally, redirect the flow according to the second step.

It is shown that when all nodes have a fixed power level (i.e., $e_{ij} = e_i, \forall j \in S_i$), the node capacities are given by

$$\sum_{j \in S_i} \sum_{c \in C} q_{ij}^{(c)} = E_i/e_i, \quad \forall i \in N, \tag{12.20}$$

and the problem is reduced to a maximum flow problem and the flow redirection algorithm leads to the optimal solution. When there are multiple power levels, the simulation results showed that the achievable lifetime is close to optimal most of the time.

12.3   Broadcast/Multicast Routing

For a given source node, the minimum power broadcast/multicast problem in MANETs is to communicate to all/some remaining nodes, such that the overall transmission power is minimized. Some heuristic algorithms for constructing the minimum power tree in wireless networks have been proposed, such as BIP/MIP (Wieselthier et al.[15]), G-REMiT (B. Wang et al.[16]), node weighted Steiner tree based heuristic (Deying Li et al.[14]), and an internal nodes based broadcasting procedure (Stojmenovic et al.[17]). While the performances of the above algorithms can certainly be compared among themselves, in the absence of any optimal solution procedure, it is impossible to judge the quality of the solutions with respect to the optimal. Thus, Das et al.[4] presented three different integer programming models that can be used for an optimal solution of the minimum power broadcast/multicast problem in MANETs.

Let V be the set of all nodes in the network and $|V| = N$, and D the set of intended destination nodes. Y_i denotes the transmitter power levels at node i and X_{ij} is a binary variable such that $X_{ij} = 1$ if the transmission $i \rightarrow j$ is used in the final solution and 0 otherwise. P_{ij} is the transmission power required from i to j.

From a Traveling Salesman aspect, implicit visitations in broadcast/multicast trees can be alternately interpreted as the salesman being allowed to make any number of actual trips outside the city, with the condition that the cost he incurs is the maximum of the individual costs of the trips he makes outside the city (that is, for each city, only the farthest trip to one of its neighboring cities has contribution to the cost).

For example, in Figure 12.1, assume that city 3 is the source, and the solid lines indicate the costliest paths outside of any city. Suppose the optimal minimum power broadcast tree for the given example is: $\{3 \rightarrow 4, 4 \rightarrow 6, 6 \rightarrow 8, 6 \rightarrow 7\}$. This solution was interpreted as follows. The salesman

1. Makes three actual trips out of city 3, to cities 1, 2, and 4. Charged only for the trip to 4.
2. Makes two actual trips out of city 4, to cities 5 and 6. Charged only for the trip to 6.
3. Makes one actual trip out of city 5, to city 7. Charged for the trip to 7.
4. Makes one actual trip out of city 6, to city 8. Charged for the trip to 8.
5. Makes no trips out of cities 1, 2, 7, and 8.

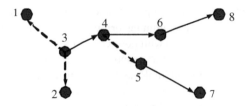

FIGURE 12.1   An example eight-node network to illustrate an alternate view of implicit visitation.

From this point of view, the minimum power broadcast/multicast problem can be formulated as follows.

Objective: Minimize the overall transmission power of a broadcast/multicast tree.

$$\min \sum_{i=1}^{N} Y_i \tag{12.21}$$

$$\text{Subject to: } Y_i - \sum_{j=1}^{N} P_{ij} X_{ij} = 0, \quad \forall i \in V, i \neq j \tag{12.22}$$

$$\sum_{j=1}^{N} X_{ij} = 1, \quad i = source, i \neq j \tag{12.23}$$

$$\sum_{j=1}^{N} X_{ij} \leq 1, \quad \forall i \in \{V \backslash source\}, i \neq j \tag{12.24}$$

$$X_{ij} - \sum_{k=1}^{N-1} X_{ijk} = 0, \quad \forall(i, j) \in V, i \neq j \tag{12.25}$$

$$\sum_{j=1}^{N} X_{ijk} = 1, \quad i = source, i \neq j, k = 1 \tag{12.26}$$

$$\sum_{i=1}^{N} \sum_{j=1}^{N} X_{ijk} = 1, \quad i \neq source, i \neq j, k = 1 \tag{12.27}$$

$$\sum_{j=1}^{N} X_{ijk} - \sum_{p=1}^{k-1} \sum_{m=1}^{N} \sum_{n=1}^{N} R_{mn(i)} X_{mnp} \leq 0, \quad \forall i \in \{V \backslash source\}, i \neq j, m \neq n, 2 \leq k \leq k^{MAX} \tag{12.28}$$

$$\sum_{m=1}^{N} \sum_{n=1}^{N} X_{mnk} \leq 1, \quad m \neq n, 2 \leq k \leq k^{MAX} \tag{12.29}$$

$$\sum_{j=1}^{N} X_{ijk} - \sum_{p=1}^{k-1} \sum_{m=1}^{N} \sum_{n=1}^{N} R_{mn(i)} X_{mnp} \leq 0, \quad \forall i \in \{V \backslash source\}, i \neq j, m \neq n, 2 \leq k \leq k^{MAX}$$

$$\tag{12.30}$$

$$\sum_{m=1}^{N} \sum_{n=1}^{N} R_{mn(i)} X_{mn} \geq 1, \quad \forall i \in D, m \neq n \tag{12.31}$$

$$X_{ijk} \geq 0, \quad integers, \forall(i, j) \in V, i \neq j, 1 \leq k \leq k^{MAX} \tag{12.32}$$

where X_{ijk} equals 1 if the kth transmission in the final solution is $i \rightarrow j$ and 0 otherwise. Note that no upper bound is required to be declared for the integer variables X_{ijk} as it is set implicitly by constraints 12.23

through 12.25. k^{MAX} denotes the maximum number of transmission in the network. For an N-node network, there can be at most $N - 1$ steps (transmissions) in the solution. $R_{mn(p)}$ is called the reward matrix and is computed as follows:

$$R_{mn(p)} = \begin{cases} 1, & \text{if } P_{mp} \leq P_{mn} \\ 0, & \text{otherwise} \end{cases} \tag{12.33}$$

The authors further proposed two more formulations for the minimum power broadcast/multicast problem from different views. One utilizes the alternate view of the traveling salesman model, while the other is built upon a network flow model.

12.4 Data Extraction and Gathering in Sensor Networks

In many sensor network applications involving environmental monitoring in remote locations, planetary exploration, and military surveillance, the efficiency of data extraction and gathering is very important. In practical applications, the information from the entire network is usually extracted *en masse* after a prolonged period of sensing and local storage. Due to the limited batteries of sensor nodes, it is often impossible to collect all the data stored in the network. How to maximize the data collection from such an energy-limited, store-and-extract wireless sensor network becomes a new challenge.

The maximum data extraction problem was discussed by Sadagopan and Krishnamachari.[2] It is an analog of the maximum lifetime problem[1,3,6] discussed in Section 12.2. However, this problem introduces an additional element of "data awareness" that must be considered in addition to "energy-awareness." Consider a scenario where some sensors that are deployed in a remote region have completed their sensing task and have some locally stored data. Let N denote the total number of sensors, and T the target (sink) to which the data is to be sent. Let $D_{\max}^j$ be the amount of data collected by sensor j, where $D_{\max}^j > 0$; $e_{\max}^j$ is the residual energy of sensor j; and d_{ij} is the Euclidean distance between sensors i and j. It is assumed that all sensors have the same communication range R. Thus, R results in a graph $G(V, E)$, where $|V| = N$, an edge $(i, j) \in E$ iff $d_{ij} \leq R$. It is further assumed that Tx_{ij} is the energy consumed in transmitting a single byte from sensor i to j and is proportional to the d_{ij}^2; that is:

$$Tx_{ij} = \beta_t d_{ij}^2, \quad \beta_t > 0 \tag{12.34}$$

Let β_r denote the energy consumed at sensor j for receiving a single byte of data from sensor i, and f_{ij} the amount of data transmitted from node i to node j.

For the ease of modeling, Sadagopan and Krishnamachari[2] added a fictitious source S with arbitrary location, such that there exists an edge from S to every other node in V, except for T. Also added was an edge from T to S. Then the new graph is $G'(V', E')$, where $V' = V \cup \{S\}$ and $E' = E \cup \{(S, i)\} \cup \{(T, S)\}$, where $i \in V, i \neq T$. Then the maximizing data extraction problem can be formulated as follows.

Objective: Maximize the data collection:

$$\max f_{T,S} \tag{12.35}$$

Subject to:

$$\sum_{j=1}^{N} f_{ji} - \sum_{j=1}^{N} f_{ij} = 0 \tag{12.36}$$

$$\sum_{(i,j) \in E} f_{ij} d_{ij}^2 + \sum_{(j,i) \in E} f_{ji} \leq e_i, \quad \forall i \neq S, T \tag{12.37}$$

$$f_{Si} \leq D_{\max}^i, \quad \forall i \neq S, T \tag{12.38}$$

$$f_{ij} \geq 0, \quad \forall (i, j) \in E' \tag{12.39}$$

where

$$\beta = \frac{\beta_t}{\beta_r} \tag{12.40}$$

$$e_i = \frac{e_{max}^i}{\beta_r}, \quad \forall i \neq S, T \tag{12.41}$$

Constraint 12.36 is the usual flow conservation constraint. Constraint 12.37 denotes that the total energy consumed of a node in receiving and transmitting cannot exceed the residual energy of the node.

By adding the fictitious source and its associated links, the problem becomes one of maximizing the circulation of the commodities from S to T and back. Subsequently, the authors gave the *dual* formulation of the above LP, and then proposed an iterative algorithm that is a modification of the Garg-Koenemann algorithm.[10]

The optimal role assignment of data gathering in sensor networks was considered by Bhardwaj and Chandrakasan.[5] There are some sources located in a specified region of observation (R). At any given instant, nodes in a sensor network can be classified as *live* or *dead*, depending on whether or not they have any energy left. By assuming different roles, live nodes collaborate to ensure that whenever a source resides in R, it is sensed using a minimum specified number of sensors and the resultant data relayed to an energy-unconstrained *base station* (B). In the collaborative model, the authors assumed that a role is composed of one or more of the following sub-roles:

- Sensor: the node observes the source.
- Relay: the node simply forwards the received data onward without any processing.
- Aggregator: the node receives two or more raw data streams and then aggregates them into a single stream.

It is assumed that a role assignment is termed *Feasible Role Assignment* (FRA) if it:

- Results in data being relayed from the minimum specified number of sensors to the base station, and,
- Has no redundancy; that is, no sub-role in any node can be deleted while still obeying the previous property.

Consider the example network in Figure 12.2. If it is assumed that node 1 is the sensor, there are four FRAs, given by:

$$F = \{f_1, f_2, f_3, f_4\}, \text{ where,}$$
$$f_1 : 1 \rightarrow B$$
$$f_2 : 1 \rightarrow 2 \rightarrow B$$
$$f_3 : 1 \rightarrow 3 \rightarrow B$$
$$f_4 : 1 \rightarrow 2 \rightarrow 3 \rightarrow B$$

There are, in fact, eight FRAs, but the self-crossing FRAs are ignored, such as $1 \rightarrow 3 \rightarrow 2 \rightarrow B$, because they are clearly sub-optimal in collinear networks. Note that there can be no aggregations due to only one sensor.

Let f denote a FRA and F the set of all feasible role assignments. The power dissipated in node i when FRA f is being sustained by the network is denoted by $p(i, f)$. Let t_i denote the time for which FRA f_i is

$B \qquad\qquad 3 \qquad\qquad 2 \qquad 1$

FIGURE 12.2   A collinear three-node network with node 1 as the assigned sensor.

sustained, and e_i the initial energy of node i. N is the number of nodes in the network. Determining the optimal collaborative strategy can be formulated as follows.

Objective: Maximize the data collection:

$$\max t = \sum_{i=1}^{|F|} t_i \qquad (12.42)$$

Subject to:

$$t_j \geq 0, \qquad 1 \leq j \leq |F| \qquad (12.43)$$

$$\sum_{j=1}^{|F|} p(i, f_j) t_j \leq e_i, \qquad 1 \leq i \leq N \qquad (12.44)$$

The authors showed that a class of role assignment problems permits a transformation to linear programming problems based on network flows that can be solved in polynomial time. But not all role assignment problems can be similarly transformed, only several ones of practical importance — pure routing, non-hierarchical and constrained hierarchical aggregation, multiple or moving sources, sources with specified trajectories — are amenable to such a transformation.

12.5   QoS Topology Control

There are many modern network applications that require QoS provisions in ad hoc networks, such as transmission of multimedia data, real-time collaborative work, and interactive distributed applications. Topology control in MANETs is to allow each node in the network to adjust its transmitting power (i.e., to determine its neighbors) so that a *good* network topology can be formed. Extensive research has been done on QoS routing[24,25] and topology control.[26-28] To the best of our knowledge, however, the construction of a network topology that can overall meet QoS requirements was first studied by Jia et al.[13]

This section focuses on the problem of QoS topology control in ad hoc wireless networks. Given a set of wireless nodes in a plane where nodes have different maximal transmitting powers and bandwidth capacities, and given QoS requirements between node-pairs, the objective is to find a network topology that can meet the QoS requirements and in which the maximal power utilization ratio of nodes is minimized. The QoS requirements of concern are traffic demands (bandwidth) and maximum delay bounds (in terms of hop counts) between end-nodes at the application level. The power utilization of a node is the actual power consumption divided by the energy capacity of the node. Minimizing the maximal power utilization of nodes would balance the power consumption of all nodes, which would avoid the situation that some nodes run out of energy faster than others. The lifetime of the network can thus be prolonged.

The network was modeled by a directed graph $G = (V, E)$, where V is the set of n nodes and E is a set of directed edges. Each node i has a bandwidth capacity B_i and a maximal level of transmitting power P_i. The bandwidth of a node is shared for both transmitting and receiving signals. That is, the total bandwidth of the outgoing edges plus the total bandwidth of incoming edges of node i shall not exceed B_i. It is also assumed that each node can adjust its transmitting power level. Let p_i denote the transmission power that node i chooses, $0 \leq p_i \leq P_i$. A directed edge $(i, j) \in E$ iff $p_i \geq d_{i,j}^\alpha$, where α is a real number typically between 2 to 4. Let $\lambda_{s,d}$ and $\delta_{s,d}$ denote the traffic demands and the maximum allowed hop-count for each node-pair (s, d), respectively. For node i, a power utilization ratio was defined by $R_i = \frac{p_i}{P_i}$. Let $R_{\max} = \max\{R_i | 0 \leq i \leq n\}$.

Then the QoS topology control problem can be formally defined as: given a node set V with their locations and each node i with B_i and P_i, and given $\lambda_{s,d}$ and $\delta_{s,d}$ for each node-pair (s, d), find the transmitting power p_i for $0 \leq i \leq n$, such that all the traffic demands can be routed within the hop-count bound, and $R_{\max}$ is minimized.

There are two cases: (1) end-to-end traffic demands are splittable (i.e., $\lambda_{s,d}$ can be routed on several different paths from s to d); and (2) end-to-end traffic demands are not splittable (i.e., $\lambda_{s,d}$ for node-pair (s, d) must be routed on the same path from s to d). Case 1 was formulated as a mixed integer linear programming problem and then solved by integrating with a variant multi-commodity flow problem. Case 2 was formulated into an integer linear programming problem. Some numeric examples are presented at the end of this section.

12.5.1   QoS Topology Control with Traffic Splittable

Because the traffic between a node-pair can be routed via several different paths in this case, there is no delay constraint in the following formulation. The QoS topology control problem with traffic splittable can be formulated as follows.

Objective: Minimize the maximum power utilization of nodes:

$$\min R_{\max} \tag{12.45}$$

Subject to:

Topology constraint: $x_{i,j} \leq x_{i,j'}, \forall i, j, j' \in V, d(i, j') \leq d(i, j)$ (12.46)

Transmission power constraint:

$$P_i \geq p_i \geq d_{i,j}^{\alpha} x_{i,j}, \quad \forall i, j \in V \tag{12.47}$$

$$R_{\max} \geq \frac{p_i}{P_i}, \quad \forall i \in V \tag{12.48}$$

Bandwidth constraint: $\displaystyle\sum_{(s,d)} \sum_j f_{i,j}^{s,d} + \sum_{(s,d)} \sum_j f_{j,i}^{s,d} \leq B_i, \quad \forall i \in V$ (12.49)

Routes constraints: $\displaystyle\sum_j f_{i,j}^{s,d} - \sum_j f_{j,i}^{s,d} = \begin{cases} \lambda_{s,d} & \text{if } s = i \\ -\lambda_{s,d} & \text{if } d = i \\ 0 & \text{otherwise} \end{cases} \quad \forall i \in V$ (12.50)

$$f_{i,j}^{s,d} \leq f_{i,j}^{s,d} x_{i,j}, \quad \forall i, j \in V, (s, d) \tag{12.51}$$

Variable constraint: $\begin{aligned} &x_{i,j} = 0, \text{ or } 1 \\ &f_{i,j}^{s,d} \geq 0, \ p_i \geq 0, \ R_{\max} \geq 0, \quad \forall i, j \in V, (s, d) \end{aligned}$ (12.52)

where $x_{i,j}$ are boolean variables and $x_{i,j} = 1$ if there is a link from node i to node j, otherwise $x_{i,j} = 0$. $f_{i,j}^{s,d}$ represents the amount of traffics of node-pair (s, d) that go through link (i, j).

Constraint 12.46 ensures that nodes have broadcast ability. That is, the transmission by a node can be received by all the nodes within its transmission range. Constraint 12.47 ensures that the transmission power of each node does not exceed its power bound. Constraint 12.48 determines the maximum power utilization ratio among all nodes. Constraint 12.49 ensures that the traffic going through a node does not exceed the bandwidth of that node. The first term on the right-hand side of inequality 12.49 represents all the outgoing traffic at node i (transmitting) and the second term represents all the incoming traffic (reception). Constraints 12.50 and 12.51 ensure that the validity of the route for each node-pair.

The QoS topology control problem with traffic splittable has now been formulated as a mixed integer programming problem. Unfortunately, the mixed integer linear programming problem cannot be solved in polynomial time. To solve the problem, first consider the Load Balancing QoS Routing problem: given a network topology, and traffic demands between node-pairs, route this traffic in the network such that the maximum bandwidth utilization $L_{\max}$ is minimized. This problem can be solved in polynomial time by transforming it to a variant multi-commodity flow problem.[7–9]

Objective: Minimize the maximum bandwidth utilization:

$$\min L_{\max} \tag{12.53}$$

$$\text{Subject to: } \sum_j f_{i,j}^{s,d} - \sum_j f_{j,i}^{s,d} = \begin{cases} \lambda_{s,d} & \text{if } s = i \\ -\lambda_{s,d} & \text{if } d = i \\ 0 & \text{otherwise} \end{cases} \quad \forall i \in V \tag{12.54}$$

$$\sum_{(s,d)} \sum_j f_{i,j}^{s,d} + \sum_{(s,d)} \sum_j f_{j,i}^{s,d} \le B_i L_{\max}, \quad \forall i \in V \tag{12.55}$$

$$\begin{aligned} f_{i,j}^{s,d} &\ge 0, \\ L_{\max} &\ge 0, \quad \forall i \in V \end{aligned} \tag{12.56}$$

Note that for $\forall (s,d)$, $f_{i,j}^{s,d} = 0$ if $(i,j) \notin E(G)$. Function 12.53 is the objective, which is to minimize the maximal node bandwidth utilization. Constraint 12.55 states that a factor (i.e., $L_{\max}$) of B_i bandwidth is actually used by node i. Notice that $L_{\max}$, obtained from solving the formulations 12.53 through 12.56, can be greater than 1. When $L_{\max} > 1$, it means that the actual bandwidth usage of some nodes must have exceeded their capacities, which violates constraint 12.49. In this case, it indicates the given topology cannot accommodate the required traffic demands. In the following QoS topology control algorithm, one needs to keep on adding more links into the topology until $L_{\max} \le 1$, which means the topology can support the required traffic (i.e., no node has the actual bandwidth usage exceeding its capacity).

The proposed algorithm to solve the QoS topology control problem with traffic splittable can be described in the following steps.

1. Sort all node-pairs with $d_{i,j}^{\alpha} \le P_i$ in ascending order according to R_{ij}.
2. Pick up the node-pair (i, j) that has the smallest R_{ij} but there is no link from i to j, and increase p_i to link j, making a new topology G.
3. Solve the load balancing QoS routing problem on G to obtain $L_{\max}$. If $L_{\max} \le 1$, then stop (a solution is found); otherwise repeat Steps 2 and 3.

In Step 2, it stops if all nodes have already reached their maximal power and an error of no solution is reported in this case. To reduce the frequency of calling the QoS routing algorithm in Step 3, the binary search method was used to find the QoS topology, instead of adding an edge each time and running the routing algorithm.

12.5.2 QoS Topology Control with Traffic Non-Splittable

The definition and the system symbol of the QoS topology control problem were described in Section 12.5.1. This section discusses the traffic non-splittable case. Because $\lambda_{s,d}$ for node-pair (s, d) is routed on only one path from s to d, the hop-count of the path cannot exceed the maximum allowed hop-count $\delta_{s,d}$. The QoS topology control problem with traffic non-splittable can be formulated as follows.

Objective: Minimize the maximum power utilization of nodes.

$$\min R_{\max} \tag{12.57}$$

Subject to:

Topology constraint: $x_{i,j} \le x_{i,j'}, \quad \forall i, j, j' \in V, d(i, j') \le d(i, j)$ \hfill (12.58)

Transmission power constraint:

$$P_i \ge p_i \ge d_{i,j}^{\alpha} x_{i,j}, \quad \forall i, j \in V \tag{12.59}$$

$$R_{\max} \ge \frac{p_i}{P_i}, \quad \forall i \in V \tag{12.60}$$

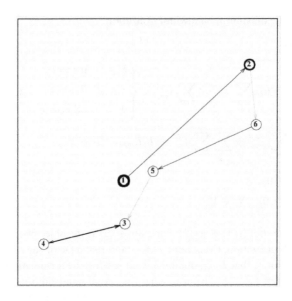

FIGURE 12.3   QoS topology for non-splittable case.

Delay constraint: $\displaystyle\sum_{(i,j)} x_{i,j}^{s,d} \le \delta_{s,d}, \quad \forall(s,d)$ (12.61)

Bandwidth constraint: $\displaystyle\sum_{(s,d)}\sum_{j} x_{i,j}^{s,d}\lambda_{s,d} + \sum_{(s,d)}\sum_{j} x_{j,i}^{s,d}\lambda_{s,d} \le B_i, \quad \forall\, i \in V$ (12.62)

Route constraint: $\displaystyle\sum_{j} x_{i,j}^{s,d} - \sum_{j} x_{j,i}^{s,d} = \begin{cases} 1 & \text{if } s = i \\ -1 & \text{if } d = i \quad \forall\, i \in V \\ 0 & \text{otherwise} \end{cases}$ (12.63)

$x_{i,j}^{s,d} \le x_{i,j}, \quad \forall i, j \in V$ (12.64)

Binary constraint: $\begin{aligned} & x_{i,j} = 0,\ or\ 1, \quad x_{i,j}^{s,d} = 0,\ or\ 1 \\ & P_i \ge 0,\ R_{\max} \ge 0, \quad \forall i, j \in V, (s,d) \end{aligned}$ (12.65)

The problem has been formulated as an integer linear programming problem. The objective and most of the constraints are the same as the traffic splittable case in Section 12.5.1. Note that the topology constructed from the above formulation is directed. To make the topology undirected (or bidirectional), it can simply add another constraint: $x_{i,j} = x_{j,i}$ for $\forall i, j \in V$.

Integer linear programming problems belong to the linear programming problems in which some or all of the variables are restricted to be integers. Integer linear programming problems are combinatorial problems and, in general, much more difficult to solve than linear programming problems. In fact, there

TABLE 12.1   Requests and Their
Routing for Figure 12.3.

s	d	$\lambda_{s,d}$	route
1	2	29.9568	$1\to2$
2	3	36.4634	$2\to6\to5\to3$
2	5	34.2944	$2\to6\to5$
3	4	29.7357	$3\to4$
4	3	35.9753	$4\to3$
6	4	33.5743	$6\to5\to3\to4$

TABLE 12.2 Requests and Their Routing for Figure 12.4.

s	d	$\lambda_{s,d}$	splitted $\lambda_{s,d}$	Route
1	2	29.9568	16.4993	$1{\to}6{\to}2$
			13.4575	$1{\to}2$
2	3	36.4634	14.3784	$2{\to}5{\to}1{\to}4{\to}3$
			11.8406	$2{\to}5{\to}3$
			10.2444	$2{\to}5{\to}1{\to}3$
2	5	34.2944	34.2944	$2{\to}5$
3	4	29.7357	15.6646	$3{\to}4$
			14.0710	$3{\to}1{\to}4$
4	3	35.9753	35.9753	$4{\to}3$
6	4	33.5743	31.0260	$6{\to}2{\to}5{\to}1{\to}4$
			2.5483	$6{\to}2{\to}5{\to}3{\to}4$

is no single algorithm that can be applied to all integer linear programming problems as the simplex algorithm that was used to effectively solve any linear programming problem.

lp_solve (ftp://ftp.es.ele.tue.nl/pub/lp_solve) and Matlab 6.5 were used to solve the problem for experimental purposes.[13] Figure 12.3 shows the topology of a network with six nodes and six requests, where node 1 is a high-power node, node 2 a medium-power node, and the rest are low-power nodes. The details of the requests and the routing computed by the *lp_solve* are given in Table 12.1. $\delta_{s,d}$ is set to 4 (consistent with the maximal hop-count for splittable case, which is 4. See Table 12.2). Details of the settings of the simulation environment and other parameters were given Jia et al.[13] For comparison purposes, the topology for the same node setting and requests for the *traffic splittable* case were also computed using the proposed algorithm in Section 12.5.1. Figure 12.4 and Table 12.2 are the resulting topology and the routing of traffics, respectively. R_{max} is 0.7517 for the non-splittable case (Figure 12.3), and 0.5965 for the splittable case (Figure 12.4). It is obvious that the topology for the splittable case has a better balanced utilization of energy because it can split the traffic into multiple routes and take the advantage of using short-distance links. From Figure 12.3 and Figure 12.4, the long-distance link $6 \to 5$ in Figure 12.3 contributes to the high R_{max}. Note that nodes 3–6 are low-power nodes and it is very costly for them to reach nodes at long distances. The topology in Figure 12.4 uses more short-distance edges for low-power nodes to carry traffic via multiple routes, which results in a low R_{max}.

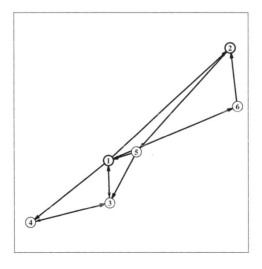

FIGURE 12.4 The QoS topology for splittable case.

12.6   Conclusion

This chapter introduced applications of LP/ILP techniques to optimization problems of energy efficiency in wireless ad hoc networks. The discussed applications include energy efficiency routing problems, broadcast/multicast routing problems, data extraction and gathering problems, and QoS topology control problems. We focused on how these problems can be formulated and solved by linear programming. We found that a majority of these problems are related to multi-commodity problems or variant multi-commodity problems.

Although linear programming problems have relatively high computational complexity — even integer linear programming problems cannot be computed in polynomial time — this mature and powerful tool will be continually used and completed as actual implementations increase.

References

1. Arvind Sankar and Zhen Liu, Maximum Lifetime Routing in Wireless Ad-hoc Networks, *IEEE INFOCOM 2004,* 2004.
2. Narayanan Sadagopan and Bhaskar Krishnamachari, Maximizing Data Extraction in Energy-Limited Sensor Networks, *IEEE INFOCOM 2004,* 2004.
3. Gil Zussman and Adrian Segall, Energy Efficient Routing in Ad Hoc Disaster Recovery Networks, *IEEE INFOCOM 2003,* 2003.
4. Arindam K. Das, Robert J. Marks, Mohamed El-Sharkawi, Payman Arabshahi, and Andrew Gray, Minmum Power Broadcast Trees for Wireless Networks: Integer Programming Formulations, *IEEE INFOCOM 2003,* 2003.
5. Manish Bhardwaj and Anantha P. Chandrakasan, Bounding the Lifetime of Sensor Networks via Optimal Role Assignment, *IEEE INFOCOM 2002,* 2002.
6. Jae-Hwan Chang and Leandros Tassiulas, Energy Conserving Routing in Wireless Ad-Hoc Networks, *IEEE INFOCOM 2000,* pp. 22–31, 2000.
7. B. Awerbuch and T. Leighton, Improved Approximation Algorithms for the Multi-Commodity Flow Problem and Local Competitive Routing in Dynamic Networks, *Proc. ACM STOC'94,* May 1994.
8. O. Gunluk, A New Min-Cut Max-Flow Ratio for Multi-Commodity Flows, *Proc. IPCO 2002, LNCS,* Vol. 2337, Springer, May 2002.
9. T. Leighton and S. Rao, Multi-Commodity Max-Flow Min-Cut Theorems and Their Use in Designing Approximation Algorithms, *J. ACM,* 46, 782–832, Nov. 1999.
10. N. Garg and J. Koenemann, Faster and Simpler Algorithms for Multicommodity Flow and Other Fractional Packing Problems, *FOCS 1998.*
11. George B. Dantzig and Mukund N. Thapa, *Linear Programming 2: Theory and Extensions,* Springer Press, 1997.
12. James P. Ignizio and Tom M. Cavalier, *Linear Programming,* Prentice Hall, 1994.
13. Xiaohua Jia, Deying Li, and Dingzhu Du, QoS Topology Control in Ad Hoc Wireless Networks, *IEEE INFOCOM 2004,* 2004.
14. Deying Li, Xioahua Jia, and Hai Liu, Energy Efficient Broadcast Routing in Ad Hoc Wireless Networks, *IEEE Trans on Mobile Computing,* 3(2), 144-151, 2004.
15. J.E. Wieselthier, G.D. Nguyen, and A. Ephremides, Energy-Efficient Broadcast and Multicast Trees in Wireless Networks, *Mobile Networks and Applications* 7, 481-492, 2002.
16. B. Wang and S.K.S. Gupta, G-REMiT: An Algorithm for Building Energy Efficient Multicast Trees in Wireless Ad Hoc Networks, *Proceeding of the Second IEEE International Symposium on Network Computing and Applications (NCA'03),* 2003.
17. Ivan Stojmenovic, Mahtab Seddigh, and Jovisa Zunic, Internal Nodes Based Broadcasting in Wireless Networks, *Proceedings of the 34th Hawaii International Conference on System Sciences,* 2001.
18. T. Cormen, C. Leiserson, and R. Rivest, *Introduction to Algorithms,* McGraw-Hill and MIT Press, 1990.

19. Javier Gomez, Andrew T. Campbell, Mahmoud Naghshineh, and Chatschik Bisdikian, Conserving Transmission Power in Wireless Ad Hoc Networks, in *Proc. of IEEE Conference on Network Protocols (ICNP'01)*, November 2001.
20. Suresh Singh, MikeWoo, and C.S. Raghavendra, Power-Aware Routing in Mobile Ad Hoc Networks, *Proc. of 4th Annual International Conference on Mobile Computing and Networking (MobiCom'98)*, pp. 181–190, 1998.
21. Suman Banerjee and Archan Misra, Minimum Energy Paths for Reliable Communication in Multi-Hop Wireless Networks, *Proc. of Workshop on Mobile and Ad Hoc Networking and Computing (MobiHoc'02)*, June 2002.
22. Archan Misra and Suman Banerjee, MRPC: Maximizing Network Lifetime for Reliable Routing in Wireless Environments, *Proc. of IEEE Wireless Communications and Networking Conference (WCNC'02)*, March 2002.
23. C.-K. Toh, Maximum Battery Life Routing to Support Ubiquitous Mobile Computing in Wireless Ad Hoc Networks, *IEEE Communications Magazine,* 39(6), June 2001.
24. C. Zhu and M.S. Corson, QoS Routing for Mobile Ad Hoc Networks, *IEEE INFOCOM'02,* 2002.
25. S. Chen and Klara Nahrstedt, Distributed Quality-of-Service Routing in Ad Hoc Networks, *IEEE Journal on Selected Areas in Communications,* 17(8), 1488–1505, 1999.
26. R. Ramanathan and R. Rosales-Hain, Topology Control of Multihop Wireless Networks Using Transmit Power Adjustment, *INFOCOM'00,* pp. 404–413, 2000.
27. R. Wattenhofer, L. Li, P. Bahl, and Y.M. Wang, Distributed Topology Control for Power Efficient Operation in Multihop Wireless Ad Hoc Networks, *INFOCOM'01,* Vol. 3, pp. 1388–1397, 2001.
28. N. Li, J. Hou, and Lui Sha, Design and Analysis of an MST-based Topology Control Algorithm, *IEEE INFOCOM'03,* 2003.

13

Wireless Networks World and Security Algorithms

Nicolas Sklavos

Nikolay A. Moldovyan

Odysseas Koufopavlou

Abstract

Wireless communication protocols have specified security layers that support encryption with high-level strength. The performance results of these integrations, in some cases, cannot meet all the wireless communications standards and specified security demands. This chapter is a survey on the advantages and trade-offs of the different implementation approaches to wireless protocol security. Alternative solutions are proposed, so that the implementation problem can be faced successfully. The different solutions to this issue are compared in terms of both performance and supported security level. Technical specifications for software and hardware implementations of the past and present are given. Future directions on both cipher designs and solutions to the problem of implementation are proposed. In particular, recommendations for wireless communications security engines are discussed.

13.1   Introduction

Wireless communications have become a very attractive and interesting sector for the provision of electronic services. Mobile networks are available almost anytime, anywhere, and user acceptance of wireless hand-held devices is high. The services offered are constantly increasing due to the different and large range of user needs. These vary from simple communications services to special-purpose and sensitive applications such as electronic commerce and digital cash. While the wireless devices are coming to the offices and houses, the need for strong, secure transport protocols seems to be one of the most important issues facing mobile standards. From e-mail services to cellular-provided applications, from secure Internet possibilities to banking operations, cryptography is an essential part of today's user needs.[1]

The standards for mobile applications and services are maturing and new specifications in security systems are being defined. This leads to a large set of possible technologies from which a service provider can choose. Although organizations and forums seem to agree as to the increasing need for secure systems with wide strength, cryptography is still a big black hole in the wireless networks because of implementation difficulties. The security layers of many wireless protocols use encryption algorithms designed in the past, which have proven unsuitable for hardware implementations, especially for wireless hand-held devices. In general, the ciphers use large arithmetic and algebraic modifications that are not appropriate for hardware integrations. That is why cipher implementations allocate many of the system resources, in hardware terms, in order to be implemented as components. So, in many cases, software applications have been developed to support the security and cryptography needs. But the software solution is not acceptable for the case of hand-held devices and mobile communications with high speed and performance specifications.

It is obvious that, in future wireless protocols and communications environments (networks), security will play a key role in transmitted information operations. This chapter summarizes key issues that should be solved for achieving the desirable performance in security implementations and focuses on alternative integration approaches for wireless communications security. It gives an overview of the current security layer of wireless protocols and presents the performance characteristics of implementations in both software and hardware. We also propose some efficient methods to implement security schemes in wireless protocols with high performance. The purpose of this chapter is to provide a state-of-the-art survey and research trends on implementations of wireless protocol security for current and future wireless communications.

13.2   Cryptography: An Overview

A cryptographic algorithm is a mathematical function used for both encryption and decryption. Of course, the strength of the privacy that an algorithm offers is not based on keeping its architecture secret. Today, the security of most algorithms is a function of two things: (1) the strength of the algorithm and (2) the length of the key.[1] With the rule key, modern cryptography means a large number or a set of large number values.

In general, every cipher has two operation modes: encryption and decryption. Both processes use the same or different keys. The strength of an encryption system has to do only with these keys and is not based on the architecture of the algorithm itself. That is, the details of the algorithm's architecture have been published and analyzed. The algorithm with the appropriately used keys, plus all possible plaintext and ciphertext, comprise the basic parts of a cryptosystem.

Encryption algorithms can be broadly categorized as follows:

Private (single, shared) key encryption algorithms. These algorithms are used for bulk encryption, as they are extremely fast, and this makes them suitable for large amounts of transport data. Well-known algorithms in this category include IDEA, Blowfish, RC5, and RIJNDAEL (AES-Proposed).[2]

Public key ciphers. In this category belong all those algorithms that use public keys. They can be used for key validation and distribution, for message authentication, and sometimes for signature operations. They are low-speed algorithms and their computational cost limits their use. Some of these algorithms are RSA, Diffie-Hellman, and elliptic curve.

Hash functions. Another fundamental primitive in modern cryptography is the cryptographic hash function, often informally called a one-way hash function. A hash function is a computationally efficient function mapping binary strings of arbitrary length to binary strings of some fixed length, called hash-values. This category includes algorithms that can compress and transform data down to a fixed sized of signing. MD5, SHA-1, and SHA-2 are the most widely used algorithms for this kind of encryption.

All the encryption systems and security layers of today's protocols are needed to support the three different kinds of cryptographic algorithms mentioned above, at the same implementation of their application.

13.3   Security and Wireless Protocols

13.3.1   GSM

GSM (Global System for Mobile Communications) is a standard for digital mobile telephones, as defined by the European Telecommunications Standards Institute (ETSI). This system supports security operations of authentication and encryption between the home network and each SIM (subscriber identity module) card. The algorithms for authentication are A3 and A8. The actual data encryption is based on the stream cipher algorithm A5. The algorithm is implemented in the end device and the network. Implementations of the above algorithms, called COMP128, exist on the Internet[3,4] but they have not been published yet as a standard. This is one basic trade-off of the security features of GSM networks, although further publications on A5 are expected.

13.3.2   WAP

The Wireless Application Protocol (WAP) is the *de facto* world standard for the presentation and delivery of wireless information and telephony services on mobile phones and other wireless terminals.[5] WAP is being defined by the WAP Forum, which is an industry group consisting of handset manufacturers, wireless service providers, infrastructure providers, and software developers. The WTLS (Wireless Transport Layer Security) is the layer of WAP dedicated to security. It provides privacy, data integrity, and message authentication for WAP users. This layer provides the transport service interface for the upper level layers. WTLS is based on the philosophy of the well-known TLS (Transport Layer Security) and introduces new characteristics to WAP. Handshake, optimized data size, and dynamic key refreshing are some of them. Classic ciphers have been chosen to support the three different cryptographic operations. Bulk encryption uses DES, IDEA, and RC5; message authentication is based on RSA, Diffie-Hellman, and the elliptic curve method; while MD5 and SHA are used as hash function algorithms. Today's equipment and technology of cellular phones allow the implementation of the security layer to support only a set of cipher suites. The WAP stack is shown in Figure 13.1. This figure also presents how the WAP stack relates to the protocols on the Internet.

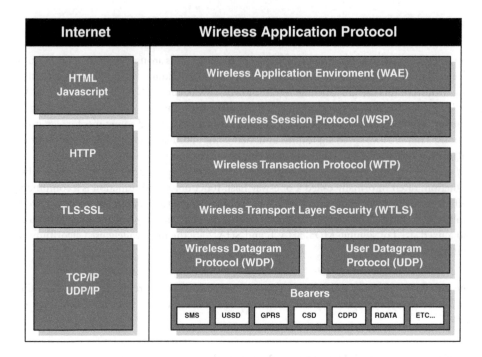

FIGURE 13.1   WAP stack architecture.

13.3.3   Bluetooth

Bluetooth is system that defines the way that portable computers, cellular telephones, and a variety of other devices can be connected using low-power, short range wireless links.[6] Bluetooth technology permits devices to communicate with each other, synchronize data, and connect to the Internet at high speeds without the need for wires or cables. Bluetooth specifications include security features. The system supports authentication and encryption processes. These features are based on a secret link key that is shared by a pair of devices. To generate this key, a pairing procedure is used when the two devices communicate for the first time. The encryption of the payloads is carried out with a stream cipher called Eo that is resynchronized for every payload (Figure 13.2). This algorithm is based on a method derived from the summation stream cipher generator attributable to Massey and Rueppel. The authentication function proposed for Bluetooth is a computationally secure authentication code, often called a MAC. The algorithm used for this kind of encryption is SAFER+. It was one of the contenders for the Advanced Encryption Standard (AES) submitted by Cylink Corp. Sunnyvale, California.

13.3.4   HIPERLAN

HIPERLAN is ETSI's wireless broadband access standard that defines the MAC sublayer, the Channel Access Control (CAC) sublayer, and the physical layer. The MAC accesses the physical layer through the CAC, which allows easy adaptation for different physical layers. The specifications of this communication protocol define an encryption-decryption part for optional use. All the HIPERLAN MAC entities use common keys for the encryption algorithms that the protocol supports. These are called the HIPERLAN key set. Every key of the set is described as unique with its own identifier. In the security part of the protocol, the ciphertext is produced by an XOR procedure over the plaintext. The encryption function, except the keys, also requires a random sequence generation. An overview of the HIPERLAN encryption-decryption scheme is shown in Figure 13.3.

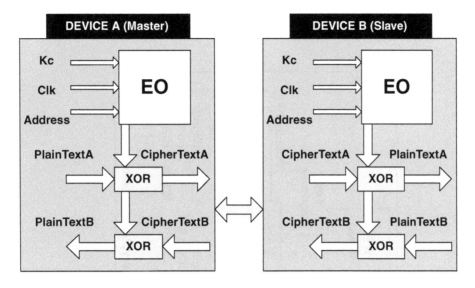

FIGURE 13.2 Bluetooth security scheme.

13.3.5 IEEE 802.11 Standard

The IEEE 802 Standards Committee formed the 802.11 Wireless Local Area Networks Standards Working Group in 1990. The IEEE 802.11 standard is the protocol for both ad hoc and client/server networks.[7] In an ad hoc network, the communications are established between multiple stations in the covered area without the use of servers and access points. The WLAN (wireless local area network) specifies an optional encryption part that is called WEP (Wireless Encryption Privacy). WEP tries to keep out unwelcome interferences in the protocol's established communications. The authentication function uses the same key with encryption (Figure 13.4). The use of the common key for both security operations imports a high level of risk for the protocol. The authentication works efficiently only if WEP is supported by the WLAN. Without the encryption mode, the authentication procedure is canceled.

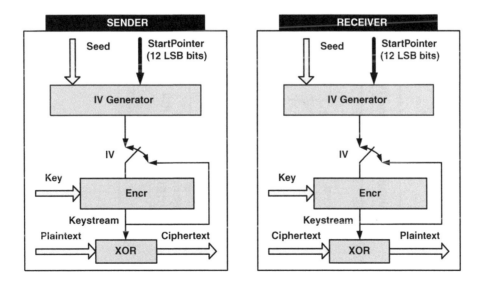

FIGURE 13.3 HIPERLAN encryption-decryption scheme.

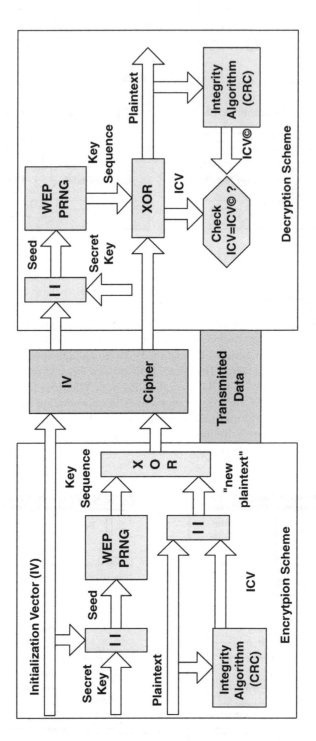

FIGURE 13.4   WEP Security in IEEE 802.11.

13.4   Wireless Network Algorithm Implementations: The Software Approach

Today, the most complicated cryptographic systems are implemented in software rather than in hardware. One major reason is the implementers' increased knowledge of software programming versus hardware design. Software tools vary, while VLSI CAD commercial tools are only of interest to large companies and specialized research groups. Individual users and class projects are restricted to software possibilities. Almost all encryption algorithms have been implemented in assembly and in other language compilers such as C++ and Java. For many years, the majority of applied development techniques were related to the sequential applications rather than those related to real-time systems. Lately, programmers have made diligent efforts to find efficient solutions to the implementation problem using different software compilers. Today, a programmer can develop a real-time system in many software languages such as Ada, Modula2, or Occam. These languages contain constructs for programming concurrency, which make them suitable for large real-time systems. At the same time, new compilers appear in the foreground, especially those with many possibilities for real-time developed systems. Erlang is a good example of such a language compiler.[8] A team of employees at the Swedish communications firm Ericsson developed this language. It is used to write huge real-time control programs for telephone exchanges and network switches. A basic advantage of Erlang is the language support code replacement in a running system. It allows execution of new versions of code functions at the same time. This is very useful in non-stop systems, telephone exchanges, air traffic control systems, etc., where the operation cannot be stopped to make software changes. In such systems, encryption algorithms and security schemes are also included, and can be easily modified and updated "on-the-fly" due to certain language specifications. Another primitive characteristic is that this language supports three constructs for detecting runtime errors.

One basic advantage of software is that the development of an operation such as encryption is a significantly easy process. Many cryptographic libraries exist and someone can prototype ciphers with no special effort and no major waste of time. On the other side, a hardware implementation needs much effort and a lot of time for designing and testing. In general, the software development performance is much slower than typical network bandwidths. Fast hardware systems are implemented and are projected to achieve speeds for encryption-decryption processes that are comparable to the network bandwidths. The software performance characteristics of the most well-known cryptographic algorithms are shown in Table 13.1.

The ciphers have been implemented in Java (using the Cryptix Library) in one run interpreted on a Pentium II/266 Linux system.[9] The encryption transformed plaintext data is 1 MB. Works from other research groups[10,11] have also been included in Table 13.1.

TABLE 13.1   Software Implementation Performance of Ciphers

Encryption Algorithm	Encryption Time (ms)	Rate (kbit/sec)
IDEA	43409	193
SAFER	41442	202
Blowfish	20506	409
Triple-DES	160807	52
Loki91	31071	269
RC2	43329	193
Square	29610	283
RC4	12945	648
DES	48629	172
CAST5	23772	352
SHA-1 [10]	—	4.23 Mbps
		41.51 Mbps
SAFER+ [11]	—	25.6

TABLE 13.2   Encryption Algorithms Software Implementation Performance Comparison

Encryption Algorithm	Clocks/Round	Number of Rounds	Clocks/Byte of Output
RC4	N.A.	N.A.	7
SEAL	N.A.	N.A.	4
Blowfish	9	16	18
Knufu/Khafre	5	32	20
RC5	12	16	23
DES	18	16	45
IDEA	50	8	50
Triple-DES	18	48	108
Rijndael [2]	369	11	32

Another comparison study of cipher software implementation has been done in C++.[12] The algorithms' performance has been analyzed in a Pentium processor and their performance characteristics are illustrated in Table 13.2. These measurements were taken for needed clock cycles per output byte.

13.5   The Hardware Solution for Mobile Devices: Architectures, Designs, and Integration

An application's increasing demand for computation power, as well as the power reduction requirements for portable devices, force us to consider that general-purpose processors are no longer an efficient solution for mobile systems. So, new hardware approaches are needed to implement some computationally heavy and power-consuming functions in order to meet current network speed requirements. Such approaches include the following.

Recent Application-Specific Integrated Circuit (ASIC) technology is the solution that created better opportunities for implementing real-time and more sophisticated systems. ASIC devices guarantee better performance, with a sufficiently small dedicated size. The reliability reaches high limits and the turnaround time is fast. The implementations in these modules are characterized by a tighter design security than any other type of device. ASICs include several custom and semi-custom hardware designs. These devices are based on programmable logic devices (PLDs), gate arrays (GAs), and standard cells (SCs). In our case, ASICs can be described as follows: custom-designed hardware, specially tailored to one particular encryption process. They require a significant initial investment in design and testing. If such a device cannot be produced in mass quantities, it is not economical for the market. ASICs seem to be more suitable for dedicated applications and not for an extended-purpose encryption system.

Between the software applications and the ASIC devices, there is some middle ground. This area is covered by the field programmable gate arrays (FPGAs) and complex programmable logic devices (CPLDs). These components provide reconfigurable logic and are commercially available at low cost. They support the benefits of customizable hardware and are software-driven implementations. Of course, these devices vary in capacity and performance. The main disadvantage is that they are not suitable for the implementation of large functions. Programmable logic has several advantages over custom hardware. It is less time-consuming, for the development and the design phase, than the custom hardware approach. These devices are more flexible than ASICs. They can be reused for the cryptanalysis of many different encryption algorithms with little extra effort.

Another solution to the implementation platform problem is smart cards. This issue has to do more with fit than with performance. In smart cards, the RAM requirements are more important than the clock's frequency. Most commodity smart card CPUs today include 128 to 256 bytes of on-board RAM. Each CPU family contains members with more RAM capacity and a correspondingly higher cost. Although some CPUs include enough RAM to hold the keys of the algorithms, it is often not realistic to assume

TABLE 13.3 Hardware Implementation Characteristics of Encryption Algorithms

Encryption Algorithm	Device Type	Area Cost	Frequency (MHz)	Throughput (Mbps)
RSA [13]	ASIC	47.61 mm^2	80	0.301
IDEA [14]	ASIC	50.01 mm^2	8	128
DES [15]	FPGA	318 CLBs	55	42.5
Elliptic Curve [16]	FPGA	1290 CLBs	45	0.031
SAFER + [17]	FPGA	6068 CLBs	50	640
Rijndael [18]	FPGA	2358 CLBs	22	259
Triple-DES [19]	ASIC	1225 mm^2	105	6.7 Gbps
Twofish [20]	ASIC	35,000 gates	66	200
Kasumi [21]	FPGA	749 CLBS	35.35	71

that such a large fraction of RAM can be dedicated solely to the encryption function. In a smart-card operating system, encryption is a minor, small part of the system and will not be able to use more than half of the available RAM. Obviously, if an encryption-decryption system does not fit on a certain CPU, with a particular configuration of its components, then the performance of the system is unrealistic. Even if an algorithm fits onto the smart card, the encryption function will not be able to use all of the RAM capacity. For example, an algorithm that needs about 100 bytes RAM seems to fit in a 128-byte smart card. Of course, this is a theoretical result because there are also RAM requirements for the control procedure that handles the total security process, and these requirements increase the need of the memory-limited capacity. It is clear that the devices in this category are not appropriate for large encryption systems with special specifications.

In general, hardware implementations have proven to be a better approach compared with the software developments, in terms of throughput and operating frequency. Of course, the covered area resources are a factor that must be taken into consideration. The covered area and the performance results of some good hardware implementation examples of ciphers are shown in Table 13.3.

For all the hardware devices there are some common factors that make the implementation of the ciphers in powerful hardware engines a very difficult process. The most critical factor is the large number of registers for key storage, which are used by most of the algorithms. As previously discussed, the security of ciphers is a function of two factors: (1) the strength of the algorithm and (2) the length of the key. While the strength of a cipher is a fixed factor because of the algorithm's definition, the key length is a parameter that can vary. Cipher introducers and cryptographers use large keys for more secure operations. This results in larger numbers of buffers and storage units and larger memory requirements for hardware integration. This event has an associated cost in the chip's covered area and sometimes in the I/O devices of the system. To face this problem, RAM blocks are mainly used in hardware implementations. However, in many cases, the availability of RAM usage is restricted. The internal memory capacity of many hardware devices is limited. The use of external RAM reduces the total system performance and increases the system's covered area. All these factors are critical items that must be given special attention by the designers. The application itself defines the impact grade of these factors.

13.6 Alternative Solutions for Security Implementations

The problem of hardware implementation is a function of two different factors: (1) cryptographic algorithm architectures and (2) their efficient integration. All forums and organizations in the wireless communication world have specified security layers/systems and have published the selected ciphers upon which these systems are based. For security with high-level strength to be ensured, three schemes of encryption must be applied in a communication handshake: (1) bulk encryption, (2) message authentication, and (3) data integrity. The wireless protocols have defined alternative ciphers in each of the above schemes. Large encryption systems have been primarily implemented only in software. In the hardware world, only separate encryption algorithms and some simple encryption schemes have been implemented.[22,23]

In the past, the hardware integration approach to the issue of security implementation was the ASIC solution. Implementations on these modules achieve high-speed performance and have proved to be confidant solutions. However, in the case of wireless protocols, this implementation aspect has proven unfeasible. The hardware integration of a set of ciphers that a protocol defines requires a very large circuit. Encryption algorithm implementations in ASICs thus far cover an area of 40 to 60 mm^2 each. For example, the WAP cipher set integration (eight algorithms in total), in one or more ASICs, needs an area about 400 to 480 mm^2, plus the space needed for the total control unit and allocated routing area. Such an ASIC device is very difficult to design and manufacture. Of course, the cost of the chip increases dramatically in this case.

Currently, a flexible encryption system that would support the operation of a set of ciphers integrated in the same module can be implemented with hardware and software cooperation. This type of cooperation could be achieved efficiently by the principles of reconfigurable computing. A proposed solution is the design of a reconfigurable cryptographic system that will support at least bulk and message authentication encryption. Reconfigurable computers are those machines that use the reconfigurable aspects of reconfigurable processing units (RPUs) and FPGAs to implement a system or algorithm. The algorithms are partitioned into a sequence of hardware-implementable objects (hardware objects). This type of object represents the serial behavior of the algorithm and can be executed sequentially. The design technique based on hardware objects offers to the developer/designer a logic-on-demand capability that is based on reconfigurable computing. The appropriate software, suitable for the application at hand, modifies the architecture of these computing platforms. This means that within the application program, a software routine has been written to download a digital circuit (chip design) directly into the RPU. The main idea of these designs is the alternating among the static and dynamic performance of the system.

Static circuitry is the part of the operation performance that remains in action between the different configurations of the system. This must take high priority when discussing design possibilities and much attention must be paid to its optimization. General-purpose blocks, such as adders, belong to this aspect of performance. Another example of static parts is the storage unit. Storage units are the parts of each system that are never changed during different operations. They always maintain the characteristics that the initialization process has set. On the other hand, there is also the dynamic circuitry, that is, the parts of the system that change during configuration. These blocks must be minimized to increase system performance. If there are no basic common parts between the selected algorithms, the dynamic circuitry is high enough and this is bad for the system operation. Dynamic circuitry increases the system demands for resources and decreases its performance. It has been shown that the selection of similar architectural algorithms is a critical factor in both design and implementation.

Of course, there are not many choices for similar cipher architectures in the technical literature. To achieve this, the designer of a powerful security system must choose one flexible algorithm for bulk encryption with the ability to operate as a hash function (data integrity). The addition of some extra parameters in the algorithm's architecture is necessary for the efficient operation of the two encryption modes. In this way, the needs of the system resources are reduced. At the same time, we have to avoid ciphers with heavily arithmetic functions, such as multiplication and modulo processes. These operations are difficult to implement in hardware devices and have no commonality.

The implementation of a security system with some common basic parts, which can be used for the implementation of the ciphers' common functions, seems to be the more sophisticated alternative solution for a large encryption engine. The term "basic parts" implies "heavy" algebraic or logical components of the algorithm architecture. In most cases, it is difficult to implement these parts in a hardware device with high-speed performance and minimized covered area. An example is the multiplication modulo that the IDEA algorithm needs.[14] The reconfigurable computing method has proven efficient enough to solve the implementation problem of encryption engines and is suitable for the different types of cipher architectures. Of course, the specifications of the application itself would prove this method as a good or best solution.

In recent years, implementations on the smart card devices have been very attractive to hardware designers. Compared with the other hardware devices such as ASICs and FPGAs, smart cards have limited computing power and minimized storage capacity. Therefore, security applications that allocate large amounts of storage or that require extensive computation power might cause conflicts.

The persistent storage of a smart card is currently limited to a few kilobytes, which prevents storing larger items on the card. This can be circumvented if the smart card delegates the storage of the item in an external environment. The smart card receives and processes the transmitted data. It encrypts the data and saves it to the sender's or receiver's external storage units (RAM, registers of general use). Later, when this data is needed again, the smart card can request it from these storage units. Using this described method, the smart card's internal storage requirements can be reduced significantly. However, we must be careful that we do not create another bottleneck; that is, the communication speed of the smart card is not very high and so we will have to handle the transmission of the same data back and forth with special care.

Another limitation of smart cards is their small processing power. The appropriate data modifications, due to encryption-decryption, may possibly exceed the computing power of a smart card. In this case, it will take an unacceptably long time to finish the appropriate data transformation. Thus, it is important to minimize the amount of computation power that the smart card has to pay for the requested tasks. For such applications, it is better for the design to be kept as simple as possible. The requested task can be divided in smaller parts with no hard processing specifications. The requested round keys for encryption-decryption can be generated in the initialization procedure and not at the same time as the encryption round transformation (on-the-fly key generation). In this way, we avoid spending extra processing power for the key expansion unit during encryption-decryption. The same methodology can be followed for appropriate specified constants generation.

13.7 New Encryption Algorithm Standards

By 1997, it had become evident that standard DES had become an out-of-date cipher because of its very small key, and the cryptographic community proposed several new block ciphers comprising many new design ideas. At the same time, theoretical research resulted in the formulation of main requirements to the block ciphers to make them secure against different variants of differential and linear attacks. Such prerequisites provided the basis for the National Institute of Standards and Technology (NIST) to initiate a process to select a block encryption algorithm to replace DES (proposed minimum requirements and evaluation criteria were discussed widely at the FSE'97 workshop[24]). The cryptographic community responded with inspiration to the call, and the leading cryptographers proposed their best designs. In 1998, NIST announced the acceptance of fifteen 128-bit block ciphers from eleven countries as candidates for the new Advanced Encryption Standard (AES). The selection of AES was intended as a public process and NIST requested the assistance of the cryptographic community in reviewing the candidates. Most well-respected cryptographers participated in the analysis of all AES candidates and numerous research results were published as a response to this appeal. Each of the fifteen candidates was studied regarding security and efficiency characteristics. The candidates were evaluated for more than two years with respect to security against known and new attacks, performance in hardware and in software implementations, and suitability for different applications. NIST reviewed the results and selected five algorithms (called AES finalists) for further consideration: MARS, RC6, Rijndael, Serpent, and Twofish. In the next stage of the AES competition after more detailed investigation of the finalists, the Rijndael algorithm was selected as the AES. After additional investigation, in October 2000, the winner was announced: the Rijndael cipher developed by Drs. Joan Daemen and Vincent Rijmen from Belgium.

Thus, the AES call has united the efforts of cryptographers all over the world and has assisted further progress in block cipher design. As a result of the AES competition, the main theoretical ideas have become practical designs. As a result, users currently have a new AES standard, in addition to many other ciphers that deserve trust. The AES finalists are the ciphers that can be efficiently embedded in different security mechanisms of information and communication systems. In general, the novelty of the finalists can be characterized as follows.

MARS comprises an innovative, heterogeneous overall structure. It incorporates a wide variety of operations, including multiplication and variable (data-dependent) rotations, and uses fast key setup. MARS possesses a large security margin. The security margin is defined as follows. If the minimal number of rounds needed for a cipher to be secure is measured as Rmin and the number of specified rounds is R, then the security margin is measured as $100\% \cdot (R - Rmin)/Rmin$.

RC6 is an algorithm that is easy to implement in both software and hardware. It represents significant improvement over its predecessor, the RC5 cipher based on data-dependent rotations (DDR). RC6 does not use substitution tables. Its security is based on combining variable rotations and multiplication operations. The global structure of RC6 corresponds to the extended Feistel network. Its key setup is fast. The security margin of RC6 is estimated as relatively low.

Rijndael is designed in the frame of some uniform Simple Permutation Networks (SPN), with the S layers comprising the 8×8 substitution boxes. Its straightforward design facilitates security analysis, which is assisted by the analysis of its predecessor, the SQUARE cipher. Rijndael is very efficient in both hardware and software implementations. The security margin of Rijndael is estimated as good.

Serpent is also designed in the framework of some uniform SPN, with the S layers comprising the 4×4 substitution boxes. It is conservative in its security margin. The designers chose to use twice as many operations as they believed secure against known attacks. As a result, Serpent's software performance is relatively low compared to the other finalists. In hardware, Serpent is comparatively efficient. Its security margin is estimated as large.

Twofish is designed in the framework of the Feistel network. It features key-dependent substitution tables. The designers believed that such tables generally offer greater security than fixed substitution tables and the key-dependent S boxes can be precomputed. Twofish can be considered a cipher descended from the Blowfish algorithm. Both these ciphers are efficient in software across a variety of platforms. Twofish possesses a large security margin.

Rijndael was selected as the AES for the following reasons:

1. It is suitable for secure implementation and provides a good security margin.
2. It is the only AES finalist that supports different block lengths (128, 192, and 256 bits).
3. It is suitable for both hardware and software implementations. (Rijndael uses the single S-box and simple linear transformation; it can be coded in a small number of bytes; a minimal dedicated hardware implementation can be built by hardwaring a single S-box and a single 32-bit to 8-bit linear transform.)

The AES competition was oriented only toward block cipher design; the practical needs for cryptographic algorithms are wider, however. In January 2000, as a response to this need, a new research project called NESSIE (New European Schemes for Signature, Integrity, and Encryption) was initiated by the Information Societies Technology (IST) Programme of the European Commission. Detailed information on the NESSIE project is available at the Web site http://cryptonessie.org. The main formulated goal of the NESSIE project was to put forward a portfolio of efficient cryptographic primitives obtained after an open call and transparent and open evaluation process concerning the security and performance aspects of the primitives. The main features of the NESSIE project against the AES competition include the following:

Many different nominations were announced: block ciphers, synchronous stream ciphers, self-synchronizing stream ciphers, message authentication codes (MACs), collision-resistant hash functions, one-way hash functions, families of pseudo-random functions, asymmetric encryption schemes, asymmetric digital signature schemes (DSSs), and asymmetric identification schemes (DISes). It was also announced that NESSIE would consider primitives designed for use in specific environments.

The main goal was to select the best ideas and designs rather than propose a new European standard based on some of the nominations. The project results (including selected primitives) should be disseminated widely. The selected primitives should be introduced in standards bodies (e.g., ISO, ISO/IEC, CEN, IEEE, IETF).

The list of NESSIE submissions included sixteen block ciphers, five stream ciphers, one hash function, two MACs, five asymmetric encryption systems, six DSSs, and one DIS. The main selection criteria included long-term security (the most important criterion); market requirements (need for a primitive, its usability, and possibility for worldwide use); efficiency (performance in different hardware and software implementations); and flexibility (suitability for use in a wide range of environments).

Although the AES and NESSIE projects initiated great progress in block cipher design, the data-dependent operations (DDOs) with large numbers of realizable modifications were not represented in the ciphers considered. The RC5, RC6, and MARS ciphers have attracted attention to the variable rotations with 32 modifications. To increase the number of modifications, some other type of operation should be used. In Moldovyan and Moldovyan,[25] data-dependent permutations were proposed to advance the DDO-based approach, the permutation networks (PNs) being used to perform variable permutations. It is interesting that PNs had been used previously in the cipher design to perform key-dependent operations.[26,27] Detailed cryptanalysis of the ciphers using key-dependent permutations has shown they are not able to survive in the speed competition with other block ciphers.[28] The use of the PNs to perform DDP has spurred new interest in the use of the PNs in cryptography. Initially, the DDP-based approach was to design fast ciphers suitable for cheap implementations. Currently, DDP-based ciphers are very slow in software, and remain so at least until a new CPU instruction is imbedded in general-purpose processors. Moldovyan et al.[29] have discussed a variant of such an instruction and its applications (cryptographic and general-purpose ones).

13.8   Future Approach Based on Data-Dependent Operations

13.8.1   Data-Dependent Permutations as a Cryptographic Primitive

A permutation network is the simplest controlled SPN. Moldovyan and Moldovyan[25] have proposed PNs to implement data-dependent permutations (DDP) in block ciphers. A security analysis of DDP-based ciphers[30–32] has shown that this method for the use of PNs is more attractive than their use to perform key-dependent permutations. Hardware implementation estimates of these ciphers confirmed that they provide fast and cheap cryptographic hardware. A critical argument against using the DDP as a cryptographic primitive is its low software performance. As a reply to this argument, controlled permutation instruction (CPI) has been proposed for embedding in general-purpose processors.[15] This is a very attractive solution because such instruction is highly desirable in a lot of practical algorithms using arbitrary permutations, which are slow. One should note some peculiarities of the cryptographic and non-cryptographic use of the CP. In the first case, the CPI should perform variable permutations at each clock, which depend on transformed data and are not required to be predefined; they should possess certain properties, however. In the second case, the CPI should perform specified bit permutations of many different types. A universal CPI should cover both cases. Designing a universal CPI represents a special interest. Let us consider a more simple case of the cryptographically oriented CPI. For example, it can be used while embedding cryptography in firmware. Some firmware-oriented DDP-based ciphers are considered below; these give a picture of the DDP-based design and can be naturally transformed into software-suitable systems if some universal CPI can be embedded in general-purpose processors.

The CP boxes $\mathbf{P}_{n/m}$ with n-bit input and m-bit control input are usually constructed using elementary switching elements $\mathbf{P}_{2/1}$ as elementary building blocks. The $\mathbf{P}_{2/1}$ box is controlled with one-bit v and forms two-bit output (y_1, y_2), where $y_1 = y_{1+v}$ and $y_2 = x_{2-v}$. A $\mathbf{P}_{n/m}$ box can be represented as a superposition:

$$\mathbf{P}_{n/m} = \mathbf{L}^{(V_1)} \circ \pi_1 \circ \mathbf{L}^{(V_2)} \circ \pi_2 \circ \ldots \circ \pi_{s-1} \circ \mathbf{L}^{(V_s)}$$

where $\mathbf{L}$ is an active layer composed of $n/2$ switching elements, $V_1, V_2, \ldots, V_s$ are controlling vectors of the active layers from 1 to s, and $\pi_1, \pi_2, \ldots, \pi_{s-1}$ are fixed permutations. The inverse CP box has the following structure:

$$\mathbf{P}_{n/m}^{-1} = \mathbf{L}^{(Vs)} \circ \pi_{s-1}^{-1} \mathbf{L}^{(Vs-1)} \circ \pi_{s-2}^{-1} \ldots \circ \pi_1^{-1} \mathbf{L}^{(V_1)}$$

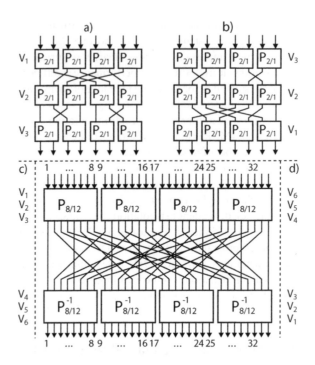

FIGURE 13.5 The CP boxes $\mathbf{P}_{8/12}$(a), $\mathbf{P}_{8/12}^{-1}$(b), $\mathbf{P}_{32/96}$ (c), and $\mathbf{P}_{32/96}^{-1}$ (d).

The components $V_1, V_2, \ldots, V_s$ comprise the controlling vector of the $\mathbf{P}_{n/m}$ box: $V = (V_1, V_2, \ldots, V_s)$. Examples are presented in Figure 13.5.

Suppose for arbitrary $h \leq n$ input bits $x_{\alpha_1}, x_{\alpha_2}, \ldots, x_{\alpha_h}$ and arbitrary h output bits $y_{\beta_1}, y_{\beta_2}, \ldots, y_{\beta_h}$ there is at least one value of the controlling vector V that specifies a CP-box permutation moving x_{α_i} to y_{β_i} for all $i = 1, 2, \ldots, h$. Such a $\mathbf{P}_{n/m}$ box is called a CP box of order h.[25] It is easy to see that the boxes $\mathbf{P}_{8/12}$, $\mathbf{P}_{8/12}^{-1}$, $\mathbf{P}_{32/96}$ (c), and $\mathbf{P}_{32/96}^{-1}$ are first order. Note that due to the symmetric structure, the mutual inverses $\mathbf{P}_{32/96}$ and $\mathbf{P}_{32/96}^{-1}$ differ only with the distribution of controlling bits over the boxes $\mathbf{P}_{2/1}$ in the same topology. When performing DDP operations with $\mathbf{P}_{32/96}$, we form a 96-bit controlling vector depending on some 32-bit data sub-block. Let L be a controlling data sub-block. Thus, bits of $L = (l_1, \ldots, l_{32})$ are used, on average, three times while defining the controlling vector. When designing respective extension boxes, it is reasonable to use the following criteria:

Criterion 1: Let $X = (x_1, \ldots, x_{32})$ be the input vector of the $\mathbf{P}_{32/96}^{(V)}$ box. Then for all L and I, the bit x_i should be permuted, depending on six different bits of L.

Criterion 2: For all i, the bit l_i should define exactly three bits of V.

Below we use the extension box $\mathbf{E}$, which provides the following relation between V and L:

$$V_1 = L_l; \quad V_2 = L_l^{>>>6}; \quad V_3 = L_l^{>>>12}; \quad V_4 = L_r; \quad V_5 = L_r^{>>>6}; \quad V_6 = L_r^{>>>12},$$

where $L_l = (l_1, \ldots, l_{16})$, $L_r = (l_{17}, \ldots, l_{32})$, and $Y = X^{>>>k}$ denotes rotation of the n-bit word X by k bits, where we have $y_i = x_{i+k}$ for $1 \leq i \leq n - k$ and $y_i = x_{i+k-n}$ for $n - k + 1 \leq i \leq n$. Due to the symmetric structure of $\mathbf{P}_{32/96}$, its modifications $\mathbf{P}_{32/96}^{(V)}$, where $V = (V_1, V_2, \ldots, V_6)$, and $\mathbf{P}_{32/96}^{(V')}$, where $V' = (V_6, V_5, \ldots, V_1)$, are mutually inverse. Such symmetry can be used to construct switchable CP boxes. This idea can be realized using a very simple transposition box $\mathbf{P}_{96/1}^{(e)}$ implemented as some single-layer CP box consisting of three parallel single-layer boxes $\mathbf{P}_{2\times16/1}^{(e)}$ (Figure 13.6). The input of each $\mathbf{P}_{2\times16/1}^{(e)}$

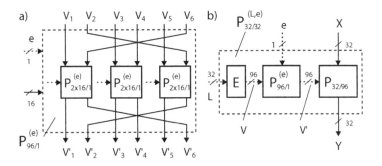

FIGURE 13.6 Switchable CP boxes $\mathbf{P}^{(e)}_{96/1}$(a) and $\mathbf{P}^{(e)}_{32/32}$(b).

box is divided into 16-bit left and 16-bit right inputs. The box $\mathbf{P}^{(e)}_{2\times 16/1}$ contains 16 parallel $\mathbf{P}^{(e)}_{2/1}$ boxes controlled with the same bit e. For example, $\mathbf{P}^{(0)}_{2\times 16/1}(U) = U$ and $\mathbf{P}^{(0)}_{2\times 16/1}(U) = U' = (U_r, U_l)$, where $U = (U_l, U_r) \in \{0, 1\}^{32}$. The left (right) inputs of the $\mathbf{P}^{(e)}_{2/1}$ boxes correspond to the left (right) 16-bit input of the box $\mathbf{P}^{(e)}_{2\times 16/1}$. If the input vector of the box $\mathbf{P}^{(e)}_{96/1}$ is $(V_1, V_2, \ldots, V_6)$, then at the output of $\mathbf{P}^{(e)}_{96/1}$ we have $V' = (V_1, V_2, \ldots, V_6)$ (if $e = 0$) or $V' = (V_6, V_5, \ldots, V_1)$ (if $e = 1$). The structure of the switchable CP box $\mathbf{P}^{(L, e)}_{32/32}$ is shown in Figure 13.6.

The firmware-suitable ciphers Cobra-F64a, Cobra-F64b, and Cobra-S128 use a 128-bit key $K = (K_1, K_2, K_3, K_4)$, where $\forall i\, K_i \in \{0, 1\}^{32}$. No secret key preprocessing is used. While performing j round transformation, subkeys are used directly as 32-bit round subkeys $Q^{(1, e)}_j$, $Q^{(2, e)}_j$, where $j = 1, \ldots, R + 1$ and $e = 0$ ($e = 1$) denotes encryption (decryption). The number of rounds is $R = 16$ for Cobra-F64a, $R = 20$ for Cobra-F64b, and $R = 12$ for Cobra-S128. Correspondence between secret key and round subkeys is defined by Table 13.4 and the following formulas for Cobra-F64a and Cobra-F64b:

$$\left(Q^{(1, 1)}_1, Q^{(1, 1)}_{R+1}\right) = \left(Q^{(1, 0)}_{R+1}, Q^{(1, 0)}_1\right)$$

$$\left(Q^{(2, 1)}_1, Q^{(2, 1)}_{R+1}\right) = \left(Q^{(2, 0)}_{R+1}, Q^{(2, 0)}_1\right)$$

$$\left(Q^{(1, 1)}_j, Q^{(2, 1)}_j\right) = \left(Q^{(2, 0)}_{R-j+2}, Q^{(1, 0)}_{R-j+2}\right), \forall j = 2, \ldots, R$$

For Cobra-S128, we have:

$$\left(Q^{(1, 1)}_j, Q^{(2, 1)}_j\right) = \left(Q^{(2, 0)}_{R-j+1}, Q^{(1, 0)}_{R-j+1}\right), \quad \forall j = 1, \ldots, R.$$

In Cobra-F64a and Cobra-F64b, the 64-bit data block X is divided into two 32-bit sub-blocks A and B. The ciphering is performed in two stages: (1) R rounds with e-dependent procedure **Crypt**$^{(e)}$ shown in Figure 13.7 and (2) final transformation. For both ciphers, the data ciphering algorithm can be represented

TABLE 13.4 Key Scheduling in Cobra-F64a, Cobra-F64a, and Cobra-S128

j	Qj(1,0)	Qj(2,0)	j	Qj(1,0)	Qj(2,0)	j	Qj(1,0)	Qj(2,0)
1	K_1	K_4	8	K_3	K_4	15	K_2	K_3
2	K_2	K_3	9	K_1	K_2	16	K_3	K_4
3	K_3	K_1	10	K_2	K_3	17	K_1	K_2
4	K_4	K_2	11	K_4	K_1	18	K_4	K_1
5	K_2	K_3	12	K_3	K_2	19	K_3	K_4
6	K_1	K_2	13	K_1	K_3	20	K_1	K_2
7	K_4	K_1	14	K_4	K_1	21	K_2	K_3

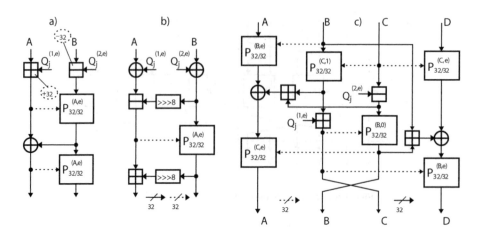

FIGURE 13.7   Procedure **Crypt**$^{(e)}$ in Cobra-F64a (a), Cobra-F64b (b), and Cobra-S128 (c).

as follows:

For $j = 1$ to $R - 1$, do: $\{(A, B): = \mathbf{Crypt}^{(e)}\left(A, B, Q_j^{(1,\, e)}, Q_j^{(2,\, e)}\right); (A, B): = (B, A).$

For $j = R$, do: $\{(A, B): = \mathbf{Crypt}^{(e)}\left(A, B, Q_j^{(1,\, e)}, Q_j^{(2,\, e)}\right)\}.$

Perform final transformation: $\{Y = (Y_l, Y_h) := (A \oplus Q_{R+1}^{(1,\, e)}, B \oplus Q_{R+1}^{(2,\, e)})$, where Y is the 64-bit output data block, for Cobra-F64b or $Y = (Y_l, Y_h) := (A -_{32} Q_{R+1}^{(1,\, e)}, B +_{32} Q_{R+1}^{(2,\, e)})$, where "$+_{32}$" ("$-_{32}$") denotes modulo 2^{32} addition (subtraction) for Cobra-F64a$\}.$

In addition to suitability to firmware, the ciphers Cobra-F64a, Cobra-F64a, and Cobra-S128 can also be efficiently implemented in cheap hardware providing high performance.

In Cobra-S128, the 128-bit input block X is divided into four 32-bit sub-blocks (A, B, C, and D) and data ciphering is performed using procedure **Crypt**$^{(e)}$ (Figure 13.7) as follows:

Perform initial transformation: $\{(A,B,C,D):= (A \oplus Q_1^{(1,\, e)}, B \oplus Q_2^{(1,\, e)}, C \oplus Q_3^{(1,\, e)}, D \oplus Q_4^{(1,\, e)})\}$

For $j = 1$ to $R - 1$, do: $\{(A,B,C,D):= \mathbf{Crypt}^{(e)}(A,B,C,D,Q_j^{(1,\, e)}, Q_j^{(2,\, e)}); (A,B,C,D):= (B,A,D,C)\}$

Do: $\{(A,B,C,D):= \mathbf{Crypt}^{(e)}(A,B,C,D, Q_R^{(1,\, e)},Q_R^{(2,\, e)})\}$

Do final transformation: $Y = (A,B,C,D):= (A \oplus Q_R^{(2,e)}, B \oplus Q_{R-1}^{(2,e)}, C \oplus Q_{R-2}^{(2,e)}, D \oplus Q_{R-3}^{(2,e)})\}$

13.8.2   FPGA Suitable Primitives

A CP box is a particular controlled SPN (CSPN) constructed using the switching elements $\mathbf{P}_{2/1}$ as standard building blocks, which is a minimum size controlled element (CE). The box $\mathbf{P}_{2/1}$ is only a particular variant of several possible minimum size controlled elements $\mathbf{F}_{2/1}$ (Figure 13.8) that can be described as two Boolean functions in three variables or as a switchable 2×2 substitution box and implements two different linear substitutions S_1 (if $v = 0$) and S_2 (if $v = 1$). Analogous to PNs, different types of the CSPNs constructed using CEs $\mathbf{F}_{2/1}$ can be applied as data-dependent operations (DDOs) suitable for designing fast hardware-oriented ciphers. Below we show that there exist 192 CEs of the $\mathbf{F}_{2/1}$-type CE that have advanced properties against $\mathbf{P}_{2/1}$. Use of the $\mathbf{F}_{2/1}$ CE provides for advancing the DDP-based approach to the design of hardware-oriented ciphers. The FPGA implementation of the $\mathbf{F}_{2/1}$ elements uses only 50 percent of the resources of two standard cells of some FPGA device. Therefore, another step to increase the efficiency of the FPGA implementation of the DDO-based ciphers is to use the $\mathbf{F}_{2/2}$ CEs controlled with two bits v and z (Figure 13.8). Elements $\mathbf{F}_{2/2}$ can be described as a pair of Boolean functions (BFs) in four variables (Figure 13.8) or as a set of four 2×2 substitutions (Figure 13.9) called modifications $\mathbf{F}_{2/2}^{(00)}$, $\mathbf{F}_{2/2}^{(01)}$, $\mathbf{F}_{2/2}^{(10)}$, and $\mathbf{F}_{2/2}^{(11)}$. Implementation of CE $\mathbf{F}_{2/2}$ consumes also two 4-bit memory cells, but uses 100 percent of

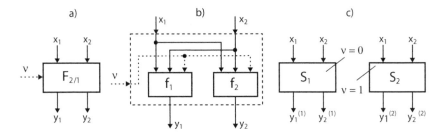

FIGURE 13.8 Element $F_{2/1}$ (a) represented as a pair of BFs in four variables (b) or as two 2×2 substitutions (c).

their resources. Two description forms of the CE are connected with the following expressions:

1. For $F_{2/1}$:

$$y1 = v(y1^{(1)} \oplus y1^{(2)}) \oplus y1^{(1)}$$

$$y2 = v(y2^{(1)} \oplus y2^{(2)}) \oplus y2^{(1)}$$

2. For $F_{2/2}$:

$$y_1 = vz\left(y_1^{(1)} \oplus y_1^{(2)} \oplus y_1^{(3)} \oplus y_1^{(4)}\right) \oplus v\left(y_1^{(1)} \oplus y_1^{(3)}\right) \oplus z\left(y_1^{(1)} \oplus y_1^{(2)}\right) \oplus y_1^{(1)}$$

$$y_2 = vz\left(y_2^{(1)} \oplus y_2^{(2)} \oplus y_2^{(3)} \oplus y_2^{(4)}\right) \oplus v\left(y_2^{(1)} \oplus y_2^{(3)}\right) \oplus z\left(y_2^{(1)} \oplus y_2^{(2)}\right) \oplus y_2^{(1)}$$

Using the notion of nonlinearity NL in sense of the distance from the set of affine BFs in the same number of variables, the following criteria have been used to select nonlinear CEs $F_{2/1}$ and $F_{2/2}$ suitable for designing efficient cryptographic DDOs[33]:

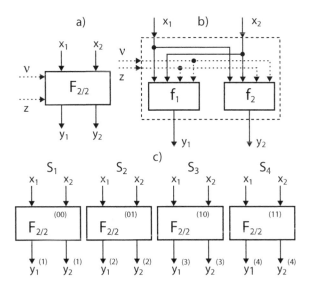

FIGURE 13.9 Element $F_{2/2}$ (a) represented as a pair of BFs in four variables (b) or as four 2×2 substitutions (c).

TABLE 13.5   Examples of $F_{2/1}$ and $F_{2/2}$ Nonlinear Controlled Involutions

$F_{2/1}$	$R_{2/1}$ $Q_{2/1}$	(e,g); (e,h); (e,i); (e,j); (f,g); (f,h); (f,i); (f,j); (g,e); (h,e); (i,e); (j,e); (g,f); (h,f); (i,f); (j,f) (g,h); (g,i); (h,g); (h,j); (i,g); (i,j); (j,h); (j,i)
$F_{2/2}$		(e,i,j,f); (e,g,h,f); (f,i,j,f); (e,h,g,e); (i,f,f,g); (e,j,i,e) (h,e,e,j); (f,g,h,f); (e,i,g,e); (e,h,g,e); (h,e,f,g); (f,j,i,e) (e,i,f,j); (e,i,e,j); (f,h,f,j); (e,i,f,g); (h,f,j,e); (e,g,f,h)

Each of two outputs of CEs should be a nonlinear BF having maximum possible nonlinearity NL for balanced BFs.

Each modification of CEs should be bijective transformation $(x_1, x_2) \rightarrow (y_1, y_2)$.

Each modification of CEs should be involution.

The linear combination of two outputs of CEs (i.e., $f = y_1 \oplus y_2$), should have maximum possible nonlinearity NL for balanced BFs.

There exist 24 different CEs $F_{2/1}$ and 2208 different CEs $F_{2/2}$ satisfying the above criteria 1 through 4. They implement only modifications shown in Figure 13.10. Table 13.5 presents all involutions $F_{2/1}$ and some examples of $F_{2/2}$.

Using the known topologies of PNs and replacing the switching element by the $F_{2/1}$ or by $F_{2/2}$ elements, one can design different CSPNs that can be used as DDO boxes while designing block ciphers suitable for FPGA implementation. Such DDO boxes, including mutually inverse and switchable ones,[34] are significantly more attractive as cryptographic primitives for the following reasons:

They are nonlinear operations, whereas the DDPs are linear ones.

Due to their good differential characteristics (DCs), the DDOs contribute to avalanche significantly better than DDP.

Figure 13.11 illustrates the DC of the $F_{2/1}$ and $F_{2/2}$ elements. Table 13.6 presents probabilities of the main DCs of nonlinear $F_{2/1}$ CEs and of the $P_{2/1}$ element. The $F_{2/1}$ elements possess two different types of DCs, $\Pr(\Delta_i^Y / \Delta_j^X, \Delta_k^V)$; therefore, they are denoted as $R_{2/1}$ and $Q_{2/1}$ elements.

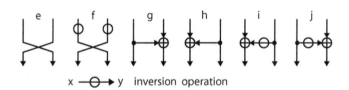

FIGURE 13.10   The 2×2 substitutions implemented by nonlinear controlled involutions.

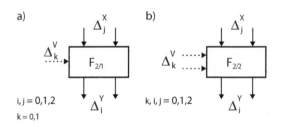

FIGURE 13.11   Differences corresponding to the DC of the $F_{2/1}$.

TABLE 13.6 Probabilities of Differential Characteristics of the $F_{2/1}$ and $P_{2/1}$ Elements

i	j	k	$Q_{2/1}$	$R_{2/1}$	$P_{2/1}$	i	j	k	$Q_{2/1}$	$R_{2/1}$	$P_{2/1}$
0	0	1	1/4	1/4	1/2	2	1	0	1/2	1/4	0
1	0	1	1/2	1/2	0	0	2	0	1	1/2	0
2	0	1	1/4	1/4	1/2	2	2	0	0	1/2	1
0	1	1	1/4	1/4	0	0	2	1	1/4	1/4	1/2
1	1	1	1/2	1/2	1	1	2	1	1/2	1/2	0
2	1	1	1/4	1/4	0	2	2	1	1/4	1/4	1/2
1	1	0	1/2	3/4	1	-	-	-	-	-	-

For the $F_{2/2}$ CEs, there exist many different types of DCs. To characterize DCs, one can use an integral parameter, called the average entropy and defined as follows:

$$H^* = \frac{\left(\sum_{j=0}^{2} \sum_{k=1}^{2} H_{jk} + \sum_{j=1}^{2} H_{j0} \right)}{8}$$

where $H_{jk} = -\sum_{i=0}^{2} p\left(\Delta_i^Y / \Delta_j^X, \Delta_k^V\right) \log_3 p\left(\Delta_i^Y / \Delta_j^X, \Delta_k^V\right)$ and $p\left(\Delta_i^Y / \Delta_j^X, \Delta_k^V\right)$ is the probability of having the output difference Δ_i^Y, if the input difference is Δ_j^X and the difference at the controlling input is Δ_k^V.

Table 13.7 presents classification of the $F_{2/2}$ involutions having maximum nonlinearity. There exist 128 different variants of the $F_{2/2}$ elements having maximum value $H^* = 0.84$.

The general structure of the DDO boxes constructed on the basis of the $F_{2/2}$ CEs is presented in Figure 13.12, where the controlled vector (V) of the box is represented as the concatenation of $2s$ components V_i and Z_i ($i = 1, \ldots, s$) having the length $n/2$ bits. Thus, the (ns)-bit controlling vector is represented as $V = (V_1, Z_1 \ldots, V_s, Z_s)$. If in the same structure the $F_{2/1}$ CEs are used as the main building blocks, then the controlling vector is $V = (V_1, V_2, \ldots, V_s)$.

Replacing the switching elements $P_{2/1}$ by CE $F_{2/1}$ or $F_{2/2}$ in the boxes $P_{32/96}$ that are internal parts of the switchable boxes $P_{32/32}^{(e)}$ and obtaining the switchable DDO boxes $R_{32/32}^{(e)}$ or $Q_{32/32}^{(e)}$, we can modernize the ciphers Cobra-F64a, Cobra-F64b, and Cobra-S128. Due to the advanced properties of the DDO boxes, such replacement provides secure reduction of rounds (see Table 13.8) and higher performance for the loop VLSI implementation architecture or lower cost for the pipeline architecture, especially while using the $F_{2/2}$ CEs. In an analogous way, one can update other known DDP-based ciphers — CIKS-1,[25] SPECTR-H64, CIKS-128,[30] Cobra-H64, and Cobra-H128[35] — in which some round keys are used to modify the controlling vector. In the last case, the DDO operations became key-dependent ones, which leads to vulnerability to some hypothetical differential related key attacks[36] because of the use of very simple key scheduling. Key-dependent DDOs are very attractive; however, the designer should pay attention to the mentioned attack while using simple key scheduling.

The Eagle-64 cipher considered below combines CSPNs with SPNs. The $F_{32/96}$, $F'_{32/96}$, and $F_{16/16}$ boxes represent the components corresponding to the CSPNs. One can note that this cipher also comprises some

TABLE 13.7 Average Entropy of the Maximum Nonlinearity Involutions $F_{2/2}$

H^*	0.84	0.834	0.815	0.813	0.812	$0.5 < H^* < 0.8$
# $F_{2/2}$	128	704	128	192	256	800

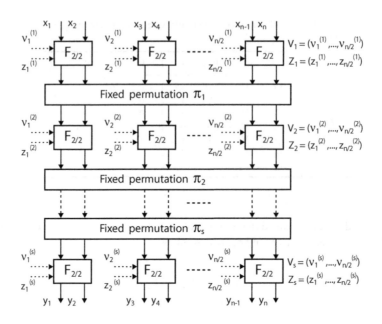

FIGURE 13.12   Controlled substitution-permutation network built up using the minimum size CEs $F_{2/2}$.

elements of the Feistel cryptoscheme: (1) the same algorithm is used to perform both the data encryption and the data decryption; and (2) in the round transformation, one of two enciphered data sub-blocks influences the transformation of the other one. A general encryption scheme and round transformation of Eagle-64 are presented in Figure 13.13. The structure of the $F_{32/96}$ and $F'_{32/96}$ boxes used in the round transformation is shown in Figure 13.13, where the boxes $F_{8/24}$ and $F_{8/24}^{-1}$ have the same topology as the corresponding boxes $P_{8/12}$ and $P_{8/12}^{-1}$. The box $F_{8/24}$ represents a cascade of eight elements $F_{2/2}$. Two mutually inverse SPNs used in the right branch are specified in Figure 13.14, where the 4×4 substitutions $S_0, \ldots, S_7$ are specified in Table 13.9 (specification of the $S_0^{-1}, \ldots, S_7^{-1}$ boxes can be easily derived from this table).

Subkeys $K_i \in \{0, 1\}^{32}$ of the 128-bit secret key $K = (K_1, K_2, K_3, K_4)$ are used directly in procedure **Crypt** as round keys Q_j (encryption) or Q'_j (decryption) specified in Table 13.10. The I_1 permutational involution is described as follows:

$$(1)(2, 9)(3, 17)(4, 25)(5)(6, 13)(7, 21)(8, 29)(10)(11, 18)(12, 26)$$
$$(14)(15, 22)(16, 30)(19)(20, 27)(23)(24, 31)(28)(32)$$

TABLE 13.8   Updating Some Firmware-Oriented DDP-Based Ciphers

Cipher	CEs		DDO Boxes		Secure Number of Rounds	
	Initial	New	Initial	New	Initial	New
Cobra-F64a	$P_{2/1}$	(i,e)	$P_{32/96}\ P_{32/32}^{(e)}$	$R_{32/96}\ R_{32/32}^{(e)}$	16	12
Cobra-F64b	$P_{2/1}$	(i,e)	$P_{32/96}\ P_{32/32}^{(e)}$	$R_{32/96}\ R_{32/32}^{(e)}$	20	16
Cobra-F64b	$P_{2/1}$	(e,i,j,f)	$P_{32/96}\ P_{32/32}^{(e)}$	$F_{32/192}\ F_{32/32}^{(e)}\ ^*$	20	12
Cobra-S128	$P_{2/1}$	(i,j)	$P_{32/32}^{(e)}$	$Q_{32/32}^{(e)}$	12	8
		(i,e)	$P_{32/32}^{(1/0)}$	$R_{32/32}^{(1/0)}$		

* A modified E box is used that forms 192-bit controlling vector.

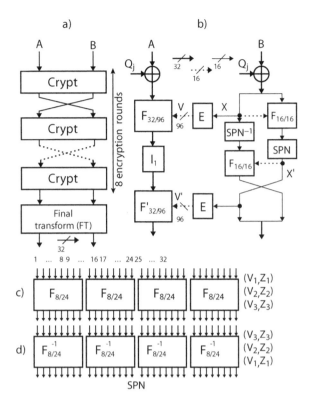

FIGURE 13.13 Eagle-64: general structure (a), one round (b) topology of the boxes.

The 96-bit controlling vectors V and V' corresponding to the $\mathbf{F}_{32/96}$ and $\mathbf{F}'_{32/96}$ boxes, respectively, are formed with the extension box $\mathbf{E}$ described as follows:

$$\mathbf{E}(X) = V = (V_1, Z_1, V_2, Z_2, V_3, Z_3)$$
$$V_1 = X; Z_1 = X^{>>>2}; V_2 = X^{>>>6}; Z_2 = X^{>>>8}; V_3 = X^{>>>10}; Z_3 = X^{>>>12}$$

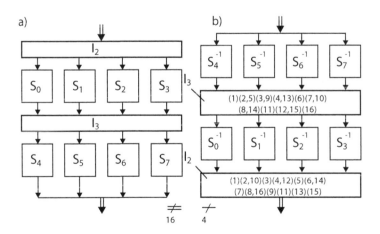

FIGURE 13.14 Mutually inverse operations.

TABLE 13.9 The 4×4 Substitution Boxes $S_0, \ldots, S_7$

S_0	14	4	13	1	2	15	11	8	3	10	6	12	5	9	0	7
S_1	3	13	4	7	15	2	8	14	12	0	1	10	6	9	11	5
S_2	10	0	9	14	6	3	15	5	1	13	12	7	11	4	2	8
S_3	1	4	11	13	12	3	7	14	10	15	6	8	0	5	9	2
S_4	10	6	9	0	12	11	7	13	15	1	3	14	5	2	8	4
S_5	11	8	12	7	1	14	2	13	6	15	0	9	10	4	5	3
S_6	10	15	4	2	7	12	9	5	6	1	13	14	0	11	3	8
S_7	1	15	13	8	10	3	7	4	12	5	6	11	0	14	9	2

where $X^{>>>b}$ denotes cyclic rotation of the word $X = (x_1, \ldots, x_{16})$ by b bits; that is, $\forall i \in \{1, \ldots, 16 - b\}$ so we have $y_i = x_{i+b}$, and $\forall i \in \{17 - b, \ldots, 16\}$ so we have $y_I = x_{i+b-16}$.

The 16-bit controlling vector (V_1, Z_1) of the $\mathbf{F}_{16/16}$ operation is described as follows:

$$V_1 = (x_1, \ldots, x_8) \quad \text{and} \quad Z_1 = (x_9, \ldots, x_{16})$$

The encryption algorithm of Eagle-64 is as follows:
For $j = 1$ to 7, do:

$$\{(L, R) \leftarrow \mathbf{Crypt}^{(e)} L, R, Q_j); (L, R) \leftarrow (R, L)\}$$

Perform transformation: $\{(L, R) \leftarrow \mathbf{Crypt}^{(e)}(L, R, Q_8)\}$
Perform FT: $\{(L, R) \leftarrow (L \oplus Q_9, R \oplus Q_9)\}$

Eagle-64 provides very fast encryption while constrained FPGA resources are utilized. Implementation comparison results show that it is significantly more efficient than conventional 64- and 128-bit block ciphers estimated with "Performance/(Cost·Frequency)" and "Performance/Cost" models.

13.9 Wireless Communications Security in the Near Future

The needs for personal wireless communications systems are growing rapidly. Coupled with this increase is telecommunication-related crime. In unwired networks, an invader with a suitable receiver can intercept transfer data. It is clear that such systems, although specified at a satisfactory level of security, are vulnerable. Security is a primary requirement of all wireless cryptographic protocols. Cryptographic algorithms are meant to provide secure communications applications. However, if the system is not designed properly, it can fail. Although there are many well-known ciphers, with different specifications and characteristics, the security of some of them remains questionable. Many works, from many different research groups, have been published in the technical literature in which cryptanalysis methods have been applied in an effort to find any existing black holes in the security strength of the encryption algorithms. From many points of view, such attempts offer valuable knowledge in the growth of and improvement in cryptography. Encryption algorithms must perform efficiently in a variety of current and future applications, doing different encryption tasks. The algorithms should be used to encrypt streaming audio and video data in real-time. They must work correctly in 32- and 64-bit CPUs. Many of today's applications run with

TABLE 13.10 Key Scheduling in Eagle-64 ($j = 9$ corresponds to final transformation)

j	1	2	3	4	5	6	7	8	9
Q_j (encryption)	K_1	K_2	K_3	K_4	K_1	K_4	K_2	K_3	K_2
Q'_j (decryption)	K_2	K_3	K_2	K_4	K_1	K_4	K_3	K_2	K_1

smart cards based on 8-bit CPUs, such as burglar alarms and pay-TV. All hardware implementations must be efficient, with less allocated area resources. This means simplicity in algorithm's architecture with sufficient "clever" data transformation components. A wireless protocol implementation demands low-power devices and fast computation components, which implies that the number and complexity of the encryption operations should be kept as simple as possible. A basic transformation in the operation of the encryption algorithms is needed, including modifications to the data blocks and key sizes.

The ciphers of the future must be key agile. Many applications need a small amount of text to be encrypted with keys that are frequently changed. Many well-known applications, such as IPSec, use this mode of algorithm operation. Although the most widely used mode of operation is encryption with the same key for all the transport data, the previous mode may also be very useful in future applications. Ciphers that require subkey precomputation have lower key agility due to computation time, and they also require extra RAM to hold the subkeys. This RAM requirement does not exist in the implementation of algorithms, which compute their keys during the encryption-decryption operation. Cellular phone technology demands hard specifications of the cryptography science. Ciphers must be compatible with the wireless devices' restricted standards in hardware resources. New mobile phones will have proper encryption parts built into them. In these devices, there is not enough room for a large integrated security layer. A solution to decrease the required hardware resources is to use ciphers for both bulk encryption and data integrity, with a simple change of their operating mode. All of the above make the effort of designing new security algorithms for wireless applications a really difficult process. However, this is the only way to change today's security wireless standards to succeed in solving the implementation problem. The AES and SHA-2 hash function standards are very good steps for the design of communications security schemes in the next few decades.

Technology's growth has led to many promises for the future of security. If the strength of the applied cryptography that is used in the wireless industry is increased enough, the security protocol would be sufficiently efficient to withstand the various attempts of attackers. Today, many ciphers can support the defense of the communications links in external invaders. On the other hand, the implementation of these is a difficult process and sometimes cannot meet the wireless network requirements. This is due to the fact that that the ciphers in use today were designed some years ago and for general cryptography reasons. They were not specifically designed for wireless communications. Security improvement needs strong, flexible encryption algorithms with efficient performance. New algorithms must be designed for this type of application. In addition, any new cipher designs will require invention. They are notoriously difficult to demonstrate or otherwise establish trust in. On the other hand, everything should be demonstrated in software before committing to hardware.

References

1. Bruce Schneier, *Applied Cryptography – Protocols, Algorithms and Source Code in C, second edition,* John Wiley & Sons, New York, 1996.
2. NIST, Advanced Encryption Standard Call, http://www.nist.gov/aes, 2003.
3. ISAAC Research Group at the University of California, Berkeley, http://www.isaac.cs.berkeley.edu/isaac/gsm.html.
4. GSM Pages at Smart Cards Developers Association, http://www.scard.org/gsm/bdy.html, 2004.
5. WAP Forum, Wireless Application Protocol Architecture Specification, www.wapforum.org, 2004.
6. Bluetooth Forum, Specification of the Bluetooth System, http://www.bluetooth.com, 2004.
7. S.H. Park, A. Gaz, and Z. Ganz, Security protocol for IEEE 802.11 wireless local area network, *Mobile Networks and Applications,* 3, 237–246, 1998.
8. J. Armstrong, R. Virding, C. Wikstrom, and M. Williams, *Concurrent Programming in Erlang, second edition,* Ericsson Telecommunications Systems Laboratories, http://www.ericsson.se/cslab/erlang.
9. L. Brown, A Current Perspective on Encryption Algorithms, presented at the UniforumNZ'99 Conference in New Zealand, April 2000.

10. M. Roe, Performance of block ciphers and hash functions-one year later, *Proc. Second Int. Workshop for Fast Software Encryption '94,* Leuven, Belgium, December 14–16, 1994.

11. J.L. Massey, G.H. Khachatrian, and M.K. Kuregian, SAFER+ Cylink Corporation's submission for the Advanced Encryption Standard, *First Advanced Encryption Standard Candidate Conference,* Ventura, CA, August 20–22, 1998.

12. B. Schneier and D. Whiting, Fast software encryption algorithms for optimal software speed on the Intel Pentium processor, *Proc. Fast Software Encryption Workshop,* 1997.

13. H. Nozaki, M. Motoyama, A. Shimbo, and S. Kawamura, Implementation of RSA algorithm based on RNS Montgomery multiplication, *Proc. CHES 2001, Lect. Notes Computer Sci.,* Vol. 2162, 2001, 364–376.

14. N. Sklavos and O. Koufopavlou, Asynchronous low power VLSI implementation of the International Data Encryption Algorithm, *Proc. 8th IEEE Int. Conf. Electronics, Circuits and Systems (ICECS'01),* Malta, September 2–5, 2001, Vol. III, 1425–1428.

15. G. Selimis, N. Sklavos, and O. Koufopavlou, Crypto processor for contactless smart cards, *Proc. IEEE Mediterranean Electrotechnical Conf. (MELECON'04),* Dubrovnik, Croatia, May 12–15, 2004.

16. K.H. Leung, K.W. Ma, W.K. Wong, and P.H.W. Leong, FPGA implementation of a microcoded elliptic curve cryptographic processor, *Proc. Field-Programmable Custom Computing Machines (FCCM'00),* 2000.

17. P. Kitsos, N. Sklavos, and O. Koufopavlou, Hardware implementation of the SAFER+ encryption algorithm for the Bluetooth system, *Proc. IEEE Int. Symp. on Circuits & Systems (ISCAS'02),* May 26–29, 2002, Vol. IV, pp. 878–881.

18. N. Sklavos and O. Koufopavlou, Architectures and VLSI implementations of the AES-Proposal Rijndael, *IEEE Trans. Computers,* 51(12), 1454–1459, 2002.

19. D.C. Wilcox, L.G. Pierson, P.J. Robertson, E.L. Witzke, and C. Gass, A DES ASIC suitable for network encryption at 10 GPS and beyond, *Proc. CHES'99,* 1999, 37–48.

20. Y.-K. Lai, L.-G. Chen, J.-Y. Lai, and T.-M. Parng, VLSI architecture design and implementation for Twofish block cipher, *Proc. IEEE Int. Symp. on Circuits & Systems (ISCAS'02),* May 26–29, 2002.

21. K. Marinis, N.K. Moshopoulos, F. Karoubalis, and K.Z. Pekmestzi, On the hardware implementation of the 3GPP confidentiality and integrity algorithms, *Proc. 4th Int. Conf. for the Information Security, ISC,* October 1-3, 2001, Malaga, Spain, pp. 248–265.

22. J. Goodman and A.P. Chandrakasan, An energy-efficient reconfigurable public-key cryptography processor, *IEEE J. Solid-State Circuits,* 36(11), 1808–1820, November 2001.

23. N. Sklavos, P. Kitsos, E. Alexopoulos, and O. Koufopavlou, Open Mobile alliance (OMA) security layer: architecture implementation and performance evaluation of the integrity unit, in *New Generation Computing: Computing Paradigms and Computational Intelligence,* Springer-Verlag, Berlin, 2004.

24. Advanced Encryption Standard, *Proc. 4th Int. Workshop, Fast Software Encryption – FSE '97, Lecture Notes in Computer Science,* Vol. 1267, E. Biham, Ed., Springer-Verlag, Berlin, 1997, pp. 83–87.

25. A.A. Moldovyan and N.A. Moldovyan, A cipher based on data-dependent permutations, *J. Cryptol.,* 15(1), 61–72, 2002.

26. M. Portz, A generallized description of DES-based and Benes-based permutation generators, *Lecture Notes in Computer Science,* 718, 397–409, 1992.

27. M. Kwan, The design of the ICE encryption algorithm, *Proc. 4th Int. Workshop, Fast Software Encryption — FSE'97, Lect. Notes Comput. Sci.,* Berlin: Springer-Verlag, Vol. 1267, 1997, pp. 69–82.

28. A. Biryukov, Methods of Cryptanalysis, research thesis, Israel Institute of Technology, Haifa, September 1999.

29. N.A. Moldovyan, N.D. Goots, P.A. Moldovyanu, and D.H. Summerville, Fast DDP-based ciphers: from hardware to software, *Proc. 46th IEEE Midwest Symp. on Circuits and Systems,* Cairo, Egypt, December 27–30, 2003.

30. N.D. Goots, B.V. Izotov, A.A. Moldovyan, and N.A. Moldovyan, Modern cryptography: protect your data with fast block ciphers, A-LIST Publishing, Wayne, 2003 (www.alistpublishing.com).

31. Ch. Lee, D. Hong, Sun. Lee, San. Lee, S. Yang, and J. Lim, A chosen plaintext linear attack on block cipher CIKS-1, *Lect. Notes Comput. Sci.*, Springer-Verlag, Berlin, Vol. 2513, pp. 456–468.

32. Y. Ko, D. Hong, S. Hong, S. Lee, and J. Lim, Linear cryptanalysis on SPECTR-H64 with higher order differential property, *Proc. Int. Workshop, Methods, Models, and Architectures for Network Security, Lect. Notes Comput. Sci.*, Springer-Verlag, Berlin, Vol. 2776, 2003, pp. 298–307.

33. A.A. Moldovyan, N.A. Moldovyan, and N. Sklavos, Minimum size primitives for efficient VLSI implementation of DDO-based ciphers, *Proc. MELECON 2004*, Dubrovnik, Croatia, May 12–15, 2004.

34. N.A. Moldovyan, M.A. Eremeev, N. Sklavos, and O. Koufopavlou, New class of the FPGA efficient cryptographic primitives, *Proc. ISCAS 2004*, Vancouver, Canada, May 23–26, 2004.

35. N. Sklavos, N.A. Moldovyan, and O. Koufopavlou, High speed networking: design and implementation of two new DDP-based ciphers, *Mobile Networks and Applications,* Kluwer Academic Publishers, Vol. 25, Issue 1–2, pp. 219–231, 2005.

36. Y. Ko, S. Hong, W. Lee, S. Lee, and J.-S. Kang, Related key differential attacks on 27 round of XTE and full-round GOST, *Proceedings of the 11th International Workshop, Fast Software Encryption – FSE '2004, Lect. Notes Comput. Sci.* Springer-Verlag, Berlin.

14

Reliable Computing in Ad Hoc Networks

Patrick Th. Eugster

14.1   Reliable Computing in Unpredictable Ad Hoc Networks

Through the advent of mobile devices and ad hoc communication between these devices, a plethora of issues have been added to the already challenging area of networked distributed computing. The extraordinary dynamism (e.g., node unavailability through mobility) in such settings makes it yet harder to reason about the success of distributed executions. In many cases, nevertheless, one would like to be able to rely on certain actions, or at least have an idea of the state of their completion. In other terms, users and components require guarantees, ideally statically insurable and verifiable, on the *interaction* (one would want *fault tolerance* to a certain amount of links, or node failures) in distributed computations, as well as on the *data* semantics (*type safety* is desirable for any interaction between two parties) of distributed applications.

Addressing this issue is already hard in wired networks, even without considering guarantees on *data*. This can be measured by the size of an entire community whose efforts, devoted to "dependable distributed computing," has been nourished over the years by issues arising from the desire to achieve

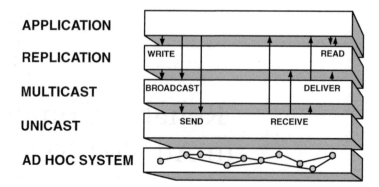

FIGURE 14.1 Building probabilistic reliable algorithms through successive layering. Overview of primitives and corresponding layers discussed in this context.

fault tolerance in distributed computations. Seminal work in that community has focused primarily on providing algorithms with "exactly once"-like guarantees, based on rigorous assumptions made on the underlying system. Driven by the desire to achieve deterministic guarantees, those assumptions appear as deterministic properties. Based on these properties, especially in the most prominent *asynchronous system model*[9] which actually only ensures few properties, costly algorithms are required to achieve the targeted guarantees.

Achieving "exactly once" guarantees in ad hoc networks is even harder, as it is already hard, if not impossible, to define a deterministic system model. Actually, not much effort has been put into providing any widely accepted accurate system model for ad hoc networks. The field is still very young, and the community does not entirely agree yet on a single, compound system model, or even target setting, and it is not even sure whether a single model would do for all purposes given the wide variety of assumptions made in certain contexts and avoided in others (e.g., node mobility, static backbone, multihop).

One way out of this shortage that might be applicable to various system models is to work with *probabilistic guarantees* and properties expressing a notion of partial success, rather than with deterministic ones. This approach has already led to interesting solutions in wired networks (e.g., *probabilistic* atomic broadcast,[7] *probabilistic* leader election[10]) when striving for scalability, and seems intuitively even more interesting in ad hoc networks, due to the aforementioned dynamic nature of such settings. Tightly coupled with the notion of probabilistic guarantees is that of *probabilistic algorithms*. These are indeed ideal candidates to provide probabilistic guarantees—guarantees of great usefulness in practice when associated with high probabilities. In particular, algorithms inspired by mother nature herself tend to fall into this category. Ideally, but not necessarily, stochastics can then be used to approximate probabilistic guarantees and tune algorithm parameters.

The goal of this chapter is to bridge the gap between the disparate views of reliability-centered distributed systems and communication-centered networking. Following the usual approach of building (reliable distributed) systems by successive layering, we investigate and argue for probabilistic guarantees and algorithms at three levels, which are (1) *unicast routing*, (2) *multicast routing*, and (3) *data replication* (see Figure 14.1). These illustrations do not include overhead measures, although the depicted algorithms have been shown to mostly incur even lower cost than most deterministic solutions to the same respective problems. The reader is referred to cited literature for such evaluations.

14.2   Modeling the System

Before delving into the presentation of the aforementioned examples, we outline our assumptions made on the system. Clearly, the goal of this chapter is not to define, or to propose, a globally valid system model. Nevertheless, a set of assumptions on the targeted environment has been necessary.

We suppose a set of uniquely identified nodes $\Pi = \{n_1, \ldots, n_s\}$ with fixed identical transmission ranges. When communication is possible between two nodes, communication manifests a latency smaller than that of any algorithm run. For presentation simplicity, each node is viewed as a single computationally active entity, meaning that the terms "process" and "node" are equivalent.

We choose a peer-based, symmetric view. Mobility and failure patterns are homogenous, meaning that they are associated with means and probabilities, respectively, identical for all nodes for a given algorithm execution. Nodes can fail by crashing, or can be deliberately switched off. When considering a node involved in a distributed computation, both kinds of events can similarly jeopardize the outcome of that computation. A *correct* node is hence one that does not crash and is not powered off. A node moving out of the transmission range of neighbor nodes and hence invalidating routing paths to it can similarly lead to an unsuccessful completion of an algorithm run. This kind of event is hence combined under the hood of a single probability P_f with the two kinds of events defining node correctness.

When drawing comparisons with wired networks, from which we usually start, the asynchronous system model[9] is assumed.

14.3 Unicast Routing

When reasoning about distributed computations in a message passing model, one of the basic ingredients is the *communication channel*, which describes the semantics of communication between two nodes.

14.3.1 Guarantees

Communication channels provide an abstraction for network communication through two primitives, SEND and RECEIVE. The two following guarantees define any useful channel:

U1 (no-creation). If a node n_2 RECEIVEs a message m, then some node n_1 has made a SEND of m to n_2.
U2 (no-duplication). Every message m sent through SEND by any node n is RECEIVEd at most once.

The thereby chareacterized *unreliable channel* provides very weak guarantees. A common channel abstraction mentioned in distributed systems literature focusing on wired networks is the *reliable channel*, a necessary assumption for solving many distributed problems (such as related to agreement). A reliable channel is characterized by U1, U2, and the following strong assumption:

U3 (reliability). If a node n_1 SENDs a message m to a node n_2, and n_2 does not crash, then n_2 eventually RECEIVEs m.

In addition, these channels are commonly assumed to provide FIFO ordering of RECEIVEd messages with respect to the invocations of SEND, in addition to being bidirectional.

In much of the literature on ad hoc networks, reliable channels are supposed *between any two nodes n_1 and n_2 in communication range*, that is, by any two nodes separated by a single hop. Often, reliable channels are furthermore supposed by transitivity *between any two nodes n_1 and n_2 at the two extremities of a chain of single hops*. This assumption provides the illusion of a homogenous (topology-oblivious) system and aids when reasoning about decentralized distributed algorithms. When considering node crash failures and particularly node movements occurring during the execution of a distributed algorithm in an ad hoc network, this assumption hardly makes sense. Also, restrictions in resources cause most unicast routing algorithms to apply more of a best-effort policy than trying until completion. In other terms, we believe reliable channels are a reasonable assumption for single hops, but cannot in the face of failures be extended to multiple hops.

A weaker assumption on communication originating from wired networks is captured by the *fair lossy channel*,[1] which is defined by U1, U2, and the following, weaker alternative to U3:

U4 (fairness). If a node n_1 SENDs an infinite number of messages to node n_2 and n_2 does not crash, then n_2 RECEIVEs an infinite subset of these messages.

Fair lossy channels ensure that, when a nodes repeatedly SENDs a message, that message will eventually be RECEIVEd at the destination node. This assumption is, in the face of node mobility, still too strong, as it might take an unbounded time span until there is a valid routing path between two nodes.

As a response to this, we propose a *probabilistic reliable channel*, whose associated probability of successful transmission relies on the model of the system and the considered unicast routing algorithm. This kind of channel is characterized by U1, U2, and the following guarantee:

U5 (probabilistic reliability). If a node n_1 SENDs a message m to node n_2, then with probability P_u, n_2 eventually RECEIVEs m.

P_u thus also depends on P_f, as a probabilistically reliable link can traverse any number of nodes. P_u can be a function of various further parameters, including even n_1 and n_2, or the moment of observation, i.e., time. The fact that the RECEIVEr node itself can crash, be switched off, or move out of range is captured by P_u as well, and hence that node does not have to be assumed to be correct (as in U3 and U4). The fact that the position and correctness of the nodes in an ad hoc network appears already in the specification of communication channels seems intuitive. Unlike in wired networks, no fixed infrastructure is provided (e.g., routers) that could help factor out hosts other than those at the origin of a message exchange and hence abstract from the network itself.

14.3.2   Algorithms

What makes this notion of probabilistic reliable channel attractive is that any kind of routing algorithm, even without a probabilistic system model, can be shown to implement the specification with some P_u. P_u can be expressed through analytical results, but can also be simply obtained from simulations or measurements. Indeed, there are a considerable number of factors influencing the reception of messages, such as the precise models for node failures, node mobility and transmission failures, network topology, percolation, and obviously also the routing algorithm itself. Some of these factors are hard if not impossible to capture analytically.

Intuitively, algorithms that are themselves non-deterministic, such as *ant-based* algorithms, seem particularly appealing for such a probabilistic notion of communication channel. Ant-based algorithms are based on the principle of *stigmergy* consisting in basing choices between several alternatives on the outcome of previous selections of these respective alternatives.[2] Before being applied for selecting routing paths, this paradigm was first successfully employed to approximate solutions to the traveling salesman problem, as well as several other NP-hard problems (e.g., sequential ordering, quadratic assignment, vehicle routing, scheduling, graph coloring, partitioning). In the context of unicast routing, ant-based algorithms appeared first in wired networks before being adapted to ad hoc networks.

Ants act as mobile agents conveying routing information, gathered at nodes in the form of tuples (*dest, nbor, prob*). Such a tuple represents the *prob*ability with which a given *neighbor* node will be chosen as next relay when routing a message toward a given *dest*ination from the current node. Probabilities are updated by ants to reflect the costs (e.g., number of hops) associated with the route choice learned from previous experiences. These probabilities are comparable to the *pheromone* concentrations left behind by ants, and usually also decrease periodically.

From there, algorithms differ substantially. Ants can be proactively sent out periodically to continuously update routing tables, or reactively in on-demand routing protocols. The probable sole concept common to all prominent representatives of this class of algorithms is *forward* ants, which are used to carry information on traversed nodes. Especially in proactive algorithms, this information can be used by these traversed nodes. Alternatively, this information can be sent back to previously passed nodes by *backwards* ants, as in most reactive algorithms inspecting the network only in the context of a route request. Forward ants can be unicast, but also multicast (e.g, single-hop broadcasts). They can be handled just like any messages at a given destination, in the sense that their next node is chosen with the probabilities from the routing tables (*regular* ants), or the next node can be chosen uniformly out of the neighbor nodes (*uniform* ants) to enforce the (re)exploration of (old) routes.

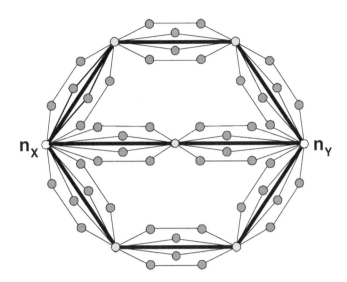

FIGURE 14.2 Sample interconnection graph of two nodes with $l_{1,1} = 0, l_{1,2} = 1, l_{1,3} = 2, l_{2,1} = 1, l_{2,2} = 2, l_{2,3} = 2$.

Algorithms relying on this concept of stigmergy are both *reliable* and *efficient*, and these two desirable properties can be traded one for the other. Eventually, choices leading to success are selected with high probability, and faulty solutions (e.g., missing links) are naturally avoided. Nevertheless, by working with probabilities, new solutions are continuously sought, and especially when combining ants with regular communication, a low overhead can be achieved. Increasing the size of the network requires only little more ants to retain the same reliability. Increasing the number of ants increases the traffic, but also reliability, and vice versa.

We next illustrate probabilistic reliable channels in the context of an abstract ant-based routing algorithm.

14.3.2.1 Ant-Based Unicast Routing

Our ant-based routing algorithm constructs tables of the form presented above, which are used in all prevalent ant-based algorithms (e.g, see Refs. 3, 11, 22). The *prob*ability of choosing a given *neighbor* node as next hop toward the message *dest*ination is proportional to the inverse of the shortest possible ("min") path achievable through that neighbor. We assume *optimistic* routing, meaning that a node receiving a message (1) acknowledges the reception to the SENDer, (2) forwards the message to the next node (different from the SENDer), and (3) forgets about the message upon reception of acknowledgment from the latter node. We make *pessimistic* assumptions about failures: if nodes crash, they do so after acknowledging messages but prior to forwarding them, that is, between (1) and (2).

When working with probabilities, the question (just like with any approximation) is the resilience. This question is particularly important with respect to ant-based algorithms, as these are hard to analyze. This seems plausible given the fact that these algorithms have been introduced precisely as practical approaches to approximate solutions for computationally hard problems. We thus start by illustrating the compliance of the above algorithm with our specifiation of probabilistic reliable channels through a topology-aware network model, before averaging to achieve more concise expressions at the expense of weaker resilience.

We suppose reliable channels (with $P_u = 1$) for every pair of nodes within transmission range. We consider a pair of nodes n_x and n_y, and assume that these are separated by $l_{1,i} \geq 0$ paths with $min = i$, $1 \leq i \leq B$. Each link in such a path can itself be subdivided; that is, each such link contains in fact $l_{2,j}$ paths with $min = j$, $1 \leq j \leq B$. To respect the minimum values of the first subdivision, $l_{2,1} \geq 1$. This subdivision can be recursively repeated. Figure 14.2 depicts an interconnection (not complete network) graph with two subdivisions.

Supposing that at the moment of observation routing information is up-to-date, the probability $p_{k,i}$ of choosing a path of min length i at a node of subdivision k when routing messages from source n_x to the destination n_y is hence given by:

$$p_{k,i} = \left(\frac{\left(\sum_j \frac{1}{l_{k,j}} \right)^{-1}}{l_{k,i}} \right)$$

The probability of successfully traversing a route of min length i at the kth subdivision, $s_{k,i}$, is recursively given by (with d subdivisions, $s_{d+1,i} = p_{d+1,i} = 1 \ \forall i$):

$$s_{k,i} = \left[\sum_j s_{k+1,j} \ p_{k+1,j} \right]^i$$

U5 is thus satisfied with:

$$P_u = \sum_i p_{1,i} s_{1,i}$$

An ant-based (or deterministic) algorithm that always selects the min path trivially has the highest probability of succeeding with $P_u = (1 - P_f)^{min}$, yet loses the low-cost inherent collection and propagation of routing information, and makes the respective path a bottleneck and a single point of failure. Several messages might be lost when a link on that path breaks before another path is chosen.

In settings that are rather homogenous, it can be interesting to average over the entire network. For any two nodes with an existing route in between them, P_u can then be approximated by observing the distribution of the routing path lengths, and weighting these by the respective probabilities of successful transmission over routes of such lengths:

$$P_u = \sum_h (1 - P_f)^h P(H = h) = E_H \left[(1 - p_f)^H \right] \tag{14.1}$$

14.4   Multicast Routing

Another basic building block of distributed computing is made up of primitives providing an abstraction of *multicast communication channel* for disseminating copies of data to multiple nodes in a group $\Delta \subset \Pi$.

14.4.1   Guarantees

The problem of *reliable broadcast*[13] is traditionally defined in wired networks by two primitives, BROADCAST and DELIVER, and the following desired guarantees ensured for nodes in Δ:

B1 (integrity). For every message m, every correct node DELIVERs m at most once, and only if m was previously BROADCAST by node *sender(m)*.

B2 (validity). If a correct node BROADCASTs a message m, then that node eventually DELIVERs m.

B3 (agreement). If a correct node n_1 DELIVERs a message m, then eventually, every other correct node n_2 DELIVERs m.

In wired networks, a crash failure-tolerant reliable broadcast is straightforwardly (yet in a costly manner) implemented based on reliable channels. Now, with weaker guarantees provided by channels, seeking a weaker specification for reliable broadcast seems natural. We define *probabilistic reliable broadcast* through B1, B2, and the following probabilistic guarantee:

B4 (probabilistic agreement). If a correct node n_1 DELIVERs a message m, then eventually, any given other node n_2 DELIVERs m with probability P_m.

In other terms, a subset of size $P_m|\Delta|$ of nodes in Δ are expected to DELIVER on average a broadcast message. Similarly to guarantee C5, nodes are not assumed to be correct for agreement, as faulty nodes are captured by the introduced probability.

Depending on the application requirements, an alternative specification could consist of combining B1 and B2 with a guarantee expressing the *probability with which all nodes* DELIVERing a given broadcast message, rather than the *average fraction of nodes* doing so:

B5 (probabilistic atomic agreement). If a correct node DELIVERs a message m, then eventually, with probability P_a, all other correct nodes DELIVER m.

This approach, however, demands a more precise definition of *which* nodes are considered to be in a multicast group (and which are not). In contrast, the first definition gives more freedom in that respect. Through its flavor of "partial coverage," B4 naturally embraces weaker (and in practice predominantly encountered) notions of membership consistency.

14.4.2 Algorithms

Just like ant-based unicast routing algorithms have been inspired by observing mother nature at her best as ants mark paths in the process of stigmergy, so-called *gossip-based* broadcast routing algorithms have been inspired by *epidemiology*, the study of the (although less benign, still instrumental for our purposes) spreading of diseases throughout populations. In gossip-based algorithms, cooperating nodes continuously share parts of their "state" (e.g., broadcast messages to be DELIVERed) with a subset of other nodes ("gossiping" to those nodes). With high probability, after a certain time, if shared/exchanged states are appropriately portioned and gossip destination nodes are adequately chosen, all nodes will have seen all relevant state information. Algorithms differ in what nodes exchange, and how such an exchange takes place. Gossips can be used to propagate the actual payload (e.g., broadcast messages), digests of RECEIVEd messages, or both. Digests can be used by gossip RECEIVErs to update gossip SENDers, but also to query gossip SENDers for further, possibly older, messages.

Gossip-based algorithms were first introduced in wired networks for replicating databases,[5] before being praised more generally as a promising alternative to deterministic broadcast algorithms. The appeal of gossip-based algorithms is threefold. First, through the limited number of destinations a node gossips to, and the limited number of times such a gossiping occurs for given information, gossip-based algorithms are *efficient* with respect to the broadcast group size (typically, "only" $O(N \log N)$ unicast transmissions are required, $N = |\Delta|$ being the size of a broadcast group). Second, the redundancy achieved through random selection of gossip destinations nevertheless makes these algorithms *reliable* despite transmission and node failures (the probability of DELIVERy of a broadcast message can come arbitrarily close to 1). Third, mathematical models from epidemiology make these algorithms *predictable*, and in particular, permit trading between efficiency and reliability by fine-tuning their parameters.

14.4.3 Gossip-Based Multicast Routing

We present an overview of a gossip-based broadcast algorithm we proposed for ad hoc networks, which builds on the notion of probabilistic reliable channel introduced in Section 14.3.1.

The Route-Driven Gossip (RDG)[17] algorithm has been inspired by a broadcast algorithm for wired networks,[6] gossiping both about membership information and broadcast messages. In contrast to the Anonymous Gossip (AG) algorithm,[4] a seminal gossip-based approach using MAODV[21] for a preliminary rough dissemination of broadcast messages, RDG is a "pure" gossip-based algorithm relying on DSR[14] *without* inherent broadcast primitives.

Periodically, gossiping nodes select F other nodes in the group to gossip to, and messages are forwarded (at most) τ times. Gossips propagate broadcast messages directly, and piggyback digests of messages

RECEIVEd by their respective source nodes, as well as a subset of the membership view (nodes in the group and routing paths for these) of these nodes. A node that RECEIVEs a gossip DELIVERs broadcast messages (*push gossip*) not seen so far, provides the gossiper node with any messages that did not appear in the digest from the latter node (*pull gossip*), and updates its membership view.

Being a pure gossip algorithm, RDG can be analytically captured.[17] Supposing synchronized rounds for analysis, $s_r \in \{1, \cdots, n\}$ and $\Delta s_r = \mathbf{E}[s_r - s_{r-1}]$ denote the number of nodes that RECEIVE a given broadcast message *after* round r and the average number of nodes that RECEIVE it *within* round r, respectively.

The probability p that a given node RECEIVEs the message at any round is a combination of the two conditions that (1) the considered node is chosen as gossip destination and (2) the message is successfully RECEIVEd (Equation 14.1 in Section 14.1):

$$p = \left(\frac{F}{N-1} \right) E_H[(1 - p_f)^H]$$

With $s_r = i$ and $\sum_{t=1}^{\tau} \Delta s_{r+1-t} = \delta$ in the current round, X_k is a random variable denoting for each of the remaining $N - i$ nodes whether it RECEIVEs the message in the next round. The probability that j nodes have RECEIVEd the message by the next round is expressed as:

$$p_{(i,j,\delta)} = P\left(s_{r+1} = j | s_r = i, \sum_{t=1}^{\tau} \Delta s_{r+1-t} = \delta \right)$$

$$= P\left(\sum X_k = j - i \right) \quad (j \geq i, \ 0 \ otherwise)$$

$$= \binom{N-i}{j-i} (1 - (1-p)^{\delta})^{j-i} (1-p)^{\delta(N-j)}$$

The distribution of s_r, v_r is then computed based on the transition matrix $\mathcal{P}_{\delta} = \{p_{(i,j,\delta)}\}$ and the initial distribution $v_0(1) = 1$:

$$v_{r+1}^T = v_r^T \mathcal{P}_{\delta} \tag{14.2}$$

where $v_r(i) = P(s_r = i)$ is the ith element of the column vector v_r, and v is the respective value after the final round.

Hence, B4 is satisfied with P_m expressed in terms of $v(i)$ as follows:

$$P_m = \frac{1}{N} \sum_{i=1}^{N} i \, v(i)$$

Luo et al.[17] furthermore investigated (analysis and simulation) the DELIVERY of streams of broadcast messages, the overhead in terms of unicasts necessary to disseminate a broadcast message, as well as an optimization (only simulation), called Topology-Aware RDG (TARDG). As its name suggests, TARDG makes use of an estimation of the topological disposition of nodes in a multicast group. Gossip destinations are chosen with probabilities inversely proportional to the number of hops required to reach them. TARDG has the same overhead as RDG, yet achieves a higher P_m. Readers may refer to Ref. 17 for pseudo code of the RDG algorithm, as well as a comparison of simulation results with analytical results.

14.5   Data Replication

While broadcast primitives enable the dissemination of data copies to a set of distributed nodes, they do not provide any means of coordinating operations performed then on individual copies. Specific primitives are (deemed) necessary for replicating data for high availability in a way such that concurrent accesses respect a precise *consistency model*.

14.5.1 Guarantees

Concurrently accessed data behaves ideally as implemented through (synchronous) shared memory. In other terms, access to such data is governed by a *strict consistency*[20] model, which states that shared data accessed by concurrent events e_i of types READ or WRITE satisfies the following guarantees:

C1 (order). There exists an ordering $\{e_i\}_i$ of events such that if any node perceives an event e_i as occurring before another event e_j, then $i < j$.

C2 (freshness). The value yielded by a READ e_i is the value of the WRITE e_j with $j < i$ such that $\nexists$ a WRITE e_k with $j < k < i$.

C3 (global sequencing). For any two events e_i and e_j, $i < j$ if and only if e_i was issued before e_j.

Here, events are considered as atomic and instantaneous. The fact is that in a distributed execution, the effect(s) of these operations take place somewhere in between the invocation of the primitive and its returning.

In any case, implementing strict consistency (more precisely, guarantee C3) requires globally synchronized clocks or logical timestamps,[15] which are both very expensive to achieve. Actually, the semantics of many shared objects do not even demand for such strong guarantees as provided by C3. The *sequential consistency*[16] model is defined by C1, C2, and the following substitute for C3:

C4 (local sequencing). If e_i and e_j are issued by the same node n, then $i < j$ if and only if e_i was issued before e_j.

In many contexts, it is sufficient to provide a single node the possibility of performing WRITE accesses to replicated data, as described by the *shared-private*[8] model. This guarantee, which strongly simplifies the implementation of any of the above consistency models when combined with it, can be formalized as follows:

C5 (single writer). The issuer of every WRITE event e_i, $issuer(e_i)$, is the same node.

Ensuring that every WRITE is perceived, and in particular, that every READ yields the last WRITE preceding it in time is hard in an ad hoc setting, when nodes but also communication can fail. We define *probabilistic shared-private sequential consistency* through C1, C4, C5, and the following guarantee:

C6 (probabilistic freshness). The value yielded by a READ e_i is the value of the WRITE e_j with $j < i$ such that with probability P_r $\nexists$ a WRITE e_k with $j < k < i$.

Just like in the case of M4, many interesting alternatives exist. One could, for instance, choose a probabilistic guarantee splitting P_r into probabilities $P_r(k)$ for obtaining the value of any of the kth last WRITEs. We are, however, interested in algorithms with probabilistic guarantees that are in practice ensured with high probabilities, meaning that $P_r(k)$ can be expected to very quickly decrease for an increasing k.

14.5.2 Algorithms

In the following we will outline an algorithm implementing this specification of probabilistic shared/private sequential consistency. The algorithm is inspired by the concept of *probabilistic quorum systems*,[19] and makes use of a subalgorithm of the probabilistic reliable broadcast algorithm presented in Section 14.4 for the dissemination of READs and WRITEs.

The principle underlying all quorum systems consists of selecting a set of nodes (*storage set*) to host replicas of data, and having clients issue READs and WRITEs only to subsets of nodes (quorums) in that set. The *access protocols* defining the intersections and interactions of the different quorums are then designed to specifically ensure the suggested consistency model. Quorum systems are particularly suited for weaker consistency models, which do not require *every node* in a storage set to perform *every operation* on the shared data, as known from the *state machine* approach,[15] or where certain inconsistencies can be tolerated or eventually repaired.

In probabilistic quorum systems, quorums intersect with a high probability (only), however, requiring access protocols with a substantially smaller complexity than in the original (strict) quorum systems. Probabilistic quorum systems have been proven to be able to tolerate $\omega(N)$ node failures, by imposing a load of $O(1/\sqrt{N})$ on every node only.[19]

14.5.3 Probabilistic Quorums

PAN (Probabilistic quorum system for Ad hoc Networks)[18] is a probabilistic quorum system implementing probabilistic shared-private sequential consistency. In contrast to the seminal approach presented in Ref. 12, PAN uses an *asymmetric* quorum construction. More precisely, the size of WRITE quorums is larger than that of READ quorums, as the former kind of events are assumed to occur less often than latter kind.

A client issues a READ or a WRITE through an *agent*, which is a "close" node in the storage set. Requests are propagated within the storage set through a subalgorithm of RDG leading to high probabilities of intersection for individual quorums.

As there is only one client node issuing WRITEs, and these events are tagged by distinct increasing timestamps, nodes in the intersection of two WRITE quorums can easily determine which value is more recent. A node issuing a READ can very well obtain an outdated value, namely if the READ quorum has no intersection with the latest WRITE quorum, or if all replies sent from nodes in the intersection are lost. Both these events are captured by the probability P_c in C6. For analytical evaluation (not presented in Ref. 12), Luo et al.[18] assume a storage set of size N, and arrivals of READs and WRITEs following independent poisson processes of intensities λ_R and λ_W, respectively. $\hat{\xi}_R$ and $\hat{\xi}_W$ represent the number of nodes that RECEIVE on average a READ and WRITE request in quorums of ξ_R and ξ_W servers, respectively.

P_r is the probability that a READ quorum intersects the most recent WRITE quorum. There exists an $\bar{r}$ for which the diffusion process of that WRITE is terminated. For analysis, the time axis after that event β is divided into $\bar{r} + 1$ intervals, as shown in Figure 14.3. A READ quorum, resulting from a query happening in between two consecutive gossip rounds r and $r + 1$ intersects a WRITE quorum of size $\hat{\xi}_W^r$ with a distribution ν_r (see Equation 14.2). To find the probability of intersection, we need to calculate the READ quorum size $\hat{\xi}_R$ (with a distribution μ) and p_r, the probability that the query event occurs in between rounds r and $r + 1$.

Denoting by $p = \mathbf{E}_H[(1 - p_f)^{2H}]$ the probability that the agent forwarding a query obtains the reply from a server belonging to the corresponding READ quorum, the distribution of $\hat{\xi}_R$, conditioned on $\xi_R = s$, is calculated as follows (considering $P(\hat{\xi}_R = 1|\xi_R = 1) = 1$):

$$\mu(k) = P(\hat{\xi}_R = k|\xi_R = s) \ (k \in [1,s], \ 0 \ otherwise)$$
$$= P(\hat{\xi}_R = k - 1|\xi_R = s - 1) \cdot p + P(\hat{\xi}_R = k|\xi_R = s - 1) \cdot (1 - p)$$

The estimation of μ is somewhat conservative because servers with relatively old versions of replicated data do not reply to a query.

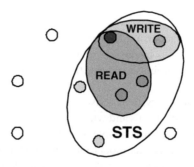

FIGURE 14.3 Incremental processes of READ and WRITE quorum sizes: $\hat{\xi}_W$ increases round by round, while $\hat{\xi}_R$ increases with the amount of queries propagated by an agent.

Given the weak assumptions on synchrony, we consider only the second READ of a data item.[18] The time interval between a WRITE and that READ is characterized by an Erlang distribution $\lambda_R^2 t e^{-\lambda_R t}$. Therefore, we have

$$
p_r = \begin{cases} \displaystyle\int_{t_r}^{t_{r+1}} \lambda_R^2 t e^{-\lambda_R t} dt & r < \bar{r} \\[2ex] \displaystyle\int_{t_r}^{\infty} \lambda_R^2 t e^{-\lambda_R t} dt & r = \bar{r} \end{cases}
$$

Now, the probability of intersection is expressed by taking an average over all possible cases. Hence, C6 is satisfied with P_c approximated as follows:

$$
P_c = \sum_{r=0}^{\bar{r}} \sum_{i=1}^{n} \sum_{j=1}^{\xi_R} \left(1 - \frac{\binom{\dot{\xi}_R}{n - \dot{\xi}_W^r}}{\binom{\dot{\xi}_R}{n}} \right) \mu(j) v_r(i) p_r
$$

The in-depth presentation of PAN in Ref. 18 includes pseudo code of the algorithms, as well as simulation results and a comparison of those with analysis.

14.6 Applications

Rather than asking *which applications* can make use of probabilistic guarantees, we believe that we should immediately ask ourselves *which guarantees* are feasible by distributed algorithms in ad hoc networks.

As pointed out, we believe that further masking the uncertainty and dynamism of interaction in ad hoc settings by strong assumptions is a limited approach, and leads to either disappointing efficiency, or to unexpected behavior should certain assumptions turn out to not always hold in practice. Strong assumptions on the underlying system lead to *hidden* probabilities of success, where algorithms behave correctly in almost all cases, but in rare cases, exhibit an arbitrary behavior that is intolerable in many applications. Instead, we opt for embracing the unpredictability of ad hoc networks by providing guarantees for distributed algorithms that make probabilistics *explicit*.

Defining which guarantees *have to be deterministic*, and which ones *can be probabilistic* then becomes one of the two main challenges, whose outcome of course depends also on targeted applications. The second challenge, as expected, is then to propose efficient algorithms that provide high probabilities in practice for all probabilistic guarantees provided to applications, and degrade gracefully in the face of an increasing dynamism of the underlying system. Ideal candidates are given by algorithms (and problems) that permit the detection of runs in which a probabilistic guarantee is not met (e.g., in subsequent runs), such as the missing of a multicast message (e.g., based on successive identifiers associated with successive broadcast messages).

One concrete application of our proposed abstractions is a service for public key sharing, built on the algorithm outlined in Section 14.5. Under the assumption of infrequent key updates, a node requesting the latest key version will yield that key with very high probability, despite node and transmission failures. Another application of that algorithm is mobility management based on a repository for storing node positions.

References

1. A. Basu, B. Charron-Bost, and S. Toueg. Simulating Reliable Links with Unreliable Links in the Presence of Failures. In *Proceedings of the 10th Int. Workshop on Distributed Algorithms (WDAG'96)*, pages 105–122, October 1996.
2. E. Bonabeau, M. Dorigo, and T. Theraulaz. *From Natural to Artificial Swarm Intelligence*. Oxford University Press, 1999.

3. D. Câmara and A.A.F. Loureiro. A Novel Routing Algorithm for Ad Hoc Networks. *Baltzer Journal of Telecommunication Systems*, 1-3(18):85–100, 2001.

4. R. Chandra, V. Ramasubramanian, and K. Birman. Anonymous Gossip: Improving Multicast Reliability in Mobile Ad-Hoc Networks. In *Proceedings of the 21st IEEE International Conference on Distributed Computing Systems (ICDCS'01)*, pages 275–283, April 2001.

5. A. Demers, D. Greene, C. Hauser, W. Irish, J. Larson, S. Shenker, H. Sturgis, D. Swinehart, and D. Terry. Epidemic Algorithms for Replicated Database Maintenance. In *Proceedings of the 6th ACM Symposium on Principles of Distributed Computing (PODC'87)*, pages 1–12, August 1987.

6. P.Th. Eugster, R. Guerraoui, S.B. Handurukande, A.-M. Kermarrec, and P. Kouznetsov. Lightweight Probabilistic Broadcast. *ACM Transactions on Computer Systems*, to appear 2003.

7. P. Felber and F. Pedone. Probabilistic Atomic Broadcast. In *Proceedings of the 21st IEEE Symposium on Reliable Distributed Systems (SRDS'02)*, pages 170–179, October 2002.

8. A.W. Fu and D.W. Cheung. A Transaction Replication Scheme for a Replicated Database with Node Autonomy. In *Proceedings of the 20th International Conference on Very Large Data Bases (VLDB'94)*, pages 214–225, September 1994.

9. M.J. Fischer, N.A. Lynch, and M.S. Paterson. Impossibility of Distributed Consensus with One Faulty Process. *Journal of the ACM*, 32(2):217–246, September 1985.

10. I. Gupta, R. van Renesse, and K.P. Birman. A Probabilistically Correct Leader Election Protocol for Large Groups. In *Proceedings of the 14th International Conference on Distributed Computing (DISC 2000)*, pages 89–103, October 2000.

11. M. Günes, U. Sorges, and I. Bouazizi. ARA—The Ant-colony Based Routing Algorithm for MANETs. In *Proceedings of the 2002 ICPP Workshop on Ad Hoc Networks (IWAHN 2002)*, pages 79–85, August 2002.

12. Z.J. Haas and B. Liang. Ad Hoc Mobility Management with Randomized Database Groups. In *Proceedings of the IEEE International Conference on Communications (ICC'99)*, Vol. 3, pages 1756–1762, June 1999.

13. V. Hadzilacos and S. Toueg. Fault-Tolerant Broadcasts and Related Problems. In S. Mullender, Editor, *Distributed Systems*, chapter 5, pages 97–145. Addison-Wesley, 2nd edition, 1993.

14. D.B. Johnson, D.A. Maltz, Y.-C. Hu, and J.G. Jetcheva. *The Dynamic Source Routing Protocol for Mobile Ad Hoc Networks (DSR)*, February 2002. Internet Draft, draft-ietf-manet-dsr-07.txt. Work in progress.

15. L. Lamport. Time, Clocks, and the Ordering of Events in a Distributed System. *Communications of the ACM*, 21(7):558–565, July 1978.

16. L. Lamport. How to Make a Multiprocessor Computer that Correctly Executes Multiprocess Programs. *ACM Transactions on Computer Systems*, 28(9):690–691, September 1979.

17. J. Luo, P.Th. Eugster, and J.-P. Hubaux. Route-Driven Gossip: Probabilistic Reliable Multicast for Ad Hoc Networks. In *Proceedings of the 22nd Annual Joint Conference of the IEEE Computer and Communications Societies (INFOCOM'03)*, March 2003.

18. J. Luo, J.-P. Hubaux, and P.Th. Eugster. PAN: Providing Reliable Storage in Mobile Ad-hoc Networks with Probabilistic Quorum Systems. In *Proceedings of the Fourth ACM International Symposium on Mobile Ad Hoc Networking and Computing (MobicHoc 2003)*, to appear June 2003.

19. D. Mahlki, M.K. Reiter, and A. Wool. Probabilistic Quorum Systems. *Information and Computation*, 170(2):184–206, November 2001.

20. C.H. Papadimitriou. The Serializability of Concurrent Database Updates. *Journal of the ACM*, 26(4):631–653, October 1979.

21. E.M. Royer and C.E. Perkins. Multicast Operation of the Ad-hoc On-demand Distance Vector Routing Protocol. In *Proceedings of the 5th Annual ACM/IEEE International Conference on Mobile Computing and Networking (MobiCom'99)*, pages 207–218, August 1999.

22. M. Shivanajay, C.K. Tham, and D. Srinivasan. Mobile Agents Based Routing Protocol for Mobile Ad Hoc Networks. In *Proceedings of the IEEE Global Telecommunications Conference 2002 (GLOBECOM'02)*, Vol. 1, pages 163–167, November 2002.

15

Medium Access Control Protocols in Mobile Ad Hoc Networks: Problems and Solutions*

Hongqiang Zhai

Yuguang Fang

*This work was supported in part by the Office of Naval Research under Young Investigator Award N000140210464 and the National Science Foundation under Faculty Early Career Development Award ANI-0093241.

15.1   Introduction

Recent advancements in wireless technologies and mankind's long-time dream of free communication are the driving forces behind the proliferation of wireless local area networks (WLANs) and the "hot" research activities in mobile ad hoc networks (MANETs). One of the most active topics is medium access control (MAC) protocols, which coordinate the efficient use of the limited shared wireless resource. However, in these wireless networks, the limited wireless spectrum, time-varying propagation characteristics, distributed multiple access control, low complexity, and energy constraints together impose significant challenges for MAC protocol design to provide reliable wireless communications with high data rates.

Among all MAC protocols, random medium access control (MAC) protocols have been widely studied for wireless networks due to their low cost and easy implementation. IEEE 802.11 MAC[10] is such a protocol that has been successfully deployed in wireless LANs and has also been incorporated in many wireless testbeds and simulation packages for wireless multihop mobile ad hoc networks. It uses four-way handshake procedures (i.e., RTS/CTS/DATA/ACK). The RTS and CTS procedures are used to avoid collisions with long data packets. The value of the NAV (network allocation vector) carried by RTS or CTS is used to reserve the medium to avoid potential collisions (i.e., virtual carrier sensing) and thus mitigate the hidden terminal problem. The ACK is used to confirm successful transmission without errors.

However, there are still many problems that IEEE 802.11 MAC has not adequately addressed. How to design a more effective transmission scheme based on the channel condition is still open and challenging. How to make full use of multiuser diversity in terms of multiple transmitters with the same receiver or the same transmitter with multiple receivers to maximize the throughput is also an interesting issue. In addition, in multihop ad hoc networks, the MAC layer contention or collision becomes much more severe than in the wireless LANs. Due to the MAC layer contention, the interaction or coupling among different traffic flows also deserves serious attention, which may limit the stability and scalability of multihop ad hoc networks.

At the MAC layer, the open shared channel imposes a lot of challenges for medium access control design. The hidden terminals may introduce collisions and the exposed terminals may lead to low throughput efficiency. In addition to these two notorious problems, the receiver blocking problem (i.e., the intended receiver does not respond to the sender with CTS or ACK due to the interference or virtual carrier sensing operational requirements due to the other ongoing transmissions) hence deserves serious consideration. In fact, this problem becomes more severe in multihop ad hoc networking environments and may result in throughput inefficiency, starvation of some traffic flows or nodes, or re-routing. Many proposed solutions actually aggravate this problem by not allowing the hidden terminal to transmit. Furthermore, how to maximize spatial reuse by allowing the hidden terminals to receive and the exposed terminals to transmit is a very interesting issue.

Higher layer network protocols may be affected by wireless MAC protocols. It has been shown in many articles that multihop ad hoc networks perform poorly with TCP traffic and heavy UDP traffic.[3,7,15,16,25] This is because all wireless links in the neighborhood share the same wireless resource. All traffic flows passing through these links need to contend for the channel before transmission. Hence, severe MAC layer contention and collision can result in the contention among traffic flows. On the other hand, MAC contention can introduce network congestion with backlogged packets, which implies that network congestion is closely coupled with MAC contention. Some researchers have already noticed this kind of coupling. Fang and McDonald[6] demonstrated that the throughput and delay can be affected by the path coupling, that is, the MAC layer contention among the nodes distributed along the node-disjoint paths. Thus, cross-layer design and optimization is necessary for MANETs.

Moreover, at the physical layer, the time-varying channel condition makes rate adaptation necessary to improve network throughput. The diversity in the link quality due to the various channel conditions could be exploited to design opportunistic packet scheduling. The MAC protocol should be designed accordingly to adapt to the varying channel conditions.

In this chapter we first discuss the identified problems and challenges at different protocol layers to the design of MAC protocol. Then, we present several recently proposed novel schemes to address MAC layer problems, traffic flow-related issues, rate adaptation, and link quality diversity, respectively, in Sections 15.3 through 15.6. Finally, Section 15.7 concludes this chapter.

15.2 Problems

This section first discusses the inherent problems of the IEEE 802.11 MAC protocol in shared wireless channel environments in MANETs, and then illustrates the impact of traffic flows and physical layer channel conditions on the performance of this MAC protocol.

15.2.1 MAC Layer Related Problems

A packet collision over the air is much more severe in multihop environments than that in wireless LANs. Packet losses due to MAC layer contention will definitely affect the performance of the high layer networking schemes such as the TCP congestion control and routing maintenance because a node does not know whether an error is due to the collision or the unreachable address. It has been shown in many articles that multihop ad hoc networks perform poorly with TCP traffic as well as heavy UDP traffic.[3,16,25]

The source of the above problems comes mainly from the MAC layer. The hidden terminals may introduce collisions and the exposed terminals may lead to low throughput efficiency. In addition to these two notorious problems, the receiver blocking problem (i.e., the intended receiver does not respond to the sender with CTS or ACK due to the interference or virtual carrier sensing operational requirements for the other ongoing transmissions) also deserves serious consideration. In fact, this problem becomes more severe in multihop environments and results in throughput inefficiency and starvation of some traffic flows or nodes. The next few subsections describe a few problems in multihop mobile ad hoc networks when the IEEE 802.11 MAC protocol is deployed.

15.2.1.1 Hidden Terminal Problem

A hidden terminal is the one within the sensing range of the receiver, but not in the sensing range of the transmitter. The hidden terminal does not know that the transmitter is transmitting, and hence can initiate a transmission, resulting in a collision at the receiving node of the ongoing transmission.

One simple example is shown in Figure 15.1, where the small circles indicate the edges of the transmission range and the large circles represent the edges of the sensing range. D is the hidden terminal to A when

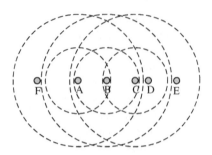

FIGURE 15.1   A simple scenario to illustrate the problems.

A is transmitting to B, and it cannot sense A's transmission but may still interfere with B's reception if D begins a transmission.

15.2.1.2   Exposed Terminal Problem

An exposed terminal is the one within the sensing range of the transmitter but not within that of the receiver. The exposed node senses the medium busy and does not transmit when the transmitter transmits, leading to bandwidth under-utilization. In 15.1, F is the exposed terminal to A when A is transmitting to B. F senses A's transmission and keeps silent, although F can transmit to other nodes outside of A's sensing range without interfering with B's reception.

In fact, in the four-way handshake procedures in IEEE 802.11 MAC, either RTS and CTS or DATA and ACK bidirectional packets are exchanged. Thus, the exposed node of one transmitter-receiver pair is also the hidden node of the other pair. So, in addition to the hidden terminal, the exposed terminal of the transmitter should not initiate any new transmission during the whole transmission process to avoid collision with the short packets ACK or CTS in IEEE 802.11 MAC. Thus, the carrier sensing strategy based on the RTS/CTS handshake will lead to a significant deficiency in spatial reuse.

15.2.1.3   Limitation of NAV Setup Procedure

IEEE 802.11 family protocols adopt short control packets (i.e., RTS/CTS) to resolve the long data packet collision and NAV setup procedures to claim the reservation for the channel for a certain period to avoid collision from the hidden terminals. This implies that any node that hears RTS/CTS correctly must set its NAV carried the received packets and keeps silent during the NAV period.

The NAV setup procedure cannot work properly when there are collisions. All kinds of packets, RTS, CTS, DATA, or ACK, can be corrupted due to collisions. For example, in Figure 15.1, A wants to send packets to B. They exchange RTS and CTS. If E is transmitting when B transmits CTS to A, B's CTS and E's transmission will collide at C, and C cannot set its NAV according to the corrupted CTS from B.

The NAV setup procedure is redundant if a node is continuously doing carrier sense. For example, in Figure 15.1, we can observe that both A's and B's transmission ranges are covered by the common area of A's and B's sensing ranges. If there is no collision, C will set NAV correctly when receiving B's CTS. However, it can also sense A's transmission, so NAV setup procedure is just redundant to prevent C from transmitting. RTS's NAV is not necessary either because any node that can receive RTS correctly can also sense B's CTS and succeeding A's DATA and B's ACK, and will not initiate new transmission to interrupt the ongoing transmission.

The NAV setup procedure does not help solve the hidden terminal problems even if there are no other collisions to prevent the CTS from setting up the neighbors' NAV. For example, in Figure 15.1, D is the hidden terminal to A. It cannot sense A's transmission and cannot receive B's CTS correctly either, because it is out of the transmission range of B. Thus, when A is transmitting a long data packet to B, D may begin to transmit a packet, which will result in a *collision* at B.

15.2.1.4 Receiver Blocking Problem

The blocked receiver is the one that cannot respond to the RTS intended for this receiver due to the other ongoing transmission in its sensing range. This may result in unnecessary retransmissions of RTS requests and subsequent DATA packet discarding. When the intended receiver is in the range of some ongoing transmission, it cannot respond to the sender's RTS according to the carrier sensing strategy in the IEEE 802.11 standard. The sender may attempt to retransmit several times if the backoff window is shorter than the long data packet. Then, the backoff window size becomes larger and larger when the RTS transmission fails and the window size is doubled, until the sender finally discards the packet. If the ongoing transmission finishes before the new sender reaches its maximum number of retransmissions allowed, the packet in the queue of an old sender will have higher priority than a new one because the old sender resets its backoff window size and is much shorter in size than that of a new one. So the old sender has a high probability of continuing to transmit and the new one continues doubling the backoff window size and discards packets when the maximum number of transmission attempts is reached. This will therefore result in serious unfairness among flows and severe packet discarding.

For example, as shown in Figure 15.1, when D is transmitting to E, A will not receive the intended CTS from B if it sends RTS to B. This is because B cannot correctly receive A's RTS due to collision from D's transmission. Thus, A keeps retransmitting and doubling the contention window until it discards the packet. If D has a burst of traffic, it will continuously occupy the channel, which will starve the flow from A to B.

The hidden terminal problem could make the receiver blocking problem worse. In the above example, even if A has a chance to transmit a packet to B, its hidden terminal D could start transmission and collide with A's transmission at B because D cannot sense A's transmission. Therefore, A almost has no chance to successfully transmit a packet to B when D has packets destined to E.

15.2.1.5 The Desired Protocol Behaviors to Achieve Maximum Spatial Reuse

The desired MAC protocol for multihop and wireless mobile ad hoc networks should at least resolve the hidden terminal problem, the exposed terminal problem, and the receiver blocking problem. Therefore, the ideal protocol should guarantee that there is only one receiver in the range of the transmitter and there is also only one transmitter in the range of the receiver. The exposed nodes may start to transmit despite the ongoing transmission. The hidden nodes cannot initiate any transmissions but may receive packets. Thus, to maximize the spatial reuse or network capacity, it should allow multiple transmitters to transmit in the range of any transmitter and multiple receivers in the range of any receiver to receive. In addition, the transmitter should know whether its intended receiver is blocked or is just outside its transmission range in case it does not receive the returned CTS to avoid packet discarding and the undesirable protocol behaviors at the higher layer, such as unnecessary rerouting requests.

15.2.1.6 Limitation of IEEE 802.11 MAC Using a Single Channel

The collisions between RTS/CTS and DATA/ACK, and that between DATA and ACK, are the culprits preventing us from achieving the aforementioned desired protocol behaviors.

The exposed terminal cannot initiate new transmission because its transmission would have prevented the current transmitter from correctly receiving the CTS or the ACK due to a possible collision.

The hidden terminal, which cannot sense the transmission or correctly receive the CTS, may initiate a new transmission, which will cause collision to the current ongoing transmission. In addition, it should not become a receiver because its CTS/ACK may collide with the current transmission. Moreover, its DATA packet reception can be corrupted by the ACK packet from the current receiver.

If the intended receiver for a new transmission is in the range of the ongoing transmission, it may not be able to correctly receive RTS and/or sense the busy medium, which prevents it from returning the CTS. Thus, the new sender cannot distinguish whether the intended receiver is blocked or out of its transmission range.

To summarize, many aforementioned problems cannot be solved if a single channel is used in the IEEE 802.11 MAC protocol.

15.2.2   Flow Level Related Problems

In wireless multihop ad hoc networks, nodes must cooperate to forward each other's packets through the networks. Due to contention for the shared channel, the throughput of each single node is limited not only by the raw channel capacity, but also by the transmissions in its neighborhood. Thus, each multihop flow encounters contentions not only from other flows that pass by the neighborhood (i.e., the *inter-flow contention*), but also from the transmissions of itself because the transmission at each hop must contend the channel with the upstream and downstream nodes (i.e., the *intra-flow contention*). These two kinds of flow contentions could result in severe collisions and congestion, and seriously limit the performance of ad hoc networks. In the following paragraphs, we discuss in detail their impacts on the performance of MANETs.

15.2.2.1   Intra-Flow Contention

The *intra-flow contention* here means the MAC layer contentions for the shared channel among nodes that are in each other's interference range along the path of the same flow. Li et al.[15] have observed that the IEEE 802.11 protocol fails to achieve optimum chain scheduling. Nodes in a chain experience different amounts of competition, as shown in Figure 15.2, where the small circle denotes a node's valid transmission range and the large circle denotes a node's interference range. Thus, the transmission of node 0 in a seven-node chain experiences interference from three subsequent nodes, while the transmission of node 2 is interfered by five other nodes. This implies that node 0 (i.e., the source) could actually inject more packets into the chain than the subsequent nodes can forward. These packets are eventually dropped at the two subsequent nodes. On the other hand, the redundant transmissions from node 0 grab the transmission opportunities of node 1 and node 2 because they cannot simultaneously transmit, and hence keep the end-to-end throughput far from the maximum value. We call this problem the *intra-flow contention* problem.

15.2.2.2   Inter-Flow Contention

In addition to the above contentions inside a multihop flow, the contentions between flows could also seriously decrease network throughput. If two or more flows pass through the same region, the forwarding nodes of each flow encounter contentions not only from its own flow, but also from other flows. Thus, the previous hops of these flows could actually inject more packets into the region than the nodes in the region can forward. These packets are eventually dropped by the congested nodes. On the other hand, the transmissions of these packets grab the transmission opportunities of the congested nodes, and hence impact the end-to-end throughput of the flows passing through the region. As shown in Figure 15.3, where there are two flows, one is from 0 to 6 and the other is from 7 to 12. Obviously, node 3 encounters the most frequent contentions and has little chance to successfully transmit packets to its downstream nodes. The packets will accumulate at and be dropped by nodes 3, 9, 2, 8, and 1. We call this problem as the *inter-flow contention* problem.

In the shared channel environments in multihop ad hoc networks, these two kinds of contentions are widespread and result in congestion at some nodes, where packets continuously accumulate, which

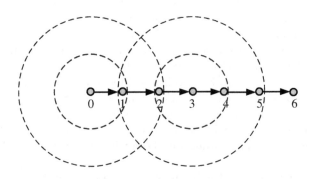

FIGURE 15.2   Chain topology.

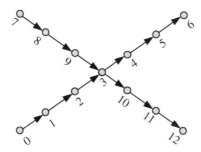

FIGURE 15.3 Cross traffic.

then aggravates the contentions and finally results in packet dropping. This not only greatly decreases the end-to-end throughput, but also increases the end-to-end delay due to the long queueing delay.

15.2.3 Physical Layer Related Issues

15.2.3.1 Time-Varying Channel Condition and Rate Adaptation

A typical wireless communication link in wireless local area networks (LANs) is time-varying, and how to more effectively design transmission schemes based on the channel condition is challenging. Many adaptive transmission schemes have been proposed in the literature to enhance throughput performance. Many of these schemes are designed by varying the data rate, the transmission power, or the packet length. One of the popular schemes is based on rate adaptation, the adaptive transmission method that employs different modulation and coding schemes to adjust the data rate based on the channel condition (in terms of the signal-to-noise ratio [SNR]). The basic idea is to employ a high-level modulation scheme when a higher SNR is detected, as long as the target error rate is satisfied. The target error rate can be characterized by the bit error rate (BER), the symbol error rate (SER), or the packet error rate (PER), as specified by the designer. For receiver-based rate-adaptation schemes, the receiver usually carries out the channel estimation and rate selection, and the selected rate is then fed back to the transmitter.

Most of these protocols are receiver-based and employ the RTS/CTS collision avoidance handshake specified in the IEEE 802.11 standard. However, these protocols have not considered the possibility of bursty transmission of fragments in the corresponding rate adaptation schemes. The fixed preamble at the physical layer and the fixed inter-frame spacing (IFS) at the MAC layer have relatively large overheads when a high data rate is used and the transmission time for the payload is relatively short. Thus, reducing the overhead at a high data rate is essential for improving protocol efficiency.

15.2.3.2 Link Diversity

One of most interesting approaches to combating scarce spectrum resources and channel variations in wireless environments is opportunistic multiuser communication. Following the philosophy of cross-layer design, opportunistic multiuser communication utilizes the physical layer information feedback from multiple users, that is, multiuser diversity, to optimize media access control, scheduling, and rate adaptation. By allowing users with good link quality to transmit data in appropriately chosen modulation schemes, system performance in terms of goodput and energy efficiency can be greatly improved.

As the counterpart of multi-downlink diversity and multi-uplink diversity in cellular networks, multiuser diversity in ad hoc networks can be characterized as *Multi-Output Link Diversity* and *Multi-Input Link Diversity*.

Multi-Output Link Diversity is the diversity of instantaneous channel quality and congestion status of output links. Multiple flows may originate from or pass through a given node and take different neighbors as the next hop forwarding nodes or destinations. After this node acquires a transmission opportunity, it can choose a link with good instantaneous quality to transmit data in the given cycle. For example, as shown in Figure 15.4, node 1 is interfered by ongoing transmission of node 5 and the link of $0 \rightarrow 2$

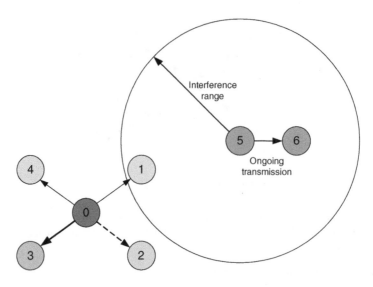

FIGURE 15.4   Multi-output link diversity.

suffers deep fading or shadowing. The link of $0 \rightarrow 4$ has instantaneous quality to support basic data rate transmission. The link quality of $0 \rightarrow 3$ happens to be "on-peak." At this time, it is better for node 0 to transmit data to node 3 or 4 rather than to node 1 or 2. Thus, the Head-of-Line blocking problem[1] can be alleviated and higher throughput can be achieved.

Multi-Input Link Diversity is the diversity of the channel quality and queue status of input links. Multiple flows originating from or passing through different neighbors take a given node as the next hop forwarding node or destination. Differences in instantaneous channel qualities of those input links form the multi-input link diversity. For example, as shown in Figure 15.5, node 1 is in the carrier sensing range of ongoing transmission of node 5. Similar to the previous example, node 3 or 4 instead of node 1 or 2 should take the opportunity to transmit packets to node 0 in this scenario.

Although diversity techniques have been widely studied and shown feasible and efficient in cellular networks, previous schemes may not apply to MANETs because they are based on an infrastructure where the base station acts as the central controller and dedicated control channels are normally available to feed back channel state periodically. To the best of our knowledge, multiuser diversity is still under investigation.

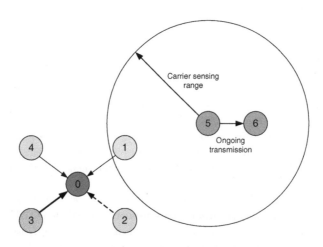

FIGURE 15.5   Multi-input link diversity.

However, there is little work that provides a comprehensive and realistic study of multiuser diversity with desired goals in protocol design of ad hoc networks.

Thus far, we have discussed a set of problems that the IEEE MAC protocol may present when we deploy it in multihop wireless ad hoc networks. The next few sections present some possible solutions that we have investigated recently to overcome or mitigate these problems.

15.3 DUCHA: A New Dual-Channel MAC Protocol

This section presents a new dual-channel MAC protocol (DUCHA) for multihop mobile ad hoc networks to mainly address the MAC layer related problems discussed above. More details can be found in Ref. 28.

15.3.1 Protocol Overview

To achieve the desired protocol behavior, we utilize two channels (dual-channel) for control packets and data packets, separately. RTS and CTS are transmitted over the control channel. Negative CTS (NCTS) is used to solve the receiver blocking problem and is also transmitted in the control channel. Data is transmitted over the data channel. An outband receiver-based busy tone[9,19] is used to solve the hidden terminal problem. The ACK is not necessary here because our protocol can guarantee that there is no collision to data packets. To deal with wireless channel error, we introduce the NACK signal, which is the continuing busy tone signal when the receiver determines that the received data packet is corrupted. The sender will not misinterpret this NACK signal because there are no other receivers in its sensing range and hence no interfering NACK signals and will conclude that the transmission is successful if no NACK signal is sensed.

Our protocol DUCHA adopts the same transmission power and capture threshold CP_{Thresh} in the control channel and the data channel. And the transmission power level for correctly receiving RX_{Thresh} is also the same for the two channels.

15.3.2 Basic Message Exchange

15.3.2.1 RTS

Any node must sense the control channel idle at least for DIFS long and sense no busy tone signal before initiating a new transmission of an RTS. If it senses the noisy (busy) control channel longer than or equal to the RTS period, it should defer long enough (at least for SIFS + CTS + 2 × max-propagation-delay) to avoid possible collision to the CTS's reception at some sender. For example, in Figure 15.1, when A finishes transmitting its RTS to B, F should wait at least long enough for A to finish receiving the possible returning CTS/NCTS from B.

15.3.2.2 CTS/NCTS

Any node correctly receiving the RTS should return CTS after SIFS spacing, regardless of the control channel status if the DATA channel is idle.

If both the control and data channels are busy, it ignores the RTS to avoid possible interference to the CTS's reception at other RTS's transmitter. If the control channel is idle for at least one CTS packet long and the data channel is busy, it returns NCTS. The NCTS estimates the remaining data transmission time in its duration field according to the difference between the transmission time of maximum data packet and the length it has sensed a busy medium in the data channel.

15.3.2.3 DATA

RTS's transmitter should start data transmission after correctly receiving the CTS if no busy tone signal is sensed. If the sender receives an NCTS, it defers its transmission according to the duration field of NCTS. Otherwise, it assumes that a collision occurred and will then double its backoff window and defer its transmission.

15.3.2.4 Busy Tone

The intended receiver begins to sense the data channel after it transmits CTS. If the receiver does not receive a signal with enough power in the data channel in the due time that the first few bits of the data packet reaches it, it will assume that the sender does not transmit data and finish the receiving procedure. Otherwise, it transmits a busy tone signal to prevent possible transmissions from hidden terminals.

15.3.2.5 NACK

The intended receiver has a timer to indicate when it should finish the reception of the data packet according to the duration field in the previously received RTS. If the timer expires and has not received the correct data packet, it assumes that the data transmission fails and sends NACK by continuing the busy tone signal for an appropriate time period. If it correctly receives the data packet, it stops the busy tone signal and finishes the receiving procedure.

The sender assumes that its data transmission is successful if there is no NACK signal sensed over the busy tone channel during the NACK period. Otherwise, it assumes that its transmission fails because of wireless channel error and then starts the retransmission procedure.

In addition, during the NACK period, in addition to the data transmission period, any other nodes in the sensing range of the sender are not allowed to become the receiver of data packets, and any other nodes in the sensing range of the receiver are not allowed to become the sender of data packets. This is to avoid confusion between NACK signals and the normal busy tone signals.

In the above message exchange, our protocol transmits or receives packets in only one channel at any time. We only use receive busy tone signals and not transmit busy tone signals. Thus it is necessary to sense the data channel before transmitting CTS/NCTS packets to avoid becoming a receiver in the sensing range of the transmitters of some ongoing data packet transmissions.

15.3.3 Solutions to the Aforementioned Problems

In the following discussions, we use examples to illustrate how our DUCHA solves those well-known problems.

15.3.3.1 Solution to the Hidden Terminal Problem

As shown in Figure 15.1, B broadcasts a busy tone signal when it receives a data packet from A. The hidden terminal of A (i.e., D) could hear B's busy tone signal and thus will not transmit in the data channel to avoid interference with B's reception. Thus, the busy tone signal from the data's receiver prevents any hidden terminals of the intended sender from interfering with the reception. Moreover, no DATA packets are dropped due to the hidden terminal problem.

15.3.3.2 Solution to the Exposed Terminal Problem

In Figure 15.1, B is the exposed terminal of D when D is transmitting a data packet to E. B could initiate RTS/CTS exchange with A although it can sense D's transmission in the data channel. After the RTS/CTS exchange is successful between B and A, B begins to transmit the data packet to A. Because A is out of the sensing range of D and E is out of the sensing range of B, both A and E could correctly receive the data packet destined to them. Thus, the exposed terminal could transmit data packets in DUCHA, which could improve the spatial reuse ratio.

15.3.3.3 Solution to the Receiver Blocking Problem

In Figure 15.1, B is the blocked receiver in the IEEE 802.11 MAC protocol when D is transmitting data packets to E. In our protocol DUCHA, B can correctly receive A's RTS in the control channel while D sends data packets in the data channel. Then B returns the NCTS to A because it senses a busy medium in the DATA channel. The duration field of NCTS estimates the remaining busy period in the data channel required to finish D's transmission. When A receives the NCTS, it defers its transmission and stops the unnecessary retransmissions. It retries the transmission after the period indicated in the duration field of NCTS. Once the RTS/CTS exchange is successful between A and B, A begins to transmit the data packet to B. B will correctly receive the data packet because there is no hidden terminal problem for receiving data packets.

15.3.3.4 Maximum Spatial Reuse

As discussed above, the exposed terminals could transmit data packets. Furthermore, in our protocol, the hidden terminal could receive data packets although it cannot transmit. In Figure 15.1, D is the hidden terminal of A when A is transmitting the data packet to B. After the RTS/CTS exchange between E and D is successful in the control channel, E can transmit data packets to D. Because D is out of A's sensing range and B is out of E's sensing range, both D and E can correctly receive the intended data packets. Thus, our protocol DUCHA can achieve maximum spatial reuse by allowing multiple transmitters or multiple receivers in the sensing range of each other to communicate. At the same time, there are no collisions for data packets or for the NACK signals because there is only one transmitter in its intended receiver's sensing range and only one receiver in its intended transmitter's sensing range.

15.3.3.5 Inherent Mechanisms to Solve the Intra-flow Contention Problem

In our DUCHA protocol, the receiver of data packets has the highest priority to access the channel for the next data transmission. When one node correctly receives a data packet, it could immediately start the backoff procedure for the new transmission, while the upstream and downstream nodes in its sensing range are prevented from transmitting data packets during the NACK period. In fact, this could achieve optimum packet scheduling for chain topology and is similar for any single flow scenario.

For example, in Figure 15.2, node 1 has the highest priority to access the channel when it receives one packet from node 0 and hence immediately forwards the packet to node 2. For the same reason, node 2 immediately forwards the received packet to node 3. Then node 3 forwards the received packet to node 4. Because node 0 can sense node 1 and 2's transmissions, it will not interfere with these two nodes. Node 0 cannot send packets to node 1 when node 3 forwards a packet to 4 because node 1 is in the interference range of node 3. When node 4 forwards a packet to node 5, node 0 may have the chance to send a packet to node 1. In general, nodes that are four hops away from each other along the path could simultaneously send packets to their next hops. Thus, the procedure could utilize 1/4 of the channel bandwidth, the maximum throughput that can be approached by the chain topology.[15]

15.3.4 Bandwidth Allocation

We split the whole bandwidth into control and data channels. While the nodes are negotiating the transmission by RTS and CTS in the control channel, there is no transmission in the data channel for these nodes. On the other hand, when the nodes are transmitting data packets in the data channel, the bandwidth in the control channel is not fully utilized. There exists an optimal bandwidth allocation for the two channels to reach the maximum throughput.

For simplicity of analysis, we assume that there are no collisions to all the packets, and the spacings between RTS, CTS, and data are fixed and can be neglected when compared to the control and data frames. The maximum throughput is determined by the packet length and the data rate of each channel. Let L_R, L_C, and L_D be lengths of RTS, CTS, and DATA, respectively; R_c and R_d be data rates of control and data channels, respectively; and BW be the total data rate (bandwidth). We observe that maximizing throughput is equivalent to minimizing the total time for a successful transmission of a packet, say, T_p. Thus, the problem is to minimize T_p under the condition $R_c + R_d = BW$. We can easily obtain

$$T_p = \frac{L_R + L_C}{R_c} + \frac{L_D}{R_d} \geq \frac{(\sqrt{L_R + L_C} + \sqrt{L_D})^2}{R_c + R_d}$$

$$T_p = \frac{(\sqrt{L_R + L_C} + \sqrt{L_D})^2}{BW}, \quad \text{when } \frac{R_c}{R_d} = \frac{\sqrt{L_R + L_C}}{\sqrt{L_D}}$$

(15.1)

In IEEE 802.11, the total time for successful packet transmission when there is no transmission error (due to collisions or channel condition) is

$$T_p' = \frac{L_R + L_C + L_D}{BW}. \tag{15.2}$$

We can observe that the bandwidth splitting sacrifices bandwidth utilization ($T_p > T_p'$). However, our protocol can eliminate the collisions to data packets and greatly improve spatial reuse, leading to performance improvement for multihop ad hoc networks.

15.4   Distributed Flow Control and Medium Access Control

This section proposes a scheme to address flow level related issues by optimizing the cross-layer interaction between the MAC layer and the higher layer. More details can be found in Refs. 26 and 27.

15.4.1   Motivation

Considering the fact that contentions are from the transmission attempts of packets at different nodes, which are generated by various traffic flows, it is natural to exploit flow control to schedule the packet transmissions to reduce the collisions and congestion at the MAC layer. The intuitive solution is to allow the downstream nodes and the congested ones to transmit packets while keeping others silent, and hence smoothly forward each packet to the destination without encountering severe collisions or excessive delay at the forwarding nodes. This motivates us to develop our scheme presented in the next subsection.

15.4.2   Solution

We present a framework of flow control over the MAC layer and queue management to address the collisions and congestion problem due to the *intra-flow contention* and *inter-flow contention*. Based on the framework, a multihop packet scheduling algorithm is incorporated into the IEEE 802.11 MAC protocol. The salient feature here is to generalize the optimum packet scheduling of chain topology, which allows nodes four hops away to transmit simultaneously to any traffic flows in general topology.

The framework includes multiple mechanisms. The *fast relay* assigns high priority of channel access to the downstream nodes when they receives packets, which reduces a lot of intra-flow contentions. The *backward-pressure congestion control* gives transmission opportunity to the congested node and keeps its upstream nodes silent. This could not only reduce excessive contentions in the congested area, but also quickly eliminate the congestion. It is also a quick method to notify the source to slow the sending rate down by exploiting the RTS/CTS of the IEEE 802.11 MAC protocol. The *receiver-initiated transmission scheme* uses a three-way handshake to resume the blocked flow at the upstream nodes when the congestion is cleared. It is a timely and economical approach with even less control overhead than the normal four-way handshake transmission in the IEEE 802.11 protocol. We discuss each of these mechanisms in detail in the next subsections.

15.4.2.1   Rule 1: Assigning High Priority of the Channel Access to the Receiver

In each multihop flow, the intermediate node on the path needs to contend for the shared channel with the upstream nodes when forwarding the received packet to the next hop. One way to avoid the first few nodes on the path to inject more packets than the succeeding nodes can forward is to assign high priority of channel access to each node when it receives a packet. This can achieve better scheduling for the chain topology.

For example, in Figure 15.2, node 1 has the highest priority when it receives one packet from node 0 and then forwards the packet to node 2. Node 2 immediately forwards the received packet from node 1 and forwards it to node 3. It is the same for node 3, which immediately forwards the received packet to node 4. Because node 0 can sense the transmissions of nodes 1 and 2, it will not interfere with these two nodes. Node 0 cannot send packets to node 1 when node 3 forwards a packet to node 4 because node 1 is in the interference range of node 3. When node 4 forwards a packet to node 5, node 0 may have the

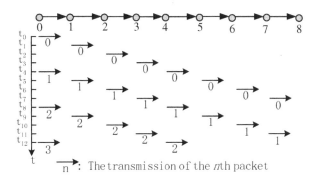

FIGURE 15.6 Optimum packet scheduling for chain topology. To simplify the illustration of how our scheme works, we use chain topology in this figure and the following ones, which is conceptually the same with any random multihop path in mobile ad hoc networks.

chance to send a packet to node 1. Similar procedures are adopted by succeeding nodes along the path. Node 0 and node 4 can simultaneously send packets to their next hops, and a similar event happens to nodes that are four hops away from each other along the path. Thus, the procedure can utilize 1/4 of the channel bandwidth, the maximum throughput that can be approached by the chain topology.[15]

To incorporate this procedure into the IEEE 802.11 MAC protocol, our scheme OPET sets the initial value of the backoff window size of each receiver to 8. When it finishes the transmission, the scheme resets its contention window size to the normal value of 32.[10] The example in Figure 15.6 shows the optimum packet scheduling for the chain topology implemented by our scheme.

Rule 1 only considers the interference in a single flow. If the next hop of the current receiver is busy or interfered by other transmission, the receiver cannot seize the channel even with the highest priority. So we introduce backward-pressure scheduling to deal with inter-flow contention.

15.4.2.2 Rule 2: Backward-Pressure Scheduling

If one flow encounters congestion, it should decrease its sending rate to alleviate contention for the shared channel. Therefore, other flows in the neighborhood can obtain more channel bandwidth to transmit their packets to achieve higher utilization efficiency of the limited channel resource.

In addition to lowering the sending rate of the source, it is necessary to prevent the node, referred to as the *restricted node* in the following discussions, from transmitting packets to its next hop if the latter has already had many packets from the same flow. This can yield the transmission opportunity to the next hop as well as alleviate the congestion status.

The backward-pressure scheduling procedure takes advantage of RTS/CTS exchange in the IEEE 802.11 MAC protocol to restrict the transmission from the upstream nodes. A negative CTS (NCTS) should respond to the RTS when the intended receiver has reached the *backward-pressure threshold* for this flow. To uniquely identify each flow, RTS for the multi-hop flows (RTSM) should include two more fields than RTS, that is, the source address and the flow ID.

Our scheme OPET sets the *backward-pressure threshold* to 1, which indicates the upper limit of the number of packets for each flow at each intermediate node. As discussed, the optimum chain throughput in the IEEE 802.11 MAC protocol is 1/4 of the chain bandwidth and therefore the optimum threshold for the backward-pressure objective is 1/4, which is similar in operations for any single path. Because 1/4 is difficult to implement in the actual protocol, we select the nearest integer 1 as the value of this threshold.

Our scheme OPET adopts the receiver-initiated transmission mechanism to resume the *restricted* transmission. It uses the three-way handshake CTS/DATA/ACK instead of the normal four-way handshake RTS/CTS/DATA/ACK, because the downstream node already knows that the restricted node has packets destined to it. The CTS to resume the transmission (CTSR) should include two more fields than CTS, the source address and the flow ID, to uniquely specify the flow. CTSR, as well as CTS, has no information

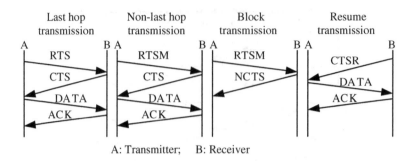

A: Transmitter;   B: Receiver

FIGURE 15.7   Message sequence for packet transmission.

about its transmitter as that in RTS. The two fields (i.e., the source address and the flow ID) are used to uniquely specify the next hop that the flow should pass through; hence, we assign different flow IDs to the flows from the same application but with different paths if multipath routing is used. The procedure of transmitting CTSR is similar to that of RTS and allows multiple retransmissions before dropping it. Different message sequences at different situations are shown in Figure 15.7.

To use the receiver-initiated transmission mechanism, we must consider that the mobility in ad hoc networks could result in link breakage followed by the transmission failure of CTSR. And CTSR may also collide several times and be dropped. The blocked node should drop CTSR after multiple retransmissions, as in the mechanism for RTS transmission. The restricted node should start a timer and begin retransmission if its intended receiver has not sent CTSR back in a long time, which we set as 1 second in our study of the proposed scheme.

One simple example to illustrate how our scheme works is shown in Figure 15.8 and Figure 15.9. To simplify the illustration, we use chain topology, which is conceptually the same as any random multihop path in mobile ad hoc networks. When node 4 has congestion and cannot forward packet 0 to its downstream node 5, as shown in Figure 15.8, the flow along the chain will accumulate one packet at each node from node 1 to node 4 and then prevent nodes 0, 1, 2, and 3 from contending for the channel in order to reduce contention at congested node 4. After eliminating the congestion at node 4, the transmission will be resumed by the congested node, as shown in Figure 15.9.

It is important to note that the control overhead of backward-pressure scheduling is relatively low. The information of backward-pressure is carried by the original message sequence RTS/CTS in IEEE 802.11. And the blocked flows are resumed by a three-way handshake procedure with less overhead than the original four-way handshake. Moreover, our scheme only maintains several short entries for each active flow with at least one packet queueing up at the considered node to indicate the *blocked* status. We observe that in a mobile ad hoc network, the number of active flows per node is restricted by the limited bandwidth and processing capability, and hence is much smaller than that in wired networks; thus, the scalability problem should not be of major concern in our scheme.

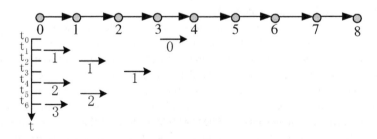

FIGURE 15.8   Packet scheduling when congestion occurs at node 4. The congestion can result from interference or contention from any crossing flow such that node 4 cannot grab the channel in time.

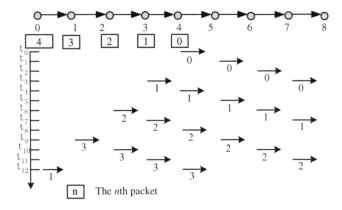

FIGURE 15.9 Packet scheduling after eliminating the congestion at node 4. After backward-pressure scheduling takes effect, the upstream nodes of this flow and all other crossing flows yield the transmission opportunity to the congested node. Thus, node 4 can quickly forward the backlogged packets and hence the congestion is eliminated.

Extensive simulation experiments are carried out to validate their performance. It turns out that our scheme can maintain stable performance with high throughput independent of traffic status, and improve the aggregated throughput by up to more than 12 times, especially for multihop flows under heavy traffic load. At the same time, it also improves the fairness among flows, and has much smaller delay and much less control overhead compared to the IEEE 802.11 MAC protocol. Moreover, it is scalable for large networks where there are more multihop flows with longer paths without incurring explosion of control packets under heavy load as the original 802.11 MAC protocol does.

15.5 Rate Adaptation with Dynamic Fragmentation

A rate-adaptive protocol with dynamic fragmentation is proposed to enhance throughput based on fragment transmission bursts and channel information. Instead of using one fragmentation threshold as in the IEEE 802.11 standard, we propose the use of multiple thresholds for different data rates so that more data can be transmitted at higher data rates when the channel is good. In our proposed scheme, whenever the rate for the next transmission is chosen based on the channel information from the previous fragment transmission, a new fragment is then generated using the fragment threshold for the new rate. This way, the channel condition can be more effectively utilized to squeeze more bits into the medium. Further details can be found in Refs. 12 through 14.

15.5.1 Fragmentation Scheme

The proposed dynamic fragmentation scheme contains the following key changes, as compared to IEEE 802.11 MAC, to enhance throughput in the time-varying wireless environment:

- The transmission durations of all fragments, except the last fragment, are set the same in the physical layer, regardless of the data rate.
- Different *aFragmentationThreshold*s for different rates are used, based on the channel condition; namely, a Rate-based Fragmentation Thresholding (RFT) scheme is employed.
- A new fragment is generated from the fragmentation process only when the rate is decided for the next fragment transmission, namely, Dynamic Fragmentation (DF).

In IEEE 802.11, with a single *aFragmentationThreshold*, the sizes of fragments are equal regardless of the channel condition. Therefore, the channel access time for a fragment varies with respect to the selected rate. For example, the channel access time for a fragment at the base rate is longer than that for a fragment at a higher rate. It is generally assumed that the channel remains unchanged during the transmission of a fragment at the base rate. Thus, more data frames can, in fact, be transmitted when a higher rate is used in the same duration provided that the SNR is high enough to support the higher rate. Due to this observation, the OAR protocol[17] proposes a multi-packet transmission scheme. However, multi-packet transmission has a higher overhead because of additional MAC headers, PHY headers, preambles in data and ACK, and SIFS idle times.

To overcome the shortcoming of multi-packet transmission, we fix the time duration of all data transmissions except for the last fragment. To generate fragments with the same time duration in a physical layer, the number of bits in a fragment should be varied based on the selected rate. Thus, it is necessary to have different *aFragmentationThreshold*s for different data rates selected by the receiver.

In the fragmentation process in IEEE 802.11 MAC, an MSDU is fragmented into equal-sized fragments that remain unchanged until all fragments in the burst are transmitted. If the channel quality is constant during the transmission of the fragment burst, the target PER (packet error rate) can be met. However, this is not guaranteed in a wireless LAN for two reasons. The first reason is that different fragments of a burst experience different channel qualities because of the time-varying nature of a wireless channel. The second reason is that after the transmission of a fragment fails, the sender contends for the channel again to transmit remaining fragments; thus, the channel quality is not guaranteed to be the same as that at the time that the first fragment is transmitted. To achieve the target PER, both the data rate and the fragment size should vary according to the changing channel condition. Moreover, to better match the varying channel, instead of generating all fragments before transmitting the first fragment, each fragment should be generated at each time when the rate is chosen for the next transmission. As a result, the fragments in a burst may not be of the same size. Figure 15.10 illustrates the process of the proposed dynamic fragmentation scheme. Notice that when the transmission of a fragment fails, the size of the retransmitted fragment may not be the same as that of the originally transmitted fragment because the channel condition may have changed.

15.5.2 Rate-Adaptive MAC Protocol for Fragment Bursts

With fragment burst transmission and rate adaptation for each fragment, data and ACK frames also participate in the rate adaptation process in the same way as RTS/CTS frames do. To support the rate adaptation process of a fragment burst, the physical layer header is modified as shown in Figure 15.11. The *signal* field in the PLCP header is divided into two 4-bit subfields, namely the current rate and the next rate subfields. The *current rate* subfield indicates the data rate of the current frame, while the *next rate* subfield indicates the selected data rate for the next incoming data frame. The values of the two subfields

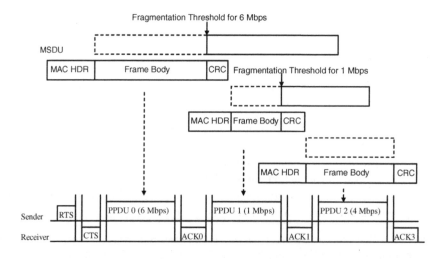

FIGURE 15.10   Dynamic fragmentation process and the timeline of data transmission.

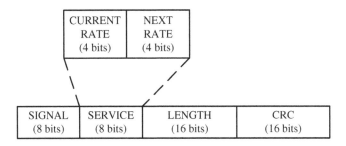

FIGURE 15.11   Physical layer header format in the proposed protocol.

in PLCP headers for RTS and data frames are the same because the *next rate* subfields in these headers indicate rates of frames transmitted from the receiver. After a sender sends an RTS frame at the base rate, a receiver estimates the channel and sends back a CTS frame to the sender with the selected rate stored in the *next rate* subfield. The sender modulates the fragment with the rate and sends a data frame to the receiver. After receiving the frame, the receiver predicts the channel condition for the next data frame and sends an ACK frame to the sender with the selected rate.

15.6   Opportunistic Media Access Control and Auto Rate Protocol (OMAR)

Due to the physical locations of various nodes in ad hoc networks, multiple nodes with various link qualities can transmit to (receive from) a common node; how to schedule the transmissions to utilize this diversity (user diversity) is challenging. The fundamental idea of OMAR is to exploit this diversity discussed in Section 15.2.3.2 through a collision avoidance process, which is necessary for CSMA/CA-based MAC. Based on sender-initiated and receiver-initiated collision avoidance, we introduce Multicast RTS (Request to Send) and Multicast RTR (Ready to Receive) to exploit multi-output link diversity and multi-input link diversity, respectively. Before the transmission of RTS (RTR) from a sender, each node selects a list of candidate receivers (senders), each with a different priority level, according to specific scheduling policy. The intended sender (receiver) multicasts a channel-probing message (i.e., RTS or RTR) to the selected group of candidate receivers (senders). Each candidate receiver (sender) evaluates the instantaneous link quality based on the received channel-probing message. The candidate receiver (sender) with the highest priority among those with channel quality better than a certain level (threshold) is granted access to the medium. The details appear in Refs. 20 through 22.

The major components adopted in our scheme are hybrid opportunistic media access, rate adaptation, and packet scheduling. We detail these mechanisms in the following sections.

15.6.1   Hybrid Opportunistic Media Access

15.6.1.1   Sender-Initiated Opportunistic Media Access

Recognizing that RTS/CTS is a common mechanism to avoid collision in sender-initiated CSMA/CA, we extend the RTS/CTS handshake procedure to probe channels and utilize multi-output link diversity. Proposed Multicast RTS and Prioritized CTS are given as follows.

15.6.1.1.1   *Multicast-RTS*

The RTS used by IEEE 802.11 is a unicast short packet in that only one receiver is targeted. In our protocol, we use multiple candidate receiver addresses in RTS and request those receivers in the receiver list to receive the RTS and measure the channel quality simultaneously. Of course, each node must use an omni-directional antenna. The targeted data rate is added to the RTS for the declaration of the data rate that the sender wants to achieve at a given directed link. We dynamically set the targeted data rate according to recently measured channel conditions among those candidate receivers in the list. Each node monitors the transmissions of its neighbors and records the received power. In addition, considering that both MPDU size and data rate are variable, we use the packet size rather than the duration into RTS for each candidate receiver so that the corresponding receiver can derive duration according to the selected data rate based on the channel condition.

Anyone except the candidate receiver that receives the MRTS should tentatively keep silent to avoid possible collision before the sender receives the CTS. After the selected qualified receiver determines the transmission duration and send back CTS, the sender sets the duration field accordingly in the MAC header of the data frame[10] for the final NAV setting. The MAC header is sent at the basic rate so that all overhearing nodes can decode.

15.6.1.1.2 *Prioritized CTS*

The candidate receivers evaluate the channel condition based on physical-layer analysis of the received RTS message. If the channel quality is better than a certain level and its NAV is zero, the given receiver is a good candidate. To avoid collision when there are two or more good candidates, different inter-frame spacings (IFSs) are employed according to the listing order of intended receivers in the RTS. For example, the IFS of the nth receiver equals to $SIFS + (n-1) * Time\_slot$. The receiver with the shortest IFS among those having the capability to receive data packet would reply to CTS first. Because all candidate receivers are within one-hop transmission range of the sender and the physical carrier sensing range is normally larger than two hops of transmission,[15] the CTS may be powerful enough for all other qualified candidate receivers to hear or sense. Suppose that busy tone[9,19] is available; it would further enhance the physical carrier sensing capability. The intended receiver turns on the busy tone upon receiving the CTS; thus, those receivers that detect CTS or busy tone from other sources would yield the opportunity to the one transmitting CTS in the first place; that is, the one with the good channel condition and the highest priority. The duration to be advertised in the CTS is set to 2*SIFS plus transmission time for DATA and ACK.

15.6.1.2 Receiver-Initiated Opportunistic Media Access

RTR, as discussed in the literature,[2,8,18,23] is also a unicast packet. To reduce the control packets (i.e., RTR to probe channel and queue status, especially when the link condition changes significantly and/or each candidate sender has no packet with high probability when the receiver polls it), we propose to use the Multicast RTR to poll the candidate senders. A candidate sender list is included in the frame of Multicast RTR. The noise power level (dB) indicates the interference and power level at the receiver. Here we assume that link gain is symmetric and RTR is sent at the default power level. The candidate receivers can derive link gains according to the receiving power. By the link gain and nomic transmit power of DATA, the sender can calculate the expected receiving power at the receiver. Finally, the candidate senders can determine the average signal to noise plus interference ratio (SINR) with known interference power level.

Upon hearing the RTR, the candidate senders estimate the channel gain. The idle candidate sender with link gain better than a certain level is allowed to access media. Similar to the sender-initiated strategy, IFS is employed to differentiate the media access priority of candidate senders. The IFS of the nth sender is equal to $SIFS + (n-1) * Time\_slot$. The sender with the highest priority among those having good link conditions transmits data first. Similar to sender-initiated opportunistic media access, we propose to incorporate busy tone to enhance carrier sensing, even if it is not necessary. Thus, other intended senders would yield the opportunity to the one transmitting DATA first, that is, the one with the good channel condition and the highest priority.

15.6.2 Rate Adaptation

According to the channel condition evaluated by the physical layer analysis of Multicast RTS or Multicast RTR, SINR is determined and the appropriate modulation scheme can be selected to efficiently use the channel, as discussed in Section 15.5.

15.6.3   Scheduling

Here we discuss how to determine the candidate receiver (sender) list, the MPDU size, and the targeted data rate of each candidate directed link, which is closely related to QoS and energy efficiency. Considering that there are many constraints such as CPU and energy consumption for portable wireless devices, one of the crucial requirements for the scheduling algorithm is simplicity. Many scheduling algorithms in the literature,[4] such as Round Robin (RR) and Earliest Timestamp First (ETF), can be tailored to our framework to achieve the desired goals. The scheduling policy in our simulation is based on the window-based weighted Round Robin. The targeted data rate is dynamically set using an algorithm similar to ARF.[11] A station is allowed to transmit multiple packets successively without contending for the media again after accessing the channel, as long as the total access time does not exceed a certain limit. We follow the thought of OAR[17] to grant channel access for multiple packets in proportion to the ratio of the achievable data rate over the basic rate so that the time-share fairness as in IEEE 802.11 can be assured. We show from our simulation study that both throughput and fairness can be significantly enhanced even by this simple scheduling method. We believe that fairness, QoS, and energy consumption can be further enhanced by more advanced scheduling algorithms.

15.7   Conclusion

This chapter discussed several important issues in MAC protocol design in IEEE 802.11 wireless LANs and mobile ad hoc networks, including severe MAC layer contention and collision in multihop environments, traffic flow contention, rate adaptation with dynamic fragmentation, multiple-input link diversity, and multiple-output link diversity. Then we proposed several novel schemes to address these issues, schemes that can greatly improve the performance of wireless networks in terms of throughput, end-to-end delay, fairness, stability, and scalability.

Acknowledgment

The authors would like to extend their appreciation to Byungseo Kim and Jianfeng Wang for their help during the preparation of this chapter.

References

1. P. Bhagwat, P. Bhattacharya, A. Krishna, and S.K. Tripathi, Enhancing Throughput over Wireless LANs Using Channel State Dependent Packet Scheduling, in *Proc. of INFOCOM'96*, 1996.
2. V. Bharghavan, A. Demers, S. Shenker, and L. Zhang, MACAW: A Media Access Protocol for Wireless LANs, in *ACM SIGCOMM '94*.
3. J. Broch, D.A. Maltz, D.B. Johnson, Y. Hu, and J. Jetcheva, A Performance Comparison of Multihop Wireless Ad Hoc Network Routing Protocols, in *Proc. of the Fourth ACM International Conference on Mobile Computing and Networking (MobiCom'98)*, pp. 85–97, Oct. 1998.
4. Y. Cao and V.O.K. Li, Scheduling Algorithms in Broad-Band Wireless Networks, in *Proc. of the IEEE*, 89(1), January 2001.
5. K. Chen, Y. Xue, and K. Nahrstedt, On Setting TCP's Congestion Window Limit in Mobile Ad Hoc Networks, in *Proc. of IEEE International Conference on Communications (ICC 2003)*, May 2003.
6. Y. Fang and A.B. McDonald, Cross-Layer Performance Effects of Path Coupling in Wireless Ad Hoc Networks: Power and Throughput Implications of IEEE 802.11 MAC, in *Proc. of 21st IEEE International Performance, Computing, and Communications Conference*, April 2002.
7. Z. Fu, P. Zerfos, H. Luo, S. Lu, L. Zhang, and M. Gerla, The Impact of Multihop Wireless Channel on TCP Throughput and Loss, in *Proc. IEEE INFOCOMM'2003*, March 2003.
8. J.J. Garcia-Luna-Aceves and A. Tzamaloukas, Reversing the Collision-Avoidance Handshake in Wireless Networks, *Mobicom'99*.

9. Z.J. Haas and J. Deng, Dual Busy Tone Multiple Access (DBTMA) — A Multiple Access Control for Ad Hoc Networks, *IEEE Transactions on Communications*, 50(6), 975–985, June 2002.

10. *IEEE standard for Wireless LAN Medium Access Control (MAC) and Physical Layer (PHY) Specifications*, ISO/IEC 8802-11: 1999(E), Aug. 1999.

11. A. Kamerman and L. Monteban, WaveLAN II: A High-Performance Wireless LAN for the Unlicensed Band, Bell Labs Technical Journal, 1997.

12. B. Kim, Y. Fang, T. Wong, and Y. Kwon, Throughput Enhancement through Dynamic Fragmentation in Wireless LANs, accepted for publication in *IEEE Transactions on Vehicolar Technology*.

13. B. Kim, Y. Fang, and T.F. Wong, Rate-Adaptive MAC Protocol in High-Rate Personal Area Networks, in *Proc. of IEEE Wireless Communications and Networking Conference 2004 (WCNC'04)*, Atlanta, GA, March 2004.

14. B. Kim, Y. Fang, T.F. Wong, and Y. Kwon, Dynamic Fragmentation Scheme for Rate-Adaptive Wireless LANs, *IEEE International Symposium on Personal, Indoor and Mobile Radio Communications (PIMRC'2003)*, Beijing, China, September 7–10, 2003.

15. J. Li, C. Blake, D.S.J. De Couto, H.I. Lee, and R. Morris, Capacity of Ad Hoc Wireless Network, in *Proc. of ACM MobiCom 2001*, July 2001.

16. C. Perkins, E.M. Royer, S.R. Das, and M.K. Marina, Performance Comparison of Two On-Demand Routing Protocols for Ad Hoc Networks, *IEEE Personal Communications*, pp. 16–28, Feb. 2001.

17. B. Sadeghi, V. Kanodia, A. Sabharwal, and E. Knightly, Opportunistic Media Access for Multirate Ad Hoc Networks, in *Proc. ACM MOBICOM' 02*, September 2002.

18. F. Talucci, M. Gerla, and L. Fratta, MACA-BI (MACA by invitation) — A Receiver Oriented Access Protocol for Wireless Multihop Networks, *PIMRC'97*.

19. F.A. Tobagi and L. Kleinrock, Packet Switching in Radio Channels. Part II- The Hidden Terminal Problem in Carrier Sense Multiple-Access and the Busy-Tone Solution, *IEEE Trans. Commun.*, Vol. COM-23, pp. 1417–1433, Dec. 1975.

20. J. Wang, H. Zhai, and Y. Fang, OMAR: Utilizing Diversity in Wireless Ad Hoc Networks, submitted for publication.

21. J. Wang, H. Zhai, Y. Fang, and M.C. Yuang, Opportunistic Media Access Control and Rate Adaptation for Wireless Ad Hoc Networks, *IEEE International Conference on Communications (ICC'04)*, Paris, France, June 2004.

22. J. Wang, H. Zhai, and Y. Fang, Opportunistic Packet Scheduling and Media Access Control for Wireless LANs and Multi-Hop Ad Hoc Networks, in *Proc. of IEEE Wireless Communications and Networking Conference (WCNC'04)*, Atlanta, March 2004.

23. Y. Wang and J.J. Garcia-Luna-Aceves, A New Hybrid Channel Access Scheme for Ad Hoc Networks, *ACM WINET*, Vol. 10, No. 4, July 2004.

24. S. Wu, Y. Tseng, and J. Sheu, Intelligent Medium Access for Mobile Ad Hoc Networks with Busy Tones and Power Control, *IEEE Journal on Selected Areas in Communications*, 18, 1647–1657, Sept. 2000.

25. S. Xu and T. Safadawi, Does the IEEE 802.11 MAC Protocol Work Well in Multihop Wireless Ad Hoc Networks?, *IEEE Communications Magazine*, pp. 130–137, June 2001.

26. H. Zhai and Y. Fang, Distributed Flow Control and Medium Access in Multihop Ad Hoc Networks, submitted for publication.

27. H. Zhai, J. Wang, and Y. Fang, Distributed Packet Scheduling for Multihop Flows in Ad Hoc Networks, in *Proc. of IEEE Wireless Communications and Networking Conference (WCNC'04)*, Atlanta, March 2004.

28. H. Zhai, J. Wang, and Y. Fang, DUCHA: A New Dual-Channel MAC Protocol for Multihop Mobile Ad Hoc Networks, in IEEE Workshop on Wireless Ad Hoc and Sensor Networks, in conjuction with IEEE Globecom 2004, Dallas, Texas, Nov. 2004.

16

On Using Ad Hoc Relaying in Next-Generation Wireless Networks

B.S. Manoj

C. Siva Ram Murthy

Until recently, the research into wireless networks concentrated only on single-hop wireless networks. The number of simultaneous users and the volume of data communicated over wireless networks have driven the need for high-capacity wireless networks. This chapter examines some of the recent architectures that have been proposed to enhance the capacity of wireless networks by improving spectrum reuse through ad hoc radio relaying. Such next-generation wireless network architectures that fuse the properties of ad hoc networking into the existing infrastructure-based wireless networks point to the fact that ad hoc networking will be an essential feature of next-generation systems. We describe the design of each architecture, identify its pros and cons, and brief the open issues that should be addressed in such architectures.

16.1   Introduction

The phenomenal growth of the Internet and wireless connectivity have caused an explosive need for higher-capacity wireless networks that can efficiently handle high network load, service highly mobile users with smooth hand-offs, manage both best-effort and real-time connections concurrently, and above all be extendible from the existing infrastructure to form the basis for the next-generation wireless systems. Recently there have been a few attempts at throughput enhancement in single-hop wireless networks such as Multi-hop Cellular Network (MCN),[1,2] Ad hoc-GSM (A-GSM),[3] Integrated Cellular and Ad hoc Relaying system (iCAR),[4] Hybrid Wireless Network (HWN) Architecture,[5] Self-Organizing Packet Radio Networks with Overlay (SOPRANO),[6] Multi-Power Architecture for Cellular Networks (MuPAC),[7,8] Throughput-enhanced Wireless in Local Loop (TWiLL),[9] and Directional throughput-enhanced Wireless in Local Loop (DWiLL).[10] The basic ingredient in these architectures has been the introduction of ad hoc network characteristics. Although these hybrid architectures[1–9] are still confined to the research arena, they are believed to be potential candidates for next-generation wireless systems. A concept group for the 3rd Generation Partnership Project (3GPP) has included Opportunity Driven Multiple Access (ODMA)[11] in the initial versions of the draft standard, although it currently appears to have been excluded in order to clear the concerns of signaling overhead and complexity, and to achieve a finalized standard. The major advantages of such multihop architectures are reduced interference, extended coverage, broadband support over an extended range, increased reliability, and support for a large number of users.

   This chapter contains the following: a survey of the existing hybrid architectures,[1–9] their classification, qualitative comparison, and finally identification of some of the open issues in this area. These architectures have hybrid properties (i.e., the multihop relaying property together with the infrastructure support of present-day wireless networks). The rest of this chapter is organized as follows. Section 16.1.1 discusses the broad classification of wireless networks. Section 16.1.2 provides a classification of hybrid wireless network architectures. Section 16.2 discusses description and a brief qualitative comparison of recent hybrid wireless network architectures. Section 16.4 summarizes the chapter.

16.1.1   Classification of Wireless Networks

Figure 16.1 illustrates the major classifications of wireless networks in terms of their dependency on fixed infrastructure for call completion. Traditional wireless networks such as cellular networks, wireless LANs, and wireless in local loop systems use a fixed base station (BS) for serving a geographical region. Because these networks depend fully on such fixed infrastructure-based stations, they are classified into infrastructure-based networks. The wireless networks that do not require any fixed infrastructure are called infrastructure-less wireless networks. An ad hoc wireless network is one example of such networks that

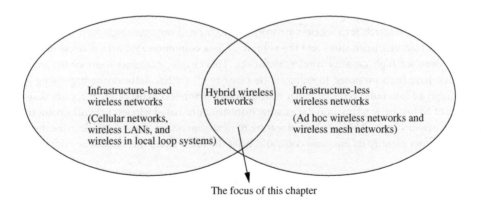

FIGURE 16.1   Classification of wireless networks.

form a communication network among a set of nodes with wireless network interfaces by self-organization. Infrastructure-less networks employ multihop wireless relaying for communication between nodes that are separated beyond their transmission ranges. Hybrid wireless networks are the result of research attempts to merge the properties of multihop relaying used in ad hoc wireless networks into the infrastructure-based wireless networks. This chapter focuses on hybrid wireless networks.

16.1.2   Classification of Hybrid Wireless Network Architectures

Figure 16.2 illustrates the classification of hybrid architectures. These architectures can be classified as the ones that use dedicated relay stations and the ones that have host-cum-relay stations. Here, the dedicated relay stations do not originate data traffic on their own; rather, they assist in forwarding on behalf of the senders. In most cases, these dedicated relay stations are fixed or stationary, whereas the host-cum-relay stations generate traffic on their own in addition to forwarding others' traffic. These host-cum-relay stations can be either stationary or mobile. These architectures can further be classified into single-mode systems and multimode systems. In single-mode systems, the mobile hosts (MHs) operate only in one mode (multihop mode), whereas in multi-mode systems, the MHs can operate either in single-hop or multihop fashion.

   The architectures that use fixed relay stations can be further divided into two categories: multi-mode systems and single-mode systems. In multi-mode systems, the mobile hosts (MHs) act either in single-hop mode or in multihop mode, depending on the architecture and routing scheme employed. In single-mode systems, MHs operate only in multihop mode. The hybrid wireless systems that use no fixed relay stations can also be divided into two categories: multi-mode systems and single-mode systems. Similar to the above description, multi-mode operation refers to the use of different modes, for example, the use of single-hop channels to communicate with the base station (BS) over a single-hop or multihop channel to communicate with the BS over multiple hops.

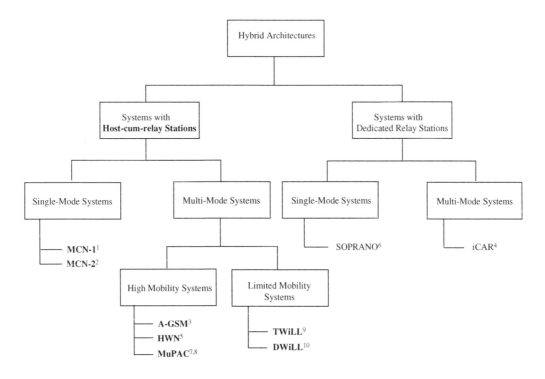

FIGURE 16.2   Classification of hybrid architectures.

16.2   Hybrid Wireless Network Architectures

This section describes some of the recently proposed architectures that use multihop relaying in the presence of infrastructure. The objectives, operation, and design philosophy of each of the architectures are discussed in this section. The treatment in this chapter excludes the lower-layer complexities of the system.

16.2.1   The MCN Architecture

Lin and Hsu[1] have proposed (this architecture is referred to as MCN-1 in this chapter) a novel cellular architecture where a connection between the source and the destination is established over a multihop path. Ananthapadmanabha et al.[2] extended this architecture and provided a unicast routing protocol (referred to as MCN-2) for best-effort and real-time traffic. The design philosophy of MCN is that the transmission power of the MHs and the base station (BS) over the data channel is reduced to a fraction $\frac{1}{k}$ (where k is referred to as reuse factor) of the cell radius, as shown in Figure 16.3. The node density expected in MCN is fairly high, hence the chances of a network partition within a cell are quite small. The throughput is expected to increase linearly with k. The analysis Lin Hsu,[1] however assumes that the straight-line path between the source and destination will be available, and moreover that the routing protocol is capable of discovering it. Hence, the actual gain will be probably lower because of the overhead of the routing protocol and the possibility of absence of relaying nodes along the straight-line path. A unicast routing protocol, a real-time scheme, and extensions to the above architecture were suggested by Manoj et al.[12] All MHs in a cell take part in the topology discovery, wherein each MH regularly sends to the BS, information about its neighbors. For best-effort communication, all the cells share a single data channel and a single control channel. While the transmission range on the data channel is kept at half of the cell radius, that on the control channel is equal to the cell radius. The value of $k = 2$ was arrived at as a compromise between increasing the spatial reuse and keeping the number of wireless hops to a minimum. The operation of

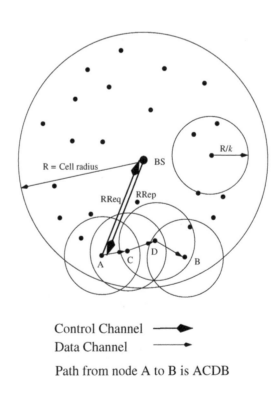

Control Channel   ⟶●

Data Channel   ⟶

Path from node A to B is ACDB

FIGURE 16.3   MCN architecture.

the unicast routing protocol is illustrated in Figure 16.3. A base-assisted-on-demand approach is used in routing for MCN. The routing protocol has a route discovery phase and a route maintenance phase. When a source A has a packet to send to a destination B for which a path is not known, it sends a route request ($RReq$) packet to the BS over the control channel. The BS responds with a route reply ($RRep$) packet containing the route, which is sent back to node A over the control channel. This route is computed by Dijkstra's shortest path algorithm. The source A, upon reception of the $RRep$ packet, transmits the data packet with the entire route information contained in it to the next node on the path, which in turn forwards it. This path is also cached in its local route cache (RC). Subsequent packets to the same destination are source routed using the same path until the RC entry times out or a route break is detected. When an intermediate node detects a break or interference in any link on the route from node A to node B, it sends a packet similar to an $RReq$ packet to the BS, which sends a new route from node A to node B, to node A (used for subsequent packets headed for node B) and a route from node C to node B, to node C (used to forward the packets that got buffered at node C due to the route break). A real-time scheme for MCN is proposed by Manoj et al.[12] The available bandwidth is split into one control channel and several data channels. These data channels are not clustered among the cells and hence can be used throughout the system area. While the transmission range of the data channels is kept at half of the cell radius, that of the control channel is equal to the cell radius. Using the channel allocation scheme described by Manoj et al.,[12] the BS also chooses the data channels to be used along each wireless hop in a route. The BS then broadcasts this information (the path and the channels to be used in every hop) over the control channel. On reception of this information, the sender, the intermediate nodes, and the destination become aware of the call setup. If the call cannot be established, the sender is appropriately informed by a unicast packet over the control channel. The channels for each hop in the wireless path obtained are allocated through a first-available-channel allocation policy. For each hop, the channels are checked in a predetermined order and the first channel that satisfies the constraints is used. At some point in the duration of the call, the wireless path may become unusable. This may happen due to either a path break or interference because of channel collision. The MH that detects this sends a packet similar to the $RReq$ packet, called a route error ($RErr$) packet, to the BS, which computes a new route and broadcasts the information to all MHs in the cell. To reduce the chance of call dropping, some channels are reserved for rerouting calls; that is, of N_{ch} channels, N_{ch}^r are reserved for handling route reconfiguration requests. The more the value of N_{ch}^r, the less the probability of dropping during reconfiguration. MCN-2 uses a new routing protocol called the Base Assisted Ad hoc Routing (BAAR) protocol.

16.2.1.1 Base Assisted Ad Hoc Routing (BAAR) Protocol

The BAAR protocol proposed by Ananthapadmanabha et al.[2] tries to assign as much responsibility to the BSs as possible. The BSs are a natural choice for all kinds of databases, including the location database and topology information. Mobile stations access these databases at the BSs through packets sent over the control channel. The computation of routes is also done at the BSs. Routing between the BS and the MS uses source routes (the overhead in the header is not an issue as the number of wireless hops is small). Here $BS(x)$ denotes the BS to which node x is registered. (x, y) is an edge between nodes x and y, and $w(x, y)$ is the edge weight of the link (x, y). The operator '.' stands for concatenation. Here, node A is the source node and Node B is the destination node.

1. If $BS(A) = BS(B)$,
   - Run the shortest path algorithm over all the wireless links in the common cell.
   - Return the path obtained.

2. Else If $BS(A)$ and $BS(B)$ are BSs for adjacent cells,
   - Get the state of links in the adjacent cell B belongs to.
   - Run the shortest path algorithm over all the wireless links in the two adjacent cells, including the link $(BS(A),\ BS(B))$ with $w(BS(A), BS(B)) = 0$.
   - Return the path obtained.

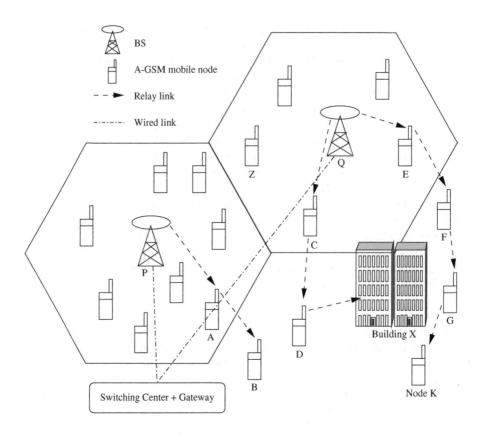

FIGURE 16.4   A typical scenario of A-GSM relaying.

3. Else,

   - Run the shortest path algorithm over all the links in A's cell and get the shortest path $p1$ from A to $BS(A)$.

   - Similarly, get the shortest path $p2$ from $BS(B)$ to B. Note that because all this happens at the BS, it has access to the location database of nodes. All the information required to find $BS(B)$ and the path from $BS(B)$ to B is available from the $BS(B)$.

   - Return $p1.(BS(A), BS(B)).p2$.

The BS runs the BAAR protocol upon request from any mobile node.

16.2.2   The A-GSM Architecture

The ad hoc GSM (A-GSM)[3] architecture was proposed by Aggelou and Tafazolli as an extension to the GSM cellular architecture for providing extended service coverage to *dead spots*.* This is a multihop-relaying-enabled extension to the GSM architecture in which service coverage can be provided to the regions that are not covered otherwise. It aims to use the existing GSM system modules and entities with minimal changes for providing compatibility with existing GSM systems. Figure 16.4 illustrates the use of

---

*These are areas within a cell where providing service coverage is very difficult due to geographical properties. Examples of dead spots include inside the basements of buildings and subways.

the A-GSM architecture. A node that is not in the coverage area of the BSs P and Q can still receive the services through relay nodes. For example, node B can make and receive calls through node A and node K can avail network services relayed through the path Q → E → F → G. Generally at the fringes of a cell, the signal strength may not be strong enough to give good coverage inside buildings located there. A-GSM can improve the service coverage in such cases, as shown in Figure 16.4 where the building X can receive better signals through the relay-enabled A-GSM node (node D). The GSM link layer protocol is modified to handle a beaconing scheme by which every A-GSM-enabled node originates beacons periodically. The important fields carried in the beacon are *Relay*, *RelayCapacity*, *LinkToBTS*, *HopsToBTS*, and *RelayToBTS*. The *LinkToBTS* flag carries information about its reachability to the BS. This flag is set to *ON* if the sender of the beacon has a direct link to the BS. If *LinkToBTS* is set to the *OFF* state, then the beacon carries information about the hop length (in *HopsToBTS*) to reach the BS and the address of the next hop field (in the *RelayToBTS*) in the beacon node to reach the BS. If a sender has no path to the BS, then it sets the *Relay* field in the beacon to −1, indicating that it has neither a direct link nor a multihop relay path to the BS. The *RelayCapacity* field carries information about the current level of resources available for the purpose of relaying. Every relayed call requires bandwidth, buffer space, and processing time. Hence, there is a limit to the total number of calls that an A-GSM node can relay. A resource manager module running in an A-GSM node executes a call admission control procedure for accepting or rejecting a call through the node. Once a call is admitted, the necessary resources are reserved. The *RelayCapacity* flag is set to *ON* to indicate that the sender node has sufficient resources to admit relay calls, and is set to *OFF* when the node is not in a position to accept any more calls. Alternatively, an A-GSM node can avoid transmitting the beacons to indicate that the node is busy, hence implicitly conveying its inability to accept calls for relaying. When an A-GSM node that has only a relay link to the BS requires to set up a call to a destination, it sends a call request to its next-hop node to the BS. The relay node would in turn relay the message to the BS. This relaying process can take place over multiple hops. Once the BS receives the call request, it forwards it to the mobile switching center (MSC), which then sets up the call and sends an acknowledgment to the caller. This is similar to the traditional GSM network. All the intermediate nodes relaying the call request and acknowledgment reserve resources for completion of the call.

16.2.3 The iCAR Architecture

Integrated Cellular and Ad hoc Relaying system (iCAR) was proposed[4] as a next-generation wireless architecture that can easily evolve from the existing cellular infrastructure. The design philosophy behind iCAR is to achieve a throughput closer to the theoretical capacity by dynamically balancing the load among different cells. In a normal single-hop cellular network (SCN; refers to the traditional single-hop cellular network), even if the network load does not reach the capacity of the entire network, several calls may be blocked or dropped in a congested cell. To counter this, the fundamental idea that iCAR deploys is to use a number of ad hoc relaying stations (ARSs) placed at appropriate locations to relay excess call traffic from a hot cell (congested cell) to relatively cooler cells (lightly loaded) around it. Thus, the excess bandwidth available in some cells can be used to accept new call requests from MHs in a congested cell. They can communicate with a BS, another ARS, or an MH through appropriate radio interfaces. The placement of ARSs is decided by the network operator and there can be limited mobility as well. The ARS radio interfaces operate at a lesser power than those of BSs, and hence only a subset of MHs can set up relay routes to BSs. The iCAR system proposes three modes of relaying (viz. primary, secondary, and cascaded relaying), which are illustrated by Figures 16.5(a), 16.5(b), and 16.5(c), respectively.

1. *Primary relaying.* In an SCN, if an MH X in a congested cell A makes a call request and no data channel (henceforth referred to as DCH) is free at BS A, the call will be blocked. Through primary relaying, it can still set up the call by establishing a relay route with an adjacent BS B through one or more ARSs. The interfaces used for communication with BSs are called C interfaces and the ones used with ARSs are called R interfaces.

2. *Secondary relaying.* Because the coverage probability of an ARS in a practical environment is not 1, not all MHs will be able to set up relay routes to neighborhood BSs. Secondary relay capitalizes on

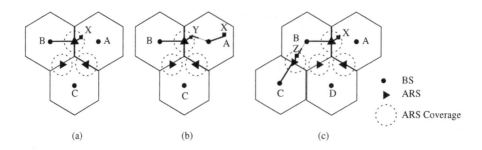

FIGURE 16.5   iCAR relaying strategies: (a) primary relay, (b) secondary relay, and (c) cascaded relay.

the fact that not all MHs with ongoing calls in the ARS coverage region would be using primary relay. The idea is to free up a DCH from the BS A for use by MH X. Assume that MH Y in the ARS coverage region has acquired a DCH from BS A to complete a call setup. On BS A's request, MH Y could switch over to the R interface and set up a relay route to a neighboring cooler cell B. The freed DCH can be used by MH X.

3. *Cascaded relaying.* It is possible that a relay path can be set up between MH X in cell A and a neighboring BS B, which is unfortunately congested. Here, secondary relaying can be deployed to free a DCH in cell B by establishing a relay path between MH Z in cell B and a neighboring cool cell C. The freed DCH can be used to connect MH X to BS B through a relay path.

16.2.4   The HWN Architecture

The Hybrid Wireless Network (HWN) architecture[5] is a novel multihop architecture having the capability of switching between multihop mode and single-hop mode of operation based on the throughput achieved. HWN has two modes of operation: the cellular mode and the ad hoc mode. In the cellular mode, nodes send packets to the BS, which forwards them to the destination (similar to SCN). Figure 16.6(a) shows the operation of HWN in the cellular mode. In this mode, the node A sends its packets to its BS B, which forwards them to the destination node C. In the ad hoc mode, nodes use the Dynamic Source Routing (DSR) protocol to discover routes. The operation of the ad hoc mode is shown in Figure 16.6(b). The transmission range of the nodes is r, where r is chosen by the BS such that all the nodes form a connected

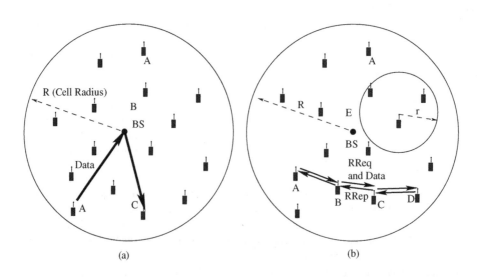

FIGURE 16.6   The HWN architecture: (a) in cellular mode and (b) in ad hoc mode.

graph. In this mode, node A uses the DSR protocol to discover routes to the destination D: it floods *RReq* packets, which reach destination D after being relayed by nodes B and C. Node D now sends a *RRep* packet, which reaches node A, which then begins to send packets along the newly discovered route.

The mode switching algorithm at the BS operates as follows. If the BS is in cellular mode, then the BS estimates the throughput if the BS has been in ad hoc mode by simulating a packet scheduling algorithm. This throughput is compared with the actual throughput achieved in the cellular mode to decide in which mode to operate. In the ad hoc mode, the BS compares the throughput achieved in ad hoc mode with $BW/2n$ (the average bandwidth achievable per user) to find out to which mode the topology is best suited (n is the number of nodes in the cell and BW is the available bandwidth per cell).

Ad hoc mode works well for dense topologies and the cellular mode is better suited to sparse topologies. An algorithm operates at the BS, which uses the network topology to decide in which mode the cell should operate, so as to maximize the throughput. This decision is broadcast to all the nodes. In the ad hoc mode, partitions are avoided by having the BS periodically check the topology and broadcasting the minimum power required to keep the network connected.

16.2.5   The SOPRANO Architecture

The Self-Organizing Packet Radio Ad hoc Networks with Overlay (SOPRANO) is a wireless multihop network overlaid on a cellular structure.[6] SOPRANO aims at providing high data rate Internet access at the fringes of a cell using inexpensive relay stations. Two separate frequency bands are assumed to carry the information, one each for up and down streams. This is a slotted packet CDMA system with fixed repeaters, where the repeaters form a hexagon or a random shape as shown in Figure 16.7. The repeaters are not expected to generate traffic on their own; rather, they help in forwarding traffic originated by other MHs. A channel assignment process is used to inform every node about the channel to be used by that node. MHs are assigned one channel while routers are expected to operate in all the channels as per BS's decision. An MH or router works either in transmit mode or in receive mode. The mode of operation is decided by the BS during the call setup/path setup phase. Scheduling the transmission slots on a path is done such that the system capacity is maximized and at the same time reduction in interference is also achieved. Two routing strategies were studied for this architecture (viz. Minimum Path Loss (MPL) and

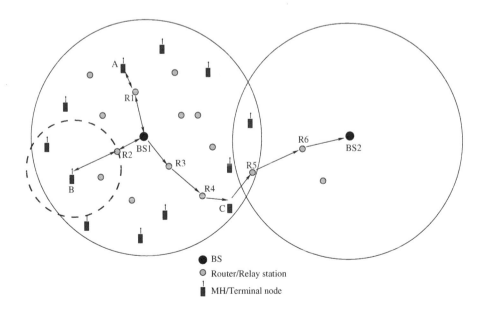

FIGURE 16.7   The SOPRANO architecture: load balancing through BS2 when BS1 is heavily loaded.

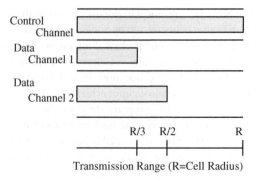

FIGURE 16.8   A two-channel MuPAC network.

Minimum path loss with Forward Progress (MFP)). In the former case, the packets are forwarded to a receiver with minimum link propagation loss; whereas in the latter case, a node sends a packet to a receiver with minimum link propagation loss along with a transmission direction toward the BS. SOPRANO can employ dynamic load balancing schemes as shown in Figure 16.7. The downlink to the node C is BS1-R3-R4-C; and if the uplink through the same path gives rise to increased interference or if it does not have enough slots to accommodate the uplink stream, then the upstream can be completed through BS2. The upstream path now is C-R5-R6-BS2.

16.2.6   The MuPAC Architecture

The Multi-Power Architecture for Cellular Networks (MuPAC) architecture[7,8] is a multichannel architecture. In an n-channel MuPAC architecture, there are $n + 1$ channels, each operating at a different transmission range. The total bandwidth is divided into a control channel using a transmission range equal to the cell radius (R) and n data channels each operating at different transmission ranges as depicted in Figure 16.8, where the channel 1 uses a transmission range of $R/3$ and channel 2 uses a transmission range of $R/2$. Figure 16.9 shows packet transmission from node A to node C. On its first hop from node A to node B, the packet is transmitted over channel 1 with transmission range $R/3$; and for the second hop from node B to Node C, channel 2 is used. The decision to use a particular channel at an intermediate node in a path is a local decision taken by the intermediate node based on the current load on each channel.

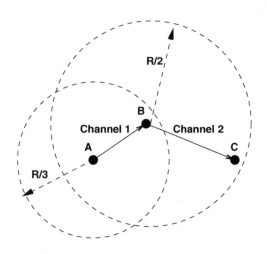

FIGURE 16.9   Packet transmission in a two-channel MuPAC network.

The bandwidth required for the channels is directly proportional to the transmission range used. With an increase in the number of data channels, the transmission ranges will be different and the corresponding bandwidths will be used. The path selection at the BS is done by assigning appropriate edge weights that are directly proportional to the approximate distances between the nodes. This approximate distance is measured using Geographical Positioning System (GPS) information. Dijkstra's shortest path algorithm is used to obtain a minimum weight path. MuPAC solves the network partition problem at low node density using the single-hop control channel for data transmission. Kumar et al.[7] propose the best-effort and also the real-time MuPAC architecture.[8] They show that the two-channel MuPAC outperforms the MCN in terms of throughput.

16.2.7 The TWiLL Architecture

The Throughput-enhanced Wireless in Local Loop (TWiLL) architecture proposed by Manoj et al.[9] is a multihop architecture for limited mobility systems such as Wireless in Local Loop (WLL). Henceforth, MH will also be used to refer to a WLL or TWiLL subscriber, be it a stationary fixed subscriber unit (FSU) or an FSU with limited mobility. The bandwidth available is split into one control channel and several data channels. TWiLL solves the problem of network partitions by allotting a channel ch in single-hop mode when there is no multihop path to the BS. That is, in TWiLL, every channel is designated as a multihop channel (MC) or a single-hop channel (SC), as illustrated in Figure 16.10. An MH transmits in the control channel and SCs with a range of R (cell radius), and in the MCs with a range of $r = R/2$, thus keeping the reuse factor $k = 2$ among the MCs. The call establishment process is similar to that in MCN.

16.2.8 The DWiLL Architecture

Directional throughput-enhanced Wireless in Local Loop (DWiLL)[10] is a high-performance architecture for wireless in local loop (WLL) systems. DWiLL uses the dual throughput enhancement strategies of multihop relaying and directional antennas. The major advantages of DWiLL include a reduction in the

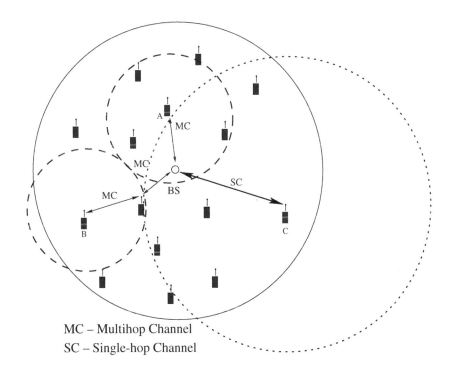

MC – Multihop Channel
SC – Single-hop Channel

FIGURE 16.10 The TWiLL architecture.

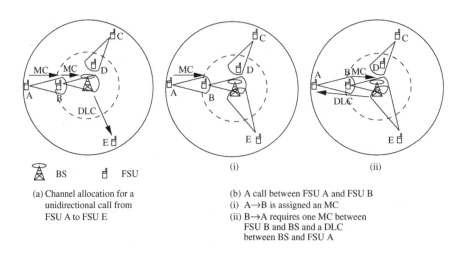

(a) Channel allocation for a
    unidirectional call from
    FSU A to FSU E

(b) A call between FSU A and FSU B
(i) A→B is assigned an MC
(ii) B→A requires one MC between
     FSU B and BS and a DLC
     between BS and FSU A

FIGURE 16.11   Illustration of call setup in DWiLL.

energy expenditure at the fixed subscriber units (FSUs)* and the ability to provide enhanced throughput when the number of subscribers becomes large. Similar to the TWiLL architecture, the spectrum is divided into a number of channels. The key difference between TWiLL and DWiLL is the use of directional relaying by the FSUs in DWiLL. DWiLL assumes that the directional antenna at the FSU is oriented in the direction of the BS. Because there is no significant requirement for the directionality to be changed, there is no need for a sophisticated electronically and dynamically steerable antenna. The FSU uses the directional antenna to transmit control information, beacon signals, and the data messages. Also, due to directionality, the wireless link-level connectivity between two nodes is not symmetric. The system works by building the topology information at the BS, as in the BAAR protocol. Each node reports the set of nodes from which it receives a beacon, along with the received power to the BS. DWiLL also designates the data channels into two categories, namely, multihop channels (MCs) and single-hop channels (SCs). The SCs are further divided into uplink channels (ULCs) and downlink channels (DLCs). MCs operate over a transmission range of r meters, where r is a fraction of the cell radius R ($r = \frac{R}{k}$ where $k = 2$). The SCs operate over a transmission range of R meters. ULCs are assigned to those nodes that do not find intermediate relaying stations to use MCs for setting up data paths to the BS. The DLCs are used by the BS for the downlink transmissions to the FSUs. Figure 16.11 shows the call setup process in DWiLL. Figure 16.11(a) shows a unidirectional call from FSU A to FSU E. It uses multihop relaying from FSU A to the BS and a DLC channel on the downlink from BS to FSU E. Another example is shown in Figures 16.11(b)(i) and 16.11(b)(ii), where a duplex path is set up between FSU A and FSU B. On the unidirectional path $A \rightarrow B$, a single MC is allotted and the unidirectional path $B \rightarrow A$ is obtained through $B \rightarrow BS \rightarrow A$ (an MC is assigned between FSU B and BS, and a DLC is assigned between BS and FSU A).

If a path cannot be obtained using MCs, then the MH is given an SC to communicate directly with the BS. The allocation of channels in single-hop mode reduces the spatial reuse of bandwidth, thus reducing the network throughput, but will also increase the number of accepted calls when the node density is less. The optimal number of SCs was calculated as a function of the required probability of blocking for the entire system, as discussed by Manoj et al.[9] As shown in Figure 16.10, MHs A and B are connected to the BS through multihop paths. Node A can reach the BS over one hop while B does so over two hops. MH C is in a partition and cannot reach the BS through a multihop path. Hence, it is allowed to use an SC to communicate with the BS in single-hop mode. The probability that a call's destination is within the same

---

cell as the call's source is defined as the locality of traffic. In TWiLL, the locality of traffic is used to improve the throughput by a technique called shortcut relaying. Since in TWiLL, users are mostly stationary or have very limited mobility, the number of path reconfigurations will be much less than in MCN, thus improving the quality of service.

16.3 A Qualitative Comparison and Open Problems in Hybrid Wireless Networks

Having classified (in Figure 16.2) and described the hybrid architectures, in this section we compare them qualitatively. iCAR and SOPRANO proposed to use fixed or semi-mobile relay stations that do not generate traffic on their own. iCAR uses a seed growing algorithm[4] to place the ARSs, which places the seed ARSs at the boundary between two cells; whereas, SOPRANO architecture places the relay stations either in layers forming a co-centric hexagonal pattern or randomly, inside the cell.

Open Issues in Hybrid Wireless Architectures

Issue	Description
Flexibility of operation	Transparent hand-off to various heterogeneous networks, possibly influenced by user preferences, is necessary to effect this.
Pricing schemes	In some hybrid wireless networks relaying is done by the intermediate nodes which expend their own resources such as battery charge and buffers. Hence those nodes should be compensated for the services rendered. New reimbursement based multihop pricing schemes are required.
Multimedia support	Support for multimedia traffic in hybrid architectures is still in a very early stage of research for packet based systems. Better multihop scheduling, with or without assistance from the BS, will be required for this purpose.
QoS at high mobility	Due to smaller transmission range and frequent path breaks caused by the instability of relaying nodes, mechanisms are required to achieve good performance at high mobility. Dynamic prediction based approaches can be used to take proactive steps to reroute calls before the actual break of the path. The path selection process can be modified to select a path which is more resilient to mobility.
Resource management	Resource allocation and management is a major issue which takes multiple dimensions. Some of the areas in which new resource management schemes are required include (i) channel allocation (time slot/code/frequency to be used at every hop), (ii) power control (the transmission power to be used at every hop), (iii) mode switching and selection (whether to work in ad hoc mode or infrastructure mode), and (iv) packet scheduling (reordering packets to introduce the property of fairness or priority).
Power control	Power control in hybrid networks is important and more complex than in cellular networks. This is because of the multihop nature, and is essential in reducing the interference and improving efficiency of CDMA systems. New efficient power control algorithms are required.
Efficient dissemination of control information	Registration and location information, neighbor topology updates, path maintenance packets, etc., need to be quickly and efficiently made available to the router/BS which is responsible for handling them. The assumption of a single hop control channel simplifies this problem but demands more resources in terms of battery and computing power. Hence efficient control protocols are required for the architectures in which the single hop control channel is not used.
Routing protocols	Efficiency of the routing protocol is key to the performance of these hybrid systems. New routing protocols with different routing constraints, like signal to noise ratio, minimum power routing, path diversity, etc. and which can make use of partial or full topological information available at the BS, are required.
Path reconfiguration mechanisms	In order to improve the performance of the system, efficient path reconfiguration mechanisms are required. In those architectures where fixed relay stations are used, base assisted pro-active reconfiguration mechanisms can be employed.
Support for multicast routing	An important requirement of next generation wireless networks is the support for multicast routing protocols. Since these architectures use multiple short range hops, efficient multicast routing protocols are required to support applications such as broadcast audio/video, multimedia multicast groups, and video conferencing. Multicast routing protocols that make best use of BSs are required.
Security	Base assisted multihop security schemes are required in these architectures.

TABLE 16.1　Comparison of Hybrid Wireless Architectures

Issue	MCN-1	MCN-2	A-GSM	iCAR	HWN	SOPRANO	MuPAC	TWiLL	DWiLL
Dedicated relay stations	No	No	No	Yes	No	Yes	No	No	No
Routing efficiency	Low	High	High	High	Low	High	High	High	High
Cost of MH	Low	High	Low	Low	High	Low	High	Low	Low
Routing complexity	High	Low	Low	Low	High	Low	Low	Low	Low
Partitions handled	No	No	Yes	Yes	Yes	No	Yes	Yes	Yes
Performance at high mobility	Low	Low	Low	Good	Low	Good	Good	Not applicable	Not applicable
Connection oriented or packet based	Packet	Packet and connection	Connection	Connection	Packet	Connection	Packet and connection	Connection	Connection
Real-time traffic support	No	Yes	Yes	Yes	Only in cellular mode	Yes	Yes[8]	Yes	Yes
Multiple interfaces	Yes	Yes	Yes	Yes	Yes	No	Yes	Yes	No
Control overhead	High	High	Low	High	Low	High	Low	Low	Low
Relay by MH	Yes	Yes	Yes	No	Only in ad hoc mode	No	Yes	Yes	Yes
New pricing schemes required	Yes	Yes	Yes	No	Yes	No	Yes	Yes	Yes
Easiness of implementation of pricing schemes	Difficult	Easy	Easy	Easy	Difficult	Easy	Easy	Easy	Easy

Routing efficiency is high in iCAR, SOPRANO, DWiLL, and TWiLL due to the presence of fixed relay stations that do not originate traffic and are stationary. Similarly, routing efficiency is high in TWiLL due to the stationary or limited mobility nature of MHs. MuPAC uses a dynamic channel switching scheme to alleviate the effects of path breaks. HWN, MCN-2, and MuPAC require increased hardware requirements in terms of multiple interfaces operating at different powers and associated power sources. The cost increases with miniaturization of hardware and hence a proportional cost increase does not occur for an MH in TWiLL. iCAR, HWN, MuPAC, TWiLL, and DWiLL handle network partitions. In iCAR, any node can either connect to the BS directly or through the nearby ARS; hence, connectivity does not depend on the presence or absence of neighbor MHs. SOPRANO assumes the density of relay stations to be high enough to avoid partitions but placing a relay station in an ideal location may not be feasible in practice. In HWN, all the nodes use transmission power that is sufficient to maintain the network connected in the ad hoc mode, but chances of transient partitions still exist. MCN-1 and MCN-2 do not handle the situation caused by the occurrence of network partitions. MuPAC, TWiLL, and DWiLL handle partitions using single-hop control channels when nodes get isolated. TWiLL and DWiLL use a novel approach of classifying channels as single-hop channels (SCs) and multihop channels (MCs). Isolated nodes are assigned a single-hop channel by the BS as and when it is required. Generally, chances of partitions are less in TWiLL because it is a limited mobility system. TWiLL is the only architecture that utilizes multihop relaying with directional antennas. Performance at high mobility is better with systems having fixed relay stations. A novel use of channels with higher transmission range in MuPAC does not degrade the performance as much as in MCNs. A-GSM, iCAR, SOPRANO, TWiLL, and DWiLL envision connection-oriented systems using TDMA, FDMA, or CDMA, whereas MCNs, HWN, and MuPAC are mainly proposed for packet-based data cellular systems. The HWN architecture proposes to use a polling mechanism in its cellular mode of operation; hence, it can support real-time traffic in that mode, whereas in the ad hoc mode it does not provide any such support. The control overhead is higher for systems that use relaying by MHs (except in the case of TWiLL and DWiLL), as the instability of MHs could cause a higher number of control packets such as route reconfiguration packets. TWiLL and DWiLL are limited mobility systems that do not suffer from this, although they use MHs for relaying. Table 16.1 summarizes a qualitative comparison of all the hybrid architectures described in this chapter.

16.4 Summary

A set of recently proposed hybrid wireless network architectures that effectively combine the benefits of using infrastructure-based systems with ad hoc radio relaying was discussed in this chapter, incuding classification, a qualitative comparison of these architectures, and a discussion of several research issues without detailing the complexities of the physical layer. These architectures reflect the importance of ad hoc radio relaying in the next-generation wireless networks where the dimension of providing service to the fast-growing user community, along with provisioning of broadband access, is considered. Several issues, such as efficient and resilient routing schemes, support for multicasting, pricing schemes, QoS at high mobility, decisions on the use of different modes of operation, real-time traffic support, and load balancing, require further work prior to the deployment of these architectures.

References

1. Y.D. Lin and Y.C. Hsu, Multi-Hop Cellular: A New Architecture for Wireless Communications, in *Proc. IEEE INFOCOM 2000*, March 2000.
2. R. Ananthapadmanabha, B.S. Manoj, and C. Siva Ram Murthy, Multi-Hop Cellular Networks: The Architecture and Routing Protocol, in *Proc. IEEE PIMRC 2001*, October 2001.
3. G.N. Aggelou and R. Tafazolli, On the Relaying Capability of Next-Generation GSM Cellular Networks, *IEEE Personal Communications Magazine*, 8(1) 40–47, February 2001.
4. H. Wu, C. Qiao, S. De, and O. Tonguz, Integrated Cellular and Ad Hoc Relaying Systems: iCAR, in

IEEE Journal on Selected Areas in Communications, 19(10), 2105–2115, October 2001.

5. H.Y. Hsieh and R. Sivakumar, Performance Comparison of Cellular and Multi-Hop Wireless Networks: A Quantitative Study, in *Proc. ACM SIGMETRICS 2001*, June 2001.

6. A.N. Zadeh, B. Jabbari, R. Pickholtz, and B. Vojcic, Self-Organizing Packet Radio Ad hoc Networks with Overlay, in *IEEE Communications Magazine*, 40(6), 140–157, June 2002.

7. K.J. Kumar, B.S. Manoj, and C. Siva Ram Murthy, MuPAC: Multi Power Architecture for Packet Data Cellular Networks, in *Proc. IEEE PIMRC 2002*, September 2002.

8. K.J. Kumar, B.S. Manoj, and C. Siva Ram Murthy, RT-MuPAC: Multi-Power Architecture for Voice Cellular Networks, in *Proc. HiPC 2002*, December 2002.

9. B.S. Manoj, D. Christo Frank, and C. Siva Ram Murthy, Throughput Enhanced Wireless in Local Loop (TWiLL) — The Architecture, Protocols, and Pricing Schemes, in *ACM Mobile Computing and Communications Review*, 7(1), January 2003.

10. V. Mythili Ranganath, B.S. Manoj, and C. Siva Ram Murthy, A Wireless in Local Loop Architecture Utilizing Directional Multihop Relaying, to appear in *Proc. IEEE PIMRC 2004*, September 2004.

11. 3GPP TR 25.833 V 1.1.0, Physical Layer Items Not for Inclusion in Release '99, April 2000.

12. B.S. Manoj, R. Ananthapadmanabha, and C. Siva Ram Murthy, Multi-Hop Cellular Networks: The Architecture and Routing Protocol for Best-Effort and Real-Time Communication, in *Proc. IRISS 2002*, Bangalore, India, March 2002.

13. K.J. Kumar, B.S. Manoj, and C. Siva Ram Murthy, On the Use of Multiple Hops in Next Generation Cellular Architectures, in *Proc. IEEE ICON 2002*, August 2002.

17

Ad Hoc Networks: A Flexible and Robust Data Communication

Mehran Abolhasan

Tadeusz Wysocki

The 1990s witnessed a rapid growth in research interests in mobile ad hoc networks (MANETs). The infrastructureless and dynamic nature of these networks demand that a new set of networking strategies be implemented to provide efficient end-to-end communication. This, along with the diverse application of these networks in many different scenarios, such as battlefield and disaster recovery, have seen MANETs being researched by many different organizations. MANETs employ the traditional TCP/IP structure to provide end-to-end communication between nodes. However, due to their mobility and the limited resources in wireless networks, each layer in the TCP/IP model requires redefinition or modifications to function efficiently in a MANET. One interesting research area in MANETs is routing. Routing in MANETs is a challenging task and has received a tremendous amount of attention from researchers. This has led to the development of many different routing protocols for MANETs. This chapter presents a discussion on a number of different routing protocols proposed for ad hoc networks and presents a number of future challenges in ad hoc and mobile ad hoc networking.

17.1 Background

17.1.1 Mobile Ad Hoc Networks

Ever since the early 1990s the use and demand for mobile wireless networks and devices has continued to grow. This is largely due to the ever-growing popularity of mobile phones. The parallel growth in popularity of the Internet has sparked new interests in providing Internet-type applications over mobile wireless networks. Traditionally, mobile wireless networks can be classified into two categories: infrastructured and infrastructureless. The infrastructured networks are coordinated by a centralized controller (also known as a base station or an access point), which directs the flow of traffic to and from each end-user node. The infrastructureless (also known as ad hoc) networks are made up of end-user nodes only. This means all nodes in the networks are capable of transmitting, receiving, and routing data to different nodes in their network without using the services of a base station (used in cellular networks). In ad hoc networks, nodes can be fixed (static) or mobile, or a mixture of the two. The ad hoc networks that have mobile nodes are commonly referred to as mobile ad hoc networks (or MANETs). In MANETs, each node is characterized by its transmission range, which is limited by the transmission power, attenuation, interference, and terrain topology. Direct communication can occur between two intermediate nodes if they are within each other's transmission range. Indirect communication can be established by determining a route through a number of intermediate nodes between the source and the destination. For example, in Figure 17.1 node A and node B can communicate directly because they are both in each other's transmission range, whereas node C and node A must establish a route through node B to be able to communicate.

17.1.2 Applications

MANETs are useful in dynamic networking environments where the topology of the network changes continuously. They are also useful in areas where a networking infrastructure cannot be easily implemented.

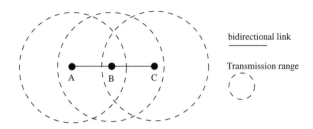

FIGURE 17.1 A mobile ad hoc network (MANET) made up of three mobile nodes.

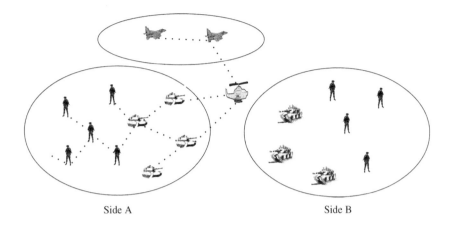

Side A   Side B

FIGURE 17.2   Military communication in a mobile ad hoc network.

Some typical applications of these networks include:

- Coordinating military operations in the battlefield
- Disaster relief operations
- Conferencing
- Sensor networking
- Vehicular networking
- Personal area networks (PANs)

In the highly dynamic battlefield environment (see Figure 17.2), efficient communications between different types of forces may give a significant advantage to one side over the other. In disaster relief operations, the search and rescue teams can coordinate their effort using MANETs to save the victims of fires, hurricanes, earthquakes, and other natural disasters. During conferences or exhibitions, where a temporary networking infrastructure may be required, MANETs can provide a more cost-effective and rapid implementation solution than wired networks. In sensor networking, MANETs can be used to control mobile devices to gather data in contaminated areas instead of sending an emergency team. In vehicular networking, MANETs can be used to control traffic in the city by providing drivers with up-to-date traffic information from the surrounding streets and intersections. Another application for MANETs is to provide communication between small devices in a personal area network (PAN) with a dynamic networking environment. For example, a number of people in a shopping mall carrying small devices such as personal digital assistants (PDAs) can interact with each other using MANETs. A MANET can be formed among PDAs equipped with Bluetooth.[6] Bluetooth is a short-range radio device that can provide communication links between mobile devices in a small networking environment.[54]

17.1.3   Challenges

The applicability of MANETs to a variety of different applications (mentioned previously) has attracted interest from many different organizations such as large companies, governments, and universities. This has made MANETs one of the most highly researched areas in wireless local area networking. The current research in MANETs ranges throughout all layers of the TCP/IP model, as the very nature of these networks demands some redesign for each layer to provide efficient end-to-end communication. Furthermore, before

MANETs can be used successfully in the scenarios described in the previous sections, a number of critical issues should be addressed:

- *Bandwidth.* The capacity of wireless networks is significantly lower than the capacity of wired networks. Route discovery and updates may cause significant bandwidth problems as the size or the density of the network grows.
- *End-to-end delay.* Data packets traveling between two nodes may experience long delays. This can be due to route recalculation as a result of broken links, queueing for gaining access to the transmission medium due to heavy channel contention, processing delays at each node, and traffic bottlenecks (intermediate nodes that receive packets from many different locations).
- *Energy (power).* Each mobile node must carry a mobile power supply (such as a battery). Periodic route updates, beaconing, and data transmission can consume significant amounts of battery power, which may require each node to frequently recharge its power supply. This means that each node must minimize processing to preserve battery; thus, lean (lightweight) protocols are desirable.
- *Security.* Because each node in a MANET broadcasts via a radio channel, there is a high security risk from eavesdropping, spoofing, denial-of-service, and other types of attacks from rogue users.[48,52]
- *Quality-of-Service (QoS).* Each node in the network may be transmitting different types of information with different levels of importance. Therefore, the available resource in the network must be distributed in such a way that each user gets different levels of access according to the level of service required. The dynamic nature of MANETs, along with the limited resources that vary with time (such as bandwidth, battery power, and storage space), makes providing QoS a challenging problem.
- *Scalability.* As the size and density of the network, in terms of the number of nodes and amount of traffic, grows, the efficiency of routing and data transmission begins to suffer. This is because access to resources such as bandwidth and the wireless medium becomes more competitive. This may result in packet loss, long delays, and the creation of traffic bottlenecks (one intermediate node may be in the path to many required destinations).

17.2   Routing in Ad Hoc and Mobile Ad Hoc Networks

The limited resources in MANETs have made designing of an efficient and reliable routing strategy a very challenging problem. An intelligent routing strategy is required to efficiently use the limited resources while at the same time being adaptable to the changing network conditions such as network size, traffic density, and network partitioning. In parallel with this, the routing protocol may need to provide different levels of QoS to different types of applications and users.

Prior to the increased interest in wireless networking, in wired networks two main algorithms were used. These algorithms are commonly referred to as the link-state and distance-vector algorithms. In link-state routing, each node maintains an up-to-date view of the network by periodically broadcasting link-state information about links to its neighboring nodes to all other nodes using a flooding strategy. When each node receives an update packet, it updates its view of the network and its link-state information by applying a shortest-path algorithm to choose the next-hop node for each destination. In distance-vector routing, for every destination x, each node i maintains a set of distances D_{ij}^x, where j ranges over the neighbors of node i. Node i selects a neighbor k to be the next hop for x if $D_{ik}^x = min_j\{D_{ij}^x\}$. This allows each node to select the shortest path to each destination. The distance-vector information is updated at each node by a periodic dissemination of the current estimate of the shortest distance to every node.[59] The traditional link-state and distance-vector algorithms are not highly scalable in large MANETs. This is because periodic or frequent route updates in large networks may consume a significant part of the available bandwidth, increase channel contention, and may require each node to frequently recharge its power supply.

To overcome the problems associated with the link-state and distance-vector algorithms, a number of routing protocols have been proposed for MANETs. These protocols can be classified into three different

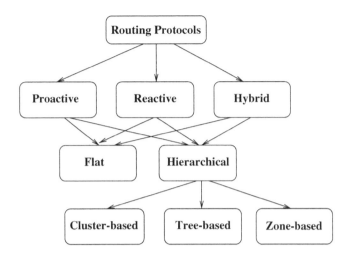

FIGURE 17.3   Classification for mobile ad hoc network routing protocols.

groups (as illustrated in Figure 17.3): (1) global/proactive, (2) on-demand/reactive, and (3) hybrid. In proactive routing protocols, the routes to all the destinations (or parts of the network) are determined from the start-up and maintained using a periodic route update process. In reactive protocols, routes are determined when they are required by the source using a route discovery process. Hybrid routing protocols combine the basic properties of the first two classes of protocols into one. That is, they are both reactive and proactive in nature. Each group has a number of different routing strategies that employ a flat or hierarchical routing structure (routing structures are described in the following section).

17.2.1   Proactive Routing Protocols

Proactive routing protocols maintain global or partial routing information. The routing information is usually kept in a number of different tables. These tables are periodically updated or when the network topology changes. The difference between these protocols exists in the way the routing information is updated, detected, and the type of information kept at each routing table. Furthermore, each routing protocol can maintain a different number of tables.

The advantage of proactive route discovery is that end-to-end delay is reduced during data transmission, when compared to determining routes reactively. Simulation studies[5,8,23] carried out for different proactive protocols show high levels of data throughput and significantly fewer delays than on-demand protocols (such as DSR[35]) for networks composed of up to 50 nodes with high traffic levels. Therefore, in small networks using real-time applications (e.g., videoconferencing), where low end-to-end delay is highly desirable, proactive routing protocols may be more beneficial. This section describes a number of different route update strategies proposed in the literature to perform proactive routing.

17.2.1.1   Global Updates

Proactive routing protocols using global route updates are based on the link-state and distance-vector algorithms, which were originally designed for wired networks. In these protocols, each node periodically exchanges its routing table with every other node in the network. To do this, each node transmits an update message every T seconds. Using these update messages, each node then maintains its own routing table, which stores the most recent or best route to every known destination. The disadvantage of global updates is that they use significant amounts of bandwidth because they do not take any measures to reduce control overheads. As a result, data throughput may suffer suffer significantly, especially as the number of nodes in the network increases. Two such protocols are DSDV[53] and WRP.[47]

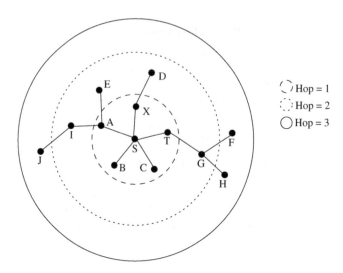

FIGURE 17.4   Illustration of the fisheye scope in FSR.

17.2.1.2   Localized Updates

To reduce the overhead in global updates, a number localized updating strategies were introduced in protocols such as GSR[12] and FSR.[23] In these strategies, route update propagation is limited to a localized region. For example, in GSR, each node exchanges routing information with its neighbors only, thereby eliminating the packet flooding methods used in global routing. FSR is a direct descendent of GSR. This protocol attempts to increase the scalability of GSR by updating the nearby nodes at a higher frequency than the nodes that are located far away. To define the nearby region, FSR introduces the fisheye scope (as shown in Figure 17.4). The fisheye scope covers a set of nodes that can be reached within a certain number of hops from the central node, as shown Figure 17.4. The update messages with greater hop counts are sent at a lower frequency. This reduces the accuracy of the routes in remote locations; however, it significantly reduces the amount of routing overhead disseminated in the network. The idea behind this protocol is that as the data packets get closer to the destination, the accuracy of the routes increases. Therefore, if the packets know approximately in what direction to travel, then as they get close to the destination, they will travel over a more accurate route and have a better chance of reaching the destination.

17.2.1.3   Mobility-Based Updates

Another strategy that can be used to reduce the number of update packets is introduced in DREAM.[5] The authors propose that routing overhead can be reduced by making the rate at which route updates are sent proportional to the speed at which each node travels. Therefore, the nodes that travel at a higher speed disseminate more update packets than the ones that are less mobile. The advantage of this strategy is that in networks with low mobility, this updating strategy can produce fewer update packets than using a static update interval approach such as DSDV. Similar to FSR, in the DREAM protocol, updates are sent more frequently to nearby nodes than the ones located far away.

17.2.1.4   Displacement-Based Updates

Displacement-based updates were introduced in Abolhasan and Wysocki.[2] This route update strategy attempts to disseminate route updates packets into the network when they are required, rather than using purely periodic updates. This is achieved by making the rate at which updates are sent proportional to the rate at which a node migrates from one location to a new location. That is, when a node changes location by a threshold distance, a route update is transmitted into the network. The required displacement can be measured using a Global Positioning System (GPS). The advantage of this strategy over purely mobility-based updates[5] is that updates are sent only when there is a chance for a topological change, which may alter the connectivity of the network.

17.2.1.5 Conditional or Event-Driven Updates

The number of redundant update packets can also be reduced by employing a conditional (also known as event-driven) update strategy.[22,53] In this strategy, a node sends an update if certain different events occur at any time. Some events that can trigger an update are when a link becomes invalid or when a new node joins the network (or when a new neighbor is detected). The advantage of this strategy is that if the network topology or conditions are not changed, then no update packets are sent. Therefore, redundant periodic update dissemination into the network is eliminated.

17.2.2   Reactive Routing Protocols

Reactive (also referred to as on-demand) routing protocols determine and maintain routing information for active for nodes that require to send data to a particular destination (i.e., active routes). Route discovery usually occurs by flooding a Route Request (RREQ) packet through the network. When a node with a route to the destination (or the destination itself) is reached, a Route Reply (RREP) is sent back to the source node using link reversal if the route request has traveled through bidirectional links or by piggy-backing the route in a route reply packet via flooding. Reactive protocols can be classified into two categories: source routing and hop-by-hop routing.

17.2.2.1   Source Routing Protocols

In source routing on-demand protocols DSR[35] and SSA,[20] each data packet carries the complete source-to-destination address. Therefore, each intermediate node forwards these packets according to the information kept in the header of each packet. This means that the intermediate nodes do not need to maintain up-to-date routing information for each active route in order to forward the packet toward the destination. Furthermore, nodes do not need to maintain neighbor connectivity through periodic beaconing messages. The major drawback to source routing protocols is that in large networks they do not perform well. This is due to two main reasons: (1) as the number of intermediate nodes in each route grows, so does the probability of route failure; and (2) as the number of intermediate nodes in each route grows, so does the amount of overhead carried in each header of each data packet. Therefore, in large networks with significant levels of multihopping and high levels of mobility, these protocols may not scale well.

17.2.2.2   Hop-by-Hop Routing Protocols

In hop-by-hop routing (also known as point-to-point routing) AODV,[16] each data packet only carries the destination address and the next-hop address. Therefore, each intermediate node in the path to the destination uses its routing table to forward each data packet toward the destination. The advantage of this strategy is that routes are adaptable to the dynamically changing environment of MANETs because each node can update its routing table when it receives fresher topology information and hence forwards the data packets over fresher and better routes. Using fresher routes also means that fewer route recalculations are required during data transmission. The disadvantage of this strategy is that each intermediate node must must store and maintain routing information for each active route, and each node may be required to be aware of its surrounding neighbors through the use of beaconing messages.

17.2.2.3   Control Overhead Reduction in On-Demand Routing

Most on-demand routing protocols employ a pure flooding approach to discovering routes. In pure flooding, an RREQ packet is flooded throughout the entire network, which can consume a significant amount of bandwidth. Therefore, for networks with a large number of nodes and high traffic, a pure flooding approach to route discovery is not scalable. A number of different on-demand routing protocols have been proposed to improve the performance of reactive routing. These include AODV,[16] LAR,[37] RDMAR,[4] ABR,[63] SSA,[20] and LPAR.[3] The characteristic features of these protocols can be distinguished by the strategies they employ to reduce the amount of control overhead during their route discovery phase. For example, in LAR and RDMR, the route history is used to control the route discovery procedure by localizing the route requests to a calculated region. Another method used to minimize the number of control packets during route discovery is through selection of stable routes (e.g., ABR and SSA). In ABR and SSA, the destination nodes select routes based on their stability. ABR also allows the use of shortest path route selection during route selection at the destination (but only secondary to stability), which means that shorter delays may be experienced in ABR during data transmission than in SSA. LPAR introduces a number of different route discovery strategies that use both the concept of route request localization and route stability to reduce the number of control packets during route discovery. Reduction in control overhead can be obtained by introducing a hierarchical structure to the network. CBRP[33] is a hierarchical on-demand routing protocol that attempts to minimize control overhead disseminated into the network by breaking the network into clusters. During the route discovery phase, cluster heads (rather than each intermediate node) exchange routing information. This significantly reduces the control overhead disseminated into the network when compared to flooding algorithms. In highly mobile networks, CBRP may introduce significant amounts of processing overhead during cluster formation and maintenance. This protocol suffers from temporary invalid routes as the destination nodes travel from one cluster to another. Therefore, this protocol is suitable for medium-sized networks with slow to moderate mobility. This protocol can also perform best in scenarios with group mobility where the nodes within a cluster are more likely to stay together.

17.2.3   Hybrid Routing Protocols

Hybrid routing protocols are a new generation of protocols that are both proactive and reactive in nature. These protocols are designed to increase scalability by allowing nodes in close proximity to work together to form some sort of backbone and to reduce the route discovery overhead. This is primarily achieved by proactively maintaining routes to nearby nodes and determining routes to far away nodes using a route discovery strategy. Most hybrid protocols proposed to date are zone based, which means that the network is partitioned or seen as a number of zones by each node. Others group nodes into trees or clusters. This section describes a number of different hybrid routing protocols proposed for MANETs. Furthermore, it provides a performance comparison between the described strategies.

17.2.3.1   Zone Routing Protocol (ZRP)

In ZRP,[25] the nodes have a routing zone that defines a range (in hops) that each node uses to maintain proactive network connectivity. Therefore, for nodes within the routing zone, routes are immediately available. For nodes that lie outside the routing zone, routes are determined on-demand (i.e., reactively), and can use any on-demand routing protocol to determine a route to the required destination. The advantage of this protocol is that it has significantly reduced the amount of communication overhead when compared to pure proactive protocols. It also has reduced the delays associated with pure reactive protocols, such as DSR, by allowing routes to be discovered faster. This is because, to determine a route to a node outside the routing zone, the routing only has to travel to a node that lies on the boundaries (edge of the routing zone) of the required destination — because the boundary node would proactively maintain routes to the destination (i.e., the boundary nodes can complete the route from the source to the destination by sending a reply back to the source with the required routing address). The disadvantage of ZRP is that if the zone radius of each routing zone is too large, then the protocol can behave like a pure proactive protocol; while if the value of the zone radius is set to be too small, then ZRP behaves like a reactive protocol. It may be important to optimize the value of the zone radius in ZRP in each specific network to best suit its characteristics (such as node density).

17.2.3.2 Zone-Based Hierarchical Link State (ZHLS)

Unlike ZRP, the ZHLS[34] routing protocol employs a hierarchical structure. In ZHLS, the network is divided into non-overlapping zones, and each node has a node ID and a zone ID, which are calculated using a GPS. The hierarchical topology is made up of two levels: (1) node level topology and (2) zone level topology, as described previously. In ZHLS, location management has been simplified. This is because no cluster head or location manager is used to coordinate data transmission. This means that there is no processing overhead associated with cluster head or location manager selection when compared to the HSR, MMWN, and CGSR protocols. This also means that a single point of failure and traffic bottlenecks can be avoided. Another advantage of ZHLS is that it has reduced the communication overhead when compared to pure reactive protocols such as DSR and AODV. In ZHLS, when a route to a remote destination is required (i.e., the destination is in another zone), the source node broadcasts a zone-level location request to all other zones. That way, it generates significantly lower overhead when compared to the flooding approach in reactive protocols. Moreover, in ZHLS, the routing path is adaptable to the changing topology because only the node ID and the zone ID of the destination are required for routing. This means that no further location search is required as long as the destination does not migrate to another zone. However, in reactive protocols, any intermediate link breakage would invalidate the route and might initiate another route discovery procedure. The disadvantage of ZHLS is that all nodes must have a preprogrammed static zone map in order to function. This may not be feasible in applications where the geographical boundary of the network is dynamic. Nevertheless, it is highly adaptable to dynamic topologies and it generates far less overhead than pure reactive protocols, which means that it may scale well to large networks.

17.2.3.3 Scalable Location Update Routing Protocol (SLURP)

Similar to ZHLS, in SLURP[66] the nodes are organized into a number of non-overlapping zones. However, SLURP further reduces the cost of maintaining routing information by eliminating a global route discovery. This is achieved by assigning a home region for each node in the network. The home region for each node is one specific zone (or region), which is determined using a static mapping function, $f(NodeID) \rightarrow regionID$, where f is a many-to-one function that is static and known to all nodes. An example of a function that can perform the static zone mapping is $f(NodeID) = g(NodeID) \bmod K$,[66] where $g(NodeID)$ is a random number generating function that uses the node ID as the seed and output a large number and k is the total number of home regions in the network. Now, because the node ID of each node is constant (i.e., a MAC address), the mapping function always calculates the same home region. Therefore, all nodes can determine the home region for each node using this function, provided they have their node IDs. Each node maintains it current location (current zone) with the home region by unicasting a location update message toward its home region. Once the location update packet reaches the home region, it is broadcast to all the nodes in the home region. Hence, to determine the current location of any node, each node can unicast a location discovery packet to the home region of the nodes (or the area surrounding the home region), in order to find its current location. Once the location is found, the source can start sending data toward the destination using the MFR (Most Forward with fixed Radius) geographical forwarding algorithm.When a data packet reaches the region in which the destination lies, then source routing* is used to get the data packet to the destination. The disadvantage of SLURP is that it also relies on a preprogrammed static zone map (as does ZHLS).

17.2.3.4 Distributed Spanning Trees Based Routing Protocol (DST)

In DST,[55] the nodes in the network are grouped into a number of trees. Each tree has two types of nodes: (1) route node and (2) internal node. The root controls the structure of the tree and whether the tree can merge with another tree, and the rest of the nodes within each tree are the regular nodes. Each node can be in one of three different states: (1) router, (2) merge, or (3) *configure*, depending on the type of task

---

*In this protocol, DSR is used as a source routing protocol.

that it tries to perform. To determine a route, DST proposes two different routing strategies: (1) hybrid tree-flooding (HTF) and (2) distributed spanning tree shuttling (DST). In HTF, control packets are sent to all the neighbors and adjoining bridges* in the spanning tree, where each packet is held for a period of time called the holding time. The idea behind the holding time is that as connectivity increases, and the network becomes more stable, it might be useful to buffer and route packets when the network connectivity is increased over time. In DST, the control packets are disseminated from the source and rebroadcast along the tree edges. When a control reaches down to a leaf node, it is sent up the tree until it reaches a certain height, referred to as the shuttling level. When the shuttling level is reached, the control packet can be sent down the tree or to the adjoining bridges. The main disadvantage of the DST algorithm is that it relies on a root node to *configure* the tree, which creates a single point of failure. Furthermore, the holding time used to buffer the packets can introduce extra delays into the network.

17.2.3.5   Distributed Dynamic Routing (DDR)

DDR[50] is also a tree-based routing protocol. However, unlike in DST, in DDR the trees do not require a root node. In this strategy, trees are constructed using periodic beaconing messages that are exchanged by neighboring nodes only. The trees in the network form a forest that is connected together via gateway nodes (i.e., nodes that are in transmission range but belong to different trees). Each tree in the forest forms a zone that is assigned a zone ID by running a zone naming algorithm. Furthermore, because each node can only belong to a single zone (or tree), the network can also be seen as a number of non-overlapping zones. The DDR algorithm consists of six phases:

1. Preferred neighbor selection
2. Forest construction
3. Intra-tree clustering
4. Inter-tree clustering
5. Zone naming
6. Zone partitioning

Each of these phases is executed based on information received in the beacon messages. During the initialization phase, each node starts in the preferred neighbor selection phase. The preferred neighbor of a node is a node that has the most number of neighbors. After this, a forest is constructed by connecting each node to its preferred neighbor. Next, the intra-tree clustering algorithm is initiated to determine the structure of the zone** (or the tree) and to build up the intra-zone routing table. This is then followed by the execution of the inter-tree algorithm to determine the connectivity with neighboring zones. Each zone is then assigned a name by running the zone naming algorithm, and the network is partitioned into a number of non-overlapping zones. Once the zones are created, a hybrid routing strategy, called the Hybrid Ad hoc Routing Protocol (HARP),[49] is used (which is built on top of DDR) to determine routes. HARP uses the intra-zone and inter-zone routing tables created by DDR to determine a stable path between the source and the destination. The advantage of DDR is that, unlike ZHLS, it does not rely on a static zone map to perform routing and it does not require a root node or a cluster head to coordinate data and control packet transmission between different nodes and zones. However, the nodes that have been selected as preferred neighbors can become performance bottlenecks. This is because they would transmit more routing and data packets than any other node. This means that these nodes would require more recharging because they will have less sleep time than other nodes. Furthermore, if a node is a preferred neighbor for many of its neighbors, many nodes may want to communicate with it. This means that channel contention would

---

* A bridge is formed when two nodes from different spanning trees are in radio range.

** The terms "tree" and "zone" are used interchangeably.

increase around the preferred neighbor, which would result in larger delays experienced by all neighboring nodes before they can reserve the medium. In networks with high traffic, this can also result in a significant reduction in throughput, due to packets being dropped when buffers become full.

17.2.3.6 Summary of Hybrid Routing

Hybrid routing protocols have the potential to provide higher scalability than pure reactive or proactive protocols. This is because they attempt to minimize the number of rebroadcasting nodes by defining a structure (or some sort of a backbone) that allows the nodes to work together to organize how to perform routing. By working together, the best or the most suitable nodes can be used to perform route discovery. For example, in ZHLS, only the nodes that lead to the gateway nodes will forward the interzone route discovery packets. Collaboration between nodes can also help in maintaining routing information much longer. For example, in SLURP, the nodes within each region (or zone) work together to maintain location information about the nodes that are assigned to that region (i.e., their home region). This can potentially eliminate the need for flooding because the nodes know exactly where to look for a destination every time. Another novelty of hybrid routing protocols is that they attempt to eliminate single points of failure and create bottleneck nodes in the network. This is achieved by allowing any number of nodes to perform routing or data forwarding if the preferred path becomes unavailable.

17.2.4 Summary of Routing Strategies

Thus far, this chapter has described three classes of routing protocols: (1) global/proactive, (2) on-demand/reactive, and (3) hybrid. The global routing protocols, which are derived mainly from the traditional link-state or distance-vector algorithms, maintain network connectivity proactively, and the on-demand routing protocols determine routes when they are needed. The hybrid routing protocols employ both reactive and proactive properties by maintaining intra-zone information proactively and inter-zone information reactively. By looking at the performance metrics and characteristics of all categories of routing protocols, a number of conclusions can be made for each category. In global routing, flat routing structure can be simple to implement, however it may not scale very well to large networks.[28] To make flat addressing more efficient, the number of routing overheads introduced into the networks must be reduced. One way to do this is to use a device such as the GPS. In DREAM, for example, nodes only exchange location information (coordinates) rather than complete link-state or distance-vector information. Another way to reduce routing overhead is by using conditional updates rather than periodic ones. For example, in STAR, updates occur based on three conditions (as described previously). The global routing schemes, which use hierarchical addressing, have reduced the routing overhead introduced into the networks by introducing a structure, which localizes the update message propagation. However, the current problem with these schemes is location management, which also introduces significant overhead into the network. In on-demand routing protocols, the flooding-based routing protocols such as DSR and AODV also have scalability problems. To increase scalability, route discovery and route maintenance must be controlled. This can be achieved by localizing the control message propagation to a defined region where the destination exists or where the link has been broken. For example, in LAR1, which also uses the GPS, the route request packets propagate in the request zone only; and in ABR, a localized broadcast query (LBQ) is initiated when a link goes down. Hybrid routing protocols such as ZHLS and SLURP can also perform well in large networks. The advantage of these protocols over other hierarchical routing protocols is that they have simplified location management by using the GPS instead of a cluster head to coordinate data transmission. As a result, a single point of failure and performance bottlenecks can be avoided. Another advantage of such protocols is that they are highly adaptable to changing topology because only the node ID and zone ID of the destination is required for routing to occur. ZRP is another routing protocol described in this section. The advantage of this protocol is that it maintains strong network connectivity (proactively) within the routing zones while determining remote route (outside the routing zone) more quickly than flooding. Another advantage of ZRP is that it can incorporate other protocols to improve its performance. For example, it can use LAR for inter-zone routing.

17.3   Future Challenges in Ad Hoc and Mobile Ad Hoc Networking

This section describes a number of other research areas in ad hoc and mobile ad hoc networking.

17.3.1   Implementation Study

Up until now, most of the research performed for ad hoc networks has been performed via simulations.[1,3,8,9,15,18,24,36,42,58,61] Recently, a few implementation solutions were undertaken to investigate the functionality of ad hoc networks.[27,56,62,65] Even with the many different routing protocols proposed for MANETs (as discussed), more work is required to investigate their behaviors in real-world scenarios. Such studies will prove very valuable to the MANET research community.

17.3.2   Scalable Multicasting

In MANETs, each user may want to communicate with a number of different users at the same time. In this case, a routing strategy is required to determine multiple routes. Routing protocols that perform this functionality are refered to as multicast routing protocols. Recently, a number of different multicasting protocols have been proposed for ad hoc and mobile ad hoc networks.[14,17,19,21,29,30–32,40,41,45,57,60,64,67,68] However, the issue of scalable multicasting still requires further research.

17.3.3   Quality of Service (QoS)

Providing QoS in a MANET environment is a challenging task, and many routing strategies proposed to date aim to provide best-effort delivery. This is because the dynamic nature of MANETs makes the available routing information less precise. Therefore, a distributed QoS routing strategy is required. Currently, a few different strategies have been proposed for MANETs to perform QoS routing.[11,13,26,43,44] However, QoS routing still remains a challenging research issue in MANETs.

17.3.4   Security

Security is another challenging research issue in mobile ad hoc networks. This is because wireless communication is susceptible to a range of attacks, such as passive eavesdropping, active interfering, denial-of-service, and break-ins.[38,39,46]

17.3.5   Sparse and Long-Range Ad Hoc Networking

With most of the attention being received by short to medium range (up to 1 km) communication in ad hoc networks, research is needed to design new strategies that can provide long-range communication in an ad hoc manner. Such networks will be useful for communities situated in very remote areas such as in the desert. Current work in this area includes efforts by Borg et al.,[7] Chambers,[10] and Nokia.[51]

References

1. M. Abolhasan, T. Wysocki, and E. Dutkiewicz. A Scalability Study of Mobile Ad Hoc Networks Routing Protocols. In *Proceedings of Sixth International DSPCS*, NSW, Australia, 2001.
2. M. Abolhasan and T. Wysocki. Displacement-Based Route Update Strategies for Proactive Routing Protocols in Mobile Ad Hoc Networks. In *The 2nd Workshop on the Internet, Telecommunications and Signal Processing (WITSP'2003)*, 2003.
3. M. Abolhasan, T. Wysocki, and E. Dutkiewicz. LPAR: An Adaptive Routing Strategy for MANETs. In *Journal of Telecommunication and Information Technology*, pages 28–37, 2/2003.

4. G. Aggelou and R. Tafazolli. RDMAR: A Bandwidth-Efficient Routing Protocol for Mobile Ad Hoc Networks. In *ACM International Workshop on Wireless Mobile Multimedia (WoWMoM)*, pages 26–33, 1999.

5. S. Basagni, I. Chlamtac, V.R. Syrotivk, and B.A. Woodward. A Distance Effect Algorithm for Mobility (DREAM). In *Proceedings of the Fourth Annual ACM/IEEE International Conference on Mobile Computing and Networking (Mobicom'98)*, Dallas, TX, 1998.

6. Bluetooth. Bluetooth wireless technology. In *http://www.bluetooth.com/*.

7. G. Borg, C.T. Chung, B. Heslop, D. Kurylowicz, R. Lam, G. Wade, J. Vedi, K. Modrak, M. Jolly, J. Harris, H. Harris, and J. McInerney. Bushlan: A VHF Wireless Local Area Network to Connect Regional Australia to the Internet. In *3rd Australian Communications Theory Workshop AusCTW*, Canberra, Australia, Feb. 4–5, 2002.

8. J. Broch, D.A. Maltz, D.B. Johnson, Y.-Ch. Hu, and J. Jetcheva. A Performance Comparison of Multi-Hop Wireless Ad Hoc Network Routing Protocols. In *Mobile Computing and Networking*, pages 85–97, 1998.

9. T. Camp, J. Boleng, B. Williams, L. Wilcox, and W. Navidi. Performance Cmparison of Two Location Based Routing Protocols for Ad Hoc Networks. In *Proceedings of IEEE INFOCOM*, pages 1678–1687, 2002.

10. B.A. Chambers. The Grid Roofnet: A Rooftop Ad Hoc Wireless Network. *Masters thesis. Massachusetts Institute of Technology*, 2002.

11. S. Chen and K. Nahrstedt. A Distributed Quality-of-Service Routing in Ad-Hoc Networks. *IEEE Journal on Selected Areas in Communications*, 17(8), August 1999.

12. T.-W. Chen and M. Gerla. Global State Routing: A New Routing Scheme for Ad-Hoc Wireless Networks. *Proc. IEEE ICC*, 1998.

13. Y.-S. Chen, T.-Y. Juang, and Y.-T. Yu. A Spiral-Multi-Path QoS Routing Protocol in a Wireless Mobile Ad-Hoc Network. In *Proceedings of IEEE ICOIN-16 2002: The 16th International Conference on Information Networking*, Cheju Island, Korea. Jan. 30–Feb. 1, 2002.

14. C.-C. Chiang, M. Gerla, and L. Zhang. Forwarding Group Multicast Protocol (fgmp) for Multihop, Mobile Wireless Networks. In *Baltzer Cluster Computing*, 1(2), 187–196, 1998.

15. S. Das, R. Castaneda, and J. Yan. Simulation Based Performance Evaluation of Mobile Ad Hoc Network Routing Protocols. In *ACM/Baltzer Mobile Networks and Applications (MONET) Journal*, pages 179–189, July 2000.

16. S. Das, C. Perkins, and E. Royer. Ad Hoc On Demand Distance Vector (AODV) Routing. In *Internet Draft, draft-ietf-manet-aodv-11.txt*, work in progress, 2002.

17. S.K. Das, B.S. Manoj, and C.S.R. Murthy. Dynamic Core Based Multicast Routing Protocol. In *In Proc. ACM Mobihoc 2002*, June 2002.

18. S.R. Das, C.E. Perkins, and E.E. Royer. Performance Comparison of Two On-Demand Routing Protocols for Ad Hoc Networks. In *INFOCOM (1)*, pages 3–12, 2000.

19. V. Devarapalli, A.A. Selcuk, and D. Sidhu. Multicast Zone Routing Protocol (MZR). In *Internet Draft*, draft-vijay-manet-mzr-01.txt, work in progress, June 2001.

20. R. Dube, C. Rais, K. Wang, and S. Tripathi. Signal Stability Based Adaptive Routing (SSA) for Ad Hoc Mobile Networks. In *IEEE Personal Communication, pp. 36–45 Feb.* 1997.

21. J.J. Garcia-Luna-Aceves and E.L. Madruga. The Coreassisted Mesh Protocol (CAMP). In *IEEE Journal on Selected Areas in Communications, Special Issue on Ad-Hoc Networks*, 17(8), 1380–1394, August 1999.

22. J.J. Garcia-Luna-Aceves and C.M. Spohn. Source-Tree Routing in Wireless Networks. In *Proceedings of the Seventh Annual International Conference on Network Protocols*, Toronto, Canada, page 273, Oct. 1999.

23. M. Gerla. Fisheye State Routing Protocol (FSR) for Ad Hoc Networks. In *Internet Draft*, draft-ietf-manet-aodv-03.txt, work in progress, 2002.

24. Z.J. Haas and M.R. Pearlman. The Performance of Query Control Schemes for the Zone Routing Protocol. In *SIGCOMM*, pages 167–177, 1998.

25. Z.J. Hass and R. Pearlman. Zone Routing Protocol for Ad-Hoc Networks. In *Internet Draft*, draft-ietf-manet-zrp-02.txt, work in progress, 1999.

26. Y.-K. Ho and R.-S. Liu. On-Demand QoS-Based Routing Protocol for Ad Hoc Mobile Wireless Networks. In *Proceedings of the Fifth IEEE Symposium on Computers and Communications*, 2000.

27. Y.-C. Hu and D.B. Johnson. Design and Demonstration of Live Audio and Video over Multihop Wireless Ad Hoc Networks. In *Proceedings of the MILCOM*, 2002.

28. A. Iwata, C. Chiang, G. Pei, M. Gerla, and T. Chen. Scalable Routing Strategies for Multi-Hop Ad Hoc Wireless Networks. In *IEEE Journal on Selected Areas in Communcations, Vol. 17, No. 8*, August 1999.

29. J.G. Jetcheva, Y.-C. Hu, D.A. Maltz, and D.B. Johnson. A Simple Protocol for Multicast and Broadcast in Mobile Ad Hoc Networks. In *Internet Draft*, draft-ietf-manet-simple-mbcast-00.txt, work in progress, June 2001.

30. J.G. Jetcheva and D.B. Johnson. Adaptive Demand-Driven Multicast Routing Protocol (ADMR). In *Internet Draft*, draft-jetcheva-manet-admr-00.txt, work in progress, June 2001, 2001.

31. L. Ji and M.S. Corson. Lightweight Adaptive Multicast Protocol (LAM). In *Proceedings of IEEE GLOBECOM'98*, Sydney, Australia, Nov. 1998.

32. L. Ji and M.S. Corson. Differential Destination Multicast (DDM) Specification. In *Internet Draft*, draft-ietf-manet-ddm-00.txt, 2000.

33. M. Jiang, J. Ji, and Y.C. Tay. Cluster Based Routing Protocol. In *Internet Draft*, draft-ietf-manet-cbrp-spec-01.txt, work in progress, 1999.

34. M. Joa-Ng and I.-T. Lu. A Peer-to-Peer Zone-Based Two-Level Link State Routing for Mobile Ad Hoc Networks. *IEEE Journal on Selected Areas in Communications*, 17(8), 1999.

35. D. Johnson, D. Maltz, and J. Jetcheva. The Dynamic Source Routing Protocol for Mobile Ad Hoc Networks. In *Internet Draft*, draft-ietf-manet-dsr-07.txt, work in progress, 2002.

36. B. Karp and H.T. Kung. GPSR: Greedy Perimeter Stateless Routing for Wireless Networks. In *Mobile Computing and Networking*, pages 243–254, 2000.

37. Y.-B. Ko and N.H. Vaidya. Location-Aided Routing (LAR) in Mobile Ad Hoc Networks. In *Proceedings of the Fourth Annual ACM/IEEE International Conference on Mobile Computing and Networking (Mobicom'98)*, Dallas, TX, 1998.

38. J. Kong, H. Luo, K. Xu, D.L. Gu, M. Gerla, and S. Lu. Adaptive Security for Multi-Layer Ad-Hoc Networks. *Special Issue of Wireless Communications and Mobile Computing*, Wiley Interscience Press, 2002.

39. J. Kong, P. Zerfos, H. Luo, S. Lu, and L. Zhang. Providing Robust and Ubiquitous Security Support for Mobile Ad Hoc Networks. In *IEEE ICNP01*, 2001.

40. H. Laboid and H. Moustafa. Source Routing-Based Multicast Protocol (SRMP). In *Internet Draft*, draft-labiod-manet-srmp-00.txt, work in progress, June 2001.

41. S. Lee, W. Su, and M. Gerla. Demand Multicast Routing Protocol in Multihop Wireless Mobile Networks. In S.-J. Lee, W. Su, and M. Gerla, *On-Demand Multicast Routing Protocol in Multihop Wireless Mobile Networks, ACM/Baltzer Mobile Networks and Applications, Special Issue on Multipoint Communication in Wireless Mobile Networks*, 2000.

42. J. Li, J. Jannotti, D. De Couto, D. Karger, and R. Morris. A Scalable Location Service for Geographic Ad-Hoc Routing. In *Proceedings of the 6th ACM International Conference on Mobile Computing and Networking (MobiCom '00)*, pages 120–130, August 2000.

43. C.R. Lin and J.-S. Liu. QoS Routing in Ad Hoc Wireless Networks. *IEEE Journal on Selected Areas in Communications*, 17(8), August 1999.

44. C.R. Lin. An On-Demand QoS Routing Protocol for Mobile Ad Hoc Networks. In *INFOCOM*, pages 1735–1744, 2001.

45. M. Liu, R.R. Talpade, A. Mcauley, and E. Bommaiah. Ad Hoc Multicast Routing Protocol (AMROUTE). In *UMD TechReport 99–8*, 1999.

46. H. Luo, P. Zefros, J. Kong, S. Lu, and L. Zhang. Self-Securing Ad Hoc Wireless Networks.

47. S. Murthy and J.J. Garcia-Luna-Aceves. A Routing Protocol for Packet Radio Networks. In *Mobile Computing and Networking*, pages 86–95, 1995.

48. R.K. Nichols and P.C. Lekkas. *Wireless Security: Models, Threats, and Solutions.* New York: London: McGraw-Hill, 2002.

49. N. Nikaein, C. Bonnet, and N. Nikaein. HARP — Hybrid Ad Hoc Routing Protocol. In *Proceedings of IST: International Symposium on Telecommunications,* Sept. 1–3, Tehran, Iran, 2001.

50. N. Nikaein, H. Laboid, and C. Bonnet. Distributed Dynamic Routing Algorithm (DDR) for Mobile Ad Hoc Networks. In *Proceedings of the MobiHOC 2000: First Annual Workshop on Mobile Ad Hoc Networking and Computing,* 2000.

51. Nokia. Nokia rooftop wireless router. In http://www.nwr.nokia.com/.

52. D. Senie and P. Ferguson. RFC2267 Network Ingress Filtering: Defeating Denial of Service Attacks Which Employ IP Source Address Spoofing. In *Internet Engineering Task Force, 1998.* Online: http://www.ietf.org/rfc/rfc2267.txt., 1998.

53. C.E. Perkins and T.J. Watson. Highly Dynamic Destination Sequenced Distance Vector Routing (DSDV) for Mobile Computers. In *ACM SIGCOMM'94 Conference on Communications Architectures,* London, U.K., 1994.

54. C.E. Perkins. *Ad Hoc Networking.* Addison-Wesley, 2000.

55. S. Radhakrishnan, N.S.V Rao, G. Racherla, C.N. Sekharan, and S.G. Batsell. DST — A Routing Protocol for Ad Hoc Networks Using Distributed Spanning Trees. In *IEEE Wireless Communications and Networking Conference,* New Orleans, 1999.

56. E.M. Royer and C.E. Perkins. An Implementation Study of the AODV Routing Protocol. In *Proc. of the IEEE Wireless Communications and Networking Conference,* Chicago, IL, September 2000.

57. E.M. Royer and C.E. Perkins. Multicast Operation of the Ad-Hoc On-Demand Distance Vector Routing Protocol. In *Proceedings of ACM/IEEE MOBICOM 99,* Seattle, WA, Aug. 1999, pages 207–218, 1999.

58. C. Santivanez, B. McDonald, I. Stavrakakis, and R. Ramanathan. The Scalability of Ad Hoc Routing Protocols. In C.A. Santivanez, B. McDonald, I. Stavrakakis, and R. Ramanathan, *On the Scalability of Ad Hoc Routing Protocols, Proc. IEEE INFOCOM 2002,* 2002.

59. A. Udaya Shankar, C. Alaettinoglu, I. Matta, and K. Dussa-Zieger. Performance Comparison of Routing Protocols using MARS: Distance-Vector versus Link-State. In *Proc. 1992 ACM SIGMETRICS and PERFORMANCE '92 Int'l. Conf. on Measurement and Modeling of Computer Systems,* page 181, Newport, RI, 1–5 1992.

60. P. Sinha, R. Sivakumar, and V. Bharghavan. MCEDAR: Multicast Core-Extraction Distributed Ad Hoc Routing. In *Proceedings of IEEE WCNC,* New Orleans, LA, pages 1313–1317, 1999.

61. Mario Gerla, Sung-Ju Lee, and Chai-Keong Toh. A Simulation Study of Table-Driven and On-Demand Routing Protocols for Ad Hoc Networks. *IEEE Network, 13(4), July/August 1999, pp. 48–54.*

62. L. Christensen, T.H. Clausen, G. Hansen, and G. Behrmann. The Optimized Link State Routing Protocol, Evaluation through Experiments and Simulation. In *Proc. of IEEE Symposium on Wireless Personal Mobile Communication,* September 2001.

63. C. Toh. A Novel Distributed Routing Protocol to Support Ad-Hoc Mobile Computing. In *IEEE 15th Annual International Phoenix Conf.,* pp. 480–86, 1996.

64. C.-K. Toh, G. Guichal, and S. Bunchua. On-Demand Associativity-Based Multicast Routing for Ad Hoc Mobile Networks (ABAM). In *Vehicular Technology Conference, 2000. IEEE VTS Fall VTC 2000. Volume 3, Page(s) 987–993,* 2000.

65. C. Tschudin and R. Gold. Simple Ad Hoc Routing with LUNAR. In *2nd Swedish Workshop on Wireless Ad Hoc Networks,* March 2002.

66. S.-C. Woo and S. Singh. Scalable Routing Protocol for Ad Hoc Networks. Accepted for Publication in Journal of Wireless Networks (WINET), 2001.

67. C.W. Wu and Y.C. Tay. Ad Hoc Multicast Routing Protocol Utilizing Increasing ID-Numbers (AMRIS). In *Proceedings of IEEE MILCOM'99,* Atlantic City, NJ, Nov. 1999.

68. H. Zhou and S. Singh. Content Based Multicast (CBM) in Ad Hoc Networks. In *Proceedings of the ACM/IEEE Workshop on Mobile Ad Hoc Networking and Computing (MOBIHOC),* Boston, MA, August 2000.

18

Adaptive Cycle-Controlled E-Limited Polling in Bluetooth Piconets

Jelena Mišić

Vojislav B. Mišić

18.1   Introduction

Bluetooth is an emerging standard for wireless personal area networks (WPANs).[17] Bluetooth devices are organized in piconets: a small, centralized network with $\mu \leq 8$ active nodes or devices.[2] One of the nodes acts as the master while the others are slaves, and all communications in the piconet take place under the master's control. All slaves listen to downlink transmissions from the master. The slave can reply with an uplink transmission if and only if addressed explicitly by the master, and only immediately after being addressed by the master.

Thus, the performance of data traffic in the Bluetooth piconet is critically dependent on the manner in which the master visits or polls its slaves — the polling or scheduling scheme. The current Bluetooth specification does not require or prescribe any specific polling scheme,[2] and a number of such schemes have been proposed.[5] However, most of these do not provide any support regarding any predefined delay bounds; only recently was a scheme proposed that can support such bounds, albeit at the expense of efficiency and/or fairness.[6] This observation has motivated us to consider the lightweight scheduling scheme described here.

In the new scheme, each slave is allocated a number of consecutive time slots, depending on its data traffic. This number is dynamically determined in each piconet cycle. Slaves that do not use all the allocated slots in one cycle are given less time in the next cycle, while slaves that have not exchanged all packets during the allocated time are given more time in the next cycle. In this manner, the number of wasted (POLL and NULL) packets is minimized and the scheme is thus able to optimize the efficiency of data transfers, especially in piconets with highly asymmetric traffic. The allocation of consecutive time slots tends to preserve the integrity of baseband packet bursts and thus leads to reduction in delays for IP packets in the piconet. Furthermore, each slave will be visited at least once during the cycle, which guarantees fairness by making sure that no slave(s) can monopolize the piconet at the expense of others. Finally, the actual algorithm is computationally simple, and the bulk of the calculations are performed only once per piconet cycle.

In cases where the piconet contains slaves with synchronous traffic, the scheme can simply be modified so as to control the duration of the piconet cycle. In this manner, the scheme is able to support piconet with both synchronous and asynchronous traffic. Owing to this modification, the scheme will be referred to as the Adaptive Cycle-controlled E-limited polling (ACE). This scheme is based on the ACLS polling scheme described by Mišić et al.[12]

We start this discussion with a detailed description of the ACE scheme and related algorithms. Section 18.3 outlines the queueing model for the master–slave pair of queues under constant maximum bandwidth allocation within the piconet cycle. Due to space limitations, the complete probability distributions for the access and end-to-end packet delays are not derived; they can be found in Mišić et al.[11] We compare the performance of the new scheme to that of other polling schemes and discuss possible modifications to improve its performance. Finally, we summarize our findings and provide some concluding remarks.

18.2   An Overview of the ACE Scheme

The ACE scheme is based on the well-known E-limited scheme.[16] In this scheme, the master stays with a slave for a fixed number M of frames ($M > 1$) or until there are no more packets to exchange, whichever comes first. Packets that arrive during the visit are allowed to enter the uplink queue at the slave and may be serviced — provided the limit of M frames is not exceeded.[16] Boundary values of $M = 1$ and ∞ correspond to the well-known 1-limited scheme, also known as (Pure) Round Robin and exhaustive service.[5,16]

We note that the Bluetooth piconet does indeed resemble a simple polling system in which multiple input queues are serviced by a single server — a well-known problem in queueing theory.[15] In such cases, exhaustive service has been shown to perform better than either 1-limited or E-limited service.[7]

However, communications in Bluetooth are bidirectional by default, the master polls the slaves using regular packets (possibly empty), and all slave–slave communications must be routed through the master. As a consequence, existing analytical results on polling schemes do not hold in Bluetooth piconets. Indeed, previous research has shown that the 1-limited scheme achieves good results at high traffic loads but tends to waste a lot of bandwidth under low and medium loads; the exhaustive service scheme achieves good results at low to medium loads but cannot guarantee fairness at high loads, especially if the traffic loads of individual slaves are not uniform.[5] In such cases, the E-limited scheme is able to achieve good results, which has been confirmed through both queueing theoretic analysis and discrete event simulations.[9]

Now the basic concept of E-limited polling can be extended to provide dynamic bandwidth allocation according to the behavior in the previous piconet cycle. To that effect, the ACE scheme (in its simpler form) allocates one of two possible values of M to each slave. Slaves that have finished their packet exchange, which is detected through the presence of a POLL-NULL sequence as is customary in Bluetooth,[4] will get less bandwidth (M_L) in the next cycle. Slaves that have undelivered packets for exchange will get more bandwidth (M_H). This scheme is described with the following pseudocode.

```
procedure ACE polling
do forever
    for all slaves
        poll slave i
        if both uplink and downlink queues are empty
```

> then set $M(i) = M_L$
>> break (i.e., move on to next slave)
> else if M packets are exchanged
> then set $M(i) = M_H$
>> break (i.e., move on to next slave)
> end if
> loop (i.e., poll the same slave again)
> end for
end do

18.2.1 Piconet Cycle Control

When the piconet contains both synchronous and asynchronous slaves, the master can modify its bandwidth allocation algorithm so as to control the duration of the piconet cycle. In this case, the available piconet cycle is divided into two parts. The first part is allocated directly to the slaves: those that did not use all the allocated time in the previous cycle will get a smaller part (analogous to M_L in the previous case), while those that did use all the allocated time and still have packets to exchange will get a larger part (analogous to the M_H in the previous case). The other part is known as the free `pool`, and slaves can use it to accommodate extra traffic they might have in the current cycle. (The percentage of total cycle duration set aside for the `pool` is an adjustable parameter.)

```
procedure ACE polling with cycle limit
do forever
    for all slaves
        poll slave i
        if both uplink and downlink queues are empty
           or slave has used allocation plus current pool
        then record used time t(i)
                return unused time (if any) to the pool
                break (i.e., move on to next slave)
        else loop (i.e., poll the same slave again)
        end if
    end for (i.e., move on to next slave)
    calculate pool, basic per-slave allocation
    for all slaves
        if the slave used more time than allocated
        then allocate twice basic per-slave time
        else allocate basic per-slave time
    end for
    restart the polling cycle
end do
```

In case some slaves have a predefined bandwidth requirement (e.g., on account of synchronous traffic or negotiated QoS), they can be excluded from both dynamic bandwidth allocation and reordering, and they can be assigned a fixed position within the piconet cycle — say, at the beginning of each cycle.

18.2.2 The Role of the Reference Slave and Slave Reordering

The fairness of ACE polling depends on the choice of the reference slave — the slave from which the piconet polling cycle starts. Any of the slaves can be chosen as the reference slave, and the master can simply poll the slaves in fixed order, as shown in Figure 18.1(a). However, this would destroy the fairness of bandwidth

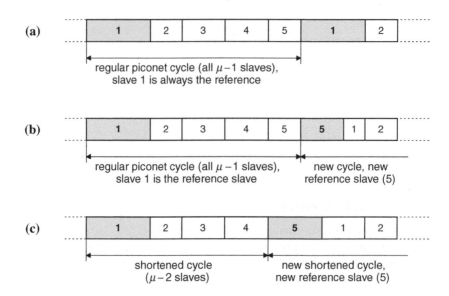

FIGURE 18.1   Pertaining to the choice of reference slave.

allocation to individual slaves. Namely, the first slave in a given cycle can use its direct allocation plus the entire `pool`. After the first slave finishes its transfer, the remaining unused time (if any) is returned to the `pool`. Each subsequent slave can use its direct allocation, plus the current value of the `pool`. In this manner, although the dynamic allocation is made on the basis of the traffic in the *previous* piconet cycle, slaves with higher values of traffic in the *current* cycle can still get extra time.

However, if the slaves are polled in fixed order, the first slave will always have the entire `pool` at its disposal, whereas the last one will seldom have more than the time obtained through direct allocation. Therefore, the scheme in this form is inherently unfair. Moreover, the worst-case cycle time and, consequently, the maximum polling interval will suffer as well.

Several possibilities exist to restore the fairness. In the deterministic approach, the ACE scheme can be modified by rotating the role of the reference slave among all the slaves. For example, the role of the reference slave could be assigned to the last slave from the previous cycle, as shown in Figure 18.1(b). In this case, the last slave will get *two* chances to exchange packets with the master, but this role is taken by each slave in turn. The last slave in one cycle will become the reference slave in the next one, so it will get slightly more time to exchange packets with the master. This slave will probably get only a short time allocation at first, and much more time immediately after that, so that any leftover traffic will be taken care of immediately. In the worst case, when there is no data to or from this slave, two frames will be wasted instead of one.

Note that with the deterministic approach, the slave that is polled last in the current cycle becomes the slave that is polled first in the coming cycle. Therefore, the polling cycle may be shortened to $\mu - 2$ slaves, as shown in Figure 18.1(c). In this case, the **newcycle()** procedure allocates bandwidth to the current set of $\mu - 2$ slaves, with the C parameter suitably modified; the queues value for the missing slave could be taken from the cycle before the last. Each slave will thus get equal attention in a super-cycle of $\mu - 1$ shorter cycles, or $\mu - 2$ normal piconet cycles. Therefore, under equally loaded slaves, the average number of slots per cycle devoted for polling one slave is equal among the slaves.

In the probabilistic approach, the fairness is achieved by choosing the next slave in the cycle with equal probability. For example, the first slave in the cycle will be chosen with probability $1/(\mu - 2)$, the second slave with probability $1/(\mu - 2)$, and so on. However, this approach is computationally more complex than the deterministic one.

The third solution — which we have chosen to implement — is a variation of a well-known concept from queueing theory. Namely, it has been shown that single-server, multiple-queue polling systems achieve best performance if the server always services the queue with the highest number of packets; this policy is known as Stochastically-Largest Queue (SLQ).[8] Again, this policy cannot be followed in Bluetooth because the master cannot know the status of *all* the slaves' queues at any given time. The best that can be done is to rearrange the polling order of the slaves in each cycle according to the length of the corresponding downlink queue; these queues are maintained by the piconet master and thus their lengths are known to the master at any given time. We refer to this policy as Longest Downlink Queue First (LDQF); as will be seen, it consistently provides a small but noticeable improvement in piconet performance. We note that this modification is particularly well suited to the situation where the piconet master acts as the access point to another network (e.g., Ethernet, as per BNEP profile[1]). In such cases, the downlink traffic can be expected to exceed the uplink one, possibly by an order of magnitude or more; and the overall performance will be primarily determined by the performance of downlink traffic.

With this modification, the ACE algorithm can be described by the following pseudocode.

```
procedure ACE polling with cycle limit and LDQF
do forever
    for all slaves according to the current sequence
        poll next slave in the current sequence
        if both uplink and downlink queues are empty
            or slave has used allocation plus current pool
        then record used time t(i)
                return unused time (if any) to the pool
                break (i.e., move on to next slave)
        else loop (i.e., poll the same slave again)
        end if
    end for
    calculate pool, basic per-slave allocation
    for all slaves
        if the slave used more time than allocated
        then allocate twice basic per-slave time
        else allocate basic per-slave time
    end for
    reorder slaves per LDQF policy
end do
```

18.3   The Queueing Model of the ACE Scheme

The performance of the ACE scheme can be assessed using a queueing theoretic approach. Consider an isolated piconet with the master and $\mu - 1$ active ACL slaves, as shown in Figure 18.2. The operation of the piconet can be described by a queueing model in which each slave maintains an uplink queue, while the master maintains a number of downlink queues, one per each active slave. All queues are assumed to

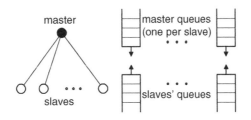

FIGURE 18.2 A single piconet and its queueing model.

have buffers of infinite size; as data traffic will be generated by the applications running on the devices themselves, packet arrival rates will be low and buffer overflows will be sufficiently rare.

Each slave generates packets in bursts that correspond to the application packets.[1] Application packet arrivals to the uplink queue of slave i follow a Poisson distribution, which has been shown to be a satisfactory approximation for the traffic of many Internet applications.[13] If the application packets themselves arrive in bursts, the probability-generating function (PGF) of the burst size of application packets should be integrated into the corresponding PGF for the baseband burst size.

The length of the burst of baseband packets is geometrically distributed, but other distributions can be easily accommodated in our model, provided the corresponding probability distribution (i.e., the PGF) is known. Assuming that all slaves use the same segmentation/reassembly mechanism (which is reasonable because this mechanism is commonly implemented in firmware), the burst length distribution will be the same for all slaves.

We assume that the traffic goes from slaves to other slaves only, which simplifies the calculations without undue loss of generality. (Any traffic generated or received by the master can be easily modeled by increasing the packet arrival rates in the downlink queues without any other changes to the analytical model.) All packets within the given burst will have the same destination node, and the distribution of destinations is assumed to be uniform. Of course, the case where the destination probabilities for individual slaves differ can be dealt with by simply recalculating the resulting downlink arrival rates may be calculated from the matrix of probabilities for slave-to-slave communication, without changing the model itself.

Data packets can last for one, three, or five slots of the Bluetooth clock (with length $T = 625 \, \mu s$),[2] with probabilities of p_1, p_3 and $p_5 = 1 - p_1 - p_3$, respectively. For simplicity, we assume that $p_1 = p_2 = p_3$ and, consequently, $\overline{L} = G'_p(1) = 3$, although other distributions can be accommodated without difficulty. The sequence of the one downlink and following uplink packet transmission will be denoted as a frame.

Using the theory of queues with vacations, we are able to derive the probability distributions for the following variables:

- *Piconet cycle time*, which is the time for the master to visit all the slaves in the piconet exactly once (note that the visit may last for several frames).

- *Slave channel service time*, which is the time the time from the moment when master polls the slave for the first time, until either an empty frame has been encountered or the maximum number of frames have been exchanged. (Note that this maximum number of frames is variable.)

- *Vacation time from the standpoint of one slave*. When the concept of vacation[16] is applied to Bluetooth networks, the piconet master acts as the server, while the client corresponds to a slave or, rather, the pair of uplink queues at the slave and the corresponding downlink queue at the master. From the viewpoint of a given client, the vacation time is the time during which the server is busy servicing other client queues and thus unavailable to service that particular client.

- *Frame length time*, which is not equal to the packet length because of the presence of empty (POLL and NULL) packets.

Using these variables, we can derive the probability distributions for the packet access delay — the time the data packet has to wait in the uplink queue of a slave before being serviced by the master — and the packet end-to-end delay — the time that elapses between the data packet enters the uplink queue at the source slave and the time it is received by the destination slave. The derivations are rather involved and thus are omitted here; the interested reader can consult a previous article[10] for details.

A similar approach is based on the so-called imbedded Markov points that correspond to the end of individual piconet cycles. Any given slave i can be in the state of low or high bandwidth allocation during any given piconet cycle, depending on their usage of allocated bandwidth during the previous piconet cycle:

- The slaves that did not use all of the allocated slots in the previous cycle will be given low bandwidth of $M_{L,i}$ frames.

- The slaves that have used all of the allocated slots will be given high bandwidth of $M_{H,i}$ frames.

For both states, we can determine the joint queue length distributions for the uplink and downlink queues corresponding to the slave i, $Q_i(z, w)$ and $\Pi_{i,k}(z, w)$, $k = 1 .. M_{L,i}$ for the low state and $k = 1 .. M_{H,i}$ for the high state. (The index i has been previously omitted for simplicity.) The transition probabilities between the states are determined according to the corresponding value of queues[i]:

$$T_{L,L,i} = \text{Prob}(\text{queues[i]} = 0 \text{ or } 1)|\text{low state}$$
$$= Q_i(1,0) + Q_i(0,1)$$
$$+ \sum_{m=1}^{M_{L,i}} (\Pi_{i,m}(1,0) + \Pi_{i,m}(0,1)) \tag{18.1a}$$

$$T_{L,H,i} = \text{Prob}(\text{queues[i]} = 2)|\text{low state}$$
$$= 1 - T_{L,L,i} \tag{18.1b}$$

$$T_{H,L,i} = \text{Prob}(\text{queues[i]} = 0 \text{ or } 1)|\text{high state}$$
$$= Q_i(1,0) + Q_i(0,1)$$
$$+ \sum_{m=1}^{M_{H,i}} (\Pi_{i,m}(1,0) + \Pi_{i,m}(0,1)) \tag{18.1c}$$

$$T_{H,H,i} = \text{Prob}(\text{queues[i]} = 2)|\text{high state}$$
$$= 1 - T_{H,L,i} \tag{18.1d}$$

In the presence of k slaves in the piconet, the number of states of the Markov chain is 2^k. One state of the chain is described by the k-tuple of states corresponding to each slave. Components of the transition probabilities between the states are given by Equation 18.1. For every state of the chain, the slot/frame allocations must be determined. The Markov chains for the piconets with one and two slaves are shown in Figures 18.3(a) and (b), respectively.

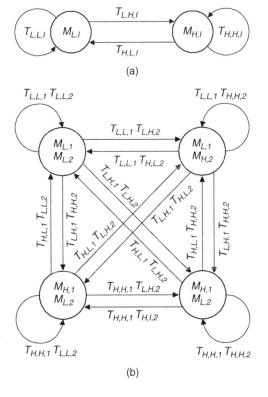

FIGURE 18.3 Markov chains that describe adaptive bandwidth allocation: (a) for one slave and (b) for two slaves.

We note that actual bandwidth allocations for low- and high-bandwidth states for one slave are not equal across the states of the Markov chain. For example, let us assume that we have two slaves, and the cycle is defined as 36 slots (which corresponds to 6 frames, or 12 packets with the mean length $\overline{L} = 3$). Also note that the reordering equally distributes the remaining bandwidth (i.e., `pool`) among the slaves. Now, $M_{L,2}$ is 12 slots (for both `queues[2]` $= 0$ or 1), and $M_{H,1} = 24$. However, for the state $M_{H,1}$, $M_{H,2}$, both slaves will get 18 slots.

When the actual bandwidth allocations are found for each slave in each state of the Markov chain, the probability generating functions for each slave, in each state, must be found; and the transition probabilities for the Markov chain must be calculated. From these, we can calculate the queue length probability distribution, as well as the resulting distributions for the access and end-to-end packet delays.

Again, the calculations are rather involved and are thus omitted; full derivations can be found in Mišić et al.[11]

18.4   Performance of the New Scheme

To evaluate the performance of the ACE scheme, we have built a simulator of the Bluetooth piconet, using the Artifex[14] object-oriented Petri Net simulation engine. The simulator operates at the MAC level, consisting of the master and up to seven slaves. The simulator operates under the assumptions listed in the beginning of Section 18.3; to avoid transition effects, the simulation results were measured after an initial warm-up period. All delays are expressed in time slots of the Bluetooth device clock, $T = 625\,\mu s$.

The performance of traditional E-limited polling, which is used as the reference against which the improvements offered by the ACE scheme can be measured, is shown in Figure 18.4. As can be seen, end-to-end packet delays generally increase with both offered load and traffic burstiness (expressed through mean burst size $\overline{B}$). At high volumes of very bursty traffic, the delays tend to increase to values well over $400T$ (which corresponds to about 0.25 s), which signals that the uplink and downlink queues become longer and longer. As Bluetooth devices are mobile, and thus are expected to run for a long time on battery power, the size of device queues will tend to be small, and buffer overflows may occur at such ranges of load and burstiness. Therefore, Bluetooth devices should probably not be operated under such high loads.

Another observation is that the cycle time is rather independent of mean burst size — it mainly depends on the offered load. This is not altogether unexpected, because longer bursts at constant load mean that the bursts will come in longer intervals. (A similar conclusion, although obtained in a slightly different manner, has been reported.[9]) However, the variance of piconet cycle time will increase because the bursts will be handled in fewer piconet cycles.

The improvements obtained through the use of adaptive E-limited polling can be seen in Figure 18.5. The upper diagram (Figure 18.5(a)) shows the ratio of end-to-end packet delay obtained under adaptive E-limited polling (with lower limit $M_L = 3$) to the similar delay obtained under the original E-limited polling with $M = 3$ (this essentially corresponds to the ACE scheme in which $M_L = M_H = 3$). As the upper limit increases, the delays decrease. This decrease is more pronounced for higher offered loads, where the adaptivity of the ACE scheme improves performance significantly. (Of course, the absolute values of end-to-end packet delay are still high.) The lower diagram (Figure 18.5(b)) shows the delay ratio under asymmetric loads, where $M_L = 3$ and $M_H = 6$. Again, the ACE scheme performs better at higher loads, and it performs better as the asymmetry between the traffic of individual slaves becomes more pronounced.

The improvement in end-to-end packet delay due to slave reordering in each cycle using the LDQF policy is shown in Figure 18.6. Figure 18.6(a) shows the delay as a function of the maximum number of packets transferred in each visit M. (Note that the value of $M = 1$ corresponds to 1-limited polling.) As can be seen, LDQF reordering reduces the delay, but also makes the minimum of the curve (which, in the case of fixed slave order, corresponds roughly to the value of $M = \overline{B}$) slightly wider. Therefore, the use of LDQF reordering not only improves the delay, but also makes the minimum delay less dependent on the mean burst size.

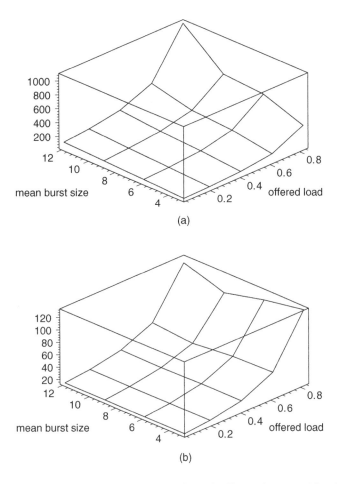

(a)

(b)

FIGURE 18.4 Performance of traditional (non-adaptive) E-limited polling with $M = 6$: (a) end-to-end packet delay as a function of offered load ρ and mean burst size $\overline{B}$; and (b) mean cycle time as a function of offered load ρ and mean burst size $\overline{B}$.

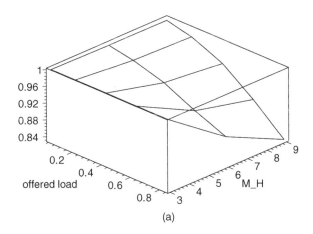

(a)

FIGURE 18.5 Delay improvement due to adaptability, as compared to the simple E-limited polling with $M = 3$: (a) performance under varying upper limt M_H; and (b) performance under asymmatric traffic.

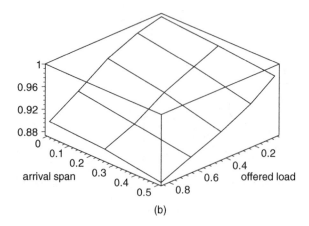

(b)

FIGURE 18.5   (*Continued*)

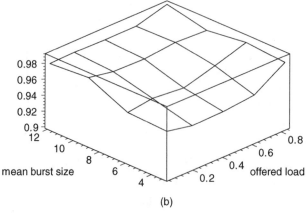

(a)

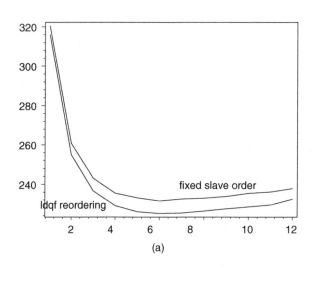

(b)

FIGURE 18.6   Pertaining to LDQF slave reordering: (a) End-to-end packet delays with and without LDQF reordering as functions of *M*, for fixed mean burst size $\overline{B} = 6$; and (b) ration of end-to-end packet delays with and without LDQF slave reordering.

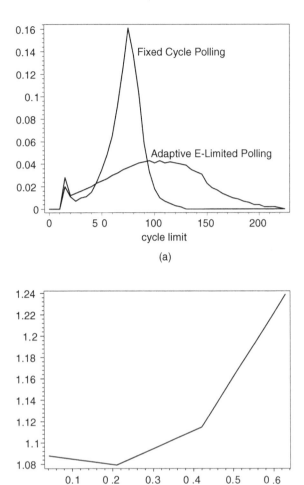

FIGURE 18.7 Performance of fixed cycle variant compared to adaptive E-limited polling: (a) cycle time distribution; and (b) Ratio of end-to-end packet delay: ACE with soft cycle control vs. plain ACE.

Finally, Figure 18.7 shows the effects of introducing the fixed cycle control. For simplicity, we have plotted the distribution of cycle durations in a sample run for both the original ACE scheme and its fixed cycle variant. The piconet has seven active ACL slaves, and the cycle time limit for the fixed cycle ACE scheme has been set to $C = 80T$. As can be seen, the fixed cycle control is much more effective in controlling the piconet cycle: the cycle time exhibits a noticeable peak close to the preset cycle limit, and the maximum cycle value (which may exceed the limit due to the presence of the free pool) is about $130T$. On the contrary, under the original adaptive E-limited scheme, the cycle times are spread over a much wider range (as there may be cycles in which most slaves get maximum bandwidth allocation), and there is no discernible peak. Note that the smaller peak at about $14T$ corresponds to the cycles in which no slave has any traffic, hence, each visit of the master takes only two time slots T.

At the same time, fixed cycle control does lead to some deterioration of delays, especially at higher loads. One should bear in mind, however, that the main goal of this scheme is to provide guaranteed polling interval for slaves with synchronous traffic, rather than optimum performance under high volumes of asynchronous traffic. For values of offered load that do not exceed $\rho = 0.4$, the deterioration of delay is below 12 percent, which appears to be an acceptable price to pay for the improvement in cycle control.

18.5   Summary and Possible Enhancements

This chapter has described a lightweight adaptive polling scheme with soft cycle control that is suitable for piconets with asymmetric traffic. We have also developed a complete queueing theoretic model of the piconet operation under dynamic bandwidth allocation within a limited piconet cycle. The new scheme is shown to perform well under a wide range of traffic loads in the piconet, while being much simpler than other comparable schemes.

The ACE scheme can be extended to support slaves with predefined QoS parameters such as bandwidth and maximum polling interval. Such requirements may stem from the presence of synchronous, CBR traffic, but also from the more common asynchronous traffic. (It should be noted that the negotiated maximum polling interval is the only mechanism to support QoS which is provided by the current Bluetooth specification[3].) Slaves with QoS requirements can be excluded from the ACE adaptive bandwidth allocation mechanism in two ways: (1) they can be polled using a fixed number of frames, and (2) they can be excluded from slave reordering in each cycle. As the piconet cycle is not fixed, slaves with CBR traffic may have less traffic in some cycles, which can easily be dealt with by ending the exchange when a POLL/NULL frame is encountered, as is common in Bluetooth.

Further extension of the ACE scheme could introduce admission control, in which slaves can be admitted in the piconet, or rejected, depending on the ability of the piconet to support the QoS requirements. These requirements can be imposed by the slaves requesting admission, but also by the slaves that are already admitted in the piconet. Some initial results in that direction, but using plain E-limited polling only, have been reported.[10]

References

 1. Bluetooth SIG. Bluetooth Network Encapsulation Protocol (BNEP) Specification. Technical Report, Revision 0.95a, June 2001.
 2. Bluetooth SIG. *Specification of the Bluetooth System*. Version 1.1, February 2001.
 3. Bluetooth SIG. *Specification of the Bluetooth System — Architecture & Terminology Overview*, Volume 1. Version 1.2, November 2003.
 4. Bluetooth SIG. *Specification of the Bluetooth System — Core System Package [Controller volume]*, Volume 2. Version 1.2, November 2003.
 5. Antonio Capone, Rohit Kapoor, and Mario Gerla. Efficient polling schemes for Bluetooth pico-cells. In *Proceedings of IEEE International Conference on Communications ICC 2001*, Volume 7, pages 1990–1994, Helsinki, Finland, June 2001.
 6. Jean-Baptiste Lapeyrie and Thierry Turletti. FPQ: a fair and efficient polling algorithm with QoS support for Bluetooth piconet. In *Proceedings Twenty-Second Annual Joint Conference of the IEEE Computer and Communications Societies IEEE INFOCOM 2003*, Volume 2, pages 1322–1332, New York, April 2003.
 7. Hanoch Levy, Moshe Sidi, and Onno J. Boxma. Dominance relations in polling systems. *Queueing Systems Theory and Applications*, 6(2):155–171, 1990.
 8. Zhen Liu, Philippe Nain, and Don Towsley. On optimal polling policies. *Queueing Systems Theory and Applications*, 11(1–2):59–83, 1992.
 9. Jelena Mišić, Ka Lok Chan, and Vojislav B. Mišić. Performance of Bluetooth piconets under E-limited scheduling. Tech. Report TR 03/03, Department of Computer Science, University of Manitoba, Winnipeg, Manitoba, Canada, May 2003.
10. Jelena Mišić, Ka Lok Chan, and Vojislav B. Mišić. Admission control in Bluetooth piconets. To appear in *IEEE Transactions on Vehicular Technology*, 53(3), May 2004.
11. Jelena Mišić, Vojislav B. Mišić, and Eric Wai Sun Ko. Fixed cycles and adaptive bandwidth allocation can coexist in Bluetooth. *Canadian Journal of Electrical and Computer Engineering*, 29(1–2), January/April 2004.

12. Vojislav B. Mišić, Eric Wai Sun Ko, and Jelena Mišić. Adaptive cycle-limited scheduling scheme for Bluetooth piconets. In *Proceedings 14th IEEE International Symposium on Personal, Indoor and Mobile Radio Communications PIMRC'2003*, Volume 2, pages 1064–1068, Beijing, China, September 2003.

13. V. Paxson and Sally Floyd. Wide area traffic: the failure of Poisson modeling. *ACM/IEEE Transactions on Networking*, 3(3):226–244, June 1995.

14. RSoft Design, Inc. *Artifex v.4.4.2*. San Jose, CA, 2003.

15. Hideaki Takagi. *Analysis of Polling Systems*. The MIT Press, Cambridge, MA, 1986.

16. Hideaki Takagi. *Queueing Analysis*, Volume 1: Vacation and Priority Systems. North-Holland, Amsterdam, The Netherlands, 1991.

17. Wireless PAN medium access control MAC and physical layer PHY specification. IEEE Standard 802.15, IEEE, New York, 2002.

19

Scalable Wireless Ad Hoc Network Simulation

Rimon Barr

Zygmunt J. Haas

Robbert van Renesse

Ongoing research into dynamic, self-organizing, multihop wireless networks, called ad hoc networks, promises to improve the efficiency and coverage of wireless communication. Such networks have a variety of natural civil and military applications. They are particularly useful when networking infrastructure is impossible or too costly to install and when mobility is desired. However, the ability to scale such networks to large numbers of nodes remains an open research problem. For example, routing and transmitting packets efficiently over ad hoc networks becomes difficult as they grow in size.

Progress in this area of research fundamentally depends on the capabilities of simulation tools and, more specifically, on the scalability of wireless network simulators. Analytically quantifying the performance and complex behavior of even simple protocols in the aggregate is often imprecise. Furthermore, performing actual experiments is onerous: acquiring hundreds of devices, managing their software and configuration, controlling a distributed experiment and aggregating the data, possibly moving the devices around, finding the physical space for such an experiment, isolating it from interference, and generally ensuring *ceteris paribus* are but some of the difficulties that make empirical endeavors daunting. Consequently, the vast majority of research in this area is based entirely on simulation, a fact that underscores the critical role of efficient simulators.

19.1 Background

Discrete event simulators have been the subject of much research into their efficient design and execution.[8,9,17,20] However, despite a plethora of ideas and contributions to theory, languages, and systems, slow sequential simulators remain the norm.[21] In particular, most published ad hoc network results are based on simulations of a few nodes, for a short time duration, and over a limited size coverage area. Larger simulations usually compromise on simulation detail. For example, some existing simulators simulate only at the packet level without considering the effects of signal interference. Others reduce the complexity of the simulation by curtailing the simulation duration, reducing the node density, or restricting mobility. At a minimum, one would like to simulate networks of many thousands of nodes.

The two most popular simulators in the wireless networking research area are ns2 and GloMoSim. The ns2 network simulator[16] has a long history with the networking community, is widely trusted, and has been extended to support mobility and wireless networking protocols. ns2 uses a clever "split object" design, which allows Tcl-based script configuration of C-based object implementations. This approach is convenient for users. However, it incurs substantial memory overhead and increases the complexity of simulation code. Researchers have extended ns2 to conservatively parallelize its event loop,[22] but this technique has proved beneficial primarily for distributing ns2's considerable memory requirements. Based on numerous published results, it is not easy to scale ns2 beyond a few hundred simulated nodes. Recently, simulation researchers have shown ns2 to scale, with substantial hardware resources and effort, to simulations of a few thousand nodes.[21]

GloMoSim[27] is a newer simulator written in Parsec,[3] a highly optimized C-like simulation language. GloMoSim has recently gained popularity within the wireless ad hoc networking community. It was designed specifically for scalable simulation by explicitly supporting efficient, conservatively parallel execution with lookahead. The sequential version of GloMoSim is freely available. The conservatively parallel version has been commercialized as QualNet. Due to Parsec's large per-entity memory requirements, GloMoSim implements a technique called "node aggregation," wherein the state of multiple simulation nodes are multiplexed within a single Parsec entity. While this effectively reduces memory consumption, it incurs a performance overhead and also increases code complexity. Moreover, the aggregation of state also renders the lookahead techniques impractical, as has been noted by the authors. GloMoSim has been shown to scale to 10,000 nodes on large, multi-processor machines.

This chapter describes the design of SWANS, a new **S**calable **W**ireless **A**d hoc **N**etwork **S**imulator. SWANS is a componentized, virtual machine-based simulator[5] built atop the JiST (**J**ava **i**n **S**imulation **T**ime) platform, a general-purpose discrete event simulation engine. SWANS significantly outperforms ns2 and GloMoSim, both in time and space. We show results with networks that are more than an order of magnitude larger than what is possible with the existing tools at the same level of simulation detail. SWANS can also, unlike any existing network simulator, efficiently embed existing network applications and run them over simulated networks.

19.2   Design Highlights

The SWANS software is organized as independent software components, called *entities*, that can be composed to form complete wireless network or sensor network simulations, as shown in Figure 19.1. Its capabilities are similar to ns2[16] and GloMoSim[27] described above. There are entities that implement different types of applications: networking, routing, and media access protocols; radio transmission, reception, and noise models; signal propagation and fading models; and node mobility models. Instances of each type of entity are shown italicized in Figure 19.1.

The SWANS simulator runs atop JiST, a Java-based discrete-event simulation engine that combines the benefits of the traditional systems-based (e.g., ns2) and languages-based (e.g., GloMoSim) approaches to simulation construction. JiST converts a standard virtual machine into a simulation platform by embedding simulation time directly into the Java object model and into the virtual machine execution semantics. Thus, one can write a simulator in a standard systems language (i.e., Java) and transparently perform optimizations and cross-cutting program transformations that are found in specialized simulation languages. JiST extends the Java object model with the notion of simulation entities. The simulation entities represent components of a simulation that can progress independently through simulation time, each encapsulating a disjoint subset of the simulation state. Simulation events are intuitively represented as

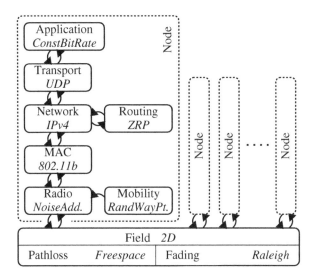

FIGURE 19.1 The SWANS simulator consists of event-driven components that can be configured and composed to form wireless network simulations.

method invocations across entities. This programming model is convenient, efficient, and flexible. We encourage the interested reader to further learn about JiST through the documentation and software available online.[4]

SWANS is able to simulate much larger networks and has a number of other advantages over existing tools. We leverage the JiST design within SWANS to (1) achieve high simulation throughput; (2) reduce its memory footprint; and (3) run existing, Java-based network applications over simulated networks. In addition, SWANS implements a technique called hierarchical binning to model wireless signal propagation in a scalable manner. The combination of these attributes leads to a flexible and highly efficient simulator. We discuss each of these concepts in turn, and then conclude with a discussion of possible directions for future work.

We limit our discussion in the following sections to techniques implemented within SWANS for maximizing *sequential* simulation performance. Others, in projects such as PDNS,[22] SWAN-DaSSF,[15] WiPPET-TeD,[12] and SWiMNet,[7] have presented algorithms and techniques to achieve scalability through distributed, concurrent, and even speculative simulation execution. These techniques can sometimes provide around an order of magnitude improvement in scale, but may require multiprocessor hardware or fast interconnects to reduce synchronization costs. More importantly, such techniques are orthogonal to the ideas presented here. A truly scalable network simulator requires raw sequential performance as well as effective distribution and parallelism.

19.3 Throughput

Conventional wisdom regarding language performance[2] argues against implementing a simulator in Java. In fact, the vast majority of existing simulators have written in C and C++, or their derivatives. SWANS, however, performs surprisingly well: aggressive profile-driven optimizations combined with the latest Java runtimes result in a high-performance system.

We selected to implement JiST and SWANS in Java for a number of reasons. First, Java is a standard, widely deployed language and not specific to writing simulations. Consequently, the Java platform boasts a large number of optimized virtual machine implementations across many hardware and software configurations, as well as a large number of compilers and languages[26] that target this execution platform. Java is an object-oriented language and it supports object reflection, serialization, and cloning, features that facilitate reasoning about the simulation state at runtime. The intermediate bytecode representation conveniently permits instrumentation of the code to support the simulation time semantics. Type-safety and garbage collection greatly simplify the writing of simulations by addressing common sources of error.

In addition, the following are some aspects of SWANS and of the underlying JiST design that contribute to its high computational throughput:

- *Dynamic compilation.* The simulator runtime, which is a standard Java virtual machine, continuously profiles running simulations and dynamically performs important code optimizations, such as constant propagation and function inlining. Dynamic optimizations provide significant performance benefits because many stable simulation parameters are not known until the simulation is running. In general, a dynamic compiler has more information available to it than a static compiler and should, therefore, produce better code. Despite virtual machine overheads for profiling, garbage collection, portability, runtime safety checks, and dynamic compilation, significant speedups can be achieved. For example, greater than $10\times$ speedups have been observed within the first few seconds of simulation execution.

- *Code inlining.* Long-running simulations often exhibit tight computation loops through only a small fraction of the code. The dynamic compiler can aggressively inline these "hot spots" to eliminate function calls in the generated code. Because both the simulator (SWANS) and the simulation kernel (JiST) are written in Java, inlining can occur both within simulation entities and also across the simulation kernel–entity boundary. For example, the dynamic Java compiler may decide to inline portions of the kernel event queuing code into hot spots within the simulation code that frequently enqueue events. Or, conversely, small and frequently executed simulation event handlers may be inlined into the kernel event loop. A similar idea was first demonstrated by the Jalapeño project.[1]

- *No context switch.* JiST provides protection and isolation for individual simulation entities from one another and from the simulation kernel. However, this separation is achieved using safe language techniques, eliminating the runtime overhead of traditional process-based isolation. A simulation event between two co-located source and target entities can be dispatched, scheduled, delivered, and processed without a single context switch.

- *No memory copy.* Each JiST entity has its own, independent simulation time and state. Therefore, to preserve entity isolation, any mutable state transferred via simulation events across entity boundaries must be passed by copy. However, temporally stable objects are an exception to this rule. These objects can safely be passed across entities by reference. To that end, JiST defines the notion of a *timeless* object as one that will not change over time. This property may either be automatically inferred through static analysis or specified explicitly. In either case, the result is zero-copy semantics and increased event throughput.

- *Cross-cutting program transformations.* The timeless property just introduced is an apt example of a cross-cutting optimization: the addition of a single tag, or the automatic detection of the timeless property, affects all events within the simulation that contain objects of this type. Similarly, the design of JiST entities abstracts event dispatch, scheduling, and delivery. Thus, the implementations of this functionality can be transparently modified. In general, the JiST bytecode-level rewriting phase that occurs at simulation load time permits a large class of transparent program transformations and simulation-specific optimizations, akin to aspect-oriented programming.[13]

High event throughput is essential for scalable network simulation. Thus, we present results showing the raw event throughput of JiST versus competing simulation engines. These measurements were all taken on a 2.0-GHz Intel Pentium 4 single-processor machine with 512 MB of RAM and 512 KB of L2 cache, running the version 2.4.20 stock Redhat 9 Linux kernel with glibc v2.3. We used the publicly available versions of Java 2 JDK (v1.4.2), Parsec (v1.1.1), GloMoSim (v2.03), and ns2 (v2.26). Each data point presented represents an average of at least five runs for the shorter time measurements.

The performance of the simulation engines was measured in performing a tight simulation event loop, using equivalent, efficient benchmark programs written for each of the engines. The results are plotted in Figure 19.2 on log-log and linear scales. As expected, all the simulations run in linear time with respect to the number of events. A counter-intuitive result is that JiST, running atop Java, outperforms all the other systems, including the compiled ones.

We therefore added a reference measurement to serve as a computational lower bound. This reference is a program, written in C and compiled with `gcc -O3`, that merely inserts and removes elements from an efficient implementation of an array-based priority queue. JiST comes within 30 percent of this reference measurement, an achievement that is a testament to the impressive JIT dynamic compilation and optimization capabilities of the modern Java runtime. Furthermore, the performance impact of these optimizations can actually be seen as a kink in the JiST curve during the first fraction of a second of the simulation. To confirm this, JiST was warmed with 10^6 events (or, for two tenths of a second) and the kink disappeared. As seen on the linear plot, the time spent on profiling and dynamic optimizations is negligible. Table 19.1 shows the time taken to perform 5 million events in each of the benchmarked systems and also those figures normalized against both the reference program and JiST performance. JiST is twice as fast as both Parsec and ns2-C, and GloMoSim and ns2-Tcl are one and two orders of magnitude slower, respectively.

SWANS builds on the performance of JiST. We benchmark SWANS running a full ad hoc wireless network simulation, running a UDP-based beaconing node discovery protocol (NDP) application. Node discovery protocols are an integral component of many ad hoc network protocols and applications.[10,11,24] Also, this experiment is representative both in terms of code coverage and network traffic; it utilizes the entire network stack and transmits over every link in the network every few seconds. However, the experiment is still simple enough to provide high confidence of simulating *exactly* the same operations across the different platforms (SWANS, GloMoSim, and ns2), which permits comparison and is difficult to achieve with more complex protocols. We simulate exactly the same network configuration across each of the simulators, measuring the overall computation time required, including the simulation setup time and the event processing overheads.

The throughput results are plotted both on log-log and on linear scales in Figure 19.3. ns2 is highly in-efficient compared to SWANS, running two orders of magnitude slower. SWANS outperforms GloMoSim by a factor of 2. However, as expected, the simulation times in all three cases are still quadratic functions of the number of nodes. To address this, we designed a scalable, hierarchical binning algorithm (discussed next) to simulate the signal propagation. As seen in the plot, SWANS-hier scales linearly with the network size.

19.4   Hierarchical Binning

In addition to an efficient simulator design, it is also essential to model wireless signal propagation efficiently because this computation is performed on every packet transmission. When a simulated radio transmits a signal, SWANS must simulate the reception of that signal at all the radios that could be affected, after considering fading, gain, and path loss. Some small subset of the radios in the coverage area will be within interference range, above some sensitivity threshold. An even smaller subset of those radios will be within reception range. The majority of the radios will not be tangibly affected by the transmission.

ns2 and GloMoSim implement a naïve signal propagation algorithm that uses a slow, $O(n)$, linear search through *all* the radios to determine the node set within reception range. This clearly does not scale as the number of radios increases. ns2 has recently been improved with a grid-based algorithm.[18] We have implemented both of these algorithms in SWANS. In addition, we have a new, more efficient and scalable algorithm that uses *hierarchical* binning. The spatial partitioning imposed by each of these data structures is depicted in Figure 19.4.

In the grid-based or flat binning approach (Figure 19.4(b)), the coverage area is subdivided into a grid of node bins. A node location update requires constant time because the bins divide the coverage area in a regular manner. The neighborhood search is then performed by scanning all bins within a given distance

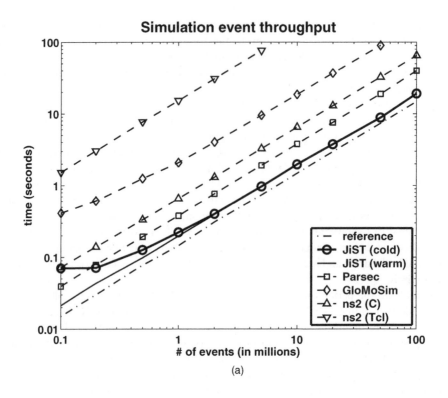

(a)

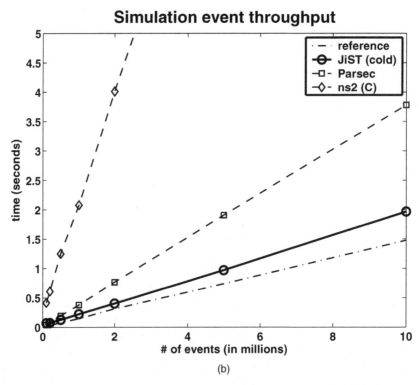

(b)

FIGURE 19.2   JiST has higher event throughput and comes within 30 percent of the reference lower bound program. The kink in the JiST curve in the first fraction of a second of simulation is evidence of JIT dynamic compilation and optimization at work.

TABLE 19.1 Time Elapsed to Perform 5 Million Events, Normalized Both against the Reference and JiST Measurements

5×10^6 Events	Time (sec)	vs. Reference	vs. JiST
Reference	0.738	1.00x	0.76x
JiST	**0.970**	**1.31x**	**1.00x**
Parsec	1.907	2.59x	1.97x
ns2-C	3.260	4.42x	3.36x
GloMoSim	9.539	12.93x	9.84x
ns2-Tcl	76.558	103.81x	78.97x

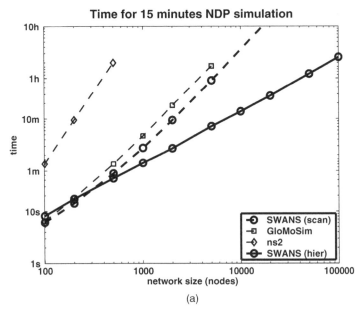

(a)

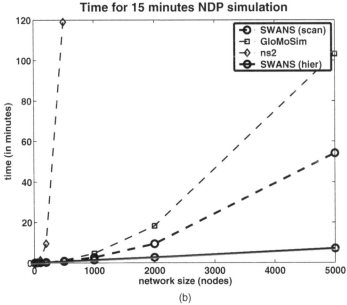

(b)

FIGURE 19.3 SWANS outperforms both ns2 and GloMoSim in simulations of NDP.

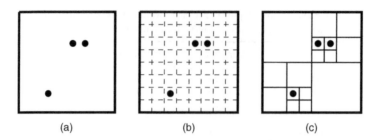

FIGURE 19.4 Hierarchical binning of radios in the coverage area allows location updates to be performed in expected amortized constant time and the set of receiving radios to be computed in time proportional to its size. (a) Linear lookup, (b) flat binning, and (c) hierarchical binning.

from the signal source. While this operation is also of constant time, given a sufficiently fine grid, the constant is sensitive to the chosen bin size: bin sizes that are too large will capture too many nodes and thus not serve their search-pruning purpose; bin sizes that are too small will require the scanning of many empty bins. A bin size that captures only a small number of nodes per bin on the average is most efficient. Thus, the bin size is a function of the local radio density and the transmission radius. However, these parameters may change in different parts of the coverage area, from radio to radio, and even as a function of time, for example, as in the case of power-controlled transmissions.

Hierarchical binning (Figure 19.4(c)) improves on the flat binning approach by dividing the coverage area recursively. Node bins are leaves of a balanced, spatial decomposition tree, which is of height equal to the number of divisions, or $h = \log_4(\frac{\text{coverage area}}{\text{bin size}})$. The hierarchical binning structure is like a quad-tree, except that the division points are not the nodes themselves, but rather fixed coordinates. A similar idea is proposed in GLS, a distributed location service for ad hoc networks.[14] Note that the height of the tree changes only logarithmically with changes in the bin or coverage area sizes. Furthermore, because nodes move only a short distance between updates, the amortized height of the common parent of the two affected node bins is constant in expectation. This, of course, is under the assumption of a reasonable node mobility model that keeps the nodes uniformly distributed and also selects trajectories uniformly. Thus, the amortized node location update cost is constant, including the maintenance of the inner node counts. When scanning for node neighbors, empty bins can be pruned as we descend. Thus, the set of receiving radios can be computed in time proportional to the number of receiving radios. Because, at a minimum, we will need to simulate delivery of the signal at each simulated radio, the algorithm is asymptotically as efficient as scanning a cached result, as proposed by Boukerche et al.,[7] even when assuming no cache misses. But, the memory overhead of the hierarchical binning scheme is minimal. Asymptotically, it amounts to $\lim_{n\to\infty}\sum_{i=1}^{\log_4 n}\frac{n}{4^i} = \frac{n}{3}$, where n is the number of network nodes. The memory overhead for function caching is also $O(n)$, but with a much larger constant. Furthermore, the memory accesses for hierarchical binning are tree structured, exhibiting better locality.

19.5 Memory Footprint

Memory is critical for simulation scalability. In the case of SWANS, memory is frequently the limiting resource. Thus, conserving memory allows for the simulation of larger network models. SWANS benefits greatly from the underlying Java garbage collector. Automatic garbage collection of events and entity state not only improves robustness of long-running simulations by preventing memory leaks, but also saves memory by facilitating more sophisticated memory protocols:

- *Shared state.* Memory consumption can often be dramatically reduced by sharing common state across entities. For example, simulated network packets are modeled in SWANS as a chain of objects that mimic the chain of packet headers added by successive layers of the simulated network stack.

Moreover, because the packets are timeless, by design, a single broadcasted packet can be shared safely among all the receiving nodes. In fact, the very same data object sent by an application entity at the top of one network stack would be received by the application entity at a receiving node. In addition to the performance benefits of zero-copy semantics, as discussed previously, this sharing also saves the memory required for multiple packet copies on every transmission. Although this optimization may seem trivial, depending on their size, lifetime, and number, packets can occupy a considerable fraction of the simulation memory footprint. Similarly, if one employs the TCP component within a simulated node stack, then the very same object that is received at one node can be referenced within the TCP retransmit buffer at the sending node, reducing the memory required to simulate even large transmission windows. Naturally, this type of memory protocol can also be implemented within the context of a non-garbage collected simulation environment. However, dynamically created objects, such as packets, can traverse many different control paths within the simulator and can have highly variable lifetimes. In SWANS, accounting for when to free unused packets is handled entirely by the garbage collector, which not only greatly simplifies the code, but also eliminates a common source of memory leaks that plague long simulation runs in other non-garbage collected simulators.

- *Soft state.* Soft state, such as various caches within the simulator, can be used to improve simulation performance. However, these caches should be reclaimed when memory becomes scarce. An example of soft state within SWANS is routing tables computed from link state. The routing tables can be automatically collected to free up memory and later regenerated, as needed. A pleasing side effect of this interaction with the garbage collector is that when memory becomes scarce, only the most useful and frequently used cached information will be retained. In the case of routing tables, the cached information would be dropped altogether, with the exception of the most active network nodes. As above, this type of memory management can also be implemented manually, although it likely would be too complex to be practical. A garbage collected environment simplifies the simulator code dramatically and increases its robustness.

To evaluate the JiST and SWANS memory requirements, we perform the same experiments as presented earlier, but measure memory consumption. The simulator memory footprint that we measure includes the base process memory, the memory overhead for simulation entities, and all the simulation data at the beginning of the simulation. Figure 19.5 shows the JiST micro-benchmark (Figure 19.5(a)) and SWANS macro-benchmark (Figure 19.5(b)) results. The plots show that the base memory footprint for each of the systems is less than 10 MB. Also, asymptotically, the memory consumed increases linearly with the number of entities, as expected. JiST performs well with respect to the memory overhead for simulation entities, because they are just small Java objects allocated on a common heap. It performs comparably to GloMoSim in this regard, which uses a technique called node aggregation specifically to reduce Parsec's memory consumption. A GloMoSim "entity" is also a small object containing an aggregation identifier and other variables similar to those found in SWANS entities. In contrast, each Parsec entity contains its own program counter and a relatively large logical process stack. In ns2, the benchmark program allocates the smallest split object possible, which duplicates simulation state in both the C and the Tcl memory spaces. JiST provides the same dynamic configuration capability using reflection, without requiring the memory overhead of a split object design.

Because JiST is more efficient than GloMoSim and ns2 by almost an order and two orders of magnitude, respectively, SWANS is able to simulate networks that are significantly larger. The memory overhead of hierarchical binning is shown to be asymptotically negligible. Also, as a point of reference, regularly published results of a few hundred wireless nodes occupy more than 100 MB, and simulation researchers have scaled ns2 to around 1500 non-wireless nodes using a 1-GB process.[19,22] Table 19.2 provides exact time and space requirements under each of the simulators for simulations of NDP across a range of network sizes.

Finally, we present SWANS with some very large networks. For these experiments, we ran the same simulations, but on dual-processor 2.2-GHz Intel Xeon machines (although only one processor was used)

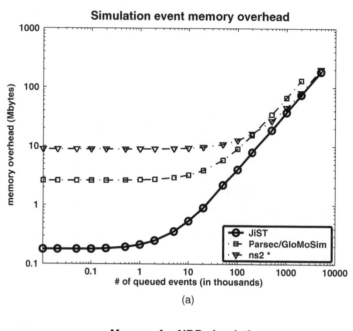

(a)

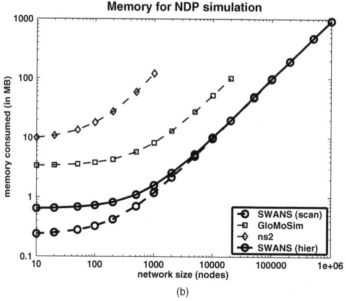

(b)

FIGURE 19.5   JiST allocates entities efficiently: comparable to GloMoSim at 36 bytes per entity and over an order of magnitude less than Parsec or ns2. SWANS can simulate correspondingly larger network models due to this more efficient use of memory and more sophisticated memory management protocols enabled by the garbage collector.

with 2 GB of RAM running Windows 2003. The results are plotted in Figure 19.6 on a log-log scale. We show SWANS both with the naïve propagation algorithm and with hierarchical binning. We observe linear behavior for the latter in all simulations up to networks of one million nodes. The 10^6 node simulation consumes just less than 1 GB of memory on initial configuration, runs with an average footprint of 1.2 GB (fluctuating due to delayed garbage collection), and completes within 5.5 hours. This size of the network exceeds previous ns2 and GloMoSim capabilities by two orders of magnitude.

TABLE 19.2 SWANS Outperforms ns2 and GloMoSim in Both Time and Space

Nodes	Simulator	Time (sec)	Memory (KB)
500	**SWANS**	**54**	**700**
	GloMoSim	82	5759
	ns2	7136	58761
	SWANS-hier	*43*	*1101*
5000	**SWANS**	**3250**	**4887**
	GloMoSim	6191	27570
	SWANS-hier	*430*	*5284*
50,000	**SWANS**	**312019**	**47717**
	SWANS-hier	*4377*	*49262*

19.6   Embedding Applications

SWANS has a unique and important advantage over existing network simulators. It can run regular, unmodified Java network applications over simulated networks, thus allowing for the inclusion of existing Java-based software, such as Web servers, peer-to-peer applications, and application-level multicast protocols, within network simulations. These applications do not merely send packets to the simulator from other processes. They operate in simulation time, embedded within the same SWANS process space, incurring no blocking, context switch, or memory copy, and thereby allowing far greater scalability.

This tight, efficient integration is achieved through a sequence of transparent program transformations, whose end result is to embed a process-oriented Java network application into the event-oriented SWANS simulation environment. The entire Java application is first wrapped within a special Java application entity harness. Because multiple instances of an application can be running within different simulated nodes, the harness entity serves as an anchor for the application context. Like the regular Java launcher, the harness entity invokes an application's `main` method to initiate it.

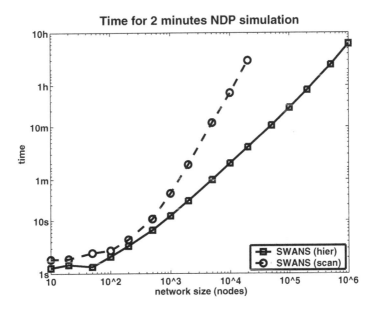

FIGURE 19.6 SWANS scales to networks of 10^6 wireless nodes. The figure shows the time for a sequential simulation of a heartbeat NDP.

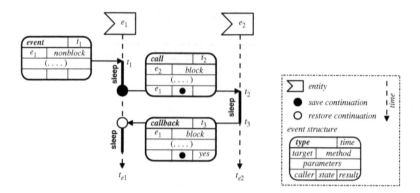

FIGURE 19.7 The addition of blocking methods allows simulation developers to regain the simplicity of process-oriented development. When a blocking entity method is invoked, the continuation state of the current event is saved and attached to a call event. When this call event is complete, the kernel schedules a callback event to the caller. The continuation is restored and the caller continues its processing from where it left off, albeit at a later simulation time.

Before loading an application, SWANS inserts a custom rewriting phase into the simulation kernel class loader. Among other, more subtle modifications, this rewriting phase replaces all Java socket calls within the original application with invocations to corresponding functionality within simulated SWANS sockets. These SWANS sockets have identical semantics to their native counterparts in the standard Java library, but send packets through the simulated network. Most importantly, the input (receive) and output (send) methods are implemented using *blocking* JiST events and simulation time continuations, which we explain next.

The semantics of a blocking events, as depicted in Figure 19.7, are a natural extension atop the existing nonblocking events. The kernel first saves the call-stack of the calling entity and attaches it to the outgoing event. When the call event is completed, the kernel notices that the event has caller information and the kernel dispatches a callback event to the caller with its continuation information. Thus, when the callback event is eventually dequeued, the state is restored and the execution continues right after the point of the blocking entity method invocation. In the meantime, however, the local entity simulation time will have progressed to the simulation time at which the calling event was completed, and other events may have been processed against the entity in the interim.

To support these blocking semantics, JiST automatically modifies the necessary application code into a continuation-passing style. This allows the application to operate within the event-oriented simulation time environment. Our design allows blocking and nonblocking entity methods to coexist, which means that event-oriented and process-oriented simulations can coexist. Unfortunately, saving and restoring the Java call-stack for continuation is not a trivial task.[23] The fundamental difficulty arises from the fact that stack manipulations are not supported at either the language, the library, or the bytecode level. Our solution draws and improves on the ideas in the JavaGoX[25] and PicoThreads[6] projects, which also save the Java stack for other reasons. Our design eliminates the use of exceptions to carry state information. This is considerably more efficient for our simulation needs because Java exceptions are expensive. Our design also eliminates the need to modify method signatures. This fact is significant because it allows our continuation capturing mechanism to function even across the standard Java libraries. In turn, this enables us to run standard, unmodified Java network applications within SWANS. A network socket read, for example, is rewritten into a blocking method invocation on a simulated socket, so that an application is "frozen" in simulation time until a network packet is delivered to the application by the simulated socket entity.

19.7 Conclusion

In summary, SWANS is a componentized wireless network simulator built according to the virtual machine-based simulator design.[5] It significantly outperforms ns2 and GloMoSim, both in time and space. We have shown results with networks that, at the same level of detail, are more than an order of magnitude larger than what is possible with existing tools. SWANS can also, unlike any existing network simulator, efficiently run existing Java-based network applications over simulated networks. In general, SWANS inherits the advantages of JiST and the Java platform.

The SWANS simulator is freely available[4] and can be extended in a number of interesting directions, including:

- *Parallel, distributed, and speculative execution.* The current implementations of JiST and SWANS have focused exclusively on sequential performance. The system, however, was explicitly designed and implemented with more sophisticated execution strategies in mind. It is possible to extend the simulation kernel to allow multiple processing threads to operate concurrently on the simulation state in order to leverage the full processing power of commodity multiprocessor machines. For distributed simulation, JiST entity separators can readily be extended to support a single system abstraction by transparently tracking entity locations as they are dynamically migrated across a cluster of machines to balance computational and network load. The JiST kernel can then be extended to support conservatively synchronized, distributed, cooperative operation with peer JiST kernels, which would increase the available simulation memory and allow larger network models to be processed. Synchronization protocols need not remain conservative. Because the JiST design can already transparently support both checkpointing and rollback of entities, and because the cost of synchronization is critical to the performance of a distributed simulator, it is worthwhile to investigate various speculative execution strategies as well. Each of these three extensions to the simulation kernel — parallel, distributed, and speculative execution — are already supported within the current JiST semantics and can therefore be implemented transparently with respect to existing SWANS components. Such extensions to the simulation kernel pose a rich space of design and research problems.
- *Declarative simulation specifications.* SWANS is a componentized simulator and therefore tends naturally to be configured as a graph of interconnected entities that can model some network with given topology and node configurations. Because these component graphs often have repetitive and highly compressible structure, it would be beneficial to construct them using short declarative specifications, rather than the current approach, driven by imperative scripts. In general, it would be interesting to develop high-level, possibly domain-specific, yet expressive, simulation configuration languages.
- *Simulation debuggers and interactive simulators.* A significant advantage of leveraging the Java platform is the ability to adopt existing Java tools, such as debuggers, often lacking in simulation environments. Event-driven programs are particularly difficult to debug, thus compounding the problem. An existing Java debugger could readily be extended to understand simulation events and other simulation kernel data structures, resulting in functionality that is unparalleled in any existing simulation environment. Because Java is a reflective language, SWANS simulations can be paused, modified in-flight, and then resumed. The appropriate tools to perform such inspection effectively (i.e., a graphical, editable view of the network and the state of its nodes, for example) would facilitate interactive simulation and present interesting research opportunities. For example, one could use the debugger to control the distributed simulation kernel and its global virtual time scheduler, not only to obtain consistent cuts of the simulation state, but also to permit stepping *backwards* in simulation time to understand root causes of a particular simulation state.

To conclude, we hope that the scalability, performance, and flexibility of SWANS, combined with the popularity of the Java language, will facilitate its broader adoption within the network simulation community.

Acknowledgment

This work has been supported in part by the DoD Multidisciplinary University Research Initiative (MURI) program administered by the Office of Naval Research under the grant number N00014-00-1-0564 and by the DoD Multidisciplinary University Research Initiative (MURI) program administered by the Air Force Office of Scientific Research under the grant number F49620-02-1-0233.

References

1. Bowen Alpern, C. Richard Attanasio, John J. Barton, Anthony Cocchi, Susan Flynn Hummel, Derek Lieber, Ton Ngo, Mark F. Mergen, Janice C. Shepherd, and Stephen Smith. Implementing Jalapeño in Java. In *Object-Oriented Programming Systems, Languages and Applications*, pages 314–324, November 1999.

2. Doug Bagley. The great computer language shoot-out, 2001. http://www.bagley.org/~doug/shootout/.

3. Rajive L. Bagrodia, Richard Meyer, Mineo Takai, Yuan Chen, Xiang Zeng, Jay Martin, and Ha Yoon Song. Parsec: A parallel simulation environment for complex systems. *IEEE Computer*, 31(10):77–85, October 1998.

4. Rimon Barr and Zygmunt J. Haas. JiST/SWANS Web site, 2004. http://www.jist.ece.cornell/edu/.

5. Rimon Barr, Zygmunt J. Haas, and Robbert van Renesse. JiST: Embedding simulation time into a virtual machine. In *Proceedings of EuroSim 2004*, September 2004.

6. Andrew Begel, Josh MacDonald, and Michael Shilman. PicoThreads: Lightweight threads in Java. Technical report, U.C. Berkeley, 2000.

7. Azzedine Boukerche, Sajal K. Das, and Alessandro Fabbri. SWiMNet: A scalable parallel simulation testbed for wireless and mobile networks. *Wireless Networks*, 7:467–486, 2001.

8. Richard M. Fujimoto. Parallel discrete event simulation. *Communications of the ACM*, 33(10):30–53, October 1990.

9. Richard M. Fujimoto. Parallel and distributed simulation. In *Winter Simulation Conference*, pages 118–125, December 1995.

10. Zygmunt Haas and Marc Pearlman. Providing ad hoc connectivity with reconfigurable wireless networks. In Charles Perkins, Editor, *Ad hoc Networks*. Addison-Wesley Longman, 2000.

11. David B. Johnson and David A. Maltz. Dynamic source routing in ad hoc wireless networks. In *Mobile Computing*. Kluwer Academic Publishers, 1996.

12. Owen Kelly, Jie Lai, Narayan B. Mandayam, Andrew T. Ogielski, Jignesh Panchal, and Roy D. Yates. Scalable parallel simulations of wireless networks with WiPPET. *Mobile Networks and Applications*, 5(3):199–208, 2000.

13. Gregor Kiczales, John Lamping, Anurag Menhdhekar, Chris Maeda, Cristina Lopes, Jean-Marc Loingtier, and John Irwin. Aspect-oriented programming. *European Conference on Object-Oriented Programming*, 1241:220–242, 1997.

14. Jinyang Li, John Jannotti, Douglas S.J. De Couto, David R. Karger, and Robert Morris. A scalable location service for geographic ad hoc routing. In *ACM/IEEE International Conference on Mobile Computing and Networking (MOBICOM)*, pages 120–130, 2000.

15. Jason Liu, L. Felipe Perrone, David M. Nicol, Michael Liljenstam, Chip Elliott, and David Pearson. Simulation modeling of large-scale ad-hoc sensor networks. In *Simulation Interoperability Workshop*, 2001.

16. Steven McCanne and Sally Floyd. ns (Network Simulator) at http://www-nrg.ee.lbl.gov/ns, 1995.

17. Jayadev Misra. Distributed discrete event simulation. *ACM Computing Surveys*, 18(1):39–65, March 1986.

18. Valeri Naoumov and Thomas Gross. Simulation of large ad hoc networks. In *ACM MSWiM*, pages 50–57, 2003.

19. David M. Nicol. Comparison of network simulators revisited, May 2002. http://www.ssfnet.org/Exchange/gallery/dumbbell/dumbbell-performance-May 02.pdf

20. David M. Nicol and Richard M. Fujimoto. Parallel simulation today. *Annals of Operations Research*, pages 249–285, December 1994.

21. George Riley and Mostafa Ammar. Simulating large networks: How big is big enough? In *Conference on Grand Challenges for Modeling and Sim.*, January 2002.

22. George Riley, Richard M. Fujimoto, and Mostafa A. Ammar. A generic framework for parallelization of network simulations. In *Symposium on Modeling, Analysis and Simulation of Computer and Telecommunication,* March 1999.

23. T. Sakamoto, T. Sekiguchi, and A. Yonezawa. Bytecode transformation for portable thread migration in Java. In *International Symposium on Mobile Agents,* 2000.

24. Prince Samar, Marc Pearlman, and Zygmunt Haas. Hybrid routing: The pursuit of an adaptable and scalable routing framework for ad hoc networks. In *Handbook of Ad Hoc Wireless Networks.* CRC Press, 2003.

25. Tatsurou Sekiguchi, Takahiro Sakamoto, and Akinori Yonezawa. Portable implementation of continuation operators in imperative languages by exception handling. *Lecture Notes in Computer Science,* 2022:217, 2001.

26. Robert Tolksdorf. Programming languages for the Java virtual machine at http://www.robert-tolksdorf.de/vmlanguages, 1996–.

27. Xiang Zeng, Rajive L. Bagrodia, and Mario Gerla. GloMoSim: a library for parallel simulation of large-scale wireless networks. In *Workshop on Parallel and Distributed Simulation,* May 1998.

II

Sensor Networks

A sensor network consists of a large number of sensor nodes that are densely deployed either inside or close to the phenomenon. In general, the positions of sensor nodes need not be engineered or predetermined. This allows for random deployment in inaccessible terrain or disaster relief operations. Sensor networks can be considered as a special type of ad hoc wireless networks, where sensor nodes are, in general, stationary. A unique feature of sensor networks is the cooperative effort of sensor nodes. Sensor nodes are usually fitted with onboard processors. Instead of sending the raw data to the nodes responsible for the fusion, they use their processing abilities to locally carry out simple computations and transmit only the required and partially processed data. A sensor system normally consists of a set of sensor nodes operating on limited energy and a base system without any energy constraint. Typically, the base station serves as the gathering point for the collected data (through fusion). The base station also broadcasts various control commands to sensor nodes.

The applications of sensor networks include health, military, and civilian. In military application, the rapid deployment, self-organization, and fault-tolerance characteristics of sensor nodes make them a promising sensing technique for command, control, communication, computing, intelligence, surveillance, reconnaissance, and targeting systems. In health care, sensor nodes can be used to monitor patients and assist disabled patients. Other applications include managing inventory, monitoring product quality, and monitoring disaster areas.

The design factors involved in sensor networks include fault tolerance, scalability, sensor network topology, data dissemination and gathering, transmission media, and power consumption. Most work is focused on sensor network topology (or topology control), data dissemination and gathering, and power consumption. As in ad hoc wireless networks, the use of protocol stacks to address various technical issues is also a popular approach in sensor networks.

There are many similarities between ad hoc wireless networks and sensor networks, such as energy constraints and dynamic network topology. However, the number of sensor nodes can be several orders of magnitude higher than the numbers of nodes in an ad hoc wireless network, and hence sensor nodes are more densely deployed. Also, sensor nodes are more prone to failures than those of an ad hoc wireless network. Therefore, the topology of a sensor network is changed mainly by the disabling or enabling of sensor nodes, whereas the topology of an ad hoc wireless network is changed by the movement of mobile hosts. In addition, sensor nodes may not have global identification because of the large amount of overhead and the large number of sensors. Finally, nodes in ad hoc wireless networks are strictly peer-to-peer, whereas nodes in sensor networks form a two-level hierarchy with the base station being the gathering point.

This group includes 16 chapters, starting with a chapter of general overview. Four chapters cover four different applications of distributed algorithms: network initialization, self-organization, self-stabilization, and time synchronization. One chapter relates routing (including broadcasting) to topology control. Three chapters are devoted to sensor coverage and deployment. Two chapters deal with surveillance and detection. One chapter compares energy conservation methods. The issues of QoS and scalability are covered in two separate chapters. One chapter presents solutions for a special security issue. The group ends with some discussion on the recent development on a new IEEE standard (IEEE 802.15.4) for sensor networks.

20

Sensor Systems: State of the Art and Future Challenges

Dharma P. Agrawal

Ratnabali Biswas

Neha Jain

Anindo Mukherjee

Sandhya Sekhar

Aditya Gupta

20.1 Introduction

Recent technological advances have enabled distributed information gathering from a given location or region by deploying a large number of networked tiny microsensors, which are low-power devices equipped with programmable computing, multiple sensing, and communication capability. Microsensor systems enable the reliable monitoring and control of a variety of applications. Such sensor nodes networked by wireless radio have revolutionized remote monitoring applications because of their ease of deployment, ad hoc connectivity, and cost effectiveness.

20.1.1 Characteristics of Wireless Sensor Networks

A wireless sensor network is typically a collection of tiny disposable devices with sensors embedded in them. These devices, referred to as sensor nodes, are used to collect physical parameters such as light intensity, sound, temperature, etc. from the environment where they are deployed. Each node (Figure 20.1) in a sensor network includes a sensing module, a microprocessor to convert the sensor signals into a sensor reading understandable by a user, a wireless interface to exchange sensor readings with other nodes lying within its radio range, a memory to temporarily hold sensor data, and a small battery to run the device. Wireless sensors typically have a low transmission data rate. A small form factor or size of these nodes (of the order of 5 cm^3) limits the size of the battery or the total power available with each sensor node. The low cost of sensor nodes makes it feasible to have a network of hundreds or thousands of these wireless sensors.

A large number of nodes enhance the coverage of the field and the reliability and accuracy of the data retrieved (Figure 20.2). When deployed in large numbers, sensor nodes with limited radio communication range form an ad hoc network. An ad hoc network is basically a peer-to-peer multi-hop wireless network where information packets are transmitted in a store-and-forward method from source to destination, via intermediate nodes. This is in contrast to the well-known single-hop cellular network model that supports the needs of wireless communications by installing base stations (BSs) as access points. Sensor nodes are attractive due to the ease of deployment and autonomous ad hoc connectivity that eliminates the need for any human intervention or infrastructure installation. Sensor networks need to fuse data obtained from several sensors sensing a common phenomenon to provide rich, multidimensional pictures of an environment that a single powerful macrosensor, working alone, may not provide.

Multiple sensors can help overcome line-of-sight issues and environmental effects by placing sensors close to an event of interest. This ensures a greater signal-to-noise ratio (SNR) by combining information from sources with different spatial perspectives. This is especially desirable in those applications where sensors may be thrown in an inhospitable terrain with the aid of an unmanned vehicle or a low-flying

FIGURE 20.1 Sensor mote. (*Source:* www.sce.umkc.edu/~leeyu/Udic/SCE2.ppt)

aircraft. Instead of carefully placing macrosensors in exact positions and connecting them through cables to obtain accurate results, a large number of preprogrammed sensor nodes are randomly dispersed in an environment. Although communication may be lossy due to the inherent unreliable nature of wireless links, a dense network of nodes ensures enough redundancy in data acquisition to guarantee an acceptable quality of the results provided by the network. These sensor nodes, once deployed, are primarily static; and it is usually not feasible to replace individual sensors on failure or depletion of their battery.

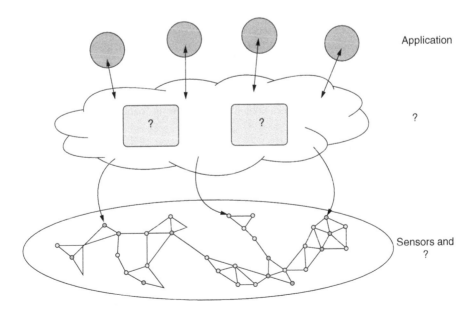

FIGURE 20.2 A generic sensor network. (*Source:* eyes.eu.org/eyes-sa.gif)

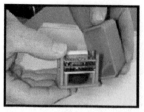  

February 2001   February 2002   February 2003

FIGURE 20.3 Progression of sensors developed by CITRIS investigators. (*Source:* www.citris-uc.org/about_citris/annual_report.html)

Each node has a finite lifetime determined by its rate of battery consumption. It is a formidable task to build and maintain a robust, energy-efficient multi-hop sensor network in an ad hoc setting without any global control. It is therefore necessary to consider the different types of sensors that are commercially available.

20.1.2 Types of Sensors

Sensor networks present unprecedented opportunities for a broad spectrum of applications such as industrial automation, situation awareness, tactical surveillance for military applications, environmental monitoring, chemical or biological detection, etc.

 Figure 20.3 shows a progression of sensors developed by CITRIS (Center for Information Technology in the Interest of Society) investigators (www.citris-uc.org). These wireless sensors can be used to sense magnetic or seismic attributes in military security networks; or sense temperature or pressure in industrial sensing networks; strain, fatigue, or corrosion in civil structuring monitoring networks; or temperature and humidity in agricultural maintenance networks. Over a period of two years, the size of these sensors has decreased from a few cubic inches to a few cubic millimeters, and they can be powered using tiny solar cells or piezoelectric generators running on the minute vibrations of walls inside buildings or vehicles.

 Microstrain Inc. (www.microstrain.com) has also developed a variety of wireless sensors for different commercial applications.

 The SG-link^{TM} Wireless Strain Gauge system (Figure 20.4) is a complete wireless strain gauge node, designed for integration with high-speed wireless sensor networks. It can be used for high-speed strain, load, torque, and pressure monitoring and finds application in sports performance and sports medicine analysis.

 The TC-Link^{TM} Wireless Thermocouple System (Figure 20.5) is a complete, cold junction compensated, multichannel thermocouple node designed to operate as part of an integrated, scalable, ad hoc wireless sensor network system. It finds application in civil structures sensing (concrete maturation), industrial

FIGURE 20.4 SG-link^{TTM} wireless strain gauge system. (*Source:* www.microstrain.com/SG-link.htm)

FIGURE 20.5   TC-Link^{TTM} wireless thermocouple system. (*Source:* www.microstrain.com/TCLink.htm)

sensing networks (machine thermal management), food and transportation systems (refrigeration, freezer monitoring), and advanced manufacturing (plastics processing, composite cure monitoring).

The miniature EmbedSense™ wireless sensor (Figure 20.6) is a tiny wireless sensor and data acquisition system that is small enough to be embedded in a product, enabling the creation of smart structures, smart materials, and smart machines. A major advantage is that batteries are completely eliminated, thereby ensuring that the embedded sensors and EmbedSense node can be queried for the entire life of the structure. EmbedSense uses an inductive link to receive power from an external coil and to return digital strain, temperature, and unique ID information. Applications range from monitoring the healing of the spine to testing strains and temperatures on jet turbine engines.

We need to consider how a query can be processed in a sensor network.

20.2   Routing in Wireless Sensor Networks

There are a few inherent limitations of wireless media, such as low bandwidth, error-prone transmissions, the need for collision-free channel access, etc. These wireless nodes also have only a limited amount of energy available to them, because they derive energy from a personal battery and not from a constant power supply. Furthermore, because these sensor nodes are deployed in places where it is difficult to replace the nodes or their batteries, it is desirable to increase the longevity of the network. Also, preferably all the nodes should die together so that one can replace all the nodes simultaneously in the whole area. Finding individual dead nodes and then replacing those nodes selectively would require preplanned deployment and eliminate some advantages of these networks. Thus, the protocols designed for these networks must strategically distribute the dissipation of energy, which also increases the average life of the overall system.

20.2.1   Query Classification in Sensor Networks

Before discussing routing protocols for sensor networks, let us first categorize the different kinds of queries that can be posed to a sensor network. Based on the temporal property of data (i.e., whether the user is

FIGURE 20.6   EmbedSense™ wireless sensor. (*Source:* www.microstrain.com/embed_sense.htm)

interested in data collected in the past, the current snapshot view of the target regions, or sensor data to be generated in future for a given interval of time), queries can be classified as follows:

Historical queries. This type of query is mainly used for analysis of historical data stored at a remote base station or any designated node in the network in the absence of a base station. For example, "What was the temperature two hours prior in the northwest quadrant?" The source nodes need not be queried to obtain historical data as it is usually stored outside the network or at a node equidistant from all anticipated sinks for that data.

One-time query. One-time or snapshot queries provide the instantaneous view of the network. For example, "What is the temperature in the northwest quadrant now?" The query triggers a single query response; hence, data traffic generated by one-time queries is the least. These are usually time critical as the user wants to be notified immediately about the current situation of the network. A warning message that informs the user of some unusual activity in the network is an example of a one-time query response that is time critical.

Persistent. Persistent, or long-running, queries are mainly used to monitor a network over a time interval with respect to some parameters. For example, "the temperature in the northwest quadrant for the next 2 hours." A persistent query generates maximum query responses in the network, depending on its duration. The purpose of the persistent query is to perform periodic background monitoring. Energy efficiency is often traded with delay in response time of persistent queries to maximize utilization of network resources, as they are usually noncritical.

20.2.2   Characteristics of Routing Protocols for Sensor Networks

Traditional routing protocols defined for wireless ad hoc networks (Broch et al., 1998; Royer and Toh, 1999) are not well suited for wireless sensor networks due to the following reasons (Manjeshwar and Agrawal, 2001, 2002:

Sensor networks are data centric. Traditional networks usually request data from a specific node but sensor networks request data based on certain attributes such as, "Which area has temperature greater than $100°F$?"

In traditional wired and wireless networks, each node is given a unique ID, which is used for routing. This cannot be effectively used in sensor networks because being data centric they do not require routing to and from specific nodes. Also, the large number of nodes in the network implies large IDs (Nelson and Estrin, 2000), which might be substantially larger than the actual data being transmitted.

Adjacent nodes may have similar data. So instead of sending data separately from each node to the requesting node, it is desirable to aggregate similar data before sending it.

The requirements of the network change with the application and, hence, it is application specific (Estrin et al. 1999). For example, some applications need the sensor nodes to be fixed and not mobile, while others may need data based only on one selected attribute (i.e., here the attribute is fixed).

Thus, sensor networks need protocols that are application specific, data centric, and capable of aggregating data and minimizing energy consumption.

20.3   Network Architecture

The two main architecture alternatives used for data communication in a sensor network are the hierarchical and the flat network architectures. The hierarchical network architecture is energy efficient for collecting and aggregating data within a large target region, where each node in the region is a source node. Hence, hierarchical network protocols are used when data will be collected from the entire sensor network. A flat network architecture is more suitable for transferring data between a source destination pair separated by a large number of hops.

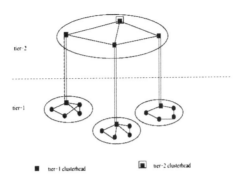

tier-1 clusterhead   tier-2 clusterhead

FIGURE 20.7 Two-tier sensor network architecture. (*Source*: A. Manjeshwar, Energy Efficient Routing Protocols with Comprehensive Information Retrieval for Wireless Sensor Networks, M.S. thesis, 2001.)

20.3.1 Hierarchical Network Architecture

One way of minimizing the data transmissions over long distances is to cluster the network so that signaling and control overheads can be reduced, while critical functions such as media access, routing, and connection setup could be improved. While all nodes typically function as switches or routers, one node in each cluster is designated as the cluster head (CH) and traffic between nodes of different clusters must always be routed through their respective CHs or gateway nodes that are responsible for maintaining connectivity among neighboring CHs. The number of tiers within the network can vary according to the number of nodes, resulting in hierarchical network architecture as shown in Figure 20.7.

Figure 20.7 shows two tiers of cluster heads where the double lines represent that CHs of tier-1 are cluster members of the cluster at the next higher level (i.e., tier-2). A proactive clustering algorithm for sensor networks called LEACH is one of the initial data-gathering protocols introduced by MIT researchers Heinzelman et al. (2000). Each cluster has a CH that periodically collects data from its cluster members, aggregates it, and sends it to an upper-level CH. Only the CH needs to perform additional data computations such as aggregation, etc., and the rest of the nodes sleep unless they have to communicate with the CH. To evenly distribute this energy consumption, all the nodes in a neighborhood take turns to become the CH for a time interval called the cluster period.

20.3.2 Flat Network Architecture

In a flat network architecture as shown in Figure 20.8, all nodes are equal and connections are set up between nodes that are in close proximity to establish radio communications, constrained only by connectivity conditions and security limitations. Route discovery can be carried out in sensor networks using flooding that does not require topology maintenance as it is a reactive way of disseminating information. In flooding, each node receiving data packets broadcasts until all nodes or the node at which the packet was originated gets back the packet. But in sensor networks, flooding is minimized or avoided as nodes could receive multiple or duplicate copies of the same data packet due to nodes having common neighbors or sensing similar data. Intanagonwiwat et al. (2000) have introduced a data dissemination paradigm called directed diffusion for sensor networks, based on a flat topology. The query is disseminated (flooded) throughout the network with the querying node acting as a source and gradients are set up toward the requesting node to find the data satisfying the query. As one can observe from Figure 20.8, the query is propagated toward the requesting node along multiple paths shown by the dashed lines. The arcs show how the query is directed toward the event of interest, similar to a ripple effect. Events (data) start flowing toward the requesting node along multiple paths. To prevent further flooding, a small number of paths can be reinforced (shown by dark lines in the figure) among a large number of paths initially explored to form the multi-hop routing infrastructure so as to prevent further flooding. One advantage of flat networks is the ease of creating multiple paths between communicating nodes, thereby alleviating congestion and providing robustness in the presence of failures.

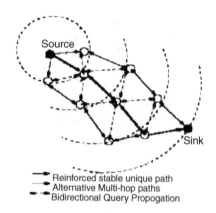

Reinforced stable unique path
Alternative Multi-hop paths
Bidirectional Query Propogation

FIGURE 20.8   A flat sensor network that uses directed diffusion for routing. (*Source:* N. Jain, Energy Efficient Information Retrieval in Wireless Sensor Networks, Ph.D. thesis, 2004.)

20.4   Classification of Sensor Networks

Sensor networks can be classified into two types based on their mode of operation or functionality and the type of target applications (Manjeshwar and Agrawal, 2001, 2002):

Proactive networks. In this scheme, the nodes periodically switch on their sensors and transmitters, sense the environment, and transmit the data of interest. Thus, they provide a snapshot of the relevant parameters at regular intervals and are well suited for applications that require periodic data monitoring.

Reactive networks. In this scheme, the nodes react immediately to sudden and drastic changes in the value of a sensed attribute and are well suited for time-critical applications.

Once we have a network, we have to come up with protocols that efficiently route data from the nodes to the users, preferably using a suitable MAC (Medium Access Control) sub-layer protocol to avoid collisions.

Having classified sensor networks, we now look at some of the protocols for sensor networks (Manjeshwar and Agrawal, 2001, 2002).

20.4.1   Proactive Network Protocol

20.4.1.1   Functioning

At each cluster change time, once the cluster heads are decided, the cluster head broadcasts the following parameters (see Figure 20.9):

Report time (T_R): the time period between successive reports sent by a node.

Attributes (A): a set of physical parameters about which the user is interested in obtaining data.

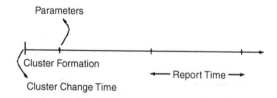

FIGURE 20.9   Timeline for proactive protocol. (*Source:* A. Manjeshwar and D.P. Agrawal, "TEEN: a routing protocol for enhanced efficiency in wireless sensor networks, in *Proc. 15th Int. Parallel and Distributed Processing Symp. (IPDPS'01) Workshop,* 2001.)

At every report time, the cluster members sense the parameters specified in the attributes and send the data to the cluster head. The cluster head aggregates this data and sends it to the base station or a higher-level cluster head, as the case may be. This ensures that the user has a complete picture of the entire area covered by the network.

20.4.1.2 Important Features

Because the nodes switch off their sensors and transmitters at all times except the report times, the energy of the network is conserved.

At every cluster change time, T_R and A are transmitted afresh and thus can be changed. By changing A and T_R, the user can decide what parameters to sense and how often to sense them. Also, different clusters can sense different attributes for different T_R.

This scheme, however, has an important drawback. Because of the periodicity with which the data is sensed, it is possible that time-critical data may reach the user only after the report time, making this scheme ineffective for time-critical data sensing applications.

20.4.1.3 LEACH

LEACH (Low-Energy Adaptive Clustering Hierarchy) is a family of protocols developed by Heinzelman et al. (www-mtl.mit.edu/research/icsystems/uamps/leach). LEACH is a good approximation of a proactive network protocol, with some minor differences.

Once the clusters are formed, the cluster heads broadcast a TDMA schedule giving the order in which the cluster members can transmit their data. The total time required to complete this schedule is called the frame time T_F. Every node in the cluster has its own slot in the frame, during which it transmits data to the cluster head. The report time T_R discussed earlier is equivalent to the frame time T_F in LEACH. However T_F is not broadcast by the cluster head but is derived from the TDMA schedule and hence is not under user control. Also, the attributes are predetermined and not changed after initial installation.

20.4.1.4 Example Applications

This network can be used to monitor machinery for fault detection and diagnosis. It can also be used to collect data about temperature (or pressure, moisture, etc.) change patterns over a particular area.

20.4.2 Reactive Network Protocol: TEEN

A new network protocol called TEEN (*Threshold sensitive Energy Efficient sensor Network*) has been developed that targets reactive networks and is the first protocol developed for reactive networks (Manjeshwar and Agrawal, 2001).

20.4.2.1 Functioning

In this scheme, at every cluster change time, in addition to the attributes, the cluster head broadcasts the following to its members (see Figure 20.10):

Hard threshold (H_T): a threshold value for the sensed attribute. It is the absolute value of the attribute, beyond which the node sensing this value must switch on its transmitter and report to its cluster head.

Soft threshold (S_T): a small change in the value of the sensed attribute that triggers the node to switch on its transmitter and transmit.

The nodes sense their environment continuously. The first time a parameter from the attribute set reaches its hard threshold value, the node switches on its transmitter and sends the sensed data. The sensed

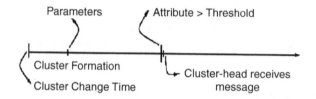

FIGURE 20.10   Timeline for TEEN. (*Source:* A. Manjeshwar and D.P. Agrawal, "TEEN: a routing protocol for enhanced efficiency in wireless sensor networks, in *Proc. 15th Int. Parallel and Distributed Processing Symp. (IPDPS'01) Workshops*, 2001.)

value is also stored in an internal variable in the node, called the *sensed value (SV)*. The nodes will next transmit data in the current cluster period but only when *both* the following conditions are true:

The current value of the sensed attribute is greater than the hard threshold.
The current value of the sensed attribute differs from SV by an amount equal to or greater than the soft threshold.

Whenever a node transmits data, *SV* is set equal to the current value of the sensed attribute. Thus, the hard threshold tries to reduce the number of transmissions by allowing the nodes to transmit only when the sensed attribute is in the range of interest. The soft threshold further reduces the number of transmissions by eliminating all the transmissions that might have otherwise occurred when there is little or no change in the sensed attribute once the hard threshold is reached.

20.4.2.2   Important Features

Time-critical data reaches the user almost instantaneously and hence this scheme is well suited for time-critical data sensing applications.
Message transmission consumes much more energy than data sensing. So, although the nodes sense continuously but because they transmit less frequently, the energy consumption in this scheme can be much less than that in the proactive network.
The soft threshold can be varied, depending on the criticality of the sensed attribute and the target application.
A smaller value of the soft threshold gives a more accurate picture of the network, at the expense of increased energy consumption. Thus, the user can control the trade-off between energy efficiency and accuracy.
At every cluster change time, the parameters are broadcast afresh and thus the user can change them as required.

The main drawback of this scheme is that if the thresholds are not reached, the nodes will never communicate. Thus, the user will not get any data from the network and will not even know if all the nodes die. Hence, this scheme is not well suited for applications where the user needs to get data on a regular basis. Another possible problem with this scheme is that a practical implementation would have to ensure that there are no collisions in the cluster. TDMA scheduling of the nodes can be used to avoid this problem. This will, however, introduce a delay in the reporting of time-critical data. CDMA is another possible solution to this problem.

20.4.2.3   Example Applications

This protocol is best suited for time-critical applications such as intrusion detection, explosion detection, etc.

20.4.3   Hybrid Networks

There are applications in which the user might need a network that reacts immediately to time-critical situations and also gives an overall picture of the network at periodic intervals to answer analysis queries. Neither of the above networks can do both jobs satisfactorily and they have their own limitations.

Manjeshwar and Agrawal (2001) have combined the best features of the proactive and reactive networks, while minimizing their limitations, to create a new type of network called a *hybrid network*. In this network, the nodes not only send data periodically, but also respond to sudden changes in attribute values. A new routing protocol (Adaptive Periodic Threshold-sensitive Energy Efficient sensor Network Protocol; APTEEN) has been proposed for such a network and uses the same model as the above protocols but with the following changes. APTEEN works as follows.

20.4.3.1   Functioning

In each cluster period, once the cluster heads are decided, the cluster head broadcasts the following parameters (see Figure 20.11):

Attributes (A): a set of physical parameters about which the user is interested in obtaining data.

Thresholds: this parameter consists of a hard threshold (H_T) and a soft threshold (S_T). H_T is a particular value of an attribute beyond which a node can be triggered to transmit data. S_T is a small change in the value of an attribute that can trigger a node to transmit.

Schedule: a TDMA schedule similar to the one used in Heinzelman et al. (2000), assigning a slot to each node.

Count time (T_C): the maximum time period between two successive reports sent by a node. It can be a multiple of the TDMA schedule length and it introduces the proactive component in the protocol.

The nodes sense their environment continuously. However, only those nodes that sense a data value at or beyond the hard threshold will transmit. Furthermore, once a node senses a value beyond H_T, it next transmits data only when the value of that attribute changes by an amount equal to or greater than the soft threshold S_T. The exception to this rule is that if a node does not send data for a time period equal to the count time T_C, it is forced to sense and transmit the data, irrespective of the sensed value of the attribute. Because nodes near each other may fall into the same cluster and sense similar data, they may try sending their data simultaneously, leading to collisions between their messages. Hence, a TDMA schedule is used and each node in the cluster is assigned a transmission slot, as shown in Figure 20.11.

20.4.3.2   Important Features

It combines both proactive and reactive policies. By sending periodic data, it gives the user a complete picture of the network, like a proactive scheme. It behaves like a reactive network also by sensing data continuously and responding to drastic changes immediately.

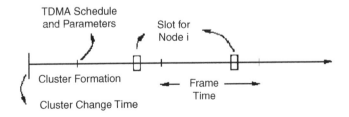

FIGURE 20.11   Timeline for APTEEN. (*Source:* A. Manjeshwar and D.P. Agrawal, APTEEN: a hybrid protocol for efficient routing and comprehensive information retrieval in wireless sensor networks, in *Proc. Int. Parallel and Distributed Processing Symp. (IPDPS'02) Workshops*, 2002.)

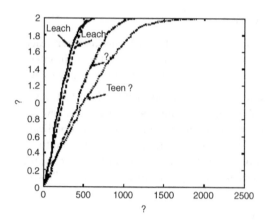

FIGURE 20.12 Comparison of average energy dissipation. (*Source:* A. Manjeshwar and D.P. Agrawal, APTEEN: a hybrid protocol for efficient routing and comprehensive information retrieval in wireless sensor networks, in *Proc. Int. Parallel and Distributed Processing Symp. (IPDPS'02) Workshops,* 2002.)

It offers a lot of flexibility by allowing the user to set the time interval (T_C) and the threshold values (H_T and S_T) for the attributes.

Energy consumption can be controlled by changing the count time as well as the threshold values.

The hybrid network can emulate a proactive network or a reactive network, based on the application, by suitably setting the count time and the threshold values.

The main drawback of this scheme is the additional complexity required to implement the threshold functions and the count time. However, this is a reasonable trade-off and provides additional flexibility and versatility.

20.4.4   A Comparison of the Protocols

To analyze and compare the protocols TEEN and APTEEN with LEACH and LEACH-C, consider the following metrics (Manjeshwar and Agrawal, 2001, 2002):

Average energy dissipated: shows the average dissipation of energy per node over time in the network as it performs various functions such as transmitting, receiving, sensing, aggregation of data, etc.

Total number of nodes alive: indicates the overall lifetime of the network. More importantly, it gives an idea of the area coverage of the network over time.

Total number of data signals received at BS: explains how TEEN and APTEEN save energy by not transmitting data continuously, which is not required (neither time critical nor satisfying any query). Such data can be buffered and later transmitted at periodic intervals. This also helps in answering historical queries.

Average delay: gives the average response time in answering a query. It is calculated separately for each type of query.

The performance of the different protocols is given in Figures 20.12 and 20.13.

20.5   Multiple Path Routing

The above protocols considered a hierarchical network; now consider multiple path routing in a flat sensor network. Multiple path routing aims to exploit the connectivity of the underlying physical networks by providing multiple paths between source destination pairs. The originating node therefore has a choice of more than one potential path to a particular destination at any given time.

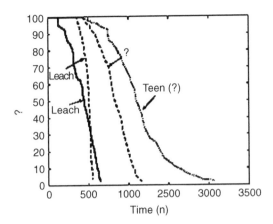

FIGURE 20.13 Comparison of the number of alive nodes, (*Source:* A. Manjeshwar and D.P. Agrawal, APTEEN: a hybrid protocol for efficient routing and comprehensive information retrieval in wireless sensor networks, in *Proc. Int. Parallel and Distributed Processing Symp. (IPDPS'02) Workshops*, 2002.)

20.5.1   The Need for Multiple Path Routing

Classical multiple path routing has been explored for two reasons. The first is *load balancing*, where traffic between the source and destination is split across multiple (partially or fully) disjoint paths to avoid congestion on any one path. The second use of multipath routing is to increase the probability of *reliable data delivery* due to the use of independent paths. To ensure reliable data delivery, duplicate copies of the data can be sent along alternate routes.

Multiple path routing is popularly used to avoid disparity in energy consumption in the network. This suggests that a multiple path scheme would be preferable when there are simultaneous active sources in the network with high traffic intensity. It is typical to have a number of overlapping source sink pairs unevenly distributed in the sensor field. It is a challenging problem to distribute traffic load evenly among a majority of the sensor nodes in the network with such random traffic conditions. Multiple path routing is cost effective in heavy load scenarios, while a single path routing scheme with a lower complexity may be more desirable when the numbers of packets exchanged between the random source sink pairs are few.

Load balancing is especially useful in energy-constrained networks because the relative energy level of the nodes does affect the network lifetime more than their absolute energy levels. With classic shortest path routing schemes, a few nodes that lie on many of these shortest paths are depleted of their energy at a much faster rate than the other nodes. As a result of these few dead nodes, the nodes in its neighborhood may become inaccessible, which in turn causes a ripple effect, leading to network partitioning. Chang and Tassiulas (2000) have proved, assuming each node to have a limited lifetime, that the overall lifetime of the network can be improved if the routing protocol minimizes the disparity in the residual energy of every node, rather than minimizing the total energy consumed in routing. In the multiple path routing protocol proposed by Jain et al. (2003a,b), the traffic is spread over the nodes lying on different possible paths between the source and the sink, in proportion to their residual energy. The rationale behind traffic spreading is that for a given total energy consumption in the network, every node should have spent the same amount of energy for data transmission. Their objective is to assign more load to underutilized paths and less load to overcommitted paths so that uniform resource utilization of all available paths can be ensured. They construct multiple paths of variable energy cost, and then design a traffic scheduling algorithm that determines the order of path utilization to enable uniform network resource utilization. They also grade the multiple paths obtained according to their quality-of-service (QoS), where the QoS metric is delay in response time (Jain et al., 2003a,b). Thus, they use a reservation-based scheme to provide a good service to time-critical applications, along with dynamic reallocation of network resources to noncritical applications to avoid underutilization of resources.

There is a need to adapt multiple path routing to overcome the design constraints of a sensor network. Important design considerations that drive the design of sensor networks are the energy efficiency and scalability (Hill et al., 2000) of the routing protocol. Discovery of all possible paths between a source and a sink might be computationally exhaustive for sensor networks because they are power constrained. In addition, updating the source about the availability of these paths at any given time might involve considerable communication overhead. The routing algorithm designed for a sensor network must depend only on local information (Ganesan et al., 2002a) or the information piggy-backed with data packets, as global exchange of information is not a scalable solution because of the sheer number of nodes.

20.5.2   Service Differentiation

Service differentiation is a basic way to provide QoS by giving one user priority over another. The data traffic is classified based on the QoS demands of the application. QoS parameters for typical Internet applications include bounds for bandwidth, packet delay, packet loss rate, and jitter. Certain additional parameters that deal with problems unique to wireless and mobile networks are power restrictions, mobility of nodes, and unreliable link quality. For a sensor network, node mobility is not high but severe power restrictions may force the packet to be routed through longer paths that have a higher residual energy than the shortest route connecting the source to the destination. INSIGNIA, a service differentiation methodology developed by Lee and Campbell (1998), employs a field in the packet to indicate the availability of resources and perform admission control. Such a testing is based on the measured channel capacity to the destination or utilization and the requested bandwidth. When a node accepts a request, resources are committed and subsequent packets are scheduled accordingly. If adequate resources are not available, then a flow adaptation is triggered to adjust the available resources on the new requested path.

Service requirements could be diverse in a network infrastructure. Some queries are useful only when they are delivered within a given timeframe. Information provided may have different levels of importance; therefore, the sensor network should be willing to spend more resources in disseminating packets carrying more important information. Service differentiation is popularly used in the Internet (Vutukury, 2001) to split the traffic into different classes based on the QoS desired by each class. In the multiple path routing protocol proposed by Jain et al. (2003b), a priority and preemption mechanism is used to control end-to-end delay for time-critical queries. Some specific examples of applications that could benefit from a sensor network supporting service differentiation are described here.

   Battlefield surveillance. Soldiers conduct periodic monitoring for situational awareness of the battlefield. If the network senses some unusual or suspicious activity in the field that requires immediate attention of the military personnel, an alarm is triggered. These warning signals must reach the end user or the soldier immediately to expedite quick decision making.

   Disaster relief operations. In case of natural calamities such as floods, wild fires, tornadoes, or earthquakes, or other catastrophes such as terrorist attacks, coordinated operation of sensor nodes could be very useful in conducting efficient rescue operations. Precise information about the location of victims or environmental parameters of risky areas could provide facts that help in the planning of rescue operations. Here, a flexibility to prioritize information retrieval could be beneficial in avoiding communication delays for time-critical responses.

   Infrastructure security. A network of sensors could be deployed in a building or a campus that needs to be secured from any intrusion detection. The sensors could be programmed to discriminate among the attacks if they occur simultaneously or associate higher priority for packets confirming intrusion, as compared to other normal data packets containing information related to the usual background monitoring of the building for parameters like light intensity, temperature, or the number of people passing by.

Environmental or biomedical applications. Monitoring presence of certain gases or chemicals in remote areas such as mines, caves, or under water, or in chambers where research experiments are carried out, or radiation levels in a nuclear plant. Alarm signals might be required infrequently to report the presence of attributes in a volume or degree at more than an expected threshold. These warning messages, if received on time, can help in accomplishing the research goals or desired monitoring operations.

20.5.3 Service Differentiation Strategies for Sensor Networks

Bhatnagar et al. (2001) discussed the implications of adapting these service differentiation paradigms from wired networks to sensor networks. They suggest the use of adaptive approaches; the sensor nodes learn the network state using eavesdropping or by explicit state dissemination packets. The nodes use this information to aid their forwarding decisions; for example, low-priority packets could take a longer route to make way for higher-priority packets through shorter routes. The second implication of their analysis is that the applications should be capable of adapting their behavior at runtime based on the current allocation, which must be given as a feedback from the network to the application.

In our work (Manjeshwar and Agrawal, 2003a,b), we aim to achieve twofold goals of the QoS-aware routing described by Chen et al. (1998). The two goals are (1) selecting network paths that have sufficient resources to meet the QoS requirements of all admitted connections, and (2) achieving global efficiency in resource utilization. In ad hoc networks, static provisioning is not enough because node mobility necessitates dynamic allocation of resources. In sensor networks, although user mobility is practically absent, dynamic changes in the network topology may be present because of node loss due to battery outage. Hence, multi-hop ad hoc routing protocols must be able to adapt to the variation in the route length and its signal quality while still providing the desired QoS. It is difficult to design provisioning algorithms that achieve simultaneously good service quality as well as high resource utilization. Because the network does not know in advance where packets will go, it will have to provision enough resources to all possible destinations to provide high service assurance. This results in severe underutilization of resources.

20.6 Continuous Queries in Sensor Networks

Multiple path routing is used for uniform load distribution in a sensor network with heavy traffic. Now consider energy optimization techniques for sensor networks serving continuous queries. Queries in a sensor network are spatio-temporal; that is, the queries are addressed to a region or space for data, varying with time. In a monitoring application, the knowledge of the coordinates of the event in the space, and its time of occurrence, is as important as the data itself. Queries in a sensor network are usually location based; therefore, each sensor should be aware of its own location (HighTower and Borreillo, 2001). When self-location by GPS is not feasible or too expensive, other means of self-location, such as relative positioning algorithms (Doherty et al., 2001), can be used. A timestamp associated with the data packet reveals the temporal property of data.

Depending on the nature of the application, the types of queries injected in a sensor network can vary. Queries posed to a sensor network are usually classified as one-time queries or periodic queries. "One-time" queries are injected at random times to obtain a snapshot view of the data attributes, but "periodic" queries retrieve data from the source nodes after regular time intervals. An example of a periodic query is, "Report the observed temperature for the next week at the rate of one sample per minute." We now concentrate on periodic queries that are long running; that is, they retrieve data from the source nodes for a substantially long duration, possibly the entire lifetime of the network. We now classify queries into three different categories based on the nature of data processing demanded by the application.

Simple queries. These are stand alone queries that expect an answer to a simple question from all or a set of nodes in the network. For example, "Report the value of temperature."

Aggregate queries. These queries require collaboration among sensor nodes in a neighborhood to aggregate sensor data. Queries are addressed to a target region consisting of many nodes in a geographically bounded area instead of individual nodes. For example, "Report the average temp of all nodes in region X."

Approximate queries. These are queries that require data summarization and rely on synopsis data structures to perform holistic data aggregation (Ganeson et al., 2002a) in the form of histograms, isobars, contour maps, tables, or plots. For example, "Report the contours of the pollutants in the region X." Such sophisticated data representation results in a tremendous reduction in data volume at the cost of additional computation at nodes. Although offline data processing by the user is eliminated, due to a lack of raw sensor data, the user may not be able to analyze it later in ways other than the query results.

Complex queries. If represented in SQL, these queries would consist of several joins nested or condition-based sub-queries. Their computation hierarchy is better represented by a query tree. For example, "Among regions X and Y, report the average pressure of the region that has higher temperature." Sub-queries in a complex query could be simple, aggregate, or approximate queries. In our proposed work, we design a general in-network query processing architecture for evaluating complex queries.

20.6.1   The Design of a Continuous Query Engine

We now present the four basic design characteristics of a continuous query processing architecture:

Data buffering. It is required to perform blocking operations (Babcock et al., 2002) that need to process an entire set of input tuples before an output can be delivered. Examples of blocking operators are GroupBy, OrderBy, or aggregation functions such as maximum, count, etc. A time-based sliding window proposed by Datar et al. (2002) is used to move across the stream of tuples and restrict the data buffered at a node at any instant of time for data processing. Time synchronization to process data also requires data buffering (Motwani et al., 2003). If the data streams arriving from different sources must be combined based on the times each tuple is generated, then synchronization between sensor nodes along the communication path is required. To ensure temporal validity of the results produced while evaluating operators, tuples arriving from the faster stream should be buffered until the tuples in the same time window belonging to the other input stream reach the QP node evaluating the operator.

Data summarization. If data is arriving too fast to be processed by the node, it will not be able to hold the incoming tuples in its limited memory. Therefore, tuples might have to be dropped or a sample of tuples could be selected from the data stream to represent the entire data set (Carney et al., 2002).

Sharing. It is important to share processing whenever feasible among multiple queries for maximizing the reuse of resources. This is used to achieve scalability with increasing workload.

Adaptability. In spatio-temporal querying, the number of sources and their data arrival rates may vary during a query's lifetime, thereby rendering static decisions ineffective. Hence, the continuous query evaluation (Madden et al., 2002) should keep adapting to changes in data properties.

20.6.2   Applications of an Adaptive Continuous Query Processing System

Distributed query processing has numerous applications in remote monitoring tasks over wireless sensor networks. Described here some of the real-world applications that will benefit from the proposed query processing architecture. These examples illustrate the range of applications that can be supported by the proposed query processing architecture. We classify potential commercial applications of wireless sensor networks into two broad classes.

Infrastructure-based monitoring. Sensor nodes are *attached to an existing physical structure* to remotely monitor complex machinery and processes, or the health of civil infrastructures such as highways, bridges, or buildings.

Field-based monitoring. A space in the environment is monitored using a dense network of *randomly scattered* nodes that are not particularly installed on any underlying infrastructure. Environmental monitoring of ecosystems, toxic spills, and fire monitoring in forests are examples of field monitoring.

We now describe infrastructure-based and field-based monitoring in detail.

20.6.2.1 Infrastructure-Based Monitoring

Commercial interest in designing solutions for infrastructure-based monitoring applications using sensor networks is growing at a fast pace. EmberNet (Ember Corporation; www.ember.com) has developed an embedded networking software for temperature sensing and heat trace control using wireless sensors. This drastically reduces the installation cost when the number of temperature monitoring points runs into thousands.

We propose that the efficiency of the monitoring task can be enhanced by deploying additional powerful nodes that are able to process the temperature readings and draw useful inferences within the network. As results occupy fewer data bytes than raw data, data traffic between the monitoring points and the external monitoring agency can be reduced. Data is routed faster and processing time at the external monitoring agency is saved, which enables quicker decision making based on the results of data monitoring. Another potential example of infrastructure-based monitoring is *detection of leakage* in a water distribution system. A self-learning, distributed algorithm can enable nodes to switch among various roles of data collection, processing, or forwarding among themselves so that the network continues retrieving data despite the failure of a few nodes. Human intervention would be deemed necessary only when a majority of the nodes in an area malfunction or run out of battery power. Some other infrastructure-based applications that might benefit from the proposed query processing and the resulting real-time decision support designed in this work are as follows:

Civil structure/machine monitoring. Continuous monitoring of civil structures such as bridges or towers yields valuable insight into their behavior under varying seismic activity. By examining moisture content and temperature, it is possible to estimate the maturity of concrete or a corrosive subsurface in structural components before serious damage occurs. Similar principles apply to underground pipes and drainage tiles. Another relevant application is monitoring the health of machines. Thousands of sensor nodes can track vibrations coming from various pieces of equipment to determine if the machines are about to fail.

Traffic monitoring on roadways. Sensors can be deployed on roads and highways to monitor the traffic or road conditions to enable quick notification of drivers about congestion or unfavorable road conditions so that they can select an alternate route.

Intrusion detection. Several sensor nodes can be deployed in buildings at all potential entrances and exits of the building to monitor movement of personnel and any unusual or suspicious activity.

Heat control and conditioning. Precise temperature control is very crucial for many industries, such as oil refineries and food industries. This can be achieved by placing a large number of sensors at specific places in the civil structure (such as pipes).

Intel (2003) has developed a heterogeneous network to improve the scalability of wireless sensor networks for various monitoring applications. Intel overlays a 802.11 mesh network on a sensor network analogous to a highway overlaid on a roadway system. Data is collected from local sensor nodes and transmitted across the network through the faster, more reliable 802.11 network. The network lifetime is therefore enhanced by offloading the communication overhead to the high-end 802.11 nodes. This network is capable of self reorganization in case any 802.11 node fails. Similarly, we propose to offload computation-intensive tasks from low-power sensor motes to high-performance nodes.

A few important inferences that can be drawn about infrastructure-based applications are as follows:

The layout (blueprint) of the infrastructure (like a machine or a freeway) where the sensor nodes are to be placed is known.

As nodes may have been manually placed or embedded on the structure at monitoring points while manufacturing it, the location of sensors may be known to the user.

It might be possible to provide a renewable source of power to some of the nodes.

The communication topology should adapt to the physical limitations (shape or surface area where nodes are placed) of underlying infrastructure.

20.6.2.2 Field-Based Monitoring

To perform field-based monitoring, the sensor nodes are randomly dispersed in large numbers to form a dense network to ensure sensor coverage of the environment to be monitored. The purpose is to observe physical phenomena spread over a large region, such as pollution levels, the presence of chemicals, and temperature levels. An example query for field monitoring would be: "Report the direction of movement of a cloud of smoke originating at location (x, y)." The purpose is to monitor the general level of physical parameters being observed — unlike infrastructure-based monitoring, which is usually applied for high-precision monitoring. The use of a heterogeneous network for field monitoring is not very obvious because node placement in the field usually cannot be controlled. But, at the expense of increasing the cost of deployment, the density of high-performance nodes can be increased to ensure that most of the low-power nodes in the network can access at least one high-performance node. For example, in a greenhouse, the same plant is grown in varying soil or atmospheric conditions, and the growth or health of the plant is monitored to determine the factors that promote its growth. The different soil beds can be considered the different target regions, and their sensor data is compared or combined with each other within the field to derive useful inferences to be sent to a remote server. Some other applications that might benefit from such an automated system of data monitoring are as follows.

Scientific experiments. Sensors can be randomly deployed in closed chambers in laboratories, or natural spaces such as caves or mines, to study the presence of certain gases, elements, or chemicals. Similarly, levels of radioactive materials can be observed to monitor toxic spills.

Examination of contaminant level and flow. The sensor nodes with chemical sensing capabilities can be used to monitor the levels and flow patterns of contaminants in the environment.

Habitat or ecosystem monitoring. Studying the behavior of birds, plants, and animals in their natural habitats using sensor networks has been employed on a small scale (Mainwaring et al., 2002).

Wild fire monitoring. This is particularly useful in controlling fires in forests by studying the variation in temperature over the areas affected, and the surrounding habitat.

20.7 Mobile Sensor Systems

In-situ sensors have been the traditional monitoring method for sensor networks. Integrating data from all these sensors is time-consuming and tedious. When data is analyzed, it is a view of what was and not what to expect, which is a major drawback. Hence, there is a need to have on-site configuration, and rapid collection and integration of data using mobile outfits. Sensors can be mounted onto mobile devices (e.g., a PDA or a robot) to enable mobility. These provide flexibility in data collection for dynamic and spatially extensive environments. Portable Palm pilots and PDAs are now replacing the traditional data collectors that have on-board storage. These are lightweight and long-standing devices that can work without GPS capabilities. At remote sites, mobile data collectors can provide precise field data and metadata information to the users.

20.7.1 Characteristics of Mobile Sensor Systems

There are some features that are inherent to a *mobile sensor network*. The mobile nodes must be amenable in a distributed setting. The network should be scalable with respect to communication and computation complexity. It is also important for the mobile nodes to be adaptive to hostile surroundings (see Figure 20.14). The nodes must also be reliable and the scheme involved should be asynchronous.

FIGURE 20.14   Sensors can be mounted on hand-held devices. (*Source:* www.ia.hiof.no/prosjekter/hoit/html/nr2_02/ grafikk/palm-pilot.jpg)

Consider a sensor network with mobile robots — a network architecture for low-power and large-scale sensor networks. This network has two types of nodes: (1) sensors and (2) mobile robots. Sensors are low-power and low-cost nodes that have limited processing and communication capability. They are deployed in large quantity, perhaps randomly through aerial drops covering the entire network. The mobile robots are powerful hardware nodes, both in their communication or processing capability and in their ability to traverse the network. Mobile robots perform information retrieval and post processing (see Figure 20.15).

These nodes can sense various physical phenomena such as light, temperature, humidity, chemical vapors, and sound. They are deployed in large quantities, perhaps randomly through an aerial drop, covering a large area and constituting a single network. Any sensor node detecting interesting data reports to a predetermined location that receives information. This location is termed a *sink*.

20.7.2   Need for Mobile Sensor Networks

In general, sensors are low-power and low-cost nodes and the energy consumed in communicating their data is much more than that for complex computations. For these reasons, the network dies when a majority of the sensors run out of energy. Robots have high energy or can be renewed periodically because they are mobile and often report to the sink. Excellent communication between robots and sensor network is desirable for precise movement and actions. The collaboration between mobile robots and sensor networks is a key factor in achieving efficient transmission of data, network aggregation, quick detection of events, and timely action by robots. Coordination between multiple robots for resource transportation has been explored for quite some time. Transporting various types of resources for different applications such as

FIGURE 20.15   Mobile sensing environment. (*Source:* www.isiindustrysoftware.com/main.html)

defense, manufacturing process, etc. has been suggested (Vaughan et al., 2000). In these schemes, the time taken to detect an event depends entirely on the trail followed by the robots. Although the path progressively gets better with the use of an ant-type algorithm (Hayes et al., 2003; Vaughan et al., 2000), the whole process must be started anew when the position of the event changes. Suitable algorithms for the detection of events at any point in the network have not been formulated. Another drawback in these systems is that there are no sensors that guide them toward the event. Two visually guided robots can simultaneously carry resources from place to place (Schenker et al., 2001). Constant information exchange is mandatory here and the use of a single or multiple robots (more than two) has been overlooked.

Enhancements in the field of robotics are paving the way for industrial robots to be applied to a wider range of tasks. Newer robots are being developed with better control, safety measures, guidance, and robust sensing. Various mechanisms for producing dexterous motion, safety, and issues related to coordinating multiple robots have also been taking giant strides in neoteric times. Thus, it seems very roseate to look in terms of merging the two fields of sensor networks and robotics.

Robots are rapidly moving from the pages of science fiction novels to everyday life. The enhancements in the field of robotics are paving the way for industrial robots to be applied to a wider range of tasks. Advances in materials and technology have made modern robots much smaller, much lighter, and more precise, which means that there can be more applications of these robots than previously envisioned. However, harnessing their full efficiency also depends on how accurately they understand their environment. To measure physical parameters of the surroundings, different sensor devices can be applied so that useful information can be procured. Also, to get a global view, each robot needs to retrieve and aggregate information from sensors, while sensors themselves can exchange information using wireless devices.

20.7.3   Applications of Mobile Sensor Networks

Gupta et al. (2004) have analyzed the possibility of coalescing sensor networks with mobile robots. Multiple mobile robots and their coexistence were examined, while the relative merits and demerits were also evaluated. The cooperation between the mobile robots and sensor networks is a key factor in achieving speedy in-network aggregation and transmission of data. The scheme (Gupta et al. 2004) is very efficient for monitoring queries. The authors claim that the frontiers of sensor networks will be significantly advanced by enabling mobile robots to herd them. Low-power sensor nodes are used to detect an event and guide mobile robots to such locations. These robots have been modeled as resource-carrying ones and are used to transport resources within the network. The symbiosis of two independently powerful spheres leads to the overall efficiency of the network. The adroitness of the mobile robots forms the backbone of our proposed scheme and their synchronization forms the communication channel of the network (see Figure 20.16).

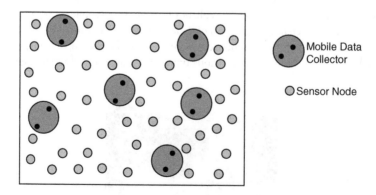

FIGURE 20.16   Heterogeneous sensor network. (*Source:* A.K. Gupta, S. Sekhar, and D.P. Agrawal, Efficient event detection by collaborative sensors and mobile robots, in *Ohio Graduate Symp. on Computer and Information Science and Engineering*, Dayton, OH, June 2004.)

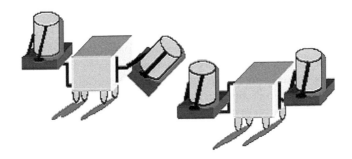

FIGURE 20.17 A resource carrying robot. (*Source:* A.K. Gupta, S. Sekhar, and D.P. Agrawal, Efficient event detection by collaborative sensors and mobile robots, in *Ohio Graduate Symp. on Computer and Information Science and Engineering,* Dayton, OH, June 2004.)

In terrains where human ingress is difficult, we use mobile robots to imitate the human's chore. These shoebox-sized robots are vibrant with energy, immune to poison, impervious to pain, vacuum resistant, hunger tolerant, invulnerable to sleep, etc. Typical resource-carrying robots are depicted in Figure 20.17.

Figure 20.18 shows a robot transferring its resource to another. These robots have the capability to carefully transport their contents and transfer their resources to another. Once depleted of their resource, they can get themselves refilled from the sink, which is a local reservoir of resources. The resource in demand could be water or sand (to extinguish fire), oxygen supply, medicine, bullets, clothes, chemicals to neutralize hazardous wastes, etc.

The choice of resource is limited only by the carrying capacity of the robots. The nature of the resource depends on the application that the network supports. Hence, one can see that there are a number of applications where sensors and robots could work together. In all these cases, the important factor is that the whole process is self-organizing without any external surveillance. Sensors detect events autonomously and the mobile robots take appropriate actions based on the nature of the event. The performance of the system considerably increases when continued assistance to the event location is needed. Coordination between the mobile robots is critical in achieving better network efficiency. These networks can be implemented in various applications, such as defense, locating a user, environmental monitoring, fire fighting, rescue operations, tele-monitoring, damage detection, traffic analysis, etc.

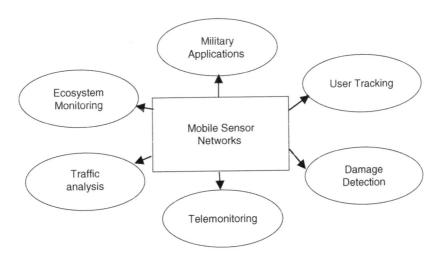

FIGURE 20.18 Applications of mobile sensor networks. (*Source:* A.K. Gupta, S. Sekhar, and D.P. Agrawal, Efficient event detection by collaborative sensors and mobile robots, in *Ohio Graduate Symp. on Computer and Information Science and Engineering,* Dayton, OH, June 2004.)

Caching on two independently powerful spheres, we propose a scheme that conjoins them. Both the entities try to complement each other and the pitfall of each field is covered by the other. This directly leads to an improved network lifetime. The possible applications of this scheme are manifold. Because the network has a stable lifetime, it could be deployed in many areas that need continuous sensing. Because the field of robotics is ever-improving, one can expect the network to perform better with time. Unlike traditional networks, data collection can be done periodically or intermittently as per the needs of the system. Loss of data is also largely alleviated. Very efficient and automated surveillance and monitoring systems will be of significant value in the near future and that is exactly what our system achieves! The adroitness of these mobile robots forms the backbone of the scheme and their synchronization forms the communication channel of the network.

20.8   Security

Before proceeding further, it should be noted that errors in sensor measurements are inevitable. Thus, to achieve high-fidelity measurements from any sensor, there is a need to map from raw sensor readings to the correct values before the decision process is invoked. However, in addition to erroneous sensor readings, there could be malicious or compromised nodes in a network and, hence, security measures are also important. Research on security mainly focuses on the creation of a trustworthy, efficient, and easy-to-manage ad hoc scenario. Any security infrastructure essentially consists of two components. The first involves intrusion prevention and the second concerns intrusion detection. We focus on both and also look at schemes for security for group communications and secure routing. The following sections look at the various aspects of research into these various domains.

20.8.1   Security for Group Communication

Security for group communication essentially requires managing keys such that common keys are shared and the rekey overhead is minimized. To this end, a multitude of schemes have been suggested.

Group communication has a few inherent issues. First, when a node joins the group for the first time, it should receive a communication key shared by each of the other members of the group. However, providing it with the past keying material would mean that the node would be able to decipher conversation to which it is not entitled. Thus, a new set of keying material must be distributed. A leave also results in a similar problem, however with magnified impact. This problem has been solved by providing a scheme (Mukherjee et al. 2004) in which the problem of rekeying is carried out at each individual step and in a level-based manner.

In Mukherjee et al. (2004), a node is assigned a "level" as soon as it joins the network. This procedure also establishes a level key between the node and its parent. The level key is the same as the one that the parent shares with its other children. The parent node passes on its level key to any node that wishes to join it. We thus have a tree in which each parent-children group shares the same common level key. In our scheme, an entire area need not be rekeyed whenever a "transfer" occurs, thereby drastically reducing the rekeying overhead. The multicast group key update takes place by repeated decryption and encryption of the multicast key at each level. We thus achieve low communication overhead (because the update message would travel only once along each link); low latency (by the assumption that the routing layer has some cross-layer information and can do this encryption-decryption process very fast without the packet going to the application layer); effective handling of user mobility (by introducing mobility parameters, as shown later); and most importantly, a security framework that is entirely distributed in nature.

Mukherjee et al. (2004) defined a parameter called *self-mobility* for a node based on the change in the neighborhood of node in a specified time period. Based on this, they calculated a combined mobility factor. This factor enables one to predict a join or a leave operation's overhead consequences with high accuracy. A communicating entity wishing to leave or join the system would make a decision to join the multicast or group tree so as to have minimal post join or leave rekey overheads. Simulation results show very positive results as well as a drastic reduction in the keying overhead.

20.8.2   Key Exchange

A major problem in tackling security issues in ad hoc networks is the absence of any central authority. Without any central body to distribute the key material, nodes joining a network need to get their key material from the network itself. A lot of assumptions are made about the structure of the network, thus leading to a multitude of solutions.

20.8.2.1   ID-Based Key Exchange

The central idea behind an ID-based key exchange process is that a node joining the network gets its keying material according to its ID. The primary assumption here is that a central authority initializes a group of communicating nodes with the keying material and these nodes provide shares for the keying material to any node that joins the network.

Deng et al. (2004) have proposed an asymmetric keying mechanism where the node's ID is considered its public key. A major issue with most asymmetric key mechanisms is the distribution of the private keys. This problem is solved by the ID itself becoming the public key. As soon a node joins a network, it sends out a share request to all its neighbors. The neighbors respond to the request by calculating a share using a master key share and the requesting node's ID. It then sends this share to the requesting node. The requesting node now combines all the shares to form a private key. The public key is the node's ID, as previously mentioned.

20.8.2.2   Group Key Mechanisms

Next, consider a group key generation mechanism. The idea here is mainly based on the Diffie-Hellman primitive. Ding et al. (2004) have defined new trust relationships.

Ding et al (2004) have suggested a novel method to carry out distributed key exchange in such a manner that each user now introduces a separate level of trust into the system for itself. Most authentication schemes treat the problem of authentication as separate from that of communication. Their contention is that for highly secure networks, a user should be able to choose a level of trust for the party he wishes to communicate with, and the authentication process should be embedded with the key exchange process.

The basic Diffie-Hellman primitive derives itself from the fact that factorization is hard in a finite field. Thus, if two parties wish to exchange a symmetric key, they exchange numbers and raise these numbers to the power of a secret number.

Ding et al. (2004) also introduced the notion of a body of servers. These servers are responsible for authenticating and distributing key materials to an incoming member. Trust assumptions are such that no node that has not been authenticated by this set of servers would be able to communicate with this node. Thus, we establish the notion of distributed trust.

As future work we can come up with schemes to reduce the number of messages required. One way in which this can be done is to have neighbor graphs to direct a node to choose servers that are close to it and not incur too much communication overhead on the network. Also, traffic analysis might be done and key messages can be piggy-backed on other data messages. A third way is to have servers maintaining multicast groups to which messages might be sent, thereby saving the overhead of unicasting to each server separately. A possible extension would be to apply the scheme to real-life scenarios involving multiple security levels for different users. Finally, we would like to make our system fool-proof by coming up with schemes to detect whether or not a server sends the shares of the secret key honestly. This can be done by either introducing intrusion detection mechanisms or by having some kind of a mechanism to see if the obtained shares satisfy a set of criteria.

We can also build robustness into the distributed keying mechanisms by building validity verifier protocols. Thus, when the key shares are generated in a distributed fashion and compromised nodes lie about the keying material, there should be a way in which the incoming nodes can detect the lies and work toward omitting the wrong values and also pinpoint the malicious node, thereby contributing to intrusion detection.

Furthermore, the symmetry in the polynomials can be used to derive symmetric keys. Thus, multivariate symmetric polynomials, partially evaluated by a central authority, can be utilized to generate pair-wise or group keys between nodes. We have exploited this fact and are in the process of making the process distributed.

Finally, polynomial-based hierarchical keying mechanisms can be considered. The evaluation of the polynomials would be such that the higher the tiers, the more information would be provided to the nodes. Thus, the lowest tier nodes would have much less information and the nodes at the upper tiers would have a lot of information.

20.8.3 Secure Routing Schemes

Secure routing schemes concern protocols in which routing messages are neither tampered with nor dropped.

In the main CBRP routing scheme, routing is carried out in a clustered manner. A central cluster head is responsible for the routing decisions in each cluster. It is our contention that this scheme can fall victim to severe routing attacks if the cluster head is compromised.

Ojha et al. (2204) and Poosarla et al. (2004) have come up with schemes to prevent this from happening. A cluster head is not just one node but a collection of nodes. These nodes together form a COUNCIL. Any routing decision that needs to be taken must be done after an agreement with this COUNCIL. The structure of the COUNCIL is such that compromise of less than t members will not compromise the system. Thus, by choosing a suitable threshold value for the COUNCIL, a robust routing algorithm can be devised.

20.8.4 Intrusion Detection

Next consider intrusion detection. Here, the primary goal is to indicate if malicious activity goes around in an ad hoc network. In considering this problem, there are two issues to solve. The first concerns the issue of selecting parameters to monitor and the second concerns getting statistics for the network.

As far as the first issue is concerned, we first identified the various forms of attacks that plague an ad hoc network. These involve blackhole, wormhole, false routes, extra data packets, gracious detour, etc. Based on these attacks, we identified parameters that might indicate anomalies in the network. Deng et al. (2003a,b,c) have mainly focused on indicating the black hole attack.

The second issue involves the actual gathering of statistics. In a practical scenario, an ad hoc node would not have enough computational power to carry out the various calculations for anomaly detection. Thus, a dimension reduction scheme is needed. Two dimension reduction schemes can be considered: (1) random projection and (2) the use of support vector machines.

20.9 Middleware Infrastructure for Sensor Networks

Middleware sits between the operating system and the application. On traditional desktop computers and portable computing devices, operating systems are well established, both in terms of functionality and systems. For sensor nodes, however, the identification and implementation of appropriate operating system primitives is still a research issue. Hence, at this early stage, it is not clear on which basis future middleware for sensor networks can typically be built (Römer et al., 2002).

The main purpose of middleware for sensor networks is to support the development, maintenance, deployment, and execution of sensing-based applications. This includes mechanisms for formulating complex, high-level sensing tasks; communicating this task to the sensor network; coordination of sensor nodes to split the task and distribute it to the individual sensor nodes; data fusion for merging the sensor readings of the individual sensor nodes into a high-level result; and reporting the result back to the task issuer.

There are already some projects to develop middleware for sensor networks. Cougar (www.cs.cornell.edu/database/cougar) adopts a database approach wherein sensor readings are treated like virtual relational database tables. The Smart Messages Project (www.rutgers.edu/sm) is based on agent-like messages containing code and data that migrate through the sensor network. NEST (www.cs.virginia.edu/nest) provides microcells that are similar to operating system tasks with support for migration, replication, and grouping. SCADDS (Scalable Coordination Architectures for Deeply Distributed Systems; www.isi.edu/dov7/scadds) is based on *directed diffusion*, which supports robust and energy-efficient delivery and in-network aggregation of sensor events. However, most of these projects are in an early stage, focusing on developing algorithms and components that might later serve as a foundation for middleware for sensor networks.

20.10 Future Challenges of Sensor Networks

Sensor networks represent a paradigm shift in computing. Traditional computing involves computers directly interacting with human operators. However, in the near future, hundreds of computers will be embedded deep in the world around us. When we are in control of hundreds or thousands of computers each, it will be impossible for us to interact directly with each one. On the contrary, the computers themselves will interact more directly with the physical world. They will sense their environments directly, compute necessary responses, and execute them directly (CSTB Publications, 2001).

For example, the networks in current cars are highly engineered systems in which each microprocessor and the overall network are carefully designed as a whole. However, as the complexity of the network and the functionality of the networked elements grow, the ability to approach the networks as single, fully engineered, closed systems becomes strained. For example, owners might want to integrate their own devices into the car viz. integrating the address book in a PDA with the navigation system in the car. The major automobile companies plan to change the car from a self-contained network (or pair of networks) into a node in a much larger network. One approach to this is General Motors' immensely successful OnStar offering. OnStar connects the car to the manufacturer, allowing the latter to monitor emergency situations and give on-demand help to the occupants of the car.

Intel (www.intel.com/research/exploratory/digital_home.htm) is contributing to the development of digital home technologies for aging in place by linking together computers and consumer electronic (CE) devices throughout the home in a wireless network. Once the digital home infrastructure is in place, any computer or CE device could be used to deliver health and wellness applications. Older adults will be able to access these applications through whatever interfaces are most familiar to them, from phones to PCs to televisions; they will not have to learn new technology. The goal is to have a variety of interfaces distributed throughout the home, all within easy reach of the person needing assistance.

Sensors 2000 has developed an easy-to-implement wireless biotelemetry system — called the Wireless On-Patient Interface for Health Monitoring (WOPI) — that can nonintrusively measure the health parameters of humans and animals in space (see Figure 20.19). The device's sensors, which are connected to miniature transceiver modules, are implanted, ingested, or attached to the body with Band-Aids. Sensors communicate with a belt-worn device that retransmits or records the data and also sends basic commands to each sensor. The device also displays a quick status of all physiological and biological parameters (see Figure 20.20). The technology also has potential applications in athletics and emergency response activities.

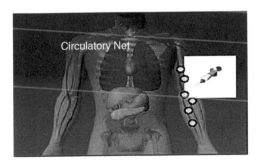

FIGURE 20.19   Biosensors. (*Source:* www-lia.deis.unibo.it/Courses/RetiLS/seminari/WSN.pdf)

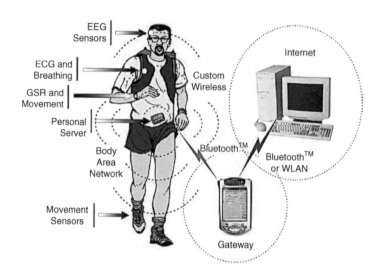

FIGURE 20.20   Wireless body area network of intelligent sensors in the telemedical environment. (*Source*: E. Jovanov, A. O'Donnell Lords, D. Raskovic, P. Cox, R. Adhami, and F. Andrasik, Stress monitoring using a distributed wireless intelligent sensor system, *IEEE Eng. in Medicine and Biology Magazine*, May/June 2003.)

References

1. D.P. Agrawal and Q. Zeng, *Introduction to Wireless and Mobile Systems*, Brooks/Cole Publishing, 436 pp., 2003.
2. R. Avnur and J.M. Hellerstein, Eddies: continuously adaptive query processing, in *Proc. 2000 ACM SIGMOD Int. Conference on Management of Data*, May 2000, pp. 261–272.
3. B. Babcock, S. Babu, M. Datar, R. Motwani, and J. Widom, Models and issues in data stream systems, in *Proceedings of the 2002 ACM Symposium on Principles of Database Systems*, June 2002, pp. 1–16.
4. S. Bhatnagar, B. Deb, and B. Nath, Service differentiation in sensor networks, in *Proc. Fourth Int. Symp. Wireless Personal Multimedia Communications*, 2001.
5. B.J. Bonfils and P. Bonnet, Adaptive and decentralized operator placement for in-network query processing information processing in sensor networks, *Second Int. Workshop, IPSN 2003*, Palo Alto, CA, April 22–23, 2003.
6. P. Bonnet, J.E. Gehrke, and P. Seshadri, Towards sensor database systems, in *Proc. Second Int. Conf. on Mobile Data Management*, Hong Kong, January 2001.
7. E. Brewer, R. Katz, and E. Amir, A network architecture for heterogeneous mobile computing, *IEEE Personal Communications Magazine*, October 1998.
8. J. Broch, D. Maltz, D. Johnson, Y. Hu, and J. Jetcheva, A performance comparison of multi-hop wireless ad hoc network routing protocols, in *Proc. 4th Annual ACM/IEEE Int. Conf. Mobile Computing (MOBICOM)*, ACM, October 1998.
9. Campbell Scientific, Inc., Measurement and Control Systems. Web page. www.campbellsci.com.
10. D. Carney, U. Cetintemel, M. Cherniack, C. Convey, S. Lee, G. Seidman, M. Stonebraker, N. Tatbul, and S.B. Zdonik, Monitoring streams—a new class of data management applications, in *Proc. 28th VLDB*, 2002.
11. A. Cerpa, J. Elson, D. Estrin, L. Girod, M. Hamilton, and J. Zhao, Habitat monitoring: application driver for wireless communications technology, *2001 ACM SIGCOMM Workshop on Data Communications in Latin America and the Caribbean*, Costa Rica, April 2001.
12. S. Chandrasekaran, O. Cooper, A. Deshpande, M.J. Franklin, J.M. Hellerstein, W. Hong, S. Krishnamurthy, S.R. Madden, V. Raman, F. Reiss, and M.A. Shah, TelegraphCQ: continuous dataflow

processing for an uncertain world, *1st Biennial Conf. Innovative Data Systems Research (CIDR 2003)*, January 2003.

13. J. Chang and L. Tassiulas, Energy conserving routing in wireless ad-hoc networks, *Proc. IEEE INFO-COM*, pp. 22–31, 2000a.

14. J. Chang and L. Tassiulas, Maximum lifetime routing in wireless sensor networks, in *Proc. Advanced Telecommunications and Information Distribution Research Program*, 2000b.

15. J. Chen, P. Druschel, and D. Subramanian, An efficient multipath forwarding method, *Proc. IEEE INFOCOM*, 1998, pp. 1418–1425.

16. S. Chen and K. Nahrstedt, Distributed quality-of-service routing in ad-hoc networks, *IEEE Journal on Special Areas in Communications*, 17(8), August 1999.

17. M. Chiang, D. O'Neill, D. Julian, and S. Boyd, Resource allocation for QoS provisioning in ad hoc wireless networks, *Proc. IEEE GLOBECOM*, San Antonio, November 2001, pp. 2911–2915.

18. Chu, H. Haussecker, and F. Zhao, Scalable information-driven sensor querying and routing for ad hoc heterogeneous sensor networks, in *Int. J. High Performance Computing Applications*, 2002.

19. I. Cidon, R. Rom, and Y. Shavitt, Analysis of multi-path routing, in *IEEE/ACM Transactions on Networking*, 7(6), 885–896, December 1999.

20. CITRIS, Center for Information Technology Research in the Interest of Society. Web Page. www.citris-uc.org.

21. Compaq iPAQ. Web page. www.compaq.com/products/iPAQ.

22. Cougar Project. Web page. www.cs.cornell.edu/database/cougar.

23. CSTB Publications, Embedded, everywhere: a research agenda for networked systems of embedded computers, CSTB Publications, 236 pages, ISBN: 0-309-07568-8, 2001.

24. Datar, A. Gionis, P. Indyk, and R. Motwani, Maintaining stream statistics over sliding windows, *Proc. Thirteenth Annu. ACM-SIAM Symp. on Discrete Algorithms*, pp. 635–644, January 6–8, 2002, San Francisco, CA.

25. H. Deng, A. Mukherjee, and D.P. Agrawal, Threshold and identity-based key management and authentication for wireless ad hoc networks, *IEEE Int. Conf. Information Technology (ITCC'04)*, April 5–7, 2004.

26. H. Deng, Q.-A. Zeng, and D.P. Agrawal, SVM-based intrusion detection system for wireless ad hoc networks, *Proc. IEEE Vehicular Technology Conf. (VTC'03)*, Orlando, October 6–9, 2003a.

27. H. Deng, Q.-A. Zeng and D.P. Agrawal, Projecting high-dimensional data for network intrusion detection, *Proc. Joint Conf. on Information Sciences (JCIS'03)*, September 26–30, 2003b, pp. 373–376.

28. H. Deng, Q.-A. Zeng, and D.P. Agrawal, An unsupervised network anomaly detection system using random projection technique, *Proc. 2003 Int. Workshop on Cryptology and Network Security (CANS'03)*, Miami, FL, September 24–26, 2003c, pp. 593–598.

29. J. Ding, A. Mukherjee, and D. Agrawal, Distributed authentication and key generation with multiple user trust level (fast abstract), *The Int. Conf. on Dependable Systems and Networks, (DSN 2004)*, June 28–July 1, 2004.

30. L. Doherty, K.S.J. Pister, and L.E. Ghaoui, Convex position estimation in wireless sensor networks, in *Proc. IEEE INFOCOM*, Alaska, April 2001, pp. 1655–1663.

31. Ember Corporation, Process Temperature Integrated Sensing and Control, www.ember.com/products/solutions/industrialauto.html.

32. D. Estrin, R. Govindan, J. Heidemann, and S. Kumar, Next century challenges: scalable coordination in wireless networks, in *Proc. 5th Annu. ACM/IEEE Int. Conf. on Mobile Computing and Networking (MOBICOM)*, 1999, pp. 263–270.

33. D. Ganesan, D. Estrin, and J. Heidemann, DIMENSIONS: why do we need a new data handling architecture for Sensor Networks?, *Proc. ACM Workshop on Hot Topics in Networks*, 2002a.

34. D. Ganesan, R. Govindan, S. Shenker, and D. Estrin, Highly resilient, energy efficient multipath routing in wireless sensor networks, in *Mobile Computing and Communications Review (MC2R)*, Vol. 1, No. 2, 2002b.

35. J. Gehrke and S. Madden, Query processing in sensor networks, in *Pervasive Computing*, 2004.

36. A.K. Gupta, S. Sekhar, and D.P. Agrawal, Efficient event detection by collaborative sensors and mobile robots, in *Ohio Graduate Symposium on Computer and Information Science and Engineering*, Dayton, OH, June 2004.

37. J. Hayes, M. McJunkin and J. Kosecka, Communication enhanced navigation strategies for teams of mobile agents, *IEEE Robotics and Automation Society*, Las Vegas, October 2003.

38. W. Heinzelman, A. Chandrakasan, and H. Balakrishnan, Energy-efficient communication protocols for wireless microsensor networks, in *Proc. Hawaaian Int. Conf. on Systems Science*, January 2000.

39. W. Heinzelman, A. Chandrakasan, and H. Balakrishnan, μAMPS ns Code Extensions, www-mtl.mit.edu/research/icsystems/uamps/leach.

40. J.M. Hellerstein, P.J. Haas, and H. Wang, Online aggregation, in *Proc. ACM SIGMOD*, Tucson, AZ, May 1997, pp. 171–182.

41. J.M. Hellerstein, W. Hong, S. Madden, and K. Stanek, Beyond average: towards sophisticated sensing with queries, in *Proc. First Workshop on Information Processing in Sensor Networks (IPSN)*, March 2003.

42. J. HighTower and G. Borreillo, Location systems for ubiquitous computing, *IEEE Computer*, 34, 57–66, August 2001.

43. J. Hill, R. Szewczyk, A.Woo, S. Hollar, D. Culler, and K. Pister, System Architecture Directions for Network Sensors, in *Proc. 9th Int. Conf. on Architectural Support for Programming Languages and Operating Systems*, November 2000, pp. 93–104.

44. T. Imielinski and B. Nath, Wireless graffiti — data, data everywhere, in *Int. Conf. on Very Large Data Bases (VLDB)*, 2002.

45. C. Intanagonwiwat, R. Govindan, and D. Estrin, Directed diffusion: a scalable and robust communication paradigm for sensor networks, in *Proc. 6th Annu. ACM/IEEE Int. Conf. on Mobile Computing and Networking (MOBICOM)*, August 2000, pp. 56–67.

46. C. Intanagonwiwat, D. Estrin, R. Govindan, and J. Heidemann, Impact of network density on data aggregation in wireless sensor networks, Technical Report 01-750, University of Southern California, November 2001.

47. Intel; Intel Exploratory Research; www.intel.com/research/exploratory/digital_home.htm.

48. Intel Research Oregon. Heterogeneous Sensor Networks, Technical report, Intel Corporation, 2003. Web Page. www.intel.com/research/exploratory/heterogeneous.htm.

49. N. Jain, Energy Efficient Information Retrieval in Wireless Sensor Networks, Ph.D. thesis, 2004.

50. N. Jain, D.K. Madathil, and D.P. Agrawal, Energy aware multi-path routing for uniform resource utilization in sensor networks, in *Proc. IPSN'03 Int. Workshop on Information Processing in Sensor Networks*, Palo Alto, CA, April 2003a.

51. N. Jain, D.K. Madathil, and D.P. Agrawal, Exploiting multi-path routing to achieve service differentiation in sensor networks, *Proc. 11th IEEE Int. Conf. on Networks (ICON 2003)*, Sydney, Australia, October 2003b.

52. N. Jain and D.P. Agrawal, Current trends in wireless sensor networks, Technical Report, CDMC, University of Cincinnati.

53. E. Jovanov, A. O'Donnell Lords, D. Raskovic, P. Cox, R. Adhami, and F. Andrasik, Stress monitoring using a distributed wireless intelligent sensor system, *IEEE Engineering in Medicine and Biology Magazine*, May/June 2003.

54. K. Kar, M. Kodialam, and T.V. Lakshman, Minimum interference routing of bandwidth guaranteed tunnels with MPLS traffic engineering applications, *IEEE J. Selected Areas in Commun.*, Vol. 18, No. 12, December 2000.

55. B. Karp and H.T. Kung, GPSR: Greedy perimeter stateless routing for wireless networks, in *Proc. ACM/IEEE MobiCom*, August 2000.

56. R. Kumar, V. Tsisatsis, and M. Srivatsava, Computation hierarchy for in-network processing, in *Proc. WSNA'03*, 2003.

57. S. Lee and A.T. Campbell, INSIGNIA: in-band signaling support for QOS in mobile ad hoc networks, in *Proc. 5th Int. Workshop on Mobile Multimedia Communications(MoMuC'98)*, Berlin, Germany, October 1998.

58. S. Lee and M. Gerla, Split multipath routing with maximally disjoint paths in ad hoc networks, *Proc. IEEE ICC, 2001*. pp. 3201–320.

59. S. Madden, The Design and Evaluation of a Query Processing Architecture for Sensor Networks, Ph.D. thesis. University of California, Berkeley, Fall 2003.

60. S. Madden, M.J. Franklin, J.M. Hellerstein, and W. Hong, The design of an acquisitional query processor for sensor networks, *ACM SIGMOD Conf.*, San Diego, CA, June 2003.

61. S. Madden, M. Shah, J.M. Hellerstein, and V. Raman, Continuously adaptive continuous queries over streams, in *ACM SIGMOD Int. Conf. on Management of Data*, Madison, WI, 2002, pp. 49–60.

62. I. Mahadevan and K.M. Sivalingam, Architecture and experimental framework for supporting QoS in wireless networks using differentiated services, *MONET* 6(4), 2001, pp. 385–395.

63. A. Mainwaring, J. Polastre, R. Szewczyk, and D. Culler, Wireless sensor networks for habitat monitoring, in *ACM Workshop on Sensor Networks and Applications*, 2002.

64. A. Manjeshwar, Energy Efficient Routing Protocols with Comprehensive Information Retrieval for Wireless Sensor Networks, M.S. thesis, 2001.

65. A. Manjeshwar and D.P. Agrawal, TEEN: a routing protocol for enhanced efficiency in wireless sensor networks, in *Proc. 15th Int. Parallel and Distributed Processing Symp. (IPDPS'01) Workshops*, 2001.

66. A. Manjeshwar and D.P. Agrawal, APTEEN: a hybrid protocol for efficient routing and comprehensive information retrieval in wireless sensor networks, in *Proc. Int. Parallel and Distributed Processing Symp. (IPDPS'02) Workshops*, 2002.

67. A. Manjeshwar, Q. Zeng, and D.P. Agrawal, An analytical model for information retrieval in wireless sensor networks using enhanced APTEEN protocol, in *IEEE Trans. Parallel and Distributed Systems*, 13(12), 1290–1302, December 2002.

68. N.F. Maxemchuk, Dispersity routing in high-speed networks, *Computer Networks and ISDN System* 25, 1993, 645–661.

69. MICA Sensor Mote. Web page. www.xbow.com/Products/Wireless Sensor Networks.htm.

70. MicroStrain Microminiature Sensors. Web page. www.microstrain.com.

71. H. Mistry, P. Roy, S. Sudarshan, and K. Ramamritham, Materialized view selection and maintenance using multi-query optimization, in *ACM SIGMOD*, 2001.

72. G.E. Moore, Cramming more components onto integrated circuits, *Electronics*, April 1965, pp. 114–117.

73. R. Motwani, J. Window, A. Arasu, B. Babcock, S. Babu, M. Data, C. Olston, J. Rosenstein, and R. Varma, Query processing, approximation and resource management in a data stream management system, in *First Annu. Conf. on Innovative Database Research (CIDR)*, 2003.

74. A. Mukherjee, M. Gupta, H. Deng, and D.P. Agrawal, Level-based key establishment for multicast communication in mobile ad hoc networks, submitted to *IEEE International Symposium on Personal, Indoor and Mobile Radio Communications, (PIMRC'04)*, September 5–8, 2004.

75. A. Nasipuri and S. Das, On-demand multipath routing for mobile ad hoc networks, *Proc. 8th Annu. IEEE Int. Conf. on Computer Communications and Networks (ICCCN)*, October 1999, pp. 64–70.

76. J. Nelson and D. Estrin. An Address-Free Architecture for Dynamic Sensor Networks, Technical Report 00-724, Computer Science Department, University of Southern California, January 2000.

77. NEST, A Network Virtual Machine for Real-Time Coordination Services. www.cs.virginia.edu/nest.

78. The Network Simulator — ns-2, www.isi.edu/nsnam/ns.

79. Ojha, H. Deng, S. Sanyal, and D.P. Agrawal, Forming COUNCIL based clusters in securing wireless ad hoc networks, *Proc. 2nd Int. Conf. on Computers and Devices for Communication (CODEC'04)*, January 1–3, 2004.

80. V.D. Park and M.S. Corson, A highly distributed routing algorithm for mobile wireless networks, in *Proc. IEEE INFOCOM*, 1997, pp. 1405–1413.

81. M.R. Pearlman, Z.J. Hass, P. Sholander, and S.S. Tabrizi, On the impact of alternate path routing for load balancing in mobile ad hoc networks, in *Proc. IEEE/ACM MobiHoc*, 2000.
82. J. Pinto, Intelligent robots will be everywhere, *Robotic Trends*, December 2003.
83. V. Raman, B. Raman, and J. M. Hellerstein, Online dynamic reordering, *The VLDB Journal*, 9(3), 2002.
84. Poosarla, H. Deng, A. Ojha, and D.P. Agrawal, A cluster based secure routing scheme for wireless ad hoc networks, *The 23rd IEEE Int. Performance, Computing, and Communications Conf. (IPCCC'04)*, April 14–17, 2004.
85. K. Römer, O. Kasten, and F. Mattern, Middleware challenges for wireless sensor networks, in *ACM SIGMOBILE Mobile Computing and Communication Review (MC2R)*, Vol. 6, No. 4, October 2002.
86. E.M. Royer and C.-K. Toh. A review of current routing protocols for ad-hoc mobile wireless networks, in *IEEE Personal Communications Magazine*, April 1999, pp. 46–55.
87. SCADDS, Scalable Coordination Architectures for Deeply Distributed Systems. www.isi.edu/div7/scadds.
88. P.S. Schenker, T.L Huntsberger, and P. Pirjanian, Robotic autonomy for space: ooperative and reconfigurable mobile surface systems, *6th Int. Symp. on Artificial Intelligence*, Montreal, Canada, June 2001.
89. L. Schwiebert, S.D.S. Gupta, and J. Weinmann, Research challenges in wireless networks of biomedical sensors, *MobiCom* 2001.
90. C. Schurgers and M.B. Srivastava, Energy efficient routing inwireless sensor networks, *MILCOM'01*, October 2001.
91. S.D. Servetto and G. Barrenechea, Constrained random walks on random graphs: routing algorithms for large scale wireless sensor networks, *in Proc. 1st ACM Int. Workshop on Wireless Sensor Networks and Applications*, September 2002, pp. 12–21.
92. R.C. Shah and J. Rabaey, Energy aware routing for low energy ad hoc sensor networks, *IEEE Wireless Communications and Networking Conference (WCNC)*, March 2002.
93. Smart Messages Project. www.rutgers.edu/sm.
94. K. Sohrabi, J. Gao, V. Ailawadhi, and G. J. Pottie, Protocols for self-organization of a wireless sensor network, *IEEE Personal Communications*, 7(5), 16–27, October 2000.
95. L. Subramanian and R. Katz, An architecture for building self-configurable systems, in *1st Annu. Workshop on Mobile and Ad Hoc Networking and Comp.*, 2000, pp. 63–78.
96. H. Suzuki and F.A. Tobagi, Fast bandwidth reservation scheme with multi-link multi-path routing in ATM networks, in *Proc. IEEE INFOCOM*, 1992.
97. TinyOS: Operating System for Sensor Networks Web Page. http://tinyos.millennium.berkeley.edu.
98. R. T. Vaughan, K. Støy, G. S. Sukhatme, and M. J. Matari'c, Blazing a trail: insect-inspired resource transportation by a robot team, *DistributedAutonomous Robotic Systems*, DARS 2000, Tennessee, October 2000, pp. 111–120.
99. S. Vutukury, Multipath Routing Mechanisms for Traffic Engineering and Quality of Service in the Internet, Ph.D thesis, March 2001.
100. H. Wang, D. Estrin, and L. Girod, Preprocessing in a tiered sensor network for habitat monitoring, in *EURASIP JASP* Special Issue on Sensor Networks, v2003(4), 392–401, March 15, 2003.
101. Xu, J. Heidemann, and D. Estrin, Geography-informed energy conservation for ad hoc routing, in *Proc. ACM/IEEE Int. Conf. on Mobile Computing and Networking*, Rome, Italy, USC/Information Sciences Institute, July 2001, pp. 70–84.
102. Y. Yu, R. Govindan, and D. Estrin, Geographical and Energy Aware Routing: A Recursive Data Dissemination Protocol for Wireless Sensor Networks, UCLA Computer Science Department Technical Report UCLA/CSD-TR-01-0023, May 2001.

21

How to Structure Chaos: Initializing Ad Hoc and Sensor Networks

Thomas Moscibroda

Roger Wattenhofer

21.1   Introduction

One of the main characteristics of ad hoc and sensor networks is that the communication infrastructure is provided by the nodes themselves. When being deployed, the nodes of such networks initially form a chaotic *unstructured radio network*, which means that no reliable and efficient communication pattern has been established yet. Before any reasonable communication can be carried out, nodes must *structure* the network; that is, they must set up a medium access scheme. The problem of initializing and structuring radio networks is of great importance in practice. Even in an ad hoc network with a small number of devices such as Bluetooth, initialization tends to be slow. In a multihop scenario with a large number of nodes, the time consumption for establishing a reasonable communication pattern increases even further. This chapter focuses on the vital transition from an unstructured to a structured network, the *initialization phase*.

One prominent approach to solving the problem of bringing structure into a multihop radio network is *clustering*.[1-4] Clustering allows the formation of virtual backbones enabling efficient routing and broadcasting,[5] it improves the usage of scarce resources such as bandwidth and energy,[6] and clustering is

crucial in realizing spatial multiplexing in non-overlapping clusters (TDMA or FDMA). Hence, computing a good initial clustering is a major step toward establishing an efficient MAC layer on top of which
higher-level protocols and applications can subsequently be built.

What is a good clustering? Depending on the specific network problem at hand, the answer to this question can vary. In light of the wireless and multihop nature of ad hoc and sensor networks, a good clustering
should satisfy at least one — and preferably a second — property. First, to allow efficient communication
between each pair of nodes, every node should have at least one cluster head in its neighborhood. These
cluster heads can act as coordination points for the MAC scheme. When we model a multihop radio
network as a graph $G = (V, E)$, this first property demands a *dominating set* (DS). A DS in a graph
$G = (V, E)$ is a subset $S \in V$ such that each node is either in S or has a neighbor in S. As it is usually
advantageous to compute a DS with few *dominators* (i.e., cluster heads), we study the well-known *minimum dominating set* (MDS) problem that asks for a DS of minimum size. The use of (connected) DS for
clustering networks has been motivated and investigated in literature.[7–12]

This identification of clustering to the notion of a DS, however, does not cover the second need arising
in ad hoc and sensor networks. As put forward by Basagni,[2,13] it is undesirable to have neighboring cluster
heads. In particular, if no two cluster heads are within each other's mutual transmission range, the task
of establishing an efficient MAC layer is greatly facilitated because cluster heads will not face interference.
This second property imposed on the clustering leads to the well-known notion of a *maximal independent
set* in a graph $G = (V, E)$. An independent set S of G is a subset of V such that $\forall u, v \in S, (u, v) \notin E$.
S is a *maximal independent set* (MIS) if any node v not in S has a neighbor in S.

Due to its additional constraint, computing an MIS is a harder problem than computing a DS. Additionally, it is worth noting that any MIS is a $4\mathcal{O} + 1$-approximation for the MDS problem on the unit
disk graph,[11] where $\mathcal{O}$ denotes the size of the optimal solution. That is, by computing the MIS, we obtain
a clustering that has all the advantages of a high-quality DS and moreover, has the property that cluster
heads do not interfere. Hence, an MIS provides an excellent *initial clustering*.[13] Note that the computation
of an MIS is also a key building block for coloring algorithms, as all nodes in an MIS can be safely assigned
the same color.

In view of our goal of setting up a MAC scheme in a newly deployed network, it is obvious that a clustering
algorithm for the initialization phase must not rely on any previously established MAC layer. Instead, we are
interested in a simple and practical algorithm that quickly computes a clustering completely from *scratch*.
Note that this precludes algorithms working under any sort of *message passing model* in which messages
can be sent to neighbors without fearing collision (see, for example, Refs. 3, 8–12). Studying clustering
in the absence of an established MAC layer highlights the *chicken-and-egg* problem of the initialization
phase. A MAC layer ("chicken") helps in achieving a clustering ("egg"), and vice versa. In a newly deployed
ad hoc/sensor network, there is no structure; that is, there are neither "chickens" nor "eggs."

Clustering algorithms in newly deployed networks must be capable of working under difficult conditions.
These conditions are captured in the following model assumptions:

- The network is a *multihop* network; that is, there exist nodes that are not within their mutual
  transmission range, resulting in problems such as the well-known *hidden terminal problem*. Some
  neighbors of a sending node can receive a message, while others are experiencing interference from
  other senders and do not receive the message.

- Our model allows nodes to "wake up" *asynchronously*. In a multihop environment, it is realistic
  to assume that some nodes wake up (e.g., become deployed, or switched on) later than others.
  Consequently, nodes do not have access to a global clock. It is important to observe the manifold
  implications of asynchronous wake-up. If all nodes started the algorithm simultaneously, we could
  easily assume an ALOHA-style MAC layer where each node sends with probability $\Theta(1/n)$. It
  is well known that this approach leads to a quick and simple communication scheme on top of
  which efficient clustering algorithms can be used. If nodes wake up asynchronously, however, the
  same approach results in an expected runtime $O(n)$ if only one single node wakes up for a long
  time.

- Nodes do not feature a reliable *collision detection* mechanism. In many scenarios (particularly when considering the lack of an established MAC protocol during the initialization phase!), not assuming any collision detection mechanism is realistic. Nodes may be tiny sensors with equipment restricted to the minimum due to limitations in energy consumption, weight, or cost. The sending node itself does not have a collision detection mechanism either; that is, a sender does not know how many (if any at all) neighbors have received its transmission correctly. Given these additional limitations, algorithms without collision detection tend to be less efficient than algorithms with collision detection.

- Nodes have only limited knowledge about the total number of nodes in the network and no knowledge about the nodes' distribution or wake-up pattern. In particular, they have no *a priori* information about the number of neighbors.

In this chapter we show that even in this harsh model, good initial clusterings can be computed efficiently. We present two randomized algorithms. The first computes an asymptotically optimal DS and the second computes an MIS, both algorithms running in polylogarithmic time. Section 21.2 provides an overview of related work. The model is introduced in Section 21.3. In Sections 21.4 and 21.5, the algorithms for DS and MIS are developed and analyzed using multiple communication channels. The subsequent Section 21.6 extends the analysis to the single-channel case. The chapter is based on Refs. 14 and 15.

21.2   Related Work

The problem of finding a minimum dominating set is NP-complete. Furthermore, it has been shown[16] that the best possible approximation ratio for this problem is $\ln \Delta$, where Δ is the highest degree in the graph, unless NP has deterministic $n^{O(\log \log n)}$-time algorithms. For Unit Disk Graphs, the problem remains NP-hard, but constant factor approximations are possible. Several distributed algorithms have been proposed, both for general graphs[9,10,12] and the Unit Disk Graph.[7,8,11] As mentioned in the introduction, all the above algorithms assume point-to-point connections between neighboring nodes; that is, they build on top of a functioning MAC layer.

Before being studied in the context of clustering ad hoc and sensor networks, the computation of an MIS was the focus of extensive research on parallel complexity. It has been shown[17] that the MIS problem is in $\mathcal{NC}$, meaning that a polylogarithmic running time is achievable on a PRAM containing a polynomial number of processors. A major breakthrough in understanding the computational complexity of MIS was the ingenious randomized algorithm by Luby,[18] achieving a runtime of $O(\log n)$ on a linear number of processors under the CRCW PRAM model of computation. Unfortunately, Luby's algorithm cannot be easily transformed to work under our model because it assumes synchronous wake-up, knowledge about the neighborhood, and collision-free communication. Recently, time lower bounds for the distributed construction of MIS were given.[19] At least $\Omega(\sqrt{\log n / \log \log n})$ and $\Omega(\log \Delta / \log \log \Delta)$ communication rounds are required to obtain an MIS, Δ being the largest degree in the network.

A model related to the one used in this chapter has been studied in the context of analyzing the complexity of broadcasting in multihop radio networks, yielding a vast and rich literature.[20] The same model has also been the focus of research on two important problems — *initialization problem** and *leader election problem* — in single-hop radio networks.[21–23] A striking difference from our model is that these

---

*Even though the naming may indicate a close relation between the *initialization problem* (also known as *processor identity problem* or *processor naming problem*) and the issues considered in this chapter, the two problems are not related that closely. In the *initialization problem*, unique identifiers in the range 1 to n must be assigned to indistinguishable nodes. This problem has been studied extensively in a variety of settings, leading to a rich and interesting literature. We refer the interested reader to [22] and to the various references therein.

algorithms consider *synchronous wake-up*; that is, nodes have access to a global clock and it is assumed that all nodes start the distributed algorithm at the same time. In the case of ad hoc and sensor networks distributed over a large geographical area, guaranteeing that all nodes start the distributed algorithm simultaneously appears to be difficult in practice. Moreover, if (sensor) nodes are deployed dispersed in time, it may even be impossible. As mentioned in the introduction, the additional difficulties imposed by asynchrony lead to new algorithmic designs.

A model featuring asynchronous wake-up and the *wake-up problem* in single-hop radio networks has been studied recently.[24,25] In comparison to our model, these articles define a much weaker notion of asynchrony. In particular, it is assumed that sleeping nodes are *woken up* by a successfully transmitted message. In a single-hop network, the problem of waking up all nodes hence reduces to successfully transmitting one single message. While this definition of asynchrony leads to interesting problems and algorithms, it does not closely reflect reality in many scenarios related to ad hoc and sensor networks.

21.3   Model

In this section we introduce our model, which will be used throughout the remainder of this chapter. We consider *multihop* radio networks *without collision detection*. Nodes are unable to distinguish between the situation in which two or more neighbors are sending and the situation in which no neighbor is sending. Further, we assume in Sections 21.4 and 21.5 that nodes have three independent communication channels: Γ_1, Γ_2, and Γ_3, which may be realized with an FDMA scheme. In Section 21.6 we show that even a single communication channel suffices to compute an asymptotically optimal DS and an MIS in polylogarithmic time.

Nodes can wake up *asynchronously* at any time. We call a node *sleeping* before its wake-up, and *awake* thereafter. Only awake nodes can send or receive messages. Sleeping nodes are *not woken up* by incoming messages. Observe that this asynchronous model is very general, encompassing the usually studied synchronous wake-up model in which all nodes start the distributed algorithm at the same time.

We consider the well-known *Unit Disk Graph* (UDG) to model the network. In a UDG $G = (V, E)$, there is an edge $(u, v) \in E$ iff the Euclidean distance between u and v is at most 1. Note that due to asynchronous wake-up, some nodes may still be asleep, while others are already sending. Therefore, at any time, there may be sleeping nodes that do not take part in the communication despite their being within the transmission range of the sender.

Nodes have only scarce knowledge about the network graph. In particular, they have no information on the number of nodes in their neighborhood or the density of nodes in the network. Nodes merely have an upper bound $\hat{n}$ for the total number of nodes $n = |V|$ in the graph. While n is unknown, all nodes have the same estimate $\hat{n}$. It has been shown[25] that without any estimate of n, even in the single-hop case, every algorithm requires at least time $\Omega(n/\log n)$ until one single message can be transmitted without collision. Hence, assuming n being completely unknown would ultimately preclude polylogarithmic clustering algorithms. In practice, is it usually possible to give a rough upper bound on the number of nodes in the network in advance.

For the sake of simplicity, we assume — for the analysis of the algorithm — that time is divided into time-slots. However, our algorithm does not rely on synchronized time-slots in any way. It is solely for the purpose of analyzing the algorithm that we assume slotted channels. This simplification of the analysis is justified by the standard trick introduced in the analysis of slotted versus unslotted ALOHA.[26] It was shown that the realistic unslotted case and the idealized slotted case differ only by a factor of 2. The basic intuition is that a single packet can cause interference in no more than two successive time-slots. Similarly, by analyzing our algorithm in an "ideal" setting with synchronized time-slots, we obtain a result that is only by a constant factor better when compared to the more realistic unslotted setting.

In each time-slot, a node can either send or not send. A node receives a message in a time-slot only if exactly one node in its neighborhood has sent a message in this time-slot. The variables p_k and q_k are the probabilities that node k sends a message in a given time-slot on channel Γ_1 and Γ_2, respectively. Unless otherwise stated, we use the term "sum of sending probabilities" to refer to the sum of sending probabilities on channel Γ_1. We conclude this section by providing two facts, the first of which has been proven[25] and the second can be found in standard mathematical textbooks.

Fact 21.1   *Given a set of probabilities $p_1 \ldots p_n$ with $\forall i : p_i \in [0, \frac{1}{2}]$, the following inequalities hold:*

$$\left(\frac{1}{4}\right)^{\sum_{k=1}^{n} p_k} \leq \prod_{k=1}^{n} (1 - p_k) \leq \left(\frac{1}{e}\right)^{\sum_{k=1}^{n} p_k}$$

Fact 21.2   *For all n, t, such that $n \geq 1$ and $|t| \leq n$,*

$$e^t \left(1 - \frac{t^2}{n}\right) \leq \left(1 + \frac{t}{n}\right)^n \leq e^t$$

21.4   Dominating Set

This section presents the first of two algorithms presented in this chapter: an algorithm for the computation of an asymptotically optimal dominating set in polylogarithmic time.

21.4.1   Algorithm

A node starts executing the dominator algorithm upon waking up. In the first phase (lines 1 to 5), nodes wait for messages (on all channels) without sending themselves. The reason is that nodes waking up late should not interfere with already existing dominators. Thus, a node first listens for existing dominators in its neighborhood before actively trying to become a dominator itself. In particular, we will choose the parameter δ to ensure that an awakening node, which is already within an existing dominator's transmission range, does not become a dominator itself.

The main part of the algorithm (starting in line 6) works in rounds, each of which contains δ time-slots. In every time-slot, a node sends with probability p on channel Γ_1. Starting from a very small value, this sending probability p is doubled (lines 6 and 7) in every round. When sending its first message, a node becomes a dominator and, in addition to its sending on channel Γ_1 with probability p, it starts sending on channels Γ_2 and Γ_3 with probabilities q and $q / \log n$, respectively. Once a node becomes a dominator, it will remain so for the rest of the algorithm. For the algorithm to work properly, we must prevent the sum of sending probabilities on channel Γ_1 from reaching too high values. Otherwise, too many collisions will occur, leading to a large number of dominators. Hence, upon receiving its first message (without collision) on any of the channels, a node becomes *decided* and stops sending on Γ_1. Being decided means that the node is covered by a dominator.

Thus, the basic intuition is that nodes, after some initial listening period, compete to become dominators by exponentially increasing their sending probability on Γ_1. Note that in light of asynchronous wake-up, this exponential increase is indispensable in order to achieve a sublinear running time. Channels Γ_2 and Γ_3 then ensure that the number of further dominators emerging in the neighborhood of an already existing dominator remains small.

The parameter δ is chosen large enough to ensure that with high probability, there is a round in which at least one competing node will send without collision. The parameter q is chosen such that during the first $\delta \cdot \lceil \log \hat{n} \rceil$ "waiting time-slots," a new node will receive a message from an existing dominator. Defining q too small or too large could lead to the undesirable situation in which nodes become dominators despite their being covered at the time of their waking up. Finally, β maximizes the probability of a successful execution of the algorithm. Specifically, the parameters are defined as follows:

$$q := \frac{1}{2^\beta \cdot \lceil \log \hat{n} \rceil} \qquad \delta := \left\lceil \frac{\log \hat{n}}{\log \left(\frac{503}{502}\right)} \right\rceil \qquad \beta := 6$$

Algorithm 21.1   Dominator Algorithm

---

decided := false;
dominator := false;
upon wake-up do:
  1: **for** $j := 1$ to $\delta \cdot \lceil \log \hat{n} \rceil$ by 1 **do**
  2:     **if** message received in current time-slot **then**
  3:         decided := true;
  4:     **fi**
  5: **od**
  6: **for** $j := \lceil \log \hat{n} \rceil$ to 0 by -1 **do**
  7:     $p := 1/\left(2^{j+\beta}\right)$;
  8:     **for** $i := 1$ to δ by 1 **do**
  9:         $b_i^{(1)} := 0; b_i^{(2)} := 0; b_i^{(3)} := 0$;
10:         **if not** decided **then**
11:             $b_i^{(1)} := 1$, with probability p;
12:             **if** $b_i^{(1)} = 1$ **then**
13:                 dominator := true;
14:             **else if** message received in current time-slot **then**
15:                 decided := true;
16:             **fi**
17:         **fi**
18:         **if** dominator **then**
19:             $b_i^{(2)} := 1$, with probability q;
20:             $b_i^{(3)} := 1$, with probability $q / \log \hat{n}$;
21:         **fi**
22:         **if** $b_i^{(1)} = 1$ **then** send message on Γ_1 **fi**
23:         **if** $b_i^{(2)} = 1$ **then** send message on Γ_2 **fi**
24:         **if** $b_i^{(3)} = 1$ **then** send message on Γ_3 **fi**
25:     **od**
26: **od**
27: **if not** decided **then**
28:     dominator := true;
29:     decided := true;
30: **fi**
31: **if** dominator **then**
32:     **loop**
33:         send message on Γ_2, with probability q;
34:         send message on Γ_3, with probability $q / \log \hat{n}$;
35:     **end loop**
36: **fi**

Correctness of the algorithm and time-complexity (defined as the number of time-slots of a node between wake-up and decision) follow immediately:

Theorem 21.1   *The algorithm computes a correct dominating set. Every node decides whether or not to become dominator in time* $O(\log^2 \hat{n})$.

Proof   Every node that has not received a message from a dominator at the end of the algorithm will decide to become a dominator in line 28. As for the running time, the first for-loop is executed $\delta \cdot \lceil \log \hat{n} \rceil$ times. The two nested loops of the algorithm are executed $\lceil \log \hat{n} \rceil + 1$ and δ times, respectively.   □

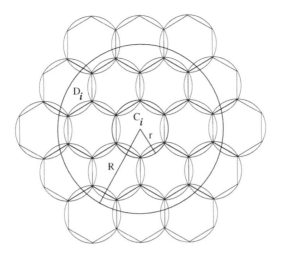

FIGURE 21.1 Circles C_i and D_i.

21.4.2 Analysis

This section shows that the expected number of dominators in the network is within $O(1)$ of an optimal solution. We cover the plane with circles C_i of radius $r = 1/2$ by a hexagonal lattice shown in Figure 21.1. Let D_i be the circle centered at the center of C_i having radius $R = 3/2$. It can be seen in Figure 21.1 that D_i is (fully or partially) covering 19 smaller circles C_j. Note that every node in a circle C_i can hear all other nodes in C_i. Nodes outside D_i are not able to cause a collision in C_i.

We first give a broad outline of the proof that contains four major steps, from Lemma 21.2 to Lemma 21.5. First, we bound the sum of sending probabilities in a circle C_i. This leads to an upper bound on the number of collisions in a circle before at least one dominator emerges. Then, we give a probabilistic bound on the number of sending nodes per collision. In the last step, we show that nodes waking up late in an already covered circle do not become dominators. More specifically, we show that all these claims hold with high probability.

Note that for the analysis, it is sufficient to assume $\hat{n} = n$ because solving minimum dominating set for $n' < n$ cannot be more difficult than for n. If it were, the imaginary adversary controlling the wake-ups of all nodes could simply decide to let $n - n'$ sleep infinitely long, which is indistinguishable from having n' nodes. Hence, we will assume $n' = n$ in this section.

Definition 21.1 Consider a circle C_i. Let t be a time-slot in which a message is sent by a node $v \in C_i$ on channel Γ_1 and received (without collision) by all other nodes in C_i. We say that circle C_i *clears* itself in time-slot t. Let t_0 be the first such time-slot. We say that circle C_i *terminates* itself in time-slot t_0. For all time-slots $t' \geq t_0$, we call C_i *terminated*.

Definition 21.2 Let $s(t) := \Sigma_{k \in C_i} p_k$ be the sum of sending probabilities on Γ_1 in C_i at time t. We define the time slot t_i^j so that for the jth time in C_i, we have $s(t_i^j - 1) < \frac{1}{2^\beta}$ and $s(t_i^j) \geq \frac{1}{2^\beta}$. We further define the Interval $\mathcal{I}_i^j := [t_i^j \ldots t_i^j + \delta - 1]$.

In other words, t_i^0 is the time-slot in which the sum of sending probabilities in C_i exceeds $\frac{1}{2^\beta}$ for the first time and t_i^j is the time-slot in which this threshold is surpassed for the jth time in C_i. The following lemma bounds the sum of sending probabilities in a circle.

Lemma 21.1 *For all time-slots $t' \in \mathcal{I}_i^j$, the sum of sending probabilities in C_i is bounded by $\Sigma_{k \in C_i} p_k \leq 3/2^\beta$.*

Proof According to the definition of t_i^j, the sum of sending probabilities $\Sigma_{k \in C_i} p_k$ at time $t_i^j - 1$ is less than $\frac{1}{2^\beta}$. All nodes that are active at time t_i^j will double their sending probability p_k exactly once in the following δ time-slots. Previously inactive nodes may wake up during that interval. There are at most n such newly active nodes and each of them will send with the initial sending probability $\frac{1}{2^\beta \hat{n}}$ in the given interval. In $\mathcal{I}_i^j$, we get

$$\sum_{k \in C_i} p_k \leq 2 \cdot \frac{1}{2^\beta} + \sum_{k \in C_i} \frac{1}{2^\beta \hat{n}} \leq 2 \cdot \frac{1}{2^\beta} + \frac{n}{2^\beta \hat{n}} \leq \frac{3}{2^\beta}.$$

$\square$

Using the above lemma, we formulate a probabilistic bound on the sum of sending probabilities in a circle C_i. Intuitively, we show that before the bound can be surpassed, C_i does either clear itself or some nodes in C_i become decided.

Lemma 21.2   *The sum of sending probabilities of nodes in a circle C_i is bounded by $\Sigma_{k \in C_i} p_k \leq 3/2^\beta$ with probability at least $1 - o(n^{-2})$. The bound holds for all C_i in G with probability at least $1 - o(n^{-1})$.*

Proof   The proof is by induction over all time-slots t_i^j in ascending order. Let $t' := t_i^0$ be the very first such time-slot in the network and let $\mathcal{I}'$ be the corresponding interval. Lemma 21.1 states that the sum of sending probabilities in C_i is bounded by $\frac{3}{2^\beta}$ in $\mathcal{I}'$. We now show that in this interval, the circle C_i either clears itself or the sum of sending probabilities falls back below $\frac{1}{2^\beta}$ with high probability.

If some of the active nodes in C_i receive a message from a neighboring node, the sum of sending probabilities may fall back below $1/2^\beta$. In this case, the sum does obviously not exceed $3/2^\beta$. If the sum of sending probabilities does not fall back below $1/2^\beta$, the following two inequalities hold for the duration of the interval $\mathcal{I}'$:

$$\frac{1}{2^\beta} \leq \sum_{k \in C_i} p_k \leq \frac{3}{2^\beta} : \quad \text{in } C_i \tag{21.1}$$

$$0 \leq \sum_{k \in C_j} p_k \leq \frac{3}{2^\beta} : \quad \text{in } C_j \in D_i, i \neq j. \tag{21.2}$$

The second inequality holds because t' is the very first time-slot in which the sum of sending probabilities exceeds $1/2^\beta$. Hence, in each $C_j \in D_i$, the sum of sending probabilities is at most $3/2^\beta$ in $\mathcal{I}'$. (Otherwise, one of these circles would have reached $1/2^\beta$ before circle C_i and t' is not the first time-slot considered).

We now compute the probability that C_i clears itself within $\mathcal{I}'$. Circle C_i clears itself when exactly one node in C_i sends and no other node in $D_i \backslash C_i$ sends. The probability P_0 that no node in any neighboring circle $C_j \in D_i, j \neq i$ sends is

$$P_0 = \prod_{\substack{C_j \in D_i \\ j \neq i}} \prod_{k \in C_j} (1 - p_k) \underset{\text{Fact 21.1}}{\geq} \prod_{\substack{C_j \in D_i \\ j \neq i}} \left(\frac{1}{4}\right)^{\Sigma_{k \in C_j} p_k}$$

$$\underset{\text{Lemma 21.1}}{\geq} \prod_{\substack{C_j \in D_i \\ j \neq i}} \left(\frac{1}{4}\right)^{\frac{3}{2^\beta}} \geq \left[\left(\frac{1}{4}\right)^{\frac{3}{2^\beta}}\right]^{18}. \tag{21.3}$$

Let P_{suc} be the probability that exactly one node in C_i sends:

$$P_{suc} = \sum_{k \in C_i} \left(p_k \cdot \prod_{\substack{l \in C_i \\ l \neq k}} (1 - p_l) \right) \geq \sum_{k \in C_i} p_k \cdot \prod_{l \in C_i} (1 - p_l)$$

$$\underset{\text{Fact 21.1}}{\geq} \sum_{k \in C_i} p_k \cdot \left(\frac{1}{4}\right)^{\Sigma_{k \in C_i} p_k} \geq \frac{1}{2^\beta} \cdot \left(\frac{1}{4}\right)^{\frac{1}{2^\beta}}.$$

The last inequality holds because the previous function is increasing in $[\frac{1}{2^\beta}, \frac{3}{2^\beta}]$.

The probability P_c that exactly one node in C_i and no other node in D_i sends is therefore given by

$$P_c = P_0 \cdot P_{suc} \geq \left[\left(\frac{1}{4} \right)^{\frac{3}{2^\beta}} \right]^{18} \cdot \frac{1}{2^\beta} \left(\frac{1}{4} \right)^{\frac{1}{2^\beta}} \underset{\beta=6}{=} \frac{2^{9/32}}{256}.$$

P_c is a lower bound for the probability that C_i clears itself in a time-slot $t \in \mathcal{I}'$. The reason for choosing $\beta = 6$ is that this value maximizes P_c.

We can now compute the probability $\overline{P_{term}}$ that circle C_i does not clear itself during the entire interval:

$$\overline{P_{term}} \leq \left(1 - \frac{2^{9/32}}{256} \right)^\delta \leq \frac{1}{n^{2.3}} \in o\left(\frac{1}{n^2} \right).$$

We have thus shown that within $\mathcal{I}'$, the sum of sending probabilities in C_i either falls back below $1/2^\beta$ or C_i clears itself.

So far, we have only shown that the lemma holds for the very first t_i^j (i.e., t'). For the induction step, we consider an arbitrary t_i^j. Due to the induction hypothesis, we can assume that all previous such time-slots have already been dealt with. That is, all previously considered time-slots $t_{i'}^{j'}$ have either led to a clearance of circle $C_{i'}$ or the sum of probabilities in $C_{i'}$ has decreased below the threshold $1/2^\beta$. Immediately after a clearance, the sum of sending probabilities in a circle C_i is at most $1/2^\beta$, which is the sending probability in the last round of the algorithm. This is true because only one node in the circle remains undecided. All others will stop sending on channel Γ_1. By Lemma 21.1, the sum of sending probabilities in all neighboring circles (both the cleared and the not cleared ones) is bounded by $3/2^\beta$ in $\mathcal{I}_i^j$ (otherwise, this circle would have been considered before t_i^j). Therefore, we know that bounds (21.1) and (21.2) hold with high probability. Consequently, the computation to show the induction step is the same as for the base case t'.

Each step of the induction only holds with high probability. But, because there are n nodes to be decided and at most n circles C_i, the number of induction steps t_i^j is bounded by n. Hence, the probability that the lemma holds for all steps is at least $(1 - o(n^{-2}))^n \geq 1 - o(n^{-1})$, which concludes the proof. $\square$

Using Lemma 21.2, we can now compute the expected number of dominators in each circle C_i. In the analysis, we will separately compute the number of dominators *before* and *after* the termination (i.e., the first clearance) of C_i.

Lemma 21.3 *Let C be the number of collisions (more than one node is sending in one time-slot on Γ_1) in a circle C_i. The expected number of collisions in C_i before its termination is $\mathrm{E}\,[C] < 6$. Further, $C < 7 \log n$ with probability at least $1 - o(\frac{1}{n^2})$.*

Proof Only channel Γ_1 is considered in this proof. We assume that C_i is not yet terminated and we define the following events.

A: exactly one node in D_i is sending
X: more than one node in C_i is sending
Y: at least one node in C_i is sending
Z: some node in $D_i \backslash C_i$ is sending

For the proof, we consider only rounds in which at least one node in C_i sends. (There will be no new dominators in C_i if no node sends.) We want to get a bound for the conditional probability $P\,[A \mid Y]$ that exactly one node in D_i is sending and this one node is located in C_i. Using $P\,[Y \mid X] = 1$ and the fact that Y and Z are independent, we get

$$P[A \mid Y] = P[\overline{X} \mid Y] \cdot P[\overline{Z} \mid Y] = P[\overline{X} \mid Y] \cdot P[\overline{Z}]$$

$$= (1 - P[X \mid Y])(1 - P[Z]) = \left(1 - \frac{P[X]\,P[Y \mid X]}{P[Y]}\right)(1 - P[Z])$$

$$= \left(1 - \frac{P[X]}{P[Y]}\right)(1 - P[Z]). \tag{21.4}$$

We can now compute bounds for the probabilities $P[X]$, $P[Y]$, and $P[Z]$:

$$P[X] = 1 - \prod_{k \in C_i}(1 - p_k) - \sum_{k \in C_i}\left(p_k \prod_{\substack{l \in C_i \\ l \neq k}}(1 - p_l)\right)$$

$$\leq 1 - \left(1 + \sum_{k \in C_i} p_k\right)\left(\frac{1}{4}\right)^{\Sigma_{k \in C_i} p_k} \tag{21.5}$$

$$P[Y] = 1 - \prod_{k \in C_i}(1 - p_k) \geq 1 - \left(\frac{1}{e}\right)^{\Sigma_{k \in C_i} p_k}. \tag{21.6}$$

The first inequality for $P[X]$ follows from Fact 21.1 and Inequality (21.4). The inequality for $P[Y]$ also follows from Fact 21.1. In the proof for Lemma 21.4, we have already computed a bound for P_0, the probability that no node in $D_i \backslash C_i$ sends. Using this result, we can write $P[Z]$ as

$$P[Z] = 1 - \prod_{C_j \in D_i \backslash C_i}\prod_{k \in C_j}(1 - p_k) \underset{\text{Eq. (21.3)}}{\leq} 1 - \left[\left(\frac{1}{4}\right)^{\frac{3}{2^\beta}}\right]^{18}. \tag{21.7}$$

Plugging Inequalities (21.5), (21.6), and (21.7) into Equation (21.4) for $P[A \mid Y]$, it can be shown that the term $P[X]/P[Y]$ is maximized for $\Sigma_{k \in C_i} p_k = \frac{3}{2^\beta}$ and therefore

$$P[A \mid Y] = \left(1 - \frac{P[X]}{P[Y]}\right)\cdot(1 - P[Z])$$

$$\geq \left(1 - \frac{1 - (1 + \frac{3}{2^\beta})(\frac{1}{4})^{\frac{3}{2^\beta}}}{1 - (\frac{1}{e})^{\frac{3}{2^\beta}}}\right)\left[\left(\frac{1}{4}\right)^{\frac{3}{2^\beta}}\right]^{18} \geq 0.18.$$

This shows that whenever a node in C_i sends, C_i terminates with constant probability at least $P[A \mid Y]$. This allows us to compute the expected number of collisions in C_i before the termination of C_i as a geometric distribution, $E[C] = P[A \mid Y]^{-1} \leq 6$. The high probability result can be derived as $P[C \geq 7\log n] = (1 - P[A \mid Y])^{7\log n} \in O(n^{-2})$. $\hfill\square$

So far, we have shown that the number of collisions before the clearance of C_i is constant in expectation. The next lemma shows that the number of *new dominators per collision* is also constant. In a collision, each of the sending nodes may already be a dominator. Hence, by assuming that every sending node in a collision is a new dominator, we obtain an upper bound for the number of new dominators.

Lemma 21.4   *Let D be the number of nodes in C_i sending in a time-slot and let Φ denote the event of a collision. Given the occurrence of a collision, the expected number of sending nodes (i.e., new dominators) is $E[D \mid \Phi] \in O(1)$. Also, $D \in O(\log n / \log\log n)$ with probability $1 - o(\frac{1}{n^2})$.*

Proof Let m, $m \leq n$, be the number of nodes in C_i and $N = \{1 \ldots m\}$. D is a random variable denoting the number of sending nodes in C_i in a given time-slot. We define $A_k := P[D = k]$ as the probability that exactly k nodes send. For example, the probability that exactly two nodes in C_i send is

$$A_2 = \sum_{k \in C_i} \left(p_k \cdot \prod_{\substack{l \in C_i \\ l \neq k}} (1 - p_l) \right).$$

Let $\binom{N}{k}$ be the set of all k-subsets of N (subsets of N having exactly k elements). We define A'_k as

$$A'_k := \sum_{Q \in \binom{N}{k}} \prod_{i \in Q} \frac{p_i}{1 - p_i}.$$

We can then write A_k as

$$A_k = \sum_{Q \in \binom{N}{k}} \left(\prod_{i \in Q} p_i \cdot \prod_{i \notin Q} (1 - p_i) \right)$$

$$= \left(\sum_{Q \in \binom{N}{k}} \prod_{i \in Q} \frac{p_i}{1 - p_i} \right) \cdot \prod_{i=1}^{m} (1 - p_i) = A'_k \cdot \prod_{i=1}^{m} (1 - p_i). \tag{21.8}$$

Fact 21.3 *The following recursive inequality holds between two subsequent A'_k:*

$$A'_k \leq \frac{1}{k} \sum_{i=1}^{m} \frac{p_i}{1 - p_i} \cdot A'_{k-1}, \quad A'_0 = 1.$$

Proof The probability A_0 that no node sends is $\prod_{i=1}^{m} (1 - p_i)$ and therefore $A'_0 = 1$, which follows directly from Equation (21.8). For general A'_k, we have to group the terms $\prod_{i \in Q} \frac{p_i}{1 - p_i}$ in such a way that we can factor out A'_{k-1}:

$$A'_k = \sum_{Q \in \binom{N}{k}} \prod_{j \in Q} \frac{p_j}{1 - p_j} = \frac{1}{k} \sum_{i=1}^{m} \left(\frac{p_i}{1 - p_i} \cdot \sum_{Q \in \binom{N \setminus \{i\}}{k-1}} \prod_{j \in Q} \frac{p_j}{1 - p_j} \right)$$

$$\leq \frac{1}{k} \sum_{i=1}^{m} \left(\frac{p_i}{1 - p_i} \cdot \sum_{Q \in \binom{N}{k-1}} \prod_{j \in Q} \frac{p_j}{1 - p_j} \right) = \frac{1}{k} \sum_{i=1}^{m} \frac{p_i}{1 - p_i} \cdot A'_{k-1}.$$

$\square$

We now continue the proof of Lemma 21.4. The conditional expected value $E[D \mid \Phi]$ is

$$E[D \mid \Phi] = \sum_{i=0}^{m} (i \cdot P[D = i \mid \Phi]) = \sum_{i=2}^{m} B_i \tag{21.9}$$

where B_i is defined as $i \cdot P[D = i \mid \Phi]$. For $i \geq 2$, the conditional probability reduces to

$$P[D = i \mid \Phi] = \frac{P[D = i]}{P[\Phi]}. \tag{21.10}$$

In the next step, we consider the ratio between two consecutive terms of Sum (21.9).

$$\frac{B_{k-1}}{B_k} = \frac{(k-1) \cdot P[D=k-1 \mid \Phi]}{k \cdot P[D=k \mid \Phi]} \underset{\text{Eq. (21.10)}}{=} \frac{(k-1) \cdot P[D=k-1]}{k \cdot P[D=k]}$$

$$= \frac{(k-1) \cdot A_{k-1}}{k \cdot A_k} = \frac{(k-1) \cdot A'_{k-1}}{k \cdot A'_k}.$$

It follows from Fact 21.3, that each term B_k can be upper bounded by

$$B_k = \frac{k A'_k}{(k-1) A'_{k-1}} \cdot B_{k-1} \underset{\text{Fact 21.3}}{\leq} \frac{k \left(\frac{1}{k} \sum_{i=1}^m \frac{p_i}{1-p_i} \cdot A'_{k-1} \right)}{(k-1) A'_{k-1}} \cdot B_{k-1}$$

$$= \frac{1}{k-1} \sum_{i=1}^m \frac{p_i}{1-p_i} \cdot B_{k-1} \leq \frac{2}{k-1} \sum_{i=1}^m p_i \cdot B_{k-1}.$$

The last inequality follows from $\forall i : p_i < 1/2$ and $p_i \leq 1/2 \Rightarrow \frac{p_i}{1-p_i} \leq 2 p_i$.

From the definition of B_k, it naturally follows that $B_2 \leq 2$. Furthermore, we can bound the sum of sending probabilities $\sum_{i=1}^m p_i$ using Lemma 21.2 to be less than $\frac{3}{2^\beta}$. We can thus sum up over all B_i recursively to obtain $E[D \mid \Phi]$:

$$E[D \mid \Phi] = \sum_{i=2}^m B_i \leq 2 + \sum_{i=3}^m \left[\frac{2}{(i-1)!} \left(\frac{6}{2^\beta} \right)^{i-2} \right]$$

$$= 2 + \frac{6}{2^\beta} + \frac{1}{3} \left(\frac{6}{2^\beta} \right)^2 + \cdots \leq 2.11.$$

The high probability result can be derived using the upper tail Chernoff bound. Let $\mu = E[D \mid \Phi]$ and $\delta = \tau \log n / \log \log n$ for some constant τ. For P_+ defined as $P[X > (1+\delta)\mu]$, it holds that $P_+ < (e^{-\delta}(1+\delta)^{1+\delta})^\mu$. Taking the logarithm of P_+, this term simplifies to

$$\log P_+ < \mu(-\delta \cdot \log e - (1+\delta) \log(1+\delta))$$

$$\leq -\frac{\mu \tau \log n}{\log \log n} \log \left(1 + \frac{\tau \log n}{\log \log n} \right)$$

$$\leq -\frac{\mu \tau \log n}{\log \log n} (\log(\tau \log n) - \log \log \log n)$$

$$\leq -\mu \tau \log n \cdot \left(1 - \frac{\log \log \log n}{\log \log n} \right) \leq -2 \log n$$

for large enough $\tau > \mu/2$. The lemma now follows from $P_+ < 2^{-2 \log n} \leq n^{-2}$.   □

The final key lemma shows that the expected number of new dominators *after* the termination of circle C_i is also constant.

Lemma 21.5   *Let A be the number of new dominators emerging after the termination of C_i. Then, $A \in O(1)$ with high probability.*

Proof   We define B and B_i as the set of dominators in D_i and C_i, respectively. Immediately after the termination of C_i, only one node in C_i remains sending on channel Γ_1 because all others will be decided. By Lemmas 21.3 and 21.4, we can bound the number of dominators in a C_i with high probability as

$|B_i| \leq \tau' \log^2 n / \log\log n$ and therefore $|B_i| \leq \tau' \log^2 n$ for a constant τ'^*. Potentially, all $C_j \in D_i$ are already terminated and therefore $1 \leq |B| \leq 19 \cdot \tau' \log^2 n$ with high probability. In the following, we write $\tau := 19 \cdot \tau'$.

We now distinguish the two cases $1 \leq |B_i| \leq \tau \log n$ and $\tau \log n < |B_i| \leq \tau \log^2 n$. We consider channels Γ_2 and Γ_3 in the first and second cases, respectively. In particular, we will show that in either case, a new node will receive a message on one of the two channels with high probability during the waiting period at the beginning of the algorithm.

First, consider case one; that is, $1 \leq |B_i| \leq \tau \log n$. The probability P_0 that one dominator is sending alone on channel Γ_2 is $P_0 = |B| \cdot q \cdot (1-q)^{|B|-1}$. This is a concave function in $|B|$. For $|B| = 1$, we get $P_0 = q = \frac{1}{2^\beta \cdot \lceil \log n \rceil}$ and for $|B| = \tau \log n$, $n \geq 2$, we have

$$P_0 = \frac{\tau \log n}{2^\beta \lceil \log n \rceil} \cdot \left(1 - \frac{1}{2^\beta \lceil \log n \rceil}\right)^{\tau \log n - 1} \geq \frac{\tau}{2^\beta} \cdot \left(1 - \frac{\tau/2^\beta}{\tau \log n}\right)^{\tau \log n}$$

$$\underset{\text{Fact 21.2}}{\geq} \frac{\tau}{2^\beta} e^{-\frac{\tau}{2^\beta}} \left(1 - \frac{(\tau/2^\beta)^2}{\tau \log n}\right) \underset{(n \geq 2)}{\geq} \frac{\tau}{2^\beta} e^{-\frac{\tau}{2^\beta}} \left(1 - \frac{\tau}{2^{2\beta}}\right) \in O(1).$$

A newly awakened node in a terminated circle C_i will not send during the first $\delta \cdot \lceil \log \hat{n} \rceil$ rounds. If during this period, the node receives a message (without collision) from an existing dominator, it will become decided and hence will not become a dominator. The probability P_{no} that such an already covered node does *not* receive any messages from an existing dominator during the first $\delta \cdot \lceil \log \hat{n} \rceil$ rounds is asymptotically bounded by

$$P_{no} \leq \left(1 - \frac{1}{2^\beta \cdot \lceil \log n \rceil}\right)^{\delta \cdot \lceil \log n \rceil} \underset{\text{Fact 21.2}}{\leq} e^{-\frac{\delta}{2^\beta}} \in O(n^{-7}). \tag{21.11}$$

This shows that the probability of new dominators emerging in C_i after the termination of C_i is very small and with high probability, the number of new dominators is bounded by a constant in this case.

The analysis in the second case follows along the same lines. For $|B| = \tau \log n$, we get

$$P_0 = \frac{\tau \log n}{2^\beta \lceil \log^2 n \rceil} \cdot \left(1 - \frac{1}{2^\beta \lceil \log^2 n \rceil}\right)^{\tau \log n - 1}$$

$$\underset{\text{Fact 21.2}}{\geq} \frac{\tau}{2^\beta \log n} e^{-\frac{\tau}{2^\beta \log n}} \left(1 - \frac{\tau}{2^{2\beta}}\right) \in O(1/\log n).$$

For $|B| = \tau \log^2 n$, it can be shown that $P_0 \in O(1)$ and, hence, the remainder of the analysis is analogous to the first case. $\qquad \square$

We are now ready to prove the following theorem.

Theorem 21.2 *The expected number of dominators in circle C_i is $E[D] \in O(1)$.*

Proof We consider a circle C_i. By Lemma 21.3, the expected number of collisions before the termination of C_i is less than 6. Lemma 21.4 states that the expected number of new dominators per collision is not higher than 2.11. Because C and $D \mid \Phi$ are independent variables, we can compute the expected number of dominators in C_i before the termination of C_i as $E[D] = E[C] \cdot E[D \mid \Phi] \in O(1)$. By Lemma 21.5, the number of dominators emerging after the termination of C_i is also constant. $\qquad \square$

---

*Taking the extra $\log\log n$ factor into account, it would be possible to shorten the waiting period to $\delta \cdot \lceil \log \hat{n} / \log\log \hat{n} \rceil$. As this does not improve the asymptotic running time of the algorithm, we will ignore this extra factor in this proof for the sake of simplicity.

As Lemma 21.2 holds only with high probability, one can argue that the expected number of dominators may be higher than the calculated value above. However, because Lemma 21.2 does *not* hold with probability less than $O(n^{-1})$ and because at most n dominators can be chosen, the expected number of dominators is bounded by $n \cdot O(\frac{1}{n}) \in O(1)$ even in this case.

Theorem 21.3 *Our algorithm computes a correct dominating set in time $O(\log^2 \hat{n})$ and achieves an approximation ratio of $O(1)$ in expectation.*

Proof Theorem 21.3 follows from Theorems 21.1, 21.2, and the fact that the optimal solution for the dominating set problem must choose at least one dominator in D_i.   □

Algorithm 21.2 MIS-Algorithm (Main-Loop)

---

state := uncovered; excited := false;
upon wake-up do:
 1: **for** $j := 1$ to $2\delta \cdot \lceil \log^3 \hat{n} / \log\log \hat{n} \rceil$ **do**
 2:     wait();
 3: **od**
 4: counter := 0;
 5: **for** $j := \lceil \log \hat{n} \rceil$ to 0 by -1 **do**
 6:     $p := 1/(2^{j+\beta})$;
 7:     **for** $i := 1$ to $\gamma \cdot \lceil \log \hat{n} \rceil$ **do**
 8:        $b := \begin{cases} 1 & \text{with probability } p \\ 0 & \text{with probability } 1 - p \end{cases}$
 9:        **if** $b = 1$ **then**
10:            **send**() on Γ_1;
11:            **start candidacy**();
12:            **stop executing main-loop**;
13:        **fi**
14:     **od**
15: **od**

Candidacy Phase():
16: **loop**
17:     $b := \begin{cases} 1 & \text{with probability } q \\ 0 & \text{with probability } 1 - q \end{cases}$
18:     **if** $b = 1$ **then**
19:        excited := true;
20:        counter := Max $\{$counter, 0$\}$;
21:        **send**(*counter*) on Γ_2;
22:     **fi**
23:     **if** excited **then**
24:        counter := counter + 1;
25:     **fi**
26:     **if** counter $= \delta \cdot \lceil \log^3 \hat{n} / \log\log \hat{n} \rceil$ **then**
27:        state := MIS;
28:        **send** on Γ_3 with probability 1/6 forever;
29:     **fi**
30: **end loop**

21.5 Maximal Independent Set

In light of our goal of establishing an efficient MAC layer in unstructured and newly deployed ad hoc and sensor networks, a clustering based on an MIS features the important advantage of non-interfering cluster heads, as compared to a mere DS clustering. Based on the DS algorithm of Section 21.4, we now present an MIS algorithm with polylogarithmic running time. Note that the additional MIS requirement only leads to an increase of the running time by a factor $O(\log n / \log \log n)$.

21.5.1 Algorithm

The algorithm consists of two main phases. The purpose of the main loop (which corresponds to the DS algorithm) is the selection of *candidates* that will subsequently compete for joining the MIS in a *candidacy phase*. More precisely, a node becomes a candidate when sending its first message on channel Γ_1 (lines 10 and 11). Once a node has become a candidate, the *Candidacy Phase()* procedure runs until termination. The main loop is designed as to bound the number of candidates simultaneously executing the candidacy phase, therefore enabling a quick election of MIS nodes. This selection in the candidacy phase takes place entirely on channel Γ_2. While Γ_1 and Γ_2 correspond to communication in the main loop and in the candidacy phase, respectively, Γ_3 is reserved for nodes having already joined the MIS.

Receive Triggers:
(Only executed if the node does not send a
message in the same time-slot.)
upon receiving msg on Γ_1 do:
    if not candidate **then**
        restart main-loop at line 1;
    fi

upon receiving msg (c') **on Γ_2 do:**
    $\Delta c := c' - \text{counter}$;
    if candidate **and** $\Delta c \geq 0$ **and** $\Delta c \leq 8 \log \hat{n}$ **then**
        counter $:= -\lceil 8 \log \hat{n} \rceil$;
    fi

upon receiving msg on Γ_3 do:
    state $:=$ covered;
    terminate();

Note that due to asynchronous wake-up, the candidacy phases of different nodes are not aligned with each other. On the contrary, just as they can start the main loop at different times, nodes can join the candidacy phase later than others. Moreover, unless a node has received a message from a neighbor, it has no knowledge of whether other nodes have previously joined the main loop or candidacy phase. In fact, overcoming the absence of any such knowledge is one of the key challenges when designing algorithms for our model.

In more detail, the algorithm works as follows. The main part of the algorithm is identical to the DS algorithm (Algorithm 21.1). A node becomes candidate (and starts executing the *start candidacy()* procedure) upon its first sending on channel Γ_1. As soon as it receives a message on Γ_1, however, it quits the current execution of the main loop and restarts at line 1. As we have seen in the analysis of the DS algorithm, the number of nodes simultaneously being candidates is bounded. In this section, we show that each time a restart occurs, some node in the 2-neighborhood will join the MIS within the required time-bounds. We call nodes in the waiting loop (lines 1–3) *inactive* and nodes in the main part of the algorithm *active*.

Having bounded the number of candidates, the candidacy phase works as follows. In each time-slot, a candidate sends on Γ_2 with probability q. After sending the first time, a node becomes *excited* and starts increasing a counter in every time-slot. This counter is attached to each message. Upon receiving a message on channel Γ_2 by another candidate, the receiver compares the sender's counter c' with its own. In case its own value is smaller *and* within $\log \hat{n}$ of the sender's counter, a node resets its own counter. This prevents two neighboring nodes from joining the MIS shortly in succession. It is interesting to note that this method of comparing counters is sufficiently powerful to avoid long cascading chains of resettings. Once a node's counter reaches $\delta \cdot \lceil \log^3 \hat{n} / \log\log \hat{n} \rceil$, the node joins the MIS and immediately starts sending on channel Γ_3 with constant probability. Because no two nodes' counter reaches the threshold within $\log \hat{n}$ time-slots, there is sufficient time for the first MIS node to inform its neighbors, thus ensuring that no two neighbors join the MIS.

The algorithm's parameter q is defined as $q := \log\log \hat{n} / \log^2 \hat{n}$ while the definition of $\beta := 6$ remains as in the DS algorithm. Intuitively, the choice of q is motivated by two contradicting aims. On the one hand, q must be large enough to make sure that some node will join the MIS within the desired runtime. On the other hand, q must be small enough as to ensure that no two neighboring nodes will join the MIS. Our choice of q results in *exactly one* node in each "neighborhood" joining the MIS. The constants δ and γ can be used to tune the trade-off between running-time (small δ and γ) and probability of success (large δ and γ).

21.5.2 Analysis

In this section we show that with high probability the algorithm computes an MIS in time $O(\log^3 \hat{n} / \log\log \hat{n})$. For the sake of clarity, we sometimes omit the ceiling signs as imposed by the algorithm; a more rigorous analysis leads to the same results.

First of all, observe that the algorithm's main loop corresponds to the DS algorithm and we can therefore make use of the results obtained in Section 21.4. Upon sending on Γ_1 in the main loop of the algorithm, a node becomes a candidate and competes for joining the MIS. In this subsection, we show that each candidate will either join the MIS or will be covered by an MIS node within time $2\delta \cdot \log^3 n / \log\log n$. Further, we know by Lemmas 21.3 and 21.4 that with high probability, there are at most $\tau \log^2 n / \log\log n$ candidates emerging in C_i *before a clearance*, for a constant τ. All but the sending node restart the main loop after a clearance, and the sending node itself stops executing the main loop altogether. Due to the waiting loop at the beginning of the algorithm, no node in C_i is going to compete for becoming a candidate during the next $2\delta \cdot \log^3 n / \log\log n$ time-slots. That is, nodes *after a clearance* do not interfere with the current candidacy phase due to their being inactive. The same holds for all $C_i \in D_i$ and hence the number of candidates within the transmission range of a node v may not exceed $19\tau \log^2 n / \log\log n$. This crucial observation allows us to *separate* candidacy phases in a circle D_i and analyze them individually because a node's candidacy phase does not take longer than $2\delta \cdot \log^3 n / \log\log n$ time-slots, as shown in the sequel.

Lemma 21.6   *Let t_m be the time-slot in which node v_m joins the MIS. The counter of all neighboring nodes v_c, $(v_m, v_c) \in E$, at time t_m is at most $c \le 2\delta \log^3 n / \log\log n - 8\log n$ with high probability.*

Proof   Let v_c be a neighboring node having counter $c > \delta \log^3 n / \log\log n - 8\log n$ by the time t_m. Assume for contradiction that v_c exists. By the definition of the algorithm, v_m must have sent in time-slot $t_m - \delta \log^3 n / \log\log n$ and v_c must have sent within the subsequent $8\log n$ time-slots. Afterward, v_c has not received a message from v_m. If it had, it would have reset its counter to $-\lceil 8\log \hat{n} \rceil$. The probability

$P_{recv}(t)$ that v_c receives a message from v_m in an arbitrary time-slot t is

$$P_{recv}(t) \geq \frac{\log\log n}{\log^2 n}\left(1 - \frac{\log\log n}{\log^2 n}\right)^{d(t)}$$

where $d(t)$ denotes the number of candidates within the transmission range of v_c at time t. We know that $d(t)$ is in the range between 1 and $19\tau \log^2 n / \log\log n$. $P_{recv}(t)$ is a monotonously decreasing function in $d(t)$ and therefore,

$$P_{recv}(t) \geq \frac{\log\log n}{\log^2 n}\left(1 - \frac{\log\log n}{\log^2 n}\right)^{\frac{19\tau \log^2 n}{\log\log n}} \in \mathrm{O}(\log\log n / \log^2 n).$$

The probability that this event does not occur in any of the $\delta \cdot \lceil \log^3 \hat{n} / \log\log \hat{n} \rceil$ time-slots following t_m can be shown to be $n^{-\nu\delta}$ for some constant ν by applying Fact 21.2. By choosing δ accordingly, this probability can be made arbitrarily small. $\qquad\square$

Let E_i denote the circle with radius $5/2$ centered at the center of C_i. Further, let t_i be the first time-slot in which a node becomes a candidate in D_i. The next lemma shows that with high probability, exactly one node in a circle E_i joins the MIS within time $t_i + 2\delta \log^3 n / \log\log n$.

Lemma 21.7 *For every candidate v_c, either v_c joins the MIS or a neighboring candidate v'_c, $(v_c, v'_c) \in E$, joins the MIS within time $t_i + 2\delta \log^3 n / \log\log n$ with high probability.*

Proof The main idea is that once a candidate v_c sends without collision at time t_c, it will either have joined the MIS by time $t_c + \delta \log^3 n / \log\log n$ or a neighboring candidate will have joined the MIS before. Let $c(v_c)$ be the value of the v_c's counter at time t_c. As the message is sent without collision, all neighboring candidates v'_c having the same counter value will set $c(v'_c) := -\lceil 8\log \hat{n} \rceil$ due to the received message on Γ_2. The sending node v_c's counter is at least D. Consequently, after time-slot t_c, $c(v_c) \neq c(v'_c)$ for all neighboring candidates v'_c. By sending the message at time t_c, v_c has become excited and, hence, $c(v_c)$ is increased in each time-slot. In the absence of neighboring candidates with equal counters within $-\lceil 8\log \hat{n} \rceil$ of $c(v_c)$, there is no way to prevent it from reaching $\delta \cdot \lceil \log^3 \hat{n} / \log\log \hat{n} \rceil$, which enables it to join the MIS.

It remains to be shown that with high probability, one candidate in D_i sends without collision in the interval $[t_i, \ldots, t_i + \delta \log^3 n / \log\log n]$, such that the above observation can conclude the proof. Let t be an arbitrary time-slot. Again, $d(t)$ denotes the number of candidates within the transmission range of the first candidate in D_i. The probability $P_{suc}(t)$ that one node sends without collision is given by

$$P_{suc}(t) = \frac{d(t)\log\log n}{\log^2 n}\left(1 - \frac{\log\log n}{\log^2 n}\right)^{d(t)-1}.$$

$P_{suc}(t)$ being a concave function in $d(t)$, we can focus attention on the two border values $d(t) \geq 1$ and $d(t) \leq 19\tau \log^2 n / \log\log n$, for all $t_i \leq t \leq t_i + \delta \log^3 n / \log\log n$. For $d(t) = 1$, $P_{suc}(t)$ simplifies to $P_{suc}(t) = \log\log n / \log^2 n$ while for $d(t) = 19\tau \log^2 n / \log\log n$, we have

$$P_{suc}(t) = 19\tau \cdot \left(1 - \frac{\log\log n}{\log^2 n}\right)^{\frac{19\tau \log^2 n}{\log\log n} - 1}$$

$$\underset{\text{Fact 21.2}}{\geq} 19\tau e^{-19\tau}\left(1 - \frac{\log\log n}{\log^2 n}\right) \geq 15\tau e^{-19\tau} \in \mathrm{O}(1).$$

Putting things together, the probability of a successful time-slot is lower bounded by

$$P_{suc}(t) \geq \min \left\{ \frac{\log \log n}{\log^2 n}, 15\tau e^{-19\tau} \right\}$$

throughout the considered time interval. The probability P_n that no candidate sends without collision in the interval $[t_i, \ldots, t_i + \delta \log^3 n / \log \log n]$ is therefore

$$P_n \leq \left(1 - \min \left\{ \frac{\log \log n}{\log^2 n}, 15\tau e^{-19\tau} \right\} \right)^{\frac{\delta \log^3 n}{\log \log n}} \leq \max \left\{ n^{-\delta}, n^{-\frac{15\tau e^{-19\tau} \delta \log^2 n}{\log \log n}} \right\}.$$

For large enough δ, this probability becomes arbitrarily small. Thus, with high probability, at least one node will send without collision within the first $\delta \log^3 n / \log \log n$ time-slots of the candidacy phase. Because the same argument can be repeated for every node, the lemma follows from the observation stated at the beginning of the proof. □

Theorem 21.4   *With high probability, no two neighboring candidates join the MIS; that is, the resulting independent set is correct.*

Proof   Let v_m be an MIS node. Assume for contradiction that $v'_m, (v_m, v'_m) \in E$, is the first node violating the MIS condition. By Lemma 21.6, v'_m joins the MIS at least $8 \log n$ time-slots after v_m. During these time-slots, v_m sends with constant probability $1/6$ on channel Γ_3. It is well-known that in a unit disk graph, v'_m can have at most six independent neighbors (i.e., MIS nodes). The probability that v'_m has received no message by v_m can thus easily be shown to be $\overline{P_{recv}} \in O(n^{-2})$. Observe that the same argument holds for nodes with area already covered by (up to 6) MIS nodes by the time of their wake-up. □

Finally, we derive the algorithm's running time. By Lemma 21.7, every node will either join the MIS or become covered within time $\delta \log^3 n / \log \log n$ upon becoming a candidate. The following observation immediately follows from the algorithm's definition.

Lemma 21.8   *Consider a circle C_i and let t_i be the time-slot in which the first node $v_c \in C_i$ executes line 5 of the main loop. With high probability, there is a node in D_i that becomes a candidate before time $t_i + \gamma \log^2 n$.*

Proof   By the definition of the algorithm, v_c sends with $p_{v_c} = 2^{-\beta}$ on Γ_1 after $\log n$ rounds (unless v_c receives a message from a neighbor, in which case the claim holds). The probability P_{no} that v_c does not send in any of this round's $\gamma \log n$ time-slots can be made arbitrarily small by choosing γ large enough; that is, $P_{no} \leq (1 - \frac{1}{2^\beta})^{\gamma \log n} \leq n^{-\gamma/2^\beta}$. □

We are now ready to prove the claimed running time of the algorithm.

Theorem 21.5   *Every node $v \in G$ either joins the MIS or becomes covered by a neighboring node joining the MIS within time $O(\log^3 n / \log \log n)$ upon waking up.*

Proof   By Lemma 21.7, we know that if a node $w \in D_i$ becomes a candidate at time t_w, it will be covered (possibly by joining the MIS itself) before $t_w + 2\delta \log^3 n / \log \log n$. This implies that there is a node $v_m \in E_i$ joining the MIS before $t_i + 2\delta \log^3 n / \log \log n$, where t_i is defined as the first time-slot a candidate emerges in D_i.

Consider an arbitrary node $v \in C_i$. By Lemma 21.8, we know that $2\delta \log^3 n / \log \log n + \gamma \log^2 n \in O(\log^3 n / \log \log n)$ time-slots after its wake-up, v will either become a candidate or there will be another candidate in D_i, from which v has received a message. In the first case, v will be covered within the next $2\delta \log^3 n / \log \log n$ time-slots by Lemma 21.7. In the latter case, at least one node in E_i joins the MIS within the same period. If this node covers v, we are done. If not, we know that the same conditions as above hold in the remaining, uncovered part of E_i because the waiting period before the main loop guarantees that a node cannot take part in the same candidacy phase twice. The above argument can thus be repeated. Each time, v either joins the MIS or becomes covered or one node in E_i joins the MIS in time $O(\log^3 n / \log \log n)$.

By Theorem 21.4, no two neighboring nodes join the MIS. Hence, the number of different nodes joining the MIS in E_i is bounded by a constant because no more than a constant number of nodes with transmission range 1 can be packed in a circle E_i of radius $5/2$ such that no two nodes are within each other's mutual transmission range. That is, at most, a constant number of repetitions are required, and it follows that node v is covered by a node in the MIS (possibly itself) within time $O(\log^3 n / \log \log n)$ upon its wake-up. The same argument holds for every node $v \in G$, which concludes the proof. $\qquad\square$

Combining Theorems 21.4 and 21.5 leads to the following corollary.

Corollary 21.1   With high probability, the algorithm computes a correct MIS such that each node is covered within time $O(\log^3 n / \log \log n)$ upon waking up.

21.6   Single Channel

Realizing independent communication channels by means of an FDMA scheme may not always be desirable or possible. In this section, we show that DS or MIS clustering can also be efficiently computed in the most basic *single-channel* setting. Intuitively, the idea is to simulate each time-slot in the multichannel model by a number of time-slots in the single-channel model. In particular, we show that the algorithm's time complexity remains polylogarithmic.

Let s and t be time-slots in the single-channel and multichannel models, respectively. We write $suc(t) = 1$ if a message is successfully transmitted in time-slot t and $suc(t) = 0$ otherwise.

Lemma 21.9   *Time-slot t can be simulated with $O(\log^3 n)$ time-slots s_i, $i \in [1 \ldots 3\alpha \log^3 n]$, for a large enough constant α such that $suc(t) = 1 \Leftrightarrow \exists i : suc(s_i) = 1$ with probability $1 - O(\frac{1}{n^2})$.*

Proof   We first investigate the critical cases by analyzing the different sending possibilities that can occur in the multichannel case (channels Γ_1, Γ_2, and Γ_3) and how they map to the single-channel model.

Γ_1	Γ_2	Γ_3	Multi	Single	Critical
0	0	0	0	0	no
0	1	0	1	1	no
0	1	1	1	0	yes
1	1	1	1	0	yes

Γ_1	Γ_2	Γ_3	Multi	Single	Critical
0	≥ 2	0	0	0	no
0	1	≥ 2	0	1	yes
1	≥ 2	1	1	0	yes
≥ 2	≥ 2	1	1	0	yes

The table shows some of the possible cases. The columns Γ_1, Γ_2, and Γ_3 denote how many senders are sending on these channels in a given time-slot. The next two columns show whether or not the transmission was successful, depending on the number of channels used. For the single-channel case, we assume that all senders sending on any channel are sending on a common channel Γ. The critical cases are those in which a node receives a message in the multichannel case but does not receive it in the single-channel case, due to a collision. When simulating three channels by a single channel, we must ensure that a message can be successfully transmitted (without collision) in these critical cases.

We write $send(t) = 1$ if a sender sends in time-slot t and $send(t) = 0$, otherwise. Further, we use the abbreviations $\lambda := \alpha \log^3 n$ and $p := 1/\log^2 n$. Each node simulates time-slot t by 3λ single-channel time-slots $s_1 \ldots s_{3\lambda}$ in the following way:

$$send(t) = 0 \Rightarrow \forall s_i \in [s_1 \ldots s_{3\lambda}]: \ send(s_i) := 0$$

$$send(t) = 1 \Rightarrow \forall s_i \in [s_1 \ldots s_{\lambda}, s_{2\lambda+1} \ldots s_{3\lambda}]: \ send(s_i) := 0$$

$$send(t) = 1 \Rightarrow \forall s_i \in [s_{\lambda+1} \ldots s_{2\lambda}]: \ send(s_i) := \begin{cases} 1, & \text{with probability } p \\ 0, & \text{with probability } 1 - p \end{cases}$$

In words, each node that sends on Γ_1, Γ_2, or Γ_3 in a time-slot t sends randomly with probability $1/\log^2 n$ in the λ time-slots $[s_\lambda \ldots s_{2\lambda}]$ on channel Γ. We call $[s_\lambda \ldots s_{2\lambda}]$ the *sending period*, $[s_1 \ldots s_\lambda]$ and $[s_{2\lambda} \ldots s_{3\lambda}]$ the *quiet periods*.

Obviously, if there is more than one sender, they may choose the same or overlapping time-slots, which will lead to collisions. Unless there is at least one time-slot in which exactly one sender is sending, the message is not transmitted successfully. Thus, there is a non-zero probability that sending a message fails in the critical cases, as defined above. We now show, however, that this probability becomes sufficiently small to make sure the algorithm works the same way as in the multichannel case.

Let T be the set of sending nodes in time-slot t. Due to asynchronous wake-up, we cannot assume that the periods $[s_1 \ldots s_{3\lambda}]$ of sending nodes $v \in T$ are aligned. It is easy to observe, however, that the probability of a successful transmission is minimized when these periods are exactly aligned. If some nodes $v \in T$ are in the sending period while others are in a quiet period, the probability of a successful transmission (exactly one node sends in a time-slot s) is larger compared to the case when all sending nodes are in the sending period at the same time. Consequently, we only have to consider the case of perfect alignment between sending periods.

By Lemma 21.4, we know that the number of sending nodes on channel Γ_1 in a given time-slot does not exceed $\log n / \log \log n$ with high probability. In combination with Lemma 21.3, we know that the number of nodes sending on Γ_2 and Γ_3 is bounded by $\tau \log^2 n / \log \log n$. In the sequel, we will ignore the $\log \log n$ factor for the sake of simplicity. Note that the result of Lemma 21.9 can easily be improved to $O(\log^3 n / \log \log n)$ if this factor was also considered.

The probability that exactly one node $v \in T$ sends in time-slot s is $P_1 \leq \frac{|T|}{\log^2 n}(1 - \frac{1}{\log^2 n})^{|T|-1}$. Because this is a concave function, we again must consider the cases $|T| = 2$ and $|T| = 2\tau \log^2 n + 1$. In the first case, the probability P_{no} that no message is successfully transmitted in the entire sending period is

$$P_{no} = (1 - P_1)^\lambda \leq \left(1 - \frac{2}{\log^2 n}\left(1 - \frac{1}{\log^2 n}\right)\right)^{\alpha \log^3 n}$$

$$= \left(1 - \frac{2\log^2 n - 2}{\log^4 n}\right)^{\frac{\alpha \log^4 n}{\log n}} \underset{\text{Fact 21.2}}{\leq} e^{-2\alpha \log n} \in O\left(\frac{1}{n^{2\alpha}}\right).$$

As for the second case, $|T| = 2\tau \log^2 n + 1$, we have

$$P_{no} = (1 - P_1)^\lambda \leq \left(1 - 2\tau\left(1 - \frac{1}{\log^2 n}\right)^{2\tau \log^2 n}\right)^{\alpha \log^3 n}$$

$$= (1 - 2\tau e^{-2\tau})^{\alpha \log^3 n} \in O\left(\frac{1}{n^{\alpha \log^2 n}}\right).$$

Because the same computation holds for all three channels, a message from each channel is successfully transmitted with high probability. $\square$

Theorem 21.6 *The DS and MIS algorithms in the single-channel model have time complexity $O(polylog(n))$. With high probability, all critical steps are executed like in the multichannel algorithm.*

Proof Time complexity follows immediately. For correctness, we compute the probability P that all critical steps are correctly simulated. Because the MIS algorithm's execution takes at most $C \cdot n \log^3 n / \log \log n$ steps (and the DS algorithm is even faster) for a constant C in the multichannel case, P is

$$P \geq \left(1 - \frac{1}{n^{2\alpha}}\right)^{\frac{Cn\log^3 n}{\log\log n}} \in 1 - O\left(\frac{\log^3 n}{n^\alpha}\right).$$

$\square$

21.7 Conclusion

How can we efficiently compute a structure completely from *nothing*? We have tried to provide an answer to this question by analyzing the initialization of multihop radio networks, that is, the transition from an unstructured to a structured network. We have shown that the problem of bringing structure into a network (i.e., organizing an efficient medium access scheme) is strongly related to the problem of clustering in the absence of an established MAC layer. We have proposed two randomized algorithms that compute clusterings in polylogarithmic time under a model featuring many of the realities of unstructured networks.

We believe that due to being fast and simple, our algorithms have practical relevance in a variety of scenarios, particularly in newly deployed ad hoc and sensor networks. Analyzing important issues such as energy efficiency in the *unstructured radio network* model is an interesting and promising field for future research.

References

1. Baker, D.J. and Ephremides, A., The architectural organization of a mobile radio network via a distributed algorithm, *IEEE Trans. on Communications,* 29(11), 1694–1701, 1981.
2. Basagni, S., Distributed clustering for ad hoc networks, *Proc. IEEE Int. Symp. Parallel Architectures, Algorithms, and Networks (I-SPAN),* Perth, pp. 310–315, 1999.
3. Chatterjee, M., Das, S.K., and Turgut, D., An on-demand weighted clustering algorithm (WCA) for ad-hoc networks, *Proc. IEEE GLOBECOM,* San Fransisco, pp. 1697–1701, 2000.
4. Gerla, M. and Tsai, J., Multicluster, mobile, multimedia radio network, *ACM/Baltzer J. of Wireless Networks,* 1(3), 255–265, 1995.
5. Stojmenovic, I., Seddigh, M., and Zunic, J., Dominating sets and neighbor elimination-based broadcasting algorithms in wireless networks, *IEEE Trans. Parallel and Distributed Systems,* 12(12), 14–25, 2001.
6. Heinzelman, W., Chandrakasan, A., and Balakrishnan, H., Energy-efficient communication protocol for wireless microsensor networks, *Proc. 33rd Annual Hawaii Int. Conf. System Sciences,* Hawaii, pp. 3005–3014, 2000.
7. Alzoubi, K., Wan, P.J., and Frieder, O., Message-optimal connected dominating sets in mobile ad hoc networks, *Proc. 3rd ACM Int. Symp. Mobile Ad Hoc Networking and Computing (MOBIHOC),* Lausanne, pp. 157–164, 2002.
8. Gao, J. et al., Discrete mobile centers, *Proc. 17th Symp. Computational Geometry (SCG),* Medford, pp. 188–196, 2001.
9. Jia, L., Rajaraman, R., and Suel, R., An efficient distributed algorithm for constructing small dominating sets, *Proc. 20th ACM Int. Symp. Principles of Distributed Computing (PODC),* Rhode Island, pp. 33–42, 2001.
10. Kuhn, F. and Wattenhofer, R., Constant-time distributed dominating set approximation, *Proc. 22nd ACM Int. Symp. Principles of Distributed Computing (PODC),* Boston, pp. 25–32, 2003.
11. Wan, P.J., Alzoubi, K., and Frieder, O., Distributed construction of connected dominating set in wireless ad hoc networks, *Proc. Infocom,* New York, 2002.
12. Wu, J. and Li, H., On calculating connected dominating set for efficient routing in ad hoc wireless networks, *Proc. of the 3rd Int. Workshop on Discrete Algorithms and Methods for Mobile Computing and Communications (DIALM),* Seattle, pp. 7–14, 1999.

13. Basagni, S., A distributed algorithm for finding a maximal weighted independent set in wireless networks, *Proc. 11th IASTED Int. Conf. Parallel and Distributed Computing and Systems (PDCS'99)*, Cambridge, pp. 517–522, 1999.

14. Kuhn, F., Moscibroda, T., and Wattenhofer, R., Initializing newly deployed ad hoc and sensor networks, *Proc. 10th Int. Conf. on Mobile Computing and Networking (MOBICOM)*, pp. 260–274, Philadelphia, 2004.

15. Moscibroda, T. and Wattenhofer, R., Efficient Computation of Maximal Independent Sets in Unstructured Multi-Hop Radio Networks, *Proc. 1st IEEE Int. Conf. on Mobile Ad-Hoc and Sensor Systems (MASS)*, Florida, pp. 51–59, 2004.

16. Feige, U., A threshold of ln n for approximating set cover, *J. of the ACM*, 45(4), 634–652, 1998.

17. Karp, R.M. and Widgerson, A., A fast parallel algorithm for the maximal independent set problem, *Proc. 16th ACM Symp. Theory of Computing (STOC)*, pp. 266–272, 1984.

18. Luby, M., A simple parallel algorithm for the maximal independent set problem, *SIAM J. on Computing*, 15, 1036–1053, 1986.

19. Kuhn, F., Moscibroda, T., and Wattenhofer, R., What cannot be computed locally!, *Proc. 23rd ACM Int. Symp. Principles of Distributed Computing (PODC)*, 2004.

20. Bar-Yehuda, R., Goldreich, O., and Itai, A., On the time-complexity of broadcast in radio networks: an exponential gap between determinism randomization, *Proc. 6th ACM Symp. Principles of Distributed Computing (PODC)*, Vancouver, pp. 98–108, 1987.

21. Hayashi, T., Nakano, K., and Olariu, S., Randomized initialization protocols for packet radio networks, *Proc. 13th Int. Parallel Processing Symp. (IPPS)*, pp. 544–548, 1999.

22. Nakano, K. and Olariu, S., Energy-efficient initialization protocols for single-hop radio networks with no collision detection, *IEEE Trans. Parallel and Distributed Systems*, 11(8), 851–863, 2000.

23. Nakano, K. and Olariu, S., A survey on leader election protocols for radio networks, *Proc. 6th International Symposium on Parallel Architectures, Algorithms, and Networks (ISPAN)*, pp. 71–78, 2002.

24. Gasieniec, L., Pelc, A., and Peleg, D., The wakeup problem in synchronous broadcast systems (extended abstract), *Proc. 19th ACM Symp. Principles of Distributed Computing (PODC)*, Portland, pp. 113–121, 2000.

25. Jurdzinski, T. and Stachowiak, G., Probabilistic algorithms for the wakeup problem in single-hop radio networks, *Proc. 13th Int. Symp. Algorithms and Computation (ISAAC)*, Vancouver, pp. 535–549, 2002.

26. Roberts, L.G., Aloha packet system with and without slots and capture, *ACM SIGCOMM, Computer Communication Review*, 5(2), 28–42, 1975.

22

Self-Organization of Wireless Sensor Networks

Manish M. Kochhal

Loren Schwiebert

Sandeep K.S. Gupta

22.1   Introduction

The continuing improvements in computing and storage technology as envisioned in *Moore's Law*, along with advances in *MEMS* (Micro-Electro-Mechanical Systems) and battery technology, have enabled a new revolution of distributed embedded computing where micro-miniaturized low-power versions of processor, memory, sensing, and communication units are all integrated onto a single board. One of the interesting applications of distributed embedded computing is an ad hoc deployed wireless sensor network (WSN)[59] that is envisioned to provide target sensing, data collection, information manipulation, and dissemination within a single integrated paradigm.

Wireless sensor networks (WSNs) have many possible applications in the scientific, medical, commercial, and military domains. Examples of these applications include environmental monitoring, smart homes and offices, surveillance, intelligent transportation systems, and many others. A WSN could be formed by tens to thousands of randomly deployed sensor nodes, with each sensor node having integrated sensors, processor, and radio. The sensor nodes then self-organize into an ad hoc network so as to monitor (or sense) target events, gather various sensor readings, manipulate this information, coordinate with each other, and then disseminate the processed information to an interested data sink or a remote base station. This dissemination of information typically occurs over wireless links via other nodes using a *multihop* path.[3,26]

The problem of self-organization (or self-configuration) has been a hot topic of research in wireless ad hoc networks, including mobile and sensor networks. Self-organization involves abstracting the communicating entities (or nodes) into an easily controllable network infrastructure. That is, when powered on, it is the ability of the nodes deployed to locally self-configure among themselves to form a global interconnected network. The resulting network organization needs to support efficient networking services while dynamically adapting to random network dynamics. The self-organization problem becomes even more exacting when the devices forming such a collaborative network are crucially constrained in energy, computational, storage, and communication capabilities. The unpredictability of wireless communication media adds a third dimension to this challenge. On top of that, the vision of having unattended and untethered network operation makes it even more complicated to provide even basic network services such as network discovery, routing, channel access, network management, etc. An example of a network that imposes such extreme demands is a wireless ad hoc sensor network.

One of the crucial design challenges in sensor networks is energy efficiency. This is because individual sensor nodes use a small battery as a power source, and recharging or replacing batteries in a remote environment is not feasible. In some cases, sensors may also use solar cells that provide limited power. Thus, to achieve a longer network lifetime, one has to tackle energy efficiency at all levels of the sensor network infrastructure. Because the wireless radio is the major energy consumer in a sensor node, systematic management of network communication becomes critical. Sensor network tasks such as routing, gathering, or forwarding sensing data to a nearby data sink or a remote base station requires network communication. To effectively coordinate these activities, one must address the problems of sensor network organization and the subsequent reorganization and maintenance. However, it should be clear that although energy is currently one of the biggest challenges, the key problems and solutions identified and summarized in this chapter will hold equally well for future generations of wireless sensor networks.[3,26] This is because in any era of technological progress, there arises a necessity to support new challenging applications that inherently magnify analogous critical technological limitations with similar trade-offs against other objectives.

The focus of this chapter is to understand the key problems and their respective solutions toward the various steps involved in self-organization for wireless sensor networks. In keeping with this approach, in this chapter we first summarize self-organization concepts as applicable to the regime of wireless sensor networks. For the sake of simplicity, we consider typical algorithmic aspects of certain elementary network organizations such as a cluster, mesh, tree, and other implicit variations. This provides the necessary platform for exploring existing protocol solutions that consider self-organization as a primary problem. Finally, we provide an elaborate comparative discussion of these self-organization approaches with respect to the metrics used for network formation and the underlying services supported.

The contributions of this chapter are thus threefold: (1) we first identify the typical characteristics of wireless sensor networks, and discuss their impact on the self organization problem; (2) we present several existing protocols that extend the basic network organization schemes to address the unique requirements of wireless sensor networks; and (3) we discuss the pros and cons of these self-organization protocols and provide some insight into their behavior. The remainder of the chapter is organized as follows: In Section 22.2, we provide a detailed overview of sensor network self-organization. The overview section includes discussions on the sensor network communication paradigm (Section 22.2.1), the impact of wireless sensor network characteristics (Section 22.2.2), and self-organization preliminaries (Section 22.2.3). We provide elementary concepts related to sensing and networking in Section 22.2.3. Section 22.3 provides a complete discussion of the currently available approaches and solutions to relevant self-organization issues. Section 22.4 provides a brief summary of the entire chapter. Finally, Section 22.5 provides a glimpse of the future research directions.

22.2 Overview

22.2.1 The Sensor Network Communication Paradigm

Ad hoc wireless networks, by definition, lack a centralized base station that could otherwise be useful for initial coordination of network start-up and self-organization. In the case of ad hoc wireless sensor networks, the remote base station is available only as an application front end rather than as a centralized arbitrator for coordinating basic network activities. Moreover, due to scalability concerns, it becomes difficult, if not impossible, for a base station to manage in fine detail the entire sensor network consisting of possibly a thousand or more sensors. On the contrary, in most sensing application scenarios, the base station at best expects only application-specific query resolution from a sensor (or a group of sensors) and it leaves the network management responsibility to individual solitary sensors that are isolated from each other by certain geographical distances. For processing on-demand application-specific queries, the sensor network adopts some distributed load-balancing heuristics to select an optimal sink (or a sensor node) to gather and process the sensing information from various sensing sources. This sink can also perform the energy-expending, long-haul query-reply to the base station. Figure 22.1 highlights this primary communication paradigm for wireless sensor networks. Figure 22.1(a) shows the basic sensing and networking services organized in a traditional layered fashion. It should be noted here that it is the self-organization protocol that engineers the critical preliminary network infrastructure support for these

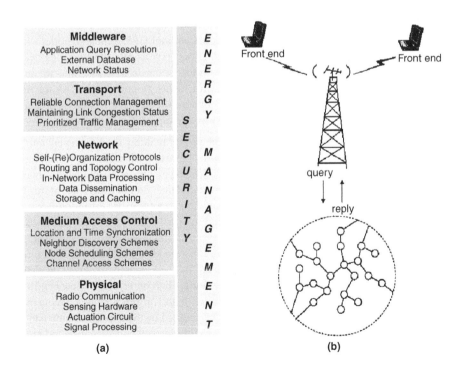

FIGURE 22.1 (a) Communication services required for sensor networks and (b) typical sensor network infrastructure.

future protocol services that may be required by the sensor network. Hence, a careful design encompassing several critical service requirements at various layers of the protocol stack should be considered in the self-organization protocol. Some of these requirements may be conflicting and, hence, a general trade-off may need to be included in the design. For example, the self-organization protocol may need to control the topology of the network by having more node neighbors, which may be at odds with the energy conservation schemes that turn off idle nodes to increase the lifetime of the network. An integrated approach with some adaptive behavior is therefore warranted in the design of self-organization protocols. In subsequent sections of the chapter, we highlight this design aspect by comparing several approaches with respect to basic self-organization requirements. Protocol cross-layering may be required to have reliable context-aware self-organization for certain application-specified tasks. Also, as shown in Figure 22.1(a), network functions such as energy management and security are inherently cross-layer in nature and must be addressed by all protocol layers in some form or other. However, protocol cross-layering[19,39] is outside the scope of this chapter.

22.2.2   Impact of Wireless Sensor Network Characteristics

In this section, we discuss some relevant characteristics of wireless sensor networks that should be incorporated into the design of self-organization protocols. These characteristics are listed below:

1. Network makeup:

    a. *Homogeneous or heterogeneous sensor devices.* The sensor network may consist of specialized nodes having special hardware and software capabilities deployed randomly or deterministically with other low-end sensor devices. This heterogeneous deployment may be required by certain applications, where placement of the sensor nodes is practical. An example of such an application may be monitoring a high-rise building for cracks and other critical hazards or faults. Another example could be some sensor nodes equipped with a GPS system that serve as position-estimating beacons for other nodes without GPS. Similarly, nodes with higher processing capability and higher battery power can serve as data sinks for their neighboring underprivileged nodes. A self-organization protocol thus needs to take into account special node capabilities by tasking these nodes with higher responsibilities. As we will see in Section 22.3, one way of doing this is to organize the network by way of assigning roles to sensor nodes depending upon their performance capabilities.[40,41]

    b. *Indoor or outdoor environments.* Sensors deployed for building monitoring fall into the category of indoor environment, whereas sensors deployed to track enemy movements in a tactical military scenario can be categorized as experiencing an outdoor radio environment. In the outdoors, there are few obstructions, so the radio signals do not experience as much attenuation due to reflections and multipath fading. This is not the case for indoor environments, where walls contribute to a drastic reduction in signal strength. Moreover, sensor network applications are driven by environmental events, such as earthquakes and fires, anywhere anytime following an unpredictable pattern. Sensor node failures are common due to these hostile environments. The radio media shared by these densely deployed wireless sensors may be subject to heavy congestion and jamming. High bit error rate, low bandwidth, and asymmetric channels make the communication highly unpredictable.[63] Frequent network monitoring and feedback should therefore be adopted by the network maintenance part of self-reorganization algorithms to provide a certain degree of fault diagnosis and repair under extreme situations.

    c. *Random or controlled node placement.* Biomedical sensor networks[74] are examples of stationary wireless sensor networks.[64] In such a network, the placement of sensor nodes is controlled and predetermined. A stationary sensor network normally has little or no mobility. One can also decide in advance the number of neighbors a node may have, depending upon application requirements and the position of the sensor (border or internal node) within the deployment.[66] In contrast, a tactical wireless sensor network deployed in a hostile area to track enemy movements in the battlefield is characteristically required to have a random deployment. In this chapter, we consider self-organization under the more challenging case of a random node deployment. For stationary sensor networks, interested readers are encouraged to refer to Ref. 64.

    d. *Node mobility.* There is an important difference between a wireless sensor network and a mobile ad hoc network. In general, wireless sensor networks have zero to limited mobility compared to the relatively high mobility in mobile ad hoc networks (MANETs[53]). Network protocols for MANET mostly address system performance for random node mobility rather than for the energy depletion[42] caused by the execution of various network protocols. However, for ad hoc sensor networks, energy depletion is the primary factor in connectivity degradation and overall operational lifetime of the network. Therefore, for WSNs, overall performance becomes highly dependent on the energy efficiency of the algorithm. However, mobile sensors make wireless networking solutions extremely challenging. A mobile sensor network essentially becomes a special research challenge in the field of mobile ad hoc networks. In this chapter, we discuss only self-organization protocols that *assume fixed or stationary sensors*. Interested readers are encouraged to extend mobility management concepts from MANETs in order to factor mobility into their design of self-organization protocols for mobile wireless sensor networks.[94]

    e. *Node density (or redundancy).* Depending on the application scale, tens of thousands of sensors can be deployed in a very large area. Examples of such an application would be deep space probing and habitat monitoring. The highly unpredictable nature of sensor networks necessitates a high level of redundancy. Nodes are normally deployed with a high degree of connectivity. With high redundancy, the failure of a single node has a negligible impact on the overall capacity of the sensor network. Network protocols for self-organization and maintenance need to adapt the topology of the network to its density and redundancy in order to have maximum energy savings. Sparser networks may need special treatment to avoid network partitions due to the existence of several orphan nodes.[72] Higher confidence in data can also be obtained through the aggregation of multiple sensor readings.[21,84] Moreover, a higher density of beacon nodes (i.e., nodes with position information) could be used to reduce localization errors[11] during position estimations by way of localized triangulations or multilaterations. A similar approach could be used to provide fine- or coarse-grained time synchronization.[23,24,29,31] A better determination of these network parameters could affect decisions for other primary network operations such as medium access or node scheduling schemes, event localization, topology control, etc.

2. *Data-centric addressing*: Data-centric addressing is an intrinsic characteristic of sensor networks. It is impractical to access sensor data by way of *ID* (or IP, as in the Internet). It is more natural to address the data through content or location. The *IDs* of the sensor nodes may not be of any interest to the application. The naming schemes in sensor networks are thus often data oriented. For example, an environmental monitoring application may request temperature measurements through queries such as "collect temperature readings in the region bounded by the rectangle $[(x_1, y_1), (x_2, y_2)]$," instead of queries such as "collect temperature readings from a set of nodes with the sensor network addresses x, y, and z." Thus, in a large-scale sensor deployment, nodes may not have unique *IDs*. However, to pursue localized control of network operations among nodes, unambiguous unique *IDs* become necessary among neighbors. Also, because the scale of a network deployment could be very large, node identification by way of unique hardware *IDs* is not feasible as it becomes an overhead on the MAC header. This may be a significant part of the packet payload and hence may represent an important source of energy consumption. Generating unique addresses within the transmission neighborhood of a node and reusing it elsewhere greatly reduces the size of the MAC

address size.[71] Auto-address configuration thus becomes one of the preliminary steps during sensor network self-organization.

3. *In-network processing:* There are essentially two options to communicate event information of interest to the base station application. One is sending individual readings of each sensor to the base station, which is impractical given the constraints discussed earlier. Another more feasible option is to gather, process, and compress neighboring correlated sensing event information within the network and then send it to the base station. This promotes energy efficiency by having a reduced data volume for long-distance transmission to the base station. It also promotes simple multihop reliability as nodes gather and process data in a localized hop-by-hop fashion as opposed to TCP's complex end-to-end reliability schemes.[67,80,85] As discussed in Section 22.2.3, self-organization algorithms usually achieve in-network processing by organizing nodes according to various aggregation patterns and network layouts or architectures.

4. *Sensing application characteristics:* Some biomedical applications[74] such as a glucose level monitoring require periodic monitoring of a patient's insulin level. In other applications, the sensor network needs to send information to the base station application only when an interesting event has been sensed. There may be other applications that desire both periodic and discrete event-triggered monitoring. A more challenging scenario could be an application requesting an on-demand organization of sensors in a certain region of network deployment for pursuing a fine- or coarse-grained monitoring of current or upcoming sensing events. Highly *resilient* self-organization protocols that (re)organize sensors either around some sensing event traffic or around a geographic region for periodic or on-demand monitoring by applications are thus warranted.

22.2.3 Self-Organization Preliminaries

In this section, we discuss the elementary concepts of network self-organization as applicable to the regime of wireless sensor networks. These concepts serve as the necessary foundation for interested readers to pursue future research in the area of sensor network self-organization. For ease of understanding, we have classified these basics into sensing and network organization concepts. We also formalize the necessary steps (or protocols) that fall under the unified umbrella of sensor network self-organization.

The *sensing concepts* (*sensing phenomenon*[41]) are concerned with the characteristics of the sensors, the events to be detected, and their topological manifestations, both in the spatial and temporal domains. For example, it is obvious that sensors in close proximity to each other should have correlated readings. A temporal dual of this observation implies that sensor readings among neighboring sensors also have some correlation within some nearby time intervals. In addition to supporting the properties associated with the sensing phenomenon, it is also necessary to support hierarchical event processing to have an incremental comprehensive global view of an area of deployment at different levels of the self-organized network hierarchy.

As mentioned previously, self-organization involves abstracting the communicating sensor nodes into an easily controlled network infrastructure. Cluster, connected dominating set (CDS), tree, grid, or mesh based organizations are typical. We provide some insight into these organizations for use in wireless sensor networks.

22.2.3.1 Elementary Networked-Sensing Concepts

The sensing phenomena mentioned previously relate to the natural property of sensors sensing events collaboratively as well as individually in a group. Figures 22.2 through 22.5 illustrate these sensing concepts of wireless sensor network organization for target detection or tracking. In the following discussion, we use the terms "sensing groups" and "sensing zones" interchangeably.

Figure 22.2 illustrates that the sensing capability of sensors sensing events collaboratively or individually in a group depends essentially on the sensitivity of the sensors with respect to the target event. The sensitivity of a sensor diminishes with increasing distance of the sensor from the target. This sensitivity

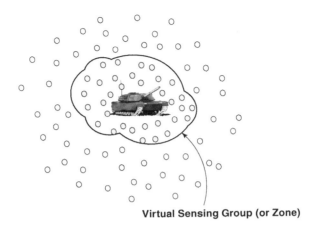

Virtual Sensing Group (or Zone)

FIGURE 22.2   Spatial group sensing concept.

can be characterized theoretically by sensor models that are based on two concepts. One is that the sensing ability (*coverage*) diminishes with increasing distance.* Second is that noise bursts diminish the sensing ability but this effect of noise can be minimized by allowing sensors to sense over longer time periods (more *exposure*). Several algorithms based on the above sensitivity model have been developed that formulate the exposure and coverage properties of sensor networks. These algorithms use traditional computational geometry-based structures such as the Voronoi diagram and the Delaunay triangulation[48,49] to compute sensing coverage and exposure. However, distributed versions of these algorithms are challenging and computationally intensive, and hence are impractical for use during the initial network organization phase.

In general, self-organization protocols usually employ the concept of *redundant sensing* to account for fault-tolerant sensing in the presence of environmental vagaries. By redundant sensing, we mean that an observation of the presence of a nearby target event (i.e., a tank* in Figure 22.2) should be supported not only by one sensor, but also by a group of neighboring sensors.[21,84] This requires selection of a group of neighboring sensors that can take sole responsibility for any event appearing within their region or group. The selection of sensors to form such a group requires quantifying relative proximity distances of each and every neighboring sensor. It also requires an intelligent discrimination between near and far sensors to avoid grouping sensors from distant locations.

Figures 22.3 and 22.4 illustrate the sensing zone dependency situation during tracking by sensing zones formed around a mobile enemy tank. Specifically, here we are discussing an initial network organization that statically forms sensing zones in anticipation of the occurrence of any future event. In the case of a random sensor deployment scenario, it is not possible to precisely control and place sensors so that they end up in groups having no overlap with neighboring sensor groups. This means that although an attempt was made to form stand-alone sensing groups (or zones) that independently take responsibility for detecting and tracking events, there are some overlapping regions where collaboration among neighboring sensing groups may be needed. However, the boundary nodes in each region can also serve as anchors for tracking events moving from one neighboring region to another. This is essentially a dichotomous scenario because,

---

*The sensing range may depend on the dimensions of the observed target; for example, a seismic sensor can detect a tank at a greater distance than it can detect a soldier on foot. For ease of discussion, we assume the sensing range to be the same for targets of similar dimensions.[77] However, in general, for an application specific sensor deployment, nodes are assumed to be preconfigured for desired targets in terms of their sensing signatures or readings. In the case of on-demand target detection and tracking, the application is free to provide respective target sensing signatures in its queries.

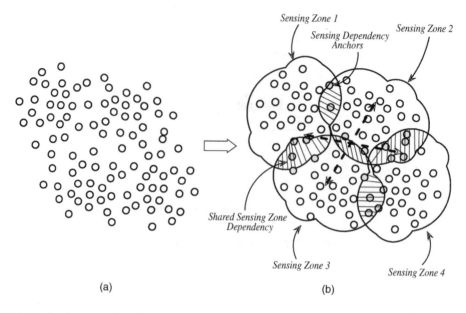

FIGURE 22.3 Sensing group dependency concept: (a) sensor nodes randomly deployed, and (b) preliminary sensing zone based organization.

on the one hand, we need independence between neighboring sensing zones but on the other hand, we also want to efficiently track events moving across neighboring sensing zone boundaries. An event monitoring and tracking algorithm that runs on top of such a self-organized network would have to analyze this dependency and utilize it to its best advantage. This can be done by either identifying neighboring dependent sensing zones and allowing collaboration among them for events moving around their neighborhood or tracking applications can dynamically specify an on-demand incremental reorganization of a new sensing zone around the moving event as it crosses the old sensing zone boundaries. The *EnviroTrack* project[1,10,25] that is currently being pursued at the University of Virginia is an initial proof-of-concept implementation that supports such application-specific sensing group network (re)organization for tracking in a physical environment.

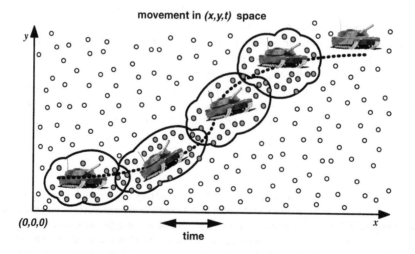

FIGURE 22.4 Tracking a mobile tank around neighboring sensing groups.

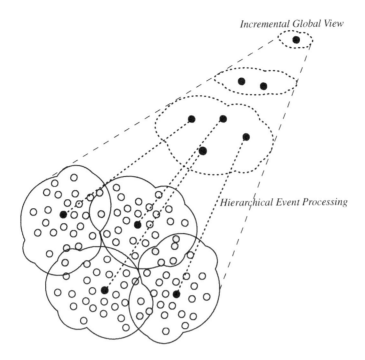

FIGURE 22.5   Hierarchical event processing for incremental global view.

Figure 22.5 illustrates event processing at various levels of a hierarchical sensor organization. It can be seen that as we go higher in the hierarchy, fewer nodes are involved in event processing. However, as we go up the hierarchy, we also lose detail about the event(s). This is because wireless communication is an overhead in terms of draining energy. Also, due to the small form factor of the sensors, memory is also a crucial resource and, hence, a lot of information cannot be maintained by an individual sensor or a small group of sensors. If we assume that the sensors selected for the upper levels of the hierarchy are powerful in terms of both communication energy as well as memory, the problem is still not resolved, due to scalability issues. However, any feedback from the sensing application about the granularity of monitoring would help in reducing overheads in information gathering and processing. In any case, hierarchical processing motivates the concept for distributed gathering, caching, and processing of sensing events where certain nodes in the hierarchy are assigned apt roles[40,41] according to their capability in the current network organization.

22.2.3.2   Elementary Network-Organization Concepts

The primary objectives of this section are to categorize several elementary network organization architectures and discuss some relevant approximation algorithms that can be extended by self-organization protocols.

Figure 22.6 shows a simple classification of various network architectures that can be employed by self-organization protocols. This classification is not complete, as there can be certain combinations of different network architectures. However, it provides the principal categories under which several current implementations can be studied and analyzed. Self-organization protocols can be either proactive or reactive. That is, protocols can organize the sensor network statically in preparation for any future event or they can dynamically configure the network around any current event of interest. Additionally, self-organizing protocols can pursue either a difficult-to-maintain hierarchical manifestation of the above network architectures or they can simply satisfy requests with their corresponding flat manifestations. Figure 22.7 provides a visual blueprint for the above elementary network architectures such as the chain,

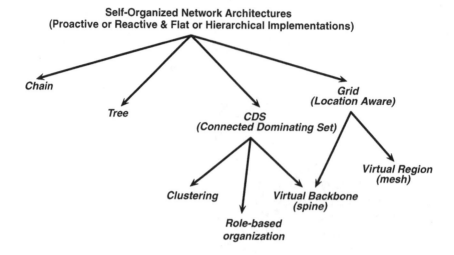

FIGURE 22.6   Self-organized network architectures.

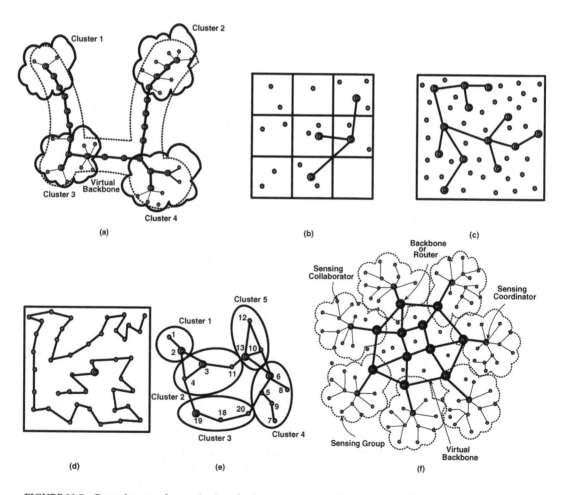

FIGURE 22.7   Example network organizations for (a) spine, (b) virtual grid, (c) tree, (d) chain, (e) clustering, and (f) role-based virtual zones.

tree, spine, virtual grid, and role-based virtual regions. We discuss the typical algorithmic aspects of these network formations for general wireless ad hoc networks. This facilitates easier comprehension and analysis of those sensor network organization protocols that extend or modify these algorithms in order to meet various sensing application requirements. However, there are certain concepts that are common across all these network organizations. In all these organizations, nodes adopt certain performance metrics for selecting neighbor(s) in their local network formation heuristics. These performance metrics might be minimum distance, minimum energy, minimum transmission power, maximum/minimum node degree, delay, bandwidth, etc. Some of these metrics can be used collectively in some particular order (depending on priority) to break ties among several eligible competitors. To have an optimally ideal neighbor selection scheme for self-organization, nodes may require complete global state information of the network. However, in an ad hoc network, nodes that execute distributed algorithms for localized self-organization do not have the luxury of gathering, maintaining, and using complete network knowledge. As mentioned previously, this is because there are trade-offs among storage capability, communication costs, computational capability, and time to completion. This effectively results in nodes maintaining network state information for only two- to three-hop neighbors. Using this information, nodes execute local decisions to select neighbors to form a global self-organized network.

The chain-based organization is one of the simplest ways of organizing network communication, where nodes farther from the base station initiate chain formation with their nearest neighbor. The idea is to gather and fuse all the data from every node by forming a chain among them. A leader is then selected from the chain to transmit the fused data to the base station. However, building a chain to minimize its total length is similar to the traveling salesman problem, which is known to be intractable. A greedy chain formation algorithm,[46] when pursued recursively for every node, results in a data-gathering chain oriented toward the base station. As we discuss later, self-organization algorithms for sensor networks usually also include certain sensing metrics in order to form an optimal organization that is efficient from both sensing[37,41] and networking perspectives.[65,75]

The tree type of network formation is similar to the chain and can be considered an extension of the chain-based mechanism. Tree type network organizations utilize the multipoint connectivity nature of the wireless medium, where one source can be heard simultaneously by several nearby receivers that act as its children. If both the sender and receiver scheduling are made collision-free, then a tree-based network organization can support both *broadcast or multicast* (i.e., dissemination of information from a central node) and *convergecast* (i.e., gathering of information toward a central node) communication paradigms across all application domains.[5] A considerable amount of research work is available for constructing multicast trees[7,27] for dynamic wireless ad hoc networks.[2,34] Algorithms for generating multicast trees typically must balance the goodness of the generated tree, execution time, and the storage requirements. Several cost metrics such as delay, communication costs, etc. are used to generate cost-optimal multicast trees. However, this gives rise to the well-known *Steiner Tree problem*, which is NP-complete.[86] Instead, heuristics are used to generate "good" rather than optimal trees. This is still an active area of research.

Cluster-based organizations (refer to Figure 22.6(e)) partition the entire network into groups called clusters. Each cluster is formed by selecting some nodes based upon some quality metric such as connectivity or distance.[4,15,81] as cluster members and a group leader, known as the cluster head, is also selected using some metric such as maximum energy to manage that cluster. These cluster heads, when connected, form a virtual backbone or spine as in Figure 22.7(b)[14,76] or a connected dominating set (*CDS*) of nodes. Related to clustering is the problem of finding a minimum connected dominating set (*MCDS*) of the nodes, which is NP-complete. An *MCDS* satisfies two properties: (1) each node is either a backbone node or is one hop from a backbone node, and (2) the backbone nodes are connected. There are several approximation algorithms[32] available in the literature that engineer virtual backbone based network configurations satisfying the *MCDS* properties. Of particular importance are Wu's[88,89] distributed and localized algorithms for constructing a hierarchical connected dominating set. This algorithm is inherently distributed and simple in nature. Ideally, it requires only local information and a constant number of iterative rounds of message exchanges among neighboring hosts. The algorithm for CDS formation involves a dominating set reduction process and some elimination rules based on quality metrics that are executed by nodes locally to identify their dominating neighbors. The dominating set reduction process, when executed recursively by an already existing set of dominating nodes, produces a domination hierarchy. Dominating nodes at any level of the hierarchy can serve as local network coordinators for nodes in the next lower level of the hierarchy. Figure 22.7(f) shows a role-based hierarchical CDS organization of nodes where the lower levels of nodes form the cluster, whereas upper levels of dominating nodes can be used for routing or hierarchical information processing. Accordingly, nodes at every level of the domination hierarchy assume respective roles in the network, depending upon the quality metrics used for role selection at that level.[40,41]

Finally, if nodes have location information (e.g., using GPS), then virtual grid based organizations[93,94] typically configure the network by partitioning the area of deployment into uniform grids also known as a mesh. From each grid, a dominating node is chosen using some selection rules. Dominating nodes from each grid, when connected, form a virtual backbone, which can then be used for gathering or forwarding information from one geographic region to another. The grid-based organization allows an implicit and simple naming system by having grids (regions) in the network be addressed by their relative geographic locations. Thus, it avoids the complex and nonscalable address generation mechanisms for individual nodes that are densely deployed in a very large area. However, the efficiency of such an organization depends critically on location accuracy and network partitioning schemes.

22.2.3.3   Steps to Sensor Network Self-Organization

Self-organization or self-configuration is one of the basic and initial steps toward an ad hoc deployment of wireless sensors. The network deployment as mentioned previously can be done deterministically or randomly. In any case, the objective is to have nodes discover their neighbors, establish their positions, and form an easily manageable network architecture. All these self-organization activities must be performed in a localized and distributed manner with high energy efficiency and little or no communication overhead. Moreover, the self-organized network infrastructure should be *adaptive* and *resilient* to being easily reorganized with respect to the ultimate traffic pattern that may run on top of it.

Following the self-organization steps as extended from Refs. 17 and 82 forms the complete basis for any self-organization algorithm for ad hoc wireless sensor networks:

1. *Network discovery or initialization phase:*
    a. Each node discovers its set of neighbors,
    b. Depends on communication transmission radius (Tx_{max}), and
    c. Random or deterministic initial channel scheduling for neighbor discovery.

2. *Coarse grained estimation phase:*
    a. Location estimation and
    b. Time synchronization.

3. *Organizational phase:*
    a. Formation of a hierarchical or flat network organization with the help of local group formations,
    b. Performing group reorganization if necessary,
    c. Generation of addresses for nodes,
    d. Generation of routing table at every node,
    e. Generation of broadcast or multicast trees and graphs within a group,
    f. Merging of broadcast trees and graphs when groups are aggregated to form hierarchical networks,
    g. Establish medium access control schemes both for intra-group and inter-group communication, and
    h. Establish key setup schemes for secure communication.

4. *Maintenance phase:*

   a. Active or passive monitoring (e.g., by "*I am Alive*" messaging),

   b. Network quality evaluation schemes such as connectivity and sensing coverage,

   c. Maintenance of routing tables,

   d. Maintenance of broadcast infrastructure,

   e. Fine-grained tuning of network parameters such as location and network time,

   f. Topology control schemes to maximize network throughput and spectral reuse or network capacity,[14,68,73] and

   g. Energy conservation schemes for increased network longevity using dynamic node scheduling.[83,93]

5. *Self-reorganizing phase:*

   a. Redeployment leading to discovery of new node neighbors and

   b. Fault detection and recovery schemes under node or link failure and group partitions.

In general, the steps listed above can also be considered as services provided by self-organization protocols for wireless sensor networks. This means that some of these services may be optional whereas some are fundamental to any algorithm that self-organizes the sensor network. Thus, Steps 1, 3, and 5 are necessary. On the other hand, Steps 2 and 4 are optional and can be developed separately. Location estimation protocols are generally referred to as self-configuring localization protocols. Network time synchronization can be considered orthogonal and implemented separately without regard to any specific network design or architecture. Similarly, the maintenance phase can be implemented separately as a suite of network management protocols.

22.3   Approaches and Solutions

In this section we discuss the current approaches and solutions along the same lines as the various self-organization stages mentioned earlier. There is much research in location estimation,[11,20,69,70] time synchronization techniques,[23,24,29,31] medium access control (MAC),[33,87,95] and key exchange schemes for secure communication[38,58,100] for ad hoc sensor networks. Discussion of these are outside the scope of this chapter.

22.3.1   Neighbor Discovery or Initialization Protocols

Neighbor discovery is one of the initial steps after node deployment toward sensor network self-organization. However, it can also be periodically used as an intermediate step after self-organization for network maintenance. For network maintenance, a node may have to listen continuously to find dying neighbors or new neighbors. Neighbor discovery at this stage can also be used to evaluate the connectivity and sensing coverage of the network before further network deployment is deemed necessary. Further deployment of nodes may be needed, which again requires additional neighbor discovery to find new nodes as neighbors. Because listening for wireless communication is also a power-consuming process that may exhaust batteries, some randomized or probabilistic node scheduling scheme can be used during the network maintenance stage.

Neighbor discovery messages are essentially broadcast messages that are transmitted at different radio transmission power levels, depending upon the desired number of neighbors for a topology. Due to the wireless medium and dense ad hoc deployment, broadcast radio signals are likely to overlap with others in a geographical area. A straightforward broadcasting by flooding is usually very costly and will result in serious redundancy, contention, and collision problems, which are collectively referred to as the *broadcast storm*[55] problem. Also, initial neighbor discovery by way of sending broadcast messages cannot use sophisticated MAC protocols. As we will see, the presence of a medium access scheme during network maintenance provides important flexibility in terms of avoiding frequent arbitrary channel collisions. This flexibility is not available immediately after initial network deployment, thus affecting the quality of the neighbor discovery process. The performance metrics for the neighbor discovery process include parameters such as the number of neighbors discovered, time to completion of the network discovery process, and energy efficiency. Other relevant network characteristics that affect neighbor discovery include the density of node deployment, radio transmission range, and efficiency of the channel arbitration schemes.

Initial neighbor discovery or network initialization protocols typically employ certain deterministic channel access schemes such as the TDMA-based scheme. Early efforts in using a TDMA-style neighbor discovery can be traced back to the formation of a link clustered architecture (LCA)[6,30,45] in ad hoc wireless networks. In this case, a base station can initiate network time synchronization and then allow nodes to communicate in TDMA fashion, where a TDMA slot is identified by a node's unique ID. However, there are several limitations to such an approach. One important problem is the convergence speed of such a neighbor discovery algorithm. As the networks become large, the time required to complete neighbor discovery will increase proportionally. Also, nodes may have to broadcast at their maximum radio transmission power, which will increase energy dissipation.

Neighbor discovery can also be done randomly or probabilistically in order to randomize listening and broadcast times for neighbor discovery messages. McGlynn and Borbash[47] propose an energy-efficient neighbor discovery protocol based on the birthday paradox. In the birthday paradox, we compute the probability that at least two people in the same room will have the same birthday. When there are as few as 23 people, the probability that at least two have the same birthday already exceeds $1/2$. This idea is borrowed for a channel access scheme for neighbor discovery. Over a period of n slots, two wireless nodes independently and randomly select k slots. The first node transmits a message on its k slots, and the second node listens on its k slots. For the remaining $n - k$ slots, the nodes are idle or in power-saving mode. The probability that a second node hears the first is nearly 1 when the ratio k/n is relatively small. The authors calculate that even with 93 percent idle time, there is a high probability that a second node is able to hear the first. For dense networks, the birthday protocol may fail to discover some links. However, by trading energy for minimal discovery loss, the authors show that their protocol is highly effective, even when a node has many neighbors.

22.3.2   Sensor Network Organization Protocols

As discussed previously, sensor network organization protocols usually employ typical organizational designs such as the chain, tree, cluster, or variations of the connected dominating set (CDS) architecture. Distributed localized algorithms for network self-organization also utilize certain quality metrics that effectively make this problem difficult or even intractable, and the solutions approximate. Although the organizational phase (i.e., Step 3) lists several key requirements for self-organization, these requirements alternatively can be looked upon as services supported by the underlying network infrastructure. Because the main goal of sensor network organization protocols is to provide uniform or balanced conservation of network energy or lifetime, the protocols essentially end up providing some or all of these services at varying degrees of QoS. All this depends on the energy-saving scheme that is integrated implicitly with the network organization, and the network parameters that are allowed to be in trade-off to gain extra energy savings.

PEGASIS (Power-Efficient GAthering in Sensor Information Systems)[46] is one of the early representatives of the chain-based design that organizes the sensor network into a chain. The main idea in PEGASIS is for each node to receive from and transmit to close neighbors and take turns being the leader for transmission to the base station (BS). The goal is to distribute the energy load evenly among the sensor nodes in the network for data-gathering services. Because forming a reduced distance chain is similar to the traveling salesman problem, a greedy approach is pursued locally by nodes first selecting its close neighbors in the chain. The chain formation is initiated at the farthest node from the base station. The greedy algorithm assumes that all nodes have global knowledge of the network. Due to the impractical nature of this assumption, the authors also suggest offloading this complex chain formation process to the base station, which then computes and broadcasts the chain organization to all the sensor nodes.

Cluster-based designs are the most favorable approach to sensor network self-organization. One of the preliminary works that uses clusters is the LEACH (Low Energy Adaptive Clustering Hierarchy) protocol.[36] LEACH is designed to support periodic remote monitoring on ad hoc sensor networks. The LEACH algorithm provides dynamic distributed cluster formation, local processing of sensing data, and randomized rotation of the cluster heads for every round of communication to the base station. All these activities are essentially targeted to support uniform energy usage and to maximize system lifetime. However, there are several limitations in the LEACH design that limit its ability to conserve energy. Because the LEACH protocol selects the cluster heads randomly rather than deterministically, this often results in suboptimal selection of cluster heads. In other words, the distance variation among cluster members in a cluster is large, and the variance becomes more pronounced for sparse network topologies. This effectively translates into nodes communicating information to their clusterhead or base station at larger distances, which results in an appreciable energy loss. Also, there are no checks to avoid large membership overheads in clusters. However, their organization infrastructure does integrate services such as data inter- and intra-cluster gathering, channel access, and network reorganization and maintenance.

Deterministic approaches to cluster-based network organization solve the above problems associated with randomized cluster formations. Krishnan and Starobinski[43] present two algorithms that produce clusters of bounded membership size and low diameter by having nodes allocate local *growth budgets* to neighbors. Unlike the expanding ring approach,[61] their algorithms do not involve the initiator (cluster head) in each round and do not violate the specified upper bound on the cluster size at any time, thus having a low message overhead compared to the expanding ring approach. Kochhal, Schwiebert, and Gupta[40,41] propose a role-based hierarchical CDS-based self-organization (RBHSO, refer Figure 22.7(f)) that deterministically groups sensors into clusters. To pursue such an organization, they develop fault-tolerant group-sensing metrics to partition the sensor network into several sensing zones (clusters). These sensing zones individually act as an aggregate consisting of sensor nodes collaborating to achieve a common sensing objective with a certain *sensing quality-of-service (sQoS)*. Routing and sensing roles are deterministically assigned to sensor nodes, depending upon their connectivity and sensing quality, respectively. Sensing coordinators are also deterministically elected that act as leaders for their respective sensing zones. The sensor coordinators play the role of systematizing collaboration among members in the sensing zone. To limit the the number of sensing zones as well their membership, the RBHSO protocol uses two specified minimum and maximum sensing zone membership limits. However, in the final stages of their algorithm, orphan nodes are allowed to join any nearest neighboring sensor coordinator or a sensing zone member. This is done to cover the maximum possible number of nodes in the organized hierarchy. Hierarchical cluster-based organization of a network of wireless sensors was also proposed by Chevallay et al.[17] Their proposal essentially builds on the hierarchical self-configuration architecture proposed by Subramanian and Katz.[82] The cluster head election is based primarily on the energy level and processing capability of each sensor node. Within each group, nodes are randomly assigned IDs drawn from numbers of a limited size. However, if this pool of numbers is large enough, then there is a high probability of the identifiers being unique both temporally and spatially. This approach is similar to RETRI (Random, Ephemeral Transaction Identifiers[22]), which also assigns probabilistically unique identifiers. However, by tagging respective group identifiers to node IDs, Chevally et al.[17] go one step further to provide a globally unique address for each sensor in the network.

Tree-based organizations are usually used to dynamically group disparate sensing sources with a particular sink node(s). Mirkovic et al.[52] organize a large-scale sensor network by maintaining a dynamic multicast tree-based forwarding hierarchy that allows multiple sinks to obtain data from a sensor. Sohrabi, Pottie et al.[18,78,79] propose several algorithms for the self-organization of a sensor network, which include the self-organizing medium access control, energy-efficient routing, and formation of ad hoc subnetworks for cooperative signal processing functions. Their self-organization technique forms an on-demand minimum-hop spanning tree to a central node or sink elected among neighboring sensors that sense environmental stimuli. They[78,79] propose a self-organization protocol for WSN that trades available network bandwidth by activating only necessary links for random topologies using TDMA-based node scheduling schemes to save energy. Their self-organizing algorithm includes a suite of protocols designed to meet various phases of network self-organization. One protocol, SMACS, forms a joint TDMA-like schedule, similar to LCA,[6] for the initial neighbor-discovery phase and the channel-assignment phase. Other protocols (such as EAR, SAR, SWE, and MWE) take care of mobility management, multihop routing, and the necessary signaling and data-transfer tasks required for local cooperative information processing.

The TTDD (Two Tier Data Dissemination) Model[94] and GAF (Geographic Adaptive Fidelity)[93] are network organization schemes that use geographic location information to partition the network into a grid. The goal of TTDD is to enable efficient data dissemination services in a large-scale wireless sensor network with sink mobility. Instead of waiting passively for queries from sinks, TTDD exploits the property of sensors being stationary and location-aware to let each data source build and maintain a grid structure. Sources then proactively propagate the existence information of sensing data globally over the grid structure, so that each sink's query flooding is confined within a local grid cell. Queries are forwarded upstream to data sources along specified grid branches, pulling sensing data downstream toward each sink. The goal of GAF[93] is to build a predefined static geographical grid only to turn off nodes for energy conservation. Similar algorithms exist that elect or schedule nodes according to certain quality metrics. These algorithms loosely come under the energy-conserving maintenance part of self-organization. Tian and Georganas[83] propose an algorithm that simultaneously preserves original sensing coverage and also autonomously turns off redundant sensor nodes in order to conserve energy. However, they need a specialized multidirectional antenna for measuring the angle of arrival (AoA) of signals to evaluate sensing coverage of neighboring areas that are geometrically modeled as sectors. They also suffer from problems due to the randomization of the backoff timer that is used to avoid having all neighbors turn themselves off, leaving a blind spot. Slijepcevic and Potkonjak[77] propose a heuristic that organizes the sensor network by selecting mutually exclusive sets of sensor nodes that together completely cover the monitored area.

In mobile ad hoc networks, self-organization essentially involves maintaining some form of network organization to support routing infrastructure in the presence of random uncontrollable node mobility. Some relevant research in this area includes the ZRP protocol[35] and the terminodes-based[9] approach. For mobility management, ZRP uses zones that are similar to clusters, whereas the terminodes-based approach uses the concept of self-organized virtual regions. Routing using either of these approaches involves two different schemes, a proactive routing scheme for nodes within a local virtual region or zone and a reactive scheme for nodes located in remote virtual regions or zones. Because in mobile ad hoc networks the availability of the network depends on each user's discretion, an incentive for cooperation by way of virtual money called nuglets is employed in terminodes.

22.3.3 Sensor Network Maintenance Protocols

Frequent network monitoring among neighbors and execution of protocols such as topology control and node scheduling form part of the maintenance phase of the self-organization protocol. The main goal of the maintenance phase is to provide increased network lifetime and also to maintain the infrastructure support for critical networking services. The topology of an ad hoc network plays a key role in the performance of networking services such as scheduling of transmissions, routing, flooding, and broadcasting. In many cases, several network links are not needed for establishing efficient sharing of the channel among neighboring nodes or the routing of data packets. Removing redundant and unnecessary topology information is usually called topology control or topology management.[8]

One way of controlling network topology is to have nodes select their neighbors based on a certain quality metric, say connectivity or available energy, or a combination of both. Nodes that are not selected remain in sleep (or power-saving) mode. In other words, the topology is controlled by having active nodes control their respective node degrees.[62] Another approach is to have nodes adjust their radio transmission power to control their neighboring topology.[54] However, the additional problem of unidirectional links must be solved when nodes are allowed to transmit at different transmission powers.[60] In general, topology control problems are concerned with the assignment of power values to the nodes of an ad hoc network so that the power assignment leads to a graph topology satisfying some specified properties; for example,

1. Energy efficiency with an emphasis on increasing network lifetime
2. Maximize network throughput
3. Provide strong network connectivity
4. Maximize wireless spectrum reuse or network capacity

Xu[90,91] proposes two topology control protocols (GAF and CEC) that extend the lifetime of dense ad hoc networks while preserving connectivity by turning off redundant nodes. Geographic Adaptive Fidelity (GAF)[93] finds redundant nodes by dividing the whole area of deployment into small "virtual grids" and turning ON only those nodes that are essential for communication among neighboring grids. The other protocol, Cluster-based Energy Conservation (CEC), forms a cluster-based network organization to directly observe radio connectivity to determine redundancy. GAF needs exact location information using GPS and an idealized radio model, which are its limitations. CEC, on the other hand, needs to only monitor network organization (in this case, a cluster) to find redundant nodes. However, this monitoring process employs random sleep and awake timers (i.e., duty cycle) to save energy. Chen et al.[14] propose a topology maintenance algorithm (Span) for energy efficiency in ad hoc wireless networks. Span is a localized power-saving protocol that removes redundant nodes in a dense ad hoc network and adaptively elects only a small number of coordinators that stay awake continuously and perform multihop packet routing. Other nodes remain in power-saving mode and periodically check if they should wake up and become a coordinator. These coordinators form a strongly connected network backbone, also known as a *connected dominating set (CDS)*. Bao and Garcia-Luna-Aceves[8] propose another topology management algorithm that constructs and maintains a backbone topology based on a minimal dominating set (MDS) of the network. Sparse Topology and Energy Management (STEM)[73] accepts delays in path-setup time in exchange for energy savings. It uses a second radio, operating at a lower duty cycle, as a paging channel. When a node needs to send a packet, it pages the next node in the routing path. This node then turns on its main radio so that it can receive the packet.

Using only transmission control techniques for controlling sensor network topology, both centralized[12] and distributed[44] approaches have been proposed. Tseng et al.[12] consider the problem of topology control by tuning the transmission power of hosts to control the structure of the network. The target topology for their protocol is 1-edge, 1-vertex, 2-edge, and 2-vertex-connected graphs. They propose two global centralized algorithms to support minimum spanning tree construction, where host transmission powers are fixed or variable during the lifetime of the network. Kubisch et al.[44] propose two distributed algorithms (LMA and LMN) that dynamically adjust the transmission power level on a per-node basis. The "local mean algorithm" (LMA) uses a predefined minimum (NodeMinThresh) and maximum (NodeMaxThresh) threshold for the number of neighbors a node can have and continues to increase its transmission power until it finds neighbors within NodeMinThresh and NodeMaxThresh. The "local mean number of neighbors algorithm" (LMN) goes a step further than LMA by considering the mean number of neighbors each of its neighbors has and increases transmission power accordingly.

Because MAC-level protocols have a very small view of the network, the main approach followed by such energy-efficient protocols has been to turn off radios that are not actively transmitting or receiving packets. Because there is a certain amount of time involved in turning radios back on when they are needed, MAC protocols typically trade off network delay for energy conservation. Energy-efficient MAC and routing protocols can be used together to increase energy conservation. AFECA[92] seeks to maintain a constant density of active nodes by periodically turning radios off for an amount of time proportional to the measured number of neighbors of a node. By following this approach, more energy can be conserved as the density increases. However, AFECA needs to be conservative in its decision to turn off radios as the density measurement is not absolute.

By doing energy conservation with *application-level* information, it is possible to save much more energy, yet the sacrifice is having a network with application-specific characteristics. ASCENT[13] measures local connectivity based on neighbor threshold and packet loss threshold to decide which nodes should join the routing infrastructure based on application requirements.

22.3.4 Self-Healing and Recovery Protocols

Self-healing or recovery might also be one of the objectives of network management. The type of network architecture, flat or hierarchical, could also be a factor in deciding the complexity of network management. Mechanisms for detecting or repairing network partitions is necessary for energy-constrained wireless sensor networks. Partitions are more prevalent when node density is low. Network partitions can also occur even in dense networks when a number of nodes are destroyed or obstructed. When such network partitions do occur, complementary mechanisms will be needed for detecting and repairing partitions. Prediction-based approaches are essentially an implicit way of recording and maintaining the vital networking stats of a node and its neighboring resources. This history information can then be used to predict network partitions (or holes) due to node failures. Salvage techniques would involve either an explicit redeployment of new nodes around the area of lost coverage or having a subset of nodes move in a controlled manner from a dense coverage area to sparser areas. The latter technique of self-healing is also known as *self-aware actuation*,[28] where the actuation ability of mobile sensors allows the network to adaptively reconfigure and repair itself in response to unpredictable runtime dynamics.

Mini et al.[50,51] present a sensor network state model for each sensor node that is modeled as a Markov chain. This model is then used to predict the energy consumption rate and to create an energy map of a wireless sensor network. The energy map is built based on a prediction approach that tries to estimate the amount of energy a node will spend in the near future. Chessa and Santi[16] present an energy-efficient fault diagnosis protocol that can identify faulty or crashed nodes in a wireless sensor network. This diagnostic information gathered by the operational sensors can then be used by the external operator for the sake of network reconfiguration or repair, thus extending network lifetime. Zhao et al.[97–99] propose a tree-based monitoring architecture to monitor wireless sensor networks with different levels of detail, and focus on the design of computing network digests. Digests represent continuously computed summaries of network properties (packet counts, loss rates, energy levels, etc.) and can serve the need for more detailed, but perhaps energy-intensive, monitoring. Finally, Zhang and Arora[96] propose a virtual hexagonal cell-based organization, similar to the virtual grid-based concept of GAF,[93] for wireless sensor networks. In their approach, they also provide self-healing protocols to contain and heal small perturbations, including node joins, leaves, deaths, state corruptions, and node movements, occurring at moderate frequencies. Their approach is similar to the tree monitoring architecture except instead of the root of the tree, the head or cluster head of the cell maintains the state of the cell.

22.4 Summary

Self-organization is the ability of the deployed nodes to locally configure among themselves in order to form a global interconnected network. In other words, self-organization involves abstracting the communicating entities (sensor nodes) into an easily manageable network infrastructure. The extreme dynamics of wireless sensor networks, along with challenging demands posed by the sensing applications, make self-organization solutions difficult and novel. Energy is one of the crucial design challenges for sensor network self-organization protocols. It is the self-organization protocol that engineers the critical preliminary network infrastructure support for sensor network services at various layers of the network protocol stack. Hence, a careful design encompassing several critical service requirements at various layers of the protocol stack must be considered in the self-organization protocol. Some of these requirements may be conflicting and hence a general trade-off may need to be included in the design. An integrated approach with some adaptive resilient behavior is therefore warranted in the design of self-organization protocols.

The design of sensor network organization protocols includes concepts that are common across all sensing applications. These include the *sensing phenomena* and elementary organizational architectures such as the chain, tree, cluster, or variations of the connected dominating set (CDS). Protocols for network self-organization also utilize certain quality metrics that effectively make the self-organization problem difficult or even intractable and the solutions approximate. The energy-conserving capability of a self-organization protocol depends on the energy-saving scheme that is integrated implicitly with the network and the network parameters that are allowed to be traded off with each other.

22.5 Future Research Directions

Integrating cross-layer requirements with the self-organization algorithm for supporting diverse concurrent applications is an open research field. Ideally, it requires maintaining and sharing cross-layer network status information across all layers of the protocol stack. It also requires a general formalization of the various QoS requirements and application policies toward network self-(re)organization and management. This also motivates the mapping of these requirements and policies[56,57] into the self-organization algorithm. A role-based self-organization approach,[40,41] where nodes in the network are assigned roles depending upon their performance capabilities, promotes easy integration with the cross-layer design mentioned earlier. It also promotes future evolution of the self-organized infrastructure in terms of new roles that meet novel services across multiple platforms and applications.

Target event characterization in terms of its dynamics and dimensions is an open field. Tracking multiple targets with different sensing signatures requires dynamic reorganization of the sensing groups or regions. As mentioned previously, this requires an explicit feedback mechanism between the event tracking protocol and the maintenance or reorganization protocol to dynamically control the topological dimension of the overlap among neighboring sensing regions. There is limited knowledge available that provides detailed specifications of the characteristics of the applications, the desired targets, and the required feedback necessary for efficient tracking.

Finally, as discussed previously, mobile sensor networks are a special research challenge that is an outcome of the merger of critical requirements of both the sensor networks and mobile ad hoc networks. However, a considerable amount of research is available in both domains, and interested readers are encouraged to apply mobility management concepts from MANETs to include mobility in their design of self-organization protocols for mobile wireless sensor networks.

Acknowledgment

This material is based upon work supported by the National Science Foundation under Grant ANI-0086020 and a summer dissertation fellowship from Wayne State University.

References

1. T. Abdelzaher, B. Blum, Q. Cao, Y. Chen, D. Evans, J. George, S. George, L. Gu, T. He, S. Krishnamurthy, L. Luo, S. Son, J. Stankovic, R. Stoleru, and A. Wood. EnviroTrack: Towards an Environmental Computing Paradigm for Distributed Sensor Networks. In *IEEE International Conference on Distributed Computing Systems*, Tokyo, Japan, March 2004.
2. F. Adelstein, G. Richard, and L. Schwiebert. Building Dynamic Multicast Trees in Mobile Networks. In *ICPP Workshop on Group Communication*, pages 17–22, September 1999.
3. I.F. Akyildiz, W. Su, Y. Sankarasubramaniam, and E. Cayirci. Wireless Sensor Networks: A Survey. *Computer Networks (Elsevier)*, March 2002.
4. A.D. Amis, R. Prakash, T.H.P. Vuong, and D.T. Huynh. Max-Min D-Cluster Formation in Wireless Ad Hoc Networks. In *Proceedings of IEEE INFOCOM'2000*, Tel Aviv, March 2000.
5. V. Annamalai, S.K.S. Gupta, and L. Schwiebert. On Tree-Based Convergecasting in Wireless Sensor Networks. In *IEEE Wireless Communications and Networking Conference*, Vol. 3, pages 1942–1947, March 2003.
6. D.J. Baker and A. Ephremides. The Architectural Organization of a Mobile Radio Network via a Distributed Algorithm. *IEEE Transactions on Communications*, COM-29(11):1694–1701, November 1981.

7. A. Ballardie, P. Francis, and J. Crowcroft. Core Based Trees (CBT). In *ACM SIGCOMM*, pages 85–95, September 1993.

8. L. Bao and J.J. Garcia-Luna-Aceves. Topology Management in Ad Hoc Networks. In *4th ACM International Symposium on Mobile Ad Hoc Networking and Computing (MOBIHOC 2003)*, Annapolis, MD, June 2003.

9. L. Blazevic, L. Buttyan, S. Capkun, S. Giordano, J. Hubaux, and J. Le Boudec. Self-Organization in Mobile Ad-Hoc Networks: The Approach of Terminodes. *IEEE Communications Magazine*, pages 166–174, June 2001.

10. B. Blum, P. Nagaraddi, A. Wood, T. Abdelzaher, S. Son, and J. Stankovic. An Entity Maintenance and Connection Service for Sensor Networks. In *The First International Conference on Mobile Systems, Applications, and Services (MobiSys)*, San Francisco, CA, May 2003.

11. N. Bulusu. Self-Configuring Localization Systems. Ph.D. thesis, University of California, Los Angeles (UCLA), 2002.

12. Y.C. Tseng, Y.-N. Chang, and B.-H. Tzeng. Energy Efficient Topology Control for Wireless Ad Hoc Sensor Networks. In *23rd International Conference on Distributed Computing Systems Workshops (ICDCSW'03)*, Providence, Rhode Island, May 2003.

13. A. Cerpa and D. Estrin. ASCENT: Adaptive Self-Configuring sEnsor Networks Topologies. In *21st International Annual Joint Conference of the IEEE Computer and Communications Societies (INFOCOM)*, June 2002.

14. B. Chen, K. Jamieson, H. Balakrishnan, and R. Morris. Span: An Energy-Efficient Coordination Algorithm for Topology Maintenance in Ad hoc Wireless Networks. In *Proceedings of MobiCom 2001*, pages 70–84, July 2001.

15. G. Chen, F. Garcia, J. Solano, and I. Stojmenovic. Connectivity Based k-Hop Clustering in Wireless Networks. In *Proceedings of the IEEE Hawaii Int. Conf. System Science*, Jan. 2002.

16. S. Chessa and P. Santi. Crash Faults Identification in Wireless Sensor Networks. *Computer Communications*, 25(14):1273–1282, Sept. 2002.

17. C. Chevallay, R.E. Van Dyck, and T.A. Hall. Self-Organization Protocols for Wireless Sensor Networks. In *Thirty Sixth Conference on Information Sciences and Systems*, March 2002.

18. L.P. Clare, G.J. Pottie, and J.R. Agre. Self-Organizing Distributed Sensor Networks. In *Proc. SPIE, Unattended Ground Sensor Technologies and Applications*, Vol. 3713, pages 229–237, 1999.

19. M. Conti, G. Maselli, G. Turi, and S. Giordano. Cross-Layering in Mobile Ad Hoc Network Design. *IEEE Computer*, 37(2):48–51, Feb. 2004.

20. L. Doherty, L.E. Ghaoui, and K.S.J. Pister. Convex Position Estimation in Wireless Sensor Networks. In *Proceedings of the IEEE INFOCOM 2001*, Anchorage, AK, April 2001.

21. R.E. Van Dyck. Detection Performance in Self-Organized Wireless Sensor Networks. In *IEEE International Symposium on Information Theory*, Lausanne, Switzerland, June 2002.

22. J. Elson and D. Estrin. Random, Ephemeral Transaction Identifiers in Dynamic Sensor Networks. In *21st International Conference on Distributed Computing Systems*, Mesa, AZ, April 2001.

23. J. Elson and D. Estrin. Time Synchronization for Wireless Sensor Networks. In *International Parallel and Distributed Processing Systems (IPDPS) Workshop on Parallel and Distributed Computing Issues in Wireless Networks and Mobile Computing*, pages 1965–1970, April 2001.

24. J. Elson and K. Römer. Wireless Sensor Networks: A New Regime for Time Synchronization. *ACM Computer Communication Review (CCR)*, 33(1):149–154, January 2003.

25. EnviroTrack: An Enviromental Programming Paradigm for Sensor Networks. EnviroTrack Homepage. http://www.cs.virginia.edu/~ll4p/EnviroTrack/.

26. D. Estrin, R. Govindan, J. Heidemann, and S. Kumar. Next Century Challenges: Scalable Coordination in Sensor Networks. In *ACM MOBICOM'99*, 1999.

27. D. Estrin et al. Protocol Independent Multicast-Sparse Mode (PIM-SM): Protocol Specification. RFC 2362, Internet Engineering Task Force, June 1998.

28. S. Ganeriwal, A. Kansal, and M.B. Srivastava. Self-Aware Actuation for Fault Repair in Sensor Networks. In *IEEE International Conference on Robotics and Automation (ICRA)*, 2004.

29. S. Ganeriwal, R. Kumar, and M. Srivastava. Timing-Sync Protocol for Sensor Networks. In *ACM Conference on Embedded Networked Sensor Systems (SenSys)*, pages 138–149, November 2003.

30. M. Gerla and J.T. Tsai. Multicluster, Mobile Multimedia Radio Network. *ACM Wireless Networks*, 1(3):255–266, October 1995.

31. L. Girod, V. Bychkovskiy, J. Elson, and D. Estrin. Locating Tiny Sensors in Time and Space: A Case Study. In *IEEE International Conference on Computer Design*, pages 214–219, 2002.

32. S. Guha and S. Khuller. Approximation Algorithms for Connected Dominating Sets. Technical Report 3660, Univ. of Maryland Inst. for Adv. Computer Studies, Dept. of Computer Science, Univ. of Maryland, College Park, June 1996.

33. C. Guo, L.C. Zhong, and J.M. Rabaey. Low Power Distributed MAC for Ad Hoc Sensor Radio Networks. In *IEEE Globecom 2001*, San Antonio, November 2001.

34. S.K.S. Gupta and P.K. Srimani. An Adaptive Protocol for Reliable Multicast in Mobile Multi-Hop Radio Networks. In *IEEE Workshop on Mobile Computing Systems and Applications*, pages 111–122, 1999.

35. Z.J. Haas, M.R. Pearlman, and P. Samar. The Zone Routing Protocol (ZRP) for Ad Hoc Networks. Internet Draft draft-ietf-manet-zone-zrp-04.txt, Internet Engineering Task Force, July 2002.

36. W. Heinzelman, A. Chandrakasan, and H. Balakrishnan. Energy-Efficient Communication Protocol for Wireless Microsensor Networks. In *International Conference on System Sciences*, January 2000.

37. M. Inanc, M. Magdon-Ismail, and B. Yener. Power Optimal Connectivity and Coverage in Wireless Sensor Networks. Technical Report 03–06, Rensselaer Polytechnic Institute, Dept. of Computer Science, Troy, NY, 2003.

38. K. Jamshaid and L. Schwiebert. Seken (Secure and Efficient Key Exchange for Sensor Networks). In *IEEE Performance Computing and Communications Conference (IPCCC)*, April 2004.

39. V. Kawadia and P.R. Kumar. A Cautionary Perspective on Cross Layer Design. Submitted to *IEEE Wireless Communication Magazine*, 2004.

40. M. Kochhal, L. Schwiebert, and S.K.S. Gupta. Role-Based Hierarchical Self Organization for Ad Hoc Wireless Sensor Networks. In *ACM International Workshop on Wireless Sensor Networks and Applications (WSNA)*, pages 98–107, September 2003.

41. M. Kochhal, L. Schwiebert, and S.K.S. Gupta. Integrating Sensing Perspectives for Better Self Organization of Ad Hoc Wireless Sensor Networks. *Journal of Information Science and Engineering*, 20(3), May 2004.

42. M. Kochhal, L. Schwiebert, S.K.S. Gupta, and C. Jiao. An Efficient Core Migration Protocol for QoS in Mobile Ad Hoc Networks. In *21st IEEE International Performance Computing and Communications (IPCCC'02)*, Phoenix, pages 387–391, April 2002.

43. R. Krishnan and D. Starobinski. Message-Efficient Self-Organization of Wireless Sensor Networks. In *IEEE WCNC 2003*, 2003.

44. M. Kubisch, H. Karl, A. Wolsz, L. Zhong, and J. Rabaey. Distributed Algorithms for Transmission Power Control in Wireless Sensor Networks. In *IEEE Wireless Communications and Networking Conference (WCNC)*, New Orleans, LA, March 2003.

45. C. Lin and M. Gerla. Adaptive Clustering for Mobile Wireless Networks. *IEEE Journal Selected Areas of Communications*, 15(7):1265–1275, September 1997.

46. S. Lindsey and C.S. Raghavendra. PEGASIS: Power Efficient GAthering in Sensor Information Systems. In *IEEE Aerospace Conference*, March 2002.

47. M.J. McGlynn and S.A. Borbash. Birthday Protocols for Low Energy Deployment and Flexible Neighbor Discovery in Ad Hoc Wireless Networks. In *2nd ACM International Symposium on Mobile Ad Hoc Networking and Computing*, Long Beach, CA, pages 137–145, 2001.

48. S. Meguerdichian, F. Koushanfar, M. Potkonjak, and M.B. Srivastava. Coverage Problems in Wireless Ad-Hoc Sensor Networks. In *IEEE INFOCOM'01*, Vol. 3, pages 1380–1387, April 2001.

49. S. Meguerdichian, F. Koushanfar, G. Qu, and M. Potkonjak. Exposure in Wireless Ad Hoc Sensor Networks. In *ACM SIGMOBILE (Mobicom)*, pages 139–150, July 2001.

50. R. Mini, B. Nath, and A. Loureiro. A Probabilistic Approach to Predict the Energy Consumption in Wireless Sensor Networks. In *IV Workshop de Comunica sem Fio e Computao Mvel*, Sao Paulo, Brazil, October 2002.

51. R. Mini, B. Nath, and A. Loureiro. Prediction-Based Approaches to Construct the Energy Map for Wireless Sensor Networks. In *21 Simpsio Brasileiro de Redes de Computadores*, Natal, RN, Brazil, 2003.

52. J. Mirkovic, G.P. Venkataramani, S. Lu, and L. Zhang. A Self Organizing approach to Data Forwarding in Large Scale Sensor Networks. In *IEEE International Conference on Communications (ICC'01)*, Helsinki, Finland, June 2001.

53. Mobile Ad hoc Networks (MANET) Charter. MANET Homepage. http://www.ietf.org/html.charters/manet-charter.html.

54. S. Narayanaswamy, V. Kawadia, R.S. Sreenivas, and P.R. Kumar. Power Control in Ad-Hoc Networks: Theory, Architecture, Algorithm and Implementation of the COMPOW Protocol. In *Proceedings of European Wireless 2002. Next Generation Wireless Networks: Technologies, Protocols, Services and Applications*, Florence, Italy, pages 156–162, Feb. 2002.

55. S. Ni, Y. Tseng, Y. Chen, and J. Chen. The Broadcast Storm Problem in a Mobile Ad Hoc Network. In *Proceedings of MOBICOM*, pages 151–162, August 1999.

56. M. Perillo and W. Heinzelman. Optimal Sensor Management under Energy and Reliability Constraints. In *Proceedings of the IEEE Wireless Communications and Networking Conference (WCNC'03)*, March 2003.

57. M. Perillo and W. Heinzelman. Sensor Management. To appear in *Wireless Sensor Networks*, Kluwer Academic Publishers, 2004.

58. A. Perrig, R. Szewczyk, V. Wen, D. Culler, and J.D. Tygar. Spins: Security Protocols for Sensor Networks. In *Proceedings of the ACM MobiCom 2001*, Rome, Italy, pages 189–199, July 2001.

59. G.J. Pottie and W. Kaiser. Wireless Sensor Networks. *Communications of the ACM*, 43(5):51–58, May 2000.

60. R. Prakash. Unidirectional Links Prove Costly in Wireless Ad Hoc Networks. In *Proceedings of ACM DIALM'99 Workshop*, Seattle, WA, pages 15–22, August 1999.

61. C.V. Ramamoorthy, A. Bhide, and J. Srivastava. Reliable Clustering Techniques for Large, Mobile Packet Radio Networks. In *Proceedings of IEEE INFOCOM'87*, pages 218–226, 1987.

62. R. Ramanathan and R. Rosales-Hain. Topology Control of Multihop Wireless Networks Using Transmit Power Adjustment. In *Proceedings of IEEE INFOCOM 2000*, Tel-Aviv, Israel, March 2000.

63. T.S. Rappaport. *Wireless Communications: Principles and Practice*, chapter Mobile Radio Propagation: Large-Scale Path Loss, Small-Scale Fading and Multipath, pages 105–248. Prentice Hall, December 2001.

64. A. Salhieh. Energy Efficient Communication in Stationary Wireless Sensor Networks. Ph.D. thesis, Wayne State University, 2004.

65. A. Salhieh and L. Schwiebert. Power Aware Metrics for Wireless Sensor Networks. In *IASTED International Conference on Parallel and Distributed Computing and Systems*, pages 326–331, November 2002.

66. A. Salhieh, J. Weinmann, M. Kochhal, and L. Schwiebert. Power Efficient Topologies for Wireless Sensor Networks. In *International Conference on Parallel and Processing*, pages 156–163, September 2001.

67. Y. Sankarasubramaniam, O.B. Akan, and I.F. Akyildiz. Esrt: Event-to-Sink Reliable Transport in Wireless Sensor Networks. In *Proceedings of ACM MobiHoc'03*, Annapolis, MD, June 2003.

68. P. Santi. Topology Control in Wireless Ad Hoc and Sensor Networks. Submitted to *ACM Comp. Surveys*, 2003.

69. A. Savvides, C. Han, and M. Srivastava. Dynamic Fine-Grained Localization in Ad Hoc Networks of Sensors. In *Proceedings of the ACM MobiCom 2001*, Rome, Italy, pages 166–179, July 2001.

70. A. Savvides, H. Park, and M.B. Srivastava. The Bits and Flops of the n-Hop Multilateration Primitive for Node Localization Problems. In *Proceedings of the 1st ACM International Workshop on Wireless Sensor Networks and Applications*, pages 112–121, September 2002.

71. C. Schurgers, G. Kulkarni, and M.B. Srivastava. Distributed On-Demand Address Assignment in Wireless Sensor Networks. *IEEE Transactions on Parallel and Distributed Systems,* 13(10):1056–1065, October 2002.

72. C. Schurgers, V. Tsiatsis, S. Ganeriwal, and M. Srivastava. Topology Management for Sensor Networks: Exploiting Latency and Density. In *Proceedings of the 3rd ACM International Symposium on Mobile Ad Hoc Networking and Computing (MOBIHOC 2002),* Lausanne, Switzerland, pages 135–145, June 2002.

73. C. Schurgers, V. Tsiatsis, and M. Srivastava. STEM: Topology Management for Energy Efficient Sensor Networks. In *IEEE Aerospace Conference,* pages 78–89, March 2002.

74. L. Schwiebert, S.K.S. Gupta, J. Weinmann, A. Salhieh, M. Kochhal, and G. Auner. Research Challenges in Wireless Networks of Biomedical Sensors. In *Proceedings of the Seventh Annual International Conference on Mobile Computing and Networking (MOBICOM 2001),* Rome, Italy, pages 151–165, July 2001.

75. S. Singh, M. Woo, and C.S. Raghavendra. Power Aware Routing in Mobile Ad Hoc Networks. In *Proceedings of MOBICOM,* pages 181–190, 1998.

76. R. Sivakumar, B. Das, and V. Bharghavan. An Improved Spine-Based Infrastructure for Routing in Ad Hoc Networks. In *IEEE Symposium on Computers and Communications '98,* Athens, Greece, June 1998.

77. S. Slijepcevic and M. Potkonjak. Power Efficient Organization of Wireless Sensor Networks. In *IEEE International Conference on Communications (ICC'01),* Helsinki, Finland, pages 472–476, June 2001.

78. K. Sohrabi, J. Gao, V. Ailawadhi, and G. Pottie. Protocols for Self Organization of a Wireless Sensor Network. *IEEE Personal Communication Magazine,* 7(5):16–27, October 2000.

79. K. Sohrabi and G. Pottie. Performance of a Novel Self-Organization Protocol for Wireless Ad Hoc Sensor Networks. In *50th IEEE Vehicle Technology Conference,* The Netherlands, September 1999.

80. F. Stann and J. Heidemann. RMST: Reliable Data Transport in Sensor Networks. In *Proceedings of the First International Workshop on Sensor Net Protocols and Applications (SNPA'03),* Anchorage, Alaska, April 2003.

81. M. Steenstrup. *Ad hoc Networking,* chapter Cluster-Based Networks, pages 75–135. Addison-Wesley, December 2000.

82. L. Subramanian and R.H. Katz. An Architecture for Building Self-Configurable Systems. In *IEEE/ACM Workshop on Mobile Ad Hoc Networking and Computing (MobiHOC 2000),* Boston, August 2000.

83. D. Tian and N.D. Georganas. A Coverage-Preserving Node Scheduling Scheme for Large Wireless Sensor Networks. In *Proceedings of ACM Workshop on Wireless Sensor Networks and Applications (WSNA'02),* Atlanta, October 2002.

84. P.K. Varshney. *Distributed Detection and Data Fusion.* Springer-Verlag, New York, 1996.

85. C. Wan, A. Campbell, and L. Krishnamurthy. PSFQ: A Reliable Transport Protocol for Wireless Sensor Networks. In *First ACM International Workshop on Wireless Sensor Networks and Applications (WSNA 2002),* Atlanta, September 2002.

86. P. Winter. Steiner Problem in Networks: A Survey. *Networks,* 17(2):129–167, 1987.

87. A. Woo and D. Culler. A Transmission Control Scheme for Media Access in Sensor Networks. In *Proceedings of the Seventh Annual International Conference on Mobile Computing and Networking (MOBICOM),* Rome, Italy, July 2001.

88. J. Wu. *Handbook of Wireless and Mobile Computing,* chapter Dominating Set Based Routing in Ad Hoc Wireless Networks, pages 425–450. John Wiley, February 2002.

89. J. Wu and H. Li. On Calculating Connected Dominating Set for Efficient Routing in Ad Hoc Wireless Networks. In *Proceedings of the 3rd International Workshop on Discrete Algorithms and Methods for Mobile Computing and Communications,* pages 7–14, August 1999.

90. Y. Xu. Adaptive Energy Conservation Protocols for Wireless Ad Hoc Routing. Ph.D. thesis, University of Southern California (USC), 2002.

91. Y. Xu, S. Bien, Y. Mori, J. Heidemann, and D. Estrin. Topology Control Protocols to Conserve Energy in Wireless Ad Hoc Networks. Technical Report 6, University of California, Los Angeles, Center for Embedded Networked Computing, January 2003. Submitted for publication.

92. Y. Xu, J. Heidemann, and D. Estrin. Adaptive Energy-Conserving Routing for Multihop Ad Hoc Networks. Technical Report TR-2000-527, University of California, Los Angeles, Center for Embedded Networked Computing, Oct. 2000.

93. Y. Xu, J. Heidemann, and D. Estrin. Geography Informed Energy Conservation for Ad Hoc Routing. In *ACM/IEEE International Conference on Mobile Computing and Networking,* pages 70–84, July 2001.

94. F. Ye, H. Luo, J. Cheng, S. Lu, and L. Zhang. A Two-Tier Data Dissemination Model for Large-Scale Wireless Sensor Networks. In *International Conference on Mobile Computing and Networking (MobiCOM),* pages 148–159, September 2002.

95. W. Ye, J. Heidemann, and D. Estrin. An Energy Efficient MAC protocol for Wireless Sensor Networks. In *IEEE INFOCOM,* New York, pages 3–12, June 2002.

96. H. Zhang and A. Arora. GS^3: Scalable Self-Configuration and Self-Healing in Wireless Networks. In *21st ACM Symposium on Principles of Distributed Computing,* July 2002.

97. Y. J. Zhao, R. Govindan, and D. Estrin. Sensor Network Tomography: Monitoring Wireless Sensor Networks. In *Student Research Poster, ACM SIGCOMM 2001,* San Diego, CA, August 2001.

98. Y.J. Zhao, R. Govindan, and D. Estrin. Residual Energy Scans for Monitoring Wireless Sensor Networks. In *IEEE Wireless Communications and Networking Conference (WCNC'02),* Orlando, FL, March 2002.

99. Y.J. Zhao, R. Govindan, and D. Estrin. Computing Aggregates for Monitoring Wireless Sensor Networks. In *The First IEEE International Workshop on Sensor Network Protocols and Applications (SNPA'03),* Anchorage, AK, May 2003.

100. L. Zhou and Z.J. Haas. Securing Ad Hoc Networks. *IEEE Networks,* 13(6):24–30, Nov. 1999.

23

Self-Stabilizing Distributed Systems and Sensor Networks

Zhenghan Shi

Pradip K. Srimani

The purpose of this chapter is to (1) provide a brief description of the basic models of self-stabilization that have been used in designing fault-tolerant distributed algorithms; (2) provide detailed descriptions of models for sensor networks,[1] P2P networks,[2] and cooperative mobile agents;[3] and (3) critically evaluate how efficiently the standard models of self-stabilization can be used to design fault-tolerant protocols for mobile environments[1–3] in the presence of mobile clients (along with their constraints, such as low battery power, unreliable communication medium, frequent disconnection, and reconnection characteristics). We show how the standard models of self-stabilization can be modified to accommodate the requirements of a sensor network and provide a couple of example protocols. We also explore designing protocols for some simple global primitives using the modified models.[4]

23.1 Distributed Systems

A distributed system essentially consists of a number of *autonomous* computers (with their own hardware and software) that are connected to each other to form a communication network. Although it may be connected to other computers, each computer can run its applications independently of the others, while the networking software (and hardware) provides the functionalities as database systems and shared storage (data and code). Typically, there is no shared memory, and communication between two connected nodes (sites) is implemented through explicit message passing. An important characteristic of the system is that it presents to the user, at any of the participating sites, a view of a *single*, highly reliable program, thus making the decentralized nature of the hardware and the software transparent to the user. The challenge of developing software, hardware, and network interfaces becomes more formidable because of the need to dynamically respond to failures and subsequent recoveries.

In the past decade we have seen excellent growth in the areas of client/server and networked personal computing systems. In the next phase of distributed computing, we will probably see a growth in the area of mobile computing systems where very close coupling (coordination) is necessary between a large number of applications running on different platforms across the globe, where the computing nodes are no longer stationary in space and frequently enter and leave the system. In addition to mobile computing systems in general, large-scale sensor networks, cooperative peer-to-peer (P2P) networks, and cooperating mobile agents are all example settings where distributed computing is indispensable. Cellular networks and sensor networks consist of a collection of a large number of identical relatively low-powered nodes, with limited communication and computing capabilities. An important charateristic is that the nodes are distributed at random, or even with some *a priori* planning, but at any time some of them may not be available, and there is only one channel for message transmission between nodes and base stations. A P2P network is a dynamic, and scalable, set of nodes (peers), where the objective is to distribute the cost of storing and sharing large amounts of data. "Each peer can join or leave the system at any moment, and can communicate with any other peer under the only hypothesis that the two peers are aware of each other. The main characteristics of peer-to-peer systems are the ability to pool together and harness a large amount of resources, self-organization, load-balancing, adaptation and fault-tolerance."[5]

The communication network architecture is a key component in a distributed system and its characteristics are fundamental to the performance of the applications running on that system. In the past, the relative performance of the communication system has lagged behind that of the computing engine, but the advent of fiber optics and gigabit networking has started to close the gap. The class of applications that can be cost effectively implemented on distributed systems has broadened, and intranets are now more popular then ever.

23.2 Fault Tolerance in Mobile Distributed Systems and Self-Stabilization

Two of the most desirable attributes of the modern mobile distributed systems are fault tolerance and scalability. We need systems designed so that they can recover spontaneously from transient faults and reconfigure (scale) themselves without any need for external intervention.

A common approach to designing fault-tolerant systems is to mask the effects of the fault. However, fault masking is not free; it requires additional hardware or software and it considerably increases the cost of the system. This additional cost may not be an economic option, especially when most faults are transient in nature, and temporary unavailability of a system service is unacceptable. *Self-stabilization* is a relatively new way of looking at system fault tolerance; it provides a "built-in-safeguard" against "transient failures" that might corrupt the data in a distributed system.

The objective of self-stabilization is (as opposed to masking faults) to recover from faults in a reasonable amount of time and without intervention by any external agent. Self-stabilization is based on two basic ideas: (1) the code executed by a node is re-entrant, incorruptible (as if written in a fault-resilient memory), and transient faults affect only data locations; (2) fault-free system behavior is usually checked by evaluating some predicate of the system state variables. The checking oracle may be complex but is not part of the self-stabilizing algorithm design. Every node has a set of local variables, whose contents specify the local state of the node. The state of the entire system, called the *global state*, is the union of the local states of all the nodes in the system. Each node is allowed to have only a partial view of the global state, and this depends on the connectivity of the system and the propagation delay of different messages. The objective in a distributed system is to arrive at a desirable global final state called a *legitimate state*.

23.2.1 Requirements of Self-Stabilization

There exist several models for self-stabilization; we present here only the basic common concepts of these models. The *state* of a node is specified by its local variables. The system state is a vector of all local states of the participating nodes. We use T to denote the set of all possible system states. A system state is either *legitimate* or *illegitimate*. The precise specification of a legitimate state depends on the algorithm; but as a general rule, when the system is in a legitimate state, it has the property required by that application. To allow system recovery after transient faults, each node repeatedly executes a piece of code. This code consists of a set of rules:

> **begin**
> > rule
> > $\vdots$
> > rule
> **end**

Each rule has the form:

> (label) [guard]: <program>;

A guard is a Boolean expression of the variables that the processor can read: its own variables and the variables of its neighbors. The program part of a rule is the description of the algorithm used to compute the new values for local variables. If the guard of a rule is true, that rule is called *enabled*. When at least one rule is enabled, the node is *privileged*. An *execution* of an enabled rule is the determination of the new node state value using the algorithm described by the program part of the rule. A *move* of a node is the execution of a nondeterministically chosen enabled rule.

That is, there is a relation $\mathcal{R} \subset T \times T$ such that if $(s_i, s_f) \in \mathcal{R}$, then (1) the states s_i, s_f differ by a single node x value; and (2) if the system is in state s_i, there is an enabled rule of node x such that after execution of the corresponding code, the system is in state s_f. A system evolution $\mathcal{E} = (s_i)_{i \in I}$ is a finite or infinite sequence of moves such that (1) if (s_i, s_{i+1}) is a consecutive pair of states in $\mathcal{E}$, then $(s_i, s_{i+1}) \in \mathcal{R}$; and (2) if the system evolution has a finite number of states, s_f being the last one, then there is no state $s \in T$ such that $(s_f, s) \in \mathcal{R}$.

To prove the correctness of a self-stabilizing algorithm, the conditions of closure and the convergence must be shown. The closure property means that when the system is in a legitimate state, the next state is always a legitimate state. The convergence property means that for any state and for any sequence of possible moves, after a finite number of moves the system is in a legitimate state. As Gouda observes,[6] self-stabilization can be, in principle, defined by a set T, a relation $\mathcal{R} \subset T \times T$, and a specification of the legitimate state set $\mathcal{L}$. Different classes of closure and convergence can be defined and general methods of proving the self-stabilization can be sketched.

One useful and elegant strategy to prove the correctness of self-stabilizing algorithms is to use bounded monotonically decreasing functions defined on global system states.[7] Some existing self-stabilizing algorithms are proved correct by defining a bounded function that is shown to decrease monotonically at every step.[8] Many self-stabilizing algorithms do not use this bounded function method because it is usually very very difficult to design such a function. Instead, they develop a different proof technique using induction on the number of nodes in the graph.

23.2.2 Implementation Issues

The stabilizing algorithms achieve fault tolerance in a manner that is radically different from traditional fault tolerance in distributed systems. The paradigm allows us to abandon failure models and a bound on the number of failures. The theory is elegant, but how practical is the concept for implementation with present-day technology? Here are some issues:

- The concept of the global state of the system requires a *common* time for all nodes. The physical time can be used as a common time but it cannot be explicitly used by the component processes (drifts in local clocks, relativity, etc.). The partial order relation (among events) generated by message exchange cannot be uniquely extended to a total order relation. If there is no global clock, global states cannot be defined and legitimate states must defined locally (i.e., based on the local state of a node and the states of its neighboring nodes and on the partial order relation associated to sending and receiving messages). This greatly complicates the correctness proof of any stabilizing algorithm.

- A *move* is a complex operation; the state of the neighbors must be read, the guards must be evaluated, and the associated code segment must be executed. Some models assume that a new move cannot start until the previous move is completed (i.e., the moves are atomic). In a real distributed system, the reading of a neighbor's state can be implemented in two ways; one option is to request every node to send its state to its neighbors periodically or whenever it changes its state. Each node caches the state received from its neighbors and moves according to the state cached in its memory. The other option is to use a query message: when a node needs to read its neighbor's state, it sends a query message and waits for a reply. In both cases, when a node moves, it uses the cached states of the neighbors instead the real states of its neighbors. Because the states of the neighbors may have already been changed, the moves are not really atomic.

- To enforce the atomicity and the serializability of the moves, Dijkstra[9] has introduced the concept of the central daemon. When multiple nodes are privileged, the central daemon arbitrarily selects one node to be active next. The concept of a central daemon is very much against the concept of a distributed system in that it serializes the moves and does not allow concurrent node executions. Proving correctness is easier for a serial execution, but there are many parallel executions that might be "equivalent" with a serial execution.[10] These executions should be allowed by an efficient move scheduler.

Different models of self-stabilization offer different prospects of cost-effective implementation of the concept and it is not clear at this point which would win.

23.2.3 Classification of Self-Stabilizing Systems

A *model* can be viewed as an *interface definition* and a set of assumptions (the algorithm designer can make) that define the behavior of the system. Unfortunately, there is no unifying model for the distributed system concept.[11] If some specific features of a system are not introduced in the model, the algorithms may be less efficient than they could possibly be; on the other hand, those features may not be general enough to be present in all systems.

Self-stabilizing algorithms have been designed for two interprocess communications paradigms: (1) shared memory and (2) message passing. Because we are interested in self-stabilizing algorithms for distributed systems, we assume the message passing paradigm. It should be observed that a lower-level protocol that sends and receives messages can transform a message passing model into a shared memory one.

23.2.3.1 Anonymous versus ID-Based Networks

The concept of *node identity* is important in designing distributed algorithms. If each node has a unique hardwired *ID,* the network is *ID*-based; otherwise, the network is *anonymous*. The anonymous network is a weaker model than an *ID*-based network. For some problems there are no deterministic algorithms in anonymous networks.[12] The impossibility stems from the lack of deterministic symmetry breaking mechanisms without unique IDs. In general, it is far more difficult to design algorithms for anonymous networks than for *ID*-based networks. The *ID*-based network is a more realistic model to design self-stabilizing algorithms, but it is necessary to maintain a database of the used *ID*s by a central authority; the addition of a new node requires a database search and the assignment of a new distinct *ID*. This concept has been used in practice for a while (Ethernet addresses and IP addresses) and has proved convenient. In principle, each node in an *ID*-based network can have global information of the topology and the state of all other nodes. Hence, a local algorithm can be used to solve the problem. This scheme may be unacceptable because the information needed to update each node state is large and takes considerable bandwidth. In addition, the system might respond too slowly to dynamic configuration changing.

23.2.3.2 Deterministic versus Probabilistic Algorithms

The self-stabilizing algorithms can also be divided into two classes: (1) *deterministic* and (2) *probabilistic* (randomized) algorithms. This criterion has nothing to do with the underlying distributed system but concerns the algorithm design strategy. Randomization is normally used to break the symmetry in anonymous networks. Many randomized algorithms succeed with probability $1 - \epsilon, \epsilon > 0$ (the success is not certain). Additionally, the random number generators used are actually pseudorandom number generators and some undesirable correlations may appear between neighboring nodes.

23.3 Sensor Network

Currently, there is increasing interest in the theory and application of sensor networks; and due to varied application domains, the sensor networks are also widely different in terms of capabilities and inherent infrastructure support. A very detailed survey of different kinds of sensor networks and their applications can be found.[13] Our purpose in this chapter is to use some kind of abstract model of sensor networks[1,14] to best suit the standard paradigms of designing self-stabilizing distributed algorithms for ad hoc networks and study some of the known protocols under the so-called *sensornet model*. For our purpose, the sensor network is built from a set of (sensor) nodes with the same computational and communication capabilities. There is no external backbone network or message repeater facility. Each node p can communicate with a subset of nodes, called neighbors of p, determined by the range of the radio signal of p. Each node uses the same radio frequency. Each node cannot send and receive concurrently; that is, the communication channels are half-duplex. Nodes do not have collision detection hardware. The network is asynchronous.

23.3.1 Unbounded Asynchronous Unison

To coordinate the need for phase synchronization in distributed systems, the problem of asynchronous unison becomes important. To present our algorithm on asynchronous unison, we define the following identifiers and concepts. Let $G = (V, E)$ be a finite, undirected, and connected graph. Let v_p be the clock of node p in the unison system. Let K be the upper bound of the unison algorithm. Let N be a positive integer that defines the *behind* relationship in the following subsection. A *state* of the system is defined by a value for every clock variable and a value for every local variable in each node program in the system. A *system transition* is a pair (r, s) of system states such that there is a rule that can be fired, the system starting from r, yields state s. A *system computation* is a maximal sequence of system states such that every pair of successive states is a transition. The legitimacy prediction for an unbounded asynchronous unison system is: $L \equiv (\forall p, q : (p, q) \in E : |(v_p - v_q)| \leq 1)$. The system stabilizes to satisfy L and every node p executes $v_p := v_p + 1$ infinitely often. Herman[1] proposed the following algorithm and proved its correctness.

Algorithm A-Unison

Rule 1. if $\forall q : (p, q) \in E, v_p \leq \boxtimes_p v_q$
   then $v_p := v_p + 1;$

In the next section, we present the algorithm for the bounded version of asynchronous unison in sensor network.

23.3.2   Bounded Unison

The legitimacy prediction for a bounded asynchronous unison system is: $L \equiv (\forall p, q : (p,q) \in E : |(v_p - v_q) \bmod K| \leq 1)$. Note that $v_p - v_q$ can be viewed as either a positive integer or a negative integer in modulo K. We always use the one with a smaller absolute value. The correctness of a unison algorithm satisfies the following three properties:

1. Liveness: every computation of the system is infinite.
2. Progress: every infinite computation of the system has an infinite suffix where every clock variable v_p is updated infinitely often and only by executing the assignment statement $v_p = v_p + 1 \bmod K$.
3. Asynchronous unison: every infinite computation of the system has an infinite suffix where each state satisfies legitimacy prediction L. For the bounded algorithm, the conduction is: for any neighboring nodes p and q, $v_p = v_q \cup v_p = v_q + 1 \bmod K \cup v_q = v_p + 1 \bmod K$.

To present the algorithm of bounded unison, we introduce the following two relationships. Relationship *behind* is an extension to the *greater equal* than in the unbounded unison. The two relationships assist to achieve the self-stabilization of bounded unison in sensor network.

Relation *behind* $\heartsuit$: Let variable x and y range over $0 \ldots K - 1$, $x \heartsuit y = ((y - x) \bmod K \leq N)$.
Relation *far* $\diamondsuit$: Let variable x and y range over $0 \ldots K - 1$, $x \diamondsuit y = \neg(x \heartsuit y) \cap \neg(y \heartsuit x)$.

In a sensor net model, the condition L becomes $C \equiv (\forall p, q : (p,q) \in E : \boxtimes_p v_q \heartsuit v_q$.
The following algorithm, which achieves an asynchronous unison system, executes at every node p in the network.

Algorithm B-Unison
Rule 1. **if** $\forall q : (p,q) \in E, v_p \heartsuit \boxtimes_p v_q$
  then $v_p := v_p + 1 \bmod K$;
Rule 2. **if** $\exists q : (p,q) \in E, v_p \diamondsuit \boxtimes_p v_q \cap v_p > v_q$
  then $v_p := 0$.

Lemma 23.1   *C is an invariant of L of any execution.*

Proof   If $\boxtimes_p v_q$ is an estimation that is behind the value of v_q, then Rule 1 may block p's progress; however, whenever $\boxtimes_p v_q$ is updated, the result of incrementing v_p satisfies L. Then, the same argument applies to the neighbors of p. If the cache update message for the increment is not correctly received by a neighbor q, Rule 1 may block that neighbor's progress.   □

The clock variables in unison systems are time-driven. They increment infinitely even if no other significant events occur in the system. Therefore, every computation of the system is infinite. We show that the proposed algorithms in asynchronous sensor networks have an infinite suffix where each state satisfies condition C.

Lemma 23.2   *With probability 1, every execution eventually satisfies $L \cap C$.*

Proof   If a node p never increments v_p, the system will eventually be deadlocked and cache coherent. This contradicts the self-stabilizing behavior of the algorithm.

For p to infinitely increment v_p in modulo K, it must forever correctly receive messages from each neighbor q containing values of v_q. In turn, each neighbor q must correctly receive updated clock values from all its neighbors; by an inductive argument, it follows that all nodes increment their clocks infinitely often. Therefore, from Rule 1 and Lemma 23.1, every execution eventually satisfies $C \cap L$. $\square$

23.4 Maximal Independent Set

Let $G = (V, E)$ be a finite, undirected, and connected graph. Let s_p be a variable of node p in the network that indicates the membership of the node in the maximal independent set. Herman[15] proposed an ID-based algorithm for leaders via maximal independent set. We propose here an algorithm that does not use IDs.

If $s_p = 1 \cap (\forall q : (p,q) \in E, s_q = 0$, refer to as L_1, we say that the node p is independent. If $s_p = 0 \cap (\exists q : (p,q) \in E, s_q = 1$, refer to as L_2, we say that the node p is dominated.

In a legitimate result, each node in the network should satisfy $L_1 \cap L_2$. In a sensor network, condition L_1 becomes $s_p = 0 \cap (\exists : (p,q) \in E, \boxtimes_p s_q = 1$, refer to as C_1. Condition L_2 becomes $s_p = 0 \cap (\exists : (p,q) \in E, \boxtimes_p s_q = 1$, refer to as C_2. For this algorithm, the weakened cache coherent model implies that the cached values may be stale.

The following algorithm, which achieves a maximal independent set, executes at every node p in the network.

Algorithm M-Independent-Set
Rule 1. if $\forall q : (p,q) \in E, s_p = 0 \cap \boxtimes_p s_q = 0$
   then $v_s := 1$;
Rule 2. if $\exists q : (p,q) \in E, s_p = 1 \cap \boxtimes_p s_q = 1$
   then $s_p := 0$.

Without the knowledge of coherency of a cache entry, the system may come to a stable state but later reconfigure due to a recognized stale cache entry. Hedetniemi et al.[16] supply the following lemma. We include it here without further detail.

Lemma 23.3 *The algorithm achieves $C_1 \cap C_2$ upon stabilization.*

Lemma 23.4 *With probability 1, $C_1 \cap C_2$ is an invariant of $L_1 \cap L_2$ of any execution.*

The following algorithm has also appeared.[15] Each node in the network has a Boolean variable L. In an initial state, the value of L is arbitrary.

Algorithm Leaders-M-Independent-Set
Rule 1. if $\forall q \in N(p) : p > q$
   then $L_p = true$;
Rule 2. if $[]q \in N(p) : \boxtimes_p L_q \cap q < p$
   then $L_p = false$;
Rule 3. if $(\exists q \in N(p) : q < p) \cap (\forall q \in N(p), q > p : \neg \boxtimes_p L_q)$
   then $L_p = true$;

This algorithm does not use randomization explicitly. However, its convergence is probabilistic because the underlying model of communication uses random delay. The article[15] shows that with probability 1, the algorithm converges to a maximal independent set with convergence time $O(1)$.

Proof The probability of a node in a sensor network with eventual coherent cache is 1. Let p be one of the nodes that contains stale cache values of $\boxtimes_p s_q$. Once the node receives a correctly updated value of s_q, it will trigger a round of updating in the network. Subsequently, each such error will trigger a round of updating. Let τ be the probability that one or more nodes failed to receive a message. If τ is less than 1, the probability of the system infinitely executing without achieving $L_1 \cap L_2$ is 0. Upon stabilization, $C_1 \cap C_2$ is equivalent to $L_1 \cap L_2$. □

23.5 Minimal Domination

Using the notation from the previous section, a legitimate state of the network will satisfy L_2. A weakened condition in a sensor network is that every node satisfies C_2.

Hedetniemi et al.[16] have proved the following lemma.

Lemma 23.5 *A set S is a minimal dominating set iff its dominating and every $u \in S$ has a private neighbor.*

We use the variable d_p to denote the pointer at node p used in the algorithm. The value of d_p is the identifier for a node, such as q. The pointer variable d_p is used to indicate the private neighbor relationship.

The following algorithm, which finds a minimal dominating set, executes at every node p in the network. Let $N(p)$ denote the open neighborhood of node p.

Algorithm M-Dominating-Set
Rule 1. if $v_p = 0 \cap \forall q \in N(p) : \boxtimes_p v_q = 0$
   then $v_p = 1$;
Rule 2. if $v_p = 1 \cap \forall q \in N(p) : \boxtimes_p d_q \neq p \cap \exists k \in N(p) : \boxtimes_p v_k = 1$
   then $v_p = 0$;
Rule 3. if $v_p = 1 \cap d_p \neq NULL$
   then $d_p = NULL$;
Rule 4. if $v_p = 0 \cap \exists! q \in N(p) : \boxtimes_p v_q = 1 \cap d_p \neq q$
   then $d_p = q$;
Rule 5. if $v_p = 0 \cap \exists$ **more than one** $q \in N(p) : \boxtimes_p v_q = 1 \cap d_p \neq NULL$
   then $d_p = NULL$;

Lemma 23.6 *With probability 1, C_2 is an invariant of L_2 of any execution and is achieved by every execution.*

Proof The probability of a node in a sensor network with eventual coherent cache is 1. Let p be one of the nodes that contains stale cache values of $\boxtimes_p s_q$. Once the node receives a correctly updated value of s_q, it will trigger a round of updating in the network. Subsequently, each such error will trigger a round of updating. Let τ be the probability that one or more nodes failed to receive a message. If τ is less than 1, the probability of the system infinitely executing without achieving $C_2 \cap L_2$ is 0. □

23.6 Neighborhood Identification

Because the sensor network is ad hoc, there is a need to configure the neighborhood knowledge of each node. At a node p, we can represent $N(p)$ by a list of identifiers learned by receiving messages from neighboring nodes. Let L be a list of nodes recording the close neighborhood of a node. Note that we do not include an aging mechanism for this algorithm because the sensor network model is not mobile. The following algorithm achieves neighborhood identification.

Algorithm Neighborhood
Rule 1. if receive message from q**, never receive any message before**
  then $\boxtimes_p L := \{q\}$;
Rule 2. if receive message from q
  then $\boxtimes_p L := \boxtimes_p L + q$;

Lemma 23.7 *The algorithm recognizes the neighborhood of each node.*

Proof With probability 1, a node p receives all the messages from its neighbors. Therefore, each node p can configure the neighborhood list correctly. □

23.7 Neighborhood Unique Naming

Let $N3_p$ denote the distance three neighborhood of node p. The algorithm finds a unique color for each node that is distinct within $N3_p$ neighborhood. Define namespace $\Delta = \lceil \delta^t \rceil$ for some $t > 3$. Let Id be the unique naming, $Id \in \{0 \ldots \Delta\}$. Let set $Cids_p = \{\boxtimes Id_q | q \in N3_p \backslash \{p\}\}$.

The following algorithm was proposed by Herman;[13] we include it here without proof.

Algorithm Leaders-M-Independent-Set
Rule 1. if $Id_p \in Cids_p$
  then $Id_p := $ **random** $(\Delta \backslash Cids_p)$;

Acknowledgement

The work reported in this chapter was supported by an NSF award ANI #0218495.

References

1. T. Herman. Models of self stabilization and sensor networks. In *Proceedings of IWDC 2003*, Vol. *LNCS 2918*, pages 205–214, Springer-Verlag, 2003.
2. A.K. Datta, M. Gradinariu, M. Raynal, and G. Simon. Anonymous publish/subscribe in P2P networks. In *Proceedings of the International Parallel and Distributed Processing Symposium (IPDPS'03)*, 2003.
3. S. Ghosh. Cooperating mobile agents and stabilization. In *Proceedings of WSS-2001*, Vol. *LNCS 2194*, pages 1–18, Springer-Verlag, 2001.
4. W. Goddard, S.T. Hedetniemi, D.P. Jacobs, and P.K. Srimani. Self-stabilizing distributed algorithm for strong matching in a system graph. In *Proceedings of HiPC 2003*, Vol. *LNCS 2913*, pages 66–73, Springer-Verlag, 2003.
5. M. Gardinariu, M. Raynal, and G. Simon. Looking for a common view for mobile worlds. In *Proceedings of the Ninth IEEE Workshop on Future Trends of Distributed Computing Systems (FTDCS'03)*, 2003.
6. M.G. Gouda. The triumph and tribulation of system stabilization. In *WDAG95 Distributed Algorithms 9th International Workshop Proceedings*, Springer-Verlag *LNCS:972*, pages 1–18, 1995.
7. J.L.W. Kessels. An exercise in proving self-stabilization with a variant function. *Inf. Processing Letters*, 29(2):39–42, 1988.
8. S.T. Huang. Leader election in uniform rings. *ACM Transactions on Programming Languages and Systems*, 15(3):563–573, July 1993.
9. E.W. Dijkstra. Self-stabilizing systems in spite of distributed control. *Communications of the ACM*, 17(11):643–644, November 1974.

10. P.A. Bernstein, V. Hadzilacos, and N. Goodman. *Concurrency Control and Recovery in Databases Systems.* Addison-Wesley, 1987.

11. S. Sahni and V. Thanvantri. Parallel Computing: Performance Metrics and Models. Technical Report TR-008, University of Florida, Gainesville, 1996.

12. D. Angluin. Local and global properties in networks of processors. In *Conference Proceedings of the Twelfth Annual ACM Symposium on Theory of Computing,* pages 82–93, 1980.

13. I.F. Akyildiz, W. Su, and E. Cayirci. A survey on sensor networks. *IEEE Communication Magazine,* pages 102–115, August 2002.

14. S.K. Kulkarni and U. Arumugam. Collison free communication in sensor networks. In *Proceedings of SSS 2003,* Vol. *LNCS 2704,* Springer-Verlag, 2003.

15. T. Herman. A distributed TDMA slot assignment algorithm for wireless sensor networks. Technical Report LRI 1370, Universite Paris Sud, September 2003.

16. S.M. Hedetniemi, S.T. Hedetniemi, D.P. Jacobs, and P.K. Srimani. Self-stabilizing algorithms for minimal dominating sets and maximal independent sets. *Computers and Mathematics with Applications,* 46:805–811, 2003.

24

Time Synchronization in Wireless Sensor Networks

Qing Ye

Liang Cheng

24.1   Introduction

Integration of recent advances in sensing, processing, and communication leads to emerging technologies of wireless sensor networks (WSNs), which interconnect a set of ad hoc deployed sensor nodes by their wireless radios. There exist a wide variety of applications such as battlefield surveillance, habitat tracking, inventory management, and disaster assistance.[1] It is envisioned that WSN will be an important platform to perform real-time monitoring tasks.

Time synchronization is essential for WSN applications. For example, data fusion[2] as a basic function in WSN may require synchronized clocks at different sensors. Consider cases where sensors are deployed in a dense fashion. When an event happens, multiple sensors may report observed phenomena at the same time. Based on synchronized timestamps, redundant messages can be recognized and suppressed to reduce the unnecessary traffic across the network. Time synchronization can also be used to realize synchronized sleeping periods for task scheduling. It is desirable to put sensor nodes into sleeping mode and wake them up to exchange information only when necessary to save battery energy. In this case, time synchronization is vital to maintain the accurate schedule among multiple sensors. Last but not least, with time synchronization, some MAC (Media Access Control) layer protocols (e.g., TDMA [Time Division Multiple Access]) can be realized.[3]

By nature, the hardware clocks at different sensors run at different speeds with different time offset to the UTC (Universal Coordinated Time) provided by NIST (National Institute of Standards and Technology). Moreover, their clocks drift differently to the environmental conditions. NTP (Network Time Protocol)[4] is the Internet standard for time synchronization, which synchronizes computer clocks in a hierarchical way by using primary and secondary time servers. Based on multiple data points, clock skew, offset, and drift can be estimated for time synchronization. However, strict resource limitations of sensor nodes make it difficult, if not impossible, to apply the well-studied NTP in WSNs. The resource constraints in WSN require lightweight design of time synchronization mechanisms that work with small storage occupation, low computation complexity, and little energy consumption.

This chapter presents multiple proposed protocols for time synchronization in WSN (Section 24.3) and our new lightweight approach (Section 24.4), which work in the sender-receiver way or a similar fashion. Section 24.2 presents a theoretical model for time synchronization based on the sender-receiver schemes of time information exchange used by these protocols. Our new approach can efficiently decrease the number of packet exchanges of time information while still maintaining certain synchronization accuracy. This chapter also compares the new approach with the conventional time synchronization mechanisms of two-way message exchanging (TWME). The simulation results in Section 24.5 also show that the new approach can achieve the time synchronization accuracy of 34 μs in one-hop WSN based on a certain delay model, and its maximum time synchronization error increases with the increment of the number of communication hops in multi-hop WSN.

24.2   Theoretical Model of Time Synchronization

The principle of the sender-receiver based time synchronization requires receivers follow the clock of a sender, which can be regarded as a time server. Consider two networked nodes where the sender S sends out a sequence of timestamped reference packets to the receiver A. There are four delay factors along the packet transmission path: (1) processing delay (P part), which includes the time spent at the node S to prepare and process the reference packet and the time for the sensor's microprocessor to transfer the packet to its networking components; (2) accessing delay (A part), the time for the packet staying in the buffer and waiting for wireless channel access; (3) propagation delay (D part), the time spent from sending the packet from the sender's wireless radio to receiving it by the receiver's wireless radio; and (4) receiving delay (R part), which is the time for the node R to retrieve the packet from the buffer, pass it to upper layer applications, and get the reading of its local clock.

We use *PADR delay* to denote these four factors. Although the propagation delay can be calculated based on the physical distance form S to R, the processing, accessing, and receiving delays are uncertain. They are related to the current work load of sensor nodes, implementation of their MAC layer, capabilities of microprocessors and communication radios, and the size of the reference packets. Based on the illustration in Figure 24.1, the time reference relationship between the sender and the receiver satisfies Equation 24.1, where a_{RS} is the relative clock skew:

$$t_r = a_{RS}t_s + t_0 + P_S A_S D_{SR} R_R + \mathrm{Drift}_R \tag{24.1}$$

Therefore, one of the essential tasks of time synchronization algorithms is to estimate the PADR delay accurately because the drift component is much smaller compared to the PADR delay and t_0 is fixed. In general, the PADR delay can be estimated by the conventional method of two-way message exchanging. Suppose the node S sends out a packet at time t_{s1}, with respect to (w.r.t.) the local clock at S, which is received by the node A at t_{a1}, w.r.t. A's local clock. Node A sends this packet back with the timestamp t_{a1} and S receives the reply at t_{s1}'. Then the sender S can get two time synchronization lines passing through

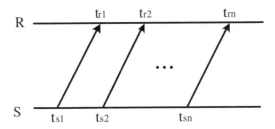

FIGURE 24.1 Referenced time synchronization.

synchronization points (t_{s1}', t_{a1}) to (t_{s2}', t_{a2}), and (t_{s1}, t_{a1}) to (t_{s2}, t_{a2}), respectively, as shown in Figure 24.2, where a_{as} is a_{RS} and assume that $Drift_R$ is 0. It has been proved that the difference between these two lines is approximately two times the PADR delay.[5]

24.3 Existing WSN Synchronization Techniques

24.3.1 Tiny-Sync and Mini-Sync

Tiny-sync and mini-sync[6] are proposed to synchronize local clocks in a pair-wise manner where the clock of the node R follows the time of node S using their bidirectional radios. Timestamped packets that contain the information of their local clocks are exchanged to estimate the time differences between R and S. Each packet stands for a time point of (t_s, t_r) in the coordinates of the two clocks. After receiving a sufficient number of packets, tiny-sync provides a method to get the upper bound and the lower bound of the time offsets between R and S from a series of time points. In this way, the node R is able to adjust its local clock to follow the node S. Mini-sync extends the idea of tiny-sync to get faster estimation speed with smaller amount of time points. It reduces the heavy traffic overheads of tiny-sync but still requires that at least four packets be exchanged between each pair at every synchronizing cycle.

Neither tiny-sync nor mini-sync keeps the global synchronized time in the WSN. For multi-hop WSNs, they assume that the sensor networks can be organized into a hierarchical structure by a certain level-discovery algorithm and each node only synchronizes with its direct predecessor. The basic idea is shown

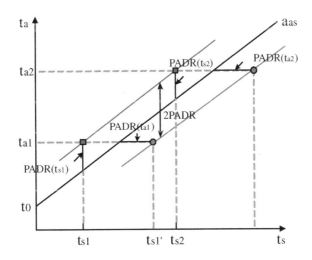

FIGURE 24.2 PADR delay estimation for time synchronization by two-way message exchanging.

in Figure 24.3. The obvious drawbacks include: (1) heavy traffic of synchronization packets will traverse the WSN because each pair has to perform its own synchronization, and (2) each node has to keep the state information of its ancestors and children. Experiment results show that mini-sync can bound the offset by 945 μs (i.e., time synchronization accuracy) in a little over an hour, corresponding to a drift of 23.3 ms in a day.[6]

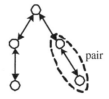

FIGURE 24.3 Time synchronization idea of tiny-sync/mini-sync.

24.3.2 Post-Facto

Post-facto[7] is a method to synchronize a local neighborhood of sensor nodes in the sender-receiver manner with the help of a "third-party" beacon node. In post-facto, the local clocks inside receivers are normally unsynchronized. When a stimulus from the sender arrives, each receiver records the arrival time using its own local clock. Immediately afterward, a "third-party" node that acts as a beacon broadcasts a synchronization pulse to all the nodes in this area. The receiving nodes then normalize their stimulus timestamps with respect to this pulse time reference. Post-facto assumes that the sender's clock is as accurate as that at the "third-party" beacon node. Therefore, each receiver knows the offset between its current local clock and the accurate time. It then adjusts its own time and waits for the next stimulus from the sender. Figure 24.4 depicts its basic idea. The solid lines represent the stimulus sent by the sender, and the dotted lines represent the references from the beacon node.

Post-facto does not require any two-way message exchanging between the sender and receivers. However, it cannot be applied to multi-hop WSN scenarios, and its synchronization range is critically limited by the communication range of the beacon node. In experiments of post-facto presented by Elson and Estrin,[7] the stimulus and sync pulse are sent and received by the standard PC parallel port, which is not wireless, and thus it is reported to achieve an accuracy of 1 μs.

24.3.3 Reference Broadcasting Synchronization

Elson et al.[8] proposed a Reference Broadcasting Synchronization (RBS) that extends the idea of the post-facto. In RBS, a node (the sender) periodically broadcasts reference beacons without an explicit timestamp to its neighbors. The receivers record the arrival timestamps based on their local clocks, and then exchange these observed time values. In this way, all the nodes get acknowledged of the offsets between each other. RBS is different from the conventional sender-receiver synchronization mechanisms because it works in a receiver-receiver manner. The receivers do not synchronize with the sender but try to keep a relative network time among the nodes that receive the same reference from the sender. Figure 24.5 illustrates the basic idea of the RBS. The solid lines represent the references from the sender while the dotted lines represent the packets exchanged among the receivers.

RBS performs well in synchronizing a neighborhood of receivers with an average accuracy of 29 μs and a worst synchronization case of 93 μs.[9] However, it is not yet applicable for multi-hop WSN. Another

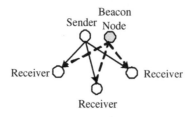

FIGURE 24.4 Time synchronization method of post-facto.

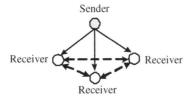

FIGURE 24.5 Time synchronization method of RBS.

disadvantage is the heavy communication overhead. To synchronize N nodes, the RBS requires at least N packets to be transmitted: one broadcast reference from the sender plus $N-1$ packets from all the receivers to exchange timestamp information. The larger the network scale, the more traffic overhead introduced by the RBS.

24.3.4 Time-Sync Protocol for Sensor Networks

A recently proposed sender-receiver time synchronization protocol is Time-sync Protocol for Sensor Networks (TPSN).[9] It is applicable for both single- and multi-hop WSNs. TPSN first organizes a randomly deployed sensor network into a hierarchical structure within a level discovery period. The new nodes that join the network after that period can get their own level values by broadcasting *level request* messages to their neighbors. Time synchronization is performed in a pair-wise fashion along the edges of this tree-like structured network, from the root to all leaf nodes, level by level. TPSN uses the conventional sender-receiver time synchronization mechanism, that is, to synchronize clocks in a pair by two-way message exchanging. Its significant contributions include that (1) it provides a good mechanism to accurately estimate the time delays along communication path between two nodes, and (2) it is implemented to be a MAC layer component of TinyOS,[11] which is the operating system for Berkeley Motes.[12] As reported by Ganeriwal et al.,[9] TPSN can achieve the one-hop time accuracy of 44 μs and get almost two times better performance when compared to the RBS.

However, TPSN does not reduce the synchronization traffic overhead, because it is based on two-way handshaking between every synchronized pair. Another disadvantage is that a receiver may have more than one sender at its upper level. In this case, there would be some redundant synchronization operations performed. Moreover, each sender at the upper level must keep all state information of its receivers at the lower level when it tries to synchronize them. Figure 24.6 shows the basic idea of TPSN.

24.3.5 Delay Measurement Time Synchronization

Ping[10] has proposed a new approach called Delay Measurement Time Synchronization (DMTS) for WSNs. DMTS organizes the sensor networks into a hierarchical structure in the same way as TPSN. It is designed to suppress the number of synchronization packets by only performing one-way message transmission

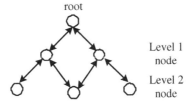

FIGURE 24.6 Time synchronization method of TPSN: two-way handshaking is needed between levels.

from the time leaders at the upper level to the receivers at the lower level, as depicted in Figure 24.7.
Thus, it synchronizes the global network time with
small traffic overhead. For example, only one broadcast packet is needed for a single-hop WSN with N
nodes, regardless of the value of N. DMTS estimates
the time delays along the communication path of
two nodes with an assumption that the distance
and bandwidth between these two nodes and the
communication packet size are known. However,

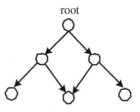

FIGURE 24.7   Time synchronization method of DMTS:
only one-way message transmission between levels.

this assumption may be impractical and system related. Different platforms of WSN may have different
communication parameters for different WSN applications. DMTS can achieve an accuracy of 32 μs in
one-hop WSNs.[10]

Note that TPSN and DMTS represent two extremes of the time synchronization schemes. TPSN is
based on pure two-way message transmissions of time information between any pairs, while DMTS
performs one-way message transmissions. In general, time synchronization via a two-way transmission
approach has better time synchronization accuracy but may not be lightweight in terms of computation
complexity, storage consumption, and traffic overhead, while a one-way transmission approach does
the opposite. There would be trade-offs between decreasing time synchronization traffic overhead and
achieving acceptable synchronization accuracy. This motivates the new design approach presented in this
chapter. Also note that current performance discussions of time synchronization protocols are based on
stable WSN scenarios; that is, no sensor node moves around.

24.4   A New Lightweight Approach for Time Synchronization

24.4.1   Time Synchronization in Single-Hop WSN

In a single-hop WSN, the sink (the sender) can synchronize a set of receivers by estimating the PADR delay
for each receiver. To decrease the number of communication packets, we provide a lightweight solution
with the basic idea of selecting an *adjuster* node from the receiver group as depicted in Figure 24.8. Only the
adjuster node is required to do two-way message exchanges with the sink and estimate the PADR delay for
the purpose of time synchronization. All the others will take the estimated PADR delay by the adjuster as
their own estimations. In this way, no matter how large the one-hop WSN is, only five packets are needed:
two references broadcast by the sender plus two replies from the adjuster, and the fifth packet from the
sender synchronizing all the receivers. And the computation and storage tasks in the sink would be simple.
Because the adjuster is selected from the receivers' group, it is expected to have properties similar to other
receivers, from the type and frequency of internal oscillators to the workload of their jobs. This expectation
should be reasonable, considering the WSN applications mentioned in Section 24.1. Another estimation
error caused by the differences of the receivers' physical distances to the sink is the communication range
of the sink divided by the speed of light, which should be a tiny value.

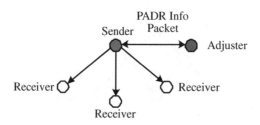

FIGURE 24.8   Basic idea of our approach: two-way handshaking only between the sender and the adjuster.

24.4.2 Time Synchronization in Multi-Hop WSN

The idea presented in the previous subsection can be extended to multi-hop WSNs. It first organizes the overall network into a hierarchical structure with different levels. Assuming that each node has a unique ID, then the level discovery can be performed in a flooding way.

The sink as the root is assigned level 0 and broadcasts a *level discovery* packet to its neighbors. The nodes that receive this packet are assigned level 1 and broadcast their own *level discovery* packets containing the new level information to other nodes. A node can receive several such packets but it can only accept the packet with the lowest level number as the one from its predecessor and takes this value plus one as its own level. Then the broadcasting operation continues. Eventually, all the sensor nodes get a unique level number assigned. A node at an upper level (i.e., a node with a smaller level number) means it is closer to the sink in terms of number of hops.

A sender selection algorithm, which is illustrated in Section 24.4.3, is used to select at least one sender per level. The communication ranges of these senders and the sink should guarantee to cover the overall WSN. Our time synchronization approach for multi-hop WSN is performed as follows:

1. The sink is selected as the sender for the nodes at level 1. It periodically broadcasts time references and performs one hop algorithm from level 0 to level 1 by randomly picking up an adjuster from level 1 nodes. Thus, level 1 is synchronized with the sink.
2. At least one sender will then be selected from the level 1 nodes as the senders for the level 2 nodes. The communication ranges of these selected senders must cover all the nodes at level 2. Then, in each sender's neighborhood, a node at the level 2 is randomly picked up to act as the adjuster. Thus, the level 2 is synchronized with the level 1.
3. In general, nodes at level n synchronized with their level $(n-1)$ predecessors may act as the senders for level $(n+1)$ nodes to perform single-hop synchronization. Only randomly picked adjusters conduct two-way message exchanging with their upper level senders.
4. This process is finished when all the nodes are synchronized. When a time reference packet is received, a node will check the level value of the source. If it is from a sender in the upper level, it accepts the time reference. Otherwise, it discards the packet silently.

24.4.3 Sender Selection Algorithm

24.4.3.1 Centralized Method

A centralized sender selection algorithm can only be performed at the sink node. After the level discovery phase, the sink can gain knowledge of the overall network topology by asking every node to report the information of its neighbors and its level value. To decrease the traffic overhead and keep the entire network synchronized, two goals should be achieved: (1) the minimum number of senders is selected, and (2) the overall network scope should be covered.

This specific network coverage problem can be transferred to a 0-1 *integer programming* problem. The leveled WSN can be envisioned as a directed graph $G(V, E)$, which consists of a set V of sensors and a set E of communication paths. Consider the node–node adjacency matrix A of G: $a_{ij} = 1$ means that there is an arc (i, j) in E with node i directly connected to node j wirelessly, and node i is exactly located at node j's upper level (i.e., $level_i = level_j - 1$), otherwise $a_{1j} = 0.a_{ij}$ also equals 0 if $i = j$. Notice that the node–node adjacency matrix A actually represents the "from-to" relationships of the sensor nodes. Take vector x as the decision variable; then the sender selection problem can be modeled as follows:

$$\text{Min} \qquad \{x_1 + x_2 + \cdots + x_n\} \tag{24.2}$$

$$\text{s.t.} \sum_{j=1}^{n} a_{ij}x_j \geq 1, \quad x_1 = 1, \quad x_j \in \{0, 1\} \forall j \in V \tag{24.3}$$

Equation 24.2 represents the optimal goal of the sender selection: to select the least number of senders possible. The model is subject to the condition expression (Equation 24.3), which means that each node should have at least one arc directed to itself from its upper level, and $x_1 = 1$ because the sink must be selected. This 0-1 integer programming problem is always solvable because there is at least one feasible solution for selecting all the non-leaf nodes in the graph G. This model can be easily solved by branch and bound algorithm with the commercial optimization solver CPLEX.[13]

24.4.3.2 Distributed Method

Although the centralized method can solve the sender selection problem, it is not efficient due to a large number of communication packets that are needed to gather the overall network topology and a huge computation task has to be done by the sink. We then design a distributed method to locally select senders at each level.

In fact, the above specific network coverage problem can be solved by a greedy select-and-prune method as follows.

1. *Report.* Each node at level n only reports its neighbors at level $(n + 1)$ up to its direct predecessor at level $(n - 1)$.
2. *Select.* The predecessor then selects the one that has the maximum number of neighbors to be the sender. This node covers its children at level $(n + 1)$.
3. *Prune.* Then the predecessor deletes all the nodes covered by the selected sender from the neighbors of those unselected nodes at level n, and goes back to Step 2 until all the nodes at level $(n + 1)$ are covered.

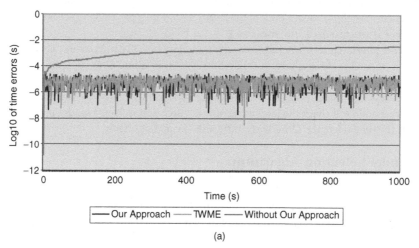

(a)

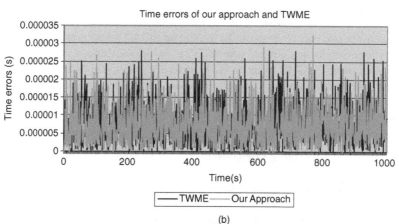

(b)

FIGURE 24.9   Time synchronization errors using different approaches in one-hop WSN scenarios.

This distributed greedy algorithm is performed in an up-down manner from level 0 to the entire network. The distributed sender selection method is able to get the same result as the centralized approach but in a more scalable and practical way.

24.5   Simulation Results

NS-2 has been used[14] to simulate our new time synchronization approach in both single- and multi-hop WSN scenarios. The PADR delay associated with each node is taken as an error factor, with the mean value of 50 μs. All simulations run for 1000 s.

In single-hop scenarios, the average time synchronization errors at the receivers are recorded every second. Under the same network topology, three cases have been studied and compared, including the pure two-way message exchanging (TWME), our approach, and a case without using any time synchronization method at all. Figure 24.9 shows the one-hop simulation results. The Figure 24.9a results demonstrate that both time synchronization schemes can effectively synchronize the receiver's local time with the sender.

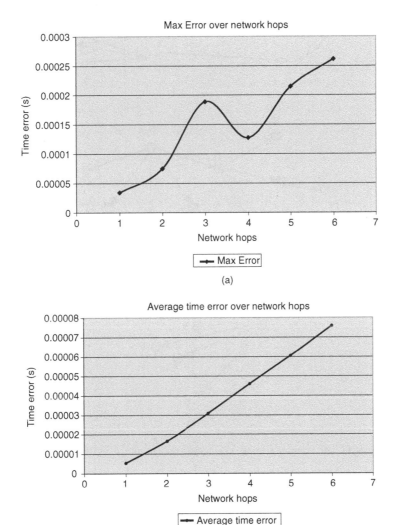

(a)

(b)

FIGURE 24.10   Maximum and average time synchronization errors from one-hop to six-hop WSN scenarios.

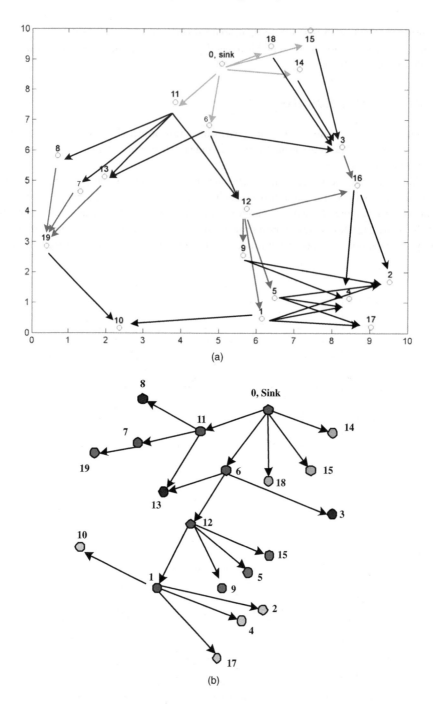

FIGURE 24.11   An example of sender selection by the centralized and distributed methods.

Figure 24.9b illustrates that the maximum time synchronization error of our approach in one-hop is 34 μs, while that of the TWME is 28 μs. This proves that our new approach can achieve a certain acceptable synchronization accuracy but it significantly decreases communication overhead by only requiring adjusters to perform two-way message exchanging.

In multi-hop WSN scenarios, the performance of our new approach in the worst case is studied, where all the senders are located in a linked list from one to six hops. The senders also act as the adjusters. Figure 24.10a

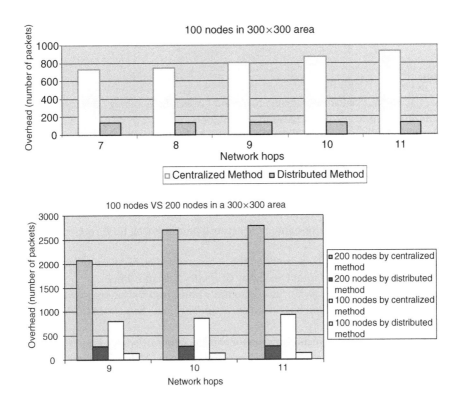

FIGURE 24.12 Overhead comparison between the centralized and distributed sender selection methods.

depicts that the maximum time synchronization error increases as the number of network hops increases. This phenomenon happens because the approach organizes WSNs into a hierarchical structure. Time errors will be propagated across levels. Figure 24.10b shows that the average time synchronization error increases almost linearly with the increase in the number of network hops.

Figure 24.11 illustrates an example to solve the sender selection problem by the centralized method. Twenty nodes are randomly deployed in a 10 × 10 area shown in Figure 24.11a with the communication range of four units, and the sender selection results by CPLEX are shown in Figure 24.11b. Using the distributed method, the same result has been generated.

The benefit of using the distributed method is that it decreases the communication overhead because every node only needs to care about its children nodes and no global information is required. The overhead comparison between the centralized and distributed methods is shown in Figure 24.12. In our simulations, either 100 or 200 sensor nodes are randomly deployed in a 300 × 300 area. The comparison results clearly show that the overheads of both the centralized method and the distributed method increases with increments in the complexity of the network topology in terms of network hops and the number of sensor nodes. However, the overhead of the distributed sender selection method is much less than that of the centralized one.

24.6 Conclusion

This chapter presented a theoretical model for time synchronization based on the sender-receiver paradigm and summarized the existing time synchronization algorithms proposed for WSN. Then a new lightweight time synchronization protocol for WSN was presented. The new approach suppresses the communication overhead in terms of the number of time synchronization packets needed by requiring only part of overall nodes in the WSN to perform two-way handshaking while still achieving acceptable time synchronization accuracy. Simulation results have validated the authors' design ideas. Moreover, the simulations reveal that the maximum and average time synchronization errors increase when communication hops increase.

References

1. Akyildiz, I.F., Su, W., Sankarasubramaniam, Y., and Cayirci, E., A survey on sensor networks, *IEEE Communications Magazine,* 40(8), 102–116, August 2002.
2. Heinzelman, W.R., Chandrakasan, A., and Balakrishnan, H., Energy efficient communication protocol for wireless microsensor network, *Proc. 33rd Hawaii Int. Conf. on System Sciences (HICSS'2000),* Vol. 8, Maui, Hawaii, January 2000.
3. Claesso, V., Lönn, H., and Suri, N., Efficient TDMA synchronization for distributed embedded systems, *Proc. 20th Symp. on Reliable Distributed Systems (SRDS'01),* New Orleans, Louisiana, October 2001, pp. 198–201.
4. Mills, D.L., Internet time synchronization: the network time protocol, *IEEE Trans. Commun.,* 39(10), 1482–1493, October 1991.
5. Ye, Q., Zhang, Y., and Cheng, L., A study on the optimal time synchronization accuracy in wireless sensor networks, under second round review by *Computer Networks,* 2004.
6. Sichitiu, M.L. and Veerarittiphan, C., Simple accurate time synchronization for wireless sensor networks, *Proc. IEEE Wireless Communiactions and Networking Conf. 2003 (WCNC'2003),* New Orleans, LA, March 2003.
7. Elson, J. and Estrin, D., Time synchronization for wireless sensor networks, *Proc. IPDPS Workshop on Parallel and Distributed Computing Issues in Wireless Networks and Mobile Computing (IPDPS'01),* San Francisco, CA, April 2001, pp. 186–191.
8. Elson, J., Girod, L., and Estrin, D., Fine-grained network time synchronization using reference broadcasts, *Proc. Fifth Symp. on Operating Systems Design and Implementation (OSDI'2002),* Boston, MA, December 2002.
9. Ganeriwal, S., Kumar, R., and Srivastava, M.B., Timing-sync protocol for sensor networks, *Proc. First ACM Int. Conf. on Embedded Networked Sensor Systems (Sensys'03),* Los Angeles, CA, November 2003.
10. Ping, S., Delay Measurement Time Synchronization for Wireless Sensor Networks, *Intel Research Berkeley Technical Report* 03-013, June 2003.
11. Levis, P., Madden, S., Gay, D. et al., The emergence of networking abstractions and techniques in TinyOS, in *Proc. First USENIX/ACM Symp. on Networked Systems Design and Implementation (NSDI'2004),* San Francisco, CA, March 2004.
12. Warneke, B.A., Ultra-Low Energy Architectures and Circuits for Cubic Millimeter Distributed Wireless Sensor Networks, Ph.D. dissertation, University of California, Berkeley, May 2003.
13. CPLEX, http://www.ilog.com/products/cplex.
14. The Network Simulator—ns-2, http://www.isi.edu/nsnam/ns/.

25

Routing and Broadcasting in Hybrid Ad Hoc and Sensor Networks

François Ingelrest

David Simplot-Ryl

Ivan Stojmenović

25.1 Introduction

In the past few years, networking technology has advanced very rapidly. Internet access is a standard commodity, and most companies use *LANs* (*local area networks*) to forward information between employees. Fiber optics deployment allowed high-speed Internet access for personal use. The next step in technological development is to provide high-quality Internet access to nomadic users who want to check their mail or keep in touch with their offices, using portable devices such as cell phones, laptops, or *PDAs* (*personal digital assistants*). WLANs (*wireless LANs*) have emerged to fill this growing demand, with the *WiFi* (*Wireless Fidelity*) technology, which provides such an access to a user who is in the physical neighborhood of an access point. These access points are being deployed at densely populated stations such as airports. Despite its advantages, this technology is still very restrictive, as users must be in the communicating range of an access point to use it. This means that a huge number of access points must be installed to have a seamless wireless network available.

To allow greater mobility, and to reduce the impact of collisions with multiple users attached to the same access point, multihop access mode is being considered. Instead of direct communication with the access point, it may be beneficial, in terms of energy efficiency, extended coverage, and bandwidth capacity, to contact the access point via other users in multihop fashion. A similar scenario also exists with cellular networks in areas of high user populations, such as a stadium during events. Multihop cellular networks are a viable alternative to direct access from mobile phone to public phone networks in such scenarios.

Wireless ad hoc networks are being considered to provide multihop communication between peers. They are formed by a set of hosts that operate in a self-organized and decentralized manner, forming a dynamic autonomous network without relying on any fixed infrastructure. Communications take place over a wireless channel, where each host has the ability to communicate directly with any other one in its physical neighborhood, which is determined by a communicating range. These networks have multiple applications in areas where wired infrastructure may be unavailable, such as battlefields or rescue areas.

These two technologies (pure ad hoc networks, fixed infrastructure) can be combined into one to better satisfy user needs. Using an ad hoc communicating mode, fewer access points are needed to cover a "crowded" area. The access points themselves can participate in ad hoc communication in addition to providing access to a fixed infrastructure. For example, some nodes in a network (possibly even mobile) could be equipped with, say, satellite access for communication among themselves and for Internet access.

Figure 25.1 illustrates how hybrid networks can be formed to replace existing single-hop access. Case (a) shows a wireless network that relies on a fixed infrastructure. To cover the whole area in this mode, two access points are needed. With the use of ad hoc communicating mode, illustrated in case (b), it is possible to use only one access point. Users that are relatively far from an access point can still access it, using other mobile users as relays. Such networks are referred to as being *hybrid ad hoc networks*. Examples of such networks include multihop cellular and wireless Internet access networks. In addition to having access to a fixed infrastructure, hybrid ad hoc networks can also provide communication between network nodes. For example, friends may look for each other at a stadium. The communication may use only ad hoc network nodes, or may, in addition, involve one or two access nodes. Access nodes have some advantages over mobile nodes. They have an "unlimited" amount of energy, and are therefore reliable nodes for receiving and transmitting messages. Access nodes can use the same transmission range as regular nodes (mobile or stable nodes in an ad hoc network, or sensors in a sensor network) to provide symmetric communication, or could have increased transmission range for one-way message transmissions. This chapter mainly focuses on the first case, with access points and regular nodes using the same transmission ranges, assuming that all messages need to be acknowledged directly for reliability. Also, the same transmission ranges are needed to establish routes from each node to the access point, as will be discussed later.

Sensor networks are composed of autonomous devices with integrated sensing, processing, and radio-communication capabilities. They have applications in many domains, such as military, health, or home environments. For example, they can be used for battlefield surveillance, monitoring equipment, or for disaster prevention and monitoring. Our context of hybrid ad hoc networks is also applicable to heterogeneous sensor networks, considered by Intel for practical applications. In addition to regular tiny sensors, bandwidth, and energy constraints, they contain some "supernodes" that have much higher bandwidth and energy (possibly even no energy limitations), and which create a high bandwidth backbone for communication between themselves and connection to the monitoring station. We assume here that these supernodes serve as access points to tiny sensors, and that the communication cost between them is negligible compared to the cost of communicating between regular sensors. These access points can all be linked by their own network to a single base station. Alternatively, some or all of them can also serve as base stations in a multi-sink structure. With such assumptions made, the heterogeneous sensor networks become a special case of hybrid ad hoc networks, considered here as a general network model.

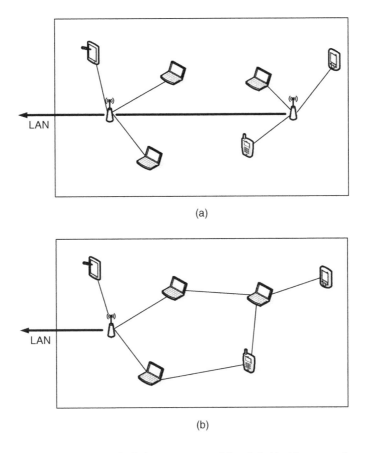

(a)

(b)

FIGURE 25.1   From single-hop access to multihop hybrid ad hoc network.

We have assumed here that all regular nodes have the same maximum transmission radius, which differs from the one used by supernodes. This is a different kind of network from the heterogeneous ad hoc networks where each node uses its own maximum transmission radius.[9]

The goal of this chapter is to consider some basic data communication problems of hybrid ad hoc networks, such as broadcasting or routing, and to propose some techniques adapted to this kind of network. Indeed, these tasks must be performed by taking advantage of the presence of access points and, as a consequence, existing algorithms for ad hoc networks must be adjusted. The organization is as follows. We first define a terminology for hybrid networks in Section 25.2 and present literature review in Section 25.3. We then propose some protocols for broadcasting in Section 25.4 and for routing in Section 25.5. We finally give a brief conclusion and ideas for future work in Section 25.6.

The preliminary version of this chapter appeared as a technical report.[6]

25.2   Preliminaries

We represent a wireless ad hoc network by a graph $G = (V, E)$ where V is the set of vertices (mobiles or access points) and $E \subseteq V^2$ the set of edges between these vertices. An edge exists between two nodes if they are able to communicate directly (1-hop communication) to each other: two nodes u and v can communicate directly if they are in the communicating radius of each other. If all nodes have the same range R, the set E is then defined as:

$$E = \{(u, v) \in V^2 \mid u \neq v \wedge d(u, v) \leq R\},$$

$d(u, v)$ being the Euclidean distance between u and v. We also define the neighborhood set $N(u)$ of the vertex u as

$$N(u) = \{v \mid (u, v) \in E\}.$$

We consider here hybrid networks, which are formed by mobile and fixed access points, denoted by P_i. Depending on their position, mobiles can be either directly connected to an access point, or constrained to use ad hoc mode if they are too distant. We assume that access points are mutually connected by a fast, high-bandwidth backbone network. It is reasonable to assume that access nodes are able to emit radio messages with a radius pR, p being a constant multiplier ≥ 1. A radio message emitted by an access point P_i will be received by every mobile u such that

$$d(P_i, u) \leq pR.$$

We use the assumption that $p = 1$, so that access points and mobiles have the same transmission radius. We denote by $\mathrm{hc}(u, v)$ the distance in hops between nodes u and v, which is simply the number of edges a message has to cross between these two nodes. We also denote by $\mathrm{AP}(u)$ the closest access point to the mobile u, in term of hops. If several access points are at the same hop distance from the node, then the identifier (id) of access points is used as a tie breaker; that is, the one with the smallest id is chosen.

For the sake of simplicity, we denote by $\mathrm{hc}(u)$ the distance in hops between u and its nearest access point:

$$\mathrm{hc}(u) = \mathrm{hc}(u, \mathrm{AP}(u)).$$

The set of mobiles attached to an access point P_i is denoted by $\mathrm{AN}(P_i)$:

$$\mathrm{AN}(P_i) = \{u \mid \mathrm{AP}(u) = P_i\}.$$

We suppose that each node u regularly emits special short messages called *HELLO* messages, containing its id, denoted by $\mathrm{id}(u)$, and the values of $\mathrm{AP}(u)$ and $\mathrm{hc}(u)$. Those messages are not relayed like broadcast messages; they are only intended for the neighborhood of a node. We suppose that a node can determine its distance to the nearest access point by first receiving the distances of its neighbors, and then setting its own distance to the smallest one-plus-one hop. If access points send their *HELLO* messages with a distance of 0, each node is thus able to recursively determine its distance to the nearest access point. This process can be avoided for some protocols that do not request nodes to have information about their closest access point, like the blind flooding (see Section 25.4.1).

Figure 25.2 shows an example of such a hybrid network. Access points P_1 and P_2 form a wired network. In this example, we have $\mathrm{AP}(a) = P_1$ while $\mathrm{AP}(e) = P_2$, $\mathrm{AN}(P_1) = \{a, b, c, d\}$, and $\mathrm{AN}(P_2) = \{e, f\}$.

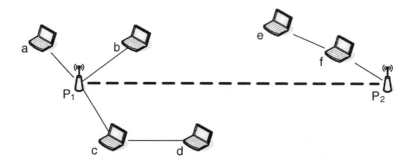

FIGURE 25.2   Example of a hybrid network with two access points.

25.3   Literature Review

25.3.1   Broadcasting

The broadcasting is defined to be a one-to-all communication, where one node sends a message that should be received by all other users in the network (provided they are connected). For further reading, an extensive review of energy-efficient broadcasting protocols for pure ad hoc networks can be found.[5] In case of hybrid networks, the sender can be one of regular nodes, or one of access points (supernodes).

The most basic broadcasting protocol is known as *Blind Flooding*, in which a source node transmits the message to all its neighbors, and then each node that receives it for the first time re-emits it. Assuming an ideal *MAC* layer, this protocol is reliable; that is, every node in the network will receive the message at least once. However, because of its simplicity, this protocol leads to a lot of duplicated packets and thus to a huge waste in energy consumption. Further, with a realistic *MAC* layer, such as IEEE 802.11, a number of almost simultaneous scheduled transmissions causes excessive message collisions; therefore, the protocol is counterproductive in dense networks.

A more intelligent protocol, called the *Neighbor Elimination Scheme* (NES), has been independently proposed.[3,11] Its principle is as follows. Each node that receives the message for the first time does not retransmit it immediately, but waits for a given duration, which can be computed or randomly generated. Then, the node starts monitoring its neighborhood and after each received copy of the broadcast message, it eliminates from its rebroadcast list neighbors that are assumed to have correctly received it. If the list becomes empty before the node decides to relay the message, the re-emission is canceled. This protocol allows some energy savings by canceling redundant emissions, while still insuring an entire coverage of the network.

Another category of protocols is based on the computation of a connected dominating set S. A set is a dominating one if each node in the graph is either in S or a neighbor of a node in S. The broadcasting step, in its simplest variant, can be described as follows. When a node receives a broadcast message for the first time, it drops it if it is not in the considered connected dominating set, or retransmits it otherwise.[3] Nodes ignore subsequent receptions of the same message. When the Neighbor Elimination Scheme is applied, some transmissions can be avoided. A node that is in the dominating set, but observes that all its neighbors have already received the same message, can also drop the packet without retransmitting it.

Connected dominating sets can be defined in several ways. A localized algorithm that computes such a set, called the *Generalized Self-Pruning Rule*, can be found.[1] In this method, each node u must be assigned a key denoted by $key(u)$, the key used in Ref. 1 being equal to $id(u)$. First, each node checks its state: it is intermediate if it has at least two neighbors that are not directly connected. Non-intermediate nodes are never dominants. Then, each intermediate node u constructs a subgraph G of its neighbors with higher keys. If G is empty or disconnected, then u is in the dominating set. If G is connected but there exists a neighbor of u that is not a neighbor of any node from G, then u is in the dominating set. Otherwise, u is covered and is not in the dominating set. In this source-independent protocol, all broadcasting tasks are always supported by the same nodes (unless NES is applied). This allows the rest of the nodes to be placed in sleeping mode without affecting network operation.

If all nodes remain active, to better balance the energy consumption, some source-dependent protocols can be used. In this category, the *Multipoint Relay* (MPR) *protocol* was proposed by Qayyum et al.[12] A node uses greedy heuristics to compute an optimal selection of its direct neighbors to act as relays, in order to reach all of its two-hops neighbors. The node forwards this selection with the broadcast packet, and only selected nodes relay it. A set of one-hop neighboring relay nodes is then selected to cover all two-hops neighbors as follows. Repeatedly, the neighbor that covers the maximal number of two-hops uncovered neighbors is selected as relay, and newly covered nodes are eliminated from the list of uncovered neighbors.

25.3.2   Routing

When two nodes want to communicate with each other, two cases can occur: (1) either they are neighbors, in which case they can communicate directly; or (2) they are too distant, in which case messages must be routed. Routing is the problem of sending a packet from a source node to a destination node. A simple solution to this problem would be to broadcast the messages to the whole network. However, such a solution uses huge network capacity and leads to network congestion after only few such tasks. For a particular communication, a path must be therefore found to utilize only nodes needed for forwarding the packets.

Dynamic Source Routing (DSR) is a reactive IETF protocol,[8] considered as a possible standard for ad hoc network routing, that uses the broadcasting process to find a route between two nodes. When a host wants to find a route to another one, it initiates a broadcast containing the *id* of the searched host. Each node that receives this message inserts its *id* in the packet, and possibly some other control overhead (depending on the particular variant of the considered protocol), and will retransmit it. Because the broadcast reaches every connected node in the network, the destination will receive it and will be able to reply to the source by following the chain of nodes traversed by the packet in the reverse order. When the reply is received, a communication route has been established between the two nodes.

While reactive protocols, such as DSR, create routes only when they are needed, a proactive protocol creates and maintains routes before their use. To do so, each host maintains routes to other ones in the network by exchanging routing tables between neighbors. These routing protocols can be also used in hybrid networks. Although proactive algorithms allow a source node to immediately have a route to a destination, they may require a large amount of data for their maintenance and therefore cause huge communication overhead. Some efficient proactive protocols have been proposed for pure ad hoc networks, including *Optimized Link State Routing* (OLSR),[7] in which *MPR* is used for route maintenance.

We identified only two routing protocols for hybrid wireless networks in the literature. Li et al.[10] proposed a system that connects public service buses to form a wireless network. Some stationary gateways must be installed along the roads for users to be able to access the Internet. Communications between buses and gateways can be done directly or in ad hoc mode, depending on the distance between them. In ad hoc mode, buses serve as relays for other ones not directly connected. The routing task is done by a top-level router, which knows the closest gateway for each bus using any proactive or reactive method. When a message must be routed, the closest gateway forwards it to the top-level router, which redirects it to the destination gateway or to the Internet, as needed.

Fujiwara et al.[4] proposed a mechanism that allows nodes to maintain their routes to the base station via multihopping if needed. If a direct link between any node and its base station is broken, the node starts monitoring communications in its neighborhood to find a node that is still connected to the base station, either directly or by multihopping. When the node finds a connected neighbor, it marks it as its "router" and sends to it the packets that must be sent to the access point. This allows nodes to always be able to connect to their base station. The authors[4] considered only the case of single access point.

Ding et al.[2] considered the problem of finding a route from a sensor to the single base station (sink) in a sensor network. Following a DSR-based strategy, the sink floods the network and sets the routes. The difference is that each sensor does not memorize the whole route, or a single pointer to predecessor sensor on the route, but instead it memorizes its hop count distance to the sink. When a packet is sent toward the sink, any neighbor at one less hop distance can forward it. Nodes can memorize few such alternatives during the setup phase and try them one by one. Alternatively, a neighbor at one less hop distance can simply retransmit, and the node can block further retransmissions by a separate blocking packet.

25.4 Broadcasting

We present here several broadcasting protocols for hybrid ad hoc networks. These protocols are either new or generalizations of existing protocols for pure ad hoc networks described in Section 25.3.1.

25.4.1 Hybrid Blind Flooding

This protocol is a very simple extension of the existing blind flooding protocol for pure ad hoc networks. In this protocol, each node, receiving a packet for the first time, will retransmit it. Subsequent copies of the same packet are ignored. If the node that received the packet (or a source node) is an access point, then all other access points receive the packet via their backbone network. Therefore, in the next step, all access points can retransmit the message.

25.4.2 Component Neighbor Elimination Based Flooding

This protocol is based on an observation that transmissions from nodes to other ones directly connected to an access point are a waste of energy. Indeed, the mobile could have received the message from the access point, which would have been done "for free" (we do not take into account energy spent by access points).

To prevent these useless transmissions and to allow access points to be the first ones to reach their neighborhood, we divide the network into components, one for each access point. Each component $C(P_i)$ is defined by:

$$C(P_i) = \{P_i \cup \text{AN}(P_i)\}.$$

We can notice that these components are connected, because there exists a path between any node in $\text{AN}(P_i)$ and P_i. To further limit energy consumption, we use a neighbor elimination scheme.

To limit the propagation inside each component, we suppose that there exists a field named P_{msg} in the broadcast packet that defines which component is going to be flooded; that is, only nodes within the component $C(P_{\text{msg}})$ relay the message. When the message is transmitted for the first time by a node s, the value of P_{msg} is set to $\text{AP}(s)$, in order to flood the component $C(\text{AP}(s))$. A node u that receives a message with $P(u) \neq P_{\text{msg}}$ does not relay it. Otherwise, if it is the first reception, it enters an *NES*, monitors its neighborhood, and relays the packet at the end of the timeout only if there exists uncovered neighbors in $P(u)$. When the access point $\text{AP}(s)$ receives the message, it relays it to all other access points. Depending on the structure of the network of access points, this can be done by direct forwarding to each of them, or by applying a corresponding broadcast protocol among access points. When an access point P_i receives the message for the first time, it changes the value of P_{msg} to its own *id* before rebroadcasting it via the radio interface to all nodes in its component.

Note that the same protocol can be applied if the source node is one of the access points. In this case, the message is first relayed to other access points via their backbone. Then, component flooding is performed independently from each access point.

25.4.3 Adaptive Flooding

The main drawback of component neighbor elimination based flooding is its increased latency, which is the elapsed time between the start of the broadcast and its end. Indeed, some nodes that could have received the message earlier from a close neighbor in another component are ignored, because they have different respective access points. *Adaptive Flooding* is designed to minimize the latency of the broadcast.

For a given node a, there are two ways to receive the message:

1. In "ad hoc mode," from the source node s, or other node (in the same or different component as s), without passing through any access point
2. In "access point mode," from the node s to $\text{AP}(s)$, from $\text{AP}(s)$ to $\text{AP}(a)$, and from $\text{AP}(a)$ to the node a. We assume that the cost of the communication (in terms of duration) between $\text{AP}(s)$ and $\text{AP}(a)$ is equal to zero.

This protocol selects the shorter path between these two modes to reduce the overall latency. When a node a receives from a node p a broadcast message initiated by a node s, two cases can happen:

1. The message has not crossed an access point. The node a decides to forward the message if there exists a node b that belongs to $N(a)\backslash N(p)$ such that $\text{hc}(s, a) + 1 < \text{hc}(s) + \text{hc}(b)$. In this case, $\text{hc}(s, a)$ can be approximated by the number of links the message has crossed from s to a and $\text{hc}(s)$ should have been written into the packet by s. These distances are illustrated in Figure 25.3. Note that the message may cross several components in this process.

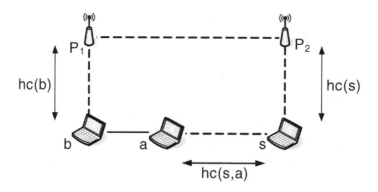

FIGURE 25.3   Illustration of the distances for adaptive flooding.

2. The message has crossed an access point. Each node relays the message if there exists a neighbor, in the same component, that would benefit from this retransmission. That is, the corresponding access point of the component is treated as the message source, neighbors from other components are ignored, and neighbor elimination based flooding is applied within the component. Note that some nodes in the same component could have received the same message by applying the first "non-crossing access point" mode, and these nodes do not participate in this mode (except in cases when they did not retransmit the message, and neighbors, not knowing about their reception, could transmit because of them).

Note that when source node is one of access points, adaptive flooding becomes equivalent to the component neighbor elimination based flooding.

25.4.4   Multipoint Relay Broadcasting Protocol

This protocol is very efficient in terms of energy savings, and can be easily generalized to hybrid networks. Regular nodes should be used as relays only if they are needed in addition to access points. When considering which neighbors should relay, access points (if any in the neighborhood) should be first added to the list of relays and then, if there remains some uncovered two-hops neighbors, an optimal selection of remaining neighboring relay nodes should be computed.

If we assume that regular nodes do not have components information, this protocol can be applied without any further modification. When an access point receives the message, it simply has to send it to the other access points to speed up the broadcasting process. However, if mobiles are aware of their component membership, and the hop count distances of the source, one-hop, or even two-hops neighbors to their corresponding access points, some transmissions could be avoided. For example, two-hops neighbors a for which $hc(s, a) > hc(s) + hc(a)$ do not need to be covered (note that current node adds two hops to its own distance to s to its estimate for $hc(s, a)$ which may not be a correct value).

When the source node is one of the access points, the method first shares the packet among all access points, and then applies MPR in each component independently. Some optimization can be considered, so that border nodes can be covered by transmissions from nodes in other components. However, this adds to the complexity of the procedure without providing significant gains.

25.4.5   Dominating Sets Based Broadcasting Protocol

The generalized self-pruning rule, as described in Section 25.3.1, is very flexible because the key can be composed of any collection of values, while still guaranteeing the construction of a connected dominating set. To adapt it to hybrid networks, we replace the *id* by two values, so that the key key(u) of a node u is defined by:

$$\text{key}(u) = \{E_u, \text{id}(u)\}, \tag{25.1}$$

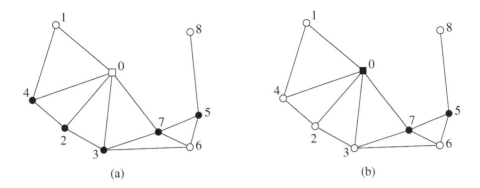

FIGURE 25.4 Generalized self-pruning rule applied to hybrid networks.

E_u being the energy level of u. The comparison between two keys is made using their primary keys; and if they are equal, then the comparison is made using the secondary key. Note that this replacement of id by node energy has been proposed.[13]

In this particular scenario, if we consider that access points have an "infinite" amount of energy, they will always be selected as dominant and thus will be part of the broadcasting process.

Figure 25.4 shows an example of the application of such key definition, square 0 being an access point and circles being regular nodes. Case (a) is the result of the generalized rule applied with ids of nodes used as keys; while in case (b), ids have been replaced by the key given in Equation 25.1. Access point 0 has been selected in the dominating set; and as a result, nodes 2, 3, and 4 are now covered (not in dominating set), so that they will not spend their energy for the broadcast process.

25.5 Routing

25.5.1 Adaptation of Existing Protocols

The simplest way to adapt routing protocols such as DSR or OLSR to hybrid networks is obviously to replace the broadcasting protocol by its adapted version as described in previous sections.

For example, the broadcasting step of DSR can use a variety of protocols, depending on the taken assumptions. If we assume that nodes do not have any information about their closest access point, one can use blind flooding, as defined in the original version of DSR. However, if we remove this assumption, some more intelligent protocols can be used. For example, using a dominating sets based protocol, the number of retransmissions will be reduced. The version adapted to hybrid networks gives top priority to access points, so that they will be used whenever it is possible for routing in DSR, saving energy of regular nodes. Adaptive flooding can also be used, which would lead to the discovery of the shortest paths, which is the property of the protocol.

Similar discussion is valid for OLSR, which can take advantage of the presence of access points if MPR is modified appropriately, as described in the previous section.

25.5.2 Hybrid Routing for Hybrid Ad Hoc Networks

This protocol allows regular nodes to communicate to each other using the faster mode between infrastructure or ad hoc communicating modes. It is a hybrid routing protocol because it combines proactive and reactive approaches. Proactive routing is used to maintain links of each ad hoc node to its access point, while reactive routing is used to find routes between two ad hoc nodes. We assume that nodes know the component memberships and that each access point P_i knows the regular nodes that are in $AN(P_i)$ and the hop distance to each of them. For example, Fujiwara's protocol[4] can be used to achieve this, when

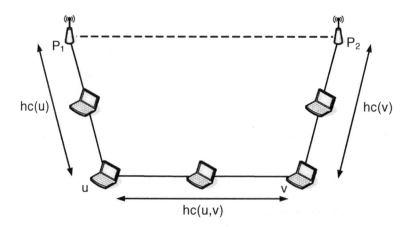

FIGURE 25.5   Ad hoc communicating mode can be faster than infrastructure mode.

applied from each access point, with nodes forwarding traffic only from their closest access points. When a node u wants to communicate with a node v, two modes can be used:

1. *The infrastructure mode.* The node u sends the packets to AP(u), which forwards them to AP(v). Access points can periodically exchange their routing tables to determine which one has to be contacted, depending on the packet that has to be routed. Finally, AP(v) will forward packets to v.
2. *The ad hoc mode.* The node u sends the packets "directly" to v, using other mobiles as relays.

As illustrated by Figure 25.5, the ad hoc communicating mode can sometimes be faster (the path is shorter) and should then be used to speed up the routing process. To determine which mode to use, the node u first asks the value of hc(v) to AP(u). If AP(u) does not have this information, it requests it from other access points in the wired network. When u retrieves hc(v), it launches a broadcast with a Time-To-Live (*TTL*) equal to hc(A) + hc(B) − 1 to find a route in pure ad hoc mode (using DSR, for example). If v is not found using this broadcast, it means that the path between them in ad hoc mode is longer than the one in infrastructure mode (i.e., hc(u, v) > hc(u) + hc(v)). In this case, the infrastructure mode will simply be used. Using this protocol, any two nodes can communicate to each other by knowing their routes and distances to access points.

It can be noticed that packets can be rerouted by any node in infrastructure mode if needed. Indeed, if a packet that has to be routed via a certain path arrives at a node that knows this route is no longer available, it can reroute the packet using its own path to the access point. The latter and the source node will then update their routing tables.

25.5.3   Routing in Heterogeneous Sensor Networks

Sensor nodes normally send their data to a specific node called a sink (or a monitoring station), which collects the requested information. Depending on the distance between a given node and its nearest monitoring station, a direct communication is not always available, and multihop routing must be used.

In the case of a network with multiple sinks (access points, base stations), each node must determine which sink is the nearest one, and its distance to it. For this purpose, the algorithm of Fujiwara et al.,[4] designed for the case of one sink (base station), can be generalized. The process is first synchronized among multiple sinks so that they start flooding at the same time. Using a variant of hybrid blind flooding, each node can retransmit the packet the first time it receives it, using the access point identifier. Further copies of the same message, coming from the same or different access node components, are ignored.

Some optimization to this procedure can be made with the help of NES. Sensors whose neighbors already received the packets from access points do not need to retransmit the message. Note that sensors that receive two messages from two different components may still need to retransmit, because they could have a third neighbor without deciding the component and without receiving any packet. This procedure sets pointers from each sensor to one or more of its neighbors with hop distance smaller by 1 from the same access points, which can be used in establishing a reporting tree for the sensor to its nearest sink.

25.6   Conclusion

In this chapter, we have considered hybrid networks, which are composed of regular ad hoc network nodes and access points, in which the ad hoc communicating mode is available to increase the flexibility, mobility of users, and reachability. We presented several algorithms for basic data communication tasks in a network, such as broadcasting and routing. These algorithms are adapted from ad hoc networks to hybrid networks to take advantage of access point as much as possible.

In our future work, we want to further improve some of these protocols and to design some experiments to obtain their respective performances, which could allow a fair comparison between them. We want to study broadcast protocols involving topology management with radius adjustment in hybrid networks. Finally, some assumptions can also be removed and their consequences studied. A particular example is the case in which access points have a constant factor p times larger transmission radius than the maximum radius of ad hoc mobile nodes.

References

1. F. Dai and J. Wu. Distributed dominant pruning in ad hoc networks. In *Proceedings of the IEEE International Conference on Communications (ICC'03)*, Anchorage, Alaska, May 2003.

2. J. Ding, K.M. Sivalingam, R. Kashyapa, and L.J. Chuan. A multi-layered architecture and protocols for large-scale wireless sensor networks. In *Proceedings of the IEEE Vehicular Technology Conference (VCT2003)*, Orlando, FL, October 2003.

3. I. Stojmenović (Ed.) and M. Seddigh. Broadcasting algorithms in wireless networks. In *Proceedings of the International Conference on Advances in Infrastructure for Electronic Business, Science, and Education on the Internet SSGRR*, L'Aquila, Italy, July 31–Aug. 6, 2000.

4. T. Fujiwara, N. Iida, and T. Watanabe. An ad hoc routing protocol in hybrid wireless networks for emergency communications. In *Proceedings of the International Workshop on Wireless Ad Hoc Networking (WWAN'04) at IEEE International Conference on Distributed Computing Systems (ICDCS'04)*, Tokyo, Japan, March 2004.

5. F. Ingelrest, D. Simplot-Ryl, and I. Stojmenović. *Resource Management in Wireless Networking*, Chapter 17, Energy-Efficient Broadcasting in Wireless Mobile Ad Hoc Networks. Kluwer, 2004. To be published.

6. F. Ingelrest, D. Simplot-Ryl, and I. Stojmenović. Routing and broadcasting in hybrid ad hoc multi-hop cellular and wireless Internet networks. Technical Report INRIA RT-291, February 2004.

7. P. Jacquet, P. Mühlethaler, T. Clausen, A. Laouiti, A. Qayyum, and L. Viennot. Optimized link state routing protocol for ad hoc networks. In *Proceedings of the IEEE International Multi-Topic Conference (INMIC'01)*, Lahore, Pakistan, December 2001.

8. D.B. Johnson, D.A. Maltz, and Y.-C. Hu. The dynamic source routing protocol for mobile ad hoc networks (DSR). Internet Draft, draft-ietf-manet-dsr-09.txt, April 2003. Work-in-progress.

9. N. Li and J. Hou. Topology control in heterogeneous wireless networks: problems and solutions. In *Proceedings of the IEEE INFOCOM 2004*, Hong Kong, March 2004.

10. T. Li, C.K. Mien, J.L.S. Arn, and W. Seah. Mobile Internet access in BAS. In *Proceedings of the International Workshop on Wireless Ad Hoc Networking (WWAN'04) at IEEE International Conference on Distributed Computing Systems (ICDCS'04)*, Tokyo, Japan, March 2004.

11. W. Peng and X.C. Lu. On the reduction of broadcast redundancy in mobile ad hoc networks. In *Proceedings of the ACM MobiHoc 2000,* pages 129–130, Boston, MA, August 2000.

12. A. Qayyum, L. Viennot, and A. Laouiti. Multipoint relaying for flooding broadcast messages in mobile wireless networks. In *Proceedings of the Hawaii International Conference on System Sciences (HICSS'02),* January 2002.

13. J. Wu, B. Wu, and I. Stojmenović. Power-aware broadcasting and activity scheduling in ad hoc wireless networks using connected dominating sets. *Wireless Communications and Mobile Computing,* Vol. 4, No. 1, June 2003, 425–438.

26

Distributed Algorithms for Deploying Mobile Sensors

Guohong Cao

Guiling Wang

Tom La Porta

Shashi Phoha

Wensheng Zhang

26.1 Introduction

Recent advances in digital electronics, microprocessors, micro-electro-mechanics, and wireless communications have enabled the deployment of large-scale sensor networks where thousands of tiny sensors are distributed over a vast field to obtain fine-grained, high-precision sensing data. Due to many attractive characteristics of sensor nodes, such as small size and low cost, sensor networks[2,11,14,19,22] have been adopted by many military and civil applications, from military surveillance to smart home,[23] from formidable remote environment monitoring to in-plant robotic control and guidance, from data collection on other planets to guarding the agricultural field.[1] Due to the limited sensing range of the sensor nodes, deploying sensors appropriately to reach an adequate coverage level is critical for the successful completion of the issued sensing tasks.[8,24]

Sensor deployment has received considerable attention recently. Most previous works[8,9,16,17] assumed that the environment is sufficiently known and under control. However, when the environment is unknown or hostile (such as remote harsh fields, disaster areas, and toxic urban regions), sensor deployment cannot be performed manually. To scatter sensors by aircraft is one possible solution. However, using this technique, the actual landing position cannot be controlled due to the existence of wind and obstacles such as trees and buildings. Consequently, the coverage may be inferior to the application requirements, no matter how many sensors are dropped. Moreover, in many cases, such as during in-building toxic-leak detection,[12,13] chemical sensors must be placed inside a building through the entrance. In such cases, it is necessary to make use of mobile sensors, which can move to the right place to provide the required coverage. Based on the work of Sibley et al.,[21] mobile sensors have already become a reality. Their mobile sensor prototype is smaller than 0.000047 m³ at a cost of less than $150 in parts.

There have been some research efforts into deploying mobile sensors, but most of them are based on centralized approaches. For example, the work of Zou and Chakrabarty[28] assumes that a powerful cluster head is available to collect the sensor location and determine the target location of the mobile sensors. However, in many sensor deployment environments such as disaster recoveries and battlefields, a central server may not be available, and it is difficult to organize sensors into clusters due to network partitions. Further, the centralized approach suffers from the problem of single point failure. Sensor deployment has also been addressed in the field of robotics,[12,13] where sensors are deployed one by one, utilizing the location information of previously deployed sensors. The deployment time of this method is long and it strongly depends on the initial sensor placement to ensure the communication between the deployed and undeployed sensors. In case of network partitions, this method may not be feasible.

To address the limitations of the previous work, we propose distributed algorithms for deploying mobile sensors. When all sensors are mobile, we propose three distributed algorithms (VOR, VEC, and Minimax) based on the principle of moving sensors from a densely deployed area to a sparsely deployed area. To achieve a good balance between sensor cost and sensor coverage, a mix of static and mobile sensors can be used. To deploy mixed sensors, bidding protocols are designed to assist the movement of mobile sensors. In the bidding protocol, mobile sensors can move in a zig-zag pattern, and move longer than the direct distance from the initial place to the final destination. To address this issue, we propose a proxy-based approach, where mobile sensors move logically instead of physically, to reduce the physical moving distance of the mobile sensors. Finally, we present some of the challenges and possible solutions for relocating mobile sensors.

The remainder of the chapter is organized as follows. Section 26.2 presents three self-deployment algorithms assuming all sensors are mobile. In Section 26.3, we present a bidding protocol to deploy a mix of static and mobile sensors. In Section 26.4, we present a proxy-based sensor deployment protocol to reduce the physical moving distance of the mobile sensors. Section 26.5 presents some of the challenges and possible solutions for relocating mobile sensors, and Section 26.6 concludes the chapter.

26.2   Deploying Mobile Sensors

In this section, we first give some technical background on the Voronoi diagram, and then present our distributed sensor deployment algorithms.

26.2.1   Technical Preliminary: Voronoi Diagram

The Voronoi diagram[3,20] is an important data structure in computational geometry. It represents the proximity information about a set of geometric nodes. The Voronoi diagram of a collection of nodes partitions the space into polygons. Every point in a given polygon is closer to the node in this polygon than to any other node. Figure 26.1(a) is an example of a Voronoi diagram, where each sensor is represented by a number and is enclosed by a Voronoi polygon. As shown in the figure, sensor s_{21} is enclosed by a Voronoi polygon. Each polygon edge of s_{21} is the vertical bisector of the line passing s_{21} and its Voronoi neighbor; that is, line AB is the bisector of $s_{21}s_9$.

Our sensor deployment algorithms are based on Voronoi diagrams. As shown in Figure 26.1(a), the Voronoi polygons together cover the target field. The points inside one polygon are closer to the sensor inside this polygon than the sensors positioned elsewhere. Thus, if this sensor cannot detect the expected phenomenon, no other sensor can detect it, and then each sensor is responsible for the sensing task in its Voronoi polygon. In this way, each sensor can examine the coverage hole (the area not covered by any sensor) locally, and only needs to monitor a small area around it. To construct the Voronoi polygon, each sensor only needs to know the existence of its Voronoi neighbors, which reduces the communication complexity.

26.2.2 Movement-Assisted Sensor Deployment Protocols

Our sensor deployment algorithm runs iteratively until it terminates or reaches the specified number of rounds. In each round, sensors first broadcast their locations and construct their local Voronoi polygons based on the received neighbor information. To construct its Voronoi polygon, each sensor first calculates the bisectors of its neighbors and itself based on the location information. These bisectors and the boundary of the target field form several polygons. The smallest polygon encircling the sensor is the Voronoi polygon of this sensor. After the Voronoi polygons have been constructed, they are examined to determine the existence of coverage holes. If any coverage hole exists, sensors decide where to move to eliminate or reduce the size of the coverage hole; otherwise, they stay. Next, we present three movement-assisted sensor deployment algorithms: VEC (VECtor-based), VOR (VORonoi-based), and Minimax, based on the principle that evenly distributed sensors can provide better coverage. For these three algorithms, VEC pushes sensors away from a densely covered area, VOR *pulls* sensors to the sparsely covered area, and Minimax moves sensors to their local center area.

26.2.2.1 The VECtor-Based Algorithm (VEC)

VEC is motivated by the attributes of electromagnetic particles: when two electromagnetic particles are too close to each other, an expelling force pushes them apart. Assume $d(s_i, s_j)$ is the distance between sensor s_i and sensor s_j. d_{ave} is the average distance between two sensors when the sensors are evenly distributed in the target area, which can be calculated beforehand because the target area and the number of sensors to be deployed are known. The virtual force between two sensors s_i and s_j will push them to move $(d_{ave} - d(s_i, s_j))/2$ away from each other. In case one sensor covers its Voronoi polygon completely and should not move, the other sensor will be pushed $d_{ave} - d(s_i, s_j)$ away. In summary, the virtual force will push the sensors d_{ave} away from each other if a coverage hole exists in either of their Voronoi polygons.

In addition to the virtual forces generated by sensors, the field boundary also exerts forces to push sensors too close to the boundary inside. The force exerted on s_i will push it to move $d_{ave}/2 - d_b(s_i)$, where $d_b(s_i)$ is the distance of s_i to the boundary. Because d_{ave} is the average distance between sensors, $d_{ave}/2$ is the distance from the boundary to the sensors closest to it when sensors are evenly distributed. The final overall force on sensors is the vector summation of virtual forces from the boundary and all Voronoi neighbors. These virtual forces will push sensors from the densely covered area to the sparsely covered area. Thus, VEC is a "proactive" algorithm that tries to relocate sensors so that they are evenly distributed.

As an enhancement, we add a *movement-adjustment* scheme to reduce the error of *virtual-force*. When a sensor determines its target location, it checks whether the local coverage will be increased by its movement. The local coverage is defined as the coverage of the local Voronoi polygon and can be calculated by the intersection of the polygon and the sensing circle. If the local coverage is not increased, the sensor should

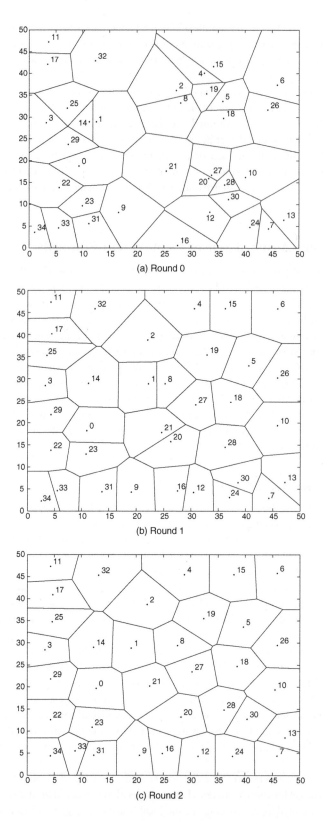

FIGURE 26.1   Snapshot of the execution of VEC.

not move to the target location. Although the general direction of the movement is correct, the local coverage may not be increased because the target location is too far away. To address this problem, the sensor will choose the midpoint between its target location and its current location as its new target location. If the local coverage is increased at the new target location, the sensor will move; otherwise, it will stay.

Figure 26.1 shows how VEC works. Round 0 is the initial random deployment of 35 sensors in a 50 × 50 m flat space, with the sensing range of 6 m. The initial coverage is 75.7 percent. After Round 1 and Round 2, the coverage is improved to 92.2 and 94.7 percent, respectively. More information on the VEC algorithm can be found.[26]

26.2.2.2 VOR and Minimax

Compared to the VEC algorithm, VOR is a *pull-based* algorithm that pulls sensors to their local maximum coverage holes. In VOR, if a sensor detects the existence of coverage holes, it moves toward its farthest Voronoi vertex. VOR is a greedy algorithm that tries to fix the largest hole. Moving oscillations may occur if new holes are generated due to sensor leaving. To deal with this problem, we add *oscillation control*, which does not allow sensors to move backward immediately. Each time a sensor wants to move, it first checks whether its moving direction is opposite to that in the previous round. If yes, it stops for one round. In addition, the movement adjustment mentioned in VEC is also applied here.

Similar to VOR, Minimax fixes holes by moving closer to the farthest Voronoi vertex, but it does not move as far as VOR to avoid the situation that the vertex which was originally close becomes a new farthest vertex. Minimax chooses the target location as the point inside the Voronoi polygon whose distance to the farthest Voronoi vertex is minimized. This algorithm is based on the belief that a sensor should not be too far away from any of its Voronoi vertices when the sensors are evenly distributed. Minimax can reduce the variance of the distances to the Voronoi vertices, resulting in a more regular shaped Voronoi polygon. Compared with VOR, Minimax considers more information and it is more conservative. Compared with VEC, Minimax is reactive; it fixes the coverage hole by moving toward the farthest Voronoi vertex. More information on VOR and Minimax can be found.[26]

26.2.3 Optimizations and Open Issues

In some cases, the initial deployment of sensors may form clusters, resulting in low initial coverage. In this case, sensors located inside the clusters cannot move for several rounds because their Voronoi polygons are well covered initially. This problem can significantly prolong the deployment time. To reduce the deployment time, we propose an optimization that detects whether too many sensors are clustered in a small area. The algorithm "explodes" the cluster to scatter the sensors apart, if necessary, which works as follows. Each sensor compares its current neighbor number to the neighbor number it will have if sensors are evenly distributed. If a sensor finds that the ratio of these two numbers is larger than a threshold, it concludes that it is inside a cluster and chooses a random position within an area centered at itself, which will contain the same number of sensors as its current neighbors in the even distribution. The explosion algorithm only runs in the first round. It scatters the clustered sensors and changes the deployment to be close to random.

26.2.3.1 Movement versus Computation versus Communication

The proposed protocols require sensors to move iteratively, eventually reaching the final destination. Other approaches can be envisioned in which the sensors move only once to their destinations to minimize the sensor movement. Two such approaches are a centralized approach and an approach using simulated movement.

Although the centralized approach may minimize the sensor movement, the computation overhead is very high. Further, because a central server may not be feasible in some applications such as the battlefield, sensors have to send their positions to the faraway server, which calculates the target positions and sends the result back through multihop routing. As a result, the delay and the communication overhead may be high.

Another alternative is to let sensors stay fixed and obtain their final destinations by simulated movement. They can calculate their target locations based on VEC/VOR/Minimax, logically move there, and exchange these locations with the sensors that would be their neighbors if they had actually moved. The real movement only happens when they know the final destinations. Although the logical movement can minimize the physical moving distance, it may incur large communication overhead. Also, it is susceptible to poor performance under network partitions that are likely to occur in a sensor deployment. If a network partition occurs, each partition may still run the movement algorithms without knowledge of the others. Consequently, the obtained final destination is not accurate and the required coverage cannot be reached. Using real movement, network partitions will be healed and all sensors will be eventually considered in the algorithm. Considering all these possibilities, we plan to further investigate the trade-offs among moving distance, computation overhead, and communication overhead.

26.3   Deploying a Mix of Mobile and Static Sensors

Although VEC, VOR, and Minimax can be used to increase the sensor coverage with low moving distance and low message complexity, to equip each sensor with a motor increases the cost and is unnecessary when the coverage requirement is not very high. To achieve a balance between sensor cost and coverage, we can deploy a mix of mobile and static sensors. In this section, we present a bidding-based protocol to deploy a mix of mobile and static sensors, and discuss some implementation issues and results.

26.3.1   Problem Statement

When a portion of deployed sensors are mobile, the deployment problem can be described as follows: given a target field covered by a number of circles (the sensing circle of the static sensors), but still having some uncovered areas, how to place a certain number of additional circles (the sensing circle of the mobile sensors) to maximize the overall coverage.

We showed[25] that this problem can be reduced to the set-covering problem, and then it is NP-hard. Although there is no optimal solution, we can still find some practical solutions to approximate the optimal solution based on heuristics.

26.3.2   General Idea of the Bidding Protocol

In the proposed bidding protocol, we use the following heuristic: mobile sensors should move to the area where the most additional coverage can be obtained. After a mobile sensor leaves its original location to cover (heal) another coverage hole, it can generate a new hole in its original location. Thus, a mobile sensor only moves to heal another hole if its leaving will not generate a larger hole than that to be healed. However, due to the lack of global information, mobile sensors may not know where the coverage hole exists. Even with the location of the coverage hole, it is still a challenge to find the target position inside the coverage hole, which can bring the most additional coverage when a mobile sensor is placed there compared to other positions. We propose to let the static sensors detect the coverage holes locally, estimate the size of these holes, and determine the target position inside the hole. Based on the properties of the Voronoi diagram, static sensors can find the coverage holes locally and provide a good way to estimate the target location of the mobile sensors.

The roles of mobile and static sensors motivate us to design a bidding protocol to assist the movement of mobile sensors. We view a mobile sensor as a *hole healing server*. Its service has a certain base price, which is the estimate of generated coverage hole after it leaves the current place. Static sensors are the bidders of the coverage hole healing services. Their bids are the estimated sizes of the holes they detect. Mobile sensors choose the highest bids and move to the target locations provided by the static sensors.

The bidding protocol runs round by round after the initialization period. During the initialization period, all static sensors broadcast their locations and identities locally. We choose the broadcast radius to be two hops, with which sensors can construct the Voronoi diagram in most cases. After the initialization period, static sensors broadcast this information again only when new mobile sensors arrive and need this information to construct their own Voronoi cells.

Each round consists of three phases: (1) **service advertising**, (2) **bidding**, and (3) **serving**. In the advertising phase, mobile sensors broadcast their base prices and locations in a local area. The base price is set at zero initially. In the bidding phase, static sensors detect coverage holes locally by examining their Voronoi cells. If such holes exist, they calculate the bids and the target locations for the mobile sensors. Based on the received information from the mobile sensors, the static sensor can find a closest mobile sensor whose base price is lower than its bid, and sends a bidding message to this mobile sensor. In the serving phase, the mobile sensor chooses the highest bid and moves to heal that coverage hole. The accepted bid will become the new base price of the mobile sensor. After the serving phase, another new round can start after the mobile sensors broadcast their new locations and new base prices. As the base price increases monotonically, when no static sensors can give out a bid higher than the base price of the mobile sensors, the protocol terminates.

26.3.3 Some Implementation Issues and Results

In the bidding message, static sensors give out the estimated coverage hole size and the target location to which the mobile sensor should move. This information is calculated based on their Voronoi cells. Static sensors construct Voronoi cells considering only static neighbors and mobile neighbors that are not likely to move. These mobile sensors are detected by examining their base prices. If the base price of a mobile sensor is zero, this sensor has not moved yet and most likely it will move to heal some coverage holes soon. Thus, when detecting coverage holes, static sensors do not consider those mobile sensors that are about to leave.

Having constructed the Voronoi cells, static sensors examine these cells. If there exists a coverage hole, the static sensor chooses the farthest Voronoi vertices as the target location of the coming mobile sensor. Inside one coverage hole, there are many positions where a mobile sensor can be located. If the mobile sensor is placed at the position farthest from any nearby sensors, the gained coverage is the highest because the overlap of the sensing circles between this new coming mobile sensor and existing sensors is the lowest.

Having determined the target location of the mobile sensor it bids, static sensors calculate the bid as: $\pi * (d - sensing\_range)^2$, where d is the distance between the bidder and the target location. Although this is not exactly the additional coverage, it can be used as an approximate to reduce the computation overhead.

Due to the limited service advertising radius, static sensors can have different knowledge about the mobile sensors. Therefore, it is possible that several static sensors independently bid different mobile sensors for the same coverage hole because the cheapest mobile sensor or the closest mobile sensor in their views are different. If several of them succeed in bidding, multiple mobile sensors can move to heal the same hole, which is not necessary. We propose a self-detection algorithm for mobile sensors to solve this problem. In the advertising phase, mobile sensors broadcast their locations and base prices. If a mobile sensor hears that another mobile sensor in its neighborhood has a higher base price than its own, it will run the detection algorithm to check whether a duplicate healing has occurred. If yes, the mobile sensor reduces its base price to zero and will likely move to cover a different hole.

To evaluate the trade-off between cost and coverage, we compare three sensor deployment algorithms[25]: (1) the *random deployment* algorithm, (2) the *VEC* algorithm, and (3) the *bidding* algorithm. Figure 26.2 shows the sensor cost of these three algorithms to reach a 95 percent sensor coverage. In this figure, random deployment is used when the percentage of mobile sensors is 0 percent, and the VEC algorithm is used when the percentage of the mobile sensors is 100 percent. The bidding protocol is used when the percentage of mobile sensors varies from 10 to 50 percent. Based on the *cost ratio* between the mobile sensor and the static sensor, the overall sensor cost of these three algorithms may be different. Intuitively, if the cost ratio is low (e.g., 1.5), increasing the percentage of mobile sensors can reduce the overall sensor cost. On the other hand, if the mobile sensors are very expensive, using only static sensors may have the lowest sensor cost (not shown in the figure). When the cost ratio is somewhere in the middle, the bidding algorithm that has a mix of mobile and static sensors can achieve the lowest sensor cost. For example, when the cost ratio is 3.5, the bidding algorithm has the lowest cost when 10 percent sensors are mobile. Based on

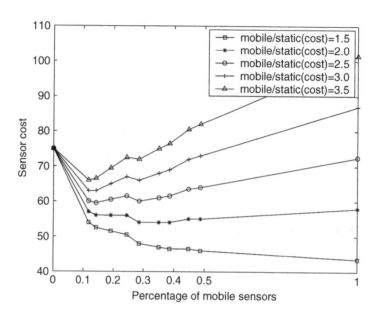

FIGURE 26.2   The cost of sensors to reach 95 percent coverage.

this figure, we can see there is a trade-off between cost and coverage. The bidding protocol can achieve a balance between these two most of the time.

26.4   Proxy-Based Sensor Deployment

Although the bidding protocol can be used to achieve high coverage, sensors may move in a zig-zag way and waste lots of energy compared to moving directly to the final destination. To address this problem, we propose a proxy-based sensor deployment protocol. Instead of moving iteratively, sensors calculate their target locations based on a distributed iterative algorithm, move logically, and exchange new logical locations with their new logical neighbors. Actual movement only occurs when sensors determine their final destinations.

26.4.1   Basic Idea: Logical Movement and Proxy Sensor

To realize the logical movement, mobile sensors should be able to identify the final destination. However, it is difficult to final the final destination (i.e., the largest hole) using a distributed algorithm. This is why mobile sensors have to identify the largest hole through iterative movement in the bidding protocol.

To identify the largest coverage holes, but not through iterative movement, we propose the idea of *logical* movement; that is, mobile sensors logically move from small holes to large holes iteratively, and only conduct a real movement when no larger holes can be detected. This *logical* movement can also solve the problem of load balance by exchanging the logical positions among mobile sensors. If the distance of one mobile sensor to its logical position is too long, it can exchange its logical position with another sensor before the move so that no sensor is penalized to move a much longer distance.

The difficulty of realizing the logical movement is to obtain the position-related information without really being placed in the logical position. In the bidding protocol, mobile sensors advertise their base prices within a certain advertising radius, collect the bidding messages from the static sensors, move to the new location (the largest hole in the advertising area), and advertise again. One possible solution for this position-related advertising and bidding is a network-wide broadcast. However, this method will significantly increase the message overhead. To obtain the location-related information with low message complexity, each mobile sensor chooses a static sensor closest to its logical position as its proxy. Proxy sensors are responsible for advertising the base prices of the mobile sensors and choosing the largest holes for them. As mobile sensors logically move, new proxies can be selected. Finally, if proxy sensors do not receive any bidding messages, they will notify their delegated mobile sensors to conduct the real movement.

26.4.2 Protocol Overview

In the proxy-based sensor deployment protocol, each mobile sensor acts as a hole-healing server, whose service has a certain base price determined by the size of the new hole to be generated by its leaving. The static sensor detects coverage holes locally, and uses the hole size as the bid to bid mobile sensors. The mobile sensor accepts the highest bid, logically moves to that hole, and delegates the winning bidder as its proxy. The winning bidder must be the static sensor closest to its logical position because the static sensor detects the coverage hole in its local area, and the largest hole it finds must be closer to it than other static sensors. Therefore, the static sensor that wins the bidding of a mobile sensor must be the sensor closest to the logical position of that mobile sensor.

Each proxy sensor advertises the service of its delegated mobile sensor, collects bidding messages, and chooses the highest bid. Then it delegates the bidder as the new proxy. The proxy sensor also examines the possible moving distance of its mobile sensor and performs *hole exchange* if needed. When no bidding message is received for a number of rounds, denoted as δ, and no hole exchange is needed, proxy sensors conclude that their delegated mobile sensors already obtained the final position, and notify them to move.

The protocol runs round by round until mobile sensors obtain their final locations and move there directly. Each round consists of four phases: the service advertising phase, bidding phase, logical movement phase, and hole-exchange phase. In the *service advertising phase*, each proxy sensor advertises the logical location, physical location, and base price for its delegated mobile sensor. In the first round, the mobile sensor does not have a proxy and advertises its physical location and base price by itself. In the *bidding phase*, each static sensor detects coverage holes, estimates the hole size, chooses a mobile sensor to bid, and sends *bidding* messages to its proxy. In the *logical movement phase*, the proxy sensor (or mobile sensors without a proxy) chooses the highest bid and sends a *delegate* message to the bidder. The bidder becomes the new proxy. The base price of the mobile sensor is updated by its new proxy. Also, the proxy sensor checks whether hole exchange is needed. If yes, it chooses the mobile sensor suitable for exchange and sends out an *exchange* request to the proxy of that mobile sensor. In the *hole-exchange phase*, the proxy sensor checks the received requests, chooses the one with the highest priority, and returns a confirm message to the requester. Then the mobile sensors delegated by these two proxy sensors exchange the hole to heal.

Figure 26.3 compares the moving distance of the proxy-based and bidding protocols. We distribute 40 sensors on a 50×50 m target field, among which 30 percent are mobile, with a sensing range of 6 m. Figure 26.3(a) shows the moving trace of the mobile sensors. In this example, the mobile sensors move 13.65 m, on average, and s_{38} moves the furthest: 27.85 m. For comparison, Figure 26.3(b) shows how mobile sensors move in the bidding protocol under the same initial distribution. The result shows that mobile sensors move 23.77 m on average, and s_{28} moves the furthest: 68.68 m. From this example, we can see that the proxy-based protocol is much more energy efficient and load balanced.

26.5 Sensor Relocation

The motion capability of sensor nodes can also be used for purposes other than sensor deployment. For example, in case of sensor failure or node malfunction, other sensors can move to replace the role of the failed node. As an event (i.e., fire, chemical spill, incoming target) occurs, more sensors should relocate to the area of the event to achieve better coverage. Compared with sensor deployment, *sensor relocation*, which relocates mobile sensors from one place to another, has many challenges. First, sensor relocation has strict time constraints. Sensor deployment is done before the network is in use, but sensor relocation is on-demand and should finish in a short time. For example, if the sensor monitoring a security-sensitive area dies, another sensor should take the responsibility as soon as possible; otherwise, some security policy

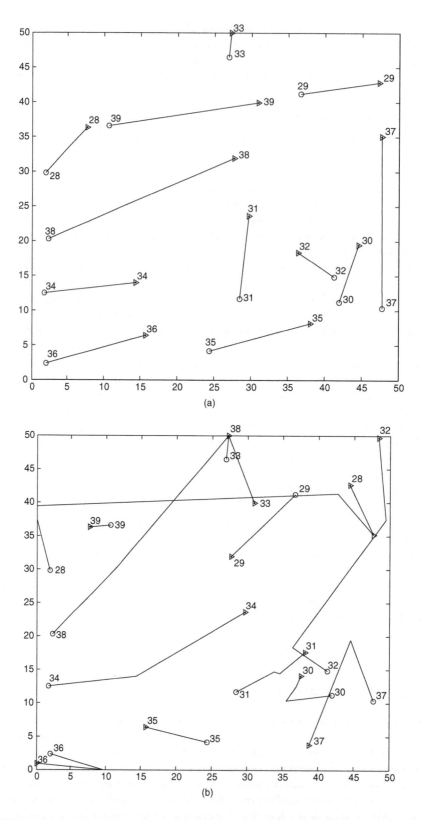

FIGURE 26.3   Comparison of the moving trace: (a) the proxy-based protocol, and (b) the bidding protocol.

may be violated. Second, relocation should not affect other missions currently supported by the sensor network, which means that the relocation should minimize its effect on the current topology. Finally, because physical movement costs much more energy than computation and communication, the moving sensor may suffer. As some nodes die due to low battery power, other nodes need to move again and cost more power. To be fair to each sensor and to prolong network lifetime, it is important to balance the trade-offs between minimizing the total energy consumption and maximizing the minimum remaining energy of the mobile sensors. Sensor relocation[4] focuses on finding the target locations of the mobile sensors based on their current locations and the locations of the sensed events. However, Butler and Rus[4] did not address the challenges of finding the relocation path under time, topology, and energy constraints.

Due to these new challenges, our deployment protocols[25,26] cannot be directly applied to senor relocation. For example, if the area covered by a failed sensor does not have redundant sensors, moving neighbor sensors can create new holes in that area. To heal these new holes, more sensors must be involved. This process continues until some area having redundant sensors is reached. During this process, sensors can move back and forth and waste lots of energy. Based on this observation, we propose to first find the location of the redundant sensors and then design an efficient relocation schedule for them to move to the target area (destination).

Using flooding to find the extra sensors can significantly increase the message overhead. To reduce message overhead, solutions similar to TTDD[27] can be used. In TTDD, the target field is divided into grids, each having a grid head. The grid head is responsible for disseminating the sensing data to other grid heads. To find the interested data, the sink floods the query, which will be served by the grid head that has the sensing data. Because the data needs to be flooded to the whole network, although only grid heads, it still has significant overhead. We propose to apply the quorum concept[5–7,10,15,18] to reduce message overhead. A quorum is a set of nodes, and any two quorums must have an intersection node. If the grid with extra sensors advertises to nodes in its quorum, any destination grid head can obtain this information by sending a query to the nodes in its quorum. A simple quorum can be constructed by choosing the nodes in a row and a column. Suppose N is the number of grids in the network. Using this quorum-based system, the message overhead can be reduced from $O(N)$ to $O(\sqrt{N})$.[7]

After obtaining information about where the extra sensors are, the grid head needs to determine how to relocate them. At one extreme, sensors can move to the destination directly. Although this solution can minimize the moving distance, the relocation time can be long, especially when the destination is far away from the source. As an alternative, the sensor can move to its neighbor grid, and ask some sensor node in that grid to move to its neighbor grid close to the destination. This process can propagate until a sensor node moves to the destination. Because the sensors can first exchange communication messages (i.e., logically move) and ask all relevant sensors to (physically) move at the same time, the relocation time is much shorter. However, the total physical moving distance of this approach can be long, and it is a challenge to make sure that the sensor coverage is maintained during sensor movement. We are currently working on various solutions to balance the trade-offs among communication overhead, relocation time, and moving distance.

26.6   Conclusion

To provide satisfactory coverage is very important in many sensor network applications such as object tracking, perimeter defense, homestead monitoring, and military surveillance. To obtain the required coverage in harsh environments, mobile sensors are helpful because they can move to cover the area not reachable by static sensors. In this chapter, we summarized our work on deploying mobile sensors. First, assuming all sensors are mobile, we presented three distributed algorithms (VOR, VEC, and Minimax) for controlling the movement of sensors that are initially randomly placed to get high coverage. These algorithms are based on the principle of moving sensors from densely deployed areas to sparsely deployed areas. Second, to achieve a good balance between sensor cost and sensor coverage, we proposed a bidding protocol to assist the movement of mobile sensors when a mix of mobile and static sensors are used. In the protocol, static sensors detect coverage holes locally using Voronoi diagrams, and bid for mobile sensors based on the size of the detected coverage hole. Mobile sensors choose coverage holes to heal based on the bid. Third, to reduce the physical moving distance of mobile sensors, we proposed a proxy-based sensor deployment protocol. Instead of moving iteratively, sensors calculate their target locations based on a distributed iterative algorithm, move logically, and exchange new logical locations with their new logical neighbors. The actual movement only occurs when sensors determine their final destinations. Finally, we presented some of the challenges and possible solutions for sensor relocation.

References

1. http://sensorwebs.jpl.nasa.gov.
2. I.F. Akyildiz, W. Su, Y. Sankarasubramaniam, and E. Cayirci, A Survey on Sensor Networks, *IEEE Communications Magazine*, pp. 102–114, August 2002.
3. F. Aurenhammer, Voronoi diagrams — A Survey of a Fundamental Geometric Data Structure, *ACM Computing Surveys*, 23, 345–406, 1991.
4. Z. Butler and D. Rus, Event-Based Motion Control for Mobile Sensor Networks, *IEEE Pervasive Computing*, 2(4), 34–43, 2003.
5. G. Cao and M. Singhal, A Delay-Optimal Quorum-Based Mutual Exclusion Algorithm for Distributed Systems, *IEEE Transactions on Parallel and Distributed Systems*, 12(2), 1256–1268, Dec. 2001.
6. G. Cao, M. Singhal, Y. Deng, N. Rishe, and W. Sun, A Delay-Optimal Quorum-Based Mutual Exclusion Scheme with Fault-Tolerance Capability, *Proc. 18th IEEE Int. Conf. on Distributed Computing Systems*, pp. 444–451, May 1998.
7. S.Y. Cheung, M.H. Ammar, and M. Ahamad, The Grid Protocol: A High Performance Scheme for Maintaining Replicated Data, *IEEE Trans. Knowl. Data Eng.*, June 1992.
8. T. Clouqueur, V. Phipatanasuphorn, P. Ramanathan, and K.K. Saluja, Sensor Deployment Strategy for Target Detection, *First ACM International Workshop on Wireless Sensor Networks and Applications*, 2002.
9. S. Dhillon, K. Chakrabarty, and S. Iyengar, Sensor Placement for Grid Coverage under Imprecise Detections, *Proceedings of the International Conference on Information Fusion*, 2002.
10. H. Garcia and D. Barbara, How to Assign Votes in a Distributed System, *Journal of ACM*, May 1985.
11. W.R. Heinzelman, J. Kulik, and H. Balakrishnan, Adaptive Protocols for Information Dissemination in Wireless Sensor Network, *ACM Mobicom*, August 1999.
12. A. Howard, M.J. Mataric, and G.S. Sukhatme, An Incremental Self-Deployment Algorithm for Mobile Sensor Networks, *Autonomous Robots, Special Issue on Intelligent Embedded Systems*, September 2002.
13. A. Howard, M.J. Mataric, and G.S. Sukhatme, Mobile Sensor Networks Deployment Using Potential Fields: A Distributed, Scalable Solution to the Area Coverage Problem, *6th International Symposium on Distributed Autonomous Robotics Systems*, June 2002.
14. C. Intanagonwiwat, R. Govindan, and D. Estrin, Directed Diffusion: A Scalable and Robust Communication, *ACM SIGMOBILE, Sixth Annual International Conference on Mobile Computing and Networking (MobiCOM '00)*, August 2000.
15. Y.-C. Kuo and S.-T. Huang, A Geometric Approach for Constructing Coteries and k-Coteries, *IEEE Trans. Parallel and Distributed Systems*, 8, 402–411, Apr. 1997.
16. S. Meguerdichian, F. Koushanfar, G. Qu, and M. Potkonjak, Exposure in Wireless Ad-Hoc Sensor Networks, *ACM Mobicom*, 2001.
17. S. Meguerdichian, F. Koushanfar, M. Potkonjak, and M.B. Srivastava, Coverage Problems in Wireless Ad-hoc Sensor Network, *IEEE INFOCOM*, April 2001.
18. D. Peleg and A. Wool, Crumbling Walls: A Class of Practical and Efficient Quorum Systems, *Distributed Computing*, 10(2), 120–129, 1997.
19. G.J. Pottie and W.J. Kaiser, Wireless Integrated Network Sensors, *Communications of the ACM*, May 2000.
20. D. Du, F. Hwang, and S. Fortune, Voronoi Diagrams and Delaunay Triangulations, *Euclidean Geometry and Computers*, 1992.

21. G.T. Sibley, M.H. Rahimi, and G.S. Sukhatme, Robomote: A Tiny Mobile Robot Platform for Large-Scale Sensor Networks, *2002 IEEE International Conference on Robotics and Automation*, 2002.

22. K. Sohrabi, J. Gao, V. Ailawadhi, and G.J. Pottie, Protocols for Self-Organization of a Wireless Sensor Network, *IEEE Personal Communication*, 7(5), 16–27, October 2000.

23. M. Srivastava, R. Muntz, and M. Potkonjak, Smart Kindergarten: Sensor-Based Wireless Networks for Smart Developmental Problem-Solving Environments, *ACM Mobicom*, 2001.

24. S. Tilak, N.B. Abu-Ghazaleh, and W. Heinzelman, Infrastructure Tradeoffs for Sensor Networks, *First ACM International Workshop on Wireless Sensor Networks and Applications*, 2002.

25. G. Wang, G. Cao, and T. La Porta, A Bidding Protocol for Deploying Mobile Sensors, *IEEE International Conference on Network Protocols (ICNP)*, pp. 315–324, November 2003.

26. G. Wang, G. Cao, and T. La Porta, Movement-Assisted Sensor Deployment, *IEEE INFOCOM*, March 2004.

27. F. Ye, H. Luo, J. Cheng, S. Lu, and L. Zhang, A Two-Tier Data Dissemination Model for Large-Scale Wireless Sensor Networks, *ACM Mobicom*, pp. 148–159, September 2002.

28. Y. Zou and K. Chakrabarty, Energy-Aware Target Localization in Wireless Sensor Networks, *Proc. IEEE Int. Conf. Pervasive Computing and Communications*, 2003.

27

Models and Algorithms for Coverage Problems in Wireless Sensor Networks

Chi-Fu Huang

Po-Yu Chen

Yu-Chee Tseng

Wen-Tsuen Chen

27.1 Introduction

The rapid progress in wireless communication and embedded micro-sensing MEMS technologies has made *wireless sensor networks* possible. Such environments may have many inexpensive wireless nodes, each capable of collecting, storing, and processing environmental information, and communicating with neighboring nodes. In the past, sensors were connected by wire lines. Today, this environment is combined with the novel ad hoc networking technology to facilitate inter-sensor communication.[20,24] The flexibility of installing and configuring a sensor network is thus greatly improved. Recently, a lot of research activities have been dedicated to sensor networks, including physical and medium access layers[22,29,32] and routing and transport layers.[2,4,6]

Because sensors can be spread in an arbitrary manner, one of the fundamental issues in a wireless sensor network is the *coverage problem*. Given a sensor network, the coverage problem is to determine how well the sensing field is monitored or tracked by sensors. In the literature, this problem has been formulated in various ways. A lot of works have been dedicated to coverage-related problems in wireless sensor networks in the past few years. These include the surveillance and exposure of sensor networks,

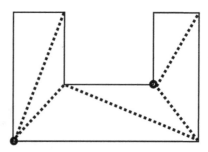

FIGURE 27.1   An example of the Art Gallery problem. Circles represent positions of cameras.

and the concerns of coverage versus connectivity issues when deploying a sensor network. Through this chapter, we intend to provide a comprehensive survey of the literature. We also discuss some application scenarios of the coverage problem. For example, to reduce sensors' on-duty time, those sensors that share the common sensing region and task can be turned off to conserve energy and thus extend the network lifetime.

This chapter is organized as follows. Section 27.2 and Section 27.3 study two relevant geometric problems. Sections 27.4 through 27.6 present works aimed at solving coverage-related problems in wireless sensor networks. Several coverage-preserving, energy-conserving protocols are then presented in Section 27.7, and Section 27.8 draws our conclusions.

27.2   The Art Gallery Problem

The *Art Gallery Problem*[18] is defined as follows. Imagine that the owner of an art gallery wants to place cameras in the gallery such that the whole gallery is thief-proof. There are two questions to be answered in this problem: (1) how many cameras are needed, and (2) where these cameras should be deployed. The first problem is related to cost, while the second is related to coverage. Every point in the gallery should be monitored by at least one camera. Cameras are assumed to have a viewpoint of 360 degrees and rotate at infinite speed. Moreover, a camera can monitor any location as far as nothing is in the middle (i.e., a line-of-sight exists). The number of cameras used should be minimized. The gallery is usually modeled as a simple polygon on a two-dimensional plane. A simple solution to this problem is to divide the polygon into non-overlapping triangles and place one camera in each of these triangles. By *triangulating* the polygon, it has been shown that any simple polygon can be guarded by $\lceil n/3 \rceil$ cameras, where n is the number of triangles in the polygon. This is also the best result in the worst case.[18] An example of triangulating a simple polygon is shown in Figure 27.1 and two cameras are sufficient to cover the gallery. Although this problem can be solved optimally in a two-dimensional plane, it is shown that its three-dimensional version is an NP-hard problem.[19]

27.3   The Circle Covering Problem

Another related problem in computational geometry is the *circle covering problem*,[28] which is to arrange identical circles on a plane that can fully cover the plane. Given a fixed number of circles, the goal is to minimize the radius of circles. This issue is discussed for covering a rectangle.[7,16,17] The coverings with less than or equal to five circles and seven circles can be done optimally.[7] For example, an optimal covering of a square by five circles is shown in Figure 27.2. Melissen and Schuur[16] have shown the coverings of six and eight circles, and present a new covering with eleven circles by a simulated annealing approach. Table 27.1 lists the minimun radius r_n to cover a unit square with n identical circles for $n = 1 \ldots 30$, as reported by Nurmela and Östergård.[17]

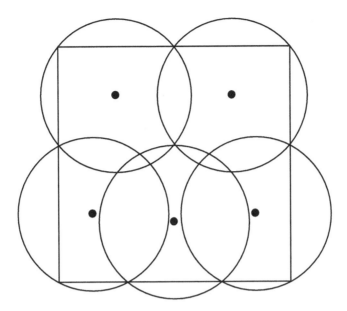

FIGURE 27.2   An example of the circle covering problem. The radius of each circle is about 0.3621605.

27.4   The Breach and Support Paths

In a sensor network, coverage can be regarded as the path between a given pair of points in the sensing field that is best or worst monitored by sensors when an object traverses along the path alone. Meguerdichian et al.[13] define the *maximal breach path* and the *maximal support path* as paths on which the distances from any point to the closest sensor are maximized and minimized, respectively. Polynomial-time algorithms are proposed to find such paths. The key idea is to use the Voronoi diagram and the Delaunay triangulation of sensor nodes to limit the search space. The Voronoi diagram is formed from the perpendicular bisectors of lines that connect two neighboring sensors, while the Delaunay triangulation is formed by connecting nodes that share a common edge in the Voronoi diagram. Examples of the Voronoi diagram and Delaunay triangulation are shown in Figure 27.3.

Because line segments in the Voronoi diagram have the maximal distance to the closest sensors, the maximal breach path must lie on the line segments of the Voronoi diagram. To find the maximal breach path, each line segment is given a weight equal to its minimum distance to the closest sensor. The proposed algorithm then performs a binary search between the smallest and largest weights. In each step, a breadth-first-search is used to check the existence of a path from the source point to the destination point using only line segments with weights that are larger than the search criterion. If a path exists, the criterion is increased

TABLE 27.1   Minimum Radius r_n to Cover a Unit Square by n Circles

n	r_n	n	r_n	n	r_n	n	r_n
1	0.7071067...	2	0.5590169...	3	0.5038911...	4	0.3535533...
5	0.3261605...	6	0.2987270...	7	0.2742918...	8	0.2603001...
9	0.2306369...	10	0.2182335...	11	0.2125160...	12	0.2022758...
13	0.1943123...	14	0.1855105...	15	0.1796617...	16	0.1694270...
17	0.1656809...	18	0.1606396...	19	0.1578419...	20	0.1522468...
21	0.1489537...	22	0.1436931...	23	0.1412448...	24	0.1383028...
25	0.1335487...	26	0.1317468...	27	0.1286335...	28	0.1273175...
29	0.1255535...	30	0.1220368...						

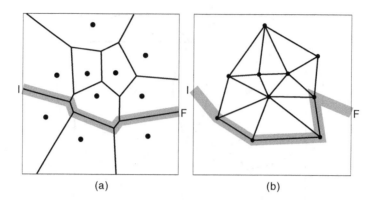

(a)   (b)

FIGURE 27.3 Examples of (a) the Voronoi diagram and the maximal breach path, and (b) the Delaunay triangulation and the maximal support path. I and F are the source and destination points, respectively.

to further restrict the lines considered in the next search iteration. Otherwise, the criterion is decreased. An example of the maximal breach path is shown in Figure 27.3(a). Similarly, because the Delaunay triangulation produces triangles that have minimal edge lengths among all possible triangulations, the maximal support path must lie on the lines of the Delaunay triangulation. To find the maximal support path, line segments are assigned weights equal to their lengths. The search part is then similar to the above case. An example is shown in Figure 27.3(b).

Li et al.[11] propose decentralized algorithms to find the maximal support path by constructing the Delaunay triangulation locally. The authors claim that the Delaunay triangulation can be replaced by a *relative neighborhood graph (RNG)* and the maximal support path can still be found. The key idea is that the RNG can be constructed locally. Given any two sensors u and v, $lune(u, v)$ is the intersection of the two disks centered at u and v, both of the same radius $\|u, v\|$, where $\|u, v\|$ is the distance between u and v. If $lune(u, v)$ does not contain any sensor, an edge is established to join u and v with a weight of $\frac{1}{2}\|u, v\|$. The RNG is constructed by such weighted edges and all sensors. Note that the RNG can be constructed distributedly by all sensors.

Figure 27.4 shows an example of constructing an RNG. There are six sensors $\{A, B, C, D, E, F\}$. In Figure 27.4(a), $lune(A, B)$ contains node C, so the link $\overline{AB}$ should not be built. On the contrary, $lune(A, C)$ and $lune(B, C)$ do not contain any other nodes, so the links $\overline{AC}$ and $\overline{BC}$ should be built. The final RNG is shown in Figure 27.4(b).

After constructing the RNG, a decentralized algorithm is used to find the maximal support path. First, the starting and ending points must connect to the sensors that are closest to them. Then, a modified Bellman-Ford algorithm is adopted to find the minimum weight path connecting the starting and ending points.

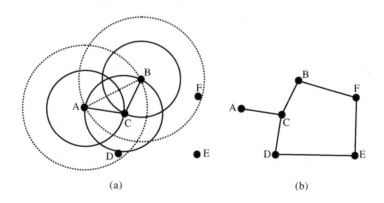

(a)   (b)

FIGURE 27.4 An RNG example.

It is possible that there are several maximal support paths. In this case, the path that consumes the least energy is selected.

It is proved[11] that the maximal support path that can be found using the Gabriel graph can also be found using the RNG. Using both the Gabriel graph and the RNG has the same complexity, $O(n \log n)$, where n is the number of sensor nodes. Thus, this conquers the problem with Delaunay triangulation that global information needs to be collected.

27.5 Exposure to Sensors

The concept of time should also be included to reflect how much a moving target is exposed to sensors. The exposure time should be accounted for. Consider the example in Figure 27.5. Suppose that s is a sensor and an object moves from point A to point B at a constant speed via three possible paths. Although path 3 is the farthest path from s, it is also the longest exposure time. In contrast, path 2 is the shortest path, but it has the strongest sensing intensity. Path 1 has neither the longest exposure time nor the strongest sensing intensity.

How to find the *minimal exposure path* is addressed by Meguerdichian et al.[14] The exposure for an object in the sensor field during an interval $[t_1, t_2]$ along a path $p(t)$ is defined as:

$$E(p(t), t_1, t_2) = \int_{t_1}^{t_2} I(F, p(t)) \left| \frac{dp(t)}{dt} \right| dt,$$

where $I(F, p(t))$ is the sensor field intensity measured at location $p(t)$ from the closest sensor or all sensors in the sensor field F, and $|\frac{dp(t)}{dt}|$ is the arc length. A numerical approximation is proposed[14] to find the minimal exposure path by dividing the sensor network region into grids and forcing the path to only pass the edges of grids and/or the diagonals of grids. Each line segment is assigned a weight equal to the exposure of this segment. Then a single-source-shortest-path algorithm is used to find the minimal exposure path.

Meguerdichian et al.[15] further discuss how to compute the exposure of a sensor network in a distributed manner. The key idea is to use the Voronoi diagram to partition the sensor field and then each sensor is responsible for the calculation of exposure in its region. Inside each region, the above grid approximation is used. Huang[10] proposes another method to find the minimal exposure path using *variational calculus*. This work first studies the sensing field with a single sensor and then discusses how to extend the result to a more general case. Veltri et al.[26] further propose a localized algorithm that can reduce the

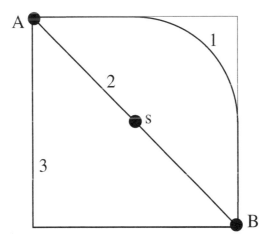

FIGURE 27.5 An example of exposure.

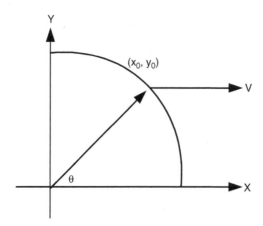

FIGURE 27.6   Moving direction of an object with respect to a sensor at the origin.

computational complexity.[15] It is also proved[26] that finding the maximal exposure path is NP-hard. Several heuristics are then provided to find the maximal exposure path.

Another important issue in sensor networks is to estimate the number of sensors required to achieve complete coverage of a desired area. Adlakha and Srivastava[1] define the *critical density thresholds* for complete coverage. The exposure is used to find the critical number of sensor nodes required to cover an area. For a sensor s, the signal received from a target decreases as the distance from the target increases. If the signal strength is less than the noise signal, the sensor cannot detect the target. The authors then investigate the influence of sensors and define two radii: *radius of complete influence* (r_1) and *radius of no influence* (r_2). Objects within the former radius are surely detected, and objects outside the latter radius are undetectable. If $r_1 = r_2$, the decision degenerates to the zero-one model.

Suppose there is a sensor located at the origin. Without loss of generality, it is assumesd[1] that the target is initially located at (x_0, y_0) on the arc at an angle θ from the x-axis, as shown in Figure 27.6, and moves in a straight line with a constant speed v parallel to the x-axis. At a period of T, the exposure value is

$$E_s = \int_0^T \frac{\lambda}{(x_0 + v * t)^2 + (y_0)^2} dt,$$

where λ is a constant value depending on the sensor property. The exposure value can be written using polar coordinates as follows:

$$E_s = \frac{\lambda}{v \sin(\theta)} \tan^{-1}\left(\frac{\delta \sin(\theta)}{r + \delta \cos(\theta)}\right),$$

where δ is the travelling distance. Let E_{th} be the object detection threshold. It is shown[1] that r_1 can be given by

$$E_{th} = \frac{\lambda}{vr_1}\left(\frac{\delta}{\delta + r_1}\right)$$

and r_2 can be given by

$$E_{th} = \frac{2\lambda}{vr_2} \tan^{-1}\left(\frac{\delta}{2r_2}\right).$$

According to Adlakha and Srivastava,[1] to cover an area A, the number of nodes required would be of the order $O(\frac{A}{r^2})$, where r is the sensing radius lying between r_1 and r_2. Via simulations, it is shown that using r_2 is a good estimation for finding the number of sensors required and the probability of detection is 98 percent or above.

27.6 Coverage and Connectivity

In this section we discuss some works that consider the coverage and connectivity of sensor networks.[5,8,21,27,33] In this kind of work, each sensor is assumed to have a fixed sensing region and a fixed communication range, both of which are modeled as disks. The goal is to achieve certain sensing coverage or communication connectivity requirements.

The coverage problem is formulated as a decision problem by Huang and Tseng.[8] Given a set of sensors deployed in a target area, the problem is to determine if the area is sufficiently *k-covered*, in the sense that every point in the target area is covered by at least k sensors, where k is a given parameter. Rather than determining the coverage of each location, the proposed approach looks at how the perimeter of each sensor's sensing range is covered, thus leading to an efficient polynomial-time algorithm. Specifically, the algorithm tries to determine whether the perimeter of a sensor under consideration is sufficiently covered. By collecting this information from all sensors, a correct answer can be obtained. The algorithm can be easily translated to a distributed algorithm.

An example of determining the perimeter-coverage of a sensor's perimeter is shown in Figure 27.7. Each sensor first determines which segments of its perimeter are covered by its neighboring nodes. As shown in Figure 27.7(a), segments $[0, a]$, $[b, c]$, and $[d, \pi]$ of sensor S_i's perimeter are covered by three of its neighboring nodes. These segments are then sorted in ascending order on the line segment $[0, 2\pi]$, as shown in Figure 27.7(b). By traversing the line segment $[0, 2\pi]$, the perimeter coverage of the sensor S_i can be determined. In this example, the perimeter coverage of S_i from 0 to b is one, from b to a is two, from a to d is one, from d to c is two, and from c to π is one. Huang and Tseng[8] prove that as long as the perimeters of all sensors are sufficiently covered, the whole area is sufficiently covered. The solution proposed in this chapter can be easily translated to a distributed protocol where each sensor only needs to collect local information to make its decision. The result can be applied to unit and non-unit disk sensing regions, and can even be extended to irregular sensing regions of sensors. How to use the results for discovering insufficiently covered areas, for conserving energy, and for supporting coverage of hot spots is also discussed.

Huang et al.[9] consider the same coverage problem but in three-dimensional space and show that tackling this problem in a three-dimensional space is still feasible within polynomial time. In a three-dimensional space, the sensing region of each sensor is modeled by a three-dimensional ball and the problem is to determine if every point in the field is within at least k balls. The proposed solution reduces the problem from a three-dimensional space to a two-dimensional space, and further to a one-dimensional space, thus leading to a very efficient solution. In essence, the solution tries to look at how the sphere of each sensor's sensing range is covered. As long as the spheres of all sensors are sufficiently covered, the whole sensing field is sufficiently covered. To determine whether each sensor's sphere is sufficiently covered, it in turn looks

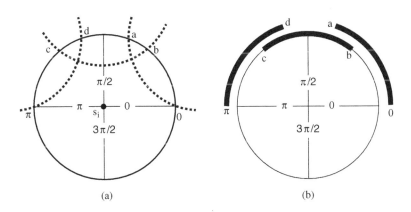

(a)   (b)

FIGURE 27.7 Determining the perimeter-coverage of a sensor S_i's perimeter.

at how each spherical cap and how each circle of the intersection of two spheres is covered. By stretching each circle on a one-dimensional line, the level of coverage can be easily determined.

For the sensor network to operate successfully, the active nodes must maintain both sensing coverage and network connectivity. Wang et al.[27] propose another solution to determine if a target area is k-covered and further study the relationship between coverage and connectivity. To determine the coverage level, this work looks at how intersection points between the sensors' sensing ranges are covered. It is claimed that a region is k-covered by a set of sensors if all intersection points between sensors and between any sensor and the boundary of this region are at least k-covered. For network connectivity, the work proves that if a region is k-covered, then the sensor network is k-connected as long as the sensors' communication ranges are no less than twice their sensing ranges.

Based on the above two properties, a Coverage Configuration Protocol (CPP) that can provide different degrees of coverage and meanwhile maintain communication connectivity is presented[27] when communication ranges are no less than twice the sensing ranges. Initially, all sensors are in the *active* state. If an area exceeds the required degree of coverage, redundant nodes will find themselves unnecessary and switch to the *sleep* state. It is not necessary for a sensor to stay active if all the intersection points inside its sensing circle are at least k-covered. A sleeping node needs to periodically wake up and enter the *listen* state. In the listen state, the sensor evaluates whether it must return to the active state.

If the communication ranges are less than twice the sensing ranges, Wang et al.[27] propose to integrate CCP with SPAN[3] to provide both sensing coverage and communication connectivity. SPAN is a connectivity-maintaining protocol that can turn off unnecessary nodes such that all active nodes are connected through a communication backbone and all inactive nodes are directly connected to at least one active node. Wang et al.[27] propose that an inactive node should become active following rules of SPAN or CCP. An active node will turn to sleep if it satisfies neither SPAN's nor CCP's wakeup rules.

How to maintain the sensing coverage and connectivity is also addressed.[33] Similar to Wang et al.,[27] Zhang and Hou[33] also show that coverage can imply connectivity if the transmission range is at least twice the sensing range. If so, we only need to focus on the coverage problem. A decentralized density control algorithm called *Optimal Geographical Density Control (OGDC)* is proposed to choose as few working nodes as possible to cover the network. Initially, all nodes are in the *UNDECIDED* state. We first find several starting nodes to enter the *ON* state. Nodes in the *ON* state can bring other *UNDECIDED*-state nodes to the *ON* state. The basic idea is to reduce the overlapping areas covered by nodes in the *ON* state. For example, in Figure 27.8, there are four sensors $\{A, B, C, D\}$ and A is a starting node. Then A selects its neighbor B to enter the *ON* state because B's distance from A is closest to $\sqrt{3}r$, where r is the sensing

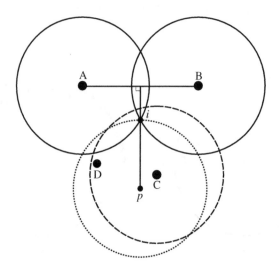

FIGURE 27.8 An example of the OGDC algorithm.

radius of each sensor. To cover the intersection point i of A's and B's circles, we then select the node whose position is closest to the optimal position p, which is on the perpendicular bisector of the line connecting A and B and is at a distance of r from i. As a result, C is selected and turned to the *ON* state. This procedure is repeated until the whole network has been covered. Note that a node in the *UNDECIDED* state can enter the *OFF* state if it finds its sensing range has been fully covered by other *ON*-state nodes.

Shakkottai et al.[21] investigate three coverage-related issues about a sensor network based on the grid-based deployment. In a unit square, $\sqrt{n} \times \sqrt{n}$ sensors are deployed in the field. However, each sensor has a probability of $p(n)$ to remain functioning, and a probability of $1 - p(n)$ to be dead. The authors show that when $p(n)r^2(n) \approx \frac{\log(n)}{n}$, it is very likely that the network will remain fully covered and connected, where $r(n)$ is the sensing and communication range of each sensor. Also, under such a condition, the network diammeter will be of the order $\sqrt{\frac{n}{\log(n)}}$.

27.7   Coverage-Preserving and Energy-Conserving Protocols

Because sensors are usually powered by batteries, sensors' on-duty time should be properly scheduled to conserve energy. If some nodes share the common sensing region and task, we can turn off some of them to conserve energy and thus extend the lifetime of the network. This is feasible if turning off such a node still provides the same "coverage." An example is shown in Figure 27.9(a). The sensor e can be put into sleeping mode because all its sensing area is covered by the other nodes. Sensor f satisfies this condition too and can go to sleeping mode. However, e and f are not allowed to be turned off at the same time; otherwise, the region in gray as shown in Figure 27.9(b) is not covered by any sensor. As a result, sensors not only need to be checked if they satisfy certain eligibility rules but also need to synchronize with each other.

Slijepcevic and Potkonjak[23] propose a heuristic to select mutually exclusive sets of sensor nodes such that each set of sensors can provide complete coverage of the monitored area. They claim that this problem is an NP-complete problem by reducing it to the minimum cover problem. The key idea of the proposed heuristic is to find out which sensors cover fields that are less covered by other sensors and then avoid including those sensors into the same set. Also targeted at turning off some redundant nodes, Ye et al.[31] propose a probe-based density control algorithm to put some nodes in a sensor-dense area to a doze mode to ensure a long-lived, robust sensing coverage. In this solution, nodes are initially in the sleeping mode. After a sleeping node wakes up, it broadcasts a probing message within a certain range and then waits for a reply. If no reply is received within a predefined time period, it will keep active until it depletes its energy. The coverage degree (density) is controlled by the sensor's probing range and wake-up rate. However, this probing-based approach has no guarantee of sensing coverage and thus blind points could appear.

A coverage-preserving scheduling scheme is presented[25] to determine when a node can be turned off and when it should be rescheduled to become active again. It is based on an eligibility rule that allows a node to turn itself off as long as other neighboring nodes can cover its sensing area. After evaluating its

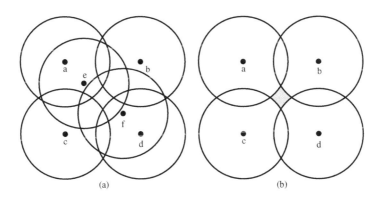

(a)   (b)

FIGURE 27.9   An example of the blind point if both sensors e and f are put into sleeping at the same time.

eligibility for off-duty, each sensor adopts a backoff scheme to prevent the appearance of blind points. If a node is eligible for off-duty, it will delay a random backoff time before actually turning itself off. During this period of time, if it receives any message from its neighbors requesting to go to sleep, it marks the sender as an off-duty node and reevaluates its eligibility. If the eligibility still holds after the backoff time, this node broadcasts a message to inform its neighbors, waits for a short period of time, and then actually turns itself off. A sleeping node will periodically wake up to check if it is still eligible for off-duty and then decide to remain sleeping or go back to on-duty.

Another node scheduling scheme is proposed.[30] In this scheme, the time axis is divided into rounds with equal duration. Each sensor node randomly generates a reference time in each round. In addition, the whole sensing area is divided into grid points that are used to evaluate whether or not the area is sufficiently covered. Each sensor has to join the schedule of each grid point covered by itself based on its reference time such that the grid point is covered by at least one sensor at any moment of a round. Then a sensor's on-duty time in each round is the union of schedules of grid points covered by the sensor. However, this scheme may suffer from the time synchronization problem in a large-scale sensor network.

A coverage-aware, self-scheduling scheme based on, probabilistic sensing model is proposed.[12] Each sensor S_j is assumed to be able to detect a nearby event happened at location P_i with a probability

$$S_j(P_i) = \frac{1}{(1 + \alpha D_{ij})^\beta},$$

where D_{ij} is the distance between sensor S_j and point P_i and constants α and β are device-dependent parameters. Thus, the level of coverage perceived by P_i contributed by all sensors can be written as

$$C(P_i) = 1 - \prod_{\forall S_j} (1 - S_j(P_i)).$$

Now suppose a sensor S_n is removed from the network. The loss of coverage at point P_i can be derived by

$$\Delta(P_i) = S_n(P_i) \prod_{\forall S_j \neq S_n} (1 - S_j(P_i)).$$

Therefore, sensor S_n's contribution to the network coverage can be defined by summing the losses over all possible points

$$SD_n = \sum_{\forall P_i \text{ within distance } R \text{ from } S_n} \Delta(P_i),$$

where R is the largest range that a sensor can detect with a predefined accuracy.

A self-scheduling scheme based on the above SD value is then proposed.[12] Periodically, each sensor S_i calculates its SD value in a distributed manner and decides whether to enter sleeping state with a hibernating probability defined as follows:

$$P_{hibernate} = (SD_{base} - SD_i)/SD_{base},$$

where SD_{base} is half of the maximum possible SD value. Therefore, a sensor with a higher SD value has a higher chance to stay active. To prevent an area from becoming uncovered due to all sensors in this area turning themselves off at the same time, a random backoff mechanism[25] is used.

27.8   Conclusion

We have discussed several coverage-related works in the literature. As to future research, distributed protocols are needed to resolve these coverage issues in a wireless sensor network. Sensing regions are typically assumed to be circles. In practice, they may be irregular in shape, or even follow a probabilistic model. In several works, the communication distance of sensors is assumed to be twice the sensing distance.

This is not necessarily true and deserves further investigation. Field studies are also important in validating the effectiveness of an algorithm.

Acknowledgments

Y.C. Tseng's research is co-sponsored by the NSC Program for Promoting Academic Excellence of Universities under grant number 93-2752-E-007-001-PAE; by Computer and Communications Research Labs, ITRI, Taiwan; by Intel Inc.; by the Institute for Information Industry and MOEA, R.O.C, under the Handheld Device Embedded System Software Technology Development Project; and by Chung-Shan Institute of Science and Technology under contract number BC93B12P.

References

1. S. Adlakha and M. Srivastava. Critical density thresholds for coverage in wireless sensor networks. In *IEEE Wireless Communications and Networking Conf. (WCNC)*, pages 1615–1620, 2003.

2. D. Braginsky and D. Estrin. Rumor routing algorithm for sensor networks. In *ACM Int. Workshop on Wireless Sensor Networks and Applications (WSNA)*, 2002.

3. B. Chen, K. Jamieson, H. Balakrishnan, and R. Morris. SPAN: an energy-efficient coordination algorithm for topology maintenance in ad hoc wireless networks. *ACM/Kluwer Wireless Networks*, 8(5):481–494, Sept. 2002.

4. D. Ganesan, R. Govindan, S. Shenker, and D. Estrin. Highly resilient, energy efficient multipath routing in wireless sensor networks. *ACM Mobile Comput. and Commun. Review*, 5(4):11–25, Oct. 2001.

5. H. Gupta, S.R. Das, and Q. Gu. Connected sensor cover: self-organization of sensor networks for efficient query execution. In *ACM Int. Symp. on Mobile Ad Hoc Networking and Computing (MobiHOC)*, pages 189–200, 2003.

6. W.R. Heinzelman, A. Chandrakasan, and H. Balakrishnan. Energy-efficient communication protocols for wireless microsensor networks. In *Hawaii Int. Conf. on Systems Science (HICSS)*, 2000.

7. A. Heppes and J.B.M. Melissen. Covering a rectangle with equal circles. *Period. Math. Hung.*, 34:65–81, 1996.

8. C.-F. Huang and Y.-C. Tseng. The coverage problem in a wireless sensor network. In *ACM Int. Workshop on Wireless Sensor Networks and Applications (WSNA)*, pages 115–121, 2003.

9. C.-F. Huang, Y.-C. Tseng, and L.-C. Lo. The coverage problem in three-dimensional wireless sensor networks. Submitted to *GLOBECOM'04*.

10. Q. Huang. Solving an Open Sensor Exposure Problem Using Variational Calculus. Technical Report WUCS-03-1, Washington University, Department of Computer Science and Engineering, St. Louis, Mo, 2003.

11. X.-Y. Li, P.-J. Wan, and O. Frieder. Coverage in wireless ad hoc sensor networks. *IEEE Trans. Comput.*, 52(6):753–763, June 2003.

12. J. Lu and T. Suda. Coverage-aware self-scheduling in sensor networks. In *IEEE Computer Communications Workshop (CCW)*, 2003.

13. S. Meguerdichian, F. Koushanfar, M. Potkonjak, and M.B. Srivastava. Coverage problems in wireless ad-hoc sensor networks. In *IEEE INFOCOM*, pages 1380–1387, 2001.

14. S. Meguerdichian, F. Koushanfar, G. Qu, and M. Potkonjak. Exposure in wireless ad-hoc sensor networks. In *ACM Int. Conf. on Mobile Computing and Networking (MobiCom)*, pages 139–150, 2001.

15. S. Meguerdichian, S. Slijepcevic, V. Karayan, and M. Potkonjak. Localized algorithms in wireless ad-hoc networks: location discovery and sensor exposure. In *ACM Int. Symp. on Mobile Ad Hoc Networking and Computing (MobiHOC)*, pages 106–116, 2001.

16. J.B.M. Melissen and P.C. Schuur. Improved coverings of a square with six and eight equal circles. *Electronic Journal of Combinatorics*, 3(1), 1996.

17. K.J. Nurmela and P.R.J. Östergård. Covering a Square with up to 30 Equal Circles. Research Report A62, Helsinki University of Technology, Laboratory for Theoretical Computer Science, Espoo, Finland, June 2000.

18. J. O'Rourke. *Art Gallery Theorems and Algorithms.* Oxford University Press, Aug. 1987.

19. J. O'Rourke. Computational geometry column 15. *Int. Journal of Computational Geometry and Applications,* 2(2):215–217, 1992.

20. G.J. Pottie and W.J. Kaiser. Wireless integrated network sensors. *Commun. ACM,* 43(5):51–58, May 2000.

21. S. Shakkottai, R. Srikant, and N. Shroff. Unreliable sensor grids: coverage, connectivity and diameter. In *IEEE INFOCOM,* pages 1073–1083, 2003.

22. E. Shih, S.-H. Cho, N. Ickes, R. Min, A. Sinha, A. Wang, and A. Chandrakasan. Physical layer driven protocol and algorithm design for energy-efficient wireless sensor networks. In *ACM Int. Conf. on Mobile Computing and Networking (MobiCom),* pages 272–287, 2001.

23. S. Slijepcevic and M. Potkonjak. Power efficient organization of wireless sensor networks. In *IEEE Int. Conf. on Communications (ICC),* pages 472–476, 2001.

24. K. Sohrabi, J. Gao, V. Ailawadhi, and G.J. Pottie. Protocols for self-organization of a wireless sensor network. *IEEE Personal Commun.,* 7(5):16–27, Oct. 2000.

25. D. Tian and N.D. Georganas. A node scheduling scheme for energy conservation in large wireless sensor networks. *Wireless Commun. and Mobile Comput. (WCMC),* 3:271–290, 2003.

26. G. Veltri, Q. Huang, G. Qu, and M. Potkonjak. Minimal and maximal exposure path algorithms for wireless embedded sensor networks. In *ACM Int. Conf. on Embedded Networked Sensor Systems (SenSys),* pages 40–50, 2003.

27. X. Wang, G. Xing, Y. Zhang, C. Lu, R. Pless, and C. Gill. Integrated coverage and connectivity configuration in wireless sensor networks. In *ACM Int. Conf. on Embedded Networked Sensor Systems (SenSys),* pages 28–39, 2003.

28. R. Williams. *The Geometrical Foundation of Natural Structure: A Source Book of Design,* pages 51–52. Dover, New York, 1979.

29. A. Woo and D.E. Culler. A transmission control scheme for media access in sensor networks. In *ACM Int. Conf. on Mobile Computing and Networking (MobiCom),* pages 221–235, 2001.

30. T. Yan, T. He, and J.A. Stankovic. Differentiated surveillance for sensor networks. In *ACM Int. Conf. on Embedded Networked Sensor Systems (SenSys),* pages 51–62, 2003.

31. F. Ye, G. Zhong, S. Lu, and L. Zhang. PEAS: A robust energy conserving protocol for long-lived sensor networks. In *Int. Conf. on Distributed Computing Systems (ICDCS),* 2003.

32. W. Ye, J. Heidemann, and D. Estrin. An energy-efficient MAC protocol for wireless sensor networks. In *IEEE INFOCOM,* pages 1567–1576, 2002.

33. H. Zhang and J.C. Hou. Maintaining sensing coverage and connectivity in large sensor networks. In *NSF Int. Workshop on Theoretical and Algorithmic Aspects of Sensor, Ad Hoc Wireless, and Peer-to-Peer Networks,* 2004.

28

Maintaining Sensing Coverage and Connectivity in Large Sensor Networks

Honghai Zhang

Jennifer C. Hou

28.1 Introduction

Recent technological advances have led to the emergence of pervasive networks of small, low-power devices that integrate sensors and actuators with limited on-board processing and wireless communication capabilities. These sensor networks have opened new vistas for many potential applications, such as civilian surveillance, environment monitoring, biological detection, and situational awareness in the battlefield.[2,6,11,14]

Because most low-power devices have limited battery life and replacing batteries on tens of thousands of these devices is infeasible, it is well accepted that a sensor network should be deployed with high density (up to 20 nodes/m^3)[16] to prolong the network lifetime. In such a high-density network with energy-constrained sensors, if all the sensor nodes operate in the active mode, sensor data collected is likely to be highly correlated and redundant; and moreover, excessive packet collision may occur as a result of sensors attempting to send packets simultaneously in the presence of certain triggering events. All these amount to energy waste. It is thus neither necessary nor desirable to have all nodes simultaneously operate in the active mode.

One important issue that arises in such high-density sensor networks is *density control* — the function that controls the density of the working sensors to a certain level.[22] Specifically, density control ensures only a subset of sensor nodes operates in the active mode, while fulfilling the following two requirements: (1) *coverage*: the area that can be monitored is not smaller than that which can be monitored by a full set of sensors; and (2) *connectivity*: the sensor network remains connected so that the information collected by sensor nodes can be relayed back to data sinks or controllers. Under the assumption that an acoustic (or light) signal can be detected with certain minimum signal-to-noise ratios by a sensor node if the sensor is within a certain range of the signal source, the first issue essentially boils down to a coverage problem: assuming that each node can monitor a disk (the radius of which is called the *sensing range* of the sensor node) centered at the node on a two-dimensional surface, what is the minimum set of nodes that should be put in the active mode to cover the entire area? How should the operational set rotate among all the sensors to maximize the operational period? Moreover, can the relationship between coverage and connectivity be well characterized (e.g., under what condition coverage may imply connectivity or vice versa), so that the connectivity issue can be studied, in conjunction with the coverage issue? Finally, due to the distributed nature of sensor networks, a practical density control algorithm should be not only distributed, but also completely localized (i.e., relies on and makes use of local information only).[6]

In this chapter, we first outline several important properties of density control in an analytical framework, and then present and categorize a few recently proposed algorithms/protocols[3,7,17–19,21–23] in the analytical framework. The analytical framework includes three parts. First, under the assumption (**A1**) that the radio range is at least twice the sensing range, it has been shown that complete k-coverage of a convex area implies k-connectivity among the set of working nodes and $2k$-connectivity of interior networks (composed of nodes whose sensing range is completely contained in the surveillance region). Note that as indicated in Tables 28.1 and 28.2, (**A1**) holds for a wide spectrum of sensor devices that recently emerge. As a result,

TABLE 28.1 Radio Transmission Range of Berkeley Motes[15]

Product	Transmission Range (m)
MPR300[a]	30
MPR400CB	150
MPR410CB	300
MPR420CB	300
MPR500CA	150
MPR510CA	300
MPR520CA	300

[a] MPR300 is a second-generation sensor, while the rest are third-generation sensors.

TABLE 28.2 Sensing Range of Several Typical Sensors

Product	Sensing Range (m)	Typical Applications
HMC1002 Magnetometer sensor[5]	5	Detecting disturbance from automobiles
Reflective type photoelectric sensor[13]	1	Detecting targets of virtually any material
Thrubeam type photoelectric sensor[13]	10	Detecting targets of virtually any material
Pyroelectric infrared sensor (RE814S)[10]	30	Detecting moving objects
Acoustic sensor on Berkeley Motes[a 5]	~1	Detecting acoustic sound sources

[a] This result is based on our own measurement on Berkeley Motes.[5]

the property allows us to focus only on the coverage problem, as complete coverage implies connectivity. Second, it is observed that for the uniform sensing range, minimizing the number of working nodes is equivalent to minimizing the common overlapping area covered by multiple sensors. This observation enables us to transform a global system attribute (the number of working nodes in the network) into a local measure that can be obtained locally. Third, we present under the ideal case that the node density is sufficiently high, a set of optimality conditions under which a subset of working nodes can be chosen for complete coverage.

The rest of the chapter is organized as follows. In Section 28.2 we present the analytical framework on maintaining coverage and connectivity. In Section 28.3 we discuss several recently proposed protocols that maintain coverage and connectivity in the analytical framework. We present a simulation study in Section 28.4 and conclude the chapter in Section 28.5.

28.2 Analytical Framework

28.2.1 Relationship between Coverage and Connectivity

The relationship between coverage and connectivity is important because devising algorithms to fulfill both conditions is, in general, more difficult than fulfilling only one of them. If one condition (coverage) can infer another (connectivity), the problem of maintaining both conditions can be simplified to that of maintaining one of them. Indeed, it has been shown[19,23] that complete coverage of a convex region infers connectivity of the underlying network if the transmission range is at least twice the sensing range. Let the sensing range and the radio transmission range be denoted as, respectively, r_s and r_t. Specifically, Zhang and Hou[23] derive the following lemma.

Lemma 28.1 *Assuming the number of sensors in any finite area is finite, the condition of $r_t \geq 2r_s$ is both necessary and sufficient to ensure that complete coverage of a convex region implies connectivity.*

The necessary condition is shown by devising a scenario in which coverage does not imply connectivity under the condition of $r_t < 2r_s$. The sufficient condition is proved by contradiction: if assuming the network is not connected under the condition of complete coverage, the resulting network must have a pair of disconnected nodes that have the shortest distance among all disconnected pairs. The proof proceeds by finding another pair of disconnected nodes with shorter distance. Although the above lemma is stated on a two-dimensional surface, both the lemma and its proof apply to three-dimensional space as well.

Wang et al.[19] go one step further and show that if $r_t \geq 2r_s$ and the coverage regions of a set of nodes k-cover a convex region, the set of nodes forms a k-connected communication graph and the interior network has $2k$-connectivity, where the interior network is composed of nodes whose coverage regions are completely contained in the surveillance region R.

As mentioned in Section 28.1, the condition $r_t \geq 2r_t$ indeed holds for most of recently developed sensors (Tables 28.1 and 28.2) Therefore, the problem of ensuring both coverage and connectivity can be reduced to that of ensuring coverage only. We will henceforth only consider the coverage problem. Zhang and Hou[23] have devised additional procedures to ensure both connectivity and complete coverage in the (rare) case that the radio ranges are smaller than twice the sensing ranges.

Given a fixed sensing range, the above findings also suggest the minimum radio range that ensures connectivity. Note also that if the radio range is too large as compared to the sensing range, the network may be subject to excessive radio interference although its connectivity is ensured. It would be desirable if the wireless devices can adjust their radio ranges to be around twice their sensing ranges.

28.2.2 Optimal Sensing Coverage in the Ideal Case

Given that the coverage area of a sensor node is a disk centered at itself, we define a *crossing* as an intersection point of two circles (boundaries of disks) or that of a circle and the boundary of region R. A crossing is said to be *covered* if it is an *interior point* of a third disk. The following lemma provides a sufficient condition for complete coverage.[8] This condition is also necessary if we assume that the circle boundaries of any three disks do not intersect at a point. (Note that the probability of the circle boundaries of three disks intersecting at a point is zero if all sensors are randomly placed in a region with uniform distribution.) Lemma 28.2 serves as an important theoretical base for the design of (near-)optimal density control algorithms.

Lemma 28.2   *Suppose the size of a disk is sufficiently smaller than that of a convex region R. If one or more disks are placed within the region R, and at least one of those disks intersects another disk, and all crossings in the region R are covered, then R is completely covered.*

Although Lemma 28.2 states the sufficient condition for complete coverage, it does not take into account the power consumption issue (e.g., the condition is satisfied by putting all the nodes in the active mode). As stated in Section 28.1, it is desirable that the set of working sensors consumes as little power as possible so as to prolong the network lifetime. If each sensor consumes the same amount of power when it is active and has the same sensing range, minimizing power consumption translates to minimizing the number of working sensors. On the other hand, if sensors have different sensing ranges (e.g., using different levels of power to sense), a minimum number of working sensors does not necessarily imply minimum power consumption.

To derive conditions that ensure complete coverage while consuming the minimal possible energy, we first define the *overlap* at a point x as the number of sensors whose sensing ranges can cover the point minus $I_R(x)$, where

$$I_R(x) = \begin{cases} 1 & \text{if } x \in R, \\ 0 & \text{otherwise.} \end{cases} \tag{28.1}$$

The overlap of sensing areas of all the sensors is then the integral of overlaps of the points over the area covered by all the sensors. In general, the larger the overlap of the sensing areas, the more redundant data will be generated and more power will be consumed.

We claim that overlap is a better index for measuring power consumption than the number of working sensors for two reasons. First, while the number of working sensors is not directly related to power consumption in the case that sensors have different sensing ranges, the measure of overlap is; that is, a larger value of overlap implies more data redundancy and power consumption. Second, as will be stated in the following lemma, minimizing the overlap value is equivalent to minimizing the number of working sensors in the case that all sensors have the same sensing ranges (i.e., the coverage disks of all sensors have the same radius r).

Lemma 28.3   *If all sensor nodes (1) completely cover a region R and (2) have the same sensing range, then minimizing the number of working nodes is equivalent to minimizing the overlap of sensing areas of all the nodes.*

Lemma 28.3 is important as it relates the total number of working sensor nodes to the common overlapping area between working nodes. Because the latter can be readily measured in a local manner, this greatly simplifies the task of designing a decentralized and localized density control algorithm.

28.2.3 Optimality Conditions under the Ideal Case

With Lemmas 28.2 and 28.3, we are now in a position to discuss how to minimize the overlap of sensing areas of all the sensor nodes. Our discussion is built upon the following assumptions:

(A1): The sensor density is high enough that a sensor can be found at any desirable point.
(A2): The region R is large enough as compared to the sensing range of each sensor node so that the boundary effects can be ignored.

Although **(A1)** may not hold in practice, as will be shown in Section 28.3, the result derived under **(A1)** still provides insightful guidance into designing distributed algorithms. Assumption **(A2)** is usually valid.

By Lemma 28.2, to totally cover the region R, some sensors must be placed inside region R and their covering areas intersect one another. If two disks A and B intersect, at least one more disk is needed to cover their crossing points. Consider, for example, in Figure 28.1, disk C is used to cover the crossing point O of disks A and B. In order to minimize the overlap while covering the crossing point O (and its vicinity not covered by disks A and B), disk C should also intersect disks A and B at the point O; otherwise, one can always move disk C away from disks A and B to reduce the overlap.

Given that two disks A and B intersect, we now investigate the number of disks needed, and their relative locations, in order to cover a crossing point O of disks A and B and at the same time minimize the overlap. Take the case of three disks (Figure 28.1) as an example. Let $\angle PAO = \angle PBO \overset{\triangle}{=} \alpha_1$, $\angle OBQ = \angle OCQ \overset{\triangle}{=} \alpha_2$, and $\angle OCR = \angle OAR \overset{\triangle}{=} \alpha_3$. We consider two cases: (1) $\alpha_1, \alpha_2, \alpha_3$ are all variables; and (2) α_1 is a constant but α_2 and α_3 are variables. Case (1) corresponds to the case where all the node locations can be chosen, while case (2) corresponds to the case where two nodes (A and B) are already fixed and the location of a third node C is to be chosen to minimize the overlap. Both of the above two cases can be extended to the general situation in which $k - 2$ additional disks are placed to cover one crossing point of the first two disks (that are placed on a two-dimensional plane), and α_i, $1 \leq i \leq k$, can be defined accordingly. Again, the boundaries of all disks should intersect at point O to reduce the overlap. In the following discussion we assume, for simplicity, that the sensing range $r = 1$. Note, however, that the results still hold when $r \neq 1$.

Case 28.1 α_i, $1 \leq i \leq k$, **are all variables.** We first state the following lemma.

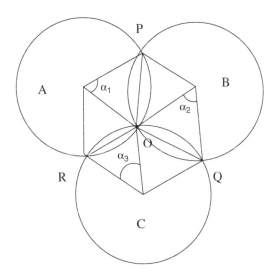

FIGURE 28.1 An example that demonstrates how to minimize the overlap while covering the crossing point O.

Lemma 28.4

$$\sum_{i=1}^{k} \alpha_i = (k-2)\pi, \tag{28.2}$$

Now the overlap between the ith and $(i \bmod k) + 1$th disks (which are called adjacent disks*) is $(\alpha_i - \sin \alpha_i)$, $1 \le i \le k$. If we ignore the overlap caused by* non-adjacent *disks, then the total overlap is $L = \Sigma_{i=1}^{k}(\alpha_i - \sin \alpha_i)$. The coverage problem can be formulated as*

Problem 28.1

$$\text{minimize} \sum_{i=1}^{k} (\alpha_i - \sin \alpha_i)$$

$$\text{subject to} \sum_{i=1}^{k} \alpha_i = (k-2)\pi. \tag{28.3}$$

The Lagrangian multiplier method can be used to solve the above optimization problem. The solution is $\alpha_i = (k-2)\pi/k, i = 1, 2, \ldots, k$ and the resulting minimum overlap using k disks to cover the crossing point O is

$$L(k) = (k-2)\pi - k \sin\left(\frac{(k-2)\pi}{k}\right) = (k-2)\pi - k \sin\left(\frac{2\pi}{k}\right).$$

Note that the overlap per disk

$$\frac{L(k)}{k} = \pi - \frac{2\pi}{k} - \sin\left(\frac{2\pi}{k}\right) \tag{28.4}$$

monotonically increases with k when $k \ge 3$. Moreover, when $k = 3$ (which means that we use one disk to cover the crossing point), the optimal solution is $\alpha_i = \pi/3$ and there is no overlap between *non-adjacent* disks. When $k > 3$, the overlap per disk is always higher than that in the case of $k = 3$, even if we ignore the overlaps between *non-adjacent* disks. This implies that using one disk to cover the crossing point and its vicinity is optimal in the sense of minimizing the overlap. Moreover, the centers of the three disks should form an equilateral triangle with edge $\sqrt{3}$. We state the above result in the following theorem.[23]

Theorem 28.1   *To cover one crossing point of two disks with the minimum overlap, only one disk should be used and the centers of the three disk should form an equilateral triangle with side length $\sqrt{3}r$, where r is the radius of the disks.*

Case 28.2 α_1 is a constant, while α_i, $2 \le i \le k$, are variables.   In this case, the problem can still be formulated as in Problem 28.1, except that α_1 is fixed. The Lagrangian multiplier method can again be used to solve the problem, and the optimal solution is $\alpha_i = ((k-2)\pi - \alpha_1)/(k-1), 2 \le i \le k$. Again, a similar conclusion can be drawn that using one disk to cover the crossing point gives the minimum overlap. We state the result in the following theorem.[23]

Theorem 28.2   *To cover one crossing point of two disks whose locations are fixed (i.e., α_1 is fixed in Figure 28.1), only one disk should be used and $\alpha_2 = \alpha_3 = (\pi - \alpha_1)/2$.*

In summary, to cover a large region R with the minimum overlap, one should ensure (1) at least one pair of disks intersects; (2) the crossing points of any pair of disks are covered by a third disk; (3) if the locations of any three sensor nodes are adjustable, then as stated in Theorem 28.1, the three nodes should form an

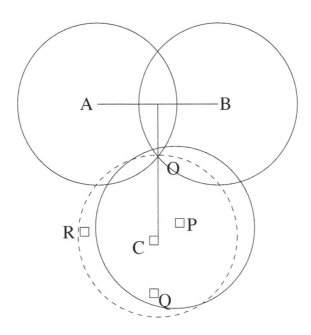

FIGURE 28.2 Although C is the optimal place to cover the crossing O of A, B, there is no sensor node there. The node closest to C, P, is selected to cover the crossing O.

equilateral triangle with side length $\sqrt{3}r$. If the locations of two sensor nodes A and B are already fixed, then as stated in Theorem 28.2, the third sensor node should be placed on the line that is perpendicular to the line connecting nodes A and B and have a distance r to the intersection of the two circles (e.g., the optimal point in Figure 28.2 is C). These conditions are optimal for the coverage problem in the ideal case in which assumptions (**A1**) and (**A2**) hold.

As mentioned above, the notion of overlap can be extended to the heterogeneous case in which sensors have different sensing ranges. Moreover, Theorems 28.1 and 28.2 can be generalized to the heterogeneous case. The following theorem establishes the theoretical base for the heterogeneous case.

Theorem 28.3 *Assuming that different nodes have different sensing ranges, to cover one crossing point O of two disks with the minimum overlap, the three disks should be placed such that $\overline{OP} = \overline{OQ} = \overline{OR}$. If disks A and B are already fixed, disk C should be placed such that $\overline{OR} = \overline{OQ}$.*

Proof See Appendix A. □

28.3 Algorithms and Protocols for Density Control

Power management protocols in wireless ad hoc networks. Minimizing energy consumption and prolonging the system lifetime has been a major design objective for wireless ad hoc networks. Many research efforts have been made in this area.[4,9,20,24,25] Some of them touch the issue of maintaining coverage and connectivity in wireless sensor networks. For example, GAF[20] assumes the availability of GPS and conserves energy by dividing a region into rectangular grids, ensuring that the maximum distance between any pair of nodes in adjacent grids is within the transmission range of each other, and electing a leader in each grid to stay awake and relay packets (while putting all the other nodes into sleep). The leader election scheme in each grid takes into account the battery usage at each node, and a sleeping node wakes up periodically to attempt to elect itself as an active node. To support mobility, in the mobile adaptation version of GAF (GAP-ma), each node estimates the time at which it expects to leave its grid (based on its current speed obtained from GPS and grid size). The simulation study shows that GAF extends the network lifetime by 30 to 40 percent. SPAN[4] defines a virtual core as a connected dominating set. Nodes make local decisions on whether they should sleep or join a forwarding backbone as a coordinator. The nodes that choose to stay awake and maintain network connectivity/capacity are called *coordinators*. The rule for electing coordinators is that if two neighbor nodes of a non-coordinator nodes can neither directly communicate with each other nor through one or two coordinators, then this node volunteers to be a coordinator. The information needed for electing oneself as a coordinator is exchanged among neighbors via HELLO messages. The coordinator announcement is broadcast, based on a delay interval reflecting the "benefit" that each neighbor will perceive and taking into account of the total energy available. The authors then investigate the performance of SPAN combined with a simple geographic routing protocol and report power savings by a factor of 3.5 or more.

The key differences between wireless ad hoc networks and sensor networks are twofold from the perspective of power savings. First, algorithms used for wireless ad hoc networks do not address the issue of sensing coverage. Second, although reducing power consumption is a common design objective, algorithms used for wireless ad hoc networks often aim to maximize the lifetime of each individual node, while those used for sensor networks aim to maximize the time interval of continuously performing certain (monitoring) functions. As long as the coverage and connectivity are maintained, a sensor network is considered to function well even if some sensors die much earlier than others.

Recall that coverage implies connectivity as long as the transmission range is twice the sensing range, and hence algorithms/protocols that ensure complete coverage also ensure connectivity. In what follows, we discuss several centralized and distributed algorithms that have been proposed for ensuring full sensing coverage (and some of them for connectivity) in sensor networks.[3,7,17–19,21,22] We roughly categorize them into centralized and decentralized algorithms. Centralized algorithms can be used to study the performance limits and provide a baseline for distributed/localized algorithms.

28.3.1 Centralized Algorithms

28.3.1.1 Critical Element Approach

Slijepcevic et al.[17] address the problem of finding the maximal number of covers in a sensor network, where a cover is defined as a set of nodes that can completely cover the monitored area. They prove the NP-completeness of this problem, and provide a centralized heuristic algorithm. The algorithm starts by organizing the points in the monitored area into disjoint fields where a field is a maximal set of points that are covered by the same set of sensors (a field is also called an element). The algorithm then selects a cover in each pass until the remaining set of nodes cannot fully cover the whole monitored region.

At the beginning of pass i, the set of unmarked fields U contains all fields and the set of available nodes V contains all sensors that are not chosen in earlier passes. In each step within the pass i, as long as there are unmarked fields in U, one member of V is chosen into the cover as follows. First, a critical field in U, which is covered by the smallest number of sensors in V, is selected from U. Then for each of the sensors that cover the critical field, a value of the objective function is calculated. The sensors with the highest objective function value is selected and included in the current cover. The set of unmarked fields U are updated by removing the fields that are covered by the selected sensor and V is updated by removing the selected sensor. The above operations of selecting a sensor into current cover repeat until all fields are marked (i.e., U becomes empty) and then a cover is completed. The objective function is designed to measure the likelihood that the current cover will redundantly cover some of the sparsely covered parts of the area. A higher value of the objective means the likelihood is lower. The proposed algorithm has been shown to approach the upper bound of the solution under most cases through simulations.

28.3.1.2 Connected Sensor Cover

Gupta et al.[7] devise both a centralized algorithm and a distributed algorithm to find a subset (called *connected sensor cover*) of nodes that ensure both coverage and connectivity. They define a *sub-element* as the maximal set of points that are covered by the same set of sensors (which is the same as the field[17]).

A sub-element is *valid* if it intersects with the query region. The centralized algorithm constructs a (near) minimal connected sensor cover M as follows. It begins by including in M an arbitrary sensor that lies in the query region. In each stage, the algorithm determines the *candidate sensors*, the set of sensors whose sensing region overlaps with that of the sensors in M. For each such *candidate sensor* C_i, the algorithm constructs a *candidate path* P_i of sensors that forms a communication path connecting C_i to some sensor in M. The candidate path P_i that covers the maximum number of uncovered valid sub-elements per sensor (defined as the *benefit* of P_i) is added to M in this stage of the algorithm. The centralized algorithm guarantees that the size of the formed subset is within $O(\log n)$ factor of the optimal size, where n is the network size.

28.3.2 Distributed and Localized Algorithms

28.3.2.1 ASCENT

Cerpa and Estrin[3] present ASCENT to automatically configure sensor network topologies. ASCENT consists of several phases. Initially, only a few nodes are operational in the multihop network. The other nodes enter the *neighbor discovery phase*, where they remain passively listening to the messages but not transmitting. During this phase, each node obtains an estimate of the number of neighbors actively transmitting messages based on local measurements. Each node also maintains an estimate of the *data message loss* — the ratio of data packets not received by any active neighbors. A node that has joined the network sends *help* messages if it detects *High data message loss*, where the level of *High* data message loss depends on applications.

Upon completion of the *neighbor discovery phase*, nodes enter the *join decision phase*, where they decide whether to join the multihop diffusion sensor network. The decision is made based on three factors: the number of active neighbors, the data message loss, and whether a node receives a help message. Table 28.3 summarizes the basic decision engine. If the number of active neighbors is high, the node does not join the network. Otherwise, if the message loss is high, the node will join the network. In the last case that both the number of active neighbors and message loss are low, the node will only temporarily join the network and enter a probe phase if it receives a help message. If the message loss is still low after the probe phase period expires, the node will join the network. Otherwise, the node will not join the network.

If a node decides to join the network, it enters the active phase and immediately sends a neighbor announcement message, which signals other passive neighbors to increase their active neighbor count. Only active nodes send and receive control and routing messages. If a node decides not to join the network, it enters the *adaptive phase*, where it turns itself off for a period of time or reduces its transmission range. In the first case, after a certain time period, it will re-enter the *neighbor discovery phase*. Based on several heuristics with tunable parameters, ASCENT does not consider the issue of completely covering the monitored region either.

28.3.2.2 Sponsored Area Approach

Tian et al.[18] devise an algorithm that ensures complete coverage using the concept of "sponsored area." A node A's sponsored area provided by node B is the maximum sector of A's coverage disk that is covered by node B. In this algorithm, time is divided into rounds. At the beginning of each round, each node broadcasts a Position Advertisement Message (PAM). Whenever a sensor node receives a PAM packet

TABLE 28.3 Join Decision Engine

Message Loss	Number of Neighbors	
	Low	High
Low	If the node receives help message(s), it temporarily joins	Do not join
High	Join	Do not join

from one of its neighbors, it calculates the sponsored area provided by the neighbor. If the union of all the sponsored areas provided by its neighbors covers the coverage disk of the node, the node is eligible to turn off.

To avoid the situation that two neighboring nodes expect each other's sponsorship and turn off themselves simultaneously, the algorithm has a back-off scheme. If a node is eligible to turn off, it will back off for a random period and then broadcast a Status Advertisement Message (SAM). When a node receives a SAM message from a neighbor, the node will remove the neighbor from its neighbor list and reevaluate its eligibility of being off. An eligible node waits for a short period after sending out its SAM message. If it is still eligible for turning off, it will then turn itself off. The operation will repeat at the beginning of each round. As will be shown in Section 28.4, this approach is less efficient than a hexagon-based GAF-like algorithm.

28.3.2.3   PEAS

Ye et al.[21,22] present a distributed, probing-based density control algorithm, called *PEAS*, for robust sensing coverage. In this work, each node has three operation modes: *Sleeping, Probing* and *Working*. Initially nodes are in the *Sleeping* mode. Each node sleeps for a duration randomly distributed according to an exponential density function $f(t) = \lambda e^{-\lambda t}$, where λ is the average probing rate. After a node wakes up, it enters the *Probing* mode, where it broadcasts a PROBE message within its probing range R_p. Any working nodes within that range should respond with a REPLY message. The REPLY message is also sent within the range R_p after waiting for a short random period. If the probing node hears a REPLY message, it will go back to the Sleeping mode for another random period of time, generated according to the same exponential distribution function. If the probing node does not hear any REPLY message, it enters the *Working* mode until it fails or consumes all its energy.

PEAS dynamically adjusts the average probing rate λ to achieve a goal that the aggregate probing rate (the sum of probing rates) of all nodes that monitor any given point is approximately constant. Notice the probing range R_p is different from (usually smaller than) the sensing range. The algorithm guarantees that the distance between any pair of working nodes is at least the probing range, but does not ensure that the coverage area of a sleeping node is completely covered by working nodes; that is, it does not guarantee complete coverage.

28.3.2.4   Distributed Version of Connected Sensor Cover

A distributed version of the connected sensor cover algorithm[7] works in a similar manner to the centralized version, except that (1) the most recently added candidate sensor C performs the operation of searching and determining the next candidate sensor to be added to M; (2) the search for the next candidate sensor is limited within $2r$ hops from the most recently added sensor, where r is the maximum number of hops between any two nodes whose sensing regions intersect. Instead of calculating the exact value of r through global search, the authors give several rules of thumb on how to choose the value of r. Specifically, in a dense network (with more than $4r_s/r_t$ sensors within a distance of $2r_s$, where r_s and r_t are respectively the sensing range and the transmission range) of area 100×100, $r = 2r_s/r_t + 1$ is used. In a not-so-dense network, $r = (2r_s/r_t + 1) \times (200/r_t)^2/n$ is used, where n is the number of sensors. The distributed algorithm is heuristic-based and does not guarantee the $O(\log n)$ factor. It is also a non-trivial task to implement the distributed algorithm in practice because it requires each node to reliably broadcast messages to all the nodes within $2r$ hops.

28.3.2.5   Integrated Coverage and Connectivity Configuration

Wang et al.[19] establish part of the analytical framework in Section 28.2 and prove that k-coverage of a convex region infers k-connectivity of the entire network and $2k$-connectivity of the interior network (that is composed of nodes whose sensing region is completely contained in the query region), if the radio range is at least twice the sensing range ($r_t \geq 2r_s$). In addition, they also prove that if all the crossing points inside a region (or disk) are k-covered, then the region (or disk) is k-covered (note that the case for $k = 1$ is stated[8]).

Based on the properties derived above, the authors then devise the *Coverage and Configuration Protocol* (CCP) in which each node collects neighboring information and then uses this information as an eligibility rule to determine whether or not it can sleep. More specifically, nodes in CCP can have three states: *SLEEP*, *LISTEN*, and *ACTIVE*. Initially, all nodes are in the *ACTIVE* state. *ACTIVE* nodes send a HELLO message periodically. When a node receives a HELLO message, it updates its neighbor list and determines if it is eligible to remain active. If it is ineligible, it starts a withdrawal timer upon the expiration of which the node broadcasts a WITHDRAW message and enters the *SLEEP* state. If the node becomes eligible as a result of receiving a WITHDRAW message from a neighbor before its own withdrawal timer expires, it cancels the timer. Nodes in the *SLEEP* state enter the *LISTEN* state after a timeout period. Nodes in the *LISTEN* state listen to the channel for a certain period T_1. During this period, a node collects its neighbor information through HELLO messages of active neighbors. If it becomes eligible to be active after T_1, the node will start a join timer upon expiration of which it broadcasts a JOIN message. If it becomes ineligible after the join timer is started (e.g., due to receipt of a JOIN message from a neighbor), it cancels the join timer. If it remains ineligible to be active after T_1, it returns to the *SLEEP* state. In the (rare) case that the radio range is less than twice the sensing range, the authors combine CCP with SPAN[4] to form a connected covering set; that is, a node is eligible to be active if and only if it is eligible in CCP or in SPAN.

28.3.2.6   Hexagon-Based GAF-Like Algorithm

GAF[20] can be straightforwardly extended to ensure full coverage in the following way. The entire region is divided into square grids and one node is selected to be awake in each grid as in GAF. To maintain coverage, the grid size must be set to be less than or equal to $r_s/\sqrt{2}$. Thus, for a large area with size $l \times l$, it requires at least $\frac{2l^2}{r_s^2}$ nodes to operate in the active mode to ensure complete coverage.

As pointed out by Kandula and Hou,[12] hexagonal cells are more "homogeneous" than square grids and thus offer more scaling benefits, (e.g., the number of working nodes is significantly smaller). To maintain coverage in hexagonal grids, the side length of each hexagon is at most $r_s/2$, and it requires at least $\frac{8l^2}{3\sqrt{3}r_s^2} \approx \frac{1.54l^2}{r_s^2}$ working nodes to completely cover a large area with size $l \times l$. The above analytical result does not take into account the boundary effect in a bounded area, which, if considered, will increase the required number of working nodes. To illustrate this point, consider a scenario where the monitored region is a square area with side length $l = 50$ m and sensing radius of each sensor is $r_s = 10$ m. In the 50×50 m^2 area, 45 hexagon cells are required to cover the entire area if the hexagon-based, GAF-like algorithm is used (Figure 28.3). In practice, there may not be sensors in some of the cells, and thus the coverage may not always be 100 percent.

28.3.2.7   Optimal Geographical Density Control (OGDC) Algorithm

OGDC[23] leverages the optimality conditions in the analytical framework (Section 28.2) to design a sub-optimal algorithm for maintaining coverage and connectivity. For clarity of presentation, we first describe OGDC under the assumptions that the radio range is at least twice the sensing range and all sensors are time synchronized. We then discuss how OGDC relaxes these assumptions.

At any time, a node is in one of the three states: "UNDECIDED," "ON," or "OFF." Time is divided into rounds. Each round has two phases: the *node selection phase* and the *steady-state phase*. At the beginning of the node selection phase, all the nodes wake up, set their states to "UNDECIDED," and carry out the operation of selecting working nodes. By the end of this phase, all the nodes change their states to either "ON" or "OFF." In the steady-state phase, all nodes keep their states fixed until the beginning of the next round. The length of each round is so chosen that it is much larger than that of the node selection phase but much smaller than the average sensor lifetime. As indicated,[23] the time it takes to execute the node selection operation for networks of size up to 1000 nodes in an area of 50×50 m^2 is usually well below 1 second, and most nodes can decide their states (either "ON" or "OFF") in less than 0.2 seconds from the time instant when at least one node volunteers to be a starting node. The interval for each round is usually set to approximately hundreds of seconds, and the overhead of density control is small (<1 percent).

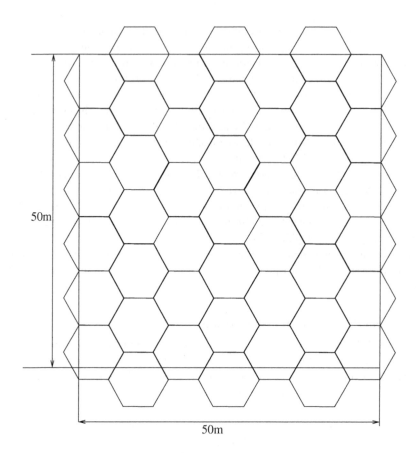

FIGURE 28.3   45 hexagons are required to cover a $50 \times 50 \, \text{m}^2$ area.

The node selection phase in each round commences when one or more sensor nodes volunteer to be starting nodes. Consider Figure 28.2, for example. Suppose node A volunteers to be a starting node. Then one of its neighbors with an (approximate) distance of $\sqrt{3}r$, say node B, will be "selected" to be a working node. To cover the crossing point of disks A and B, the node whose position is closest to the optimal position C (e.g., node P in Figure 28.2) will then be selected, in compliance with Theorems 28.1 and 28.3 (Section 28.2), to become a working node. The process continues until all the nodes change their states to either "ON" or "OFF," and the set of nodes with state "ON" forms the working set. As a node volunteers itself to be a starting node with the probability that is proportional to its remaining power in each round, the set of working sensor nodes is not likely to be the same in each round, thus ensuring uniform (and minimum) power consumption across the network, as well as complete coverage and connectivity.

In the case that the transmission range is less than twice the sensing range, coverage may not imply connectivity, and OGDC is extended as follows: a node should go to the "OFF" state only if its coverage disk is completly covered by its neighbors and its neighbors are connected without it. The authors also discuss how to relax the assumption of time synchronization. In the first round, a sensor node is designated to be the starting node. When the starting node sends a power-on message, it includes in its power-on message a duration δT after which the receivers should wake up for the next round. When a non-starting node broadcasts a power-on message, it reduces the value of δT by the time elapsed since it received the last power-on message and includes the new value of δT in its power-on message. In this fashion, all the nodes get "synchronized" with the starting node and will all wake up at the beginning of the next round. To overcome the small clock drifting over the network lifetime, when a node wakes up, it needs to wait for a short time ($\geq$ the maximum clock drifting) before it starts to send any message.

28.4 Performance Evaluation

28.4.1 Simulation Environment Setup

To evaluate the different algorithms/protocols for maintaining coverage and connectivity in sensor networks, we have conducted a simulation study using *ns-2* with the CMU wireless extension in a $50 \times 50 \text{ m}^2$ region where up to 1000 sensors are uniformly randomly distributed.[1] Each data point reported below is an average of 20 simulation runs.

28.4.1.1 Schemes for Comparison

We evaluate the PEAS algorithm proposed by Ye et al.,[22] the CCP algorithm by Wang et al.,[19] the hexagon-based GAF-like algorithm discussed in Section 28.3, and the OGDC algorithm proposed by Zhang and Hou.[23] We observe that, under the same scenario, at least 47 nodes are required to operate in the active mode using the "sponsored area" approach proposed[18] to ensure complete coverage, according to the simulation results given. In addition, as the number of sensor nodes in the sensor network increases, the "sponsored area" algorithm requires more nodes to cover the entire area. But because the sponsored area algorithm performs worse than the hexagon-based GAF-like algorithm, we do not include it in our simulation study.

28.4.1.2 Parameters Used

We use the energy model of Ye et al.,[22] where the power consumption ratio for transmitting, receiving (idling), and sleeping is 20:4:0.01. We define one unit of energy (power) as that required for a node to remain idle for 1 second. Each node has a sensing range of $r_s = 10$ m, and a lifetime of 5000 seconds if it is idle all the time. The system parameters, such as the initial energy of a node, the radio transmission rate, and the energy consumption rate, are the same for all nodes.

28.4.1.3 Performance Metrics

The performance metrics of interest are (1) the percentage of coverage (i.e., the ratio of the covered area to the entire area to be monitored); (2) the number of working nodes required to provide the percentage of coverage in (1); and (3) α-lifetime, defined as the total time during which at least α portion of the total area is covered by at least one node. The conventionally defined network lifetime is then 100 percent-lifetime. Note that the lifetime defined here is slightly different from that of Ye et al.,[21] where the lifetime is defined as the time interval until coverage falls below a predetermined percentage and never comes back again.

The way coverage is measured is as follows. We divide the area into 50×50 square grids, and a grid is considered covered if the center of the grid is covered; and coverage is defined as the ratio of the number of grids that are covered by at least one sensor to the total number of grids. In the first part of the simulation, we assume the transmission range is at least twice the sensing range (which is set to 20 m) so that we can focus on coverage alone. In the second part of the simulation, we study the cases in which the transmission range is smaller than twice the sensing range.

28.4.2 Simulation in the Cases of Sufficient Transmission Ranges

28.4.2.1 Number of Working Nodes and Coverage

Figure 28.4 shows the number of working nodes, and coverage, versus the number of sensor nodes deployed in the network. Both metrics are measured after the density control process is completed. In most cases, OGDC takes less than 1 second to perform density control in each round, while PEAS[22] and CCP[19] can take up to 100 seconds. As shown in Figure 28.4 OGDC needs only half as many nodes to operate in the active mode as compared to the hexagon-based GAF-like algorithm, but achieves almost the same coverage (in most cases OGDC achieves more than 99.5 percent coverage). As the PEAS algorithm can control the number of working nodes using different probing ranges, we tried two different probing ranges: 8 m and 9 m.

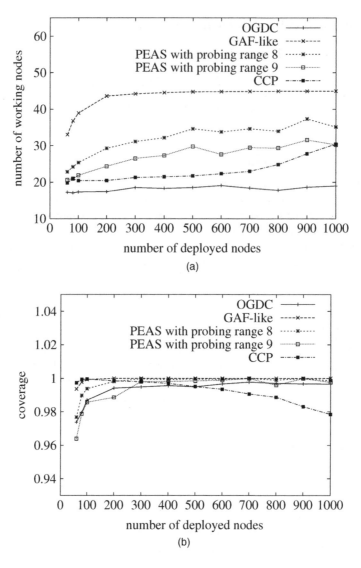

FIGURE 28.4   Number of working nodes (a) and coverage (b) versus number of sensor nodes in a 50×50 m^2 area.

(Using a probing range of 10 m leads to insufficient coverage, the result of which is thus not reported here.) As shown in Figure 28.4, use of a smaller probing range results in more working nodes. With a probing range of 9 m, the resulting coverage is less than that achieved by OGDC, while the number of working nodes is up to 50 percent more than that of OGDC. Moreover, the number of working nodes required under OGDC modestly increases with the number of sensor nodes deployed, while both PEAS and CCP incur a 50 percent increase in the number of working nodes when the number of sensor nodes deployed in the network increases from 100 to 1000. We also observe that when the number of working nodes becomes very large, the coverage ratio of CCP actually decreases. This is because a large number of message exchanges are required in CCP to maintain neighbor information. When the network density is high, packets incur collision more often and the neighbor information may be inaccurate. In contrast, in OGDC, each working node sends out at most one power-on message in each round; and as a result, the packet collision problem is not so serious. The result of CCP reported here is a little different from that which is reported by Ye et al.[19] because the latter results[19] are obtained assuming error-free channel conditions.

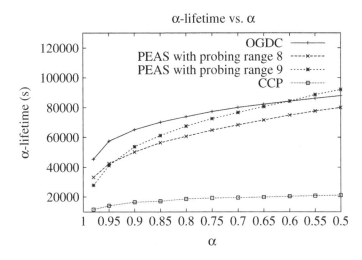

FIGURE 28.5 Comparison of α-lifetime versus α under OGDC, PEAS, and CCP.

28.4.2.2 α-Lifetime

Figure 28.5 compares the α-lifetime achieved by OGDC, PEAS, and CCP in a sensor network of 300 nodes, where α varies from 98 to 50 percent. For the PEAS algorithm, we again tried two different probing ranges: 8 m and 9 m. As shown in Figure 28.5, for a reasonably large α, the α-lifetime of PEAS is much shorter than that of OGDC. Only when α is less than 60 percent is the lifetime of PEAS with the use of the probing range 9 m longer than that of OGDC. This is because with a relatively small probing range, PEAS requires an excessive number of nodes to operate simultaneously. Hence, its lifetime is consistently shorter than OGDC. On the other hand, with a large probing range of 9 m, PEAS only guarantees that no two working nodes are in each other's probing range and does not ensure complete coverage. Moreover, when a node dies, it may take more than 100 seconds for another node to wake up to take its place. During that transition period, the network is not completely covered. As a result, the low percentage lifetime is prolonged in PEAS. A nice property of OGDC is that during most of the lifetime, the monitored region is covered with a high percentage. It is clear that OGDC is preferred to PEAS no matter what probing range is used, unless the desired coverage percentage is very low (i.e., less than 60 percent). Although CCP uses less working nodes than PEAS in most cases, its lifetime is much shorter than both PEAS and OGDC. This is due to two reasons. First, CCP needs to periodically broadcast HELLO messages, the operation of which consumes energy. Second, in CCP when a node wakes up from the *SLEEP* mode, it must stay awake and wait until it receives HELLO messages from a sufficient number of neighbors and determines if it can return to the *SLEEP* mode.

Figure 28.6 shows the 98 percent-lifetime and 90 percent-lifetime under OGDC, CCP, and PEAS with a probing range of 9 m, when the number of sensor nodes deployed in a network varies from 100 to 800. The α-lifetime scales linearly as the number of sensors deployed increases for both OGDC and PEAS algorithms. However, OGDC achieves nearly 100 percent more 98 percent-lifetime and 40 percent more 90 percent-lifetime than PEAS. Again, CCP achieves a much shorter lifetime than OGDC and PEAS.

28.4.3 Simulation in the Cases of Insufficient Transmission Range

We now investigate the effect of small transmission ranges on coverage and connectivity. Because PEAS does not consider the connectivity issue, we only compare OGDC against CCP. Figure 28.7 shows the number of working nodes versus the number of sensor nodes deployed with respect to different radio transmission ranges r_t under OGDC and CCP. As shown in Figure 28.7, OGDC uses a much smaller number of working nodes than CCP, especially when the radio range is small. Due to wireless channel errors, the sensor network may not always be connected in the case of small radio ranges, even if all the sensor nodes are powered on. Hence, instead of using the coverage of the network as the performance

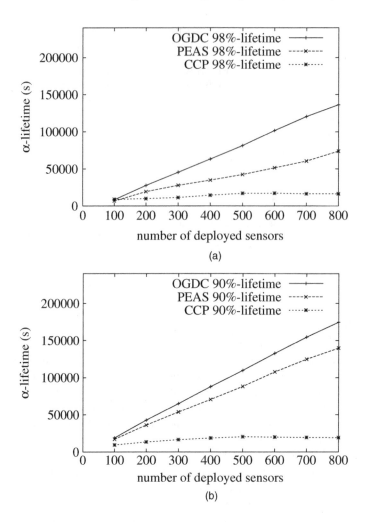

FIGURE 28.6   Comparison of α-lifetime versus number of sensor nodes under OGDC, PEAS (with probing range 9 m), and CCP: (a) 98 percent-lifetime and (b) 90 percent-lifetime.

index, we measure the coverage of the largest connected component and plot the result in Figure 28.8. The coverage of the largest connected component is very close to 1 under both algorithms, except in the cases that the number of sensor nodes deployed and the radio range are both small (e.g., $n = 100$ and $r_t = 5$). As a matter of fact, in the case of $n = 100$ and $r_t = 5$, the sensor network with all the sensor nodes active is not connected, and has more than 18 connected components with a 45 percent coverage for the largest connected component on average. In general, we observe that as the radio range decreases, the coverage increases slightly and the number of nodes also increases. This is the cost of maintaining connectivity.

28.5   Conclusions and Suggested Research Topics

In this chapter we have investigated the issues of maintaining coverage and connectivity by keeping a minimal number of sensor nodes operating in the active mode in wireless sensor networks. We begin with the discussion on the relationship between coverage and connectivity, and state that if the radio range is at least twice the sensing range, complete coverage implies connectivity. That is, if the condition holds, we only need to consider coverage. Then, we discuss, under the ideal case in which node density is

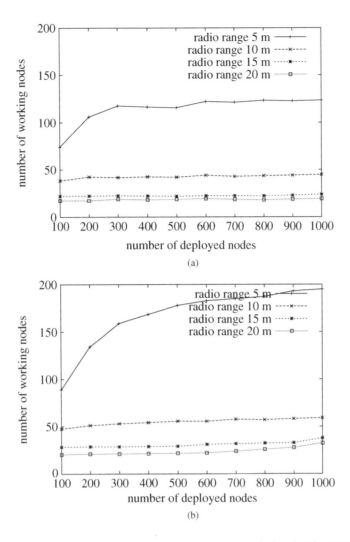

FIGURE 28.7 Number of working nodes versus number of sensor nodes deployed with respect to different radio ranges under OGDC (a) and CCP (b) (the sensing range is fixed at 10 m).

sufficiently high, a set of optimality conditions under which a subset of working sensor nodes can be chosen for complete coverage. Following that, we discuss several recently proposed algorithms and protocols for maintaining coverage and connectivity in wireless sensor networks in the analytical framework. We also conduct ns-2 simulation to evaluate these algorithms and protocols, and show that OGDC outperforms the PEAS algorithm,[22] the CCP algorithm,[19] the hexagon-based GAF-like algorithm, and the sponsor area algorithm[18] with respect to the number of working nodes needed and the network lifetime (with up to 50 percent improvement), and achieves almost the same coverage as the best algorithm.

Despite all the research effort, there are several issues that are not fully addressed. First, most of the existing solutions require that each node knows its own location. It would be desirable to relax this requirement to, for example, that each node knows its relative location to its neighbors. Second, the issue of k-coverage and its impact on fault tolerance has not been fully investigated. Finally, most existing algorithms attempt to maintain complete coverage, while in reality it might be sufficient to cover a certain percentage of the surveillance region. It would be interesting to investigate to what extent power consumption can be further reduced (and hence the network lifetime further extended) if an algorithm or protocol only attempts to maintain the predetermined percentage of coverage.

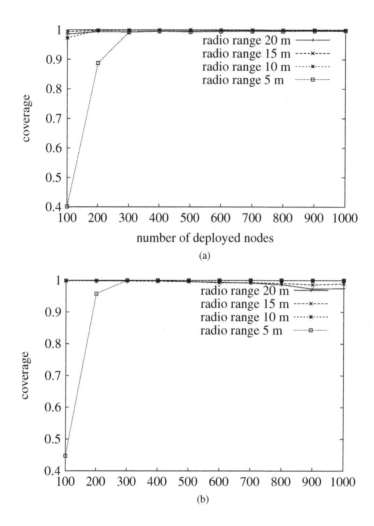

FIGURE 28.8   Coverage of the largest connected component versus the number of sensor nodes deployed with respect to different radio ranges under OGDC (a) and CCP (b) (the sensing range is fixed at 10 m).

References

1. ns-2 network simulator. http://www.isi.edu/nsnam/ns.
2. I.F. Akyildiz, W. Su, Y. Sankarasubramaniam, and E. Cayirci. Wireless sensor networks: a survey. *Computer Networks*, 38(4):393–422, March 2002.
3. A. Cerpa and D. Estrin. Ascent: adaptive self-configuring sensor networks topologies. In *Proc. of INFOCOM 2002*.
4. B. Chen, K. Jamieson, H. Balakrishnan, and R. Morris. SPAN: an energy-efficient operation in multihop wireless ad hoc networks. In *Proc. of ACM MobiCom'01*, 2001.
5. Inc. Crossbow Technology. http://www.xbow.com/support/support_pdf_files/mts-mda_series_user_manual_revb.pdf.
6. D. Estrin, R. Govindan, J.S. Heidemann, and S. Kumar. Next century challenges: scalable coordination in sensor networks. In *Proc. of ACM MobiCom'99*, Washington, August 1999.
7. H. Gupta, S. Das, and Q. Gu. Connected sensor cover: self-organization of sensor networks for efficient query execution. In *Proc. of Mobihoc 2003*, 2003.
8. P. Hall. *Introduction to the Theory of Coverage Processes*. John Wiley & Sons, 1988.

9. C. Hu and J.C. Hou. LISP: a link-indexed statistical traffic prediction approach to impoving IEEE 802.11 PSM. In *Proc. of IEEE Int. Conf. on Distributed Computing Systems (ICDCS 2004)*, 2004. An enhanced version will appear in *Ad Hoc Networks Journal*.

10. I. Infrared Sensor. http://www.interq.or.jp/japan/se-inoue/e_pyro.htm.

11. J.M. Kahn, R.H. Katz, and K.S.J. Pister. Next century challenges: mobile networking for "smart dust." In *Proc. of ACM MobiCom'99*, August 1999.

12. S. Kandula and J.C. Hou. The Case for Resource Heterogeneity in Large Sensor Networks. Technical report, University of Illinois at Urbana, Champaign, August 2002. An abridged version appeared in *IEEE Milcom 2004*.

13. I. KEYENCE America. http://www.keyence.com/products/sensors.html.

14. A. Mainwaring, J. Polastre, R. Szewczyk, and D. Culler. Wireless sensor networks for habitat monitoring. In *First ACM International Workshop on Wireless Workshop in Wireless Sensor Networks and Applications (WSNA 2002)*, August 2002.

15. MOTES. http://www.xbow.com/products/wireless_sensor_networks.htm.

16. E. Shih, S. Cho, N. Ickes, R. Min, A. Sinha, A. Wang, and A. Chandrakasan. Physical layer driven protocol and algorithm design for energy-efficient wireless sensor networks. In *Proc. of ACM MobiCom'01*, Rome, Italy, July 2001.

17. S. Slijepcevic and M. Potkonjak. Power efficient organization of wireless sensor networks. In *ICC 2001*, Helsinki, Finland, June 2001.

18. D. Tian and N.D. Georganas. A coverage-preserving node scheduling scheme for large wireless sensor networks. In *First ACM International Workshop on Wireless Sensor Networks and Applications*, Atlanta, GA, 2002.

19. X. Wang, G. Xing, Y. Zhang, C. Lu, R. Pless, and C. Gill. Integrated coverage and connectivity configuration in wireless sensor networks. In *ACM Sensys'03*, Nov. 2003.

20. Y. Xu, J. Heidemann, and D. Estrin. Geography-informed energy conservation for ad hoc routing. In *Proc. of ACM MOBICOM'01*, Rome, Italy, July 2001.

21. F. Ye, G. Zhong, S. Lu, and L. Zhang. Energy Efficient Robust Sensing Coverage in Large Sensor Networks. Technical report, UCLA, 2002.

22. F. Ye, G. Zhong, S. Lu, and L. Zhang. PEAS: a robust energy conserving protocol for long-lived sensor networks. In *The 23nd International Conference on Distributed Computing Systems (ICDCS)*, 2003.

23. H. Zhang and J.C. Hou. Maintaining sensing coverage and connectivity in large sensor networks. In *International Workshop on Theoretical and Algorithmic Aspects of Sensor, Ad hoc Wireless and Peer-to-Peer Networks*, Feb. 2004. Also a technical report with reference number UIUCDCS-R-2003-2351 in the Department of Computer Science, University of Illinois at Urbana-Champaign.

24. R. Zheng, J.C. Hou, and L. Sha. Asynchronous wakeup for ad hoc networks. In *4th ACM International Symposium on Mobile Ad Hoc Networking and Computing (MobiHoc'03)*, 2003.

25. R. Zheng and R. Kravets. On-demand power management for ad hoc network. In *Proc. of IEEE INFOCOM*, 2003.

Appendix A: Proof of Theorem 28.3

We only prove the first part of the theorem where the location of all three disks can change. To prove the second part when node A and B are fixed we only need to take the variable x_1 below as a fixed value.

Refer to Figure 28.9. Let r_1, r_2; and r_3 denote the radii of disks A, B, and C; let $x_1 = \overline{OP}/2, x_2 = \overline{OQ}/2, x_3 = \overline{OR}/2$; and let $\alpha_1 = \angle OAP, \alpha_2 = \angle OBP, \alpha_3 = \angle OBQ, \alpha_4 = \angle OCQ, \alpha_5 = \angle OCR, \alpha_6 = \angle OAR$. Notice that if $r_1 = r_2 = r_3$, then $\alpha_1 = \alpha_2, \alpha_3 = \alpha_4, \alpha_5 = \alpha_6$. The angles $\alpha_i, 1 \le i \le 6$, can be expressed as:

$$\alpha_1 = 2\arcsin(x_1/r_1), \quad \alpha_2 = 2\arcsin(x_1/r_2),$$
$$\alpha_3 = 2\arcsin(x_2/r_2), \quad \alpha_4 = 2\arcsin(x_2/r_3),$$
$$\alpha_5 = 2\arcsin(x_3/r_3), \quad \alpha_6 = 2\arcsin(x_3/r_1), \tag{28.5}$$

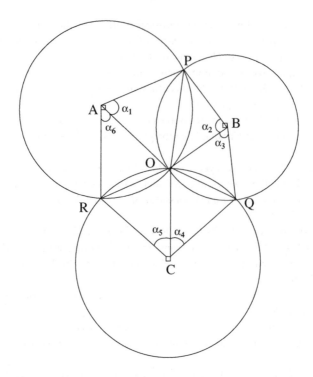

FIGURE 28.9 Minimizing the overlap while covering the crossing point O when each node has a different sensing range.

and the total overlap can be written as

$$\frac{1}{2} \sum_{i=1}^{6} r_i^2(\alpha_i - \sin \alpha_i). \tag{28.6}$$

Now the problem is to minimize Equation (28.6) subject to the same constraint as in Lemma 28.4:

$$\sum_{i=1}^{6} \alpha_i = 2\pi. \tag{28.7}$$

Now we apply Lagrangian multiplier theorem with the Lagrangian function

$$L = \frac{1}{2} \sum_{i=1}^{6} \left(r_i^2(\alpha_i - \sin \alpha_i) \right) + \lambda \left(\sum_{i=1}^{6} \alpha_i - 2\pi \right). \tag{28.8}$$

Note that the variables α_i are not independent; for example, both α_1 and α_2 depend on x_1. Hence, we have to apply the Lagrangian multiplier theorem on the independent variables x_i and regard α_i as $\alpha_i(x_j)$, where x_j is one of the x_k's that α_i depends on. First we apply the first order necessary condition on x_1.

$$\frac{\partial L}{\partial x_1} = \sum_{i=1}^{6} \frac{\partial L}{\partial \alpha_i} \cdot \frac{\partial \alpha_i}{\partial x_1}$$

$$= \left(2x_1^2 + \lambda \right) \left(\frac{1}{\sqrt{r_1^2 - x_1^2}} + \frac{1}{\sqrt{r_2^2 - x_1^2}} \right) = 0 \tag{28.9}$$

If $x^* = (x_1^*, x_2^*, x_3^*)$ and λ^* satisfy the first-order Lagrangian necessary condition, we have $2x_1^{*2} = -\lambda^*$. Applying the same necessary condition on x_2 and x_3 renders $2x_2^{*2} = 2x_3^{*2} = -\lambda^*$. Thus, $x_1^* = x_2^* = x_3^*$ satisfies the first-order necessary conditions. To show it also satisfies the second-order sufficient conditions, it suffices to verify that

$$\frac{\partial L^2(x^*, \lambda^*)}{\partial x_i \partial x_j} = 0 \text{ for } i \neq j, \tag{28.10}$$

and

$$\frac{\partial L^2(x^*, \lambda^*)}{\partial x_i^2} > 0 \text{ for all } i \tag{28.11}$$

to show the Hessian matrix of the Lagrangian is positive definite. That is, (x_1^*, x_2^*, x_3^*) is a local minimum. Because there is only one local minimum, it is also a global minimum. Hence, (x_1^*, x_2^*, x_3^*) minimizes the Equation (28.6) subject to constraint Equation (28.7), and $\overline{OP} = \overline{OQ} = \overline{OR}$ minimizes the overlap.

29

Advances in Target Tracking and Active Surveillance Using Wireless Sensor Networks

Yi Zou

Krishnendu Chakrabarty

29.1   Introduction

Wireless sensor networks that are capable of observing the environment, processing data, and making decisions based on these observations have recently attracted considerable attention.[A] These networks are important for a number of applications, such as coordinated target detection and localization, surveillance, and environmental monitoring. Breakthroughs in miniaturization, hardware design techniques, and system software, such as advances in microelectromechanical systems (MEMS), have made it possible to design a single wireless sensor node as a stand-alone system small in size and low in cost yet with full sensing and communication functionalities.[B]

A sensor node integrates hardware and software for sensing, data processing, and communication; the basic structure of a node is shown in Figure 29.1. Wireless sensor nodes are deployed in large numbers and organized in an ad hoc manner,[C] where nodes collect and exchange data about the environment and higher-level decisions can be made based on the integrated sensor data to achieve the required goal. Target tracking highlights the enormous potential of wireless sensor networks, especially in cases such as enemy intrusion detection in combat scenarios. Some practical examples of sensor networks[D] include AWAIRS at UCLA/RSC,[1] Smart Dust at UC Berkeley,[24] USC-ISI network,[16] SensIT systems/networks,[12] the ARL Advanced Sensor Program,[47] and the DARPA Emergent Surveillance Plexus (ESP).[48]

Research on centralized target tracking has been carried out for quite some time, originating from early work on target tracking by radar during World War II.[14] General centralized tracking techniques such as Kalman filtering and Bayesian methods[19] are beyond the scope of this chapter. Detailed discussions on this topic can be found.[E] The constraints of wireless sensor networks, such as limited energy due to battery-based power supply, limited storage for time-series data, scalability for network management, and distributed sensing and data processing, have posed many new challenges for target tracking. This is especially true because wireless sensor networks are made up of large numbers of cheap sensor nodes with limited capabilities.[F]

Recent research on target tracking in wireless sensor networks[G] focuses on collaborative sensing,[H] energy-efficient routing and management,[I] and sensor nodes deployment.[J] An extensive discussion on all these topics is beyond the scope of this chapter; here we limit ourselves to collaborative sensing and energy-efficient network management for target tracking in wireless sensor networks.

Collaborative sensing and signal processing provide raw sensory data from the low-level sensing unit on sensor nodes. In many cases, cheap sensors such as omnidirectional acoustic sensors[K] are used because other sensors such as CCD cameras generally require more resources for power, memory, bandwidth, and computation capability. Although the target information from a single node is generally limited, more useful information can be obtained by data exchange and aggregation between multiple nodes, based upon which higher-level strategic decisions can be made.[23] Targets in this case are the acoustical wave sources such as vehicle engine noise, which provide inputs to acoustic sensor nodes. Detailed discussion and analysis are provided by Chen et al.,[14] where source localization, direction-of-arrival (DOA) and beamforming techniques[14,27] are used to carry out target tracking.

Another important issue for target tracking in wireless sensor networks is energy management because battery-driven sensor nodes are severely energy-constrained. Considerable research has been recently

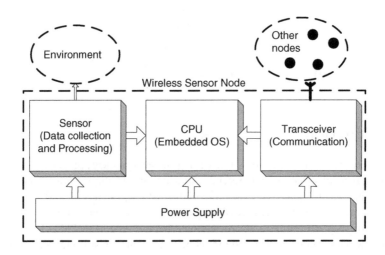

FIGURE 29.1 Sensor node architecture.

carried out in an effort to make sensor networks energy-efficient. Techniques published in the literature include the power-saving MAC-layer protocol PAMAS,[31] the cluster-based energy-efficient routing protocol LEACH,[20] PEGASIS as an improvement over LEACH,[26] SPAN[13] with off-duty and on-duty cycles for sensor nodes, and rumor routing.[8] Additional techniques include geography-information-based routing,[43] the energy-efficient topology management protocol STEM,[32] connected-dominating-set (CDS) based routing and querying,[42] the TEEN protocol,[28] dynamic power management,[35] energy-saving strategies for the link layer,[34] IDSQ for energy-efficient target tracking,[L] energy-efficient management for target localization,[46] and coverage-preserving node organization.[M] Representatives among these as typical examples to achieve energy-efficiency without sacrificing the goals of target localization and tracking are an energy-efficiency management strategy[46] and information-driven sensor query (IDSQ).[15,44] Energy saving is achieved by reducing the communication used for transmitting or receiving redundant sensory data in the target localization procedure.[46] Under considerations of collaborative sensing and signal processing, an information-driven routing protocol achieves energy efficiency without losing any useful target information.[15,44]

Note that target tracking in wireless sensor networks requires the integration of multidisciplinary technologies from many areas. The following issues provide an overview of the challenges involved in target tracking using wireless sensor networks.

- How to achieve collaborative sensing among multiple sensor nodes with sufficient tracking quality and time latency. The nodes in wireless sensor networks are physically distributed,[A] as opposed to many existing multiple sensor integration techniques where exchange of sensor data is carried out in a centralized manner. The tracking quality may have to be traded off between latency and energy concerns due to the limited resources available.[44]

- How to aggregate sensor data more efficiently in the sense of both quality-of-service (QoS) and energy savings. A good example is the directed diffusion routing protocol that minimizes communication distance between data sources and data sinks, resulting in energy efficiency compared with traditional methods. However, for applications with tighter demands (e.g., enemy intrusion), a local decision is necessary based on local sensor data to guarantee an event-reaction from the sensor network with low latency.

- Target tracking itself requires a large amount of communication, data processing, and caching of historical data, which is in conflict with the reality that only limited resources are available to wireless sensor networks. Therefore, it is necessary to manage the wireless sensor nodes to prolong the operational lifetime without undermining the goal of target tracking. Nodes can be designed to operate in different states such as the off-duty and on-duty scheduling methods.[B] The sensing coverage should be maintained at all times, for any possible target events, with sensing units being able to join the collaborative sensing for a better tracking quality whenever necessary.[46]

- Other issues range from networking to human–computer interfaces[44] include node mobility, data naming, data abstraction, query data browsing, search and visualization, network initialization and discovery, time synchronization, location services, fault tolerance, and security. Research into many of these problems for wireless sensor networks is still in its infancy. Due to the lack of systematic theoretical support, most work has focused on heuristics that are applicable to specific application scenarios.

This chapter is organized as follows. In Section 29.2 we discuss known collaborative sensing technologies for target tracking using acoustic sensors in wireless sensor networks. In Section 29.3, an energy-saving target localization strategy proposed by the authors is introduced with experimental results presented as case studies. Section 29.4 concludes this chapter by presenting future directions of research.

29.2   Collaborative Sensing in Wireless Sensor Networks Using Acoustic Sensors

Sensor data processing for target tracking is generally referred to as collaborative signal processing (CSP).[C] Acoustic sensors have been widely used in wireless sensor networks for target tracking and localization.[D] A typical application is military surveillance, where engine noises of wheeled or tracked vehicles generate acoustic waves, referred to as wideband signals.[14] The use of acoustic sensors is relatively easy and it entails less hardware cost; thus, these sensors can be deployed in large numbers. However, because the amount of information provided by a single sensor node is limited and inaccurate,[39] the potential overhead for signal processing and communication can be very high. In this section we discuss recent representative research that shows how target tracking is achieved in acoustic sensor based wireless sensor networks.

29.2.1   Background

For the simplified and ideal case, based on the traveling principle of sound waves, the distance from the sound source to the sensor can be easily measured.[25] However, in the real world, it is much more complicated because of energy loss due to the actual environment, source separation when multiple acoustic sources are close to each other, etc.[14] A detailed discussion of acoustic signal processing is beyond the scope of this chapter. In the following we focus on recent research on the use of acoustic sensors in wireless sensor networks for target tracking. Some of these tracking methods use the technique of time-difference of arrival (TDOA).[7] The general acoustic sensor model can be expressed by the following equation,[25] where the acoustic intensity attenuates at a rate that is inversely proportional to the distance between the source target and the sensor.

$$p = \frac{a}{||\vec{x} - \vec{z}||^\alpha} + p_e \qquad (29.1)$$

where p is the measurement from the sensor, a is the near-source sound energy or amplitude, α is the attenuation factor, and p_e is the noise. $\vec{x}$ is the target state variable, where in a 2D or 3D coordination system, $\vec{x}$ corresponds to target location. $\vec{z}$ is the sensor location in the same coordination system. Normally a and α are known for a certain environment, and p_e is assumed as Gaussian, that is, $p_e \sim N(0, \sigma^2)$. Note that the acoustic signals are additive, corresponding to the case of multiple targets where the received signal at the sensor side is the sum of acoustic signals from all source targets. In this case, the received acoustic signal is expressed as

$$p = \sum_{i=1}^{k} \frac{a}{||\vec{x}(i) - \vec{z}||^\alpha} + p_e \qquad (29.2)$$

where target i is located at $\vec{x}(i)$. Note that the current p_e is used to approximate the total background noise in the received sound signal but it is not necessarily Gaussian. Also note that the above two equations only represent the sensor measurement at a single time instant. For mobile targets, mobile sensor nodes, or harsh environments, they must be time-varying.

Common techniques to achieve localization and tracking include triangulation, least-square source localization, maximum likelihood source localization, beamforming, blind beamforming, time difference of arrival, and direction of arrival. A detailed discussion on these techniques can be found in Ref. 14. In the application of these techniques, concerns such as communication bandwidth, and energy conservation should be taken into consideration.

29.2.2   Target Localization Using Triangulation

We first discuss the triangulation technique because it is simple and easy to implement. Triangulation is carried out based on the time delays of acoustic signal measured from multiple sensor nodes. Assume the acoustic source of the target is generated at time t and detected by three sensor nodes at t_1, t_2, and t_3, respectively. Denote the locations of these three sensor nodes as $\vec{z}_1, \vec{z}_2, \vec{z}_3$, and target location at time t as $\vec{x}(t)$; then the triangulation is shown by the following equations:[39]

$$||\vec{z}_1 - \vec{x}(t)||^{\frac{1}{2}} == (t - t_1) \cdot v \tag{29.3}$$

$$||\vec{z}_2 - \vec{x}(t)||^{\frac{1}{2}} == (t - t_2) \cdot v \tag{29.4}$$

$$||\vec{z}_3 - \vec{x}(t)||^{\frac{1}{2}} == (t - t_3) \cdot v \tag{29.5}$$

where v is the propagation speed of acoustic waves. Obviously $\vec{x}(t)$ can be obtained from the above equations. Note that for the pathological case where nodes are equally far away from a static target (i.e., we have correlation in the above equations), the triangulation approach will fail. Second, a minimum node density is required such that all cluster heads must have at least three (2D) or four (3D) neighbor nodes whose relative locations to the current target are uncorrelated with each other. Another issue is that errors in sensor nodes locations will have a direct impact on the results of target location estimation. Also, according to Wang et al.,[39] other problems such as the coarse-grained acoustic sampling data of low-performance sensors and non-trivial errors generate in cross-correlation have made it too impractical for actual implementation. However, triangulation is easy to implement and it does not require much computation. Techniques based on triangulation can still achieve good results for target location estimation.[33]

29.2.3   TDOA-Based Energy-Efficient Collaborative Sensing

Energy-efficient collaborative sensing strategy using the principle of difference of arrival for acoustic sensors has been proposed by Boettcher.[7] Accordingly, energy consumption in data processing per instruction is considerably low compared with the energy consumption for communication per bit. The technique of acoustic time-difference of arrival (TDOA) requires small communication bandwidth, which motivates the authors to limit distributed processing of sensor data as much as possible unless absolutely necessary. Only essential information is extracted from sensor nodes and sent out to the sensor network, which results in communication reduction and consequently a smaller communication latency.

Instead of transmitting time-series data from one node to another, the relative time-difference of arrival is estimated using the dominant frequency of the acoustic spectrum as a feature, which is then used for location estimation calculation.[7] The estimation procedure is initiated by an initiating node that shares its frequency estimate with its neighbor nodes. Neighbor nodes calculate the delay estimates and send them back to the initiating node, which subsequently calculates the estimate of bearing to target from the received estimated time delays.[7] Because all nodes are assumed to have GPS installed, target location is then available from the bearing estimates and with time-series estimation fed to sink nodes, target tracking is achieved. Based on the results of Boettcher et al.,[7] the estimation error of bearing is within $\pm 15°$ in most cases.

The above approach gives a good example of distributing collaborative sensing using acoustic sensors for target tracking without engaging a large amount of sensor data processing and communication. However, when the assumption of GPS being available to each sensor node does not hold, which is quite common in wireless sensor networks, accurate location discovery algorithms must be used to reduce the errors in nodes locations; otherwise, performance degradation can be expected. To compensate for this problem, the potential overhead of computation and communication may have to be tolerated.

29.2.4   Energy-Saving Target Tracking Architecture

Wang et al.[39] proposed an acoustics-based wireless sensor network architecture for target tracking using UC Berkeley mica motes with energy efficiency concerns. The architecture contains two subsystems, namely the acoustic target tracking subsystem and the communication subsystem. The target location is obtained by a maximum likelihood estimator.[39] Assuming that a total of n sensor nodes are participating in the collaborative sensing, the target location is obtained by

$$\vec{x} = \arg \min_{\vec{z}_i} d(\vec{t}(\hat{\vec{x}}), \vec{t}) \tag{29.6}$$

where $\vec{x}$ is the target location vector, $\hat{\vec{x}}$ is the actual target location, $\vec{z}_i$ is the location vector for node i, $\vec{t}$ is the measured acoustic signal propagation delays from hypothetical acoustic source (target) to the n nodes, that is, $\vec{t} = \{t_1, \cdots, t_n\}$, $\vec{t}(\hat{\vec{x}})$ is the theoretical or real acoustic signal propagation delays from hypothetical acoustic source to the n nodes (i.e., $\vec{t}_r = \{\tilde{t}_1, \cdots, \tilde{t}_n\}$), and d is an algorithm specific difference measurement function.[39] (Note that we have rewritten the original equation to make the notations consistent.) $\vec{t}(\hat{\vec{x}})$ is obtained by exhaustive calculation over the sensing area of all nodes; however, this implies that an infinite number of points must be considered. Therefore, the surrounding field is discretised by dividing it into an N-by-N-grid,[39] where computation limitation is traded off with estimation accuracy. This is similar to the detection probability used in the energy-saving target localization approach introduced in the next section.[46]

The acoustic target tracking subsystem contains a cluster of sensor nodes, with cluster head as the central unit for acoustic sensor data processing and member nodes as acoustic sensor data collection points. The cluster-based design differentiates the role of sensing and computation where the cluster head is assumed to be equipped with more computation capabilities than its member nodes. Wang et al.[39] used a PC instead of a mica mote as the cluster head. In the communication subsystem, the authors make use of the redundancy in target tracking data from multiple sensors nodes to improve the throughput and reduce the latency, resulting in energy savings. Different priorities are assigned based on the quality of tracking reports where high-quality tracking data is delivered first with low-quality tracking data suppressed or dropped. This is referred to as quality-driven redundancy suppression.[39]

This approach bears some similarity to the approach proposed by Boettcher et al.[7] because both depend on the inherent redundancy of sensor data to achieve energy conservation, that is, to avoid transmitting sensor data that is not going to help significantly in target tracking. The initiating node in Ref. 7 acts as a cluster head node in Ref. 39 except that the sensor network needs no clusterization initially and the cluster head node in Ref. 39 does not really participate in any sensing tasks. Another similar assumption is that sensor nodes location information is assumed to be accurate enough that no significant errors are caused in target tracking.

29.2.5   Collaborative Signal and Information Processing by Information-Driven Sensor Query

We next review the information-driven query approach (IDSQ) proposed by Zhao[A] which contains detailed discussions on both single and multiple target tracking scenarios in wireless sensor networks. Based on the canonical tracking optimization problem[B] the authors first formulate the target tracking problem T_r in wireless sensor network as[44]

$$T_r = <S_n, T_g, S_m, Q, O, C> \tag{29.7}$$

where S_n is the deployed wireless sensor nodes set with node state information (e.g., nodes locations, speed, active or sleeping, etc.). T_g is the state variable that describes the target of interest. S_m is a parameter defining the sensing signal propagation and attenuation. Q defines the query initiated in the ad hoc sensor network requesting the tracking data or other information. O is the objective function that essentially defines the quality of service (i.e., tracking quality in the case of target tracking), and finally, C is the constraints set such as energy stringency, latency, bandwidth, etc.[15,44]

The IDSQ approach is based on centralized sequential Bayesian filtering.[C] Denote sensor measurements up to time t as $\bar{\mathbf{z}}^{(t)} = \{\vec{z}^{(0)}, \cdots, \vec{z}^{(t)}\}$. The posterior belief at time $t+1$ for target state $\vec{x}^{(t+1)}$ can be determined based on the new measurement $\vec{z}^{(t+1)}$ and the existing belief $p(\vec{x}^{(t)}|\bar{\mathbf{z}}^{(t)})$ from time t.

$$p(\vec{x}^{(t+1)}|\bar{\mathbf{z}}^{(t+1)}) \propto p(\vec{z}^{(t+1)}|\vec{x}^{(t+1)}) \cdot$$
$$\int p(\vec{x}^{(t+1)}|\vec{x}^{(t)}) \cdot p(\vec{x}^{(t)}|\bar{\mathbf{z}}^{(t)}) d\vec{x}^{(t)} \tag{29.8}$$

where $p(\vec{z}^{(t+1)}|\vec{x}^{(t+1)})$ is the observation model and $p(\vec{x}^{(t+1)}|\vec{x}^{(t)})$ is the target state dynamics model.

When the new sensor measurement is to be integrated based on the above equation, then among available sensor detection reports, which of them should be selected and which should be selected first? The IDSQ approach is proposed to solve such problems by formulating the tracking problem as a general distributed constrained optimization problem that maximizes information gain of sensors while minimizing communication and resource usage. This achieves an optimal selection of a subset of sensor nodes that can provide the most useful information for the query of target tracking data.[15,44] The evaluation of the information from sensor nodes is achieved by balancing the information usefulness with communication cost using weighting factors in the objective function O. Suppose U is the set of sensor nodes whose measurements have already been incorporated that produces the current belief as $p(\vec{x}|\{\vec{z}_i\}_{i \in U})$. For the sensor measurement $\vec{z}_j^{(t)}$ from a node j in the set of remaining nodes, denoted as A, the objective function is expressed as[15]

$$O\left(p\left(\vec{x}|\vec{z}_j^{(t)}\right)\right) = \alpha\phi\left(p\left(\vec{x}|\vec{z}_{j-1}^{(t)}, \vec{z}_j^{(t)}\right)\right) - (1-\alpha)\psi\left(\vec{z}_j^{(t)}\right) \qquad (29.9)$$

ϕ measures the information utility of incorporating the measurement $z_j^{(t)}$, ψ is the cost of communication and other resources, and α is the relative weighting of the utility and cost. The optimal selection criterion that selects the best sensor given the current state $p(\vec{x}|\{\vec{z}_i\}_{i \in U})$ is then given as[44]

$$\hat{j} = \arg\max_{j \in A} O(p(\vec{x}|\{\vec{z}_i\}_{i \in U} \cup \vec{z}_j)) \qquad (29.10)$$

The above objective function and selection criterion can be varied to suit different application scenarios where different aspects in the operation of sensor network have different requirements.

29.2.6 Summary

Based on the above discussion, we first note that there is an inherent redundancy in the sensor data when a target is detected in the wireless sensor network. Sensor nodes are usually deployed in large numbers (i.e., redundantly). The existing redundancy in the sensor data can be used to achieve energy savings without losing the target tracking quality. However, this requires reliable evaluation metrics on the sensor data to decide whether the data would contribute to collaborative sensing and how much contribution is expected. Otherwise, it is possible that useful information may be lost, which may lead to undesirable situations such as false alarms or missed targets. This is sometimes possible with technologies such as GPS, and techniques such as location discovery and distributed calibration. In the absence of such features or techniques, alternatives such as blind beamforming[14] and GPS-less localization[11] can also be used. Nevertheless, the robustness of location-dependent methods can be improved greatly when assumptions of location information are not necessary.

29.3 Energy-Efficient Target Localization

In this section, we introduce an energy-saving target localization strategy for wireless sensor networks.[46] In general, a sensor network has an almost constant rate of energy consumption if no target activities are detected in the sensor field.[6] The minimization of energy consumption for an active sensor network with target activities is more complicated because target detection involves collaborative sensing and communication involving different nodes. The transmission of detailed target information consumes a significant amount of energy due to the large volume of raw data, as supported by the sensor data redundancy reduction approaches proposed by Boettcher et al.[7] and Wang et al.[39] Contention for the limited bandwidth among the shared wireless communication channels causes additional delay in relaying detailed target information to the cluster head. We attempt to prolong the sensor network lifetime by an *a posteriori* energy-aware target localization strategy, as discussed in the following.

29.3.1   Preliminaries

The sensor response is expressed as a probability-based value corresponding to different signal-to-noise ratios (SNRs), where the closer the target is to the sensor node, a higher SNR is expected. The sensor field is represented by an n-by-m 2D grid where a total of k sensors nodes, denoted by a set S, with a detection range r are deployed. Let $d_{xy}(s_k)$ be the distance from grid point (x, y) to node s_k. Denote the coverage probability of a target at (x, y) being detected by a node s_k at (x_k, y_k) as $c_{xy}(s_k)$. Hence, we use the following exponential function to represent the confidence level in the received sensing signal, as a simplified model corresponding to the omnidirectional acoustic sensors.

$$c_{xy}(s_k) = \begin{cases} e^{-\alpha d_{xy}(s_k)}, & \text{if } d_{ij}(s_k) \leq r_s \\ 0, & \text{otherwise} \end{cases} \tag{29.11}$$

where α is a parameter representing the physical characteristics of the sensing unit. For evaluating energy consumption, assume three basic energy consumption types, namely, sensing, transmitting, and receiving, exist on sensor nodes. Denote them as ψ_s, ψ_t, and ψ_r, respectively (in units of j/s). Assume at time instant t, there are $k(t)$ sensors that have detected the target, where $k(t) \leq k$. Therefore, the energy for sensing activities in the wireless network, denoted as $E_s(k(t))$, is

$$E_s(k(t)) = k(t)\psi_s T_s \tag{29.12}$$

where T_s is the time duration that a sensor node is involved in sensing. For a fixed time interval, E_s is a constant if all sensor nodes are assumed homogenous. The energy used for communication between nodes and the cluster head can be categorized into two types, E_b and E_c. The parameter E_c is the energy consumed by a sensor node for communication with the cluster head. This includes the energy for transmitting data and the energy for receiving data. The parameter E_b is the energy needed for broadcasting data from the head to the nodes. Both E_b and E_c are functions of T and $k(t)$, where T is the time required for either retrieving data from a sensor node or the broadcasting of data from the cluster head, and $k(t)$ is the number of sensors involved in this communication at time instant t. We define E_c and E_b as follows:

$$E_c(k(t), T) = (\psi_t T + \psi_r T)k(t) \tag{29.13}$$

$$E_b(k(t), T) = \psi_t T + \psi_r T k(t) \tag{29.14}$$

The parameter T is directly proportional to the volume of data involved in the communication. In this work, T can be one of three values: T_d for raw target data, T_e for target event reporting, or T_q for query request. They satisfy the relationship $T_e \leq T_q \ll T_d$ because raw data collected by a sensor node can be up to hundreds of bytes in size. We assume that target detection and localization are discrete processes that are derived from a discrete sampling of target activities in the sensor network. Also, because the sensor network is designed to track target activities, T_s, T_e, T_q, and T_d are assumed to be less than the granularity of the time t. Thus, for the case that a target is moving in the sensor field during the time interval $[t_{start}, t_{end}]$, the corresponding instantaneous energy consumption $E(t)$ and total energy consumption E in the wireless sensor network can be expressed as

$$E(t) = E_s(k(t)) + E_c(k(t), T_d) \tag{29.15}$$

$$E = \sum_{t=t_{start}}^{t_{end}} E(t) \tag{29.16}$$

29.3.2   Target Localization Procedure

The proposed energy-saving target localization approach contains a two-step communication protocol between the cluster head and the sensors reporting the target detection events. When a node detects a target, it sends an event notification to the cluster head. To conserve power and bandwidth, the message from the sensor to the cluster head is kept very small (e.g., a single bit is sufficient). Detailed target information is first stored locally on the node and is available upon subsequent queries. Based on the information received from the sensors within the cluster, the cluster head executes a probabilistic localization algorithm to determine candidate target locations, and it then queries the sensor(s) in the vicinity of the target.

The *a posteriori* side of this method is based on the detection probability table created by the cluster head, which contains entries for all possible detection reports from those sensors that can detect a target at all grid points. Assume (x, y) is covered by a set of k_{xy} sensors, denoted as $S_{xy} \subseteq \S$. The probability table is built on the power set of S_{xy}, including the event that none of the sensors detect anything (represented by the binary string as "00...0") as well as the event that all of the sensors (represented by the binary string as "11...1"). Thus, the probability table for grid point (x, y) then contains $2^{k_{xy}}$ entries, defined as

$$p\_table_{xy}(i) = \prod_{s_j \in S_{xy}} p_{xy}(s_j, i) \qquad (29.17)$$

where $0 \leq i \leq 2^{k_{xy}}$, and $p_{xy}(s_j, i) = c_{xy}(s_j)$ if s_j detects a target at (x, y); otherwise $p_{xy}(s_j, i) = 1 - c_{xy}(s_j)$.

Suppose $S_{rep}(t)$ is the set of sensors that have reported the detection of an object at time t, $S_{rep,xy}(t)$ is the set of sensors that can detect a target at point (x, y) and have also reported the detection of an object at time t, and the set of the sensors selected by the cluster head for querying at time t is denoted as $S_q(t)$. Their cardinalities are given as $k_{rep}(t)$, $k_{rep,xy}(t)$, and $k_q(t)$, respectively. Obviously, $S_{rep,xy}(t) = S_{rep}(t) \cap S_{xy}$ and $S_q(t) \subseteq S_{rep}(t)$. We use an inference method based on the established probability table. To save both communication energy and bandwidth, at any time instant t during the target localization process, the cluster head uses the probability table to determine $k_q(t)$ most suitable sensors out of the reported $k_{rep}(t)$ sensors to be queried for more detailed information. The score of the grid point (x, y) at time instant t is calculated as follows:

$$SCORE_{xy}(t) = p\_table_{xy}(i(t)) \times w_{xy}(t) \qquad (29.18)$$

where $i(t)$ is the index of the p_table_{xy} at time t. The parameter $i(t)$ is calculated from S_{xy} and $S_{rep,xy}$. The parameter $p\_table_{xy}(i(t))$ corresponds to the conditional probability that the cluster head receives this event information given that there was a target at (x, y). The weight $w_{xy}(t)$ reflects the confidence level in this reporting event for this particular grid point and it refines the grid point scores to narrow down grid points that are most probably close to the current target location. $w_{xy}(t)$ is defined as

$$w_{xy}(t) = \begin{cases} 0 & \text{if } S_{rep,xy}(t) = \{\phi\} \\ 4^{-\Delta k_{rep,xy}(t)} & \text{otherwise} \end{cases} \qquad (29.19)$$

where $\Delta k_{rep,xy}(t)$ measures the degree of difference in the set of sensors that reported and those sensors that can detect point $P(x, y)$ at time instant t. $\Delta k_{rep,xy}(t)$ is given as:

$$\Delta k_{rep,xy}(t) = |k_{rep}(t) - k_{rep,xy}(t)| \\ + |k_{rep}(t) - k_{xy}| \qquad (29.20)$$

In Equation (29.20), the first term $|k_{rep}(t) - k_{rep,xy}(t)|$ represents the absolute difference between the number of nodes that have reported (i.e., $k_{rep}(t)$) and the number of nodes that have reported and can also detect a target at point (x, y) (i.e., $k_{rep,xy}$). The second term $|k_{rep}(t) - k_{xy}|$ represents the absolute difference between the number of nodes that have reported (i.e., $k_{rep}(t)$) and the number of nodes that can detect the grid point (x, y) from the pre-calculated detection probability table (i.e., k_{xy}). The parameter w_{xy} is a decaying factor that is 1 only if $S_{rep}(t) = S_{xy}$. The number 4 in Equation (29.19) was chosen empirically through simulations. We are using $w_{xy}(t)$ to filter out grid points that are not likely to be close to the actual target location. The score is based on both the probability value from the probability table and the current relationship between $S_{rep}(t)$, $S_{rep,xy}(t)$, and S_{xy}.

Assume that the maximum number of sensors allowed to report an event is k_{max}. To select the sensor to query based on the event reports and the localization procedure, we first note that for time instant t, if $k_{max} \geq k_{rep}(t)$, then all reported sensors can be queried. Otherwise, we select sensors based on a score-based ranking. The sensors selected correspond to the ones that have the shortest distance to those grid points with the highest scores. This selection rule is defined as

$$S_q(t) : d(S_q(t), P_{MS}) = min\{d(s_i, P_{MS})\} \tag{29.21}$$

where $S_i \in S_{rep}(t)$, and P_{MS} denotes the set of grid points with the highest scores. Note that it is possible that there are multiple grid points that have the maximum score. When this happens, we calculate the score concentration by averaging the scores of the current grid point and its eight neighboring grid points. The grid point with the highest score (or the score concentration) is the most likely current target location. Therefore, selecting sensors that are closest to this point guarantees that the selected sensors can provide the most detailed and accurate data in response to the subsequent queries. Note that target identification is not possible as at this stage because the cluster head has no additional information other than $S_{rep}(t)$. However, the selected sensors provide enough information in the subsequent stage to facilitate target identification. We evaluate the accuracy of this target localization procedure by calculating the distance between the grid point with the highest score and the actual target location. From Equation (29.12) to Equation (29.16), we evaluate the energy consumption using the above target localization method as follows.

$$E^*(t) = E_s(k_{rep}(t)) + E_c(k_{rep}(t), T_e)$$
$$+ E_b(k_q(t), T_q) + E_c(k_q(t), T_d) \tag{29.22}$$

$$E^* = \sum_{t=t_{start}}^{t_{end}} E^*(t) \tag{29.23}$$

Therefore, let $k(t) = k_{rep}(t)$ in Equations (29.2) and (29.3); then the difference in energy consumption, $\Delta E = E - E^*$, can be expressed as:

$$\Delta E(t) = (k_{rep}(t) - k_q(t))(\psi_t + \psi_r)T_d$$
$$- (k_q(t)\psi_r + \psi_t)T_q$$
$$- k_{rep}(t)(\psi_t + \psi_r)T_e \tag{29.24}$$

$$\Delta E = \sum_{t=t_{start}}^{t_{end}} \Delta E(t) \tag{29.25}$$

The last two terms in Equation (29.24) indicate the overhead for the proposed target localization procedure. Because $T_d \gg T_e$ and $T_d \gg T_q$, the overhead is small. As $k_q < k_{max}$, with k_{max} properly selected, from Equation (29.24) and Equation (29.25), energy consumption is greatly reduced with the passage of time.

29.3.3 Case Studies

We present results for a case study carried out using MatLab. The simulation is done on a 30-by-30 sensor field grid with 20 sensors randomly placed in the sensor field, with $r = 5$. We choose energy consumption model parameters as $\psi_r \approx 400$ nJ/sec, $\psi_t \approx 400$ nJ/sec, and $\psi_s \approx 1000$ nJ/sec, assuming the sensing rate of 8 bits/sec.[6,21] We have no physical data available for T_d and T_e; however, their values do not affect the target localization procedure, and therefore we only need to set them manually to satisfy the relationship $T_d \gg T_e$ and $T_d \gg T_q$. In this case, $T_d = 100$ ms, $T_e = 2$ ms, and $T_q = 4$ ms.

The layout of the sensor field is given in Figure 29.2, with a target trace randomly generated in the sensor field. The target travels from the position marked as "Start" to the position marked as "End." We assume the target locations are updated at discrete time instants in units of seconds, and the granularity of time is long enough for sampling by two neighboring locations in the target trace with negligible errors. We have evaluated the algorithm for $k_{max} = 1$, $k_{max} = 2$, and $k_{max} = 3$. Figure 29.3 presents the instantaneous and cumulative energy saving results respectively, as the target moves along its trace in the sensor field. The energy savings are compiled relative to the base case when all sensors report complete target information in one step everywhere.

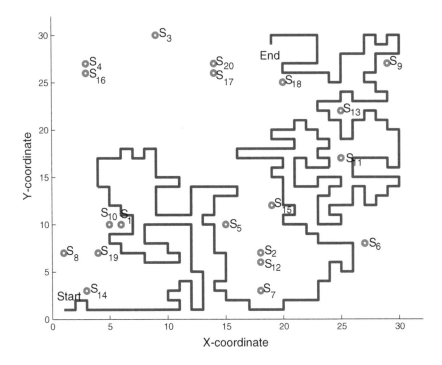

FIGURE 29.2   Sensor field layout with target trace.

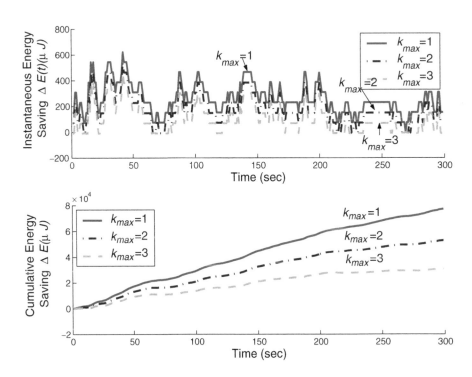

FIGURE 29.3   Instantaneous and cumulative energy saved during target localization relative to the "always report" one-step base case.

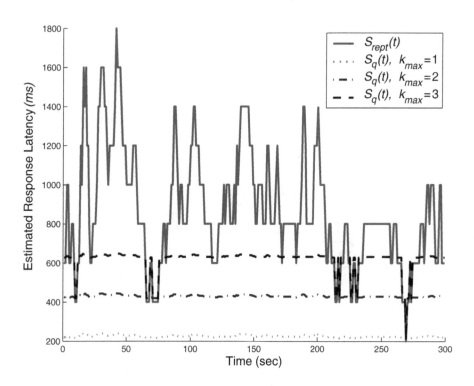

FIGURE 29.4   Latency in the localization of a target by the cluster head.

From Figure 29.3 we note that a large amount of energy is saved during target localization. Note that when k_{max} approaches $k_{rep}(t)$, the savings is less apparent due to the additional communication overhead of the two-stage query protocol. Nevertheless, there is still a considerable amount of energy saved in target localization, even in the case that $k_{max} = 3$. With an appropriate selection of k_{max}, the proposed algorithm performs exceptionally well.

Next we consider the latency in the localization of a target by the cluster head. By latency, we refer to the time that it takes for the cluster head to collect the detailed target information from sensor nodes from the time sensor nodes detect an event, assuming that the wireless sensor network uses the Time Division Multiple Access (TDMA) protocol.[30] The results are shown in Figure 29.4. The latency is reduced here compared to the base case using a "report once" strategy, because a large amount of communication for transmitting raw data has been reduced to a smaller amount of data sent by a selected set of sensors. This is an added advantage to the proposed energy-aware target localization procedure.

Because the selection of sensors for querying is based on both the detection probability table and the distance of sensors from the estimated high-score points, the proposed *a posteriori* approach offers another important advantage. It provides a substantial amount of built-in false-alarm filtering. Figure 29.5 illustrates the false-alarm filtering ability of the proposed approach. We manually generated false alarms reported by some malfunctioning sensors, which are during $t \in [18, 22]$ by s_4, during $t \in [138, 142]$ by s_{16}, and during $t \in [239, 241]$ by s_8. We calculate the distance d of the target from the sensor in $S_{rep}(t)$ that is farthest from it, as well as the distance d^* of the target from the sensor in $S_q(t)$ that is farthest from it. The difference $d - d^*$ is used as a measure of the built-in filtering ability. Figure 29.5 shows the variation of $d - d^*$ with time. Note the fact that prior to querying, the cluster head only knows which sensors have reported the detection of a target, and there is no information available to the cluster head about any detailed information of the target. We find that the proposed approach successfully narrows down the sensors that are the closest to the real target location, and selects them for detailed information querying. As shown in Figure 29.5, the three spikes present the fact that the false alarms from the sensor (which in this case is the

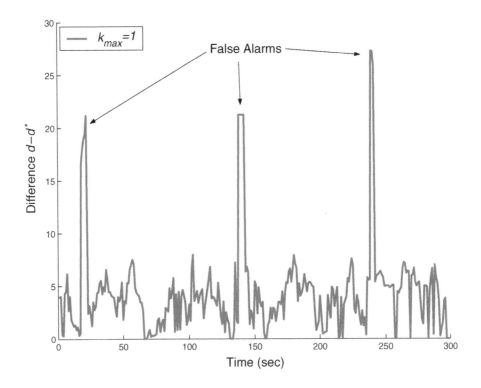

FIGURE 29.5   Results on localization error in the presence of false alarms.

furthest sensor from the actual target location) have been filtered out as the proposed target localization procedure is still able to select the most appropriate sensors to query for detailed target information.

We have described an energy-saving target localization procedure for wireless sensor networks as a two-step communication protocol and a probabilistic localization algorithm. We have shown that this approach reduces energy consumption, decreases the latency for target localization, and provides a mechanism for filtering false alarms. This approach also uses the inherent redundancy in sensor data to achieve energy savings. However, other than the approaches by Boettcher et al.[7] and Wang et al.,[39] data processing and integration are done only for certain nodes selected by the cluster head.

29.4   Conclusion

In this chapter, we discussed the problem of target tracking in wireless sensor networks, especially using acoustic sensors. Existing techniques for centralized target tracking cannot be directly used for wireless sensor networks because the sensor nodes are physically distributed and are resource-constrained. To achieve high-quality target tracking without excessive use of network resources, special care must be taken in the design of protocols and tracking algorithms Such techniques should be robust enough to handle unreliable or inaccurate sensor data from sensing units, the redundancy of sensor data due to a large number of nodes deployed, wireless communication on limited bandwidth, application-centered routing for sensed and aggregated data, the efficient management of collaborative sensing, etc. These issues provide a rich set of problems for future research in this area.

Acknowledgments

This research was supported in part by ONR under grant no. N66001-00-1-8946. It was also sponsored in part by DARPA, and administered by the Army Research Office under Emergent Surveillance Plexus MURI Award No. DAAD19-01-1-0504.

References

1. J.R. Agre, L.P. Clare, G.J. Pottie, and N.P. Romanov, Development platform for self-organizing wireless sensor networks, *Proc. SPIE,* Vol. 3713, pp. 257–268, Mar. 1999.
2. I.F. Akyildiz, W. Su, Y. Sankarasubramaniam, and E. Cayirci, A survey on sensor networks, *IEEE Communications Magazine,* pp. 102–114, August 2002.
3. Y. Bar-Shalom and X.R. Li, *Multitarget-Multisensor Tracking: Principles and Techniques,* Yaakov Bar-Shalom Publication, 1995.
4. Y. Bar-Shalom, *Multitarget/Multisensor Tracking: Applications and Advances – Volume III,* Artech House Publishers, 2000.
5. S. Blackman and R. Popoli, *Design and Analysis of Modern Tracking Systems,* Artech House Publishers, 1999.
6. M. Bhardwaj, T. Garnett, and A.P. Chandrakasan, Bounding the lifetime of sensor networks via optimal role assignments, *Proc. IEEE Infocom,* Vol. 3, pp. 1380–1387, 2001.
7. P.W. Boettcher, J.A. Sherman, and G.A. Shaw, Energy-constrained collaborative processing for target detection, tracking, and geolocation, *Proc. Information Processing in Sensor Networks,* pp. 254–268, 2003.
8. D. Braginsky and D. Estrin, Rumor routing for sensor networks, *Proc. ACM Int. Workshop on Wireless Sensor Networks and Applications,* pp. 22–31, 2002.
9. R.R. Brooks and S.S. Iyengar, *Multi Sensor Fusion: Fundamentals and Applications with Software,* Prentice Hall Publication Co., New Jersey, 1997.
10. R.R. Brooks, C. Griffin, and D. Friedlander, Distributed target classification and tracking in sensor networks, *Proceedings of IEEE,* pp. 1163–1171, 2003.
11. N. Bulusu, J. Heidemann and D. Estrin, GPS-less low-cost outdoor localization for very small devices, *IEEE Personal Communication Magazine,* Vol. 7, No. 5, pp. 28–34, 2000.
12. A.P. Chandrakasan et al., Design consideration for distributed microsensor system, *Proc. IEEE Custom Integrated Circuits,* pp. 279–286, 1999.
13. B. Chen, K. Jamieson, H. Balakrishnan, and R. Morris, SPAN: an energy-effecient coordination algorithm for topology maintenance in ad hoc wireless networks, *Proc. ACM/IEEE MobiCom,* pp. 85–96, 2001.
14. J.C. Chen, K. Yao, and R.E. Hudson, Source localization and beamforming, *IEEE Signal Processing Magazine,* Vol. 19, pp. 30–39, 2002.
15. M. Chu, H. Haussecker, and F. Zhao, Scalable information-driven sensor querying and routing for ad hoc heterogeneous sensor networks, *Int. J. of High-Performance Computing Applications,* 16(3):90–110, 2002.
16. D. Estrin, R. Govindan, J. Heidemann, and S. Kumar, Next century challenges: scalable coordination in sensor networks, *Proc. IEEE/ACM MobiCom,* pp. 263–270, 1999.
17. D. Estrin, L. Girod, G. Pottie, and M. Srivastava, Instrumenting the world with wireless sensor networks, *Proc. International Conference on Acoustics, Speech, and Signal Processing,* Vol. 4, pp. 2033–2036, 2001.
18. H. Gupta, S.R. Das, and Q. Gu, Connected sensor cover: self-organziation of sensor networks for efficient query execution, *Proc. IEEE/ACM MobiHoc,* pp. 189–200, 2003.
19. D.L. Hall and J. Llinas, *Handbook of Multisensor Data Fusion,* CRC Press, 2001.
20. W.R. Heizelman, A. Chandrakasan, and H. Balakrishnan, Energy efficient communication protocol for wireless micro sensor networks, *Proc. IEEE Int. Conf. System Sciences,* pp. 1–10, 2000.
21. W.B. Heinzelman, A. Chandrakasan, and H. Balakrishnan, An application-specific protocol architecture for wireless microsensor networks, *IEEE Transactions on Wireless Communications,* Vol. 1, pp. 660–670, 2002.
22. C. Intanagonwiwat, D. Estrin, and R. Govindan, Directed diffusion: a scalable and robust communication paradigm for sensor networks, *Proc. IEEE/ACM MobiCom,* pp. 56–67, 2000.
23. S.S. Iyengar and R.R. Brooks, *Handbook on Distributed Sensor Networks,* Chapman and Hall/CRC Press, July 2003.

24. J.M. Kahn, R.H. Katx, and K.S.J. Pister, Next century challenges: mobile networking for smart dust, *Proc. Mobicom,* 1999, pp. 483–492.

25. L.E. Kinsler et al., *Fundamentals of Acoustics,* John Wily & Sons Inc., New York, 1982.

26. S. Lindsey and C.S. Raghavendra, PEGASIS: power-efficient gathering in sensor information systems, *Proc. IEEE Aerospace Conference,* Vol. 3, pp. 1125–1130, 2002.

27. J. Liu, J. Reich, and F. Zhao, Collaborative in-network processing for target tracking, *EURASIP, J. Applied Signal Processing,* Vol. 23, No. 4, pp. 378–391, 2003.

28. A. Manjeshwar and D.P. Agrawal, TEEN: a routing protocol for enhanced efficiency in wireless sensor networks, *Proc. 15th International Parallel and Distributed Processing Symposium,* pp. 2009–2015, 2001.

29. G.J. Pottie and W.J. Kaiser, Wireless integrated network sensors, *Communications of the ACM,* Vol. 43(5), pp. 51–58, 2000.

30. T. Rappaport, *Wireless Communications: Principles & Practice,* Prentice-Hall Inc., New Jersey, 1996.

31. C.S. Raghavendra and S. Singh, PAMAS: power-aware multi-access protocol with signaling for ad hoc networks, *ACM Communications Review,* pp. 5–26, 1998.

32. C. Schurgers, V. Tsiatsis, and M.B. Srivastava, STEM: topology management for energy-efficient sensor networks, *Proc. IEEE Aero Conference,* pp. 135–145, 2002.

33. X.H. Sheng and Y.H. Hu, Energy based acoustic source localization, *Proc. Information Processing in Sensor Networks,* pp. 285–300, 2003.

34. E. Shih, B.H. Calhoun, H.C. Seong, and A.P. Chandrakasan, An energy-efficient link layer for wireless micro Sensor networks, *Proc. IEEE Computer Society Workshop on VLSI,* pp. 16–21, 2001.

35. A. Sinha and A. Chandrakasan, Dynamic power management in wireless sensor networks, *IEEE Design and Test of Computers,* Vol. 18, pp. 62–74, 2001.

36. D. Tian and N.D. Georganas, A node scheduling scheme for energy conservation in large wireless sensor networks, *Wireless Comm. Mob. Comput.,* Vol. 3, pp. 271–290, 2003.

37. S. Tilak, N.B. Abu-Ghazaleh, and W.B. Heinzelman, A taxonomy of wireless micro-sensor network models, *ACM Mobile Computing and Communications Review,* Vol. 6, No. 2, 2002.

38. M. Tubaishat and S. Madria, Sensor networks: an overview, *IEEE Potentials,* Vol. 22, pp. 20–23, 2003.

39. Q. Wang, W.P. Chen, R. Zheng, K. Lee, and L. Sha, Acoustic target tracking using tiny wireless sensor devices, *Proc. Information Processing in Sensor Networks,* pp. 642–657, 2003.

40. G. Wang, G. Cao, and T. La Porta, Movement-assisted sensor deployment, *Proc. IEEE InfoCom,* 2004.

41. X.R. Wang, G.L. Xing, Y.F. Zhang, C.Y. Lu, R. Pless, and C. Gill, Integrated coverage and connectivity configuration in wireless sensor networks, Proc. *Proc. ACM SenSys,* pp. 28–39, 2003.

42. J. Wu, Extended dominating-set-based routing in ad hoc wireless networks with unidirectional links, *IEEE Trans. Parallel and Distributed Computing,* Vol. 22(1–4), pp. 327–340, 2002.

43. Y. Xu, J. Heidemann, and D. Estrin, Geography-informed energy conservation for ad hoc routing, *Proc. IEEE/ACM MobiCom,* pp. 70–84, 2001.

44. F. Zhao, J. Liu, J. Liu, L. Guibas, and J. Reich, Collaborative signal and information processing: an information directed approach, *Proceedings of IEEE,* pp. 1199–1209, 2003.

45. Y. Zou and K. Chakrabarty, Sensor deployment and target localization based on virtual forces, *Proc. IEEE InfoCom,* pp. 1293–1303, 2003.

46. Y. Zou and K. Chakrabarty, Energy-aware target localization in wireless sensor networks, *Proc. IEEE International Conference on Pervasive Computing and Communications,* pp. 60–67, 2003.

47. ARL Federated Laboratory Advanced Sensor Programm, *http://www.arl.army.mil/alliances.*

48. Emergent Surveillance Plexus (ESP): A Multidisciplinary University Research Initiative (MURI), http://strange.arl.psu.edu/ESP.

30

Energy-Efficient Detection Algorithms for Wireless Sensor Networks

Caimu Tang

Cauligi S. Raghavendra

30.1 Introduction

Technological advances have led to the development of small, low-cost, and low-power devices that integrate microsensing with on-board processing and wireless communications. A node can typically have multiple sensors for collecting data, such as temperature, pressure, acoustic signals, video, etc. A large number of sensor nodes can be deployed or dropped in an area, and these nodes can self-organize to form a sensor network to perform useful computations on the sensed data in their environment. Sensor networks can be deployed in a variety of scenarios to collect information from the field, for example, target detection and tracking, military surveillance, building security, habitat monitoring, and for scientific investigations on other planets.[2,3,12,21]

Sensor networks can perform distributed computations in the network for a number of interesting applications, including signal and image processing to detect object signatures, tracking detected objects in the field, and performing statistical computations. However, sensor nodes have limited processing capability, memory resource, and communication bandwidth. Because they operate in the field using battery power, sensor nodes are severely constrained by energy resources. To operate a sensor network for a long period, energy resources must be carefully managed. There has been significant research progress in developing techniques for power-aware and energy-efficient computations and communications. These techniques include the use of voltage scaling, turning nodes off when not in use, operating with different power levels, turning on and off selective parts, novel energy efficient protocols, etc. In addition, application-specific techniques can also be developed to reduce energy consumption in sensor networks.

An important and common application of a sensor network is automatic detection of targets (ATR) using acoustic signals. In this ATR application, a set of sensor nodes collects acoustic signals, performs FFT, and combines their results using the beamforming algorithm for target detection and tracking, as is shown in Figure 30.1. This beamforming operation is essentially summing the outputs of different sensors to determine the line of bearing of the target in the sensor field. When the target is moving, a different set of sensor nodes will be cued to track this object. In an army ATR application, seven sensor nodes are used in a cluster, with one node at the center and the remaining nodes placed around in a circle of diameter four to six feet. Such a cluster of sensor nodes collects 1024 acoustic samples per second, and can detect and compute the line of bearing of a military vehicle target at a distance of about 1.5 km.

Another critical issue with the ATR application is false alarms. For such an application in the field, it turns out that the number of events happening in a day is very few. From the application requirements, it is critical to maximize the probability of target detection while reducing false alarms. With a cluster of sensors collecting acoustic data using microphones, there is a possibility of false alarm due to other noise sources. Targets can be detected when there is only background ambient noise or due to sources other than intended targets. Identifying false alarms due to other than intended sources requires classification using models and acoustic signatures, and is beyond the scope of this chapter. However, we will consider false alarms occurring due to background noise, and identifying such alarms early can save significant amounts of processing by sensor nodes, and hence energy.

The focus of this chapter is on energy-efficient detection of military vehicle targets using acoustic signals in sensor networks. The ATR application of interest is as described above, where a cluster of nodes collaborates to detect a target and its line of bearing. One can use low-power hardware to reduce the energy of a sensor network used in this application. However, our interest is in the architecture and algorithmic techniques to save energy with a given sensor hardware. A classic approach is to duty cycle sensor nodes to detect events. Because vehicle targets appear in the field infrequently, sensor nodes can go up and down

FIGURE 30.1   Target tracking using line-of-bearing.

periodically and not miss targets. However, false alarms will still be a problem and can cause unnecessary signal processing and communications to detect targets when there are none.

Another approach to energy-efficient target detection is to use lightweight sensor nodes (tripwires for short) to form a monitoring network and wake up processing sensor nodes (sensor nodes for short). The idea is to use a two-tier network architecture, where tripwires perform simple computations to distinguish target detection from false alarms and wake up sensor nodes for additional processing only when there is a target. When there are infrequent target events, this approach will have significant energy savings by keeping the sensor nodes off most of the time. Figure 30.2 illustrates this approach by showing both sensor networks with and without tripwires. In the left panel, all sensor processing nodes are always active or on duty-cycle to ensure no target is missed. In the right panel, selected nodes only are woken up by tripwires when a target is near that cluster.

Clearly, with no tripwires, there is wastage of energy and savings can only come from duty-cycle operation of sensor nodes. On the other hand, with tripwires, greater energy savings for this ATR application is possible. The success of this two-tier approach depends on a simple and energy-efficient technique to distinguish targets from background noise. A classic method for tripwires to detect events is the so-called energy detection using the received signal strength indication (RSSI) information. Only when the energy of tripwires sensor readings exceeds a predefined threshold can one determine that a target is in the field and an alarm can be issued to wake up sensor nodes. Using the receiver operating characteristics (ROC), a constant false alarm rate detector can be devised. In this chapter, we also discuss a simple false alarm detection algorithm based on energy distribution among subbands using wavelet transform. This algorithm can be implemented with simple processing and limited memory requirements and, therefore, is suitable for running on tripwires.

In general, tripwires do not have sufficient signal processing capability compared to sensor nodes, and the processing required for detection should be simple and efficient both in time and in space. In a sensor field, because more than one tripwire can observe an event, it is possible for a group of tripwires to collaborate in detecting a target. In general, distributed detection can reduce memory requirements and processing time required as compared to detection by a single node. Well-known distributed detection approaches include: (1) Neyman-Pearson based model[30,31] and (2) Bayesian.[13] In distributed target detection in a sensor network, fusion is also required. There are two types of fusion in this context, namely, data fusion and decision fusion.[8,9,11,17] Due to the high communication cost in a wireless sensor network, decision fusion is more favorable. Therefore, in this chapter, algorithms using decision fusion are emphasized.

This chapter is organized as follows. In the next section we describe the details of ATR application using sensor networks and some known techniques for energy savings. In Section 30.3, we discuss the false alarm

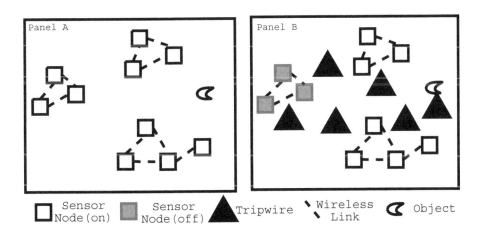

FIGURE 30.2 Sensor field without (A) and with (B) tripwire.

problem and then present details of two-tier sensor network architecture for energy-efficient target detection. In Section 30.4, we present distributed algorithms for target detection using collaborating tripwires. In Section 30.5, we show significant energy savings with our techniques using extensive simulations with experimental field data. Finally, we end this chapter with a summary and concluding remarks.

30.2   ATR Using Sensor Networks

Detection of targets is done using a cluster having several homogeneous sensors, with each sensor equipped with a microphone and radio card for wireless communications. One sensor is elected the head, which aggregates the data from the members, and the head sensor node also observes the same event. Figure 30.3 shows the details of one cluster with various processing of sensor readings. With acoustic sensors to detect military vehicles, the sampling rate is typically 1024 samples per second. Therefore, in one second, a frame of data needs to be processed by the sensor cluster.

A single sensor node can detect a target. However, to know more about the position and direction of motion, several sensor nodes collaborate. The standard approach is to compute the line of bearing using multiple sensor readings. This is usually accomplished using a beam-forming algorithm that combines computations from different sensors. It is essentially delay summing to compute the line of bearing (LOB). In the simplest approach, all nodes send their sensed data to the cluster head. The cluster head performs all the FFT and LOB computations and it will expend more energy than other nodes. One way to reduce the total energy cost for this application is by distributing FFT computations to all nodes as shown in Figure 30.3. Here, each node computes FFT and sends those coefficients to the cluster head, which performs the LOB computation. With this distribution of FFT computations, nodes can use dynamic voltage scaling to expend lesser energy as there is more time available for computing.

In this ATR application, there is a significant amount of data communication among sensor nodes, leading to high energy cost. We briefly explained above a technique for distributing computations to reduce the energy cost. However, to further reduce the energy cost for this application, we need to reduce the number of bits transmitted and received between sensors. Because sensor nodes are normally deployed in a terrain of close proximity that they are observing the same signal source(s), the signal sample correlation across nodes is high. Furthermore, because sensors observe an event for a few tens of seconds, there is also high temporal correlation. Therefore, it should be possible to achieve high compression ratios for sensor data by exploiting these correlations.

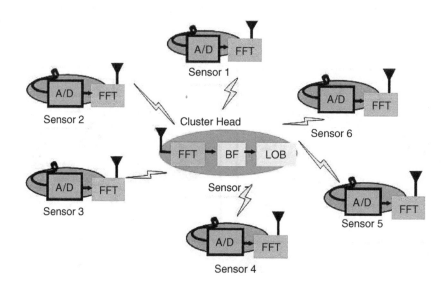

FIGURE 30.3   ATR processing in a cluster.

One can apply standard data compression algorithms or employ more sophisticated algorithms, such as distributed source coding using Syndrome (DISCUS)[19] and spatio-temporal compression ESPIHT.[28] In general, distributed source coding is based on the achievability of the Slepian-Wolf theorem.[24] For the ATR application, the achievable rate is simply $nH(Y|X)$ for two i.i.d X (beamforming node) and Y (sensing node) sources, where n is the frame length and $H(Y|X)$ is the conditional entropy because the beam-forming node also observes the common event. Therefore, any scheme aimed at approximating the achievable rate of the Slepian-Wolf theorem is directly applicable to an ATR application.

The idea in the DISCUS[19] scheme is to send only a few coded bits representing coding syndromes and the recipient will infer the correct codewords through transmitted syndromes. Trellis-based decoding is employed in DISCUS at the decoder to achieve data compression. DISCUS has been shown to be effective when sensors are densely deployed, and it is well-suited for data collection types of applications because encoding complexity is very simple while decoding is done in a node with sufficient power supply. This technique only exploits spatial correlation. Additional gains can be achieved by exploiting both spatial and temporal correlation. It is especially useful when temporal correlation in sensor networks is high, which is the case in ATR applications.

We have developed a scheme, called ESPIHT,[28] that exploits both spatial and temporal correlations present in sensor networks to reduce the communication energy cost. ESPIHT seamlessly embeds a distributed source coding (DSC) scheme with a SPIHT-[23] based iterative set partitioning scheme to achieve high efficiency. Instead of being a generic compression scheme, ESPIHT is coupled with the application, and therefore, it can adapt to application fidelity requirements so that the application-level false alarm rate is kept low. ESPIHT reduces the data rate at least by a factor of 8 and maintains a signal-to-noise ratio (SNR) gain of 20 dB or better on ATR field datasets. The coding/decoding processing is simple and takes on the order of 10 msec using iPAQs.

The major steps of the ESPIHT scheme are as follows: (1) a wavelet transformation on a frame (i.e., a fixed number of samples) to obtain subbands; and (2) compression on the coefficients with quantization using the Trellis Coded Quantization (TCQ) Technique and source coding (DSC). The wavelet transformation and wavelet packet decomposition[33] on a frame of data result in wavelet coefficients. After the transformation, the signal energy is concentrated into a few subbands, which facilitates compression. Note that this subband transformation does not change the spatial correlation present in the data. ESPIHT separates coefficients into two categories: significant and insignificant. To code significant coefficients, distributed source coding is employed to further exploit the spatial correlation. Some insignificant coefficients may be promoted to significant coefficients in successive iterations and the remaining insignificant coefficients are simply discarded. By this way, coefficients contributing to more SNR gains are encoded first and those coefficients contributing less are coded next, and finally, least SNR gain contributing coefficients are discarded to reduce communication energy cost.

One noticeable characteristic of ATR applications is that events may happen infrequently, and it could waste a substantial amount of node energy resources if all nodes are kept active at all times. One simple way to improve this is to duty-cycle nodes, that is, to put a node in sleep mode periodically based on either a predefined fixed duty-cycle rate or a dynamic duty-cycle rate adapting to field activities. However, the duty-cycling approach is only an approximation of matching node on/off periods with field activities. Next we present another approach to wake up nodes only when events are happening and keep nodes in sleep mode when no events are happening in their coverage region.

30.3 False Alarm Detection in ATR Applications

In ATR applications, a critical issue is the false alarms due to noise in the environment. Before detailed processing of an event, it would be useful to identify whether or not there is a target. A simple alarm detection based on energy threshold suffers from false alarms. The *False Alarm Problem* (FAP) refers to a situation in which noise strength is large enough and triggers an alarm when no event is actually happening. There are many sources of false alarms in sensor networks and applications: (1) ambient noise, (2) measurement thermal noise, (3) saturation of analog-digital converter (ADC), (4) scatter and/or reflection to sensing signals, and (5) multi-path and multi-source interference. Sources (4) and (5) usually cause false positives. These factors are not negligible in sensor network applications, especially when a sensor deployment is constrained by the physical limitations of sensors (e.g., acoustic sensor is very close to the ground), and these limitations could cause severe ground reflection and further degrade the algorithmic performance of event identification.

To eliminate unnecessary processing of false alarms by sensor nodes, we need to efficiently determine if an alarm from an energy detector is an event or not. A low-power tripwire detection algorithm was devised to identify false alarms. Because energy in noise spreads out in the Nyquist frequency spectrum and the energy distribution of sensor readings has a large variance when an event is present, this low-power detector computes the variance of energy distribution in sensor readings and uses it to determine if an event is present. This detector uses wavelet subband decomposition to filter the signal into different frequency subbands and then computes the variance of the energy distribution of these subbands.

One approach of detection is based on Windowed FFT. In general, this approach can detect an event from false alarms with high confidence. The drawback of this approach is mainly the processing overhead, which renders this approach inappropriate for running in tripwires. Our wavelet-based approach has an efficient implementation that uses only integer arithmetic operations; furthermore, it requires much less memory than that of the FFT-based approach. The detection performance of our detection algorithm is almost the same as that of the FFT approach.

The idea of using wavelet transformation for false alarm detection stems from the feature extraction and signal classification using wavelets packet decomposition.[16] It has been shown to be promising with efficient processing and accurate results. To show the benefits of our approach, consider the field data at two different times as shown in Figure 30.4. Figure 30.4(a) shows a plot of an acoustic signal amplitude curve with a real object (truck) moving in the test field over a period of 148 seconds, and Figure 30.4(b) shows field noise while the average acoustic energy is maintained the same as that of Figure 30.4(a). The threshold and false alarm detection period are in the first few seconds shown by the black box. The target has to be detected at the event entering stage (so called *early detection*) in order for the object to be tracked in time.

Our approach uses wavelet decomposition on the signal and iteratively checks the energy distribution of subbands derived from the decomposition to detect false alarms. When an event enters a sensor field, the energy distribution in terms of energy ratios (i.e., normalized energy difference between low and high subband) is altered. For the case of white noise, when an event enters, some energy peak corresponding to the event signal spectrum will show up in the energy distribution map. In Figure 30.5, the signal energy as shown in dark is concentrated in some frequency range and the noise energy is spread across the Nyquist

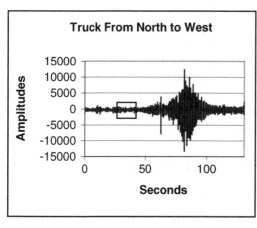

(a) Event Case

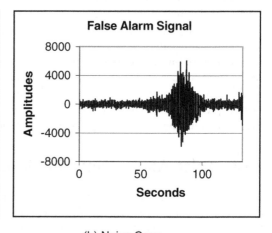

(b) Noise Case

FIGURE 30.4 Early detection of acoustic signal.

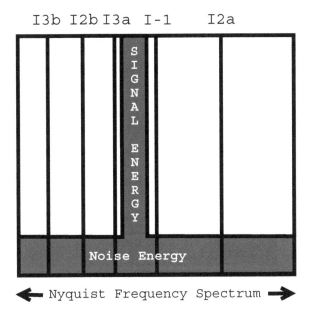

FIGURE 30.5   Recursive detection of an event of interest using wavelets.

frequency spectrum. The signal can be identified as an event of interest using recursive detection by noticing the significant difference between two subbands in iteration I3a (three iterations in Figure 30.5 as labeled by the thin vertical lines).

30.3.1   Two-Tier Sensor Networks with Detection

Sensor nodes are constrained by energy and must operate for a long time in ATR applications. Therefore, it would be useful to conserve energy for detecting targets. For this purpose, we proposed a two-tier architecture for sensor networks to detect targets. In this approach, there are two kinds of nodes, sensor nodes and tripwires. The operating power for a sensor node is about an order of magnitude higher than that for a tripwire. Tripwire[4,7,20,22] is one type of lightweight microsensor node that can perform some simple processing tasks, and it also has limited communication capability. Tripwires can be deployed along with sensor nodes, and form a network to monitor a sensor field so that sensor nodes can be kept in sleep mode to conserve energy. Tripwires will wake up sensor nodes for various signal processing tasks only when necessary. Using tripwires, significant amounts of energy can be saved because sensor nodes consume energy only for periods when events are happening. A network is made energy aware using different operating powers based on field activities.

For energy conservation, we define two network operating modes as follows: (1) low-power monitoring mode: sensor nodes are turned off, and tripwires remain in sleep mode and switch to active mode on a regular basis to check field status; and (2) high power processing mode: sensor nodes are woken up and process signals. The operating power difference between these two modes is significant, and it is at least an order of magnitude. Generally, it is a waste of energy to keep sensor nodes up longer than necessary. An important issue is how accurately to match network operating modes to field activities without any loss of detection.

For a simple two-tier architecture, each sensor node has a physical connection to a tripwire. The early detection of events is performed by the tripwire. The challenge is that the detection algorithm should be able to function correctly and detection results should be generated in a timely manner. The wavelet-based detection scheme discussed earlier fits well with this architecture. With this two-tier architecture, sensor nodes are off most of the time. In a typical scenario of a few events per day, the energy savings can be significant with this approach.

30.4   Distributed Detection Processing and Decision Fusion

Because there are many tripwire nodes, it is possible to achieve false alarm detection through collaboration. For distributed detection,[15,25,32] the processing proceeds as follows: (1) source signals are sampled by individual sensors and each is assigned to process an overlapped segment; (2) each sensor applies windowing to source readings and then performs FFT on the windowed source data; (3) each sensor generates a predicate within its frequency partition with respect to a given threshold; (4) a fusion sensor collects these predicates and generates a final detection predicate. This method utilizes a group of sensors for distributed detection and it uses decision fusion to achieve more robust results.

A two-stage detection scheme is also presented: threshold detection and false alarm detection.[27] Threshold detection uses energy as the criterion for comparison with a predefined threshold and is performed individually, while false alarm detection is done collaboratively.

30.4.1   Two-Stage Local Detection

Initially, each sensor node performs threshold detection on a frame-by-frame basis to determine if there is a possibility of an event. Next, sensors collaborate to discern if there is a false alarm. Threshold detection is performed on a frame-by-frame basis. During the k-th frame period, a tripwire computes a frame energy denoted by $W(k)$ to the end of this frame. It then compares the energy with a predefined threshold T_f. If $W(k) \leq T_f$, the tripwire continues to monitor the field; otherwise, it enters into the false alarm detection stage.

To explain our technique for false alarm detection, let u be a non-leaf node in a preassigned wavelet sub-tree for a tripwire, and let us also denote two corresponding high-pass and low-pass subbands transformed from u as u_2 and u_1, respectively. We define the energy ratio as follows:

$$\mathcal{R}(u) = \frac{|\gamma_1 W(u_1) - \gamma_2 W(u_2)|}{|\gamma_1 W(u_1) + \gamma_2 W(u_2)|}, \tag{30.1}$$

where $\gamma_1 W(u_1)$ and $\gamma_2 W(u_2)$ are the energy for low-pass and high-pass subbands, respectively, of node u; γ_1 and γ_2 are the normalization factors for a given transform for low-pass and high-pass subbands, respectively. When a signal has mean of zero, by the Parseval theorem, the denominator in Equation (30.1) actually equals to the energy of subband u. The normalization factors only depend on the basis and how the transform is actually performed (for S-Transform, γ_1 is $\sqrt{2}$ and γ_2 is $0.5\sqrt{2}$).

False alarm detection has five phases, as follows:

1. *Selection.* Select reference based on received signal strength indication (RSSI) information.
2. *Transform.* Transform the source signal or a subband to obtain the next node in the decomposition tree.
3. *Elimination.* Remove a node if the energy ratio associated with this node is close to that of the reference.
4. *Recursion.* Based on different local decision rule (i.e., using $\mathcal{L}^\infty$ (summation of absolute energy ratios) or $\mathcal{L}^1$ (maximum of absolute energy ratios)), select next node to process based on predefined tree traversal order or stop when either the threshold is exceeded or all nodes are either processed or eliminated in the *Elimination* step.
5. *Predication.* Generate a detection predicate based on the *Recursion* results.

This process is followed by a predicate fusion and the final predicate is used by a tripwire to wake up sensor nodes.

30.4.2   Distributed Detection Processing

After detection in individual nodes, the detection result (in a form of predicate) is sent to a designated tripwire. The designated tripwire fuses decisions using a Boolean function on received predicates. The

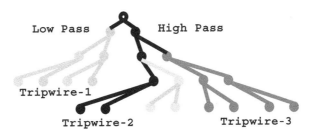

FIGURE 30.6 Partitioning for three sensor nodes.

final predicate will indicate a false alarm when all predicates indicate a false alarm, and the final predicate indicates positive detection of an event if at least one predicate indicates an event.

To process the detection in a distributed fashion to reduce individual node processing delay and memory requirement, we need to assign each node to a subtree of the wavelet packet decomposition tree. One such assignment is shown in Figure 30.6, where three sensors are assigned to different subtrees of a wavelet decomposition tree. In Figure 30.6, the light gray part on the right is not processed by any tripwires. We assume that nodes are sequentially indexed. We use a pyramidal wavelet decomposition scheme because underlying signals could be of low-pass, high-pass, or band-pass type. The tree structure corresponding to a wavelet packet basis[33] of a pyramidal wavelet decomposition scheme is predefined based on application signal characteristics, and partition of a tree is also predefined based on various application requirements (e.g., delay, accuracy). As an example of packet basis selection, for a case of vehicle acoustic tracking, a basis selection may have to take into account the type of engine (e.g., Turbo or non-Turbo). In the case of Turbo engines, the basis should include more low-pass subbands. With the partition information made available to a tripwire, each tripwire selects a subtree to process based on its index.

As a general guideline on tree partition, all nodes in a wavelet packet decomposition tree should be covered by at least one tripwire when there is no *a priori* knowledge of underlying signals available to tripwires, and nodes in a given frequency band should be covered by at least one tripwire when the underlying signal is most possibly in that frequency band (*a priori* knowledge). As for basis selection, an evenly balanced binary wavelet packet decomposition tree should be employed when no *a priori* knowledge of the underlying signal is available. However, coarser decomposition (i.e., more levels of decomposition) in a frequency band should be performed when the underlying signal is most possibly in that frequency band, and the corresponding tree is unbalanced.

At the end of detection processing in each node, a detection predicate indicating a false alarm or an event is generated corresponding to a node's subtree. Each node broadcasts a message containing its predicate and index. The message contains one bit for a predicate and $\lceil \log(N) \rceil$ bits for the node index, where N is the total number of nodes in a field. A fusion node, which is also determined based on its index (e.g., its index is a multiple of 3 for the case of three tripwires per detection), collects these messages and produces a final detection predicate using a Boolean function. For fault tolerance, we can assign two or more tripwires to process overlapped subtrees in a wavelet decomposition tree.

30.5 Experimental Results

In this section, we present some simulation and experimental results for ATR applications. Our simulation uses extensive field datasets to show improvement in energy consumption for ATR applications and the benefits of our two-tier approach.

The left plot and right plot of Figure 30.7 show time domain representations of an event signal and a false alarm signal (ambient noise), respectively, for a period of 20 seconds. The left and right plots of Figure 30.8 show the energy per frame of the event signal and false alarm signal, respectively, for the same period. In this experiment, threshold settings are simply determined using similar field datasets. Once signal energy in a frame (256 samples, i.e., quarter-second resolution for threshold detection) is greater

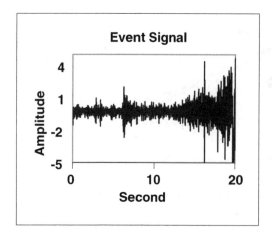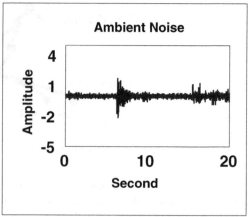

FIGURE 30.7 Data sources: event signal versus ambient noise in time domain.

than -30 dB ($T_f = -30$ dB), tripwires are alarmed and they start false detection processing. From the right plot of Figure 30.8, false alarms happen from around the 6th second untill the 9th second.

In this experiment, one predicate is generated per second. Figure 30.9 shows distributed detection plots with three tripwires for the same two cases as explained above. In false alarm detection, for the case that an event is present (see left plot of Figure 30.9), tripwire T-1 can correctly detect it; then the final detection predicate is an event predicate by the Boolean fusion function. For the case in which no event is present, tripwires are idle from start time to the 5th second and from the 10th second to the 19th second, and all three tripwires generate false alarm predicates from the 6th second to the 9th second, in which case the final detection predicate is a false alarm predicate. The false alarm threshold in this experiment can be any value between 0.1 and 0.35. Dynamic update of false alarm thresholds and energy thresholds based on field dynamics (e.g., noise floor and SNR) is possible. Further improvement can be performed by checking for a fixed number of successive frames, and only when all predicates from these frames are events, should a sensor node be activated.

Table 30.1 shows the average processing delay and energy consumption per detection and stand-alone detection using a single tripwire (shown in last row). The communication time is for the duration when

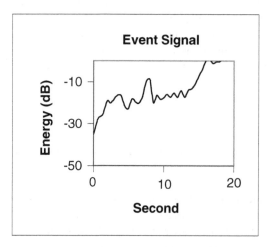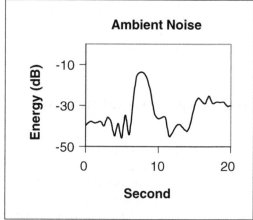

FIGURE 30.8 Time domain signal energy.

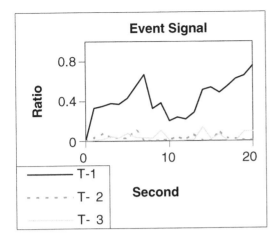

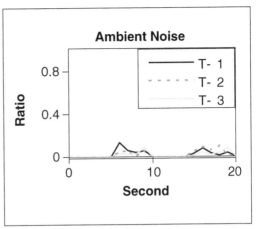

FIGURE 30.9 Distributed detection processing.

the transceiver is on, and the actual transmission time is less than 1 ms using an RFM TR1000 radio. It is clear that the total time needed is much less than that needed for stand-alone detection. With added communication cost, the overall energy is still comparable to that of stand-alone detection. We must note that the overall delay of each tripwire is much less than that of a single tripwire under stand-alone detection.

We have conducted experiments using a sensor node developed in the PASTA project.[6] A PASTA node consists of PXA255 as a sensor signal processor and a microcontroller for tripwire processing. We have studied the energy dissipation on sensor nodes and tripwires. These power profiles are shown in Table 30.2, where power consumption data of the PXA255 CPU and PASTA radio are itemized because they are not active at the same time. Tripwires are dissected into three parts — MCU, tripwire beacon radio, and ADC — where MCU and radio are not always active. ADC on a tripwire or a sensor node is active as long as the host sensor node or tripwire is on. The application used in this experiment is time-domain line-of-bearing (LOB) beamforming. The energy per LOB computation, including transmission, is 245 mJ, the duty-cycle energy per detection is 1.5 mJ, and tripwires are turned on periodically, where a tripwire

TABLE 30.1 Energy and Time Comparisons on CYGNAL C8051 F020 μC

	T. (ms)	Comm. T. (ms)	Energy (mJ)
T-1	80	45	1.35
T-2	90	43	1.33
T-3	130	57	1.97
Total	300	145	4.7
Single	370	0	4.5

TABLE 30.2 Power Profiles of Sensor Node and Tripwire

mW	CPU	Radio	MCU/Radio	Sensor
Idle	185	126	18	n/a
Active	635	171	30	0.9
Sleep	67	38	0.1	n/a

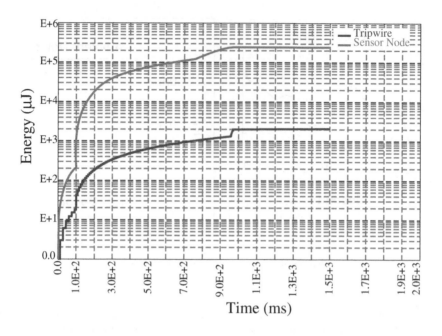

FIGURE 30.10   Energy comparison of detection and tracking.

radio is on during the first 100 ms of every second and off during the remaining period of a second. However, tripwire synchronization cost is omitted because it is required very infrequently. An energy dissipation comparison is shown in Figure 30.10. Figure 30.11 shows the power consumptions of a tripwire and sensor node. From Figure 30.11, with cueing protocol and detection at 80 percent false alarm rate and 1 percent false positive rate, the duty-cycle power consumption can be brought down to within a 10 mW region.

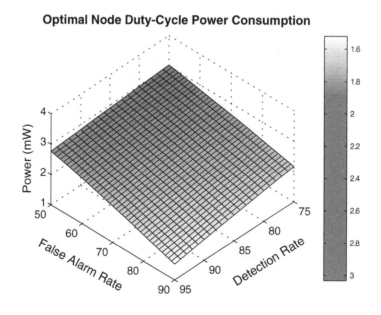

FIGURE 30.11   Power consumption of duty-cycling detection.

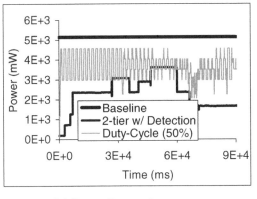 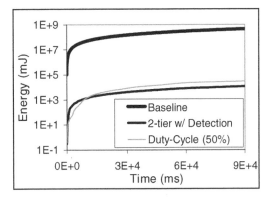

(a) Power Comparison (b) Energy Comparison

FIGURE 30.12 Power and energy comparisons.

Figure 30.12(a) shows the power consumption for a 90-second simulation with two simultaneous targets. Figure 30.13 shows the field and network configuration and the Legends except that star objects for tripwires follow those in Figure 30.1. The comparison is on three different network configurations: (1) two-tier networks with sensor nodes and tripwires, (2) baseline networks in which all sensor nodes are active all the time and there are no tripwires, and (3) duty-cycled networks in which sensor nodes are on duty-cycle at rate 50 percent and there are no tripwires. The power dissipation of the two-tier network is significantly smaller than those of baseline and duty-cycled networks. The energy dissipation is shown in Figure 30.12(b), and there is also significant energy savings using the two-tier network configuration with detection as compared to the other network configurations. These simple simulation results[29] illustrate the potentially great energy savings of a two-tier network configuration.

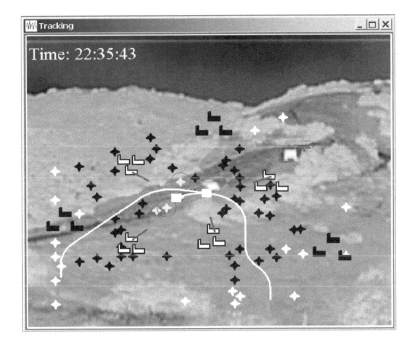

FIGURE 30.13 A two-target sensor field configuration.

30.6   Conclusions and Final Remarks

In this chapter we discussed energy efficient detection using wireless sensor networks. We first briefly reviewed applications and issues in sensor network on energy efficiency, and then discussed a two-tier sensor network architecture for energy efficiency with details on detection of events. In general, distributed detection processing with Boolean fusing is introduced, and we discussed how to distribute processing to a group of sensor nodes to improve algorithm performance and processing efficiency for the wavelet-based detection scheme. Experimental results on energy dissipation of detection and energy savings from detection are also presented.

Acknowledgments

This research is partially supported by DARPA under contract F33615-02-2-4005 in the Power Aware Computing and Communications Program (PAC/C). The views and conclusions contained herein are those of the authors and should not be interpreted as necessarily representing the official policies or endorsements, either expressed or implied, of the Defense Advanced Research Projects Agency (DARPA), the Air Force Research Laboratory, or the U.S. Government.

References

1. DARPA PAC/C Project (ongoing), Power Aware Computing and Communication [online], Available: http://www.darpa.mil/ipto/programs/pacc.
2. DARPA SensIT Project (past), Sensor Information Technology [online], Available: http://www.sainc.com/sensit.
3. NASA JPL, Sensor Systems for Space Exploration and Research [online], Available: http://www.jpl.nasa.gov.
4. Crossbow Technology Inc., Mica Motes [online], Available: http://www.xbow.com.
5. University of California at Berkeley, Habitat Monitoring on Great Duck Island, Maine [online], Available: http://www.greatduckisland.net, 2002.
6. University of Southern California—Information Systems Institute, PASTA Microsensor Node [online], Available: http://pasta.east.isi.edu, 2003.
7. Abrach, H., Bhatti, S., Carlson, J., Dai, H., Rose, J., Sheth, A., Shucker, B., Deng, J., and Han, R., MANTIS: System Support for MultimodAl Networks of *In-situ* Sensors, *2nd ACM International Workshop on Wireless Sensor Networks and Applications (WSNA)*, 50, 2003.
8. Blum, R.S., Kassam, S.A., and Poor, H.V., Distributed Detection with Multiple Sensors. II. Advanced Topics, *Proc. of the IEEE*, 85, 64, 1997.
9. Brooks, R., Ramanathan, P., and Sayeed, A., Distributed Target Classification and Tracking in Sensor Networks, *Proc. of the IEEE*, 91, 1163, 2003.
10. Calderbank, A.R., Daubechies, I., Sweldens, W., and Yeo, D., Wavelet Transforms that Map Integers to Integers, *Applied and Computation al Harmonic Analysis*, 5, 332, 1998.
11. Clouqueur, T., Ramanathan, P., Saluja, K.K., and Wang, K.-C., Value-Fusion versus Decision-Fusion for Fault-Tolerance in Collaborative Target Detection in Sensor Networks, in *Proc. of the 4th Annu Conf. on Information Fusion*, 2001.
12. Govindan, R., Wireless Sensor Networks: Cutting through the Hype [online], Available: http://cs.usc.edu/~ramesh/talks/byyear.html, 2001.
13. Hoballah, I.Y. and Varshney, P.K., Distributed Bayesian Signal Detection, *IEEE Trans. on Information Theory*, 35, 995, 1989.
14. Harris, F.J., On the Use of Windows for Harmonic Analysis with the Discrete Fourier Transform, *Proc. of the IEEE*, 66, 51, 1978.
15. Kenefic, R.J., An Algorithm to Partition DFT Data into Sections of Constant Variance, *IEEE Trans. Aerospace and Electronic Systems*, 34, 789, 1998.

16. Learned, R.E. and Wilsky, A.S., A Wavelet Packet Approach to Transient Signal Classification, *Applied and Computational Harmonic Analysis,* 2, 265, 1995.

17. Lexa, M.A., Rozell, C.J., Sinanović, S., and Johnson, D.H., To Cooperate or Not to Cooperate: Detection Strategies in Sensor Networks, in *Proc. of Int. Conf. on Acoustics, Speech, and Signal Processing (ICASSP),* 2004.

18. Li, D., Wong, K.D., Hu, Y.H., and Sayeed, A.M., Detection, Classification, and Tracking of Targets, *IEEE Signal Processing Magazine,* 19, 17, 2002.

19. Pradhan, S.S. and Ramchandran, K., Distributed Source Coding Using Syndromes (DISCUS): Design and Construction, *IEEE Transactions on Information Theory,* 49, 626, 2003.

20. Rabaey, J., Ammer, J., Karalar, T., Li, S., Otis, B., Sheets, M., and Tuan, T., PicoRadios for Wireless Sensor Networks: The Next Challenge in Ultra-Low-Power Design, in *Proc. of the International Solid-State Circuits Conference,* 2002.

21. Raghavendra, C.S., Sivalingam, K.M., and Znati, T. (eds.), *Wireless Sensor Networks,* Kluwer Academic Publishers, 2004.

22. Roundy, S., Wright, P.K., and Rabaey, J., *Energy Scavenging for Wireless Sensor Networks with Special Focus on Vibration.* Kluwer Academic Publishers, 2003.

23. Said, A. and Pearlman, W.A., A New, Fast, and Efficient Image CODEC Using Set Partitioning Hierarchical Trees, *IEEE Transactions on Circuits and Systems for Video Technology,* 6, 243, 1996.

24. Slepian, D. and Wolf, J.K., Noiseless Coding of Correlated Information Sources, *IEEE Transactions on Information Theory,* IT-19, 471, 1973.

25. Swanson, D., *Signal Processing for Intelligent Sensor Systems,* Marcel Dekker, 2000, chap. 11.

26. Tang, C. and Raghavendra, C.S., Detection and cueing for energy efficiency in wireless sensor network, submitted 2004.

27. Tang, C. and Raghavendra, C.S., Providing Power Awareness to Wireless Microsensor Networks via Tripwires, in *Proc. IEEE/CAS Symposium on Emerging Technologies: Frontiers of Mobile and Wireless Communication,* 2004.

28. Tang, C., Raghavendra, C.S., and Prasanna, V.K., Power Aware Coding for Spatio–Temporally Correlated Wireless Sensor Data in Proc. IEEE International Conference on Mobile Ad Hoc Sensor Systems, 2004.

29. Tang, C. and Raghavendra, C.S., Simulation Study of Two-Tier Wireless Sensor Network with Detection on Energy Savings, unpublished manuscript, 2004.

30. Thomopoulos, S.C., Viswanathan, R., and Bougoulias, D.K., Optimal Distributed Decision Fusion, *IEEE Trans. Aerospace and Electronic Systems,* 25, 761, 1989.

31. Viswanathan, R. and Varshney, P.K., Distributed Detection with Multiple Sensors. I. Fundamentals, in *Proc. of the IEEE,* 85, 54, 1997.

32. Wang, Z., Willett, P., and Streit, R., Detection of Long-Duration Narrowband Processes, *IEEE Trans. Aerospace and Electronic Systems,* 38, 211, 2002.

33. Wicherhauser, M.V., INRIA Lectures on Wavelet Packet Algorithms [online], http://www.math.wustl.edu/~victor/papers/lwpa.pdf, 1991.

31

Comparison of Cell-Based and Topology-Control-Based Energy Conservation in Wireless Sensor Networks

Douglas M. Blough

Mauro Leoncini

Giovanni Resta

Paolo Santi

In this chapter we compare the effectiveness of two popular energy conservation strategies, namely topology control protocols and cooperative cell-based approaches, to reduce the energy consumption, and thus extend the lifetime, of wireless sensor networks. To this end, we define a realistic (although necessarily idealized in some respects) and unified framework of investigation. Using this framework, we prove lower bounds on network lifetime for cell-based cooperative strategies with node densities that are known to be sufficient to ensure a connected network. We also perform a number of simulation experiments, under a traffic model specifically designed for sensor networks, to evaluate and compare cell-based approaches with topology control. This is the first attempt at a comprehensive understanding of the conditions under which one of these approaches outperforms the other. Indeed, our study reveals a number of properties of the techniques investigated, some of which are not at all obvious. As expected, cell-based cooperative approaches, which completely power down network interfaces of certain nodes for extended time periods, produce longer network lifetimes when node density is very high. However, even with moderately high density of nodes, cell-based approaches do not significantly extend lifetime when it is defined in terms of connectivity. While it is not surprising that cell-based approaches do not extend lifetime under low density conditions, we find that topology control techniques *can* significantly increase network lifetime under those conditions and, in fact, they substantially outperform cooperative approaches in this respect. More details on the precise findings can be found in Section 31.7.

31.1   Introduction

Recent advances in MEMS, mixed-mode signaling, RF, and low-power circuit design have enabled a new type of computing device, namely a battery-powered intelligent sensor with wireless communication capability.[22] Applications for networks of these new sensor devices are many and potential applications are virtually unlimited. Examples of applications include monitoring of environmental change in areas such as forests, oceans, and deserts; detection of chemical, biological, and nuclear contamination; monitoring of traffic conditions; and surveillance. The issue of energy conservation is paramount in sensor networks due to the reliance on power-intensive RF circuitry, the use of battery power, and a limited ability to replace or recharge batteries in many sensor environments.

Cooperative strategies and topology control protocols have been recently proposed in the literature as major approaches to reducing energy consumption in wireless networks. Cooperative strategies are motivated by the fact that, for many wireless transceivers, there is little difference between the energy consumed by the interface during transmission, reception, and listening, while significant energy savings can be achieved if the radio is shut down completely. These strategies therefore operate by having neighboring nodes coordinate times during which they can turn off their interfaces, potentially resulting in large energy savings if the coordination overhead is limited. The goal of a good cooperative strategy is to achieve energy savings while not significantly reducing the network capacity. Topology control protocols, on the other hand, try to take advantage of the at least quadratic reduction in transmit power that results from reducing the transmission range, while still attempting to guarantee that each node can reach a sufficient number of neighbors for the network to remain connected. In addition to reducing energy consumption, topology control protocols have the additional advantage of increasing network capacity, due to the reduced interference experienced by the nodes.[10]

Cooperative techniques and topology control protocols can be considered orthogonal techniques. Cooperative strategies take advantage of a "dense" communication graph, in which some nodes can turn off their radios without impairing network connectivity; the goal of topology control, on the other hand, is to produce a relatively sparse communication graph, where only energy-efficient links are used to communicate.

Despite the common goal of extending network lifetime through energy conservation, cooperative strategies and topology control protocols have not, to date, been directly compared in a unified framework. Thus, it is not clear under what conditions each technique is more effective than the other. In this chapter, we make a first step toward answering this question. We consider a cell-based cooperative strategy and a number of topology control algorithms, and we define an idealized framework for comparison. Using this framework, we compare the performances of the various protocols (in terms of network lifetime extension) considering various definitions of network lifetime and varying traffic loads, under a traffic model specifically designed to represent sensor networks.

It is our opinion that the results presented here represent novel contributions toward a better understanding of:

- The relative performance of the two techniques under different conditions (e.g., load, node density, and node number)
- The effect of the lifetime definition on the results
- The relative performance of the two approaches under a realistic sensor network traffic model

31.2 Related Work and Motivation

Measurements on the Medusa II sensor node's wireless transceiver have shown sleep:idle:receiving: transmitting ratios of 0.235:1:1.008:1.156.[23] Cooperative strategies exploit this by shutting down the interface most of the time. Clearly, the node's sleeping periods must be carefully scheduled, because otherwise the network functionality may be compromised. Typically, the awake nodes form a connected backbone, which is used by the other nodes as a communication infrastructure. To establish the backbone, some classes of "routing equivalent" nodes are identified, and only one node in every class (called the *representative*, or *leader*, node) is left active. Periodically, the set of active nodes is changed to achieve a more uniform energy consumption or to deal with mobility.

Cooperative strategies are based on the following idea. Assume that a given set S of nodes provides a functionality F to the rest of the system; instead of keeping all the nodes in S operative, a representative node u can be selected, and the remaining nodes can be turned off in order to save energy. The representative node selection is obtained as the result of a negotiation protocol executed by nodes in S, which is repeated when u dies, or after a certain wake-up time. This way, considerable energy savings can be potentially achieved. Observe that the exact definition of the cooperative strategy depends on the kind of functionality in which we are interested. For example, F could be defined as the capability of nodes in S to relay messages on behalf of the remaining nodes, without compromising network connectivity (i.e., nodes are equivalent from the point of view of a routing protocol). In the case of sensor networks, F can also be defined as the capability of sensing a given sub-region of the deployment region R.

Several routing cooperative strategies have been recently proposed in the literature. Xu et al.[34] introduce the GAF strategy, which is based on a subdivision of the network deployment region into an appropriate number of nonoverlapping square cells. The cells are used to identify equivalent nodes: all the nodes lying in the same cell are proved to be routing equivalent. An election algorithm is periodically executed to elect the leader in every cell. The cell-based approach is also used by Santi and Simon.[30] In this work, the focus is on sensor networks, and it is assumed that a loose time synchronization mechanism is available to the nodes (which is quite likely to occur in sensor networks). Given this loose synchronization, the authors present and analyze deterministic and randomized leader election algorithms. In the

SPAN algorithm proposed by Chen et al.,[7] nodes decide whether to be active or not based on two-hop neighborhood information: a node is eligible as leader if two of its neighbors cannot reach each other either directly or via one or two leaders. Eligible nodes decide whether to become leaders based on a randomized algorithm, where the probability of being elected depends on the *utility* of the node (which can be informally understood as the number of additional pairs of nodes that would be connected if the node were to be elected leader) and on the amount of energy remaining in the node's battery. The CPC strategy[32] is based on the construction of a connected dominating set: nodes in the dominating sets are the representatives, and coordinate the sleeping periods of the other units. Routing equivalence is ensured by the fact that, by the definition of dominating set, every node in the network has at least one leader node as an immediate neighbor, and by the fact that the nodes in the dominating set are connected.

Cooperative strategies rely on a homogeneity assumption on the node transmitting ranges, namely that all the nodes have the same range r, which can be intended as the maximum range. This assumption is vital in the GAF strategy and in the protocols presented by Santi and Simon[30] because the value of r is used to determine the size of the cells used to partition the deployment region. SPAN and CPC are based on neighborhood relationships between nodes and, in principle, could work also if the homogeneity assumption does not hold. However, it should be observed that routing cooperative strategies are effective only when the communication graph is very dense, and many disjoint paths exist between source–destination pairs. If this is not the case, several nodes will remain active for a long time,* thus reducing considerably the energy savings achieved by the strategy. Hence, the assumption that all nodes transmit at the maximum range r is also essential for SPAN and CPC in practice.

Contrary to cooperative strategies, topology control techniques leverage the capability of nodes to vary their transmitting ranges dynamically, in order to reduce energy consumption. In fact, the power needed to transmit data depends on the sender–receiver distance. More precisely, the power p_i required by node i to correctly transmit data to node j must satisfy inequality

$$\frac{p_i}{\delta_{i,j}^{\alpha}} \geq \beta, \tag{31.1}$$

where $\alpha \geq 2$ is the *distance-power gradient*, $\beta \geq 1$ is the *transmission quality* parameter, and $\delta_{i,j}$ is the Euclidean distance between the nodes. While β is usually set to 1, α is in the interval $[2,6]$, depending on environmental conditions.[26] A simple (but widely adopted) energy model can then be devised by simply turning Equation (31.1) into an equality. According to this, if a node transmits at distance, say, $\frac{r}{2}$ and $\alpha = 4$, it consumes $\frac{1}{16}$ of the energy required to transmit at full power.

Observe that Equation (31.1) refers to the transmit power only, and does not account for the power consumption of other components of the wireless transceiver. Thus, the energy savings achieved in a realistic setting might be considerably different than predicted by Equation (31.1). This point is discussed in detail in the remainder of the chapter.

Several topology control protocols have been recently proposed in the literature[2,3,12,15,16,25,27,33]. They differ in the type of topology generated and in the way it is constructed. Typically, the goal of a topology control algorithm is to build a connected and relatively sparse communication graph that can be easily maintained in the presence of node mobility. Having a sparse communication graph has the further advantage of making the task of finding routes between nodes easier because there are relatively few paths between source–destination pairs. Topology control protocols have been surveyed by Rajaraman.[24]

---

*We recall that the goal of cooperative strategies is to reduce energy consumption *while preserving capacity (hence, connectivity) as much as possible.*

Energy-efficient routing protocols[4–6,11,31] can be used in conjunction with either cooperative approaches or topology control to produce further energy savings. Most of these works are targeted toward ad hoc networks, while Heinzelman et al.[11] consider sensor networks. In our traffic model used to compare the two approaches, we assume the most energy-efficient paths are used without specifically considering the underlying routing algorithm. This follows our philosophy of attempting to evaluate the network lifetime extensions possible with these approaches under the best possible conditions.

31.3 Overview of System Model

We consider a set of n nodes distributed uniformly at random over a fixed-size region. In Section 31.4, where we derive an analytical lower bound on network lifetime for cell-based approaches, we consider one-, two-, and three-dimensional regions, but we make the simplifying assumption that network traffic is uniformly distributed over the cells in the region. Beginning with Section 31.5, we focus on two-dimensional regions, which are the most common scenario, and we consider a more realistic traffic model. For this situation, we assume that sensor nodes supply data to external entities through a set of k gateway nodes, also referred to as data collection sites, which are distributed uniformly at random across the boundary of the region. The traffic model consists of all sensor nodes communicating an equal amount of traffic to their closest (in an energy cost sense) data collection site.

We assume that the nodes have uniform transceivers, each of which has a maximum communication range r. We assume that the energy-intensive RF communication is the dominant energy cost incurred by the nodes. When analyzing cooperative approaches, we assume that nodes have the capability to shut down their transceivers; and for topology control protocols, we assume the ability to vary the transmit power used by the transceivers. Because the energy used when no energy conservation measures are employed is highly dependent on r, we compare both approaches against the situation where r is chosen *a priori* as small as possible while still guaranteeing that the resulting network is connected with high probability. Such a choice can be considered a simple static form of topology control, which is already a crude mechanism for energy conservation. Thus, the lifetime extensions we report for the two approaches are improvements over and above those achieved by this simple energy conservation mechanism.

This system model corresponds to a class of sensor networks where the network is active at all times or at least is periodically active. That is, there is a continuous or regular quantity to be measured by the network, and this requires regular communication across the network. Another class of sensor networks are event driven; that is, they monitor for specific events and might only need to communicate information when those events occur. In event-driven sensor networks, other energy conservation approaches are likely to be more effective than those studied herein (e.g., using ultra-low-power wake-up channels to wake up nodes and turning on their higher-power radios only when absolutely necessary).

More details on our precise assumptions are provided in the next section.

31.4 A Lower Bound to Network Lifetime for an Idealized Cell-Based Energy Conservation Approach

Measurements of the energy consumption of the Medusa II sensor node have shown that the energy required to sense the channel is only about 15 percent less than the energy required to transmit.[23] In nodes such as these, approaches to energy conservation that completely shut down the network interface during idle periods therefore seem to have the most promise in terms of network lifetime extension. In this section, we analyze an idealized version of a popular class of these approaches, which we refer to as cooperative cell-based energy conservation.

31.4.1   A Model for Idealized Cell-Based Energy Conservation

The effectiveness of a cooperative strategy depends heavily on the node density. Intuitively, if node density is low, almost all the nodes must stay up all the time, and no energy saving can be achieved. Considering the overhead required for coordination of nodes, the actual network lifetime could actually be reduced with respect to the case where no cooperative strategy is used. Conversely, if node density is high, consistent energy savings (and, consequently, extension of network lifetime) can be achieved. This behavior is displayed by the GAF protocol,[34] while the energy savings achieved by SPAN does not increase with node density. This is due to the fact that the overhead required for coordination with SPAN tends to "explode" with node density, and thus counterbalances the potential savings achieved by the increased density. For this reason, in the following we focus attention primarily on a class of approaches based on the GAF protocol, that is, cell-based cooperative energy conservation techniques. Several existing protocols fall into this category.[30,34]

Observe that the positive effect of an increased node density on network lifetime could be counterbalanced by its detrimental effect on the network capacity. In fact, it is known that, for stationary networks, the network capacity does not scale with node density, and the end-to-end throughput achievable at each node goes to 0 as the density increases.[10] Furthermore, increasing node density entails a higher network cost. On the other hand, node density cannot be too low because, otherwise, network connectivity would be impaired. Hence, the trade-off between node density, network lifetime, and capacity/cost must be carefully evaluated. As a first step in this direction, in the following we investigate the relation between the expected benefit of cooperative strategies and the node density, under the following simplifying hypotheses:

a1. n nodes are distributed uniformly and independently at random in $R = [0, l]^d$, with $d = 1, 2, 3$.

a2. Nodes are stationary.

a3. Nodes can communicate directly when they are at distance at most r; that is, there exists the bidirectional link (i, j) in the communication graph if and only if $\delta_{i,j} \leq r$.

a4. r is set to the minimum value of the transmitting range that ensures that the communication graph is connected with high probability.

a5. F is defined as the capability of nodes in S to relay messages on behalf of the remaining nodes, without compromising network connectivity. To this end, R is divided into nonoverlapping d-dimensional cells of equal side $\frac{r}{2\sqrt{d}}$.[*1] The total number of cells is then $N = \frac{k_d l^d}{r^d}$, where $k_d = 2^d d^{d/2}$.

a6. We consider an ideal cooperative strategy, in which the overhead needed to coordinate nodes amongst themselves is zero. Hence, the energy savings derived in the following can be seen as the best possible a cooperative strategy can achieve.

a7. Network lifetime is defined in terms of connectedness.

a8. The traffic is balanced over all cells.

a9. The energy consumed by other components of a node is negligible compared to the energy consumption of its transceiver.

Observe that, by Assumption *a8*, all cells are subject to the same load. If this load must be handled by a single node, it will die at the *baseline time* T; however, if a cell contains h nodes, the load can be evenly divided among them, and it follows from Assumptions *a6* and *a9* that the last node in the cell will die at time hT. Hence, a lower bound to the network lifetime can be obtained by evaluating the probability distribution of the minimum number of nodes in a cell, and occupancy theory[13] can be brought to bear on the problem.

---

[*]This ensures that a node in one cell can communicate with all nodes in the complete neighborhood of cells surrounding it. Note that the side of the cell as defined here is slightly different from that used in the GAF protocol,[34] which ensures that nodes residing in a cell can communicate with all the nodes in the upper, lower, left, and right cell. However, this slight difference does not impair the validity of our analysis for the GAF protocol.

31.4.2 Analysis of the Lower Bound

In this section we will use the standard notation regarding the asymptotic behavior of functions, which we recall. Let f and g be functions of the same parameter x. We have:

$f(x) = O(g(x))$ if there exist constants C and x_0 such that $f(x) \leq C \cdot g(x)$ for any $x \geq x_0$

$f(x) = \Omega(g(x))$ if $g(x) = O(f(x))$

$f(x) = \Theta(g(x))$ if $f(x) = O(g(x))$ and $f(x) = \Omega(g(x))$. In this case, we also use the notation $f(x) \approx g(x)$

$f(x) = o(g(x))$ if $\frac{f(x)}{g(x)} \to 0$ as $x \to \infty$

$f(x) \ll g(x)$ or $g(x) \gg f(x)$ if $f(x) = o(g(x))$

The probability distribution of the minimum number of nodes in a cell can be evaluated using some results of occupancy theory,[13] which studies properties of the random independent allocations of n balls into N urns* when $n, N \to \infty$. Let $\eta(n, N)$ be the random variable denoting the minimum number of nodes in a cell. The form of the limit distribution (i.e., of the probability distribution of $\eta(n, N)$, when $n, N \to \infty$) depends on the asymptotic behavior of the ratio $\frac{\alpha}{\ln N}$, where $\alpha = \frac{n}{N}$. The following theorem holds:[13]

Theorem 31.1   *If $\frac{\alpha}{\ln N} \to 1$ as $n, N \to \infty$ and $h = h(\alpha, N)$ is chosen so that $h < \alpha$ and $N p_h(\alpha) \to \lambda$, where $p_h(\alpha) = \frac{\alpha^h}{h!} e^{-\alpha}$ and λ is a positive constant, then:*

$$P(\eta(n, N) = h) \to 1 - e^{-\lambda}$$
$$P(\eta(n, N) = h + 1) \to e^{-\lambda}$$

Theorem 31.1 states that if $\frac{\alpha}{\ln N} \to 1$ as $n, N \to \infty$, then $\eta(n, N)$ is either h or $h+1$ asymptotically almost surely** (a.a.s. for short), where h is such that $N p_h(\alpha) \to \lambda$, for some positive constant λ. Similar results[13] determine the limit distribution of $\eta(n, N)$ when $\frac{\alpha}{\ln N} \to x$, for some $x > 1$, or when $\frac{\alpha}{\ln N} \to \infty$. However, for our purposes, it is sufficient to note that denoting by $\eta_1(n, N)$, $\eta_x(n, N)$ and $\eta_\infty(n, N)$ the value of $\eta(n, N)$ for the three asymptotic cases, it is $\eta_1(n, N) \leq \eta_x(n, N) \leq \eta_\infty(n, N)$, a.a.s. This follows immediately by the fact that in the three asymptotic cases, either a strictly increasing number of nodes are distributed into the same number of cells, or the same number of nodes is distributed in a smaller number of cells.

Observe that Theorem 31.1 gives very precise information on the asymptotic value of $\eta(n, N)$, but gives no explicit value of h. To determine the value of h, we have to do some hypotheses on the relative magnitude of $r, n,$ and l.

Assume $n = dk_d \frac{l^d}{r^d} \ln l$, where $d = 1, 2, 3$ and $k_d = 2^d d^{d/2}$. Further assume that $r \ll l$. With these hypotheses, we have $\lim_{l \to \infty} n = \infty$. Let $N = k_d \frac{l^d}{r^d}$ be the number of cells into which the deployment region $R = [0, l]^d$ is divided. Under the assumption $r \ll l$, we have $\lim_{l \to \infty} N = \infty$. That is, by setting n and N as above, we can use the results on the asymptotic distribution of the minimally occupied urn derived in the occupancy theory. In particular, we want to apply Theorem 31.1. To this end, we have to check that $\lim_{n, N \to \infty} \frac{\alpha}{\ln N} = 1$.

In the hypotheses above we have:

$$\lim_{n, N \to \infty} \frac{\alpha}{\ln N} = \lim_{l \to \infty} \frac{\ln l^d}{\ln \frac{k_d l^d}{r^d}}. \tag{31.2}$$

It is immediate to see that the limit above equals 1 under the further assumption that $r^d = \Theta(1)$ (we remark that this assumption does not contradict assumption $r \ll l$).

---

*For consistency, in the following we will use the words *node* and *cell* instead of *ball* and *urn*, respectively.

**We say that an event E_m, describing a property of a random structure depending on a parameter m, holds *asymptotically almost surely* if $P(E_m) \to 1$ as $m \to \infty$.

We are now in the hypotheses of Theorem 31.1. What it is left to do is to determine h such that $h < \alpha$ and $Np_h(\alpha) \to \lambda$, where λ is a positive constant.

Define $h = \alpha - \epsilon$, where ϵ is an arbitrarily small, positive constant. It is immediate to see that $h < \alpha$. Let us now consider $\lim_{n,N\to\infty} Np_h(\alpha)$. Under the hypotheses above, the limit can be rewritten as:

$$\lim_{l\to\infty} \frac{k_d (\ln l^d)^h}{r^d h!} \tag{31.3}$$

Because $r^d = \Theta(1)$ by assumption, $(31.3) \approx \frac{(\ln l^d)^h}{h!}$. Taking the logarithm, we have:

$$\ln \frac{(\ln l^d)^h}{h!} = h \ln (\ln l^d) - \ln h! \approx h \ln (\ln l^d) - h \ln h = h \ln \frac{\ln l^d}{h}$$

Plugging $h = \alpha - \epsilon$ into the equation above, we obtain:

$$(\ln l^d - \epsilon) \ln \frac{\ln l^d}{\ln l^d - \epsilon} = (\ln l^d - \epsilon) \ln \left(1 + \frac{\epsilon}{\ln l^d - \epsilon}\right)$$

Because $\epsilon/(\ln l^d - \epsilon) \to 0$ for $l \to \infty$, we can use the first term of the Taylor expansion and write:

$$(\ln l^d - \epsilon) \ln \left(1 + \frac{\epsilon}{\ln l^d - \epsilon}\right) \approx (\ln l^d - \epsilon) \frac{\epsilon}{\ln l^d - \epsilon} = \epsilon > 0.$$

Thus, we have proved that, defining h as above, we have $Np_h(\alpha) \to \lambda$ for some positive constant λ. We can then conclude that when $n = dk_d \frac{l^d}{r^d} \ln l$, $r \ll l$, and $r^d \in \Theta(1)$, the minimum number of nodes in a cell is at least $h = \ln l^d - \epsilon$ a.a.s. Note that this condition implies that every cell contains at least one node a.a.s., which in turn implies that the network is connected a.a.s. So, in the conditions above, we also have a.a.s. connectivity.

What happens if n is set as above but $r^d \gg 1$? In this case, we can prove by a scaling argument (i.e., by dividing r, n, and l by r) that a similar result holds under the assumption $r \in O(l^\beta)$, for some $0 < \beta < 1$. However, in this case we are able to prove a less precise result, that is, that $h \in \Omega(\ln l)$. Finally, what happens if $n > dk_d \frac{l^d}{r^d} \ln l$? For any possible setting of r (intended as a function of l) with $r \in O(l^\beta)$, we have just proved that the minimum cell occupancy is at least $h \in \Omega(\ln l)$ when $n = dk_d \frac{l^d}{r^d} \ln l$. So, if we increase n with respect to this value, we can only increase the cell occupancy (recall that N does not depend on n) and the statement still holds. We can then conclude with the following theorem.

Theorem 31.2   *Assume that n nodes with transmitting range r are distributed uniformly and independently at random in $R = [0,l]^d$, for $d = 1,2,3$, and assume that $n \geq d \cdot k_d \frac{l^d}{r^d} \ln l$, where $k_d = 2^d d^{d/2}$. Further assume that $r \in O(l^\beta)$, for some $0 < \beta < 1$. If a cooperative strategy is used to alternately shut down "routing equivalent" nodes, then $P(NL_l \geq hT) \to 1$ as $l \to \infty$, where NL_l is the random variable denoting network lifetime and $h \in \Omega(\ln l)$.*

It turns out that the condition on n stated in Theorem 31.2 is also a sufficient condition for obtaining a.a.s. connectivity (see Theorem 8 in Ref. 29). Thus, Theorem 31.2 gives a lower bound to an ad hoc network's lifetime under a sufficient condition for the network to be connected with high probability.

It should be observed that in the optimal case, that is, when the n nodes are evenly distributed into the N cells, the network lifetime is exactly $d \cdot k_d \ln l$. The result stated in Theorem 31.2 therefore implies that, in the case of nodes distributed uniformly at random, the network lifetime differs from the optimal at most by a constant factor.

31.4.3   Limitations of the Analysis

The lower bound of Theorem 31.2 on network lifetime is mainly of theoretical interest for $d = 2, 3$. This is because reported simulations[30] have shown that, while the sufficient condition for a.a.s. connectedness is tight for $d = 1$, it becomes looser for two- and three-dimensional networks. Thus, the transmitting range r specified in the theorem might actually be substantially larger (or equivalently, the node density might be substantially greater) than what is necessary for connectedness for higher-dimensional regions. Hence, a careful evaluation of network lifetime under true minimum density conditions is still necessary. Identifying the true minimum density condition would also allow us to study the scalability of network lifetime with node density. These would provide a more accurate picture of the lifetime extension possible for cell-based approaches for $d = 2, 3$. Simulations investigating this issue were reported in our earlier work.[1] The remainder of this chapter focuses on $d = 2$ and compares cell-based approaches against topology control.

To make the preceding analysis tractable, we made use of Assumption *a8*, which states that traffic is balanced over all cells. In a realistic environment, this balanced traffic assumption will not hold. This assumption was also used in the simulation results.[1] In the next section, we present a more realistic traffic model that is designed for sensor networks.

The next section complements the analysis of this section by identifying (through simulation) the true minimum density conditions for $d = 2$ and formulating a more realistic traffic model. These are then used in simulations to perform a comprehensive evaluation and comparison of cell-based approaches and topology control.

31.5 A Framework for Comparison of Topology Control and Cell-Based Approaches

In this section, we define a framework for the comparison of cell-based and topology control techniques. The framework is *unified* because we make the same assumptions (to the extent to which this is possible*), especially on what concerns the traffic and the energy model. This framework builds on the idealized model presented in Section 31.4 for cell-based strategies.

31.5.1 The Cooperative Cell-Based Approach

As in Section 31.4, we focus on cell-based techniques because they appear to be the most promising of the cooperative approaches. In the remainder of the chapter, we adopt Assumptions *a1* through *a9* with the following changes. We restrict ourselves to the two-dimensional unit square region, as is common in ad hoc and sensor networks research, and we adopt a realistic traffic model, which is discussed in Section 31.5.3.

In view of Assumption *a6*, the energy savings derived in the following can be seen as the best possible any cell-based cooperative strategy can achieve. However, we want to remark that at least in the case where a loose synchronization mechanism is available to the nodes, energy savings very close to this ideal value can be achieved in practice.[30] That is, at least under certain conditions, the message overhead required for node coordination can actually be considered negligible.

Assumption *a4* is motivated by the fact that, in principle, the nodes' transmitting range should be set to the minimum value that makes the communication graph connected, known as the *critical transmitting range*. When the transmitting range is set to the critical value, the network capacity is maximal and the node energy consumption is minimal.[8,10,14,17] Setting the transmitting range to the critical value can be seen as a limited form of topology control, in which all the nodes are forced to use the same transmit power. That is, the results presented in this chapter, both referring to cooperative strategies and topology control protocols, must be interpreted as relative lifetime extensions *with respect to a simple energy saving approach*. Thus, we believe that even relatively small lifetime extensions (say, in the order of 30 percent) can be regarded as significant in this setting.

Although computing the critical transmitting range in a distributed manner is feasible,[17] it is quite common to characterize it using a probabilistic approach, that is, determining the value r of the transmitting

---

*Some assumptions will be simply not relevant for the topology control setting.

range that ensures connectivity with high probability (w.h.p.).[9,28,29] In this chapter, we will follow this probabilistic approach (see Section 31.6 for details).

31.5.2   Topology Control Protocols

Ideally, a topology control protocol should build a connected and relatively sparse communication graph. Furthermore, the protocol should preserve energy-efficient links between nodes, while avoiding energy inefficient multihop paths. Another desirable feature of the communication graph is a small node degree, which in principle implies reduced interference between nodes. Finally, a topology control protocol should be able to build the communication graph using local information only, and exchanging as few messages as possible. Clearly, designing a protocol with all these features is not an easy task. Among the many protocols proposed in the literature, in this chapter we consider the CBTC protocol presented by Li et al.[15] and Wattenhofer et al.[33] and the KNeigh protocol.[2] CBTC is one of the most popular topology control protocols, and is based on the idea of connecting each node to the closest node "in every direction," that is, in all nonoverlapping cones of a certain angle centered at the node. In KNeigh, every node is connected to the k closest neighbors, where k is a properly tuned network parameter. For details on these protocols, see Refs. 2, 15, and 33. The choice of considering these two protocols is motivated by the fact that they show good performance in terms of energy savings, and that they can be implemented exchanging relatively few messages. In particular, KNeigh requires that any node sends only two messages. Thus, our simplifying assumption of negligible message overhead (i.e., the counterpart of Assumption $a6$) can be regarded as quite realistic (at least in the case of stationary networks).

31.5.3   Multihop Traffic Generation

A typical strategy for generating multihop data traffic in WSN is the following: there are one or more fixed data collection sites (also called base stations) that are usually situated on the boundary of the deployment region; the (sensor) nodes generate CBR traffic directed to the collection sites. In this case, multihop traffic is generated when a node cannot reach a collection site directly.

In our experiments we have considered a variation of the above strategy. The basic idea is to compute the average load generated on every node when a (possibly) multihop message circulates in the network. We first compute, for every sensor node s, the most energy-efficient path $P_{sd} = \{s, u_1, \ldots, u_k, d\}$ that connects s to some data collection site d. If there are multiple energy optimal paths between a certain s, d pair, we randomly select one of them. Under the assumption that transmissions are perfectly scheduled (i.e., that no collision to access the wireless channel occurs), exactly $k + 1$ transmissions are needed to communicate a message from s to d. Using this information, the total number S_i of send operations and the total number R_i of receive operations performed by node i can be determined. The average load of node i when C multihop messages are transmitted in the network is then set equal to $C \frac{S_i}{n}$ send and $C \frac{R_i}{n}$ receive operations.

We stress that the above scheme (that uses energy-efficient routes) is only used to generate average node loads and does not prevent our approach from being used with any routing protocol.

31.5.4   The Energy Model

Our energy model for WSN is based on the measurements reported by Raghunathan et al.,[23] which refers to a Medusa II sensor node. That article reported the *sleep:idle:rx:tx_{min}:tx_{max}* ratios to be 0.235:1:1.008: 0.825:1.156, where tx_{min} corresponds to a transmit power of 0.0979 mW, and tx_{max} to a power of 0.7368 mW. The measurements refer to a data rate of 2.4 Kb/sec. Note that in this model, the energy consumed to transmit at minimum power is lower than that consumed to receive, or to listen to the wireless channel (idle state). This means that nodes that reduce their transmitting range significantly will actually consume substantially less energy when transmitting than when receiving or listening. Thus, nodes that transmit a lot might actually live longer than nodes that do not transmit often. This point, which is quite counter-intuitive, will be carefully investigated in our simulations.

TABLE 31.1 Values of the Critical Transmitting Range r for Different Values of n

n	r	n	r
10	0.65662	250	0.15333
25	0.44150	500	0.10823
50	0.32582	750	0.08943
75	0.27205	1000	0.07656
100	0.23530		

31.6   Simulation Setup

In this section we discuss some details of our simulation setup, including (1) node placement, (2) routing, (3) traffic generation, (4) energy consumption, (5) simulation time period, and (6) network lifetime definition.

31.6.1   Node Placement and Communication Graph

The simulator distributes n nodes in the deployment area $R = [0, 1]^2$ uniformly at random. Then it distributes $k = 4k'$ data collection sites in the region's boundary according to the following rule: for each side of R, k' base stations are located uniformly at random on that side. The transmitting range of a node depends on the particular scenario. In case of no energy saving (NES) or cell-based (CB) approach, this is set to the critical value r for connectivity (see Assumption *a4*). The values of the critical transmitting range are taken from Santi and Blough,[29] and are reported in Table 31.1. These values of r are also used to define the desired cell size in the CB scenario.

Under topology control, the transmitting range depends on the node placement itself and the particular protocol considered (CBTC or KNeigh). At the end of the execution of a given protocol,* a node u in the network is assigned a certain transmitting range r_u, with $r_u \leq r$.

For a given placement of the nodes, we then build the communication graph as follows:

- In the NES and CB scenario, we insert a bidirectional edge (i, j) whenever nodes i and j are at distance at most r (see Assumption *a3*).
- In case of topology control, we put a bidirectional edge between nodes u and v if and only if $\delta(u, v) \leq \min\{r_u, r_v\}$, where $\delta(u, v)$ denotes the Euclidean (two-dimensional) distance between u and v.

Note that, in all cases, we only consider bidirectional communication links.

The communication graphs generated by the various topology control protocols for a particular random instance are shown in Figure 31.1(b) and (c). Figure 31.1(a) shows the communication graph on the same set of nodes in the NES/CB scenario.

31.6.2   Routing

We assume that messages are routed along the most energy-efficient path from a sensor node to any data collection site. Note that in the NES/CB scenario in general, there exist several paths with minimal energy consumption for a given sensor node. In fact, the nodes use the same transmitting power, and any path connecting s to a collection site with minimum hop count is optimal from the energy efficiency

---

*In case of CBTC, the cone width is set to $2/3\pi$ (see Ref. 33 for details); the value of k in KNeigh is chosen according to the data reported by Blough et al.[2]

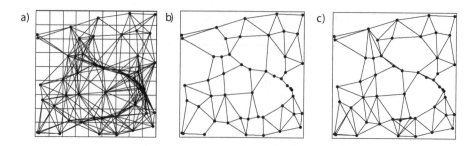

FIGURE 31.1   Communication graphs in the NES/CB scenario (a), and with CBTC (b) and KNeigh (c) topology control.

point of view. Thus, we must define a strategy to route the messages from the sensor node to the closest collection site in this scenario. In our simulation, we have implemented a cell-based routing algorithm, which can be seen as an instance of the widely studied class of geographical routing algorithms.

The cell-based routing algorithm is defined as follows. Given the communication graph G, we build a *cell graph* CG by inserting one node for every non-empty cell. Then we insert the bidirectional edge (i, j) in CG if and only if there exist nodes u, v in G such that $u \in C_i$, $v \in C_j$, and $(u, v) \in G$, where C_i (resp., C_j) denotes the cell to which node i (resp., j) corresponds. For any node x, let i_x denote the node in CG that corresponds to the cell to which x belongs. Now, for any source s, we compute one possible single source shortest path tree rooted at i_s, by selecting at random the node to be expanded next, among those eligible, in the classical Dijkstra's algorithm. We then use such tree for all the messages originating at s.

Note that the cell-based routing algorithm defined above does not specify completely the path from the source to the collection site: nodes in the same cell are perfectly equivalent from the routing algorithm's point of view, thus capturing the intuition upon which the cell-based strategy is founded. Indeed, the routing algorithm, as defined here, does not account for possible intra-cell communications needed to transmit/receive the message to/from the current leader node. That is, we make the following assumption, which is coherent with our idealized setting: node states and transmissions are perfectly scheduled; that is, when a node is scheduled to transmit/receive, it is also active.

31.6.3   Traffic Load

For the cell-based approach, we first compute the cell load by assigning counters S and R (see Section 31.5.3) to every cell in CG. For every sensor node, we select the path in CG that connects the corresponding cell to a collection site using the routing algorithm specified above and update the counters accordingly. The load of each cell C_i is then equally subdivided among the $n_i \geq 1$ nodes in it. That is, we assume that the routing algorithm is "smart," and that it perfectly balances the load between the nodes in the same cell. The assumption of perfect balance is again coherent with our idealized setting.

In the case of topology control, the evaluation of the per-node load is simpler because the minimum energy path that connects any sensor node to a data collection site is unique.

In both cases, the load is then normalized with respect to the total number of sensor nodes, and multiplied by n (i.e., we assume that n multihop messages, one for each node, are generated in the network). An example of the load generated in the NES/CB and with KNeigh topology control when $n = 25$ is shown in Figure 31.2. The numbers refer to the values of the sum of the send and receive counters. In the case of CB, the counters are referred to the cell, while in the case of topology control they are referred to the nodes. As expected, when an equal number of multihop messages circulate in the network, the traffic generated with topology control is higher than in the NES/CB scenario. This is due to the fact that the average hop distance between nodes and the closest base station is larger in case of topology control. For example, in Figure 31.2, a total of 33 send/receive operations are performed in the NES/CB scenario, while 65 operations are performed in the case of topology control.

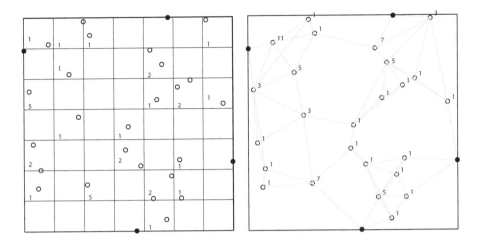

FIGURE 31.2 Load generated when $n = 25$ multihop messages are circulating in a network composed by n nodes in the NES/CB scenario (left) and with KNeigh topology control (right). Nodes (cells in the NES/CB scenario) are labeled with the sum of their S and R counters.

31.6.4 Energy Consumption

Let us first consider the NES scenario. Let $L_i = S_i + R_i$ be the overall load of cell C_i, and let n_i be the number of nodes in C_i. Assuming no contention, cell C_i will take L_i time units to handle the traffic. Let us consider a period of time T, with $T > L_i$ (we discuss how to set T later). If the traffic load in each cell is evenly distributed, every node in C_i spends $\frac{L_i}{n_i}$ units of time sending or receiving messages. The amount of energy consumed by a node $u \in C_i$ in this period is $\frac{rx \cdot R_i + tx_{max} \cdot S_i}{n_i}$, where rx is the energy needed to receive a message and tx_{max} is the energy needed to send a message at distance r.* In the remaining time $T - \frac{L_i}{n_i}$ of the period, the node remains idle, consuming one unit of energy per time unit. Thus, the energy consumption of node u during the period T is $E_u^{NES} = T - \frac{L_i}{n_i} + \frac{rx \cdot R_i + tx_{max} \cdot S_i}{n_i}$. Dividing this value by T, we obtain the average energy consumption per time unit of node u during the period T. Assuming that the traffic load is time invariant, the average energy consumption per time unit can then be used to compute the lifetime of node u.

Let us now consider the CB scenario. Every node in cell C_i sleeps for a fraction $(1 - \frac{1}{n_i})$ of the time period T. During this time, any node $u \in C_i$ has an energy consumption of $sleep \cdot (1 - \frac{1}{n_i})T$. In the fraction $\frac{T}{n_i}$ of the period, node u is active and spends $\frac{L_i}{n_i}$ units of time sending/receiving, consuming $\frac{rx \cdot R_i + tx_{max} \cdot S_i}{n_i}$ units of energy. In the remaining $\frac{T - L_i}{n_i}$ units of time, the node is idle, consuming $\frac{T - L_i}{n_i}$ units of energy. Thus, the energy consumption of node u during the period T is $E_u^{CB} = sleep \cdot (1 - \frac{1}{n_i})T + \frac{rx \cdot R_i + tx_{max} \cdot S_i}{n_i} + \frac{T - L_i}{n_i}$. As in the previous case, we divide this value by T, obtaining the average energy consumption per time unit which will be used to evaluate the lifetime of node u.

Finally, let us consider the case of topology control. Let again S_u (respectively, R_u) denote the number of send (respectively, receive) operations performed by node u (its load), and let us consider a period of time T as above. The node requires $L_u = S_u + R_u$ time units to handle the traffic, consuming $rx \cdot R_u + f(r_u) \cdot S_u$ units of energy. Here, r_u is the transmitting range of node u as set by the topology control protocol, and $f(r_u)$ is defined as follows: $f(r_u) = tx_{min} + (tx_{max} - tx_{min}) \cdot (r_u/r)^2$. With this definition, we have $f(0) = tx_{min}$ (energy consumed to send a message at minimum transmit power), and $f(r) = tx_{max}$ (when the transmitting range is set to the critical value r, the unit consumes energy tx_{max} to send a

---

*Without loss of generality, we assume that the power needed to transmit a message at distance r is the maximum transmit power.

message). In the remaining time $T - L_u$, node u remains idle, consuming $T - L_u$ units of energy. Thus, the energy consumption of node u during the period T is $E_u^{TC} = rx \cdot R_u + f(r_u) \cdot S_u + T - L_u$.

Note that in order to reduce the complexity of our experiments, we made some simplifying assumptions: no contention, traffic load evenly distributed in the cell, and time invariant traffic load. Admittedly, these assumptions may affect the significance of the *absolute* figures obtained. While this is clearly a point that deserves more investigations (see also Section 31.8), we observe that here we are primarily interested in the *relative* values of network lifetime extensions.

31.6.5   Simulation Time Period

The value of T is set as follows. Let $\bar{L}_i$ be the average per-cell load in the NES/CB scenario; T is set to $k \cdot \bar{L}_i$, where k is a parameter that accounts for the network load. For example, setting k to 10 corresponds to considering a network in which (on the average) a cell spends 10 percent of the time in communications.

31.6.6   Network Lifetime

We have used three different definitions of network lifetime, which are representative of notions of lifetime commonly used in the literature and which account for connectivity and the number of alive nodes.

Definition 31.1   Network lifetime is defined as:

1D: the time for the first node to die

DISC: the time to network disconnection

90LC: let N_C be the set of nodes that are connected to at least one data collection site; 90LC is defined as the time for $|N_C|$ to drop below $0.9n$

Before concluding this section, we give a numerical example of the node energy consumption computed according to the traffic and energy models described above. Recall that the power ratios we assume are *sleep:idle:rx:tx$_{min}$:tx$_{max}$* equal to 0.235:1:1.008:0.825:1.156. Suppose node u belongs to cell C_i, with $n_i = 3$, and assume $S_i = R_i = 5$. Let us set k to 20, which corresponds to a 5 percent average load scenario. We have $T = 100$, and the energy consumption per time unit is 1.002 in the NES scenario and 0.492 in the CB scenario. Suppose that the load of node u in the KNeigh scenario is $S_u = R_u = 12$ and that $r_u = \frac{r}{2}$. With these settings and the same value for T, we obtain an energy consumption per time unit of 0.989. Setting $k = 5$ (20 percent average load) yields an energy consumption per time unit of 1.011, 0.501, and 0.959 in the NES, CB, and KNeigh scenarios, respectively. Note that the energy consumption in the KNeigh is comparable to that in the NES scenario, and that considerable savings are achieved only with the cell-based approach. It is interesting to note that, in the KNeigh scenario, an increased traffic load results in decreased energy consumption.

31.7   Simulation Results

The goal of our simulation experiments is to answer the following questions:

- Given the same "minimal" node density for connectivity, which one between cell-based and topology control is more effective in extending network lifetime?
- How does the lifetime extension achieved by the two approaches scale with the node density?

We have performed experiments considering three values of the average load (5, 10, and 20 percent), three definitions of lifetime (1D, DISC, and 90LC), and different numbers of base stations (4, 8, and 12). Due to space limitations, we report only the most representative results. In particular, we have observed very similar performance in the cases of 1D and DISC lifetime definitions. In the case of CB, this is due to the fact that nodes tend to die in groups: first those that are alone in their cell, then those that are in a cell with two nodes, and so on. Because in the minimum density scenario cells on the average contain relatively few nodes,[1] the death of the nodes in the first group is very likely to disconnect the network. In the case of topology control, the graphs generated are quite sparse (see Figure 31.1), so the death of the first node is quite likely to disconnect the network also. Given this observation, in the following we report only the results obtained with the 1D and 90Lc lifetime definitions.

The results obtained with CBTC and KNeigh are practically indistinguishable in all the experiments. For this reason, in the graphics we report only the plots relative to the KNeigh protocol.

All the results reported are averaged over 1000 experiments.

31.7.1 Lifetime with Minimum Density

In the first set of experiments, we evaluated network lifetime in the NES, CB, and topology control scenarios under the hypothesis that the node density is the minimal for connectivity: given a value of n, we have set the maximum transmitting range to the critical value, as reported in Table 31.1.

The results of our simulations in the case of four data collection sites and 5 percent average load when network lifetime is defined as 1D are reported in Figure 31.3. The results of the same experiments with 20 percent average load are reported in Figure 31.4. Figures 31.5 and 31.6 report the results of similar experiments obtained with the 90Lc lifetime definition.

A few remarks are in order:

- Topology control outperforms the CB strategy in all the experiments performed. However, the relative advantage of topology control with respect to CB depends heavily on the definition of lifetime: it is very evident with the 1D definition, while it is only marginal in the case of the 90Lc definition. This indicates that topology control significantly postpones the times of first node death and loss of connectivity, but does not postpone the time at which large numbers of nodes begin to die, relative to the static and homogeneous range assignment policy embodied by NES.

- In absolute terms, topology control achieves considerable savings with respect to the NES scenario, as high as 370 percent for large networks with high load and 1D lifetime definition.

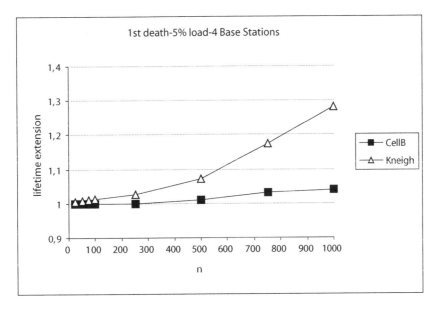

FIGURE 31.3 Lifetime extension for the CB strategy and KNeigh topology control for increasing values of n. The lifetime extension is expressed as a multiple of the lifetime in the NES scenario. Four base stations are used to collect data, and the average load is 5 percent.

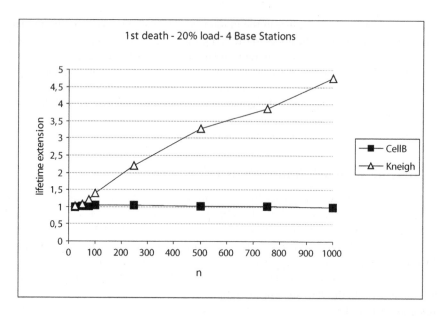

FIGURE 31.4 Lifetime extension for the CB strategy and KNeigh topology control for increasing values of n. The lifetime extension is expressed as a multiple of the lifetime in the NES scenario. Four base stations are used to collect data, and the average load is 20 percent.

The fact that topology control provides relatively better performance with high network load deserves some discussion as well. This better performance is due to the fact that nodes in NES transmit at maximum power. Thus, the large number of send operations generated in the 20 percent load scenario induces a better relative performance of topology control.

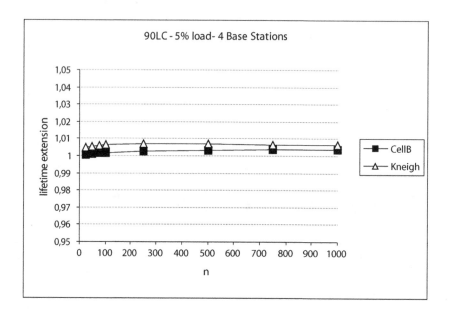

FIGURE 31.5 Lifetime extension for the CB strategy and KNeigh topology control for increasing values of n. The lifetime extension is expressed as a multiple of the lifetime in the NES scenario. Four base stations are used to collect data, and the average load is 5 percent.

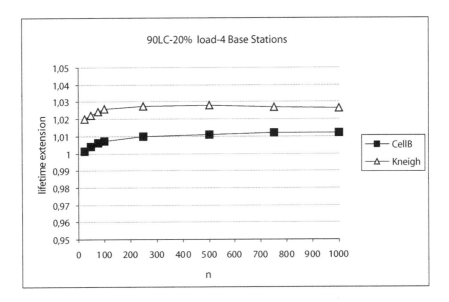

FIGURE 31.6 Lifetime extension for the CB strategy and KNeigh topology control for increasing values of *n*. The lifetime extension is expressed as a multiple of the lifetime in the NES scenario. Four base stations are used to collect data, and the average load is 20 percent.

We have repeated a similar set of experiments varying the number of data collection sites (4, 8, and 12 sites). The only difference with respect to the previous results is a somewhat decreased performance of topology control with respect to NES. For example, Figure 31.7 reports the results of the simulations with 12 base stations, the 1D definition of lifetime, and 20 percent load.

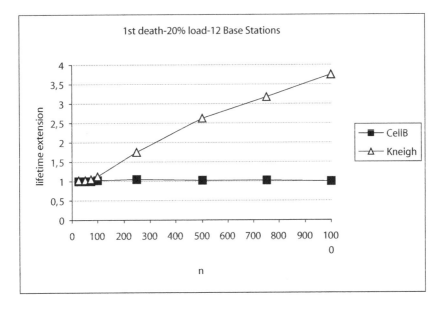

FIGURE 31.7 Lifetime extension for the CB strategy and KNeigh topology control for increasing values of *n*. The lifetime extension is expressed as a multiple of the lifetime in the NES scenario. Twelve base stations are used to collect data, and the average load is 20 percent.

31.7.2   Lifetime for Increasing Density

In the second set of experiments, we investigated how the lifetime extension achieved by CB and topology control scales with node density, relative to the minimum density scenario with $n = 50$ (hence with $r = 0.32582$, see Table 31.1). We increased node density by distributing ρn nodes and leaving the value of the transmitting range, and consequently also the cell size, unchanged. As the expected number of nodes in a cell is ρ times that of the minimum density scenario, the CB strategy is likely to achieve better energy savings as ρ increases. In the case of topology control, we simply computed the various topologies on the set of nodes as in the previous experiments. In this situation, the advantage of having a higher node density is that the transmitting range r_u of a generic node u is likely to be smaller than in the minimum density scenario. Thus, because $r = 0.32582$ independently of the node density, the node lifetime should, in principle, be longer than in the case of minimum density. However, whether or not the relative savings of topology control with respect to the NES scenario increase with density is not clear, due to the effect of multihop data traffic. This effect can be explained as follows. In the NES scenario, when the density is very high, the average hop-length between node pairs tends to be quite small, because the transmitting range is fixed. On the other hand, in the case of topology control the nodes' transmitting ranges are adjusted in order to build a sparse graph even when the node density is very high. Because the absolute number of nodes in the network increases, the average hop-length between node pairs increases as well. Because there is a considerable fixed energy cost associated with any one-hop communication, it could be the case that the high average hop-count in the case of topology control counterbalances the potential benefits of the reduced transmitting ranges.

The results of our experiments for values of ρ ranging from 1 to 10 (i.e., n ranging from 50 to 500) are reported in Figures 31.8 and 31.9, which refer to a 20 percent load with 1D and 90LC lifetime definitions, respectively. The number of base stations is set to 4. Contrary to the previous set of experiments, we have obtained very similar results for low load (5 percent), and for different numbers of data collection sites. For this reason, the results of these experiments are not reported.

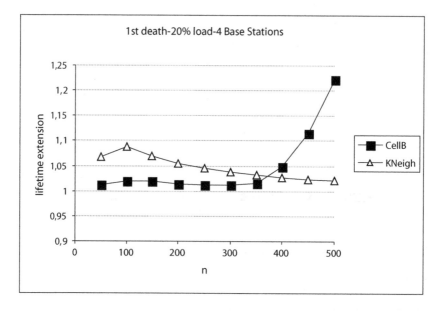

FIGURE 31.8   Lifetime extension for increasing values of the node density ρ. Network lifetime is defined as the time for the first node to die, and is expressed as a multiple of the lifetime in the NES scenario. The number of base stations is set to 4, and the load is 20 percent.

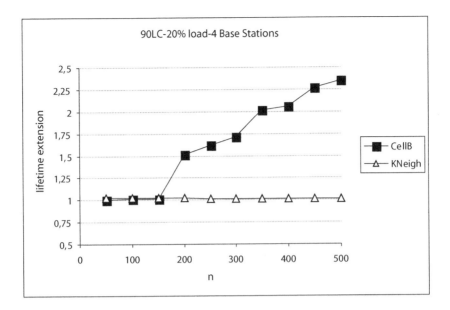

FIGURE 31.9 Lifetime extension for increasing values of the node density ρ. Network lifetime is defined as 90LC, and is expressed as a multiple of the lifetime in the NES scenario. The number of base stations is set to 4, and the load is 20 percent.

We remark that:

- The situation is reversed with respect to the case of minimum node density: with high node density, CB performs considerably better than topology control, especially when lifetime is defined as 90LC. In this case, CB can extend network lifetime by as much as 130 percent as compared to NES. Note, however, that when 1D lifetime is considered, node density must be *very* high before the benefits of CB become clear. For example, even when 3.5 times the minimum number of nodes are present, the lifetime extension is only about 1 percent. The benefits of CB become clear much earlier when considering the 90LC lifetime definition.

- The potential benefits of topology control due to the reduced transmit power consumption and the negative effect of the increased average hop-length balance almost perfectly. Almost independently of the node density, topology control protocols achieve basically the same savings as in the NES scenario. We believe this result is quite interesting and should not be underestimated. In fact, topology control has the potential to increase considerably the network capacity in very dense networks. So, increasing capacity while saving as much energy as a very simple energy saving technique is a very good feature in this scenario.

- As in the previous case, the definition of lifetime has a strong influence on the investigated energy saving techniques. For example, if we consider the 1D definition, CB extends lifetime by as much as 23 percent with respect to NES; with the 90LC definition, the relative extension can be as high as 130 percent. Note, however, that differently from the minimum density scenario, the better savings are achieved with the 90LC lifetime definition.

31.8 Discussion and Future Work

Our study has yielded a number of interesting insights into the issue of network lifetime and the studied approaches to energy conservation. First, we have seen that the lifetime definition has a significant effect on the results, both at minimum node density and at higher densities. We found that, in minimum density conditions, topology control was effective at postponing the times until first node death and loss of connectivity, but it was not effective at delaying the time at which large numbers of nodes die. Conversely, at moderately high densities, we found that the cell-based cooperative approach was able to significantly increase the lifetime when it is defined in terms of a large number of node deaths, but it did not significantly delay the times to first node death and connectivity loss. At very high densities, the cell-based approach significantly extends network lifetime according to all of the definitions we considered.

In comparing the cell-based approach with topology control, we can summarize the results as follows. When node density is near the minimum necessary for connectivity and the number of nodes is large, topology control performs better. However, the cell-based cooperative approach has a clear performance advantage over topology control at higher node densities. It is worth repeating that for lifetime definitions of first death and loss of connectivity, we have found that node density must be very high (at least four times more nodes than necessary for connectivity) to achieve significant benefits.

There are several points worthy of additional research.

- *Consideration of the relative performances of the two approaches in ad hoc networks.* Both the energy consumption parameters as well as the traffic model must be altered to consider an ad hoc network environment. It is not at all clear that the results we report herein will translate to this different setting.

- *Accounting for channel contention.* In general, topology control increases network capacity because of reduced contention. This fact has not been considered herein, where contention is nonexistent due to the perfect scheduling assumption. Indeed, achieving perfect scheduling in CB, especially with moderate load, could be very difficult. Refining our analyses to take collisions into account is the subject of future work.

- *Investigating the combined benefits of topology control in terms of energy reduction and capacity improvement.* In this chapter, we focused solely on evaluation of energy conservation. However, as noted earlier, topology control has the added benefit of reducing channel contention and therefore increasing overall network capacity. Quantifying the capacity improvement attained by topology control, and combining this with its energy reduction benefits to present a complete picture of its capabilities, is a worthy endeavor.

References

1. D.M. Blough and P. Santi, Investigating Upper Bounds on Network Lifetime Extension for Cell-Based Energy Conservation Techniques in Stationary Ad Hoc Networks, *Proc. ACM Mobicom 02*, pp. 183–192, 2002.
2. D.M. Blough, M. Leoncini, G. Resta, and P. Santi, The k-Neighbors Protocol for Symmetric Topology Control in Ad Hoc Networks, *Proc. ACM MobiHoc 03*, Annapolis, MD, pp. 141–152, June 2003.
3. S.A. Borbash and E.H. Jennings, Distributed Topology Control Algorithm for Multihop Wireless Networks, *Proc. IEEE Int. Joint Conference on Neural Networks*, pp. 355–360, 2002.
4. T. Brown, H. Gabow, and Q. Zhang, Maximum Flow-Life Curve for a Wireless Ad Hoc Network, *Proceedings of the ACM International Symposium on Mobile Ad Hoc Networking and Computing*, pp. 128–136, 2001.
5. J. Chang and L. Tassiulas, Routing for Maximum System Lifetime in Wireless Ad Hoc Networks, *Proceedings of the 37th Annual Allerton Conference on Communication, Control, and Computing*, 1999.
6. J. Chang and L. Tassiulas, Energy Conserving Routing in Wireless Ad Hoc Networks, *Proceedings of INFOCOM*, pp. 22–31, 2000.
7. B. Chen, K. Jamieson, H. Balakrishnan, and R. Morris, SPAN: An Energy-Efficient Coordination Algorithm for Topology Maintenance in Ad Hoc Wireless Networks, *Proc. ACM Mobicom 01*, pp. 85–96, 2001.

8. M. Grossglauser and D. Tse, Mobility Increases the Capacity of Ad Hoc Wireless Networks, *Proc. IEEE INFOCOM 01,* pp. 1360–1369, 2001.

9. P. Gupta and P.R. Kumar, Critical Power for Asymptotic Connectivity in Wireless Networks, *Stochastic Analysis, Control, Optimization and Applications,* Birkhauser, Boston, pp. 547–566, 1998.

10. P. Gupta and P.R. Kumar, The Capacity of Wireless Networks, *IEEE Trans. Information Theory,* Vol. 46, No. 2, pp. 388–404, 2000.

11. W. Heinzelman, A. Chandrakasan, and H. Balakrishnan, Energy-Efficient Routing Protocols for Wireless Microsensor Networks, *Proceedings of the 33rd Hawaii International Conference on System Sciences,* 2000.

12. Z. Huang, C. Shen, C. Srisathapornphat, and C. Jaikaeo, Topology Control for Ad Hoc Networks with Directional Antennas, *Proc. IEEE Int. Conference on Computer Communications and Networks,* pp. 16–21, 2002.

13. V.F. Kolchin, B.A. Sevast'yanov, and V.P. Chistyakov, *Random Allocations,* V.H. Winston and Sons, 1978.

14. J. Li, C. Blake, D.S.J. De Couto, H. Imm Lee, and R. Morris, Capacity of Ad Hoc Wireless Networks, *Proc. ACM Mobicom 01,* pp. 61–69, 2001.

15. L. Li, J.H. Halpern, P. Bahl, Y. Wang, and R. Wattenhofer, Analysis of a Cone-Based Distributed Topology Control Algorithm for Wireless Multi-Hop Networks, *Proc. ACM PODC 2001,* pp. 264–273, 2001.

16. N. Li, J. Hou, and L. Sha, Design and Analysis of an MST-based Topology Control Algorithm, *Proc. IEEE INFOCOM'03,* 2003.

17. S. Narayanaswamy, V. Kawadia, R.S. Sreenivas, and P.R. Kumar, Power Control in Ad Hoc Networks: Theory, Architecture, Algorithm and Implementation of the COMPOW Protocol, *Proc. European Wireless 2002,* pp. 156–162, 2002.

18. E.M. Palmer, *Graphical Evolution,* John Wiley & Sons, 1985.

19. P. Panchapakesan and D. Manjunath, On the Transmission Range in Dense Ad Hoc Radio Networks, *Proc. IEEE SPCOM,* 2001.

20. T.K. Philips, S.S. Panwar, and A.N. Tantawi, Connectivity Properties of a Packet Radio Network Model, *IEEE Trans. Information Theory,* Vol. 35, pp. 1044–1047, 1989.

21. P. Piret, On the Connectivity of Radio Networks, *IEEE Trans. Information Theory,* Vol. 37, pp. 1490–1492, 1991.

22. G. Pottie and W. Kaiser, Wireless Integrated Network Sensors, *Communications of the ACM,* Vol. 43, pp. 51–58, May 2000.

23. V. Raghunathan, C. Schurgers, S. Park, and M. Srivastava, Energy-Aware Wireless Microsensor Networks, *IEEE Signal Processing Magazine,* Vol. 19, No. 2, pp. 40–50, 2002.

24. R. Rajaraman, Topology Control and Routing in Ad Hoc Networks: A Survey *SIGACT News,* Vol. 33, No. 2, pp. 60–73, 2002.

25. R. Ramanathan and R. Rosales-Hain, Topology Control of Multihop Wireless Networks Using Transmit Power Adjustment, *Proc. IEEE INFOCOM 00,* pp. 404–413, 2000.

26. T.S. Rappaport, *Wireless Communication Systems,* Prentice Hall, 1996.

27. V. Rodoplu and T.H. Meng, Minimum Energy Mobile Wireless Networks, *IEEE Journal Selected Areas in Comm.,* Vol. 17, No. 8, pp. 1333–1344, 1999.

28. M. Sanchez, P. Manzoni, and Z.J. Haas, Determination of Critical Transmitting Range in Ad Hoc Networks, *Proc. Multiaccess, Mobility and Teletraffic for Wireless Communications Conference,* 1999.

29. P. Santi and D.M. Blough, The Critical Transmitting Range for Connectivity in Sparse Wireless Ad Hoc Networks, *IEEE Transactions on Mobile Computing,* Vol. 2, No. 1, pp. 1–15, January–March 2003.

30. P. Santi and J. Simon, Silence is Golden with High Probability: Maintaining a Connected Backbone in Wireless Sensor Networks, to appear in *Proc. European Workshop on Wireless Sensor Networks,* Berlin, Jan. 2004.

31. S. Singh, M. Woo, and C.S. Raghavendra, Power-Aware Routing in Mobile Ad Hoc Networks, *Proceedings of the International Conference on Mobile Computing and Networking,* 1998.

32. C. Srisathapornphat and C. Shen, Coordinated Power Conservation for Ad Hoc Networks, *Proc. IEEE ICC 2002,* pp. 3330–3335, 2002.

33. R. Wattenhofer, L. Li, P. Bahl, and Y. Wang, Distributed Topology Control for Power Efficient Operation in Multihop Wireless Ad Hoc Networks, *Proc. IEEE INFOCOM'01,* pp. 1388–1397, 2001.

34. Y. Xu, J. Heidemann, and D. Estrin, Geography-Informed Energy Conservation for Ad Hoc Routing, *Proc. ACM Mobicom 01,* pp. 70–84, 2001.

QoS Support for Delay Sensitive Applications in Wireless Networks of UAVs

Ionut Cârdei

32.1   Introduction

For the last few years, advancements in wireless technology have brought high-speed wireless networking closer to critical mass for civilian applications. Wireless Ethernet networks are now pervasive in airports, urban wireless hotspots, and office buildings, while start-up companies[1] develop products for ad hoc wireless networks that will provide ubiquitous connectivity to mobile stations.

Developments in wireless technology will also have an impact on how the military and homeland defense operate. DARPA's Future Combat Systems program[2] targets development and demonstration of future network-centric systems consisting of highly mobile robotic sensors and weapons platforms, supported by unmanned air vehicles (UAVs) and unmanned ground vehicles (UGVs). These assets are connected by multi-megabit ad hoc wireless networks with links having low probability of interception and detection, providing real-time data delivery with quality-of-service (QoS) constraints for C4ISR applications. Communications QoS (e.g., availability, bandwidth, jitter, and delay) is key for network support of mission-critical applications. Because wireless links typically have large variations in link quality and are prone to interference and fading, effective Medium Access Control (MAC) is key for providing QoS in the context of limited resources and multiuser scenarios.

This chapter presents the design and performance evaluation of a QoS-enabled MAC protocol for ad hoc wireless networks, called Receiver-initiated Access Control with Sender Scheduling (RACSS), specifically designed to address the communication and mission requirements of networks of Micro Air Vehicles (MAVs) and other FCS assets. Primary applications for MAVs include situational awareness, surveillance, and target monitoring. The MAV communication system must be able to support QoS for both sensor-generated high-rate data streams (video, IR, acustic) and delay-critical short messages carrying mission control, telemetry, or inter-vehicle coordination commands.

The proposed protocol uses contention-based MAC techniques and employs receiver-initiated sender scheduling for effective link-layer bandwidth allocation. Support for low latency transmissions is made possible by an adaptive RTS lead-time mechanism and priority-based frame scheduling. The basic principle behind a receiver-initiated MAC protocol, such as MACA-BI,[3] is that data frame transmissions are controlled by the receiver, by means of a control frame handshake for relaying invitations to senders. We present a sender scheduling mechanism, Burst Mode Priority Sender Scheduling (BMPSS), that allows traffic receivers to schedule data frame bursts for effective implementation of link-layer bandwidth allocation between active neighboring senders.

This chapter continues in Section 32.2 with a presentation of recent developments in QoS-aware MAC protocols. Section 32.3 describes the project objectives and the detailed protocol design. Performance analysis, evaluation, and simulation results for the RACSS MAC protocol follow in Section 32.4.

32.2   MAC Protocols and Quality-of-Service in Wireless Networks

An approach for QoS in wireless networks must span the protocol stack, starting with the link layer. To handle QoS from the network or transport layer only, considering the inherent dynamic nature of wireless networks in terms of RF link quality, mobility, or resource availability, would require an overly complex solution.

Design of Medium Access Control (MAC) protocols with QoS is an active research area. Most solutions use either a contention-free, scheduled access control, such as Time Division Multiple Access (TDMA) or a contention-based medium access control. With TDMA, a centralized controller divides transmission time into frames and allocates frame subdivisions, called time slots, to network nodes that have data to transmit. All nodes in the network are synchronized and each node is allocated one or more time slots to transmit its data. This simple mechanism is very effective for controlling bandwidth allocation and for providing low and deterministic communication latency. TDMA wireless protocols are widely used in cellular telephony by GSM networks,[4] satellite communications,[5] and in personal area networking Bluetooth[6] and Wireless Personal Area Networking — IEEE 802.15.[7]

The second class of MAC protocols uses contention-based access. Senders compete for transmission time and two or more frames may overlap, causing collisions at receivers. Therefore, most contention-based MAC protocols (except the first one, Aloha[8]) use methods for collision avoidance. Well-known contention-based MAC protocols include MACA,[9] MACAW,[10] and IEEE 802.11.[11]

Without additional transmission scheduling and coordination between transmitters, contention-based MAC protocols are not capable of supporting QoS. On the other hand, the centralized nature and strict time synchronization requirements of TDMA MAC protocols hinder efficient and simple support for ad hoc network topologies with hidden terminals. Apparently, any successful QoS MAC protocol for ad hoc networks from one class borrows design principles from the other class, as one can see in the examples below.

The work by Zhu and Corson[12] introduces the Five-Phase Reservation Protocol (FPRP) for MANETs. FPRP is basically a TDMA protocol in which nodes use a contention-based mechanism to reserve transmission slots within TDMA broadcast schedules. The authors also propose a multi-hop Aloha policy with a pseudo-Bayesian algorithm to speed up the convergence of the reservation phase. FPRP is a fully distributed protocol that supports concurrent and independent slot reservations for nodes at least two hops apart. FPRP also handles the hidden terminal problem with minimal collision probability.

A centralized TDMA MAC protocol designed for asymmetric connections to an aerial UAV backbone network is C-ICAMA.[13] This MAC protocol runs a slotted Aloha contention-based reservation protocol during the reservation subframe located at the end of each periodic frame. Reservation requests received correctly at the central controller are queued. For the next period, the controller sends a FRAME-START packet that indicates successful reservations for each time slot and availability for the reservation minislots. The protocol supports asymmetric links by adapting the number of downlink transmission slots during a frame. In addition, the controller handles congestion by changing the number of available reservation slots from the FRAME-START packet. C-ICAMA does not support ad hoc topologies and is strictly a backbone-access MAC protocol. Its centralized operation gives total QoS control to the switching node.

Most recent QoS research efforts on contention-based MAC solutions are extensions to IEEE 802.11. The IEEE 802.11e standard[14] proposes QoS support for wireless LANs with new Enhanced Distributed Coordination Function (EDCF) and Hybrid Coordination Function (HCF), a polling scheme based on EDCF. EDCF introduces traffic differentiation with Traffic Categories (TCs). The MAC protocol maintains one backoff instance and a separate transmission queue for each TC. Before transmission, a station has to sense the medium idle for a time interval equal to the Arbitration Inter Frame Space (AIFS) associated with the corresponding Traffic Category. Shorter AIFSs promote higher probability of medium capture for higher-priority traffic. In addition, IEEE 802.11e uses TC-specific backoff parameters such as the Persistence Factor and the minimum/maximum contention window, CWmin/CWmax. The HCF provides a good traffic prioritization method, but with no fine control over bandwidth allocation.

The work by Kanodia et al.[15] contributes two interesting QoS extensions to IEEE 802.11. First, a distributed priority scheduling technique piggybacks the priority of the head-of-line packet on control and data frames. Each node monitors transmitted packets and maintains a local scheduling table used to compare the node's priority level with other nodes. Transmitting packets ordered by priority practically allows implementation of selected scheduling policies, such as EDF. Priority-based access is provided by a modified backoff mechanism. The second contribution consists of a multi-hop coordination scheme that enables downstream nodes to adjust the priority of packets in transit to compensate for upstream delays.

Sheu and Sheu[16] propose extensive changes to the collision avoidance algorithm of IEEE 802.11 to implement a sophisticated bandwidth allocation mechanism, called DBASE, that supports both CBR and VBR traffic.

MAC-layer QoS solutions do not scale well end to end in ad hoc wireless networks. Resource management protocols at the network and transport layers are more effective and more compatible with application representation for QoS. QoS protocols that have been used in ad hoc wireless networks include Insignia,[17] Mobiware,[18] and RSVP. For best performance in ad hoc wireless networks, IP QoS models should be integrated with MAC layer QoS mechanisms wherever possible. DiffServ[19] traffic differentiation would fit in naturally with some of the MAC protocols mentioned above, while IntServ[20] could benefit from low-latency and efficient bandwidth allocation schemes, such as DBASE, FPRP, or the one we describe in this chapter.

32.3 Protocol Design

The next few paragraphs describe the applicability of QoS in applications with micro air vehicles (MAVs) and the motivation for a new MAC protocol approach.

Primary applications for MAVs include remote sensing, surveillance, and battlefield situational awareness for small combat units (Figure 32.1). MAVs will deliver video, infrared, and other sensor information

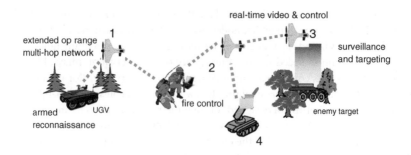

FIGURE 32.1   Micro air vehicle mission scenario.

to portable ground stations. Supporting competing QoS demands with a variety of attributes requires a vertical cross-layer approach for resource management and QoS, especially because MAVs have very limited on-board resources. Adverse channel effects (e.g., multipath fading, jamming, and co-channel interference) impair the effective transmission range when operating in dense foliage environments or in urban areas. In these scenarios, multiple vehicles will be used to form a multi-hop network to relay packets over line-of-sight links while maintaining the expected delay and bandwidth requirements.

One major design factor of our MAC QoS approach is the fact that the expected load on communication links is close to capacity whenever payload applications, especially video and radar, are active. The network must accommodate the high bandwidth requirements of payload data and the delay-sensitive constraints raised by transmissions from navigation and collision avoidance.

Existing MAC protocols for ad hoc wireless networks with support for QoS do not fully address the specific requirements for extended range, throughput, overhead, and delay constraints of MAV systems. Most MAC protocols have been built to suit generic scenarios with random mobility patterns and trade-off overhead for fair medium access.

32.3.1   Design Factors and MAC Protocol Features

We have analyzed a set of MAV networking scenarios and derived the set of traffic characteristics and requirements shown in Table 32.1.

We designed the RACSS as a receiver-initiated MAC protocol with QoS support and reliable transmissions based on link-layer retransmissions. Receiver-initiated medium access was first proposed in the MACA-BI protocol.[13] The main QoS features backing QoS are receiver-initiated sender scheduling for effective link-layer bandwidth allocation and support for low-latency transmissions enabled by an adaptive RTS lead-time mechanism and priority-based frame scheduling. The main MAC protocol features are summarized in Table 32.2.

TABLE 32.1   Traffic Characteristics and Requirements for MAV Networks

Traffic Characteristics	Communication Requirements
High rate data streams usually flow toward the base station	Mission-critical command/control transmissions must preempt any other communication
Transmission range can be extended ad hoc with multi-hop relay networks	Support for multi-hop ad hoc topologies
Multiple high data rate streams compete for the same network resources (bandwidth, relay buffer space)	High-priority traffic must be differentiated
Bandwidth and delay requirements change over time	Network must adapt and continue providing QoS to critical streams
Independent missions can operate in overlapping areas	Multi-channel capability
Variable link quality	Reliable link-layer transmissions

TABLE 32.2 RACSS Protocol Features

Functional Protocol Features	QoS Support Features
1 Mbps link speed, 10 km range	Reliable communication (ARQ with link-level acknowledgments and retransmissions)
Receiver-initiated transmission	Burst communication mode for high data rate streams and low overhead
Contention-based medium access with packet sensing collision avoidance	Low overhead under heavy load
Binary exponential backoff for contention resolution, with adaptive RTS lead time	Receiver-controlled transmitter scheduling
Broadcast and multicast	Maximum time-to-live for delay-sensitive packet filtering
Multi-channel operation for supporting independent missions	Packet priorities and priority transmission scheduling
Simple implementation; few resources required; no expensive carrier sensing	

32.3.2 MAC with Receiver-Initiated Transmission

MACA-BI[3] introduced first the concept of collision avoidance with receiver-initiated transmission. In contrast to IEEE 802.11,[11] MACA,[9] and MACAW,[10] receiver-initiated MAC protocols rely on receivers to decide which node transmits, when, and how much. This principle fits very well with communication patterns in MAV networks. For typical applications, high data rate demanding media/sensor streams flow down the network from remote vehicles to a central node for processing, with short delay-sensitive messages being transmitted between all mission participants.

RACSS frame types derived from MACA-BI:

Data — data frame. Fields: payload, length, priority, reliable (ACK is required from destination) and backlog — length of next frame for current destination.

Ready-to-receive (RTR) — control frame: receiver computes a transmission schedule and sends RTR invitations to neighbor nodes for a burst transmission. The RTR frame piggybacks acknowledgment information in frame bursts.

Acknowledgement (ACK) — control frame: upon receiving the last data frame without errors from a burst, a node replies with an ACK frame. If the sender does not get back an ACK within a certain time, it will attempt retransmission.

Request-to-send (RTS) — control frame: if a sender does not get an invitation (RTR) in time from a receiver, it will transmit an RTS frame. Upon reception of the RTS, the receiver can reply with an RTR invitation, or can defer and send an invitation to a different node.

Simulations have shown that the bandwidth overhead incurred from using ACKs to provide reliable communications is just 1 percent, while the additional delay is, on average, 700 microseconds. The relatively small overhead from ARQ pays well, especially in environments where FEC cannot provide sufficient immunity to noise and interference.

To relieve the overhead from the RTS—*RTR*—DATA—*ACK* handshake when a sender transmits consecutive frames to the same destination node, we introduced support for burst communication. After each incoming data frame, the receiver replies with an RTR, inviting the sender to continue transmitting data frames. When the receiver determines that the burst from a sender should be interrupted, it replies with a final ACK after the last burst frame. Starting with the sender's initial RTS, a burst follows this frame sequence for reliable transmissions: RTS—*RTR*—DATA—*RTR*—DATA—…—DATA—*ACK*, where *italicized* frames come from the receiver. RTR frames play two roles. First, they acknowledge correct data frame reception, and second, they keep the sender data frame flow open. In our current protocol implementation, if a data frame is received with unrecoverable errors, the receiver will transmit no RTR or ACK frame. The sender will timeout waiting for an acknowledgment and will reattempt transmission, starting with a new RTS frame.

For the purpose of discussing frame timing, we assume that processing overhead for frame transmission and reception is negligible compared to the actual transmission time, discounting also the radio transceiver turn-around time, necessary to switch the radio from transmit to receive mode.

After completing transmission of an RTS frame, the sender blocks, waiting up to a time T_{bc} for an RTR reply from the receiver:

$$T_{bc} = 2T_p + T_{RTR}$$

where T_p is the maximum propagation delay and T_{RTR} is the transmission time for an RTR frame. If the sender timeouts waiting for RTR, it assumes its RTS did not reach the receiver and will perform Binary Exponential Backoff (see Section 32.3.3) before retrying a new RTS to the receiver. Upon reception of an RTS frame, the receiver node transmits an RTR invitation to the sender and starts waiting for a time T_{bd} for the data frame, whose length is known from the RTS header:

$$T_{bd} = 2T_p + T_{DATA}$$

If the data frame does not arrive within T_{bd}, or is received with unrecoverable errors, the receiver switches to the idle state. In case the frame has been received correctly, the receiver replies with a new RTR, if the burst may continue, or with an ACK, to finish the burst for reliable data transmissions. The sender will therefore wait time T_{bc} for the reception of the ACK or RTR, before it timeouts and restarts the transmission process with an RTS frame, after performing backoff. For our RACSS implementation $T_p = 50$ μs, considering a maximum signal propagation distance under ideal conditions of 15 km and $T_{RTR} = T_{RTS} = 256$ μs, based on the control frame total size of 32 bytes, including an 8-byte preamble.

Transmission of broadcast and multicast frames in the RACSS protocol is performed similarly to MACAW and 802.11.

32.3.3   Collision Avoidance and Contention Resolution

For contention resolution, the RACSS MAC protocol uses a Binary Exponential Backoff (BEB) mechanism applied to packet sensing. Carrier sensing requires supplemental hardware resources and consumes extra power to keep the decoding stages active on the carrier frequency, even when the radio is idle. Therefore, the RACSS MAC protocol depends on the RTS—RTR handshake to prevent collisions through packet sensing.

The work done for MACA-BI and MACAW indicates that with error-free channels, there are no collisions between data frames, but only between control frames. Because RACSS control frames are very small (32 bytes, including preamble) compared to data frames, we expect the extra overhead from collisions between control frames to be low.

To reduce the probability of repeated overlapping RTS transmissions, the RACSS protocol uses a Binary Exponential Backoff scheme. The time unit for backoff delays (time slot) is determined by the time needed to receive an invitation, including the RTS:

$$T_{slot} = 2T_p + T_{RTS} + T_{RTS} = 612\mu s$$

In contrast with the 802.11 BEB version, our implementation resets the backoff counter to 0 when any of these events occurs while waiting for the blocking period to expire:

- Sender receives any kind of frame intended for itself (control or data)
- Sender receives the final ACK for other node's burst
- Maximum retry count is reached and the frame is dropped

IEEE 802.11 and its variations reset the backoff counter (named "contention windows" in IEEE 802.11 documents) to 0 when a sender receives a CTS reply or successfully completes a DATA transmission and an ACK is received. Our implementation of BEB resets the backoff counter to 0 after completion of a multi-frame burst of any node, not just the sender, reducing the medium access delay for all nodes to the penalties incurred by the variable RTS lead-time algorithm.

32.3.4 Design for Quality-of-Service

Our MAC protocol has an innovative combination of several mechanisms for providing support for link-layer QoS and traffic differentiation. We start our description with two frame differentiation mechanisms, and then continue with a receiver-driven sender scheduling mechanism called Burst-Mode Priority Sender Scheduling (BMPSS).

32.3.4.1 Variable RTS Lead Time

The RACSS MAC protocol implements a technique based on variable RTS lead times to provide fair medium access for non-critical frames and low-delay access to critical frames. Similar in concept to the 802.11 Inter Frame Space, the RTS lead time is the time a sender must wait when the medium is available before taking initiative and sending an RTS to the frame destination node. In our MAC protocol, most contention occurs between competing RTS frames. The original BEB scheme reduces contention by delaying nodes with higher backoff counters. An RTS frame with a lower lead time has greater probability of being received at the receiver. We propose an adaptive scheme for the RTS lead time that takes into consideration the criticality of the data frame and medium access fairness.

RACSS defines three values for the RTS lead time (Figure 32.2). The critical RTS lead time, equal to the maximum propagation delay T_p, is used for time-critical transmissions, such as for inter-aircraft collision avoidance coordination or weapon commands, $T_{RTSwait0} = T_p$. The regular RTS lead time, for normal (non-time-critical) traffic, is a randomly distributed number of time slots T_{slot}. More time slots are used when the sender has completed a successful transmission in the recent past ($T_{w1} = 10$ ms):

$$T_{RTSwait1} = U(0, rtsWaitSlots) \cdot T_{slot}$$

If the sender has not communicated within T_{w1}, the RTS lead time is set to

$$T_{RTSwait2} = U(0, rtsWaitSlots/2) \cdot T_{slot}$$

where *rtsWaitSlots* is a protocol parameter. Thus, nodes that have been waiting longer for access are more likely to be assigned a lower number of time slots than nodes that have been communicating recently. RTS frames for time-critical data are assigned the minimum lead time equal to T_p, preempting all other transmissions. This scheme can be easily extended for more priority levels, considering also fairness.

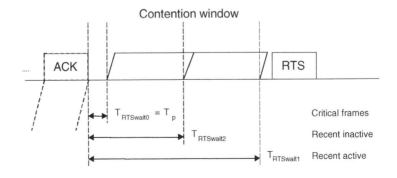

FIGURE 32.2 RTS frame lead time.

32.3.4.2   Priority Frame Scheduling

RACSS uses a priority number to select internally the frame to transmit from its internal queues. To break a tie, the oldest frame is selected. Priority-based transmission scheduling provides frames that carry critical information access to the medium with less queuing delay, reducing their total communication latency. The 8-bit priority field from the MAC frame header supports up to 256 priority levels and can be integrated with priority schemes from a higher layer, such as IP DiffServ.[19]

32.3.4.3   Delay Frame Filtering

For sensor information communications, frames received after their deadline expires are useless. For real-time video display and analysis, only the most recent frames are of interest. Similarly, voice or acustic data frames lose their value after a maximum delay or when more recent frames become available. The RACSS MAC header has a *time-to-live* field that indicates the remaining lifetime for a data frame in milliseconds. The *ttl* for a frame is set by higher layers, according to mission QoS information. Whenever a frame arrives at the MAC layer (from a higher layer), the *ttl* is reduced by the expected queuing time and the transmission time. If *ttl* < 0, the frame will be dropped immediately, because it would not reach its destination in a timely manner anyway. Right before a frame is sent to the physical layer, its *ttl* is updated again to reflect the actual time spent in the MAC frame queue. A frame with negative *ttl* will be dropped before transmission. This filtering mechanism saves buffer space and medium time, effectively improving QoS for data frames that can be delivered in time or have no deadline. Setting the *ttl* field to 0 indicates to the scheduler that filtering is not required for a frame.

32.3.4.4   Sender Scheduling

In this section, we propose an efficient mechanism for link-layer bandwidth allocation based on Burst Mode Priority Sender Scheduling. In the RACSS protocol, immediate medium access is arbitrated by the RTS lead time and the BEB mechanisms, while priorities determine the frame ordering. In centralized wireless networks, bandwidth allocation and medium access can be tightly managed with a TDMA link-layer protocol. TDMA does not perform as well in environments with hidden terminals that require support for ad hoc topologies with multi-hop routes. In these scenarios, TDMA MAC protocols must employ a coordination protocol that implements distributed time slot allocation. The FPRP MAC protocol[12] uses such an approach for slot reservation. A bandwidth allocation solution for contention-based MAC, implemented for IEEE 802.11, is described at http://grouper.ieee.org/groups/802/11/.[11]

With our contention-based MAC protocol with receiver-initiated transmission, we introduce a lightweight and simple solution for bandwidth allocation based on sender scheduling that supports ad hoc topologies and delay-sensitive transmissions. The mechanism for directing sender scheduling is based on the control frame handshake, centered on the Ready-To-Receive frame. If node A knows that nodes B and C have data frames queued up for node A, and that C's frames are more urgent, A will send an RTR invitation to node C first. The information on pending traffic and its QoS attributes would come from higher layers.

The sender scheduling algorithm is initiated with a list of bandwidth weights for each neighbor. A weight defines the fraction of total channel capacity that should be allocated to a specific sender. A mechanism derived from the Deficit Weighted Round Robin IP packet scheduling[21] is used to compute the sender bursts. A frame burst consists of several back-to-back data frame transmissions from the same sender. After each data frame reception, the receiver sends a new invitation with an RTR frame and piggybacks acknowledgment information. When the receiver decides to terminate the current burst, it replies with an ACK message instead of another invitation. The receiver can select to switch to another sender (including itself) or just relinquish medium control and let other nodes transmit.

During initialization, or whenever the bandwidth allocation changes, the receiver computes a new sender schedule $S = < t_i, b_i >$, consisting of medium transmission time t_i and the number of bursts allocated to each sender b_i. The receiver selects the next sender to be the node with the highest priority traffic pending.

The priority for a sender is considered the most recent received. If more senders have the same high priority, the receiver cycles through them in a round-robin way.

In typical scenarios with multi-hop paths, local traffic forwarders act as sinks for upstream nodes. Their bandwidth allocation is coordinated through *per-node* time division.

$$D = < f_j >$$

where $j = 1, \ldots, n$; and n is the number of "local sink nodes" (receivers) in the network.

The allocation f_j for each receiver is computed such that interference is kept low and spatial reuse is maximized. How to compute $<f_j>$ is outside the scope of this chapter.

The transmission time at a receiver node is further divided into equally sized scheduling cycles. f_j gives the actual time in a second that a particular receiver can schedule sender transmissions. Sender i is thus allocated a time t_i for transmission in each second by a receiver node j:

$$t_i = w_i f_j$$

During a scheduling cycle, all recent traffic participants are invited to transmit in order of their priority until they exhaust the maximum burst transmission time. For example, in Figure 32.1, the ground station node allocates 50 percent of transmission time to in-range relays 1 and 2, and 2 further allocates another 45 percent of transmission time to vehicle 3, and 5 percent to mobile node 4.

The transmission order is not strict. If a receiver gets an RTS from a node that is out of order, or without an allocation, it will allow the sender to complete a burst, and it will record its transmission time. During the next scheduling cycle that node will have its share reduced accordingly so that, overall, all nodes use their fraction of the allocated transmission time.

The schedule setup algorithm is listed in Figure 32.3.

If the total weighted sum of time allocations is less than 1, the remaining medium time is left unused by the receiver. The BMPSS algorithm includes in its schedule transmissions from the local node.

The sender schedule is activated at a receiver when a frame is received from a neighbor. The receiver updates the time the sender spent during the current scheduling cycle. In case the sender has more data frames intended for the receiver and has not exceeded its cycle time, the receiver will reply with a new RTR invitation. When the burst time or the cycle time is exceeded, the receiver will stop sending a new RTR to that sender. Instead, it will select another node (or itself) that has time left for transmissions in the current scheduling cycle. Another case where the receiver switches to a different sender is when the incoming packet indicates that the sender has no more data to transmit. The steps executed by BMPSS for an incoming DATA frame are shown in Figure 32.4:

The BMPSS sender selection is shown in Figure 32.5.

```
proc ScheduleSetup(w: bandwidth fraction array,
  sc: schedule cycle, f: receiver time fraction
  maxBS :maximum burst size)

  for each neighbor node i with w[i] > 0:
    // time allocated to sender I
    // during one scheduling cycle:
    cycleTime[i] = w[i] * sc * f
    // burst count for sender i
    // during a cycle:
    burstCount[i] = floor(cycleTime[i] / maxBS)
  endfor

  return cycleTime, burstCount
```

FIGURE 32.3 Sender scheduler setup.

```
eventhandler EventDATAReceived(frm)
//update scheduler state and select next
//sender:
  nextSender = ScheduleOnReception(frm)

  if nextSender = this node
    initiate transmission from this node
    //select frame from out queue and send RTS
    return
  else if nextSender != frm.source
    //burst interrupted, send ACK, if required:
    if frm.reliable == 1
      send ACK to frm.source
    endif
    if nextSender != NULL
      send RTR invitation to nextSender
    else
      defer transmission,
      let other nodes transmit
    endif
  endif
endif
```

FIGURE 32.4   Scheduler activation by the DATA reception event handler.

After executing ScheduleOnReception(), the receiver sends an RTR to the nextSender node. If nextSender is the receiver node itself, the receiver will initiate a transmission with an RTS, while keeping the receiver schedule updated just as it does for other nodes.

The SelectSender() function selects the highest priority sender with a non-zero allocation that has transmission time left available in the current scheduling cycle. If a node does not reply to an RTR invitation, the receiver will skip it for the current scheduling cycle. This prevents precious medium time being wasted on waiting frames from nodes that have gone out of range or have no other data pending for the current receiver node.

The BMPSS mechanism is opportunistic in the sense that a sender node transmitting to an idle receiver will be allowed to transmit an entire burst, even if it is out of order. The sender schedule is activated on a node when it receives data frames. Thus, when a network is lightly loaded, receivers do not waste medium time sending unnecessary invitations just to keep with the round-robin sender ordering.

Scheduling senders ordered on priority of their data provides additional support for low-latency communication for critical data.

32.4   Protocol Performance Evaluation

To evaluate the performance of the RACSS MAC protocol, we implemented the protocol engine and several scenarios for the OpNet network simulator.[22] Our initial measurements focused on protocol performance in static scenarios with variable load and in multi-hop topologies with hidden terminals. A topology with ten wireless nodes was defined, with all nodes in the same collision domain. The MAC protocol parameters are listed in Table 32.3.

The first series of measurements describe protocol performance depending on traffic load. The aggregate source traffic load is varied from 10 to 110 percent, where 100 percent corresponds to the total channel capacity (1 Mbps). The generated traffic mimics an application with MAVs transmitting sensor data to a ground-station node, and the GS sending short periodic control frames to each vehicle.

```
function ScheduleOnReception(frm: incoming frame,f: receiver time)
  nextSender = NULL

  //time left to schedule tx ?
  if receiver time fraction f depleted
                  return
  //yes, attempt to schedule sender:
  if frm.backlog > 0
    //sender has more data for this receiver
    //update time left for sender in current cycle:
    cycleTime[frm.source] =
      cycleTime[frm.source] - frm transm.time
    //update time used by sender in current burst:
    burstTime[frm.source] = burstTime[frm.source] + frm transm.time

    //transmission time for the next frame from sender:
    nextFrmTime = transmission time for a frame with frm.backlog bytes

    //if enough time left in this cycle and burst:
    if cycleTime[frm.source] - nextFrmTime > 0
        AND burstTime[frm.source] + nextFrmTime < maxBurstSize

      //current sender can keep transmitting:
      nextSender = frm.source
  endif
  endif

  if nextSender == NULL
    //select next sender among nodes with nonzero allocation
    //(including receiver node):
    nextSender = SelectSender()

    //setup a new burst:
    burstTime[nextSender] = 0
  endif
  return nextSender
```

FIGURE 32.5   Sender scheduling code.

Figure 32.6 shows the achieved goodput, throughput, and the relative overhead depending on the aggregate traffic source load. The network reaches its maximum goodput (80.6 percent) when the sources generate more than 900 kbps (90 percent capacity). The overall network throughput, including goodput and all control frames, is also limited at 91 percent of total channel capacity.

A key metric for MAC protocol performance is medium access delay. This measures the time spent by a frame within MAC buffers before transmission starts. It is the sum of the queuing delay and the channel

TABLE 32.3   MAC Simulation Parameters

Parameter	Value	Parameter	Value
Data rate	1 Mbps	Queue size	8 kB
MTU	1000 B	RTSWait	6 slots
Max. range	10 km	Schedule cycle	200 ms
Max. retry count	16	Max. burst time	20 ms

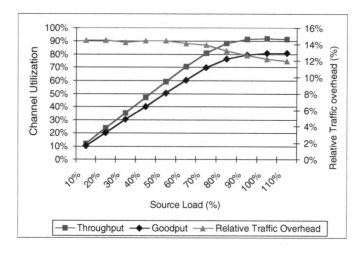

FIGURE 32.6   Goodput, throughput, and relative overhead, depending on source traffic load.

reservation delay. The medium access delay dependency on the aggregated offered goodput is illustrated in Figure 32.7. The maximum achievable goodput in the network is 80.6 percent from channel capacity (1 Mbps). As expected in networks with contention-based MAC, the delay rises at high load (after the goodput exceeds 70 percent from capacity). The eight-frame MAC transmission queue capacity and the round-robin transmission scheduling cause the average delay to exceed 0.5 s before the traffic load reaches the 80.6-percent plateau.

The overhead is measured relative to the total network throughput and decreases slightly, from 14.5 down to 12 percent, when the source load grows from 10 to 110 percent. Because the measured overhead includes transmission for all control frames and does not grow with network load, our numbers indicate that contention and collisions do not penalize the medium access delay as the network load increases.

The effectiveness of the sender scheduling mechanism is shown in Figure 32.8. The ground station node assigns each node an allocation weight that defines the fraction of channel capacity a sender is supposed to use for its transmissions. We measured the effective node transmission throughput for the (10%, 5%, 5%, 5%, 10%, 10%, 10%, 10%, 5%, 30%) allocation and the deviation from allocation. The ground station shows the best compliance to the desired schedule, only 0.02 percent, because it does the transmitter scheduling. The highest deviation from schedule (5.49 percent) is experienced by the sender with the

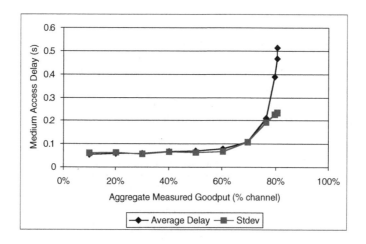

FIGURE 32.7   Evolution of the medium access delay with the offered aggregate network goodput.

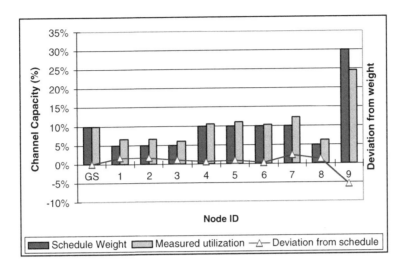

FIGURE 32.8   Schedule allocation weights, measured utilization, and deviation from schedule.

highest allocation (30 percent), achieving lower throughput than scheduled. Meanwhile, other senders with lower allocations consistently receive more capacity from the ground station. The deviation is caused by the source traffic definitions. Node 9 generates 10,000-byte-long messages that are fragmented in 1000-byte frames. The 20-ms maximum burst size and the preemptive nature of sender scheduling introduces more delay for node 9's transmissions, while contending with all other nodes.

The performance of priority-based sender scheduling is illustrated in Table 32.4. We ran the ten-node scenario with one vehicle node with critical priority. The table shows how medium access delay improves compared with the scenario with best effort priority. The average maximum delay and the jitter are more than twice shorter for higher-priority traffic. Also, the goodput improves 2.17 times for critical frame transmissions. The flexible sender scheduling mechanism does not strictly enforce the initial allocation, thus allowing capacity to be diverted to critical nodes on demand.

32.4.1 MAC Performance with Hidden Terminals

MAC protocol performance was evaluated in two scenarios with hidden terminals (Figure 32.9). The transmitter data rate is 1 Mbps in both scenarios and the MAC parameters are the same as in the previous simulations. In the first scenario, four nodes are placed in a typical hidden terminal topology. Node 1 transmits packets to node 2 and node 3 to 4, with a variable load factor. Transmissions from node 3 cause interference on the $1 \rightarrow 2$ link.

Both transmitters are assigned the same load, ranging from 10 percent of link capacity (aggregate 100 kbps) to 110 percent. To provide fair access in this asymmetric configuration, node 2 assigns 50 percent transmission capacity to node 1.

TABLE 32.4   Delay Performance for Best Effort and Critical Frame Transmission

	Best Effort	Critical	Improvement
Ave. (ms)	493.549	235.445	2.10
Min. (ms)	2.900	0.568	5.11
Max. (ms)	1391.911	595.472	2.34
Std. Dev. (ms)	298.434	127.308	2.34

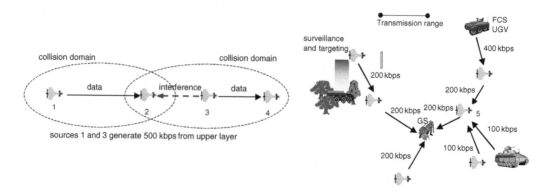

FIGURE 32.9   Scenarios with hidden terminals.

The charts in Figure 32.10 illustrate the mean aggregate goodput, throughput, overhead, and medium access delays when the transmitter aggregate load factor varies for nodes 1 and 3 from 10 to 100 percent of link capacity. The total achievable goodput exceeds 850 kbps, with a 10 percent overhead at highest load — performance comparable to a scenario without hidden terminal. The receiver scheduler active at node 2 allows node 1 to get a fair share of the medium time without excessive contention. The average delay measured at node 2 for transmissions from node 1 is, at maximum, 19 percent higher than the

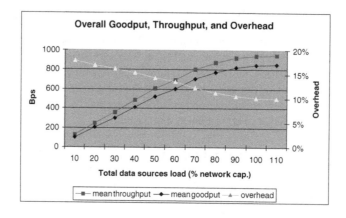

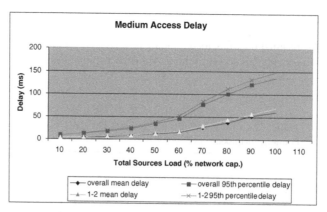

FIGURE 32.10   MAC performance for the four-node hidden terminal scenario.

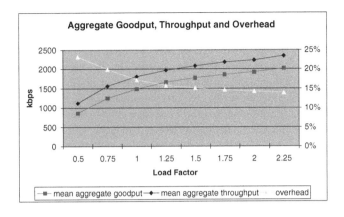

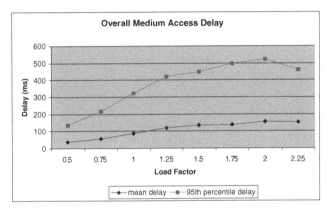

FIGURE 32.11 MAC performance for the nine-node hidden terminal scenario.

delay experienced at node 4. Also, 95 percent of all packets are transmitted within 175 ms, even when the wireless channel is saturated at capacity. Although transmissions from hidden node 3 interfere with node 1's packets, the RTS/RTR handshake and the bandwidth scheduling mechanism provide fair access for challenged node 1. The DATA frame retransmission rate for node 1 is 1.28/s and the sustained goodput is 407 kbps from an aggregate of 853 kbps.

The second scenario with hidden terminals places the nodes in a multi-hop topology and sets a reference load for point-to-point transmissions as indicated in Figure 32.9, with a total goodput of 1600 kbps. Nodes 5 and GS have a schedule assigned for transmitter scheduling to improve access fairness.

Performance measurements for the nine-node multi-hop scenario are shown in Figure 32.11. Maximum aggregate goodput reaches 2.02 Mbps when the load factor is 2.25 and the average medium access delay is 152 ms. The maximum retransmission rate for DATA frames does not exceed 2.68/s, which indicates that the MAC is effective in reducing contention from hidden terminals. Similar to other contention-based protocols with RTS/CTS handshakes, RACSS is susceptible to the exposed terminal problem, with nodes unnecessarily deferring transmission in scenarios where interference between DATA frames would not be an issue. The exposed terminal problem can be mitigated through additional sender synchronization and limiting the scope of control frame handshakes that cause interference.

32.5 Conclusion

This chapter presented the design of a Medium Access Control (MAC) protocol for multi-hop wireless networks of micro unmanned air vehicles (MAVs). Envisioned MAV applications, such as surveillance, sensing, and video monitoring, have demands for communication quality-of-service (QoS) with an emphasis on low overhead and low delay under heavy transmission load. The drivers for our MAC protocol design are support for multi-hop ad hoc networking, range, data rate, and constraints on volume and processing resources. We designed a new contention-based MAC protocol that implements a sender scheduling mechanism for effective and low overhead link-layer bandwidth allocation for nodes within communication range. Starting from a channel time allocation for neighbor nodes specified from a higher layer, receiving nodes generate a schedule for senders within communication range. The receiver orders senders and invites them to send data frame bursts, so each sender gets close to the specified channel time for each scheduling cycle. The scheduling mechanism is flexible and adapts when some senders become inactive or go out of range. We envision that MAC sender scheduling will be coordinated across a multi-hop topology by employing a network-layer configuration protocol integrated in an IntServ/DiffServ QoS architecture.

Further support for MAC layer QoS comes from priority-based frame scheduling with delay-sensitive frame filtering and a variable RTS lead-time scheme integrated with the Binary Exponential Backoff algorithm. We have evaluated the MAC protocol performance with simulations in OpNet and have shown promising results.

References

1. http://www.meshnetworks.com.
2. http://www.darpa.mil/fcs/index.html, http://www.boeing.com/fcs.
3. Fabrizio Talucci and Mario Gerla, MACA-BI (MACA by invitation). A Wireless MAC Protocol for High Speed Ad Hoc Networking, *Proceedings of IEEE ICUPC'97*.
4. Joerg Eberspaecher, Hans-Joerg Voegel, and Christian Bettstetter. *GSM Switching, Services, and Protocols, 2nd edition*, John Wiley & Sons, 2001.
5. Gérard Maral and Michel Bousquet, *Satellite Communications Systems: Systems, Techniques and Technology, 4th edition*, John Wiley & Sons, 2002.
6. http://www.bluetooth.com.
7. http://www.ieee802.org/15/.
8. N. Abramson, The Throughput of Packet Broadcasting Channels, *IEEE Transactions on Communications*, 25(1), 117–128, 1977.
9. Phil Karn, MACA — a new channel access method for packet radio, RRL/CRRL *Amateur Radio 9th Computer Networking Conference*, April 1990, pp. 134–140.
10. V. Bharghavan, A. Demers, S. Shenker, and L. Zhang, MACAW: a media access protocol for wireless LANs, *SIGCOMM'94*, September 1994, pp. 212–225.
11. http://grouper.ieee.org/groups/802/11/.
12. Chenxi Zhu and M.S. Corson, A Five-Phase Reservation Protocol (FPRP) for mobile ad hoc networks, in *Wireless Networks*, Kluwer Academic Publishers, Vol. 7, Issue 4 (August 2001).
13. D.L. Gu, H. Ly, G. Pei, M. Gerla, and X. Hong, C-ICAMA, A centralized intelligent channel assigned multiple access for the acess net of aerial mobile backbone networks, *Proc. IEEE WCNC 2000*, Chicago, IL, September 2000.
14. http://grouper.ieee.org/groups/802/11/Reports/tge_update.htm.
15. Vikram Kanodia, Chengzhi Li, Ashutosh Sabharwal, Bahareh Sadeghi, and Edward Knightly, Distributed priority scheduling and medium access in ad hoc networks, in *Wireless Networks*, Kluwer Academic Publishers, Vol. 8, Issue 5, September 2002.
16. Shiann-Tsong Sheu and Tzu-Fang Sheu, A bandwidth allocation/sharing/extension protocol for multimedia over IEEE 802.11 ad hoc wireless LANs, *IEEE J. Selected Areas in Communications*, 19(10), 2065–2080, 2001.
17. Seoung-Bum Lee, Gahng-Seop Ahn, Xiaowei Zhang, and Andrew T. Campbell, INSIGNIA: an IP-based quality of service framework for mobile ad hoc networks, *J. Parallel and Distributed Computing*, 60(4), 374–406, April 2000.

18. A. Campbell, Mobiware: QoS-aware middleware for mobile multimedia communications, *Proc. IFIP 7th Int. Conf. on High Performance Networking,* White Plains, NY, April 1997.

19. http://www.ietf.org/html.charters/diffserv-charter.html.

20. http://www.ietf.org/html.charters/intserv-charter.html

21. M. Shreedhar and G. Vargese, Efficient fair queuing using deficit round robin, *Proc. ACM SICOGMM'95,* 1995, pp. 231–242.

22. http://www.opnet.com.

33

A Scalable Solution for Securing Wireless Sensor Networks

Ashraf Wadaa

Kennie Jones

Stephen Olariu

Larry Wilson

Mohamed Eltoweissy

33.1   Introduction

Recent advances in nano-technology make it technologically feasible and economically viable to develop low-power, battery-operated devices that integrate general-purpose computing with multiple sensing and wireless communications capabilities. It is expected that these small devices, referred to as sensor nodes, will be mass-produced, making production costs negligible. Individual sensor nodes have a nonrenewable power supply and, once deployed, must work unattended. For most applications, we envision a massive random deployment of sensor nodes, numbering in the thousands or tens of thousands. Aggregating sensor nodes into sophisticated computation and communication infrastructures, called sensor networks, will have a significant impact on a wide array of applications, including military, scientific, industrial, health, and domestic, as illustrated in Figure 33.1. The fundamental goal of a wireless sensor network is to produce, over an extended period of time, meaningful global information from local data obtained by individual sensor nodes.[1,4,9,11,21]

It is expected that networking unattended wireless sensors will have a significant impact on the efficiency of many military and civil applications, such as combat field surveillance, intrusion detection, and disaster management. Wireless sensor networks process data gathered by multiple sensors to monitor events in an area of interest. For example, in a disaster management scenario, a large number of sensors can be dropped from a helicopter. Networking these sensors can assist rescue operations by locating survivors, identifying risky areas, and making the rescue crew aware of the overall situation. On the military side, the use of wireless sensor networks can limit the need for personnel involvement in the usually dangerous reconnaissance missions. Homeland security applications — including law enforcement, remote reconnaissance, monitoring, surveillance, and security zones ranging from persons to borders — can benefit enormously from the use of an underlying wireless sensor network. Similarly, medical, biological, scientific, and industrial applications can exploit wireless sensor networks to complement traditional data gathering, data fusion, and pattern matching capabilities. Wireless sensor networks also are key to the emerging "smart spaces," which will include numerous sensor nodes interacting with the physical world to provide information services almost everywhere and at all times.[14,27]

However, a wireless sensor network is only as good as the information it produces. In this respect, perhaps the most important concern is *information* security. Indeed, in most application domains, sensor networks will constitute an *information source* that is a mission-critical system component and thus require commensurate security protection. Security must be provided although sensor nodes are unattended and vulnerable to tampering.[5]

Due to the fact that individual sensor nodes are anonymous and communication is via wireless links, sensor networks are highly vulnerable to security attacks. It is clear that if wireless sensor networks are to

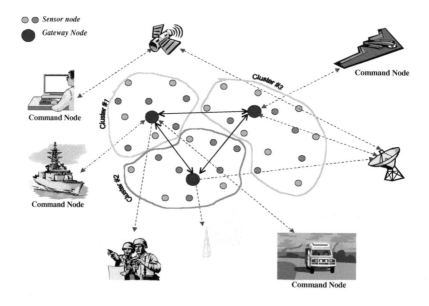

FIGURE 33.1 Applications of wireless sensor networks.

play a role in homeland security and other application areas, they must offer a high degree of security. If an adversary can thwart the work of the network by perturbing the information produced, stopping production, or pilfering information, then the usefulness of sensor networks will be drastically curtailed. Thus, security is a major issue that must be resolved if the potential of wireless sensor networks is to be fully exploited. The task of securing wireless sensor networks is complicated by the fact that the sensors are mass-produced anonymous devices with a severely limited energy budget and initially unaware of their location.[1,9,11,16,22]

Recently, the problem of securing ad hoc networks has received a great deal of well-deserved attention in the literature.[4,10,12,13,23,25] Many wireless sensor networks are mission-oriented, must work unattended, and espouse data-centric processing. Consequently, they are significantly different in their characteristics from ad hoc networks; security solutions designed specifically for wireless sensor networks are therefore required. Quite recently, a number of solutions for securing wireless sensor networks have been proposed in the literature.[16,19,23] Somewhat surprisingly, none of these solutions address the problem of jamming. Furthermore, all assume sensors with unique identities. Section 33.3 examines some of these solutions in more detail.

The main contribution of this work is to propose a novel solution to the problem of securing wireless sensor networks. Specifically, we show that by a suitable enhancement, the classic frequency hopping strategy[7,26] provides a lightweight and robust mechanism for securing wireless sensor networks. A significant advantage of our solution is that it is readily applicable to networks having anonymous nodes that are initially unaware of their location after deployment. It is worth noting that our solution supports a *differential security service* that can be dynamically configured to accommodate changing application and network state.

The remainder of this chapter is organized as follows. Section 33.2 introduces the wireless sensor network model assumed in the chapter, along with a surprisingly lightweight protocol for organizing the network into clusters. Section 33.3 provides the parameters of the security service that we propose, as well as the motivation, background, and state-of-the-art in securing wireless sensor networks. Section 33.4 presents the details of our proposed security solution. Section 33.5 evaluates our solution in terms of well-known security goals. Finally, Section 33.6 offers concluding remarks and maps out directions for future investigations.

33.2   The Sensor Network Model

We assume a class of wireless sensor networks consisting of a *sink node* and a large number of individual *sensors nodes* randomly deployed within the transmission range of the sink.

33.2.1   The Sensor Node Model

We assume that individual sensor nodes operate subject to five fundamental constraints. First, sensor nodes are anonymous; that is, they do not have either fabrication-time or pre-deployment runtime identities. Second, each sensor has a modest, nonrenewable power budget. Third, the sensor nodes are in sleep mode most of the time, waking up at random points in time for short intervals under the control of a timer. Fourth, upon deployment, the nodes must work unattended, because we assume human intervention is both impractical and undesirable. Fifth, the sensor nodes have a modest transmission range, perhaps a few meters, with the ability to send and receive a wide range of frequencies.

The range constraint implies that outbound messages transmitted by a sensor node can reach only the sensors in its proximity, typically a small fraction of the sensors in the entire network. As a consequence, the sensor network must be multi-hop and only a limited number of the sensor nodes count the sink among their one-hop neighbors. For reasons of scalability, it is assumed that no sensor node knows the topology of the network.

Sensor nodes possess three basic capabilities — sensing, computation, and communication — as illustrated in Figure 33.2. At any point in time, a sensor node will be engaged in performing one of a

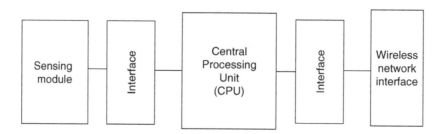

FIGURE 33.2   The anatomy of a sensor node.

finite set of possible operations, or will be idle (asleep). The unit of activity of a sensor node is an operation. Examples include sensing (data acquisition), routing (data communication; sending or receiving), computing (e.g., data aggregation), and roaming (e.g., receiving control data). We assume that each operation performed by a sensor node consumes a fixed amount of energy that may vary according to the operation type and that a sleeping sensor performs no operation and consumes essentially no energy. Also note that the cost per node of communication operations is significantly higher than that of computation.

33.2.2   The Sink Node Model

We assume that the sensor network interfaces to the outside world through a sink node. The sink has a full range of computational capabilities, can send long-range directional broadcasts to all sensors on a wide range of frequencies, can receive messages from nearby sensors, and has a steady power supply. The sink may or may not be collocated with the sensor network, and may be stationary or mobile (e.g., an LEO satellite, an airplane, or a helicopter).[9,11,20] This chapter assumes for simplicity that the sink is collocated with the sensor network. Importantly, the sink is in charge of performing any necessary training and maintenance operations involving the sensor network. To mitigate the risk of a collocated sink being a single point of failure, the role of the sink can be redundantly implemented using a collection of devices.[1,20]

33.2.3   Acquiring Location Awareness

We refer to *training* as the task of endowing individual sensor nodes with *coarse-grain* location awareness in the area of deployment. Training must be contrasted with localization, the stated goal of which is to provide sensor nodes with *approximate* location information.[1]

Briefly stated, the goal of training is to provide rough location information, establish clusters, and organize a network for node-to-sink multi-hop communications. Figure 33.3a features an untrained sensor network immediately after deployment. For simplicity, we assume that the sink node is centrally placed, although this is not necessary. Training imposes a *coordinate system* onto the sensor network in such a way that each sensor belongs to exactly one *sector*. Referring to Figure 33.3b, the coordinate system divides the sensor network area into equiangular wedges. In turn, these wedges are divided into sectors by means of *coronas* centered at the sink and whose radii are determined to optimize the transmission efficiency of sensors-to-sink transmission.[20] Note that we can decompose the network into clusters as a by-product of training at no additional cost. This is accomplished by using the collection of nodes in each sector as a cluster.

The intuition for establishing coronas and wedges is simple and is summarized below. The full technical details are rather complex and can be found in Wadaa et al.[20]

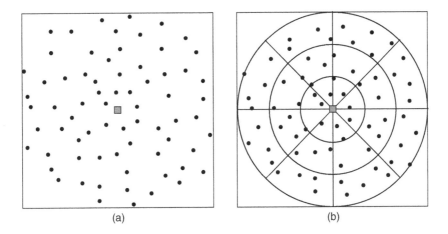

FIGURE 33.3   (a) An untrained sensor network with a central sink node; and (b) a trained sensor network.

Distance coronas. The sink starts out by transmitting a beacon at the lowest power level. All the sensors that receive the beacon above a certain threshold belong to the first *corona*. This is repeated at progressively higher power levels until every sensor in the network determines the identity of the *corona* to which it belongs. This process is termed "corona training."

Angular wedges. Using a reduced angular aperture, the sink will transmit a beacon to a small wedge of the sensor nodes. All the sensor nodes that receive the signal above a certain threshold belong to the first wedge. This is then continued for the second wedge and so on. This process is termed "wedge training."

Referring to Figure 33.3b, at the end of the training period, each sensor node has acquired two coordinates: (1) the identity of the corona in which it lies and (2) the identity of the wedge to which it belongs. The locus of all the sensor nodes that have the same coordinates is a cluster. It is interesting to note that clusters are very similar to sectors on a hard disk. The entire training process consumes on the order of log the number of sectors units of energy per sensor node. For the details of the training process the interested reader should refer to Wadaa.[20]

33.2.4   Routing in a Trained Sensor Network

Recall that sensor network communication paths are multi-hop from node to sink. Thus, for the sensing information to be conveyed to the sink node, routing is necessary. Our cluster structure allows a very simple routing scheme in that the information is routed within one wedge along a virtual path joining the outermost sector to the sink; a hop constitutes a direct transmission from one sector to the adjacent wedge sector that is closer to the sink, as illustrated in Figure 33.4.

33.3   Network Security: Motivation and Background

This section defines the parameters supported by our proposed solution to securing wireless sensor networks. We begin the discussion of these parameters by briefly reviewing some fundamentals of network security. We then go on to describe the principles underlying our application of these fundamentals.

33.3.1   Network Security Fundamentals

Encryption and denial of access to the physical layer are two basic techniques for securing communications in wireless networks.

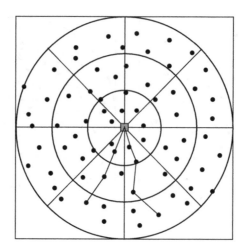

FIGURE 33.4   Illustrating the multi-hop routing paths.

33.3.1.1   Encryption

It is intuitively clear and confirmed by recent work[5,16] that due to the modest energy budget and limited on-board storage capabilities at the individual sensor nodes, public key cryptography is not an option for securing wireless sensor networks. Indeed, increasing the ratio of the total number of bits transmitted to the effective data bits (a result of encryption) increases the total number of bits transmitted and, thus, the energy consumed. Key management poses problems in sensor networks; for example, how are keys generated and disseminated? How can compromised keys be securely revoked and rekeyed with an acceptable delay time? Humans are not available at each sensor, distribution and modification of keys are difficult, and the sensor (and thus, embedded keys) is physically at risk. Perrig et al.[16] describe techniques for securing sensor networks primarily using encryption and delayed distribution of keys. While these techniques are novel and substantial, their overhead can be excessive. For example, they report that the energy overhead associated with transmission of security-related information in their scheme is more than a *quarter* of the entire initial energy budget of a node.

33.3.1.2   Denial of Physical Layer Access

33.3.1.2.1   *Frequency Hopping*

Frequency hopping can provide this service to wireless sensor networks. Given that techniques are known to discover a hopping sequence by monitoring transmissions, security can only be provided if the design modifies the hopping sequence in less time than that required to discover the sequence. Parameters in the specification of frequency hopping determine the time required to discover the sequence:

Hopping set: the set of frequencies available for hopping.
Dwell time: the time interval per hop.
Hopping pattern: the sequence in which frequencies in the hopping set are visited.

A dynamic combination of these parameters can improve security with little expense in terms of memory, computation, and power. As frequency hopping requires events to happen simultaneously for both senders and receivers, all must maintain a synchronized clock.

33.3.1.2.2   *Resistance to Physical Tampering*

Along with other workers,[2,3,5,22] we assume that the form factor and low cost of individual sensor nodes makes for minimal tamper resistance and protection. Indeed, rudimentary tamper protection can be obtained by blanking out memory if the sensor is pried open. However, the protection offered by this solution is far from adequate. One of our contributions in this chapter is to propose a lightweight solution to the tampering problem that does not rely on the use of sophisticated tamper-resistant hardware.

33.3.2 Guiding Principles of Holistic Security in Wireless Sensor Networks

We view this chapter as an initial contribution toward developing a holistic solution for securing wireless sensor networks. Our solution provides security not only for the various individual layers of the system, but also for the entire system in an integrated fashion.

We now propose a set of four *guiding principles* for addressing the problem of securing wireless sensor networks. A solution in the context of these principles supports a *differential security service* that can be dynamically configured to cope with changing network state (e.g., a detected state change in security risks or energy content in the network). Reconfiguration of dynamic security service can potentially minimize the energy cost of security over the network lifetime.

The four guiding principles for securing wireless sensor networks include:

The security of a network is determined by the security over all layers. For example, provisioning confidentiality, two-party authentication, and data freshness typically addresses the security of the link layer. Referring to Figure 33.5, we note that securing the link layer confers the layers above it some security; however, it does not address security problems in the physical layer below, most notably jamming. In general, an insecure physical layer may render the entire network insecure, even if the layers above are secure. This is especially true in the wireless sensor network environment because basic wireless communication is inherently not secure.

In a massively distributed network, security measures should be amenable to dynamic reconfiguration and decentralized management. Given the basic characteristics of wireless sensor networks, a security solution must work without prior knowledge of the network configuration after deployment. Also, the security solution should work with minimal or no involvement of a central node to communicate globally (or regionally) shared information.

In a given network, at any given time, the cost incurred due to the security measures should not exceed the benefit provided by the security. A sensor network is expected to experience risks of varying magnitude over its possibly long lifetime. Thus, security services should adapt to changes in assessed levels of risk. This suggests that a cost model for both security provisioning and risk be an integral part of the security model.

If the physical security of nodes in a network is not guaranteed, the security measures must be resilient to physical tampering with nodes in the field of operation. For example, a wireless sensor network deployed in a battlefield should exhibit graceful degradation if some individual sensor nodes are captured.

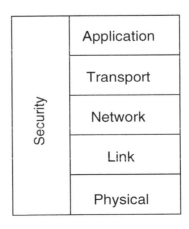

FIGURE 33.5 A holistic view of security.

33.3.3   Related Work

Recently, the problem of securing ad hoc networks has received a great deal of well-deserved attention in the literature.[10–13,23,25] However, because wireless sensor networks are different in their characteristics from ad hoc networks, security solutions designed specifically for the former do not apply to the latter. Quite recently, a number of solutions for securing wireless sensor networks have been proposed in the literature.[16,19,22] We now examine some of these solutions from the viewpoint of the guiding principles proposed above.

Carman et al.[5] surveyed a wide array of protocols for key agreement and key distribution in wireless sensor networks. They analyzed the overhead of these protocols on a variety of hardware platforms but did not address issues related to denial-of-service and solutions to tampering.

Perrig et al.[16] proposed SPINS, a general security infrastructure for sensor networks. The infrastructure consists of an encryption primitive, SNEP, and an authenticated streaming broadcast primitive, *micro TESLA*. These primitives constitute building blocks that can be used to construct higher-level security services. SPINS, however, does not address security in the physical layer. In addition, SPINS supports a *binary* security model — either *no security* or *maximum security*. In addition, the base station acts as a central authority for establishing trust between any pair of nodes, departing, to some extent, from the principle of decentralized security management. Further, the key management model in SPINS does not scale in a massively deployed network because it prescribes a unique key per node (called master key) to be loaded into the node prior to deployment. Physical tampering is not addressed in SPINS.

TinySec[19] proposes a link layer security mechanism for sensor networks based on an efficient symmetric key encryption protocol. Similar to SPINS, TinySec does not address security in the physical layer, and is based on a binary security model. TinySec presents an assessment of the trade-offs between security on one hand, and energy consumption and RAM utilization on the other hand. However, this cost model is not integrated into their security model and is not a factor in configuring the security service supported. In fact, the security service supported by TinySec is not dynamically reconfigurable. Another limitation of TinySec is that it is tightly coupled with the Berkeley TinyOS radio stack[21] and is therefore not applicable to a general sensor network model.

33.4   Our Solution

We propose a solution for securing wireless sensor networks that adheres closely to the guiding principles stated previously. The proposed solution uses parameterized frequency hopping and cryptographic keys in an integrated framework to provide differential security services for the network. Parameters are used to define a *configuration space* for the security service. In general, different configurations of the security service are characterized by the energy cost assessed and the amount of security afforded.

It should be noted that frequency hopping in radio communications is not a new idea and has been explored previously in the context of sensor networks.[7,22,26] However, these explorations use frequency hopping as a means of implementing frequency diversity and interference averaging in a non-hostile context. Typically, these efforts offered little security.

33.4.1   Basic Problems to Address

Our solution will address the fundamental problems or tampering, authentication, path security, and clock synchronization.

33.4.1.1   Tamperproofing Nodes

The most obvious tamper-resistance strategies are hardware based and involve special hardware circuits within the sensor node to protect sensitive data, special coatings, or tamper seals.[5] However, hardware solutions to the tampering problem require extra circuitry that increases the cost and hardware complexity of sensor nodes. Worse yet, the additional hardware is very likely to consume valuable energy, already in short supply. Also, special coatings and seals may offer protection against some — but certainly not all — tampering attempts. Indeed, it is assumed that a sufficiently capable adversary can extract confidential information, thus compromising the sensor node. Thus, not surprisingly, tamper resistance or tamper protection is not found in present-day sensor nodes.[5,22] Because wireless sensor networks must function unattended, the potential for physical tampering attacks is significant. It is worth noting that while *pre-deployment* tamper detection may be worthwhile, *post-deployment* tamperdetection is of little use in wireless sensor networks because, in the vast majority of applications, inspecting individual sensor nodes is not an option. Also, physical tampering can compromise only the node attacked (ideal), the immediate neighborhood of the node attacked, or the entire network. To cope with these conditions, our solution subscribes to the notion of "self-guarding" in that each sensor node should be able to detect, independently, physical tampering and should react such that the impact of the attack is minimal. Our solution to endow individual sensor nodes with tamper resistance does require additional or more sophisticated hardware.

To set the stage for discussing our solution, we note that the tampering *threat model* assumes that the adversary is (1) either trying to force open an individual sensor node in place or (2) physically removing the sensor node from the deployment area. We guard against the first threat by blanking out the memory. We guard against the second threat by relying on local data that the sensor can collect, thus establishing a unique signature of its neighborhood that is difficult to forge. To be more specific, immediately after deployment, each sensor transmits, during its wake time, on a specified set of frequencies, using a frequency hopping sequence established prior to deployment. This allows individual sensor nodes to collect an array of signal strengths from the sensors in their locale. It is important to recall that sensors do not have identities and that, consequently, the array of signal strengths is the only data available to the sensor node. This array establishes, in the obvious way, a signature of the neighborhood of the node. For this reason, the array will be referred to as the node's *signature array* (SA). If the node is removed from the area of deployment, it will notice changes in the signals received when compared to its SA and erase its own memory to prevent the tampering agent from gaining access to information secret to the sensor network. Note also that tampering attempts that involve the simultaneous removal of several sensor nodes will also be defeated because some node in the set of removed nodes will notice changes in its SA and can alert the others.

33.4.1.2 Authentication of Neighbors

The node *authentication* problem is one of the key problems in securing wireless networks. Our solution to node authentication relies on the signature array discussed above. Specifically, neighboring nodes exchange learned SA information, creating a matrix of signatures. When a node wishes to communicate with a neighbor, it establishes an authenticated connection by identifying itself with its own SA. Upon receiving the SA, the target node only needs verify that the corresponding signature is valid. The scheme can be made even more secure by storing several previous instances of the matrix of SAs. With this, the authentication dialogue can ask, for example, for the "second to the last SA."

33.4.1.3 Trusted Communication Paths

In our solution, all communication uses a frequency hopping mechanism, as we explain shortly. The *security measure* we employ for paths is a *randomization process* defined on the frequency hopping mechanism and driven by a secret shared by the path from sender to receiver. Our solution scales in the number of nodes and provides security for an entire *path*, as opposed to *hop or link* security.

33.4.1.4 Synchronization Provided

For communication to occur using frequency hopping, both the sender and receiver must be in sync. We propose a lightweight synchronization solution scalable in the number of nodes that enables a sender and a receiver sharing a common secret to synchronize.

In Section 33.4.2, we leverage frequency hopping to provide significant security for wireless sensor networks. In Section 33.4.3, we extend our solution to further enhance this security for a wireless sensor network that has been trained as described in Section 33.2.

33.4.2   Frequency Hopping in a Wireless Sensor Network

33.4.2.1   Secure Communications

We assume that the sender and receiver are mutually trusting, and also synchronized. Our solution works as follows. We assume that time is ruled into epochs. For a given sender s and receiver r at time epoch t, s transmits (and r receives) following a *hopping pattern* across a set of frequencies, called the *hopping set* for t. We assume that, for each epoch, the hopping set is drawn from a designated frequency space (band) that provides the set of all possible frequencies that can be used (e.g., ISM band).

The key idea is that the shared secret between s and r is used to drive a randomization of the frequency-hopping process. Specifically, the shared secret enables the *epoch length*, the hopping pattern, and the size and membership of the hopping set for each epoch to be changed according to random number sequences. Let both s and r be in sync at epoch t_i. Seeded by the shared secret, a random number generation scheme is used in both s and r to generate the successive epoch lengths, hopping sets, and hopping patterns, for the epochs $t_i, t_{i+1}, t_{i+2}, \ldots$. To an observer, however, successive epoch lengths, hopping sets, and hopping patterns appear as the product of an unknown random process.

The cost incurred by the network is a function of the configuration of the security parameters; each of the epoch length, frequency set, and frequency pattern can be dynamically configured. This gives rise to a differential security service that potentially incurs differential cost. On the one hand, a constant epoch length, hopping pattern, and hopping set correspond to minimal security and minimal cost incurred. However, randomized epoch length, hopping pattern, and hopping set correspond to maximum security provided and cost incurred.

Synchronization is an important concern in any frequency-hopping scheme. In our solution, we propose a scalable and lightweight synchronization scheme to enable arbitrary pairs of sender and receiver nodes that can exchange messages directly to synchronize. This scheme is detailed next.

33.4.2.2   Synchronization

The main goal of this subsection is to spell out the details of a scalable synchronization protocol that underlies our new security paradigm for wireless sensor networks. Our protocol achieves synchronization in a probabilistic sense. The natural way for nodes to synchronize is by following the master clock running at the sink node. Thus, the sink node here is the sender and the node that wants to synchronize is the receiver.

We assume that the sink dwells τ microseconds on each frequency in the hopping sequence. It is clear that determining the epoch and the position of the sink in the hopping sequence corresponding to the epoch is sufficient for synchronization.

For the purpose of showing how synchronization is effected, recall that time is ruled into epochs t_1, $t_2, \ldots, t_n$. For every $i, (1 \leq i \leq n)$, we let l_i stand for $\left\lceil \frac{t_i}{\tau} \right\rceil$; thus, epoch t_i involves a hopping sequence of length l_i. Further, with epoch $t_i (1 \leq i \leq n)$, we associate a set of n_i frequencies and a corresponding hopping sequence $\lambda_1, \lambda_2, \ldots, \lambda_{l_i}$. We can think of the epoch $t_i (1 \leq i \leq n)$ as being partitioned into l_i slots $s_1, s_2, \ldots, s_{l_i}$ such that in slot $j, (1 \leq j \leq l_i)$, the sink is visiting frequency λ_j.

We assume that, just prior to deployment, the sensor nodes are synchronized. However, due to natural clock drift, resynchronization is necessary. Our synchronization protocol is predicated on the assumption that clock drift is bounded, as we are about to explain. Specifically, assume that whenever a sensor node wakes up during its *local* time epoch t_i, the master clock is in one of the time epochs t_{i-1}, t_i, or t_{i+1}. The sensor node knows the *last* frequencies $\lambda_{l_{i-1}}, \lambda_{l_i}, \lambda_{l_{i+1}}$ on which the sink will dwell in the time epochs t_{i-1}, t_i, and t_{i+1}. Its strategy, therefore, is to tune in, cyclically, to these frequencies, spending r/3 time units on each of them. It is clear that, eventually, the sensor node meets the sink node on one of theses frequencies. Assume, without loss of generality, that the node meets the sink on frequency λ_{i+1} in some (unknown) slot s of one of the epochs t_{i-1}, t_i, or t_{i+1}. To verify the synchronization, the node will attempt to meet the sink in slots $s + 1, s + 2$, and $s + 3$ at the start of the next epoch. If a match is found, the node declares itself synchronized. Otherwise, the node will repeat the above process. We note that even if the sensor node declares itself synchronized with the sink, there is a slight chance that it is not. The fact that the node has not synchronized will be discovered quickly and it will again attempt to synchronize. There are ways in which we can make the synchronization protocol deterministic. For example, the hopping sequence can be designed in such a way that the last frequency in each epoch is unique and it is not used elsewhere in the epoch. However, this entails less flexibility in the design of the hopping sequence and constitutes, in fact, an instance of a *differential security service* where the level of security is tailored to suit the application or the power budget available.

33.4.3   Trusted Paths in a Trained Sensor Network

33.4.3.1   How Do a Sender and Receiver Establish a Trusted Communication Path (i.e., Acquire a Shared Path Secret)?

(Precondition: at pre-deployment, the entire network shares a secret for an initial post-deployment frequency-hopping phase [this is a condition for the proper operation of the network even if secret paths are not used.] A network without node identifiers receives training, for example, as described in Section 33.2. On the average, nodes are distributed equally among wedges.)

We establish secure paths between a sender and a receiver as follows. Pre-deployment sensors are loaded with a set of m keys that are selected at random from a set of k keys. The number of keys k is chosen such that two random subsets of size |m| overlap in at least one key with probability p. Post-deployment, a link can be established between neighboring sensors on the path to the sink if a key of their selected sets of m keys overlaps. It is to be noted that the number of overlapping keys can be a parameter in the security solution. On the one hand, increasing the number of overlapping keys needed to establish a link will reduce the number of paths between nodes, which will make it more difficult to eavesdrop; on the other hand, it may limit the existence of paths that might otherwise be selected due to other criteria, for example, energy budget. Similar approaches to the one used here for secure path determination have been proposed by Eschenauer and Gligor[28] and Chan et al.[29] to use encryption keys to construct a connection graph for distributed security over an entire wireless sensor network. The solutions proposed limit the size of the network.

33.4.3.2   Path Determination

Using the shared frequency-hopping secret (FHS), the sink endows each wedge Wi with a unique wedge key WKi and (possibly) a new wedge FHS, WFHSi, for $1 \le i \le NW$, where NW is the number of wedges. This, in effect, creates a firewall at wedge boundaries. This process can initially take place during the training phase. Note that the time to start using the new WFHSi should also be broadcast to the wedge.

Using encryption with WKi or a hopping set with seed WFHSi or both, depending on the level of security required, a source sensor broadcasts indices (or puzzles) to its set of m keys. If an overlap is detected with a neighbor, a link can be established.

All neighbors in the direction of the sink with established links (i.e., with overlapping keys), in turn, broadcast indices (or puzzles) to their sets of m keys. Again, a link can be established between neighbors with overlapping sets in the direction of the sink.

The process continues all the way to the sink.

The source node generates a path key PKj and sends it along each of the established j paths within the same wedge. If either the frequency-hopping set based on WFHSi is used or PKj is encrypted with WKi, then only nodes on the established paths that are within the same wedge will know the path key.

The path key is used to send: (a) new path FHS, PFHSj, that is generated by either the source (or the sink) for each path j, and (b) the time to start running the FH algorithm with the new seed PFHSj.

33.4.3.3   Remarks

After the initial frequency-hopping phase, the secure operation of the network will not depend on the entire network sharing a common secret.

To limit the probability of compromising the wedge key and the wedge frequency-hopping secret, the sink can randomly update their values one at a time. Also, once a path key is established, nodes can purge their stored wedge key and wedge frequency-hopping secret.

If a node is compromised, its impact will be limited to the paths in which it can participate. If anomalous behavior along a path is detected, either the source or the sink can purge that path.

Our solution allows for graceful degradation; it is possible to start by purging paths within a wedge, then purging wedges, etc.

33.5   Evaluation of the Proposed Solution

It is widely recognized that the task of securing a network involves achieving the following important goals[16]:

Availability: ensures the survivability of network services despite denial-of-service attacks.
Confidentiality: ensures that certain information is not disclosed to unauthorized entities.
Integrity: guarantees that a message being transferred is never corrupted by an attack.
Authentication: enables a node to ensure the identity of the peer node with which it communicates.
Non-repudiation: ensures that the origin of a message cannot deny having sent the message.

The main goal of this section is to evaluate our proposed solution to securing wireless sensor networks in terms of achieving the goals above. As already mentioned, just prior to deployment, the individual sensor nodes are synchronized and are injected with *genetic material* consisting, essentially, of a program capable of generating the random sequences defined in this chapter.

Availability. We note that in our solution it is highly probable that the adversary cannot infiltrate the system other than by physically tampering with the individual sensor nodes. Furthermore, our solution provides protection against jamming, unlike other schemes discussed by Hill et al.[9]

Confidentiality and integrity. Our solution provides both confidentiality and message integrity because the adversary does not have the time to learn our hopping sequence in any given epoch. Indeed, our assumption that individual sensor nodes are tamper-proof, combined with the process of securely migrating between various sets of frequencies and between various frequency-hopping sequences from one time epoch to the next, makes the task of breaking into the system extremely difficult.

Authentication and non-repudiation. Due to the fact that sensor nodes are anonymous, the classic definitions of authentication and non-repudiation do not apply to wireless sensor networks presented in this chapter. These two goals are extremely important. We presented a context-based solution that uses signature arrays from neighboring nodes to authenticate a node within a neighborhood. There is ongoing work investigating the non-repudiation issue. We also are investigating solutions that endow the sensor nodes with temporary IDs.

33.6   Conclusion and Directions for Future Work

It is anticipated that in the near future, wireless sensor networks will be employed in a wide variety of applications establishing ubiquitous networks that will pervade society. It is widely recognized that sensor network research is in its infancy.[1,2,4−6,9,11−14] In particular, there is precious little known about how to get sensor networks to self-organize in a way that maximizes the operational longevity of the network and that guarantees a high level of availability in the face of potential security attacks. However, given the characteristics of sensor networks, security protocols developed for wired, cellular, or ad-hoc networks do not apply.[2,9,14]

We have presented a new solution to the problem of securing wireless sensor networks. Specifically, we showed that suitable enhancement to the classic frequency-hopping strategy would provide a lightweight and scalable mechanism for securing wireless sensor networks. Our solution supports a *differential security service* that can be dynamically configured to accommodate changing application and network system state.

We view this chapter, in part, as an initial contribution toward developing a holistic solution for securing sensor networks. Research is underway toward a solution that provides security not only for the various individual layers of the system, but also for the entire system in an integrated fashion. This will necessitate addressing open problems such as non-repudiation and dynamic security configuration.

Acknowledgments

This work was supported in part by a grant from the Commonwealth of Virginia Technology Research Fund (SE 2001-01) through the Commonwealth Information Security Center.

References

1. I.F. Akyildiz, W. Su, Y. Sankarasubramanian, and E. Cayirci, Wireless sensor networks: a survey, *Computer Networks,* 38(4), 393–422, 2002.
2. R. Anderson, Security Engineering: A Guide to Building Dependable Distributed Systems, Wiley, New York, 2001.
3. R. Anderson and M. Kuhn, Tamper resistance — a cautionary note, *Proc. 2nd Usenix Workshop on Electronic Commerce,* Berkely, CA, 1996, pp. 1–11.
4. P. Bahl, W. Russell, Y.-M. Wang, A. Balachandran, G.M. Voelker, and A. Miu, PAWNs: satisfying the need for ubiquitous secure connectivity and location services, *IEEE Wireless Communications,* 9(1), 40–48, 2002.
5. D.W. Carman, P.S. Kruus, and B.J. Matt, Constraints and Approaches for Distributed Sensor Network Security, Technical Report #00-010, NAI Labs, 2000.
6. Cross Bow Technologies, http://www.xbow.com/.
7. A. Ephremides, J. Wieselthier, and D. Baker, A design concept for reliable mobile radio networks with frequency hopping signaling, *Proc. of the IEEE,* 75(1), 56–73, 1987.
8. J.C. Haartsen, The Bluetooth radio system, *IEEE Personal Communications,* 2000, 28–36.
9. J. Hill, R. Szewczyk, A. Woo, S. Hollar, D. Culler, and K. Pister, System architecture directions for networked sensors, *Proc. 9th Int. Conf. on Architectural Support for Programming Languages and Operating Systems (ASPLOS'2000),* Cambridge, MA, November 2000.
10. J.-P. Hubaux, L. Buttyan, and S. Capkun, The quest for security in mobile ad-hoc networks, *Proc. ACM Symp. on Mobile Ad-Hoc Networking and Computing (MobiHoc'01),* Long Beach, CA, October 2001, pp. 146–155.
11. J.M. Kahn, R.H. Katz, and K.S.J. Pister, Mobile networking for Smart Dust, *Proc. 5th Annu. Int. Conf. on Mobile Computing and Networking (MOBICOM'99),* Seattle, WA, August 1999.
12. J. Kong, P. Zerfos, H. Luo, S. Lu, and L. Zhang, Providing robust and ubiquitous security support for mobile ad-hoc networks, *Proc. 9th Int. Conf. on Network Protocols,* Riverside, CA, 2001, pp. 251–260.
13. S. Marti, T.J. Giuli, K. Lai, and M. Baker, Mitigating routing misbehavior in mobile ad-hoc networks, *Proc. 6th Annu. Int. Conf. on Mobile Computing and Networking (MOBICOM'00),* Boston, MA, August 2000.
14. National Research Council, *Embedded, Everywhere: A Research Agenda for Systems of Embedded Computers,* Committee on Networked Systems of Embedded Computers, for the Computer Science and Telecommunications Board, Division on Engineering and Physical Sciences, Washington, D.C., 2001.
15. NIST. FIPS PUB 140-1, Security Requirements for Cryptographic Modules, National Institute of Standards and Technology, January 1994.
16. A. Perrig, R. Szewczyk, V. Wen, D. Culler, and J.D. Tygar, SPINS: security protocols for sensor networks, *Proc. 7th Annu. Int. Conf. on Mobile Computing and Networking (MOBICOM'01),* Rome, Italy, August 2001, pp. 189–199.
17. A. Pfitzmann, B. Pfitzmann, and M. Waidner, Trusting mobile user devices and security modules, *IEEE Computer,* 30(2), 61–68, 1997.
18. P. Saffo, Sensors, the next wave of innovation, *Communications of the ACM,* 40(2), 93–97, 1997.
19. TinySec http://www.cs.berkeley.edu/-nks/tinysec/.

20. A. Wadaa, S. Olariu, L. Wilson, K. Jones, and Q. Xu, On training wireless sensor networks, *MONET*, January 2005, to appear. A preliminary version of this paper has appeared in *Proc. 3rd International Workshop on Wireless, Mobile and Ad Hoc Networks (WMAN'03)*, Nice, France, April 2003.

21. B. Warneke, M. Last, B. Leibowitz, and K. Pister, SmartDust: communicating with a cubic-millimeter computer, *IEEE Computer,* 34(1), 44–51, 2001.

22. A.D. Wood and J.A. Stankovic, Denial of service in sensor networks, *IEEE Computer,* 35(4), 54–62, 2002.

23. Y. Zhang and W. Lee, Intrusion detection in ad-hoc networks, *Proc. 6th Annu. Int. Conf. on Mobile Computing and Networking (MOBICOM'00)*, Boston, MA, August 2000.

24. V.V. Zhirnov and D.J.C. Herr, New frontiers: self-assembly and nano-electronics, *IEEE Computer,* 34(1), 34–43, 2001.

25. L. Zhou and Z.J. Haas, Securing ad-hoc networks, *IEEE Network,* 13(6), 24–30, 1999.

26. J. Zyren, T. Godfrey, and D. Eaton, Does frequency hopping enhance security? http://www.packetnexus.com/docs/20010419_frequencyHopping.pdf.

27. U. Hansmann, L. Merk, M. Nicklous, and T. Strober, *Foundations of Pervasive Computing*, Springer, 2003.

28. L. Eschenauer and V. Gligor, A key management scheme for distributed sensor networks, *Proc. 9th ACM Conf. on Computing and Communication Security,* November 2002.

29. H. Chan, A. Perrig, and D. Song, Random key pre-distribution schemes for sensor networks, *Proc. IEEE Symp. on Security and Privacy,* Berkeley, CA, May 2003.

30. K. Jones, A. Wadaa, S. Olariu, L. Wilson, and M. Eltoweissy, Towards a new paradigm for securing wireless sensor networks, *Proc. New Security Paradigms Workshop (NSPW'2003)*, Ascona, Switzerland, August 2003.

34

Antireplay Protocols for Sensor Networks

Mohamed G. Gouda

Young-ri Choi

Anish Arora

34.1 Introduction

A sensor consists of a sensing board (that can sense magnetism, sound, heat, etc.), a battery-operated small computer, and an antenna. Sensors in a network can communicate in wireless fashion by broadcasting messages over radio frequency; and due to the limited range of their radio transmission, the network is usually multihop. One of the challenging problems in designing sensor networks is to secure the network against adversarial attacks. This problem has received considerable attention in recent years.[4,8,11,12]

Two common examples of adversarial attacks are called *message insertion attacks* and *message replay attacks*. In a message insertion attack, the adversary inserts arbitrary messages into the message stream from a process p executing on one sensor to a process q executing on a second sensor. Such an attack can be thwarted by attaching a digest[10] to each legitimate message in the message stream from p to q. The digest attached to a message is computed using the different fields of the message and a secret key that is shared only between processes p and q. The adversary does not know the shared key between p and q and thus cannot compute the digest of any arbitrary message that the adversary wishes to insert into the message stream from p to q.

To define message replay attacks on the message stream from p to q, we need to introduce the following notation. Let d.i denote the i-th data message, where i = 0, 1, . . . , that process p sends to process q. Thus, a message stream, that consists of seven messages, from p to q can be represented as follows:

Stream 0: d.0, d.1, d.2, d.3, d.4, d.5, d.6

In a replay attack on the message stream from p to q, the adversary makes copies of some data messages in this stream and inserts these copies anywhere in the stream. For example, if the adversary makes two copies of the data message d.4 in Stream 0 and inserts the first copy after d.1 and the second copy after the original d.4, then the resulting message stream can be represented as follows:

$$\text{Stream 1: d.0, d.1, d.4, d.2, d.3, d.4, d.4, d.5, d.6}$$

Copy insertion (by the adversary) can conspire with natural transmission errors to make the task of antireplay protocols difficult. Three types of transmission errors can befall a message stream: (1) message loss, (2) message corruption, and (3) message reorder. We define these three types of errors next.

A message d.i is said to be *lost* from the message stream from process p to process q if d.i is removed from the stream. For example, if message d.1 and the third occurrence of message d.4 are lost from Stream 1, then the resulting stream is as follows:

$$\text{Stream 2: d.0, d.4, d.2, d.3, d.4, d.5, d.6}$$

A message d.i in the message stream from process p to process q is said to be *corrupted* if the values of some fields in d.i are changed such that the digest of d.i becomes inconsistent with the values of the other fields of d.i. A corrupted message is denoted as C in its message stream. For example, if message d.3 in Stream 2 is corrupted then the resulting stream is as follows:

$$\text{Stream 3: d.0, d.4, d.2, C, d.4, d.5, d.6}$$

An uncorrupted message d.i is said to be *reordered* if the stream has a preceding uncorrupted message d.j where $j > i$. For example, message d.2 is the only reordered message in Stream 3.

Each message in the message stream from p to q (after this stream is subjected to copy insertion, message loss, corruption, and reorder) can be classified into one of three disjoint classes: fresh, replayed, and corrupted. Corrupted messages are defined above; it remains to define fresh and replayed messages.

An uncorrupted message d.i in the message stream from p to q is called *fresh* if the stream has no preceding d.i messages. Message d.i is called *replayed* if the stream has at least one preceding d.i message. For example, the messages d.0, d.2, d.5, and d.6 in Stream 3 are fresh. Also in Stream 3, the first d.4 message is fresh and the second d.4 message is replayed.

The sending process p and the receiving process q use an *antireplay protocol* to allow process q to receive the messages of the message stream from p to q, one by one, and to correctly classify each received message into one of the three classes: fresh, replayed, or corrupted. Process q accepts each fresh message and discards each replayed or corrupted message. Perhaps the most well-known antireplay protocol in the Internet is the one developed for IPSec.[5–7]

As discussed below, the used antireplay protocol between the sending process p and the receiving process q can be made more efficient if the "degrees" of loss and reorder, for the message stream from p to q, have relatively small values. Next we define the degrees of loss and reorder for a message stream.

The *degree of loss* for the message stream from p to q is a nonnegative integer dl that satisfies the following two conditions: (1) The first fresh message d.i in the message stream is such that $i \leq dl$. (2) For every pair of fresh messages d.i and d.j where d.i precedes d.j in the message stream and $i < j$: If every other fresh message d.k that occurs between d.i and d.j in the message stream is such that $i > k$, then $i + dl + 1 \geq j$.

The *degree of reorder* for the message stream from p to q is a nonnegative integer dr that satisfies the following condition for every pair of fresh messages d.i and d.j where d.i precedes d.j in the message stream and $i > j$: If every other fresh message d.k that occurs between d.i and d.j in the message stream is such that $i > k$, then $i \leq j + dr$.

Note that if both dl and dr have the value 0, then the next fresh message that can occur in the message stream from p to q after a fresh message d.i is d.$(i + 1)$. Note also that if only dr has the value 0, then the next fresh message that can occur after a fresh message d.i is d.j, where $i < j < i + dl + 1$.

The antireplay protocols presented in the following sections are specified formally using the Abstract Protocol Notation.[1]

34.2   A Perfect Antireplay Protocol

In this section we describe a protocol that can perfectly overcome any message replay attack, no matter how unlikely this attack is to occur. As discussed below, this protocol is expensive: it requires that the sending process attaches unbounded sequence numbers to sent messages and that the receiving process maintains an infinite Boolean array. In the next sections, we describe more efficient protocols that can overcome only those replay attacks that are more likely to occur in sensor networks.

In our perfect antireplay protocol, the sending process p and the receiving process q share a secret key sk. Each data message sent from process p to process q has three fields as follows:

$$data(x, s, m)$$

Field x denotes the text of the data message, field s denotes the unique sequence number of the data message, and field m denotes the digest of the data message that is computed (by process p) as follows:

$$m := MD.(x|s|sk)$$

where MD is a message digest function, the "." denotes the function application operator, and the "|" denotes the integer concatenation operator.

The sending process p in our perfect antireplay protocol can be specified as follows:

```
process p

const   sk      :       integer                 /shared key/

var     x       :       integer,                /message text/
        s       :       integer,                /initially 0/
        m       :       integer                 /message digest/

begin
        true    -->     /send the next data message/
                        x := any;
                        m := MD.(x|s|sk);
                        send data(x, s, m);
                        s := s+1
end
```

The receiving process q in this protocol maintains an infinite Boolean array named rcvd. At each instant, the value of an element rcvd[s] in array rcvd is false iff process q has not yet received a (fresh) data message whose sequence number is s. Thus, the value of every element in array rcvd is initially false. The receiving process q is specified as follows:

```
process q

const   sk      :       integer                 /shared key/

var     rcvd    :       array [integer] of boolean,
                                                /initially false/
        x       :       integer,                /message text/
        s       :       integer,                /sequence number/
        m       :       integer                 /message digest/
```

```
begin
        rcv data(x, s, m) -->
                if   rcvd[s] --> skip                /message is not fresh/
                [] !rcvd[s] -->
                        if m   = MD.(x|s|sk) --> /message is fresh/
                                                 rcvd[s] := true
                        [] m != MD.(x|s|sk) --> /message is corrupted/
                                                 skip
                        fi
        fi
end
```

It is straightforward to show that this protocol satisfies the following three properties:

1. *Corruption detection.* If process q receives a corrupted message, then q detects that the message is not fresh and discards it. The protocol satisfies this property assuming that the degrees of loss and reorder, for the message stream from p to q, have any values.
2. *Replay detection.* If process q receives a replayed message, then q detects that the message is replayed and discards it. The protocol satisfies this property assuming that the degrees of loss and reorder, for the message stream from p to q, have any values.
3. *Freshness detection.* If process q receives a fresh message, then q detects that the message is fresh and accepts it. The protocol satisfies this property assuming that the degrees of loss and reorder, for the message stream from p to q, have any values.

These three properties are the most that one can hope for from an antireplay protocol. Thus, any antireplay protocol that satisfies these properties is regarded as "perfect." Perfect antireplay protocols are usually expensive, and thus they are not suitable for sensor networks where processes have limited resources.

In particular, the perfect antireplay protocol discussed in this section is expensive in two ways. First, the sending process p attaches unbounded sequence numbers, namely the s sequence numbers, to the data messages before sending the message to q. Second, the receiving process q maintains an infinite array rcvd to keep track of all the fresh messages that q has received in the past. In the following sections, we discuss several antireplay protocols that are less expensive than the perfect protocol in this section.

34.3   An Explicit Sequencing Protocol

In this section we discuss a second antireplay protocol for transmitting data messages from a sending process p to a receiving process q. The receiving process q in this protocol, unlike the one in the perfect protocol in Section 34.2, does not maintain an infinite array. As a result, the freshness detection property for this protocol is weaker than that for the perfect antireplay protocol. However, as discussed below, this weakening of the freshness detection property does not hinder the effective use of this new protocol in sensor networks.

In the new protocol, the sending process p attaches a unique sequence number s to each data message before it sends the message to the receiving process q. Hence, we refer to this protocol as an explicit sequencing protocol.

The sending process p in the explicit sequencing protocol is identical to the one in the perfect antireplay protocol in Section 34.2, and thus we do not need to specify process p in this section.

The receiving process q in the explicit sequence protocol maintains an integer variable named exp. The value of variable exp is the sequence number of the next data message that process q expects to receive from p. Initially, the value of variable exp equals the initial value of variable s in process p.

When process q receives a data(x, s, m) message, q compares its own variable exp with the sequence number s of the received message. This comparison yields one of two outcomes. First, exp > s, and in this case q concludes, possibly wrongly, that the message is not fresh and discards the message. Second, exp < s, and in this case q checks digest m of the received message and accepts the message iff m matches the digest MD.(x|s|sk) that q computes for the message.

The receiving process q in the explicit sequencing protocol is specified as follows:

```
process q

const   sk      :           integer                 /shared key/

var     x       :           integer,                /message text/
        exp, s  :           integer,                /init. exp.q = s.p = 0/
        m       :           integer                 /message digest/

begin
        rcv data(x, s, m) -->
                if   exp >  s -->  skip              /message is not fresh/
                []   exp =< s -->
                        if m  = MD.(x|s|sk) --> /message is fresh/
                                               exp := s+1
                        [] m != MD.(x|s|sk) --> /message is corrupted/
                                               skip
                        fi
                fi
end
```

The explicit sequencing protocol satisfies three properties: (1) the corruption detection property in Section 34.2, (2) the replay detection property in Section 34.2, and (3) the following freshness detection property.

4. *Freshness detection with 0 reorder.* If process q receives a fresh message, then q detects that the message is fresh and accepts it. The protocol satisfies this property assuming that the degree of loss for the message stream from p to q has any value, and the degree of reorder for the same stream is 0.

To show that the detection of fresh messages fails when the degree of reorder for the message stream from p to q is more than 0, consider the following scenario. Assume that the protocol starts from the initial state, where variable s in p and variable exp in q have the value 0. Then process p sends two messages, data(x, 0, m) followed by data(x', 1, m'), to process q. Assume also that due to an occurrence of message reorder, process q first receives the second message then the first message. On receiving the data(x', 1, m') message, q concludes correctly that the message is fresh, accepts it, and assigns value 2 to its own variable exp. On receiving the data(x, 0, m) message, q concludes incorrectly that this message is not fresh and discards it. This scenario shows that any message reorder causes q to discard fresh messages.

The fact that message reorder causes the receiving process in the explicit sequencing protocol to discard fresh messages should not be terribly alarming if this protocol is used in sensor networks. This is because the probability of message reorder in sensor networks is usually very small.

Nevertheless, there are explicit sequencing protocols that can correctly detect fresh messages and accept them, even if these messages are received out of order. Examples of these protocols are presented by Gouda et al.[2] and Huang and Gouda.[3]

An explicit sequencing protocol still has the problem that unbounded sequence numbers are attached to all sent messages. We solve this problem next.

34.4 An Implicit Sequencing Protocol

In this section we discuss a third antireplay protocol where the data messages carry no explicit sequence numbers, unlike the above antireplay protocols. We call this protocol an implicit sequencing protocol. (For another example of an implicit sequencing protocol, the reader is referred to Ref. 9.)

To be exact, the sending process p in our implicit sequencing protocol does compute a sequence number h for each data message before the message is sent to the receiving process q. However, the sequence number of a message is not attached to the message; it is merely used in computing the digest m of this message. (Note that in this protocol, the sequence number of a data message is denoted h to signify that this sequence number is hidden and not explicitly attached to the message when it is sent from process p to process q.)

Each data message in the implicit sequencing protocol has only two fields as follows:

$$\text{data}(x, m)$$

Field x denotes the text of the data message and field m denotes the digest of the message that is computed (by process p) as follows:

$$m := \text{MD}.(x|h|sk)$$

where h is the sequence number of the message and sk is the shared key between processes p and q.

In this protocol, each of the two processes p and q has an integer variable h. Variable h in process p stores the sequence number of the next message to be sent by p, and variable h in process q stores the sequence number of the next expected message to be received by q. Clearly, the initial values of these two variables are 0.

Process p in the implicit sequencing protocol is specified as follows.

```
process p

const   sk      :       integer                 /shared key/

var     x       :       integer,                /message text/
        h       :       integer,                /hidden seq. #/
        m       :       integer                 /message digest/

begin
        true    -->     /send the next data message/
                        x := any;
                        m := MD.(x|h|sk);
                        send data(x, m);
                        h := h+1
end
```

To allow the receiving process q to detect fresh messages, we need to assume that the degree of loss for the message stream from p to q is some known value dl and that the degree of recorder for the same stream is 0. That is, we assume that if consecutive messages in the message stream from p to q are lost or corrupted, then the number of these messages is no more than dl. Moreover, we assume that messages cannot be reordered in the message stream from p to q.

Based on these assumptions, when process q receives a data(x, m) message, q knows that this message is fresh iff the hidden sequence number of the received message is one of the following:

$$\{h, h + 1, h + 2, \ldots, h + dl - 1, h + dl\}$$

where h is the current value of variable h in process q. Note that if the received message is fresh, then its sequence number is h+d, for some d in the range 0..dl, and each of the messages, whose sequence numbers are h, h + 1, ..., h + d − 1, is lost and will not be received. Also, if each of the messages, whose sequence numbers are h, h + 1, ..., h + dl − 1, is lost, then the message whose sequence number is h + dl cannot be lost because the degree of loss for the message stream from p to q is dl.

It follows from this discussion that when q receives a data(x, m) message, then q checks whether the sequence number of this message is some h + d, where h is the current value of variable h in q and d is some value in the range 0..dl. If the sequence number of the received message is indeed some h + d, then q concludes that the message is fresh and accepts it. Otherwise, q concludes that the message is not fresh and discards it. Process q checks that the sequence number of the received data(x, m) message is some h + d by checking that the digest m in the received message matches the digest MD.(x|h + d|sk) that q computes for the received message.

The receiving process q in the implicit sequencing protocol is specified as follows:

```
process q

const   sk      :       integer,        /shared key/
        dl      :       integer         /degree of loss/

var     x       :       integer,        /message text/
        h       :       integer,        /init. h.q = h.p/
        m       :       integer,        /message digest/
        d       :       0..dl,
        match   :       boolean
begin
        rcv data(x, m)  -->
                d := 0;
                match := false;
                do d < dl and   !match -->
                        if m  = MD.(x|h+d|sk)   -->  match := true
                        [] m != MD.(x|h+d|sk)   -->  d := d+1
                        fi
                od;
                if  match or  m =MD.(x|h+d|sk) -->  /message is fresh/
                                                    h := h+d+1
                [] !match and m!=MD.(x|h+d|sk) -->  /message is not fresh/
                                                    skip
                fi
end
```

The implicit sequencing protocol satisfies three properties: (1) the corruption detection property in Section 34.2, (2) the replay detection property in Section 34.2, and (3) the following freshness detection property:

5. *Freshness detection with dl loss and 0 reorder.* If process q receives a fresh message, then q detects that the message is fresh and accepts it. The protocol satisfies this property assuming that the degree of loss for the message stream from p to q is dl and the degree of reorder for the same stream is 0.

The implicit sequencing protocol exhibits a new problem: the receiving process computes a possibly large number (up to dl + 1) of digests for each received data message. We solve this problem next.

34.5   A Mixed Sequencing Protocol

The reason that process q in the previous section computes up to dl+1 digests for each received message is that q needs to check whether the sequence number of the received message is h+d, where h is the current value of variable h in q, and d is a value in the range 0..dl. Thus, to allow process q to compute only one digest for each received message, we modify the antireplay protocol in the previous section such that p includes the corresponding value d in each message that p sends to q.

In the new antireplay protocol, the sequence number of each message is a pair of two components:

$$(s, h)$$

Component s is in the range 0..dl; it is included as a field in the message and is used in computing the message digest. Component h is a nonnegative integer; it is not included as a field in the message but is used in computing the message digest. Because component s can be viewed as providing explicit sequencing and component h can be viewed as providing implicit sequencing, we refer to the new antireplay protocol as a *mixed sequencing protocol*.

Each data message sent from process p to process q in the mixed sequencing protocol has three fields as follows:

$$data(x, s, m)$$

Field x denotes the message text, field s is the explicit or clear sequence number of the message, and field m is the message digest computed (by process p) as follows:

$$m := MD.(x|s|h|sk)$$

where h is the current value of variable h, which stores the implicit or hidden sequence number of the message, and sk is the shared secret between processes p and q.

After process p sends a data(x, s, m) message whose sequence number is (s, h), p increments this sequence number by one to get the sequence number of the next message to be sent. Process p increments the sequence number (s, h) by one as follows:

$$(s, h) + 1 = (s + 1, h) \qquad \text{if } s < dl$$
$$(0, h + 1) \qquad \text{if } s = dl$$

It follows from this definition that a sequence number (s, h) is less than a sequence number (s', h') iff either h is less than h', or h equals h' and s is less than s'.

The sending process p in the mixed sequencing protocol is specified as follows:

```
process p

const   sk      :       integer,                /shared key/
        dl      :       integer                 /degree of loss/

var     x       :       integer,                /message text/
        s       :       0..dl,                  /clear seq. #/
        h       :       integer,                /hidden seq. #/
        m       :       integer                 /message digest/

begin
        true    -->     /send the next data message/
                        x := any;
                        m := MD.(x|s|h|sk);
                        send data(x, s, m);
                        s := s+1 mod (dl+1)
                        if s   = 0 --> h := h+1
                        [] s != 0 --> skip
                        fi
end
```

Process q in the mixed sequencing protocol has two variables, exp and h. Variable exp is used to store the explicit or clear sequence number of the next message that q expects to receive; the value of exp is in the range 0..dl. Variable h is used to store the implicit or hidden sequence number of the next message that q expects to receive; the value of h is a nonnegative integer. Clearly, the two variables s in p and exp in q have the same initial value (which is 0), and the two variables h in p and h in q have the same initial value.

Process q also has an integer variable g. When q receives a data(x, s, m) message, it computes the implicit or hidden sequence number of that message and stores the result in variable g. Clearly, the resulting g satisfies $g = h$ or $g = h + 1$, where h is the current value of variable h in process q.

To compute the variable of g when a data(x, s, m) message is received, process q uses a Boolean function BET.(u, v, w) whose three arguments are in the range 0..dl. The value of BET.(u, v, w) is true iff one of the following two conditions holds:

1. $u = v = w$
2. $u \neq w$ and v is an element of the following set of nonnegative integers:

```
{u          ,
 u+1   mod (dl+1),
 u+2   mod (dl+1),
 . . .
 w-1   mod (dl+1),
 w                 }
```

The receiving process q in the mixed sequencing protocol can be specified as follows.

```
process q

const   sk      :       integer,        /shared key/
        dl      :       integer         /degree of loss/

var     x       :       integer,        /message text/
        exp, s  :       0..dl,          /init. exp.q = s.p = 0/
        h, g    :       integer,        /init. h.q = h.p/
        m       :       integer         /message digest/

begin
        rcv data(x, s, m) -->
                if exp  = 0 or   !BET.(exp, 0, s) --> g := h
                [] exp != 0 and   BET.(exp, 0, s) --> g := h+1
                fi;
                if m  = MD.(x|s|g|sk) --> /message is fresh/
                                          h := g;
                                          exp := s+1 mod (dl+1);
                                          if exp  = 0 --> h := h+1
                                          [] exp != 0 --> skip
                                          fi
                [] m != MD.(x|s|g|sk) --> /message is not fresh/
                                          skip
                fi
end
```

The mixed sequencing protocol satisfies the same three properties (namely, the corruption detection property in Section 34.2, the replay detection property in Section 34.2, and the freshness detection property in Section 34.4) satisfied by the implicit sequencing protocol in Section 34.4. However, the mixed sequencing protocol is better than the implicit sequencing protocol because it requires only one message digest to be computed per received message.

34.6   An Antireplay Sensor Protocol

The only problem with the mixed sequencing protocol in Section 34.5 is that its freshness detection property holds only under the assumption that the degree of reorder for the message stream from p to q is 0. In this section, we modify this protocol so that the freshness detection property of the modified protocol holds under the assumption that the degree of reorder has a (small) value dr that is not necessarily 0. We refer to this modified protocol as an *antireplay sensor protocol*.

In the antireplay sensor protocol, if the receiving process q is waiting to receive a data message whose sequence number is (s, h), then process q may receive a data message whose sequence number is (s', h') such that

```
either    (s',h') = (s,h) + u     where 0 ≤ u ≤ dl
or        (s',h') = (s,h) − v     where 1 ≤ v ≤ dr+1
```

(This is because the degree of loss for the message stream from the sending process p to the receiving process q is dl and the degree of reorder for the same stream is dr.)

When process q expects to receive a data message whose sequence number is (s, h), but receives a data(x', s', m') message, then q uses s' to deduce the sequence number of the received message from the following set of candidate sequence numbers:

```
{ (s,h)  - dr - 1,
  (s,h)  - dr    ,
  ...
  (s,h)  - 1     ,
  (s,h)          ,
  (s,h)  + 1     ,
  ...
  (s,h)  + dl    }
```

Because this set has dl + dr + 2 elements, s' should have dl + dr + 2 distinct values. Thus, the value of the explicit or clear sequence number of a message in the antireplay sensor protocol is in the range 0..dl + dr + 1.

The sending process p in the antireplay sensor protocol is specified as follows:

```
process p

const   sk    :       integer,          /shared key/
        dl    :       integer,          /degree of loss/
        dr    :       integer           /degree of reorder/

var     x     :       integer,          /message text/
        s     :       0..(dl+dr+1),     /clear seq. #/
        h     :       integer,          /hidden seq. #/
        m     :       integer           /message digest/

begin
        true    -->   /send the next data message/
                      x := any;
                      m := MD.(x|s|h|sk);
```

```
                              send data(x, s, m);
                              s := s+1 mod (dl+dr+2);
                              if s  = 0 --> h := h+1
                              [] s != 0 --> skip
                              fi
end
```

The receiving process q in the antireplay sensor protocol can be specified as follows. (Note that to simplify this specification we have omitted mod (dl + dr + 2) from several mathematical expressions that involve variables exp or s. Thus, the reader should read the expression "exp-dr-1" as "exp-dr-1 mod (dl + dr + 2)," and read the expression "s + 1" as "s + 1 mod (dl + dr + 2)," etc.)

```
process q

const    sk       :          integer,                 /shared key/
         dl       :          integer,                 /degree of loss/
         dr       :          integer                  /degree of reorder/

var      x        :          integer,                 /message text/
         exp, s   :          0..(dl+dr+1),            /init. exp.q = s.p =0/
         h, g     :          integer,                 /init. h.q = h.p/
         m        :          integer,                 /message digest/
         rcvd     :          array[0..(dl+dr+1)] of boolean

                                                      /init.
                                                      rcvd[exp] = false and
                                                      rcvd[exp+1] = false and
                                                      ...
                                                      rcvd[exp+dl] = false and
                                                      rcvd[exp−1] = true and
                                                      rcvd[exp−2] = true and
                                                      ...
                                                      rcvd[exp−dr−1] = true/

begin
        rcv data(x, s, m)  −>
             if BET.(exp−dr−1, s, exp−1)  −>
                    if  BET.(s+1, 0, exp)  −> g := h−1
                    [] !BET.(s+1, 0, exp)  −> g := h
                    fi;
                    if !rcvd[s] and m  = MD.(x|s|g|sk)  −>
                            /msg is fresh/ rcvd[s] := true
                    []   rcvd[s] or  m != MD.(x|s|g|sk)  −>
                            /msg is not fresh/ skip
                    fi
             [] BET.(exp, s, exp+dl)   −>
                    if !BET.(exp+1, 0, s)   −> g := h
                    []  BET.(exp+1, 0, s)   −> g := h+1
                    fi;
                    if m != MD.(x|s|g|sk) −> /msg is not fresh/
                            skip
                    [] m  = MD.(x|s|g|sk) −> /msg is fresh/
```

```
                                   h := g;
                                   do exp != s ->
                                              rcvd[exp-dr-1] := false;
                                              exp := exp+1
                                   od;
                                   rcvd[exp-dr-1] := false;
                                   rcvd[exp] := true;
                                   exp := exp+1;
                                   if exp  = 0 -> h := h+1
                                   [] exp != 0 -> skip
                                   fi
                      fi
            fi
end
```

This antireplay protocol for sensor networks satisfies three properties: (1) the corruption detection property in Section 34.2, (2) the replay detection property in Section 34.2, and (3) the following freshness detection property:

6. *Freshness detection with dl loss and dr reorder.* If process q receives a fresh message, then q detects that the message is fresh and accepts it. The protocol satisfies this property assuming that the degree of loss for the message stream from p to q is dl and the degree of reorder for the same stream is dr.

Note that array rcvd in process q has $dl + dr + 2$ elements. It is possible to modify process q such that array rcvd has only dr elements.

34.7   Conclusion

The antireplay sensor protocol presented in Section 34.6 can be implemented as follows. First, the degree of loss for the message stream should be a relatively large value, say $dl = 128$. Second, the degree of reorder for the message stream should be a relatively small value, say $dr = 16$. Third, from these values of dl and dr, we conclude that the range of explicit or clear sequence numbers that are attached to sent messages is as follows:

$$0 .. dl + dr + 1 = 0 .. 128 + 16 + 1$$
$$= 0 .. 145$$
$$\sim 0 .. 255$$

Therefore, one byte in each sent message is sufficient to store the explicit or clear sequence number of that message.

Both the sending and receiving processes can have say two bytes to store the implicit or hidden sequence numbers of sent messages. Assuming that the sending process sends continuously one message per second, the sequence numbers are exhausted in about six months. When the sequence numbers are exhausted, the sending and receiving processes are reset and supplied with a new shared secret, and the cycle repeats.

Finally, array rcvd in the receiving process should have dr bits. Because dr is assumed to be 16, only two bytes are needed to implement array rcvd in the receiving process.

References

1. M.G. Gouda. *Elements of Network Protocol Design.* John Wiley & Sons, Inc, New York, 1998.
2. M.G. Gouda, C.-T. Huang, and E. Li. Anti-Replay Window Protocols for Secure IP. In *Proceedings of the 9th IEEE International Conference on Computer Communications and Networks,* Las Vegas, October 2000.

3. C.-T. Huang and M.G. Gouda. An Anti-Replay Window Protocol with Controlled Shift. In *Proceedings of the 10th IEEE International Conference on Computer Communications and Networks*, Scottsdale, AZ, October 2001.

4. C. Karlof and D. Wagner. Secure Routing in Sensor Networks: Attacks and Countermeasures. To appear in Elsevier's *AdHoc Networks Journal, Special Issue on Sensor Network Applications and Protocols*.

5. S. Kent and R. Atkinson. IP Authentication Header. RFC 2402, November 1998.

6. S. Kent and R. Atkinson. IP Encapsulating Security Payload (ESP). RFC 2406, November 1998.

7. S. Kent and R. Atkinson. Security Architecture for the Internet Protocol. RFC 2401, November 1998.

8. J. Kong, P. Zerfos, H. Luo, S. Lu, and L. Zhang. Providing Robust and Ubiquitous Security Support for Mobile Ad-Hoc Networks. In *Proceedings of IEEE 9th International Conference on Network Protocols (ICNP)*, pages 251–260, 2001.

9. A. Perrig, R. Szewczyk, V. Wen, D.E. Culler, and J.D. Tygar. SPINS: Security Protocols for Sensor Networks. In *Proceedings of the 7th Annual International Conference on Mobile Computing (MobiCom 2001)*, pages 189–199, New York, 2001.

10. R. Rivest. The MD5 Message-Digest Algorithm. RFC 1321, April 1992.

11. A. Wood and J. Stankovic. Denial of Service in Sensor Networks. *IEEE Computer*, 35(10):54–62, 2002.

12. L. Zhou and Z.J. Haas. Securing Ad Hoc Networks. *IEEE Networks Special Issue on Network Security*, November/December, 1999.

Low Power Consumption Features of the IEEE 802.15.4 WPAN Standard

Edgar H. Callaway, Jr.

35.1 Introduction to IEEE 802.15.4^{TM}/ZigBee^{TM}

The Institute of Electrical and Electronics Engineers (IEEE) 802.15.4 Low-Rate Wireless Personal Area Network (WPAN) standard[1] is part of the IEEE 802 family of digital communication standards. The standard was published in October 2003, after three years' development, and is now available by free download from the "Get IEEE 802^{TM}" Web site.[2] It is designed for low-cost, low-power applications that require relatively low data throughput (down to an average of less than 1 bps). It is not designed for wireless local area network (WLAN) service, nor is it optimized for multimedia, TCP/IP, or other applications that require a specific quality of service (QoS). It is also differentiated from IEEE 802.15.1^{TM}(Bluetooth^{TM}) in several respects; it does not support isochronous voice, as Bluetooth does, for example, while it natively supports multihop networks, something Bluetooth does not do.

In a relationship not unlike that between the IEEE 802.11[TM] Local Area Network standard and Wi-Fi[TM], the IEEE 802.15.4 standard has an industry consortium, called the ZigBee Alliance[TM].[3] The ZigBee Alliance specifies higher layers of the protocol not otherwise defined, and performs marketing and compliance certification of that specification. As an industry consortium, the ZigBee Alliance is neither an open standards-defining organization (SDO), nor is it officially associated with the IEEE. It also does not specify lower layers of the communication stack.

Unlike many other wireless standards, the IEEE 802.15.4/ZigBee standard was designed to meet the needs of a wide variety of applications, from industrial control and monitoring to home automation, security, personal computer peripherals, intelligent agriculture, asset tracking, toys, and military sensing. These applications require a protocol capable of varying trade-offs between message latency, device duty cycle, and power consumption, while still being simple enough to be competitive in the market with dedicated application solutions. That is, a design goal of the standard was to have the marginal cost of any unused features in a given application be subsidized by the overall increased volumes used by all applications, so that the net product cost of an implementation would be lower than available alternatives.

Despite the long list of desired applications, there are several features that most of these have in common, including:

Low power consumption. These applications typically require lifetimes of months or years with very small batteries — coin cells or AAA cells — and many prefer the use of energy-scavenging techniques so that batteries need not be used at all. This requirement leads to an average power consumption specification of less than 100 μW in many cases. Many applications require networks of relatively high order (hundreds or thousands of nodes), making battery replacement impractical in any case.

Low cost. Most applications require that the devices be very inexpensive — a few dollars, at most — and some require that the nodes be disposable. As a result, the nodes must have very limited communication and computation resources; a typical device might use an 8-bit, 16-MHz microcomputer with 60 kB ROM and 8 kB RAM, often embedded with the RF transceiver in a single system-on-a-chip (SoC).

Low offered message throughput. An assumption that has far-reaching consequences for communication protocol design is that the system has a low rate of offered message throughput. Individual nodes are not expected to produce significant amounts of traffic, on average, although they may have significant bursts of traffic in certain applications (for example, in a security system during an emergency).

Large network order. As noted above, wireless sensor networks must support networks of large order. The IEEE 802.15.4 standard has a 16-bit address field, supporting networks of up to 65k devices.

Few QoS guarantees. To simplify (and therefore reduce the implementation cost of) the protocol, IEEE 802.15.4/ZigBee provides few QoS guarantees. Link layer acknowledgment is available, but message latency limits, jitter, and connection-based channels in multihop networks are not supported. (However, in star networks, guaranteed time slots (GTSs) are available to reserve channel access for low message latency applications.) Some features, such as end-to-end message acknowledgment, can still be performed at the application layer.

Selectable levels of security. Security is a difficult technical challenge in ad hoc, multihop wireless sensor networks, which as already noted face stringent cost and power limitations.[4] To complicate matters further, different applications have different security requirements. To address these issues, IEEE 802.15.4 offers variable levels of security, each based on the Advanced Encryption Standard[5] with 128-bit keys (AES-128). IEEE 802.15.4 supports the use of (1) no security; (2) privacy (encryption) only; (3) message integrity and sender authentication only, with 32-, 64-, or 128-bit message integrity codes appended to the message; or (4) a combination of (2) and (3). The multiple available levels of security meet the needs of a wide variety of applications, while the use of a common security engine for all helps minimize cost and complexity.

IEEE 802.15.4/ZigBee has two available physical layers. In the 2.4-GHz band, it supports a data rate of 250 kbps; there are 16 available channels, centered at 2405 + 5*k* MHz, where $0 \leq k \leq 15$. The other physical layer is a regional one, covering the 868.0- to 868.6-MHz band available in Europe and the 902- to 928-MHz band available in much of the Americas. There is a single channel in the 868-MHz band, centered at 868.3 MHz, with a BPSK data rate of 20 kbps; the standard supports a BPSK data rate of 40 kbps in the 902- to 928-MHz band, with channels centered at 906 + 2*k* MHz, where $0 \leq k \leq 9$.

To meet its wide range of potential applications, IEEE 802.15.4/ZigBee supports star, mesh, and tree networks, the latter two being multihop networks.

The standard specifies carrier sense multiple access with collision avoidance (CSMA-CA) channel access; however, as noted above, support for low-latency applications in star networks is provided in the form of optional guaranteed time slots, which reserve channel time for specific devices such as wireless keyboards and joysticks.

To meet its low average power consumption goals, IEEE 802.15.4/ZigBee is capable of extremely low duty cycles — below 10 ppm. The standard also supports beaconless operation — an asynchronous, asymmetrical mode supporting unslotted CSMA-CA channel access for star networks — that enables devices other than the (constantly) receiving central node to remain asleep for indefinite periods, thus reducing their average power consumption still further.

35.2   Low Power Features

Because low power operation is important, each layer of the IEEE 802.15.4/ZigBee protocol stack has been optimized with that goal in mind.

35.2.1   Physical Layer

As noted previously, low average power is achieved with a low overall system duty cycle. However, low duty cycle must be achieved with low peak power consumption during active periods because most of the target power sources (coin cells, energy-scavenging techniques) have limited current sourcing capabilities and low terminal voltage, and it is not desired to burden implementations with complex power conditioning systems for cost and efficiency reasons.[6] At the physical layer, the need for low duty cycle yet low active power consumption implies the need for a high data rate (to finish active periods quickly and return to sleep), but a low symbol rate (because signal processing peak power consumption is more closely tied to the symbol rate than the data rate). These simultaneous requirements imply the need for multilevel signaling (or *m*-ary signaling, with $m > 2$), in which multiple information bits are sent per transmitted symbol. However, simple multilevel signaling, such as 4-FSK, results in a loss of sensitivity that may defeat the low-power goal: to recover the needed link margin (i.e., range) without resorting to directive antennas, the transmitted power must be increased or the receiver noise figure must be reduced, both of which can increase power consumption significantly.

To resolve this dilemma, the IEEE 802.15.4 2.4-GHz physical layer uses a 16-ary quasi-orthogonal signaling technique — trading signal bandwidth to recover sensitivity with coding gain.[7] A particular 32-chip, pseudo-random (PN) sequence is used to represent four bits. Information is placed on the signal by cyclically rotating or conjugating (inverting chips with odd indices) the PN sequence.[8] The PN sequence is rotated in increments of four chips: symbols 0 through 7 represent rotation without conjugation; and symbols 8 through 15 represent the same rotations, but with conjugation. In this way, four bits are placed on each transmitted symbol and, because transmitted symbols are related by simple rotations and conjugations, receiver implementations can be greatly simplified over other orthogonal signaling techniques that employ unrelated PN sequences.

Half-sine shaped, offset-Quadrature Phase Shift Keying (O-QPSK) is employed, in which the chips of even index are placed on the I-channel and the chips of odd index are delayed one-half chip period and then placed on the Q-channel. The chip rate on either the I- or the Q-channel is 1 Mchip/s, so the overall chip rate is 2 Mchip/s. The symbol rate is 62.5 ksymbols/s, leading to a data rate of 250 kbps.

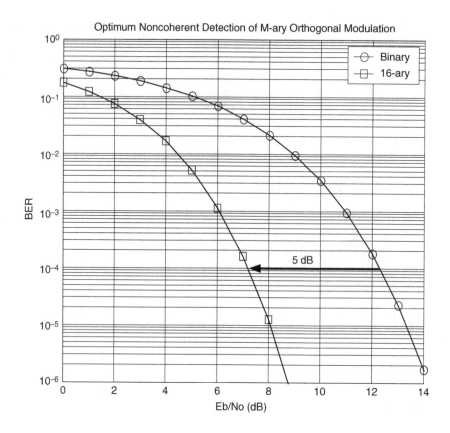

FIGURE 35.1   Performance of binary and 16-ary orthogonal modulation.

The bit error rate performance of this scheme, compared to binary signaling, is shown in Figure 35.1. The 16-ary quasi-orthogonal technique provides a 5-dB *increase* in sensitivity over binary signaling at the BER $= 10^{-4}$ level. The technique has the additional advantage that it can be implemented digitally, so it can shrink in size and cost with future semiconductor process improvements.

A second area in which the IEEE 802.15.4 physical layer has been designed to minimize power consumption is warm-up power loss. Because the active periods of IEEE 802.15.4 devices can be very short (an acknowledgment frame is only 352 μs long), significant power can be lost if the transceiver warm-up time is long.[9,10]

Warm-up time can be dominated by the settling of transients in the signal path, especially the transients of channel selectivity filters, which are usually active filters, integrated to minimize external parts count. Wideband techniques, such as the quasi-orthogonal signaling approach used in IEEE 802.15.4, have an advantage in that their wide channel filters have inherently short settling times. Wideband systems also have a larger channel spacing; this enables their frequency synthesizers to employ higher reference frequencies, which reduces their lock time. The IEEE 802.15.4 2.4-GHz physical layer specifies a 5-MHz channel spacing. These features enable implementations with greatly shortened warm-up periods.

The IEEE 802.15.4 physical layer has several other features that enable low-power operation. As noted above, half-sine shaped O-QPSK is employed; this type of modulation has a constant envelope, which simplifies the transmitter and reduces active current in the transmitter's power amplifier. Duplex operation is not supported; this reduces peak power consumption by eliminating simultaneous operation of the transmitter and receiver, and by eliminating the insertion loss of a duplexer.

The receiver blocking specification is reduced in IEEE 802.15.4, compared to many other wireless protocols, so that the power consumption of the receiver front end (low-noise amplifier and mixer) can be reduced. This is a significant savings because the receiver front end often has the highest power consumption of any functional block in the transceiver. Finally, the standard supports the use of low transmit output power: it reads that compliant devices must be "capable" of transmitting -3 dBm output power; however, the amount of power actually used in service is not specified. In practice, devices are free to use the minimum amount of power necessary to achieve communication.

It is interesting to note that there is little point in reducing transmitted output power below 0 dBm, to reduce power consumption. Zero dBm equates to one milliwatt, and the fixed power needed to operate the transmitter circuits (oscillator, modulator, etc.) is often 10 mW or more, independent of the output power. Reducing the transmitted output power below 1 mW therefore will have a negligible effect on overall power consumption. In fact, in these systems, the receiver often consumes more power than does the transmitter, because it has more signal processing circuits active. This can have a significant effect on low power consumption strategies because it implies that it is more power efficient to blindly transmit than to blindly receive, for an equal period of time. For example, in many applications, such as wireless door locks, one device must discover another, and one device is mains powered and the other battery powered. In these applications, it is better to have the mains-powered device constantly receive, and the battery-powered device transmit beacons, than to have the mains-powered device constantly transmit beacons and have the battery-powered device periodically receive.

35.2.2 Data Link Layer

A typical application of IEEE 802.15.4/ZigBee is a wireless lamp switch. In this application, the lamp is always connected to the ac mains. The switch, however, is desired to be wireless so that it can be placed anywhere in a room the user desires, or even carried in a pocket or purse. Further, it is desired to have a very long battery life so that the user is not inconvenienced. Finally, the communication latency between switch and lamp should be undetectable to the user. These criteria are best met by the IEEE 802.15.4 standard's beaconless mode. As noted above, the beaconless mode of operation is suitable for unslotted CSMA-CA channel access. In this case, the lamp is connected to the ac mains so it can employ a receiver that is constantly active. The switch, however, is battery powered. To minimize its power consumption in this application, the switch remains in sleep mode until the user toggles the switch. The switch transmits a frame to the lamp announcing the change of state in the switch, and then listens for an acknowledgment. The lamp receives the frame, sends the acknowledgment, and toggles the lamp power. Once the switch receives the acknowledgment, it can return to sleep mode. If it does not receive the acknowledgment within a timeout period, it waits a random period of time and then retransmits the frame; this process repeats until an acknowledgment is received or a predetermined number of attempts is exceeded.

This scheme takes advantage of the asymmetrical (one-way) nature of the communication needed in this application, and the asymmetrical nature of the available power supplies, to optimize the battery life of the wireless switch. Many applications, however, require more symmetrical communication links and do not have ac mains power available. For these applications, IEEE 802.15.4/ZigBee has an optional superframe structure, as shown in Figure 35.2.

With the superframe, networked devices transmit beacons periodically. Following beacon transmission, there is a contention access period (CAP) during which the receiver of the beaconing device is active, and other devices in range can attempt to communicate with it via a slotted CSMA-CA protocol. Following the CAP, the beaconing device can enter a low-power sleep mode until it is time to transmit the next beacon.

The beacon period for IEEE 802.15.4 in the 2.4-GHz band is defined as 15.36×2^{BO} ms, $0 \leq BO \leq 14$. The minimum beacon period is therefore 15.36 ms, and the beacon period can be expanded to a maximum of 251.65824 s, or more than four minutes, by adjustment of the beacon order (BO) parameter. As shown in Figure 35.2, the active period of the beaconing device can be similarly adjusted using the superframe order (SO) parameter, with the restriction that SO $\leq$ BO. The device duty cycle decreases as SO becomes smaller relative to BO; in the extreme, when SO $= 0$ and BO $= 14$, the device duty cycle reaches approximately 61 ppm.

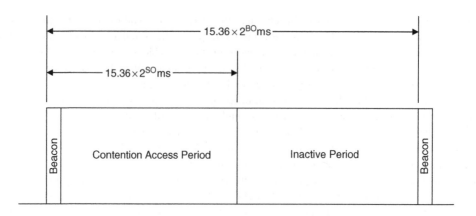

FIGURE 35.2   The IEEE 802.15.4 superframe.

CSMA-CA was selected as the channel access method for IEEE 802.15.4 for a number of reasons, including the fact that it was desired to support multihop networks without a central device coordinating channel access. Nevertheless, the comparison between polling, used by IEEE 802.15.1 (Bluetooth), and CSMA-CA as low-power channel access methods is quite interesting. For low offered traffic rate applications, transmission and reception activity associated with polling can create a lower bound on attainable duty cycle and, therefore, power consumption. However, in CSMA-CA, most power consumption is due to the receiver alone, due to the long monitoring periods required to support operation during high offered traffic periods. This seemed incongruous to the needs of IEEE 802.15.4, so a "battery life extension" (BLE) mode was incorporated into the standard for applications expecting a low offered traffic rate. In the BLE mode (announced via a flag in the beacon), the beaconing device limits the length of its CAP to only six backoff period (twoslots, a total of 120 symbols). Devices attempting to transmit to the beaconing device detect the BLE flag in the beacon and limit their CSMA-CA backoff exponent to the values 0, 1, or 2. This scheme utilizes the assumption of a low offered traffic rate to enable the beaconing device to return to sleep rapidly. Devices employing the BLE mode are capable of duty cycles nearing 10 ppm and, because they achieve a lower duty cycle for given beacon and superframe orders than devices that do not employ it, they can achieve a better trade between power consumption and message latency.

35.2.3   A Note about Batteries

Many, if not all, types of batteries exhibit a recovery effect: when current is drawn from them in pulses, batteries have a greater capacity than when a continuous current of equal average value is drawn from them.[11,12] The shorter the pulse, and the more rest between pulses, the longer the battery life is extended. This is a fortunate situation in which the condition that minimizes the power consumed by the load (low duty cycle) also maximizes the power available from the source. This effect has been proposed for use in route determination in multihop networks.[13]

35.3   Low Power Features Compatible with ZigBee

The ZigBee platform is very flexible and there are many opportunities for creative use of the ZigBee platform to minimize power consumption in ways that are compatible with the ZigBee specification but not contained in it. One of these opportunities is in the data link layer synchronization of devices in a multihop network.

35.3.1 The Mediation Device[14]

One of the fundamental problems in an ad hoc, multihop network of battery-powered devices is the need for synchronization. The lower the duty cycle of network devices, the better their battery life, yet the more difficult is the problem of synchronizing any given nodes that need to communicate, so that the proper one is receiving when the proper one is transmitting. Global synchronization would be useful, yet is difficult to achieve in ad hoc networks of low-cost devices while retaining the goals of low power and self-organizing operation. However, constant synchronization of all devices in a network at all times is not required — especially because the offered data throughput rate is assumed to be low and devices are expected to be sleeping a large fraction of the time they are in operation. At any given time, synchronization is only needed between devices that need to communicate with each other.

One approach to achieve this is to consider an analogy with ordinary telephone communication. The desired recipient of a telephone call does not know when a caller will make a call, so he does not know when to be near his telephone. Similarly, the caller does not know when to place his call because he does not know when the recipient will be available to receive it. To solve this problem, a telephone answering machine can be used. In the operation of a telephone answering machine, the caller leaves a message, saying when he will be near his telephone and available to receive a return call. Sometime later, the recipient replays the message and, at the appointed time, returns the call. At this point, the two are (temporarily) synchronized and communication takes place. Following the call, the caller and the recipient desynchronize and go their separate ways. This type of synchronization, which occurs only when needed, has been termed "dynamic synchronization."

Dynamic synchronization can be employed in a ZigBee network by requiring all network devices to periodically transmit beacons at a regular interval, set by the beacon order, and by employing the BLE mode. Because these devices can be enabled at random times, employ low-cost time bases, and have low duty cycles, they will be unsynchronized. With reasonable device densities, however, the probability of beacon collision can be very low — minimal-length IEEE 802.15.4 beacons last only 544 μs, and the beacon period can be many seconds. Ordinarily, the beacon is a "query," or "hello," beacon, containing only the identity of the beaconing device. However, when a device has a message for another network device, it advertises this fact in its beacon by changing it to a Request To Send (RTS) type that includes the address of the desired recipient.

The synchronization problem is, of course, that the desired recipient likely does not receive the RTS beacon because it is sleeping. Similarly, the network device attempting to send the message does not know when the recipient's receiver will be active. To solve this problem, a *Mediation Device* is introduced. A Mediation Device is a third device in the network that has its receiver active for at least one beacon period. It receives the beacons of both the desired sender and the desired receiver and records the timing of each beacon. It then notifies the desired recipient of the existence of the pending message and gives it the time offset between the sender and recipient beacons. The recipient then waits until the sender's receiver is active (following one of the sender's beacons) and transmits a "Clear To Send" (CTS) message. Now synchronized, the sender and recipient transfer the message. At the conclusion of their communication, the recipient returns to his original beacon phase (so his beacon will not interfere with that of the sender) and the process is complete. This process is shown in Figure 35.3.

If the Mediation Device detects the sender's RTS beacon after the desired recipient, the Mediation Device can optionally contact the sender so that it can suppress transmission of further RTS beacons while the Mediation Device contacts the desired recipient. Note that the Mediation Device does not require that the query and RTS beacons be sent in any particular order; it is merely necessary that the Mediation Device receive both. This does, however, illustrate another point of the Mediation Device protocol: the Mediation Device will attempt to synchronize the sender and recipient nodes, even if they are out of range of each other. In IEEE 802.15.4 networks, this situation is avoided by requiring the sending device to send messages only to those devices from which it has already received beacons, that is, those devices in its neighborhood. Messages destined for other devices are routed via the ZigBee routing protocol, which identifies which device in the sender's neighborhood can best be used to relay the message.

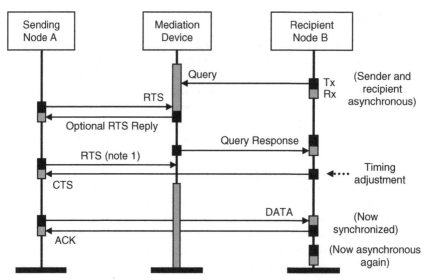

Note 1: Not sent if optional RTS Reply message is received from mediation device.

FIGURE 35.3   Mediation device operation.

35.3.2   The Distributed Mediation Device

The Mediation Device protocol as described above has the disadvantage that a dedicated device is necessary for network communication to take place. Placement of such devices so that one is within range of all network devices is contrary to the self-organizing, ad hoc nature desired of wireless sensor networks. To address this, instead of using a dedicated Mediation Device, one can distribute the Mediation Device functionality among all network devices. Rather than keeping their receivers (almost) constantly active, however, each device is required to periodically stop its beaconing and to operate as a Mediation Device only for the duration of a single beacon period. At the conclusion of this time, if no RTS beacons were received, the device resumes its normal beaconing. If an RTS beacon were received, however, the device would continue to function as a Mediation Device, dynamically synchronize the appropriate devices, and then return to its normal beaconing.

To maintain its low overall duty cycle, a network node cannot enter "mediation mode" too often, because when in that mode it must activate its receiver for an entire beacon period (strictly speaking, an entire beacon period plus a beacon duration). However, in a network of reasonable density, there will likely be several devices within range of the sending and recipient nodes, any one of which can serve as the Mediation Device for the transaction. This greatly reduces the average period of time a sending node must wait for a Mediation Device to hear its RTS beacon.

35.4   Power-Conscious Implementation

35.4.1   The Effect of Time Base Errors

The IEEE 802.15.4 standard is capable of very low duty cycle operation. However, in practical implementations, the lowest attainable duty cycle is often limited by the effect of time base errors in both the sending and receiving devices.

Consider a receiving node receiving from a sending node, ideally sending beacons T_b seconds long every T_c seconds. The ideal duty cycle is therefore simply T_b/T_c. The receiving node should turn on its receiver every T_c seconds; however, the sending node has a time base stability specification ε_t and the recipient node has a time base stability specification ε_r. Because the receiving node cannot know the time base difference between the two devices *a priori*, in order to be sure that it receives the beacon, it must turn on its receiver $|\varepsilon_t + \varepsilon_r| \times T_c$ seconds early, leading to a duty cycle of $[T_b + |\varepsilon_t + \varepsilon_r| \times T_c]/T_c$. The time base stability therefore sets a lower bound on the attainable duty cycle, no matter how short T_b is made or how long T_c is made.

It can be pointed out that the receiving device can learn about the value of $|\varepsilon_t + \varepsilon_r|$ from beacons received earlier, and this is indeed the case. However, IEEE 802.15.4 devices are physically very small and can react quickly to changes in temperature; further, beacon periods can be longer than one minute. It is therefore possible for the time base to vary significantly over a single beacon period; one can reduce the early receiver activity period somewhat but the risk of missing the next beacon transmission increases as this is done.

35.4.2   Security

Like all wireless systems, security is an important component of IEEE 802.15.4 implementations. As noted above, IEEE 802.15.4 supports seven different security levels: encryption only; message integrity only, with one of three message integrity code (MIC) lengths (4, 8, or 16 bytes); or encryption plus message integrity, again with one of three MIC lengths.

The use of security increases the power consumed by IEEE 802.15.4 devices. There are two components of this increase: (1) communication and (2) computation. Additional communication power consumption is due to additional packet length, caused by the auxiliary security header and any MIC appended to the packet. Additional computational power consumption is due to the need to perform the security algorithms themselves.

To evaluate the power cost of security in IEEE 802.15.4, one needs first to determine which security level to consider. If security is used at all, the standard requires that at least the encryption-plus-8-byte-MIC level (ENC-MIC-64) be available; this therefore seems a reasonable level to evaluate.

First evaluate the energy needed to transmit (but not receive) an unsecured frame. Now make the following assumptions:

The RF transmitter active power $P_{act} = 50$ mW.
The data payload is 16 bytes; this is added to the 6-byte PHY and 13-byte MAC headers.
Short (one-byte) logical addresses are used in the MAC header.
The 2.4-GHz-band is used, implying a 250-kbps (or 32 μs/byte) raw data rate.
Warm-up energy is ignored.

With these assumptions, the energy needed to transmit an unsecured frame is

$$(6 + 13 + 16)\text{bytes} \times 32\,\mu\text{s/byte} \times 50\text{ mW} = \underline{56\mu\text{J}}.$$

To transmit this frame securely, using the ENC-MIC-64 security level, requires an additional 5 bytes for the auxiliary security header, plus the 8-byte MIC appended to the end of the frame. The additional energy needed to transmit the frame, above that required to send it unsecured, is

$$(5 + 8)\text{bytes} \times 32\,\mu\text{s/byte} \times 50\text{ mW} = \underline{21\mu\text{J}}.$$

The computational power needed to perform the security algorithm must now be determined. First consider the energy consumed when performing the security algorithm in software. One can make the following assumptions, noting in addition that the AES algorithm must be performed seven times for ENC-MIC-64 on a packet of this length:

The Motorola MC9S08GB60 MCU is used, drawing 4.8 mA at 16 MHz from 2 V (9.6 mW).[15]
The AES algorithm requires 8390 cycles (524 μs at 16 MHz) to perform.[16]

TABLE 35.1 Energy Consumed by the Transmission of an Unsecured Frame, the Same Frame Secured with a Software AES Implementation, and Secured with a Hardware AES Implementation

		Secured	
	Unsecured (μJ)	Software (μJ)	Hardware (μJ)
Data frame Tx	56	56	56
Additional security Tx	0	21	21
AES calculations	0	35	0.0875
Total	56	112	77.0875

The calculation energy used for security operations, when performed in software, is therefore

$$524\,\mu s \times 7 \times 9.6\,\text{mW} = \underline{35\mu J}.$$

Alternatively, the security operations can be performed using a dedicated AES engine in hardware. One can make the following assumptions about the hardware implementation, noting in advance that the values given are both implementation-specific and conservative:

When in operation, the AES hardware engine draws 1 mA from 2 V at 16 MHz (2 mW).

100 cycles (6.25 ms at 16 MHz) are needed per output.

As before, seven AES runs are needed for the secured frame.

The calculation energy used for security operations, when performed in hardware, is therefore

$$6.25\,\mu s \times 7 \times 2\,\text{mW} = \underline{87.5\,\text{nJ}}.$$

A summary of these results is in Table 35.1.

As Table 35.1 shows, the transmission of the frame secured with a software AES implementation consumes twice the energy as transmission of the same frame unsecured. Use of a hardware AES implementation drops this penalty to about 37 percent [i.e., $(77.0875 - 56)/56$], almost all of which is due to the additional length of the frame because the energy consumed by the hardware AES engine itself is trivial. Use of a hardware AES engine is clearly the lower-power solution.

However, it must be kept in mind that in low duty cycle systems, it is often not the active power consumption that determines battery life: because the transceiver is so rarely active, it is often the standby power consumption that dominates — or perhaps even the shelf life of the battery itself. For example, if an environmental application is contemplated that transmits only a single secured message per day ("It rained today"), the additional power consumed by a software AES implementation may be quite acceptable, because it would not materially affect the average power consumed by the device. Because the development or purchase of a dedicated hardware engine can be avoided, this may be a lower cost solution as well.

35.4.3　Power Conditioning

Power sources for wireless sensor network nodes are critical to their practical implementation. Unfortunately, for a variety of reasons, the current and voltage sourcing capabilities of most available power sources do not match particularly well with the current and voltage requirements of most IEEE 802.15.4/ZigBee implementations. Hence, some type of power conditioning (voltage conversion or charge storage) is often required to maximize the performance of the device.

Power conditioning is an inherently lossy process because power is lost through voltage conversion stages or leaked away when charge is stored; thus, it is an important part of the design of low-power IEEE 802.15.4 devices. There is an inherent contradiction in the design of such devices because the manufacturer typically wants to maximize the number of different power sources on which his device will operate, to maximize his total available market, while the efficiency of power conditioning is often maximized when the power conditioning circuit is designed for a single pair of sources and loads.

Because even annual battery replacement for networks with many hundreds of devices is impractical, an area of great research interest at present is energy scavenging, or the extraction of energy from the environment of the network device. Of principal interest is the extraction of energy from vibration. Because the ambient power (energy per unit time) available in the environment is often very low, energy-scavenging techniques require well-designed power conditioning circuits to be effective. Often, these must be designed for specific applications; the efficient extraction of energy from external vibration, for example, requires a vibration-to-electricity converter that is tuned to the frequency of the vibration.[17]

Many power sources, especially scavenging sources, produce power at voltages significantly less than 1 V. Examples of these include solar cells, vibration-to-electricity converters, and microbial fuel cells. To avoid power conditioning losses, the optimum design of future scavenging-friendly IEEE 802.15.4 devices should be able to operate directly from low supply voltages.

35.5　Conclusion

The IEEE 802.15.4 WPAN standard has many features that enable very low-power implementations. At the physical layer, these include a signaling scheme with a high data rate but relatively low symbol rate, a short warm-up capability, and constant-envelope modulation. In the data link layer, the standard incorporates a beaconless mode for asymmetrical communication applications, an optional superframe with low duty cycle capability, and an optional battery life extension mode for even lower duty cycle operation in low-offered-throughput applications.

The standard, including the upper layers specified by the ZigBee Alliance, also supports low-power dynamic synchronization, as performed by the Mediation Device protocol. Operation of the security suite can also consume little power, especially if a hardware AES engine is employed. Along with the design of low-voltage, low-power transceiver circuits, the design of efficient low-voltage power conversion systems will be critical for the success of IEEE 802.15.4/ZigBee systems in applications requiring very large networks, especially those employing energy-scavenging techniques.

References

1. Institute of Electrical and Electronics Engineers, IEEE Std. 802.15.4-2003, IEEE Standard for Information Technology — Telecommunications and Information Exchange between Systems — Local and Metropolitan Area Networks — Specific Requirements — Part 15.4: Wireless Medium Access Control (MAC) and Physical Layer (PHY) Specifications for Low Rate Wireless Personal Area Networks (WPANs). New York: IEEE Press, 2003.
2. http://standards.ieee.org/getieee802/.
3. http://www.zigbee.org.
4. Tom Messerges et al., A Security Design for a General Purpose, Self-Organizing, Multihop Ad Hoc Wireless Network, *ACM Workshop on Security of Ad Hoc and Sensor Networks,* October 2003.
5. U.S. Department of Commerce, National Institute of Standards and Technology, Information Technology Laboratory, Specification for the Advanced Encryption Standard (AES). Federal Information Processing Standard Publication (FIPS PUB) 197. Springfield, VA: National Technical Information Service. 26 November 2001. http://csrc.nist.gov/CryptoToolkit/aes/.
6. Edgar H. Callaway, Jr., *Wireless Sensor Networks.* Boca Raton, FL: Auerbach Publications, 2003, Chap. 7.
7. Jose A. Gutierrez, Edgar H. Callaway, Jr., and Raymond L. Barrett, Jr., *Low-Rate Wireless Personal Area Networks: Enabling Wireless Sensors with IEEE 802.15.4^{TM}.* New York: IEEE Press, 2003, Chap. 4.
8. Isao Okazaki and Takaaki Hasegawa, Spread spectrum pulse position modulation and its asynchronous CDMA performance — a simple approach for Shannon's limit, *Proc. IEEE Sec. Int. Symp. Spread Spectrum Techniques and Applications,* 1992, pp. 325–328.

9. Eugene Shih et al., Physical layer driven protocol and algorithm design for energy efficient wireless sensor networks, *Proc. MOBICOM,* 2001, pp. 272–287.
10. Andrew Y. Wang et al., Energy efficient modulation and MAC for asymmetric RF microsensor systems, *IEEE Int. Symp. Low Power Electronics and Design,* 2001, pp. 106–111.
11. S. Okazaki, S. Takahashi, and S. Higuchi, Influence of rest time in an intermittent discharge capacity test on the resulting performance of manganese-zinc and alkaline manganese dry batteries, *Progress in Batteries & Solar Cells,* 6, 106–109, 1987.
12. C.F. Chiasserini and R.R. Rao, A model for battery pulsed discharge with recovery effect, *Proc. Wireless Commun. Networking Conf.,* 2, 636–639, 1999.
13. C.F. Chiasserini and R.R. Rao, Routing protocols to maximize battery efficiency, *Proc. MILCOM,* 1, 496–500, 2000.
14. Qicai Shi and Edgar H. Callaway, An ultra-low power protocol for wireless networks, *Proc. 5th World Multi-Conference on Systemics, Cybernetics and Informatics,* IV, 321–325, 2001. See also Callaway, Jr., *ibid.,* pp. 74–84.
15. http://e-www.motorola.com/files/microcontrollers/doc/data_sheet/MC9S08GB60.pdf.
16. Joan Daemen and Vincent Rijmen, *AES submission document on Rijndael,* Version 2, September 1999. Section 6.1.2. http://csrc.nist.gov/CryptoToolkit/aes/rijndael/Rijndael-ammended.pdf.
17. Shad Roundy, Paul Kenneth Wright, and Jan M. Rabaey, *Energy Scavenging for Wireless Sensor Networks.* Boston: Kluwer Academic Publishers, 2004.

Peer-to-Peer
Networks

Peer-to-peer (P2P) computing refers to technology that enables two or more peers to collaborate spontaneously in a network of equals (peers) by using appropriate information and communication systems without the necessity for central coordination. The P2P network is dynamic, in the sense that peers come and go (i.e., leave and join the group) for sharing files and data through direct exchange.

The most frequently discussed applications include popular file-sharing systems, such as early Napster. In addition to file-sharing collaborative P2P service, grid computing and instant messaging are key applications of P2P. P2P systems offer a way to make use of the tremendous computation and storage resources on computers across the Internet. Unlike sensor networks and ad hoc wireless networks, P2P networks are overlay networks operated on infrastructured (wired) networks, such as the Internet. However, P2P networks are also highly dynamic. Users join and leave the network frequently. Therefore, the topology of the overlay network is dynamic.

Most current research in this field focuses on location management, which is also called the *lookup problem*. Specifically, how can we find any given data item in a large P2P network in a scalable manner? In serverless approaches, flooding-based search mechanisms are used; these include DFS with depth limit D (as in Freenet) or BFS with depth limit D (as in Gnutella), where D is the system-wide maximum TTL of a message in hops. There are several efficient alternatives; these include iterative deepening to slowly increase the flooding rings, random walks instead of blind flooding to reduce the flooding space, and dominating-set-based searching to reduce the searching scope. In server-based approaches, Napster uses a centralized approach by maintaining a central database. KaZaA uses a hierarchical approach based on the notion of supernode.

Data in P2P networks are sometimes structured to facilitate an efficient searching process through the use of a *distributed hash table* (DHT). Each node acts as a server for a subset of data items. The operation *lookup(key)* is supported, which returns the node ID storing the data item with that key. The values of the node could be data items or pointers to where the data items are stored. Each data item is associated with a key through a hashing function. Nodes have identifiers, taken from the same space as the keys. Each node maintains a routing table consisting of a small subset of nodes in the system. In this way, an overlay network is constructed that captures logical connections between nodes. Usually, the logical network is a regular network such as a ring, tree, mesh, or hypercube. When a node receives a query for a key for which it is not responsible, the node routes the query to the neighbor that makes the most "progress" (defined in terms of "distance" between source and destination) towards resolving the query. A promising approach is limited server-based approach, in which location information is limited to a limited region of nodes.

Other issues related to P2P systems include network control, security, interoperability, metadata, and cost sharing. Some open problems include operation costs, fault tolerance and concurrent changes, proximity routing, malicious nodes, and indexing and keyword searching.

Among 12 chapters in this group, one chapter gives a general overview of P2P networks. Two chapters deal with searching techniques, including one on semantics search. One chapter provides an overview of structured P2P networks. The next three chapters are devoted to three specific aspects of structured P2P networks: distributed data structure, state management, and topology construction. One chapter deals with overlay optimization. Reliability and efficient issues are covered in two chapters. The discussion on security issues is given in one chapter. The group ends with the application of the peer-to-peer concept in ad hoc wireless networks.

Peer-to-Peer: A Technique Perspective

Weimin Zheng

Xuezheng Liu

Shuming Shi

Jinfeng Hu

Haitao Dong

36.1 Introduction

Peer-to-peer (P2P) systems have drawn much attention from end Internet users and the research community in the past few years. From Napster,[1] the earliest P2P system, appearing in 1999, to popular systems such as Gnutella,[2] Freenet,[3] KaZaA,[4] and BitTorrent,[5] more and more P2P file sharing systems come to fame. In research community, P2P has become one the most active research fields. Many universities and research institutes have research groups focusing on P2P techniques. And P2P has become one of the hottest topics at many conferences and workshops on distributed systems or networking.

Different from traditional client/server architectures, each node in a peer-to-peer system acts as both a producer (to provide data to other nodes) and a consumer (to retrieve data from other nodes). Key features of P2P systems include large-scale self-organization, self-scaling, decentralization, fault tolerance, and heterogeneity. In a P2P system, an increase in the number of peers also adds capacity of the system. There is a large heterogeneity in the properties (including capacity, availability, bandwidth, etc.) of peers in most P2P systems. A typical P2P system often has no central servers to which users connect. Instead, the whole system is self-organized by the interaction of peers. The above features of P2P systems make them suitable for file sharing. In addition to file sharing, there are also some proposals that use P2P to build large-scale, fault-tolerant storage systems,[6–9] do content distribution,[10] or even as a replacement of current networks.[11]

The earliest P2P system was Napster, which maintains a centralized directory to facilitate file sharing. Files shared by peers are indexed in a centralized directory. To search a file, peers first refer to the centralized directory and results are returned after local lookup; and then result files are downloaded directly from peers. Relying on this kind of centralized-lookup and decentralized-download mode, Napster became one of the most popular applications in just a few months. However, Napster was forced to shut down because of legal issues. Then there came decentralized architectures where the search and download processes are both distributed among peers. Existing decentralized P2P networks can be partitioned into two groups: unstructured and structured. Unstructured P2P systems (e.g., Gnutella and Freenet) have few constraints on overlay topology and the mapping of data to peers. Whereas, for structured P2P systems (e.g., CAN,[12] Pastry,[13] Chord,[14] Tapestry,[15] etc.), the topology of overlay and the mapping of data items to nodes are strictly controlled.

Because of the heterogeneous and dynamic properties of peers, it is a challenge for peers to self-organize themselves to build a large-scale P2P system in a decentralized way. Many researchers and system designers have been involved in solving key problems of P2P systems in the past several years.

This chapter analyzes the design issues and applications of P2P systems from a technique perspective. It focuses on key design issues of P2P systems and studying how these techniques are used to support practical applications.

36.2   Design Issues

This section discusses the fundamental issues of P2P system design. It is very difficult to build a large-scale distributed system and combine enormous weak, dynamic participants to provision powerful and persistent services, especially in the Internet environment which is prone to failures. The system should scale to large numbers of peers, tolerate unpredictable failures, fully utilize peers' capacities and heterogeneities, guarantee strong availability of data or services, and efficiently provide diverse functionalities.

Therefore, there are many challenging problems to address. First, peers should be perfectly organized so as to facilitate communication and cooperation. Thus, we need to construct an efficient routing infrastructure in P2P networks, an infrastructure that is scalable and fault tolerant, and also exploits network proximity (Section 36.2.1). Second, on top of the underlying routing, we need to deploy services in P2P networks. Thus, we should answer how the services and data are placed and supplied, so as to achieve data availability and full utilization of peers' capacities (Section 36.2.2). In addition, the difficulty of looking for the most suitable service from numerous provided services leads us to build efficient search mechanisms in P2P applications (Section 36.2.3). Finally, for the special cases of group communication (e.g., network meeting, media streaming, and bulk data multiple delivery), only unicast communication and one-to-one data transmission remain far from sufficient. Thus, we want to implement application-level multicast on top of overlay and P2P networks (Section 36.2.4).

36.3 Routing

Scalability is an extraordinarily important issue for the P2P system. Unfortunately, the original designs for P2P systems have significant scaling problems. For example, Napster introduces a centralized directory service, which becomes a bottleneck for the millions of users' access; Gnutella employs a flooding-based search mechanism that is not suitable for large systems.

To solve the scaling problem, several approaches have been simultaneously but independently proposed, all of which support a distributed hash table (DHT) functionality. Among them are Tapestry,[15] Pastry,[13] Chord,[14] and Content-Addressable Networks (CAN).[12] In these systems, which we call DHTs, files are associated with a key, which is produced, for instance, by hashing the filename, and each node in the system is responsible for storing a certain range of keys. There is one basic operation in these DHT systems: lookup(key), which returns the identity (e.g., the IP address) of the node storing the object responsible for that key. By allowing nodes to put and get files based on their key with such operation, DHTs support the hash-table-like interface. This DHT functionality has proved a useful substrate for large distributed systems, which is promising to become an integral part of the future P2P systems.

The core of these DHT systems is the routing algorithm. The DHT nodes form an overlay network with each node having several other nodes as neighbors. When a lookup(key) operation is issued, the lookup is routed through the overlay network to the node responsible for that key. Then, the scalability of these DHT algorithms depends on the efficiency of their routing algorithms. Each of the proposed DHT systems listed above—Tapestry, Pastry, Chord, and CAN—employs a different routing algorithm. Although there are many details that are different between their routing algorithms, they share the same property that every overlay node maintains $O(\log n)$ neighbors and routes within $O(\log n)$ hops (n is the system scale).

Researchers have made inroads in the following issues related to routing: state–efficiency trade-off, resilience to failures, routing hotspots, geography, and heterogeneity, and have forwarded various algorithms to improve the performance of the initial DHT systems.

This section is organized as follows. First we review the history of research for routing algorithms in P2P system; then we introduce the representative DHT systems; and finally we discuss several issues related to DHT routing algorithms.

36.3.1 $O(\log N)$ DHT Overlay

This section reviews some representative routing algorithms that were introduced in the early period. All of them take, as input, a key and, in response, route a message to the node responsible for that key. The keys are strings of digits of some length. Nodes have identifiers, taken from the same space as the keys (i.e., same number of digits). Each node maintains a routing table consisting of a small subset of nodes in the system. When a node receives a query for a key for which it is not responsible, the node routes the query to the neighbor node that makes the most "progress" toward resolving the query. The notion of progress differs from algorithm to algorithm but in general is defined in terms of some distance between the identifier of the current node and the identifier of the queried key.

36.3.1.1 Plaxton et al.

Plaxton et al.[16] developed perhaps the first routing algorithm that could be scalably used by DHTs. While not intended for use in P2P systems, because it assumes a relatively static node population, it does provide very efficient routing of lookups. The routing algorithm works by "correcting" a single digit at a time: if node number 47532 received a lookup query with key 47190, which matches the first two digits, then the routing algorithm forwards the query to a node that matches the first three digits (e.g., node 47603).

To do this, a node needs to have, as neighbors, nodes that match each prefix of its own identifier but differ in the next digit. For a system of n nodes, each node has on the order of $O(\log n)$ neighbors.

Because one digit is corrected each time the query is forwarded, the routing path is at most $O(\log n)$ overlay (or application-level) hops. This algorithm has the additional property that if the $O(n^2)$ node-node latencies (or "distances" according to some metric) are known, the routing tables can be chosen to minimize the expected path latency and, moreover, the latency of the overlay path between two nodes is within a constant factor of the latency of the direct underlying network path between them.

36.3.1.2 Tapestry

Tapestry[15] uses a variant of the Plaxton et al. algorithm. The modifications are to ensure that the design, originally intended for static environments, can adapt to a dynamic node population. The modifications are too involved to describe in this short review. However, the algorithm maintains the properties of having $O(\log n)$ neighbors and routing with path lengths of $O(\log n)$ hops.

36.3.1.3 Pastry

Each node in the Pastry[13] peer-to-peer overlay network is assigned a 128-bit node identifier (nodeID). The nodeID is used to indicate a node's position in a circular nodeID space, which ranges from 0 to 2128. The nodeID is assigned randomly when a node joins the system. It is assumed that nodeIDs are generated such that the resulting set of nodeIDs is uniformly distributed in the 128-bit nodeID space. As a result of this random assignment of nodeIDs, with high probability, nodes with adjacent nodeIDs are diverse in geography, ownership, jurisdiction, network attachment, etc.

Assuming a network consisting of n nodes, Pastry can route to the numerically closest node to a given key in less than $O(\log n)$ steps under normal operation and this is a configuration parameter with typical value. Despite concurrent node failures, eventual delivery is guaranteed unless L/2 nodes with adjacent nodeIDs fail simultaneously (L is a configuration parameter with a typical value of 16 or 32). For the purpose of routing, nodeIDs and keys are thought of as a sequence of digits with base 2b. Pastry routes messages to the node whose nodeID is numerically closest to the given key. To support this routing procedure, each node maintains routing state with length of $O(\log n)$. In each routing step, a node normally forwards the message to a node whose nodeID shares with the key a prefix that is at least one digit (or b bits) longer than the prefix that the key shares with the present node's ID. If no such node is known, the message is forwarded to a node whose nodeID shares a prefix with the key as long as the current node, but is numerically closer to the key than the present node's ID. Figure 36.1 is an example of Pastry node's routing table.

Node ID 10233102			
Leaf set			
10233021	10233033	10233120	10233122
Suffix set			
-0-2212102	**1**	-2-2303203	-3-1203203
0	1-1-301233	1-2-230203	1-2-021022
10-0-31203	10-1-32102	2	10-3-23302
102-0-0230	102-1-1302	102-2-2302	3
1023-0-322	1023-1-000	1023-2-121	3
10233-0-01	1	10233-2-32	
0		102331-2-0	
		2	

FIGURE 36.1 State of a hypothetical Pastry node with nodeID 10233102, b = 2, and l = 4. All numbers are in base 4. The top row of the routing table is row zero. The shaded cell in each row of the routing table shows the corresponding digit of the present node's nodeID. The nodeIDs in each entry have been split to show the *common prefix with 10233102–next digit–rest of nodeID*. The associated IP addresses are not shown. (*Source:* From Reference 13.)

36.3.1.4 Chord

Chord[14] also uses a one-dimensional circular keyspace. The node responsible for the key is the node whose identifier most closely follows the key (numerically); that node is called the key's successor. Chord maintains two sets of neighbors. Each node has a successor list of k nodes that immediately follow it in the keyspace. Routing correctness is achieved with these lists. Routing efficiency is achieved with the finger list of $O(\log n)$ nodes spaced exponentially around the keyspace. Routing consists of forwarding to the node closest to, but not past, the key; path lengths are $O(\log n)$ hops.

36.3.1.5 CAN

CAN[12] chooses its keys from a d-dimensional toroidal space. Each node is associated with a hypercubal region of this keyspace, and its neighbors are the nodes that "own" the contiguous hypercubes. Routing consists of forwarding to a neighbor that is closer to the key. CAN has a different performance profile than the other algorithms; nodes have d neighbors and path lengths are $O(dn^{1/d})$ hops. Note, however, that when $d = O(\log n)$, CAN has neighbors and $O(\log n)$ path lengths like the other algorithms.

36.3.2 Improvements

36.3.2.1 State–Efficiency Trade-Off

The most obvious measure of the efficiency of these routing algorithms is the resulting path length. Most of the algorithms have path lengths of $O(\log n)$ hops, while CAN has longer paths of $O(dn^{1/d})$. The most obvious measure of the overhead associated with keeping routing tables is the number of neighbors. This is not just a measure of the state required to do routing, but is also a measure of how much the state needs to be adjusted when nodes join or leave. Given the prevalence of inexpensive memory and the highly transient user populations in P2P systems, this second issue is likely much more important than the first. Most of the algorithms require $O(\log n)$ neighbors, while CAN requires only $O(d)$ neighbors.

Xu et al.[17] studied this fundamental trade-off, which is shown in Figure 36.2. We can see that, in a network consisting of n nodes, when n neighbors are maintained at each node, the search cost is $O(1)$, but to maintain so large a routing table will introduce heavy maintenance costs because of the frequent joins and leaves of the P2P nodes. When each node only maintains one neighbor, the search cost is $O(n)$, which incurs intolerable network delay. This plots two endpoints on the trade-off curve shown in Figure 36.2. They also point out that there exist algorithms that achieve better trade-offs than existing DHT schemes, but these algorithms cause intolerable levels of congestion on certain network nodes.

Gupta et al.[18] argue that it is reasonable to maintain complete information at each node. They propose a well-defined hierarchy system to ensure a notification of membership change events (i.e., joins and

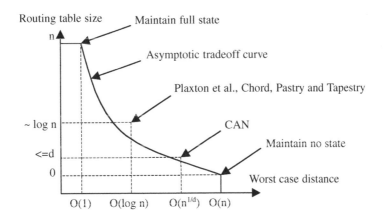

FIGURE 36.2 Asymptotic trade-off curve between routing table size and network diameter.

leaves) can reach every node in the system within a specified amount of time (depending on the fraction of failed queries) and with reasonable bandwidth consumption. There is a distinguishing character in their approach: a node in the system need not probe all the items in its routing table. It only watches a small fraction of nodes; and when a membership change event occurs, the node responsible for this event will initial an event multicast in the whole system.

36.3.2.2 Proximity

The efficiency measure used above was the number of application-level hops taken on the path. However, the true efficiency measure is the end-to-end latency of the path. Because the nodes could be geographically dispersed, some of these application-level hops could involve transcontinental links, and others merely trips across a LAN; routing algorithms that ignore the latencies of individual hops are likely to result in high-latency paths. While the original "vanilla" versions of some of these routing algorithms did not take these hop latencies into account, almost all of the "full" versions of the algorithms make some attempt to deal with the geographic proximity of nodes. There are (at least) three ways of coping with geography.

36.3.2.2.1 *Proximity Routing*

Proximity routing is when the routing choice is based not just on which neighboring node makes the "most" progress towards the key, but also on which neighboring node is "closest" in the sense of latency. Various algorithms implement proximity routing differently but all adopt the same basic approach of weighing progress in identifier space against cost in latency (or geography); usually, when choosing the next hop, the node selects, among the possible next hops, the one that is closest in the physical network or one that represents a good compromise between progress in the ID space and proximity. Simulations have shown that this is a very effective tool in reducing the average path latency. But the construction of overlay does not consider physical networks, and its performance is largely dependent on the number of alternative next hops.

36.3.2.2.2 *Proximity Neighbor Selection*

This is a variant of the idea above, but now the proximity criterion is applied when choosing neighbors, not just when choosing the next hop. During the construction of the overlay, nodes choose routing table entries to refer to the topologically nearest among all nodes with nodeID in the desired portion of the ID space. So, its performance depends on the degree of freedom an overlay protocol has in choosing routing table entries without affecting the expected number of routing hops. As mentioned, if the node-pair distances (as measured by latency) are known, the Plaxton/Tapestry algorithm can choose the neighbors so as to minimize the expected overlay path latency. This is an extremely important property that is (so far) the exclusive domain of the Plaxton/Tapestry algorithms. Its mechanism depends on satisfying triangle inequality, which may not hold for the Internet.

36.3.2.2.3 *Geographic Layout*

In most of the algorithms, the node identifiers are chosen randomly (e.g.,hash functions of the IP address, etc.) and the neighbor relations are established based solely on these node identifiers. One could instead attempt to choose node identifiers in a geographically informed manner. An initial attempt to do so in the context of CAN was reported by Ratnasamy et al.[12] This approach was quite successful in reducing the latency of paths. There was little in the layout method specific to CAN but the high dimensionality of the keyspace may have played an important role. Recent work[19] suggests that latencies in the Internet can be reasonably modeled by a d-dimension geometric space with $d \geq 2$ This raises the question of whether systems that use a one-dimensional key set can adequately mimic the geographic layout of the nodes. However, this may not matter because the geographic layout may not offer significant advantages over the two proximity methods.

Moreover, these geographically informed layout methods may interfere with the robustness, hotspot, and other properties mentioned in previous sections.

36.3.2.3 Heterogeneity

All the algorithms start by assuming that all nodes have the same capacity to process messages and then, only later, add on techniques for coping with heterogeneity. However, the heterogeneity observed in current P2P populations[20] is quite extreme, with differences of several orders of magnitude in bandwidth. One can ask whether the routing algorithms, rather than merely *coping* with heterogeneity, should instead use it to their *advantage*. At the extreme, a star topology with all queries passing through a single hub node and then routed to their destination would be extremely efficient, but would require a very highly capable hub node (and would have a single point of failure). But perhaps one could use the very highly capable nodes as mini-hubs to improve routing. Chawathe et al.[21] argue that heterogeneity can be used to make Gnutella-like systems more scalable.

It may be that no sophisticated modifications are needed to leverage heterogeneity. Perhaps the simplest technique to cope with heterogeneity, and one that has already been mentioned in the literature, is to *clone* highly capable nodes so that they could serve as multiple nodes; that is, a node that is ten times more powerful than other nodes could function as ten virtual nodes. When combined with proximity routing and neighbor selection, cloning would allow nodes to route to themselves and thereby "jump" in keyspace without any forwarding hops.

36.4 Data Placement

The DHT (distributed hash table)[12–15] is a basic manner of data placement in peer-to-peer systems. In such structured systems, a unique identifier is associated with each data item as well as each node in the system. The identifier space is partitioned among the nodes and each node is responsible for storing all the items that are mapped to identifiers within its portion. Thus, the system provides two basic functions: (1) put (ID, item), which stores an item with its identifier ID, and (2) get(ID), which retrieves the item. How to find the appropriate node for a given ID is by the routing algorithms mentioned in Section 36.2.1.

Many DHT-based systems[6–8] assume that item IDs and node IDs are all randomly chosen by a consistent hashing function.[22] This offers a significant advantage for the system; that is, items are evenly distributed among all the nodes.

To improve the availability of the items, replication is desired. A simple and useful replication method is putting one item on k nodes whose node IDs are closest to the item's ID, where k is a constant value. For example, PAST[6] deploys Pastry as its routing infrastructure and a node replicates its data on the k closest nodes in its leaf set. Because node IDs are chosen randomly, these k nodes are scattered around the world with high probability, thus reducing the probability of simultaneous replica failures.

Other replication strategies on DHT include simultaneously using k different hashing functions for data items and simultaneously running k different routing infrastructures.[12]

However, all these methods cannot escape from a significant issue: unbalanced load on DHT nodes. This can come from the following reasons:

There is a $\Box(\log N)$ imbalance factor in the number of items stored at a node because it is impossible to achieve perfect randomicity.

Items are not of identical size.

Nodes are heterogeneous. This is a serious and unavoidable reality in the peer-to-peer world.[20] Some applications associate semantics with item IDs, which makes distribution of item IDs skewed. For example, a peer-to-peer database might stores tuples using their primary keys as tuple IDs.

In PAST,[6] items are first intended to be stored on k contiguous nodes in Pastry's leaf set. But once a node in the leaf set is overloaded, items on it must be moved to the other nodes in the leaf set, leaving corresponding pointers. Once all the nodes in the leaf set are overloaded, a node out of one leaf set must then undertake the responsibility of item storing. How to maintain the pointers, as well as the pointers to the pointers, significantly increases the complexity of the system. Therefore, other systems tend toward another choice, performing load balancing in a direct way:

An overloaded node claims its status and lets a light node take over part of its data.[23,24] This work first involves a concept of *virtual servers*. A virtual server works as a peer in the P2P infrastructure, while a physical node can simultaneously host many virtual servers. When a node is overloaded, it attempts to move one or many virtual servers to other nodes.

Then the key problem turns to how to find a light partner, in the peer-to-peer environment, for a heavy node to transfer its load. Methods can be classified into three types of schemes: (1) one-to-one, (2) one-to-many, and (3) many-to-many, from the simplest to the most complicated.

36.4.1   One-to-One

Each light node periodically picks a random ID and then routes to the node that is responsible for that ID. If that node is a heavy node, then a transfer can take place between the two nodes.

36.4.2   One-to-Many

This scheme is accomplished by maintaining *directories* that store load information about a set of light nodes in the system. We use the same DHT system to store these directories. Assume that there are d directories in the system. A light node l is hashed into a directory using a well-known hash function h' that takes values in the interval $[0, d)$. A directory i is stored at the node that is responsible for the identifier $h(i)$, where h is another well-known hash function. A light node l will periodically advertise its target load and current load to the node $i = h(h'(l))$, which is responsible for directory $h'(l)$. In turn, the heavy nodes will periodically sample the existing directories. A heavy node n picks a random number $k \in [0, d)$ and sends the information about its target load and the loads of all its virtual servers to node $j = h(h'(k))$. Upon receiving such a message, node j looks at the light nodes in its directory to find the best virtual server that can be transferred from n to a light node in its directory. This process repeats until all the heavy nodes become light.

36.4.3   Many-to-Many

The many-to-many scheme is similar to the one-to-many scheme. The difference is that all the heavy nodes, as well as the light ones, advertise their target load and current load to the directories, and the nodes in charge of the directory determine how to transfer virtual servers, by a heuristic algorithm that is more complex than in the one-to-many scheme. Details are provided by Rao et al.[23]

Godfrey et al.[24] have examined the many-to-many scheme in a dynamic P2P environment and the results show that a such mechanism achieves load balancing for system utilization as high as 90 percent while moving only about 8 percent of the load that arrives into the system. Moreover, in a dynamic system, less than 60 percent of the load is moved due to node arrivals and departures. Compared to a similar centralized algorithm, such a distributed algorithm is only negligibly worse.

Although these balancing algorithms make the data placement fit for the heterogeneous peer-to-peer systems, another serious problem remains unresolved, that is, *the bandwidth cost*. Peer-to-peer systems maintain an invariant of k replicas, even in the dynamic environment. So when a node leaves, all its data must be reproduced on some other nodes, which is equivalent to transferring this data. Unfortunately, in peer-to-peer storage, nodes store a large amount of data and enter and leave the system frequently. This makes data always in transfer, which needs a very high bandwidth that is out of practice in today's Internet world. Blake and Rodrigues[25] argue that such a scenario will get worse and worse, not better and better, over time.

So, DHT-based storage can be eployed only in a relatively stable system, such as a data center. However, if the nodes are stable, DHT (or say, structured routing overlay) is not a necessity because full-connection topology is then practical.

A compromise method is not directly storing items by DHT, but instead letting users determine where and how to replicate their data by themselves, and only employing DHT to store the replica lists.[26] A replica list is much smaller than physical items, so moving it when nodes change will not cost too much bandwidth.

36.5   Data Lookup and Search

As more and more data is stored in P2P systems, upper applications demand that the infrastructure provides search capacity. There are two kinds of P2P search functionalities: (1) data lookup and (2) keyword searching. Data lookup takes the object ID as an input to get the corresponding data item or its hosting peer(s). Keyword searching means feeding one or some keywords into the system and getting some data items that contain the keywords. Data lookup and keyword searching are nearly the same for unstructured P2P networks because they process queries by adopting flooding or random walk[27] approaches. For structured P2P networks, the two search functionalities are quite different. As nearly all structured P2P systems presented by now actually implement distributed hash tables (DHTs), the data lookup functionality can be naturally supported. However, keyword searching is not supported directly by DHTs. In this section we review some data lookup and search techniques proposed in recent years.

36.5.1   Search in Unstructured P2P Networks

36.5.1.1   Basic Search Techniques

Flooding and random walk are two basic searching mechanisms in unstructured networks. Gnutella[2] is the most famous unstructured P2P system and uses flooding to deal with queries. To locate a file using flooding, a node sends a query to all its neighbors, which in turn propagate the query to their neighbors, and so forth until the query reaches all the peers within a certain radius of the query initiator. With random walk, a query is forwarded to a randomly chosen neighbor at each step until sufficient query results are found. Random walk is recommended by Lv et al.[27] as a replacement strategy for flooding.

36.5.1.2   Search Optimizations

Peer flooding and random walk have proved inefficient in answering search requests. To make the search process more efficient and scalable, some optimization mechanisms must be adopted. Search in unstructured P2P networks can be boosted in three directions. The first one is content aggregation, that is, peers summarizing the content from other peers. With content aggregation, peers know more information about the system, and therefore there is hope for a boost in search efficiency. The second direction is content clustering. With content clustering, all contents in the system are clustered (maybe implicitly) by semantic or user interest, and clustering information is used by peers to reduce the number of nodes that must be accessed in searching. As the third direction, peer heterogeneity can be exploited.

36.5.1.2.1   *Directed Flooding and Biased Random Walk*

Directed flooding means that each peer sends query messages to a subset of its neighbors that hopefully have more high-quality search results than other neighbors. Biased random walk means that in random walk, rather than forwarding incoming queries to randomly chosen neighbors, each peer selects a neighbor that has the highest probability of having query results. Compared with pure flooding and random walk, directed flooding and biased random walk not only reduce the number of nodes bothered to process a query, but also improve the quality of results. To intelligently select one or some neighbors, a peer must maintain statistics on its neighbors. Statistical information of a node can be collected based on its basic information (degree, capacity, availability, latency, bandwidth, etc.) and the contents of it. Adamic et al.[28] suggest that queries are forwarded to high-degree nodes. A number of neighbor-selection heuristics are listed by Yang and Garcia-Molina,[29] including the number of results for previous queries, the largest number of messages forwarded, etc. Routing indices[30] is another way to collect node statistics and enable directed flooding and biased random walk. Different from other statistics collecting mechanisms, routing indices collect statistics of nodes in a "direction" rather than from a specified neighbor.

36.5.1.2.2   *r-Hop Replication*

In the *r*-hop replication technique, a node maintains an index over the data of all nodes within *r* hops of itself. When a node receives a query message, it can answer the query on behalf of every node within *r* hops of itself. In this way, search overhead can be dramatically reduced. r-hop replication (especially one-hop replication) is adopted as a building block in many search boosting proposals.[21,29]

36.5.1.2.3 *Super-Nodes*

A super-peer network[4,31] is a P2P network in which all super-nodes are connected as a pure P2P network and each super-peer is connected to a set of ordinary peers. A super-node, which typically has high capacity, is a node that operates as a centralized server to a set of ordinary peers. Each ordinary peer connects to one or several super-nodes. For each query initiated from an ordinary peer, the ordinary peer submits the query to one of its super-nodes and receives query results from it. Every super-node indexes the contents of all its attached ordinary peers and then answers queries on behalf of them. In a super-peer network, queries are only processed by super-nodes, and both flooding and random walk mechanisms can be adopted. Because queries are processed only by super-nodes that have relatively higher capacities than ordinary peers, searches in super-peer networks are much more efficient than in ordinary unstructured P2P networks. Yang and Garcia-Molina[31] studied the behavior of super-peer networks and presented an understanding of their fundamental characteristics and performance trade-offs.

36.5.1.2.4 *Guide Rules and Interest-Based Shortcuts*

These kinds of optimizations are in the "content clustering" direction, assuming that contents and peers in the system have semantic locality. A guide rule is a set of peers that satisfies some predicate. A group of peers belonging to a certain guide rule should contain contents that are semantically similar. Cohen et al.[32] proposed to build associative overlays based on guide rules and adopt two algorithms, RAPIER and GAS, for improving search efficiency. Sripanidkulchai et al.[33] observed the phenomenon of interest-based locality; that is, if a peer has a data item that one is interested in, then it is likely that it will have other data items that one is also interested in. All peers in the system are clustered by creating interest-based shortcuts between peers sharing similar interests. And search efficiency is improved by using these shortcuts.

36.5.1.3 Case Study: Gia

Gia[21] is a decentralized and unstructured P2P file-sharing system that combines several existing search optimization techniques in its design. The four boosting schemes used in Gia include (1) a dynamic topology adaptation protocol, (2) an active flow control scheme, (3) a one-hop replication mechanism, and (4) a search protocol based on biased random walks.

The dynamic topology adaptation protocol ensures that high-capacity peers are indeed the ones with high degrees, and that low-capacity peers are within short reach of high-capacity ones. Each peer maintains a level of satisfaction value, determined by its capacity and neighbors. Each node tends to gather more neighbors to improve its satisfaction level until it is up to 1. Each node also tends to use high-capacity nodes to replace low-capacity neighbors without dropping already poor-connected neighbors. Dynamic topology adaptation is a key protocol for Gia to explicitly make use of node heterogeneity to speed up search efficiency.

The active flow control scheme is used for avoiding overloaded hot spots. It explicitly acknowledges the existence of heterogeneity and adapts to it by assigning flow-control tokens to nodes based on available capacity. Each peer periodically assigns flow-control tokens to its neighbors. A node can send a query to a neighbor only if it has received a token from that neighbor, thus avoiding overloaded neighbors. To provide an incentive for high-capacity nodes to advertise their true capacity, Gia peers assign tokens in proportion to the neighbor's capacities, rather than distributing them evenly between all neighbors.

To improve search efficiency, Gia uses a one-hop replication mechanism; that is, each peer maintains an index of the contents of all its neighbors. Indices are incrementally updated between neighbors periodically.

Gia uses a biased random walk search protocol in which each peer selects the highest-capacity neighbor for which it has flow-control tokens and sends the query to that neighbor.

The combination of dynamic topology adaptation, one-hop replication, and biased random walk generates a similar effect to that of the super-node mechanism. However, compared with the super-node network, Gia provides a more adaptive mechanism. The active flow-control scheme makes Gia one of few proposals to consider node capacity constraints in search optimization. Refer to Chawathe et al.[21] for more details on Gia.

36.5.2   DHT-Based Keyword Searching

Structured P2P networks actually implement a distributed hash table (DHT) layer by firmly controlling the topology of overlay networks and the mapping from data items to peers. Upper applications can insert a <key, value> pair into the system and retrieve a data item by its key. As DHT can only directly support search by an opaque key, to support keyword searching, some mechanism is needed to map keyword queries to unique routing keys of DHT.

Most existing DHT-based keyword searching proposals use the inverted index, the most widely used indexing structure in full-text searching systems, as the basic index structure. An inverted index comprises many inverted lists, one for each word. An inverted list for a word contains all the identifiers of documents in which the word appears. In a P2P keyword searching system, the logically global inverted index must be partitioned and placed at certain peers. There are two basic P2P index partitioning and placing strategies: (1) local indexing and (2) global indexing. With local indexing, each host maintains a local inverted index of the documents for which it is responsible. Using this strategy, each query must be flooded to all peers. Unstructured P2P networks use this strategy by default. The global indexing strategy assigns each keyword undivided to a single node, and each node maintains the inverted lists of some keywords. Global indexing is the basic strategy used by structured P2P networks.

36.5.2.1   Global Indexing and its Optimizations

The most obvious keyword searching scheme in structured P2P networks is global indexing, that is, partition the index by keyword. By this scheme, each keyword (and its corresponding inverted list) is mapped by hashing to a unique key of DHT layer and thus to a unique peer of the system. To answer a query consisting of multiple keywords, the query is sent to peers responsible for those keywords. Their inverted lists are transmitted over the network and intersected to get a list of documents that contain the keywords. Using this strategy, for a query that contains k keywords, at most k nodes need to be contacted. However, as keyword intersection requires that one or more inverted lists be sent over the network, bandwidth consumption and transmission time can be substantial. To alleviate this problem, some optimizations have been proposed. Here are some of them.

36.5.2.1.1   *Compression*

To reduce communication overhead of the inverted list intersection, data can be transferred in a compressed way. Bloom filters[34] are the most frequently used compression methods in existing optimization proposals.[35,36] A bloom filter is a hashing-based data structure that can summarize the contents of a set with a relatively small space, at the cost of a small probability of false positives. For two nodes whose inverted lists are to intersect using bloom filters, one node compresses its inverted list and sends the bloom filter to another node. The receiving node intersects the bloom filter and its inverted list, and sends back the result list of documents. The original sender then removes false positives and sends final query results to end users. Li et al.[35] employed four rounds of inverted list intersections using compressed bloom filters and obtained a compression ratio of roughly 50. Li et al.[35] also tried other compression techniques (e.g., gap compression, adaptive set intersection, document clustering, etc.).

36.5.2.1.2   *Caching*

Caching has been widely used in many sub-fields of computer science to improve efficiency. For P2P systems with global indexing, caching schemes can be divided into two categories: (1) inverted list caching and (2) result caching. Inverted list caching means that peers cache the inverted lists sent to them for some queries to avoid receiving them again for future queries. Result caching, on the other hand, caches query results rather than inverted lists. Because keyword popularity in queries roughly follows a Zipf distribution, it is hoped that a small cache of inverted lists or query results can reduce communication overhead and improve search efficiency remarkably. A data structure called *view tree* is proposed[37] to efficiently store and retrieve prior results for result caching. Reynolds and Vahdat[36] [efficiently] analyzed the effect of inverted list caching and proposed to cache bloom filters instead of inverted lists. This has two effects: (1) storage space is saved because bloom filters are more compact than inverted lists; and (2) caching bloom filters spends more communication overhead in answering queries than caching inverted lists. That is because, by caching bloom filters, the caching node needs to send the intersection results of its inverted list and the bloom filter to another node. Query results can be returned to end users directly by caching inverted lists.

36.5.2.1.3   *Pre-Computation*
Using pre-computation, the intersection of some inverted lists can be computed and stored in the system in advance. Li et al.[35] tried to choose 7.5 million term pairs (about 3 percent of all possible term pairs) from the most popular terms of their dataset and pre-compute intersections. As a result, they obtained an average communication reduction of 50 percent. Gnawali[38] proposed a keyword-set search system (KSS) in his thesis. In KSS, the logically global index is partitioned by sets of keywords rather than keywords; and each query is divided into sets of keywords accordingly. In answering a query, the document list for each set of keywords is retrieved and intersected if necessary. KSS can be viewed as a kind of pre-computation. The effects of adopting pre-computation are twofold. Pre-computation reduces query processing overhead, but on the other hand, it also dramatically improves index building overhead and storage usage. Pre-computation shares some similar features with caching. Their relationship and the comparison of their effects remain to be studied.

36.5.2.1.4   *Incremental Intersection*
In most keyword searching systems, users nearly always need a few most relevant documents rather than all query results. This allows adopting the incremental intersection of inverted lists for reducing communication overhead and improving efficiency. Incremental intersection means that, in the intersection of two (or more) inverted lists, instead of transferring an inverted list as a whole to another node, the list is partitioned into multiple blocks and transferred incrementally. The intersection process terminates before the whole inverted list is transferred when some conditions are satisfied. Reynolds and Vahdat[36] combine the bloom filter technique and incremental intersection by dividing an inverted list into chunks and sending bloom filters of each chunk incrementally. To generate precise results when adopting incremental intersection, some algorithms must be used to guarantee that top k (k indicates the number of results users required) results have been generated when the intersection process terminates. One of the most frequently used algorithms is Fagin's algorithm.[39] Fagin's algorithm has some constraints to ranking functions that judge the relevance between each document and a given query. Unfortunately, some important ranking functions (e.g., term proximity) are not applicable to Fagin's algorithm. We face three choices in this circumstance. The first choice is compromising the quality of search results, either by abandoning the ranking functions not supported by Fagin's algorithm, or including such ranking functions while terminating early without the guarantee of generating top k results. As a second choice, we can negatively cease using incremental intersection to guarantee query result quality. The third choice may be the best choice; however, it needs our effort; that is, discover algorithms other than Fagin's algorithm that can support existing important ranking functions.

There also have been other optimizations. P-VSM[40] improves search efficiency by sacrificing query quality to some extent. In P-VSM, each document is represented by a vector by using the vector space model. And then the m most important components (keywords) of the document are indexed in the P2P system. Compared with the basic global indexing scheme, P-VSM reduces the overhead of index building and query processing by only indexing a small portion of keywords. However, the omission of some terms may affect its search precision. Another problem is that the IDF (inverse document frequency) values for terms needed in the vector space model must be globally computed.

36.5.2.2 Local Indexing

36.5.2.2.1 *Partition by Document*

P-LSI takes CAN[12] as its P2P substrate to implement keyword searching. The basic idea of P-LSI is to control the placement of documents such that documents stored close to each other in CAN are also close in semantics. Each document has a semantic vector that can be generated by LSI (latent semantic indexing), which uses SVD (singular value decomposition) to transform and truncate a matrix of documents and terms. To build the index, the index of each document is inserted into CAN using the semantic vector of the document as the key. Query processing in P-LSI follows a two-phase process. In answering a query, as the first phase, the semantic vector of the query (note that each query can be regarded as a, although short, document) is computed and the query is routed to a peer using its semantic vector as the key. In the second phase, the destination peer floods the query only to peers in a predefined radius r. Because the index of documents is stored by semantics, the search mechanism of P-LSI can retrieve highly relevant documents with high probability, while bothering only a small fraction of peers.

P-LSI is similar to the guide rules[32] and interest-based shortcuts[33] proposals in unstructured P2P networks. P-LSI uses information retrieval techniques to discover the latent semantic between documents, while the latter two proposals are comparatively more ad hoc.

36.5.2.3 Hybrid Indexing Schemes

Global indexing and local indexing each have their advantages and drawbacks. There have been some combinations of the two for the purpose of achieving good trade-offs between them.

36.5.2.3.1 *eSearch*

The primary index partitioning and placement strategy used in eSearch[41] is global indexing; that is, an inverted index is distributed based on keywords, and each node is responsible for maintaining the inverted lists of some keywords. To avoid transmitting large amounts of data in query processing, eSearch also replicates document information on nodes that are responsible for at least one keyword of the document, called *metadata replication*. Metadata replication is implemented by having each node store the complete lists of terms for documents in its inverted lists. It is the combination of global index and document replication that makes the index structure "hybrid." With this hybrid index structure, a multi-keyword query need only be sent to any one of the nodes responsible for those terms and do a local search. Refer to Figure 1 in Ref. 41 for an intuitive description of the index structure of eSearch.

The hybrid index structure used by eSearch actually trades storage space for communication cost and search efficiency. However, its communication cost in index publishing is much more than that of the more commonly proposed global indexing scheme. To overcome this drawback, eSearch enhances the above naïve hybrid scheme by two optimizations. We call the first optimization *metadata selection*, which means that only the most important terms for each document are published. To implement metadata selection, each document is represented as a vector using the vector space model (VSM). And then, the top elements (or their corresponding terms) are selected. The second optimization involves adopting of overlay source multicast. Overlay source multicast provides a lightweight way to disseminate metadata to a group of peers and reduces the communication cost compared with other data disseminating schemes.

36.5.2.3.2 *MLP*

Multi-level partitioning (MLP)[42] is a P2P-based hybrid index partition strategy. The goal of MLP is to achieve a good trade-off among end-user latency, bandwidth consumption, load balance, availability, and scalability.

The definition of MLP relies on a node group hierarchy. In the hierarchy, all nodes are logically divided into k groups (not necessarily of equal size), and nodes of each group are further divided as k sub-groups. Repeat this process to obtain a node hierarchy with l levels. Groups on level l are called leaf groups. Given the hierarchical node groups, the logically global inverted index is partitioned by document among groups; that is, each group maintains the local inverted index for all documents belonging to this group. For each leaf group, the index is partitioned by keywords among nodes; that is, a global index is built inside each leaf group.

MLP processes queries by the combination of broadcasting and inverted list intersection. Take a query (initiated from a node A) containing keyword w_1 and w_2 as an example. The query is broadcast from node A to all groups of level 1, and down to all groups of next levels, until level l is reached. For each group in level l, the two nodes that contain the inverted lists corresponding to keywords w_1 and w_2, respectively, are responsible for answering the query. In each group, the two inverted lists are intersected to generate the search results of the group. And then, the search results from all groups are combined level by level and sent back to node A. The implementation of MLP on top of SkipNet[43] is demonstrated by Shi et al.[42]

MLP has two features: (1) uniformity and (2) latency locality. Uniformity implies that all nodes must have roughly the same number of inverted lists on them. This is required for the balance of load and storage among peers in the system. Latency locality infers that intra-group latency should be smaller than inter-group latency on each level of the node hierarchy. This is needed for reducing end-user latency and bisection bandwidth consumption. With latency locality, all nodes in a group are roughly in the same sub-network. Therefore, bisection backbone bandwidth is saved by confining inverted list intersection inside each sub-network.

Compared with local indexing, MLP avoids flooding a query to all nodes by having only a few nodes in each group to process the query. MLP can dramatically reduce bisection backbone bandwidth consumption and communication latency when compared with the global indexing scheme.

36.5.2.3.3 *Hybrid Search Infrastructure*

It well known that flooding-based techniques (with local index structure) are effective for locating popular items, but are not suitable for finding rare items. On the other hand, DHT-based global indexing techniques excel at retrieving rare items, but are inefficient in processing queries containing multiple popular terms. Loo et al.[44] proposed a simple hybrid index building and search mechanism, in which global indexing is used to index rare data items and local indexing with flooding is used for indexing and retrieving popular data items. Based on this hybrid mechanism, a hybrid search infrastructure is built. A key problem in hybrid index building and query processing is the identification of rare items. The hybrid search infrastructure utilizes selective publishing techniques that use some heuristics to identify rare items.

36.6   Application-Level Multicast

In some cases, we meet the requirements of data transmission or communication among more than two end-hosts (i.e., the group communication). In group communication, many end-hosts take part in a common task and exchange data simultaneously and cooperatively to implement more complicated applications beyond only one-to-one connection. For example, in considering an application for a network meeting with many attendant people, the system needs to transmit the voice or picture of the talker to all other persons. Other examples for group communication include online video stream and efficient delivery for bulk contents. In these applications of group communication, the *multicast* is a key component that facilitates efficient data delivery from one source to multiple destinations. Although network-level IP multicast was proposed over a decade (e.g., network architectures for multicast[45–47] and reliable protocols[48,49]), the use of multicast in applications has been very limited because of the lack of widespread deployment and the issue of how to track group membership. To alleviate this problem, several recent proposals have advocated an alternative approach, termed "*application layer multicast*" or "*end-host multicast*," that implements multicast functionality at the application layer using unicast network-level services only, forming an overlay network between end-hosts.

Application-level multicast has a number of advantages over IP multicast. First, because it requires no router support, it can be deployed incrementally on existing networks. Second, application-level multicast is much more flexible than IP multicast and can adapt to diverse requirements from the application. The end-to-end argument[50] suggests that functions placed at low levels of a system (e.g., the network level) may be redundant or of little value when compared with the cost of providing them at that low level, partly because of insufficiency in low levels to accomplish the function alone and partly because of incomplete knowledge about an upper level. For example, to deal with the usual problem that various links in the distribution tree have widely different bandwidth, a commonly used strategy is to decrease the fidelity of the content over lower-bandwidth links. Although such a strategy has merit for video streaming that must be delivered live, there exist content types that require bit-for-bit integrity, such as software. Obviously, such application-aimed diversity can only be supported in the overlay layer, using application-level multicast rather than IP multicast in the network layer.

As a result, the overlay-based application-level multicast has recently become a hot category in research, and many algorithms and systems are being proposed for achieving scalable group management and scalable, reliable propagation of messages.[51–56] For such systems, the challenge remains to build an infrastructure that can scale to and tolerate the failure modes of the general Internet, while achieving low delay and effective use of network resources. That is, to realize an efficient overlay-based multicast system, there are three important problems to be addressed.

In particular, multicast systems based on overlay should mainly consider the following problems:

What is the appropriate multicast structure? In addition, how can it be constructed in the (mostly decentralized) overlay networks?

How can overlay networks manage the numerous participating hosts (users), especially when there is huge diversity between users, either in their capacities or in behaviors?

How can the multicast algorithm and overlay networks adapt themselves to an Internet environment whose links are prone to frequent variations, congestion, and failures?

The first problem often reduces to how to construct and maintain an efficient, fault-tolerant spanning tree on top of overlay networks, because a spanning tree is the natural structure for multicast. Beyond single tree structures, recent research on peer-to-peer networks further takes advantage of "perpendicular links" between siblings of the tree to overcome some of the drawbacks of tree structure. The second problem is often called "group management" in IP multicast; and in overlay and peer-to-peer networks, it becomes even more difficult because of the large system scale and decentralization. Indeed, sometimes the overlay is made up of only volunteer hosts, and we are unable to count on a dedicated, powerful host to take charge of the entire group management. In this case, it is helpful to employ suitable self-organized infrastructures presented in previous sections to perform decentralized group management. The third problem is typically solved by actively measuring the Internet and multicast connections, and using adaptive approaches for data transmission. Overlay networks usually consider the network connections between pairs of overlay nodes as direct, independent network links, regardless that some connections may actually share the same network link between routers. Thus, by measuring the underlying "black box" (mainly the connectivity and available bandwidth of overlay connections), the overlay network will become aware of the runtime status and may further perform self-adaptation to improve the performance.

The following sections explain how existing designs and approaches address the above three fundamental problems of overlay-based multicast.

36.6.1 Construction of Multicast Structures

To perform multicast on overlay, the primary problem lies in constructing the multicast structure, that is, deciding the paths for data flows. Consider the scenario that there is one source and a number of destinations, and the goal is to deliver data from the source to all destinations as quickly as possible. Most applications of group communication can be simplified to such a scenario. To deal with numerous destinations, some destination hosts act as intermediate nodes in the data transmission path and forward received data to other destinations. Thus, many approaches employ a multicast tree as the basic structure for one-source-multiple-destination application. Theoretically, the problem now becomes one of constructing a minimum spanning tree (MST) rooted at the source, given the overlay connective graph $G = (V, E)$ with weighted edges, where V represents all end hosts and E characterizes the overlay connections. Although there is an efficient algorithm for constructing an MST,[57] in practice the one-source-multiple-destination problem is not that easy, because it is usually impracticable to get the entire overlay connective graph G in the Internet, and because the resulting MST usually contains nodes with high degree that which will be overloaded in multicast. So, in practice many multicast algorithms do not seek to find the optimal tree solution, but turn to construct the near-optimal spanning tree that has bounded node degree and is easily built based on only partial knowledge of connective graphs. Therefore, the considerations of a multicast structure are mainly the following: (1) improvement of latency and bandwidth; (2) fault tolerance; and (3) ease of construction, especially in a decentralized manner.

Overcast[62] is motivated by real-world problems to deliver bandwidth-intensive content using the Internet, which is faced with content providers. An overcast system is an overlay consisting of a central source, any number of internal Overcast nodes (standard PCs with permanent storage) sprinkled throughout a network fabric, and clients with standard HTTP connection. Overcast organizes the internal nodes into a distribution tree rooted at the source and using a simple tree-building protocol. The goal of Overcast's tree algorithm is to maximize bandwidth to the root for all nodes, which is achieved by placing a new node as far away from the root as possible without sacrificing bandwidth to the root. Once a node initializes, it begins from the root a process of self-organization with other nodes. In each round, the new node considers its bandwidth to the current node as well as the bandwidth to the current node *through each of the current node's children*. If the bandwidth through any of the children is as high as the direct bandwidth to the current node, then one of these children becomes current and a new round commences. In this way, the new node can locate itself further away from the root while also guaranteeing available bandwidth. Therefore, the tree-building protocol is inclined to locate a node to a parent that is very near the node, mostly within a localized Internet area. Thus, Overcast takes advantage of the multicast tree to maximize available bandwidth and the power for delivering content to large-scale hosts.

Bayeux[54] utilizes a prefix-based routing scheme that it inherits from an application-level routing protocol (i.e., the Tapestry) and thus it combines randomness for load balancing and scales to arbitrarily large receiver groups, with the properties of locality and tolerance to failures in routers and network links. Bayeux follows the idea of using the "natural multicast tree" associated with DHT-based routing schemes. Because we can easily derive a tree structure from most DHT-based routing schemes, it is beneficial to use such a tree for overlay multicast and thus leave the failure recovery and proximity issues to overlay's routing layer, as Bayeux and other systems do.[63] In Bayeux, the Tapestry overlay assists efficient multipoint data delivery by forwarding packets according to suffixes of listener node IDs, and the node ID base defines the fan-out factor used in the multiplexing of data packets to different paths on each router. A multicast packet only needs to be duplicated when the receiver node identifiers become divergent in the next digit. In addition, the maximum number of overlay hops taken by such a delivery mechanism is bounded by the total number of digits in the Tapestry node IDs. Therefore, Bayeux has a bounded multicast approach; that is, both hops (critical to latency) and host's out-degree (critical to bandwidth) are guaranteed to be small.

However, tree-based multicast has some inherent drawbacks and is sometimes insufficient for full use of available resources in a cooperative environment. The reason is that in any multicast tree, the burden of duplicating and forwarding multicast traffic is carried by the small subset of the hosts that are interior nodes in the tree. The majority of hosts are leaf nodes and contribute no resources. This conflicts with the expectation that all hosts should share the forwarding load. The problem is further aggravated in high-bandwidth applications, such as video or bulk file distribution, where many receivers may not have the capacity and availability required of an interior node in a conventional multicast tree. In addition, in tree structures, the bandwidth is guaranteed to be monotonically decreasing moving down the tree. Any loss high up the tree will reduce the bandwidth available to receivers lower down the tree. Despite a number of techniques being proposed to recover from losses and hence to improve the available bandwidth in an overlay tree (e.g., Birman et al.[51] and Byers et al.[58]), fundamentally, the bandwidth available to any host is limited by the bandwidth available from that node's single parent in the tree.

To overcome the inherent limitations of the tree, recent techniques propose to use "perpendicular links" to augment the bandwidth available through the tree (e.g., Bullet[59]) or to construct a multicast *forest* in place of a single tree (e.g., SplitStream[60]). In essence, these techniques share the same idea: that we should make use of every host in the overlay to enhance data transmission, rather than use only interior nodes of a single tree.

The key idea in SplitStream[60] is to split the content into k stripes and to multicast each stripe using a separate tree. Hosts join as many trees as there are stripes they wish to receive, and they specify an upper bound on the number of stripes they are willing to forward. The challenge is to construct this forest of multicast trees such that an interior node in one tree is a leaf node in all remaining trees and the bandwidth constraints specified by the nodes are satisfied. This ensures that the forwarding load can be spread across all participating hosts. For example, if all nodes wish to receive k stripes and they are willing to forward k stripes, SplitStream will construct a forest such that the forwarding load is evenly balanced across all nodes while achieving low delay and link stress across the system. Figure 36.3 illustrates the stripes in SplitStream and the multicast forest.

The challenge in the design of SplitStream is to construct such a forest of interior-node-disjoint multicast trees in a decentralized, scalable, efficient, and self-organizing manner. A set of trees is said to be "interior-node disjoint" if each node is an interior node in at most one tree and a leaf node in the other trees. SplitStream exploits the properties of Pastry's routing and Scribe's group member management (Scribe[61]; see details of Scribe in the section Group Member Management) to facilitate construction of interior-node-disjoint trees. Recall that Pastry normally forwards a message toward nodes whose nodeIDs share progressively longer prefixes with the message's key. Because a Scribe tree is formed by the routes from all members to the groupID (see below), the nodeIDs of all interior nodes share some number of digits with the tree's groupID. Therefore, it can be ensured that k Scribe trees have a disjoint set of interior nodes simply by choosing groupIDs for the trees that all differ in the most significant digit. Figure 36.3 illustrates the construction. The groupID of a stripe group is also called the stripeID of the stripe. To limit a node's out-degree, the "push-down" method, which is a built-in mechanism in Scribe, is used. When a node that has reached its maximal out-degree receives a request from a prospective child, it provides the prospective child with a list of its current children. The prospective child then seeks to be adopted by the child with lowest delay. This procedure continues recursively down the tree until a node is found that can take another child. Moreover, to fully use the spare capacity in some powerful hosts, these hosts are organized in an independent group of Scribe, called the spare capacity group. All SplitStream nodes that have less children in stripe trees than their forwarding capacity limit are members of this group. When there is a forwarding demand that the current node cannot afford, SplitStream will seek an appropriate node in the spare capacity group to help the fully loaded node in supplying stripe.

Bullet[59] aims at high-bandwidth multicast data dissemination, for example, delivering bulk content to multiple receivers via the Internet. Rather than sending identical copies of the same data stream to all nodes, Bullet proposes that participants in a multicast overlay cooperate to strategically transmit *disjoint* data sets

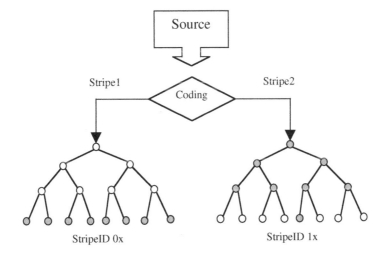

FIGURE 36.3 Stripes and forest of multicast trees in Splitstream.

to various points in the network. Nodes still receive a set of objects from their parents, but they are then responsible for locating hosts that hold missing data objects. Bullet has a distributed algorithm that aims to make the availability of data objects uniformly spread across all overlay participants. In this way, it avoids the problem of locating the "last object" and achieves persistent high-bandwidth data dissemination.

36.6.2 Group Member Management (Scribe)

In addition to the multicast structure, there are always requirements for managing participants, especially organizing hosts into many different groups according to the hosts' properties and also delivering messages to all members belonging to a certain group. Group member management is a basic overlay service and usually a necessary component for many multicast applications.

For this purpose, Scribe[61] is proposed as a decentralized application-level multicast infrastructure, aimed at member management for a large number of groups. Scribe is built on top of Pastry and takes advantage of Pastry's remarkable properties in network proximity, fault tolerance, and decentralization to support simultaneously a large number of groups. Any Scribe node can create a group; other nodes can then join the group, or multicast messages to all members of the group (provided they have the appropriate credentials). Each Scribe group has a unique groupID, and the Scribe node with a nodeID numerically closest to the groupID acts as the rendezvous point for the associated group. The rendezvous point is the root of the multicast tree created for the group. When a Scribe node wishes to join a group g, it asks Pastry to route a JOIN message with the g's groupID as the key. This message is routed towards the g's rendezvous point. Each node along the route checks its list of groups to see if it is currently a forwarder of g; if so, it accepts the node as a child, adding it to the children table. Otherwise, it creates an entry for the g, adds the source node as a child, and becomes a new forwarder for g by sending a JOIN message to the next node along the route from the joining node to the rendezvous point. So, the membership management mechanism is efficient for groups with a wide range of memberships, varying from one to all Scribe nodes. Pastry's randomization properties ensure that the tree is well balanced and that the forwarding load is evenly balanced across the nodes. This balance enables Scribe to support large numbers of groups and members per group. Furthermore, joining requests are handled locally in a decentralized fashion. In particular, the rendezvous point does not handle all joining requests. Therefore, Scribe can be used as a basic component for overlay-based multicast applications, and the notable example is SplitStream, which uses Scribe to form interior-disjoint multicast trees and organize hosts' spare capacities.

CoopNet[64] focuses on the problem of distributing "live" streaming media content from a server to a potentially large and highly dynamic population of interested clients. CoopNet uses multicast through clients to alleviate the heavy load in a streaming server and help a server tide over crises such as flash crowds, rather than replace the server with a pure peer-to-peer system. So, CoopNet benefits from a centralized management, which is much more efficient than a decentralized approach. There is a dedicated manager (a workstation) for managing the joins and leaves of clients. The manager stores the entire structure of the multicast tree (or trees) in its memory. When a client begins to receive live streaming, the client contacts the manager for JOIN operation. The manager then chooses an appropriate link place in the multicast tree from its memory, and responds with the designated parent to the client. In this way, the JOIN operation is very efficient and only needs one round of messaging. Although centralized management is not self-scaling, it is shown in CoopNet that such management is a rather lightweight task, and that a laptop with a 2-GHz Mobile Pentium 4 processor can keep up with about 400 joins and leaves per second. So, centralized management is also feasible and scales to large systems in some practical scenarios.

36.6.3 Overlay-Based Measurement and Adaptation for Multicast

In addition to multicast structure construction and group member management, an overlay application must continually adapt itself to the variational Internet environment. As is well known, the Internet is a highly dynamic environment and prone to unpredictable partitions, congestion, and crowd flashes. So, the multicast structure well-constructed at early time may become very inefficient after a period of time.

Consequently, overlay networks should be able to learn Internet variations by means of repetitive overlay-based measurement and re-estimation of its connections, and also adaptively transform the multicast structure in response.

Many probing-based techniques to measure overlay connections have been proposed[65–70]; they use lightweight message probing to estimate the latency and available bandwidth between host pairs. In these techniques, the RTT (round-trip time) and 10-KB TCP probing are commonly utilized. In addition, some approaches[71] manage to estimate the bottleneck bandwidth. Although probing messages and lightweight estimations cannot fully characterize the property of connections, they are usually helpful in failure detection and path selection. In general, overlay networks regard the connections between end-host pairs as "black-boxes," and the measurement out-of-the-box is sufficient for overlay-based multicast systems.

After detecting variations, overlay networks must adapt themselves and multicast structures to improve performance. Many overlay designs employ localized adjustments of multicast structures, allowing hosts to dynamically change to a better service node. In Overcast,[62] to approximate the bandwidth that will be observed when moving data, the tree protocol measures the download time of 10 kbytes. This approach gives better results than approaches based on low-level bandwidth measurements such as using ping. In addition, a node periodically reevaluates its position in the tree by measuring the bandwidth to its current siblings (an up-to-date list is obtained from the parent), parent, and grandparent. Just as with the initial building phase. a node will relocate below its siblings if that does not decrease its bandwidth back to the root. The node checks bandwidth directly to the grandparent as a way of testing its previous decision to locate under its current parent. If necessary, the node moves back up in the hierarchy to become a sibling of its parent. As a result, nodes constantly reevaluate their position in the tree and the Overcast network is inherently tolerant of non-root node failures and Internet congestion.

36.7 Application

In recent years, P2P applications and practical systems have become a large category. A practical P2P system always needs integrated considerations in many aspects and combines many of the above approaches into one design. In this section we review a typical P2P application (i.e., P2P-based network storage) and see how the design issues that were previously discussed are implemented in practice.

36.7.1 Network Storage

Storage is a main application based on peer-to-peer architecture, in which storage spaces of all the nodes are combined and offered to users as a whole storage service for data storing, replication, backup, and other uses. Major projects on P2P storage systems include OceanStore,[8] PAST,[6] CFS,[7] Farsite,[72] Freenet,[3] and Granary.[26] As a practical application, peer-to-peer storage systems face many technical problems, such as routing, data location, data access control, replication, cache, update manner, service model, etc. Not all projects focus on all these problems, and different projects choose different resolutions. Figure 36.4 provides a brief summary of the major aspects of current projects.

In the following we introduce OceanStore as a tangible example, as its research almost covers all the areas. The key design parameters of other projects will also be mentioned when necessary.

OceanStore is designed to be a very large storage pool, spanning the globe and offering reliable and efficient storage service to end users, especially to mobile clients, such as ubiquitous computing devices. Users can approach their data in OceanStore anytime, anywhere, and through any kind of device. Figure 36.5 illustrates the basic architecture of OceanStore, showing that the system is composed of a multitude of highly connected *pools*. Most of these pools are dedicated servers, provided by a confederation of companies, and any organization can be invited to participate in the system, even airports or small cafés, as long as they offer some storage servers. Users pay their fees for occupied spaces in the system, perhaps monthly, and all their data would be safe and never be leaked to other users or even to the system managers. Of course, users are permitted to grant data access authority to other users.

	Routing & Location	Access Control	Replication	Cache
OceanStore	Tapestry	✓	✓	✓
CFS	Chord	–	✓	✓
Farsite	Broadcast	✓	✓	–
Freenet	DFS	✓	–	–
PAST	Pastry	✓	–	–
Granary	Tourist	✓	✓	–

	Update Manner	Service Model	Others
OceanStore	on block & archival	API & facades	Introspection
CFS	whole file replace	file system	load balance
Farsite	whole file replace	file system	
Freenet	whole file replace	file system	
PAST	file create & reclaim	file system	smart card
Granary	on object level	object storage	object query

FIGURE 36.4   Brief comparison of P2P storage projects.

To accomplish such goals, many technical problems should be tackled, and these are introduced in the following one by one.

36.7.1.1   Service Model

Different storage systems are based on different data units, including block, file, object, etc. The atomic data unit in OceanStore is "*object*," a piece of data with a global unique identifier. However, such a data form shows no substantial difference with the *file*, which is widely chosen by many other projects (e.g., CFS, Farsite, Freenet, and PAST).

OceanStore offers a set of its own APIs, which provides full access to OceanStore functionality in terms of sessions, session guarantees, updates, and callbacks. Applications with more basic requirements are supported through facades to the standard APIs. A facade is an interface to the APIs that provides a traditional, familiar interface. For example, a transaction facade would provide an abstraction atop the OceanStore API so that the developer could access the system in terms of traditional transactions.

Another object-based system, Granary, uses a more complex object model, which consists of a collection of user-defined attributes, just like the object concept in the OOP. This enables data query service for Granary, a significant functionality for many applications.

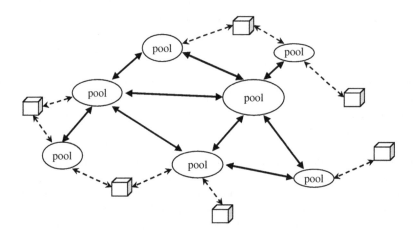

FIGURE 36.5   Architecture of OceanStore.

36.7.1.2 Data Location and Routing

PAST and CFS directly use DHT to store their data, which cannot avoid the trouble of massive data movement. OceanStore and Granary only use DHTs to store the metadata, mainly consisting of the replica lists. This greatly reduces the amount of data stored by DHTs and leave sufficient space for replication-algorithm design.

The basic routing layer in OceanStore is Tapestry,[15] which was introduced in Section 36.2.1. Moreover, a modified Bloom filter algorithm, called an attenuated Bloom filter, was combined. The attenuated Bloom filter[73] summarizes object items on a node's neighbors, in which whether a neighbor does not hold a given item can be judged accurately, while whether it does hold the item can be judged with high probability.

36.7.1.3 Access Control

OceanStore supports two primitive types of access control: *reader restriction* and *writer restriction*. More complicated access control policies, such as working groups, are constructed from these two.

To prevent unauthorized reads, all data is encrypted and the encryption key is distributed to those users with read permission. To revoke read permission, the owner must request that replicas be deleted or re-encrypted with the new key.

To prevent unauthorized writes, OceanStore requires that all writes be signed so that well-behaved servers and clients can verify them against an access control list (ACL). The owner of an object can securely choose the ACL for an object by providing a signed certificate. The specified ACL may be another object or a value indicating a common default. An ACL entry extending privileges must describe the privilege granted and the signing key, but not the explicit identity, of the privileged users. Such entries publicly are readable so that servers can check whether a write is allowed.

Peer-to-peer storage systems run on the open Internet, so the service provider must keep the data encrypted and unknown to other users or the system managers. Thus, almost every project involves access control design, and most are similar in spirit.

Farsite employs an interesting method—convergent encryption—to encrypt the data: first hashing the data content and then using the hashing result as the key to encrypt the data, by a symmetric encryption algorithm. When granting read authority to another user, the owner delivers the symmetric key, encrypted by the receiver's public key. In this way, the same files will always generate the same keys and then produce the same encrypted result, even if they belong to different owners. Therefore, the system can directly judge whether two files are identical, without knowing their content. This allows the system to reduce redundant storing of the same files. A following backup project Pastiche[74] also employs such an algorithm.

36.7.1.4 Replication

Objects in OceanStore are replicated and stored on multiple servers. This replication provides availability in the presence of network partitions and durability against failure and attack. A given replica is independent of the server on which it resides at any one time, and thus is called a floating replica.

Because object written authority can be granted to others, simultaneously updating must be coped with; that is, all the update requests should be uniformly ordered before executing. OceanStore separates all the replicas of a given object into two tiers, the primary tier performing a Byzantine agreement protocol to commit the updates and the secondary tier propagating the updates among themselves in an epidemic manner.

Most other projects do not assume such sophisticated replication. They only let the owner have the write permission and detour the replica consistence design.

36.7.1.5 Archival Storage

OceanStore objects exist in both active and archival forms. An active form of an object is the latest version of its data, together with a handle for update. An archival form represents a permanent, read-only version of the object. Archival versions of objects are encoded with an erasure code and spread over hundreds or thousands of servers; as data can be reconstructed from any sufficiently large subset of fragments, the result is that nothing short of a global disaster could ever destroy information.

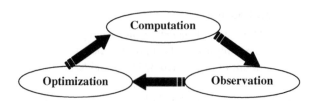

FIGURE 36.6   Introspection in OceanStore.

Erasure coding[75,76] is a process that treats input data as a series of fragments (say, n) and transforms these fragments into a greater number of fragments (say, $2n$ or $4n$). The essential property of the resulting code is that any n coded fragments are sufficient to construct the original data. In that OceanStore spans the globe, and all the fragments are identified by hashing and distributed in a DHT manner, all the fragments would be dispersed around the world, thus greatly improving the availability of archival data.

36.7.1.6   Introspection

OceanStore consists of millions of servers with varying connectivity, disk capacity, and computational power. Servers and devices will connect, disconnect, and fail sporadically. Server and network load will vary from moment to moment. Manually tuning a system so large and varied is prohibitively complex. Worse, because OceanStore is designed to operate using the utility model, manual tuning would involve cooperation across administrative boundaries.

To address these problems, OceanStore employs introspection, an architectural paradigm that mimics adaptation in biological systems. As shown in Figure 36.6, introspection augments a system's normal operation (computation) with observation and optimization. Observation modules monitor the activity of a running system and keep a historical record of system behavior. They also employ sophisticated analyses to extract patterns from these observations. Optimization modules use the resulting analysis to adjust or adapt the computation.

OceanStore takes advantage of introspection mainly in two ways: (1) cluster recognition, which detects clusters of strongly related objects; and (2) replica management, which adjusts the number and location of floating replicas in order to service access requests more efficiently.

References

1. Napster. http://www.napster.com.
2. Gnutella. http://gnutella.wego.com.
3. I. Clarke, O. Sandberg, B. Wiley, and T.W. Hong. Freenet: a distributed anonymous information storage and retrieval system. In *Proceedings of the Workshop on Design Issues in Anonymity and Unobservability*, Berkeley, CA, June 2000. http://freenet.sourceforge.net.
4. KaZaA. http://kazaa.com.
5. BitTorrent. http://bitconjurer.org/BitTorrent/.
6. A. Rowstron and P. Druschel. Storage management and caching in PAST, a large-scale, persistent peer-to-peer storage utility. In *Proceedings of ACM SOSP'01*, 2001.
7. F. Dabek, M.F. Kaashoek, D. Karger, et al. Wide-area cooperative storage with CFS. In *Proc. ACM SOSP'01*, Banff, Canada, Oct. 2001.
8. J. Kubiatowicz, D. Bindel, Y. Chen, et al. OceanStore: an architecture for global-scale persistent storage. In *Proceedings of the Ninth international Conference on Architectural Support for Programming Languages and Operating Systems (ASPLOS 2000)*, 2000.
9. A. Muthitacharoen, R. Morris, T.M. Gil, and B. Chen. Ivy: A read/write peer-to-peer file system. In *Proceedings of the 5th Symposium on Operating Systems Design and Implementation*, 2002.

10. M. Castro, P. Druschel, A.-M. Kermarrec, et al. SplitStream—high-bandwidth multicast in cooperative environments. *SOSP'03,* 2003.

11. J. Eriksson, M. Faloutsos, and S. Krishnamurthy. PeerNet: pushing peer-to-peer down the stack. In *Proceedings of the 2nd International Workshop on Peer-to-Peer Systems (IPTPS'03),* 2003.

12. S. Ratnasamy, P. Francis, M. Handley, R. Karp, and S. Shenker. A scalable content-addressable network. *Annual Conference of the Special Interest Group on Data Communication (SIGCOMM 2001),* August 2001.

13. A. Rowstron and P. Druschel. Pastry: scalable, distributed object location and routing for large-scale peer-to-peer systems. *International Conference on Distributed Systems Platforms (Middleware 2001),* November 2001.

14. I. Stoica, R. Morris, D. Karger, M.F. Kaashoek, and H. Balakrishnan. Chord: a scalable peer-to-peer lookup service for Internet applications. *Annual Conference of the Special Interest Group on Data Communication (SIGCOMM 2001),* August 2001.

15. B. Zhao, Jo. Kubiatowicz, and A. Joseph. Tapestry: An Infrastructure for Fault-Tolerant Wide-Area Location and Routing. Technical Report UCB/CSD-01-1141, Computer Science Division, University of California, Berkeley. April 2001.

16. C. Plaxton, R. Rajaraman, and A. Richa. Accessing nearby copies of replicated objects in a distributed environment. In *Proceedings of the ACM SPAA,* Newport, RI, June 1997, pp. 311–320.

17. J. Xu, A. Kumar, and X. Yu. On the fundamental tradeoffs between routing table size and network diameter in peer-to-peer networks, *IEEE Journal on Selected Areas in Communications,* 22(1), 151–163, January 2004.

18. A. Gupta, B. Liskov, and R. Rodrigues. One hop lookups for peer-to-peer overlays. In *Proceedings of the Ninth Workshop on Hot Topics in Operating Systems,* Lihue, Hawaii, May 2003.

19. E. Ng, and H. Zhang. Towards global network positioning. In *Proceedings of ACM SIGCOMM Internet Measurement Workshop 2001,* November 2001.

20. S.Saroiu, P.K. Gummadi, and S.D. Gribble. A measurement study of peer-to-peer file sharing systems. In *Proceedings of Multimedia Computing and Networking 2002 (MMCN'02),* January 2002.

21. Y. Chawathe, S. Ratnasamy, L. Breslau, N. Lanham, and S. Shenker. Making Gnutella-like P2P systems scalable. In *Proceedings of ACM SIGCOMM 2003,* Karlsruhe, Germany, August 2003.

22. D. Karger, E. Lehman, T. Leighton, M. Levine, D. Lewin, and R. Panigrahy. Consistent hashing and random trees: distributed caching protocols for relieving hot spots on the World Wide Web. In *29th ACM Annual Symposium on Theory of Computing (STOC 1997),* May 1997.

23. A. Rao, K. Lakshminarayanan, S. Surana, R. Karp, and I. Stoica. Load balancing in structured P2P systems. In *2nd International Workshop on Peer-to-Peer Systems (IPTPS'03),* February 2003.

24. B. Godfrey, K. Lakshminarayanan, S. Surana, R. Karp, and I. Stoica. Load balancing in dynamic structured P2P systems. In *The 23rd Annual Joint Conference of the IEEE Computer and Communications Societies (INFOCOM 2004),* March 2004.

25. C. Blake and R. Rodrigues. High availability, scalable storage, dynamic peer networks: pick two. In *9th Workshop on Hot Topics in Operating Systems (HOTOS IX),* May 2003.

26. Granary project. Homepage at: http://166.111.68.166/granary/.

27. Q. Lv, P. Cao, E. Cohen, K. Li, and S. Shenker. Search and replication in unstructured peer-to-peer networks. In *Proc. ACM ICS 2002,* 2002.

28. L.A. Adamic, R.M. Lukose, A.R. Puniyani, and B.A. Huberman. Search in power-law networks. *Physical Review E 64,* 2001.

29. B. Yang and H. Garcia-Molina. Efficient search in peer-to-peer networks. In *ICDCS'02,* 2002.

30. A. Crespo and H. Garcia-Molina. Routing indices for peer-to-peer systems. In *ICDCS'02,* 2002.

31. B. Yang and H. Garcia-Molina. Designing a super-peer network. In *ICDE'03,* 2003.

32. E. Cohen, A. Fiat, and H. Kaplan. Associative search in peer to peer networks: harnessing latent semantics. In *INFOCOM'03,* 2003.

33. K. Sripanidkulchai, B. Maggs, and H. Zhang. Efficient content location using interest-based locality in peer-to-peer systems. In *INFOCOM'03,* 2003.

34. B.H. Bloom. Space/time trade-offs in hash coding with allowable errors. *Communications of the ACM*, 13(7):422–426, 1970.

35. J. Li, B.-T. Loo, J.M. Hellerstein, M.F. Kaashoek, D.R. Karger, and R. Morris. On the feasibility of peer-to-peer Web indexing and search. In *Proceedings of the 2nd International Workshop on Peer-to-Peer Systems (IPTPS'03)*, 2003.

36. P. Reynolds and A. Vahdat. Efficient peer-to-peer keyword searching. In *Middleware'03*, 2003.

37. B. Bhattacharjee, S. Chawathe, V. Gopalakrishnan, P. Keleher, and B. Silaghi. Efficient peer-to-peer searches using result-caching. In *Proceedings of the 2nd International Workshop on Peer-to-Peer Systems (IPTPS'03)*, 2003.

38. O.D. Gnawali. A Keyword Set Search System for Peer-to-Peer Networks. Masters thesis, Massachusetts Institute of Technology, June 2002. UCB/CSD-01-1141, University of California, Berkeley, April 2001.

39. R. Fagin, A. Lotem, and M. Naor. Optimal aggregation algorithms for middleware. In *Symposium on Principles of Database Systems*, 2001.

40. C. Tang, Z. Xu, and M. Mahalingam. Peersearch: efficient information retrieval in peer-to-peer networks. In *Proceedings of HotNets-I, ACM SIGCOMM*, 2002.

41. C. Tang and S. Dwarkadas. Peer-to-peer information retrieval in distributed hashtable systems. In *USENIX/ACM Symposium on Networked Systems Design and Implementation (NSDI'04)*, March 2004.

42. S. Shi, G. Yang, D. Wang, J. Yu, S. Qu, and M. Chen. Making peer-to-peer keyword searching feasible using multi-level partitioning. In *Proceedings of the 3rd International Workshop on Peer-to-Peer Systems (IPTPS'04)*, 2004.

43. N.J.A. Harvey, M.B. Jones, S. Saroiu, M. Theimer, and A. Wolman. SkipNet: a scalable overlay network with practical locality properties. In *USITS'03*, 2003.

44. B.-T. Loo, R. Huebsch, I. Stoica, and J.M. Hellerstein. The case for a hybrid P2P search infrastructure. In *Proceedings of the 3rd International Workshop on Peer-to-Peer Systems (IPTPS'04)*, 2004.

45. S. Deering and D. Cheriton. Multicast routing in datagram internetworks and extended LANs, *ACM Transactions on Computer Systems*, 8(2), May 1990.

46. S.E. Deering. Multicast Routing in a Datagram Internetwork, Ph.D. thesis, Stanford University, December 1991.

47. S. Deering, D. Estrin, D. Farinacci, V. Jacobson, C. Liu, and L.Wei. The PIM architecture for wide-area multicast routing, *IEEE/ACM Transactions on Networking*, 4(2), April 1996.

48. S. Floyd, V. Jacobson, C.G. Liu, S. McCanne, and L. Zhang. A reliable multicast framework for lightweight sessions and application level framing, *IEEE/ACM Transaction on Networking*, 5(4):784–803, December 1997.

49. J.C. Lin and S. Paul. A reliable multicast transport protocol. In *Proc. of IEEE INFOCOM'96*, 1996, pp. 1414–1424.

50. J.H. Saltzer, D.P. Reed, and D.D. Clark. End-to-end arguments in system design, *ACM Transactions on Computer Systems*, 4(2): 277–288, November 1984.

51. K.P. Birman, M. Hayden, O. Ozkasap, Z. Xiao, M. Budiu, and Y. Minsky. Bimodal multicast, *ACM Transactions on Computer Systems*, 17(2): 41–88, May 1999.

52. P. Eugster, S. Handurukande, R. Guerraoui, A.-M. Kermarrec, and P. Kouznetsov. Lightweight probabilistic broadcast. In *Proceedings of The International Conference on Dependable Systems and Networks (DSN 2001)*, Gothenburg, Sweden, July 2001.

53. L.F. Cabrera, M.B. Jones, and M. Theimer. Herald: achieving a global event notification service. In *HotOS VIII*, Schloss Elmau, Germany, May 2001.

54. S.Q. Zhuang, B.Y. Zhao, A.D. Joseph, R.H. Katz, and J. Kubiatowicz. Bayeux: an architecture for scalable and fault-tolerant wide-area data dissemination. In *Proceedings of the Eleventh International Workshop on Network and Operating System Support for Digital Audio and Video (NOSSDAV 2001)*, Port Jefferson, NY, June 2001.

55. Y.-H. Chu, S.G. Rao, and H. Zhang. A case for end system multicast. *Proceedings of ACM Sigmetrics*, Santa Clara, CA, June 2000, pp. 1–12.

56. P.T. Eugster, P. Felber, R. Guerraoui, and A.-M. Kermarrec. The Many Faces of Publish/Subscribe, Technical Report DSC ID:2000104, EPFL, January 2001.

57. H. Gabow. Two algorithms for generating weighted spanning trees in order. *Siam J. Computing,* 6:139–150, 1977.

58. J.W. Byers, J. Considine, M. Mitzenmacher, and S. Rost. Informed content delivery across adaptive overlay networks. In *Proceedings of ACM SIGCOMM,* August 2002.

59. D. Kostic, A. Rodriguez, J.R. Albrecht, and A. Vahdat: Bullet: high bandwidth data dissemination using an overlay mesh. In *Proceedings of the 19th ACM Symposium on Operating System Principles,* October 2003, pp. 282–297.

60. M.Castro, P. Druschel, A.-M. Kermarrec, A. Nandi, A. Rowstron, and A. Singh. Splitstream: high-bandwidth content distribution in cooperative environments. In *Proceedings of the 19th ACM Symposium on Operating System Principles,* October 2003.

61. M. Castro, P. Druschel, A.-M. Kermarrec, and A. Rowstron. SCRIBE: a large-scale and decentralized application-level multicast infrastructure. *IEEE JSAC,* 20(8), October 2002.

62. J. Jannotti, D.K. Gifford, and K.L. Johnson. Overcast: reliable multicasting with an overlay network. In *USENIX Symposium on Operating System Design and Implementation,* San Diego, CA, October 2000.

63. S. Ratnasamy, M. Handley, R. Karp, and S. Shenker. Application-level multicast using content-addressable networks. In *Proceedings of the 3rd International Workshop on Networked Group Communication,* November 2001.

64. V.N. Padmanabhan, H.J. Wang, and P.A. Chou. Resilient peer-to-peer streaming. In *11th IEEE International Conference on Network Protocols (ICNP'03),* November 4–7, 2003, Atlanta, GA.

65. T.S.E. Ng, Y.-H. Chu, S.G. Rao, K. Sripanidkulchai, and H. Zhang. Measurement-based optimization techniques for bandwidth-demanding peer-to-peer systems. In *IEEE INFOCOM'03,* 2003.

66. R.L. Carter and M. Crovella. Server selection using dynamic path characterization in wide-area networks. In *Proceedings of IEEE INFOCOM,* 1997.

67. M. Sayal, Y. Breitbart, P. Scheuermann, and R. Vingralek. Selection algorithms for replicated Web servers. In *Proceedings of the Workshop on Internet Server Performance,* 1998.

68. Z. Fei, S. Bhattacharjee, E.W. Zegura, and M.H. Ammar. A novel server selection technique for improving the response time of a replicated service. In *Proceedings of IEEE INFOCOM,* 1998.

69. K. Obraczka and F. Silva. Network latency metrics for server proximity. In *Proceedings of IEEE Globecom,* December 2000.

70. K. Hanna, N. Natarajan, and B.N. Levine. Evaluation of a novel two-step server selection metric. In *Proceedings of IEEE ICNP,* November 2001.

71. K. Lai and M. Baker. Nettimer: a tool for measuring bottleneck link bandwidth. In *Proceedings of the 3rd USENIX Symposium on Internet Technologies and Systems,* March 2001.

72. W. Bolosky, J. Douceur, D. Ely, and M. Theimer. Feasibility of a serverless distributed file system deployed on an existing set of desktop PCs. In *2000 ACM SIGMETRICS International Conference on Measurement and Modeling of Computer Systems (Sigmetrics 2000),* June 2000.

73. S.C. Rhea and J. Kubiatowicz. Probabilistic location and routing. In *The 21st Annual Joint Conference of the IEEE Computer and Communications Societies (INFOCOM 2002),* June 2002.

74. L.P. Cox, C.D. Murray, and B.D. Noble. Pastiche: making backup cheap and easy. In *5th Symposium Operating Systems Design and Implementation (OSDI'02),* December 2002.

75. M. Luby, M. Mitzenmacher, M. Shokrollahi, D. Spielman, and V. Stemann. Analysis of low density codes and improved designs using irregular graphs. In *30th ACM Annual Symposium on Theory of Computing (STOC 1998),* May 1998.

76. J. Plank. A tutorial on Reed-Solomon coding for fault tolerance in RAID-like systems. *Software Practice and Experience,* 27(9):995–1012, September 1997.

Searching Techniques in Peer-to-Peer Networks

Xiuqi Li

Jie Wu

37.1 Introduction

There has been a growing interest in peer-to-peer networks since the initial success of some very popular file-sharing applications such as Napster[53] and Gnutella.[15] A P2P network is a distributed system in which *peers* employ distributed resources to perform a critical function in a decentralized fashion. Nodes in a P2P network normally play *equal roles*; therefore, these nodes are also called peers. A typical P2P network often includes computers in unrelated administrative domains. These P2P participants join or leave the P2P system frequently; hence, P2P networks are *dynamic* in nature. P2P networks are overlay networks, where nodes are end systems in the Internet and maintain information about a set of other nodes (called neighbors) in the P2P layer. These nodes form a virtual overlay network on top of the Internet. Each link in a P2P overlay corresponds to a sequence of physical links in the underlying network. Examples of P2P applications are distributed file-sharing systems, event notification services, and chat services.[1,3–5]

P2P networks offer the following benefits:[1,3]

- They do not require any special administration or financial arrangements.
- They are self-organized and adaptive. Peers can come and go freely. P2P systems handle these events automatically.
- They can gather and harness the tremendous computation and storage resources on computers across the Internet.
- They are distributed and decentralized. Therefore, they are potentially fault tolerant and load balanced.

P2P networks can be classified based on the control over data location and network topology. There are three categories: *unstructured, loosely structured*, and *highly structured*.[7] In an unstructured P2P network such as Gnutella,[15] no rule exists that defines where data is stored and the network topology is arbitrary. In a loosely structured network such as Freenet[34] and Symphony,[31] the overlay structure and the data location are not precisely determined. In Freenet, both the overlay topology and the data location are determined based on hints. The network topology eventually evolves into some intended structure. In Symphony, the overlay topology is determined probabilistically but the data location is defined precisely. In a highly structured P2P network such as Chord,[16] both the network architecture and the data placement are precisely specified. The neighbors of a node are well-defined and the data is stored in a well-defined location.

P2P networks can also be classified into *centralized* and *decentralized*.[7,11,12] In a centralized P2P such as Napster,[53] a central directory of object location, ID assignment, etc. is maintained in a single location. Peers find the locations of desired files by querying the central directory server. Such P2Ps do not scale well and the central directory server causes a single point of failure. Decentralized P2Ps adopt a distributed directory structure. These systems can be further divided into *purely decentralized* and *hybrid*.[11,12] The difference between them lies in the role peers play. In purely decentralized systems such as Gnutella and Chord, peers are totally equal. In hybrid systems, some peers called *dominating nodes*[2] or *superpeers*[25] serve the search request of other regular peers. Peers in a P2P system are often heterogeneous in computation power, stability, and connectivity. Purely decentralized systems cannot take advantage of this heterogeneity while hybrid systems can. However, dominating nodes and superpeers must be carefully selected to avoid single points of failure and service bottlenecks.

P2P systems can also be classified into *hierarchical* and *nonhierarchical*, based on whether or not the overlay structure is a hierarchy. Most purely decentralized systems have flat overlays and are nonhierarchical systems. All hybrid systems and few purely decentralized systems such as Kelips[23] are hierarchical systems. Nonhierarchical systems offer load balancing and high resilience. Hierarchical systems provide good scalability, the opportunity to take advantage of node heterogeneity, and high routing efficiency.

There are many research issues in P2P computing. This chapter focuses on searching techniques in P2P networks. Searching means locating desired data. Most existing P2P systems support the simple object lookup by key or identifier. Some existing P2P systems can handle more complex keyword queries, which find documents containing keywords in queries. More than one copy of an object can exist in a P2P system. There may be more than one document that contains desired keywords. Some P2P systems are interested in a single data item; others are interested in all data items or as many data items as possible that satisfy a given condition. Most searching techniques are forwarding based. Starting with the requesting node, a query is forwarded (or routed) node to node until the node that has the desired data (or a pointer to the desired data) is reached. To forward query messages, each node must keep information about some other nodes called neighbors. The information of these neighbors constitutes the routing table of a node.

The desired features of searching algorithms in P2P systems include high-quality query results, minimal routing state maintained per node, high routing efficiency, load balance, resilience to node failures, and support of complex queries. The quality of query results is application dependent. In general, it is measured by the number of results and relevance. The routing state refers to the number of neighbors each node maintains. The routing efficiency is generally measured by the number of overlay hops per query. In some systems, it is also evaluated using the number of messages per query. Different searching techniques make different trade-offs between these desired characteristics.

Searching in highly structured systems follows the well-defined neighboring links. For this reason, highly structured P2P systems provide guarantees on finding existing data and bounded data lookup efficiency in terms of the number of overlay hops; however, the strict network structure imposes high overhead for handling frequent node join-leave. Unstructured P2P systems are extremely resilient to node join-leave, because no special network structure needs to be maintained. Searching in unstructured networks is often based on flooding or its variation because there is no control over data storage. The searching strategies in unstructured P2P systems are either blind search or informed search. In a blind search such as iterative deepening,[6] no node has information about the location of the desired data. In an informed search such as routing indices,[8] each node keeps some metadata about the data location. To restrict the total bandwidth consumption, data queries in unstructured P2P systems can be terminated prematurely before the desired existing data is found; therefore, the query may not return the desired data even if the data actually exists in the system. An unstructured P2P network cannot offer bounded routing efficiency due to lack of structure. Searching in a loosely structured system depends on the overlay structure and how the data is stored. In Freenet, searching is directed by the hints used for the overlay construction and the data storage. In Symphony, the data location is precisely defined but the overlay structure is probabilistically formed. Searching in Symphony is guided by reducing the numerical distance from the querying source to the destination node where the desired data is located. The loosely structured systems can offer a balanced trade-off if they are properly designed.

This chapter provides a survey of state-of-the-art searching schemes in different types of P2P systems. The survey focuses on searching schemes in unstructured P2Ps. The chapter is organized as follows. Section 37.2 explores searching in various unstructured systems. Section 37.3 investigates searching in strictly structured systems. The discussion in this section focuses on hierarchical DHT P2Ps and non-DHT P2Ps. Non-hierarchical DHT P2Ps are briefly overviewed here because a survey of searching in such systems has been performed by Balakrishnan.[1] Searching in loosely structured systems is examined in Section 37.4, and a summary appears in Section 37.5.

37.2 Searching in Unstructured P2Ps

In an unstructured P2P system, no rule exists that strictly defines where data is stored and which nodes are neighbors of each other. To find a specific data item, early work such as the original Gnutella[15] used flooding, which is the Breadth First Search (BFS) of the overlay network graph with depth limit D. D refers to the system-wide maximum TTL of a message in terms of overlay hops. In this approach, the *querying node* sends the query request to all its neighbors. Each neighbor processes the query and returns the result if the data is found. This neighbor then forwards the query request further to all its neighbors except the querying node. This procedure continues until the depth limit D is reached. Flooding tries to find the maximum number of results within the ring that is centered at the querying node and has the radius D-overlay-hops. However, it generates a large number of messages (many of them are duplicate messages) and does not scale well.

Many alternative schemes have been proposed to address the problems of the original flooding. These works include iterative deepening,[6] k-walker random walk,[7] modified random BFS,[10] two-level k-walker random walk,[52] directed BFS,[6] intelligent search,[10] local indices-based search,[6] routing indices-based search,[8] attenuated bloom filter-based search,[9] adaptive probabilistic search,[11] and dominating set-based search.[2] They can be classified as BFS based or Depth First Search (DFS) based. The routing indices based-search and the attenuated bloom filter-based search are variations of DFS. All the others are variations of BFS. In the iterative deepening and local indices, a query is forwarded to all neighbors of a forwarding node. In all other schemes, a query is forwarded to a subset of neighbors of a forwarding node.

The searching schemes in unstructured P2P systems can also be classified as deterministic or probabilistic. In a deterministic approach, query forwarding is deterministic, while in a probabilistic approach, query forwarding is probabilistic, random, or is based on ranking. The iterative deepening, local indices based search, and the attenuated bloom filter based search are deterministic. The others are probabilistic.

Another way to categorize searching schemes in unstructured P2P systems is regular-grained or coarse-grained. In a regular-grained approach, all nodes participate in query forwarding. In a coarse-grained scheme, the query forwarding is performed by only a subset of nodes in the entire network. Dominating set based search is coarse-grained because query forwarding is performed only by the dominating nodes in the CDS (Connected Dominating Set). All the others are regular-grained.

Another taxonomy is blind search or informed search.[11,12] In a blind search, nodes do not keep information about data location. In an informed search, nodes store some metadata that facilitates the search. Blind searches include iterative deepening, k-walker random walk, modified random BFS, and two-level k-walker random walk. All the others are informed search.

37.2.1 Iterative Deepening

Yang and Garcia-Molina[6] borrowed the idea of iterative deepening from artificial intelligence and used it in P2P searching. This method is also called *expanding ring*. In this technique, the querying node periodically issues a sequence of BFS searches with increasing depth limits $D_1 < D_2 < \ldots < D_i$. The query is terminated when the query result is satisfied or when the maximum depth limit D is reached. In the latter case, the query result may not be satisfied. All nodes use the same sequence of depth limits called *policy P* and the same time period W between two consecutive BFS searches.

For example, assume that $P = \{3, 5, 8\}$ and $W = 6$ seconds. The query node S first sends a BFS search with depth limit 3 to all its neighbors via a *query message*. This BFS search message will reach all nodes within three-hops distance from S. These nodes will process this BFS message and store (*freeze*) that message for a time period ($\geq W$) when they receive it. If any desired data is located on these nodes, the data will be sent back to S. If the query is satisfied within W ($=6$) seconds following the first BFS search, S will terminate the query and will not continue. Otherwise, S will initiate the second BFS search with depth limit 5 via a *resend message*. The resend message carries the same query ID as in the corresponding query message. Any node within two-hops distance from S will simply forward the resend message to all its neighbors after receiving it. The nodes at three-hops distance from S will drop the resend message and then unfreeze the stored query message with the matching query ID. "Unfreeze" means forwarding the respective stored query message with a new depth limit of 2 ($= 5 - 3$) to all its neighbors. This unfrozen query message will be processed similarly to the query message in the first BFS search. If the resend message with maximum depth limit 8 is sent by the querying node, nodes within eight-hops distance from S will not store (freeze) this query message. The querying node will not issue another resend message with a larger depth limit.

Iterative deepening is tailored to applications where the initial number of data items returned by a query is important. However, it does not intend to reduce duplicate messages and the query processing is slow.

37.2.2 k-Walker Random Walk and Related Schemes

In the *standard random walk* algorithm, the querying node forwards the query message to one randomly selected neighbor. This neighbor randomly chooses one of its neighbors and forwards the query message to that neighbor. This procedure continues until the data is found. Consider the query message as a walker.

The query message is forwarded in the network the same way a walker randomly walks on a network of streets. The standard random walk algorithm uses just one walker. This can greatly reduce the message overhead but causes longer searching delay.

In the k-walker random walk algorithm,[7] k walkers are deployed by the querying node. That is, the querying node forwards k copies of the query message to k randomly selected neighbors. Each query message takes its own random walk. Each walker periodically "talks" with the querying node to decide whether that walker should terminate. Nodes can also use soft states to forward different walkers for the same query to different neighbors. The k-walker random walk algorithm attempts to reduce the routing delay. On average, the total number of nodes reached by k random walkers in H hops is the same as the number of nodes reached by one walker in kH hops. Therefore, the routing delay is expected to be k times smaller.

A similar scheme is the *two-level random walk*.[52] In this scheme, the querying node deploys k_1 random walkers with the TTL being l_1. When the TTL l_1 expires, each walker forges k_2 random walkers with the TTL being l_2. All nodes on the walkers' paths process the query. Given the same number of walkers, this scheme generates less duplicate messages but has longer searching delays than the k-walker random walk.

Another similar approach, called the *modified random BFS*, was proposed by Kalogeraki et al.[10] The querying node forwards the query to a randomly selected subset of its neighbors. On receiving a query message, each neighbor forwards the query to a randomly selected subset of its neighbors (excluding the querying node). This procedure continues until the query stop condition is satisfied. No comparison to the k-walker random walk was given by Kalogeraki et al.[10] It is expected that this approach visits more nodes and has a higher query success rate than the k-walker random walk.

The work by Lv et al.[7] and Cohen et al.[41] also address the data replication issue in unstructured P2P systems. The question studied is: assuming the fixed amount of total storage space in the P2P system, what is the optimal number of copies for each object in terms of the average search overhead per successful query? Three replication strategies were analyzed: (1) uniform, (2) proportional, and (3) square-root replication. In uniform replication, the same number of copies is created for each object, regardless of the query distribution. In proportional replication, the number of copies for each object is proportional to its query distribution. The higher the query rate of an object, the higher the number of copies for that object. In square-root replication, the number of copies per object is proportional to the square root of the query rate. The performance measures are the average search size (i.e., the average number of nodes probed) and the utilization rate of a copy (i.e., the rate of queries that a copy serves). The search size reflects the query efficiency. The utilization rate indicates the load balance. The k-walker random walk is used as the searching scheme in the evaluation.

The analysis and simulation results show that uniform replication and proportional replication achieve the same average search size, and this search size is larger than that of the square-root replication. As for the utilization rate, the proportional replication has the same rate for all objects; the uniform replication has the rate proportional to the query rate, and the square-root replication has a varying utilization rate per object. However, the square-root replication has much smaller variances than uniform and proportional replication in the two performance measures. In summary, the square-root replication has the best query efficiency and the proportional replication achieves the best load balance. In practice, the square-root replication is implemented by replicating copies proportional to the number of sites probed.

The work by Lv et al.[7] and Cohen et al.[41] also studies where to replicate an object. Three approaches are considered and evaluated using k-walker random walk: *owner replication*, *path replication*, and *random replication*. All three schemes replicate the found object when a query is successful. The owner replication replicates an object only at the requesting node. The path replication creates copies of an object on all nodes on the path from the providing node to the requesting node. The random replication places copies on the p randomly selected nodes that were visited by the k walkers. The path replication implements the square-root replication. The random replication has slightly less overall search traffic than the path replication, because path replication intends to create object copies on the nodes that are topologically along the same path. Both path replication and random replication have less overall search traffic than owner replication.

37.2.3 Directed BFS and Intelligent Search

The basic idea of the directed BFS approach[6] is that the query node sends the query message to a subset of its neighbors that will quickly return many high-quality results. These neighbors then forward the query message to all their neighbors just as in BFS.

To choose "good" neighbors, a node keeps track of simple statistics on its neighbors, for example, the number of query results returned through that neighbor and the network latency of that neighbor. Based on these statistics, the best neighbors can be intelligently selected using the following heuristics:

- The highest number of query results returned previously
- The least hop-count in the previously returned messages (i.e., the closest neighbors)
- The highest message count (i.e., the most stable neighbors)
- The shortest message queue (i.e., the least busy neighbors)

By directing the query message to just a subset of neighbors, directed BFS can reduce the routing cost in terms of the number of routing messages. By choosing good neighbors, this technique can maintain the quality of query results and lower the query response time. However, in this scheme, only the querying node intelligently selects neighbors to forward a query. All other nodes involved in a query processing still broadcast the query to all their neighbors as in BFS. Therefore, the message duplication is not greatly reduced.

A similar approach called *intelligent search* was presented by Kalogeraki et al.[10] The query type considered in this work is the *keyword query*: a search for documents that contain desired keywords listed in a query. A query is represented using a keyword vector. This technique consists of four components: (1) a search mechanism, (2) a profile mechanism, (3) a peer ranking mechanism, and (4) a query-similarity function.

When the querying node initiates a query, it does not broadcast the query to all its neighbors. Instead, it evaluates the past performance of all its neighbors and propagates the query only to a subset of its neighbors that have answered similar queries before and therefore will most likely answer the current query. On receiving a query message, a neighbor looks at its local datastore. If the neighbor has the desired documents, it returns them to the querying node and terminates. Otherwise, the neighbor forwards the query to a subset of its own neighbors that have answered similar queries previously. The query forwarding stops when the maximum TTL is reached.

The *cosine similarity model* is used to compute the query similarity. Based on this model, the similarity between two queries is the cosine of the angle between their query vectors. To determine whether a neighbor answered similar past queries, each node keeps a profile for each of its neighbors. The profile for a neighbor contains the most recent queries answered by that neighbor. The profile is created and updated using two schemes. In one scheme, each peer continuously monitors the query and query response message. Queries answered by a neighbor are stored in the profile for that neighbor. In the second scheme, the peer that replies to a query message broadcasts this information to all its neighbors.

Neighbors are ranked to facilitate the selection. The rank of a neighbor P_i of the peer P_j in terms of the query q is determined by the following formula:

$$R_{P_j}(P_i, q) = \sum_{q_l \in A_i} (Q_{sim}(q_l, q))^\alpha$$

In this formula, A_i denotes the set of queries among the K most similar ones that were answered by peer P_i; α is a configurable parameter used to add more weight to more similar queries. The ranking formula aggregates the similarities of K most similar past queries answered by a neighbor.

37.2.4 Local Indices-Based Search

The *local indices*[6] intend to get the same number of query results as scoped-flooding with less number of nodes processing a query. In local indices, each node keeps indices of data on all nodes within k-hop distance from it. Therefore, each node can directly answer queries for any data in its local indices without resorting to other nodes. All nodes use the same policy P on the list of depths at which the query should be processed. The nodes whose depths are listed in P check their local indices for the queried data and return the query result if the sought data is found. These nodes also forward the query message to all their neighbors if their depths are not the maximum depth limit. All other nodes whose depths are not listed in P just forward the query message to all their neighbors once they have received it and do not check their local indices. For example, assume that $P = \{0, 3, 6\}$. To route a query, the querying node processes the query because its depth, 0 (i.e., the depth from itself is 0), is listed in P. The querying node then forwards the query message to all its neighbors at depth 1. Because their depth 1 is not listed in P, these nodes will not process the query. They will simply forward the query message to all their neighbors at depth 2. For the same reason, all nodes at depth 2 will simply forward the query message to all their neighbors at depth 3. All nodes at depth 3 will process the query because their depth is listed in P. These nodes then forward the query to their neighbors at depth 4. This procedure continues until the query message is forwarded to all nodes at depth 6. These nodes will process the query. However, they will not forward the query because their depth is the maximum depth in P. At this point, the query is terminated even if the query result is not satisfied. Note that all nodes in a P2P system organized using local indices play equal roles.

The local indices are updated when a node joins, leaves, or modifies its data. A node Y joins the network by sending a join message with a TTL of r. This join message contains the metadata (indices) about the data collection in Y. All nodes within r-hop distance from Y will receive this join message. If a node X receives the join message from Y, it replies with another join message that includes the metadata over its own data collection. X sends this replied join message directly to Y over a temporary connection. Then, both X and Y add each other's metadata into their own local indices.

A new node Y can add a new path of length k or less between two other nodes A and B. These two nodes can discover this new path in a number of ways without introducing additional messages. One way to achieve this is through periodic ping-pong messages. Nodes constantly send *ping* messages to all nodes within a depth D. Every node replies with a *pong* message. If A receives a pong message from B that is, at most, k hops away and A does not contain indices about B's data collection, then A learns that there is a new path between A and B. A will inform B about its data collection by sending a join message directly to B. B will reply directly to A with another join message containing the indices of its own data collection.

When a node Z gracefully leaves the network or fails, other nodes will detect this event after a timeout. If these nodes index Z's data collection, they will remove those index entries. When the data collection on a node Z is modified, Z will send a short update message with a TTL of r to all its neighbors. This update message includes information about all affected data elements and how they are affected: inserted, deleted, or updated. Any node that receives such a message and contains index entries for those affected elements will update their local indices accordingly.

The local indices approach is similar to iterative deepening. Both broadcast the query message based on a list of depths; however, in iterative deepening, all nodes within the maximum depth limit process the query. In local indices, only nodes whose depths are listed in the policy P process the query. In addition, the iterative deepening approach spreads the query message iteratively with increasing TTL; the local indices approach spreads the query message once with the maximum TTL.

37.2.5 Routing Indices-Based Search

Routing indices[8] is similar to directed BFS and intelligent search in that all of them use the information about neighbors to guide the search. Directed BFS only applies this information to selected neighbors of the querying source (i.e., the first hop from the querying source). The rest of the search process is just as that of BFS. Both intelligent search and routing indices guide the entire search process. They differ in the information kept for neighbors. Intelligent search uses information about past queries that have been answered by neighbors. Routing indices stores information about the topics of documents and the number of documents stored in neighbors.

Routing indices consider content queries, queries based on the file content instead of file name or file identifier. One example of such a content query is: a request for documents that contain the word "networks." A query includes a set of subject topics. Documents may belong to more than one topic category. Document topics are independent. Each node maintains a local index of its own document database based on the keywords contained in these documents.

The goal of a *routing index* (RI) is to facilitate a node to select the "best" neighbors to forward queries. An RI is a distributed data structure. Given a content query, the algorithms on this data structure compute the top *m* best neighbors. The goodness of a neighbor is application dependent. In general, a good neighbor is the one through which many documents can be quickly found.

A routing index is organized based on the single-hop routes and document topics. There is one index entry per route (i.e., per neighbor) per topic. An RI entry, (*networks, B*), at node *A* stores information about documents in the topic: *networks* that may be found through the route $(A \rightarrow B)$. This entry gives *hints on the potential query result* if *A* forwards the query to *B* (i.e., the route $A \rightarrow B$ is chosen) — hence the name *routing index*. A routing index entry is very different from a regular index entry. If (*networks, B*) were the regular index entry, it would mean that node *B* *stores* documents in the topic: *networks*. By organizing the index based on neighbors (routes) instead of destinations (indexed data locations), the storage space can be reduced.

Three types of RIs — compound RI, hop-count RI, and exponentially aggregated RI — are proposed. They differ in RI entry structures. A compound RI (CRI) stores information about the number of documents in each interesting topic that might be found if a query is forwarded to a single-hop neighbor. A sample CRI at a node *B* is shown in Table 37.1. Each row in the table describes the number of documents along a specific path and the number of documents on each interesting topic along that path. For example, the first row in the table indicates that if *B* forwards the query to *A*, 1000 documents can be found. Among those documents, 100 are database (DB) documents, 200 are network documents, 400 are theory documents, and there are no language documents.

The goodness of a neighbor for a query in CRI is the number of desired documents that can be found through that neighbor. This can be estimated by the following formula:

$$ND \times \prod_i \frac{CRI(t_i)}{ND}$$

In the formula above, t_i refers to the subject topic that appears in both the query and the CRI table; $CRI(t_i)$ denotes the value in the intersection of the row for a path and the column for the topic t_i; and *ND* represents the value at the column *#docs* for the path considered. Use the CRI example for node *B* in Table 37.1. Assume that *B* receives a query for documents on "networks" and "theory." The goodness of each neighbor for the query is:

$$A: \quad 1000 \times (200/1000) \times (400/1000) = 80$$
$$E: \quad 300 \times (0/300) \times (200/300) = 0$$
$$F: \quad 800 \times (100/800) \times (160/800) = 20$$

Therefore, *B* will select *A* to forward the query because its goodness score is the highest.

Figure 37.1 shows a partial P2P network and some CRI values. An additional row is added into the CRI at each node to summarize the local indices in that node. For example, the summary at node *B* indicates that there are 200 documents at *B*; 50 of them are related to database, 60 of them are about theory, and 20

TABLE 37.1 An Example of a Compound RI at Node *B*

Path	# Docs	Documents in Topics Database (DB)	Networks (N)	Theory (T)	Languages (L)
A	1000	100	200	400	0
E	300	60	0	200	100
F	800	0	100	160	200

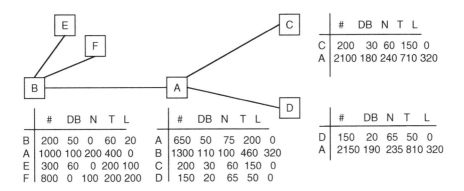

FIGURE 37.1 A partial P2P with CRI indices.

of them are about languages. B does not store documents about networks. The CRIs at node B, A, C, and D show that node B can access 200 network documents via A; 75 of them are at A, 60 at C, and 65 at D.

The following shows an example of searching using routing indices. Suppose that the node B initiates a query for the documents about "networks" and "theory." B first looks up its local database for the desired documents. If not enough documents are found, it calculates the goodness scores of all its neighbors: A: 80; E: 0; and F: 20. A is then chosen as the best neighbor to forward the query. After receiving the query, A first checks its local database and returns all desired documents to B. If the query result is not satisfied, A will then calculate the goodness scores of its neighbors C, D (B is excluded): C: 45, D: 23. A then selects C as the best neighbor to forward the query. C then processes the query and returns all desired data along the query path. C does not have any other neighbor to forward the query. If the query stop condition is not satisfied, C will return the query back to A. A then forwards the query to its second best neighbor D. This process continues until the desired number of documents is found.

The CRIs are expanded as follows. When a new connection is established between nodes A and D, A will add up its RI vectors (rows) and then send this aggregated RI vector to D. In the meantime, D also sums up its RI vectors (excluding A's entry if it exists) and sends the aggregated RI to A. When either party receives the other's aggregated RI, it will create a new entry in its RI for the other party. After this, both A and D inform their other neighbors about this change in a similar fashion. The CRI entry deletion and update are handled similarly. RI entry aggregation reduces the bandwidth overhead.

The compound RI (CRI) does not consider the number of hops required to reach documents of a specific topic. However, we can modify the CRI to incorporate the hop count. We can store a CRI for each hop up to a maximum hop limit H at each node. H is called the *horizon* of an RI. This modified CRI is called *hop-count routing index*. The hop-count RI contains information about the noncumulative number of documents that can be found along a path at one-hop distance, at two-hops distance, . . . , and at H-hops distance. The goodness of a neighbor with respect to a query in the hop-count RI is the number of desired documents per message. It considers both the document counts and the number of messages to reach those documents. The goodness score is computed using the regular-tree cost model.

The limitation of the hop-count RI is that it does not have information about documents at hop-distances beyond the horizon. The *exponentially aggregated RI (ERI)* solves this problem at the cost of some potential loss in accuracy. The ERI entries store the result of applying the regular-tree cost formula to a corresponding hop-count RI for the topics of interest.

37.2.6 Attenuated Bloom Filter-Based Search

The attenuated bloom filter-based search[9] assumes that each stored document has many replicas spread over the P2P network; documents are queried by names. It intends to quickly find replicas close to the query source with high probability. This is achieved by approximately summarizing the documents that likely exist in nearby nodes. However, the approach alone fails to find replicas far away from the query source.

Bloom filters[50] are often used to approximately and efficiently summarize elements in a set. A bloom filter is a bit-string of length m that is associated with a family of independent hash functions. Each hash function takes as input any set element and outputs an integer in $[0, m)$. To generate a representation of a set using bloom filters, every set element is hashed using all hash functions. Any bit in the bloom filter whose position matches a hash function result is set to 1. To determine whether an element is in the set described by a bloom filter, that element is hashed using the same family of hash functions. If any matching bit is not set to 1, the element is definitely not in the set. If all matching bits in the bloom filter are set to 1, the element is *probably* in the set. If the element indeed is not in the set, this is called a *false positive*.

Attenuated bloom filters are extensions to bloom filters. An attenuated bloom filter of depth d is an array of d regular bloom filters of the same length w. A level is assigned to each regular bloom filter in the array. level 1 is assigned to the first bloom filter and level 2 is assigned to the second bloom filter. The higher levels are considered attenuated with respect to the lower levels. Each node stores an attenuated bloom filter for each neighbor. The i-th bloom filter in an attenuated bloom filter (depth: d; $i \leq d$) for a neighbor B at a node A summarizes the set of documents that will probably be found through B on all nodes i-hops away from A. Figure 37.2 illustrates an attenuated bloom filter for neighbor C at node B. "File3" and "File4" are available at two-hops distance from B through C. They are hashed to $\{0, 5, 6\}$ and $\{2, 5, 8\}$, respectively. Therefore, the second bloom filter contains 1 at bits 0, 2, 5, 6, and 8.

To route a query for a file, the querying node hashes the file name using the family of hash functions. Then the querying node checks level 1 of its attenuated bloom filters. If level 1 of an attenuated bloom filter for a neighbor has 1s at all matching positions, the file will probably be found on that neighbor (one-hop distance from the query source). We call such a neighbor a candidate. The querying node then forwards the query to the closest one among all candidates. If no such candidate can be found, the querying node will check the next higher level (level 2) of all its attenuated bloom filters similarly to checking level-1. If no candidate can be found after all levels have been checked at the query source, this indicates that definitely no nearby replica exists. On receiving the query, a neighbor of the querying node looks up its local data store. If the data is found, it will be returned to the query source. If not, this neighbor will check its attenuated bloom filters similarly. During the query processing, if a false positive is found after d (the depth of the attenuated bloom filter) unsuccessful hops, the attenuated bloom filter-based search terminates with a failure. No backtracking is allowed.

To ease the filter update operation, for any two neighboring nodes A and B, node A keeps a copy of the attenuated bloom filter at B for the link $B \rightarrow A$. Node B also keeps a copy of the attenuated bloom filter at A for the link $A \rightarrow B$. If a new document is inserted at node A, it calculates the changed bits in the attenuated bloom filters of its own and of its neighbors. A then sends the changes to the corresponding neighbors. When A's neighbor B receives such a message, B will attenuate the changed bits one level and check changes in the attenuated bloom filters that its neighbors maintain. B will also inform its neighbors

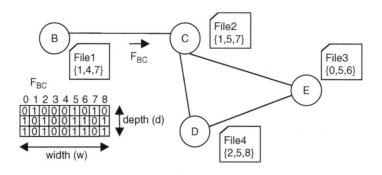

FIGURE 37.2   An example of an attenuated bloom filter.

about the changes. Thus, each update is spread outward from the update source. The duplicate update messages can be suppressed by either the source node or the destination node with the help of update message IDs.

The attenuated bloom filter approach can be combined with any structured approach to optimize the searching performance. We can use the attenuated bloom filters to try locating nearby replicas. If no nearby replica exists, we switch to the structured approach to continue the lookup. The hop-count RI is similar to the attenuated bloom filter approach. Both summarize the documents at some distance from the querying source. There are two differences between them. One is that the attenuated bloom filter is a probabilistic approach while the hop-count RI is a deterministic approach if omitting the document change. The other is that the attenuated bloom filter provides information about a specific file while the hop-count RI provides the number of documents on each document category but not a specific file.

37.2.7   Adaptive Probabilistic Search

In the *Adaptive Probabilistic Search (APS)*,[11,12] it is assumed that the storage of objects and their copies in the network follows a replication distribution. The number of query requests for each object follows a query distribution. The search process does not affect object placement or the P2P overlay topology.

The APS is based on k-walker random walk and *probabilistic* (not *random*) forwarding. The querying node simultaneously deploys k walkers. On receiving the query, each node looks up its local repository for the desired object. If the object is found, the walker stops successfully. Otherwise, the walker continues. The node forwards the query to the best neighbor that has the highest probability value. The probability values are computed based on the results of the past queries and are updated based on the result of the current query. The query processing continues until all k walkers terminate either successfully or fail (in which case the TTL limit is reached).

To select neighbors probabilistically, each node keeps a local index about its neighbors. There is one index entry for each object that the node has requested or forwarded requests for through each neighbor. The value of an index entry for an object and a neighbor represents the relative probability of that neighbor being selected for forwarding a query for that object. The higher the index entry value, the higher the probability. Initially, all index values are assigned the same value. Then the index values are updated as follows. When the querying node forwards a query, it makes some guess about the success of all the walkers. The guess is made based on the ratio of the successful walkers in the past. If it assumes that all walkers will succeed (*optimistic approach*), the querying node proactively increases the index values associated with the chosen neighbors and the queried object. Otherwise (i.e., *pessimistic approach*), the querying node proactively decreases the index values. Using the guess determined by the querying node, every node on the query path updates the index values similarly when forwarding the query.

The index values are also updated when the guess for a walker is wrong. Specifically, if an optimistic guess is made and a walker terminates with a failure, then the index values for the requested object along that walker's path are decreased. The last node on the path sends an update message to the preceding node. On receiving the message, the preceding node decreases the index value for that walker and forwards the update message to the next node on the reverse path. This update procedure continues on the reverse path until the querying node receives an update message and decreases the index value for that walker. If the pessimistic approach is employed and a walker terminates successfully, the index values for the requested object on the walker's path are increased. The update procedure is similar. To remember a walker's path, each node appends its ID in the query message during query forwarding and maintains a soft state for the forwarded query. If a walker A passes by a node which another walker B stopped by before, the walker A terminates unsuccessfully. The duplicate message is discarded.

Figure 37.3 illustrates how the search process works. Peer A issues a query for an object stored on peer F. Two walkers are deployed. Peer A made an optimistic guess. The initial values of all index entries for this object are 30. One walker w_1 takes the path $A \rightarrow B \rightarrow F$. The other one w_2 takes the path $A \rightarrow C \rightarrow D \rightarrow E$. During the search, each node except the last node on the query paths increases the

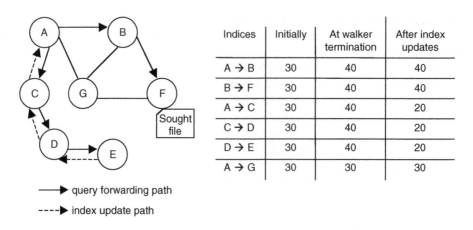

Indices	Initially	At walker termination	After index updates
A → B	30	40	40
B → F	30	40	40
A → C	30	40	20
C → D	30	40	20
D → E	30	40	20
A → G	30	30	30

⟶ query forwarding path

---▶ index update path

FIGURE 37.3   An example of adaptive probabilistic search.

index value(s) for this object and the chosen neighbor(s) by 10. Because the optimistic approach is employed and w_2 fails, the index values on the query path for w_2 will be decreased by 20 so that the final index values are smaller than the initial index values. When the subsequent request for the same object is initiated at or forwarded to A, the neighbor B will be chosen with the probability $4/9\ (=40/(20+30+40))$, C with the probability $2/9$, and G with the probability $3/9$.

Compared to the k-walker random walk, the APS approach has the same asymptotic performance in terms of the message overhead. However, by forwarding queries probabilistically to most promising neighbor(s) based on the learned knowledge, the APS approach surpasses the k-walker random walk in the query success rate and the number of discovered objects.

Two performance optimizations of the APS were also proposed.[11,12] The APS uses the same guess for all objects. This imprecision causes more messages. The *swapping-APS* (s-APS) therefore constantly observes the ratio of successful walkers for each object and swaps to a better update policy accordingly. The *weighted-APS* (w-APS) includes the location of objects in the probabilistic selection of neighbors. A distance function is embedded in the stored path of the query and is used in the index update. When the pessimistic guess is made for a walker and the walker succeeds, the index values for neighbors closer to the discovered object are increased more than those for distant neighbors.

37.2.8   Dominating Set-Based Search

The *dominating set-based search* scheme was proposed by Yang and Wu.[2] In this approach, routing indices are stored in a selected set of nodes that form a connected dominating set (CDS). A CDS in a P2P network is a subset of nodes that are connected through direct overlay links. All other nodes that are not in the CDS can be reached from some node in the CDS in one hop. Searching is performed through a random walk on the dominating nodes in the CDS.

The construction of the CDS uses solely the local information: a node's one-hop and two-hops neighbors. The construction consists of two processes: (1) marking followed by (2) reduction. The marking process marks each node in the P2P system as either a dominating node or a nondominating node. The marker T represents a dominating node while the marker F represents a nondominating node. A node is marked using T if two of its neighbors are not directly connected (i.e., these two neighbors are not neighbors of each other). At the end of the marking process, all nodes with marker T form the CDS. To reduce the size of the CDS, two reduction rules are applied during the reduction process. Each node in the CDS is assigned a one-hop ranking value. This ranking value is the sum of the number of documents on a node and the number of documents of the node's neighbor that has the most documents. The first reduction rule specifies that if the neighbors of a node A in the CDS are a proper subset of neighbors of another node *B* in the CDS and the node *A* has a smaller one-hop ranking value than node *B*, then remove node *A* from the CDS. The second reduction rule states that a node *C* is removed from the CDS if the following three conditions are satisfied:

1. Two neighbors *A* and *B* of the node *C* are also dominating nodes.
2. The neighbor set of *C* is a proper subset of the union of the neighbor sets of *A* and *B*.
3. The node *C* has a one-hop ranking value that is smaller than the values of both *A* and *B*.

Searching is conducted on the CDS as follows. If the querying source is not a dominating node, the source forwards the query to its dominating neighbor with the highest one-hop ranking value. If the querying source is a dominating node, it forwards the query to its dominating neighbor with the highest one-hop ranking value. This querying source also forwards the query to a nondominating neighbor if that neighbor has the most documents among all neighbors of the querying source. On receiving a query request, a dominating node looks up its local database for the searched document and performs the query forwarding similarly to a querying source that is a dominating node. On receiving a query request, a nondominating node only looks up the local database and does not forward the query any further. All found documents are returned from the hosting nodes to the querying source along the reverse query paths. The query stops when the TTL limit is reached or a node is visited a second time.

The dominating set-based approach intends to get the most number of documents by forwarding queries primarily on dominating nodes that are well-connected and have many documents themselves or whose neighbors have many documents. The construction of the CDS does not incur more overlay links, as often occurs in superpeer approaches (discussed in Section 37.3.2).The cost of creating and maintaining the CDS is lower than that of routing indices.

37.3 Searching in Strictly Structured P2Ps

In a strictly structured system, the neighbor relationship between peers and data locations is strictly defined. Searching in such systems is therefore determined by the particular network architecture. Among the strictly structured systems, some implement a distributed hash table (DHT) using different data structures. Others do not provide a DHT interface. Some DHT P2P systems have flat overlay structures, while others have hierarchical overlay structures.

A DHT is a hash table whose table entries are distributed among different peers located in arbitrary locations. Each data item is hashed to a unique numeric key. Each node is also hashed to a unique ID in the same keyspace. Each node is responsible for a certain number of keys. This means that the responsible node stores the key and the data item with that key or a pointer to the data item with that key. Keys are mapped to their responsible nodes. The searching algorithms support two basic operations: (1) *lookup(key)* and (2) *put(key)*. The *lookup(k)* operation is used to find the location of the node that is responsible for the key *k*, and *put(k)* is used to store a data item (or a pointer to the data item) with the key *k* in the node responsible for *k*. In a distributed storage application using a DHT, a node must publish the files that were originally stored on it before these files can be retrieved by other nodes. A file is published using *put(k)*.

In this section, searching in nonhierarchical (flat) DHT P2Ps is briefly overviewed. Then, searching in hierarchical DHT P2Ps and non-DHT P2Ps are discussed in detail. More about nonhierarchical DHT P2Ps can be found in a comprehensive survey.[1]

37.3.1 Searching in Nonhierarchical DHT P2Ps

Different nonhierarchical DHT P2Ps use different flat data structures to implement the DHT. These flat data structures include ring, mesh, hypercube, and other special graphs such as the de Bruijn graph. Chord uses a ring data structure.[16] Node IDs form a ring. Each node keeps a finger table that contains the IP addresses of nodes that are half of the ID ring away from it, one fourth of the ID ring away, one eighth of the ID ring away, etc., until its immediate successor. A key is mapped to a node whose ID is the largest number that does not exceed that key. During the searching for *lookup(k)*, a node *A* forwards the query for *k* to *successor(k)*, which is another node in *A*'s finger table with the highest ID that is not larger than *k*. In this way, the query for *k* is forwarded through the successor list until the node responsible for *k* is reached. The finger table speeds up the lookup operation. In case of the failure of *successor(k)*, a node forwards the query to its immediate successor node. Chord achieves $O(\log^N)$ routing efficiency at the cost of $O(\log^N)$ routing state per node. *N* refers to the total number of nodes in the system. The work by Xuan et al.[22] extends Chord by adding different kinds of reverse edges to Chord so that the modified Chord is resilient to routing attacks.

Pastry uses a tree-based data structure that can be considered a generalization of a hypercube.[17] The node ID is 128-digit in base 2^b, and *b* is typically 4. Each node *A* keeps a leaf set *L*, where *L* consists of the set of |L|/2 nodes whose IDs are closest to and smaller than *A*'s ID and the set of |L|/2 nodes whose IDs are closest to and larger than *A*'s ID. This leaf set guarantees the correctness of routing. To shorten the routing latency, each Pastry node also keeps a routing table of pointers to other nodes in the ID space. Each node keeps $(2^b - 1)$ entries for each prefix of its node ID. An entry for a prefix of length *i* stores the location of some node whose ID shares that prefix and whose $(i + 1)th$ digit is different.

The searching in Pastry is done as follows. Given a query for the key *k*, a node *A* forwards the query to a node whose ID is numerically closest to *k* among all nodes known to *A*. The node *A* first tries to find a node in its leaf set. If such node does not exist, node *A* tries to find a node in its routing table whose ID shares a longer prefix with *k* than *A*. If such a node does not exist, node *A* forwards the query to a node whose ID has the same shared prefix as *A* but is numerically closer to *k* than *A*. Network proximity can be considered using heuristics during query forwarding in Pastry. Each Pastry node maintains $O(\log^N)$ routing state to achieve the routing latency $O(\log^N)$. The algorithms in Tapestry[18] and Kademlia[19] are similar to Pastry.

A *d*-dimensional toroidal space is used to implement the DHT in CAN.[20] The space is divided into a number of zones. Each zone is a hyper-rectangle and is taken care of by a node. The zone boundaries identify the node responsible for that zone. A key *k* is hashed to a point *p* in the *d*-dimensional space. The node whose zone covers *p* stores the hash table entry for *k*. Each node's routing table consists of all its neighbors in the *d*-dimensional space. A node *A* is considered a neighbor of another node *B* if *B*'s zone shares a $(d - 1)$-dimensional hyperplane with *A*'s zone. Given a query for the data item with key *k*, a node forwards the query to another node in its routing table whose zone is closest to the zone of the node responsible for the key *k*. Ties are broken arbitrarily. Each CAN node maintains $O(d)$ states to achieve $O(d \sqrt[d]{N})$ routing efficiency, where *N* refers to the total number of nodes in the P2P.

The Koorde,[21] Viceroy,[46] and Cycloid[47] systems have overlays with constant degrees. Koorde embeds a de Bruijn graph on the Chord ring for forwarding lookup requests. A routing efficiency of $O(\log^N)$ can be achieved with $O(1)$ states per node. The overlay of Viceroy is an approximate butterfly network. The node ID space is [0, 1). The butterfly level parameter of a node is selected according to the estimated network size. Viceroy also achieves $O(\log^N)$ routing efficiency with $O(1)$ neighbors per node. Cycloid integrates Chord and Pastry and imitates the cube-connected-cycles (CCC) graph routing. It has a routing efficiency of $O(d)$, with a routing state per node of $O(1)$. Simulation results[47] show that Cycloid performs better than Koorde and Viceroy in large-scale and dynamic P2P systems.

37.3.2   Searching in Hierarchical DHT P2Ps

All hierarchical DHT P2Ps organize peers into different groups or clusters. Each group forms its own overlay. All groups together form the entire hierarchical overlay. Typically, the overlay hierarchies are two-tier or three-tier. They differ mainly in the number of groups in each tier, the overlay structure formed by each group, and whether or not peers are distinguished as regular peers and *superpeers/dominating nodes*. Superpeers/dominating nodes generally contribute more computing resources, are more stable, and take more responsibility in routing than regular peers. The discussion in this subsection focuses on Kelips and Coral.

37.3.2.1 Kelips

Kelips[23] is composed of k virtual *affinity groups* with group IDs in $[0, k-1]$. The IP address and port number of a node n are hashed to a group ID of the group to which the node n belongs. The consistent hashing function SHA-1 provides a good balance of group members with high probability. Each filename is mapped to a group using the same SHA-1 function. Inside a group, a file is stored in a randomly chosen group member, called the file's *homenode*. Thus, Kelips offers load balance in the same group and among different groups.

Each node n in an affinity group g keeps in memory the following routing state:

- *View of the belonging affinity group g.* This is the information about the set of nodes in the same group. The data includes the round-trip time estimate, the heartbeat count, etc.
- *Contacts of all other affinity groups.* This is the information about a small constant number of nodes in all other groups. The data for each contact is the same as that of an intra-group node.
- *File tuples.* This is the intra-group index about the set of files whose homenodes are in the same affinity group. A file tuple consists of a filename and the IP address of the file's homenode. A heartbeat count is also associated with a file tuple.

The total number of routing table entries per node is $N/k + c*(k-1) + F/k$, where N refers to the total number of nodes, c is the number of contacts per group, F is the total number of files in the system, and k is the number of affinity groups. Assume that F is proportional to N, and c is fixed. With the optimal k, the complexity of the routing state is $O(\sqrt{N})$.

To look up a file f, the querying node A in the group G hashes the file to the file's belonging group G'. If G' is the same as G, the query is resolved by checking the node A's local data store and local intra-group data index. Otherwise, A forwards the query to the topologically closest contact in group G'. On receiving a query request, the contact in the group G' searches its local data store and local intra-group data index. The IP address of f's homenode is then returned to the querying node directly. In case of a file lookup failure, the querying node retries using different contacts in the group G', a random walk in the group G', or a random walk in the group G. The query is processed in $O(1)$ time with $O(1)$ message complexity.

To insert a file f, the *origin node* hashes the filename to the belonging group G. After looking up the routing table, the origin node sends an insert request to the topologically closest contact in the group G. A node in the group G is randomly chosen by this contact to be the homenode of the file f. This contact forwards the insert request to the chosen homenode. The file is then transferred from the origin node to the homenode. A new file tuple for the file f is created and added to the states of other nodes in group G. The failure of a file insertion is handled similarly to a file lookup failure. The file insertion is also done in $O(1)$ time with $O(1)$ message overhead.

All existing routing states are periodically updated using the spatially weighted gossip scheme within a group and across groups. Any timed-out entries are deleted. An update such as the heartbeat count for a file tuple starts at the responsible node. This node gossips the update for a number of fixed time intervals. During each time interval, the update message is multicast to a small constant number of gossip target nodes. The target nodes are chosen using a weighted scheme based on the round-trip time estimates. The preferences are given to those that are topologically closer in the network.

When a new node joins Kelips, it contacts a well-known introducer node (or group). The new node then uses the introducer's routing table to create its own routing table. The new node then announces its presence through gossiping. Contacts can be replaced either proactively or reactively, taking into account node distance and accessibility. Currently, a proactive approach is used to replace the farthest contact.

37.3.2.2 Coral and Related Schemes

Coral is an indexing scheme.[24] It does not dictate how to store or replicate data items. The objectives of Coral are to avoid hot spots and to find nearby data without querying distant nodes. A distributed sloppy hash table was proposed to eliminate hot spots. In DHT, a key is associated with a single value that is a data item or a pointer to a data item. In a DSHT, a key is associated with a number of values that are pointers to replicas of data items. DSHT provides the interface: *put(key, value)* and *get(key)*. The *put(key,value)* stores a value under a key; and *get(key)* returns a subset of values under a key. There is a quota on the number of values associated with a particular key stored per node. When this quota is exceeded, the additional values are distributed across multiple nodes on the lookup path.

Specifically, when a file replica is stored locally on a node A, node A hashes the filename to a key k and inserts a pointer *nodeaddr* (A's address) to that file into the DSHT by calling *put(k,nodeaddr)*. During the processing of *put(k,nodeaddr)*, node A finds the first node whose list of values under the key k is full or the first node that is closest to key k. If a node with a full-list is found, then node A goes back one hop on the lookup path. This previous node appends the pointer *nodeaddr*, together with a timestamp to the end of its list under the key k. To query for a list of values for a key k, *get(k)* is forwarded in the identifier space until the first node storing a list for the key k is found. The requesting node can then download data from the list of nodes obtained. The unique "spill-over" scheme in Coral inserts pointers along the lookup path for popular keys. The hot spots are removed because the load is balanced during pointer insertion and retrieval and data downloading.

To find nearby data without going through distant nodes, Coral organizes nodes into a hierarchy of clusters and puts nearby nodes in the same cluster. Coral consists of three levels of clusters. Each cluster is a DSHT. In the lowest level, level 2, there are many clusters that cover peers located in the same region and have the cluster diameter (round-trip time) 30 ms. In the next-higher level, level 1, there are multiple clusters that cover peers located in the same continent and have the cluster diameter 100 ms. The highest level, level 0, is a single cluster for the entire planet and the cluster diameter is infinite. Each cluster is identified by a cluster ID. Coral's hierarchy is built on top of Chord. Each cluster is a Chord ring composed of a different set of peers. The cluster at level 0 is the original Chord ring. Each node belongs to one cluster at each level and has the same node ID in all clusters to which it belongs.

A node inserts a key/value pair into Coral by performing a *put* on all of its clusters. To retrieve a key k, the querying node A first looks in its lowest-level cluster. If the query fails in this level, the node B in the same cluster whose ID is closest to the key k is reached. The node B returns its routing information in level 1 to A. The node A then continues the search on its level-1 cluster, starting with the closest level-1 node C in B's routing table. If the query fails again, A will continue the search in the global cluster, beginning with the closest level-0 node E in C's routing table. The query latency is therefore reduced by resolving a query from nearby nodes to distant nodes. The query hop count is still $O(\log^N)$, where N is the total number of nodes in the system.

A node only joins *acceptable* clusters. A cluster is acceptable to a node if its latency (round-trip time) to 90 percent of the nodes in the cluster is below the cluster diameter. If a node cannot find such clusters, it forms its own cluster. A node first joins a lowest-level cluster. Then the node inserts itself into its higher-level cluster under the hash key of the IP addresses of its gateway routers. When a node switches to a new cluster, its information is still kept in the old cluster. When old neighbors contact this node, it replies with the new cluster information. These members in the old cluster found the new cluster with more nodes and the same diameter. They will then switch to this larger cluster. The cluster split is implemented by guiding the split into two directions. Some node in the cluster c is chosen as the *cluster center*. The nodes that are close to c form one cluster. The nodes far away from c form another cluster.

The HIERAS algorithm[38] is similar to Coral. The two differ in three aspects. First, HIERAS supports DHT while Coral supports DSHT. Second, a HIERAS node joins the P2P hierarchy from the top level to the lowest level while a Coral node joins the hierarchy in the opposite way. Third, HIERAS employs distributed binning to determine nodes in each Chord ring while Coral uses ping-pong messages to get latencies for determining peers in the same cluster (a Chord ring).

Another work similar to Coral was proposed.[48] The overlay is also a hierarchy of Chord-like rings. The hierarchy emulates the nodes' real-world organization. Each Chord ring corresponds to an administrative domain. It requires that each node knows its own position in the hierarchy, and two nodes are able to compute their common ancestor in the hierarchy. The overlay hierarchy is formed in a bottom-up manner. All nodes in each leaf domain form their own overlay, a Chord ring. The overlay for a domain in the next higher-level is formed by merging the overlays for its child domains. The merging of two Chord rings is conducted as follows. Each node keeps all neighbors in its original Chord ring. In addition, each node A in one Chord ring adds another node B in the other Chord ring into its neighbor set if the following two conditions are satisfied: (1) B is the closest node that is at least 2^k away from A in the node ID space, where $0 \leq k \leq m$ (node IDs are m-bit numbers); and (2) B is closer to A than A's immediate successor in A's original Chord ring. The query routing in [48] is performed from the bottom level to the higher level in the hierarchy, which is similar to Coral.

37.3.2.3 Other Hierarchical DHT P2Ps

In Kelips and Coral, all peers play equal roles in routing. The differences among peers, such as processing power and storage capacity, are not considered. The work by Mizrak et al.[25] takes into account peer heterogeneity such as CPU power and storage capacity. The nodes with more contributed resources are called *superpeers*. Otherwise, they are called *peers*. A superpeer can be demoted to a peer. A peer can also become a superpeer. The system architecture consists of two rings: an outer ring and an inner ring. The outer ring is a Chord ring and consists of all peers and all superpeers. The inner ring consists of only superpeers. Each superpeer is responsible for an arc in the outer ring. Each superpeer sp maintains a peer table and a superpeer table. The peer table contains the node ID and address of each peer in the sp's managed arc. The superpeer table stores the node ID and managed arc range of each superpeer. The routing state is on the order of $O(\log^N)$, where N is the total number of nodes in the system.

To look up a document with the key k, the querying node first sends the query to its superpeer. If the key k is in the superpeer's managed arc, this superpeer locates and returns the successor of k to the querying node. Otherwise, the superpeer checks its superpeer table and forwards the query to another superpeer whose arc includes k. This second superpeer then looks up its peer table and returns the successor of k to the querying node. The lookup cost is $O(1)$.

To support the superpeer selection, the system uses a *volunteer service* to keep track of resources each node is willing and able to contribute to the system. Each new node registers its resources with the volunteer service. The volunteer service is provided as a black box.

A new node first joins the outer ring, just as in Chord,[16] and obtains its superpeer from its immediate neighbor. The new node then informs its superpeer to add a new entry for itself into the peer table. Unless selected as a superpeer later, this new node remains as a peer and stays in the outer ring in its lifetime. The peer failure is detected through periodic keep-alive messages between peer neighbors. The neighbor peer detecting a peer failure notifies its superpeer to remove the corresponding entry from the peer table. Superpeer failures are detected similarly. In case of superpeer failure, the load of the failed superpeer can be taken over by newly created superpeers or existing neighbor superpeers. The actual load failover scheme is determined by the arc range of the failed superpeer. All changes to the inner ring topology are distributed to all superpeers.

The work by Garces-Erice et al.[26] also considers peer heterogeneity. However, the criterion for the superpeer selection is different. The selection primarily considers nodes with longer uptime and better connection, and secondarily, CPU power and network bandwidth. The hierarchy is also somewhat different. It contains two tiers. Peers form disjoint groups in the lower tier based on the network latency. Each group has its own overlay structure, similar to Chord or CAN. A small number of peers in each group are chosen as superpeers for that group. All superpeers in the system form a separate overlay: a Chord ring in the top tier. Each "node" in the top-tier ring refers to all superpeers in a group and is represented by a vector. Given a query for the key k, the querying node first tries to look up the key in the lower tier. If the key is not found, the querying node sends the query to one of the superpeers in its group. This superpeer routes the query on the top-tier overlay toward the group that is responsible for the key k. After passing one or more superpeers, the query reaches one superpeer in the responsible group. This superpeer routes the query to the node closest to k in its own group in a similar way to the routing in a regular Chord ring.

KaZaA[27] also employs a two-tier hierarchy. It chooses nodes with the fastest Internet connections and best CPU power as *supernodes*. A supernode indexes the files in its managed groups. In the literature, it is not clear what type of structure is formed by supernodes. In Brocade,[28] all peers in the system form an overlay. Some peers in this overlay that have significant processing power, minimal number of IP hops to the wide area network, and high-bandwidth outgoing links are chosen as supernodes. Each supernode acts as a landmark node for a network domain. Each supernode keeps a list of nodes in its managed domain. All supernodes form a Tapestry overlay on top of the base overlay. During query routing, the supernode of the querying source determines whether or not the query can be resolved in the local domain. If not, the supernode will route the query on the supernode overlay to the supernode of the node responsible for the sought key.

37.3.3   Searching in Non-DHT P2Ps

The non-DHT P2Ps try to solve the problems of DHT P2Ps by avoiding hashing. Hashing does not keep data locality and is not amenable to range queries. This section introduces three kinds of non-DHT P2Ps: (1) SkipNet,[29] (2) SkipGraph,[30] and (3) TerraDir.[32] SkipNet is designed for storing data close to users. SkipGraph is intended for supporting range queries, while TerraDir is targeted for hierarchical name searches. Searching in such systems follows the specified neighboring relationships between nodes.

37.3.3.1   SkipNet and SkipGraph

DHTs balance load among different nodes. However, hashing destroys data locality. The work by Harvey et al.[29] introduces *content locality* and *path locality*. Content locality refers to the fact that a data item is stored close to its users, and the nodes in a given organization store their data items inside the same organization. Path locality means that routing between the querying node and the node responsible for the queried data is within their organization if these two nodes belong to the same organization. The overlay SkipNet supports these two data localities using a hierarchical naming structure.[29]

SkipNet is based on SkipList. A SkipList is a sorted linked list where some nodes have pointers that skip over varying numbers of list elements in the increasing sort order. In a perfect SkipList, all elements that have pointers skipping 2^h elements form the level h. The highest level of an element is called its height. In a probabilistic SkipList, node heights are determined probabilistically. A SkipList can also be considered a hierarchy of sorted linked lists that are increasingly sparse.

SkipNet is a modification of SkipList. The data in the SkipNet are peer names (name IDs). The linked list is changed to a doubly-linked ring for path locality. All SkipNet nodes have the same $2 \log^N$ number of pointers, where N denotes the number of peers in the P2P. All pointers of a peer constitute its routing table. Figure 37.4 shows the peer name order and the routing tables for the peer A. The corresponding perfect SkipNet is shown in Figure 37.5. All peers are part of the *root ring* at level 0. The root ring is divided into two disjoint rings at level 1. The pointer of each peer at level 1 traverses two peers. Each ring at level 1 is divided into two disjoint rings at level 2. The pointers at level 2 traverse four peers. This procedure continues until at level 3, each peer forms a ring containing just itself. The pointers at level 3 traverse eight (i.e., all) peers.

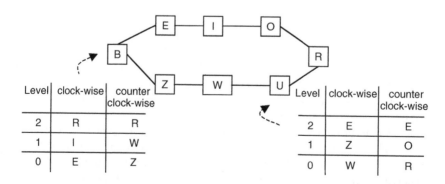

FIGURE 37.4   The peer name order and sample routing tables.

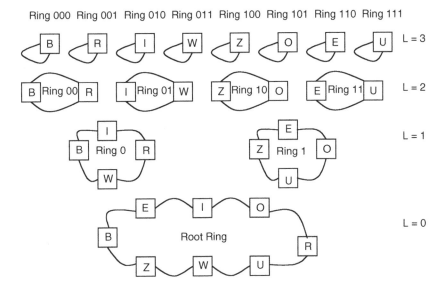

FIGURE 37.5 A sample of perfect SkipNet.

To ease efficient node insertions and deletions, the probabilistic SkipNet is used in practice. In such a probabilistic design, each ring at level i is still split into two rings at level $i + 1$. However, the peers in the two rings at level $i + 1$ are randomly and uniformly selected from the peers in the corresponding ring at level i. With such probabilistic design, a pointer at level i traverses an *expected* 2^i number of peers. The routing efficiency is $O(\log^N)$ with high probability, where N is the number of peers in the P2P. SkipNet generates a random binary bit vector for each peer. These random bit vectors are used to determine the random ring memberships of peers. A ring at level i consists of all peers whose random vectors have the same i-bit prefix. For example, the vectors for A and D are 000 and 001, respectively. Both A and T are in the same ring 0 at level 1 and the same ring 00 at level 2. The random bit vector is also used as the *numeric ID* of a peer.

A file is stored in the node whose name ID is closest to the file name. To provide content locality, the node name is used as the prefix of the file name. For example, a file *cert9i.html* stored in the node *education.oracle.com* can be named *education.oracle.com/cert9i.html*. To search for a file named *fname*, the querying node A forwards the query along its highest level pointers until the node B whose name ID is closest to but is not greater than *fname* is reached. The node B continues the search along its next lower level pointers until the node C whose name ID is closest to but is not greater than *fname* is reached. This procedure continues until the node E whose name ID is closest to *fname* among all levels is reached. If the node E stores the sought file, the query succeeds. Otherwise, the query fails. To provide path locality, the DNS name with reversed components is used as the prefix of the file name. For example, nodes in the domain *oracle.com* can be named *com.oracle.node1*, *com.oracle.node2*, etc. In summary, searching by name ID visits nodes whose name IDs share a non-decreasing prefix of the desired filename.

In SkipNet, searching can also be done by numeric ID. It is similar to searching by name ID. However, the querying node starts the search from the lowest-level (level 0). In level 0, the search stops at the node whose numeric ID matches the first bit in the desired numeric ID. This node then continues the search in its level-1 ring and stops at the node with the first two matching bits. This procedure continues until the longest prefix is found in a ring at level h. The search continues in this ring and terminates at the node that is numerically closest to the desired numeric ID.

SkipNet supports *constrained load balance*, wherein loads are balanced among peers in a constrained range such as an organization. This is implemented by dividing the file-name into two parts: a prefix and a suffix. The prefix specifies the domain where load balance should occur. The suffix is hashed uniformly to the peers in that domain.

DHTs do not support range queries very well because hashing destroys the ordering on hash keys such as filenames. The overlay SkipGraph[30] is tailored for range queries. SkipGraph is very similar to SkipNet. There are three differences. First, SkipNet is designed for providing data locality while SkipGraph is designed for supporting range queries. Second, each node in SkipNet is a computer and the node name is the computer name, while each node in a SkipGraph is a resource and the node name is the resource name. Third, SkipNet is a hierarchy of doubly-linked sorted rings while SkipGraph is a hierarchy of doubly-linked sorted lists. Searching for a specific resource in SkipGraph is similar to searching by node name in SkipNet. A range query is resolved by first locating the range boundary and then traversing the linked list in the lowest level.

37.3.3.2 TerraDir

TerraDir[32,33] is a general distributed directory service for searching data by hierarchical names such as UNIX filenames. The hierarchical namespace consists of meta-information about the data stored in the P2P system. The TerraDir directory structure is a rooted graph. Each node in this graph has one single canonical name and may have other names. All canonical names form a rooted tree. They are used to avoid cycles in wildcard queries and are also used for failure recovery. Users can query the data using any node name. Each TerraDir directory node has a single owner. The owner is a peer that permanently maintains information for a TerraDir directory node. The owner is in charge of the replication of the owned node. Only the owner can make modifications to the owned node. Many directory nodes can be owned by the same owner. An owner keeps the following information (*state*) for each owned directory node: a label, a set of incoming edges, a set of outgoing edges, a set of attributes, a record, and some bookkeeping information. The incoming and outgoing edges contain the information about the peers that own or replicate the parent nodes and children nodes, respectively. The attributes are metadata about the node and are represented using (type, value) pairs. The record is the actual data represented by this node. The bookkeeping information is used for failure recovery. Each peer also permanently maintains the metadata for all nodes replicated on it.

To reduce the routing latency, the owner of each node on the query path caches the partial query path from that node to the sought node. The querying peer (i.e., the owner of the starting node) caches the entire query path. The cache entry for a cached node includes information about that node, its parent and children, its owning peer, and a digest of the nodes permanently hosted by its owning peer. The node owner replicates an owned node in randomly selected peers. The number of replicas of an owned node is $k * h$, where k is a configurable constant and h is the level of that node in the TerraDir directory tree. Level 1 consists of all leaf nodes, while level 2 consists of all parents nodes of the leaf nodes. With this replication scheme, the average replication overhead per node is a constant. The network addresses of peers that have replicas of a node A are also part of the state that A's parent maintains for A.

Searching in TerraDir is conducted as follows. Assume that a peer A is forwarding a query toward the peer that owns, replicates, or caches the target node t. Peer A proceeds in the following order:

It generates a list L of prefixes of node names it knows. L includes the target t, names of the nodes A owns, replicates, or caches. The entire node name is also considered a prefix of that node name.

It sorts all elements in L in the increasing order of the distance (on the namespace tree) between the prefixes and the target t. This sorted list is called a candidate list.

It searches the candidate list for the first prefix whose owning peer or replicating peer B is known to A. This best prefix is closest to the target t.

It forwards the query to B.

The peer failures are handled as follows. If the peer storing the best prefix fails, the next best prefix in the candidate list is tried. If all peers storing the prefixes in the candidate list fail, the query is retried on a replica of the current node that is available and has not yet been visited. If such a replica does not exist, the query is retried on a replica of the directory root that is available and has not yet been visited. If no such replica exists, the query fails.

37.4   Searching in Loosely Structured P2Ps

In loosely structured P2Ps, the overlay structure is not strictly specified. It is either formed based on hints or formed probabilistically. In Freenet[34] and Phenix,[40] the overlay evolves into the intended structure based on hints or preferences. In Symphony[31] and the work by Zhang et al.,[35] the overlay is constructed probabilistically. Searching in loosely structured P2P systems depends on the overlay structure and how the data is stored. In Freenet, data is stored based on the hints used for the overlay construction. Therefore, searching in Freenet is also based on hints. In Phenix,[40] the overlay is constructed independent of the application. The data location is determined by applications using Phenix. Therefore, searching in Phenix is application dependent. In Symphony,[31] the data location is clearly specified but the neighboring relationship is probabilistically defined. Searching in Symphony is guided by reducing the numerical distance from the querying source to the node that stores the desired data.

37.4.1   Freenet

Freenet[34] is one loosely structured decentralized P2P designed for protecting the anonymity of data sources. It supports the DHT interface. Each node maintains a local datastore and a dynamic routing table. The routing table of a node contains addresses of some other nodes and the keys possibly stored on these nodes. Because of the storage capacity, both the datastore and the routing table are managed using the LRU algorithm.

The query routing in Freenet is similar to DFS. Given a query for a file with a key k, the querying node A first looks up its local datastore. If the file is in the local datastore, the query is resolved. Otherwise, A forwards the query to the node B in its routing table whose key is nearest to k. On receiving the query, B performs the similar computation. If the file is not stored on B, then B forwards the query to the neighbor in its routing table that has the nearest key to k. This forwarding procedure continues until the query terminates. During query routing, some node may not forward the query to the neighbor with the nearest key because that neighbor is down or a loop may be detected. In such cases, this node tries the neighbor with the second nearest key. If the node cannot forward to all its neighbors, the node reports a failure back to its upstream node. This upstream node will try its second best choice. A TTL limit is specified to restrict the number of messages in query routing. When the file is found, the file is returned to the querying node hop by hop along the reverse of the query path. Each node except the last one on the query path caches the found file and creates an entry in the routing table for the key k.

To provide anonymity, each node except the last one on the query path can change the reply message and claim itself or another node as the data source. Figure 37.6 depicts a querying routing example. Node A starts the query for the file f with the key k stored on node F. It first sends the query to node B with the nearest key. Node B then forwards to its best neighbor C. C reports the failure back to node B because C does not have any other neighbor. B then forwards the query to its second best neighbor D. D forwards the query to its best neighbor F. The file is found in F. F then returns the file to A through the path

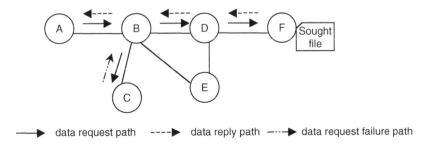

━━▶ data request path   ╌╌▶ data reply path   ━╌▶ data request failure path

FIGURE 37.6   A sample of querying processing in Freenet.

$F \rightarrow D \rightarrow B \rightarrow A$. After this query, nodes A, B, and D all have the file f in their datastores and entries for the key k in their routing tables.

To insert a file f with key k, the inserting node first issues a query for the key k to avoid duplicate keys. This search request for k is processed as a regular query. If an existing file with key k is found, the file is returned to the inserting node. All nodes except the last one on the search path cache the file and add an entry for the key k in their routing tables. The inserting node will retry the insertion with a different key. If the search fails, the last node on the search path sends a "no-collision" message back to the inserting node. The inserting node then sends an insert message to its neighbor on the path established by the initial search for duplicate keys. On receiving the insert message, the neighbor adds the new file into its own datastore and a new entry for the new file into its routing table. The neighbor then forwards the insert message to the next node on the initial search path. The insert message is normally propagated along the initial search path. If some node on the search path fails, the upstream node forwards the insert message to the neighbor with the second nearest key, etc. The insertion terminates successfully when a TTL limit is reached and fails when the message is backtracked to the original inserter. To provide data anonymity, any node on the insert path may change the insert message and claim itself or another node as the data source. Freenet does not support persistent data storage. Unpopular files with few requests are purged from the datastores by the LRU algorithm.

Freenet's query routing improves over time as more queries are answered and more files are inserted. Nodes tend to locate collections of similar keys. Once a node is listed in the routing table under a key, it will receive many requests for keys similar to that key. As the node builds up its routing table, it gets more informed about the locations of similar keys. Nodes also tend to store files with clusters of keys. Forwarding a successful query causes a node to keep a copy of the sought file with key k. File insertion normally follows the query path. Therefore files with similar keys are inserted along the same query path. The nodes on the query path accumulate files of similar keys in their datastores. Consequently, the routing tables of these nodes also contain entries with similar keys.

A new node joins the Freenet by sending messages to some existing nodes discovered through out-of-band means. The new node announces its presence using a random walk. The key associated with the new node is the *XOR* of the random seeds generated by all nodes on the random walk. A cryptographic protocol is used during the random walk for verifying the truth of each random seed.

37.4.2   Searching the Power-Law Graph Overlay

In a power-law graph, node degrees follow a power-law distribution.[14] This means that the probability of a node with the degree k, P_k, is

$$P_k = ck^{-\alpha}$$

where α is a positive integer and c is a constant. In many networks, α is close to 2. Freenet[34] and Gnutella[15] tend to evolve into a power-law graph with a power-law exponent close to 2. An efficient strategy for searching power-law overlays is proposed by Adamic et al.[14] The scheme tries to utilize high-degree nodes. First, each node forwards the query to a neighbor with a higher degree. This rule continues until the node with the highest degree is reached. After the highest-degree node is accessed, the node of approximately second highest degree will be selected. Following this procedure, the searching algorithm approximately visits nodes with degrees in the decreasing order across the entire graph.

The approach of Wouhaybi and Campbell[40] makes a power-law overlay "organically" emerge by guiding the node-join process through preferences. The new node prefers connecting to existing nodes with high degrees. Specifically, when a new node joins the system, it gets a list of live nodes using a rendezvous mechanism. Then, the new node divides this node list into two sets: *random* neighbors and *friend* neighbors. Next, the new node sends a TTL-1 ping message to each friend. Each friend returns its own neighbors in a pong message to the new node. Each friend also forwards the ping message to its own neighbors. On receiving the ping message, the friend's neighbors add the new node to their own *special lists*. The new node's friends and the friends' neighbors form a *candidate list* of the new node. The candidate list is then sorted based on the decreasing order of the number of node appearances. The top c number of nodes in this list is selected as the *preferred list* of the new node. The new node then creates connections to all nodes in its random list and preferred list. The new node also periodically contacts the neighbors in its preferred list for possible backward connections. A preferred neighbor increases its counter each time it is contacted by the new node. If the counter reaches a constant value r, the preferred neighbor decreases the counter by r and adds a backward connection to the new node. The work by Wouhaybi and Campbell[40] is intended for a generic overlay topology independent of the application. Searching on the resulting power-law overlay can be any scheme suitable for the specific application.

37.4.3 Searching the Small-World Model Overlay

A small-world graph is a graph in which each node has many local connections and a few random long-range connections. The diameter of a small-world graph is $O((\log^N)^2)$.[49] Symphony[31] employs the small-world graph model to implement the DHT. A data item with a key k is mapped to a node whose ID is numerically closest to k. Each node has two *short-distance links* to its immediate predecessor and successor in the ID ring and m *long-distance links* to other distant nodes in the ID ring. These distant nodes are chosen probabilistically. Specifically, when selecting a long-distance neighbor, a node A draws a random number r based on the probabilistic distribution function: $p_N(r) = 1/(r \ln^N)$, where r is in $[1/N, 1]$ and N is the current number of nodes in the P2P. Then node A finds another node B that is responsible for the number r. This node B is selected as one long-distance neighbor of node A. m is determined experimentally. N is estimated based on the sum of segment lengths managed by a set of distinct nodes.

Symphony has a unidirectional routing protocol and a bi-directional routing protocol. In the unidirectional routing protocol, each node on the query path forwards the query to one of its immediate or distant neighbors that is *clockwise* closest to the sought key. In the bi-directional routing protocol, the query is forwarded to one immediate or distant neighbor that has the *absolutely shortest* distance (*clockwise or counterclockwise*) from the responsible node. The *look-ahead approach* is proposed to reduce the query latency even further. In this approach, each node looks ahead at its neighbors' neighbors during query forwarding. For example, in the 1-lookahead approach, each node forwards a query to the neighbor whose neighbor is closest to the sought key.

In Freenet, the cache replacement policy LRU for datastores can destroy the key clustering in both the datastores and the routing tables when the number of files stored is huge. The work by Zhang et al.[35] solves this problem by incorporating a small-world model in the datastore cache replacement policy. Routing tables are tailored by datastores. Integrating a small-world model in the datastore cache replacement policy makes the routing tables emulate a small-world model overlay. Specifically, each new node selects a random seed from the keyspace. When a new data item with a new key k comes in and the datastore is full, the node first compares the new key with the existing key k' in its datastore that is farthest from its seed. If the new key k is closer to the seed than k', then the node removes the file with key k', stores the new file with key k, and adds a new entry for k in the routing table. (This is intended for emulating short links to close neighbors in a small-world model.) Otherwise, the node probabilistically removes k', caches k, and adds a new routing table entry for the new key k. (This is designed for emulating a small number of random long links to distant neighbors in a small-world model.) This scheme enforces clustering of keys around the random seed at each node's datastore and routing table.

37.5 Conclusion

This chapter discussed various searching techniques in peer-to-peer networks (P2Ps). First, the concept of peer-to-peer networks was introduced and different schemes were classified. Next, searching strategies in unstructured P2P systems, strictly structured P2P systems, and loosely structured P2P systems were presented. The strengths and weaknesses of these approaches were addressed.

Clearly, significant progress has been made in the P2P research field. However, there are still many issues that remain unresolved. First, good benchmarks should be developed to evaluate the actual performance of various techniques. The work by Rhea et al.[36] and Lin and Wang[37] made such an initial attempt. Second, schemes amenable to complex queries supporting relevance ranking, aggregates, or SQL are needed to satisfy the practical requirements of P2P users.[3] An initial effort[51] addressed the top-K query, the query supporting relevance ranking of query results. It uses a global index and ranks the result using the term frequency and inverse document frequency. This scheme is not amenable to large-scale systems. Third, security issues have not been addressed by most current searching techniques. Some initial work by Marti et al.[45] adds security by preferring to forward queries to friends obtained through third-party services such as instant messenger service. Fourth, P2P systems are dynamic in nature. Unfortunately, existing searching techniques cannot handle concurrent node join-leave gracefully. Fifth, good strategies are needed to form overlays that consider the underlying network proximity. Some initial effort has been made in Coral, HIERAS, and other approaches.[42–44] Both Coral and Hieras consider network proximity in the initial construction of overlays. Coral uses ping-pong messages to estimate round-trip times between nodes. HIERAS employs distributed binning to estimate the proximity between nodes. The works in Refs. 42 through 44 try to modify the existing overlay to match the underlying network. The modification is conducted by deleting inefficient overlay links and adding efficient ones. Sixth, almost all existing techniques are forwarding-based techniques. Recently, a study on non-forwarding techniques[39] was performed. More effort is required to develop good non-forwarding techniques and to compare non-forwarding techniques to various forwarding techniques.

Acknowledgments

This work was supported in part by U.S. National Science Foundation grants CCR 9900646, CCR 0329741, ANI 0073736, and EIA 0130806.

References

1. H. Balakrishnan, M.F. Kaashoek, D. Karger, R. Morris, and I. Stoica, Looking up data in P2P systems, *Commun. ACM,* Vol. 46, No. 2, 2003.
2. C. Yang and J. Wu, A dominating-set-based routing in peer-to-peer networks, *Proc. 2nd Int. Workshop on Grid and Cooperative Computing Workshop (GCC'03),* 2003.
3. N. Daswani, H. Garcia-Molina, and B. Yang, Open problems in data-sharing peer-to-peer systems, *Proc. 9th Int. Conf. on Database Theory (ICDT'03),* 2003.
4. D.S. Milojicic, V. Kalogeraki, R. Lukose, K. Nagaraja, J. Pruyne, B. Richard, S. Rolins, and Z. Xu, Peer-to-Peer Computing, HP Lab technical report, HPL-2002-57, 2002.
5. D. Barkai, Technologies for sharing and collaborating on the net, *Proc. of the 1st Int. Workshop on Peer-to-Peer Systems (IPTPS'02),* 2002.
6. B. Yang and H. Garcia-Molina, Improving search in peer-to-peer networks, *Proc. 22nd IEEE Int. Conf. on Distributed Computing (IEEE ICDCS'02),* 2002.
7. Q. Lv, P. Cao, E. Cohen, K. Li, and S. Shenker, Search and replication in unstructured peer-to-peer networks, *Proc. 16th ACM Int. Conf. on Supercomputing (ACM ICS'02),* 2002.
8. A. Crespo and H. Garcia-Molina, Routing indices for peer-to-peer systems, *Proc. 22nd Int. Conf. on Distributed Computing (IEEE ICDCS'02),* 2002.
9. S.C. Rhea and J. Kubiatowicz, Probabilistic location and routing, *Proc. 21st Annu. Joint Conf. of the IEEE Computer and Communications Societies (INFOCOM'02),* 2002.
10. V. Kalogeraki, D. Gunopulos, and D. Zeinalipour-yazti, A local search mechanism for peer-to-peer networks, *Proc. 11th ACM Conf. on Information and Knowledge Management (ACM CIKM'02),* 2002.
11. D. Tsoumakos and N. Roussopoulos, Adaptive probabilistic search in peer-to-peer networks, *Proc. 2nd Int. Workshop on Peer-to-Peer Systems (IPTPS'03),* 2003.
12. D. Tsoumakos and N. Roussopoulos, Adaptive Probabilistic Search in Peer-to-Peer Networks, Technical Report, CS-TR-4451, 2003.

13. D. Tsoumakos and N. Roussopoulos, A comparison of peer-to-peer search methods, *Proc. 2003 Int. Workshop on the Web and Databases,* 2003.

14. L.A. Adamic, R.M. Lukose, A.R. Puniyani, and B.A. Huberman, Search in power-law networks, *Physical Review,* Vol. 64, 2001.

15. The gnutella protocol specifications VO.4.CLip2 distributed search solutions, http://www.CLip2.com.

16. I. Stoica, R. Morris, D. Liben-Nowell, D.R. Karger, M.F. Kaashoek, F. Dabek, and H. Balakrishnan, Chord: a scalable peer-to-peer lookup service for Internet applications, *Proc. 2001 ACM Annu. Conf. of the Special Interest Group on Data Communication (ACM SIGCOMM'01),* 2001.

17. A. Rowstron and P. Druschel, Pastry: scalable, distributed object location and routing for large-scale peer-to-peer systems, *Proc. 18th IFIP/ACM Int. Conf. of Distributed Systems Platforms,* 2001. www.cs.rice.edu/CS/systems?Pastry.

18. K. Hildrum, J. Kubiatowicz, S. Rao, and B.Y. Zhao, Distributed object location in a dynamic network, *Proc. 14th ACM Symp. on Parallel Algorithms and Architectures (SPAA),* 2002.

19. P. Maymounkov and D. Mazieres, Kademlia: a peer-to-peer information system based on the XOR metric, *Proc. 1st Int. Workshop on Peer-to-Peer Systems (IPTPS'02),* Springer-Verlag version, 2002.

20. S. Ratnasamy, P. Francis, M. Handley, R. Karp, and S. Shenker, A scalable content-addressable network, *Proc. 2001 ACM Annu. Conf. of the Special Interest Group on Data Communication (ACM SIGCOMM'01),* 2001

21. M.F. Kaashoek, and D.R. Karger, Koorde: a simple degree-optimal distributed hash table, *Proc. 2nd Int. Workshop on Peer-to-Peer Systems (IPTPS'03),* 2003.

22. D. Xuan, S. Chellappan, and M. Krishnamoorthy, RChord: an enhanced chord system resilient to routing attacks, *Proc. 2003 Int. Conf. in Computer Networks and Mobile Computing,* 2003.

23. I. Gupta, K. Birman, P. Linga, A. Demers, and R.V. Renesse, Kelips: building an efficient and stable P2P DHT through increased memory and background overhead, *Proc. 2nd Int. Workshop on Peer-to-Peer Systems (IPTPS'03),* 2003.

24. M.J. Freedman and D. Mazieres, Sloppy hashing and self-organized clusters, *Proc. 2nd Int. Workshop on Peer-to-Peer Systems (IPTPS'03),* 2003.

25. A.T. Mizrak, Y. Cheng, V. Kumar, and S. Savage, Structured superpeers: leveraging heterogeneity to provide constant-time lookup, *Proc. IEEE Workshop on Internet Applications (WIAPP'03),* 2003.

26. L. Garces-Erice, E.W. Biersack, P.A. Felber, K.W. Ross, and G. Urvoy-Keller, Hierarchical peer-to-peer systems, *Parallel Processing Letters,* Vol. 13, 2003.

27. KaZaA, http://www.kazaa.com.

28. B.Y. Zhao, Y. Duan, L. Huang, A.D. Joseph, and J.D. Kubiatowicz, Brocade: landmark routing on overlay networks, *Proc. 1st Int. Workshop on Peer-to-Peer Systems (IPTPS'02),* 2002.

29. N.J.A. Harvey, M.B. Jones, S. Saroiu, M. Theimer, and A. Wolman, SkipNet: a scalable overlay network with practical locality properties, *Proc. 4th USENIX Symp. on Internet Technologies and Systems (USITS'03),* 2003.

30. J. Aspnes and G. Shah, Skip Graphs, technical report, Yale University, 2003.

31. G.S. Manku, M. Bawa, and P. Raghavan, Symphony: distributed hashing in a small world, *Proc. 4th USENIX Symp. on Internet Technology and Systems (USITS'03),* 2003.

32. B. Silaghi, B. Bhattacharjee, and P. Keleher, Query routing in the TerraDir Distributed Directory, *Proc. SPIE ITCom'02,* 2002.

33. B. Bhattacharjee, P. Keleher, and B. Silaghi, The design of TerraDir, technical report, CS-TR-4299, University of Maryland, College Park, 2001.

34. I. Clarke, O. Sandberg, B. Wiley, and T.W. Hong, Freenet: a distributed anonymous information storage and retrieval system, *Proc. ICSI Workshop on Design Issues in Anonymity and Unobservability,* 2000.

35. H. Zhang, A. Goel, and R. Govindan, Using the small-world model to improve Freenet performance, *Proc. 22nd Annu. Joint Conf. of the IEEE Computer and Communications Societies (INFOCOM'03),* 2003.

36. S. Rhea, T. Roscoe, and J. Kubiatowicz, Structured peer-to-peer overlays need application-driven benchmarks, *Proc. 2nd Int. Workshop on Peer-to-Peer Systems (IPTPS'03)*, 2003.
37. T. Lin and H. Wang, Search performance analysis in peer-to-peer networks, *Proc. 2nd Int. Workshop on Peer-to-Peer Systems (IPTPS'03)*, 2003.
38. Z. Xu, R. Min, and Y. Hu, HIERAS: a DHT based hierarchical P2P routing algorithm, *Proc. 32nd Int. Conf. on Parallel Processing (ICPP'03)*, 2003.
39. B. Yang, P. Vinograd, and H. Garcia-Molina, Evaluating GUESS and non-forwarding peer-to-peer search, *Proc. 24th IEEE Int. Conf. on Distributed Computing Systems (IEEE ICDCS'04)*, 2004.
40. R.H. Wouhaybi and A.T. Campbell, Phenix: supporting resilient low-diameter peer-to-peer topologies, *Proc. 23rd Annu. Joint Conf. of the IEEE Computer and Communications Societies (INFOCOM'04)*, 2004.
41. E. Cohen and S. Shenker, Replication strategies in unstructured peer-to-peer networks, *Proc. ACM Annu. Conf. of the Special Interest Group on Data Communication (ACM SIGCOMM'02)*, 2002.
42. Y. Liu, X. Liu, and L. Xiao, Location-aware topology matching in P2P systems, *Proc. 23rd Annu. Joint Conf. of the IEEE Computer and Communications Societies (INFOCOM'04)*, 2004.
43. S. Ren, L. Guo, S. Jiang, and X. Zhang, SAT-Match: a self-adaptive topology matching method to achieve low lookup latency in structured P2P overlay networks, *Proc. 18th IEEE Int. Parallel & Distributed Processing Symp. (IPDPS'04)*, 2004.
44. Y. Liu, L. Xiao, and L.M. Ni, Building a scalable bipartite P2P overlay network, *Proc. 18th IEEE Int. Parallel & Distributed Processing Symp. (IPDPS'04)*, 2004.
45. S. Marti, P. Ganesan, and H. Garcia-Molina, DHT routing using social links, *Proc. 3rd Int. Workshop on Peer-to-Peer Systems (IPTPS'04)*, 2004.
46. D. Malkhi, M. Naor, and D. Ratajczak, Viceroy: a scalable and dynamic emulation of the butterfly, *Proc. of Principles of Distributed Computing (PODC) 2002*, 2002.
47. H. Shen, C.-Z. Xu, and G. Chen, Cycloid: a constant-degree and lookup-efficient P2P overlay network, *Proc. 18th IEEE Int. Parallel & Distributed Processing Symp. (IPDPS'04)*, 2004.
48. P. Ganesan, K. Gummadi, and H. Garcia-Molina, Canon in G major: designing DHTs with hierarchical structure, *Proc. 24th IEEE Int. Conf. on Distributed Computing Systems (IEEE ICDCS'04)*, 2004.
49. J. Kleinberg, The small-world phenomenon: an algorithmic perspective, *Proc. 32nd ACM Symp. on Theory of Computing*, 2000.
50. B. Bloom, Space/time trade-offs in hash coding with allowable errors, *Commun. ACM*, Vol. 13(7), 1970.
51. F.M. Cuenca-Acuna, C. Peery, R.P. Martin, and T.D. Nguyen, PlanetP: using gossiping to build content addressable peer-to-peer information sharing communities, *Proc. 12th IEEE Int. l Symp. on High Performance Distributed Computing (HPDC'03)*, 2003.
52. I. Jawhar and J. Wu, A two-level random walk search protocol for peer-to-peer networks, *Proc. 8th World Multi-Conf. on Systemics, Cybernetics and Informatics*, 2004.
53. http://www.napster.com.

38

Semantic Search in Peer-to-Peer Systems

Yingwu Zhu

Yiming Hu

38.1   Introduction

A recent report[18] has shown that 93 percent of information produced worldwide is in digital form. The volume of data added each year is estimated at more than one terabyte (i.e., 10^{18} bytes) and is expected to grow exponentially. This trend calls for a scalable infrastructure capable of indexing and searching rich content such as HTML, music, and image files.[28]

One solution is to build a search engine such as Google. Such a solution, however, needs to maintain an enormous centralized database about all the online information. Also, for such search engines to appear to be "scalable," they need a very large and expensive infrastructure to support their operations (e.g., Google uses tens of thousands of computers). The costs of hardware and software, as well as maintenance and utilities, are very high.[2] The centralized database approach also poses a single point of failure problem. Moreover, newly created or modified information often is not indexed into the search database for weeks. Similarly, search results often contain stalled links to files that have been removed recently.

On the other hand, as P2P (peer-to-peer) systems gain more interest from both the user and research communities, building a search system on top of P2P networks is becoming an attractive and promising alternative for the following reasons:

- *High availability*. Centralized search systems are vulnerable to distributed denial-of-service attacks. However, P2P search tends to be more robust than centralized search as the demise of a single node or some nodes is unlikely to paralyze the entire search system. Furthermore, it is not easy for an attacker to bring down a significant fraction of geographically distributed P2P nodes. Recent work[9] has shown that the failure of a reasonable portion of P2P nodes will not prevent a P2P system from functioning as a whole.
- *Low cost and easy of deployment*. As discussed above, a centralized search engine requires a huge amount of investment in both hardware and software, as well as in maintenance. A P2P search system, however, is virtually free by pooling together slack resources in P2P nodes and can be deployed incrementally as new nodes join the system.
- *Data freshness*. In centralized search systems, it usually takes weeks for newly updated data to enter the data center that hosts the search database, due to the fact that it takes time for robots or crawlers to collect such information into the search database as well as the bandwidth constraints between the data center and the Internet. Therefore, there is no freshness guarantee on the index maintained in the centralized database (i.e., weeks delay). On the other hand, P2P nodes can publish their documents immediately once they appear, and the publishing traffic goes to geographically distributed nodes, thereby avoiding the bandwidth constraints imposed by centralized search systems.
- *Good scalability*. A recent study[13] has shown that no search engine indexed more than 16% of the indexable Web. The exponentially growing data added each year would be beyond the capability of any search engine. However, the self-organizing and scalable nature of P2P systems raises a hope to build a search engine with very good scalability.

The purpose of this chapter is to give an overview of P2P search techniques and present two semantic search systems built on top of P2P networks. The remainder of this chapter is structured as follows. We review the search systems built on top of unstructured P2P networks in Section 38.2. Section 38.3 provides a survey of the search systems built on top of structured P2P networks. Section 38.4 provides the necessary background. We present two representative search systems in Sections 38.5 and 38.6, respectively. Finally, we conclude with Section 38.7.

38.2   Search in Unstructured P2P Systems

In unstructured P2P systems such as Gnutella, the unstructured overlay organizes nodes into a random graph and uses flooding on the graph to retrieve relevant documents for a query. Given a query, each visited node evaluates the query locally on its own content and then forwards the query to all its neighbors.

Arbitrarily complex queries therefore can be easily supported on such systems. Although this approach is simple and robust, it has the drawback of the enormous cost of flooding the network every time a query is issued.

Improvements to Gnutella's flooding mechanism have been studied along three dimensions: *random walks*, *guided search*, and similar content group-based search.

38.2.1   Random Walks

Random walks represent the recommended search technique proposed by Lv et al.[15] It is used to address the scalability issue posed by flooding on unstructured P2P systems like Gnutella. Given a query, a random walk is essentially a blind search in that at each step, the query is forwarded to a randomly chosen node without considering any hint of how likely the next node will have answers for the query. Two techniques have been proposed to terminate random walks: TTL (time-to-live) and "checking." TTL means that, similar to flooding, each random walk terminates after a certain number of hops, while "checking" means a walker periodically checks with the query originator before walking to the next node.[15]

38.2.2   Guided Search

Guided search represents the search techniques that allow nodes to forward queries to neighbors that are more likely to have answers, rather than forward queries to randomly chosen neighbors or flood the network by forwarding queries to all neighbors.[8]

Crespo and Garcia-Molina[8] introduced the concept of *routing indices* (RIs), which give a promising "direction" toward the answers for queries. They present three RI schemes: the compound, the hop-count, and the exponentially routing indices. The basic idea behind guided search is that a distributed index mechanism maintains indices at each node. These distributed indices are small (i.e., compact summary) and they give a "direction" toward the document, rather than its actual location. Given a query, the RI allows a node to select the "best" neighbors to which to send a query. An RI is a data structure (and associated algorithms) that, given a query, returns a list of neighbors, ranked according to their *goodness* for the query. The notion of goodness generally reflects the number of relevant documents in "nearby" nodes.

38.2.3   Similar Content Group-Based Search

The basic idea of similar content group-based search[3,7,25,31] is to organize P2P nodes into similar content groups on top of unstructured P2P systems such as Gnutella. The intuition behind this search technique is that nodes within a group tend to be relevant to the same queries. As a result, this search technique will guide the queries to nodes that are more likely to have answers to the queries, thereby avoiding a significant amount of flooding.

The search in SETS[3] uses a topic-driven query routing protocol on a topic-segmented overlay built from Gnutella-like P2P systems. The topic-segmented overlay is constructed by performing node clustering* at a single designated node, and each cluster corresponds to a topic segment. Therefore, SETS partitions nodes into topic segments such that nodes with similar documents belong to the same segment. Given a query, SETS first computes R topic segments that are most relevant to the query and then routes the query to these segments for relevant documents. However, the designated node is potentially a single point of failure and performance bottleneck.

Motivated by research in data mining, Cohen et al.[7] introduced the concept of *associative overlays* into Gnutella-like P2P systems. They use *guide rules* to organize nodes satisfying some predicates into associative overlays, and each guide rule constitutes an associative overlay. A guide rule is a set of nodes that satisfies some predicate; each node can participate in a number of guide rules; and for each guide rule it participates

---

*The node is represented by a node vector that summarizes a node's documents.

in, it maintains a small list of other nodes belonging to the same guide rule. The key idea of guide rules is that nodes belonging to some guide rule contain similar data items. As a result, guided search restricts the propagation of queries to be within some specified guide rules, that is, some associative overlays, instead of flooding or blind search.

Sripanidkulchai et al.[25] propose a content location approach in which nodes are organized into an interest-based overlay on top of Gnutella by following the principle of *interest-based locality*. The principle of interest-based locality is that if a node has a piece of content in which one is interested, then it is likely that it will have other pieces of content in which one is also interested. Therefore, nodes that share similar interests create shortcuts to one another, and interest-based shortcuts form the interest-based overlay on top of Gnutella's unstructured overlay. Given a query, the interest-based overlay serves as a performance enhanced layer by forwarding the query along shortcuts. When shortcuts fail, nodes resort to the underlying Gnutella overlay.

ESS[31] is another example of efficient search on Gnutella-like P2P systems, by leveraging the state-of-the-art information retrieval (IR) algorithms. The key idea is that ESS employs a distributed, content-based, and capacity-aware topology adaptation algorithm to organize nodes into semantic groups. Thereby, nodes with similar content belong to the same semantic group. Given a query, ESS uses a capacity-aware, content-based search protocol based on semantic groups and selective one-hop node vector replication, to direct the query to the most relevant nodes responsible for the query, thereby achieving high recall while probing only a small fraction of nodes. We defer the detailed discussion of ESS to Section 38.5.

38.3   Search in Structured P2P Systems

Following the first-generation P2P systems such as Gnutella and KaZaA, structured (or DHT-based) P2P systems,[19,22,26,30] generally called second-generation P2P systems, have been proposed to provide scalable replacement for unscalable Gnutella-like P2P systems.

Such DHT-based systems are adept at *exact-match* lookups: given a key, the system can locate the corresponding document with only $O(\log N)$ hops (N is the number of nodes in system). In structured P2P systems, replication (and caching) has been exploited to improve data availability and search efficiency. Rhea and Kubiatowicz[21] proposed a probabilistic location algorithm to improve the location latency of existing DHT deterministic lookups, if the replica of a requested document exists close to query sources. Their approach is based on *attenuated bloom filters*, a lossy distributed index structure constructed on each node.

However, supporting complex queries such as *keyword search* and *semantic/content search* on top of DHTs is a nontrivial task. The following sections provide a survey of keyword search and semantic search on top of DHTs.

38.3.1   Keyword Search

In keyword search, a query contains one or more keywords (or terms) (e.g., $Q = K_1$ AND K_2 AND K_3, where Q is a query that contains three unique keywords K_1, K_2, and K_3), and the search system returns a set of documents containing all the requested keywords for the query. Basically, three indexing structures have been proposed to support keyword search in structured P2P systems: *global indexing*,[14,20] *hybrid indexing*,[27] and *optimized hybrid indexing*.[27] Figure 38.1 illustrates these three indexing structures, each of which distributes metadata for three documents (D, E, and F) containing terms from a small vocabulary (a, b, and c) to three nodes.

In global indexing (as shown in Figure 38.1(1)), the system as a whole maintains an inverted index that maps each potential term to a set of documents containing that term. Each P2P node stores the complete *inverted list* of those terms that are mapped into its responsible DHT identifier region. An inverted list $a \rightarrow D, F$ indicates that term a appears in documents D and F. To answer a query containing multiple terms (e.g., a and b), the query is routed to nodes responsible for those terms (e.g., nodes 1 and 2). Then, their inverted lists are intersected to identify documents that consist of all requested terms. Although

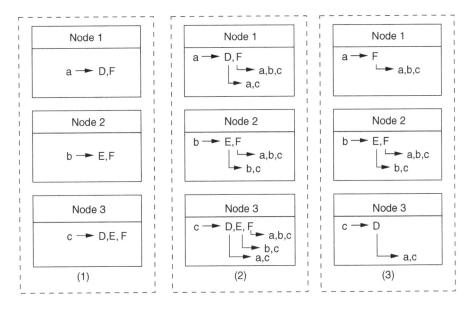

FIGURE 38.1 Three indexing structures on top of DHTs: (1) global indexing; (2) hybrid indexing; (3) optimized hybrid indexing. a, b, and c are terms; D, E, and F are documents. Forward list $D \rightarrow a, c$ indicates that document D contains terms a and c. Inverted list $a \rightarrow D, F$ means that term a appears in documents D and F.

global indexing involves only a small number of nodes for a query (i.e., proportional to the number of terms in the query), it has the drawback of requiring communication in the intersection operation for multiple term conjunctive queries. The communication cost grows proportionally with the length of the inverted lists.

In hybrid indexing (as shown in Figure 38.1(2)), each P2P node maintains the complete inverted list of those terms mapped into its responsible DHT identifier region. In addition, for each document (say, D) in the inverted list for some term t, the node also maintains the complete forward list for document D (a forward list $D \rightarrow a, c$ indicates document D contains terms a and c). Given a multiple keyword query, the query is routed to nodes responsible for those terms. Each of these nodes then performs a local search without contacting others, because they have the complete forward list for each document in their respective inverted lists. This hybrid indexing achieves search efficiency at the cost of publishing more metadata, requiring more communication and storage.

To address the associated cost in hybrid indexing, Tang and Dwarkada[27] proposed an optimized hybrid indexing scheme (as shown in Figure 38.1(3)). The basic idea behind the optimized hybrid indexing is that the metadata for a document is published under the document's top terms,* rather than all of its terms. Figure 38.1(3) illustrates such an optimization. For example, document D containing terms a and c publishes its forward list only at node 3 (responsible for term c), due to the fact that only term c is a top term in D. Given a query containing terms a and c, node 3 can still determine that document D is the answer because it stores complete forward lists for documents in its inverted lists. However, the quality of search results may be degraded because optimized hybrid indexing only publishes the metadata for a document under its top terms. Tang and Dwarkada[27] proposed to adopt automatic query expansion techniques[16] to address this problem. More details can be found in Ref. 27.

---

*Top terms are defined as those terms central to a document. In the IR algorithms such as vector space model (VSM), terms central to a document are automatically identified by a heavy weight.

38.3.2　Semantic Search

Semantic search is a content-based, full-text search, where queries are expressed in natural language instead of simple keyword match. When a query is issued by a user, a query representative is first derived from its full text, abstract, or title, and then presented to the information retrieval system. For example, when a user issues a query such as "find files similar to file F," a query representative is derived from its full text. Then the query representative is presented to the information retrieval system for those files that are similar to F. Semantic search presents a challenging problem for structured P2P systems: given a query, the system either has to search a large number of nodes or miss some relevant documents. Some semantic search systems[28,32] have been proposed on top of structured P2P systems. One important feature of such search systems is to extend the state-of-the-art information retrieval (IR) algorithms, such as Vector Space Model (VSM) and Latent Semantic Indexing (LSI),[4] to the P2P environment.

pSearch[28] introduces the concept of *semantic overlay* on top of a DHT (i.e., CAN) to implement semantic search. The semantic overlay is a logical network in which documents are organized under their semantic vectors* such that the distance (e.g., routing hops) between two documents is proportional to their dissimilarity in semantic vectors.

Two basic operations are involved in pSearch: *indexing* and *searching*. Whenever a document D enters the system, pSearch performs the indexing operations as follows:

1. Use LSI to derive D's semantic vector V_d.
2. Use rolling-index to generate a number p of DHT keys ($k_i, i = 0, \ldots, p - 1$) from V_d.
3. Index D into the underlying DHT using these DHT keys.

Whenever a query Q is issued, pSearch performs the search operations as follows:

1. Use LSI to derive Q's semantic vector V_q.
2. Use rolling-index to generate a number p of DHT keys ($k_i, i = 0, \ldots, p - 1$) from V_q.
3. Route Q to the destination nodes that are responsible for these DHT keys.
4. Upon reaching the destination, Q is either flooded to nodes within a radius r or forwarded to nodes using content-directed search.
5. All nodes that receive the query do a local search using LSI and return the matched documents to the query originator node.

More details of pSearch can be found in Ref. 28, and we leave the discussion of the work by Zhu et al.[32] to Section 38.6.

38.4　VSM and Locality-Sensitive Hashing

38.4.1　VSM

We provide an overview of VSM.[4] In VSM, each document or query is represented by a vector of terms. The terms are stemmed words that occur within the document. In addition, stop words** and highly frequent words*** are removed from the term vector. Each term in the vector is assigned a weight. Terms with a relatively heavy weight are generally deemed central to a document. To evaluate whether a document is relevant to a query, the model measures the relevance between the query vector and the document vector.

---

*A semantic vector is a vector of terms. VSM or LSI represents documents and queries as semantic vectors.

**Stop words are those words that are considered non-informative, like function words (*of, the*, etc.), and are often ignored.

***Words appear in a document very frequently, but are not useful to distinguish the document from other documents. For example, if a term t appears in a document very frequently, it should not be included in its term vector. Generally, stop words and high frequency words are removed by using a *stop list* of words.

Typically, VSM computes the relevance between a document D and a query Q as (suppose the term vectors of D and Q have been already normalized):

$$REL(D, Q) = \sum_{t \in D,Q} d_t \cdot q_t \qquad (38.1)$$

where t is a term appearing in both D and Q, q_t is term t's weight in query Q, and d_t is term t's weight in document D. Documents with a high relevance score are identified as search results for a query.

A number of term weighting schemes have been proposed for VSM, among which *tf-idf* is a scheme in which the weight of a term is assigned a high numeric value if the term is frequent in the document but infrequent in other documents. The main drawback of *tf-idf* is that it requires global information (i.e., *df*, the document frequency, which represents the number of documents where a term occurs) to compute a term's weight which is not an easy task in the P2P environment. To avoid such a global information requirement, a "dampened" *tf* scheme is proposed, wherein each term is assigned a weight in the form of $1 + \log d_t$ (d_t is the term frequency in a document). Previous work[24] has shown that this scheme not only does not require global information, but also produces higher-quality document clusters.

38.4.2 Locality-Sensitive Hashing (LSH)

A family of hash functions $\mathcal{F}$ is said to be a locality sensitive hash function family corresponding to similarity function $sim(A, B)$ if for all $h \in \mathcal{F}$ operating on two sets A and B, we have:

$$\mathbf{Pr}_{h \in \mathcal{F}}[h(A) = h(B)] = sim(A, B)$$

where **Pr** is the probability, and $sim(A, B) \in [0, 1]$ is some similarity function.[6,11]

38.4.2.1 Min-Wise Independent Permutations

Min-wise independent permutations[5] provide an elegant construction of such a locality sensitive hash function family with the Jaccard set similarity measure $sim(A, B) = \frac{|A \cap B|}{|A \cup B|}$, where sets A and B each represents a set of integers.

Let π represent a random permutation on the integer's universe U, $A = \{a_1, a_2, \ldots, a_n\} \subseteq U$, and $B = \{b_1, b_2, \ldots, b_n\} \subseteq U$. The hash function h_π is defined as $h_\pi(A) = min\{\pi(A)\} = min\{\pi(a_1), \pi(a_2), \ldots, \pi(a_n)\}$; that is, the hash function $h_\pi(A)$ applies the permutation π on each integer component in A and then takes the minimum of the resulting elements. Then, for two sets A and B, we have $x = h_\pi(A) = h_\pi(B)$ if and only if $\pi^{-1}(x) \in A \cap B$. That is, the minimum element after permuting A and B matches only when the inverse of the element lies in both A and B. In this case, we also have $x = h_\pi(A \cup B)$. Because π is a random permutation, each integer component in $A \cup B$ is equally likely to become the minimum element of $\pi(A \cup B)$. Hence we conclude that $min\{\pi(A)\} = min\{\pi(B)\}$ (or $h_\pi(A) = h_\pi(B)$) with probability $p = sim(A, B) = \frac{|A \cap B|}{|A \cup B|}$. We refer readers to Refs. 5 and 10 for further details.

38.5 Case Study on Unstructured P2P Systems

We present ESS,[31] an architecture for efficient semantic search on unstructured P2P systems, leveraging the state-of-the-art IR algorithms such as the VSM and relevance ranking algorithms.

38.5.1 Overview

The design goal of ESS is to improve the quality of search (e.g., high recall*) while minimizing the associated cost (e.g., the number of nodes visited for a query). The design philosophy of ESS is that we improve search efficiency and effectiveness while retaining the simple, robust, and fully decentralized nature of Gnutella.

---

*Recall is defined as the number of retrieved relevant documents, divided by the number of relevant documents.

In ESS, each node has a *node vector*, a compact summary of its content (as shown in Section 38.5.2). And each node can have two types of links (connections), namely *random links* and *semantic links*. Random links connect irrelevant nodes, while semantic links organize relevant nodes* into semantic groups. The topology adaptation algorithm (as discussed in Section 38.5.3) is first performed to connect a node to the rest of the network through either random links, or semantic links, or both.** The goal of the topology adaptation is to ensure that (1) relevant nodes are organized into the same semantic groups through semantic links, and (2) high capacity nodes have high degree and low capacity nodes are within short reach of higher capacity nodes.

Given a query, ESS's search protocol (as discussed in Section 38.5.5) first quickly locates a relevant semantic group for the query, relying on selective one-hop node vector replication (as shown in Section 38.5.4) as well as its capacity-aware mechanism. Then ESS floods the query within the semantic group to retrieve relevant documents. ESS will continue this search process until sufficient responses are found. The intuition behind the flooding within a semantic group is that semantically associated nodes tend to be relevant to the same query.

The main contributions of ESS include the following:

- We propose a distributed, dynamic, and capacity-aware topology adaptation algorithm to organize nodes into semantic groups for efficient search.
- We propose a capacity-aware, content-based search protocol based on semantic groups and selective one-hop node vector replication, to direct queries toward the most relevant nodes that are responsible for the queries, thereby achieving high recall at very low cost.
- Our findings suggest that an appropriate size of node vectors is a very good design choice in both search efficiency and effectiveness, and justify that a good node vector size plays a very important role in system design.
- We introduce automatic query expansion into our system to further improve the quality of search results in both recall and precision. To the best of our knowledge, this is the first work to employ the automatic query expansion technique on Gnutella-like P2P systems.
- We show that ESS's capacity-aware abilities can exploit heterogeneity to make search even more efficient.

38.5.2   Node Vectors

A node vector is a representation of the summary of a node's content. It is derived from a node's locally stored documents as follows. First, each document is represented as a term vector using VSM. The terms in a term vector are the stemmed words that occur within the document. Stop words and highly frequent words are also removed from the term vector. Each term t in the term vector is assigned a weight d_t, the term frequency. Then, all term vectors of a node's documents are summed up and we get a new vector, in which each term component t has a weight d'_t. For each term vector t, we replace its weight d'_t with $1 + \log d'_t$. Finally, we normalize the new vector, and the normalized vector is called the node vector.

As described above, the node vector characterizes a node's content. They are used to determine the relevance of two nodes (say, X and Y) according to the Equation (38.2).

$$REL(X, Y) = \sum_{t \in X, Y} w_{X,t} \cdot w_{Y,t} \qquad (38.2)$$

where t is a term appearing in both X and Y; $w_{X,t}$ is term t's weight in X; and $w_{Y,t}$ is term t's weight in Y. If the relevance score is less than a certain relevance threshold, nodes X and Y are deemed irrelevant; otherwise, they are deemed relevant.

---

*Nodes with similar contents are considered relevant.

**If a node cannot find a relevant node, it connects itself to the rest of the network *only* through random links.

Node vectors are also used to determine the relevance of the node X and a query vector (say, Q) according to Equation (38.3), as shown later in biased walks during search:

$$REL(X, Q) = \sum_{t \in X, Q} w_{X,t} \cdot w_{Q,t} \qquad (38.3)$$

Note that in the IR community, a number of term weighting schemes have been proposed. The motivation for why ESS uses the "dampened" *tf* scheme instead of $1 + \log tf$ in its design is as follows. First and most importantly, unlike other schemes such as *if-idf*, this term weighting scheme does not require global information such as *df*. A term weighting scheme requiring global information contradicts its design philosophy that we retain the simple, robust, and fully decentralized nature of Gnutella. Second, it has been shown by Schutze and Silverstein[24] that such a term weighting scheme works very well and can produce higher quality clusters.

38.5.3 Topology Adaptation Algorithm

The topology adaptation algorithm is a core component that connects a node to the rest of the network and, more importantly, connects the node to a semantic group (if it can find one) through semantic links.

When a node joins the system, it first uses a bootstrapping mechanism in Gnutella to connect to the rest of the network. However, it may not have any information about other nodes' content (i.e., node vectors) as well as semantic groups. Its attempt to gain such information is achieved through *random walks*. A random walk is a well-known technique in which a query message is forwarded to a randomly chosen neighbor at each step until sufficient responses are found. In ESS, the duration of a random walk is also bound by a TTL (time-to-live).

A random walk query message contains a node's node vector, a relevance threshold *REL_THRESHOLD*, the maximum number of responses *MAX_RESPONSES*, and TTL. The random walk returns a set of nodes. In ESS's implementation, a node actually periodically issues two queries: one requesting nodes whose relevance is lower than *REL_THRESHOLD*, and the other requesting nodes whose relevance is higher than or equal to *REL_THRESHOLD*. Note that the relevance score is computed using the Equation (38.2).

The returned nodes are added to the query initiator node's two *host caches* — *random host cache* and *semantic host cache*, respectively — according to their relevance scores. Each entry of host caches consists of *a node's IP address, port number, node capacity, node degree, node vector*,* and *relevance score*. These two caches are maintained throughout the lifetime of the node. Each cache has a size constraint and uses FIFO as replacement strategy.

The goal of topology adaptation is to ensure that, on the one hand, a node is connected to the most relevant nodes through semantic links (thereby forming semantic groups); and on the other hand, its capacity-aware mechanism makes a node connected to higher capacity nodes** through random links. To achieve this goal, each node periodically checks its two caches for random and semantic neighbor addition or replacement.

To add a new *semantic neighbor* (a neighbor node connected by a semantic link), a node (say, X) chooses a node from its semantic cache that is not dead and not already a neighbor and with the *highest* relevance score. Node X then uses a three-way handshake protocol to connect to the chosen neighbor candidate (say, Y). During the handshake, each node decides independently whether or not to accept the other node as a new semantic neighbor based upon its own *MAX_SEM_LINKS* (the maximum number of semantic neighbors), *SEM_LINKS* (the number of current semantic neighbors), and the new node (i.e., the relevance score). If *SEM_LINKS* is less than *MAX_SEM_LINKS*, the node automatically accepts this new connection. Otherwise, the node must check if it can find an appropriate semantic neighbor to drop and replace with the new neighbor candidate. X always accepts Y and drops an existing semantic neighbor

---

*The semantic host cache does not contain node vectors.
**We assume a node's capacity is a quantity that represents its CPU speed, bandwidth, disk space, etc.

if Y's relevance score is higher than all of X's current semantic neighbors. Otherwise, it makes a decision whether or not to accept Y, as follows. From all of X's semantic neighbors whose relevance scores are lower than that of Y and which are not poorly connected,* X chooses the neighbor Z that has the lowest relevance score. Then it drops Z and adds Y to its semantic neighbors.

To add a new *random neighbor* (a neighbor node connected by a random link), node X chooses a node from its random cache that is not dead and not already a neighbor, and has a capacity greater than its own capacity (e.g., the highest capacity node is preferred). If no such candidate node exists, X randomly chooses a node. X then initiates a three-way handshake to the chosen random neighbor candidate (say, Y). During the handshake, each node independently decides whether or not to accept the other node as a new random neighbor upon the capacities and degrees of its existing random neighbors and the new node. If X's *RND_LINKS* (the number of random neighbors) is less than *MAX_RND_LINKS* (the maximum number of random neighbors, and it dynamically changes as the *SEM_LINKS* changes), the node automatically accepts this new node as a random neighbor. Otherwise, the node must check if it can find an appropriate random neighbor to drop and replace with the new node. X always drops an existing random neighbor in favor of Y if Y has capacity higher than *all* of X's existing random neighbors. Otherwise, it decides whether or not to accept Y as follows. From all of X's random neighbors that have capacity less than or equal to that of Y, X chooses the neighbor Z that has the highest degree. Z will be dropped and replaced with Y *only* if Y has lower degree than that of Z. This ensures that ESS does not drop already poorly connected neighbors and avoids isolating them from the rest of the network.

38.5.3.1 Discussion

The goal of the topology adaptation is to ensure that (1) relevant nodes are organized into semantic groups through semantic links, by periodically issuing random walk queries for semantic neighbors and performing semantic neighbor addition, and (2) high capacity nodes have high degree and low capacity nodes are within short reach of higher capacity nodes, by periodically issuing random walk queries for random neighbors and performing random neighbor addition. Note that in ESS, a *direct* semantic link connects two *most* relevant nodes; while in other systems like SETS, a *local* link does not necessarily mean that two connected nodes are most relevant within a topic segment.

For random walk queries requesting relevant nodes, ESS can actually do an optimization as follows. When a query arrives at a node (say, Y) that is deemed to be relevant to the initiator node (say, X), Y can first choose other relevant nodes from its semantic host cache for responses with some probability. If the *MAX_RESPONSES* has been reached, the query reply is routed back to X. Otherwise, Y biased walks the query through one of its semantic links with some probability. Further, relevant nodes within a semantic group can exchange the content of their host caches. Currently, ESS does not adopt such optimizations. Each node also continuously keeps track of the relevance scores of both semantic links and random links. If the relevance score of a semantic link drops below the *SEM_THRESHEOLD* due to dynamically changing documents in either node (and thus changing node vectors), ESS simply drops the semantic link and adds the neighbor information into the random host cache. Similarly, if the relevance score of a random link rises above the *SEM_THRESHOLD*, ESS simply drops the random link and adds the neighbor information into the semantic host cache. As a result, the topology adaptation process performed thereafter can adapt to the dynamically changing node vectors of each node's existing neighbors.

38.5.4 Selective One-Hop Node Vector Replication

To allow ESS to quickly locate the relevant semantic group for a query during biased walks (as shown in Section 38.5.5), each node maintains the node vectors of *all* of its random neighbors in memory. Note that ESS does not maintain those of its semantic neighbors. This is why we call it *selective* replication.

---

*Each node has a minimum degree constraint, and a typical value is 3. If a node's degree is less than or equal to the minimum constraint value, this node is identified as a poorly connected node.

When a random connection is lost, either because the random neighbor node leaves the system or due to topology adaptation, the node vector for this neighbor gets flushed from memory. A node periodically checks the replicated node vectors with each random neighbor in case a neighbor node might add or remove documents. This allows nodes to adapt to dynamically changing node vectors (due to dynamically changing of documents) and keep replicated node vectors up-to-date and consistent.

38.5.5   Search Protocol

The combination of the topology adaptation algorithm (whereby relevant nodes are organized into semantic groups and high capacity nodes have more neighbors [i.e., high degree]) and selective one-hop node vector replication (whereby nodes maintain the node vectors of their random neighbors) have paved the way for ESS's content-based, capacity-aware search protocol.

Given a query, the search protocol is performed as follows. First, ESS uses *biased* walks, rather than random walks, to forward the query through *random links*. During biased walks, each node along the route looks up its locally stored documents for those satisfying the query; each document is evaluated using Equation (38.1) and a relevance score is computed. If the relevance score is higher than or equal to a certain relevance threshold, this document is identified as a relevant document for this query. If any such a relevant document is identified, then the node (say, X) is called the *semantic group target node*, where the query terminates biased walks and starts flooding. Otherwise, X selects a neighbor (say, Y) from its *random neighbors* whose node vector is *most* relevant to the query vector according to the Equation (38.3), and forwards the query to Y. The biased walks are repeated until a semantic group target node is identified.

The target node then floods the query along its *semantic links*: each semantic neighbor evaluates the query against its documents and then floods the query along its own semantic links. During flooding, we can allow the query to probe all the nodes within a semantic group or only a fraction of nodes by imposing a flooding radius constraint from the target node (called *controlled flooding*). The relevant documents found within the semantic group are directly reported to the target node. Note that each query contains a *MAX_RESPONSES* parameter. The target node aggregates the relevant documents, reports them directly to the query initiator node (which will present highest relevance ranking documents to the user), and decreases *MAX_RESPONSES* by the number of relevant documents. If *MAX_RESPONSES* becomes less than or equal to zero, the query is simply discarded; otherwise, the query starts biased walks from the target node again and repeats the above search process until sufficient responses are found.

During both biased walks and flooding, ESS uses bookkeeping techniques to sidestep redundant paths. In ESS, each query is assigned a unique identifier *GUID* by its initiator node. Each node keeps track of the neighbors to which it has already forwarded the query with the same *GUID*. During biased walks, if a query with the same *GUID* arrives back at a node (say, X), it is forwarded to a different random neighbor with the highest relevance score among those random neighbors to which X has not yet forwarded the query. This reduces the probability that a query traverses the same link twice. However, to ensure forward progress, if X has already sent the query to all its random neighbors, it flushes the bookkeeping state and starts reusing its random neighbors. On the other hand, during flooding, if a query with the same *GUID* arrives back at the node, the query is simply discarded. Note that ESS nodes treat query messages with the same *GUID*s differently during biased walks and flooding.

The search protocol discussed thus far does not consider node capacity heterogeneity. Now we incorporate the capacity-aware mechanism into the search protocol to make search more efficient in the system where node capacities are heterogeneous. Due to the fact that the topology adaptation algorithm takes into account the heterogeneity of node capacities *only* in random link construction, the search protocol only needs to adapt the biased walk while retaining the flooding part untouched. During biased walks, each node makes a decision how to forward a query based on *the query vector, its own capacity, the capacities of all of its random neighbors,* and *the node vectors of all of its random neighbors.* If the node (say, X) is a supernode (whose capacity is higher than a certain threshold), it forwards the query to a random neighbor whose node vector is most relevant to the query vector. Otherwise, X must check all its random neighbors and chooses the neighbor (say, Y) with the highest capacity. If Y is a supernode, X forwards the query to Y, hoping that high capacity nodes can typically provide useful information for the query. Otherwise,

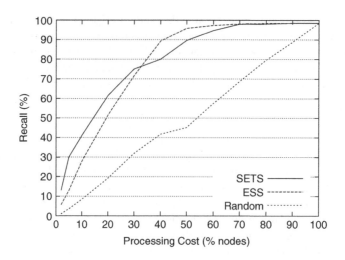

FIGURE 38.2   Recall versus processing cost for ESS, SETS, and Random.

X forwards the query to a random neighbor whose node vector is most relevant to the query vector. The biased walks are repeated until a target node if reached.

38.5.5.1 Discussion

In summary, the query flooding within a semantic group is based on the intuition that semantically associated nodes tend to be relevant to the same queries and can provide useful responses for them. The biased walk, taking advantage of the heterogeneity of node capacities and selective one-hop node vector replication, forwards a query either to a supernode neighbor with the hope that high capacity nodes can typically provide useful information for the query, or to the *most* relevant random neighbor if no such a supernode exists. Note that *biased walks direct a query along one of a node's random links, while flooding forwards the query along all of the node's semantic links.* In ESS, in addition to *MAX_RESPONSES*, each query is also bound by the *TTL* parameter. Note that flooding within semantic groups keeps us from exactly keeping track of the *TTL*. For simplification, in ESS's implementation, the *TTL* is decreased by one at each step *only* during biased walks. Once *TTL* hits zero, the query message is dropped and no longer forwarded.

38.5.6 Experimental Results

We present part of experimental results here and refer readers to Zhu and Hu[31] for more results.

The data used in the experiments is *TREC-1,2-AP*.[1] The TREC corpus is a standard benchmark widely used in the IR community. *TREC-1,2-AP* contains AP Newswire documents in TREC CDs 1 and 2. The queries used in the experiments are from TREC-3 ad hoc topics (151–200). The query vector was derived from the *title* field using VSM. These 50 queries each come with a query-relevant judgment file that contains a set of manually identified relevant documents.

Figure 38.2 plots evaluation results comparing the performance of ESS against SETS and Random.* The vertical axis is the recall, and the horizontal axis is the query processing cost in the fraction of nodes involved in a query.

Several observations can be drawn from this figure:

1. ESS and SETS outperform Random substantially, achieving higher recall at smaller query processing cost.
2. Compared to ESS, SETS achieves higher recall when exploring less than 30 percent of the nodes. This is explained by the fact that SETS takes advantage of knowing the global *C* (=256) topic

---

*Random represents the random walk technique.

TABLE 38.1 Recall Improvements with Respect to Query Processing Cost Made by ESS(1000+heter) on SETS(full)

	Processing Cost (% nodes)						
	2	*5*	*10*	*20*	*30*	*40*	*≥ 50*
ESS(1000+heter) : SETS(full)	63.8	8.3	16.1	17.9	13.3	18.5	≤ 7.4

Note: ESS(1000+heter) represents ESS, which uses an appropriate node vector size of 1000 and considers heterogeneity; "full" represents the full node vector size.

segments and therefore can quickly and precisely locate the most relevant topic segments to look up relevant documents. ESS instead has to use biased walks to locate a target node and then floods the query within the corresponding semantic group for relevant documents. If the target node is not a right one (which actually does not contain relevant documents, although some of its documents have a relevance score high enough to be deemed relevant), some irrelevant nodes are unavoidably probed. The overhead of locating a right target node hurts the performance of ESS, especially when probing *only* a small fraction of nodes. However, ESS still achieves about 71.6 percent recall by probing *only* 30 percent of the nodes.

3. ESS outperforms SETS when exploring more than 30 percent of the network. It achieves 89.3 percent recall by visiting only 40 percent of the nodes, while SETS achieves 80 percent recall in this case. We give the following explanations. First, the overhead of locating a right target node (and thus the semantic group) is amortized by exploring more nodes. Second, the nature of ESS's topology adaptation connects the *most* relevant nodes through *direct* semantic links, and it ensures that a query probes the *most* relevant nodes first along semantic links. However, SETS does not distinguish the relevance between nodes within a topic segment, and local links do not necessarily reflect that the *most* relevant nodes have *direct* connections. Therefore, some irrelevant nodes within topic segments are unavoidably visited when flooding the query within topic segments.

4. When exploring the whole network, the recall achieved by all three systems is 98.5 percent. This is because queries are short on average, with only 3.5 terms in the experiments. Some relevant documents could not be identified because their relevance scores computed using Equation (38.1) are 0. During query evaluation, they are mistakenly deemed irrelevant due to such a low relevance score. In other words, with such short queries, the maximum recall achieved by a centralized IR system is 98.5 percent.

Table 38.1 summarizes the recall improvements made by ESS(1000+heter) on SETS(full), which does not consider capacity heterogeneity in its design. The node capacities are based on a Gnutella-like profile that was derived from the measured bandwidth distribution for Gnutella.[23] Table 38.1 shows that, with an appropriate node vector size and capacity-aware mechanism, ESS outperforms SETS.

38.6 Case Study on Structured P2P Systems

We present an efficient, LSH-based semantic search system[32] built on top of DHTs. Leveraging the state-of-the-art IR algorithms such as VSM, this system aims to provide efficient semantic indexing and retrieval capabilities for structured P2P systems.

38.6.1 Overview

Figure 38.3 illustrates the system architecture. To support semantics-based access, we must add two major components into an existing P2P system: (1) *a registry of semantic extractors* and (2) *semantic indexing and locating utility.*

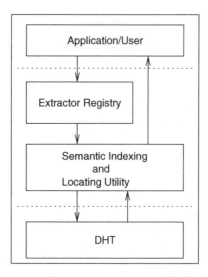

FIGURE 38.3   Major components of the system architecture.

The functionality of semantic indexing is to index each object automatically according to its semantic vector (SV) whenever an object is created or modified. The functionality of semantic locating is to find similar documents for a given query.

The index table is fully distributed. When an object is created or modified, its SV is extracted. The system then hashes the SV to an integer number called *semID*. The DHT uses this *semID* as a key to put an index entry (a pointer to the original object) into the P2P system. Note that the original locations of documents are not affected. Given a query, the system generates a *semID* based on the query's semantic vector. The semantic locating utility then uses the *semID* to locate the indices of similar documents stored in the P2P systems.

The key here is to make sure that two semantically close documents (which have similar semantic vectors) will be hashed to the same *semID* so that the underlying DHT can locate the indices. However, this is not possible in many traditional hashing functions that try to be uniformly random. As a result, two documents that are similar but slightly different (e.g., different versions of the same document) will generate different hashing results. Our system, on the contrary, relies on a very special class of hashing functions called Locality-Sensitive Hashing. If two documents are similar, then it is likely that they will generate the same hashing result. Moreover, the higher the similarity between the two files, the higher the probability that the hashing results are the same.

However, LSH cannot guarantee that two similar documents will always have the same hashing result. To increase the probability, we use a group of n LSH functions to generate n semIDs (n is a small number, about 5 to 20). If the probability of generating a matching result from a single LSH function is p, then the probability of generating at least one matching result from n LSH functions will be $1 - (1 - p)^n$. The locating utility then uses the resulting n semIDs to search the DHT. Our initial results indicate that with n set to about 10 to 20, our system can find almost 100 percent of semantically close documents. As a result, our system is very efficient: instead of sending the query to tens of thousands of nodes in the system, we only need to send it to n nodes.

38.6.1.1 Semantic Extractor Registry

The semantic extractor registry consists of a set of semantic extractors for each known file type. A semantic extractor is an external plug-in module. It is a file-type specific filter that takes as input the content of a document and outputs the corresponding semantic vector (SV) of that document.

A SV is a vector of file-type specific features extracted from the file content. For example, the VSM extracts the term frequency information from text documents, and Welsh et al.[29] have derived frequency, amplitude, and tempo feature vectors from music data.

Leveraging the state-of-the-art IR algorithms such as VSM and LSI, the fundamental functionality of the semantic extractor registry is to represent each *document* and *query* as a semantic vector where each dimension is associated with a distinct term (or keyword).

Semantically close documents or queries are considered to have similar SVs. The similarity between documents or queries can be measured as the cosine of the angle[4] or the Jaccard set similarity measure between their vector representations.

Whenever a user or application on a node X wants to store a document D in the system, the semantic extractor registry on X is responsible for deriving an SV for D. The resulting SV is then used to produce a small number of *semIDs* (semantic identifiers) as the DHT keys for D. As a result, the document D can be indexed into the DHT according to its semantic presentation, that is, with the resulting *semIDs* as the DHT keys. Note that a document in our system has two kinds of DHT keys: (1) *docID*, produced by SHA-1 hash of its content or name (which is commonly used in current P2P systems for file storage and retrieval); and (2) *semID*, derived from its semantic vector (which is used for semantic indexing, as discussed later).

38.6.1.2 Semantic Indexing and Locating Utility

The semantic indexing and locating utility provides semantics-based indexing and retrieval capabilities. The functionality of semantic indexing is to index each document or query automatically according to its semantic vector whenever a document or query is created or modified. The functionality of semantic locating is to locate semantically close documents for a given query.

Given a document, the semantic indexing and locating utility hashes its SV into a small number of *semIDs* using locality sensitive hashing functions (LSH). By having these *semIDs* as the DHT *keys* and *docID* and SV as the *objects* in the DHT's interface of *put(key, object)*, it indexes the document into the underlying DHT in the form of tuple of $< semID, docID, SV >$.

When locating semantically close files for a given query, the semantic indexing and locating utility first hashes the query's SV into a set of *semIDs*. It then interacts with the DHT to retrieve the indices of those files that satisfy the query, by having these *semIDs* as the DHT *keys* in the DHT's interface of *get(key)*. The result of a successful query will return a list of *docIDs* that satisfy the query.

The semantic indexing and locating utility also generates materialized views of query results, and allows users to reuse these materialized views as regular objects to save the expensive processing cost of popular queries. We refer readers to Zhu et al.[32] for more details.

38.6.2 LSH-Based Semantic Indexing

The objective of semantic indexing is to cluster the indices of semantically close files or queries to the same peer nodes with high probability. Without loss of generality, our focus here is on documents. Queries can also apply the same indexing procedure.

Given a document's semantic vector A, semantic indexing hashes A into a small number of *semIDs* using LSH. This process can be described as follows:

1. For each vector component t of A, convert it into a 64-bit integer* by taking the first 64 bits of t's SHA-1 hash. Therefore, A is converted into A', which is a set of 64-bit integers.
2. Using a group of m min-wise independent permutation hash functions, we derive a 64-bit *semID* from A'. Therefore, applying n such groups of hash functions on A' can yield n *semIDs* (as shown in Algorithm 38.1).

---

*Because we evaluate our system on the Pastry simulator of 64-bit identifier space, we here convert t into a 64-bit integer.

Algorithm 38.1   Semantic indexing procedure using n groups of m LSH functions.

Require: $g[1..n]$, each of which has m hash functions $h[1..m]$
1: Convert A into A', which is a set of integers
2: **for** $j = 1$ to n **do**
3:     $semID[j] = 0$
4:     **for** each $h[i] \in g[j]$ **do**
5:         $semID[j] \wedge = h[i](A')$ /* $\wedge$ is a XOR operation */
6:     **end for**
7: **end for**
8: **for** each $semID[j]$ **do**
9:     insert the index $<semID[j], docID, A>$ into the DHT by having $semID[j]$ as the DHT key
10: **end for**

Note that for two SVs A and B, their similarity is $p = sim(A, B) = sim(A', B')$. This is because the SHA-1 hash function is supposed to be collision resistant and the above process would not change the similarity p.

Let A denote the SV of a document with a *docID*, as shown in Algorithm 38.1. A *semID* is produced by XORing m 64-bit integers that are produced by applying a group of m hash functions on A'. Thus, applying n such groups of hash functions on A' yields n *semIDs*. By having the resulting *semIDs* as the DHT keys, the file is indexed into the DHT in the form of $< semID, docID, A >$. We expect that such semantic indexing could have the indices of semantically close files hashed to the same peer nodes with *high* probability.

Probability Analyses. What is the probability achieved by the procedure as shown in Algorithm 38.1? We here offer probability analyses. Consider a group of $g = \{h_1, h_2, \ldots, h_m\}$ of m hash functions chosen uniformly at random from a family of locality sensitive hash functions. Then the probability that two SVs A and B are hashed to the same 64-bit integer for all m hash functions is $\mathbf{Pr}[g(A) = g(B)] = p^m$ (where $p = sim(A, B) = sim(A', B')$). Now, for n such groups $g_1, g_2, \ldots, g_n$ of hash functions, the probability that A and B cannot produce the same integer for all m hash functions $\in g_i$ is $1 - p^m$. And the probability that this happens for all n groups is $(1 - p^m)^n$. So, the probability that A and B can produce the same integer for all m hash functions of *at least* one of n groups is $1 - (1 - p^m)^n$. That is, the probability that A and B can produce the same *semID* for at least one of n groups is $1 - (1 - p^m)^n$.

For example, if two SVs A and B have a similarity $p = 0.7$, the probability of hashing these two SVs to at least one same *semID* is 0.975 ($m = 5, n = 20$). Semantically close files are considered to have similar SVs. We ensure that the indices of semantically close files could be hashed to the same *semIDs* with very high probability (nearly 100 percent) by carefully choosing the values of m and n.

Note that each node in a P2P system is responsible for a portion of the DHT's identifier space. Close identifiers could be mapped into the same node. So even if two SVs A and B cannot be hashed to one same identifier *semID* by applying n such groups of hash functions, it is still possible that the semantic indices of these two corresponding files might be hashed to the same node if both SVs are hashed into some *semIDs* that are close together in the identifier space. This implies that the probability of hashing these two SVs to at least one same node is greater than or equal to $1 - (1 - p^m)^n$.

As a result, we can improve the performance of our probabilistic approach by adjusting the parameters n and m. Ideally, we would like the probability $1 - (1 - p^m)^n$ to approach 100 percent. By assuming that semantically close files have a relatively high similarity value (e.g., ≥ 0.7), the probability is now dependent on n and m. Recall that the probability is 97.5 percent when p, n, and m are chosen to be 0.7, 20, and 5, respectively. If we increase n to 30 while keeping p and m fixed, the probability is 99.6 percent; if we reduce m to 1 while keeping p and n fixed, the probability is nearly 100 percent. Therefore, either a relatively big n or a relatively small m would dramatically improve the performance of our probabilistic approach. However, a big n would *increase the load of indexing and querying as well as storage cost*, and a small m might *cluster the indices of those files that are not very semantically close (with low similarity) to the same nodes with non-negligible probability.* Another interesting fact is that we could use a small m if in a system those files with a relatively low similarity ($p \leq 0.5$) are also regarded as semantically close files. For example, the probability is nearly 100 percent if p, n, and m are 0.3, 20, and 1, respectively. In summary, all these problems need further exploration in our future work.

38.6.3 LSH-Based Semantic Locating

We now discuss the issue of how to locate semantically close documents that satisfy a query, given the fact that all documents in the system are automatically indexed according to their SVs in response to operations such as document creation or modification. The goal of semantic locating is to answer a query by consulting only a small number of nodes that are most responsible for the query.

Given the semantic indexing scheme described previously, we show that this goal can be easily achieved. For example, let V_q be a query Q's semantic vector. Suppose Q wants to locate those documents whose SVs are similar to A (with certain similarity degree). The semantic locating procedure (as shown in Algorithm 38.2) produces n *semIDs* from V_q for Q using the same set of hash functions (used in the semantic indexing procedure). So, if a document D satisfies such a query Q, it will be retrieved by Q with very high probability. Note that the SVs of document D and query Q could be hashed to the same *semIDs* with high probability (i.e., $1 - (1 - p^m)^n$). Thus, by having these *semIDs* as the DHT *keys* in the DHT's interface of *get(key)*, Q is able to retrieve semantically close documents from the peer nodes that are responsible for these *semIDs*. n is very small (e.g., 20) in the system, which implies that a query can be answered by consulting only a small number n of nodes.

As shown in Algorithm 38.2, upon a request, each destination peer (at most n) locally checks the list of tuples $<semID, docID, SV>$ and finds the *docIDs* such that their associated SVs are similar to the query's SV with certain similarity threshold, and sends the list of *docIDs* to the requesting node. Then, the requesting node merges the replies from all destination peers, generates a materialized view of the query result, and indexes the query according to its SV.

Actually, each destination peer can organize its own tuples in such a way that these tuples are clustered locally according to their SVs using *data clustering* techniques such as k-means clustering.[12] It should be pointed out that we do not use LSH to perform local matching because LSH could hash the indices of dissimilar files into the same *semIDs* with some probability. Each destination peer uses k-means to cluster the tuples into collections according to the semantic vector until the variance inside a collection falls below a certain threshold. Managing the tuples in the unit of collections allows each peer to narrow the search range (within a single collection instead of the whole indices) upon a request, thereby making the search efficient and fast.

Algorithm 38.2   Semantic locating procedure using n groups of m LSH functions.

Require: $g[1..n]$, each of which has m hash functions $h[1..m]$

  1: Convert V_q into V_q', which is a set of integers

  2: **for** $j = 1$ to n **do**

  3:   $semID[j] = 0$

  4:   **for** each $h[i] \in g[j]$ **do**

  5:     $semID[j] \wedge = h[i](V_q')$ /* $\wedge$ is a XOR operation */

  6:   **end for**

  7: **end for**

  8: **for** each $semID[j]$ **do**

  9:   Send the request to the destination node which is responsible for $semID[j]$

10: **end for**

11: Get replies from all the destination nodes

12: Merge the *docIDs* that satisfy the query from all replies

13: Create a materialized view of the query result asynchronously

14: Index the query according to its semantic vector asynchronously

---

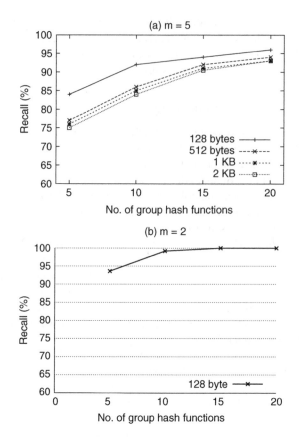

FIGURE 38.4 Performance of semantic locating, where n is the number of LSH function groups and m is the number of LSH functions in each group.

38.6.4 Experimental Results

We present part of experimental results here and refer readers to Zhu et al.[32] for more details.

The results reported here are based on a P2P file system built on top of Pastry.[22] The data used in the experiments consists of 205 unique C++ program files from a CVS repository. Each of these 205 program files consists of three different versions, on average. Hence, there are 615 files in total (about 10 MB). We divided each single file into a list of variable-sized chunks using the technique suggested in LBFS.[17] As a result, each file can be represented as a list of chunk fingerprints, each of which is a 64-bit integer, by taking the first 64 bits of the chunk's SHA-1 hash.[17] We started the experiment with an empty P2P file system and indexed each file into the system by applying the semantic indexing procedure. Then we issued a set of queries to locate different versions for each unique C++ program file, because here we consider different versions of a program file semantically close due to similar fingerprints.

Figure 38.4(a) shows the recall for different minimum chunk size limits,* including 128 bytes, 512 bytes, 1 KB, and 2 KB. The x-axis represents the number of group hash functions n used in the semantic indexing and locating procedures, while the y-axis represents the recall. As expected, the recall increases with the number n, increasing from 5 to 20. Moreover, as the minimum chunk size varies from 128 bytes to 2 KB, the recall decreases. This is because chunks with a smaller minimum chunk size limit are able to identify more similarity between different versions of a file. But even when the minimum chunk size limit

---

*When dividing a file into chunks, we impose a minimum chunk size limit like in LBFS.[17]

is 128 bytes and n is 20, our semantic locating approach was still unable to find all the files. This is because some files are very small; even a minimum chunk size limit of 128 bytes could not make the similarity high enough between different versions of a file. According to $(1 - (1 - p^m)^n)$, if p is small (say, ≤ 0.7), our locating approach might fail to find *all* semantically close files with such a small similarity (because it cannot guarantee a 100 percent probability). Further, a small m could dramatically improve the recall according to $(1 - (1 - p^m)^n)$. Figure 38.4(b) shows the result for a minimum chunk size limit of 128 bytes with $m = 2$. When n is 15, the recall is 100 percent.

38.6.5 Top Term Optimization

By realizing the fact that in a document, a small number (e.g., 30) of top terms are much more important than other terms,[31] we propose a top term-based optimization for the basic LSH-based semantic indexing and locating approach. The intuition behind this optimization is that, according to the similarity computing Equation 38.1, top terms with heavy weight tend to contribute most in similarity score in comparison to those terms with light weight in a document. Therefore, we can represent a document by a *compact term vector*, which consists of only those top terms. We believe this could reduce the cost of both LSH computation in semantic indexing and location and storage (as shown by Zhu et al.[32]). Furthermore, top term-based optimization might help to identify the most relevant documents for the user's query, thus preventing the system from returning too many documents (most of them may be irrelevant) beyond the user's capability to deal with.

38.7 Summary

This chapter addressed the advantages of constructing a P2P search system and reviewed current search systems built on top of both unstructured and structured P2P systems. It also discussed in detail two search systems built on top of unstructured P2P networks and structured P2P networks, respectively.

Structured P2P systems are adept at exact-match lookups: given a key, the system can locate the corresponding document with only $O(\log N)$ hops. However, as discussed previously, extending exact-match lookups to support keyword/semantic search[14,20,27,32] on DHTs is nontrivial. The main problem facing these search techniques is the high maintenance cost in both overlay structure and document indices due to node churn in P2P networks. To counter this problem, we expect to construct the search engine using a subset of P2P nodes that are stable and have good connectivity.

Unstructured P2P systems, on the other hand, automatically support arbitrarily complex queries. In addition, node churn causes little problem for them. However, the main problem facing the search systems built on top of unstructured P2P networks is the search inefficiency — a query might probe a very large fraction of nodes to be answered. As a result, a number of search techniques[3,7,8,15,25,31] have been proposed to improve search efficiency on unstructured P2P systems.

References

1. Text retrieval conference (trec). http://trec.nist.org.
2. L.A. Barroso, J. Dean, and U. Holzle. Web search for a planet: the google cluster architecture. *IEEE Micro*, 23(2):22–28, 2003.
3. M. Bawa, G. Manku, and P. Raghavan. SETS: Search enhanced by topic segmentation. In *Proceedings of the 26th Annual International ACM SIGIR Conference*, pages 306–313, Toronto, Canada, July 2003.
4. Michael W. Berry, Zlatko Drmac, and Elizabeth R. Jessup. Matrices, vector spaces, and information retrieval. *SIAM Review*, 41(2):335–362, 1999.
5. Andrei Z. Broder, Moses Charikar, Alan M. Frieze, and Michael Mitzenmacher. Min-wise independent permutations. *Journal of Computer and System Sciences*, 60(3):630–659, 2000.
6. Moses S. Charikar. Similarity estimation techniques from rounding algorithms. In *Proceedings of the 34th ACM Symposium on Theory of Computing (STOC)*, pages 380–388, Quebec, Canada, May 2002.

7. E. Cohen, H. Kaplan, and A. Fiat. Associative search in peer to peer networks: harnessing latent semantics. In *Proceedings of IEEE INFOCOM*, Volume 2, pages 1261–1271, San Francisco, CA, April 2003.

8. Arturo Crespo and Hector Garcia-Molina. Routing indices for peer-to-peer systems. In *Proceedings of the 22nd IEEE International Conference on Distributed Computing Systems (ICDCS)*, pages 23–32, Vienna, Austria, July 2002.

9. K. Gummadi, R. Gummadi, S. Gribble, S. Ratnasamy, S. Shenker, and I. Stoica. The impact of DHT routing geometry on resilience and proximity. In *Proceedings of ACM SIGCOMM*, pages 381–394, Karlsruhe, Germany, August 2003.

10. A. Gupta, D. Agrawal, and A. El Abbadi. Approximate range selection queries in peer-to-peer systems. In *Proceedings of the First Biennial Conference on Innovative Data Systems Research (CIDR)*, Asilomar, CA, January 2003.

11. Piotr Indyk and Rajeev Motwani. Approximate nearest neighbors: towards removing the curse of dimensionality. In *Proceedings of the 13th Annual ACM Symposium on Theory of Computing (STOC)*, pages 604–613, Dallas, TX, May 1998.

12. A.K. Jain, M.N. Murty, and P.J. Flynn. Data clustering: a review. *ACM Computing Surveys*, 31(3):264–323, 1999.

13. S. Lawrence and C.L. Giles. Accessibility of information on the web. *Nature*, 400:107–109, 1999.

14. J. Li, B. Loo, J. Hellerstein, F. Kaashoek, D. Karger, and R. Morris. On the feasibility of peer-to-peer web indexing and search. In *Proceedings of the 2nd International Workshop on Peer-to-Peer Systems(IPTPS)*, pages 207–215, Berkeley, CA, February 2003.

15. Qin Lv, Pei Cao, and Edith Cohen. Search and replication in unstructured peer-to-peer networks. In *Proceedings of 16th ACM Annual International Conference on Supercomputing (ICS)*, pages 84–95, New York, NY, June 2002.

16. Mandar Mitra, Amit Singhal, and Chris Buckley. Improving automatic query expansion. In *Proceedings of ACM SIGIR*, pages 206–214, Melbourne, Australia, 1998.

17. Athicha Muthitacharoen, Benjie Chen, and David Mazieres. A low-bandwidth network file system. In *Proceedings of the 8th ACM Symposium on Operating Systems Principles (SOSP)*, pages 174–187, Banff, Canada, October 2001.

18. C.D. Prete, J.T. MacArthur, Richard L. Villars, I.L. Nathan Redmond, and D. Reinsel. Industry developments and models, disruptive innovation in enterprise computing: storage. In *IDC*, February 2003.

19. S. Ratnasamy, P. Francis, M. Handley, R. Karp, and S. Shenker. A scalable content-addressable network. In *Proceedings of ACM SIGCOMM*, pages 161–172, San Diego, CA, August 2001.

20. P. Reynolds and A. Vahdat. Efficient peer-to-peer keyword searching. In *Proceedings of ACM/IFIP/USENIX International Middleware Conference (Middleware)*, pages 21–40, Rio de Janeiro, Brazil, June 2003.

21. Sean C. Rhea and John Kubiatowicz. Probabilistic location and routing. In *Proceedings of IEEE INFOCOM*, Volume 3, pages 1248–1257, New York, NY, June 2002.

22. Antony Rowstron and Peter Druschel. Pastry: scalable, decentralized object location, and routing for large-scale peer-to-peer systems. In *Proceedings of the 18th IFIP/ACM International Conference on Distributed System Platforms (Middleware)*, pages 329–350, Heidelberg, Germany, November 2001.

23. Stefan Saroiu, P. Krishna Gummadi, and Steven D. Gribble. A measurement study of peer-to-peer file sharing systems. In *Proceedings of Multimedia Computing and Networking(MMCN)*, San Jose, CA, January 2002.

24. H. Schutze and C. Silverstein. A comparison of projections for efficient document clustering. In *Proceedings of ACM SIGIR*, pages 74–81, Philadelphia, PA, July 1997.

25. Kunwadee Spripanidkulchai, Bruce Maggs, and Hui Zhang. Efficient content location using interest-based locality in peer-to-peer systems. In *Proceedings of IEEE INFOCOM*, Volume 3, pages 2166–2176, San Francisco, CA, March 2003.

26. I. Stoica, R. Morris, D. Karger, M. Kaashoek, and H. Balakrishnan. Chord: a scalable peer-to-peer lookup service for Internet applications. In *Proceedings of ACM SIGCOMM*, pages 149–160, San Diego, CA, August 2001.

27. Chunqiang Tang and Sandhya Dwarkada. Hybrid global-local indexing for efficient peer-to-peer information retrieval. In *Proceedings of First Symposium on Networked Systems Design and Implementation (NSDI)*, pages 211–224, San Francisco, CA, March 2004.

28. Chunqiang Tang, Zhichen Xu, and Sandhya Dwarkadas. Peer-to-peer information retrieval using self-organizing semantic overlay networks. In *Proceedings of ACM SIGCOMM*, pages 175–186, Karlsruhe, Germany, August 2003.

29. M. Welsh, N. Borisov, J. Hill, R. von Behren, and A. Woo. Querying Large Collections of Music for Similarity. Technical Report UCB/CSD00-1096, U.C. Berkeley, November 1999.

30. B.Y. Zhao, J.D. Kubiatowicz, and A.D. Joseph. Tapestry: An Infrastructure for Fault-Tolerance Wide-Area Location and Routing. Technical Report UCB/CSD-01-1141, Computer Science Division, University of California, Berkeley, April 2001.

31. Yingwu Zhu and Yiming Hu. ESS: Efficient Semantic Search on Gnutella-Like P2P Systems. Technical report, Department of ECECS, University of Cincinnati, March 2004.

32. Yingwu Zhu, Honghao Wang, and Yiming Hu. Integrating semantics-based access mechanisms with P2P file systems. In *Proceedings of the 3rd International Conference on Peer-to-Peer Computing*, pages 118–125, Linkping, Sweden, September 2003.

39

An Overview
of Structured P2P
Overlay Networks

Sameh El-Ansary

Seif Haridi

39.1   Introduction

39.1.1   Historical Background

The term "peer-to-peer" (P2P) is used in many contexts to mean different things. It was recently used to refer to the form of cooperation that emerged with the appearance of the music file sharing application Napster.[31] With that application, music files where exchanged between computers (peers), relying on a central directory for knowing which peer has which file. Napster ceased operation due to legal rather than technical reasons and was followed by a number of systems such as Gnutella[17] and Freenet,[13] in which the central directory was replaced with a flooding process where each computer connects to random members in a P2P network and queries his neighbors who act similarly until a query is resolved. The random graph of such peers proved a feasible example of an overlay network, that is, an application-level network on top of the Internet transport with its own topology and routing.

39.1.2   The Motivating Problem

The simultaneous "beauty" and "ugliness" of random overlay networks has attracted academic researchers from the networking and the distributed systems communities. The "beauty" lies in the simplicity of the solution and its ability to completely diffuse central authority and legal liability. From a computer science point of view, this elimination of central control is very attractive for — among other things — eliminating single points of failure and building large-scale distributed systems. The "ugliness" lies in the huge amount of induced traffic that renders the solution unscalable.[25,38] The problem of having a scalable P2P overlay network with no central control became a scientifically challenging problem, and the efforts to solve it resulted in the emergence of what is known as "structured P2P overlay networks," referred to also by the term "distributed hash tables" (DHTs).

39.1.3   The General Solution

The main approach introduced by the academics to build overlay networks was to let the set of cooperating peers act as a distributed data structure with well-defined operations, namely, a distributed hash table with the two primitive operations `Put(key,value)` and `Get(Key)`. The `Put` operation should result in the storage of the value at one of the peers such that any of the peers can perform the `Get` operation and reach the peer that has the value. More importantly, both operations need to take a "small" number of hops. A first naive solution would be that every peer knows all other peers, and then every `Get` operation would be resolved in one hop. Apparently, that is not scalable. Therefore, a second constraint is needed. Each node should know a "small" number of other peers. From a graph-theory point of view, this means that a directed graph of a certain known "structure" rather than a random graph must be constructed with scalable sizes of both the outgoing degree of each node and the diameter of the graph.

39.2   Definitions and Assumptions

Values. The set of values $\mathcal{V}$ such as files, directory entries, etc. Each value has a corresponding key from the set *Keys*($\mathcal{V}$). If a value is a file, the key could be, for instance, its checksum, a combination of owner, creation date, and name, or any such unique attribute.

   Nodes. The set $\mathcal{P}$ of machines/processes also referred to as nodes or peers. *Keys*($\mathcal{P}$) is the set of unique keys for members of $\mathcal{P}$, usually the IP addresses or public keys of the nodes.

   The Identifier Space. A common and fundamental assumption of all DHTs is that the keys of the values and the keys of the nodes are mapped into one range using a hashing function. For instance, the IP addresses of the nodes and the checksums of files are hashed using SHA-1[12] to obtain 128-bit identifiers. The term "identifier" is used to refer to hashed keys of items and of nodes. The term "identifier space" refers to the range of possible values of identifiers and its size is usually referred to by N. We use *id* as an abbreviation for identifier most of the time.

Items. When a new value is inserted in the hash table, its key is saved with it. We use the term "item" to refer to a key-value pair.

Equivalence of Nodes. The operations of adding a value, looking up a value, adding a new node (join), and removing an existing node (leave) are all possible through any node $p \in \mathcal{P}$.

Autonomy of Nodes. The addition or removal of any node is a decision taken locally at that node and there is a distinction between graceful removals of nodes (leaves) and ungraceful removals (failures).

The First Contact. Another fundamental assumption in all DHTs is that to join an existing set of peers that already form an overlay network, a new peer must know some peer in that network. This knowledge in many systems is assumed to be acquired by some out-of-band method. Some systems discuss the possibility of obtaining the first contact through IP multicast; however, it is an orthogonal issue to the operation of any DHT.

Ambiguous Terms. Because we are forced to use a different terminology to refer to the same logical entities in different contexts, we try to resolve those ambiguities early by introducing the following equalities. Nodes = peer = contact = reference, overlay network = overlay graph, identifier = id, edge = pointer, "point to" = "be aware of" = "keep track of", routing table = outgoing edges, diameter = lookup path length, lookup = query, routing table size = outgoing arity. Also, sometimes, letters like n, s, t, and x are used to refer to nodes and values as well as their identifiers but the meaning should be clear from the context.

39.3 Comparison Criteria

The Overlay Graph. This is the main criteria that distinguishes systems from each other. For each overlay graph, we want to know what the graph looks like and what is the outgoing arity of each node in the graph.

Mapping Items onto Nodes. For a given overlay graph, we want to know the relation between node *id*s and item *id*s (i.e., at which node should an item be stored?).

The Lookup Process. A tightly coupled property with the overlay graph is how lookups are performed and what is the typical performance.

Joins, Leaves, and Maintenance. How a new node is added to the graph and how a node is gracefully deleted from the graph. Joins and leaves make the graph change constantly, and some maintenance process is usually required to cope with such changes, so how does this process take place, and what is its cost?

Replication and Fault Tolerance. In addition to graceful removal of nodes, dealing with failures is usually more difficult. Replication is a tightly coupled property because it can be a technique to overcome failure's effect or a method of improving efficiency.

Upper Services and Applications. When applicable, we enumerate some of the applications and services developed using a certain system.

Implementation. Because many systems are of a completely theoretical nature even for their services and applications, we try to give an idea about any available implementations of a system.

39.4 DHT Systems

39.4.1 Chord

The Overlay Graph. Chord[42,43] assumes a circular identifier space of size N. A Chord node with identifier u has a pointer to the first node following it clockwise on the identifier space $(Succ(u))$ as well as the first node preceding it, $(Pred(u))$. The nodes therefore form a doubly linked list. In addition to those, a node keeps $M = \log_2(N)$ pointers called fingers. The set of fingers of node u is $F_u = \{(u, Succ(u + 2^{i-1}))\}, 1 \le i \le M$, where the arithmetic is modulo N.

The intuition of that choice of edges is that a node perceives the circular identifier space as if it starts from its *id*. The edges are, then, chosen such as to be able to partition the space into two halves, partition one of the halves into two quarters, and so forth.

In Figure 39.1(a), we show a network with an *id* space $N = 16$. Each node has $M = \log_2(N) = 4$ edges. The network contains nodes with *id*s 0, 3, 5, 9, 11, and 12. The general policy for constructing routing tables

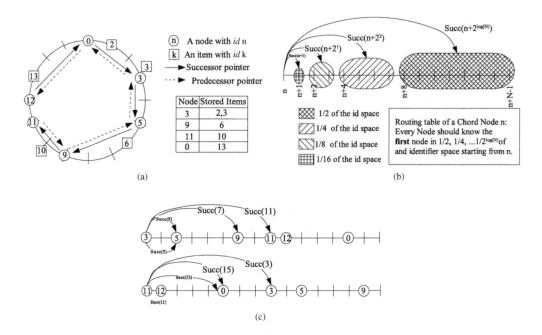

FIGURE 39.1 (a) A chord network with $N = 16$ populated with 6 nodes and 5 items; (b) the general policy for Chord's routing tables; and (c) example routing tables for nodes 3 and 11.

is shown in Figure 39.1(b). Node n chooses its pointers by positioning itself at the start of the identifier space. It chooses to have the pointers to the successors of the *ids* $n + 2^0$, $n + 2^1$, $n + 2^2$, and $n + 2^3$. The last pointer, $n + 2^3$, divides the space into two halves. The one before it, $n + 2^2$, divides the first half into two quarters, and so forth. However, there may not exist a node at the desired position so its successor is taken instead. Figure 39.1(c) shows the routing entries of node 3 and 11.

Mapping Items onto Nodes. As shown in Figure 39.1(a), an item is stored at the first node that follows clockwise on the circular identifier space. If items with *ids* 2, 3, 6, 10, and 13 are to be stored in the network given above, then {2,3} will be stored at 3; {6} at 9; {10} at 11; and {13} at 0.

The Lookup Process. The lookup process comes as a natural result of how the *id* space is partitioned. Both the insertion and querying of items depend on finding the successor of an *id*. For example, assume that node 11 wants to insert a new item with *id* 8. The lookup is forwarded to node 3, which is the closest preceding finger — from the point of view of 11 — to the *id* 8. Node 3 will act similarly and forward the query to node 5 because node 5 is the closest preceding finger for 8 from the point of view of 5. Node 5 finds that 8 is between itself and its successor 9, and therefore returns 9 as an answer to the query through the reverse path.[*] In all cases, upon getting the answer, node 11's application layer should contact node 9's application layer and ask for the storage of some value under the key 8. Any node looking for the key 8 can act similarly and in no more than M hops,[**] a node will discover the node at which 8 is stored. In general, under normal conditions, a lookup takes $O(\log_2(N))$ hops.

Joins, Leaves, and Maintenance. To join the network, a node n performs a lookup for its own *id* through some first contact in the network and inserts itself in the ring between its successor s and the predecessor

---

[*]This is known as the recursive method. Another suggested approach in the Chord papers is an iterative method where all the answers path by the node at which the lookup originated; that is instead of the path being $11 \to 3 \to 5 \to 3 \to 11$, in an iterative lookup the path will be $11 \to 3 \to 11 \to 5 \to 11$. A third approach adopted in other systems[2] would be to continue to the destination and send the result to the origin of the lookup (i.e., $11 \to 3 \to 5 \to 9 \to 11$).

[**]Chord counts a remote procedure call and the response to it as one hop.

of s using a periodic stabilization algorithm. Initialization of n's routing table is done by copying the routing table of s or letting s look up each required edge of n. The subset of nodes that need to adjust their tables to reflect the presence of n will eventually do that because all nodes run a stabilization algorithm that periodically goes through the routing table and looks up the value of each edge. The last task is to transfer part of the items stored at s; namely, items with id less than or equal to n need to be transferred to n, and that is also handled by the application layers of n and s.

Graceful removals (leaves) are done by first transferring all items to the successor and informing the predecessor and successor. The rest of the fingers are corrected by virtue of the stabilization algorithm.

Replication and Fault Tolerance. Ungraceful failures have two negative effects. First, ungraceful failures of nodes cause loss of items. Second, part of the ring is disconnected, leading to the inability to look up certain identifiers — let alone if a set of adjacent nodes fails simultaneously. Chord tackles this problem by letting each node keep a list of the $\log_2(N)$ nodes that follow it on the circle. The list serves two purposes. First, if a node detects that its successor is dead, it replaces it with the next entry in its successor list. Second, all the items stored at a certain node are also replicated on the nodes in the successor list. For an item to be lost or the ring to be disconnected, $\log_2(N) + 1$ successive nodes must fail simultaneously.

Upper Services and Applications. A few applications, such as a cooperative file-system,[9] a read/write file system,[29] and a DNS directory,[8] were built on top of Chord. As a general-purpose service, a broadcast algorithm was also developed for Chord.[10]

Implementation. The main implementation of Chord is that by its authors in C++,[44] where a C++ discrete-event simulator is also available. Naanou[19] is a C# implementation of Chord with a file-sharing application on top of it.

39.4.2 Pastry

The Overlay Graph. The overlay graph design of Pastry,[39] in addition to aiming at achieving logarithmic diameter with a logarithmic node state, also tries to target the issue of locality. In general, as a result of obtaining the node ids by hashing IP numbers/public keys, nodes with adjacent node ids may be farther apart geographically. Differently said, two machines in one country would communicate through a machine on another continent just because the hash of their ids will be far apart in the id space.

Pastry assumes a circular identifier space and each node has a list containing $\frac{L}{2}$ successors and $\frac{L}{2}$ predecessors known as the leaf set. A node also keeps track of M nodes that are close according to another metric other than the id space, such as, for instance, network delay. This set is known as the neighborhood set and is not used during routing but is used for maintaining locality properties. The third type of node state is the main routing table. It contains $\left\lceil \log_{2^b}(N) \right\rceil$ rows and $2^b - 1$ columns. L, M, and b are system parameters.

Node ids are represented as a string of digits of base 2^b. In the first row, the routing table of a node contains node ids that have a distinct first digit. Because the digits are base 2^b, a node needs to know $2^b - 1$ nodes for each possible digit except its own.

The second row of a node with id n contains $2^b - 1$ nodes that share the first digit with n but differ in the second digit. The third row contains nodes that share the first and second digit of n but differ in the third, and so forth. We stress that -1 in $2^b - 1$ is because in each row the node itself would be the best match for one of the columns; therefore, we do not need to keep an address of it. Figure 39.2 illustrates how the the id space is partitioned using this prefix matching scheme.

As one can observe, for each of the constraints about the node ids contained in a routing table, there exist many satisfying nodes. Therefore, the node with the lowest network delay, or the best according to some other criteria, is included in the routing table.

Mapping Items onto Nodes. An item in Pastry is stored at the node that is numerically closest to the id of the item. Such a node will have the longest matching prefix.

The Lookup Process. To locate the closest node to an id x, a node n checks first if x falls within the range of node ids covered by its leaf set. If so, it is forwarded to such node. Otherwise, the lookup is forwarded to the node in the interval that x belongs to, that is, to a node that shares more digits than the

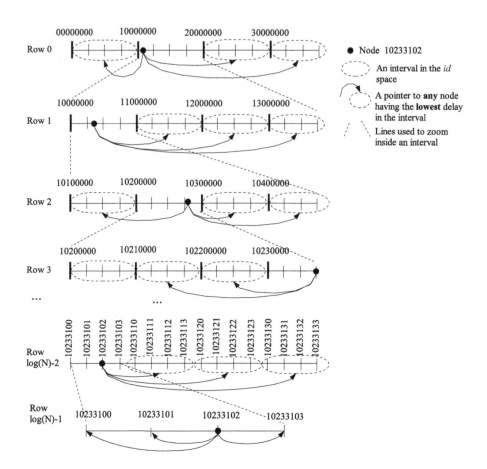

FIGURE 39.2 Illustration of how the Pastry node 10233102 chooses its routing edges in an identifier space of size $N = 2^{128}$ and encoding base $2^b = 4$.

shared prefix between n and x. If no such node is found in n's routing table, the lookup is forwarded to the numerically closest node to x. The latter case does not happen very often, provided the *id*s are uniformly distributed. With the matching of one digit of the sought *id* in each hop, after $\log_{2^b}(N)$ hops, a lookup is resolved.

Joins, Leaves, and Maintenance. When a node n joins the network through a node t, t is usually in the proximity of n and thus the neighborhood set of t is suitable for n. Due to the construction of the routing tables in Pastry, n performs a lookup for its own *id* to figure out the numerically closest node s to n. It can take the ith row from the ith node on the path from t to s and use those rows in initializing its routing table. Moreover, the leaf set of s is a good initialization for the leaf set of n. Finally, n informs every node in its neighborhood set, leaf set, and routing table of its presence. The cost is about $3 \times 2^b \log_{2^b} N$.

Node departures are detected as failures and repaired in a routing table by asking a node in the same row of the failed node for its entry on the failed position.

Replication and Fault Tolerance. Pastry replicates an item on the k closest nodes in its leaf set. This serves in saving an item after a node loss and, in the meantime, the replicas act as cached copies that can contribute to finding an item more quickly.

Upper Services and Applications. A number of applications and services were developed on top of Pastry, such as SCRIBE[7] for multicasting and broadcasting; PAST,[40] an archival storage system; SQUIRREL,[20] a cooperative Web caching system; and SplitStream,[6] a high-bandwidth content distribution.

Implementation. FreePastry[14] is an open-source Java implementation of the Pastry system.

39.4.3 Tapestry

Tapestry[46] is one of the earliest and largest efforts on structured P2P overlay networks. Like Pastry, it is based on the earlier work of a Plaxton[35] mesh. We do not describe the details of Tapestry here due to the large similarity with Pastry. However, we must point out that as software, it is probably one of the most mature implementations of a structured overlay network. In addition to network simulation, Tapestry has been evaluated using a more realistic environment, namely PlanetLab,[34] a globally distributed platform with machines all over the world that is used for testing large-scale systems.

Tapestry is a cornerstone project in the larger Oceanstore[22] project for global-scale persistent storage. Other applications based on Tapestry include the steganographic file system Mnemosyne;[18] Bayeux,[48] an efficient self-organizing application-level multicast system; and SpamWatch,[47] a decentralized spam-filtering system.

39.4.4 Kademlia

The Overlay Graph. The Kademlia[28] graph partitions the identifier space exactly like Pastry. However, it is presented in a different way, where node *id*s are leaves of a binary tree with each node's position being determined by the shortest unique prefix of its *id*. Each node divides the binary tree into a series of successively lower subtrees that do not contain the node *id* and keeps at least one contact in each of those subtrees. For example, a node with *id* 3 has the binary representation 0011 in an identifier space of size $N = 16$. Because its prefix of length 1 is the digit 0, it needs to know a node whose first digit is 1. Because its prefix of length 2 is 00, it needs to know a node with prefix 01. Because its prefix of length 3 is 001, it needs to know a node with prefix 000. Finally, because its prefix of length 4 is 0011, it needs to know a node with a prefix 0010. This policy is illustrated in Figure 39.3, which results in a space division exactly like Pastry with the special case of a binary encoding of the digits.

Kademlia does not keep a list of nodes close in the identifier space like the leaf set or the successor list in Chord. However, for every subtree or interval in the identifier space, it keeps k contacts rather than one contact if possible, and calls a group of no more than k contacts in a subtree a k-bucket.

Mapping Items onto Nodes. Kademlia defines the notion of the distance between two identifiers as the value of the bitwise exclusive or XOR of the two identifiers. An item is stored at the node whose XOR difference between the node *id* and the item *id* is minimal.

The Lookup Process. To increase robustness and decrease response time, Kademlia performs lookups in a concurrent and iterative manner. When a node looks up an *id*, it checks to see which subtree the *id* belongs to and forwards the query to α randomly selected nodes from the k-bucket of that subtree. Each node possibly returns a k-bucket of a smaller subtree closer to the *id*. From the returned bucket, another α randomly selected nodes are contacted and the process is repeated until the *id* is found. When an item is inserted, it is also stored at the k closest nodes to its *id*. Because of the prefix matching scheme, similar to Pastry, a lookup is also resolved in $O(\log(N))$ hops.

Joins, Leaves, and Maintenance. A new node finds the closest node to it through any initial contact and uses it to fill its routing table by querying about nodes in different subtrees. If it happens that a k-bucket

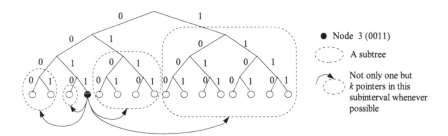

FIGURE 39.3 The pointers of node 3 (0011) in Kademlia. The same partitioning of the identifier space as in Pastry with binary-encoded digits.

is filled due to exposure to lots of nodes in a particular subtree, a least-recently-used replacement policy is applied. However, Kademlia makes use of statistics taken from existing peer-to-peer measurements studies, which indicate that a node that stayed for a longer time in the past will probably stay connected longer in the future. Therefore, Kademlia can discard the knowledge of new nodes if it knows many other stable nodes in a given subtree.

Maintenance of the routing tables after joins and leaves depends on a technique that is different from the stabilization in Chord or the deterministic update of Pastry. Kademlia maintains the routing tables using the lookup traffic. The XOR metric results in every node receiving queries from the nodes contained in its routing table (which is not the case in a system like Chord). Consequently, the reception of any message from a certain node in a certain subtree is essentially an update of the k-bucket for that subtree. This approach clearly minimizes the maintenance cost. However, it is not deeply analyzed.

Another maintenance task is that upon receiving multiple queries from the same subtree, Kademlia updates the latencies of the nodes in a particular k-bucket. This improves the choice of nodes used for doing lookups and one could say that by doing that, Kademlia also takes into consideration network delay and locality.

Replication and Fault Tolerance. Because leaves are not deeply discussed, we assume that they are treated as failures. Kademlia fault tolerance depends mainly on strong connectivity because it keeps k contacts per subtrees and not only one, and this makes the probability of a disconnected graph low.

Also, as mentioned above, Kademlia stores k copies of an item on the k closest nodes to its *id*. The nodes are also republished periodically. The policy for republishing is that any node that sees itself closer to an item *id* than all the nodes it knows about, gives it to $k - 1$ other nodes.

Applications and Implementation. Kademlia is probably the one DHT that received relatively wider non-academic adoption by being used in two file-sharing applications, namely Overnet[32] and Emule.[11]

39.4.5 HyperCup

While it has been mentioned many times in the literature that systems such as Chord and Pastry are approximations of hypercubes, those works were not presented that way by their authors. HyperCup[41] is a system that presents a way to construct and maintain hypercubes in a dynamic setting. The performance of HyperCup is similar to the many other DHTs with logarithmic order for both the routing table size and the lookup path length under particular uniformity assumptions. HyperCup also defines a broadcast algorithm based on the concept of a spanning tree of all nodes. A distinguishing feature of HyperCup is that it addresses semantic search based on ontological terms. Nodes with similar ontologies are clustered together such that a search by a certain ontological term is achieved as a localized broadcast within a cluster.

39.4.6 DKS

The Overlay Graph. DKS[2] can be perceived as an optimal generalization of Chord to provide shorter diameter with larger routing tables. As well, DKS can be perceived as a meta-system from which other systems can be instantiated. DKS stands for Distributed k-ary Search and was designed after perceiving that many DHT systems are instances of a form of k-ary search. Figure 39.4 shows the division of the space done in DKS. One can see that, in common with Chord, each node perceives itself as the start of the space. And like Pastry, each interval is divided into k rather than two intervals.

Mapping Items onto Nodes. Along with the goal of DKS to act as a meta-system, mapping items onto nodes is left as a design choice. A Chord-like mapping is valid as a simple first choice. However, different mappings are also possible.

The Lookup Process. A query arriving at a node is forwarded to the first node in the interval to which the *id* of the node belongs. Therefore, a lookup is resolved in $\log_k(N)$ hops.

Joins, Leaves, and Maintenance. Unlike Chord, DKS avoids any kind of periodic stabilization, for the maintenance of the successors, the predecessor, and the routing table. Instead, it relies on three principles: local atomic actions, correction-on-use, and correction-on-change. When a node joins, a form of atomic

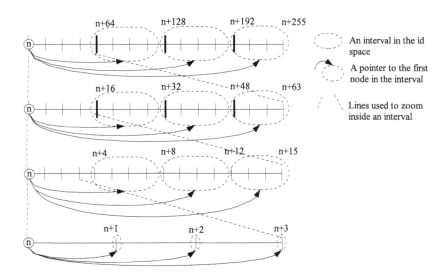

FIGURE 39.4 Illustration of how a DKS node divides the space in an identifier space of size $N = 2^8 = 256$.

distributed transaction is performed to insert it on the ring. Routing tables are then maintained using the correction-on-use technique, an approach introduced in DKS. Every lookup message contains information about the position of the receiver in the routing table of the sender. Upon receiving that information, the receiver can judge if the sender has an updated routing table. If correct, the receiver continues the lookup; otherwise, the receiver notifies the sender of the corruption of his routing table and advises him about a better candidate for the lookup according to the receiver's knowledge. The sender then contacts the candidate and the process is repeated until the correct node for the routing table of the sender is used for the lookup.

By applying the correction-on-use technique, a routing table entry is not corrected until there is a need to use it in some lookup. This approach reduces the maintenance cost significantly. However, the number of joins and leaves is assumed to be reasonably less than the number of lookup messages. In cases where this assumption does not hold, DKS combines it with the correction-on-change technique.[4] Correction-on-change notifies all nodes that need to be updated upon the occurrence of a join, leave, or failure.

Replication and Fault Tolerance. In early versions of DKS, fault tolerance was handled similar to Chord, where replicas of an item are placed on the successor pointers. In later developments,[16] DKS tries to address replication more on the DHT level rather than delegating most of the work to the application layer. Additionally, to avoid congestion in a particular segment of the ring, replicas are placed in dispersed, well-chosen positions and not on the successor list. In general, for the correction-on-use technique to work, an invariant is maintained where the predecessor pointer must always be correct, and that is provided by the atomic actions on the circle.

Upper Services and Applications. General-purpose broadcast[15] and multicast[3] algorithms were developed for DKS.

39.4.7 P-Grid

P-Grid[1] is a system based on randomized algorithms assuming that there will be random interactions between nodes of the overlay. P-Grid ensures that those peers are arranged in a graph most similar to the overlay graph of Pastry. A unique assumption of P-Grid is that nodes do not have constant *ids*; instead, *ids* change over time so that the identifier space can be partitioned fairly among them. This property is not only used for decreasing the lookup path length, but also for balancing items among nodes. A file-sharing application with the same name is implemented in Java and available in Ref. 33.

39.4.8   Koorde

The Overlay Graph. Koorde[21] is based on the DeBruijn graph.[27] Koorde stresses the point that a constant number of outgoing edges per node is enough to have a logarithmic lookup length. The DeBruijn graph is an example capable of doing that. The significance of a constant number of edges is that the maintenance overhead is lower compared to a logarithmic number, as is the case in all the previous DHTs we have shown thus far. In Figure 39.5(a), we show the pointers of all the nodes of a Koorde graph of eight nodes. A node with *id* n has edges to nodes $2n$ and $2n + 1$ in a circular identifier space like Chord. We denote the first and the second edge of node as $n\ E_n \circ 0$ and $E_n \circ 1$, respectively.

    Mapping Items onto Nodes. Exactly like Chord.

    The Lookup Process. When a node n needs to look up an *id* x represented as a string of binary digits $d_1 d_2 \ldots d_{\log_2(N)}$, it takes the top bit d_1; if it is a 0, it forwards the query to $E_n \circ 0$, otherwise to $E_n \circ 1$. The second node looks at the remaining string $d_2 \ldots d_{\log_2(N)}$ and acts similarly. After, at most, $\log_2(N)$ hops, a query is resolved. Figure 39.5(b) shows what paths nodes 1, 3, and 4 take to reach any node in the network. The Koorde article[21] also elaborates on an algorithm to handle networks where not all the nodes are present in the *id* space. Each node tries to locally traverse imaginary hops for nodes that do not exist.

    Joins, Leaves, and Maintenance. Exactly like Chord. In fact, the authors say that Koorde could be perceived as a Chord system with a constant instead of a logarithmic number of fingers. Stabilization is also the basic mechanism for maintenance.

    Replication and Fault Tolerance. To realize fault tolerance, an out-degree less than $\log(N)$ nodes must be maintained; otherwise, a node will lose all its contacts very easily. This makes the advantage of a constant node state invalid. However, because with k edges, Koorde provides a $\log_k(N)$ diameter. Then, with $\log_k(N)$ edges, it provides $\log_{\log_k(N)}(N) = \frac{\log(N)}{\log(\log(N))}$ diameter, which is an advantage over the logarithmic class of DHTs.

    Load Balancing. The load balancing of items onto nodes will depend on the uniform distribution, exactly as in Chord. However, another load-balancing issue arises, that is, the load of message passing on

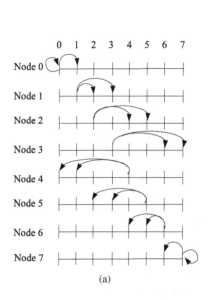

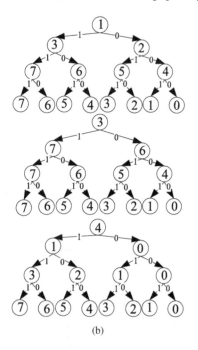

(a)   (b)

FIGURE 39.5   (a) The pointers of all the nodes in a complete Koorde network where $N = 8$. Every node n points to nodes of *ids* $2n$ and $2n + 1$. (b) Examples of how nodes 1, 3, and 4 reach other nodes by matching the destination *id* digit by digit, starting from the most significant bit.

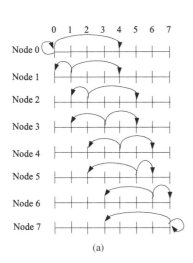

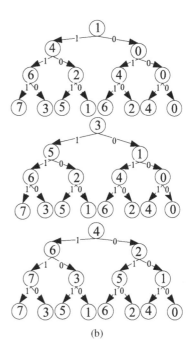

FIGURE 39.6 (a) The pointers of all the nodes in a complete Distance-Halving network where $N = 8$. (b) Examples of how nodes 1 reaches other nodes by matching the destination *id* digit by digit, starting from the least significant bit.

each node. In a DeBruijn graph, some nodes will have more traffic than others by a factor of $O(\log(N))$ of the average traffic load. For example, in the network illustrated in Figure 39.5, if every node sends a message to every other node in the network, not all the nodes will endure the same number of messages; 12 messages will be routed via a node like 7, while 21 messages will be routed via a node like 3.

39.4.9 Distance Halving

The Overlay Graph. The Distance Halving $(DH^*)^{30}$ distributed hash table is another system based on the DeBruijn graph, like Koorde. However, the way of building the graph is somewhat different. The DH is based on an approach called the continuous-discrete approach for building graphs. To build a DeBruijn graph with this approach, the identifier space is normalized into a continuous space represented by the interval $[0, 1]$. Nodes are points in that interval. Each node y has two edges, a left edge and a right edge denoted $\ell(y)$ and $r(y)$, respectively, where $\ell(y) = \frac{y}{2}$ and $r(y) = \frac{y}{2} + \frac{1}{2}$. Given the set of points and their edges, a discretization step is done to build the graph. The set of points are denoted by $\vec{x}$. The points of $\vec{x}$ divide the space into n segments. The segment of a point x_i, $S(x_i) = [x_i, x_{i+1})$, $(i = 1 \dots n - 1)$ and $S(x_n) = [x_{n-1}, 1) \cup [0, x_1)$. If a node y has an edge that belongs to the segment of some node z, then there is an edge in the discrete graph between y and z. One can also notice that the segments are defined in a way that realizes a circular identifier space.

The intuition behind that graph is that every node divides the space into two intervals and keeps a pointer to a node that is in the middle of the left interval and a pointer in the middle of the right interval. Figure 39.6(a) shows the pointers of all the nodes in a DH network of size $N = 8$. Figure 39.6(b) shows the paths to all possible destinations, starting from node 1.

---

*Do not confuse this abbreviation with the abbreviation of a Distributed Hash Table (DHT).

Mapping Items onto Nodes. Exactly like Chord and Koorde. In DH terminology, an item is stored at node y where the *id* of the item belongs to $S(y)$.

The Lookup Process. The lookup process, similar to many other DHTs, is done by the prefix matching of the sought *id*, digit by digit. The lookup is forwarded to the node pointed to by the left edge for matching a 0 digit and to the right edge for matching a 1 digit. The lookup path length is thus $O(\log_2(N))$.

Joins, Leaves, and Maintenance. A new node n joins a DH network by looking up the node s such that n belongs to $S(s)$; n then uses s to look up its left and right edges. By the construction of DH, a node can easily know the nodes that are pointing to it. Therefore, a node can easily compute the nodes that need to be updated and notifies them of n's existence. Updating of others upon a leave is done the same way. The transfer of the items upon a join or a leave is also similar to Chord.

Replication and Fault Tolerance. DH recognizes the problem of failures that can lead to a disconnected graph and advocates an additional state of $O(\log(N))$ pointers. That is in agreement with Koorde's reasoning and emphasizes that the main advantage of having a constant-degree graph will be compromised if fault tolerance is to be considered. However, with a logarithmic degree, those types of graphs can offer a diameter of $\frac{\log(N)}{\log(\log(N))}$.

Other Comments. The formal analysis of the DH graph and the continuous discrete approach are both useful tools that help gain a better understanding of the properties of a DHT system. The discussion of the *smoothness* of the graph, which is a term used by the authors to quantify the uniformity of distribution of the *ids*, is quite unique. It was noted in other DHT systems that uniform distribution could affect the performance; but in DH, an analysis of the magnitude of that effect is provided.

39.4.10   Viceroy

The Overlay Graph. Viceroy[24] is based on the Butterfly[26] network. Like many other systems, it organizes nodes into a circular identifier space and each node has successor and predecessor pointers. Moreover, nodes are arranged in $\log_2(N)$ levels numbered from 1 to $\log_2(N)$. Each node apart from nodes at level 1 has an "up" pointer and every node apart from the nodes at the last level has two "down" pointers. There is one short and one long "down" pointer. Those three pointers are called the Butterfly pointers. All nodes also have pointers to successors and predecessors pointers on the same level. In that way, each node has a total of seven outgoing pointers.

Figure 39.7(a) shows the "down" pointers of a network of $N = 16$ nodes where all nodes are present. Figure 39.7(b) shows the "up" pointers of all nodes. For simplicity, the successor pointers of the ring and the levels are not illustrated.*

Mapping Items onto Nodes. Exactly like Chord.

The Lookup Process. To look up an item x, a node n follows its "up" pointer until it reaches level 1. From there, it starts going down using the down links. In each hop, it should traverse a pointer that does not exceed the target x. For example, if node 1 is looking up the *id* 10, first it will follow its "up" pointer and reach 4, which is at level 1. At node 4, there are two choices: either to use the short pointer to 5 or the long pointer to 13; because 5 precedes the target 10, the pointer to 5 is followed. At node 5, there is a direct pointer to 10. In another example, for reaching *id* 15 from node 3, the path will be $3 \xrightarrow{up} 6 \xrightarrow{up} 9 \xrightarrow{up} 12 \xrightarrow{down} 13 \xrightarrow{down} 14 \xrightarrow{down} 15$. From this example, we can see that in a worst case, we can traverse all the levels up and down, that is, $2 \times \log(N)$ hops. Needless to say, the example includes a simplified network where all the nodes are present. When the graph is sparse, the reasoning is slightly more complicated; however, the expected lookup path length is still $O(\log(N))$.

Joins, Leaves, and Maintenance. To join, a node looks up its successor s, fixes the ring pointers, and takes the required items from s. After that, it selects a level based on the estimation of the number of

---

*In fact, some of them coincide with the Butterfly pointers.

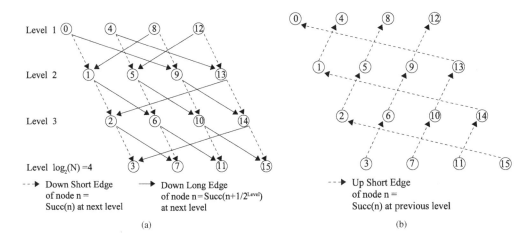

Level 1 / Level 2 / Level 3 / Level $\log_2(N) = 4$

--→ Down Short Edge
of node n =
Succ(n) at next level

——→ Down Long Edge
of node n = Succ(n+1/2^{Level})
at next level

--→ Up Short Edge
of node n =
Succ(n) at previous level

(a)   (b)

FIGURE 39.7 The Butterfly edges of a complete Viceroy network with $N = 16$ nodes; (a) the down edges; (b) the up edges.

nodes. It finds, by a combination of lookups and stepping on the ring, the rest of the pointers (successor and predecessor at the selected level, "up" and "down" pointers).

To leave, a node disconnects all its pointers; the concerned nodes consequently are aware and look up for replacements. Additionally, the stored items are transferred to the successor.

Replication and Fault Tolerance. Viceroy does not delve into ungraceful failures nor replication, but refers to Lynch et al.[23] for a general approach for handling failures in DHTs.

Implementation. There exists a Java implementation of Viceroy.[45] This homepage includes also a visualization applet that can illustrate the main topology, lookups, joins, and leaves in Viceroy.

Other Comments. While the intuitive analysis might lead to thinking that nodes at higher levels endure more lookup traffic, Viceroy's analysis shows that the congestion is not that bad; however, such a proof is beyond the scope of this chapter.

39.4.11 Ulysses

Ulysses is another system based on the Butterfly graph. It achieves the known limits of routing table and lookup path length, $O(\log(N))$ and $\frac{\log(N)}{\log(\log(N))}$, while accounting for joins, leaves, and failures. In that sense, it agrees with the conclusions of Koorde, Distance-Halving, and Viceroy and shows a second way of building a Butterfly network. Ulysses also depends on periodic stabilization for maintenance of the graph. Like Distance-Halving, it discusses the elimination of congestion. Ulysses also has an interesting discussion on the optimization of the average lookup path length.

39.4.12 CAN

The Overlay Graph. CAN[36] is in a class of its own. The design of the graph is based on a d-dimensional coordinate space. Like all other systems, the nodes and items are mapped onto a virtual space using a uniform hashing function, but the hashing is applied d times to get the d coordinates. For example, in a two-dimensional discrete coordinate space, an IP address or key of a file would be hashed once to obtain an x value and another time to obtain a y value. The coordinate space is dynamically partitioned among all the nodes in the system such that every node "owns" its distinct zone within the overall space. Figure 39.8 shows a discrete coordinate space of 16×16 partitioned among five nodes.

Mapping Items onto Nodes. An item with key k is stored at the node that owns the zone onto which k is mapped. Two nodes are neighbors; that is, they have pointers to each other if their zones have common sides.

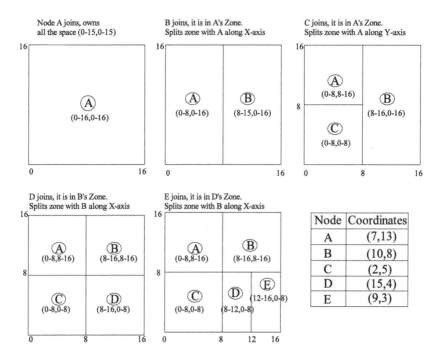

FIGURE 39.8   The process of five nodes joining a CAN network.

The Lookup Process. A lookup is achieved using the straight-line path through the Cartesian space from source to destination.

Joins, Leaves, and Maintenance. A new node w joins by selecting a random point P; it sends to its initial contact in the network u a JOIN message containing P. Node u consequently routes the message to the node v that owns the zone in which P lies. The zone of v is then split between v and w. Zones are split along the x-axis first, then along the y-axis. Upon a split, the new node learns its neighbors from the previous owner. Neighbors of a new node are neighbors of the previous owner, plus the previous owner itself. The new node informs its neighbors of the change. The cost of join in that way is $O(d)$. Finally, items that belong to the new node are obtained from the previous owner.

The leave process is the reverse; a node informs its neighbors of its leaving and merges its zone with a neighbor to produce a valid zone. If no valid zone can be formed, the items are transferred to a neighbor owning the smallest zone.

Under normal conditions, a node sends periodic updates to each of its neighbors, giving them its zone coordinates. Additionally, there is a background zone-balancing process that tries to reconfigure zones after a series of joins and leaves.

Replication and Fault Tolerance. There are two ways of detecting failures in CAN. The first occurs if a node tries to communicate with a neighbor and fails; it then takes over that neighbor's zone. The second way of detecting a failure is by not receiving the periodic update message after a long time. In the second case, the failure would probably be detected by all the neighbors, and all of them would try to take over the zone of the failed node; to resolve this, all nodes send to all other neighbors the size of their zone, and the node with the smallest zone takes over.

Replication in CAN is achieved in two ways. The first way is to use α hashing functions to map an item to α points. When retrieving an item, α queries are sent and α responses are received. The second way is to create multiple instances of the coordinate space. Each instance is called a "reality." If a node storing an item is dead in one reality, the item can be retrieved from one of the other realities because the item would be stored at other nodes in the other realities.

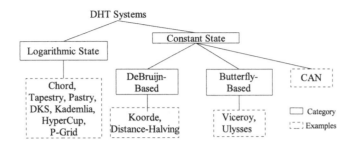

FIGURE 39.9 A classification of DHT systems based on the size of the node state and underlying graph.

Latency. Every node in CAN keeps the round-trip-time (RTT) of its neighbors. When selecting a path for a lookup, a CAN node forwards to the neighbor with maximum ratio of progress to RTT. CAN also has a mechanism for nodes to choose their points so as to make points near in the IP network also near in the Cartesian space; the technique uses root DNS servers as landmarks from which a node can approximate to which other nodes it is near in the IP network.

Upper Services. A multicast protocol is available for CAN.[37] Some work has also been done on richer queries, such as range queries.[5]

39.5 Summary

39.5.1 The Overlay Graph

We summarize the different overlay graphs by providing a classification based on the size of the node state, as shown in Figure 39.9. The first category is for systems that keep a logarithmic number of routing entries. Most DHT systems are in that category. A common property in that category is the logarithmic order lookup path length. The second category includes systems that use a constant number of routing entries. CAN is in a class of its own as it provides a polynomial order lookup path length. Other systems in the same category include the DeBruijn- and Butterfly-based DHTs, and such systems offer a logarithmic path length. Naturally, one can instead set the constant of the constant-state systems to a value logarithmic in the number of nodes and get a shorter lookup length. Table 39.1 summarizes those performance trade-offs.

39.5.2 Mapping Items onto Nodes

Four ways of assigning items to *ids* are identified and summarized in Table 39.2. In all those scenarios, the fair (load-balanced) assignment of items onto nodes relies on the uniform distribution of the hashing function. This is apart from P-Grid, where the network is in constant trial to load balance the items between nodes, irrespective of the distribution of identifiers.

TABLE 39.1 Summary of Node State and Lookup Path Length for the Different Categories of Systems

Category	Node State	Lookup Path Length
Logarithmic state	$O(\log(N))$	$O(\log(N))$
DeBruijn and Butterfly (per se)	$O(k)$	$O(\log(N))$
DeBruijn and Butterfly $(k = O(\log(N)))$	$O(\log(N))$	$\frac{O(\log(N))}{O(\log(\log(N)))}$
CAN (per se)	$O(k)$	$O(kN^{1/k})$
CAN $(k = O(\log(N)))$	$O(\log(N))$	$O(\log(N)N^{1/\log(N)})) = O(\log(N))$[a]

[a] Because $N^{1/\log(N)}$ is a constant factor.

TABLE 39.2   The Different Policies for Mapping Items onto Nodes

Assignment Policy	Example Systems
Item assigned to successor on the ring	Chord, DKS, Koorde, Viceroy, DH, Ulysses
Item assigned to numerically closest node	Pastry, Tapestry
Item assigned to XOR closest node	Kademlia
Item assigned to zone owner	CAN

39.5.3   The Lookup Process

The lookup process is a direct result of the node state. Increasing nodes decreases the lookup path length but increases the maintenance cost.

In some systems, such as Pastry, Tapestry, Kademlia, and CAN, overlay hops are not the sole optimized metric; additionally, network latency is addressed.

Congestion is a tricky issue related to the lookup process. Not only should the lookup path length be optimized, but it should not be the case that some nodes endure more traffic than others, which is the case in the DeBruijn and Butterfly graphs. However, authors of systems that suffer from congestion try to adapt those graphs to eliminate congestion.

39.5.4   Joins, Leaves, and Maintenance

Joins and leaves jeopardize the desired properties of any good graph, and different systems have adopted different techniques to bring back the overlay graph to its ideal state. Table 39.3 enumerates those techniques.

Stabilization is the most common technique, wherein routing table entries are periodically looked up and corrected. The use of traffic is adopted in Kademlia, where the graph structure makes a node receive lookups from the same nodes it is pointing to. Pastry also depends on the structure of the graph, where a new node can inform all the other nodes that need to be informed about it. Periodic activity is still needed, however, for collecting latency information. The correction-on-use introduced in DKS relies on the presence of traffic as well, but a receiving node can correct a sending node and no periodic activity is used. Where not sufficient alone, correction-on-use is complemented with a more deterministic technique, namely, correction-on-change. P-Grid has a unique correction mechanism, where the random interaction between peers can lead to the change of their *id*s in a way that causes eventual optimality of the graph.

39.5.5   Replication and Fault Tolerance

Replication is an essential tool for recovering items stored at failed nodes. The choice of nodes for replication is tightly coupled with the policy for mapping items to nodes. Local vicinity is most frequently chosen; for example, the successors on the circle or the k numerically closest nodes.

Fault tolerance is one of the most challenging and open areas in structured overlay networks. Some systems can cope with the failure of a small number of nodes at a time. However, dealing with a large

TABLE 39.3   Different Policies Overlay Graph Maintenance Policies

Maintenance Policy	Example Systems
Stabilization	Chord, Koorde, Viceroy, CAN
Use of traffic	Kademlia
Determinism + stabilization	Pastry, Tapestry
Correction-on-use + correction-on-change	DKS
Lazy + randomized	P-Grid

number of simultaneous failures is more difficult. A constant-state routing table is an advantage that must be given up if a large number of simultaneous failures must be tolerated. Nodes will have to keep their node state to at least logarithmic order to be able to cope with $N/2$ randomly distributed nodes failing simultaneously or the failure of $O(\log(n))$ adjacent node *ids* simultaneously.

39.6   Open Problems and Other Issues

Some of the open issues related to the aspects discussed in this chapter include the following:

- *Reducing maintenance cost.* While we have seen different techniques for dealing with the maintenance of the overlay structure, any optimization in that aspect is important for the overall performance of a DHT.
- *State-performance trade-off.* The trade-off between node state and lookup path length is fundamental. The current known limit is that a constant node state can provide logarithmic path length; but if fault tolerance is to be addressed, more state is required. It is still an open question as to whether a constant state suffices for a fault-tolerant system while preserving the logarithmic path length and without introducing congestion.
- *Performance of existing DHTs.* At the time of writing of this chapter, while many systems have been introduced, less work has been dedicated to measuring their performance in different aspects and for different applications.
- *Heterogeneity.* Many of the desirable DHT properties depend on some kind of uniform distribution. In practice, aspects such as connection time or bandwidth are evidently uniformly distributed for a given set of peers. Therefore, supporting heterogeneity is another open issue from a practical point of view.
- *Search and indexing.* All DHTs have the simple "put–get" interface. Consequently, the knowledge of the key of the sought item is assumed. However, an efficient listing of all items that have some common property is rather challenging in a DHT.
- *Grid computing.* Because both Grid computing and P2P computing are both forms of resource sharing, it is a current hot topic to investigate how results of both areas can be mutually beneficial.

Other unrelated active and challenging research topics include security, trust, anonymity, denial-of-service attacks, malicious node behavior, reputation, and incentives.

Acknowledgments

We thank the following members of the distributed systems lab at the Swedish Institute of Computer Science: Per Brand, Luc Onana Alima, Ali Ghodsi, and Erik Aurell, for the numerous discussions, insights, and the several points that they explained and clarified to us in the area of DHTs. This work was funded by the Swedish Vinnova PPC project and the European projects PEPITO and EVERGROW.

References

1. Karl Aberer. P-Grid: a self-organizing access structure for P2P information systems. In *Proceedings of the Sixth International Conference on Cooperative Information Systems (CoopIS'2001)*, Trento, Italy, 2001.
2. Luc Onana Alima, Sameh El-Ansary, Per Brand, and Seif Haridi. DKS(N; k; f): a family of low communication, scalable and fault-tolerant infrastructures for P2P applications. In *The 3rd International Workshop on Global and Peer-to-Peer Computing on Large Scale Distributed Systems (CCGRID'2003)*, Tokyo, Japan, May 2003. http://www.ccgrid.org/ccgrid2003.
3. Luc Onana Alima, Ali Ghodsi, Per Brand, and Seif Haridi. Multicast in DKS(n, k, f) overlay networks. In *The 7th International Conference on Principles of Distributed Systems (OPODIS'2003)*. Springer-Verlag, 2004.

4. Luc Onana Alima, Ali Ghodsi, and Seif Haridi. A framework for structured peer-to-peer overlay networks. In *LNCS Volume of the Post-Proceedings of the Global Computing 2004 Workshop*. Springer-Verlag, 2004.

5. Artur Andrzejak and Zhichen Xu. Scalable, efficient range queries for grid information services. In *2nd International Conference on Peer-to-Peer Computing*, pages 33–40, Linkping, Sweden, September 2002. IEEE Computer Scociety. ISBN-0-7695-1810-9.

6. Miguel Castro, Peter Druschel, Anne-Marie Kermarrec, Animesh Nandi, Antony Rowstron, and Atul Singh. Splitstream: high-bandwidth multicast in a cooperative environment. In *19th ACM Symposium on Operating Systems Principles (SOSP'03)*, 2003.

7. Miguel Castro, Peter Druschel, Anne-Marie Kermarrec, and Antony Rowstron. Scribe: a large-scale and decentralized application-level multicast infrastructure. *IEEE Journal on Selected Areas in Communication (JSAC)*, 20(8), October 2002.

8. Russ Cox, Athicha Muthitacharoen, and Robert Morris. Serving DKS using Chord. In *Proceedings of the 1st International Workshop on Peer-to-Peer Systems (IPTPS)*, Cambridge, MA, March 2002.

9. Frank Dabek, M. Frans Kaashoek, David Karger, Robert Morris, and Ion Stoica. Wide-area cooperative storage with CFS. In *Proceedings of the 18th ACM Symposium on Operating Systems Principles (SOSP'01)*, Chateau Lake Louise, Banff, Canada, October 2001.

10. Sameh El-Ansary, Luc Onana Alima, Per Brand, and Seif Haridi. Efficient broadcast in structured P2P networks. In *2nd International Workshop on Peer-to-Peer Systems (IPTPS'03)*, Berkeley, CA, February 2003.

11. Emule. The emule file-sharing application homepage, 2004. http://www.emule-project.net/.

12. FIPS 180-1. Secure Hash Standard. U.S. Department of Commerce/NIST, National Technical Information Service, Springfield, VA, April 1995.

13. FreeNet, 2003. http://freenet.sourceforge.net.

14. FreePastry. The FreePastry homepage, 2004. http://www.cs.rice.edu/CS/Systems/Pastry/FreePastry.

15. Ali Ghodsi, Luc Onana Alima, Sameh El-Ansary, Per Brand, and Seif Haridi. Self-correcting broadcast in distributed hash tables. In *Series on Parallel and Distributed Computing and Systems (PDCS'2003)*, Calgary, 2003. ACTA Press.

16. Ali Ghodsi, Luc Onana Alima, and Seif Haridi. A Novel Replication Scheme for Load-Balancing and Increased Security. Technical Report TR-2004-11, SICS, June 2004.

17. Gnutella, 2003. http://www.gnutella.com.

18. Steven Hand and Timothy Roscoe. Mnemosyne: peer-to-peer steganographic storage. *Lecture Notes in Computer Science (IPTPS'02)*, pages 130–140, 2002.

19. Clint Heyer. Naanou home page, 2004. http://naanou.sourceforge.net.

20. Sitaram Iyer, Antony Rowstron, and Peter Druschel. Squirrel: a decentralized peer-to-peer web cache. In *12th ACM Symposium on Principles of Distributed Computing (PODC'2002)*, 2002.

21. Frans Kaashoek and David R. Karger. Koorde: a simple degree-optimal distributed hash table. In *2nd International Workshop on Peer-to-Peer Systems (IPTPS'03)*, Berkeley, CA, February 2003.

22. John Kubiatowicz, David Bindel, Yan Chen, Patrick Eaton, Dennis Geels, Ramakrishna Gummadi, Sean Rhea, Hakim Weatherspoon, Westly Weimer, Christopher Wells, and Ben Zhao. OceanStore: an architecture for global-scale persistent storage. In *Proc. of ASPLOS*, ACM, Nov. 2000.

23. Nancy Lynch, Dahlia Malkhi, and David Ratajczak. Atomic data access in content addressable networks. In *The 1st International Workshop on Peer-to-Peer Systems (IPTPS'02)*, 2002. http://www.cs.rice.edu/Conferences/IPTPS02/.

24. D. Malkhi, M. Naor, and D. Ratajczak. Viceroy: a scalable and dynamic emulation of the Butterfly. In *Proceedings of the 21st ACM Symposium on Principles of Distributed Computing (PODC'02)*, August 2002.

25. E.P. Markatos. Tracing a large-scale peer to peer system: an hour in the life of Gnutella. In *The Second International Symposium on Cluster Computing and the Grid*, 2002. http://www.ccgrid.org/ccgrid2002.

26. MathWorld. The Butterfly graph, 2004. http://mathworld.wolfram.com/Butterfly-Graph.html.

27. MathWorld. The De Bruijn graph, 2004. http://mathworld.wolfram.com/deBruijn-Graph.html.
28. Petar Maymounkov and David Mazires. Kademlia: a peer-to-peer information system based on the XOR metric. In *The 1st International Workshop on Peer-to-Peer Systems (IPTPS'02)*, 2002. http://www.cs.rice.edu/Conferences/IPTPS02/.
29. Athicha Muthitacharoen, Robert Morris, Thomer M. Gil, and Benjie Chen. Ivy: a read/write peer-to-peer file system. In *Proceedings of the 5th Symposium on Operating Systems Design and Implementation*, Boston, MA, December 2002. USENIX Association.
30. Moni Naor and Udi Wieder. Novel architectures for P2P applications: the continuous-discrete approach. In *Proceedings of SPAA 2003*, 2003.
31. Napster. Open source napster server, 2002. http://opennap.sourceforge.net/.
32. Overnet. The Overnet file-sharing application homepage, 2004. http://www.overnet.com.
33. P-Grid. The P-Grid homepage, 2004. http://www.p-grid.org.
34. Planet-Lab. The Planet-Lab homepage. http://www.planet-lab.org.
35. C. Greg Plaxton, Rajmohan Rajaraman, and Andrea W. Richa. Accessing nearby copies of replicated objects in a distributed environment. In *ACM Symposium on Parallel Algorithms and Architectures*, pages 311–320, 1997.
36. Sylvia Ratnasamy, Paul Francis, Mark Handley, Richard Karp, and Scott Shenker. A scalable content addressable network. In *Proceedings of the ACM SIGCOMM '01 Conference*, Berkeley, CA, August 2001.
37. Sylvia Ratnasamy, Mark Handley, Richard Karp, and Scott Shenker. Application-level multicast using content-addressable networks. In *Third International Workshop on Networked Group Communication (NGC'01)*, 2001. http://www-mice.cs.ucl.ac.uk/ngc2001/.
38. M. Ripeanu, I. Foster, and A. Iamnitchi. Mapping the Gnutella network: properties of large-scale peer-to-peer systems and implications for system design. *IEEE Internet Computing Journal*, 6(1), 2002.
39. Antony Rowstron and Peter Druschel. Pastry: scalable, distributed object location and routing for large-scale peer-to-peer systems. In *IFIP/ACM International Conference on Distributed Systems Platforms (Middleware)*, 329–350, 2001.
40. Antony Rowstron and Peter Druschel. Storage management and caching in PAST, a large-scale, persistent peer-to-peer storage utility. In *18th ACM Symposium on Operating Systems Principles (SOSP'01)*, 188–201, 2001.
41. Mario Schlosser, Michael Sintek, Stefan Decker, and Wolfgang Nejdl. HyperCup — hypercubes, ontologies and efficient search on peer-to-peer networks, May 2003. http://www-db.stanford.edu/schloss/docs/HyperCuP-LNCS2530.ps.
42. Ion Stoica, Robert Morris, David Karger, M. Frans Kaashoek, and Hari Balakrishnan. Chord: a scalable peer-to-peer lookup service for Internet applications. In *Proceedings of the ACM SIGCOMM'01 Conference*, pages 149–160, San Diego, CA, August 2001.
43. Ion Stoica, Robert Morris, David Liben-Nowell, David Karger, M. Frans Kaashoek, Frank Dabek, and Hari Balakrishnan. Chord: a scalable peer-to-peer lookup service for Internet applications. *IEEE Transactions on Networking*, 11, 2003.
44. The Chord Project Home Page, 2003. http://www.pdos.lcs.mit.edu/chord/.
45. Viceroy. Java implementation and visualization applet, 2004. http://www.ece.cmu.edu/atalmy/viceroy.
46. Ben Y. Zhao, Ling Huang, Sean C. Rhea, Jeremy Stribling, Anthony D Joseph, and John D. Kubiatowicz. Tapestry: a global-scale overlay for rapid service deployment. *IEEE J-SAC*, 22(1):41–53, January 2004.
47. Feng Zhou, Li Zhuang, Ben Y. Zhao, Ling Huang, Anthony D. Joseph, and John D. Kubiatowicz. Approximate object location and spam filtering on peer-to-peer systems. In *Proc. of Middleware*, pages 1–20, Rio de Janeiro, Brazil, June 2003. ACM.
48. Shelley Q. Zhuang, Ben Y. Zhao, Anthony D. Joseph, Randy H. Katz, and John D. Kubiatowicz. Bayeux: an architecture for scalable and fault-tolerant wide-area data dissemination. In *Proc. of NOSSDAV*, pages 11–20. ACM, June 2001.

Distributed Data Structures for Peer-to-Peer Systems

James Aspnes

Gauri Shah

40.1 Introduction

Peer-to-peer (P2P) networks are distributed systems without any central authority that are used for efficient location of shared resources. Such systems have become very popular for Internet applications in a short period of time. A survey of recent P2P research yields a slew of desirable features for a peer-to-peer network such as decentralization, scalability, fault tolerance, self-stabilization, data availability, load balancing, dynamic addition and deletion of peer nodes, efficient and complex query searching, incorporating geography in searches, and exploiting spatial as well as temporal locality in searches.

Early "unstructured" systems, such as Napster,[29] Gnutella,[17] and Freenet,[7] did not support most of these features and were clearly unscalable, either due to the use of a central server (Napster) or to high message complexity from performing searches by flooding the network (Gnutella). These early peer-to-peer systems implemented loose organization protocols that did not guarantee any performance from the network.

This chapter surveys some data structures that have been proposed to build peer-to-peer systems that do not have these drawbacks. In particular, we give a broad overview of Distributed Hash Tables (DHTs),[27,34,35,39,40] as well as other organized peer-to-peer systems that do not use DHTs, such as censorship resistant networks[8,12,30] and TerraDir.[6,20,36] We then focus on a data structure called a *skip graph*[5] that overcomes some of the limitations of DHTs, such as lack of spatial locality and lack of support for range queries.

40.2   Distributed Hash Tables

The first-generation peer-to-peer systems inspired the development of more sophisticated second-generation ones such as CAN,[35] Chord,[39] Pastry,[34] and Tapestry.[40] Although these systems appear vastly different, there is a recurrent underlying theme: each resource is assigned a location in a virtual *metric space* by hashing its key, and each machine in the system is assigned responsibility for some region of the metric space. The machine maintains either the addresses of the resources that hash to this region, or it can host the resources themselves. Routing is done *greedily* by forwarding packets from each machine to its neighbor *closest* to the target node with respect to the metric distance. This inherent common structure leads to similar results for the performance of such networks.

We now give a brief overview of such distributed hash table systems, highlighting some of their major design aspects.

40.2.1   Content-Addressable Network (CAN)

CAN[35] partitions a d-dimensional coordinate space into *zones* that are owned by nodes. Each resource key is mapped to a point in the coordinate space using a uniform hash function, and then stored at the node that owns the zone in which this point is located. Each node maintains an $O(d)$ state about its neighbors that abut with the node only in one dimension. Greedy routing, by forwarding messages to the neighbor closest to the target zone, takes $O(dm^{1/d})$ time with m nodes in the network. With $d = \log m$, each node uses $O(\log m)$ space and routing takes $O(\log m)$ time.

To join the network, a new node first picks a random point in the coordinate space and then contacts the node that owns the zone in which this random point lies. The two nodes then split the zone between them, each taking responsibility for half of the previous zone. The dimension along which the split occurs is chosen in a fixed round-robin fashion among all possible dimensions. Node failures are handled in the same way by merging zones of failed neighbors. Various additional design improvements include multiple coordinate spaces to improve data availability, better routing metrics such as forwarding to the neighbor with the smallest round-trip latency, multiple peers per zone, multiple hash functions for fault tolerance, etc.

40.2.2   Tapestry and Pastry

Tapestry[40] and Pastry[34] use the PRR algorithm[32] as the core routing algorithm, with modifications to the other features of the design to make it suitable for use in a distributed environment.

The PRR algorithm, which uses *suffix-based hypercube routing*, is designed for accessing shared resources in an overlay network while using small-sized routing tables at each node in the network. Each node has a unique t-bit identifier divided into r levels of $w = t/r$ bits each; let w bits represent a *digit*. A node x

will have r sets of 2^w neighbors each, such that each set i, $0 \leq i < r$, will have nodes with identifiers as follows: i common digits with x's identifier, followed by all possible 2^w values for the $(i + 1)$st digit, and any of the 2^w possible values for each of the remaining digits.

Let $b = 2^w$. Then with m nodes in the network, the size of the routing table is $(b - 1) \cdot \lceil \log_b m \rceil$ and routing takes $\lceil \log_b m \rceil$ time. This was probably the first routing algorithm introduced that was scalable for peer-to-peer systems. However, in the original scheme, each resource was associated with a unique *root* node that maintained the address of the resource. This has two disadvantages: (1) it requires global knowledge to map resources to their root nodes, and (2) it is not fault tolerant.

To incorporate fault tolerance, Tapestry[40] maintains multiple neighbors per entry in the routing table to route around failures. It also eliminates the need for a central entity for mapping resources to root nodes, by assigning a resource to the node whose identifier matches the hash value of the resource key in the maximum number of trailing bits. *Surrogate routing* is used to locate a resource by also storing it at a node closest to its chosen root node, if the latter is absent. There are many more optimizations, such as dynamic node addition and deletion, soft-state publishing, and supporting mobile resources, that are too involved to describe here, and the interested reader is referred to Zhao et al.[40]

Pastry[34] uses *prefix*-based routing instead of suffix-based routing as in Tapestry. Each node maintains a *neighborhood* set and a *leaf* set in addition to the routing table, as per the PRR algorithm. The neighborhood set of a node consists of a group of nodes that are closest to it, as per the proximity metric; they help the node choose the closest nodes that satisfy the routing table criteria. The leaf set L is the set of nodes with the $|L|/2$ numerically closest larger identifiers, and the $|L|/2$ numerically closest smaller identifiers relative to the node's own identifier. If the routing table node is not accessible, the current node will forward the message to one of the leaf set nodes whose identifier prefix matches the target in the same number of bits as the current node, but whose identifier is numerically closer to that of the target. This procedure ensures that the routing always converges, although not necessarily efficiently.

40.2.3 Chord

Chord[39] provides hash-table functionality by mapping m machines (using consistent hashing[23]) to identities of $\log m$ bits placed around an *identifier circle*. Each node x stores a pointer to its immediate *successor*, that is, the closest node in the clockwise direction along the circle. In addition, it also maintains a *finger table* with $\log m$ entries such that the i-th entry stores the identity of the first node at or after position $x + 2^{i-1}$ on the identifier circle.

Each resource is also mapped using hashing onto the identifier circle and stored at the first node at or after the location that it maps to. Routing is done greedily to the farthest node in the routing table that does not appear beyond the target, and it is not hard to see that this gives a total of $O(\log m)$ routing hops per search. Further, consistent hashing ensures that no node is responsible for maintaining much more than its fair share of resources.

For fault tolerance, each node maintains a *successor list*, which consists of the next several successor nodes, instead of a single one. Theoretical results about the performance of Chord in the presence of concurrent joins and involuntary departures can be found in Stoica et al.[39]

40.2.4 Fault-Tolerant Routing

Aspnes et al.[3] use *small-world* routing techniques described by Kleinberg[22] to build a peer-to-peer system. The resources are embedded in a one-dimensional real line based on the hash values of their identifiers. Each node in the metric space is connected to its immediate neighbors to the left and right. In addition, each node chooses k long-distance neighbors with probability inversely proportional to the distance between itself and those neighbors. Similar to other DHTs, routing is done greedily.

The authors[3] prove that the routing time is $O(\log n / k)$ with n nodes and k long-distance neighbors. They also prove upper bounds on the number of hops in the presence of link and node failures. These bounds are inversely proportional to the probability of failure. They also present a lower bound of $\Omega(\frac{\log^2 n}{k \log \log n})$ on the number of hops, assuming certain uniformity conditions on the link distribution. Using similar techniques in higher dimensions and proving tolerance against Byzantine failures were left as open questions in the article.[3]

40.3   Censorship-Resistant Networks

Some DHT systems are partly resilient to random node failures, but their performance may be badly impaired by adversarial deletion of nodes. Fiat and Saia[12] introduced censorship-resistant networks that are resilient to adversarial deletion of a constant fraction of the nodes,* such that most of the remaining live nodes can access most of the remaining data items. With m nodes in the network, the network topology is based on a butterfly network of depth $(\log m - \log \log m)$, where each node represents a set of peers and each peer can be a member of multiple nodes. An expander graph is maintained between any two connected nodes of the network. Multiple copies of the data items are stored in the nodes in the bottom-most level, and searches are run in parallel for any copy of a data item. Each search takes $O(\log m)$ time, $O(\log^2 m)$ messages, and $O(\log m)$ storage per node.

The authors[12] also give a *spam-resistant* variant of this scheme that requires additional storage space and messages. In this variant, the adversary can take control of up to half of the nodes and yet be unable to generate false data items.

The main drawback of the basic censorship-resistant network is that it can survive only a static attack, and cannot be dynamically maintained as more nodes join the network. Some extensions of this result for dynamic maintenance can be seen in Saia et al.[37] and Datar.[8] However, the construction of such networks is very complex, and it is still an open question to dynamically maintain such a network when it grows enough to need a new level of butterfly linkage. A simpler design of a DHT that is provably fault tolerant to *random faults* was recently given by Naor and Wieder.[30]

40.4   Other Systems

The common feature of distributed hash tables is *namespace virtualization*, where the original key of each resource is mapped to a *virtual key* by a hash function. There are other peer-to-peer systems that either eliminate namespace virtualization or adopt a hybrid approach that involves a less rigid structure compared to DHTs, while still providing better search efficiency compared to first-generation systems such as Gnutella. We explore some of these designs in the following sections.

40.4.1   Non-Virtualized Systems

These systems do not hash keys, but instead structure the system in whole or in part based on the structure of keys, giving a tree-like organization in most cases. A particular case of a non-virtualized system *not* described here is a *skip graph*, which is described in more detail in Section 40.6.

40.4.1.1   TerraDir

TerraDir[6,20,36] is a recent system that provides a directory lookup for *hierarchical namespaces*, such as DNS names or UNIX file system names, as opposed to the flat namespaces in DHTs. Caching and replication are heavily used for both fault tolerance and reduction of query latencies. Nodes at height i in the hierarchical namespace tree are replicated $O(i)$ times, so that the root is replicated $O(\log m)$ times (with m nodes in

---

*The article[12] describes a network that is robust up to deletion of half of the nodes, but it can be generalized to any arbitrary fraction of deleted nodes.

the network), while the leaves have just one replica. Each node is replicated randomly at other nodes, and each parent maintains information about all the replicas of each of its children. Routing is performed by going "up" in the tree until the longest common prefix of the source and destination is reached, and then going "down" until the destination is reached.

Each node also calculates the distance to the target from all possible replicas that it maintains and chooses the closest, live one to improve routing. As the namespace is not virtualized, spatial locality is maintained, which supports non-point queries such as /J/*. But this support is only partial because if a request is serviced from some replica and not the original node, then it may be possible that the address for the next data item is not known. Further, there are as yet no provable guarantees on load balancing and fault tolerance for this network.

40.4.1.2 Distributed Trie

Freedman and Vingralek[13] use a distributed trie[10,11,25] to support non-point queries; the keys are maintained in a trie, which is distributed among the peers depending on their access locality. Updates are done lazily by piggybacking information on query traffic. This algorithm reduces the message traffic compared to flooding but can degenerate to a broadcast for all nodes with stale views of the network.

40.4.1.3 SkipNet

SkipNet is another non-virtualized system, very similar in spirit to the skip graphs described in detail in later sections. It was developed independently of skip graphs by Harvey et al.[19]

SkipNet builds a trie of circular, singly linked skip lists to link the *machines* in the system. The machines' names are grouped by associated *domains* represented by DNS domain names and corresponding to organizational boundaries (for example, www.ibm.com for machines owned and operated by IBM). Pointers between machines in different organizations are built using a structure essentially identical to a skip graph. Within an organization, the machines are also linked using a DHT, and the resources are uniformly distributed over all the machines using hashing. A search consists of two stages: (1) the search locates the domain in which a resource lies using a search operation similar to that in a skip graph; and (2) once the search reaches some machine inside a particular domain, it uses greedy routing as in the DHT to locate the resource within that domain.

The SkipNet design ensures *path locality*; that is, the traffic within a domain traverses other nodes only within the same domain. Further, each organization hosts its own data, which provides *content locality* and some security against snooping or malicious competitors. Finally, using the hybrid storage and search scheme provides *constrained load balancing*, where the cost of storing resources is spread evenly within each domain.

However, as the name of the data item includes the domain in which it is located, transparent remapping of resources to other domains is not possible, thus giving a very limited form of load balancing. Another drawback of this design is that it does not give full-fledged spatial locality. For example, if the resources are document files, sorting according to the domain on which they are served gives no advantage in searching for related files compared to DHTs.

40.4.2 Hybrid Systems

Hybrid systems mix features from distributed hash tables into unstructured networks. Two examples are given below.

40.4.2.1 Yappers

Yappers[18] has an unstructured network like Gnutella but introduces hashing for storing keys. Each machine x is assigned one of b *colors* based on its IP address: $\text{color}(x) = \text{hash}(\text{IP}(x)) \bmod b$. Each node maintains information about an *immediate neighborhood*, which comprises all nodes within a constant number of routing hops from itself. Resources are stored at nodes whose color matches the hash value of the resource key, so searches are only forwarded to nodes of the same color. While this improves the performance of the Gnutella search, it does not give any provable guarantees on routing time or load balancing.

40.4.2.2 Kelips*

Kelips* is a hybrid system between a DHT and a super-peer network, where m nodes are clustered into $O(\sqrt{m})$ *affinity groups* using hashing.[14] Each node resembles a super-peer by maintaining information about all the nodes and resources in its own group, and one contact from every other group. Space requirement per node is $O(\sqrt{m})$ and routing takes $O(1)$ time. Updated network information is propagated using gossip protocols given by Kempe et al.[21] For larger peer-to-peer systems, the space and maintenance costs are prohibitively high, and routing may fail due to stale routing tables.

40.5   The Purpose and Price of Hashing

The purpose of the hash function in DHTs is twofold:

1. Hashing maps resources to nodes uniformly so that no node gets much more than its fair share of resources to manage. In Chord with m machines and n resources, using a base hash function such as SHA-1[1] guarantees with high probability that each node is responsible for at most $(1 + \log m) \cdot n/m$ keys. Thus, hashing provides a natural load balancing among the nodes.
2. In addition, hashing the node identifiers (for example, their IP addresses) distributes the nodes uniformly in the overlay metric space. As a result of that, it does not need any additional mechanism to balance the tree that is maintained for routing in the network.

However, hashing results in scrambling the resource keys and, in turn, destroys *spatial locality*. Spatial locality is the property where related resources are located close to each other in the data structure. For example, if the resources are Web pages, then if the network maintained spatial locality `http://www.cnn.com` and `http://www.cnn.com/weather` would be located close to each other in the data structure. If a search for any resource $i + 1$ is immediately preceded by a search for resource i, the network would be able to use the information from the first search to improve the performance of the second search. This functionality helps applications such as prefetching of Web pages and smart browsing. In a DHT, hashing will map these resources to random locations in the overlay network, and the two searches will be completely independent of each other.

Also, as hashing destroys the ordering on the keys, DHT systems do not easily support complex queries such as near matches to a key, keys within a certain range, or approximate queries. Such systems cannot be easily used for applications such as versioning and user-level replication without adding another layer of abstraction and its associated maintenance costs. In contrast, a data structure that does not destroy the key ordering can be used to provide all these features, allowing for a simple underlying architecture. At the same time, we can continue to use the load-balancing properties of the DHTs by using hashing, independent of the mechanism for resource location.

40.6   Skip Graphs

We now describe a model for a peer-to-peer network based on a distributed data structure called a *skip graph*.[5] Before going on to explain skip graphs in detail, we give an overview of the advantages and disadvantage of this data structure. Resource location and dynamic node addition and deletion can be done in time logarithmic in the size of the graph, and each node in a skip graph requires only logarithmic space to store information about its neighbors. More importantly, there is no hashing of the resource keys, so related resources are present near each other in a skip graph. As explained earlier, this may be useful for certain applications such as prefetching of Web pages, enhanced browsing, and efficient searching.

Skip graphs also support *complex queries* such as range queries, that is, locating resources whose keys lie within a certain specified range*.

Skip graphs are resilient to node failures: a skip graph tolerates removal of a large fraction of its nodes chosen at random without becoming disconnected, and even the loss of an $O(1/\log n)$ fraction of the nodes chosen by an adversary still leaves most of the nodes in the largest surviving component. Skip graphs can also be constructed without knowledge of the total number of nodes in advance. In contrast, DHT systems such as Pastry and Chord require *a priori* knowledge about the size of the system or its keyspace. Although these systems initially choose a very large keyspace that cannot be exhausted easily (e.g., 2^{128} identifiers), no requirement about the knowledge of the size of the keyspace is still an interesting property.

A disadvantage of skip graphs is that they do not necessarily group resources in the same key range together on the same machine. This may lead to many pointers between machines that will need to be maintained and repaired at the cost of high message traffic. The underlying issue is that skip graphs by themselves do not include any policy for placement of resources on machines, and take responsibility only for routing between resources. The issue of how to place resources on machines to optimize a skip graph is still open; we discuss some possibilities in Section 40.6.2.

40.6.1 Structure of a Skip Graph

A skip graph is a generalization of the *skip list*, a randomized balanced tree data structure invented by Pugh.[33] We begin by describing skip lists and then describe the modifications needed to construct a skip graph.

A skip list is organized as a tower of increasingly sparse linked lists. Level 0 of a skip list is a linked list of all nodes in increasing order by key. For each i greater than 0, each node in level $i - 1$ appears in level i independently with some fixed probability p. In a doubly linked skip list, each node stores a predecessor pointer and a successor pointer for each list in which it appears, for an average of $\frac{2}{1-p}$ pointers per node. The lists at the higher level act as "express lanes" that allow the sequence of nodes to be traversed quickly. Searching for a node with a particular key involves searching first in the highest level, and repeatedly dropping down a level whenever it becomes clear that the node is not in the current level. Considering the search path in reverse shows that no more than $\frac{1}{p}$ nodes are searched on average per level, giving an average search time of $O(\frac{\log n}{p \log(1/p)})$ with n nodes at level 0. Skip lists have been extensively studied;[9,24,26,31,33] and because they require no global balancing operations, they are particularly useful in parallel systems.[15,16]

We would like to use a data structure similar to a skip list to support typical binary tree operations on a sequence whose nodes are stored at separate locations in a highly distributed system subject to unpredictable failures. A skip list alone is not enough for our purposes, because it lacks redundancy and is thus vulnerable to both failures and congestion. Because only a few nodes appear in the highest-level list, each such node acts as a single point of failure whose removal partitions the list and forms a hot spot that must process a constant fraction of all search operations. Skip lists also offer few guarantees that individual nodes are not separated from the rest, even with occasional random failures. Because each node is connected on average to only $O(1)$ other nodes, even a constant probability of node failures will isolate a large fraction of the surviving nodes.

These considerations lead to the expansion of a skip list to form a skip graph. As in a skip list, each of the n nodes in a skip graph is a member of multiple linked lists. The level 0 list consists of all nodes in sequence. Where a skip graph is distinguished from a skip list is that there may be many lists at level i, and every

---

*Skip graphs support complex queries along a single dimension (i.e., for one attribute of the resource, for example, its name key).

node participates in one of these lists until the nodes are splintered into singletons after $O(\log n)$ levels on average. A skip graph supports search, insert, and delete operations analogous to the corresponding operations for skip lists; indeed, we show in Lemma 40.1 that algorithms for skip lists can be applied directly to skip graphs, as a skip graph is equivalent to a collection of n skip lists that happen to share some of their lower levels.

Because there are many lists at each level, the chances that any individual node participates in some search is small, eliminating both single points of failure and hot spots. Furthermore, each node has $\Theta(\log n)$ neighbors on average, and with high probability no node is isolated. In Section 40.6.4 we observe that skip graphs are resilient to node failures and have an expansion ratio of $\Omega(\frac{1}{\log n})$ with n nodes in the graph.

We now give a formal definition of a skip graph. Precisely which lists a node x belongs to is controlled by a *membership vector* $m(x)$. We think of $m(x)$ as an infinite random word over some fixed alphabet, although in practice only an $O(\log n)$ length prefix of $m(x)$ needs to be generated on average. The idea of the membership vector is that every linked list in the skip graph is labeled by some finite word w, and a node x is in the list labeled by w if and only if w is a prefix of $m(x)$.

To reason about this structure formally, we will need some notation. Let Σ be a finite alphabet, let Σ^* be the set of all finite words consisting of characters in Σ, and let Σ^ω consist of all infinite words. We use subscripts to refer to individual characters of a word, starting with subscript 0; a word w is equal to $w_0 w_1 w_2 \ldots$. Let $|w|$ be the length of w, with $|w| = \infty$ if $w \in \Sigma^\omega$. If $|w| \geq i$, write $w \upharpoonright i$ for the prefix of w of length i. Write ϵ for the empty word. If v and w are both words, write $v \preceq w$ if v is a prefix of w; that is, if $w \upharpoonright |v| = v$. Write w_i for the i-th character of the word w. Write $w_1 \wedge w_2$ for the common prefix (possibly empty) of the words w_1 and w_2.

Returning to skip graphs, the bottom level is always a doubly linked list S_ϵ consisting of all the nodes in order as shown in Figure 40.1. In general, for each w in Σ^*, the doubly linked list S_w contains all x for which w is a prefix of $m(x)$, in increasing order. We say that a particular list S_w is part of level i if $|w| = i$. This gives an infinite family of doubly linked lists; in an actual implementation, only those S_w with at least two nodes are represented. A skip graph is precisely a family $\{S_w\}$ of doubly linked lists generated in this fashion. Note that because the membership vectors are random variables, each S_w is also a random variable.

We can also think of a skip graph as a random graph, where there is an edge between x and y whenever x and y are adjacent in some S_w. Define x's left and right neighbors at level i as its immediate predecessor and successor, respectively, in $S_{m(x) \upharpoonright i}$, or $\bot$ if no such nodes exist. We will write $x L_i$ for x's left neighbor at level i and $x R_i$ for x's right neighbor, and in general will think of the R_i as forming a family of associative composable operators to allow writing expressions like $x R_i R_{i-1}^2$ etc. We write x.maxLevel for the first level ℓ at which x is in a singleton list, that is, x has at least one neighbor at level $\ell - 1$.

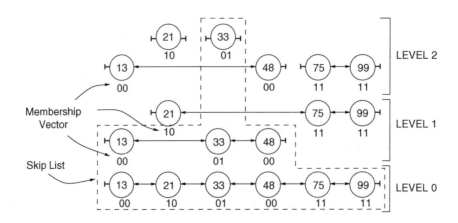

FIGURE 40.1 A skip graph with $n = 6$ nodes and $\lceil \log n \rceil = 3$ levels.

If we think of a skip list formally as a sequence of random variables $S_0, S_1, S_2, \ldots$, where the value of S_i is the level i list, then we have:

Lemma 40.1 *Let $\{S_w\}$ be a skip graph with alphabet Σ. For any $z \in \Sigma^\omega$, the sequence $S_0, S_1, S_2, \ldots$, where each $S_i = S_{z \restriction i}$, is a skip list with parameter $p = |\Sigma|^{-1}$.*

We omit all proofs in this chapter due to lack of space. The interested reader is referred to Shah.[38]

For a node x with membership vector $m(x)$, let the skip list $S_{m(x)}$ be called the *skip list restriction* of node x.

40.6.2 Implementation

Each node in a skip graph is typically assumed to be a resource. This is in contrast to DHTs, which construct a data structure on machines and assign many resources to each machine. Skip graphs are agnostic about resource placement, and do not require any particular placement policy for searches to work. But a cost of this in practice is that a straightforward implementation of a skip graph may contain many more pointers between machines than a similar DHT. Although the space used to store these extra pointers is likely to be small compared to the space used to store the data the skip graph indexes, having many pointers between machines will add to the expense of detecting and repairing damage from machine failures.

Because most systems are likely to contain more resources than machines, an implementation of a skip graph will need to adopt some policy for assigning resources to machines, to optimize trade-offs between efficiency, security, the desire to minimize inter-machine pointers, and load balancing. Some possibilities are:

1. As in the World Wide Web, each resource is hosted by some originating machine typically owned by the provider of the resource. The skip graph acts as a directory for these resources.
2. As in a DHT, each resource (or a pointer to it) is assigned to one of a set of server machines in some uniform fashion to ensure even distribution of load. Because searching a skip graph does not depend on the physical location of the resources, any load-balancing policy that ensures an even distribution can be applied.
3. To reduce the number of pointers between machines, resources with nearby keys can be stored in the same machine, and the skip graph can be limited to only a small number of "sample" resources from each machine. A search for a particular target now looks for a machine that contains a nearby sample resource; the actual target will be located on this machine or one of its immediate neighbors. This approach retains the range-query capabilities of skip graphs while reducing the number of inter-machine pointers to a level similar to DHTs; however, it requires a separate resource-migration mechanism to manage load balancing while preserving node order. An example of such a mechanism has recently been described by Aspnes et al.[4]

40.6.3 Algorithms for a Skip Graph

In this section we describe the search, insert, and delete operations for a skip graph. For simplicity, we refer to the key of a node (e.g., x.key) with the same notation (e.g., x) as the node itself. It will be clear from the context whether we refer to a node or its key. In the algorithms we denote the pointer to x's successor and predecessor at level ℓ as x.neighbor$[R][\ell]$ and x.neighbor$[L][\ell]$, respectively. We define $x R_\ell$ formally to be the value of x.neighbor$[R][\ell]$, if x.neighbor$[R][\ell]$ is a non-nil pointer to a non-faulty node, and $\perp$ otherwise. We define $x L_\ell$ similarly.

40.6.3.1 The Search Operation

The search operation (Algorithm 40.1) is identical to the search in a skip list with only minor adaptations to run in a distributed system. The search is started at the topmost level of the node seeking a key and it proceeds along each level without overshooting the key, continuing at a lower level if required, until

---

Algorithm 40.1   Search for node v

---

1   upon receiving $\langle$searchOp, startNode, searchKey, level$\rangle$:
2   **if** *(v.key = searchKey)* **then**
3 | **send** $\langle$foundOp, $v\rangle$ to startNode
4   **if** *(v.key < searchKey)* **then**
5 | **while** *level* ≥ 0 **do**
6 | **if** *((v.neighbor[R][level].key < searchKey)* **then**
7 | **send** $\langle$searchOp, startNode, searchKey, level$\rangle$ to v.neighbor[R][level]
8 | **break**
9 | **else** level $\leftarrow$level-1
10   **else**
11 | **while** *level* ≥ 0 **do**
12 | **if** *((v.neighbor[L][level]).key > searchKey)* **then**
13 | **send** $\langle$searchOp, startNode, searchKey, level$\rangle$ to v.neighbor[L][level]
14 | **break**
15 | **else** level $\leftarrow$level-1
16   **if** *(level < 0)* **then**
17 | **send** $\langle$notFoundOp, $v\rangle$ to startNode

---

it reaches level 0. Either the address of the node storing the search key, if it exists, or the address of the node storing the largest key less than the search key is returned.

Lemma 40.2   *The search operation in a skip graph S with n nodes takes expected $O(\log n)$ messages and $O(\log n)$ time.*

The performance shown in Lemma 40.2 is comparable to the performance of distributed hash tables (e.g., Chord).[39] With n resources in the system, a skip graph takes $O(\log n)$ time for one search operation. In comparison, Chord takes $O(\log m)$ time, where m is the number of machines in the system. As long as n is polynomial in m, we get the same asymptotic performance from both DHTs and skip graphs for search operations.

Skip graphs can support *range queries*, in which one is asked to find a key $\geq x$, a key $\leq x$, the largest key $< x$, the least key $> x$, some key in the interval $[x, y]$, all keys in $[x, y]$, and so forth. For most of these queries, the procedure is an obvious modification of Algorithm 40.1 and runs in $O(\log n)$ time with $O(\log n)$ messages. For finding all nodes in an interval, we can use a modified Algorithm 40.1 to find a single element of the interval (which takes $O(\log n)$ time and $O(\log n)$ messages). With r nodes in the interval, we can then broadcast the query through all the nodes (which takes $O(\log r)$ time and $O(r \log n)$ messages). If the originator of the query is capable of processing r simultaneous responses, the entire operation still takes $O(\log n)$ time.

40.6.3.2   The Insert Operation

A new node u knows some *introducing* node v in the network that will help it to join the network. Node u inserts itself in one linked list at each level until it finds itself in a singleton list at the topmost level. The insert operation consists of two stages:

1. Node u starts a search for itself from v to find its neighbors at level 0, and links to them.
2. Node u finds the closest nodes s and y at each level $\ell \geq 0$, $s < u < y$, such that $m(u)\lceil (\ell+1) = m(s)\lceil (\ell+1) = m(y)\lceil (\ell+1)$, if they exist, and links to them at level $\ell + 1$.

Because each existing node v does not require $m(v)_{\ell+1}$ unless there exists another node u such that $m(v)\lceil(\ell+1) = m(u)\lceil(\ell+1)$, it can delay determining its value until a new node arrives asking for its value; thus, at any given time, only a finite prefix of the membership vector of any node needs to be generated. Detailed pseudo-code for the insert operation can be found in Shah.[38]

Lemma 40.3   *The insert operation in a skip graph S with n nodes takes expected $O(\log n)$ messages and $O(\log n)$ time.*

Inserts can be trickier when we have to deal with concurrent node joins. Before u links to any neighbor, it verifies that its join will not violate the order of the nodes. So if any new nodes have joined the skip graph between u and its predetermined successor, u will advance over the new nodes if required before linking in the correct location.

40.6.3.3   The Delete Operation

The delete operation is very simple. When node u wants to leave the network, it informs its predecessor node at each level to update its successor pointer to point to u's successor. It starts at the topmost level and works its way down to level 0. Node u also informs its successor node at each level to update its predecessor pointer to point to n's predecessor. If u's successor or predecessor is being deleted as well, they pass the message on to their neighbors so that the nodes are correctly linked. A node does not delete itself from the graph as long as it is waiting for some message as a part of the delete operation of another node.

Lemma 40.4   *The delete operation in a skip graph S with n nodes takes expected $O(\log n)$ messages and $O(1)$ time.*

40.6.4   Fault Tolerance

In this section we describe some of the fault-tolerance properties of a skip graph with alphabet $\{0, 1\}$. Fault tolerance of related data structures, such as augmented versions of linked lists and binary trees, has been well-studied and some results can be seen in Refs. 2 and 28. We are interested in the number of nodes that can be separated from the primary component by the failure of other nodes, as this determines the size of the surviving skip graph after the repair mechanism finishes.

We consider two fault models: (1) a random failure model in which an adversary chooses random nodes to fail, and (2) a worst-case failure model in which an adversary chooses specific nodes to fail after observing the structure of the skip graph For a random failure pattern, our experiments have shown that for a reasonably large skip graph, nearly all nodes remain in the primary component until about two thirds of the nodes fail; and that it is possible to make searches highly resilient to failures, even without using the repair mechanism, by the use of redundant links. These results are shown in Figure 40.2. For searches, the fact that the average search involves only $O(\log n)$ nodes establishes trivially that most searches succeed as long as the proportion of failed nodes is substantially less than $O(\frac{1}{\log n})$. By detecting failures locally and using additional redundant edges, we can make searches highly tolerant to small numbers of random faults. Figure 40.3 shows what proportion of searches (with non-faulty source and target nodes) fail as a function both of the proportion of failed nodes and the number of additional links made to successors at each level. With a fivefold increase in the number of links, the proportion of failed searches becomes quite small, even with 20 percent node failure rate.

For a worst-case failure pattern, experimental results are less useful because the worst-case pattern may be exponentially improbable. Here, theoretical results show that even a worst-case choice of failures causes limited damage. With high probability, a skip graph with n nodes has an $\Omega(\frac{1}{\log n})$ expansion ratio, implying that at most $O(f \cdot \log n)$ nodes can be separated from the primary component by f failures. The formal statement of this result is:

Theorem 40.1   *Let $c \geq 6$. Then a skip graph with n nodes and alphabet $\{0, 1\}$ has an expansion ratio of at least $\frac{1}{c \log_{3/2} n}$ with probability at least $1 - \alpha n^{5-c}$, where the constant factor α does not depend on c.*

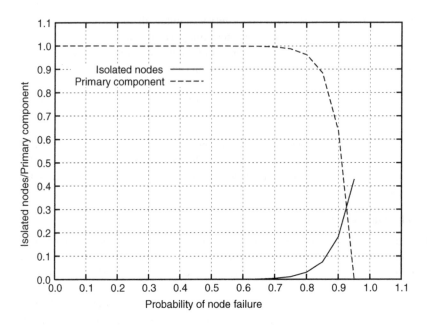

FIGURE 40.2   The number of isolated nodes and the size of the primary component as a fraction of the surviving nodes in a skip graph with 131,072 nodes.

40.6.5   Congestion

In addition to fault tolerance, a skip graph provides a limited form of congestion control, by smoothing out hot spots caused by popular search targets. The guarantees that a skip graph makes in this case are similar to the guarantees made for survivability. Just as a node's continued connectivity depends on the

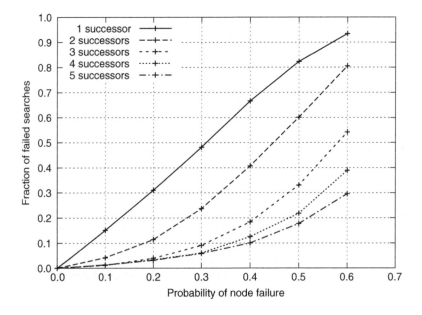

FIGURE 40.3   Fraction of failed searches in a skip graph with 131,072 nodes and 10,000 messages. Each node has up to five successors at each level.

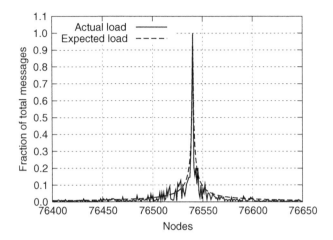

FIGURE 40.4 Actual and expected congestion in a skip graph with 131,072 nodes with the target $= 76{,}539$. Messages were delivered from each node to the target and the actual number of messages through each node was measured. The bound on the expected congestion is computed using Theorem 40.2. Note that this bound may overestimate the actual expected congestion.

survival of its neighbors, its message load depends on the popularity of its neighbors as search targets. However, this effect drops off rapidly with distance; nodes that are far away from a popular target in the bottom-level list of a skip graph get little increased message load on average.

For a single search, the probability that a particular search uses a node between the source and target drops off inversely with the distance from the node to the target. Formally:

Theorem 40.2 *Let S be a skip graph with alphabet $\{0, 1\}$, and consider a search from s to t in S. Let u be a node with $s < u < t$ in the key ordering (the case $s > u > t$ is symmetric), and let d be the distance from u to t, defined as the number of nodes v with $u < v \le t$. Then the probability that a search from s to t passes through u is less than $\frac{2}{d+1}$.*

This theoretical upper bound closely matches experimental data, as shown in Figure 40.4.

Theorem 40.2 is of small consolation to some node that draws a probability $\frac{2}{d+1}$ straw and participates in every search. Fortunately, such disasters do not happen often. Define the *average congestion* L_{tu} imposed by a search for t on a node u as the probability that an $s - t$ search hits u conditioned on the membership vectors of all nodes in the interval $[u, t]$, where $s < u < t$ or, equivalently, $s > u > t$.[*] Note that because the conditioning does not include the membership vector of s, the definition in effect assumes that $m(s)$ is chosen randomly. This approximates the situation in a fixed skip graph where a particular target t is used for many searches that may hit u, but the sources of these searches are chosen randomly from the other nodes in the graph.

Theorem 40.2 implies that the expected value of L_{tu} is no more than $\frac{2}{d+1}$. In the following theorem, we show the distribution of L_{tu} declines exponentially beyond this point.

Theorem 40.3 *Let S be a skip graph with alphabet $\{0, 1\}$. Fix nodes t and u, where $u < t$ and $|\{v : u < v \le t\}| = d$. Then for any integer $\ell \ge 0$, $\Pr[L_{tu} > 2^{-\ell}] \le 2e^{-2^{-\ell}d}$.*

---

[*]It is immediate from the proof of Theorem 40.2 that L_{tu} does not depend on the choice of s.

40.7   Remarks

Many peer-to-peer systems have been proposed that overcome many drawbacks of the early systems such as Napster and Gnutella. The new breed of systems displays an inherent trade-off between support for complex queries and minimization of system state. It would be interesting to design a system that provides the space efficiency of DHTs and the support for range queries of skip graphs.

References

1. FIPS 180-1. Secure Hash Standard. In *U.S. Department of Commerce/NIST, National Technical Information Service,* Springfield, VA, April 1995. http://www.itl.nist.gov/fipspubs/fip180-1.htm.
2. Yonatan Aumann and Michael A. Bender. Fault Tolerant Data Structures. In *Proceedings of the Thirty-Seventh Annual Symposium on Foundations of Computer Science (FOCS),* pages 580–589, Burlington, VT, October 1996.
3. James Aspnes, Zoë Diamadi, and Gauri Shah. Fault-Tolerant Routing in Peer-to-Peer Systems. In *Proceedings of the Twenty-First ACM Symposium on Principles of Distributed Computing (PODC),* pages 223–232, Monterey, CA, July 2002. Submitted to *Distributed Computing.* Available at http://arXiv.org/abs/cs/0302022.
4. James Aspnes, Johnathan Kirsch, and Arvind Krisnamurthy. Load balancing and locality in range-queriable data structures. In *Proceedings of the Twenty-Third ACM Symposium on Principles of Distributed Computing (PODC),* Newfoundland, Canada, July 2004. To appear.
5. James Aspnes and Gauri Shah. Skip Graphs. In *Proceedings of the Fourteenth Annual ACM-SIAM Symposium on Discrete Algorithms (SODA),* pages 384–393, Baltimore, MD, January 2003. Submitted to a special issue of *Journal of Algorithms* dedicated to select papers of SODA 2003.
6. Bobby Bhattacharjee, Pete Keleher, and Bujor Silaghi. The Design of TerraDir. Technical Report CS-TR-4299, University of Maryland, College Park, College Park, MD, October 2001.
7. Ian Clarke, Oskar Sandberg, Brandon Wiley, and Theodore W. Hong. Freenet: A Distributed Anonymous Information Storage and Retrieval System. In Hannes Federrath, Editor, *Designing Privacy Enhancing Technologies: International Workshop on Design Issues in Anonymity and Unobservability,* Volume 2009 of *Lecture Notes in Computer Science,* pages 46–66, Berkeley, CA, July 2000. http://www.freenet.sourceforge.net.
8. Mayur Datar. Butterflies and Peer-to-Peer Networks. In Rolf Möhring and Rajeev Raman, Editors, *Proceedings of the Tenth European Symposium on Algorithms (ESA),* Volume 2461 of *Lecture Notes in Computer Science,* pages 310–322, Rome, Italy, September 2002.
9. Luc Devroye. A limit theory for random skip lists. *The Annals of Applied Probability,* 2(3):597–609, 1992.
10. Rene de la Briandais. File Searching Using Variable Length Keys. In *Proceedings of the Western Joint Computer Conference,* Volume 15, pages 295–298, Montvale, NJ, 1959.
11. Edward Fredkin. Trie Memory. *Communications of the ACM,* 3(9):490–499, September 1960.
12. Amos Fiat and Jared Saia. Censorship Resistant Peer-to-Peer Content Addressable Networks. In *Proceedings of the Thirteenth Annual ACM-SIAM Symposium on Discrete Algorithms (SODA),* pages 94–103, San Francisco, CA, January 2002. Submitted to a special issue of *Journal of Algorithms* dedicated to select papers of SODA 2002.
13. Michael J. Freedman and Radek Vingralek. Efficient Peer-to-Peer Lookup Based on a Distributed Trie. In Peter Druschel, Frans Kaashoek, and Antony Rowstron, Editors, *Proceedings of the First International Workshop on Peer-to-Peer Systems (IPTPS),* Volume 2429 of *Lecture Notes in Computer Science,* pages 66–75, Cambridge, MA, March 2002.
14. Indranil Gupta, Kenneth Birman, Prakash Linga, Al Demers, and Robbert Van Renesse. Kelips*: Building an Efficient and Stable P2P DHT through Increased Memory and Background Overhead.

In *Proceedings of the Second International Workshop on Peer-to-Peer Systems (IPTPS)*, Berkeley, CA, February 2003.

15. Joaquim Gabarró and Xavier Messeguer. A Unified Approach to Concurrent and Parallel Algorithms on Balanced Data Structures. In *Proceedings of the Seventeenth International Conference of the Chilean Computer Society (SCCC)*, pages 78–92, Valparaíso, Chile, November 1997.

16. Joaquim Gabarró, Conrado Martínez, and Xavier Messeguer. A Top-Down Design of a Parallel Dictionary Using Skip Lists. *Theoretical Computer Science*, 158(1–2):1–33, May 1996.

17. Gnutella. http://gnutella.wego.com.

18. Prasanna Ganesan, Qixiang Sun, and Hector Garcia-Molina. YAPPERS: A Peer-to-Peer Lookup Service over Arbitrary Topology. In *Proceedings of the Twenty-Second Annual Joint Conference of the IEEE Computer and Communications Societies (INFOCOM)*, San Francisco, CA, March 2003.

19. Nicholas J.A. Harvey, Michael B. Jones, Stefan Saroiu, Marvin Theimer, and Alec Wolman. Skip-Net: A Scalable Overlay Network with Practical Locality Properties. In *Proceedings of the Fourth USENIX Symposium on Internet Technologies and Systems (USITS)*, pages 113–126, Seattle, WA, March 2003.

20. Pete Keleher, Bobby Bhattacharjee, and Bujor Silaghi. Are Virtualized Overlay Networks Too Much of a Good Thing? In Peter Druschel, Frans Kaashoek, and Antony Rowstron, Editors, *Proceedings of the First International Workshop on Peer-to-Peer Systems (IPTPS)*, Volume 2429 of *Lecture Notes in Computer Science*, pages 225–231, Cambridge, MA, March 2002.

21. David Kempe, Jon M. Kleinberg, and Alan J. Demers. Spatial Gossip and Resource Location Protocols. In *Proceedings of the Thirty-Third Annual ACM Symposium on Theory of Computing (STOC)*, pages 163–172, Crete, Greece, July 2001.

22. Jon Kleinberg. The Small-World Phenomenon: An Algorithmic Perspective. In *Proceedings of the Thirty-Second Annual ACM Symposium on Theory of Computing (STOC)*, pages 163–170, Portland, OR, May 2000.

23. David Karger, Eric Lehman, Tom Leighton, Matthew Levine, Daniel Lewin, and Rina Panigrahy. Consistent Hashing and Random Trees: Distributed Caching Protocols for Relieving Hot Spots on the World Wide Web. In *Proceedings of the Twenty-Ninth ACM Symposium on Theory of Computing (STOC)*, pages 654–663, El Paso, TX, May 1997.

24. Peter Kirschenhofer, Conrado Martínez, and Helmut Prodinger. Analysis of an Optimized Search Algorithm for Skip Lists. *Theoretical Computer Science*, 144(1–2):119–220, June 1995.

25. Donald E. Knuth. *The Art of Computer Programming: Sorting and Searching*, Volume 3. Addison-Wesley, Reading, MA, 1973.

26. Peter Kirschenhofer and Helmut Prodinger. The Path Length of Random Skip Lists. *Acta Informatica*, 31(8):775–792, 1994.

27. Dahlia Malkhi, Moni Naor, and David Ratajczak. Viceroy: A Scalable and Dynamic Emulation of the Butterfly. In *Proceedings of the Twenty-First ACM Symposium on Principles of Distributed Computing (PODC)*, pages 183–192, Monterey, CA, July 2002.

28. J. Ian Munro and Patricio V. Poblete. Fault Tolerance and Storage Reduction in Binary Search Trees. *Information and Control*, 62(2/3):210–218, August 1984.

29. Napster. http://www.napster.com.

30. Moni Naor and Udi Weider. A Simple Fault-Tolerant Distributed Hash Table. In *Proceedings of the Second International Workshop on Peer-to-Peer Systems (IPTPS)*, Berkeley, CA, February 2003.

31. Thomas Papadakis, J. Ian Munro, and Patricio V. Poblete. Analysis of the Expected Search Cost in Skip Lists. In J.R. Gilbert and R.G. Karlsson, Editors, *Proceedings of the Second Scandinavian Workshop on Algorithm Theory (SWAT 90)*, Volume 447 of *Lecture Notes in Computer Science*, pages 160–172, Bergen, Norway, July 1990.

32. C. Greg Plaxton, Rajamohan Rajaraman, and Andrea W. Richa. Accessing Nearby Copies of Replicated Objects in a Distributed Environment. *Theory of Computing Systems*, 32(3):241–280, 1999.

33. William Pugh. Skip Lists: A Probabilistic Alternative to Balanced Trees. *Communications of the ACM*, 33(6):668–676, June 1990.

34. Antony Rowstron and Peter Druschel. Pastry: Scalable, Distributed Object Location and Routing for Large-Scale Peer-to-Peer Systems. In *Proceedings of the IFIP/ACM International Conference on Distributed Systems Platforms (Middleware)*, pages 329–350, Heidelberg, Germany, November 2001.

35. Sylvia Ratnasamy, Paul Francis, Mark Handley, Richard Karp, and Scott Shenker. A Scalable Content-Addressable Network. In *Proceedings of the ACM Symposium on Communications Architectures and Protocols (SIGCOMM)*, pages 161–172, San Diego, CA, August 2001.

36. Bujor Silaghi, Bobby Bhattacharjee, and Pete Keleher. Query Routing in the TerraDir Distributed Directory. In Victor Firoiu and Zhi-Li Zhang, Editors, *Proceedings of the SPIE ITCOM 2002*, Volume 4868 of *SPIE*, pages 299–309, Boston, MA, August 2002.

37. Jared Saia, Amos Fiat, Steven Gribble, Anna Karlin, and Stefan Saroiu. Dynamically Fault-Tolerant Content Addressable Networks. In Peter Druschel, Frans Kaashoek, and Antony Rowstron, Editors, *Proceedings of the First International Workshop on Peer-to-Peer Systems (IPTPS)*, Volume 2429 of *Lecture Notes in Computer Science*, pages 270–279, Cambridge, MA, March 2002.

38. Gauri Shah. Distributed Data Structures for Peer-to-Peer Systems. Ph.D. thesis, Yale University, 2003.

39. Ion Stoica, Robert Morris, David Liben-Nowell, David R. Karger, M. Frans Kaashoek, Frank Dabek, and Hari Balakrishnan. Chord: A Scalable Peer-to-Peer Lookup Service for Internet Applications. *IEEE/ACM Transactions on Networking*, 11(1):17–32, February 2003.

40. Ben Y. Zhao, John Kubiatowicz, and Anthony D. Joseph. Tapestry: An Infrastructure for Fault-Tolerant Wide-Area Location and Routing. Technical Report UCB/CSD-01-1141, University of California, Berkeley, Berkeley, CA, April 2001.

41

State Management in DHT with Last-Mile Wireless Extension

Hung-Chang Hsiao

Chung-Ta King

Most *peer-to-peer* (P2P) overlays based on *distributed hash tables* (DHTs) focus on stationary Internet hosts. However, when nodes in the last-mile wireless extension are also allowed to join the overlay, we face immediately the problem of *peer mobility*. When a peer moves to a new location in the network, most existing overlays treated them as if they had left the network and joined as a new node. Often, the peer needs to use a new hash ID after it changes its network attachment point. All the state information regarding the old hash ID will be discarded by the overlay. This results in inefficiency in message delivery. In addition, other nodes cannot access the mobile peer and the information stored on it through the old ID. This chapter discusses a DHT-based overlay, called Bristle, which supports peer mobility and manages their states in the last-mile wireless extension. A mobile peer can disseminate its location information through Bristle, and other nodes can learn its movement or query its location through

Bristle. Bristle forms a state management overlay for DHT-based P2P systems with last-mile wireless extension.

41.1 Introduction

A *peer-to-peer* (P2P) overlay intends to provide a universal utility space that is distributed among the participating end computers (peers) across the Internet. All peers assume equal roles and there is no centralized server in the space. Often-cited features of P2P overlays include scalability, dynamic adaptation, fault tolerance, and self-configuration. The key to realizing these features lies in an effective way of naming and locating peers and the utility objects they host. Without loss of generality, we will refer to a utility object as a *data item*. An exciting strategy of organizing peers and data items is the use of *distributed hash tables* (DHT),[7,22,24,31,35] which represents peers and data items with unique hash keys to control the topology and data placement in the overlay network. Because a DHT-based overlay (abbreviated DHT) relies on a logic structure (e.g., a ring in Chord[31] and a torus in CAN[22]) to organize the interconnection of the peers, a DHT-based overlay is also called a *structured* P2P overlay.

A common strategy for naming and routing in DHTs is to store a data item with a hash key k in a peer node whose hash key (or ID) is the closest to k. To fetch the data item, the request message is routed through the intermediate peers whose hash keys are closer and closer to k. This requires that each peer node keeps some state information for routing. The state information may appear as a list of entries; each is referred to as a *state* in this chapter. A state associates, among other things, the hash key of a known peer and its network attachment point (e.g., the IP address and port number). A state thus gives the location information of a peer node in the network, which allows the local node to communicate with that node directly. We will call the tuple <hash key, network address> a *state-pair*. When a node receives a message, it forwards the message to the next peer according to the states that it maintains. Note that a state may contain other information, such as the quality of the communication path to the corresponding peer, the capability of that peer, the amount of data items on that peer, etc. Obviously, states in a DHT-based P2P overlay are distributed, and a robust and scalable DHT relies on sophisticated state management.

Previously proposed DHTs, including CAN,[22] Chord,[31] Pastry,[24] Tapestry,[35] and Tornado,[7] assume that the participating peers are stationary. This is because DHTs were originally proposed for Internet-based wide-area applications. However, a growing number of devices are now connected to the Internet through wireless links.[26] Wireless devices may want to join a DHT-based overlay and access its services or resources. For example, a wireless device may want to join a multicast group, where all members are DHT nodes,[2] in order to receive messages addressed to that group. The problem is that when nodes in the last-mile wireless extension are allowed to join a DHT overlay, they may move and we will have the *peer mobility* problem. The state-pairs regarding this node that are distributed across the P2P overlay will become stale.

A straightforward solution is to treat that node as leaving the DHT and then joining as a "new" peer in the new location. For example, a peer in CFS[3] chooses its hash key using its IP address as an input to the hash function $\mathcal{H}$. The hash key changes when the IP address of a node changes. To preserve the freshness of state information, peers are required to periodically update their states to the overlay. Old states associated with a mobile peer will be removed gradually from the system when its states expire. Such a scheme will be referred to as a Type **A** DHT. This approach may impair the performance (i.e., the number of hops for sending a message) of the system due to mobility. Routing and system maintenance operations in a DHT overlay rely on healthy states. In addition, this approach does not guarantee the end-to-end semantic. This is because an application implemented on top of a DHT-based overlay refers to a peer using the peer's hash key (or hash ID). If peers can change their IDs due to mobility, the application may not work properly.

A second approach is to let a peer to use its original ID when it rejoins the system. This approach may also be harmful to the performance of the system. This is because states associated with the mobile peer may not be updated quickly in time. However, this approach can guarantee the end-to-end semantic. We call this type of DHT a Type **B** DHT.

TABLE 41.1   A Summary of Various Design Choices

	A	B	C	Bristle
Infrastructure	IP	IP	Mobile IP	IP
Scalability	Good	Good	Poor	Good
Reliability	Good	Good	Poor	Good
Efficiency	Fair	Fair	Poor	Good
Deployment	Easy	Easy	Sophisticated	Easy
End-to-end semantic	No	Transparent	Transparent	Transparent

Alternatively, it is possible to deploy a DHT-based overlay over a mobile IP infrastructure[20] (called a Type **C** DHT). Mobile IP provides a transparent view of the underlying network to a DHT. A DHT thus does not need to consider mobility. However, mobile IP assumes that home and foreign agents are reliable. These agents may also introduce critical points of failure and performance bottlenecks. Perhaps the most serious problem with mobile IP is the triangular route it introduces. Although mobile IPv6[21] solves the problem by allowing a host to cache the binding of the address of the mobile node and the address of its network attachment point, it requires that the correspondent host is mobile-IPv6 capable and that the mobile IPv6 infrastructure is available. In addition, it still relies on a reliable home agent to resolve unknown network addresses of mobile hosts.

Table 41.1 summarizes the design choices for handling node mobility in DHTs. Types **A** and **B** DHTs need only the IP infrastructure support and can be more scalable, reliable, efficient, and deployable than Type **C** DHTs, by not relying on auxiliary or central agents. However, Type **A** DHTs do not guarantee the end-to-end semantic for applications.

By scalability we mean that the location management of mobile nodes does not depend on a central point such as the home agent in mobile IP. Types **A** and **B** DHTs do not manage the location of a mobile node, which eliminates any central point in the DHT. Efficiency denotes whether (1) a message can be forwarded to its destination in a timely manner, and (2) a DHT can be maintained at low cost. We thus are interested to know whether *it is possible to develop an efficient DHT to maintain up-to-date and consistent states for mobile nodes.* We discuss an extension to DHT, called Bristle,[6] which supports peer mobility and manages their states in the last-mile wireless extension.

A number of issues should also be addressed when designing an overlay such as Bristle. First, a DHT should not rely on the wireless devices to help forward a message due to the mobility of the mobile devices. DHT nodes that are in the wired Internet are preferred. Second, a wireless communication environment is bandwidth-limited. It may not be appropriate to eagerly probe the aliveness of wireless devices for maintaining their states, because this may increase the network traffic.

Bristle is based on the P2P infrastructure Tornado.[7] We evaluate Bristle via extensive simulation. In particular, we identify two design options for Bristle: namely, invalidation-based and invalidation-update-based approaches. The simulation results show that the invalidation-based and invalidation-update-based approaches considerably outperform the Type **B** DHT in terms of the number of hops for sending a message, and the percentage timeouts of performing a lookup and the maintenance bandwidth. Bristle can take advantage of stationary peers and for managing a DHT with mobile nodes. Bristle also takes advantage of peer heterogeneity. Furthermore, we analyze the worst case for routing in Bristle. A straightforward routing scheme in Bristle requires $O(\log N)^2$ hops to send a message from its source to the destination, where N is the number of nodes in the system. A route between two nodes may involve mobile nodes to help message forwarding; this thus may require additional $O(\log N)$ address resolution operations. The worst-case analysis not only helps identify the worst case of performance in Bristle, but also provides a hint for improving performance. We then propose a simple clustered naming scheme, in which message routing between two nodes does not need the help of mobile nodes. This boosts the routing performance to $O(\log N)$. We also evaluate the clustered naming scheme in simulation. The clustered naming scheme significantly outperforms the naming scheme without differentiating between stationary and mobile nodes.

41.2   Background

Because Bristle is implemented on top of Tornado,[7] we introduce Tornado first. Tornado is a DHT-based P2P that addresses node heterogeneity with the notion of *virtual homes*. For the purpose of this chapter, we need only know how peers in the overlay keep their states and how messages are routed. Therefore, we do not differentiate virtual homes from physical nodes. Each peer hosts one virtual home.

Similar to other works,[22,24,31,35] Tornado assigns each data item and each peer a unique hash key using a hash function $\mathcal{H}$. A data item is stored by a peer whose hash key (or ID) is "numerically closest" to that of the data item. If $\mathcal{H}$ is a uniform function, the peers will be allocated almost the same number of data items. To access a data item, the request is sent to the peer with an ID numerically closest to the key of the requested data item. By interconnecting the peers into a ring according to their hash IDs, the numerically closest peer can always be reached along the ring. Of course, this operation will take $O(N)$ hops, where a hop is an end-to-end communication between two peers and N is the total number of peers in the overlay.

To improve the above routing performance, most DHT-based overlays add extra links between peers at different parts of the ring so that a message can reach a far side of the ring in fewer hops. These extra links form a structure over the peers and thus give rise to the name "structured P2P." Following this idea, Tornado adopts a set of rules to form its own structure. Consider a peer with a hash ID x. The peer keeps a list (i.e., the states defined in Section 40.1) of other peers with information for reaching them. The connections to these peers form the "links" of the Tornado structure. The first peer in the list, referred to as the level-1 *neighbor*, has a hash ID l_1 that is any value "smaller than or equal to" $u_1 = (x + \frac{\mathcal{R}}{2}) \bmod \mathcal{R}$, where $\mathcal{R} - 1$ is the maximum key value that can be assigned to a peer. The second neighbor has a hash ID l_2 that is smaller than or equal to $u_2 = (x + \frac{\mathcal{R}}{4}) \bmod \mathcal{R}$, and so on. This list of peers is kept in a table called the *routing table*. In Tornado, a node must issue $O(\log N)$ messages for constructing its routing table.

Now suppose that x wants to send a message to fetch a data item v, where $v > (x + \frac{\mathcal{R}}{2}) \bmod \mathcal{R}$. Because x does not keep the state of v in its routing table, it has no idea of how to send the message toward the peer hosting v. What it can do is forward the message to the peer whose hash key is the closest to but smaller than v, say, peer l_i in the routing table. Using the routing table, x knows how to send a message to l_i. Peer l_i, after receiving the message, calculates $v - l_i$ and checks its own routing table to determine which of its neighbors should forward the message. In this way, the message is routed closer and closer to the peer hosting v and reaches it finally. In this way, Tornado sends a message in $O(\log_d N)$ hops, where d is a constant number. When $d = 2$, Tornado implements a Chord-like[31] routing algorithm. However, Tornado differs from Chord in the techniques of exploiting network locality and node heterogeneity. Due to space limitations, we do not elaborate on these points here.

41.3   Bristle

41.3.1   Overview

The Bristle architecture is depicted in Figure 41.1. There are two types of nodes: *stationary nodes* and *mobile nodes*. The stationary nodes form a stationary layer, and have fixed locations. On the other hand, mobile nodes are in the last-mile wireless extension (i.e., nodes in the mobile layer) and can change their network attachment points. The stationary layer is, in fact, a DHT-based overlay, which can be CAN,[22] Chord,[31] Pastry,[24] Tapestry,[35] or Tornado.[7] It is responsible for resolving network addresses for mobile nodes and handles their mobility. However, from the application's perspective, nodes in both stationary and mobile layers form a single DHT. Bristle provides GET(*id*) and PUT(*id, data*) APIs to access a data item with a hash key *id*. Conceptually, the data item is stored in the DHT on top of both stationary and mobile layers.

Consider the node **X** in Figure 41.1, which is in the mobile layer. Suppose it wants to send a message to node **Y**, whose state is kept in the routing table of **X**. In the normal case, the message can reach **Y** directly using the state information. However, if **Y** moves, its state changes and the state stored in **X** is stale. The message cannot reach **Y** according to the stale state information. The system now needs to resolve **Y**'s

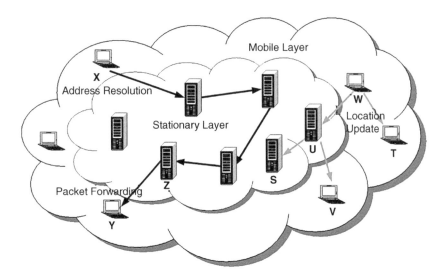

FIGURE 41.1 The Bristle architecture.

current address and then forward the message *if all* **X**'s *neighbors fail*.* With Bristle, **X** only needs to inject the message into the stationary layer to have **Y**'s network address resolved. As shown in Figure 41.1, the message is routed to a node **Z** that knows the current location of **Y** and how to forward the message to **Y**. This is done by requiring **Y** to update (or publish via PUT) its new state to the stationary layer. To be exact, the new state is published to the peer **Z** in the stationary layer whose hash key is the closest to that of **Y**.

In Bristle, nodes in the mobile layer can proactively "subscribe" to the state of other mobile nodes. When a node changes its network attachment point, the new state will be advertised to update or invalidate its subscribers. The advertisement operation is scalable by distributing only to $O(\log N)$ nodes in the system. For example, in Figure 41.1, the mobile node **W** updates its location to mobile nodes **T** and **V** and stationary nodes **S** and **U**.

We assume that when a mobile node moves to a new network domain, it can automatically configure its network address via protocols such as the DHCP.[4] The mobile node communicates with the Internet hosts using TCP/IP or a tailored version for wireless networks.[8]

41.3.2 Routing

As aforementioned, a node's state-pair is represented as <hash key, network address>. The network address field may be invalid (denoted by *null*) if the network address is unreachable. Suppose a node **X** wants to send a message and it needs to reference the state with a state-pair $<k, null>$. It then seeks the help of the stationary layer by sending a *discovery* message with the key k to a node, say **Z**, which can resolve the network address of the node with the hash ID k. Once **Z** determines the current network address of k, it replies the resolved network address to **X**, which in turn updates the local state-pair. In the meantime, **Z** forwards the message toward the destination.

There are a number of ways to determine whether or not a network address corresponding to a state is reachable. One approach is to use probing. If a peer does not receive an acknowledgment after a number of probes, it simply concludes that the probed peer has left the specified network attachment point. A second method is to let each peer periodically send alive messages to those peers that keep the state of that peer.

---

*Note that a practical implementation of a DHT is to let each node additionally maintain backup neighbors. This can reduce the need of address resolution. In this study, we evaluate a DHT where each node maintains a number of additional backup neighbors.

---

route (node i, key j, payload d) **begin**
   // does a node closer to the designated key j exist?
   if $\exists p \in state[i]$ such that $p.key$ is closer to j
      // is the network address valid?
      if $p.addr = null$ and not exist an alternative neighbor
         // resolve address for $p.key$ and forward packets
         // by a node appearing in location management layer
         $p.addr =$ **discovery** ($p.key$);
      else
         // node i forwards packets to node $p'.key$
         p' is p (if p is reachable); otherwise, p' is an
         alternative neighbor (if available) of i;
         route ($p'.addr, j, d$);
end.

FIGURE 41.2   The routing algorithm in which node i forwards a packet toward a node with the hash key closest to j.

Note that this approach is not supported by existing DHTs because a DHT overlay is asymmetric, in which a peer does not know who caches its state. If a peer does not hear the alive messages, it nullifies the state of the corresponding peer.

A design issue with the above two approaches is to determine how often one needs to probe the nodes or to disseminate the alive message. A node may require different refresh periods for each of its neighbors. However, this is out of this study's scope. Bristle mixes the probing and the update/invalidation mechanisms. It has a novel way of tracking the states of other nodes of interest without requiring the node to explicitly determine the refresh periods in the case when a node changes its location. Details are given later. Our study shows that the probability that a node needs the stationary layer to resolve the location of a mobile node is very small. However, the stationary layer is still needed because it can be used as the last resort for location resolution.

Similar to other DHTs, each state in Bristle is associated with a *time-to-live* (TTL) value. When the TTL of a state expires, Bristle will treat that node as if it had left the network attachment point. However, it may be the case that the state refreshment message is delayed or blocked due to an intermittent wireless connection or network congestion. In this case, the up-to-date state information of that node can always be recovered from the stationary layer because a mobile node will periodically publish its state to the stationary layer. Another complication arises when a node has changed to a new location but the location update messages have not reached the interested peers and the stationary layer. Other nodes using the outdated state might assume that node to remain at the same location and try to reach it there. Bristle uses an acknowledgment and time-out mechanism to determine whether or not the message has reached that node. If not, the message will be rerouted through another node. This takes advantage of multiple communication paths between any source and destination in a DHT. In the meantime, the current address of that node can be resolved through the stationary layer. If the address still is not resolved, that node is then removed from the local routing table.

Figure 41.2 shows the routing operations in the Bristle mobile layer. A node with the hash key i (denoted as node i) requests a data item with hash key j, where $state[i]$ denotes the set of state-pairs maintained by node i. Note that different DHTs have different definitions for the "closeness in the hash space." Also, a *discovery* message is only handled by the stationary layer.

41.3.3   Location Management

Bristle supports four operations — *register*, *update*, *join*, and *leave* — for location management of mobile nodes. As mentioned in Section 41.3.1, if a node **X** is interested in the state of a mobile node **Y**, it "registers"

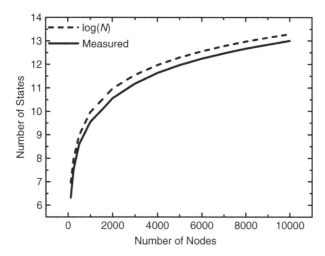

FIGURE 41.3 The number of states published per peer, on average (the results are based on a DHT that does not include backup neighbors).

its interest with **Y**. **X** is called a *registry node* of **Y**. When **Y** moves to a new network attachment point, **Y** "updates" its new network address to the stationary layer and informs **X** of the change.

Bristle organizes those nodes interested in the location of a given mobile node into a *location advertisement tree* (denoted by LDT) in the mobile layer. The tree supports multicast communication for state advertisement. This allows the mobile node to send an address update message to all the nodes in the tree. LDTs have two important features:

1. Each mobile node is associated with an LDT.
2. The number of members in a LDT is $O(\log N)$. This is because, in Tornado, each peer publishes its states to $O(\log N)$ nodes. Therefore, there will be $O(\log N)$ registry nodes of each peer. Figure 41.3 shows the average number of states each peer publishes for the system size, N, varied from 1 to 10,000. Consequently, when the size of a DHT increases, the number of LDT members increases in a logarithmic fashion.

There are two design alternatives for an LDT: *member-only* or *hybrid*. A member-only LDT consists only of the mobile node **Y** and its registry nodes (see Figure 41.4(a)). A hybrid LDT may contain other nodes in addition to **Y** and its registry nodes (see Figure 41.4(b)). This is similar to the multicast tree constructed in IP-multicast[18] and Scribe,[2] which incorporates all the nodes along the routes from the leaves to the root.

In the case of Bristle, we use the member-only LDT. The reasons are the following. Consider the worst-case scenario of the hybrid LDT protocol shown in Figure 41.4(c). The number of nodes (denoted by $S(\tau)$) in a hybrid LDT τ has an order that is the number of levels in the tree times the number of tree leaves (i.e., the registry nodes). In Tornado, $S(\tau) = O(\log N) \cdot O(\log N)$. Assuming a Bristle system with M mobile nodes ($M < N$), the total number of nodes in all LDTs (i.e., M LDTs) is thus $O(M \cdot S(\tau)) = O(M \cdot (\log N)^2)$. We can reduce this number by reusing non-member nodes as much as possible. That is, we let the non-member nodes be elected from the other $N - M$ nodes in the stationary layer. As a result, each stationary node should handle $O(\frac{M}{N-M} \cdot (\log N)^2)$ LDT nodes on average. This parameter is called the *responsibility* of the node. For a member-only approach, the responsibility is reduced to $O(\frac{M}{N-M} \cdot \log N)$.

Figure 41.5 shows the responsibility values for the member-only and hybrid approaches, where N is set to 1,048,576. It shows that when $\frac{M}{N}$ is increased linearly (i.e., the number of mobile nodes is linearly increased), the responsibility value of the hybrid approach increases exponentially. On the other hand, the member-only approach can drastically reduce the responsibility for managing nodes in LDTs.

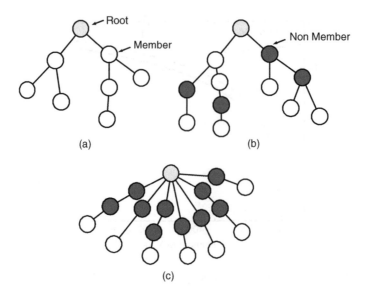

FIGURE 41.4 Alternatives to structuring an LDT tree: (a) the member-only approach, (b) the hybrid approach, and (c) a worst case of a hybrid approach.

41.3.3.1 Register and Update/Invalidate

A node **X** in Bristle sets up and maintains a list of states in two situations: (1) **X** newly joins Bristle and collects a set of states from other nodes in the system for constructing its own states, or (2) **X** receives a neighbor discovery message advertised periodically by other nodes. In either case, **X** registers itself with nodes whose state-pairs are in the routing table of **X**. As mentioned, **X** needs to register with $O(\log N)$

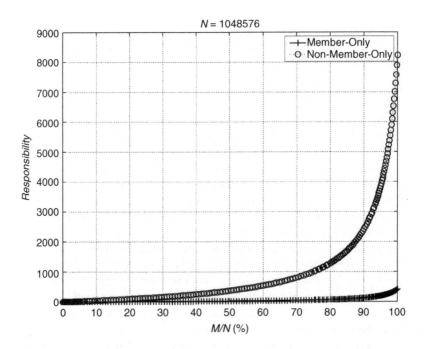

FIGURE 41.5 The responsibility value by varying M and N for the member-only and hybrid approaches.

nodes. If M of N nodes are mobile, **X** can register itself with those mobile nodes only and the number of registrations becomes $O(\frac{M}{N} \cdot \log N)$. Consequently, the total number of registrations issued from all nodes is $O(N \cdot \frac{M}{N} \cdot \log N) = O(M \cdot \log N)$. Each mobile node thus updates (and/or invalidates) its new network attachment point to $O(\frac{M}{N} \cdot \log N)$ nodes; that is, each mobile node only maintains $O(\frac{M}{N} \cdot \log N)$ members for the update. Note that $O(\frac{M}{N} \cdot \log N) < O(\log N)$ because $M < N$.

When **X** registers itself with a mobile node, it also reports its *capacity* (denoted by C_X) to that node. The capacity represents a node's ability, which is the maximum number of network connections that the node can accommodate. This parameter can be obtained from the node's computational power, network bandwidth, etc. Scheduling the update/invalidation notifications to the registry nodes is based on the capacity values of the nodes.

Figure 41.6 illustrates the location update algorithm for a mobile node i. The node updates (and/or invalidates) its new network address to its registry nodes, denoted by $R(i)$. The idea is that i first estimates its remaining capacity. If it is overloaded (e.g., running out of its network connections), it only notifies the registry node with the highest capacity. Then, that registry node performs state updates on behalf of the mobile node to other registry nodes. Otherwise, i advertises its state to the k nodes with the highest capacities based on its available capacity, denoted by $Avail_i$, such that $k \cdot v \le Avail_i < (k+1) \cdot v$, where v is the unit cost to send an update message. Note that each of the k nodes also receives a disjoint subset comprised of the nodes registering with i. The algorithm guarantees that the numbers of registry nodes of different disjoint subsets are nearly equal.

---

advertise (node i) **begin**
    // sort $R(i)$ according to nodes' capabilities order
    $list(i) = \mathbf{sort}\ R(i)$ according to C_t's $\forall t \in R(i)$;
    // the remaining capacity of i
    $Avail_i = C_i - Used_i$;
    // node i is overloaded?
    if $Avail_i - v \le 0$
        // report node i's location to the registry node having
        // the maximum capacity
        $h = list(i).head$;
        send $(h, i$'s network address, $list(i) - \{h\})$;
    else
        // equally partition $list(i)$ to $\lfloor \frac{Avail_i}{v} \rfloor$ lists
        while $list(i)$ is not empty
            for $k = 1; k \le \lfloor \frac{Avail_i}{v} \rfloor; k = k + 1$
                $h = list(i).head$;
                $partition(k) = partition(k) \cup \{h\}$;
                $list(i) = list(i) - h$;
        // update node i's location to $\lfloor \frac{Avail_i}{v} \rfloor$ registry nodes
        // having the maximum capacity
        for $k = 1; k \le \lfloor \frac{Avail_i}{v} \rfloor; k = k + 1$
            $h = partition(k).head$;
            send $(h, i$'s network address, $partition(k) - \{h\}))$;
end.

FIGURE 41.6 The state advertisement algorithm, where $Used_i$ denotes the present workload of the node i, $partition(k) \subseteq list(i)$ for $1 \le k \le \lfloor \frac{Avail_i}{v} \rfloor$, and $\bigcup_k partition(k) = list(i)$. (Note that instead of updating a node's state to its registry node, a node can be based on the same algorithm to invalidate its states stored in the DHT overlay.)

The LDT constructed by the algorithm shown in Figure 41.6 has the following features:

- It exploits the heterogeneity of nodes through a node's capacity.
- An LDT associated with a mobile node is dynamically structured based on the workload of the participating nodes. The workload depends on the consumption of a node's local resources, such as the network bandwidth and the memory used by other processes running on that node.
- To prevent an LDT from skewing, a tree node should try to advertise states to the registry nodes evenly. This can greatly reduce the height of an LDT. Thus, disseminating a state can use a minimal number of network hops.

Ideally, if an LDT is a k-ary complete tree, then performing a state advertisement takes $O(\log_k \log N)$ hops.

In addition to advertising the current state to the registry nodes, a mobile node also publishes its state to the stationary layer (this is not shown in Figure 41.6). Publishing states in the stationary layer allows a registry node to query the up-to-date states of the mobile node (see below).

We want to note first that, in Bristle, before a node changes its network attachment point, it needs to invalidate its state stored in its registry nodes (the algorithm is similar to the one shown in Figure 41.6).* When the node receives a new network address after reattaching the network, it updates its new location to its registry nodes. We denote such an approach as the *invalidation-update-based* location advertisement. Second, another design option is not to let a node to update its new network address to its registry nodes. The node only invalidates its states that appeared in its registry nodes. We call such an approach an *invalidation-based* approach.

41.3.3.2   State Discovery

Although a mobile node a in Bristle periodically broadcasts its state to the registry nodes, a registry node b may not receive the updated state. This is because (1) b may also change its network attachment point and cannot inform its registry nodes in time, or (2) the network fails to transmit the update/invalidation message. In Bristle, a registry node can thus issue a discovery message to the stationary layer to resolve the network address of the mobile node. The stationary layer constitutes a last resort.

To discover the current network address of a node **Y**, a message with the hash ID of **Y** is issued to the stationary layer (see the *discovery* operation in Figure 41.2). The stationary layer can adopt any DHT, such as CAN, Chord, Pastry, Tapestry, or Tornado. These structured P2P overlays provide the following features. (1) For *scalability*, each node in a DHT takes $O(\log N)$ memory overhead to maintain the states, except that it is $O(D)$ in CAN, where D is a constant and each node needs to maintain $2 \cdot D$ neighbors. (2) For *reliability*, each node in a DHT can periodically monitor its connectivity to other $O(\log N)$ nodes in the system, except that a CAN node only uses $2 \cdot D$ connections. In addition, a route toward a destination can have many different paths (e.g., $(\log N)!$ paths in Chord and Tornado). (3) For *availability*, a data item published to a DHT can be replicated to the k nodes whose hash keys are the closest to the key of that data item.[3,25] When a node fails, the requested data item can be accessed from the remaining $k - 1$ nodes. (4) For *responsiveness*, discovering a data item takes $O(\log N)$ hops in these systems; CAN, however, needs $O(N^{1/D})$ hops.

41.3.3.3   Join and Leave

When a Bristle node joins and leaves the system, the overlay geometry structure changes. Consider the case when a node i joins Bristle. It publishes its state to $O(\log N)$ nodes and these nodes return their

---

*We believe that most operating systems have provided a set of utilities to collect the statistics information for wireless network interfaces. For example, in Linux, a DHT code can determine whether or not to invalidate its register nodes by accessing the `/proc/net/wireless` file. The file contains the information about the quality (e.g., the signal strength) of the wireless link. If the signal strength is under a certain level, the node then sends an invalidation message to its registry nodes. A similar technique can also be applied by the Type **B** DHT.

```
join (node i) begin
    // each k in the path traveled by i's joining message
    for (each node k visited)
        // can i become k's neighbor?
        if ∃p ∈ state[k] such that i.key is closer to k than p.key
        and delay(p, k) ≥ delay(i, k)
            // i becomes k's neighbor
            state[k] = state[k] ∪ {i};
            // i registers itself to k
            register (i, k);
        // can each of k and state[k] become i's neighbor?
        for each r ∈ {k ∪ state[k]}
            if ∃q ∈ state[i] such that r.key is closer to i than q.key
            and delay(q, i) ≥ delay(r, i)
                // r becomes i's neighbor
                state[i] = state[i] ∪ {r};
                // r registers itself to i
                register (r, i);
end.
```

FIGURE 41.7 The node joining algorithm, where $delay(a, b)$ denotes the network latency from b to a.

registrations to **Y**. Note that the LDT with the root i is not yet constructed in this phase if i is a node in the mobile layer. At most, $2 \cdot O(\log N)$ messages* are sent and received by i. Figure 41.7 presents the algorithm.

The joining node i will collect the states maintained by each visited node and determine whether each of these collected states (i.e., r in Figure 41.7) should be included in the set of states it maintains. This depends on the closeness of r's and i's keys in the hash table space, and the network distance between r and i. The network distance can be the measured as the network latency of sending a packet between r and i. This shows that Bristle can incorporate network locality in its design. Maintaining states of geographically adjacent nodes allows a node to exploit the network proximity. This can also avoid selecting mobile nodes with a slow wireless network. Consequently, a node can help forward a message to a geographically adjacent node that is likely a stationary node.

Because a node can leave the system at any time, it needs to periodically refresh its state to the associated nodes to maintain the reliability of the system. This is accomplished by adopting the operations similar to those shown in Figure 41.7. The traffic overhead (i.e., $2 \cdot O(\log N)$ messages) will be added to the root of each LDT.

41.4 Worst-Case Analysis and Optimization

Note from the above discussions that the stationary layer is the last resort for resolving the address of a mobile node. The routing performance in this layer is an important issue. In addition, when a route enters into the stationary layer, it is better to remain in that layer. Otherwise, additional address resolution for mobile nodes may be required.

---

*The number of messages becomes $O(\log N) + 1$ if Pastry[24] or Tapestry[35] is used. A joining node in these systems simply sends a joining message to sequentially visit $O(\log N)$ nodes.

Consider a route between two stationary nodes in Bristle. The route normally takes $O(\log(N - M))$ hops if no mobile node is involved to forward a message, where N is the total number of nodes and M is the number of nodes in the mobile layer. Unfortunately, in Bristle, the number of hops between two stationary nodes can be up to $O(\log(N - M) \cdot \log N)$, because each intermediate node in each hop may be a mobile node and needs network address resolution. Notably, this can occur if the DHT overlay experiences an extremely high churn, and thus leads to no alternative node being able to help forward the message. Frequent "back-and-forth" between the stationary and mobile layers will greatly degrade Bristle's performance.

Two potential optimizations can be performed to boost the performance of a route in Bristle. The first is to reduce the routing overhead of each hop by exploiting the network proximity. This is accomplished by forwarding the route to a neighboring node whose hash key is closer to the destination and the cost of the network link to the neighbor is minimal; the technique is called the *route selection*.[5,7] Although this optimization still requires $O(\log(N - M))$ hops to discover a requested state, each hop may follow the network link with the minimal cost. However, studies[5] show that this optimization improves the routing performance modestly.

The second optimization is to avoid using nodes in the mobile layer for communication between nodes in the stationary layer. In Bristle, this is done by clustering the hash IDs of the nodes in the stationary layer. Figure 41.8 illustrates an example in which a stationary node $a.b.c.d$ wants to communicate with another stationary node $w.x.y.z$. If the hash keys are assigned randomly, the route may frequently need state discovery for resolving network addresses of mobile nodes. Figure 41.8(a) shows two state discovery operations. However, if the hash keys are assigned strategically, a route can better utilize the paths comprised of the stationary nodes (see Figure 41.8(b)).

Therefore, in Bristle, a stationary node is assigned a hash key k_s such that $0 < L \leq k_s \leq U < \mathcal{R}$, where L and U are the predefined system parameters. On the other hand, a mobile node is assigned a key k_m such that $k_m < L$ or $k_m > U$. Let $|U - L| = \nabla \cdot \mathcal{R}$. We assume that a uniform hash function such as SHA-1 is adopted. Then,

$$\nabla = \frac{|U - L|}{\mathcal{R}} \approx \frac{N - M}{N}. \tag{41.1}$$

If ∇ decreases, the probability of a routing path comprising of mobile nodes will be increased and a route thus cannot exploit stationary nodes in its routing path. However, if ∇ is increased, a route will try to exploit nodes in the stationary layer to help the forwarding.

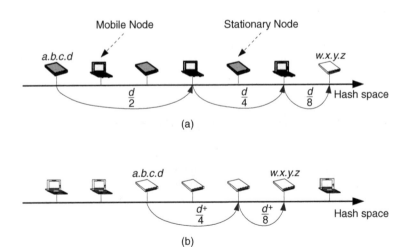

FIGURE 41.8   (a) The random and (b) the clustered naming schemes.

Consider that a stationary node x_1 sends a message to another stationary node x_2. Note that $\mathcal{H}(x)$ represents the hash key of x. If $\mathcal{H}(x_1) < \mathcal{H}(x_2)$, then the route must visit the nodes in the stationary layer and thus there does not need to be any address resolution. Otherwise, if the route can be forwarded by x_2, then we have

$$\left(\left(\mathcal{H}(x_1) + \frac{\mathcal{R} - |\mathcal{H}(x_1) - \mathcal{H}(x_2)|}{2}\right) \mod \mathcal{R}\right) \geq L. \qquad (41.2)$$

Equation (41.2) can be rewritten as $\mathcal{H}(x_1) + \mathcal{H}(x_2) \geq 2 \cdot L + \mathcal{R}$. Because $\mathcal{H}(x_1) \leq U$ and $\mathcal{H}(x_2) \leq U$, we can have

$$\frac{U - L}{\mathcal{R}} \geq \frac{1}{2}. \qquad (41.3)$$

And because $\nabla = \frac{U-L}{\mathcal{R}} \approx \frac{N-M}{M}$ (Equation 41.1), we thus have

$$\frac{N - M}{M} \geq \frac{1}{2}. \qquad (41.4)$$

This concludes that when the number of stationary nodes is greater than or equal to the number of mobile nodes, a message sent from a stationary node to another stationary node can use stationary nodes only. However, the number of stationary nodes may not be greater than or equal to the number of mobile nodes. This can be simply solved using virtual nodes.[3] We can create a sufficient number of virtual nodes in the stationary layer by having each stationary node host a number of virtual nodes. In this way, the number of virtual nodes is greater than or equal to the number of nodes in the mobile layer. However, this requires estimating the number of mobile nodes in the system. We leave the study of utilizing the virtual nodes for future work.

41.5   Performance Evaluation and Results

We evaluate the performance of various deigns of DHTs that mixes stationary and mobile nodes via simulation. We are particularly interested in the performance of the Type **B** DHT and Bristle, because both designs have the same assumption regarding the Internet infrastructure. On the other hand, the Type **A** DHT requires that a mobile node migrates its data items to another peer if it changes its network attachment point, while the Type **C** DHT needs the support of mobile IP infrastructure.

We simulate a Tornado DHT in which a peer joins the DHT every one second on average. After the DHT contains up to 1000 nodes (the 1000 nodes have uniform hash IDs), it stabilizes for one minute (see Figure 41.9(a)). Each node in the DHT pings its level-i neighbors every 30 seconds, where i is from 1 to 9 for a system with 1000 nodes. Thus, a node will maintain at least nine neighbors. However, for improving system reliability, each node caches another 20 backup neighbors. Similar to Chord, a node in Tornado also maintains a few neighbors whose hash IDs are the closest. In the simulation, a node maintains four closest neighbors. It probes each of the closest neighbors every 500 milliseconds. If there is no response from the probed node within the 500 milliseconds (denoted as the "timeout" value), the probing node issues another probe to the probed peer. After the probing node consecutively issues four ping messages without hearing any response, the probing node starts to find a new neighbor immediately.

After the one-minute stabilization when the simulated DHT comprises up to 1000 nodes, we insert 1000 data items with uniform hash keys into the system every 100 milliseconds on average. We assume that the peers have infinite storage space. After 1000 data items are inserted into the system, we let the simulated DHT stabilize for another minute. We then issue 100,000 queries every 100 milliseconds on average. Each query for a randomly selected data item is initiated by a random peer. Consequently, a simulated DHT runs for up to two hours.

We did not simulate the details (e.g., the congestion) of the Internet. Instead, we adopted the simulation technique used by Chord that models the delay of sending a packet as an exponential distribution, where the mean delay is 50 milliseconds. However, we also report the simulation results based on a topology generated by GT-ITM.[33] In addition, our simulator did not model the wireless communication environment.

TABLE 41.2 Parameters Used in the Simulation

Parameter	Value
Number of DHT nodes (i.e., N)	1000
Number of closest neighbors	4
Number of level-i neighbors	9
Number of backup neighbors	20
Stabilization period for closest neighbors	500 milliseconds
Stabilization period for non-closest neighbors	30 seconds
Number of mobile peers (i.e., M)	100, 250, 500, and 700
Stationary period	50 minutes to 8 hours; the mean is 4 hours
Peer moving period	5 minutes on average

Table 41.2 summarizes the parameters used in our simulator. The number of mobile peers simulated is 100, 250, 500, and 700.* We let the numbers of peers in the stationary layer be 900, 750, 500, and 300, respectively. We simulated the random waypoint mobility model,[10] in which a mobile node changes its network attachment point every "stationary period." The stationary period is uniformly distributed, which is from 50 minutes to 8 hours.** The mean stationary period is four hours. A mobile node takes five minutes (denoted as the "movement period"), on average, to move to a new location. The movement period is exponentially distributed. During the movement period, we assume that a moving node cannot communicate with other nodes. We note that a node starts to change its network attachment point in the simulation after the stabilization process that follows the insertion of data items.

The performance metrics we measured are as follows:

- *Number of hops.* The metric is defined as the number of hops required for routing a lookup request. The number of hops reported here is averaged over those lookups performed every 100 seconds.
- *Percentage of timeouts.* A node a may not be able to forward a lookup request if it cannot successfully forward the request to a peer b in its next hop. This is because a does not detect the failure of b before a performs its next stabilization process. Node a thus needs to forward the request to an alternative peer. The percentage of timeouts is defined as the ratio of the number of timeouts in forwarding a lookup request to the total number of hops for sending that request. Similarly, we measure and average the percentages of timeouts for lookups every 100 seconds.
- *Maintenance bandwidth.* As mentioned above, the simulated DHT performs stabilization for its neighbors and closest neighbors every 30 seconds and 500 milliseconds, respectively. The maintenance bandwidth per node is the bandwidth required for a node performing its stabilization process. Note that during stabilization, a node that detects the failure of one of its neighbors requires additional four probing messages. We refer to the message format in Gnutella.[11] The message simulated has 91 bytes in size, which comprises a 54-byte header for MAC, IP, and TCP, and a 37-byte header (including the ID filed and the message type)*** for a DHT packet.

By default, the simulated DHT implements clustered naming. We assume that each node can create infinite connections for updating or invalidating states stored in its registry peers. However, we also report the effect of varying the number of connections for each peer later.

---

*We simulated 500 mobile nodes to mimic the results of Tang and Baker,[32] which conclude that in a metropolitan wireless network, 42 percent of nodes are stationary.

**It has been shown[32] that 64 percent of mobile nodes change their location once every weekday. In addition, the study[1] shows that a mobile node has a session time of less than ten minutes. We thus used a stationary period between the values reported in Refs. 32 and 1.

***See the PING and PONG messages defined by Gnutella.

41.5.1 Overall Performance

Figure 41.9 shows the results for $M = 500$, $M = 250$, and $M = 100$, respectively. The results show that when M is large, the invalidation-based approach can significantly reduce the percentage of timeouts and the maintenance bandwidth. It performs better than the invalidation-update-based approach and leave-join approach, that is, the Type **B** DHT that uses join and leave operations to handle the movement of a node. For the simulated DHT, the invalidation-based approach can reduce a modest number of hops for routing a lookup request while introducing nearly zero percent timeouts. This is because nodes in the simulated DHT maintain backup neighbors. Backup neighbors help forward a lookup request without significantly stretching the path length of sending a message.

Similarly, the invalidation-update-based approach can significantly reduce the percentage of timeouts and the maintenance bandwidth when compared with the Type **B** DHT. However, the invalidation-update-based approach cannot outperform the invalidation-based approach because the former will create more registry nodes for a mobile node. This increases the possibility that a node will forward a message to a mobile node and thus increases the percentage of timeouts for sending a message. If the mobile node moves to a new location, the invalidation-update-based approach requires more traffic to update the new location of a mobile node. However, the invalidation-update-based approach outperforms the invalidation-based one in terms of the number of hops for sending a lookup message. This is because the invalidation-update-based approach can introduce a greater number of "healthy" neighbors to each peer although this may introduce more traffic to the system.

When M is decreased, the invalidation-based, invalidation-update-based, and Type **B** DHT have similar numbers of hops for sending a lookup request and nearly identical maintenance bandwidth. However, both invalidation-based and invalidation-update-based approaches significantly outperform the Type **B** DHT in terms of percentage of timeouts.

Our simulation study shows that a lookup message requires less help from the stationary layer. This is because each simulated DHT node aggressively maintains its closest neighbors (four such neighbors are maintained by each node) every 500 milliseconds. Thus, if all non-closest neighbors fail, a node can ask for help from one of its closest neighbors to forward its message. However, the stationary layer is still demanded, which can improve the routing performance. We discuss this later.

41.5.2 Taking Advantage of Stationary Neighbors

We investigate whether a DHT can route a message to stationary nodes instead of relying on mobile nodes. Figure 41.10 presents the simulation results, which only include the invalidation-based approach and the Type **B** DHT. The percentage of stationary peers per node in the figure means the ratio of the number of stationary neighbors to the number of mobile neighbors a node maintains in its routing table. We conclude that a DHT node simply implementing periodic probing can maintain up to 90 percent of stationary peers in its routing table. This is because the timeout mechanism can finally identify mobile nodes that are unreachable due to their movements. However, as discussed above, the timeout mechanism can create redundant traffic (in the simulation, a node is required to send an additional four probing messages to detect the failure of a node) in order to identify an unreachable peer. The results in Figure 41.10 reaffirm that the invalidation-based approach outperforms the Type **B** DHT.

Interestingly, the results in Figure 41.10 reveal a U-shaped curve. This can explained as follows. When DHT nodes experience the movement of neighbor nodes to new locations for a while, they will gradually remove those unreachable neighbors from their routing tables and include more stationary nodes.

41.5.3 Random versus Clustered Naming

Figure 41.11 depicts the invalidation-based approach with the random and clustered naming schemes. We vary the number of mobile nodes from 100 to 700. When the number of mobile nodes is small (e.g., $M = 100$), the clustered naming scheme (denoted as "C" in the figure) can outperform the random

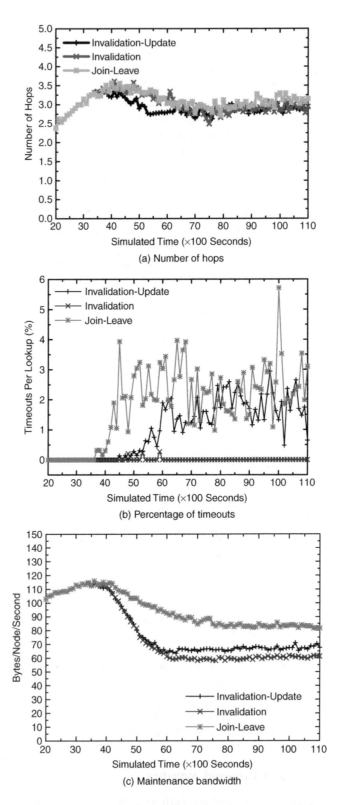

FIGURE 41.9 $M = 500$ and $N = 1000$ in (a), (b), and (c); $M = 250$ and $N = 1000$ in (d), (e), and (f); $M = 100$ and $N = 1000$ in (g), (h), and (i).

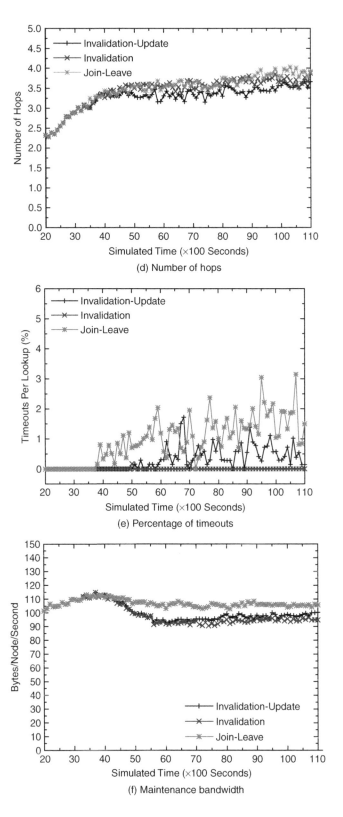

(d) Number of hops

(e) Percentage of timeouts

(f) Maintenance bandwidth

FIGURE 41.9 (*Continued*)

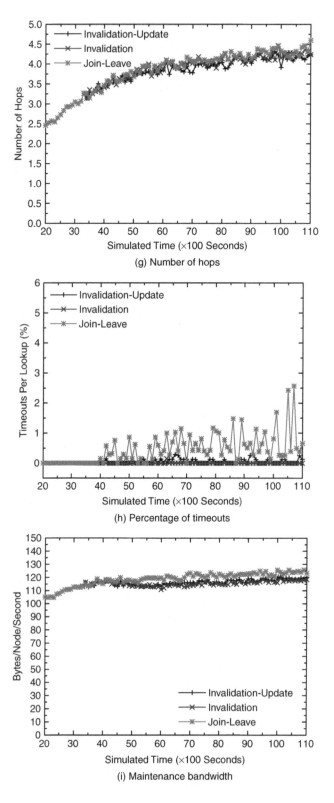

(g) Number of hops

(h) Percentage of timeouts

(i) Maintenance bandwidth

FIGURE 41.9   *(Continued)*

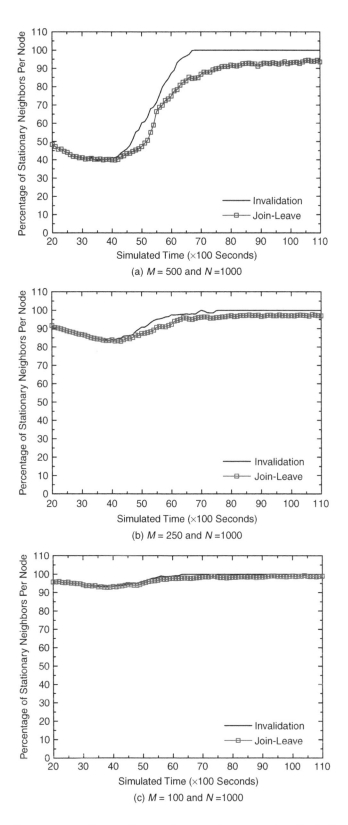

(a) *M* = 500 and *N* = 1000

(b) *M* = 250 and *N* = 1000

(c) *M* = 100 and *N* = 1000

FIGURE 41.10 The percentage of the number of stationary nodes that a node maintains in its routing table.

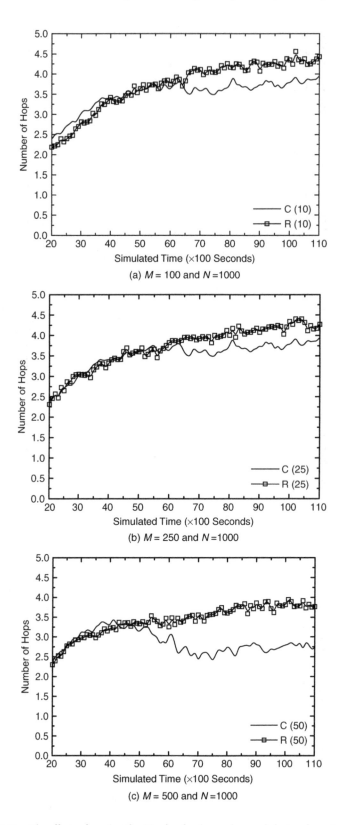

FIGURE 41.11 The effects of varying the *M* value for the random and clustered naming schemes.

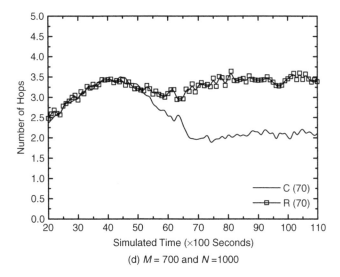

(d) $M = 700$ and $N = 1000$

FIGURE 41.11 (*Continued*)

naming scheme (denoted as "R"). As mentioned, the clustered naming scheme allows a route to be for-warded by stationary nodes if a route is forwarded into the stationary layer. We note that because each simulated DHT node maintains additional backup neighbors, a route in a DHT with the random scheme or the clustered naming scheme will experience nearly zero percent timeouts. The clustered naming scheme can outperform the random naming scheme because a node in the stationary layer can maintain a rel-atively large number of stationary neighbors in its routing table. A stationary node thus has relatively more selection when routing a message. Therefore, once a route is forwarded to the stationary layer, a stationary node can forward the message to another node that is closer to the destination in terms of hash ID space.

When M becomes enlarged, the clustered naming scheme can significantly outperform the random naming scheme. This is because a node in a DHT implementing the random naming scheme has very few stationary peers maintained in its routing table. This consequently leads to an increase in the number of hops required for sending a message. In particular, nodes frequently rely on their closest neighbor to help forward messages. The messages are thus very likely to move along the ring of the hash space. We note that similar to other DHTs, Tornado in where Bristle is built on also maintains a ring logical structure among peers.

41.5.4 Effects of Network Topology and Proximity

We model the Internet topology using the topology generator GT-ITM.[33] The topology we study consists of four transit domains. Each transit domain has five nodes on average. Each transit domain links to four stub domains on average. Each stub domain consists of ten nodes. Consequently, the topology nearly has 1000 nodes. The average packet delay of the simulated topology is 125 milliseconds (note that a node detects the failure of a neighbor node if it experiences a 500-millisecond timeout). The DHT nodes are randomly placed in the simulated topology.

Because the results for the number of hops for sending a lookup request and the maintenance band-width are similar to those discussed in Section 41.5.1, we only present the percentage of timeouts in Figures 41.12(a), (b), and (c) for $M = 500$, $M = 250$, and $M = 100$, respectively. The results show that the invalidation-based and invalidation-update-based approaches outperform the Type **B** DHT. The invalidation-based approach is superior.

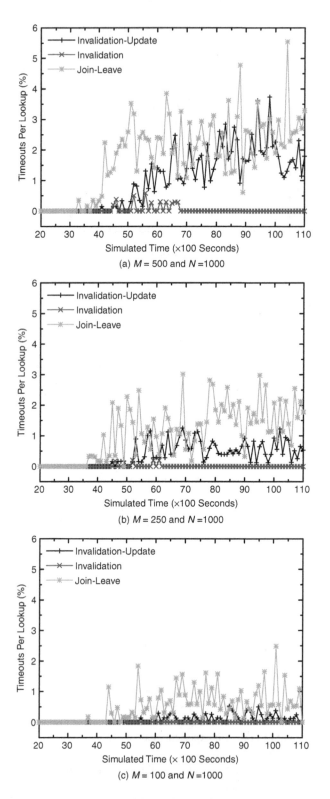

FIGURE 41.12 The percentage of timeouts per lookup request with the ITM topology [(a), (b), and (c)] and the effects of exploitation of network proximity [(d), (e), and (f)].

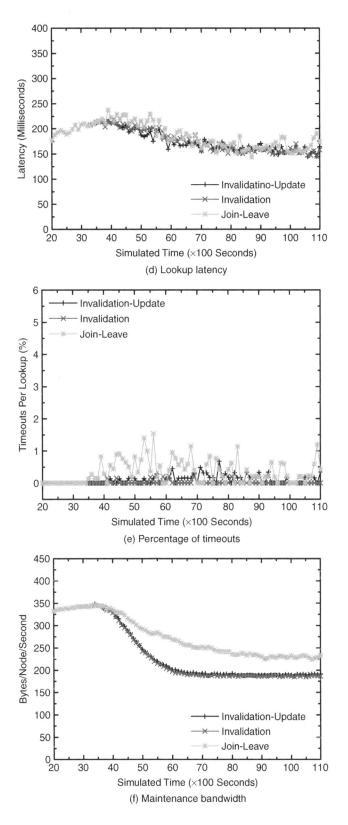

(d) Lookup latency

(e) Percentage of timeouts

(f) Maintenance bandwidth

FIGURE 41.12 (*Continued*)

We also investigate how the network proximity selection mechanism[5,7] can affect the performance of various designs. As mentioned previously, a node selects the geographically close nodes as its neighbors when it joins the system or when it sends or receives periodic neighbor discovery messages. We implement that a node periodically discovers its level-i neighbor by randomly generating the discovery message with a random hash key for the destination within the hash subspace $(x + \frac{R}{2^{i+1}}, x + \frac{R}{2^i}]$. To prevent introducing burst traffic, a DHT node simulated generates a neighbor discovery request every one second to discover a candidate of its level-i neighbor in a round-robin fashion, where $i = 1, 2, 3, \ldots, \frac{R}{2}$. When a node discovers a candidate of a level-i neighbor, it removes the distant one.

Figures 41.12(d), (e), and (f) depict the results for $M = 500$ (again, the results for $M = 250$ and $M = 100$ are similar). Basically, the invalidation-based, the invalidation-update-based, and the Type **B** DHT approaches take roughly 170 milliseconds to process a lookup request. Interestingly, when a DHT implements the proximity neighbor selection mechanism, the percentage of timeouts for performing a lookup can be significantly reduced (Figure 41.12(e)). This is because the proximity neighbor selection we implement requires periodically discovering geographically close neighbors. This helps discover healthy neighbors. However, this comes with the cost of increasing the maintenance bandwidth [from ∼60 bytes/node/second in Figure 41.9(c) to ∼200 in Figure 41.12(f) in the case of the invalidation-based approach].

41.5.5   Effects of the Heterogeneity of Nodes

A recent study[28] analyzed the traffic appearing in multiple border routers across a number of ISP networks over a period of three months. The traffic included those contributed by three popular P2P systems, namely, KaZaA, Gnutella, and DirectConnect. The study concluded that a peer-to-peer network highly relies on peers that are relatively powerful (e.g., in KaZaA, these peer are called *super-nodes*), where less than 10 percent of peers contribute around 99 percent of traffic. Another study[27] reached similar conclusions.

The simulation results discussed so far are based on the assumption that each DHT node can have an infinite number of network connections. This allows a DHT node to send its update and invalidation messages to its registry nodes without relying on other nodes to help relay. We also study how the heterogeneity of nodes affects the design of Bristle. Based on the previous analysis for real P2P systems, we thus model the capabilities (in terms of the number of simultaneous network connections) of nodes having the exponential distribution.

Figure 41.13 illustrates the results for the invalidation-based approach (the invalidation-update-based approach shows similar results). In Figure 41.13, "capability" denotes the mean number of connections a node can have. The results reveal that the capabilities of nodes have a great impact on the percentage of timeouts for performing a lookup request when the mean capability value is small (Figure 41.13(a)). If the design bears the capability (i.e., the algorithm shown in Figure 41.6) in mind when a node only has a few connections to broadcast its state, then it can greatly reduce the percentage of timeouts a lookup request can experience. Notably, the "capability-unaware" scheme is to let a DHT node randomly advertise its state to nodes randomly selected from its registry ones without taking advantage of node heterogeneity.

41.5.6   Effects of the Stabilization Period

Finally, we report the effect of different stabilization periods in Figure 41.14. In contrast to the default of stabilization period, which is 30 seconds, for non-closest neighbors, we lengthen the stabilization period to one minute [Figures 41.14(a), (b), and (c)] and two minutes [Figures 41.14(d), (e), and (f)]. Apparently, the longer the stabilization period is, the less maintenance traffic is introduced [comparing ∼60 bytes/node/second in Figure 41.9(c), ∼30 bytes/node/second in Figure 41.14(c), and ∼20 bytes/node/second in Figure 41.14(f) for the invalidation-based approach]. However, this may increase the number of stale neighbors to a node. For the Type **B** DHT, this can significantly increase the percentage of timeouts. However, the invalidation-based approach can perform quite well.

We want to note that when the stabilization period is increased, the invalidation-update-based approach can obviously outperform the Type **B** DHT [Figures 41.14(a) and (d)].

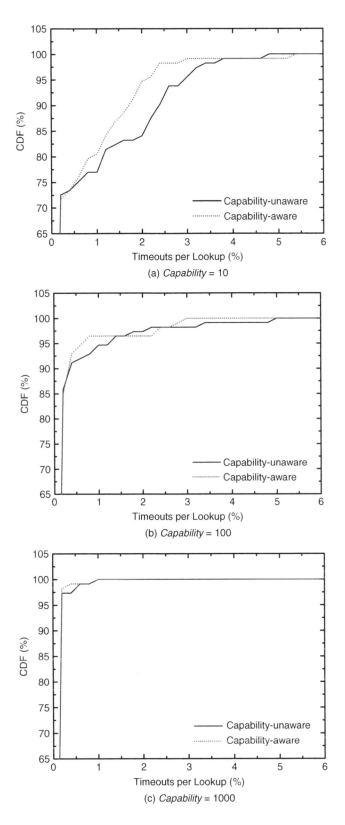

FIGURE 41.13 The percentage of timeouts by varying nodes' capabilities ($M = 500$ and $N = 1000$).

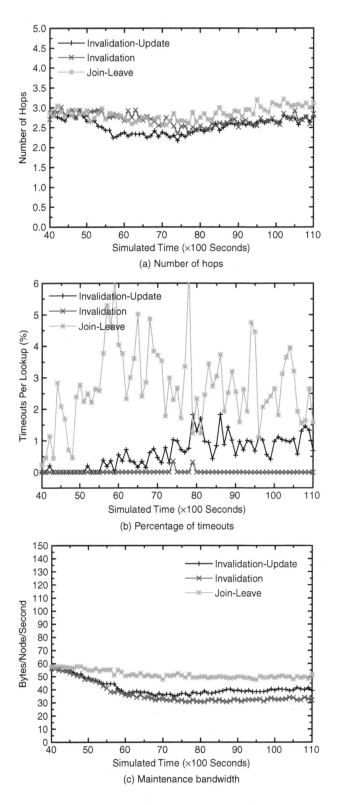

FIGURE 41.14   The one-minute stabilization period [(a), (b), and (c)] and the two-minute stabilization period [(d), (e), and (f)] for $M = 500$ and $N = 1000$.

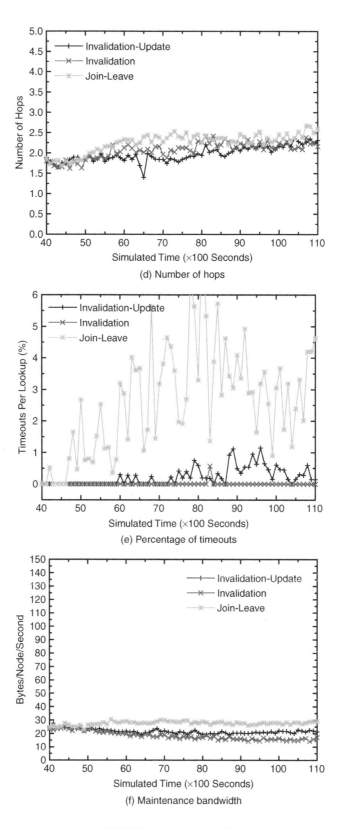

(d) Number of hops

(e) Percentage of timeouts

(f) Maintenance bandwidth

FIGURE 41.14 (*Continued*)

41.6   Related Work

Mysore and Bharghavan[18] proposed the use of the IP multicast infrastructure to handle Internet host mobility. A mobile host is uniquely identified by a multicast IP address. Packets destined to the mobile host are routed to the address using the IP multicast infrastructure. Bristle can, however, adopt the IP multicast infrastructure to support node mobility for a DHT. The scalability of a DHT is restricted by the available multicast IP addresses. In addition, due to lack of wide deployment of IP multicast, a DHT cannot rapidly support node mobility based on IP multicast. In contrast, to rely on IP multicast to provide DHTs with mobility support, Bristle distributes location management by exploiting resources in end hosts. This allows Bristle to be rapidly and incrementally deployed using the ubiquitous Internet infrastructure but without the sophisticated support of IP multicast.

Rather than handling host mobility in the IP network layer, Maltz and Bhagwat[17] proposed to do mobility management in transport, while Snoeron and Balakrishnan[29] implemented mobility management in the application layer. Handling mobility in transport and application layers allows distinguishing packets from various sessions; that is, each packet will not be identically treated. This feature enables an application to further perform additional operations such as filtering over various communication streams. However, Maltz and Bhagwat need to modify the standard TCP protocol while Snoeron and Balakrishnan[29] rely on the tailored DNS infrastructure. Bristle, however, manages mobility in the application layer with the rudimentary networking protocols (i.e., TCP/IP). Because Bristle is implemented in the application layer, similar techniques such as the filtering implemented by Snoeron and Balakrishnan[29] can also be realized in Bristle.

Li et al.[13] proposed an architecture called *Grid* to provide GPS location information for aiding routes in a wireless ad hoc networking environment. Grid preorganizes geographical location regions in a hierarchical fashion and distributes mobile nodes' locations to each mobile node in the system. To maintain location information for each mobile node in a dynamic mobile environment, Grid makes the following assumptions: (1) each mobile node is GPS-capable; (2) each node is aware of its location in the hierarchy; (3) the hierarchical region where a mobile node appears should be identified in advance; and (4) each node is aware of the identities of other nodes appearing in the same geographical "unit"-region. These four assumptions allow a mobile node to update its location and perform the location query to the mobile nodes in the geographical regions following the hierarchy. In contrast, Bristle does not structure nodes participating in the location management layer via auxiliary information such as GNP,[19] and it poses no assumptions that each node participating in the location management layer is aware of its position in the network. In particular, Bristle is based on the context of distributed hash tables.

Recently, an infrastructure $i3$[30] implemented on top of Chord[31] was proposed, which generalizes point-to-point communication using rendezvous abstraction to provide services such as multicast, anycast, and mobility for the Internet. To send a packet to a mobile node, the sender simply injects the packet to $i3$ toward a rendezvous point that helps forward the packet to the receiver. Once the receiver changes its network attachment point, it updates its location to the rendezvous point. Note that the sender and receiver are not the constituent members of $i3$. The mobility support by $i3$ is primarily for non-$i3$ members (i.e., peers that do not participate in the P2P overlay). Unlike $i3$, Bristle intends to extend a DHT with last-mile wireless extension; that is, mobile nodes in the last-mile wireless network are DHT members. Similar to $i3$, *Warp* is an architecture that aims to handle mobile crowds[34] by taking advantage of the *local convergence* property[5] inherent in a DHT. In Warp, mobile crowds are treated as "data items" that are not assembling parts of the overlay infrastructure. Manipulating a mobile crowd is to PUT and GET the corresponding data item that is represented to the crowd.

Hu et al.[9] recently proposed an implementation, called *DPSR*, of a DHT over an ad hoc wireless network. Unlike a DHT implemented on top of the Internet, each node in DPSR must maintain the routing path for each of its neighbors. To discover the routing path for a given neighbor, each DPSR node presently implements dynamic source routing.[10] With local convergence and scalability features inherent in a DHT, an ad hoc wireless network implementing DPSR can scale to a large network size. Unlike DPSR, Bristle intends to include wireless devices into a DHT implemented for a wire-line environment. Bristle assumes that the underlying communication infrastructure (e.g., wireless TCP)[8] for wireless nodes is available.

Liben-Nowell et al.[15] analyzed the traffic maintenance cost for Chord's stabilization traffic. The analysis, however, was based on the fail-stop model; that is, when a node fails, it becomes silent.

During the revision of the chapter, we became aware that Mahajan et al.[16] had proposed a mechanism to detect a failure node. When a node detects that the failure rates of its neighbors increase, the node reactively increases the probing rates to its remaining neighbors. Bristle complements such a mechanism. However, to include such a mechanism in Bristle, a node must first identify whether its neighbors are stationary or mobile nodes. Increasing the probing rate to a mobile node is inappropriate because this may increase the traffic to the bandwidth-limited wireless networks. Our study also complemented the study by Li et al.,[14] who presented a framework for performance analysis of DHTs under high churn. They concluded that the most effective parameter for a DHT is the stabilization period. The chosen stabilization period affects various DHTs (e.g., Chord, Tapestry, etc.) in different ways. A recent study by Rhea et al.[23] emulated a DHT in a cluster environment which can model the network queuing delay. Rhea et al. concluded that heavy background traffic can exacerbate a DHT to handle even a lower level of churn. Finally, a complementary study by Lam and Liu[12] proposed a mechanism of rapidly discovering a health neighbor in order to replace a failure neighbor.

41.7 Conclusion

In this chapter we discussed a design for a DHT with mobile peers, called Bristle. Bristle allows nodes to dynamically change their network attachment points. Applications implemented on top of Bristle can transparently manipulate mobile nodes without tracking their locations.

In Bristle, a mobile node is associated with an LDT that can rapidly update its state to those nodes that are interested in its movements. Nodes can also reactively discover missing states with the help of the stationary nodes. We investigated the performance of various design options using extensive simulation. The results show that the invalidation-based and the invalidation-update-based approaches outperform the Type **B** DHT, which relies on the leave and join operations to handle the movement of a node.

41.8 Further Studies

In this study, we assume that stationary and mobile nodes are installed with different versions of DHT codes. However, a mobile node might stay at a particular network location for a certain period of time. The mobile node might behave like a stationary node during that period. If we can identify such a case, the mobile node will not need the costly operations to update its location to nodes that are interested in its movement. This requires a mechanism for a DHT to take advantage of a mobile node that does not frequently change its network attachment point. Further, Bristle has not addressed the issue of when a group of DHT nodes moves simultaneously. We will extend Bristle to handle the movement of a group of DHT nodes in the future. Our simulator cannot model the wireless communication environment and the details of mobile communication protocols such as the mobile IP. We are, however, also interested in the performance of a DHT that is deployed on top of the mobile IP infrastructure. Finally, as presented in this chapter, there are several parameters that affect the performance of a DHT, and thus Bristle. Selecting best parameters is a challenging task. We will study how Bristle adaptively selects appropriate values for the parameters during its runtime.

References

1. A. Balachandran, G.M. Voelker, P. Bahl, and P.V. Rangan. Characterizing user behavior and network performance in a public wireless LAN. *ACM SIGMETRICS Performance Evaluation Review,* 33(1):195–205, June 2002.
2. M. Castro, P. Druschel, A.-M. Kermarrec, and A. Rowstron. Scalable application-level anycast for highly dynamic groups. *Lecture Notes in Computer Science,* 2816:47–57, September 2003.

3. F. Dabek, M.F. Kaashoek, D. Karger, R. Morris, and I. Stoica. Wide-area cooperative storage with CFS. In *Proceedings of the Symposium on Operating Systems Principles*, pages 202–215. ACM Press, October 2001.

4. R. Droms. Automated configuration of TCP/IP with DHCP. *IEEE Internet Computing*, 3(4):45–53, July/August 1999.

5. K. Gummadi, R. Gummadi, S. Gribble, S. Ratnasamy, S. Shenker, and I. Stoica. The impact of DHT routing geometry on resilience and proximity. In *Proceedings of the International Conference on Applications, Technologies, Architectures, and Protocols for Computer Communications*, pages 381–394. ACM Press, August 2003.

6. H.-C. Hsiao and C.-T. King. Bristle: a mobile structured peer-to-peer architecture. In *Proceedings of the International Parallel and Distributed Processing Symposium*. IEEE Computer Society, April 2003.

7. H.-C. Hsiao and C.-T. King. Tornado: a capability-aware peer-to-peer storage overlay. *Journal of Parallel and Distributed Computing*, 64(6):747–758, June 2004.

8. F. Hu and N.K. Sharma. Enhancing wireless Internet performance. *IEEE Communications Surveys*, 4(1):2–15, 2002.

9. Y.C. Hu, S.M. Das, and H. Pucha. Exploiting the synergy between peer-to-peer and mobile ad hoc networks. In *Proceedings of the International Workshop on Hot Topics in Operating Systems*. USENIX, May 2003.

10. D.B. Johnson and D.A. Maltz. *Dynamic Source Routing in Ad Hoc Wireless Networks*. Kluwer Academic, 1996.

11. T. Klingberg and R. Manfredi. The Gnutella 0.6 Protocol Draft. http://rfc-gnutella.sourceforge.net/.

12. S.S. Lam and H. Liu. Failure recovery for structured P2P networks: protocol design and performance evaluation. In *Proceedings of ACM SIGMETRICS*, June 2004.

13. J. Li, J. Jannotti, D. De Couto, D.R. Karger, and R. Morris. A scalable location service for geographic ad hoc routing. In *Proceedings of the International Conference on Mobile Computing and Networking*, pages 120–130. ACM Press, August 2000.

14. J. Li, J. Stribling, T.M. Gil, R. Morris, and M.F. Kaashoek. Comparing the performance of distributed hash tables under churn. In *Proceedings of the International Workshop on Peer-to-Peer Systems*. Springer Press, February 2004.

15. D. Liben-Nowell, H. Balakrishnan, and D. Karger. Analysis of the evolution of peer-to-peer systems. In *Proceedings of the Symposium on Principles of Distributed Computing*, pages 233–242. ACM Press, July 2002.

16. R. Mahajan, M. Castro, and A. Rowstron. Controlling the cost of reliability in peer-to-peer overlays. In *Proceedings of the International Workshop on Peer-to-Peer Systems*. Springer Press, February 2003.

17. D.A. Maltz and P. Bhagwat. Msocks: an architecture for transport layer mobility. In *Proceedings of IEEE INFOCOM*, pages 1037–1045, March 1998.

18. J. Mysore and V. Bharghavan. A new multicasting-based architecture for Internet host mobility. In *Proceedings of the International Conference on Mobile Computing and Networking*, pages 161–172. ACM Press, September 1997.

19. T.S. Eugene Ng and H. Zhang. Predicting Internet network distance with coordinates-based approaches. In *Proceedings of IEEE INFOCOM*, pages 170–179, June 2002.

20. C.E. Perkins. IP Mobility Support. RFC 2002, October 1996.

21. C.E. Perkins and D.B. Johnson. Mobility support in IPv6. In *Proceedings of the International Conference on Mobile Computing and Networking*, pages 27–37. ACM Press, November 1996.

22. S. Ratnasamy, P. Francis, M. Handley, R. Karp, and S. Shenker. A scalable content-addressable network. In *Proceedings of the International Conference on Applications, Technologies, Architectures, and Protocols for Computer Communications*, pages 161–172. ACM Press, August 2001.

23. S. Rhea, D. Geels, T. Roscoe, and J. Kubiatowicz. Handling churn in a DHT. In *Proceedings of the USENIX Annual Technical Conference*, June 2004.

24. A. Rowstron and P. Druschel. Pastry: scalable, distributed object location and routing for large-scale peer-to-peer systems. *Lecture Notes in Computer Science*, 2218:161–172, November 2001.

25. A. Rowstron and P. Druschel. Storage management and caching in past, a large-scale, persistent peer-to-peer storage utility. In *Proceedings of the Symposium on Operating Systems Principles,* pages 188–201. ACM Press, October 2001.

26. S. Saha, M. Jamtgaard, and J. Villasenor. Bringing the wireless Internet to mobile devices. *IEEE Computer,* 34(6):54–58, June 2001.

27. S. Saroiu, P.K. Gummadi, and S.D. Gribble. Measurement study of peer-to-peer file sharing systems. In *Proceedings of Multimedia Computing and Networking.* January, 2002.

28. S. Sen and J. Wang. Analyzing peer-to-peer traffic across large networks. *ACM/IEEE Transactions on Networking,* 12(2), April 2004.

29. A.C. Snoeren and H. Balakrishnan. An end-to-end approach to host mobility. In *Proceedings of the International Conference on Mobile Computing and Networking,* pages 155–166. ACM Press, August 2000.

30. I. Stoica, D. Adkins, S. Zhuang, S. Shenker, and S. Surana. Internet indirection infrastructure. In *Proceedings of the International Conference on Applications, Technologies, Architectures, and Protocols for Computer Communications,* pages 73–86. ACM Press, August 2002.

31. I. Stoica, R. Morris, D. Karger, M.F. Kaashoek, and H. Balakrishnan. Chord: a scalable peer-to-peer lookup service for Internet applications. In *Proceedings of the International Conference on Applications, Technologies, Architectures, and Protocols for Computer Communications,* pages 149–160. ACM Press, August 2001.

32. D. Tang and M. Baker. Analysis of a local-area wireless network. In *Proceedings of the International Conference on Mobile Computing and Networking,* pages 1–10. ACM Press, August 2000.

33. E.W. Zegura, K. Calvert, and S. Bhattacharjee. How to model an internetwork? In *Proceedings of IEEE INFOCOM,* pages 594–602, March 1996.

34. B.Y. Zhao, L. Huang, A.D. Joseph, and J.D. Kubiatowicz. Rapid mobility via type indirection. In *Proceedings of the International Workshop on Peer-to-Peer Systems.* Springer Press, February 2004.

35. B.Y. Zhao, L. Huang, J. Stribling, S.C. Rhea, A.D. Joseph, and J.D. Kubiatowicz. Tapestry: a resilient global-scale overlay for service deployment. *IEEE Journal on Selected Areas in Communications,* 22(1):41–53, January 2004.

Topology Construction and Resource Discovery in Peer-to-Peer Networks

Dongsheng Li

Xicheng Lu

Chuanfu Xu

42.1 Introduction

In recent years, peer-to-peer (P2P) computing has attracted significant attention in both industry and academic research (Clark, 2001; Milojcic et al., 2002; Schoder and Fischbach, 2003). Many peer-to-peer networks have been deployed on the Internet, and some of them have become the most popular Internet applications. Unlike traditional client/server systems, each node (peer) in pure P2P networks can act as both client and server with equal capability. Peers can exchange information and services directly with each other to perform a critical function coordinately in a decentralized manner.

P2P computing is a promising technique to take advantage of vast number of resources (e.g., storage, CPU cycles, content, human presence) on the Internet. Applications of P2P networks range from distributed computing (e.g., SETI@home), file sharing (e.g., Napster and Gnutella), instant messaging (e.g., ICQ,

Jabber), collaboration (e.g., Groove), cooperative Web-caching (e.g., Squirrel), persistent data storage (e.g., Oceanstore) to application-level multicast (e.g., Scribe), etc. Numerous interesting research topics have been developed within these areas. The general problem for a P2P network is how to make it scalable, robust, secure, etc.

42.1.1   Characteristics of P2P Networks

Generally, P2P networks exhibit many common characteristics, such as large-scale, dynamic, geographical distribution, heterogeneity, etc. This section gives a short review of these characteristics and introduces their impact on topology construction and resource discovery in P2P networks.

Many P2P networks are large-scale. For example, the number of users of Napster, the famous P2P file-sharing network, had reached 50 million by January 2001; and the number of registered users of Kazaa reached 150 million in 2003. Large numbers of users led to vast resources sharing. It has been reported that in the Morpheus file-sharing network, over 470,000 users shared a total of 36 petabytes of data on October 26, 2001. Measurements from CAIDA, a famous international statistic organization for Internet traffic, has also shown that the traffic of P2P applications had accounted for more than 40 percent of the total traffic on the backbone network in 2002. Such a large number of users and vast resources not only bring about valuable applications, but also make the efficient and scalable topology construction and resource discovery in P2P networks a challenging problem. The mechanisms for topology construction and resource discovery in P2P networks should be scalable and able to support large-scale users and resources.

P2P networks are strongly dynamic. Peers may join or leave P2P networks due to various reasons (peer or link failures, etc.) at any time. Some research indicates that the average MTTF (mean time to failure) of computers on the Internet is 13 hours; that is, in a P2P network with 10,000 peers, at least one peer would go down every two minutes. One important reason for dynamic is that peers in P2P networks are highly autonomous. A great number of peers can join or leave the network according to their individual wills, which causes highly changing P2P topology and increases the difficulty of resource discovery. Measurements (Saroiu et al., 2002) on the typical P2P networks — Napster and Gnutella — have shown that the frequency of peer joining or leaving is high, and the average online time of a peer is no more than one hour. Therefore, the topology construction and resource discovery mechanisms in P2P networks should be fault-tolerant and adapt well to the changing peer population.

P2P networks are highly distributed. There are numerous users and resources that participate in a P2P network, and these users and resources are geographically distributed. Therefore, P2P systems often adopt a decentralized topology. The highly distributed characteristic helps make the network robust and load balanced, but it also makes the topology construction and resource discovery become very complex.

P2P networks are often heterogeneous. One reason for the heterogeneity across peers is that peers have different capabilities, such as computing power, storage capacity, network bandwidth, etc. Another reason for heterogeneity lies in that fact that peers are highly autonomous. For example, many fashionable P2P file-sharing software programs allow users to set the download and upload bandwidth, the number of links with other peers, the number of shared resources, etc. Measurements (Saroiu et al., 2002) on the Gnutella network have shown that the bandwidth of peers varies over a large range (from 10 Kbps to more than 10 Mbps). Any topology construction and resource discovery mechanisms that ignore the heterogeneity across peers would form an inefficient P2P network, and those that can take advantage of the heterogeneity could improve the efficiency of P2P networks.

Security, trust, and incentive are problems that P2P networks must face. The individual will of autonomous peers affects the abilities and actions (e.g., online time, etc.) of peers participating in the network; therefore, it not only affects the dynamic and heterogeneity characteristics of P2P networks, but also brings about the security and incentive problems. In addition to the general security problems (e.g., identity authentication, authorization, data encryption, integrality, etc.), P2P networks should be able to keep away from the malicious behaviors of partial malicious peers. Research has indicated that there are a lot of trick behaviors and unreliable service quality in many P2P networks. Measurements have shown that there is a "free riding" phenomenon in the Gnutella network (Adar and Huberman, 2000): 70 percent of Gnutella peers share no files and 90 percent of the peers answer no searches. The topology construction and resource discovery mechanisms should adopt effective security, trust, and incentive policies to help the P2P network grow well and improve its stability, usability, and performance.

In conclusion, the main design goals of the mechanism for topology construction and resource discovery in P2P networks include:

Scalability. It should scale to millions of, or even more, peers and resources.

Performance. It should support efficient topology construction and resource discovery in large-scale P2P networks.

Adaptability. It should adapt well to the dynamic environment in which peers join and leave freely.

Resilience. It should be fault-tolerant toward peer or link failures.

Security and incentive. It can operate correctly and effectively in an untrustworthy environment.

42.2 Topology of P2P Networks

The P2P system builds a virtual overlay network at the application level (i.e., the P2P overlay network). The P2P overlay network has its own routing mechanisms that can forward messages to discover desired resources or perform other functions. The topology and its resource discovery mechanism are the two basic elements of P2P networks, and they have a significant impact on the performance, scalability, reliability, and other properties of P2P applications.

There are many classification methods for topologies of P2P networks. From the view of "degree of decentralized," P2P networks can be divided into three categories: (1) centralized topology (e.g., Napster), in which there is a central server to coordinate the interaction of peers; (2) decentralized topology (e.g., Gnutella, Chord), in which all peers act as both server and client equally; and (3) partially decentralized topology (e.g., FastTrack, Brocade), in which there exist some supernodes or superpeers that play a more important role than others.

From the view of "coupling of topology," P2P networks can also be divided into three categories: (1) unstructured topology (e.g., Gnutella), which is freely formed by peers; (2) structured topology (e.g., Chord), which is precisely controlled by determined algorithm; and (3) loosely structured topology (e.g., Freenet), in which the topology is freely formed by peers but the placement of data in the P2P network is controlled. The loosely structured P2P network utilizes a policy between unstructured and structured P2P networks: the network topology is arbitrary, as it is in unstructured schemes; however, the placement of content is controlled, as in structured schemes. Freenet and Yappers belong to this category. Many research topics in the loosely structured P2P networks are somewhat similar to those in unstructured P2P networks. For convenience, we include the loosely structured topology in the unstructured topology and divide P2P topologies into four categories: (1) centralized topology, (2) decentralized unstructured topology, (3) decentralized structured topology, and (4) partially decentralized topology. We briefly introduce the four topologies, along with related research.

42.2.1 Centralized Topology

The centralized topology is based on a central index server (or servers) that coordinates or schedules the resources on individual registered peers. Generally the central server maintains central directories of the resources on the peers in the P2P network and coordinates the interaction between peers. Sometimes, the central server can also act as a dispatcher that assigns tasks to appropriate peers. In some cases, the central index server can be a cluster of servers to improve the scalability, robustness, and load balance.

The centralized P2P network has two features: (1) centralized index and (2) distributed work. Once peers join or leave the P2P network, the resources of the peers would be registered into or removed from the centralized index. When a peer wants to discover certain particular resources, it submits a resource search request to the central server, which checks its directories and returns a list of resources satisfying the request to the requesting peer. The centralized server only provides the directory service, and the critical functions

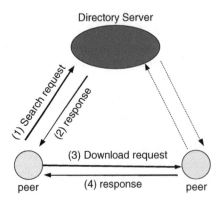

FIGURE 42.1   Napster architecture.

of the system (e.g., file download or distributed computing) are performed by distributed individual peers. Thus, these systems are still peer-to-peer systems, not pure but hybrid P2P systems. Napster, SETI@home, and certain other P2P networks adopt the topology.

42.2.1.1 Case Studies

42.2.1.1.1 *Napster*

Napster (2001) was the first famous P2P system; it enables millions of users to share music files over the Internet. Users can exchange mp3 files through Napster. Napster was founded by Shawn Fanning and Sean Parker in May 1999. After that, it gained significant success and has become one of the most popular Internet applications with more than 50 million users. In December 1999, the RIAA (Recording Industry Association of America) filed a lawsuit against Napster for its copyright infringement, and Napster was forbidden to use for exchanging copyrighted music files by the Court in February 2001. It is estimated by the RIAA that the total income of the music industry decreased by 30 percent each quarter since Napster was developed (see Figure 42.1).

Napster uses a central index server to maintain metadata information of all music files shared in the current Napster community. While an individual peer joins the system through Napster client software for the first time, the client software registers the list of music files shared on the peer to the server. When a peer wants to find a certain music file, it submits its search request to the index server. The index server receives and handles the request, returns a list of matched peers to the peer, and then the peer can select a most appropriate peer and download the music file from it directly (without the help of central server again).

Napster is not a pure P2P system but a hybrid P2P system with a central directory, but it is peer-to-peer in nature, as peers connect to each other to download files directly.

42.2.1.1.2 *BitTorrent*

BitTorrent is a protocol designed by Bram Cohen (2003a) that debuted at CodeCon 2002. It has become a very popular P2P file-sharing protocol on the Internet. The key philosophy of BitTorrent is that peers should upload to each other at the same time they are downloading. BitTorrent can work better as the number of people interested in a certain file increases. The more downloaders there are at the same time, the more efficient the downloading.

BitTorrent also uses a central server called a tracker to coordinate the action of all participating peers for the same file. But unlike Napster, the tracker only manages connections and does not have any knowledge of the contents of the files being distributed. Therefore, a large number of users can be supported with relatively limited tracker bandwidth and loads. When multiple downloads of the same file happen concurrently, the downloaders upload to each other, making it possible for the file source to support very large numbers of downloaders with only a modest increase in its load.

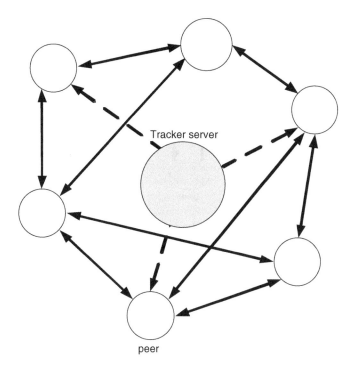

FIGURE 42.2 BitTorrent architecture.

To publish a file, the peer first cuts the file into pieces of fixed size, typically a quarter megabyte, and then a static metafile with the extension. torrent is created and published on a Web site. The .torrent metafile contains information about the file, including its length, name, hashing information of each piece, and the URL of a tracker. To make a file available, a seed peer that has the complete file already must be started.

The peer that would like to download the file needs to get the metafile from a Web site at first, and then with the help of the tracker it can know the seed peer and other peers that are downloading. It connects to them and downloads different pieces of the file from various peers concurrently to improve downloading performance. When it has downloaded a complete piece of the file, it can upload the piece to others. To keep track of which peers have what, each downloader reports to all the other peers downloading the same file as to what pieces it has. To verify data integrity, the SHA-1 hashes of all the pieces are included in the torrent file, and peers do not report that they have a piece until they have checked the hash (see Figure 42.2).

BitTorrent adopts a choking algorithm to utilize all available resources, provide reasonably consistent download rates for everyone, and be somewhat resistant to peers only downloading and not uploading. BitTorrent is fit for big and popular files sharing.

42.2.1.1.3 *SETI@home*

SETI@home is a scientific research project announced in 1998 that aimed at building a virtual supercomputer consisting of large numbers of Internet-connected computers to analyze the signal data received from space and collected by the giant Arecibo telescope (Anderson et al, 2002). It is one of the projects from SETI (Search for Extraterrestrial Intelligence), which focused on discovering alien civilizations. Before SETI@home, radio SETI projects used special-purpose supercomputers to perform data analysis. SETI@home has acquired a raw processing power of several dozens of Tflops from more than three million computers over the Internet.

SETI@home has two major components: the data server and the client. The data server splits the signal data into fixed-size work units and distributes the work units to client programs running on numerous computers on the Internet. The server also collects results after processing is done. Each peer that would like

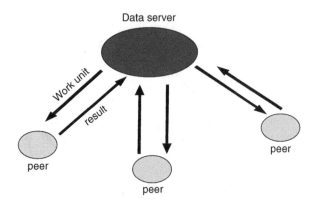

FIGURE 42.3   SETI@home architecture.

to participate in the project installs client-side software as a screen saver. The client program repeatedly gets a work unit from the data server, computes a result, and returns it to the server. There is no communication between clients. In SETI@home, each work unit is processed by multiple clients to detect and discard results from faulty processors and from malicious users.

The SETI@home client software adopts a checkpoint mechanism. The program periodically writes its state to a disk file, reading the file on start-up. Once a SETI@home computation is interrupted, the computation will resume from the last saved state and proceed. This simple mechanism helps increase the reliability of the system. The single server may be a scalability and performance bottleneck, but real-world experience shows that the system can handle more than three million clients (see Figure 42.3).

Many projects similar to SETI@home have been deployed on the Internet, such as the Great Internet Mersenne Prime Search (GIMPS), which searches for prime numbers; Distributed.net, which performs brute-force decryption; protein folding (folding@home) at Stanford University; etc.

42.2.1.2   Advantages and Disadvantages

Simplicity is an important advantage of centralized topology. Through the central index, the resource management is somewhat easy, and resource discovery is quick and efficient. As the resource search is performed on a central directory, it can be very flexible. The central directory updates the index periodically; thus, resources that peers find through their searches are immediately available.

While the centralized topology makes the efficient and comprehensive resource search possible, it may introduce single points of failure, hotspots in the network, lawsuits, and other problems. The centralized topology exposes vulnerability to technique failures or malicious attacks because it depends on a small number of central servers. The system might completely collapse if one or some of the servers failed for some reason. When the scale of peers or search requests increases, the central server may become the bottleneck of the system and thus limit the scalability of the P2P network. The central server also makes some censorship possible.

42.2.2   Decentralized Unstructured Topology

In decentralized unstructured P2P networks, there is neither a centralized directory nor any precise control over the network topology or resource placement. When a new peer joins the P2P network, it forms connections with other peers freely (e.g., selecting some random peers as neighbors). If a peer wants to publish some resources, it usually just stores them locally or places them on randomly selected peers. In general, unstructured P2P networks have loose guarantees for resource discovery; that is, the resource in such systems may not be found even though it exists and the search efficiency cannot be guaranteed. Decentralized unstructured topology is quite suitable for environments composed of highly autonomous peers, such as file-sharing environments in which a wide range of users who come from many different organizations share files with each other and strangers are unwilling to perform much additional work for others. Gnutella, Freenet, Mojo Nation, and NureoGrid are typical decentralized unstructured P2P networks.

42.2.2.1 Case Studies

42.2.2.1.1 *Gnutella*

Gnutella (2001) is not a specific system but a file-sharing protocol used to search and share files among users. Gnutella was introduced in March 2000 by employees of AOL's Nullsoft division. To avoid music copyright infringement, AOL stopped the development of Gnutella. However, the remaining online open-source program produced many clones of the Gnutella protocol and now various versions of the original Gnutella servant have developed, such as Limewire and BearShare. The Gnutella protocol is changing fast and we only introduce the popular version 0.4 here.

Unlike Napster, Gnutella is based on a purely distributed architecture and designed to provide a purely distributed file-sharing solution without a central server. There is no "client" or "server" in Gnutella. Each peer is equal and called a "servant." When a new peer joins, it initially connects to one of several known peers by out-of-bounds ways (e.g., a well-known Web site where some Gnutella peers are collected). Once it connects to a known peer, the peer will send it a list of peers it knows. Then the new peer adds more connections to some peers chosen from the list. To cope with the dynamic environment in which peers often join or leave, each peer periodically exchanges information with its neighbors to discover other active peers (see Figure 42.4).

Gnutella uses the flooding mechanism with a predetermined TTL (time-to-live) to search files. When a peer wants to search a file (based on keywords), it will broadcast the search request to all neighboring peers to which it is connected. Those peers perform a text match of the keywords and may send a response back to the requesting peer if a match is found; and they will also broadcast the request to their neighboring peers and decrease the TTL by 1. The search proceeds until TTL reaches 0. When the requesting peer looks through the file list returned and would like to download a file, it connects to the peer that reported the match and starts the download. If the destination peer cannot be connected (e.g., because of a firewall), it invokes a download request. The request travels the same way as the original search and eventually arrives at the destination peer that reported the match. Then the destination peer connects back to the requesting peer and begins to upload.

From a user's view, Gnutella is simple and effective because the hit rate of searches is reasonably high. Gnutella adapts well to dynamically join or leave peers. However, flooding limits the scalability of the

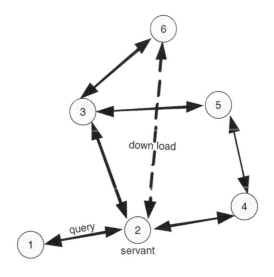

FIGURE 42.4 Gnutella architecture.

Gnutella system and consumes too much network bandwidth. Assuming a typical flooding depth of 7 and an average of 4 connections per peer, one search message can produce 26,240 messages in the Gnutella network. Recent experiments show that in a real-world setting, this accumulates up to 3.5 Mbps (Clip2.com, 2001). Research has also shown that most Gnutella peers do not contribute to the network and simply download files from a minority of contributing peers (i.e., the "free riding" phenomenon). This goes against the original peer-to-peer ideal of Gnutella, and shows the need for some incentive mechanisms in future peer-to-peer networks.

42.2.2.1.2 *Freenet*

Freenet is a distributed anonymous information storage and retrieval system initially designed by Ian Clarke in 1999 (Clarke et al., 2000). Freenet is a completely decentralized P2P system, and its primary mission is to provide a certain degree of privacy and anonymity for producers, consumers, and storers of data files. Each peer in Freenet maintains a dynamic routing table containing the addresses of other peers and the keys that they are thought to hold. Data files are replicated adaptively and dynamically.

Each file in Freenet is identified by a key generated by the hash SHA-1 function. Each peer in Freenet attempts to store files with similar keys. When a peer wants to publish or search certain files, it uses the key to locate the proper location of the file. To search a file, a peer sends a request message specifying the file key. Each request is given a hops-to-live (HTL) limit, which is decreased at each peer. Each request is also assigned a pseudo-unique random identifier; thus, peers can avoid loops by discarding requests they have seen before. When the request is received by a peer that stores the desired file, it returns the file as a successful result. If the peer does not store the file, it looks up its "routing table" and the request is forwarded to the neighbor peer corresponding to the most similar key in the routing table. The search process continues until the request is satisfied or its HTL limit is reached. If the file is found successfully, the file is replicated at each peer in the search path and the file's key is inserted into the routing tables. In this way, popular files become highly replicated and files with similar keys are clustered in the same peer. When all shared disk space of a peer is consumed, files are replaced using the least recently used (LRU) replacement strategy.

Research has shown that Freenet is scalable and its path length scales approximately logarithmically to the network size. Freenet is also surprisingly robust against quite large failures. The scalability and fault-tolerance characteristics of Freenet can be interpreted by the "small-world" model, as the Freenet network exhibits a strong "small-world" property.

42.2.2.2 Advantages and Disadvantages

Decentralized unstructured P2P networks are widely deployed and predominant on the Internet in real life for their simplicity and usability. Such networks are fault tolerant toward peers or links failures because each part of the P2P network is somewhat independent. Recently, some researches have shown that Gnutella-style unstructured P2P networks exhibit the "power-law" property (Jovanovic et al., 2001; Ripeanu et al., 2002; Albert and Baraba, 2002); that is, the number of peers with L links is proportional to L^{-k}, where k is a network-dependent constant). In a power-law P2P network, most peers have few links and only a few peers have many links. Such a network remains connected even if a large part of the network has failed. This can help explain the generally stable and resilient structures of the Gnutella network while random failures occur frequently. The power-law property also makes it difficult to attack by random malicious attacks. However, the power-law network is vulnerable to specific denial-of-service attacks: If a few high-degree peers are attacked, the P2P network might collapse.

Unstructured P2P networks adapt well to the dynamic of peers. When peers join or leave, the P2P network is easy to maintain and adapt to the changing peer population. Unstructured P2P networks can also support richer searches than just search by identifier, including keyword search with regular expressions, range search, etc.

However, unstructured P2P networks provide no guarantee of the quality and performance of resource discovery. The search results returned may not be complete and the search procedure might be very slow. Some searches fail even if the desired resources exist. Flooding, random walk, or selective forwarding is often used for resource discovery in such systems, but current techniques are often not efficient or scalable. For example, Gnutella uses flooding with limited time-to-live (TTL) for resource discovery, but flooding consumes too much bandwidth and limits its scalability.

42.2.2.3 Current Research

Performance and scalability are two important open problems on unstructured P2P networks. There is a lot of research going on about unstructured P2P networks and much of it focuses on improving performance and scalability. This section briefly describes several related techniques, especially techniques related to resource discovery (i.e., search) in unstructured P2P networks.

42.2.2.3.1 *Blind Search*

Flooding is one example of a blind search method used in unstructured P2P networks. Gnutella uses blind flooding with limited time-to-live (TTL), but flooding produces too many search messages in the network. A simple approach to reduce the flooding traffic is to set a low TTL on initial search messages. If the results returned are not enough after a timeout period, the search message is resent with a higher TTL. Expanding ring (Lv et al., 2002) and iterative deepening (Yang and Molina, 2002) techniques follow this idea and they increase the flooding radius in a slow way to decrease bandwidth consumption. Random walks are also suggested to take the place of flooding in many researches (Lv et al., 2002; Gkantsidis et al., 2004). Gkantsidis et al. (2004) adopted a deep result of stochastic processes, which indicates that samples taken from consecutive steps of a random walk can achieve statistical properties similar to independent sampling, and analyzed the advantages of random walks over flooding, in some cases theoretically.

42.2.2.3.2 *Search with Hints*

Some research has suggested that each peer in a P2P network maintains some kind of metadata that can provide "hints" to guide the search direction. Adamic et al. (2001) proposed algorithms utilizing local information such as the identities and connectedness of a peer's neighbors and forward search messages to high-degree neighbors to utilize the power-law property of Gnutella-style P2P networks. A directed BFS technique (Yang and Molina, 2002) was proposed to forward search messages to only a subset of its neighbors according to which peers with more quality results might be reached. Yang and Molina (2002) used local indices where peers index the content of other peers in the system. In NeuroGrid, Joseph (2002) proposed that peers maintain a knowledge base that stores associations between keywords and other peers, facilitating the search by forwarding queries to a subset of peers that it believes might possess matches to the search query. Crespo and Molina (2002) built summaries of content that is reachable via each neighbor of the peer in different topics. Query routing (LimeWire, 2003) used a well-known hashing function to hash all the keywords representing their shared files into a hash table and allowed search messages to be routed only to hosts that are likely to hold the specified item. Cohen (2003b) exploited associations inherent in human selections to steer the search process to peers that are more likely to have an answer to the query. These techniques are different in the content of hints and the policy to forward search messages, and thus lead to different performance and cost characteristics.

42.2.2.3.3 *Replication and Caching*

Replication and caching are two important methods to improve search performance. In Freenet, data is proactively replicated at each peer in the path where the search message passes although it is not requested by the peer. Sripandidkulchai (2001) studied the characteristics of search messages and found that the frequency of the search keyword in Gnutella networks follows the zipf-like distribution. Based on that observation, a query cache policy was proposed to reduce message cost in Gnutella-style P2P systems. Markatos (2002) studied the traffic characteristics of Gnutella networks and proposed some caching strategies that take the TTL factor into account to improve search performance. Cohen and Shenker (2002) proposed and analyzed three replication strategies for blind search ((1) a uniform strategy in which each data has the same number of replicas; (2) a proportional strategy in which the number of replicas is proportional to the search rate of the data; and (3) a square-root strategy in which the number of replicas is proportional to the square root of the search rate), and proved that the square-root strategy can minimize the search size.

42.2.2.3.4 *Topology Construction and Optimization*

Some researchers have designed distributed algorithms to construct and maintain unstructured topologies with good connectivity properties for searching. Pandurangan et al. (2001) suggested building a low-diameter P2P system with high connectivity; however, they did not discuss how to find the desired data in such a system. Sripanidkulchai et al. (2003) exploited the interest-based locality principle (i.e., if a peer has a particular piece of content that one is interested in, it is very likely to have other items that one is interested in as well), and built interest-based shortcuts among peers to improve the search performance. Phenix (Wouhaybi and Campbell, 2004) supports low-diameter operations by creating a topology of peers whose degree distribution follows a power-law, while its implementation is fully distributed.

Some research focuses on mapping the P2P overlay efficiently to the underlying Internet network topology. Clustering techniques (Mathy et al., 2002) were proposed that allow selection of topologically close hosts, which helps map overlays more efficiently to the underlying network.

42.2.2.3.5 *Downloading Issues*

In many peer-to-peer systems, data is replicated on multiple peers. When acquiring the multiple locations of a desired file, a peer must choose from where to download to get good downloading properties. Bernstein et al. (2003) proposed an approach based on a machine learning technique to construct good peer selection strategies from past experience. In their approach, decision tree learning was used for rating peers based on low-cost information, and Markov decision processes were used for deriving a policy for switching among peers. Maymounkov et al. (2003) presented a novel algorithm based on rateless erasure codes that are locally encodable and linear-time decodable to download large files from multiple sources in peer-to-peer networks.

Many other research topics have been extensively studied for unstructured P2P systems, including security, heterogeneity, trust and reputation management, etc. For example, systems such as Mojo Nation seek to develop effective incentive mechanisms for users to contribute their resources. In a word, performance, scalability, security, and incentive are still important open problems in unstructured P2P systems.

In many research efforts about unstructured P2P networks, the "small-world" network and "small-world" phenomenon are often mentioned. Thus, we introduce the small-world network below.

42.2.2.3.6 *Small-World Network*

The "small-world" phenomenon (popularly known as six degrees of separation), the principle that most people on the planet are linked by a chain of at most six acquaintances, was first discovered by Harvard professor Stanley Milgram in the late 1960s (Milgram 1967). Milgram performed a social experiment in which he randomly chose 60 people who lived in Omaha, Nebraska. He asked each of them to pass one letter to a given destination person who worked as a stockbroker in Boston, Massachusetts. In the experiment, each person was to pass his letter to a friend whom he thought might bring the letter closest to the destination; the friend would then pass it on to another friend, and so on until the letter reached someone who knew the destination person and could give it to him. When the experiment finished, 42 letters reached the destination person via a median number of just 5.5 intermediaries, while the population of the United States was 200 million at that time. This experiment demonstrated the small-world effect for the first time.

Recently, research (see Albert and Baraba, 2002) has shown that the "small-world" phenomenon also exists in many biological, technological, and social networks, such as power grid, actors network, the Internet, the WWW, etc. These networks are called "small-world" networks, by analogy with the small-world phenomenon. Watts and Strogatz (1998) modeled the small-world phenomenon and used the concepts of clustering coefficients and characteristic path lengths to analyze the characteristics of small-world networks. The characteristic path length L is defined as the number of edges in the shortest path between two vertices, averaged over all pairs of vertices. The clustering coefficient C is defined as follows:

Suppose that a vertex v has k_v neighbors; then there are, at most, $k_v(k_v - 1)/2$ edges between them; clustering coefficient C_v of vertex v is the result of dividing the number of actual edges between the neighbors by $k_v(k_v - 1)/2$; clustering coefficient $C(G)$ of a graph G is defined as the average of C_v over all vertices. Watts and Strogatz (1998) showed that, compared with corresponding random networks with the same number of peers and average degree per peer, the small-world network exhibits two features: its characteristic path length is close to that of corresponding random networks, but its clustering coefficient is much higher (i.e., $L_{small-world} \geq L_{random}$ and $C_{small-world} >> C_{random}$). Thus, the characteristic path length of the small-world network scales logarithmically with the size of the network because it follows the path length of a random network.

Research has also shown that many unstructured P2P networks are small-world networks. Freenet exhibits a strong small-world property, and Gnutella (Jovanovic et al., 2001; Ripeanu et al., 2002) has a somewhat smaller small-world effect. The trace of actual Gnutella networks [Ripeanu et al., 2002] has also shown that 95 percent of the shortest paths among peers in Gnutella are less than seven hops and 50 percent of those are less than five hops.

Many research efforts have used the small-world property to improve the performance and scalability of unstructured P2P networks. For example, Zhang et al. (2002) studied the impact of a cache replacement policy on the performance of Freenet and proposed an enhanced clustering cache replacement scheme. This replacement scheme forces the routing tables to resemble neighbor relationships in a small-world network and thus dramatically improves the hit ratio of the search cache.

42.2.3   Decentralized Structured Topology

In decentralized structured P2P networks, there are no central directories but there is tight control over the P2P network topology. The P2P topology should be reconstructed along with the join or leave of peers. There is close coupling between the network topology and resource location information. The resource (or its metadata) is placed not on local or random peers but on specified peers by certain determined algorithms. Decentralized structured P2P networks can guarantee that any existing resource can be located within predetermined bounded hops. It can also achieve good scalability and performance and thus has become a hot research topic. The core component of many structured P2P networks is the distributed hash table (DHT) scheme (Ratnasamy et al., 2002; Balakrishnan et al., 2003) that uses a hash table-like interface to publish and look up data objects. The topology of a P2P network is determined by the DHT scheme and the resource discovery is also performed based on the DHT algorithm.

42.2.3.1   DHT Schemes

In DHT schemes, each data object is hashed into a namespace and assigned a uniform identifier *key* according to its metadata (e.g., the filename) by some public hash function. Each peer takes charge of a small part of the namespace and is also assigned a uniform peerID. In general, the data object with *key O* is stored on a certain peer whose peerID has some mapping relationship to *key O* according to the DHT scheme. When peers join or leave, the responsibility is reassigned among the peers to maintain the hash table structure.

The DHT scheme must be able to determine which peer is responsible for storing the data with any given key and adapt well to the dynamics of peers. The basic operation in DHT scheme is lookup(*key*), which returns the information (e.g., the IP address) of the peer storing the data object with that *key*. For example, if a peer would like to find the file "student.doc," it first converts "student.doc" to a key, say "123456789," and then looks up "123456789" in the DHT scheme to find the location of the file "student.doc." If a peer publishes a data object with the identifier *key*, it also uses the operation lookup(*key*) to find the peer that should store the data object, and stores the data object on the peer.

To implement the lookup operation, each peer in a DHT scheme has a "forwarding table" that maintains mappings between the keys and information (e.g., IP address) of a small number of other peers ("neighbors") in the network. The P2P overlay network is formed by the neighborhood and the lookup messages are forwarded in the overlay based on the forwarding tables. The lookup operation in the DHT scheme is somewhat like routing in the IP (Internet Protocol) networks: When a peer receives the lookup request, it first checks whether it can satisfy the request by itself; if it cannot, it forwards the request to a neighbor peer based on the forwarding table; then the lookup request is handled by the neighbor peer and the lookup procedure continues. But the functions and requirements of DHT schemes are much different from that of IP routing, as DHT schemes are designed to scale to large numbers of potentially faulty peers and objects and support the strong dynamics of peers.

Chord, CAN, Tapestry, and Pastry are the most well-known DHT schemes. We give a brief introduction to some of them.

42.2.3.2 Case Studies

42.2.3.2.1 *Chord*

Chord (Stoica et al., 2001) is a distributed lookup protocol for Internet application developed as a part of MIT's SFS project. Chord supports one operation: given a key, it maps the key onto a peer. Chord uses a variant of consistent hashing (Karger, 1997) to assign keys to peers. Theoretical analysis and simulation results show that Chord adapts efficiently as peers join and leave the system, and can answer queries even if the system is continuously changing.

Both peers and keys in Chord are assigned m-bit identifiers based on a public hash function (such as SHA-1), and thus Chord forms a unidimensional, circular identifier space. Chord assigns keys to peers as follows: Key k is assigned to the first peer whose identifier is equal to or follows (the identifier of) k in the identifier space (which is called the successor of key k). Each peer maintains a "finger table" with (at most) m entries. The i-th entry of the finger table at peer n contains the identity of the first peer, s, that succeeds n by at least $2^i - 1$ on the identifier circle; that is, $s = \text{successor}(n + 2^i - 1)$, where $1 \le i \le m$ (and all arithmetic is modulo 2^m). The lookup(id) operation of peer n is as follows: If id falls between n and n's successor, the lookup process is finished and peer n returns its successor. Otherwise, n consults its finger table for the peer p, whose ID most immediately precedes id, and then the lookup request if forwarded to peer p. This process continues until the destination id is found.

When a peer n joins the network through a gateway peer n', it will ask for n' to build its finger table. To ensure that lookups execute correctly as the set of participating peers changes, Chord must ensure that each peer's successor pointer is up-to-date. It achieves this using a basic "stabilization" protocol. Chord also adopts a replication and redundancy policy to obtain highly available safe storage; each object can be stored at some number of successive peers.

In an N-peer Chord network, each peer maintains routing information for only about $O(\log N)$ other peers, and resolves all lookups via $O(\log N)$ messages to other peers. Other attractive advantages of Chord include its simplicity, provable correctness, and provable performance even in the face of concurrent peer arrivals and departures.

Chord can be used in many distributed applications as an infrastructure to locate the peer that stores a specific data object. CFS is a wide-area cooperative read-only file system based on Chord and project TWINE also uses Chord as a distributed lookup algorithm.

42.2.3.2.2 *CAN*

CAN (Content-Addressable Network; Ratnasamy et al., 2001) is a DHT scheme developed at the University of California, Berkeley. CAN maintains a virtual d-dimensional Cartesian coordinate space that is completely logical and has nothing to do with any physical coordinate systems or underlying network topology. The virtual space is dynamically partitioned into many small d-dimensional zones and each peer "owns" a zone. Each data object with unique key K is assigned a d-dimensional coordinate $P = \text{hash}(K)$ that corresponds to a point in the virtual space by a hash function; then the data object is stored on the peer that owns the zone containing the point P.

Each peer in CAN maintains a routing table that holds the information (e.g., IP address) of its neighbors in the d-dimensional coordinate space. Two peers are neighbors if their coordinate spans overlap along $d - 1$ dimensions and about along one dimension. Using its routing table, a peer routes a lookup message toward its destination by simple greedy forwarding to the neighbor with coordinates closest to the destination coordinates: For a given message lookup(key), key's destination position will be calculated first; then, starting from the initial peer, the lookup message will be iteratively passed through the neighbors until the destination peer is found. In a d-dimensional space, CAN's average routing path length is $(d/4)(n^{1/d})$ hops and the number of its neighbors is $2d$.

Every peer in CAN periodically sends refresh messages with its currently assigned zone to its neighbors. CAN uses soft-state style updates to support dynamic peer joining or leaving and ensure that all neighbors learn about the changes quickly to update their routing tables accordingly. When a peer joins the CAN network, it needs to discover a bootstrap peer currently in the system, and then randomly selects a point P in the space and sends a JOIN request destined for point P via the bootstrap peer. The destination peer that owns the zone containing point P splits its zone in half and assigns one half to the new peer. Afterward, the new peer learns its neighbors and initiates its routing table from the destination peer. When a peer leaves, it explicitly hands over its zone and the associated data objects to one of its neighbors. A takeover algorithm that ensures one of the failed peer's neighbors takes over the zone is also used to detect and recover peer failure automatically.

Currently, CAN is used in application-layer multicast and large-scale storage systems. Famous projects include PIER (a Peer-to-Peer Infrastructure for Information Exchange and Retrieval), CAN, etc.

42.2.3.2.3 *Tapestry*

Tapestry (Zhao et al., 2001, 2002) was designed as a fault-tolerant wide-area location and routing layer in the Oceanstore project. Tapestry can provide location-independent routing of messages directly to the closest copy of an object without centralized resources. Tapestry was modeled after the Plaxton (Plaxton et al., 1997) scheme (e.g., naming, structuring, locating, and routing), but it provides adaptability, fault-tolerance against multiple faults, and introspective optimizations.

Peers participating in the Tapestry overlay are assigned *nodeIDs* and objects are assigned *Globally Unique IDentifiers* (GUIDs). Both nodeIDs and GUIDs are uniformly distributed in the same identifier space of 160-bit values with a globally defined radix (e.g., hexadecimal). Each peer in Tapestry has a neighbor map that is organized into routing levels. For a peer with nodeID M, a given level of its neighbor map contains a number of entries equal to the base of the nodeID, where the i-th entry in the j-th level is the nodeID and location of the closest peer that ends in "i"+suffix(M, $j-1$). For example, the 8th entry of the 3rd level for peer 168D (based on hexadecimal) is the peer closest to 168D in network distance that ends in 88D. Each peer also maintains a backpointer list that points to peers, where it is referred to as a neighbor. The lists are used to generate the appropriate neighbor maps for a new peer. Tapestry peer uses neighbor maps to incrementally route overlay messages to the destination nodeID digit by digit. Tapestry's routing method guarantees that any existing unique peer in the system will be found within (at most) $\log_b N$ logical hops, in a system with an N-size namespace using IDs of base b, and the size of the neighbor map is $b\log_b N$.

To meet the fault-tolerant requirement, each entry in the neighbor map maintains two backup neighbors in addition to the closest (primary) neighbor. When the primary neighbor fails, Tapestry turns to the alternate neighbors in order. Each object is stored on multiple peers (called its root peer) to improve availability. To accomplish this, Tapestry concatenates a small, globally constant sequence of "salt" values (e.g., 1, 2, 3) to each object ID, then hashes the result to identify the appropriate root peers. The primary benefit of Tapestry over Chord and CAN is that it efficiently exploits network locality and actively reduces the latency of each P2P hop in addition to reducing the number of hops taken during a search.

Tapestry is self-administrating, fault-tolerant, and resilient under load, and can be a fundamental component for large-scale distributed systems. In addition to Oceanstore, other systems such as Bayeux, an application-layer multicasting system, are also based on Tapestry.

42.2.3.3 Advantages and Disadvantages

Decentralized structured P2P networks and DHT schemes have attracted much attention in academic research because of their desirable characteristics, such as scalability, robustness, self-management, and generality. Structured P2P networks have strong guarantees for resource discovery; that is, the resource in such systems can be found as long as it exists and the lookup efficiency can be guaranteed. Any existing resources can be located within predetermined hops. Thus, decentralized structured topology is well suited to environments that require a strong guarantee for resource discovery, such as persistent storage systems (e.g., Oceanstore). DHT schemes used in structured P2P networks are designed gracefully to achieve good scalability, performance, and fault-tolerance. They provide a general-purpose interface for location-independent naming on which many kinds of applications can be built. For example, the DHT scheme Pastry has been used in archival storage systems (e.g., PAST), cooperative Web cache (e.g., Squirrel), content distribution systems (e.g., SplitStream), etc.

Structured P2P networks and DHT schemes are quite prevalent topics in the research literature, but there are only few large-scale structured P2P networks deployed on the Internet. In DHT schemes, each peer needs to maintain some state information ("forwarding table") that should be updated whenever peers join or leave. This make the maintenance of DHT schemes somewhat complex and the P2P network may churn when there is an extreme change in the population of peers. DHT schemes are designed for exact-match searching and can currently only support searching by object identifier. The flexible search needed in many applications should be further developed on top of DHT schemes and it may decrease the advantages of DHT schemes. The matching from the logical P2P overlay constructed by DHT schemes to the underlying physical network also remains an open problem.

Despite these problems, the DHT scheme is still a very valuable research topic with a bright future for various applications.

42.2.3.4 Current Research

DHT schemes have been extensively studied. State–efficiency trade-off, load balance, resilience, incorporating geography, flexible search, security, and heterogeneity (Ratnasamy et al., 2002) are the main research topics. This section only gives a brief introduction to the related research on the state-efficiency trade-off topic and the other topics are discussed in other chapters of the book.

42.2.3.4.1 *State–Efficiency Trade-Off*

Two important measures of DHT schemes are degree (the size of the routing table to be maintained on each peer) and diameter (the number of hops a query needs to travel in the worst case). The trade-off between the degree and the diameter has been referred to as the "state–efficiency trade-off" of the DHT scheme (Ratnasamy et al., 2002). In many existing DHT schemes, such as Chord, Tapestry, and Pastry, both the degree and the diameter tend to $O(\log N)$ (where N is the total number of peers in the network), while in CAN the degree and the diameter are $O(d)$ and $O(dN^{1/d})$, respectively.

An open problem posed by Ratnasamy et al. (2002) is whether there exists a DHT scheme with $O(1)$ degree and $O(\log N)$ diameter. Recent work has shown that there are DHT algorithms to achieve $O(\log N)$ diameter with $O(1)$ degree. Kaashoek and Koorde (2003) is the constant degree and $O(\log N)$ diameter. D2B (Fraigniaud and Gauron, 2003) and Viceroy (Malkhi et al., 2003) are DHT schemes to achieve an expected constant degree and expected $O(\log N)$ diameter. The expected degree of D2B is constant but its high probability bound is $O(\log N)$; that is, a few unlucky peers would be of degree $\Omega(\log N)$. The expected diameter of Viceroy is about $3\log_2 N$; however, its $O(\log N)$ diameter is achieved not with certainty but with "high probability."

Xu et al. (2004) systematically studied the degree–diameter trade-off of DHT schemes and have shown that there are straightforward routing algorithms that can achieve $O(\log_2 N)$ and $O(\frac{\log_2 N}{\log_2(\log_2 N)})$ diameters when the degrees are $O(d)$ and $O(\log_2 N)$, respectively. For example, Figure 42.5 shows a sample balanced tree with out-degree $(\log_2 N)/2$. In such a tree, each node other than the root has a directed edge to its parent. The maximum distance from the root node to any other node is (at most) $\log_{(\frac{\log_2 N}{2})} N = O(\frac{\log_2 N}{\log_2(\log_2 N)})$. Thus, each node can reach any other node through their lowest common ancestor by (at most) $2*\log_{(\frac{\log_2 N}{2})} N = O(\frac{\log_2 N}{\log_2(\log_2 N)})$ steps.

Obviously, such a routing algorithm can cause serious congestion on the high-level nodes of the tree. Xu et al. (2004) defined the concept of *congestion-free* and then clarified the role that *congestion-free* plays in

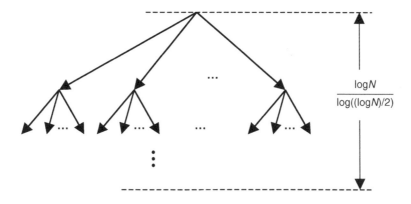

FIGURE 42.5 A balanced tree with out-degree $(\log_2 N)/2$.

the degree–diameter trade-off. A conjecture posed by Xu et al. (2004) is that "when the network is required to be c-congestion-free for some constant c, $\Omega(\log N)$ and $\Omega(N^{1/d})$ are the asymptotic lower bounds for the diameter when the degree is no more than $O(\log N)$ and d, respectively." This conjecture is true for a category of DHT algorithms known as *uniform* (Xu et al., 2004) but it is negative for general DHT schemes. For example, Ulysses (Xu et al., 2004) is a congestion-free DHT scheme that can achieve $O(\frac{\log_2 N}{\log_2(\log_2 N)})$ diameter with $O(\log_2 N)$ degree (however, the degree of Ulysses is $O(\log_2 N)$ with high probability), and FissionE (Li et al., 2002, 2003) is a $(1 + o(1))$-congestion-free DHT scheme that can achieve $O(\log_2 N)$ diameter with constant degree. FissionE is introduced in Section 42.3.

DHT schemes with constant diameter also have been proposed. For example, the Kelips scheme (Gupta et al., 2003) is of $O(\sqrt{N})$ degree and suffices to resolve lookups with $O(1)$ time and message complexity.

42.2.3.5 Topology-Theoretic Analysis of DHT Schemes

Many proposed DHT schemes are based on some traditional interconnection topology: Chord is based on the ring topology; Tapestry and Pastry are based on the hypercube topology; CAN is based on d-torus topology; Koorde and D2B are based on the de Bruijn graph; and Viceroy and Ulysses are based on the Butterfly topology. The underlying topology has a significant impact on the performance, resilience, and other properties of DHT schemes.

Loguinov et al. (2003) examined the topology theoretic properties of existing DHT schemes. They first studied the distribution of the routing length, average routing length, graph expansion, and clustering properties of the Chord, CAN, and de Bruijn topologies by theoretical analysis and simulation, and then examined bisection width, path overlap, and several other properties that affect the routing and resilience of DHT schemes. They showed that the de Bruijn topology is a good topology for DHT schemes and proposed a new DHT scheme ODRI that is based on the de Bruijn topology. However, the details of ODRI are still under investigation. Gummadi et al. (2003) studied how basic geometric approaches (i.e., interconnection topologies) affect the resilience and proximity properties of DHT schemes and showed that the ring topology allows great flexibility and achieves better resilience and proximity performance than a hypercube, tree-like topology and butterfly networks.

Compared with the hypercube, butterfly, de Bruijn graph, and torus topologies, the Kautz graph has some better properties (e.g., optimal network diameter) (Panchapakesan and Sengupta, 1999). FissionE (Li et al., 2002, 2003) was the first DHT scheme based on the Kautz graph. Table 42.1 shows a comparison of the different topologies.

42.2.4 Partially Decentralized Topology

The partially decentralized topology combines elements of both the centralized topology and decentralized topology. In partially decentralized topology, there are some superpeers that own more powerful

TABLE 42.1 Comparison of Different Topologies

Topology	Degree	Diameter	Average Path Length	Ref.
de Bruijn (ODRI)	d	$\log_d N$	$\log_d N - 1/(d-1)$	Loguinov et al., 2003
Hypercube (Chord)	$\log_2 N$	$\log_2 N$	$1/2 \log_2 N$	
d-torus (CAN)	$2d$	$1/2 dN^{1/d}$	$1/4 dN^{1/d}$	
Butterfly	d	$2 \log_d N(1-o(1))$	About $3/2 \log_d N$	Loguinov et al., 2003
Kautz (FissionE)	d	$D = \log_d N - \log_d(1+1/d)$	$D - 1/(d+1)$	Li et al, 2002

capability (processing power, storage capacity, bandwidth, etc.) than normal peers. Each superpeer acts as a centralized resource for a fraction of normal peers, and keeps an index over the data on them. A pure decentralized P2P network is formed among the superpeers. The superpeers perform searches on behalf of the normal peers for which they are responsible. If a normal peer p wants to discover some resource, it first submits the search request to its superpeer S, and then the search request is processed among the superpeers. The superpeer S acquires the search results and returns the results to peer p. However, all peers are equal in terms of file downloading.

Partially decentralized topology can be two-layer or multiple-layer; that is, there can be superpeers of superpeers in different layers. A typical partially decentralized topology is FastTrack, and the latest Gnutella also adopts this topology.

42.2.4.1 Case Studies

42.2.4.1.1 *FastTrack*

Considering heterogeneity across peers in the P2P network, the FastTrack (2002) protocol selects peers with high bandwidth and processing power as superpeers. The FastTrack protocol is implemented in two popular P2P file-sharing systems — Kazaa and Morpheus — and the users of Kazaa have reached more than 100 million.

FastTrack classifies all peers into two categories based on their bandwidth and processing power: (1) super-peers that connect to the Internet through high-speed links (e.g., Cable/xDSL or even higher-speed links), and (2) normal peers that connect to the Internet through slower links. The architecture of FastTrack has two layers: (1) the first layer comprises all superpeers that form a relatively reliable backbone, and (2) the second layer includes all normal peers. When a normal peer joins, it first finds a superpeer as its index server and then connects to it. The superpeer gathers the file location information from the attached normal nodes and becomes a central index server for normal peers that have registered to it. Searching in FastTrack is still performed by flooding among the superpeers. When a peer would like to search some file, it sends the search request to its superpeer. Then the superpeer broadcasts the request to its neighboring superpeers, and the request will be forwarded until it is satisfied or the TTL reaches 0. Suppose each superpeer is responsible for 100 normal peers; then the number of peers searched in FastTrack is 100 times more than that in a Gnutella network of the same search size. Thus, a search in FastTrack is more efficient than that in Gnutella (see Figure 42.6).

42.2.4.1.2 *Brocade*

In P2P overlay networks, messages are often routed across multiple autonomous systems (ASs) and administrative domains before reaching their destinations. Each overlay hop across an AS often incurs significantly longer latencies than that within multiples ASs. To reduce latency and network hops for a given message, Brocade (Zhao et al., 2002) was proposed in the Oceanstore project. Brocade can determine the network domain of the destination and route directly to that domain. Brocade selects a superpeer as a landmark for each network domain based on the status of the underlying network. Superpeers maintain information (e.g., peerID) of local peers and can have independent names in the Brocade overlay. They take the peerIDs of normal peers as a particular kind of *key* stored on them and still locate peers through the Tapestry algorithm. While locating a data object, the peer first contacts its local superpeer, and then routes

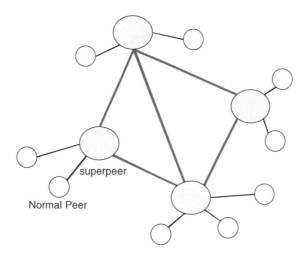

FIGURE 42.6 FastTrack architecture.

the message to the superpeer responsible for the destination peer and finally arrives at the destination peer.

42.2.4.2 Advantages and Disadvantages

The partially decentralized topology adopts a policy between the centralized topology and the decentralized topology, and it has the potential to combine the efficiency of the centralized topology with resilience, scalability, and load balance of the decentralized topology. Because superpeers act as centralized servers for normal peers, the search requests can be processed more efficiently than those in a decentralized topology. Furthermore, there are relatively more superpeers in the partially decentralized topology; thus, the problems (such as bottleneck, single point of failure, etc.) faced in a centralized topology might be avoided. The partially decentralized topology can also take advantage of the heterogeneity across peers to improve performance.

However, the partially decentralized topology might also face similar problems in both the centralized topology and decentralized topology. Superpeers play an important role in the network and the failures of a few superpeers near the top of the hierarchy might have a serious impact on the entire network. The superpeer network formed among super-peers is also a pure decentralized P2P network and can be a decentralized unstructured topology or decentralized structured topology; thus, it also has the problems mentioned in Sections 42.2.2 and 42.2.3.

42.2.4.3 Current Research

Because the network among superpeers is also of decentralized topology, many research topics in partially decentralized topology are similar to those in a decentralized unstructured topology or a decentralized structured topology. This section ignores the similar topics and focuses on its particular problems.

Despite the fact that the partially decentralized topology has been adopted in many real systems, such as KaZaA and Morpheus, the research into it is relatively minimal. Actually, there are many problems that should be solved; for example (Yang and Garcia-Molina, 2003), how are the superpeers selected? How many clients should a superpeer take charge of to maximize the efficiency? How should superpeers connect to each other? How can superpeers be made more reliable? Yang and Garcia-Molina (2003) studied the fundamental characteristics and performance trade-offs of superpeer networks in detail, and presented some practical guidelines and a general procedure for the design of an efficient superpeer network. Xu et al. (2003) proposed two approaches for constructing an auxiliary expressway network to take advantage of the inherent heterogeneity of peers (such as different connectivity, forwarding capacities, and availabilities) to speed up routing.

42.3   FissionE: A Constant Degree and Low-Congestion DHT Scheme

This section gives a brief introduction of a new DHT scheme — FissionE — that is based on the Kautz graph and is constant degree and $(1 + o(1))$-congestion-free.

The Kautz graph is of good topology with optimal diameter and optimal fault tolerance (Chiang and Chen, 1994). The Kautz graph $K(d, k)$ of degree d has connectivity d and there are d disjoint paths between any two nodes. In addition, the Kautz graph $K(d, k)$ can achieve better load-balancing characteristics than the de Bruijn graph, as shown by Panchapakesan and Sengupta (1999).

We have also proved that the Kautz graph is constant congestion (almost congestion-free) when using longest path routing. Thus, the Kautz graph is a good static topology to construct a DHT scheme.

FissionE uses Kautz graph $K(2, k)$ as its static topology. The identifiers of nodes in Fission are Kautz strings, and the neighborhood among nodes is built according to their identifiers. When peers join or leave, the peers' identifiers change dynamically. A "split large and merge small" policy is utilized to achieve load balance and the entire identifier space is dynamically partitioned among the peers. The degree of FissionE ranges from 3 to 6 and the average degree of FissionE is 4. For an N-peer FissionE system, its diameter is less than $2^* \log_2 N$ and the average routing path is less than $\log_2 N$.

FissionE is a good DHT scheme with many good properties, and the details of FissionE can be referred to Li et al. (2002, 2003).

Acknowledgments

We wish to thank Professors Wu Jie and Xu Jun for their reviews and suggestions for the FissionE scheme and the writing of the chapter.

References

1. L.A. Adamic, B. Huberman, R. Lukose, and A. Puniyani. Search in power law networks. *Phys. Rev. E,* 64, 46135–46143, 2001.
2. E Adar and B.A. Huberman. Free riding on Gnutella. *First Monday,* Vol. 5, No. 10, October 2000.
3. R. Albert and A.-L. Baraba. Statistical mechanics of complex networks. *Rev. Mod. Phys.,* 74, 48–94, 2002.
4. D.P. Anderson, J.Cobb, E. Korpela, M. Lebofsky, and D. Werthimer. SETI@home: an experiment in public-resource computing. *Commun. ACM,* Vol. 45, No. 11, November 2002.
5. H. Balakrishnan, M. Frans Kaashoek, D. Karger, et al. Looking up data in P2P systems. *Commun. ACM,* 46(2), 43–48, 2003.
6. D. Bernstein, Z. Feng, B. Levine, and S. Zilberstein. Adaptive peer selection. In *Proc. IPTPS'03,* Berkeley, CA, February 2003.
7. W.G. Bridges and S. Toueg. On the impossibility of directed Moore graphs. *J. Combinatorial Theory, Ser. B,* 29, 330–341, 1980.
8. W. Chiang and R.-J. Chen. Distributed fault-tolerant routing in Kautz networks. *J. Parallel and Distributed Computing,* 20, 99–106, 1994.
9. D. Clark. Face-to-face with peer-to-peer networking. *IEEE Computer,* 34(1), 18–21, 2001.
10. I. Clarke, O. Sandberg, B. Wiley, and T. Hong. Freenet: a distributed anonymous information storage and retrieval system. In *Proc. ICSI Workshop on Design Issues in Anonymity and Unobservability,* Berkeley, CA, June 2000.
11. Clip2.com. Gnutella: To the Bandwidth Barrier and Beyond. Available at http://www.clip2.com/gnutella.html, 2001.
12. E. Cohen and S. Shenker. Replication strategies in unstructured peer-to-peer networks. In *Proc. ACM SIGCOMM 2002,* 2002.
13. B. Cohen. Incentives build robustness in BitTorrent. In *Proc. Workshop on Economics of Peer-to-Peer Systems,* June 2003.

14. E. Cohen. Associative search in peer to peer networks: harnessing latent semantics. In *Proc. INFO-COM2003*, 2003.

15. A. Crespo and H. Garcia-Molina, Routing indices for peer-to-peer systems. In *Proc. 22nd IEEE Int. Conf. on Distributred Computing Systems (ICDCS)*, Vienna, Austria, July 2002.

16. FastTrack. FastTrack Product Description. http://www.fasttrack.nu/index_int.html, 2002.

17. P. Fraigniaud and P. Gauron. The Content-Addressable Network D2B. Technical Report 1349, CNRS University Paris-Sud, France , 2003.

18. C. Gkantsidis, M. Mihail, and A. Saberi. Random walks in peer-to-peer networks. In *Proc. INFO-COM'2004*, 2004.

19. The Gnutella home page, gnutella.wego.com, 2001.

20. K. Gummadi, R. Gummadi, S. Gribble, S. Ratnasamy, S. Shenker, and I. Stoica. The impact of DHT routing geometry on resilience and proximity. In *Proc. of SIGCOMM'03*, 2003.

21. I. Gupta, K. Birman, P. Linga, A. Demers, and R. van Renesse. Kelips: building an efficient and stable P2P DHT through increased memory and background overhead. In *Proc. IPTPS'03*, February 2003.

22. F. Harrell, Y. Hu, G. Wang, and H. Xia. Survey of Locating & Routing in Peer-to-Peer Systems. Technical Report, University of California, San Diego, 2002.

23. S.R.H. Joseph. NeuroGrid: semantically routing queries in peer-to-peer networks. In *Proc. Int. Workshop on Peer-to-Peer Computing*, Pisa, Italy, 2002.

24. M. Jovanovic, F.S. Annexstein, and K.A. Berman. Modeling peer-to-peer network topologies through "small-world" models and power laws. In *TELFOR*, Belgrade, Yugoslavia, November, 2001.

25. F. Kaashoek and D.R. Karger. Koorde: a simple degree-optimal hash table. In *Proc. 2nd Int. Workshop on Peer-to-Peer Systems (IPTPS 2003)*, 2003.

26. D. Karger, E. Lehman, F. Leighton, et al. Consistent hashing and random trees: distributed caching protocols for relieving hot spots on the World Wide Web. In *Proc. 29th Annu. ACM Symp. on Theory of Computing*, El Paso, TX, pp. 654–663, May 1997.

27. D. Li, et al. FissionE: A Scalable Constant Degree and Constant-Congestion Peer-to-Peer Network. Technical Report PDL-2002-147, National University of Defense Technology, December 2002.

28. D. Li, X. Fang, Y. Wang, X. Lu, et al. A scalable peer-to-peer network with constant degree. In *Proc. APPT'2003 (5th Int. Workshop on Advanced Parallel Processing Technologies), Lecture Notes in Computer Science*, 2834, 414–425, September 2003.

29. LimeWire Company. Query Routing for the Gnutella Network, http://www.limewire.com/developer/query_routing/keyword%20routing.htm, 2003.

30. D. Loguinov, A. Kumar, V. Rai, et al. Graph-theoretic analysis of structured peer-to-peer systems: routing distances and fault resilience. In *Proc. of ACM SIGCOMM'2003*, Karlsruhe, Germany: ACM Press, pp. 395–406, 2003.

31. Q. Lv, P. Cao, E. Cohen, K. Li, and S. Shenker. Search and replication in unstructured peer-to-peer networks. In *Proc. 16th ACM Int. Conf. on Supercomputing (ICS'02)*, New York, June 2002.

32. D. Malkhi, M. Naor, and D. Ratajczak. Viceroy: a scalable and dynamic lookup network. In *Proc. 21st ACM Symp. on Principles of Distributed Computing (PODC)*, Monterey, CA, 2002.

33. E. P. Markatos. Tracing a large-scale peer to peer system: an hour in the life of Gnutella. In *Proc. 2nd IEEE/ACM Int. Symp. on Cluster Computing and the Grid 2002*, 2002.

34. L. Mathy, R. Canonico, S. Simpson, and D. Hutchison. Scalable adaptive hierarchical clustering. *IEEE Commun. Lett.*, 6(3), 117–119, March 2002.

35. P. Maymounkov and D. Mazieres. Rateless codes and big downloads. In *Proc. IPTPS'03*, Berkeley, CA, 2003.

36. S. Milgram. The small world problem. *Psychol. Today*, No. 2, 60–67, 1967.

37. D.S. Milojicic, V. Kalogeraki, R. Lukose, K. Nagarajael, J. Pruyne, B. Richard, S. Rollins, and Zh. Xu. Peer-to-Peer Computing. Technical Report HPL-2002-57, HP Laboratories Palo Alto, CA, March 2002.

38. Napster. The Napster home page. http://www.napster.com, 2001.

39. G. Panchapakesan and A. Sengupta. On a lightwave network topology using Kautz digraphs. *IEEE Trans. Computers*, 48(10), 1131–1138, 1999.

40. G. Pandurangan, P. Raghavan, and E. Upfal, Building low diameter peer-to-peer networks. In *Proc. 42nd Annu. IEEE Symp. on the Foundations of Computer Science (FOCS)*, 2001.

41. C.G. Plaxton, R. Rajaraman, and A.W. Richa. Accessing nearby copies of replicated objects in a distributed environment. In *Proc. ACM SPAA 1997*, ACM Press, June 1997.

42. S. Ratnasamy, P. Francis, M. Handley, et al. A scalable content-addressable network. In *Proc. ACM SIGCOMM'2001*. New York: ACM Press, pp. 149–160, 2001.

43. S. Ratnasamy, S. Shenker, and I. Stoica. Routing algorithms for DHTs: some open questions. In *Proc. 1st Int. Workshop on Peer-to-Peer Systems (IPTPS'02)*, 2002.

44. M. Ripeanu, I. Foster, and A. Iamnitchi. Mapping the Gnutella network: properties of large-scale peer-to-peer systems and implications for system design. *IEEE Internet Computing*, 6(1), January–February 2002.

45. A. Rowstron and P. Druschel. Pastry: scalable, distributed object location and routing for large-scale peer-to-peer systems. In *Proc. IFIP/ACM Middleware 2001*. Heidelberg, Germany, pp. 329–350, 2001.

46. S. Saroiu, P.K. Gummadi, and S.D. Gribble. A measurement study of peer-to-peer file sharing systems. In *Proc. Multimedia Computing and Networking (MMCN'02)*, 2002.

47. D. Schoder and K. Fischbach. Peer-to-peer prospects. *Commun. ACM*, 46(2), 27–29, 2003.

48. K. Sripanidkulchai. The Popularity of Gnutella Queries and Its Implications on Scalability. White paper, Carnegie Mellon University, available online at http://www.cs.cmu.edu/~kunwadee/research/p2p/gnutella.html, March 2001.

49. K. Sripanidkulchai, B. Maggs, and H. Zhang. Efficient content location using interest-based locality in peer-to-peer systems. In *Proc. INFOCOM2003*, 2003.

50. I. Stoica, R. Morris, D. Karger, et al. Chord: a scalable peer-to-peer lookup service for Internet applications. In *Proc. ACM SIGCOMM 2001*. New York: ACM Press, pp. 160–177, 2001.

51. D.J. Watts and S.H. Strogatz. Collective dynamics of "small-world" networks. *Nature*, 393, 440–442, 1998.

52. R.H. Wouhaybi and A.T. Campbell. Phenix: supporting resilient low-diameter peer-to-peer topologies. In *Proc. INFOCOM 2004*, 2004.

53. Zh. Xu, M. Mahalingam, and M. Karlsson. Turning heterogeneity into an advantage in overlay routing. *IEEE INFOCOM 2003*, 2003.

54. J. Xu, A. Kumar, and X. Yu. On the fundamental tradeoffs between routing table size and network diameter in peer-to-peer networks. *IEEE J. Selected Areas in Communications (JSAC)*, January 2004.

55. B. Yang and H. Garcia-Molina. Efficient search in peer-to-peer networks. In *Proc. 22nd IEEE Int. Conf. on Distributred Computing Systems (ICDCS)*, Vienna, Austria, July 2002.

56. B. Yang and H. Garcia-Molina. Designing a super-peer network. In *Proc. ICDE 2003*, 2003.

57. H. Zhang, A. Goel, and R. Govindan. Using the small-world model to improve Freenet performance. In *Proc. INFOCOM 2002*, 2002.

58. B.Y. Zhao, J. Kubiatowicz, and A.D. Joseph. Tapestry: An Infrastructure for Fault-Tolerant Wide-Area Location and Routing. Technical Report UCB//CSD-01-1141, 2001, University of California, Berkeley Computer Science Division, April 2001.

59. B.Y. Zhao, Y. Duan, L. Huang, A.D. Joseph, and J.D. Kubiatowicz. Brocade: landmark routing on overlay networks. In *Proc. First Int. Workshop on Peer-to-Peer Systems (IPTPS'02)*, 2002.

43

Peer-to-Peer Overlay Optimization

Yunhao Liu

Li Xiao

Lionel M. Ni

43.1 Introduction

Popularized by Napster, peer-to-peer (P2P) has become a well-known term recently. Large amounts of files are shared in these P2P systems, such as Gnutella, KaZaA, and BitTorrent, with most of the contents being provided by the P2P users themselves. Today, millions of users (peers) join the network by connecting to some of the active peers in the P2P overlay network.

There are mainly three different architectures for P2P systems: (1) centralized, (2) decentralized structured, and (3) decentralized unstructured.[1] In a centralized model such as Napster,[2] central index servers are used to maintain a directory of shared files stored on peers so that a peer can search for the whereabouts of a desired content from an index server. However, this architecture creates a single point of failure, and its centralized nature of the service also makes systems vulnerable to denial-of-service attacks.[3] Decentralized P2P systems have the advantages of eliminating reliance on central servers and providing greater freedom for participating users to exchange information and services directly between each other. In decentralized structured models, such as Chord,[4] Pastry,[5] Tapestry,[6] and CAN,[7] the shared data placement and topology characteristics of the network are tightly controlled based on distributed hash functions. Although there are many discussions on this model, decentralized structured P2P systems are not practically in use in the Internet.

This chapter focuses on decentralized unstructured P2P systems such as Gnutella[8] and KaZaA.[9] File placement is random in these systems, which has no correlation with the network topology.[10] Unstructured P2P systems are most commonly used in today's Internet. The most popular search mechanism in use is to blindly "flood" a query to the network among peers (such as in Gnutella) or among supernodes (such as in KaZaA). A query is broadcast and rebroadcast until a certain criterion is satisfied. If a peer receiving the query can provide the requested object, a response message will be sent back to the source peer along the inverse of the query path, and the query will not be further forwarded from this responding peer. This mechanism ensures that the query will be "flooded" to as many peers as possible within a short period of time in a P2P overlay network. A query message will also be dropped if the query message has previously visited the peer.

Studies by Sen and Wang[11] and Saroiu et al.[12] have shown that P2P traffic contributes the largest portion of the Internet traffic based on their measurements of some popular P2P systems, such as FastTrack (including KaZaA and Grokster),[13] Gnutella, and DirectConnect. Measurements performed by Ripeanu et al.[14] have shown that although 95 percent of any two nodes are less than seven hops away and the message time-to-live (TTL = 7) is preponderantly used, the flooding-based routing algorithm generates 330 TB/month in a Gnutella network with only 50,000 nodes. A large portion of the heavy P2P traffic caused by inefficient overlay topology and the blind flooding is unnecessary, which makes the unstructured P2P systems being far from scalable.[15] One of the important reasons for this problem is that the mechanism of a peer randomly choosing logical neighbors without any knowledge about the underlying physical topology causes topology mismatch between the P2P logical overlay network and the physical underlying network.

In a P2P system, all participating peers form a P2P network over a physical network. A P2P network is an abstract, logical network called an overlay network. Maintaining and searching operations of a Gnutella peer are specifically described.[16] When a new peer wants to join a P2P network, a bootstrapping node provides the IP addresses of a list of existing peers in the P2P network. The new peer then tries to connect with these peers. If some attempts succeed, the connected peers will be the new peer's neighbors. Once this peer connects into a P2P network, the new peer will periodically *ping* the network connections and obtain the IP addresses of some other peers in the network. These IP addresses are cached by this new peer. When a peer leaves the P2P network and then wants to join the P2P network again (no longer the first time), the peer will try to connect to the peers whose IP addresses have already been cached. This mechanism, although simple to implement, causes mismatch problems. Take a look at the example shown in Figure 43.1, where solid lines represent physical links and dotted lines represent logical links.

When peer P inserts a query message to the network, in the case of an efficient P2P overlay as shown in Figure 43.1a, there are no message duplications. However, when a topology mismatch problem occurs, as shown in Figure 43.1b, a message from peer P will incur many unnecessary message duplications, especially if the physical distance between peer Q and M is very far and both the traffic cost and the query response time will be increased significantly.

Aiming at alleviating the mismatch problem and reducing the unnecessary traffic, many methods have been introduced. This chapter provides an overview of some traditional approaches, discusses the reason why these approaches do not work, and then introduces several recently proposed algorithms.

43.2   Traditional Approaches

There are several traditional topology optimization approaches. End system multicast, Narada, as proposed by Chu et al.,[17] first constructs a rich connected graph on which to further construct shortest path spanning trees. Each tree is rooted at the corresponding source using well-known routing algorithms. This approach

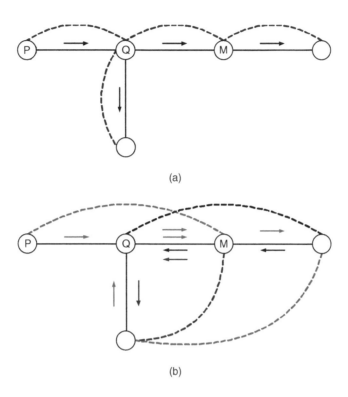

(a)

(b)

FIGURE 43.1 Topology mismatch problem.

introduces large overhead of forming the graph and trees in a large scope, and does not consider the dynamic joining and leaving characteristics of peers. The overhead of Narada is proportional to the multicast group size. This approach is infeasible to large-scale P2P systems.

Researchers have also considered clustering close peers based on their IP addresses (see, for example, Krishnamurthy and Wang[18] and Padmanabhan and Subramanian.[19]) We believe there are two limitations for this approach. First, the mapping accuracy is not guaranteed by this approach. Second, this approach may affect the searching scope in P2P networks.

Recently, Xu et al.[20] proposed to measure the latency between each peer to multiple stable Internet servers called "landmarks." The measured latency is used to determine the distance between peers. This measurement is conducted in a global P2P domain and needs the support of additional landmarks. Similarly, this approach also affects the search scope in P2P systems.

Using an example shown in Figure 43.2, one can explain why these proximity-based approaches will shrink query search scopes. In Figure 43.2, peers A, B, C, and D locate in the same AS, and peers E, F and H, G, K belong to other ASs, respectively. It is safe to assume that the physical distance between A and B or E and F is much smaller than that of K and A or C and F, as illustrated in Figure 43.2. Using the above discussed approaches, when peers successfully obtain or estimate the distance between each pair of them, and the optimization policy for each node is to connect the closest peers while retaining the original number of logical neighbors, a connected graph can be broken down into three components. As a result, before optimization, queries can visit all of the peers, while after optimization, all queries can only visit a small group of live peers in the system, and the search scope of queries is significantly reduced.

Gia[21] introduced a topology adaptation algorithm to ensure that high-capacity nodes are indeed the ones with high degree, and low-capacity nodes are within short reach of high-capacity nodes. It addresses a different matching problem in overlay networks but does not address the topology mismatch problem between the overlay and physical networks.

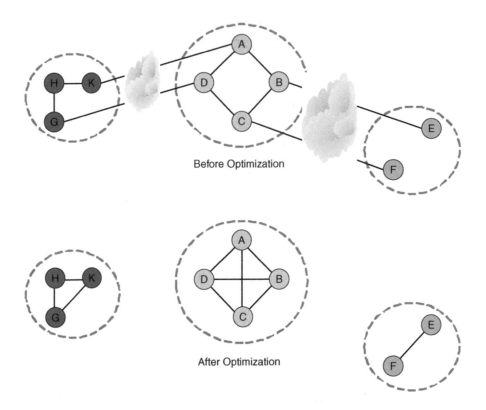

FIGURE 43.2   Before optimization, queries can visit all of the peers; while after optimization, all queries can only visit a small group of live peers.

43.3   Distributed Approaches to the Topology Mismatch Problem

This section briefly introduces three recently proposed distributed approaches to topology mismatch problems: namely, (1) Adaptive Connection Establishment (ACE), (2) Scalable Bipartite Overlay (SBO), and (3) Location-aware Topology Matching (LTM).

43.3.1   Adaptive Connection Establishment (ACE)

Optimizing inefficient overlay topologies can fundamentally improve P2P search efficiency. The first proposed approach, Adaptive Connection Establishment (ACE), includes three phases,[22] which are *neighbor cost table construction and exchanging, selective flooding, and overlay optimization.*

43.3.1.1   Phase 1: Neighbor Cost Table Construction and Exchanging

Network delay between two peering nodes is used as a metric for measuring the cost between peers. ACE modifies the Limewire implementation of Gnutella 0.6 P2P protocol by adding one routing message type. Each peer probes the costs with its immediate logical neighbors and forms a *neighbor cost table*. Two neighboring peers exchange their neighbor cost tables so that a peer can obtain the cost between any pair of its logical neighbors. Thus, a small overlay topology of a source peer and all its logical neighbors is known to the source peer. If one uses N(S) to denote the set of direct logical neighbors of peer S, each peer S has the information to obtain the overlay topology, including S itself and N(S), as illustrated in Figure 43.3a.

A critical issue in this phase is how often peers exchange their neighbor's cost table. There are two basic operations in this phase: peers first probe the cost to their neighbors and construct the cost table;

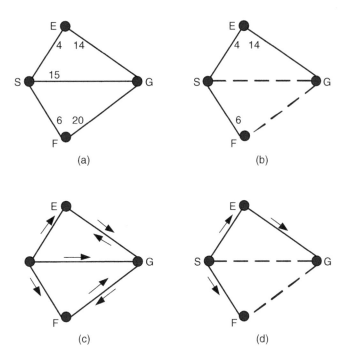

FIGURE 43.3 Selective flooding.

they then exchange the table with direct neighbors. In this design, there are two ways for each single peer to decide when to conduct neighbor probing and reporting, namely, periodic and event-driven. In the periodic approach, each peer conducts neighbor distance probing at every certain period of time q. After probing the distances to all the neighbors, a peer sends the cost table to its neighboring peers. The value q is a critical factor for the performance of the periodic approach. In the event-driven approach, a peer produces and sends an updated cost table to its neighboring peers when there is a change in its logical connections with its neighbors, such as on a neighbor's leaving or on a peer's joining as its new neighbor.

43.3.1.2 Phase 2: Selective Flooding (SF)

Based on obtained neighbor cost tables, a minimum spanning tree (MST) among each peer S and its immediate logical neighbors $(S \cup N(S))$ can be built by simply using an algorithm such as PRIM,[23] which has a computation complexity of $O(m^2)$, where m is the number of logical neighbors of the source node. A computed MST is shown in Figure 43.3b. Now the message routing strategy of a peer is to select the peers that are the direct neighbors in the MST to send its queries, instead of flooding queries to all neighbors. We thus call peer S's direct neighbors in its MST *flooding neighbors* of S, and call those that are not direct neighbors in S's MST *non-flooding neighbors*. The connections between S and its flooding neighbors are defined as *forwarding connections*. See the example shown in Figure 43.3. In Figure 43.3c, the traffic incurred by node S's flooding of messages to its direct neighbors E, F, and G is:

$$4 + 14 + 14 + 15 + 6 + 20 + 20 = 93$$

After SF, the *forwarding connections* are changed as shown in Figure 43.3d and the total traffic cost becomes:

$$6 + 4 + 14 = 24$$

In Figure 43.3d, peer S sends a message only to peers E and F, and expects that peer E will forward the message to peer G. Note that in this phase, even peer S does not flood its query message to peer G

anymore; S still retains the connections with G and keeps exchanging the neighbor cost tables with G. In this example, peer G is a *non-flooding neighbor* of peer S, which is the direct neighbor potentially to be replaced in phase 3. Peers E and F are *flooding neighbors* of S.

43.3.1.3   Phase 3: Overlay Optimization

This phase reorganizes the overlay topology. Note that each peer has a neighbor list that is further divided into *flooding neighbors* and *non-flooding neighbors* in phase 2. Each peer also has the neighbor cost tables of all its neighbors. In this phase, a peer tries to replace those physically far away neighbors by physically close by neighbors, thus minimizing the topology mismatch traffic. An efficient method to identify such a candidate peer to replace a far away neighbor is critical to system performance. Many methods can be proposed. In ACE, a non-flooding neighbor may be replaced by one of the non-flooding neighbor's neighbors.

The basic concept of phase 3 is illustrated in Figure 43.4. In Figure 43.4a, peer S is probing the distance to one of its *non-flooding neighbor* G's neighbors, for example, H. If SH is smaller than SG, as shown in Figure 43.4b, connection SG will be cut. If SG is smaller than SH but S finds that the cost between nodes G and H is even larger than the cost between peers S and H, as shown in Figure 43.4c, S will keep H as a new neighbor. Because the algorithm is executed in each peer independently, S cannot allow G to remove H from its neighbor list. However, as long as S keeps both G and H as its logical neighbors, one can expect that peer H will become a *non-flooding* neighbor to peer G after G's phase 2 because peer G expects S to forward messages to H to reduce unnecessary traffic. Then G will try to find another peer to replace H as its direct neighbor. After knowing that H is no longer a neighbor to G from periodically exchanged neighbor cost tables from node G (or from node H), S will cut connection SG, although S has already stopped sending query messages to G for a period of time, ever since the spanning tree has been built for S. Obviously, if SH is larger than SG and GH, as shown in Figure 43.4d, this connection will not be built and S will keep probing G's other direct neighbors.

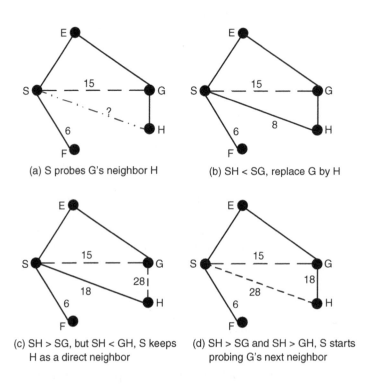

(a) S probes G's neighbor H

(b) SH < SG, replace G by H

(c) SH > SG, but SH < GH, S keeps H as a direct neighbor

(d) SH > SG and SH > GH, S starts probing G's next neighbor

FIGURE 43.4   Overlay optimization.

Let C_{ij} represent the cost from peer i to peer j. The following pseudo code describes the algorithm of phase 3, overlay optimization for a given source peer i:

```
For each j in i's non-flooding neighbors
  Replaced = false;
  List = all j's neighbors excluding i;
  While List is not empty and Replaced = false
    randomly remove a peer h from List;
    measure Cih;
    if Cih< Cij{replace j by h in i's neighbor list;
      Replaced = true;}
        else if Cih < Cjh{add h to i's neighbor list;
          remove j from i's neighbor list right after i
finds out jh is disconnected;
          Replaced = true;}
      End While;
End For;
```

43.3.2 Scalable Bipartite Overlay (SBO)

Although ACE is proved an effective algorithm and easy to follow, experimental results show that the convergent speed of ACE is not fast. One reason is that each node only works with one-hop away neighbors. SBO employs a more efficient strategy to select query forwarding path and logical neighbors.[24] Using a bipartite overlay, every node in SBO builds a two-hop MST with an even lower overhead compared with ACE. SBO consists of four phases: bootstrapping a new peer, neighbor distance probing and reporting, forwarding connections computing, and direct neighbor replacement.

43.3.2.1 Phase 1: Bootstrapping a New Peer

A typical unstructured P2P system provides several permanent well-known bootstrap hosts to maintain a list of recently joined peers so that a new incoming peer can find an initial host to start its first connection by contacting the bootstrap hosts. In the design of SBO, when a new peer is joining the P2P system, it will randomly take an initial color: red or white. A peer should keep its color until it leaves, and again randomly select a color when it rejoins the system. Thus, each peer has a color associated with it, and all peers are separated into two groups: red and white. In SBO, a bootstrap host will provide the joining peer with a list of active peers with color information. The joining peer then tries to create connections to the different color peers in the list. Figure 43.5 illustrates a new peer's joining process. In such a way, all the

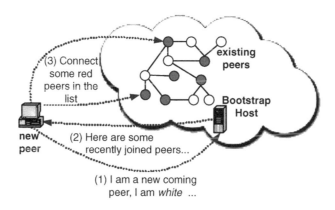

FIGURE 43.5 Bootstrapping a new peer.

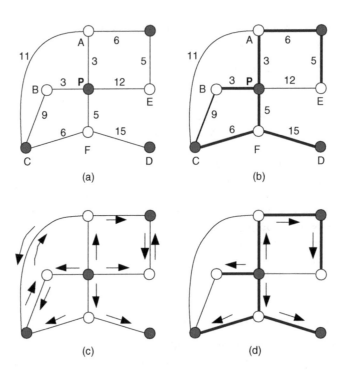

FIGURE 43.6   A red peer P has a small overlay topology of $N(P)$ and $N^2(P)$, and computes the efficient forwarding paths.

peers form a bipartite overlay in which a red peer will only have white peers as its direct neighbors, and vice versa.*

Once a peer has joined the P2P system, it will periodically ping the network connections and obtain the IP addresses of other peers in the network, which will be used to make new connections for the peer's rejoining or in the case that the peer loses some of the connections with its neighbors due to the neighbors' departure or failure, or the faults in the underlying networks.

43.3.2.2   Phase 2: Neighbor Distance Probing and Reporting by White Peers

In this phase, each white peer probes the costs with its immediate logical neighbors and forms a neighbor cost table, and then sends this table to all its neighbors who are all red peers.

We use $N^2(S)$ to denote the set of peers being two hops away from S. Because each red peer P receives the cost table from its white neighbors about its all red neighbors, the red peer P has the information to obtain the overlay topology including P itself, $N(P)$, and $N^2(P)$, as illustrated in Figure 43.6a. Note that in SBO the overlay forms a bipartite topology, so there are no connections between any pairs of peers in $N^2(P)$. Thus, we only require all the white peers to probe the costs to their neighbors and send out the cost tables. There is no need for the red peers to probe the distance.

43.3.2.3   Phase 3: Forwarding Connections Computing by Red Peers

Based on obtained neighbor cost tables, an MST can be built by each red peer, such as P in Figure 43.6b. Because a red peer builds an MST in a two-hop diameter, a white peer does not need to build an MST.

---

*Current implementation of NTP version 4.1.1 in the public domain can reach the synchronization accuracy down to 7.5 milliseconds.[25] Another approach is to use distance to measure the communication cost, such as the number of hops weighted by individual channel bandwidth.

The thick lines in the MST are selected as forwarding connections (FC), while the rest of the lines are non-forwarding connections (NFC). Queries are only forwarded along FCs. For example, in Figure 43.6b, P will send/forward queries to A, B, and F, but not E. Peer P also informs E that E is a non-forwarding neighbor. This information will be used by E in phase 4, that is, direct neighbor replacement.

Figure 43.6c illustrates how the query message from P is flooded along the connections based on Figure 43.4a. One can see many message duplications, that is, the RK problem. The total traffic cost incurred by the query is:

$$3 + 6 + 5 + 5 + 12 + 3 + 5 + 6 + 9 + 9 + 15 + 11 + 11 = 100$$

After FC computing in Figure 43.6b, the traffic cost incurred by this query becomes:

$$3 + 6 + 5 + 3 + 5 + 6 + 15 = 43$$

as shown in Figure 43.6d.

Although FC computing can reduce a lot of traffic while retaining the same search scope, as described earlier, the price is to scarify query response time, or the query latency. For example, P issues a query and E has the desired data. The response time in Figure 43.6c is $2 \times 12 = 24$. After FC computing, the response time becomes $2 \times (3 + 6 + 5) = 28$. Based on this observation, one can further improve the FC selecting algorithm (later in this section).

43.3.2.4 Phase 4: Direct Neighbor Replacement by White Peers

This operation is only conducted by white peers. After computing an MST among the peers within two hops, a red peer P is able to send its queries to all the peers within two hops. Some white peers become non-forwarding neighbors, such as E in Figure 43.6. In this case, for peer E, P is no longer its neighbor. In the phase of direct neighbor replacement, a non-forwarding neighbor E will try to find another red peer being two hops away from P to replace P as its new neighbor.

Peer P will send the neighbor cost tables it collected from A, B, and F to the non-forwarding neighbor E so that E has enough information to find another neighbor to form a more efficient topology. Having received the cost tables, E can obtain the overlay topology among P and the peers $N(P)$ and $N^2(P)$. In the design of SBO, E will probe the round-trip times (RTTs) to all the red peers in $N^2(P)$ and sort the red peers according to their RTTs. Peer E then selects the one with the smallest RTT (e.g., peer D in Figure 43.7a). There are three cases for peer E who finds D as its nearest red peer.

Case 1: The delay of ED is smaller than that of EP. The connection of ED will be created and D becomes E's direct logical neighbor. The connection EP will be put into E's *will-cut list*, that is, a list of connections to be cut later. A connection in a will-cut list will be disconnected when it has been in the list for a certain period of time. A peer will not send or forward any queries to the connections in its will-cut list.

The reason for E not to disconnect EP immediately is that some query responses might be sent back along the overlay path EP for some earlier queries. Disconnecting non-forwarding connections, such as EP, immediately may cause a serious response loss problem. Figure 43.7b is the topology after E connects with D, and disconnects with P after a timeout period.

Case 2: The delay of ED is larger than that of EP but smaller than the larger one of PF and FD. For example, if ED = 13 in Figure 43.7c, 12 < ED < 15. In this case, E will create the connection of ED and treat D as its direct neighbor. Peer E will not put connection EP into its will-cut list until it sends its neighbor cost table to D so that D still thinks the connection of EP exists. Note that the algorithm is fully distributed. Thus, when red peer D conducts the FC computing, F will become D's non-forwarding neighbor. The white peer F will conduct the same operations as what peer E has done, and may try to find a better red peer to replace node D as its neighbor.

Case 3: If ED has the largest delay among EP, PF, and FD, peer E will pick the second-nearest peer in $N^2(P)$, such as C in Figure 43.7d, and repeat the above process until it finds a better node to replace P as its neighbor, or until it has tried all the peers in $N^2(P)$.

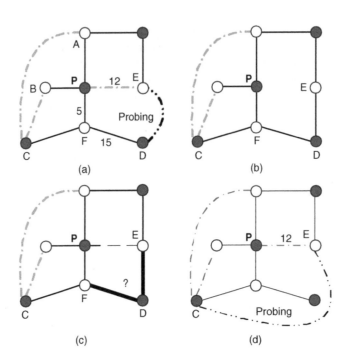

FIGURE 43.7   An example of neighbor replacement.

The first three operations are relatively straightforward, so we do not provide the detailed pseudo code, and the pseudo code of *direct neighbor replacement* operation is as below:

```
For a white peer i
For each peer j in white peer i's non-forwarding neighbor list
  Replaced = false;
  List = all the two hope away red neighbors of j, N²(j);
  Peer i pings all the peers in List;
  Add peers' RTT information to List;
  While List is not empty and Replaced = false
    remove the peer h with smallest RTT from List;
    if RTTᵢₕ < RTTᵢⱼ{replace j by h in i's neighbor list;
      Put j into will-cut list;
      Replaced = true;}
    else
      Commonlist = all common neighbors of peer j and h;
      While Commonlist is not empty and Replace = false
        Randomly remove a peer k from Common_list;
          RTTₖ = max{RTTₖⱼ, RTTₖₕ};
          if RTTᵢₕ < RTTₖ
              {add h to i's neighbor list;
              remove j from i's neighbor list right after
              i finds out jk or kh is disconnected;
          Replaced = true;};
      End While;
    End While;
End For
```

TABLE 43.1 TTL2-Detector Message Body

	Source	IP Address	Source	Timestamp				
Byte offset	0	3	4	7				
	Source	IP Address	Source	Timestamp	TTL1	IP Address	TTL1	Timestamp
Byte offset	0	3	4	7	8	11	12	15

43.3.3 Location-Aware Topology Matching (LTM)

In LTM, each peer is aware of the location of other peers within a distance of two hops.[25] The major advantage of LTM is that it is not only able to timely match the logical topology with the physical topology to significantly improve the search efficiency, but it also guarantees to retain the search scope. Three operations are defined in LTM: TTL2 detector flooding, slow connection cutting, and source peer probing.

43.3.3.1 TTL2-Detector Flooding

Based on the Gnutella 0.6 P2P protocol, a new message type called TTL2-*detector* is designed. In addition to Gnutella's unified 23-byte header for all message types, a TTL2-detector message has a message body in two formats, as shown in Table 43.1.

The short format is used in the source peer, which contains the source peer's IP address and the timestamp to flood the detector. The long format is used in a one-hop peer that is a direct neighbor of the source peer, which includes four fields: Source IP Address, Source Timestamp, TTL1 IP Address, and TTL1 Timestamp. The first two fields contain the source IP address and the source timestamp obtained from the source peer. The last two fields are the IP address of the source peer's direct neighbor who forwards the detector, and the timestamp to forward it. In the message header, the initial TTL value is 2. The payload type of the detector can be defined as 0×82.

Each peer floods a TTL2-detector periodically. We use $d(i, S, v)$ to denote the TTL2-detector that has the message ID of i with TTL value of v, and is initiated by S. A TTL2-detector can only reach peers in $N(S)$ and $N^2(S)$. The clocks in all peers can be synchronized by current techniques in an acceptable accuracy. By using the TTL2-detector message, a peer can compute the cost of the paths to a source peer. As an example, in Figure 43.8a, when peer P receives a $d(i, S, 1)$, it can calculate the cost of link

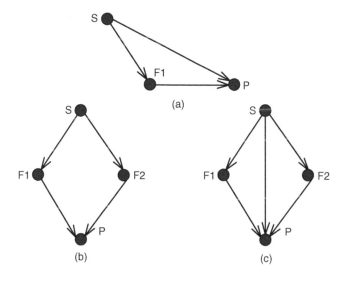

FIGURE 43.8 Peer P receives d(i, S, v) multiple times.

SP from Source Timestamp and the time P receives the $d(i, S, 1)$ from S. When P receives a $d(i, S, 0)$, it can calculate the cost of link SN_1 from TTL1 Timestamp and Source Timestamp, and N_1P from TTL1 Timestamp and the time P receives the $d(i, S, 0)$ from N_1. As one can see in an inefficient overlay topology, the peers in set $N^2(S)$ can receive $d(i, S, v)$ more than once, such as peer P in Figure 43.8a,b,c. If a peer receives $d(i, S, v)$ multiple times, it will conduct the operations in the second step of LTM, slow connection cutting.

43.3.3.2   Slow Connection Cutting

There are three cases for any peer P who receives $d(i, S, v)$ multiple times.

Case 1: P receives both $d(i, S, 1)$ and $d(i, S, 0)$ as shown in Figure 43.8a. In this case, $d(i, S, 1)$ comes from path SP, while $d(i, S, 0)$ comes from SN_1P. The costs of SP, SN_1, and N_1P can be calculated from the timestamps recorded in $d(i, S, 0)$ and $d(i, S, 1)$. If SP or N_1P has the largest cost among the three connections, P will put this connection into its *will-cut list*, that is, a list of connections to be cut later. If SN_1 has the largest cost, P will do nothing. Note that LTM is fully distributed and all peers do the same LTM operations. In the case of SN_1 having the largest cost, N_1 will put this connection into N_1's will-cut list. A peer will not send or forward queries to connections in its will-cut list, but these connections have not been cut in order for query responses to be delivered to the source peer along the inverse search path.

Case 2: P receives multiple $d(i, S, 0)$s from different paths as shown in Figure 43.8b. In LTM, P randomly takes two of the paths, such as SN_1P and SN_2P in Figure 43.8b, to process at each time. Other paths, if any, will be handled in the next round of optimization. Thus, one important factor to affect the performance of LTM is the frequency for each peer to issue TTL2-detector messages. Peer P can calculate the costs of SN_1, SN_2, N_1P and N_2P. If PN_1 or PN_2 has the largest cost, P will put it into its will-cut list. If SN_1 or SN_2 has the largest cost, P will do nothing. As discussed above, SN_1 or SN_2 having the largest cost will be cut by one of the other three nodes.

Case 3: P receives one $d(i, S, 1)$ and multiple $d(i, S, 0)$s as shown in Figure 43.8c. In this case, P will process the path receiving $d(i, S, 1)$ and one path randomly selected from the multiple paths of $d(i, S, 0)$s forming a scenario of *Case 1*.

43.3.3.3   Source Peer Probing

For a peer P that receives only one $d(i, S, 0)$ during a certain time period (e.g., 10 seconds), and $P \in (N^2(S) - N(S))$, it will try to obtain the cost of PS by checking its cut list first. If S is not in the list, P will probe the distance to S (see Figure 43.9). After obtaining the cost of PS, P will compare this cost with the costs of SN_1 and PN_1. If PS has the largest cost, P will not keep this connection. Otherwise, this connection will be created. In the Internet, the cost of SP and the cost of PS may not be the same. We use the cost of PS to estimate the cost of SP.

43.4   Summary

Peer-to-peer (P2P) computing has emerged as a popular model aiming at further utilizing Internet information and resources, complementing the available client/server services. Without assuming any knowledge of the underlying physical topology, the conventional P2P mechanisms are designed to randomly choose logical neighbors, which causes a serious topology mismatch problem between the P2P overlay network and the underlying physical network. This mismatch problem incurs a great stress in the Internet infrastructure and adversely restrains the performance gains from the various search or routing techniques. To alleviate the mismatch problem and reduce the unnecessary traffic and response time, three schemes are recently proposed, namely, Adaptive Connection Establishment (ACE), Scalable Bipartite Overlay (SBO), and Location-aware Topology Matching (LTM) techniques. All achieve the above goals without bringing any noticeable extra overheads. Moreover, these techniques are scalable because the P2P overlay networks are constructed in a fully distributed manner where global knowledge of the network is not necessary. ACE is the simplest one, but the convergent speed is relatively slow. SBO, with the same overhead, has a better

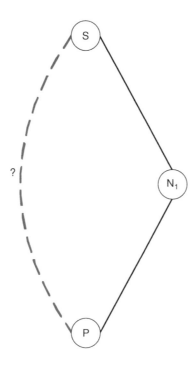

FIGURE 43.9 Source peer probing.

performance than ACE. The convergent speed of LTM is the fastest but it needs the support of NTP[26] to synchronize the peering nodes.

References

1. Q. Lv, P. Cao, E. Cohen, K. Li, and S. Shenker, Search and replication in unstructured peer-to-peer networks, *Proc. 16th ACM Int. Conf. on Supercomputing,* 2002.
2. Napster, http://www.napster.com.
3. O.D. Gnawali, A Keyword-Set Search System for Peer-to-Peer Networks, Master's thesis, Massachusetts Institute of Technology, June 2002.
4. I. Stoica, R. Morris, D. Karger, F. Kaashoek, and H. Balakrishnan, Chord: a scalable peer-to-peer lookup service for Internet applications, *Proc. ACM SIGCOMM,* 2001.
5. A. Rowstron and P. Druschel, Pastry: scalable, distributed object location and routing for large-scale peer-to-peer systems, *Proc. Int. Conf. on Distributed Systems Platforms,* 2001.
6. B.Y. Zhao, J.D. Kubiatowicz, and A.D. Joseph, Tapestry: An Infrastructure for Fault-Resilient Wide-Area Location and Routing, Technical Report UCB//CSD-01-1141, U.C. Berkeley, 2001.
7. S. Ratnasamy, P. Francis, M. Handley, R. Karp, and S. Shenker, A scalable content-addressable network, *Proc. ACM SIGCOMM,* 2001.
8. Gnutella, http://gnutella.wego.com/.
9. KaZaA, http://www.kazaa.com.
10. B. Yang and H. Garcia-Molina, Efficient search in peer-to-peer networks, *Proc. 22nd Int. Conf. on Distributed Computing Systems (ICDCS),* 2002.
11. S. Sen and J. Wang, Analyzing peer-to-peer traffic across large networks, *Proc. ACM SIGCOMM Internet Measurement Workshop,* 2002.
12. S. Saroiu, K.P. Gummadi, R.J. Dunn, S.D. Gribble, and H.M. Levy, An analysis of Internet content delivery systems, *Proc. 5th Symp. on Operating Systems Design and Implementation,* 2002.

13. Fasttrack, http://www.fasttrack.nu.
14. M. Ripeanu, A. Iamnitchi, and I. Foster, Mapping the Gnutella network, *IEEE Internet Computing*, 2002.
15. Ritter, Why Gnutella Can't Scale. No, Really, http://www.tch.org/gnutella.html.
16. The Gnutella Protocol Specification 0.6, http://rfc-gnutella.sourceforge.net.
17. Y. Chu, S. G. Rao, and H. Zhang, A case for end system multicast, *Proc. ACM SIGMETRICS*, 2000.
18. B. Krishnamurthy and J. Wang, Topology modeling via cluster graphs, *Proc. SIGCOMM Internet Measurement Workshop*, 2001.
19. V.N. Padmanabhan and L. Subramanian, An investigation of geographic mapping techniques for internet hosts, *Proc. ACM SIGCOMM*, 2001.
20. Z. Xu, C. Tang, and Z. Zhang, Building topology-aware overlays using global soft-state, *Proc. 23rd Int. Conf. on Distributed Computing Systems (ICDCS)*, 2003.
21. Y. Chawathe, S. Ratnasamy, L. Breslau, N. Lanham, and S. Shenker, Making Gnutella-like P2P systems scalable, *Proc. ACM SIGCOMM*, 2003.
22. Y. Liu, Z. Zhuang, L. Xiao, and L.M. Ni, A distributed approach to solving overlay mismatch problem, *Proc. 24th Int. Conf. on Distributed Computing Systems (ICDCS)*, 2004.
23. T. Cormen, C. Leiserson, and R. Rivest, *Introduction to Algorithms*, McGraw-Hill, New York, 1990.
24. Y. Liu, L. Xiao, and L.M. Ni, Building a scalable bipartite P2P overlay network, *Proc. 18th Int. Parallel and Distributed Processing Symposium (IPDPS)*, 2004.
25. Y. Liu, X. Liu, L. Xiao, L.M. Ni, and X. Zhang, Location-aware topology matching in unstructured P2P systems, *Proc. IEEE INFOCOM*, 2004.
26. NTP: The Network Time Protocol, http://www.ntp.org/.

44

Resilience of Structured Peer-to-Peer Systems: Analysis and Enhancement

Dong Xuan

Sriram Chellappan

Xun Wang

44.1 Introduction

Peer-to-peer (P2P) systems are a new paradigm of communication systems in which all computers are treated as equals, or peers, in the network. Individual computers may share hard drives, CD-ROM drives, other storage devices, files, etc. with the other computers on the network. This is different from a client/server set-up in which most of the computers (clients) tend to share resources provided by one main computer (the server). P2P systems have several advantages, as follows and more. The system model is inherently distributed. There is no need for huge servers and enormous indexing. Single point of failures are eliminated due to the absence of dependencies on any specific server. Sharing of available content does not require a publication step such as creating a Web page or uploading to a server. Anonymizing the communication between the requester and provider is easier. The systems can be implemented with localized security considerations and it is flexible enough to modify them.

Broadly speaking, there are two classes of P2P systems, namely unstructured and structured. Unstructured systems such as Gnutella,[1] KaZaA,[2] and Freenet[3] are constructed on-the-fly, without any regularization on the connectivity among peers or the searching mechanism. Such networks are easier to build and so is the maintenance (or the lack of it). Typically, new nodes randomly connect to existing alive nodes in the network, and the searching process for resources is flooding (called breadth first search) or iterative searching (called depth first search).

Another class of P2P systems is structured P2P systems,[4–7] where the system (comprising of peers) follows a predetermined structure. This structure must be maintained by participating peer nodes for correctness. Structured P2P systems typically use distributed hash table (DHT) functionality to assign identifiers to nodes and the resources. Structured P2P systems enjoy efficient lookup mechanisms that increase data availability and bound the time and messages required to find available data.

It is envisaged that eventually millions of users will demand access to these systems. In such a situation, it will be prohibitive to validate or constrain membership to such networks. This raises serious questions about the trustworthiness of nodes that join these networks, as well as the dynamics associated with the membership (in terms of node joins and leaves). The P2P system should be able to operate in the presence of malicious nodes and dynamics. That is, the system should be *resilient*. Resilience of P2P systems can be thought of as being impacted in two ways. The first is the resilience of P2P systems under node failures. The second is the resilience of the P2P systems under attacks by malicious nodes.

This chapter focuses on the resilience of structured P2P systems in the presence of both node dynamics and attacks. We first provide a brief introduction to the design features and performance metrics of structured P2P systems, followed by threats associated with structured P2P systems. We then describe approaches to analyze the resilience of structured P2P systems, followed by enhancements to the design features of these systems to make them more resilient. Our discussions on the analysis and enhancements will provide a survey of related work, followed by our approaches to do the same.

44.2　Structured P2P Systems and Their Threats

44.2.1　Structured P2P Systems

Structured P2P systems, as the name suggests, contain peer nodes arranged in a fixed structure. Structured P2P systems use distributed hash tables (DHTs) for resource management and searching. Typically, DHTs associate hash values (keys) with the content or resources present in the system. Participants in the system each store a small section of the contents of the hash table, improving load balancing and searching. Secure hashes such as MD5[8] and SHA-1[9] are commonly used to assign collision-free identifiers to nodes and resources. What follows is a very brief background on four popular structured P2P systems. For more details on these systems, interested readers can read the corresponding literature.

44.2.1.1　CAN

CAN[5] resembles a hypercube geometry. CAN uses a d-torus that is partitioned among nodes such that every node owns a distinct zone within the space. As explained,[5] a CAN node's identifier is a binary string representing its position in the space. For an N node system, when $d = \log N$ dimensions,* the neighbor sets in CAN are exactly those of a $\log N$-dimensional hypercube. Each node has $\log N$ neighbors; neighbor

---

*Unless otherwise specified, $\log N$ denotes $\log_2 N$.

i differs from the given node on only the i-th bit. The distance between two nodes is the number of bits on which their identifiers differ, and routing works by greedy forwarding to reduce this distance. Thus, routing is effectively achieved by correcting bits on which a forwarding node differs from the destination. The path length in CAN scales as $O(\frac{d}{4}n^{\frac{1}{d}})$.

44.2.1.2 Chord

Chord[4] is a distributed hash table storing key-value pairs. Keys are hashed into a circular N-bit identifier space $[0; 2^N)$ (identifiers on the circle are imagined as being arranged in increasing order clockwise). Each node is assigned a unique identifier (ID) drawn uniformly at random from this space, and the distance from a node m to a node k is the clockwise distance on the circle from m's ID to k's ID. Each node m maintains a link to the closest node k that is at least distance 2^i away, for each $0 \le i < N$. The set of nodes forming a Chord network is known as a Chord ring. In Chord, a node can route to an arbitrary destination in $O(\log N)$ hops because the routing semantics in Chord at each hop cuts the distance to the destination by half.

44.2.1.3 Pastry

In Pastry,[6] nodes are responsible for keys that are the closest numerically (with the keyspace considered as a circle similar to Chord). The neighbors consist of a *Leaf Set, L*, which is the set of L closest nodes (half larger, half smaller) for correctness in routing. To achieve more efficiency, Pastry has another set of neighbors spread in the key space. Routing consists of forwarding the query to the neighboring node that has the longest shared prefix with the key. In the event of node failures, Pastry uses a repairment algorithm.[6]

44.2.1.4 Tapestry

Tapestry[7] is an overlay location and routing infrastructure where the nodes maintain the routing table called *neighbormaps*. It routes along the nodes following the longest prefix routing similar to the one in CIDR IP address allocation architecture. The Tapestry location mechanism is very similar to the Plaxton location scheme[10] but has more semantic flexibility.

A host of other structured P2P systems have been designed. The Viceroy[11] algorithm emulates the operation of a traditional Butterfly network. A constant routing state is maintained at each node and routing is achieved in three stages, each in $O(\log N)$ steps toward the destination. The PRR[10] scheme employs a tree-based structure. Routing works greedily toward the destination by successively correcting the highest-order bits on which the forwarding node differs from the destination. The Kademlia[12] system uses the numerical value of the Exclusive OR (*XOR*) as the routing distance between two nodes and routing is through greedy forwarding.

The performance of any P2P system can be evaluated based on three metrics: (1) data availability, (2) resource consumption, and (3) messaging overhead. Data availability refers to the ability of the system to locate the requested data correctly. One choice to quantify it is the hit ratio, defined as the ratio of the number of successful queries to the total number of queries generated. Resource consumption during routing is estimated by measuring hop length and latency. Path length refers to the number of peer-to-peer hops (not the number of hops at the network layer) taken by the system to locate a resource. Most designs employing DHT functionality can locate available data within $O(\log N)$ hops for an N-node system. The latency from the request to the response is the actual delay in time seen by the user. Systems try to route messages based on proximity estimates to other peers in an effort to minimize the latency.

The third metric is the messaging overhead associated with system repair due to the inherent dynamics associated with the nodes in terms of joining and leaving the network frequently. The system must update pointers that otherwise incorrectly point to lost or new data, update routing tables reflecting the new changes, and correctly route current queries. Structured P2P systems must do more messaging in this realm to maintain the structure, whereas unstructured P2P systems do not handle dynamics with the intention of repairing the system.

44.2.2   Threats

This section provides an overview of the threats associated with P2P systems. To clarify for the reader, a threat when implemented or executed by a node (typically malicious) becomes an attack. Threats can be classified into two types: passive and active. Passive threats occur due to inherent threats in the system and not to malicious participants, while active threats involve malicious nodes deliberately compromising the system.

Passive threats in P2P systems are typically due to dynamics caused by frequent node failures and joins:

- *Node failures.* If preplanned node failures occur, content can be distributed across other nodes. But if failures occur unexpectedly, existing content may be unobtainable until the node restarts or some replication mechanisms are in place.
- *Node joins.* Although node joins are useful in the sense that content can be made more available and more effectively distributed, it comes at the cost of maintenance overhead, which should be minimized. This overhead depends on the number of nodes, the geometry of the system, and the maintenance protocol.

On the other hand, active threats are due to the presence of malicious nodes. Such threats can be further classified as:

- *Routing attacks.* Malicious nodes present in the system can disrupt operations by not obeying the routing logic. They can route to nodes farther than the destination with the intention of increasing latencies or path length; they can drop queries compromising data availabilities; and they can advertise false routing table information to other nodes.
- *Sybil attacks.* Malicious nodes may indulge in Sybil attacks,[13] where a node masquerades itself with the identity of other nodes. If node identities are not certified, it is very easy for malicious nodes to execute Sybil attacks. If some popular content is replicated, then the redundancy in the system can be seriously compromised under Sybil attacks.
- *Freeloading.* In the realm of file-sharing systems, several works[14] have shown the problem of freeloading. A user (or node) in a P2P system is a freeloader when it obtains content from other users but refuses to share its content with others. This results in a significant imbalance in long-term operation and defeats the basic tenet of all users (or nodes) being peers.
- *Miscellaneous attacks.* There are several other types of attacks possible. Malicious nodes can launch distributed denial-of-service (DDoS) attacks by bombarding a peer with queries. The intention is to prevent that peer from providing service to other benign peers in the system. A malicious node can also declare that a particular node X is the holder of a resource, when in fact node X may not possess the resource. That is, trusting one node to speak out for another node can be taken advantage of by malicious nodes.

All of the above attacks can be orchestrated separately by one or more nodes. However, when individual attackers coordinate to conduct these attacks, the impacts can be severe. Typical scenarios of coordinated attacks include the following. Since all structured P2P systems follow a rigid structure, if the malicious nodes locate themselves at strategic locations, performance can be significantly compromised. When a new node joins a network, it can inadvertently contact a malicious node first. If a set of malicious nodes forms a parallel network among themselves, then this newly joined node can neither contribute nor enjoy resources.

44.3   Analyzing the Resilience of Structured P2P Systems

44.3.1   A Survey

In broad terms, the resilience of a P2P system is its ability to maintain performance levels under the presence of malicious nodes and dynamics. The better the performance, the more resilient the system. Gummadi et al.[15] have defined the term "static resilience" as the ability of the P2P system to locate resources in the period of transition between node failures and full recovery by the system. The importance of static resilience lies in a measure of how quickly the recovery algorithm must work. Systems with low static resilience require faster recovery algorithms than those with higher levels of static resilience. Although static resilience is an important metric, it does not quantify the resilience in the presence of malicious nodes that other works try to address.

Typically, there are three approaches to analyze P2P resilience. Some works focus on a general description of threats and attacks, their impacts on system resilience, and optimizations to the system features. Such works[15,16] use representative case studies or simulations to study resilience. Another class of works employs graph-theoretic approaches to study resilience.[15,18,19,27] All structured P2P systems can be modeled as graphs, and the richness of graph theory is employed to dissect the system's behavior and study its resilience. A third approach uses analytical methods to study resilience. Such works[17] mathematically model the P2P system behavior to obtain insights into its performance.

An introductory analysis of the impact of geometry of P2P systems on resilience and proximity is presented by Gummadi et al.[15] The authors dissect popular P2P systems in terms of their geometry, flexibility in neighbor and route selection, and the actual routing algorithms used to locate resources. Based on formal arguments and simulations, the authors highlight the importance of flexibility in route selection; the presence of sequential neighbors (neighbors to which one can route and be sure of making progress toward all destinations); and building routing tables based on proximity to enhance overall system resilience. In the same vein, Sit and Morris[16] analyze security issues in P2P systems by specifying several possible attack scenarios and countermeasures.

Graph-theoretic approaches to analyze resilience are given by Loguinov et al.[18] and Aspnes et al.[19] The P2P systems are modeled as graphs and certain graph-theoretic properties such as connectivity and diameter are analyzed. Based on findings, the authors propose P2P systems based on classes of graphs that do possess desirable properties. Some instances of such graphs are de Bruijn graphs[18,20] and random graphs.[19]

44.3.2 A Markov-Chain-Based Analytical Approach

This section first describes our general approach to analyzing resilience of structured P2P systems. Following that, using a case study, we discuss how to apply this approach to P2P systems under node failures.

As we know, in P2P systems during the data lookup process, a node forwards a query to the next node based on its current routing table. The routing process can be modeled as a discrete absorbing *Markov chain*:

- Define a stochastic process $\{X_h : h = 0, 1, \ldots\}$, where random variable X_h is the state of a message forwarding during a routing process. Specifically, X_h can be a "failure" state or the ID of the node where the message is located after it is forwarded to the h-th hop.*

- The message is forwarded to the destination or dropped finally ("destination state" and "failure state," respectively). These two states are absorbing and the others are transient.**

In the following, we use the feature of the Markov chain to compute the average hit ratio and the average path length from any source to one specific destination node. We assume that, in a P2P system, each node has a unique ID. In a system with n nodes, the nodes are named, starting from 0 to $n - 1$. Without loss of generality, we consider node 0 as a destination, and all other nodes $1, 2, \ldots, n - 1$ as sources. We define the state i ($i = 0, 1, \ldots, n - 1$) as the state when the message is at node i. We denote n as the "failure" state; that is, if the message cannot be forwarded to any nodes in the system, it will be dropped and virtually put into n. Obviously, $1, 2, \ldots, n - 1$ are transient states and $0, n$ are absorbing states.

---

*Here, we slightly abuse the notation of node ID and the state.

**If the message can retry after previous trials fail, we can model it as a discrete Markov chain with only one absorbing state (destination state). The analysis is very similar except that we must remove state n and replace $p_{i,i}$ with $p_{i,n}$ on computing transition probabilities.

We define the *transition probability* $p_{i,j} = \Pr[X_h = j \mid X_{h-1} = i]$. The matrix of transition probabilities $P = (p_{i,j})_{(n+1) \times (n+1)}$ can be written as:

$$P = \begin{pmatrix} 1 & 0 \cdots 0 & 0 \\ U & Q & V \\ 0 & 0 \cdots 0 & 1 \end{pmatrix}, \tag{44.1}$$

where Q is an $(n-1) \times (n-1)$ matrix that delineates the transition rates between the transient states $1, 2, \ldots, n-1$. U and V are $(n-1) \times 1$ column matrices that delineate the transition rates between the transient states and the absorbing states 0 and n, respectively.

We define the *hit ratio* $a_{i,j}$ as the probability that a message is successfully forwarded from a transient state i ($i = 1, 2, \ldots, n-1$) to an absorbing state j ($j = 0, n$), and the *average path length* $m_{i,j}$ as the average number of steps for the corresponding successful forwarding. Computing these two values are nothing but the absorbing probability problem and the mean first passage time problem, respectively. In particular, for the state 0, if we define $A = (a_{1,0}, a_{2,0}, \ldots, a_{n-1,0})^\perp$ and $M = (m_{1,0}, m_{2,0}, \ldots, m_{n-1,0})^\perp$, by the analysis of Ravindran,[21] we have*

$$A = (I - Q)^{-1} U, \quad M = ((I - Q)^{-1} A) \div A. \tag{44.2}$$

From Equation (44.2) we can obtain the hit ratio and the average path length from transient state i ($i = 1, 2, \ldots, n-1$) to destination state j ($j = 0, n$). Assuming that a source is uniformly randomly chosen, we have:

Theorem 44.1 *The average hit ratio $\bar{a}$ and the average hit path length $\bar{m}$ for the algorithm sending message from any source to the destination 0 are, respectively,*

$$\bar{a} = \frac{1}{n}(\pi^0 A + 1), \quad \bar{m} = \frac{1}{n}\pi^0 M, \tag{44.3}$$

where A, M are defined in Equation (44.2) and $\pi^0 = (1, 1, \ldots, 1)$.

Based on Equation (44.3), we can compute the hit ratio and the average path lengths to node 0 from any other node. Recall that we made no assumption on node 0, and the derivation of the above formulae does not use any particular feature of this node. Hence, the above formulae are applicable to any node i as the destination. If the system is symmetric such that all nodes are equivalent in terms of looking up, the results of the formulae are automatically the hit ratio and the average path length of the system. If the system is not symmetric, the overall performance can be obtained based on the above formulae and the specific features of the system.

In the following, we demonstrate how to use the above approach to analyze the resilience of Structured P2P systems. Due to space limitations, we only discuss the case of node failures; refer to Wang and Xuan[22] for a discussion of resilience under attacks.

We can model node failures as follows. At some instant in time, to one node, say node i, its neighbors can be up or down. Departure or failure of nodes will make some items currently in the routing tables invalid. As defined by Wang et al.,[17] there are two main kinds of neighbors: *special neighbors* (such as the successor list in Chord and the leaf set in Pastry) and *finger neighbors*. To one node, we define τ_1 and τ_2 as the probabilities that a special neighbor and a finger neighbor are in failure state in terms of this node at some point in time, respectively (generally, $\tau_1 < \tau_2$).

The key to apply the Markov-chain-based approach is computing the transition probability $p_{i,j}$ from node i to node j for a given structured P2P system. For each node i, we need to compute the individual probability that node i will forward the message directly to node j. Obviously, if node j is not a neighbor or

---

*Define $Z = X \div Y$ as $z_{i,j} = x_{i,j}/y_{i,j}$ for any i, j.

itself, the probability of forwarding is 0. For all other nodes, the following steps must be taken to compute the transition probabilities:

1. The set of neighbors, including special neighbors and finger neighbors, is determined first. Recall that these two types of neighbors have different failure probabilities.
2. Then, the routing semantics specific to the given P2P system are followed to determine the probability of forwarding to each neighbor. Different P2P systems have different routing policies. Care is taken to make sure that the routing policy is honored.

In the following, we use CAN as an example to describe how to compute the transition probabilities. Sylvia et al.[5] proposed CAN as a distributed infrastructure that provides hash table-like functionality on Internet-like scales. It models the participating peers as zones in a d-dimensional toroidal space. Each node is associated with a hypercubal region of this keyspace, and has only special neighbors, which are the nodes associated with the adjoining hybercubes. Routing consists of forwarding to a neighbor that is closer to the key (in the toroidal space). Let us compute the transition probability $p_{i,j}$ from node i to node j:

1. Given node i, first determine the neighbor set of node i. Recall that the destination is defined as 0. Define $l(i)$ as the lattice distance from i to destination 0 and $l(i,j)$ as the lattice distance from i to j. Define $N(i) = \{i' : l(i') = l(i) - 1 \wedge l(i,i') = 1\}$; $N(i)$ is the neighbor set of node i. All nodes in $N(i)$ are *special neighbors*, and each is operational with probability τ_1. Their lattice distances from node i are the same under our assumption.
2. Having determined the neighbor set, let us discuss how to follow the routing semantics of CAN to determine the forwarding probability of different neighbors. In the CAN system, node i can uniformly choose one of the operational neighbors, and each node in $N(i)$ will be chosen as the next hop with probability $(1 - \tau_1^{|N(i)|})\frac{1}{|N(i)|}$. Now, if all the neighbors are down, the message will be dropped at node i, with probability $p_{i,n} = \tau_1^{N(i)}$.

Therefore, for $i = 1, 2, \ldots, n - 1$, we have

$$
p_{i,j} = \begin{cases} \left(1 - \tau_1^{|N(i)|}\right)\frac{1}{|N(i)|}, & j \in N(i) \\ \tau_1^{N(i)}, & j = n \\ 0, & \text{otherwise.} \end{cases} \tag{44.4}
$$

We apply the above formulae to compute the average path length and the hit ratio for CAN, Chord, Pastry, and Tapestry systems, using MATLAB. We also run simulations for the systems, with the number of nodes ranging from 64 to 4096. Here we only report the data of CAN and Pastry. Note that d is the dimension number for CAN and l is the leaf set size for Pastry. Note that there are no finger neighbors for the CAN system but there exist $\mathcal{O}(\log n)$ finger neighbors for Pastry.

Figure 44.1 illustrates the sensitivity of the average path length and the hit ratio to the failure of the special neighbors and the finger neighbors for CAN and Pastry, respectively.* We have the following observations:

- *Resilience to failures of finger neighbors:* The average path length is very sensitive to the failure of finger neighbors, but the average hit ratio is not, independent of the system model and the number of special neighbors used. Hence, the finger neighbors have significant impact on resilience to failures in terms of the average path length. For example, in the CAN system in Fig. 44.1(a), both the hit ratio and the average path length remain constant as τ_2 changes for different dimensions, since there are no finger neighbors in CAN; for Pastry, given different sizes of the leaf sets, the average hit ratio is not sensitive to the change of τ_2, unlike the average path length, which is very sensitive to such a change.

---

*Generally speaking, $\tau_1 < \tau_2$. For the purpose of illustration, we also show the data in the case that $\tau_1 > \tau_2$.

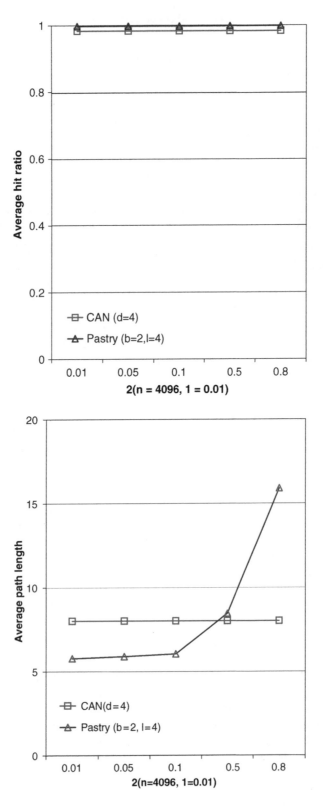

FIGURE 44.1 Performance evaluation of resilience to failure.

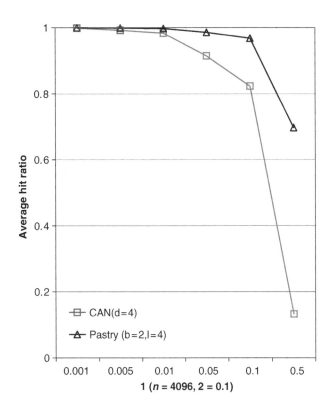

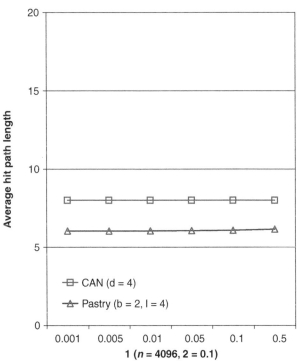

FIGURE 44.1   *(Continued)*

- *Resilience to failures of special neighbors*: The average hit ratio is very sensitive to the failure of special neighbors, but not the average path length, independent of the system model and the number of special neighbors used. Hence, the special neighbors have significant impact on resilience to failures in terms of the hit ratio. For example, in Fig. 44.1(b), as τ_1 increases, the average hit ratio for both CAN and Pastry decreases for different numbers of special neighbors (*i.e.*, $2d$ in CAN and L in Pastry), but the average path length remains almost constant.

44.3.3   Discussion

An important advantage in using a Markov-chain-based approach is its ease of usage in the realm of structured P2P systems. Once the transition probabilities are obtained, it is straightforward to calculate the average path length and hit ratio. Another benefit is the potential ease with which other stochastic theories can be incorporated to analyze other features of P2P systems.

A limitation of the Markov-chain-based approach is that, currently, it can only be applied to structured P2P systems that are in a stable state. A P2P system may or may not be in a stable state. By a *stable* system, we mean a system that obeys the following two criteria: (1) nodes are uniformly distributed in the system; and (2) the number of nodes in the system does not change much over time. It is true that a system can be in an unstable state due to an intensive attack on the system, or when nodes leave and join the system frequently. In most cases, a P2P system with DHTs is stable for the following reasons:

1. To achieve load balancing, the P2P system attempts to uniformly distribute the nodes in the system. A uniform hash function is used by most systems to achieve this goal.
2. We know that in a P2P system, individual nodes can dynamically leave and join. However, after the initial establishment phase, the number of leaving nodes will match the number of joining nodes, in which case the total number of nodes in the system will not change much.

We emphasize here that our work is only a first step toward a generalized analytical approach to analyze the resilience of P2P systems. One interesting issue is how to extend our approach to analyze P2P systems that are dynamic (absence of a stable state).

44.4   Enhancing the Resilience of Structured P2P Systems

44.4.1   A Survey

This section aims to provide a broad overview of several approaches, each of which aims to improve system resilience under failures and attacks. Some approaches exclusively focus on one performance metric to improve, while being oblivious to other metrics. Other approaches aim to optimize multiple performance metrics at the same time. Two standard approaches (or methodologies) and some newly emerging approaches that enhance P2P system resilience are described in the following.

44.4.1.1   Redundancy

Enhancing a system with redundancy is critical to the resilience of a P2P system. This includes replication of content, redundant control (or maintenance) query messages, and availability of multiple routing choices.

The importance of redundancy to improve system resilience is highlighted in Refs. 16, 23–26. The problem of efficient storage mechanisms for the data to provide high reliabilities has been addressed.[24,26] The authors propose a set of dedicated servers with replication to store content to improve availability. They validate their claim by demonstrating that eventually bandwidth, and not disk space, will be the bottleneck for the P2P applications. Similarly, Sit and Morris[16] highlight the importance of content replication under the threat of freeloading and DDoS attacks. Sufficient replication and avoiding single points of failure is their solution. The authors argue for randomizing the replication, both virtually and geographically. Such randomization will also alleviate the effects of coordinated attacks on the system.

The resilience of Tapestry and Pastry is shown to significantly increase under attacks and failures when there are redundancies in control messages and routing paths.[23] Specifically, the authors show that for an N node system, by maintaining an additional $\log N$ overhead in routing table and an additional $\log^2 N$ communication overhead (apart from the state currently maintained), data availability significantly increases under failures and malicious nodes. The attack model in that article typically involves fail stop models, where nodes fail to respond abruptly (node failures). Apart from this, malicious nodes can also generate "incorrect" messages (routing attacks).

Castro et al.[25] advocate the use of redundant queries to different replicas to increase data availability in the presence of malicious nodes. The idea is that multiple paths taken by multiple queries increase the likelihood that at least one of the nodes chosen during each hop is non-malicious. Also the presence of multiple routing paths naturally allows more choices for routing. For example, nodes can route to the nearest node in terms of network delay, thus decreasing latency. Obviously, in the P2P model, multiple paths naturally translate to increases in content availability as one dead end does not mean that the message cannot be routed correctly.

44.4.1.2 Graph-Theoretic Approaches

Graph-theoretic approaches to analyze and enhance resilience have been proposed.[15,18,19,27] The idea here is to introduce the "good" features of graphs to the structures of P2P systems to improve their resilience. The following discusses graph-theoretic approaches that have been proposed to enhance structured P2P system resilience.

Topologies such as ring (e.g., Chord[4]) and hybrid (e.g., Pastry[6]) naturally possess the feature of sequential neighbors (successor list in Chord and leaf set in Pastry, respectively). Sequential neighbors are neighbors to which one can route and be sure of making progress toward all destinations. Gummadi et al.[15] show the importance of adding sequential neighbors to enhance resilience. The addition of these to the various geometries that can accommodate this feature significantly increases the hit ratio.

A new P2P structure based on optimal-diameter de Bruijn graphs is presented by Loguinov et al.[18] De Bruijn graphs are those that are close to achieving the optimal diameter for a given graph. Typically, de Bruijn graphs can achieve a diameter of $D = \log_k N$, where N is number of nodes and k is the fixed node degree.[20] This drastic reduction in diameter achievable by de Bruijn graphs is the key motivation of Ref. 18. For a system with $N = 10^6$ nodes, Chord has a node degree and diameter equal to $\log N = 20$. On the other hand, with a fixed node degree $k = 20$, de Bruijn graphs have diameter $\log_k N = 5$. This implies improved routing efficiency and better resilience properties.

Massouli et al.[27] designed a localizer algorithm that attempts to design the P2P system in such a way that it is reflective of the underlying topology to minimize load at the network layer; it maintains connectivity among alive nodes in the system even in the presence of faults and dynamics; and the cost of using the overlay is distributed evenly among all participants. In a long-term execution of this localizer algorithm, the resultant topology is representative of the classical Erdos and Renyi model,[28] having properties that this article addresses apart from the proximity requirement.

Saia et al.[29] defined a notion of ϵ-dynamically strong fault tolerance to describe the robustness of CAN under massive targeted attacks in highly dynamic environments. A CAN network initially containing N peers is ϵ-dynamically strong fault-tolerant if, with high probability, all but an ϵ fraction of the peers in the CAN system can access all but an ϵ fraction of the data items during a period when a certain number of peers polynomial in N are removed by an adversary. They then designed a virtual CAN that is ϵ-dynamically strong fault-tolerant. The virtual CAN includes totally N virtual nodes. The virtual nodes and the data items are mapped on the super-nodes in a Butterfly network of depth $\log N - \log(\log N)$ by a set of hash functions.

An approach based on the construction of *random graphs*, where points represent resource identifiers, is presented by Aspnes.[19] In this graph, the probability of a connection between two nodes depends only on the distance between the nodes. The key principle employed by the authors is that in random graphs, failures result in a smaller random graph that can be used to locate existing resources.

Zhang et al.[30] have applied the small-world model[31] to enhance the routing performance of Freenet, which is an unstructured P2P system. Inspired by this work, we augmented the resilience of a structured P2P system (i.e., CAN) with this model.[17] A network is said to exhibit the small-world phenomenon if, roughly speaking, any two nodes in the network are likely to be connected through a short sequence of intermediate nodes. One network construction that gives rise to small-world behavior is one in which each node in the network knows its local neighbors, as well as randomly chosen remote nodes.

44.4.1.3   Miscellaneous Solutions

Apart from the above, other novel approaches using cryptography and biological principles have also been proposed to improve resilience.

Sit and Morris[16] discuss the use of authentication and cryptography techniques to enhance system resilience. They propose many attack scenarios. Their approaches to defend against attacks include assignment of keys to nodes to be done in a verifiable, trusted manner; the design of strong invariant properties that must be maintained and verified during dynamics; authenticating the joining process by a trusted authority; and keeping track of the node activity by means of signatures and maintaining a report of a node's behavior to check for maliciousness.

Castro et al.[25] discuss the importance of having a secure assignment of node identifiers; secure routing table maintenance; and secure message forwarding. For securing the assignment of identifiers to nodes, the authors propose a heavy-weight and a light-weight solution in the form of using a centralized server to assign identifiers or ask individual nodes to solve crypto puzzles before being permitted into the system. This approach also increases the security in forwarding. Stronger constraints on neighborship and the addition of trusted bootstrapping nodes will increase security in routing.

A recent paradigm of computation makes use of biological mechanisms[32,33] to address problems in computer science. In this framework, a biological framework for the realization of resilient P2P systems is explored by Montresor et al.[34] Here, the system model is distributed computing rather than purely file searching. In this context, the authors model the system, deriving inspiration from the *complex adaptive systems* (CAS) paradigm. The authors provide preliminary discussions on a very simple approach, the core idea of which is as follows. The resources (computational) are modeled as nests, and queries to find out resources are modeled as ants. In a purely random fashion, ants traverse the network to find desirable nests. They may also move computation from an overloaded nest to a non-overloaded one. In such a manner, the system is inherently fault tolerant, resilient to random failures of nests, and self-organizing. However, many implementation details must be analyzed before such models can be applied for making P2P systems resilient. Nevertheless, biological mechanisms do provide an interesting area to explore in this regard.

44.4.2   Enhancing Resilience of CAN and Chord

The following describes our work in enhancing the resilience of two structured P2P systems — namely, CAN and Chord.

44.4.2.1   CAN-SW

We apply the small-world model in CAN to form CAN-SW.[17] In the basic CAN-SW system, there are three key issues in the lookup system: (1) *CAN-SW construction*, (2) *routing mechanism*, and (3) *CAN-SW maintenance*. In CAN-SW, we still follow the greedy routing mechanism. In each step, the current message holder chooses a neighbor (either a local one or a remote one) as the next hop a node that is as close to the destination as possible, in the sense of lattice distance. The maintenance mechanisms are almost the same as those in the CAN system. The only difference is that remote neighbor nodes must be taken into account in the same manner as local neighbors. In the following, we focus on CAN-SW construction.

In the original design of the CAN system, each node maintains as its set of neighbors the IP addresses of those nodes that hold coordinate zones adjoining its own zone. In our system, in addition to local neighbors, each node will maintain an additional remote neighbor. The remote neighbor v of node u is chosen randomly with probability $\alpha_{i,j} = \frac{l^{-d}(i,j)}{\Delta}$, where $\Delta = \Sigma_{i' \neq i} l^{-d}(i,i')$. To choose remote neighbors, we need to know the size of the system (i.e., the number of nodes/zones) and the topology of the base

TABLE 44.1 Average Path Length for CAN and CAN-SW ($d = 2$)

	$n = 64$	$n = 256$	$n = 1024$	$n = 4096$
CAN	4	8	16	32
CAN-SW	3.65	6.36	10.74	17.12

CAN. Due to the dynamics of the construction of CAN, the topology of the base CAN is not regular. For simplicity, we assume that the topology is a d-torus. We can use the volume of individual zones to measure the system size. For example, for node i associated with zone Z_i, we define $[Z_i]$ as its volume. We know that the overall system volume is 1; so, the system size is estimated to be $n \approx m^d$, where $m^d \leq \frac{1}{[Z_i]} < (m+1)^d$. Once n is known, based on node i, a remote point v can be generated with probability $\alpha_{i,j}$. Using the routing and lookup mechanism, the remote zone where j is located can be found.

Note that the addition of a new node affects only a small number of existing nodes in a very small locality of the coordinate space. The number of neighbors a node maintains depends only on the dimension d of the coordinate space and the number of remote neighbors,* and is independent of the total number of nodes in the system. Thus, node insertion affects only $\mathcal{O}(d)$ existing nodes, which is important for CAN-SW with huge numbers of nodes.

We apply the Markov-chain based approach to analyze the resilience of CAN-SW.[17] We first follow the step described previously to compute the transition probabilities of CAN-SW and compute the average path length and the hit ratio.

To compute $p_{i,j}$, we first consider the neighbors, especially the remote neighbors. Although the small-world model considers all remote neighbors as being generated initially at random, we invoke the "Principle of Deferred Decisions" — a common mechanism for analyzing randomized algorithms[14] — and assume that the remote neighbors of a node i are generated only when the message first reaches j. We know that the probability that i chooses j as the remote neighbor is $\alpha_{i,j} = l^{-d}(i,j)\frac{1}{\Delta}$, where $\Delta = \Sigma_{i' \neq i} l^{-d}(i,i')$. Define the set of remote neighbors that can be the next hop as $B(i) = \{i' : l(i') \leq l(i) - 1\} \wedge l(i,i') > 1\}$. $N(i) = \{i' : l(i') = l(i) - 1 \wedge l(i,i') = 1\}$ is defined as the set of local neighbors that can be the next hop. Each node $j \in B(i)$ will be chosen as the next hop with probability $(1 - \tau_2)\alpha_{i,j}$. If no remote neighbor is available, a local neighbor must be chosen. Therefore, each node $j \in N(i)$ will be chosen as the next hop with probability $(1 - \tau_1^{|N(i)|})\frac{1}{|N(i)|}(1 - (1 - \tau_2)\Sigma_{i' \in B(i)}\alpha_{i,i'})$. As we know, if all neighbors are down, the message will be dropped at node i; hence, $\tau_1^{|N(i)|}(1 - (1 - \tau_2)\Sigma_{i' \in B(i)}\alpha_{i,i'})$. Therefore, for $i = 1, 2, \ldots, n - 1$, we have

$$
p_{i,j} = \begin{cases}
(1 - \tau_2)\alpha_{i,j}, & j \in B(i) \\
\left(1 - \tau_1^{|N(i)|}\right)\frac{1}{|N(i)|} \\
\quad \cdot \left(1 - (1 - \tau_2)\sum_{i' \in B(i)} \alpha_{i,i'}\right), & j \in N(i) \\
\tau_1^{|N(i)|}\left(1 - (1 - \tau_2)\sum_{i' \in B(i)} \alpha_{i,i'}\right), & j = n \\
0, & \text{otherwise.}
\end{cases}
\tag{44.5}
$$

We apply the above formula to compute the average path length and the hit ratio for CAN-SW and compare it with CAN.

Table 44.1 illustrates the comparison between CAN-SW and CAN in terms of the average path length when there is no failure in the system. We can clearly see that CAN-SW outperforms CAN. In particular, the

---

*So far, we assume it is 1. In the extended version, it may be a constant number larger than 1.

improvement turns out to be more significant as the number of nodes increases. We expect that the impact will increase as the number of remote neighbors increases; however, it will cost more maintenance overhead.

Figure 44.2 illustrates the sensitivity of the average path length and the hit ratio to the failure probabilities of the local neighbors (the special neighbors) (τ_1) and the remote neighbors (the finger neighbors) (τ_2) of different systems. It confirms the observation in Section 44.3.2 that the average path length is sensitive to the remote neighbors, while the hit ratio is sensitive to the local neighbors. Once $\tau_2 = 1$, CAN-SW becomes CAN. Figure 44.2(a) (where τ_1 is fixed, and $\tau_2 \rightarrow 1$) shows that CAN-SW outperforms CAN both in terms of hit ratio and the average path length.

44.4.2.2   RChord

Chord is susceptible to hostile routing attacks due to its uni-directional routing mechanism. CAN and Pastry support bi-directional routing. Hence, even if a query overshoots the destination, a benign node will be able to correctly route it by backtracking to the destination in the case of CAN and Pastry, but it is not possible in Chord. This prompts us to believe that Chord's resilience can be greatly improved by incorporating bi-directional routing. To enable bi-directional routing, we propose[35] to add reverse edges to Chord. The new Chord system is called RChord. Two important issues were addressed: (1) how to add the reverse edges to the routing table, and (2) how to do routing with reverse edges.

One simple approach to adding reverse edges is to include the nodes that are the *mirror nodes* for the nodes present in the routing table of a normal Chord node. Obviously, more reverse edges will lead to better performance; however, it would also introduce more overhead and redundancy among the edges. We believe that significant performance improvement can be achieved only by adding a few reverse edges. We focus on adding a constant number of reverse edges and propose several approaches to add them. We broadly classify them as *deterministic* and *randomized* algorithms.

With deterministic algorithms, the nodes that the reverse edges point to are determined by some system parameters, for example, the system size. The way the current Chord system adds edges is deterministic. That is, each node has a fixed set of neighbors located at predetermined distances relative to the nodes. We design several deterministic algorithms:

1. *Mirror Algorithm (M).* With this algorithm, the reverse edges of a given node are mirrors of the fingers (the clockwise edges in the original Chord system) of that node in the system.
2. *Uniform Algorithm (U).* Assume the number of reverse edges is a constant number, say R. With this algorithm, the R reverse edges are distributed uniformly in the space of log scale. The whole space is logged into R sections.
3. *Local-Remote Combination Algorithm (L-R).* This algorithm attempts to follow the idea of cooperation between local edges (L) and remote edges (R). We add some edges close to the current node, while adding others at a remote distance. They are chosen alternatively.

With the randomized algorithm, the local reverse neighbors are still selected following the same way as algorithm *L-R*; however, the remote reverse neighbors are selected in random fashion. Specifically, the probability of one node to be selected as the remote reverse neighbor is proportional to the distance between the current node and that node. With this algorithm, the node that is far away from the current node has a higher probability of being chosen as a remote node. In this sense, the algorithm is similar to the above deterministic Local-Remote combination algorithm. However, some nodes that are closer to the current node also have some probability of being selected and thus we expect the performance to be different from the deterministic one.

The formal descriptions of the above algorithms are given in Table 45.2.

We need to extend Chord's routing algorithm to consider the reverse edges. Chord's routing algorithm is greedy. Upon receiving a request, a peer will pick a node among its neighbors that is closest to the destination in the clockwise direction. The algorithm is simple and robust. A simple extension is picking the neighbor among the forward and reverse neighbors that is closest to the destination in any one of the two directions. While the algorithm is simple, and also compatible with the original algorithm, it is not efficient.[35] The basic reason lies in the asymmetry between the clockwise edges and the reverse edges.

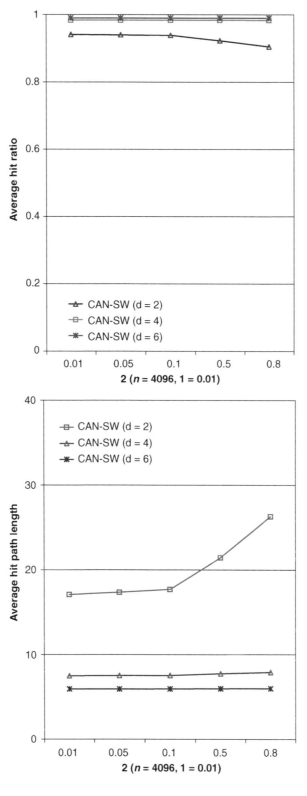

FIGURE 44.2 Resilience to failure of remote and local neighbors for CAN-SW.

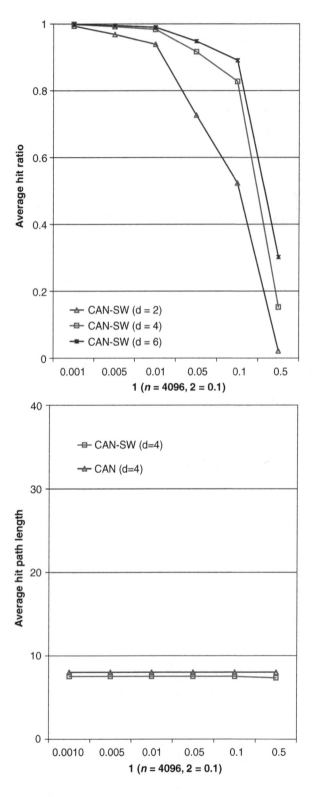

FIGURE 44.2 (*Continued*)

TABLE 44.2 Algorithms for Adding Reverse Edges to the Chord System

Input: the system size N ($m = \log N$), the number of reverse edges R, the current nodeID i
Output: the reverse neighbors r_k ($k = 0, \ldots, R - 1$)
Mirror Algorithm:
   $r_k = (i - 2^k) \bmod 2^m$, where $k = 0, \ldots, R - 1$.
Uniform Algorithm
   $r_k = (i - 2^k) \bmod 2^m$, where $k = \lfloor \frac{(p+1)m}{R+1} \rfloor$, $p = 0, \ldots, R - 1$.
Local-Remote Combination Algorithm
   $r_k = (i - 2^k) \bmod 2^m$, where $k = (m - 1 - \lfloor \frac{p}{2} \rfloor)(p \bmod 2) + \lfloor \frac{p}{2} \rfloor(1 - p \bmod 2)$, $p = 0, \ldots, R - 1$.
Local-Remote Random Algorithm
   $r_k = (i - 2^k) \bmod 2^m$, when k is even and less than R.
   The probability of a node j to be r_k (where k is odd and less than R) is
   $Prob(j) = (i - j) \bmod 2^m / \Sigma_{q=1}^{n/2-1} q \bmod 2^m$.

We conclude that our extension should consider the asymmetry between the forward edges and the reverse edges. Based on the above consideration, we design a routing algorithm that (1) is compatible with the original Chord routing algorithm, (2) considers the cooperation between the forward and reverse edges, as well as the difference in the routing capacity of the forward and reverse edges.

Our analysis results[35] show that the RChord system is much more resilient to routing attacks than the original Chord system. Due to space limitations, we do not report data here.

44.4.3 Discussion

The two approaches we have proposed to enhance resilience are basically inspired by the ideas of redundancy and the richness of graph theory.

The remote edges in the CAN-SW system and the reverse edges in the RChord system (both point to neighbors added to the routing table at a node) are fundamentally added to the existing neighbor list at nodes in the system. That is, both increase the redundancy in terms of routing choices at each node when forwarding a message. Under node failures or attacks, the *new* neighbors help in successfully routing messages, thus improving the system resilience in terms of hit ratio.

However, while adding these new neighbors, we apply ideas in graph theory with the objective of significantly improving system resilience. In CAN-SW, the remote edges are added randomly following the small-world model; while in RChord, a randomized remote edge adding algorithm was designed. Randomly selected reverse edges will make it difficult even for a highly destructive attack to cause extensive damage.

There are several directions to extend our work. For CAN-SW, the remote neighbor(s) can be added considering several factors:

1. If each node constructs multiple remote neighbors independently, the performance would improve due to better *load balancing* in the routing. However, this may increase the overhead in system maintenance.
2. Considering the issue of the *popularity* and *importance of a node*, we can adjust the number of remote neighbors for each node by assigning relative weights to it.

It will be an interesting exercise to theoretically prove the features of our reverse edge adding algorithms in our RChord system and hence derive an algorithm to add these edges that will give optimum performance.

44.5   Final Remarks

This chapter provided a detailed summary of the resilience of structured P2P systems in the presence of node dynamics and attacks. We first provided a brief introduction to the design features and performance metrics of structured P2P systems, followed by the threats associated with structured P2P systems.

We then described approaches to analyze the resilience of structured P2P systems. Our discussions on analysis provided a survey of related work. We then proposed our analytical approach to analyze the resilience of structured P2P systems under failures. The approach is Markov-chain based and can be applied to systems with a relatively stable size and uniform distribution of nodes.

Finally, we discussed several enhancements that have been proposed for the design features of structured P2P systems to make them more resilient. We then discussed our approaches to enhance the design features of two popular structured P2P systems — namely, CAN and Chord. Both our enhancements employed redundancy and graph-theoretic approaches for the enhancements. Specifically, the new edges (or the neighbors) that were added to our CAN-SW and RChord systems increased redundancy, which translates to more routing choices. This again translates to improved resilience under failures or attacks. For adding the edges, we employed the principles of small world and random graphs, such that even under highly destructive attacks, the impacts are minimized.

Acknowledgments

This work was partially sponsored by the National Science Foundation (NSF) under contract number ACI-0329155. Any opinions, findings, and conclusions or recommendations expressed in this material are those of the authors and do not necessarily reflect the views of the National Science Foundation.

References

1. Gnutella, http://www.gnutellaforums.com/.
2. KaZaA, http://www.kazaa.com.
3. Freenet, http://freenet.sourceforge.net/.
4. Ion Stoica, Robert Morris, David Karger, M. Frans Kaashoek, and Hari Balakrishnan, Chord: a scalable peer-to-peer lookup service for Internet applications, in *Proceedings of the 2003 ACM Special Interest Group on Data Communication (SIGCOMM)*, August 2001.
5. Sylvia Ratnasamy, Paul Francis, Mark Handley, Richard Karp, and Scott Shenker, A scalable content-addressable network, in *Proceedings of the 2003 ACM Special Interest Group on Data Communication (SIGCOMM)*, August 2001.
6. A. Rowstron and P. Druschel, Pastry: scalable, distributed object location and routing for large-scale peer-to-peer systems, in *IFIP/ACM International Conference on Distributed Systems Platforms (Middleware)*, November 2001.
7. Ben Y. Zhao, Ling Huang, Jeremy Stribling, Sean C. Rhea, Anthony D. Joseph, and John Kubiatowicz, Tapestry: a resilient global-scale overlay for service deployment, in *IEEE Journal on Selected Areas in Communications*, January 2004.
8. R. Rivest, The MD5 message-digest algorithm, Internet RFC, April 1992.
9. P. Jones, Us secure hash algorithm 1 (SHA1), Internet RFC, September 2001.
10. C.G. Plaxton, R. Rajaraman, and A.W. Richa, Accessing nearby copies of replicated objects in a distributed environment, in *Proceedings of ACM Symposium on Parallel Algorithms and Architecture (SPAA)*, June 1997.
11. Dahlia Malkhi, Moni Naor, and David Ratajczak, Viceroy: a scalable and dynamic emulation of the Butterfly, in *Proceedings of the Twenty-First ACM Symposium on Principles of Distributed Computing (PODC)*, July 2002.
12. Petar Maymounkov and David Mazires, Kademlia: a peer-to-peer information system based on the XOR metric, in *Proceedings of the First International Workshop on Peer-to-Peer Systems (IPTPS)*, March 2002.
13. John R. Douceur, The Sybil attack, in *Proceedings of First International Workshop on Peer-to-Peer Systems (IPTPS)*, March 2002.
14. Dennis Heimbigner, Adapting publish/subscribe middleware to achieve gnutella-like functionality, in *Proceedings of the 2001 ACM Symposium on Applied Computing (SAC)*, March 2001.

15. Krishna P. Gummadi, Ramakrishna Gummadi, Steven D. Gribble, Sylvia Ratnasamy, Scott Shenker, and Ion Stoica, The impact of DHT routing geometry on resilience and proximity, in *Proceedings of the ACM SIGCOMM Symposium on Network Architectures and Protocols,* August 2003.

16. Emil Sit and Robert Morris, Security considerations for peer-to-peer distributed hash tables, in *Proceedings of First International Workshop on Peer-to-Peer Systems (IPTPS),* March 2002.

17. Shengquan Wang, Dong Xuan, and Wei Zhao, On resilience of structured peer-to-peer systems, in *IEEE Global Telecommunications Conference (GLOBECOM),* December 2003.

18. Dmitri Loguinov, Anuj Kumar, Vivek Rai, and Sai Ganesh, Graph-theoretic analysis of structured peer-to-peer systems: routing distances and fault resilience, in *Proceedings of the 2003 ACM Special Interest Group on Data Communication (SIGCOMM),* August 2003.

19. James Aspnes, Zoe Diamadi, and Gauri Shah, Fault-tolerant routing in peer-to-peer systems, in *Proceedings of the Twenty-Second ACM Symposium on Principles of Distributed Computing (PODC),* July 2003.

20. M. Imase and M. Itoh, Design to minimize diameter on building-block network, *IEEE Trans. on Computers,* Vol. 30, 1981.

21. A. Ravindran, Don T. Phillips, and James J. Solberg, *Operations Research: Principles and Practice, 2nd edition,* John Wiley & Sons, January 1987.

22. Shengquan Wang and Dong Xuan, A Markov-Chain Based Approach to Resilience of Structured P2P Systems under Failures and Attacks, Technical Report, Dept. of Computer Science and Information, Ohio-State University, February 2003.

23. Kirsten Hildrum and John Kubiatowicz, Asymptotically efficient approaches to fault-tolerance in peer-to-peer networks, in *Proceedings of 17th International Symposium on Distributed Computing (DISC),* October 2003.

24. Rodrigo Rodrigues, An agenda for robust peer-to-peer storage, in *First Infrastructure for Resilient Internet Systems (IRIS) Student Workshop,* August 2003.

25. Micuel Castro, Peter Druschel, Ayalvadi Ganesh, Antony Rowstron, and Dan S. Wallach, Secure routing for structured peer-to-peer overlay networks, in *Proceedings of the Fifth Symposium on Operating Systems Design and Implementation (OSDI),* December 2002.

26. Charles Blake and Rodrigo Rodrigues, High availability, scalable storage, dynamic peer networks: pick two, in *Workshop on Hot Topics in Operating Systems (HotOS),* May 2003.

27. Laurent Massouli, Anne-Marie Kermarrec, and Ayalvadi J. Ganesh, Network awareness and failure resilience in self-organising overlays networks, in *Proceedings of the 22nd International Symposium on Reliable Distributed Systems (SRDS),* October 2003.

28. Hongsuda Tangmunarunkit, Ramesh Govindan, Sugih Jamin, Scott Shenker, and Walter Willinger, Network Topologies, Power Laws, and Hierarchy, Technical Report TR01-746, University of Southern California, June 2001.

29. Jared Saia, Amos Fiat, Steve Gribble, Anna R. Karlin, and Stefan Saroiu, Dynamically fault-tolerant content addressable networks, in *Proceedings of First International Workshop on Peer-to-Peer Systems (IPTPS) Workshop,* March 2002.

30. Hui Zhang, Ashish Goel, and Ramesh Govindan, Using the small-world model to improve Freenet performance, in *Proceedings of the IEEE Conference on Computer Communications (INFOCOM),* June 2002.

31. Jon Kleinberg, The small-world phenomenon: an algorithmic perspective, in *Proceedings of the 32nd ACM Symposium on Theory of Computing (STOC),* 2000.

32. Jeffrey O. Kephart, A biologically inspired immune system for computers, in *Proceedings of the Fourth International Workshop on Synthesis and Simulation of Living Systems (ALIFE),* 1994.

33. Michael Wang and Tatsuya Suda, The bio-networking architecture: a biologically inspired approach to the design of scalable, adaptive, and survivable/available network applications, in *Proceedings of the 1st IEEE Symposium on Applications and the Internet (SAINT),* January 2001.

34. Alberto Montresor, Heing Meling, and Ozalp Babaoglu, Towards self-organizing, self-repairing and resilient large-scale distributed systems, in *Proceedings of the 1st International Workshop on Future Directions in Distributed Computing*, June 2002.

35. Dong Xuan, Sriram Chellappan, and Muralidhar Krishnamoorthy, RChord: an enhanced Chord system resilient to routing attacks, in *Proceedings of IEEE International Conference on Computer Networks and Mobile Computing (ICCNMC)*, October 2003.

45

Swan: Highly Reliable and Efficient Networks of True Peers

Fred B. Holt

Virgil Bourassa

Andrija M. Bosnjakovic

Jovan Popovic

45.1   Introduction

The demands are increasing for interaction over the Internet. At the application layer, significant progress is being made in providing collaboration for office work, engineering work, and distance learning. The distributed teams and large distributed datasets involved in today's industrial, commercial, academic, scientific, and recreational projects introduce a need to distribute the computing resources. These resources include processing power, data, and bandwidth. Beyond distributing them physically, good programming will distribute these resources logically as well. Logical distribution supports modularity, portability, and simplicity for the resulting applications.

Over the years, several approaches to distributed applications have been offered. Each solution has its strengths and weaknesses. Client/server solutions lack reliability, and the bandwidth demands at the hub grow linearly in the number of participants; the server farm and its costs grow at least linearly with the number of participants, allowing no economy of scale. A simple peer-to-peer solution, connecting everyone to everyone else, introduces this scalability problem at every node. Setting the connectivity layer up as a tree can provide logarithmic efficiency in latency,[1,2] but trees have low reliability and they degrade to the lowest performance of any interior node. The solution presented in this chapter introduces fundamentally new concepts and gains many advantages over existing solutions while introducing very few drawbacks.

Small-world wide area networking ("Swan" in this chapter) is a new approach to providing the connectivity layer for distributed applications. Swan relies on the "small-world effect,"[3,4] in which, by creating a world in which everyone has only a few direct acquaintances, information can flow to everyone reliably and quickly. Making direct use of the existing networking infrastructure, Swan coordinates multiple simultaneous communication links to allow large numbers of participants to interact online in real-time. Instead of a tree, a star, or a complete graph, the logical connections between the nodes in a Swan session form a regular graph. Using regular graphs removes the problem of having a single point of failure, and these graphs have minimal numbers of hops for messages.

Swan is a self-contained software library written in C++, currently available for different versions of Windows and UNIX operating systems, and accessed by a simple and intuitive publish-and-subscribe application programmer's interface (API). This interface fits naturally into the demanding event-handling paradigm of commercial applications employing graphical user interfaces (GUIs).

Efficient multicast is an enabling technology for multiplayer online games, collaborative work, distance learning, and emergency networks. Swan provides an optimal but generic solution that is entering use in each of these areas.

45.2   Multicast for Online Interaction

There have been several types of human-to-computer interaction over the years, and the most important among them are briefly discussed in the following paragraphs, as some of those directly influence online interaction. As the need for better support for human-to-human interaction over a computer network grew, a number of solutions have been implemented. Broadcast (one-to-all) and unicast (one-to-one) solutions have matured in support of applications that require these types of interaction. At the cultural and application levels, we are just beginning to expect high levels of interaction among groups of participants. To date, the existing broadcast and unicast solutions have been pressed into service to handle an essentially multicast (any-to-all) problem. Multicast technologies[5–7] are not as mature as broadcast or unicast, due in part to the historical lack of demand and in part to the technical difficulties in providing reliable multicast. This section identifies the challenges in providing multicast, considers existing solutions, and introduces the Swan multicast technology.

When discussing the relationship between computers and users, several major trends can be taken into account. At first, only mainframe computers existed, and people considered them a scarce and precious resource, due to their high prices and low numbers. Therefore, only large teams of experts operated these computers, using them for sophisticated computations. Advancements in both hardware and software led to the rapid expansion of affordable personal computers, which in turn lowered the level of expertise required to operate a computer. This increase in the number of computers, and their increased interconnections, introduced the possibility of distributed and parallel processing. On the other end, the rapid development of a global computer network—the Internet—led to a change in the human-to-computer relationship. With a global computer network available, users could use computers to communicate with other computers or other users; they could find needed data and share their ideas, regardless of physical distance.

This computer communication at a distance evolved into increasing levels of interactivity and expanding numbers of participants. These demands increase the need for greater speed, security, and reliability. Connectivity solutions are often tuned to the application, satisfying the most important requirements while sacrificing the others. Organizing users and data began to follow standard patterns, from powerful servers and terminals with small processing power on one side, to large communities of peers having enough local processing power to deal with the most complex tasks.

Most applications now require the use of diverse resources, such as services, printers and other peripheral devices, databases, and multimedia data, that are not always locally available. Even if most of the resources needed for one application can be made available, they are not used all the time, which locally leads to high costs and low resource utilization. One can mitigate these local effects by distributing access to these resources among a large number of users. In this way, an entire online community can use the same resources when desired, without needing to replicate them. This sharing of resources has been extended to sharing processing resources for demanding computational problems. These shared resources set requirements on bandwidth and throughput.

Although gathering computer resources for shared use is undoubtedly important, it is more effective when the people using them establish a flow of experience, knowledge, and ideas. Therefore, supporting human collaboration is one of the most important aspects of shared applications.

An online session involves participants, logical networks, and the Internet. In addition to people, participants may also include autonomous computers such as servers, as illustrated in Figure 45.1. An increasing number of applications require large groups to work at the same time on some project. The Internet provides the opportunity to collaborate with people across the world, in different time zones, hundreds and thousands of miles away. One common requirement is maintaining consistency across the local views of the user community. Among the applications with this requirement for consistency are:

- Computer games
- Internet lectures
- Collaborative engineering

The need has been increasing in collaborative workflow, in Internet communities, and in distributed applications for an effective means to allow scalable and reliable sharing of information across multiple processes. To enable worldwide digital collaboration, programmers need a software mechanism allowing dozens, hundreds, or perhaps thousands of participating computer processes to share information easily,

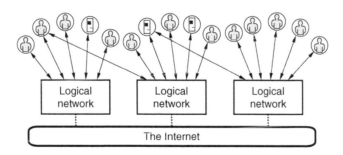

FIGURE 45.1   Three basic parts of an online session. The Internet includes the physical layer and low-level protocols. The logical network is the set of software connections among the participants, which are involved in diverse types of interaction. In practice, the flexibility and resilience of the participant level and of the Internet are lost in relying on a fragile logical network.

quickly, and reliably across the world. This demand for more interaction among communities on the Internet has been addressed primarily by client/server communications infrastructures. One thrust is to offer a true multicast capability at the router level[5-7] as part of the Internet protocol; Swan provides a multicast solution at the session level, using ubiquitous protocols such as TCP/IP for point-to-point connections.

People want to see the interactivity of chat combined with the rich features of their other software. Large enterprises have been pushing toward collaborative online work environments for a decade; given their size and geographic scope, these large companies often run into technology pains before the general markets. Computer games have provided pressure from another direction. In the last five years, game players have pushed for greater online multiplayer action. The early solutions for these two domains are now being echoed in the general markets. From enterprise software and multiplayer games to distance learning, business meetings, supply-chain management, emergency response teams, and other online communities, people are reaching for rich interaction over the Internet.

45.2.1 Encapsulating the Connectivity Layer

Today, one of the major challenges in distributed application design is implementing efficient communications for interactive online collaboration. Although the Internet and its protocols are mature enough to support most of these demands at the transport layer, the performance ultimately depends on the connectivity layer, the organization of logical connections among the participants. As this connectivity layer becomes more important, modular solutions at this layer provide the programmer with a software library with a simple API.

An API, which encapsulates the connectivity layer, should provide an intuitive abstraction, such as shared objects or the publish-and-subscribe paradigm. Hidden from the programmer is how the actual communication takes place. The underlying communication infrastructure may make use of a multicast network protocol, a graph of point-to-point network protocols, or a combination of the two. The details of the transport layer are hidden from the application programmer, and this modularity allows the same application code to work across heterogeneous operating systems and transport layer configurations.

There are four conceptual steps in using a publish-and-subscribe API. The first step concerns session management: finding and joining a session, getting a handle on the platform's event dispatcher, subscribing to events of interest, and eventually leaving the session. The second step specifies the schema of the events that will be shared. To multicast each type of application event to the group, the programmer must set the data fields that suffice to replicate this application event at the receiving end. In the third step, the programmer handles the publication of local events. This is usually a simple step, repacking local data according to the event schema and then making the library call. Finally, a programmer must decide how to handle events received by other participants. Typically, the programmer will write a callback function for this application event data and, with a library call, tie this callback function to the platform's event dispatcher. In Section 45.4, the Swan API provides an explicit example of the publish-and-subscribe paradigm.

Resources that provide entry points into an online session are designated as *rendezvous points (RPs)*. An RP helps new processes become part of a session. When a new process arrives, it needs to contact some of the participants that are already in the session, so it can establish communications with the group and receive sufficient state information to participate fully. Rendezvous points are represented as a list of Internet Protocol (IP) addresses or valid uniform resource locator (URL) addresses.

45.2.2 Requirements for Efficient Connectivity

Given the types of applications that are pressing for real-time interaction, a multicast solution must meet the following design requirements.

- *Low message latencies.* The latencies for event-message traffic must scale acceptably with the number of participants.
- *High throughput.* The solution should make efficient use of available bandwidth at every node, as reflected in the effective throughput available at the application layer.

- *Robust messaging.* The message traffic is not disrupted by routine departures of nodes. These departures may be graceful or ungraceful (consistent with Banerjee and Bhattacharjee,[8] an *ungraceful* departure occurs when a node leaves without sending the proper control messages).
- *Reliability.* The connectivity of the session is not lost due to the departure of any node, whether the departure is graceful or ungraceful.
- *Massively scalable.* The connectivity layer does not introduce an inconvenient threshold on the number of participants. The session support costs do not become prohibitive with increased participation.
- *Secure.* The solution supports strong authentication and encryption, and it respects the enforcement of data-authority, whether for treasures in multiplayer games or for proprietary engineering datasets.

Beyond the requirements, the following design goals make a multicast solution more compelling for adoption, both from a technical and commercial point of view:

- *Compatible with existing network devices.* No special hardware is presumed available. The solution can be used on most of the installed base of network devices.
- *Self-healing.* No human administration is required when a process joins or leaves the session. All the required properties are maintained automatically.
- *Easily adopted.* The solution has a simple API that respects modular programming.
- *True peer.* At the communications level, *every* node is equal. The multicast capability does not depend on hidden servers or a preferred class of nodes. Only local knowledge about the communications topology is stored.

These requirements and goals were identified in pursuit of a multicast problem from industry. In 1997, a global manufacturing company wanted to hold collaborative design reviews among its engineering teams. These reviews would initially involve up to a dozen geographically dispersed sites, but the number of sites could rapidly increase as team members joined from their desks rather than from conference rooms. The solution would have to work on the installed base of computers. Not all participants were guaranteed to arrive on time or to stay throughout the review. Any participant would have to be able to lead the review at any time. The requirement was to deliver a multicast capability in which computer processes were the nodes. The majority of the message traffic would be application events. The performance timescales and tolerances were set by applications instead of humans. Regardless of who initiated a review event, the collaborative performance should be comparable to the single-user performance of existing engineering software tools. Using the Swan multicast technology, a collaborative solution went into production use in 1999. Although developed to solve this problem, Swan itself was written as a general-purpose multicast solution.

45.2.3 Existing Distributed Architecture Solutions

This section describes a variety of existing solutions and measures them against the design goals and requirements presented above.

45.2.3.1 Client/Server

The central feature of the client/server model is the existence of one or more servers that together act as a hub for network communications. This hub mediates all network traffic for the session. In this way, the server decides whether to provide resources, determines the official application state and official clock, and provides a single point for addressing security and consistency issues. Server-based solutions are attractive due to server transparency to the client. Clients do not need to know anything about the server characteristics, and often the clients have significantly lower processing burdens than the server.

If an application needs to support a large number of participants, no small-scale solution is acceptable. Any solution with only one server is also limited because of the rapid degradation in performance as the number of clients overwhelms the server's capabilities. As a result, in client/server solutions, the logical server is often a federation of hardware servers. Instead of one server, a group of servers is coordinated to handle the message traffic. These federated servers can sit in a single rack, the same warehouse, or in different cities. This server partitions participants into groups that a single hardware server can handle, so the number of connections to the particular server is reduced. User requests are queued at the server, the server updates the application state, and it prepares the relevant updates for each participant. Each server in a server farm handles a smaller group of clients, so both the response time is acceptably short and the throughput is increased. This segmenting of the participants balances the performance demands across the servers in the server farm. In this way, server farms are not as obvious a bottleneck as single servers definitely are.

On the requirements for multicast, server-based solutions score well on security, throughput, and performance; however, they struggle with reliability and fail on scalability. For security, the server provides a central site from which to control authentication, monitor the flow of information, and restrict access to resources. All resources can be protected at the server side, so concurrent requests for resources are easy to resolve. Confidential information, such as passwords, bank accounts, and credit card numbers, is stored on the server. It is very difficult for a user to access the resources without the required access rights, which prevent malicious user behavior.

A server does not depend on a client's ability to process information. It responds to its clients' requests as quickly as possible and does not need to know what kind of processing the clients will perform. The server can work either with thin clients that have only the ability to send requests and display the results, or with fat clients that can themselves perform complex processing. On the other hand, clients depend entirely on the server. If the server goes down, this single point of failure destroys the entire session.

With server-based solutions, one has a built-in way to synchronize the participants' views of the application state. The server is the sole authority for the application state. As such, the server's view must encompass the views of all the participants. One allows the server to be the arbiter of what happens in what order. Relying on the server to synchronize the participants' views only buys a semblance of fairness and order. The server also does not suffice to keep the participants in lock step; interpolation and dead reckoning are used to guess what is going on until updates arrive from the server. The server is not sufficient for transaction synchronization because a three-phase commit is required to ensure the transaction state in the event of a communications failure. Although it provides a veneer of fairness, a server solves none of the hard issues in synchronization.

Servers maintain acceptable performance by centralizing resources and restricting the number of participants. All resources required by a client can be favorably deployed by the system architect to maintain high performance and throughput. Data access and transformation are performed on the server side, transparently to the client. It is therefore very easy to change the back-end implementation of access methods or business logic without affecting the client. In this central role, the server's resources become the limiting factor to scalability.

The inefficiencies at the hub of client/server middleware make it generally unsuitable for interactive online collaboration. To provide interactive behavior and to maintain consistency between the clients, the server establishes logical connections with all of them. As clients become more powerful, both in processing and in bandwidth, system resources are generally not well used in client/server solutions. Client resources tend toward underutilization, while the resources of the servers are in excessive demand. Each client conveys information directly to the server, and possibly, after some preprocessing and filtering, the server disseminates the information either through routing lists or by being polled by the clients. While the demands on a client's resources remain constant, the server's bandwidth and processing power are consumed in proportion to the number of participants.

This one-to-all connectivity limits the server-farm market for online communities. One reason for this is that the costs of the server farms must somehow be passed on to the end users. Typically, this is either through a higher initial price for the software or through subscriptions to online services. As the number of subscription models proliferates, people will have to choose which subscriptions they will continue to pay. Already, portals offer bundled subscriptions to a number of online applications, competing for the customers' subscriptions. This marketplace problem, as a reflection of the underlying capital demands and technical challenges, ultimately limits the scalability of server-based solutions. A new generation of Internet communication tools, described in this section, frees these online communities from the server farm.

45.2.3.2 All-to-All

To increase the reliability and throughput of client/server solutions, some connectivity strategies introduce point-to-point connections between the participants. The simplest of these solutions is *all-to-all* connectivity, in which each node is connected to every other node. In all-to-all systems, each node is connected with $N - 1$ other nodes, where N is the total number of nodes in a session. In an all-to-all system, the number of connections on every single node is equal to the number of connections on the server in a client/server solution with $N - 1$ clients and one server. The primary strengths of this approach are that every pair of participants is separated by only one logical network hop and that no participant is a single point of failure.

Using point-to-point connections directly does not scale easily as the number of participants grows. The resources at each node are now consumed at the rate of the hub in a client/server system, and linearly in the number of participants. Every node is subject to considerable network traffic overhead. In addition, joining to the point-to-point system is very complex. When newcomers want to join the group, they must find all other participants and join them. As the number of participants increases, establishing connections to all of them requires a lot of time and effort. If the number of participants changes frequently, it is possible for the newcomer to receive incorrect information while joining.

45.2.3.3 Multicast Trees

A variety of small-scale solutions has been developed over the years to provide multicast communications for small groups. Some examples of these solutions are shown in Figure 45.2. Among the most primitive of

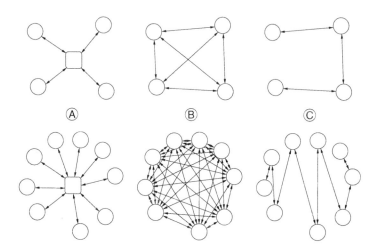

FIGURE 45.2 The most common small-scale solutions in cases with negligible and notable number of participants. This figure shows the logical networks that usually support small-scale applications: (a) hub-and-spoke, (b) all-to-all, and (c) daisy chain. If a server is present, it usually is the hub. The upper portion shows a case with four online participants. All these models have drawbacks and linear performance in the number of participants, which limits their scalability, as presented in lower portion of the picture. Hub-and-spoke has the hub as the bottleneck and the single point of failure; all-to-all has an enormous number of connections per participant, thus raising the level of bandwidth required; in the end, daisy chain suffers extreme latencies when leftmost and rightmost nodes are communicating.

these small-scale solutions are the all-to-all and client/server networks discussed separately above. Another important class of small-scale solutions is the multicast trees.

Well-designed trees can offer some efficiency in distributing messages; however, these efficiencies are easily lost in practice. Because trees rely on the non-leaf nodes to forward messages, the performance degrades to the lowest performance of all its non-leaf participants. Top performance is achieved by a balanced tree with the weakest participants as leaves and with stronger participants closer to the root. These balanced trees provide latencies that scale logarithmically in the number of participants. Note that the resulting optimization problem is a global one, so any late arrivals or early departures could trigger extensive reconnections. The network is considered optimized when the maximum latency among all pairs of nodes is minimized, which in turn leads to increase in performance.

In practice, the optimization problem is frequently ignored. The hub-and-spoke and daisy chain networks are common but degenerate examples of trees. If the connections are created by the participants or by human administrators, as in the T.120 standard, then the resulting logical networks usually end up close to the hub-and-spoke or daisy-chain networks, and far from the balanced trees. Both the hub-and-spoke and daisy chain lose the logarithmic efficiency of the balanced tree. The daisy chain has linear latency in its forwarding. The hub-and-spoke has linear efficiency as the hub sends each message to all other participants and, as seen above, the resources at the hub are consumed linearly in the number of participants. Without solving the global optimization problem to achieve logarithmic performance, trees easily degrade to linear performance instead.

If the connectivity layer is a tree, then the ungraceful departure of any non-leaf node will prune from the session the entire subtree below this node. Every non-leaf node is a single point of failure and a possible performance bottleneck.

45.2.3.4   IP Multicast

A multicast networking protocol[5,6] allows selective broadcast of messages to multiple recipients. It retains the complexity of direct network communication mentioned above, but is a natural choice for digital collaboration. Currently, multicast is available for UDP messages but virtually all UDP multicast traffic is limited to a single local area network or, at most, a small set of connected local area networks. UDP multicast, in its current implementation, could easily flood the Internet, as it would have to saturate the Internet with each message until it finds all possible participants.

Several wide area multicast networking protocols have been proposed,[9,10] and some, such as IP Multicast, are in limited commercial or research deployment. Most of these solutions require special router hardware or software to achieve data sharing without overwhelming the participating networks. Even if a standard solution were selected today, it would take years, maybe even decades, before the entire Internet infrastructure could be completely retrofitted with the new technology.

Additionally, the solutions proposed in this area, in an attempt to conserve bandwidth, are not constructed with reliability as a concern. Using spanning trees among the routers involved, any node failure can temporarily partition the collaborative session.

45.2.4   Swan: Regular Random Graphs for Multicast

The Swan multicast solution uses a connectivity of random regular graphs, as shown in Figure 45.3. A regular graph is a graph whose nodes have a fixed number of neighbors. Each node in a Swan system is logically connected with exactly r nodes, with r being set as a parameter with minimum value 4. The central innovation in Swan is weaving participating processes together into regular random graphs[11,12] using generally available point-to-point network protocols. The resulting logical connectivity is called *Swan fabric*. Swan fabrics are families of random regular graphs generated by the addition and deletion of single nodes with minimal disruption.

Swan was developed to meet all of the requirements and goals laid out in Section 45.2.2. As particular families of random regular graphs, the Swan fabrics have near-optimal performance. The message latency grows as $O(\log_{r-1} N)$ in the number of participants N. This logarithmic latency makes sessions with

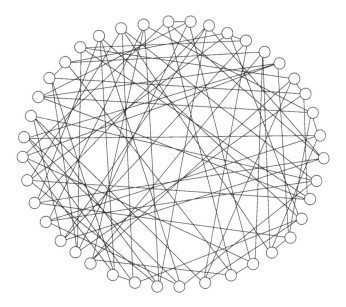

FIGURE 45.3 A Swan fabric on 40 nodes. Swan Heralds are represented as nodes. Connections to other participants are represented as edges. In this example, each participant has four connections to other participants. This type of distributed logical network is necessary to optimally support the synchronous components for online interaction. These fabrics are as flexible and resilient as the communities of participants they support are.

hundreds or thousands of participants feasible. Because messages are forwarded over multiple routes, there are no bottlenecks to degrade the effective session throughput. The multiple routes and multiple connections provide high robustness and reliability compared to trees or centralized solutions. Swan supports strong authentication and encryption. Using TCP/IP for the point-to-point connections, Swan can be readily adopted by both existing and emerging platforms. The Swan administration is completely distributed among the participants, which may join, depart, or even fail, at any time and in any order, without affecting the others.

Several additional innovations are made in Swan to provide easy, quick, and reliable multicast for highly interactive and widely distributed applications. Using existing Internet protocols in noninvasive ways, no special system administration, no operating system modifications, and no special hardware are required. All computers can participate, without requiring root access, daemons, kernel modifications, or the addition of "well-known" port numbers. Swan provides a session-level API with an intuitive publish-and-subscribe paradigm, hiding as much as possible the administration and session management.

45.3 The Swan Technology

This section provides a description of the concepts on which Swan is relying, requirements that must be met in order to start using it, as well as an introduction to the Swan Application Programmer's Interface (Swan API). Also included in this section are some notes on programming with Swan.

45.3.1 Swan Concepts

Swan avoids synchronization difficulties by making use of the "small-world effect,"[3,4] using only local knowledge to maintain global properties of reliability and scalability. The small-world effect has many advantages for large-scale Internet communications (also known as wide area networking). Furthermore, each program needs only to have a small, constant number of these direct connections to other participating programs. In addition, there is no limit to how many programs can participate in this "world."

Joining an information-sharing session is automatic. Processes can join and depart at any time. Although openly accessible, joining a Swan session is restricted to those processes sharing the Swan code base and aware of the correct channel designation. Swan is data-agnostic and network-agnostic, and thus supports additional security features such as virtual private networks and encryption.

Each node in the graph represents a *Swan Herald*, an instantiation of the top-level class in this publish-and-subscribe paradigm. Each edge in the graph represents a point-to-point connection between Swan Heralds. Participating Swan processes instantiate Swan Herald objects to act as their communication agents. Swan Heralds exchange messages regarding application events with other Swan Heralds on the same *logical channel*. These peer Swan Heralds then publish these messages within their own processes.

The graph cannot become r-regular until it has at least $r + 1$ members. The graph is referred to as being in the *small regime* as long as it has fewer than $r + 1$ members. In the small regime, all participants are completely connected. Swan Heralds make the transition seamlessly back and forth between the small and large regimes.

Swan fabrics use random r-regular graphs for the communication topology, with r even and $r \geq 4$. Recall that a graph is r-regular if every node is incident to exactly r edges. Each node participating in the Swan session uses $r + 1$ sockets, r being internal to the fabric, and "1" being the *shingle*, a socket for handling control messages external to the fabric. Currently, Swan is implemented with the parameter $r = 4$. The Swan fabrics are extended to include new participants and repaired around departing participants with minimal and local disruptions. Swan participants are true peers — self-contained members of a completely self-sufficient communication fabric.

45.3.1.1 Flexible Session Management

Swan fabrics provide reliable and easy-to-use logical networks. The nodes of the fabrics, the Swan Heralds, are very flexible. They can coexist in great numbers on a single device, and they can even coexist in the same application without any interference. If they share the same logical channel, the Swan Heralds are woven into the same fabric. Otherwise, if their logical channels are not identical, two Swan Heralds belong to different fabrics and are completely unaware and entirely independent of each other.

Two 32-bit keys designate the logical channel for a Swan Herald. The first key is the *channel type*, used to distinguish among applications or versions of applications. This key indicates compatibility for understanding the event messages being shared. The second key is the *channel instance*, which separates concurrent exercises of the application, such as two distinct sessions of the same online meeting software. Thus, a single application can support multiple fabrics for different communities or different levels of control, and multiple applications of various types can run orthogonal to each other on a single device without any collisions.

With Swan, the rendezvous points are provided as an external list, which can be arbitrarily long and can be edited. The length of the RP list is less important than the probability that some device high on the list has a Swan Herald currently participating on the right channel. Unlike the systems surveyed by Banerjee and Bhattacharjee,[8] Swan allows every node to act as a rendezvous point for the session.

45.3.1.2 Extending and Repairing the Swan Fabrics

To extend a Swan fabric in the general case, there is an existing r-regular graph on N nodes, and an additional node arrives. The newcomer needs to contact any of the existing nodes that are listed as rendezvous points (RPs). The newcomer sends to the RP's shingle a request to join. Once contacted on its shingle, the RP initiates a random search for $r/2$ disjoint edges. The newcomer is interposed on each of these disjoint edges. The result is an r-regular graph having $N + 1$ nodes. Visually, it looks like the newcomer has clamped the $r/2$ edges together. This process is illustrated in Figure 45.4.

How can Swan find random edges with only local knowledge? The newcomer has contacted one Swan Herald on its shingle. This Swan Herald initiates $r/2$ drunk walks through the existing graph. If these walks are long enough, then the nodes at the ends of these walks are almost uniformly distributed across the graph. Each of these $r/2$ nodes thus selected picks one of its edges at random and offers it to the newcomer,

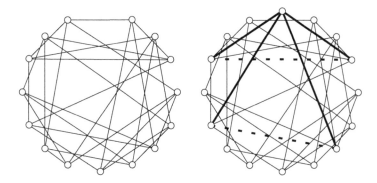

FIGURE 45.4 The process of extension of a Swan fabric. This example illustrates a sixteenth node joining a Swan fabric consisting of 15 nodes. The two dotted edges are selected at random and pinned together at the new node, shown in bold. This modification uses only local knowledge and is minimally disruptive. The resulting fabric remains r-regular. Its time diameter grows logarithmically following the number of participants.

whose address was part of the message, sent on the drunk walk. The newcomer checks that the edges are disjoint.

In fact, each half-edge corresponds to one of the internal sockets maintained by some Swan Herald. This Swan Herald disconnects this socket from its current neighbor and reconnects to the newcomer. Extending the fabric in this way involves only local operations, and it is minimally disruptive—introducing a newcomer into an r-regular graph requires a minimum of r disconnections and reconnections.

Swan accommodates both *graceful* and *ungraceful* departures. In a graceful departure, a Swan Herald, instructed by its application to disconnect from the fabric, alerts its neighbors to the impending repair. In an ungraceful departure, a Swan Herald dies for any number of reasons but without sending any control messages. In either case, the remaining Swan Heralds work to repair the fabric quickly.

When disconnecting gracefully, a Swan Herald sends a message to each of its neighbors, alerting them to its departure. This message lists the departing Herald's neighbors. To facilitate smooth reconnection, the first pair of Swan Heralds in the list attempts to connect directly to one another, while the second pair does the same. Barring successful reconnection, the first and third, and second and fourth, attempt direct connection.

When a Swan Herald departs ungracefully, it fails to alert its neighbors. Instead, the loss is discovered by its neighbors either immediately or, at the latest, when they next attempt to send information to the departed Herald. Upon discovering the loss of their neighbor, each of these Swan Heralds will broadcast a general connection request through the graph. These Heralds, which are lacking a connection, will contact each other and pair up for the repair. Although it requires more effort than in a graceful departure, the nodes lacking a connection will quickly find each other and reconnect properly.

Swan fabrics are reliable in the face of multiple simultaneous communication link and process failures. With high probability, the fabrics are r-connected,[13] meaning that the simultaneous loss of at least r nodes is required to separate some set of nodes from the session. For example, if any node loses all r of its neighbors, then this node will be separated from the session. If sets of nodes are deleted at random, it is in fact much more difficult to partition the session. In some tests, a quarter of the participants were deleted instantaneously without notice, and the session remained connected and was able to repair itself.

45.3.2 Swan Requirements

Currently, Swan employs TCP/IP network connections, an accepted Internet standard that provides reliable and ordered delivery along each connection. The details of TCP/IP are encapsulated in a telephone abstraction, therefore enabling easier maintenance and modular protocol replacement by other point-to-point protocols, such as UDP. Requirements for this technology are met in the majority of the installed base of computers. Most users already have TCP/IP-compliant network devices installed, and are most likely using some version of either UNIX or MS Windows. As for developers, Swan libraries are compliant with existing C/C++ standards and integrate seamlessly with existing code.

45.3.3   Notes from Swan-Enabled Applications

In general, programmers do not yet have much experience in writing completely distributed applications. The communications architecture of client/server is often reflected in the application architecture; the server becomes a processing hub in addition to its role as a communications hub. To benefit from Swan's efficiencies, the application must minimize these processing chokepoints. The server side of an application has provided a convenient place for addressing real challenges in supporting interactive online communities.

Swan has forced the distinctions among the roles that servers traditionally play: the communications server, the processing server, and the content server. The content server can participate as a node in the Swan fabric. Swan removes the communications server. To make best use of Swan, an application should also eliminate the processing server. To eliminate the processing server, the real challenges must be resolved in a distributed instead of centralized fashion. Despite the current disposition toward centralized strategies, reinforced by the client/server culture, the underlying challenges will yield to distributed strategies as well. This section offers notes on a variety of programming experiences with Swan.

45.3.3.1   Data Authority and Persistent Data

Although the flurry of interaction in a distributed application can be generated by all the nodes, some one participant will have particular responsibility for some aspect of the data. The content server for an online game must maintain the status of ranks and treasures; the lead engineer is responsible for the results of an online design review. Traditionally, this authority resides with the server. In a Swan session, a content server, sitting as one additional node in the fabric, can enforce this same functionality.

45.3.3.2   Managing the Level of Interaction

Different types of online interaction have different levels of interaction, from the sole leadership during a presentation to the free-for-all of an online game. Leadership in a Swan session can be managed through tokens, designed by the application developer. For an online game, there may be no leadership token. In contrast, in an application for online meetings, the token may have three settings: leader, follower, and independent. The leader has the authority to set the current state (view, annotation, etc.). A follower has a view of the state set by the leader. A participant can wander from the current state, for example, changing pages and annotations; he is independent to modify his local state. If he then becomes a follower, he returns to the leader's state; but if this independent node becomes the leader, then his local state becomes the official state for the application.

Control of the leadership token can be closely held, as during a presentation, or it can be made available, for example, to handle questions at the end of the presentation. The users of an online application tend to develop their own social rules, and these can be accommodated in the application-level design of leadership.

45.3.3.3   Access Security

Swan provides four controls on access to a Swan session. The list of rendezvous points is one control—to join a Swan fabric, one must find the Swan fabric. The second and third controls are the two channel keys. Exactly these two keys are required to find and to join a Swan fabric. Finally, if the intruding process does not include the Swan libraries and the schema for this application, it will be unable to participate in the fabric and its requests for connections will be rejected.

For confidential sessions, the list of rendezvous points and the channel instance key should be tightly controlled. If all sessions for this application are confidential, the channel type key can also be tightly controlled. For public sessions, the rendezvous points and channel instance key can be posted on a Web site.

At this time, if a participant has the application, a valid list of rendezvous points, and the channel instance key, then he can join the session. His participation may still be unwanted, as in a spammer on a publicly available demonstration program. There is currently no mechanism for denying access to unwelcome newcomers. Further authorization could be implemented at the application layer so that current participants could remove abusive participants from the current session.

45.3.3.4 Message Security

Swan also provides security in its messaging. First, Swan reveals an `Encrypt()`/`Decrypt()` pair that can be overloaded by the application developer. Additionally, the messages are multicasts of application events such that small samples of the message traffic will provide very little information. The messages are not snapshots of the application state; they are small modifications to the state. Without the broader context, many of these messages will be almost meaningless. To interpret an intercepted message, even after successful decryption, a snooper would have to infer the schema for this message. The schemas are coded into the Swan applications and are not transmitted along with the event data.

45.3.3.5 Consistency and Its Enforcement

In interactive online applications, disputes arise when two or more nodes issue conflicting events. For example, in an online game, suppose three players reach for the same treasure. In a server-based game, the server decides which player receives the treasure and broadcasts this result to all the players. With Swan, these conflicts must be resolved in a distributed but fair way. A robust and flexible approach is to use random signatures as tiebreakers. Each node attaches a random number to its event asserting that it is reaching for the treasure; every node in the fabric has a consistent way of evaluating these random numbers to determine which player will get the treasure. For a window of time that is just longer than the session latency, the treasure remains unclaimed. Because every node has a consistent way to decide between conflicting assertions, cheating is easy to detect.

To monitor the integrity of the application state, processes can participate in the session to check on the events asserted by the participants. In a client/server model, the server checks the validity of assertions before propagating the results to the clients. With Swan, we use a referee model. In this model, interaction is allowed to continue while the referee processes the validity of the assertions. These referees can propagate corrections for small inconsistencies, and they can suppress further participation for major infractions. This referee paradigm again underscores that although the server has been relieved of its communication duties, it can still play a distinguished role in the application.

45.3.3.6 Finding Sessions: Rendezvous Points

Potential participants must be able to find an active session. *Rendezvous points* direct newcomers to active sessions of the application. These rendezvous points (RPs) need not participate in the session themselves. Web sites often act as RPs. The Web site matches the newcomer to a session and leaves the newcomer the responsibility for further session management. Portals behave similarly, although they may keep a light presence in the session just to monitor participation, whether for subscriptions, advertisement, or refereeing. RPs can also be transmitted less publicly through e-mail, file attachment, or GUI form.

45.4 Programming with Swan

This section describes the steps a developer needs to take to establish the flow of event data between peers in a fabric. Simplified code taken from one application is used to illustrate the process. The programming steps needed to establish communication with other peers include defining the schema for objects representing shared events, creating the *Registry* and registering *Items* corresponding to the shared events, and preparing to subscribe by creating the Swan Herald. In addition, methods are written to prepare event data for sending

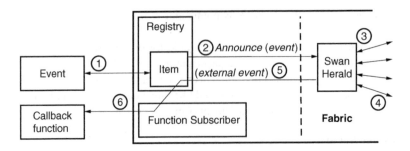

FIGURE 45.5 An illustration of the Swan API inside one client. The central feature is the Registry — a topical collection of items. Interested code may subscribe to registry items of interest by becoming subscribers to those items. These items can be updated by announcing new values of events for them, causing interested subscribers to be alerted to the updates. A typical scenario of interaction starts locally: (1) application changes its state and sets attributes of the appropriate Event; (2) Item announces this Event; (3) Swan Herald broadcasts the Announcement to the fabric. The latter three actions happen on the other end of the fabric; (4) Swan Herald receives Announcement from the fabric, (5) extracts relevant Event data and passes it over the FunctionSubscriber, (6) thus calling the predefined callback function.

and to respond upon receiving event data from other nodes. A brief overview of the publish-and-subscribe interface is depicted in Figure 45.5; Figures 45.6 and 45.7 illustrate relevant segments of the complete process.

45.4.1 The Swan API

The Swan API was developed as a general-purpose publish-and-subscribe interface so it can be used for online games, collaborative work, distance learning, and emergency networks. Both developing an application with this API and using it are very simple because its network communication is completely transparent with virtually no administration required. Furthermore, the Swan API is real-time responsive and it self-optimizes network data flow to achieve the most efficient organization available.

Swan uses a number of components to provide a flexible communication mechanism. Underneath its publish-and-subscribe API, Swan uses asynchronous message passing. Participants in any online session

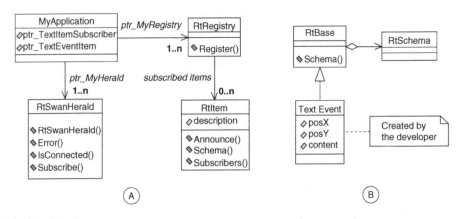

FIGURE 45.6 Crucial segments of Swan-enabled application architecture. (a) Application creates one or more instances of the RtSwanHerald class, used to communicate with other participants present in the session. Also created are one or more instances of the RtRegistry class, used to register all Items that might be of any interest to the user. (b) Every event needed for the application is derived from the RtBase class; hence, it has a schema and a unique type identifier. Developer decides on event naming, methods, and attributes.

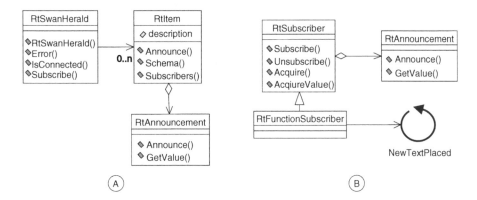

FIGURE 45.7 Segments of application architecture needed to announce and to process an Event. (a) RtSwanHerald is subscribed to some or all of the registered RtItem class instances. When other participants need to be informed about a change of an item, a call to the RtItem class' Announce() method creates an RtAnnouncement and sends it to the fabric. (b) Once an Announcement related to some of user-defined classes is received, it is passed to a callback function subscribed to handle that particular type.

have a number of topics of mutual interest. They can subscribe to any of the topics. Any action related to the subscribed topics is described using an *Event*, which contains all the information relevant to that action. In most cases, the information contained in events provides incremental changes to the current application state.

All of the topics that a participant has subscribed to are kept locally as distinct *Items* in the *Registry*. Each Item is associated with both an Event and the callback function for handling incoming events of this type. When a user triggers a shared application event, the data for this event is recorded in an Event object and published through *Announcing* the corresponding Item. All members of the group who are subscribed to this Item will receive this *Announcement* and invoke the corresponding callback function.

An application developer writes the class for a given Event and the callback function that implements action performed to handle this Event once it is acquired from the Announcement. A *FunctionSubscriber* provides a handle on the callback function.

All network communications for the multicast are in the hands of *Swan Herald*, an instantiation of the top-level class in the Swan API. Participating Swan processes instantiate Swan Herald objects to act as their communication agents.

45.4.1.1 Registries

A Registry is a named topical collection of named Items. A module of code accesses registries by *registering* them by name. The first call to register a registry creates an instance of the registry in the process' global namespace. That and subsequent calls to register the registry return a pointer to the registry of the given name. In this manner, several logically separate sections of code can attach to a registry without synchronization.

Each registry item may or may not have a corresponding value. For an item with a value, a globally accessible copy of the most recently announced value is available, referred to as the item's *state*.

Once access to a registry is obtained, items can be registered to the registry in a similar manner. Each item can have data, whose format is described by its *Schema*. The first section of code to register an item with its schema defines the characteristics of that value (memory size, etc.) and can optionally provide an initial state value. Each registry item with a schema maintains a state value reflecting the latest announced value for the item.

A special item created automatically for each registry is named "All." These items cannot be announced directly. Instead, subscribers to a registry's "All" item receive all announcements made for any of the registry's items.

45.4.1.2 Announcements

Announcements of registry items send data and control instances to the item subscribers. By analogy, if a registry can be likened to a newspaper, its items correspond to names of the newspaper's regular columns, and each announcement corresponds to a new issue of the article. Unlike a newspaper, new issues of the columns are not coordinated into an issue of the newspaper; rather, all of the articles are issued independently. The values for each update announcement are also available by inspecting the queue of announcements provided for each subscriber.

An announcement can be made in one of two ways. The simple way is to announce a registry item, providing a value if the item has one. The more expressive way is to create an announcement for a registry item and then announce the announcement. The former way is easier and handles most situations well. The second method allows timestamps and other designations to be attached to the announcement, and it allows announcements to be grouped into atomic sequences called *reports*. Either method allows specifying a single subscriber to exclude when delivering the announcement. This allows a logical section of code that subscribes to and authors an item to filter out its own announcements, so that its callback function is not invoked.

45.4.1.3 Subscribers

To receive announcements for a set of registry items, a section of code must subscribe to those items by instantiating a *subscriber* and instructing it to subscribe to the desired items. The subscriber can subscribe to as many items as it likes, from as many registries as it likes. Subscribing to a registry's "All" item causes delivery of announcements for all of the registry's items.

Each subscriber has a queue for receiving announcements. Each announcement can be examined for meta-information, such as the registry, item, schema, etc. corresponding to it, and for the announcement-specific information, such as the value or timestamp.

Announcement delivery takes place during the event-loop. The subscriber determines when and how to deal with the announcements. The subscriber has a callback method that is called when each announcement is delivered. The subscriber's callback function may choose to examine the announcement queue when called or leave the announcement for later. Alternatively, the subscriber can disable the callback invocation and examine the queue, perhaps periodically. The values for each update announcement are also available by inspecting the queue of announcements provided for each subscriber.

Announcements remain in a subscriber's queue until the subscriber explicitly *acquires* them. Once acquired by all interested subscribers, an announcement is garbage-collected, in which it is deleted and its memory footprint freed. The following assumption is made: once a function has completed an acquire action, the announcement is no longer available. If some information might be needed later, this should be copied from the announcement. When several announcements are placed at one time in a subscriber's queue, the subscriber's callback function is generally called once for each announcement. A subscriber can choose, however, to acquire as many announcements currently on its queue as it likes.

45.4.1.4 Function Subscriber

The base subscriber class defines a method, called `Response()`, as the callback to handle the arrival of announcements. The default response method, while not abstract, is useless—it removes announcements from the queue and ignores them. It is primarily a placeholder for derived classes to overload with a more useful callback.

The *function subscriber* is just such a derived class. Upon creation, each instance of a function subscriber is given its own pointer to a function to use for the callback and an optional pointer to context information. This context information could be used, for example, to distinguish differing roles handled by the same function. Note that this also allows a function subscriber to use a C function for its callback.

45.4.1.5 Herald Subscriber

Heralds are the key to performing interprocess communication. By creating a herald subscriber, announcements to the registry items it subscribes to are transparently mirrored to other processes with cooperating heralds. Swan Heralds are invisible to each other unless they belong to the same fabric. If they belong to different fabrics, the Swan Heralds are entirely independent of each other. A single application can support multiple fabrics for different communities or different levels of control, and multiple applications of various types can run orthogonal to each other on a single device.

To cooperate, the heralds must agree on an *implementation* and a *channel*. The implementation corresponds to an underlying communication protocol, which is specified by the type of herald selected. For example, one might choose to use an RtUdpHerald, which communicates over local area networks using a reliable broadcast datagram (UDP) protocol. Currently, for wide area network broadcast, the RtSwanHerald uses the TCP network protocol.

The channel is specified by two integers: the channel *type* and the channel *instance*. The channel type is compiled in, and is used to distinguish between applications. The channel instance is agreed upon by the processes while exercising the application. The channel instance is passed as a runtime parameter from the application to Swan. The channel instance separates one group of cooperating processes from an identical group of cooperating processes sharing the same implementation and channel type. For example, one group of players in a multiplayer game can coexist with another group of players on the same network by having different channel instances.

Once a herald is created, it will automatically share announcements about the items to which it subscribes. The announcements will be shared with all its peer heralds of the same implementation and channel. Peer heralds, when receiving an announcement through their communication pathway, announce the announcement directly to the same registry item in their own process space. A herald receiving an announcement for a registry item that does not yet exist in its process space will register the registry and registry item as necessary, subscribe to the item, then locally post the announcement related to the newly subscribed item. Thus, the registry items shared among a group of peer heralds are the union of all such items subscribed to by any of the heralds.

Another important consideration for heralds is that the schema definitions for all types shared among the heralds must be available to all of the heralds to allow proper creation of registry items. The schema type identifiers are communicated and used to discover the proper schema instance within the process' global memory.

45.4.2 Programming Example

This application enables users, among other things, to place some text at a desired position on the screen, and the same text will appear on the other participants' screens at the same position. Although the code presented here deals with plaintext-related data, the sequence of programming steps is fundamentally the same for any other type of event.

The main module instantiates the RtSwanHerald class, which corresponds to a node in the Swan fabric. This instance is responsible for all communication activities: joining the group, sending and receiving the events. The main module also instantiates an RtRegistry, which is a collection of the items that are declared and used in the application.

```
RtSwanHerald *ptr_MyHerald;//pointer to the Swan Herald
RtRegistry *ptr_MyRegistry;//pointer to the Registry of Items
```

Every Event tracked in the system must have an Item/Subscriber pair associated with it. As stated above, in this example, text-related Event is used. Both Item and Subscriber are declared, respectively, as instances of the RtItem and RtSubscriber classes.

```
RtItem *ptr_TextEventItem;//item correspondent to TextEvent class
RtSubscriber *ptr_TextItemSubscriber;//subscriber for ptr_TextEventItem
```

45.4.2.1 Creating the Event

The first step in developing a Swan-based application is to define the events that will be announced. Events contain all the information related to changes in application state that must be passed to other participants. The events are coded as C++ classes, derived from the RtBase class that provides iterators, schemas, etc. Swan provides streaming back and forth over the common data types. The streaming is platform independent, allowing Swan to work across diverse operating systems. In this example, an event that describes entering text to be displayed to other participants is used. This text-related event contains data fields for the author's ID, the text content, and coordinates, as shown in the following code fragment.

```
RT_TYPE(TextEvent, UNIQUE_NUMBER) // index TextEvent for broadcast
class TextEvent: public RtBase // inheriting schema macros and methods
{
public:
...
RT_STANDARD(TextEvent)// this macro declares the RtBase methods
...
private:
int authorID; // identify the creator of the text message
// Properties: 2-d position and contents
int posX; // X coordinate of the text
int posY; // Y coordinate of the text
string contents; // contents of the text
...
};
```

The Macro RT_TYPE is used to define unique type for event defined. UNIQUE_NUMBER is a number set by the programmer, and must differ for all event classes in a Swan-based system. Events must be derived from the RtBase class so they can be sent using the Swan API. Macro RT_STANDARD is used to declare RtBase methods for the class RtBase.

45.4.2.2 Creating the Registry

The Registry contains schema/name pairs for all types of events to be shared. Swan's RtRegistry class provides a conditional constructor, which finds a Registry by name and creates it, if it does not already exist.

```
ptr_MyRegistry = RtRegistry::Register("MySwanApplication");
```

Once the registry is created, everything that is of interest for the application should be registered as an Item in the Registry. The Registry creates an RtItem for each of application-event classes created in the first step. To include this event in our Registry, Swan again provides a conditional constructor, which finds the event if it has already been registered or otherwise creates it.

```
ptr_TextEventItem = ptr_MyRegistry->RegisterItem
    ("New text placed", &TextEvent::Schema());
```

The Schema() method, inherited from Swan's RtBase class, is used to record the data field types for a TextEvent with the corresponding Item. Used as the second argument in this conditional constructor, this method defines the schema for a new event and checks consistency for events that were already created by other nodes. The return value belongs to Swan's RtItem class. This class matches the event class in the client with the Swan Herald and the FunctionSubscriber, as subsequently described.

45.4.2.3 Preparing to Subscribe

The next step is to define the callback function that will process received events. In this case, the callback function is a function that will be called when a new instance of the TextEvent arrives. This function has two parameters. The first parameter is the address of the object that is capable of handling instances of this particular event type. This example uses the imaginary class type called `Painter` that, among others, has a method called `DrawString`. The second parameter is a reference to the subscriber that contains the new event in its queue. Our prototype of the function has a void pointer to a context, for flexibility in processing the given type and address of RtSubscriber.

```
void NewTextPlaced(void *context, RtSubscriber &subscriber);
```

We next create the callback mechanism, preparing to handle events that are published. Swan's RtFunctionSubscriber class matches published events to the callbacks that handle them in each node.

```
Painter *pPainter = Painter::Instance();
ptr_MyTextEventSubscriber = new RtFunctionSubscriber
   (NewTextPlaced, pPainter);
ptr_MyTextEventSubscriber->Subscribe( *ptr_TextEventItem);
```

An assumption made in this part of the code is that the application needs only one `Painter` instance; hence, the class is treated as if it were Singleton design pattern compliant,[14] and that instance is accessed through the pointer returned by the static `Instance()` method.

The `NewTextPlaced()` function handles the event data from other nodes, as returned by the Swan Herald. We registered this callback with our FunctionSubscriber above.

45.4.2.4 Creating the Swan Herald: Subscribing to the Items

Swan Heralds are the nodes in the communication fabric, as shown in Figures 45.3 and 45.4. In this example, one Swan Herald is created. The constructor takes as arguments a set of 32-bit keys and a set of rendezvous points. The keys separate Swan fabrics from each other and provide the fabric with some security. Joining an existing fabric requires this same set of keys as the one used by the other participants.

```
ptr_MyHerald = new RtSwanHerald(FabricKeys, RendezvousPoints);
```

As mentioned above, the rendezvous points are a set of IP addresses or resolvable Internet addresses where this client should look to join the fabric. The list of rendezvous points can be as short or long as the programmer desires. Shorter lists help concentrate the search for a fabric; longer lists provide more places to look for a fabric. After a successful return from the constructor, the Swan Herald has joined the fabric. Next, the Swan Herald subscribes to the Items that it wants to share with others. Typically, it will subscribe to all Items in a given Registry. Fetching a pointer to all of the Items in a Registry is accomplished through a single invocation of the `AllItem()` method of the Registry class.

```
ptr_MyHerald->Subscribe( *ptr_MyRegistry->AllItem());
```

45.4.2.5 Publishing and Reacting

The above steps initialize Swan. Once initiated, Swan works with the event-loop handler. Local events are published and incoming events trigger the corresponding callback. When there is a need to send a new value for an Item, we create an instance of corresponding Event class (if not available already), set the data values, and announce it. For example, to publish newly entered text and its position, our client fills in local data for the `TextEvent` and invokes the `Announce()` method of the corresponding Item. This method belongs to the RtItem object returned when we registered `TextEvent` as an event class in the Registry.

```
TextEvent *pTextEvent = new TextEvent;
// set pTextEvent's attributes before announcing
...
ptr_TextEventItem->Announce(pTextEvent, ptr_TextItemSubscriber);
```

The argument pTextEvent is a TextEvent containing the data for the text being published. Among the attributes mentioned above, it includes the author's ID. The second parameter is the subscriber that will not receive this event. Although this parameter can be omitted, it is typically given the value of the local subscriber to the item corresponding to the event being announced; this prevents the local application from reprocessing local events. Methods inherited from the RtBase class handle the streaming and unstreaming of this data.

On the other end of the fabric, once an Announcement related to some of the subscribed items is received, the callback function is invoked, with the application context and the event data passed as parameters.

```
void NewTextPlaced(void *context, RtSubscriber &subscriber)
{
TextEvent *pTextEvent = (TextEvent*)subscriber.Acquire()→GetValue();
// re-casting context argument
Painter *pPainter = (Painter*)context;
// invoking Painter's method for handling the text received
pPainter→DrawString(pTextEvent→content, pTextEvent→posX,
  pTextEvent→posY);
}
```

Avid readers will note that using the Swan API elegantly encapsulates all network-related issues in a few function calls. Any message related to a change in values of shared data is delivered to all participants of a session, and the callback function that handles that type of data is automatically invoked. All of this makes for comfortable adoption of this multicast solution.

45.5 Swan Performance

For the purpose of sharing information across a communicating group, the time it takes for a message starting from an arbitrary participant to reach all other participants is the *latency* of that message. The worst-case latency among all possible pairs of initiating and receiving participants is called the *time diameter* of the group. In an abstract model of Internet communications, the time diameter is represented by the maximum of the number of hops separating pairs of nodes. This graph-theoretic diameter is the maximum number of point-to-point Internet transmissions required for message delivery to every node.

The diameter for a Swan session varies with the number of participants in the session. In the small regime, where all nodes are directly connected to each other, the diameter is 1, meaning that messages are immediately exchanged. For r-regular random graphs on N nodes, there is high probability that the following bounds[11] are applied to the diameter d:

$$\lfloor \log_{r-1} N \rfloor + \lfloor \log_{r-1} \ln N - \log_{r-1}(6r/(r-2)) \rfloor + 1 \le d$$
$$d \le \lceil \log_{r-1} N + \log_{r-1} \ln N + \log_{r-1} 2r \rceil + 1 \tag{45.1}$$

The first line in Equation 45.1 defines the value for the lower bound, while the second defines the value for the upper bound. For the case when $r = 4$, formulas for upper and lower bounds of the time diameter, given in Equation 45.1, can be simplified to:

$$\lfloor \log_3 N \rfloor + \lfloor \log_3 \ln N - \log_3 12 \rfloor \le d$$
$$d \le \lceil \log_3 N + \log_3 \ln N + \log_3 8 \rceil + 1 \tag{45.2}$$

Because the r-regular graphs underlying the Swan fabrics are nearly random,[13] these bounds should apply, and experimental data confirms that the diameters of Swan fabrics fall toward the middle of the given range. That is, treating all Internet connections as approximately equal, Swan fabrics provide latencies that grow logarithmically in the number of participants. This means that for every additional hop that the application's performance can tolerate, the number of participants can more than double.

TABLE 45.1   Upper and Lower Bound for the
Time Diameter

Nodes	100	1K	10K	100K	1M	10M
Upper	9	11	14	16	18	21
Lower	4	6	8	10	13	15

Representative values for the upper and lower bounds on the diameter, for the base case of $r = 4$, are shown in Table 45.1. A sample fabric having a notable number of nodes and a low latency is shown in Figure 45.8.

The Swan fabrics are extended and repaired randomly. This approach is selected because these random graphs greatly outperformed all graphs created using deterministic local rules. This was verified on a number of occasions during Swan development. The deterministic rules tended to be more disruptive and led to much higher latencies.

Bandwidth is the limiting resource for a Swan fabric. The relevant constraint at the node labeled i is:

$$c_i \mu(t) N \leq B_i(t) \tag{45.3}$$

where c_i is the number of active connections at this node, $\mu(t)$ is the average volume of message traffic per node per second, N is the number of nodes, and $B_i(t)$ is the bandwidth available to this node.

The order of the left-hand side, in terms of the number N of nodes, depends in part on the average message traffic volume $\mu(t)$. If, as in multiplayer games, the application has an approximately constant average traffic volume per node, then local bandwidth is consumed in proportion to $c_i N$. For applications such as online classrooms, the average message traffic volume per node does not vary significantly with the number of participants; and thus for this class of application, the demand on local bandwidth is dominated by c_i and by temporal variations in the traffic volume. These applications, games and classrooms, mark two extremes on a spectrum of *leadership*, as discussed in Section 45.3.3.2. Regardless of the concentration of leadership in the application, the message traffic can be punctuated by bursts of activity, due to firefights in games or delivery of richer content to the classroom. Local bandwidth consumption will depend on how these bursts of activity correlate in time, which we earmark by writing average message traffic volume as a function of time $\mu(t)$. For demanding applications such as computer games, there has also been a lot of effort put into message culling, reducing the amount of information $\mu(t)$ that must be sent to the i-th participant. When a node sending or forwarding messages detects that one of its neighbors has fallen

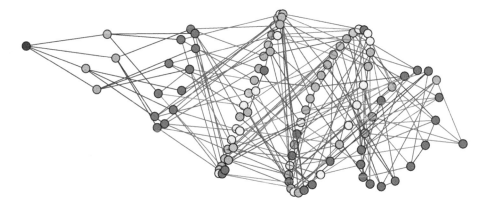

FIGURE 45.8   Wire-frame model representing a 100-node Swan fabric. Nodes are visually distinguished by broadcast distance from the leftmost node. Time diameter of this fabric is six (meaning that the two farthest nodes are six edges away one from another). For the sake of clarity, all of the graph nodes are presented identical in size, while some of the graph edges are intentionally omitted.

behind, the sending node uses priority flags or expiration tags to remove messages from the queue of outgoing messages for this lagging neighbor.

Depending on the concentration of leadership, the consumption of local bandwidth varies from c_i to $c_i N$. For servers, c_i equals $N - 1$, and bandwidth is consumed at the server on the order of N to N^2. In contrast, for every node in a Swan fabric, c_i equals $r - 1$, and the bandwidth consumption decreases to order 1 to N for this same range of applications. The constraint in Equation 45.3 makes explicit the cost of redundancy and encourages us to keep $r = 4$ for Swan fabrics.

The average traffic $\mu(t)$ and available bandwidth $B_i(t)$ are instantaneous values. These vary over windows of time. In measuring these, one would choose a window of time that represents an acceptable latency for this application. For computer games with intensive interaction, this window would be smaller than the window for an online classroom. The constraint in Equation 45.3 indicates a trade-off between the size of a session and the level of interactivity provided by the application, in the face of fixed bandwidth. As more participants join, their average participation must decrease. Conversely, if the component requires a high level of activity by each participant, then the number of participants must remain low.

On the other hand, as local bandwidth $B_i(t)$ increases, the constraint eases, allowing higher levels of participation by larger groups. Depending on the connectivity layer, a local increase in bandwidth may not help the session-level performance. Participants with low bandwidth may slow the entire session, or they may be prohibited from joining. For some synchronous components, the design process fixes the minimum level of interactivity and the maximum number of participants. As a result, a participant must meet given bandwidth requirements to participate. Some logical networks slow down to the lowest common bandwidth. This is true of the daisy chain, hub and-spoke, and other trees. If some node, such as the hub, serializes the communications from the other nodes, then the session performance will slow to the lowest bandwidth present.

This constraint also quantifies the cost of additional connections. For a server-based solution or a hub-and-spoke solution, the number of connections is $N - 1$ for the hub and "1" for all other participants. We can see that the bandwidth at the hub is used up on the order of N^2 in the number of participants. This is why server farms automatically partition sessions into small groups; thus, maybe 20,000 participants are online but they may interact only in groups of 20.

In the completely connected small-scale solution, every participant has $N - 1$ active connections to other participants. Therefore, as with the hubs, local bandwidth is consumed on the order of N^2 and the constraint in Equation 45.3 is fatal. Completely connected sessions support very small groups.

The cost of using the Swan multicast strategy is the message-traffic multiplier (c_i in constraint 45.3) at every node.

45.6 Horizons for Further Research

The Swan technology has many advantages, at the cost of a few limitations. Swan gains most of its advantages by combining r-regular connectivity with redundant message flooding. Together, these provide a high degree of reliability and optimal performance. By accepting an up-front cost of replicating message traffic at every node, Swan is able to provide an order of magnitude improvement in bandwidth consumption for the sender. However, as an optimal multicast solution, Swan is not optimized for broadcast or one-to-one transactions. Given a suitable application, Section 45.2 discussed programming challenges. This section discusses some shortcomings in the current implementation of the underlying technology.

Because multiple copies of each message flow through every node in a Swan fabric, efficiencies in creating messages directly impact performance. Already the more demanding online applications manage the size and frequency of messages. These techniques are immediately applicable for Swan fabrics. For highly interactive applications, such as online games, the messages contain the small differences that happened at one node, so the others can update their states. For example, when a player changes position, only his new position data will be sent over the network. All recipients will update this player's position with information received in the message, while other attributes, such as size and color, will be retained from the previous state. Sending only the differences makes efficient use of the bandwidth for the active participants.

However, when a newcomer joins in, he needs to obtain complete information about the current state. These "cold starts" must be managed outside the Swan fabric.

Swan provides multicast for arbitrarily large groups of peer applications. The cold-start problem touches on two general issues: (1) transferring large blocks of data to participants and (2) sharing information not intended for all participants. Swan is not optimized for sending large blocks of data to the peer community. The current implementation of Swan sends data out in single message transfer units (MTUs). After the message header, Swan packs data in the rest of the MTU. Sharing larger blocks of data is not a limitation of the underlying technology, just a boundary of the current programming implementation.

To share a data block larger than a single MTU, three solutions are possible. Currently, the application has to split the data block into MTU-size pieces, transmit these separately, and combine those pieces on the receiving end. This technique for splitting and recombining large messages is not yet implemented in the Swan API, so this issue must be solved at the application layer. This first solution allows new data to be passed in a balanced way among all participants; however, redundant copies of the data block are transmitted, which can compound the difficulties for any node experiencing bandwidth shortages. A second solution is to transmit only a URL for the data block and have the receivers retrieve this data independently; this creates localized $O(N)$ bandwidth demand at the server for this data block. The third and most elegant solution involves creating an overlay network to distribute the data block through a tree created on-the-fly through the peer community. This solution retains balanced distribution of the content without creating redundant copies of the data.

Often in interactive applications, there arises the need for one-to-one transactions as well. These temporary one-to-one transactions include the cold-start data for a newcomer, the financial transaction at the end of an auction, and consistency enforcement from a referee process. Swan provides multicast; when one node sends a message through Swan, all other nodes will receive it. The one-to-one transactions must be handled separately, by creating a temporary direct link between these two nodes. In this way, two nodes can exchange information without affecting other nodes and without overloading the entire multicast network. Although Swan provides this point-to-point support as one of its primitives, this capability is not fully integrated into the encapsulated API. Currently, the sender and receiver have to exchange addresses and apply some socket-based communication.

The RtBase currently handles flat data types. For broader use, Swan should be extended to handle richer schema. Most applications today use flexible data types such as indefinite arrays, linked lists, and structures; because one is passing the event data, one is only concerned with passing the data between objects at each node. As with large data blocks, programmers currently have to map these data types manually to the flat Swan schema, a process that must soon be automated.

There are some hard technical challenges inherent in supporting interactive online communities. The performance demands on the system increase with higher levels of interaction, with richer shared data, and with larger communities of participants. As the performance demands increase, there arises a tension between reliable messaging and NACK implosion.[15] The consistency of the application state is at stake. Consistency is easier to maintain if each participant reliably receives every message. When no packets are dropped, consistency can be addressed directly through the variation in latencies among pairs of participants. Once the flow of packets is disrupted, the potentially large time delays involved in resending messages can completely disrupt the underlying rhythm of the application. Perceived latencies become skewed, and the perturbation in rhythm is transmitted directly into any dead reckoning or other time-series based procedures. Worse, local catastrophes in reliable message propagation tend to propagate; as a sender tries to resend to a blocked receiver, it decreases its own effective throughput for the application. On the one hand, as the reliability of the Internet continues to improve, the probability of dropped messages decreases; however, applications are expected to work seamlessly across extensions to the Internet (currently WiFi and wireless[16,17], and these young infrastructures suffer from much more severe packet loss rates.

Further subtle refinements to the Swan multicast technology will come as the result of real-world feedback. It is not possible to anticipate all the complications and nuances for distributed applications, particularly when the network and the online communities are growing and changing so rapidly.

Studies of real-world behaviors[18,19] and tools for monitoring the local performance of distributed systems[20] will reveal new challenges to be addressed.

45.7   Conclusion

Swan is a significant advance in the infrastructure for synchronous online collaboration. It provides reliable and high-performance sharing of information, which scales to hundreds or thousands of asynchronous cooperating processes. It does this using generally available standard Internet protocols and requiring no system modifications or administration. Currently implemented to grow random r-regular graphs of TCP/IP connections, Swan provides latencies that grow logarithmically as the number of participants increases. It uses redundant messaging to accommodate processes with slow connections, and to provide reliability for the message stream when nodes depart. Swan has been in production use since 1999 and is the subject of five U.S. patents pending. Swan represents an enabling technology. With such an infrastructure, one can envision a new generation of applications for digital collaboration, for work or play. An airplane could be designed and developed by hundreds of engineers in a massive, long-term shared CAD session. Command, control, communication, and intelligence applications could coordinate thousands of independent software agents. Multiplayer games, with participants constantly joining and leaving, could run indefinitely and without the expense of server farms. Swan goes beyond the "pull" and "push" technologies currently seen on the Internet, giving us the ability to share and collaborate interactively.

Acknowledgments

Due to the introductory character of this chapter, some of the material that provides detailed information relevant to the topic of the chapter is not included. Readers interested to know more about matter described here are directed to conference papers by Fred B. Holt and Virgil Bourassa:

- Swan: Small-World Wide Area Networks, *SSGRR 2003*, L'Aquila, Italy, July 23–August 3, 2003.
- Load-Balancing in a Swan Network of True Peers, *IPSI 2003*, Sveti Stefan, Montenegro, October 4–11, 2003.
- Supporting Online Communities: The Untethered Approach, *SSCCII 2004*, Amalfi, Italy, January 29–February 1, 2004.

Both these conference papers and more technical information on Swan are available from Panthesis' Web site: www.panthesis.com.

References

1. Maihoefer, C. and Rothermel, K., A delay analysis of tree-based reliable multicast protocols, *Proc. 10th IEEE-ICCCN*, 2001, pp. 274–281.
2. Yoon, W., Lee, D., Youn, H.Y., Lee, S., and Koh, S.J., A combined group/tree approach for scalable many-to-many reliable multicast, *IEEE InfoCom*, 2002, pp. 1336–1345.
3. Watts, D.J. and Strogatz, S.H., Collective dynamics of 'small-world' networks, *Nature*, 393(4), 440–442, 1998.
4. Korniss, G. et al., Suppressing roughness of virtual times in parallel discrete-event simulations, *Science*, 299, 677–679, 2003.
5. Li, V.O.K. and Zhang, Z., Internet multicast routing and transport control protocols, *Proc. IEEE*, 90(3), 360–391, 2002.
6. Almeroth, K.C., The evolution of multicast: from the Mbone to interdomain multicast to Internet2 deployment, *IEEE Network*, Jan./Feb. 2000, pp. 10–20.
7. Egger, S. and Braun, T., Multicast for small conferences: a scalable multicast mechanism based on IPv6, *IEEE Commun. Mag.*, Jan. 2004, pp. 121–126.

8. Banerjee, S. and Bhattacharjee, B., A Comparative Study of Application Layer Multicast Protocols, manuscript, Dept. of Computer Science, U. of Maryland, April 2003.

9. Obraczka, K., Multicast transport protocols: a survey and taxonomy, *IEEE Commun. Mag.,* Jan. 1998, pp. 94–102.

10. Vinge, V., *True Names and the Opening of the Cyberspace Frontier,* J. Frenkel (Ed.), Tor Books, New York, 2001.

11. Bollobas, B., *Random Graphs,* Academic Press, London, 1985.

12. Bollobas, B. and Fernandez de la Vega, W., The diameter of random regular graphs, *Combinatorica,* 2(2), 125–134, 1982.

13. Cooper, C., Dyer, M., and Greenhill, C., Sampling Regular Graphs and a Peer-to-Peer Network, manuscript, School of Mathematics, University of New South Wales, Sydney, 2004.

14. Gamma, E. et al., *Design Patterns,* Addison-Wesley Publishing Company, Dec. 1995, pp.127–134.

15. Chen, S., Yener, B., and Ofek, Y., Performance trade-offs in reliable group multicast protocols, *INFO-COM 2,* IEEE, 1999, pp. 982–989.

16. Lott, M., Sitalov, A., Linsky, E., and Li, H., Performance analysis of multicast transmission in WLAN, *Veh. Tech. Conf.,* Spring 2003, IEEE VTC-2003, Vol 2. pp. 1223–1227.

17. Xylomenos, G. and Polyzos, G.C., TCP and UDP performance over a wireless LAN, *Proc. IEEE InfoCOM 1999,* pp. 439–446.

18. Ilie, D., Erman, D., Popescu, A., and Nilsson, A.A., Measurement and Analysis of Gnutella Signaling Traffic, manuscript, Dept. of Telecom Systems, Blekinge Inst. of Tech., Karlskrona, Sweden, 2004.

19. Chandra, P., Gambhire, P., and Kshemkalyani, A.D., Performance of the optimal causal multicast algorithm: a statistical analysis, *IEEE Trans. Parallel and Dist. Systems,* 15(1), 40–52, Jan. 2004.

20. Steigner, C. and Wilke, J., Isolating performance bottlenecks in network applications, *Proc. IPSI 2003,* Sveti Stefan, Montenegro, Oct. 4–11, 2003, Paper No. 74, pp. 1–7.

Scalable and Secure
P2P Overlay Networks

Haiying Shen

Aharon S. Brodie

Cheng-Zhong Xu

Weisong Shi

46.1   Introduction

Since the inception of Napster in the late 1990s, the peer-to-peer (P2P) computing model has grown dramatically. In the literature, however, there is little consensus about its definition. An Intel P2P working group defines it as "the sharing of computer resources and services by direct exchange between systems."[84] Veytsel defines P2P as "the use of devices on the internet periphery in a nonclient capacity."[108] Shirky[101] presents a more detailed definition as "a class of applications that takes advantage of resources—storage, cycles, content, human presence—available at the edges of the Internet. Because accessing these decentralized resources means operating in an environment of unstable connectivity and unpredictable IP addresses, P2P nodes must operate outside the DNS system and have significant or total autonomy from central servers." The author summarizes two criteria to judge if a system is P2P or not: (1) Does it treat variable connectivity and temporary network addresses as the norm? and (2) Does it give the nodes at the edges of the network significant autonomy?

Essentially, a P2P system is a logical network on top of physical networks in which peers are organized without any centralized coordination. A peer (or node) is an abstract notion of participating entities. It can be a computer process, a computer, an electronic device, or a group of them. Each peer has equivalent responsibilities, and offers both client and server functionalities to the network for resource sharing.

P2P systems are a complement to the client/server computing model. Although it is popular in distributed computing, the client/server model has weak scalability and availability because it is prone to bottleneck traffic and failures at the servers. On the other hand, client machines on the network edge have seen a steady increase in processing networking and storage capacities. P2P computing exploits the potential of distributed resources by shifting the service away from central servers, outward to the edge, and allowing clients to share data and other resources such as spare cycles, disk space, and network bandwidth directly.

P2P systems bring about many benefits, such as aggregating resources, cost sharing/reduction, utilizing spare resources, enhancing scalability/reliability, assuring anonymity/privacy in resource sharing, and being adaptive to dynamic/ad hoc environments.[76] These benefits are demonstrated in a wide range of applications, including distributed computing, file sharing, multicasting, collaboration, platforms, search engines, agent-based systems, awareness systems, mirror systems, naming systems, etc.[50,76]

P2P systems can be traced back to ARPANET in the late 1960s. ARPANET connected hosts as equal computing peers, which had equal rights in sending and receiving packets. The decentralized control model of early Usenet bears much resemblance to today's P2P applications; Domain Name System (DNS) combines the P2P concepts with a hierarchy of information ownership. The P2P model gets momentum with the popularity of Napster for music sharing on the Web.[80] An important novel feature introduced by Napster is that the concepts of ownership and distribution of information are differentiated so that people can distribute information, which they do not own, in a free and pseudo-anonymous manner.

Although the Napster Web site was forced to shut down due to a lawsuit for copyright infringement, research on the P2P computing model has never stopped. Many Napster offsprings have been proposed over the past few years. Representatives include Gnutella,[37] KaZaA,[53] and Freenet.[16] Gnutella inspired Sun's Infrasearch for P2P information retrieval. Freenet provides an anonymous method for storing and retrieving information. That is, a user can make requests for data without anyone being able to determine who is making these requests. Likewise, if a user stores a file in the system, it should be impossible to determine who placed the file into the system. Finally, the operator of a Freenet node should have no knowledge of what data is stored on the local disk. In Napster and its offsprings, peers are connected in a simple unstructured network, but interact in a fairly complicated protocol. Their complex interaction directly affects the overall system performance and network scalability, and limits the usefulness of the networks for new applications beyond traditional file sharing.

On another track altogether, there are studies on distributed hash tables (DHTs) that organize P2P systems in a structured network. A DHT maps document keys to network peers based on a consistent hashing function.[52] Representatives of the DHTs include CAN,[90] Chord,[105] Pastry,[94] Tapestry,[122] Kademlia,[74] and Cycloid.[99] They organize the peers in various ways for efficient management of distributed data.

Two primary concerns in the development of P2P systems are scalability and security. Scalability refers to the capability of a system to maintain consistently acceptable levels of performance, while peers are added. A highly scalable P2P system can grow to several million concurrent participants without significant performance degradation. Security represents the system's ability to resist different kinds of attacks and avoid malicious peers so as to provide a safe distributed processing environment for participants. Making a P2P system secure is a challenge because the system is open and each participant is potentially unreliable. With the P2P model quickly emerging as a computing paradigm of the future, there is an ever-increasing need for distributed algorithms that would allow P2P applications to scale to a large community of users, as well as security methods to make P2P systems secure.

The remainder of this chapter is structured as follows. Section 46.2 presents an overview of the characteristics of P2P overlay systems. Section 46.3 and Section 46.4 review scaling techniques and security measures in both structured and unstructured networks. Section 46.5 concludes this chapter with remarks on open issues.

46.2   P2P Overlay Network Characteristics

This section presents an overview of the distinguishing characteristics of P2P systems: decentralization with ad hoc connectivity and self-organization, dynamics and churn, scalability, security, performance, fault resilience, and interoperability. All have a major impact on the effectiveness and deployment of P2P systems.

46.2.1   Decentralization and Self-Organization

As its name implies, P2P upends the traditional client/server model by decentralization. In a P2P system, each peer has control over its data and resources independently. P2P computing makes new services available to end users in a novel way by making use of their machines as active participants in computing processes rather than just resource requestors. This decentralization characteristic offers each node in P2P systems a self-organization function. Self-organization is required for reliability and for the support of ad hoc connectivity of the peers. By ad hoc connectivity, we mean that the connections between peers are improvised instead of preexisting, due to continuous and fast peer joining, leaving, and failure. The objective of reliability requires P2P systems to adapt to such system membership changes rapidly and keep system consistency at a low cost. Self-organization has a twofold objective: self-maintenance and self-repair. For self-maintenance, node joining and leaving is handled in a distributed manner, without requiring the member change information to be propagated through the entire network for reorganization. When a node joins a system, it builds connections with other nodes to become an active participant. When a node leaves, it is removed from the whole system without leaving any trace of its existence. For example, in Pastry,[94,95] self-organization handles the arrival and departure of nodes by information exchange within the overlay network.

When a node fails without warning, its connected nodes still regard it as alive. In this case, a self-repair function helps them to update their routing and neighboring information. In Chord,[105] stabilization implements self-organization for node failures to keep the Chord ring intact.

46.2.2   Dynamics and Churn

Churn represents a situation where a great number of nodes join, leave, and fail continually and rapidly. Previous work shows that an ideal overlay can still locate an object efficiently even after some fraction of the nodes are failed. Because churn has the potential to increase P2P system workload and reduce the object location efficiency, an ideal overlay should perform well under churn. Liben-Nowell et al.[64] indicate that P2P systems' "maintenance bandwidth" depends on the rate at which nodes tend to enter and leave the system. They prove that the stabilization protocol of Chord is near optimal for churn. Saia et al. propose[96] a virtual CAN butterfly content addressable network that routes efficiently in the face of churn. Pandurangan et al.[85] present a centralized algorithm to ensure connectivity when many nodes fail. Ledlie et al.[62] propose a self-organizing hierarchically based P2P system. Mahajan et al.[70] describe a new technique to reduce the maintainance cost while providing high reliability and performance. Linga et al.[65] regard churn as a security threat and propose a churn-resistant P2P Web caching system based on the Kelips system[43] to ensure high cache hit rate without stabilization protocol. Li et al.[63] provide a unified framework for evaluating cost and performance under churn in different DHT protocols. Rhea et al.[92] propose Bamboo, a DHT protocol for dealing with network high churn efficiently and gracefully. Castro et al.[9] present an optimized Pastry, MSPastry, to deal with consistent routing under churn with low overhead. Based on Tapestry, Hildrum et al.[47] propose algorithms for adapting to arriving and departing nodes and provide the necessary failure recovery mechanisms.

46.2.3   Scalability

P2P computing is an alternative to the client/server computing for greatly improved scalability. However, the problems, such as low bandwidth and high latency on peer connections combined with large amounts of traffic, pose serious barriers to the scalability of P2P networks. Napster, as a centralized network, attacked the scalability problem by having the peers directly download music files from the peers that possess the requested document. However, its scalability depends entirely on the scalability of the central server. In unstructured P2P systems, such as KaZaA and Gnutella, a huge amount of bandwidth is required for routing control messages like "I'm alive" and other network chatter before any actual data messages are transferred. Many architectures and scaling techniques, approaches such as increasing ring search and random walks, are proposed for reducing congestion and enhancing the scalability of Gnutella. In addition, structured P2P overlay networks such as Chord, CAN, and Pastry appeared recently letting the P2P systems scale well to potentially billions of keys, stored on hundreds or millions of nodes. They use consistent mapping between the key and the hosting node to ensure the availability of the object when the hosting node is alive. A comprehensive review of unstructured and structured P2P systems is presented in Section 46.3.

46.2.4   Security

P2P systems provide security challenges because they are not structured in the traditional client/server model. The client/server model allows the server to authenticate users, provide measures to validate data, and poses smaller security risks because the communication has no intermediaries. P2P systems, on the other hand, especially today's popular decentralized models, lack an authoritative entity that can authenticate and verify users and data. Moreover, communication passes through foreign nodes on the network that can be malicious. P2P systems raise security issues in four aspects:

1. *Authenticity*. To authenticate data on today's P2P networks, an alternative to centralized verification systems must be implemented. Because data is replicated on the network, it is further necessary to verify existing copies as well. Different aspects of authentication will be discussed, from originality of data to ensuring copies have not been modified. Different methods proposed to solve these problems will be introduced, such as reputation systems and file hashes.

2. *Anonymity*. An important feature in P2P networks is user anonymity. Table 46.1 presents a summary for the anonymity discussed by Molnar et al.[77] Users do not need to log in, or identify themselves,

TABLE 46.1   Types of Anonymity

Types of Anonymity	Difficult for Adversary to Determine
Author	Which users created which documents?
Server	Which nodes store a given document?
Reader	Which users access which documents?
Document	Which documents are stored at a given node?

upon entering the network. In certain P2P implementations, anonymity is achieved by indirect communication, and the client and server communicate through other nodes. While this is an attractive feature for users, it complicates other aspects of the P2P system. Because a node does not know with whom it is communicating, great measures must be taken to protect itself, both in terms of security and data validity. Moreover, in certain implementations, network usability degrades. Anonymous implementations will be discussed, along with their advantages and disadvantages.

3. *Attacks.* The insecurities of P2P networks have brought forth numerous attacks. These attacks include denial of service, node masquerading, and data tampering.[24] Attacks discussed in this chapter are denial of service, which virtually disconnects a node from the network, and byzantine and sybil attacks, which can lead to malicious nodes masquerading as others, as well as communication and data tampering. Daswani and Garcia-Molina[23] and Lynch[69] provide ways to minimize the effect of such attacks. Examples of these attacks and their attempted solutions will be further discussed.

4. *Secure routing.* One of the basic requirements of a P2P network is to provide reliable and efficient routing of communication between nodes. Reliability focuses on ensuring messages arrive at their destination despite legitimate, although sometimes unforseen, circumstances. Secure routing, however, deals with malicious nodes actively attempting to disrupt network communication. This is a crucial and difficult task; although nodes will actively misroute messages or segregate nodes from the network, the network would still attempt for the messages to reach their destination. The three crucial tasks a P2P network needs to implement securely exist in different aspects of the protocol: securely assigning node IDs, updating the routing table,[35] and ensuring successful message propagation. These are further discussed, along with possible attacks and their solutions.

More details about the security requirements and current security measures are discussed in Section 46.4.

46.2.5 Performance

Recall that a P2P overlay network is a logical structure constructed upon physical networks. The shortest path (the least hop count routing) according to the routing protocol is not necessarily the shortest physical path. The logical structure of the overlay should take into account the physical structure of the underlying network. Techniques to exploit topology information in overlay routing include geographic layout, proximity routing, and proximity-neighbor selection.[11]

1. *The geographic layout method* maps the overlay's logical ID space to the physical network so that neighboring nodes in the ID space are also close in the physical network. Topologically aware CAN[91] adopts this approach. But this method can lead to an uneven distribution of nodes in the overlay, thus increasing the chances of overloading nodes and rendering the maintenance cost formidable. Xu et al.[115] claim their study shows that for a typical 10,000-node topologically-aware CAN, five nodes can occupy 85 to 98 percent of the entire cartesian space, and some nodes have to maintain 450–1500 neighbors. Castro et al.[11] indicate that this method is not suitable for a one-dimensional ID space and neighboring nodes in the ID space are more likely to suffer correlated failures.

2. *In proximity routing*, the logical overlay is constructed without considering the underlying physical topology. In a routing, the node with the closest physical distance to the object key is chosen among the next hop candidates in the routing table. This method has been applied to Chord[105] and CAN.[90] However, the benefits of this method are dependent on the determination of an appropriate number of the hop candidates.

3. *Proximity neighbor selection* is a middle-ground approach between the above methods. It selects the routing table entries pointing to the topologically nearest among all nodes with nodeId in the desired portion of the ID space. Castro et al.[11] implement this method in Pastry and Tapestry with low overhead. It maintains system load balancing and robustness with a comparable delay stretch compared to Geographic layout method. Xu et al.[115] point out some limitations of this application and propose a way to build topology-aware overlays using global soft-state. It combines

landmark clustering and RTT measurement to generate proximity information and stores the system information (such as proximity and load information) as objects on the system itself, which is easy to update and retrieve. In addition, it uses publish/subscribe functionality that allows nodes to subscribe to the relevant soft-state and get notified as the state changes necessitate neighbor re-selection.

Waldvogel and Rinaldi[110] propose an efficient topology-aware overlay network, Mithos. In contrast to other approaches, Mithos does not require full topology knowledge; even the forwarding and routing information is minimum. At the same time, Mithos provides a close conceptual integration between geographic layout and proximity routing, as well as a powerful addressing scheme directly suitable for use in DHTs. It provides locality-aware connectivity, thereby ensuring that a message reaches its destination with minimal overhead.

46.2.6 Fault Resilience

A good P2P system should still work well in the face of failure problems such as disconnections/unreachability, partitions, and node failures. The most vital issue of fault resilience is to ensure that all nodes in the P2P system are alive and there is no disconnection. Therefore, techniques that detect, manage, and recover from failed nodes or disconnections are reviewed.[76]

- *Detection.* Chord uses stabilization, in which each node sends probe queries to its successor, to maintain an intact ring. Sending an "I'm alive" message periodically is a method to detect failed nodes and disconnection. Pastry can periodically perform an expanding ring multicast search for other Pastry nodes in their vicinity to check the disconnected overlay. Gnutella and Freenet resort to resource-intensive naive broadcast queries to tolerate against node failures.

- *Management.* File replication is a way to solve the problem when the object hosting node has failed. P2P networks such as Napster and Gnutella have a passive and an uncontrolled replication mechanism based only on the file's popularity. Freenet and Publius[109] use controlled replication to provide persistence guarantees. OceanStore maintains a two-layered hierarchy of replicas. In contrast to promiscuous replication,[39] OceanStore does not bind floating replicas to specific machines, and it does not replicate all objects at each server.

- *Recover.* In a P2P system, how can the connected nodes still collaborate well when a failure happens? How can the disconnected node join in the ongoing computation when it reappears? Previous works primarily focused on providing resiliency to servers and network failures for distributed file systems. For example, Coda[97] adopted server replication and disconnected operation. Currently, some other methods are proposed. One is to let a "relay" node substitute for the destination temporarily, like Groove.[83,97] Second is to let the source wait with the messages until the destination reappears, as in Magi.[8] Third, Grid computing solutions, such as Legion,[41] deal with such problems by restarting computations on different nodes.

46.2.7 Interoperability

Different P2P networks have their advantages and disadvantages. For example, Freenet provides a certain degree of security but cannot ensure object location, and each node cannot control the files stored in itself. By contrast, structured P2P networks such as Chord can ensure object availability. Gnutella enables each node to hold what it desires. Therefore, it is desirable to combine the different kinds of P2P networks as a whole to take advantage of each network's benefits. Because different P2P networks deploy distinct protocols with different characteristics, the most important issue is to let the P2P systems interoperate. Lui and Kwok[66] propose a framework to integrate various P2P file-sharing protocols using P2P gateways to maximize the benefit of the P2P file-sharing application. The community of P2P developers gather together in the P2P Working Group[84] to build common standards that would enable a common understanding among P2P developers. The JXTA project[49] is Sun's foray into providing a utility application substrate for building P2P applications across platforms, OSs, and programming languages. A gateway project underway is World Free Web (WFW), which is to combine Freenet and the World Wide Web. Wiley[112] explains the characteristics of five popular networks—Freenet, Gnutella, Mojo Nation, Free Haven, and Publius—and evaluates the strengths of each network that can be offered to the all-encompassing OmniNetwork. P2P interoperability also has commercial significance. For instance, new file-swapping applications, such as StreamCast's test version of Morpheus,[78] are bridging separate networks, promising to improve service and content offerings.

46.3 Scaling Techniques

An ideal scalable P2P network is that when increasing the number of nodes, aggregate storage space and file availability should grow linearly, response time should remain constant, and search throughput should remain high or grow. The following are the main criteria for a system's scalability:

- *Decentralization.* In centralized systems, the central nodes easily become hot spots that quickly become overloaded as the network grows, and their limited bandwidth prevents systems from unlimited growth. Distribution of queries in networks decreases the node's average bandwidth and workload, which directly gain the speed of serving and overall system throughput.

- *Nodes' storage overhead.* In P2P systems, nodes need to contribute some storage space for other users' data or routing information that are the indices of other nodes. However, when a large amount of storage is required, some nodes cannot afford it, especially when the nodes grow to very large numbers, leading to poor system scalability.

- *Object location cost.* Generally, with increasing numbers of nodes, the object location cost (i.e., the number of hops traversed) will increase because searching is performed among a larger number of nodes. Therefore, low object location cost leads to decreased query traffic and increased system throughput, resulting in good scalability.

- *Workload: bandwidth and processing cost.* When a query passes through a number of nodes, each node uses bandwidth resources because it sends and receives query and response messages.[119] Low bandwidth and processing cost of each node, and of whole system, are essential for improving scalability.

- *System fault resilience.* A scalable system can handle node joins, leaves, and failures with little negative effect.

In the following, we present a classification of P2P systems. It is followed by a comprehensive survey of scaling techniques in unstructured and structured P2P systems.

46.3.1 Classification of P2P Systems

Napster is a typical example of centralized P2P, in which the elements of both pure P2P and client/server systems coexist. Napster clients download a file directly from another Napster client's machine and the Napster servers answer a list of matching files and locations and brokering client connections. Other centralized models include Aimster,[3] Magi,[8] Softwax,[103] and iMesh.[48] As a centralized network, it attacks the scalability problem by having the peers directly download music files from the peers that possess the requested document. However, because file location is done from the central server, the cost incurred on the central server is so high that it makes it very likely that the central server will become a bottleneck. Yang and Garcia-Molina[118] have studied issues and trade-offs in designing a scalable hybrid P2P, including chained, replication, and hash architectures.

The poor scalability of these centralized P2P systems motivates researchers to explore various methods for robust and scalable distributed P2P systems. As a result, unstructured P2P, structured P2P, and hybrid P2P (a combination of unstructured and structured P2P) are developed. Figure 46.1 shows a classification of P2P systems of Napster offspring. Unstructured P2P overlay networks such as Gnutella and Freenet do not have strict control over the topologies, and they do not assign responsibility for data to specific nodes.

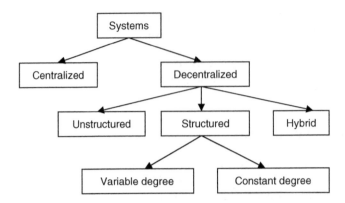

FIGURE 46.1   Classification of P2P overlay networks.

Nodes join and leave the network according to some loose rules. On the contrary, structured overlay networks such as Chord, Pastry, Tapestry, and CAN have strictly controlled topologies, and the data placement and lookup algorithms are precise. According to the node degree, that is, the number of entries in a node's routing table, structured P2P can be classified into variable-degree P2P and constant-degree P2P.

An unstructured system has improved system stability owing to the removal of central servers. Its advantages include overall simplicity, its ability to endure a high rate of churn, and its support for keyword searches. However, it cannot ensure the object location and too much message forwarding can easily result in network congestion. A structured system was developed to improve system scalability by reducing the amount of forward messaging and node storage for routing indices. It guarantees object location within a bounded lookup cost even if the system is in a continuous state of change. Because of their potential efficiency, robustness, scalability, and deterministic data location, structured networks have been studied intensively in recent years. But these systems also have some disadvantages. First, DHT is less adept at supporting keyword searches because it is designed for exact-match query. Second, it incurs substantial repair operations for high churn rate. Third, hot spots are generated for too frequently accessed files. People are realizing each system's benefits and drawbacks from the theoretical work, experiments, and practice, and are trying to develop improved scalable systems. The hybrid network is such an example; it uses the advantages of both unstructured and structured P2Ps. Representative hybrid networks include Yappers[34] and distributed ties-based DHT.

46.3.2   Unstructured P2P

Gnutella is a typical example of an unstructured P2P. Gnutella uses flooding-based search to perform object locations. That is, to find a file, a node sends a query message to each node connected to it. The nodes then rebroadcast this query in turn to every node they are connected to, except for the original node, and so on. Each query contains a time-to-live (TTL) and will be forwarded until the TTL is expired. If a node has the desired file, it will respond and the file will return along the route the query comes from. In a sense, Gnutella improves the scalability of Napster because it does not need to depend on the central server. However, the number of queries and the number of potential responses increase exponentially with each hop. For example, if each node has two connections to other nodes and the TTL is 7, then the number of queries sent will be 128 and the number of responses will depend on the file's popularity. As a result, the load on each node grows linearly with the total number of queries, the query number increases with the system size, and large systems quickly become overwhelmed by the query-induced load. So, Gnutella does not scale very well, due to a large amount of traffic. In addition, without consideration for heterogeneity, some nodes with limited capabilities become bottlenecks when the network size surges, resulting in slower response time and less availability. In the following we present a survey of key techniques for improving unstructured P2P scalability.

46.3.2.1 Gnutella TTL Control

TTL selection is very important in Gnutella; too high a value generates much network traffic, whereas too low a value will result in no search results. Lv et al.[67] propose increasing the ring search to help solve this problem. Increasing the ring achieves savings at the expense of a slight increase in delay to find the object, which improves Gnutella's scalability to a certain degree. However, the method may cause the problems of same-value TTL, query reprocessing, and message duplication, which may lead to more redundant network traffic. These problems can be solved by an iterative deepening search method proposed by Yang and Garcia-Molina.[119] Based on Gnutella, this method iteratively sends the query with different TTL with a mechanism to avoid query reprocessing until the query is satisfied. It greatly improves Gnutella's scalability by reducing the number of lookup hops.

46.3.2.2 Intelligent Peer Selection

Gnutella's poor scalability is primarily due to its flooding-based search. Some proposals suggest to reduce the number of query receivers to improve the system's scalability; that is, a query is sent to a part of a node's neighbors rather than to all its neighbors. Ripeanu and Foster[93] measured the Gnutella network, found that the network follows a power-law pattern, and estimated the generated traffic via the percentage of each kind of message with the crawler program. Lv et al.[67] proposed random-walkers search, which greatly reduces the load produced by a query. Random walks forward k query messages, and each query is sent to a randomly chosen neighbor at each step until the object is found. It also adopts active replication, which produces lower overall query load than non-active node-based replication. In addition, it uses a uniform random graph building algorithm instead of a power-law random graph[2] to reduce the likelihood of very-high degree nodes. Portmann and Seneviratne[87] studied the cost of Gnutella broadcasting and proposed the use of rumor mongering or gossip protocol as an alternative routing method to reduce the cost via simulation. Yang and Garcia-Molina[119] claim that Gnutella uses breadth-first traversal (BFS) over the network with depth limit TTL. They proposed direct breadth-first traversal (DBFS) to overcome the time-consuming problem of multiple iterations. DBFS reduces the number of nodes that a query reaches, as well as maintains the quality of responses. Yang and Garcia-Molina[119] also proposed a local indices technique, in which a node maintains an index over the data of each node within r hops of itself, where r is a system-wide variable. Queries are then processed by a smaller set of nodes than without this technique, while yielding as many results. Crespo and Garcia-Molina[22] proposed routing indices (RI) schemes, including compound RI, hop-count RI, and exponential RI. An RI is a data structure that returns a list of neighbors for a query. Simulations show that RIs can improve performance by one or two orders of magnitude compared to a flooding-based system, and by up to 100 percent compared to a random forwarding system.

46.3.2.3 Heterogeneity Consideration

Heterogeneity is an important factor to consider for scalability. Heterogeneity means that different nodes have different capabilities, particularly in terms of bandwidth, but presumably also in terms of other resources such as CPU, memory, and disk. Attention to node heterogeneity means to avoid overloading any of the nodes by taking into account the nodes' capacity constraints. In an ideal scalable system, nodes with more capability should be responsible for more work, while nodes with less capability should have less work. Flow control and topology adaptation are such methods proposed by Lv et al.,[68] based on their previous work. These methods are used to (1) restrict the flow of queries into each node so they do not become overloaded, and (2) dynamically evolve the overlay topology so that queries flow toward the nodes that have sufficient capacity to handle them. The random-walks-based search, combined with TTL and "state-keeping" algorithms,[2,67] lets the query be forwarded to high-capacity nodes, and thus makes it more possible to find the targeted files.

Chawathe et al.[14] indicate the DHT's disadvantages, and propose Gia to make Gnutella more scalable. Gia has the ability to deal with much higher aggregate query rates and functions well even with growing system size. It modifies Gnutella's algorithms to include dynamic topology adaptation, flow control, one-hop replication, and attention to node heterogeneity. Simulations show that Gia improves Gnutella's system capacity by three to five orders of magnitude.

46.3.2.4   Super-Peer Topology

FastTrack[28] is one of the super-peer topologies. Currently, two file-sharing applications, KaZaA and Morpheus, are based on the FastTrack protocol. For example, the KaZaA application has had upward of 20 million downloads and it can have anywhere up to 800,000 users connected at any one time. The Morpheus multimedia file-sharing system reported over 470,000 users sharing a total of 0.36 petabytes of data as of October 26, 2001.

The FastTrack architecture consists of super-nodes with fast connections and nodes with slower connections. A super-node connects with other super-nodes and some nodes at the same time. A node connects with a super-node. Consequently, a two-tier topology is generated in which the nodes at the center of the network are faster and therefore produce a more reliable and stable backbone. This allows more messages to be routed than if the backbone were slower, and therefore allows greater scalability.

Routing on FastTrack is almost the same as on Gnutella, but broadcasting is between the super-nodes only. A node sends its query to its super-node only. Every super-node that broadcasting reaches searches an index that contains all the files of its connected nodes. The fast-connection super-nodes lessen the workload produced by a large amount of messages. In addition, FastTrack use the nodes' heterogeneity to its advantage as the nodes with limited capabilities are shielded from query processing and traffic. By combining the P2P and client/server models, this topology provides a useful balance between the efficiency of centralized search and the autonomy, load balancing, and robustness of distributed search.

Recently, it was proposed to apply the hierarchical designs of FastTrack to the Gnutella network.[36] However, its scalability has neither been measured nor analyzed. Yang and Garcia-Molina[120] present practical guidelines and a general procedure for the design of an efficient super-peer network that is almost the same as FastTrack. They also point out a potential drawback of the super-peer node network; that is, when a super-node fails or simply leaves, all its clients (the nodes in the same cluster with that super-node) become temporarily disconnected until they find a new super-node to connect to. To address this problem, the method of super-peer redundancy is proposed. Instead of a single super-node in a cluster (the super-node with its clients), there are some redundant super-node partners with the same responsibility in a round-robin manner. To reduce the individual query load, each partner is connected with half of the clients.

46.3.2.5   Caching and Content Placement

Caching is a short-term solution to increasing the scalability of Gnutella. Sripanidkulchai[104] finds that the popularity of search strings follows a Zipf-like distribution. He proves that caching a small number of query results significantly decreases the amount of network traffic. He also finds that caching at one Gnutella node can result in up to a 3.7-time reduction in traffic while using only a few megabytes of memory. As more nodes implement caching, more traffic is reduced. Markatos[73] proposes another caching mechanism that caches Gnutella query results for a limited amount of time to reduce the query traffic. Simulations show that it improves performance by as much as a factor of two.

FastTrack's[28] proactive replication is far from optimal, as the relative index capacity dedicated to each item in the super-nodes is proportional to the number of copied nodes. Cohen and Shenker[17] studied the optimal way to replicate data, given fixed constraints on per-probe capacity. They show that the Uniform and Proportional strategies constitute two extreme points of a large family of strategies that lie "between" the two, and that any other strategy in this family has a better expected search size. They then show that one of the strategies in this family, square-root replication, minimizes the expected search size on successful queries.

46.3.2.6   Clustering/Arrange Overlay Topology

Pandurangan et al.[85] propose techniques for constructing Gnutella-like networks in a distributed fashion. In their heuristic proposed protocol, newly arriving nodes decide which network nodes to connect to, and existing nodes decide when and how to replace lost connections. The proposal results in connected networks of constant degree and logarithmic diameter. This method focuses on pure-search networks without copying content indices. CAP[56] is a cluster-based architecture for P2P systems that uses a network-aware clustering technique[55] to group hosts into clusters. Each cluster has one or more delegate nodes that act as directory servers for the objects stored at nodes within the same cluster. A central server tracks existing clusters and their delegates. To reduce query latency, delegate nodes maintain a cache of recently received queries. CAP would also be more stable because clusters join and leave the network less frequently than individual clients.

In Gnutella, the network topology is unrelated to data placement, and the nodes receiving a query are unrelated to the content of the query. A node does not have any information about which other nodes might best be able to resolve the query. Some proposals improve Gnutella by reorganizing network topology based on content location, that is, to cluster peers based on property similarity or results availability to avoid query messages flooding, reduce traffic, gain the number of query hit messages, and save resources in handling irrelevant queries in the P2P network. In these approaches, instead of blind flooding in Gnutella, the queries are forwarded to those nodes that are most likely to have the desired data to improve query performance.

Ramanathan et al.[89] propose a mechanism, using only local knowledge, to let a peer connect to the peer that frequently provides good results. This leads to clusters of peers with similar interests, and in turn allows us to limit the depth of searches required to find good results. Schlosser et al.[98] address the scalability problems by imposing a deterministic shape on P2P networks, called HyperCuP, which relies on the concept of prefix-based routing and allows for very efficient broadcast and search. They describe a broadcast algorithm that exploits the topology to reach all nodes in the network with the minimum number of messages possible. They also provide an efficient topology construction and maintenance algorithm that requires neither a central server nor super-nodes in the network. Moreover, they show how their scheme can be made even more efficient using a global ontology to determine the organization of peers in the graph topology. Hang and Sia[46] propose a method for clustering peers that share similar properties. They also propose a new intelligent query routing strategy, the firework query model. Crespo and Garcia-Molina[22] propose semantic overlay networks (SONs) based on the semantic relations among peers. SONs establish semantic relations between a query and peers and route the query directly to relevant peers. Furthermore, layered SONs are proposed to increase the availability of results.

The SON supports only limited metadata, such as simple filenames. On the other hand, currently, schema-based networks such as the Edutella[1,6,45,60,81,82] provide complex query facilities but no sophisticated means for semantic clustering of peers, and their broadcasting does not scale well. In this condition, a new class of P2P system that combines SON, super-peer topology, and schema-based networks has emerged.[59,61] Cooper and Garcia-Molina[19,20] present a Search/Index Link (SIL) model for analyzing and visualizing P2P search networks. They suggest reducing or eliminating redundant work of the network to improve network efficiency. The SIL model describes existing networks such as super-node networks and global index networks, while also yielding novel organizations such as parallel index networks and parallel index cluster networks. Based on SIL, Cooper and Garcia-Molina[21] examine how to reduce the load on nodes by allowing peers to self-organize into a relatively efficient network, and then self-tune to make the network even more efficient. Unlike their previously studied architectures, they propose "ad hoc, self-supervising" networks that avoid restrictions on whom a node can connect to or what information can be exchanged. This network is optimized for a heterogenous network, as well as for a more homogenous network where all nodes have roughly similar capabilities. Their results indicate that their ad hoc networks are more efficient than popular super-peer topologies for several important scenarios.

46.3.2.7 New Protocols

Gnutella2[38] is a reworking of the Gnutella protocol. It uses the same peer-based network as Gnutella but drops all of the old Gnutella protocol except for the connection handshake and adopts an entirely new and complex system. Gnutella2 uses UDP rather than the TCP/IP network protocol for searches, an extensible binary XML-like packet format, and includes many extensions such as intelligent query routing, SHA-1 checksums, parallel downloading in slices (swarming), etc. Preceded by their previous work, iterative deepening search, DBFS, super-peer network, SON, and SIL, which are based on Gnutella, Yang et al.[117] propose a non-forwarding architecture (NFA) in the context of the GUESS protocol. In GUESS, peers directly probe each other with their own query message, instead of relying on other peers to forward the message. Unlike the flooding-based search in Gnutella, in NFA the messages are not forwarded and peers have complete control over message receivers and message receiving time. As a result, NFA greatly reduces the message forwarding overhead.

46.3.3 Structured P2P

The different proposals for improving the scalability of Gnutella have resulted in significant progress toward better scalability. At the same time, another class of P2P, structured P2P systems, has developed with more scalable architecture. These systems have no central directory server, and thus are decentralized. "Structured" means that the P2P network topology is tightly controlled and that files are placed not at random nodes but at specified locations that will make subsequent queries easier to satisfy. These structured P2P systems include Chord, Pastry, Tapestry, CAN, etc., which support a DHT functionality. Each object is stored at one or more nodes selected deterministically by a uniform hash function. Although these DHT systems have great differences in implementation, they all support a hash table interface of put(key,value) and get(key) either directly or indirectly, and queries for the object will be routed incrementally to the node. Although hash functions can help locate content deterministically, they lack the flexibility of keyword searching—a useful operation to find content without prior knowledge of exact object names. These structured systems are highly scalable as they make very large systems feasible; lookups can be resolved in a bounded number of overlay routing hops.

In the following we review and compare some of the structured DHTs by focusing on their topological aspects within variable-degree and constant-degree P2P systems.

46.3.3.1 Variable-Degree Structured P2P

In a variable-degree structured P2P, the routing table size is related to the network size, which typically is $O(\log n)$. As a result, the routing table size increases with an increasing number of nodes in the system.

- *Hypercube based.* Plaxton et al.[86] developed perhaps the first routing algorithm that could be scalably used for P2P systems. Tapestry and Pastry use a variant of this algorithm. The approach of routing based on address prefixes, which can be viewed as a generalization of hypercube routing, is common to all theses schemes. The routing algorithm works by correcting a single digit at a time in left-to-right order. If node number 12345 received a lookup query with key 12456 that matches the first two digits, then the routing algorithm forwards the query to a node that matches the first three digits (e.g., node 12467). To do this, a node needs to have, as neighbors, nodes that match each prefix of its own identifier but differ in the next digit. For each prefix (or dimension), there are many such neighbors (e.g., node 12467 and node 12478 in the above case) because there is no restriction on the suffix (i.e., the rest bits right to the current bit). This is the crucial difference from the traditional hypercube connection pattern and provides abundance in choosing cubical neighbors and thus a high fault resilience to node absence or node failure. In addition to such cubical neighbors spreading out in the keyspace, each node in Pastry also contains a leaf set L of neighbors, that is, the set of $|L|$ numerically closest nodes (half smaller, half larger) to the present node ID, and a neighborhood set M, that is, the set of $|M|$ geographically closest nodes to the present node.

- *Ring based.* Chord uses a one-dimensional circular keyspace. The node responsible for a key is the node whose identifier most closely follows the key numerically; that node is called the key's successor. Chord maintains two sets of neighbors. Each node has a successor list of k nodes that immediately follow it in the keyspace and a finger list of $O(\log n)$ nodes spaced exponentially around the keyspace. The i-th entry of the finger list points to the node that is 2^i away from the present node in the keyspace, or to that node's successor if that node is not alive. Thus, the finger list is always fully maintained without any null pointer. Routing correctness is achieved with two such lists. A lookup(key) is, except at the last step, forwarded to the node closest to, but not past, the key. The path length is $O(\log n)$ because every lookup halves the remaining distance to the home.

- *Mesh based.* CAN chooses its keys from a d-dimensional toroidal space. Each node is identified by a binary string and is associated with a region of this keyspace, and its neighbors are the nodes that own the contiguous regions. Routing consists of a sequence of redirections, each forwarding a lookup to a neighbor that is closer to the key. CAN has a different performance profile than the other algorithms; nodes have $O(d)$ neighbors and path lengths are $O(dn^{1/d})$ hops. Note that when $d = \log n$, CAN has $O(\log n)$ neighbors and $O(\log n)$ path length like the other algorithms. This actually provides another way to deploy the hypercube as an overlay network.

eCAN,[116] which means expressway CAN, augments CAN's routing capacity with routing tables of larger span to achieve logarithmic routing performance. To build the expressways, CAN's entire Cartesian space is partitioned into zones of different spans with the smallest zones corresponding to the CAN zones. Therefore, each node owns a CAN zone and is also a resident of the expressway zones that enclose its CAN zone. In this way, the message can be traversed between not only CAN neighboring zones, but also neighboring expressway zones.

- *Tree based.* Maymounkov and Mazires[74] propose a novel XOR-based metric topology, Kademlia. It effectively treats nodes as leaves in a binary tree, with each node's position determined by the shortest unique prefix of its ID. For each node, the binary tree is divided into a series of successively lower subtrees that do not include this node, which knows at least one node in each of its subtrees. Kademlia defines the distance between two nodes as the XOR of their IDs. Its routing table is a binary tree whose leaves are $O(\log n)$ neighbors, where the i-th neighbor is a node within an XOR distance of $[2^i, 2^{i+1}]$. Kademlia employs a recursive algorithm for node lookups, which forwards the lookup message to a constant number of nodes in the current node's routing table in parallel. This can avoid timeout delays from failed nodes. The number of hops per lookup is $O(\log n)$. By increasing the routing table's size to $2^b \log_{2^b} n$, the lookup hops can be reduced to $\log_{2^b} n$. PRR's algorithm[86] uses a tree-like structure. It is used to maintain and distribute object locations information to let access requests locate the addresses of copies of objects in order to ensure fast access and efficient network resource utilization.

- *Butterfly based.* Fiat and Saia[29] build a Butterfly network of depth $\log n - \log \log n$ that can tolerate massive adversarial node failures and random object deletions. Each search for a data item in the network takes $O(\log n)$ time and requires, at most, $O(\log_2 n)$ messages. This network is censorship resistant in the sense that even after adversarial removal of an arbitrarily large constant fraction of the nodes in the network, all but an arbitrarily small fraction of the remaining nodes can obtain all but an arbitrarily small fraction of the original data items. The network can be created in a fully distributed fashion. It requires only $O(\log n)$ memory in each node. A variant of the scheme is also given that has the property that it is highly spam resistant: an adversary can take over complete control of a constant fraction of the nodes in the network and yet remains unable to generate spam. Saia et al.[96] created another highly fault-resilient CAN based on the Butterfly structure with $O(\log^3 n)$ state at each node and $O(\log^3 n)$ per message routing overhead. Xu et al.[113] present Ulysses. The namespace of Ulysses k-Butterfly consists of k-level and k-dimensional cuboids. Each node with randomly assigned level l represents a zone that is a subcuboid of l-level cuboid. By maintaining routing tables no larger than $O(\log n)$, Ulysses achieves a lookup cost of $O(\frac{\log n}{\log \log n})$ with high probability.

- *Random graphs based.* Thus far we have discussed the deterministic graphs; another direction for building DHTs relies on properties of random graphs to achieve logarithmic-time routing. The $\mathbb{H}$-graph[58] network composed of d Hamilton cycles is proposed. In a layered construction of such random expander networks, any node can be located in $O(\log n)$ time with $O(\log n)$ degree, a join operation takes $O(\log n)$ time and $O(\log n)$ messages, while a leave operation takes $O(1)$ time and $O(\log n)$ messages. Aspnes et al.[4] propose a P2P system where nodes are embedded at grid points in a one-dimensional real line. Each node is connected to its immediate neighbors and to multiple long-distance neighbors. They examine greedy routing in the random graph with degree $l + 1$ (l is a polylogarithmic value) and derive upper and lower bounds on the expected routing distance, and prove that greedy routing can be a nearly optimal mechanism for searching even in the presence of many faults. Their results show that both bounds are proportional to $\frac{\log^2 n}{l \log \log n}$.

- *Hybrid structured based.* Actually, some P2P network structures employ different geometric models. Take Pastry[94] as an example. Pastry's structure is the combination of a hypercube and a ring. Each node's routing table has the node's information that shares the current node's nodeID in the first n digits, but whose $n + 1$ digit has one of the $2^b - 1$ possible values other than the $n + 1$ digit in the present node's ID, which is used in the hypercube routing. Each node's leaf set has the information of numerically closest node IDs, which is used in the ring routing when the hypercube routing is failed. As mentioned[42] sequential neighbors can make progress toward all destination identifiers and the ring is the only geometry that supports sequential neighbors. Several designs such as Chord[105] and CAN[90] have incorporated sequential neighbors to improve proximity and resilience. Ganesan and Manku[33] propose optimal routing algorithms for Chord. The standard Chord routing algorithm uses edges in only one direction. These algorithms exploit the bi-directionality of edges for optimality. Given that Chord is a variant of the hypercube, the optimal routes possess a surprising combinatorial structure.

46.3.3.2   Constant-Degree Structured P2P

In contrast to variable-degree P2P, in which the routing table size increases with the growing number of nodes in the system, constant-degree DHTs let each node maintain the same size routing table no matter how large the system scales. As a result, storage space is saved and the bookkeeping required to respond to system membership changes remains small, thus enabling the network to scale to a large system.

- *Ring based.* Inspired by Kleinberg's small worlds,[54] Symphony[72] is proposed. It arranges all nodes along a ring. Each node manages the ring segment between its own ID and its predecessor, and it has two short links with its immediate neighbors and k long-distance links. With $k = O(1)$, the lookup path length is proved as $O(\frac{1}{k} \log^2 n)$.

- *Mesh based.* Kleinberg[54] constructs a two-dimensional grid where every point maintains four links to each of its closest neighbors and just one long-distance link to a node chosen from a suitable probability function. He shows that the small world can route a message by greedy routing in $O(\log^2 n)$ hops with constant degree. Barriere et al.[5] study Kleinberg's construction and prove its optimality under certain conditions.

- *Butterfly based.* Viceroy[71] maintains a connection graph with a constant-degree logarithmic diameter, approximating a Butterfly network. Each Viceroy node in Butterfly level l has seven links to its neighbors, including pointers to its predecessor and successor pointers in a general ring, pointers to the next and previous nodes in the same level ring, and Butterfly pointers to its left, right nodes of level $l + 1$, and up node of level $l - 1$, depending on the node location. In Viceroy, every participating node has two associated values: its identity $\in [0, 1)$ and a Butterfly level index l. The node ID is independently and uniformly generated from a range $[0, 1)$, and the level is randomly selected from a range of $[1, \log n_0]$, where n_0 is an estimate of the network size. A node's ID is fixed, but its level may need to be adjusted during its lifetime in the system. Viceroy routing involves three steps: (1) ascending to a level 1 node via uplinks, (2) descending along the downlink until a node is reached with no downlinks, and (3) traversing to the destination via the level ring or ring pointers. Viceroy takes $O(\log n)$ hops per lookup request.

  Xu et al.[113] study the fundamental trade-off between the routing table size and the network diameter and prove that the Tapestry, Pastry, Chord, and CAN schemes are indeed asymptotically optimal as uniform algorithms. They found that $O(\log n)$ is the routing table size threshold that separates the trade-off region dominated by congestion and the region dominated by reachability for uniform algorithms. They proposed a graph based on a modified static Butterfly that can achieve $O(\frac{\log n}{\log \log n})$ and $O(\log_d n)$ lookup cost when the routing table size is $\log n$ and d, respectively.

- *Random graphs based.* Pandurangan et al.[85] propose a random DHT graph with a constant degree and logarithmic diameter; however, the article does not provide an efficient routing algorithm for the proposed structure that can deterministically explore the low diameter of the graph.

- *De Bruijn graph based.* Koorde[51] combines Chord with de Bruijn graphs. Like Viceroy, it looks up a key by contacting $O(\log n)$ nodes with $O(1)$ neighbors per node. As in Chord, a Koorde node and a key have identifiers that are uniformly distributed in a 2^d identifier space. A key k is stored at its successor, the first node whose ID is equal to or follows k in the identifier space. Node $2^d - 1$ is followed by node 0. Due to the dynamic nature of P2P systems, they often contain only a few of the possible 2^d nodes. To embed a de Bruijn graph on a sparsely populated identifier ring, each participating node maintains knowledge about its successor on the ring and its first de Bruijn node. To look up a key k, the Koorde routing algorithm must find the successor of k by walking down the de Bruijn graph. Because the de Bruijn graph is usually incomplete, Koorde simulates the path taken through the complete de Bruijn graph, passing through the immediate real predecessor of each imaginary node on the de Bruijn path. Koorde can also be a variable-degree P2P. With $O(\log n)$ neighbors per node, it needs $O(\log n/\log \log n)$ hops per lookup request.

The Distance Halving DHT network proposed by Naor and Mieder[79] is isomorphic to the r-dimensional de Bruijn graph. The network is a discretization of a continuous graph based on a dynamic decomposition of the underlying Euclidean space into cells where each node is responsible for a cell. It consists of 2^r nodes and 2^{r+1} directed edges, and each node has 2 out-degree and 1 in-degree. Its routing has two phases; the first phase is routing to an almost random destination, and the second phase is from the random destination to the target. It allows logarithmic routing and load, while preserving constant degree. In a continuous graph with the edges $f_i(y) = \frac{y}{c} + \frac{i}{c}$ ($i = 0, 1, \ldots, c - 1$), when $c = \log n$ or $c = n^\varepsilon$ (ε is a constant), the lookup path length can be reduced to $O(\frac{\log n)}{\log \log n})$ or $O(1)$.

Another network structure based on the de Bruijn graph is D2B.[30] Its lookup cost is $O(\log n)$ with $O(1)$ degree. The expected number of keys managed by a node of an n-node D2B network is $|K|/n$ (K is the node representation binary string length), and is, with high probability, at most $O(|K|\log n/n)$. It achieves a trade-off between the latency for node joining or leaving the network, and the latency of a lookup. Also, a large node degree increases the connectivity of the network, and thus its robustness against processor crashes.

- *Comb based.* Considine and Florio[18] develop Chord to a constant degree comb structure with logarithmic time searches. It has many similarities with Viceroy. In the minimal topology, there are all $m \cdot 2^m$ nodes that are grouped with rank. Each node of rank i has two links: the short pointer points to the node's clockwise closest neighbor on the circle and another is called the jump pointer, which points to the node of rank i that is 2^i groups away moving clockwise. In the searching, the jump pointer is followed if it does not exceed the target; otherwise, the short pointer is followed. More links can be added for maintenance and robustness purposes. However, this comb structure needs more updates when nodes join and leave, which is within $O(\log^2 n)$ hops, and it needs to estimate network size to properly construct the structure.

- *CCC based.* Cycloid is a combination of Chord and Pastry. It emulates a Cube-Connected-Cycles (CCC) graph in the routing of lookup requests between the nodes. We explain the details in Section 46.3.5.

- *Hybrid structured based.* Like the variable-degree P2P system, some systems are actually hybrid structure-based in a more accurate sense. For example, Koorde, Viceroy, and Cycloid are actually the combination of the ring and their own main topology. All systems regard all nodes on a ring, and each node keeps its predecessor and successor indices on the ring.

We summarize the architectural characteristics of various P2P systems in Table 46.2. Their topological properties in variable-degree and constant-degree P2P systems, respectively, are shown in Table 46.3.

46.3.4 Hybrid P2P

So far, we have discussed unstructured and structured P2P systems. Another class of P2P is the hybrid network, which is the combination of unstructured and structured P2P systems. It brings together the advantages of unstructured and structured P2P systems, and avoids their disadvantages, to achieve higher scalability.

TABLE 46.2 A Comparison of Representative DHTs: $d = \log n$ in CAN and $n = d \cdot 2^d$ in Cycloid

	Systems	Base Network	Lookup Cost	Routing Table Size
Variable degree	Pastry[94]	Hypercube	$O(\log n)$	$O(|L|) + O(|M|) + O(\log n)$
	Tapestry[122]	Hypercube	$O(\log n)$	$O(\log n)$
	Chord[105]	Ring	$O(\log n)$	$O(\log n)$
	CAN[90]	Mesh	$O(d n^{1/d})$	$O(d)$
	eCAN[116]	Mesh	$O(\log n)$	$O(d)$
	Kademlia[74]	Tree	$O(\log n)$	$O(\log n)$
	PRR[86]	Tree	$O(\log n)$	$O(q \log^2 n)$
	ZIGZAG[107]	Tree	$O(\log_k n)$	k^2
	Ulysses[113]	Butterfly	$O(\frac{\log n}{\log \log n})$	$\log n$
	Censorship Resistant P2P[29]	Butterfly	$\log n$	$\log n$
	Dynamically Fault-Tolerant CAN[96]	Butterfly	$\log n$	$\log^3 n$
	$\mathbb{H}$-graph[58]	Random graph	$O(\log n)$	$O(\log n)$
	Koorde[51]	de Bruijn	$O(\log n / \log \log n)$	$O(\log n)$
Constant degree	Symphony[72]	Ring	$O(\frac{1}{k} \log^2 n)$	$k = O(1)$
	Small Worlds[54]	Mesh	$O(\log^2 n)$	$O(1)$
	Viceroy[71]	Butterfly	$O(\log n)$	7
	Low-Diameter[85]	Random graph	$O(\log n)$	$O(1)$
	Distance Halving[79]	de Bruijn	$O(\log_d n)$	$O(d)$
	Koorde[51]	de Bruijn	$O(\log n)$	≥ 2
	D2B[30]	de Bruijn	$O(\log_d n)$	$O(1)$
	Constant State Indexing based on Chord[18]	Comb	$O(\log n)$	≥ 2
	Cycloid[99]	CCC	$O(d)$	7

The recently proposed Yappers[34] is an example of a hybrid network. It has no explicit control of the overlay network. Its keyspace is partitioned into a small number of buckets, in which the DHTs are used, and intelligent forwarding is used outside the buckets. Compared to Gnutella, its lookup cost is reduced by an order of magnitude. Freedman and Vingralek[32] propose a DHT based on distributed tries. It employs a dynamic-assignment approach, in which peers replicate lookup structure partitions that they frequently access and organize those partitions as a trie. The distributed trie converges to an accurate network map

TABLE 46.3 Structured P2P Systems Summary

Base Network	Variable-Degree P2P	Constant-Degree P2P
Hypercube	Pastry[94]	
	Tapestry[122]	
Ring	Chord[105]	Symphony[72]
Mesh	CAN[90]	Small worlds[54]
	eCAN[116]	
Tree	Kademlia[74]	
	PRR[86]	
	ZIGZAG[107]	
Butterfly	Ulysses[113]	Viceroy[71]
	Censorship Resistant P2P[29]	
	Dynamically Fault-Tolerant CAN[96]	
Random graph	$\mathbb{H}$-graph[58]	Low-diameter[85]
de Bruijn	Koorde[51]	Distance halving[79]
		Koorde[51]
		D2B[30]
Comb		Constant state indexing based on Chord[18]
CCC		Cycloid[99]

over time. Its lookup algorithm shares many similarities with the algorithm proposed in Ref. 88. The lookup node first checks its local storage for the value corresponding to the lookup key. If not present, then it initiates a distributed lookup process.

46.3.5 Cycloid: An Example of Structured P2P DHTs

Cycloid[99] is a constant-degree structured P2P overlay network, which is part of the authors' work. It combines Pastry with CCC graphs. In a Cycloid system with $n = d \cdot 2^d$ nodes, each lookup takes $O(d)$ hops with $O(1)$ neighbors per node. Like Chord, it employs consistent hashing to map keys to nodes. A node and a key have identifiers that are uniformly distributed in a $d \cdot 2^d$ identifier space.

46.3.5.1 Structure and Key Allocation

A d-dimensional CCC graph is a d-dimensional cube with replacement of each vertex by a cycle of d nodes. It contains $d \cdot 2^d$ nodes of degree 3 each. Each node is represented by a pair of indices $(k, a_{d-1}a_{d-2}\ldots a_0)$, where k is a cyclic index and $a_{d-1}a_{d-2}\ldots a_0$ is a cubical index. The cyclic index is an integer, ranging from 0 to $d - 1$, and the cubical index is a binary number between 0 and $2^d - 1$. Figure 46.2 shows the three-dimensional CCC. In a Cycloid system, each node keeps a routing table and two leaf sets with a total of seven entries to maintain its connectivity to the rest of the system. Table 46.4 shows a routing state table for node (4,10111010) in an eight-dimensional Cycloid, where x indicates an arbitrary binary value, the inside leaf set maintains the node's predecessor and successor in the local cycle, and the outside leaf set maintains the links to the preceding and the succeeding remote cycles.

In general, a node $(k, a_{d-1}a_{d-2}\ldots a_k\ldots a_0)$ $(k \neq 0)$ has one cubical neighbor $(k - 1, a_{d-1}a_{d-2}\ldots \overline{a}_k xx\ldots x)$, where x denotes an arbitrary bit value, and two cyclic neighbors $(k - 1, b_{d-1}b_{d-2}\ldots b_0)$ and $(k - 1, c_{d-1}c_{d-2}\ldots c_0)$. The cyclic neighbors are the first larger and smaller nodes with cyclic index $k - 1$ mod d, and their most significant different bit with the current node is no larger than $k - 1$. That is,

$$(k - 1, b_{d-1}\ldots b_1 b_0) = \min\{\forall (k - 1, y_{d-1}\ldots y_1 y_0)|y_{d-1}\ldots y_0 \geq a_{d-1}\ldots a_1 a_0\}$$
$$(k - 1, c_{d-1}\ldots c_1 c_0) = \max\{\forall (k - 1, y_{d-1}\ldots y_1 y_0)|y_{d-1}\ldots y_0 \leq a_{d-1}\ldots a_1 a_0\}$$

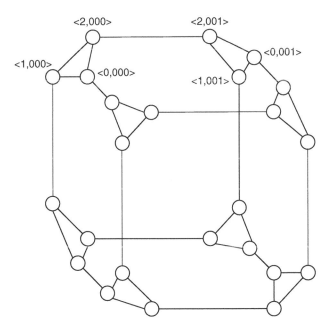

FIGURE 46.2 A three-dimensional Cube-Connected-Cycles.

TABLE 46.4 Routing Table of a
Cycloid Node (4,101-1-1010)

NodeID(4,101-1-1010)
Routing table
Cubical neighbor: (3,101-0-xxxx)
Cyclic neighbor: (3,101-1-1100)
Cyclic neighbor: (3,101-1-0011)
Leaf sets (half smaller, half larger)
Inside leaf set
(3,101-1-1010)   (6,101-1-1010)
Outside leaf set
(7,101-1-1001)   (6,101-1-1011)

The node with a cyclic index $k = 0$ has no cubical neighbor and cyclic neighbors. The node with cubical index 0 has no small cyclic neighbor, and the node with cubical index $2^d - 1$ has no large cyclic neighbor. The nodes with the same cubical index are ordered by their cyclic index mod d on a local cycle. The left inside leaf set node points to the node's predecessor and the right inside leaf set node points to the node's successor in the local cycle. The largest cyclic index node in a local cycle is called the primary node of the local cycle. All local cycles are ordered by their cubical index mod 2^d on a large cycle. The left outside leaf set node points to the primary node in the node's preceding remote cycle and the right outside leaf set node points to the primary node in the node's succeeding remote cycle in the large cycle. Our connection pattern is resilient in the sense that even if many nodes are absent, the remaining nodes are still capable of being connected. The routing algorithm is heavily assisted by the leaf sets. The leaf sets help improve the routing efficiency, check the termination condition of a lookup, and wrap around the key space to avoid the target overshooting. How the routing table and leaf sets are initialized and maintained is explained later. The Cycloid DHT assigns keys onto its ID space using a consistent hashing function. For a given key, the cyclic index of its mapped node is set to its hash value modulated by d, and the cubical index is set to the hash value divided by d. If the target node of a key's ID $(k, a_{d-1} \ldots a_1 a_0)$ is not a participant, the key is assigned to the node whose ID is first numerically closest to $a_{d-1} a_{d-2} \ldots a_0$ and then numerically closest to k.

46.3.5.2 Cycloid Routing Algorithm

Cycloid routing algorithm emulates the routing algorithm of CCC[88] from source node $(k, a_{d-1} \ldots a_1 a_0)$ to destination $(l, b_{d-1} \ldots b_1 b_0)$, incorporating the resilient connection pattern of Cycloid. The routing algorithm involves three phases, assuming that MSDB is the most significant different bit of the current node and the destination.

1. *Ascending.* When a node receives a request, if its k < MSDB, it forwards the request to a node in the outside leaf set sequentially until cyclic index $k \geq$ MSDB.
2. *Descending.* In the case of $k \geq$ MSDB, when $k =$ MSDB, the request is forwarded to the cubical neighbor; otherwise, the request is forwarded to the cyclic neighbor or inside leaf set node, whichever is closer to the target, in order to change the cubical index to the target cubical index.
3. *Traverse cycle.* If the target ID is within the leaf sets, the request is forwarded to the closest node in the leaf sets until the closest node is the current node itself.

Figure 46.3 presents an example of routing a request from node (0,0100) to node (2,1111) in a 4-D Cycloid DHT. The MSDB of node (0,0100) with the destination is 3. As (0,0100) cyclic index $k = 0$ and k < MSDB, it is in the ascending phase. Thus, the node (3,0010) in the outside leaf set is chosen. Node (3,0010)'s cyclic index 3 is equal to its MSDB; then in the descending phase, the request is forwarded to its cubical neighbor (2,1010). After node (2,1010) finds that its cyclic index is equal to its MSDB 2, it forwards the request to its cubical neighbor (1,1110). Because the destination (2,1111) is within its leaf sets, (1,1110) forwards the request to the closest node to the destination (3,1111). Similarly, after (3,1111) finds that the destination is within its leaf sets, it forwards the request to (2,1111) and the destination is reached.

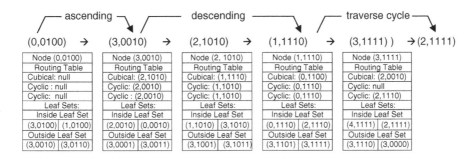

FIGURE 46.3 An example of routing phases and routing table states in Cycloid.

Each of the three phases is bounded by $O(d)$ hops; hence, the total path length is $O(d)$. The key idea behind this algorithm is to keep the distance decrease repeatedly. Also, the routing algorithm can be easily augmented to increase fault tolerance. When the cubical or the cyclic link is empty or faulty, the message can be forwarded to a node in the leaf sets.

46.3.5.3 Self-Organization

Cycloid deals with node joining and leaving in a distributed manner, without requiring that hash information be propagated through the entire network. When a new node joins, it needs to initialize its routing table and leaf sets, and inform other related nodes of its presence, which will update their neighbors. Before a node leaves, it needs to notify its inside leaf set nodes. The need to notify the nodes in its outside leaf set depends on whether the leaving node is a primary node. Upon receiving a leaving notification, the nodes in the inside and outside leaf sets update themselves. In addition, the nodes in the outside leaf set need to notify other nodes in their local cycle one by one, which will take at most d steps. Updating cubical and cyclic neighbors is the responsibility of system stabilization, as in Chord. Undoubtedly, low-degree P2P networks perform poorly in failure-prone environments, where nodes fail or depart without warning. Usually, the system maintains another list of nodes to handle such problems, such as the successor list in Chord[105] and the bucket in Viceroy.[71] In our proposal, we assume that nodes must notify others before leaving, as the authors of Koorde argued that the fault-tolerance issue should be handled separately from routing design.

Simulation results show that Cycloid delivers a higher location efficiency in the average case, it has a much shorter path length per lookup request in the average case than Viceroy and Koorde, another two constant-degree overlay networks. Cycloid distributes keys and query load more evenly between the participating nodes than Viceroy. Also, Cycloid is more robust as it continues to function correctly and efficiently with frequent node joins and leaves.

46.4 Security Concerns

Currently, most P2P systems assume that the nodes involved are cooperative or trustful. However, because of P2P systems' openness feature and decentralization, it is not desirable to constrain the membership of a P2P system. As a result, P2P systems are faced with threats such as denial of service, masquerading, and tampering. In these situations, the system must be able to operate even though some participants are malicious. In the following, we explain security issues in P2P systems from four aspects: authenticity, anonymity, attacks, and secure routing.

46.4.1 Authenticity

Searching for data on a P2P network is a straightforward process. A node on the network issues a request for a file, and the request propagates the network until nodes having that file are found. Finally, these nodes send back a reply informing they have the file, and the file transfer can begin. While this might seem simple, there is an important question that complicates this process. How can a node know it is downloading the exact file it was searching for? This file could have been altered or be a different file altogether.

There are a number of definitions for an authentic file, and a number of approaches to verify authenticity. If we view an authentic file as being the original file on the network, then a time-stamping system would be helpful in determining which of the search results is the oldest file available. An example is Mammoth,[26] a file system that presents objects with unique identifiers made up of the creator's node address and a timestamp. Another is PAST,[95] which assigns unique identifiers to files and storage nodes. These identifiers are used to determine which nodes will store replicas of the file. If we were to have trusted nodes that kept signatures of files uploaded by users, nodes would be able to compare signatures of files they downloaded with signatures on these servers. Identical signatures would verify that a file is identical to the original file uploaded and has not been modified. An example of such a signature is the MD5 checksum, which creates a virtually unique signature for every file. Even the smallest modification to the file would create a completely different signature and the change would be detected. Finally, there is the reputation-based approach, in which nodes deemed more trustworthy than others have a bigger influence on the decision of a query. This type of approach would require the system to maintain and propagate node reputation throughout the network. There has been significant research in this area, such as deploying a reputation system for the Gnutella network.[44] There are, however, many inherent problems with reputation systems on P2P networks. For example, nodes might not be eligible to vote on other nodes. This could be because of their location on the network or the tasks that they are required to perform. This issue has been addressed by introducing the need for different types of reputations for different scenarios on the same network.[111] Another proposal was to add reputation policies that would determine the pairs of nodes eligible to interact.[106]

Because P2P networks consist of nodes that can join the network without being verified as legitimate, these nodes cannot be trusted to have authentic data or to make a valid vote. Also, because P2P networks have taken a decentralized approach, there is no central main node that can be used to verify nodes or files. A solution to these problems was created; however, it is not without its drawbacks. P2P programs such as Freenet[15,16] attempt to answer these problems by introducing a variation of DHT. This is a unique hash made, in this case, of a filename. Unlike other P2P programs such as KaZaA[53] or Gnutella,[37] where a search is constructed of a partial or complete filename, a file search on Freenet is composed of a DHT. A user searching for a file on the network would have to know its exact name in order to compose the correct DHT. If a filename is different by even one character, hashing it would result in a different DHT and the search would be unsuccessful. Figure 46.4 illustrates an example of a DHT file search where only one letter is different by being uppercase. This search is unsuccessful because the DHT constructed does not match the original.

A few questions arise about Freenet and its DHT approach:

1. How can a user know the exact name of a file if he or she cannot search for it to begin with?
2. Using a DHT search, is the user not missing search results of identical files that are differently named?
3. What are the benefits of using a restrictive DHT search instead of a standard name search?

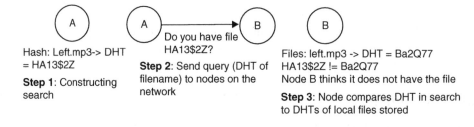

FIGURE 46.4 An example of a DHT file search.

The answers to these questions lie with the solution Freenet has developed on how to find files on its network. To publicize files on Freenet, users publish SHA hashes on their Freenet sites. Users retrieve this information from these sites and search for the acquired hash on the network. This ensures that the user is downloading the exact file requested, although other copies of the file named differently could exist. Because the download is limited to a file a user has information about, the two main authenticity issues are resolved. First, tampering with data in the file could easily be detected by adding a file integrity hash to the site on which the file is published. Second, file renaming, where a different file is named like the one a user is searching for, can also be easily detected by publishing file integrity hashes. This hash could be compared to the hash of a file just downloaded to verify its authenticity. The clear benefits of this approach are that because the user will have identifying information about the file being downloaded, it will be simple to ensure it is authentic. On the other hand, performance considerably degrades because nodes that do have this file under a different name will not return positive search results. A more important feature disabled by this approach is the partial name search. Having to know the exact filename beforehand makes it more difficult to find a file. Because nodes are not authenticated, and there is no authoritative entity to verify data, so far this has been the only completely successful approach to circumvent these issues.

46.4.2 Anonymity

Anonymity has been an important issue in P2P networks. Most implementations, such as KaZaA, Napster, or Gnutella, offer no anonymity and have given rise to numerous lawsuits by companies seeking to protect their copyright material. The main focus of anonymous peer networks is not only to allow sharing or retrieving of any data desired, but also to allow individuals who live under Internet-censure laws to freely browse the Internet and express their ideas.

There are different ways to achieve anonymity on a network, depending on the user's actions.[31] A user who wishes to upload or share a file but remain anonymous will need to use a peer network that keeps the originating IP address or node ID hidden. Freenet[16] is one such example of an anonymous P2P program that ensures author anonymity, while Free Haven[25] provides server anonymity. Once a file is available on the network, users may want to retrieve it anonymously. Nodes retrieving a file will also need to have their IP addresses or node IDs hidden.

Many P2P programs increase file availability by replicating files on various nodes on the network. This can be done by path replication,[16] where copies of files are stored along the path they are retrieved, or by replicating files based on calculations such as distance.[7] This brings forward another form of anonymity, one that protects a user from its own data. Because a user's P2P program might unknowingly be replicating data that is illegal, such as copyrighted material, it is necessary to protect that user from being held responsible for it. This can be accomplished by encrypting the data stored locally on a node and only allowing the P2P program to decrypt it when it is being accessed.[7,109] Another solution would be to split the file into chunks and allow a node to only store a part of the file. This, combined with data encryption, would prevent a user from tampering with data it is replicating on the network.

To hide the source and destination of a communication on a P2P network, it is necessary to form an anonymous path between the two nodes and ensure that nodes on the path do not know where the endpoints are. This is implemented with an overlay network that restricts a node's view only to its neighbors. A node on a message path knows only the previous and next step in the path, two of its neighbors. Figure 46.5 depicts a tunnelled communication between nodes A and E. The path of the communication is not directly

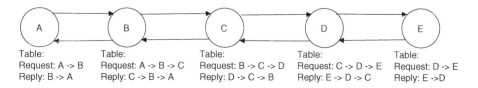

FIGURE 46.5 A tunnelled communication between nodes A and E.

between the two nodes; rather, it travels through three other nodes. Each node on the path does not know of the source and final destination of the messages transferred. For example, node C does not know if the communication is between nodes B and D or if they are passing the messages elsewhere.

Providing anonymity on the network has been shown to affect network efficiency. The number of messages on the network increases significantly because extra nodes are used to relay messages instead of using direct communication. And most importantly, download and upload speeds decrease because data must be routed through a prolonged path.

46.4.3   Attacks on the Network

The lack of security on P2P networks has not gone unnoticed. Many attacks have been introduced in the past few years, most of which deny access to the network. These range from being aimed at a single node, to attacks that can cause a large branch of the network to fail. The three most powerful attacks are introduced here: Sybil, Byzantine, and DoS attacks.

The Sybil attack[27] is named after Sybil Dorsett, who suffered from a multiple personality disorder. This attack involves one node on the network that masquerades as multiple identities. This can lead to many situations that are harmful to the network. With this attack it would be possible to actively monitor and control communication between nodes. A malicious node can impersonate all the different routes to another node and control all of its communication. Even worse, a malicious node can segment off a part of the network if it positions itself accordingly. Figure 46.6 shows an example of such an attack. In this example, node 4 controls communication between two network segments. It can monitor, alter, or disrupt communication passing between the segments, and can even separate them, effectively disabling all communication between them. A more direct attack involves masquerading as a node and performing actions on its behalf. If illegal actions are undertaken, such as downloading illegal content, an innocent user can be prosecuted.

Another form of attack capable of disrupting a node's network communication is called a denial-of-service attack. A node can cause a DoS attack against another node in a number of ways. The traditional DoS attack involves the malicious node sending messages to the victim node at a high rate. On P2P networks, this is accomplished by forwarding all requests a malicious node receives to the victim node. However, today, with available high-speed Internet and fast processing speeds at the node, such attacks originating from only one node do not generate messages fast enough to cause a real threat. The victim node will most likely be able to process the incoming messages before they fill up the message queue and cause legitimate messages to be dropped. A more efficient way to successfully construct a DoS attack is for a malicious node to reply to every request it receives stating the victim node as a source for the file. Therefore, all the nodes whose queries reach the malicious node will be directed to contact the victim node for files it does not have. On large P2P networks such as the ones deployed today on the Internet, thousands of nodes query the network every few minutes. If the victim receives a large number of requests in a short time, it will be bombarded with more messages than it can process and will be unable to send or receive legitimate communications.

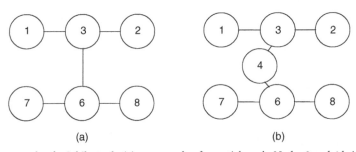

FIGURE 46.6 An example of a Sybil attack: (a) an example of a partial mesh. Nodes 3 and 6 bridge the two parts of the network; (b) attacker (node 4) positions itself between nodes 3 and 6. Each assumes node 4 has been placed legitimately.

Finally, the Byzantine attack raises the problem of reaching a consensus among nodes when an unknown number of them may be traitors.[57] In a P2P network where decisions can be based on neighboring nodes' responses, it is important to act on the legitimate ones, for example, P2P implementations that use a voting system to decide from which node to download a file, or those that determine the message path based on nodes' input of their network load. In these cases, decisions based on data sent from malicious nodes can be devastating to the network. The Byzantine attack involves multiple nodes collaborating to disrupt the network. This attack is more difficult to set up than the Sybil attack because multiple nodes must be set up on the network instead of one. The malicious possibilities a Byzantine attack can cause are similar to those that a Sybil attack can achieve. However, after malicious nodes have been placed on the network, Byzantine attacks become significantly more powerful and are easier to execute than a Sybil attack. Figure 46.7 illustrates a Byzantine attack. A Sybil attack would not be able to segment the network efficiently because one node is not present in different segments of the network. In this attack, however, the malicious user has placed nodes in different segments of the network, taking control of each one. There has been significant work in solving the Byzantine problem. One method proposed an algorithm for Byzantine fault-tolerant (BFT) state machine replication[13] that ensures that eventually clients receive correct replies.

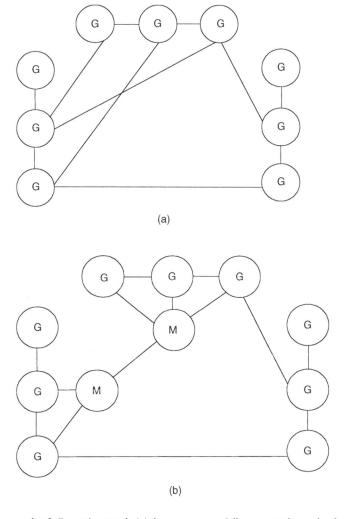

FIGURE 46.7 An example of a Byzantine attack: (a) three groups partially connected to each other; and (b) two malicious nodes M1 and M2 have been placed where they control all network communication between groups G1 and G2.

Another approach separates agreement from execution on a BFT state machine.[121] *Agreement* is what orders requests and *execution* is what processes requests; these have been coupled together in other BFT state machine architectures.

Because P2P networks are insecure by design, these attacks have remained an open problem. The solutions discovered by research so far have not managed to eliminate the threat—only to minimize it.

46.4.4   Secure Routing

Secure routing implementations are necessary to ensure that a message will be delivered despite malicious node activity such as message corruption or misrouting. There are three crucial tasks a P2P protocol must implement securely to ensure node connectivity.[10]

First, node IDs must be securely assigned to nodes on the network. This will prevent attackers from choosing node IDs. If a malicious node were to choose its node ID, it could choose where to place itself on the network. This is especially dangerous in P2P networks using ordered IDs.[105] Attackers placing malicious nodes in strategic places on the network can control the traffic of messages and data; connections can be manipulated or blocked altogether. Figure 46.8 shows an example of malicious use of a node's ability to choose its location on the network. This attack can be prevented if node IDs are related to the node's IP address,[102] as is the case of Pastry.[94] The assumption is that a malicious node cannot gain a large number of diverse addresses and position itself in multiple places on the Internet.

Another task that must be securely implemented concerns the constant updates to the routing table. This involves ensuring that the fraction of faulty nodes in the table will be lower than the fraction of faulty nodes in the network. A large number of faulty nodes inserted in the routing table can cause nodes to depend on a route that consists of malicious nodes.

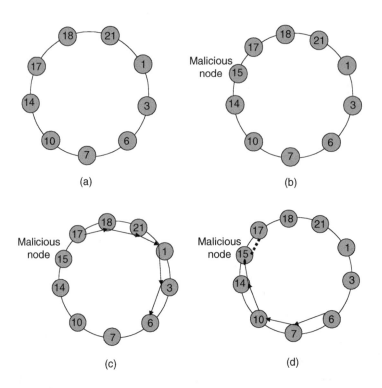

FIGURE 46.8 Malicious use of a node's ability to choose its location on the network: (a) an ordered P2P network; (b) a malicious node is interested in disrupting the communication of node 17. It inserts itself between nodes 14 and 17; (c) node 17 requests a file from node 6; (d) node 15 does not forward file to node 17. (Optional: node 15 forges message from node 6 stating it does not have file.)

Finally, message propagation must be securely implemented. Messages must be able to reach nodes that will further replicate them on the network. It is not required for all copies of a message to reach message-replicating nodes; only a minimum of one is required. Although the rate of replication and amount of results will be lower, if a node's routing table is poisoned with invalid entries, the desire is not to keep the connection efficient; it is to keep it connected.

Assigning node IDs securely without the use of a central authority remains an open problem. There have been attempted solutions but they have not been successful. First, an introduction of cryptographic puzzles[75] that need to be solved before a node can join. This proposal only slows down the malicious users from connecting; it does not prevent them from doing so. Assuming a malicious user would obtain enough resources to solve the puzzle faster, this would be ineffective altogether. Also, because connecting to the network is done anonymously, a node can connect as many times as it wants. Another proposal suggested sending a message on the network and waiting for a reply, confirming that the route is active and correct. However, because malicious nodes can act as intermediaries of a communication, they can choose to forward the confirmation message untouched while modifying the remaining data transferred. Also, a malicious node can masquerade as the destination node and confirm that the route is correct. This would be more difficult to accomplish if nodes would have predetermined keys to encrypt the communication. However, it is unrealistic to require every communication to use a unique predetermined key. P2P networks are dynamic: searching and retrieving files from nodes on the network requires no previous communication or prior knowledge of them. All solutions that exist to encrypt the communication or verify file integrity with checksums are prone to failure. A malicious node on the path can intercept the key exchange in the beginning of the communication and modify the checksum as well as the data being transferred. Because central authorities on today's decentralized P2P networks are not a feasible option, a possible solution is to use the PGP keys available on public servers worldwide. This would require all users on the network to create such keys and keep them secure. The main problem with this solution, however, is that user anonymity will not be available.

46.5   Concluding Remarks

In this chapter we have presented an overview of P2P overlay networks with a focus on scalability and security issues. P2P systems can be classified into structured, unstructured, and hybrid classes. Representatives of the systems and related scaling techniques in each class were reviewed. P2P systems provide security challenges in four aspects: authenticity, anonymity, attacks, and secure routing. The requirements and current security measures were discussed.

We conclude this chapter with discussion of several of open issues for building P2P cooperative computing applications, to include:

- *Heterogeneity.* With the increasing emergence of various end devices equipped with networking capability, coupled with the diverse network technology development, the heterogeneity of participating peers of a practical widely deployed P2P system is pervasive in the Internet domain. Their instinct properties, including computing ability, differ a lot and deserve serious consideration for the construction of a real efficient, widely deployed application.

- *Data integrity.* Due to the heterogeneity of end devices, we envision that the transformation is a very likely happening in any distributed cooperative applications, including image transcoding, personalized Web pages, etc. It is easy for this transformation to provide data integrity for static files. However, for dynamic generated content (e.g., dynamic Web content by content assembly,[100] transcoded content by adaptation[40]), it is really a challenge for both content providers and clients to validate the correctness of the content. By taking advantage of other peers' efforts to leverage the content (including transcoded content and dynamic pages) transferring burden, we realize that data integrity becomes a more and more serious problem.

- *Application-specific behavior.* This is neglected in most existing applications, but we believe it will play an important role in the future. In a distributed cooperative application, only peers showing application-specific interest would carry out the application-specific actions. Otherwise, there is no obligation for peers to serve the heavy burden of those non-interested applications. For example, to build an application-layer multicast tree, asking a peer that is not interested in the specific multicast streaming content to contribute its computing power to help the construction of the multicast tree seems reasonable. However, routing the high bandwidth streaming content is too high a price for that peer. Therefore, we envision that application-specific behavior should be considered in partner discovery.

- *Proximity.* Mismatch between logical proximity abstraction, derived from DHTs, algorithms, and physical proximity information in reality is a big obstacle for the deployment and performance optimization issues for P2P applications.

- *Reconfiguration semantics and overhead.* Using dynamic reconfiguration to adapt to dynamic environment changing and/or peer failures has been widely proposed in several systems.[12,114,123] However, few of these previous works take the reconfiguration overhead into consideration. We argue that the dynamic reconfiguration should consider the application-related semantics and also the overhead of reconfiguration.

References

1. K. Aberer, P. Cudrè-Mauroux, and M. Hauswirth. The Chatty Web: emergent semantics through gossiping. In *Proc. of the 12th International World Wide Web Conference,* 2003.
2. L.A. Adamic, B.A. Huberman, R.M. Lukose, and A.R. Puniyani. Search in power law networks. In *Physical Review E,* 64, 46135–46143, 2001.
3. Aimster. http://computer.howstuffworks.com/question587.htm.
4. J. Aspnes, Z. Diamadi, and G. Shah. Fault-tolerant routing in peer-to-peer systems. In *Proc. of the 12th ACM Symposium on Principles of Distributed Computing (PODC),* 2002.
5. L. Barriere, P. Fraigniaud, E. Kranakis, and D. Krizanc. Efficient routing in networks with long range contacts. In *Proc. of the 15th International Conference on Distributed Computing (DISC),* pages 270–284, 2001.
6. P.A. Bernstein, F. Giunchiglia, A. Kementsietsidis, J. Mylopoulos, L. Serafini, and I. Zaihrayeu. Data management for peer-to-peer computing: a vision. In *Proc. of the 5th International Workshop on the Web and Databases,* 2002.
7. S. Blackheath. The Grapevine Project. http://grapevine.sourceforge.net/.
8. G. Bolcer. Magi: an architecture for mobile and disconnected workflow. In White paper (www.endeavors.com), 2001.
9. M. Castro, M. Costa, and A. Rowstron. Performance and dependability of structured peer-to-peer overlays. Technical Report, Microsoft Research, 2003.
10. M. Castro, P. Druschel, A. Ganesh, A. Rowstron, and D.S. Wallach. Secure routing for structured peer-to-peer overlay networks. In *Proc. of the 5th USENIX Symposium on Operating Systems Design and Implementation,* December 2002.
11. M. Castro, P. Druschel, Y.C. Hu, and A. Rowstron. Topology-aware routing in structured peer-to-peer overlay networks. In *Future Directions in Distributed Computing,* 2003.
12. M. Castro, P. Druschel, A. Kermarrec, and A. Rowstron. SCRIBBLE: a large-scale and decentralized application-level multicast infrastructure. *IEEE Journal on Selected Areas in Communications,* 20(8), 2002.
13. M. Castro and B. Liskov. Practical Byzantine fault tolerance. *Proc. of the 3rd Symposium on Operating Systems Design and Implementation,* pages 173–186, 1999. USENIX Association.
14. Y. Chawathe, S. Ratnasamy, L. Breslau, N. Lanham, and S. Shenker. Making Gnutella-like P2P systems scalable. In *Proc. of ACM SIGCOMM,* 2003.
15. I. Clarke, T.W. Hong, S.G. Miller, O. Sandberg, and B. Wiley. Protecting free expression online with Freenet. *IEEE Internet Computing,* 6(1), 40–49, 2002.

16. I. Clarke, O. Sandberg, B. Wiley, and T.W. Hong. Freenet: a distributed anonymous information storage and retrieval system. In *Proc. Int. Workshop on Design Issues in Anonymity and Unobservability,* pages 46–66, 2001.

17. E. Cohen and S. Shenker. Replication strategies in unstructured peer-to-peer networks. In *Proc. of ACM SIGCOMM,* 2002.

18. J. Considine and T.A. Florio. Scalable peer-to-peer indexing with constant state. Technical Report, CS Department, Boston University, 2002.

19. B. Cooper and H. Garcia-Molina. Studying search networks with SIL. In *Proc. of the 2nd International Workshop on Peer-to-Peer Systems (IPTPS),* 2003.

20. B. Cooper and H. Garcia-Molina. SIL: Modeling and Measuring Scalable Peer-to-Peer Search Networks, 2004. In *Proc. of the International workshop on Databases, Information systems and Peer-to-Peer Computing,* 2003.

21. B.F. Cooper and H. Garcia-Molina. Ad hoc, self-supervising peer-to-peer search networks. Technical Report, CS Department, Stanford University, 2003.

22. A. Crespo and H. Garcia-Molina. Routing indices for peer-to-peer systems. In *Proc. of the 22nd International Conference on Distributed Computing Systems (ICDCS),* 2002.

23. N. Daswani and H. Garcia-Molina. Query-flood dos attacks in Gnutella. In *Proc. of the Ninth ACM Conference on Computer and Communications Security,* 2002.

24. N. Daswani, H. Garcia-Molina, and B. Yang. Open problems in data-sharing peer-to-peer systems. In *Proc. of the 9th International Conference on Database Theory,* 2003.

25. R. Dingledine, M.J. Freedman, and D. Molnar. The Free Haven Project: distributed anonymous storage service. In *Proc. of the Workshop on Design Issues in Anonymity and Unobservability,* July 2000.

26. B. Dmitry, B. Alex, P. Jody, G. Shihao, F. Michael, and H. Norman. Using file-grain connectivity to implement a peer-to-peer file system. In *21st IEEE Symposium on Reliable Distributed Systems (SRDS),* 2002.

27. J. Douceur. The Sybil attack? In *Proc. of the 1st International Workshop on Peer-to-Peer Systems (IPTPS),* 2002.

28. FastTrack product description, 2001. www.fasttrack.nu/index_int.html.

29. A. Fiat and J. Saia. Censorship resistant peer-to-peer content addressable networks. In *Proc. of the Symposium on Discrete Algorithms,* 2002.

30. P. Fraigniaud and P. Gauron. The content-addressable network D2D. Technical Report, Univerisity Paris-Sud, 2003.

31. M.J. Freedman, E. Sit, J. Cates, and R. Morris. Introducing Tarzan, a peer-to-peer anonymizing network layer. In *Proc. of the 1st International Workshop on Peer-to-Peer Systems (IPTPS),* 2002.

32. M.J. Freedman and R. Vingralek. Efficient peer-to-peer lookup based on a distributed trie. In *Proc. of the IPTPS02,* 2002.

33. P. Ganesan and G.S. Manku. Optimal routing in Chord. In *Proc. of the Symposium on Discrete Algorithms (SODA),* 2004.

34. P. Ganesan, Q. Sun, and H. Garcia-Molina. YAPPERS: a peer-to-peer lookup service over arbitrary topology. In *Proc. of IEEE Conference on Computer Communications (INFOCOM),* 2003.

35. A. Ganesh, A. Rowstron, M. Castro, P. Druschel, and D. Wallach. Security for structured peer-to-peer overlay networks. In *Proc. of the 5th USENIX Symposium on Operating Systems Design and Implementation,* 2002.

36. Gnutella development forum. The Gnutella Ultrapeer Proposal, 2002.

37. Gnutella home page. http://www.gnutella.com.

38. Gnutella2 home page. http://www.gnutella2.com.

39. J. Gray, P. Helland, P. ONeil, and D. Shasha. The dangers of replication and a solution. In *Proc. of ACM SIGMOD,* pages 173–182, 1996.

40. S.D. Gribble and et al. The Ninja architecture for robust internet-scale systems and services. *Journal of Computer Networks,* 35(4), 2001.

41. A. Grimshaw, W. Wulf, J. French, A. Weaver, and P. Reynolds Jr. Legion: the next logical step toward a nation-wide virtual computer. Technical Report, Department of Computer Science, University of Virginia, 1994.

42. K. Gummadi, R. Gummadi, S. Gribble, S. Ratnasamy, S. Shenker, and I. Stoica. The impact of DHT routing geometry on resilience and proximity. In *Proc. of ACM SIGCOMM*, August 2003.

43. I. Gupta, K. Birman, P. Linga, A. Demers, and R. van Renesse. Kelips: building an efficient and stable P2P DHT through increased memory and background overhead. In *Proc. of the 2nd International Workshop on Peer-to-Peer Systems (IPTPS)*, pages 81–86, 2003.

44. M. Gupta, P. Judge, and M. Ammar. A reputation system for peer-to-peer networks, In *Proc. of the 13th Intl. Workshop on Network and Operating Systems Support for Digital Audio and Video*, 2003.

45. A.Y. Halevy, Z.G. Ives, P. Mork, and I. Tatarinov. Piazza: data management infrastructure for semantic Web applications. In *Proc. of the 12th International World Wide Web Conference*, 2003.

46. C. Hang and K.C. Sia. Peer clustering and firework query model. In *Proc. of the 12th International World Wide Web Conference*, 2003.

47. K. Hildrum, J. Kubiatowicz, S. Rao, and B.Y. Zhao. Distributed object location in a dynamic network. In *Proc. of ACM SPAA*, 2002.

48. iMesh. http://www.imesh.com/.

49. JXTA 2001. JXTA home page: www.jxta.org.

50. F. Kaashoek. Peer-to-peer computing: a new direction in distributed computing. In *Proc. of Principles of Distributed Computing (PODC)*, 2002.

51. M.F. Kaashoek and R. Karger. Koorde: a simple degree-optimal distributed hash table. In *Proc. of the 2nd International Workshop on Peer-to-Peer Systems (IPTPS)*, 2003.

52. D. Karger, E. Lehman, T. Leighton, M. Levine, D. Lewin, and R. Panigrahy. Consistent hashing and random trees: distributed caching protocols for relieving hot spots on the World Wide Web. In *Proc. of the 29th Annual ACM Symposium on Theory of Computing (STOC)*, pages 654–663, 1997.

53. KaZaA, 2001. KaZaA home page: www.kazaa.com.

54. J. Kleinberg. The small-world phenomenon: an algorithmic perspective. In *Proc. of the 32nd ACM Symposium on Theory of Computing*, pages 163–170, 2000.

55. B. Krishnamurthy and J. Wang. On network-aware clustering of Web clients. In *Proc. of ACM SIG-COMM*, 2000.

56. B. Krishnamurthy, J. Wang, and Y. Xie. Early measurements of a cluster-based architecture for P2P systems. In *Proc. of SIGCOMM Internet Measurement Workshop*, 2001.

57. L. Lamport, R. Shostak, and M. Pease. The Byzantine generals problem. *ACM Transactions on Programming Languages and Systems*, 4(3), 382–401, 1982.

58. C. Law and K.-Y. Siu. Distributed construction of random expander graphs. In *Proc. of IEEE Conference on Computer Communications (INFOCOM)*, 2003.

59. A. Löser, F. Naumann, W. Siberski, W. Nejdl, and U. Thaden. Semantic overlay clusters within super-peer networks. In *Proc. of the International Workshop on Databases, Information Systems and Peer-to-Peer Computing in Conjunction with the VLDB*, 2003.

60. A. Löser, W. Nejdl, M. Wolpers, and W. Siberski. Information integration in schema-based peer-to-peer networks. In *Proc. of the 15th International Conference of Advanced Information Systems Engieering (CAiSE)*, 2003.

61. A. Löser, M. Wolpers, W. Siberski, and W. Nejdl. Efficient data store discovery in a scientific P2P network. In *Proc. of the Workshop on Semantic Web Technologies for Searching and Retrieving Scientific Data, International Semantic Web Conference (ISWC)*, 2003.

62. J. Ledlie, J. Taylor, L. Serban, and M. Seltzer. Self-organization in peer-to-peer systems. In *Proc. of the ACM SIGOPS European Workshop*, 2002.

63. J. Li, J. Stribling, T.M. Gil, R. Morris, and F. Kaashoek. Comparing the performance of distributed hash tables under churn. In *Proc. of the 2nd International Workshop on Peer-to-Peer Systems (IPTPS)*, 2004.

64. D. Liben-Nowell, H. Balakrishnan, and D. Karger. Observations on the dynamic evolution of peer-to-peer networks. In *Proc. of the 1st International Workshop on Peer-to-Peer Systems (IPTPS)*, 2002.

65. P. Linga, I. Gupta, and K. Birman. A churn-resistant peer-to-peer Web caching system. In *Proc. of the 2nd International Workshop on Peer-to-Peer Systems (IPTPS)*, 2004.

66. S.M. Lui and S.H. Kwok. Interoperability of peer-to-peer file sharing protocols. *ACM SIGecom Exchanges*, 3(3), 25–33, 2002.

67. Q. Lv, P. Cao, E. Cohen, K. Li, and S. Shenker. Search and replication in unstructured peer-to-peer networks. In *Proc. of ACM International Conference on Supercomputing (ICS)*, 2001.

68. Q. Lv, S. Ratnasamy, and S. Shenker. Can heterogeneity make Gnutella scalable? In *Proc. of the 1st International Workshop on Peer-to-Peer Systems (IPTPS)*, 2002.

69. N.A. Lynch. *Distributed Algorithms*. Morgan Kaufmann Publisher, 1996.

70. R. Mahajan, M. Castro, and A. Rowstron. Controlling the cost of reliability in peer-to-peer overlays. In *Proc. of the 2nd International Workshop on Peer-to-Peer Systems (IPTPS)*, 2003.

71. D. Malkhi, M. Naor, and D. Ratajczak. Viceroy: a scalable and dynamic emulation of the Butterfly. In *Proc. of Principles of Distributed Computing (PODC)*, 2002.

72. G.S. Manku, M. Bawa, and P. Raghavan. Symphony: distributed hashing in a small world. In *Proc. of the 4th USENIX Symposium on Internet Technologies and Systems (USITS)*, 2003.

73. E.P. Markatos. Tracing a large-scale peer-to-peer system: an hour in the life of Gnutella. In *Proc. of the 2nd IEEE/ACM International Symposium on Cluster Computing and the Grid*, 2002.

74. P. Maymounkov and D. Mazires. Kademlia: a peer-to-peer information system based on the XOR metric. In *Proc. of the 1st International Workshop on Peer-to-Peer Systems (IPTPS)*, 2002.

75. R.C. Merkle. Secure communications over insecure channels. In *Communications of the ACM*, pages 294–299, 1978.

76. D. Milojicic, V. Kalogerai, R. Lukose, K. Nagaraja, J. Pruyne, B. Richard, S. Rollins, and Z. Xu. Peer-to-peer computing. Technical Report, HP Laboratories, Palo Alto, CA, 2002.

77. D. Molnar, R. Dingledine, and M. Freedman. *Chapter 12: Peer-to-Peer: Harnessing the Power of Disruptive Technologies*. O'Reilly & Associates, 2001.

78. Morpheus home page: Chapter 12: Freehave, http://www.musiccity.com.

79. M. Naor and U. Mieder. Novel architectures for P2P applications: the continuous-discrete approach. In *Proc. of ACM SPAA*, 2003.

80. Napster. http://computer.howstuffworks.com/napster2.htm.

81. W. Nejdl, W. Siberski, M. Wolpers, and C. Schmnitz. Routing and clustering in schema-based super peer networks. In *Proc. of the 2nd International Workshop on Peer-to-Peer Systems (IPTPS)*, 2003.

82. W. Nejdl, M. Wolpers, W. Siberski, A. Löser, I. Bruckhorst, M. Schlosser, and C. Schmitz. Super-peer-based routing and clustering strategies for RDF-based peer-to-peer networks. In *Proc. of the 12th International World Wide Web Conference*, 2003.

83. Groove Networks. Groove networks product backgrounder, groove networks white paper, 2001. www.groove.net/pdf/groove_product_backgrounder.pdf.

84. Peer-to-Peer Working Group, 2001. www.p2pwg.org.

85. G. Pandurangan, P. Raghavan, and E. Upfal. Building low-diameter P2P networks. In *IEEE Symposium on Foundations in Comp. Sci.*, 2001.

86. C. Plaxton, R. Rajaraman, and A. Richa. Accessing nearby copies of replicated objects in a distributed environment. In *Proc. of ACM SPAA*, 1997.

87. M. Portmann and A. Seneviratne. The cost of application-level broadcast in a fully decentralized peer-to-peer network. In *ISCC02*, 2002.

88. F.P. Preparata and J. Vuillemin. The Cube-Connected Cycles: a versatile network for parallel computation. *CACM*, 24(5), 300–309, 1981.

89. M.K. Ramanathan, V. Kalogeraki, and J. Pruyne. Finding good peers in peer-to-peer networks. In *Proc. of International Parallel and Distributed Processing Symposium (IPDPS)*, 2002.

90. S. Ratnasamy, P. Francis, M. Handley, R. Karp, and S. Shenker. A scalable content-addressable network. In *Proc. of ACM SIGCOMM*, pages 329–350, 2001.

91. S. Ratnasamy, M. Handley, R. Karp, and S. Shenker. Topologically-aware overlay construction and server selection. In *Proc. of IEEE Conference on Computer Communications (INFOCOM)*, 2002.

92. S. Rhea, D. Geels, T. Roscoe, and J. Kubiatowicz. Handling churn in a DHT. Technical Report UCB/CSD-3-1299. Technical report, UC Berkeley, Computer Science Division, 2003.

93. M. Ripeanu and I. Foster. Mapping the Gnutella network. *IEEE Internet Computing Special Issue on Peer-to-Peer Networking*, 6(1), 50–57, 2002.

94. A. Rowstron and P. Druschel. Pastry: scalable, decentralized object location and routing for large-scale peer-to-peer systems. In *Proc. of the 18th IFIP/ACM International Conference on Distributed Systems Platforms (Middleware)*, 2001.

95. A. Rowstron and P. Druschel. Storage management and caching in past, a large-scale persistent peer-to-peer storage utility. In *Proc. of the 18th ACM Symp. on Operating Systems Principles (SOSP-18)*, 2001.

96. J. Saia, A. Fiat, S. Gribble, A.R. Karlin, and S. Saroiu. Dynamically fault-tolerant content addressable networks. In *Proc. of the 1st International Workshop on Peer-to-Peer Systems (IPTPS)*, 2002.

97. M. Satyanarayanan, J. Kistler, P. Kumar, M. Okasadi, E. Siegel, and D. Steere. Coda: a highly available file system for a distributed workstation environment. In *IEEE Transactions on Computers*, pages 447–459, 1990.

98. M. Schlosser, M. Sintek, S. Decker, and W. Nejdl. Hypercup—hypercubes, ontologies and efficient search on P2P networks. In *Proc. of the Workshop on Agents and P2P Computing*, 2002.

99. H. Shen, C. Xu, and G. Chen. Cycloid: a constant-degree and lookup-efficient P2P overlay network. In *Proc. of International Parallel and Distributed Processing Symposium (IPDPS)*, 2004.

100. W. Shi and V. Karamcheti. CONCA: an architecture for consistent nomadic content access. In *Proc. of the Workshop on Cache, Coherence, and Consistency (WC3)*, 2001.

101. C. Shirky. What is P2P . . . and what isn't? *The O'Reilly Network*, 2000.

102. E. Sit and R. Morris. Security considerations for peer-to-peer distributed hash tables. In *Proc. of the 1st International Workshop on Peer-to-Peer Systems (IPTPS)*, 2002.

103. SoftWax. http://www.softwax.com/.

104. K. Sripanidkulchai. The popularity of Gnutella queries and its implications on scalability. In *Proc. of the O'Reilly Peer-to-Peer and Web Services Conference*, 2001.

105. I. Stoica, R. Morris, D. Liben-Nowell, D.R. Karger, M.F. Kaashoek, F. Dabek, and H. Balakrishnan. Chord: a scalable peer-to-peer lookup protocol for Internet applications. In *IEEE/ACM Trans. on Networking*, 2002.

106. G.P. Thanasis and D.S. George. Effective use of reputation in peer-to-peer environments. In *Proc. of the International Workshop on Global and P2P Computing (GP2PC)*, 2004.

107. D. Tran, K. Hua, and T. Do. Zigzag: an efficient peer-to-peer scheme for media streaming. In *Proc. of IEEE Conference on Computer Communications (INFOCOM)*, 2003.

108. A. Veytsel. There is no P-to-P market . . . but there is a market for P-to-P. In Aberdeen Group presentation at the *P2PWG*, 2001.

109. M. Waldman, A. Rubin, and L. Cranor. Publius: a robust, tamper-evident, censorship-resistant Web publishing system. In *Proc. of the USENIX Security Symposium*, 2000.

110. M. Waldvogel and R. Rinaldi. Efficient topology-aware overlay network. In *Proc. of the 1st ACM Workshop on Hot Topics in Networks (HotNets-I)*, 2002.

111. Y. Wang and J. Vassileva. Trust and reputation model in peer-to-peer networks. In *Proc. of the 3rd International Conference on Peer-to-Peer Computing (P2P)*, September 2003.

112. B. Wiley. Chapter 19: Interoperability Through Gateways, *Peer-to-Peer: Harnessing the Power of Disruptive Technologies*. O'Reilly & Associates, 2001.

113. J. Xu, A. Kumar, and X. Yu. On the fundamental tradeoffs between routing table size and network diameter in peer-to-peer networks. In *IEEE JSAC*, 2004.

114. Z. Xu, C. Tang, S. Banerjee, and S. Lee. RITA: receiver initiated just-in-time tree adaptation for rich media distribution. In *Proc. of the 13th International Workshop on Network and Operating Systems Support for Digital Audio and Video (NOSSDAV)*, 2003.

115. Z. Xu, C. Tang, and Z. Zhang. Building topology-aware overlays using global soft-state. In *Proc. of the 23rd International Conference on Distributed Computing Systems (ICDCS)*, 2003.

116. Z. Xu and Z. Zhang. Building low-maintenance expressways for P2P systems. Technical Report, *Hewlett-Packard Labs: Palo Alto*, 2001.

117. B. Yang, P. Vinograd, and H. Garcia-Molina. Evaluating GUESS and non-forwarding peer-to-peer search. In *Proc. of the 24th International Conference on Distributed Computing Systems (ICDCS)*, 2004.

118. B. Yang and H. Garcia-Molina. Comparing hybrid peer-to-peer systems. In *Proc. of the 27th Int. Conf. on Very Large Databases*, 2001.

119. B. Yang and H. Garcia-Molina. Improving search in peer-to-peer networks. In *Proc. of the 22nd International Conference on Distributed Computing Systems (ICDCS)*, pages 5–14, 2002.

120. B. Yang and H. Garcia-Molina. Designing a super-peer network. In *Proceedings of ICDE*, 2003.

121. J. Yin, J. Martin, A. Venkataramani, L. Alvisi, and M. Dahlin. Separating agreement from execution for Byzantine fault tolerant services. In *Proc. of the 19th ACM Symp. on Operating Systems Principles (SOSP-19)*, 2003.

122. B. Zhao, J. Kubiatowicz, and A. Joseph. Tapestry: an infrastructure for fault-tolerant wide-area location and routing. Technical Report UCB/CSD-01-1141, Computer Science Division, UC Berkeley, 2001.

123. S. Zhuang, B. Zhao, A. Joseph, R. Kotz, and J. Kubiatowicz. Bayeux: an architecture for scalable and fault-tolerant wide-area data dissemination. In *Proc. of the 11th International Workshop on Network and Operating Systems Support for Digital Audio and Video (NOSSDAV)*, 2001.

47

Peer-to-Peer Overlay Abstractions in MANETs

Y. Charlie Hu

Saumitra M. Das

Himabindu Pucha

47.1   Introduction

A mobile ad hoc network (MANET) consists of a collection of wireless mobile nodes dynamically forming a temporary network without the use of any existing network infrastructure or centralized administration. In such a network, nodes operate as both end hosts and routers, forwarding packets for other mobile nodes that may not be within the direct transmission range of each other. MANETs are formed with the key motivation that users can benefit from collaborations with each other. The applications enabled by such collaborations often require services for locating resources, sharing information and data, multicasting, etc. In addition, due to the infrastructure-less environment, these applications should be designed to operate in a decentralized manner.

Recently, peer-to-peer (P2P) systems consisting of a dynamically changing set of nodes connected via the Internet have gained tremendous popularity. While initially conceived and popularized for the purpose of file sharing (for example, Gnutella and Freenet,[1,15]). P2P has emerged as a general paradigm for the construction of resilient, large-scale, distributed services and applications in the Internet. We broadly define P2P systems as self-organizing, decentralized distributed systems that consist of potentially untrusted, unreliable nodes with symmetric roles. The self-organization, decentralization, diversity, and redundancy inherent in the approach are relevant to a large class of applications beyond file sharing, anonymity, and anti-censorship.

The P2P paradigm has largely adopted a layered approach. A P2P overlay network built on top of the Internet provides a general-purpose substrate that provides many common properties desired by distributed applications, such as self-organization, decentralization, diversity, and redundancy. Such an overlay shields distributed application designers from the complexities of organizing and maintaining a secure overlay, tolerating node failures, balancing load, and locating application objects.

While largely developed independently of each other, P2P overlay networks in the Internet and mobile wireless ad hoc networks share many key characteristics such as self-organization and decentralization due to the common nature of their distributed components: a P2P overlay network consists of a dynamically changing set of nodes connected via the Internet, and a mobile ad hoc network consists of mobile nodes communicating with each other using multi-hop wireless links. These common characteristics lead to further similarities between the two types of networks: (1) both have a flat and frequently changing topology, caused by node join and leave in P2P overlays and MANETs and additionally terminal mobility of the nodes in MANETs; and (2) both use hop-by-hop connection establishment. Per-hop connections in P2P overlays are typically via TCP links with physically unlimited range, whereas per-hop connections in MANETs are via wireless links, limited by the radio transmission range.

The common characteristics shared by P2P overlays and MANETs also dictate that both networks are faced with the same fundamental challenge, that is, to provide connectivity in a decentralized, dynamic environment. Thus, there exists a synergy between these two types of networks in terms of the design goals and principles of their routing protocols and applications built on top: both P2P and MANET routing protocols and applications have to deal with dynamic network topologies due to membership changes or mobility. The common characteristics and design goals between P2P overlays and mobile ad hoc networks point to a new research direction in networking, that is, to exploit the synergy between P2P overlays and mobile ad hoc networks to design better routing protocols and applications.

47.1.1   Scope of Chapter

We use Figure 47.1 to define the scope of this chapter. Figure 47.1 depicts the network protocol design space for the Internet and MANETs as well as example protocols for each design subspace.

The protocols developed for the Internet can be broadly classified into two categories:

1. Protocols that have been developed using a router-assisted approach. Examples of such protocols include Internet routing protocols such as RIP,[25] OSPF,[45] and BGP[54] and IP multicast protocols such as MOSPF[44] and DVMRP.[21]

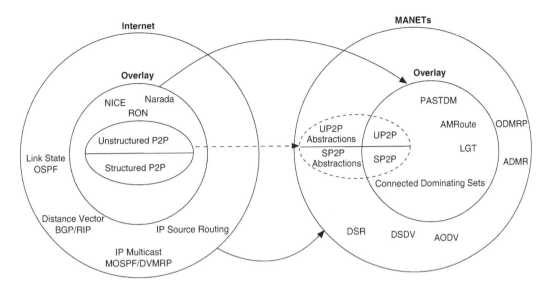

FIGURE 47.1 Scope of this chapter: how to map unstructured and structured P2P overlays in the Internet to MANETs. S-P2P refers to structured P2P and U-P2P refers to unstructured P2P.

2. Protocols that have been developed for an overlay of nodes (depicted by the inner circle in Figure 47.1), which typically refer to end hosts on the edge of the network.

P2P research in the Internet has concentrated on overlay networks of end hosts. The set of overlay-based protocols includes unstructured P2P protocols (e.g., Gnutella,[1] KaZaA[33]) and structured P2P protocols (e.g., Chord,[61] Tapestry,[71] Pastry,[56] CAN[51]). Examples of other overlay-based protocols include RON,[4] Narada,[26] and NICE.[5]

Similarly, MANETs have protocols parallel to those in the Internet domain. In MANETs, some approaches involve all the nodes in the network. These include routing protocols such as DSR[30] and AODV[50] as well as multicast protocols such as ADMR[28] and ODMRP.[37] On the other hand, some protocols (e.g., overlay-based multicast such as AMRoute,[69] virtual infrastructures[17,40]) have been developed to operate on an overlay (subset) of nodes in MANETs.

The scope of this chapter is on mapping the unstructured and structured P2P protocols developed for the Internet to the MANET protocol design space. Such a mapping raises several key questions:

- Are there any benefits of this mapping for designing new protocols and applications in MANETs?
- Should these P2P overlay protocols be mapped to an overlay or a non-overlay (the whole network) in MANETs?
- How should these P2P overlay protocols be adapted for MANETs?

In the remainder of the chapter, we survey recent studies on mapping unstructured and structured P2P overlay protocols in the Internet to MANETs. We first survey studies investigating how to efficiently support P2P overlay abstractions in MANETs. We then discuss how and when applications in MANETs can benefit from the use of P2P abstractions and examine the trade-offs involved in their operations.

We emphasize that due to the peer-to-peer nature of MANETs, all protocols designed for MANETs are inherently peer-to-peer. Examples include multi-hop routing protocols (e.g., DSR and AODV), multicast routing protocols (e.g., ADMR and ODMRP), overlay based multicast routing protocols (e.g., AMRoute and PAST-DM[23]), as well as numerous gossip-based data dissemination protocols (e.g., 7DS[48] and iFlow[39]). The scope of this chapter is focused on the P2P overlay protocols and applications that have been recently developed for the Internet and evaluates their usability and applicability for MANETs, as shown by the dashed arrow and dashed oval in Figure 47.1.

The rest of the chapter is organized as follows. Section 47.2 gives a brief overview of peer-to-peer overlay networks developed for the Internet. Section 47.3 discusses the motivation and challenges in supporting P2P overlay abstractions in MANETs. Section 47.4 and Section 47.5 survey recent work on supporting unstructured and structured P2P overlay abstractions in MANETs, respectively. Section 47.6 discusses potential applications of P2P overlay abstractions in the MANETs. Section 47.7 briefly summarizes other overlay-based techniques in MANETs. Section 47.8 concludes with a set of challenges and problems requiring further research.

47.2   Background on P2P Overlay Networks

The numerous P2P overlay networks for the Internet that have been proposed in the past few years can be broadly classified into two categories:

1. *Unstructured.* Unstructured P2P overlay networks as exemplified by Gnutella[1] do not have precise control over the overlay topology. The network is typically formed by nodes joining the network following some loose rules, for example, a node joining a Gnutella network starts by connecting to nodes in a *host cache* file that stores Gnutella nodes learned from the last time the node was part of a Gnutella network,[2] and a Gnutella node typically specifies a default maximal number of neighbors in the Gnutella overlay. The resulting network topology follows certain patterns, for example power-law-like,[42] but the placement of an object or a file is not based on any knowledge about the topology. Furthermore, the overlay is often not network-proximity-aware; that is, neighboring nodes in the overlay may be far away from each other in the underlying Internet topology. The typical way of locating an object in an unstructured overlay is to flood the network in which a query is propagated to overlay neighbors within a controlled radius. We note that more efficient ways of locating objects in unstructured overlays exist (for example, Ref. 12) and a detailed discussion is beyond the scope of this survey. While the lack of proximity-awareness and flooding-based object location are inefficient, the consequent advantage is that unstructured overlay networks and the companion object location mechanisms that do not rely on any precise structure of the topology are highly resilient to frequent node join and departure.

2. *Structured.* To overcome the inefficiency with object location in unstructured networks, structured overlay networks have been proposed to combine the inherent self-organization, decentralization, and diversity of unstructured P2P overlays with a scalable and efficient routing algorithm that can reliably locate objects in a bounded number of routing hops, typically logarithmic in the network size, while exploiting proximity in the underlying Internet topology. Numerous structured P2P overlays have been proposed, such as CAN,[51] Chord,[61] Pastry,[56] and Tapestry.[71] The routing of such structured P2P overlays effectively implements scalable and fault-tolerant *distributed hash tables* (DHTs): each node in the network has a unique node identifier (nodeID) and each data item stored in the network has a unique key, nodeIDs and keys live in the same namespace, and a message with a key is routed (mapped) to a unique node in the overlay.* Thus, DHTs allow data to be inserted without knowing where it will be stored and requests for data to be routed without requiring any knowledge of where the corresponding data items are stored. To maintain efficient routing, nodes in a structured overlay must maintain neighboring nodes that satisfy certain criteria in the namespace. As a result, structured overlays are conceptually less resilient to frequent node join and departure.

---

*In the rest of the chapter, we use DHTs and structured P2P overlays interchangeably whenever appropriate for this reason.

47.3 Supporting P2P Overlay Abstractions in MANETs

In this section we first elaborate the difference between supporting P2P overlay abstractions and P2P overlays in MANETs. We then discuss the motivations for and challenges in supporting P2P overlay abstractions in MANETs. Finally, we give a taxonomy of different design approaches to supporting P2P overlay abstractions in MANETs.

47.3.1 Why P2P Overlay Abstractions and not P2P Overlays?

Although an overlay consisting of a subset of the nodes could be constructed in an ad hoc network, similarly to in the Internet, all studies of P2P overlays in MANETs have assumed the involvement of all nodes. This is because an ad hoc network is typically formed of nodes that collaborate with each other to enable communication among all the nodes. Because all nodes are involved, the notion of overlay in the Internet (i.e., consisting of a subset of end hosts) is no longer precise. What the existing studies really support is effectively a *P2P overlay abstraction*, that is, borrowing the topologies and objection location techniques (for example, DHTs) of P2P overlays developed in the Internet and supporting them in MANETs. In the rest of the chapter, we focus on P2P overlay abstractions in MANETs, shown as "U-P2P abstractions" and "S-P2P abstractions" in Figure 47.1.

Because overlays in the Internet are built to circumvent the fact that router-assisted approaches are not feasible, message routing is done in the application-layer overlay. In contrast, because nodes in MANETs are end hosts as well as routers, all nodes in MANETs are effectively involved in supporting P2P overlay abstractions, and thus P2P overlay abstractions in MANETs have the option of being implemented either at the network layer or above, that is, at the application layer. These design options are discussed further in Section 47.3.4.

47.3.2 Why Supporting P2P Overlay Abstractions in MANETs?

The motivation for supporting P2P overlay abstractions in MANETs is that MANETs and P2P overlays share many essential characteristics such as decentralization that make P2P applications developed in the Internet potential candidates for deployment in MANETs. For example, P2P file-sharing applications such as Gnutella are designed with a serverless architecture, which makes them potentially well suited to the infrastructure-less MANET environment.

Structured P2P overlays developed for the Internet have been shown to provide a general substrate for building a variety of scalable and robust distributed applications for the Internet, such as distributed storage systems,[16,55] application-level multicast,[10,11,52,70,72] and content-based full-text search.[64] A DHT abstraction implemented by these structured P2P overlays shields many difficult issues, including fault tolerance, locating objects, scalability, availability, load balancing, and incremental deployment from the distributed application designers. The motivation for supporting the DHT abstraction in MANETs is similar to its counterpart in the Internet. Due to its support for many properties common to distributed applications, a DHT abstraction, if deployed in MANETs, could similarly shield many complexities in constructing distributed applications from the application designers. For example, applications such as file sharing and resource discovery could benefit from the distributed insert/lookup convergence provided by DHTs.

47.3.3 Challenges in Supporting P2P Overlay Abstractions in MANETs

Many fundamental differences between the Internet and a mobile ad hoc network pose challenges to implementing P2P overlay abstractions in MANETs, including:

- *Bandwidth limitation.* Unlike the wired Internet, MANETs have lower network capacity due to the use of wireless channels. This limits the usability of P2P protocols that have high message overhead.

- *Multi-access interference.* Multiple access techniques such as CSMA/CA are required for nodes in a MANET to acquire the wireless channel and transmit data. Because no central coordination point exists in a MANET, collisions and delays occur in acquiring the wireless channel. These problems can be aggravated further by P2P applications that rely on high message overhead mechanisms such as frequent pings for delay estimation.

- *Node mobility.* In the Internet, the topology of a P2P overlay changes at a large time scale. On the other hand, in a MANET, limited transmission range and node mobility results in frequent topology changes. This places pressure on P2P applications constructed in MANETs to update the overlay topology much more frequently to maintain the matching between the overlay topology and the underlying network topology.

    Topology maintenance in P2P overlays is achieved by periodically probing current and candidate neighbors and selecting closer nodes as new neighbors. In the Internet, such an approach is feasible because routes between overlay neighbors change rarely; the maintenance is more for the purpose of checking whether a neighbor is alive. Even if the routes do change, there is little overhead for discovering a new route to an overlay neighbor in the Internet. In MANETs, due to limited link capacity and multi-access interference, probing an overlay neighbor can be much more costly.

- *Churn.* In the Internet, structured P2P protocols are particularly affected by "churn" (frequent node joining and leaving the network). This occurs in the Internet primarily because most nodes in the systems are end-user desktops rather than "always-on" servers. If structured P2P protocols are used in MANETs, they could potentially suffer poor performance due to high churn caused by not only the transience of node in terms of being "on" and "off," but also network partition (un-reachability) caused by node mobility.

- *Lack of infrastructure.* Certain P2P protocols make use of some infrastructure components in their designs. For example, a P2P routing protocol may assign node identifiers based on locations determined from static landmarks to improve routing performance.[51] These techniques may not be usable in MANETs due to the lack of any static infrastructure.

- *Limited energy.* Most P2P applications in the Internet are not designed to operate with minimum message transmissions. In an energy-limited environment such as a MANET, it may be very important for nodes to reduce the number of message transmissions while keeping the performance acceptable. For example, most P2P protocols use proactive maintenance of state information (e.g., periodically exchanging and probing routing table entries to maintain the proximity of routing tables in Pastry[9]). In a MANET, reactive approaches exemplified by protocols such as AODV[50] and DSR,[30] in which the protocol state is refreshed only when required, may be more effective.

- *State-efficiency trade-off.* Structured P2P routing protocols trade increased number of routing hops (in the overlay) for dramatically reduced state in order to scale to millions of nodes. In a MANET, it is an open question whether such a large number of nodes can ever effectively be organized into one infrastructure-less network. Because MANETs are likely to be much smaller than overlays in the Internet and have much smaller capacity compared to the Internet, it may be more effective to keep more state at each node if it can reduce the number of hops traveled by messages.

- *Addressing.* Nodes in a MANET are likely to disconnect and reconnect to the network many times. Although no specific addressing architecture has been standardized for MANETs, it is plausible to assume that nodes will have changing IP addresses over time. This could challenge structured P2P protocols that store logical to physical address (nodeID-to-IP) mappings in their routing tables. In some cases, nodeIDs are assigned as hashed IP addresses, and thus the nodeIDs of MANET nodes will continuously change. This can result in consistency issues for structured P2P protocols. If nodeIDs are assigned by hashing static MAC addresses, then the nodeID-to-IP mappings of nodes could change over time.

- *Namespace versus physical space routing.* Structured P2P protocols route packets in a logical namespace, that is, based on nodeIDs. The routing in the logical namespace could be tree-like (prefix based)[56,71] or skiplist-like.[61] Although prefix-based routing allows for natural inclusion of topology awareness, that is, selecting physically nearby logical nodes with matching prefix nodeIDs in the namespace at each routing step, the routes going through the overlay are unavoidably longer than the direct routes in the underlying network. Thus, this "route stretch," along with the inflexibility in the selection of intermediate overlay nodes, could cause inefficiency for certain applications in MANETs, such as multicast.

47.3.4 A Taxonomy of Design Approaches

The various design approaches for applying peer-to-peer to MANETs can be classified based on the nature of the P2P overlays (structured versus unstructured). In addition, as discussed in Section 47.3.1, because nodes in MANETs are end hosts as well as routers and all nodes in MANETs are effectively involved in supporting P2P overlay abstractions, the P2P overlay abstraction in MANETs has the option of being implemented either at the network layer or above, that is, at the application layer. If a protocol is layered over an existing routing protocol for MANETs (DSR,[30] AODV,[50] etc.), we classify it as the *layered design*. If a protocol is integrated with a MANET routing protocol at the network layer, we classify it as the *integrated design*.

The layered design allows P2P applications developed in the Internet to be easily ported to MANETs. Furthermore, it decouples functionalities of the application layer (P2P application) and the network layer (routing), which enables independent development of protocols at the two layers. However, MANETs are a limited resource environment where the performance can be more important than portability and separation of functionalities. In fact, a multitude of cross-layer techniques have been proposed for MANETs for this reason. Thus, for supporting P2P overlay abstractions in MANETs, it is important to study the potential benefits of cross-layering, for example, in an integrated design.

The four design approaches to supporting P2P overlay abstractions in MANETs are:

1. *Layered and unstructured.* In this design, off-the-shelf unstructured P2P protocols (for example, Gnutella) are operated on top of an existing MANET routing protocol. This design is similar to the approach in the Internet, which layers a P2P protocol on top of the existing IP infrastructure.
2. *Integrated and unstructured.* In this design, the operation of an unstructured P2P protocol is closely integrated with the operation of a MANET routing protocol to support unstructured P2P APIs.
3. *Layered and structured.* In this design, a structured P2P protocol such as Pastry, CAN, Chord, or Tapestry runs as an application over a MANET routing protocol.
4. *Integrated and structured.* In this design, a structured P2P protocol is integrated with the operations of a MANET routing protocol to provide a distributed hash table abstraction.

47.4 Unstructured P2P Overlay Abstractions in MANETs

47.4.1 Layered Design

The work of Oliveira et al.[47] studies the performance of an unstructured P2P application in a MANET running over three existing MANET routing protocols. Specifically, the authors studied the relative performance of DSR, AODV, and DSDV when supporting an unstructured P2P application based on Gnutella. The results observed show that the performances of the protocols are different from those observed for typical unicast applications assumed in previous studies.[8,20] In addition, the packet delivery ratios observed are lower than those observed for typical unicast applications.

The disadvantage of the layered unstructured design is explained by considering a query search for a data item (e.g., as in Gnutella). Similar to in the Internet, in this design, Gnutella operating at the application layer will contact the overlay neighbors to resolve the query. These overlay neighbors will further forward the query to their neighbors if the decremented TTL of the query is greater than zero. However, due to node mobility, these overlay neighbors may not reflect the current physical topology of the ad hoc network, and thus may need a multi-hop route to be reached. As a result, each such overlay hop required by Gnutella at the application layer could result in a costly flooding-based route discovery by the multi-hop routing protocol. This suggests strong motivation for continuous update of the list of overlay neighbors to reflect the current physical neighbors so that neighbors can be contacted via one-hop physical broadcast rather than multi-hop routing.

47.4.2   Integrated Design

The work by Klemm et al.[34] proposes integrating an unstructured Gnutella-like P2P application into the network layer and compares it to a layered design similar to that of Oliveira et al.[47] The proposed file-sharing application, ORION, allows for the setup of on-demand overlay connections that closely match the physical topology of the underlying MANET. ORION integrates the query process required for the P2P operation with the routing process done by the routing protocols in MANETs. Specifically, ORION combines the P2P operation with routing techniques from AODV[50] and the Simple Multicast and Broadcast Protocol.[27] When a query for a data item arrives, ORION employs one-hop broadcast to contact all its physical neighbors in one transmission. These nodes, in turn, contact their physical neighbors if they are unable to resolve the query. Nodes that can resolve the query stop further propagation of the query (i.e., one-hop broadcast) and reply to the initiator of the query. Note that this may require multi-hop routing. However, such a route has already been set up as the query is propagated in the network, similar to that in AODV.

47.4.3   Comparison

The advantage of the integrated design of ORION for a file-sharing application was experimentally observed by Klemm et al.[34] ORION is implemented in the ns-2 simulator[7] and compared to an off-the-shelf design that layers Gnutella over DSR. The simulations are carried out for a range of network sizes up to 60 nodes using a random waypoint mobility model in which nodes move at a maximum speed of 2 m/s with a pause time of 50 seconds. The search accuracy (i.e., the fraction of received unique files out of all the files actually in the network that match the search query) is measured for both designs. The results show that as the network size is increased from 0 to 60 nodes, the search accuracy of the integrated design increases and is always higher than that of the layered design. In fact, the accuracy for the layered design decreases at large network sizes due to increased overhead. The layered design experiences a large fraction of total packet transmissions related to control traffic.

In summary, the primary factor for the worst performance of the layered design is not the inefficiency of the search mechanism, but rather the overhead of maintaining static overlay neighbor connections that are not adapted to the dynamic physical topology. The results of the study indicate that the integrated unstructured design has significantly lower overhead compared to the layered design while achieving better performance according to application-specific metrics.

47.5   Structured P2P Overlay Abstractions in MANETs

We separate the discussion of supporting DHTs in wireless ad hoc networks with and without GPS support. With GPS support (or some other type of position services), each node in the network can find out its geographic location. Previously, this location information was exploited to improve the efficiency and scalability of multi-hop routing protocols[43] or support other types of services such as geocast.[46] Such location information can also be potentially used to improve the efficiency of DHTs.

47.5.1   DHTs in MANETs without GPS Support

Both the layered design and the integrated design in supporting a DHT abstraction in highly dynamic mobile ad hoc networks are explored in the work of Pucha et al.[19] In particular, the simple design of directly overlaying a DHT on top of an existing multi-hop routing protocol for MANETs is compared with Ekta, which integrates a DHT with a multi-hop routing protocol at the network layer.

47.5.1.1 Layered Design

In the layered design[19], a proximity-aware DHT, Pastry,[9,56] is directly layered on top of a multi-hop routing protocol, DSR,[30] with minimum modifications to the routing protocol. Pastry maintains its leaf set and routing table entries without source routes and DSR maintains source routes passively as per the demand of the Pastry routing state.

However, a straightforward layering is not pragmatic, and three modifications are made to accommodate the shared medium access nature of MANETs: (1) Pastry's node joining process is modified to use expanding ring search in locating a bootstrap node to join the network; (2) the original Pastry uses an expensive "ping" mechanism with a delay metric to measure and maintain the proximity of nodes in its routing tables; Pastry is modified to use a hop count metric for proximity because in MANETs, delay is affected by many factors and has a high variability; (3) to reduce the cost of this proximity probing, DSR is modified to export an API that allows Pastry to inquire about the proximity values for nodes in which it is interested. DSR can then use its cache to reply to "pings" from Pastry if there is a cached path to the node being pinged.

In summary, the layered design is similar to implementing a DHT in the Internet; it leverages the existing routing infrastructure for MANETs to the fullest extent. This design, while consistent with the layered principle of the ISO model of networking, makes it difficult to exploit many optimization opportunities from the interactions between the DHT protocol and the underlying multi-hop routing protocol. For example, when the routing protocol is Dynamic Source Routing (DSR),[30] which uses caching to reduce the routing overhead, it is difficult for the routing structures of the DHT and the route cache of DSR to coordinate with each other to optimally discover and maintain source routes.

47.5.1.2 Integrated Design

In the integrated design, called Ekta by Das et al.,[19] the functions performed by the Pastry DHT protocol operating in a logical namespace are fully integrated with the MANET multi-hop routing protocol DSR operating in a physical namespace. The key idea of the integration is to bring the DHT routing protocol of Pastry to the network layer of MANETs via a one-to-one mapping between the nodeIDs of the mobile nodes in the namespace and their IP addresses. With this integration, the routing structures of a DHT and of a multi-hop routing protocol (e.g., the route cache of DSR) are integrated into a single structure that can maximally exploit the interactions between the two protocols to optimize the routing performance.

- *Node addressing.* Ekta assigns unique nodeIDs to nodes in a MANET by hashing the IP addresses of the hosts using collision-resistant hashing functions such as SHA-1.[22]
- *Node state.* The structures of the routing table and the leaf set stored in each Ekta node are similar to those in Pastry. The difference lies in the content of each leaf set and routing table entry. Because there is no underlying routing infrastructure in MANETs, each entry in the Ekta leaf set and routing table stores a nodeID and a source route to reach the designated node. As in Pastry, any routing table entry is chosen such that it is physically closer than the other choices for that routing table entry. This is achieved by making use of the vast amount of indirectly received routes (from overhearing or forwarded messages) as described below in optimizations.

  For efficiency, each routing table entry stores a vector of source routes to one or more nodes that match the prefix of that entry. Similarly, each leaf set entry stores multiple routes to the designated node. The replacement algorithm used in each leaf set or routing table entry is Least Recently Discovered (LRD), disregarding whether a route is discovered directly or indirectly. When looking up a route from a leaf set or routing table entry, the freshest among the shortest routes in that vector entry is returned.

- *Routing.* In Ekta, a message with a key is routed using Pastry's prefix-based routing procedure and delivered to the destination node whose nodeID is numerically closest to the message key. When a route lookup for the next logical hop returns a next-hop node from the leaf set for which a source route does not exist, Ekta initiates route discovery to discover a new source route. On the other hand, if the node selected as the next hop is from the routing table, a modified prefix-based route discovery is performed to discover routes to any node whose nodeID matches the prefix for that routing table entry. Note that each hop in the Ekta network is a multi-hop source route, whereas each hop in a corresponding Pastry network is a multi-hop Internet route.

- *Optimizations.* Ekta inherits all of the optimizations on route discovery and route maintenance used by the DSR protocol. In addition, Ekta updates its routing table and leafset using routes snooped while forwarding and overhearing packets, thus constantly discovering fresh and low-proximity routes for the leaf set and the routing table entries. In addition to the "prefix-based view" of the routing table and the "neighbor-node view" of the leaf set, the Ekta routing structures can be viewed as two caches of source routes. These can be used to support unicast routing by Ekta whenever required by the application. For example, an *Insert* operation in a DHT-based application can travel over multiple hops in the nodeID space while an acknowledgment to the *Insert* could be efficiently unicast back to the originator.

47.5.1.3   Comparison

A detailed simulation study[19] was also performed to compare the layered design with the integrated design. The results show that the integrated design of supporting the DHT abstraction in MANETs used by Ekta is superior to the layered design in terms of the number of data packets successfully delivered and the average delay in delivering the packets while incurring comparable routing overhead. These results suggest that integrating the functionalities of the DHT into the routing layer is much more efficient than having two independent layers with minimal interactions in supporting a DHT abstraction in MANETs.

47.5.2   DHTs in MANETs with GPS Support

The geographic location system (GLS) in GRID[38] is a scalable location service that performs the mapping of a node identifier to its location. The implementation of GLS effectively provides a DHT abstraction: it routes a message with a message key Y to a node whose nodeID is closest to Y. However, the implementation of GLS itself requires both GPS support as well as building a distributed location database.

The Geographic Hash Table (GHT)[53] is inspired by DHTs in the Internet but proposed to support the data-centric storage model in sensor networks. GHT hashes keys into geographic coordinates, and stores a key-value pair at the sensor node geographically nearest the hash of its key. That is, different from the objection location and routing in the namespace abstraction of DHT, GHT provides the distributed hashing abstraction directly in the geographic location space. Such an abstraction suits the data-centric storage model, in which a data object is stored on the unique node determined based on the object's name.

To implement GHT routing, GHT requires GPS support. In fact, GHT is implemented by extending the GPSR[32] geographic forwarding protocol with the notions of home node (the node nearest a hashed location) and home perimeter, which contains nodes enclosing a hashed location. To ensure persistence and consistency of stored key-value pairs in the presence of node failure and mobility, the stored key-value pairs are replicated among nodes on the home perimeter and are periodically refreshed.

47.6   Application of P2P Overlay Abstractions in MANETs

In this section we revisit distributed applications that have been built on top of P2P overlays in the Internet, and discuss whether it is suitable to build them on top of the P2P overlay abstractions supported in MANETs.

The design trade-offs involved when porting applications for unstructured P2P overlays depend on how the overlay topology maintained by the applications maps to the physical one. The most important modification that will be required of any unstructured P2P application deployed in MANETs is to reconcile the overlay neighbors with the physical neighbors continuously. Thus, the unstructured P2P applications would need to incorporate mechanisms for choosing physically nearby nodes as overlay neighbors, as well as techniques to gradually change the set of overlay neighbors as a node or its neighbors move.

There are subtle trade-offs involved when using a structured P2P (DHT) substrate for applications in MANETs. An efficient DHT substrate in MANETs such as Ekta can greatly ease the construction of distributed applications and services in MANETs by shielding many common and difficult issues such as fault tolerance, object location, load balancing, and incremental deployment from the developer. These issues are especially challenging in a wireless, mobile environment. Thus, providing a DHT substrate in MANETs is significant because it removes the need for developers to optimize each individual application in order to perform efficiently in MANETs. However, due to node mobility in MANETs, any DHT substrate based on on-demand routing can trigger repeated flooding-based route discoveries to discover and maintain routes. That is, an application built on top of DHT in MANETs may also experience many floodings of messages. Thus, the real question is whether applications built on top of the DHT substrate can be as efficient as or more efficient than those individually optimized to operate in MANETs.

47.6.1 File Sharing

File sharing is the most popular P2P application currently in use in the Internet. Due to the collaborative nature envisioned for MANETs, file sharing could potentially be one of the important applications for ad hoc networks. However, as discussed in Section 47.4, although Gnutella is widely used in the Internet, directly running applications like Gnutella over a routing protocol for MANETs could result in poor performance due to the mismatch between the overlay topology and the underlying network topology.

The work of Klemm et al.[34] develops an integrated unstructured P2P Protocol (ORION) that works efficiently in MANETs. As discussed in Section 47.4, ORION uses one-hop broadcast flooding when searching for files in order to use the current physical neighbors instead of overlay neighbors. ORION also maintains a file-based routing table similar to AODV, with the only difference being that it stores next hops to reach files instead of other nodes in the network. Similar to a unicast MANET routing protocol, ORION caches information about the ownership of files learned from forwarded messages in its file routing table to improve routing performance. When a query for a file reaches a node, it can send a response if it stores that file or if it knows of a route to the file through its file routing table. Thus, responses to queries performed by a node result in widespread caching of file information throughout the network. This information can then be used to respond to future requests.

Additionally, the authors[34] use their own file transfer protocol different from TCP. The motivation for this approach is that the current sender of a file could change with network conditions, making an end-to-end approach like TCP inefficient. A file is split into equal-sized blocks prior to transfer. A file is fetched block by block by the querying node. This allows for parts of files to be fetched from different nodes based on the current network conditions. Because TCP is not used, ORION incorporates its own packet scheduling and loss-recovery mechanisms. File blocks can arrive out of order as long as one copy for each block is received.

The integrated design of a P2P file-sharing application (ORION) is experimentally compared to an off-the-shelf approach that layers Gnutella over DSR and AODV.[34] The simulations are carried out for a range of network sizes up to 65 nodes using a random waypoint mobility model in which nodes move at a maximum speed of 2 m/s with a pause time of 50 seconds. The number of successful file transfers is measured for 20,000 search queries, each for a file of size 3 MB (the size of a typical MP3 file), for both designs. The results show that ORION is able to complete up to 40 percent more file transfers while using lower bandwidth compared to the layered design.

In principle, and similar to how the PAST[55] storage system is implemented on top of the Pastry DHT, a storage system in MANETs can be constructed on top of Ekta. However, the trade-offs involved and comparison with an unstructured integrated approach remain open issues.

47.6.2 Resource Discovery

Because wireless ad hoc networks are typically comprised of a wide variety of heterogeneous devices with varying energy resources, capabilities, and services to offer, such systems tend to rely on peer cooperation to efficiently use each other's resources. Examples of resource discovery include discovering nodes with GPS devices so other nodes can approximate their own locations, collecting sensed environmental information from mobile sensors, contacting location and directory servers, and locating people with specific capabilities in disaster relief and battlefield networks. For example, a resource lookup in a platoon of soldiers or in a team of coordinating disaster relief personnel can be of the form *"Find the closest medic."* These examples show that efficient discovery of resources that meet certain requirements in an ad hoc network is of great importance in building a variety of distributed applications in ad hoc networks.

The work by Pucha et al.[19] first formally defines the resource discovery problem in mobile ad hoc networks and then presents two alternative designs for resource discovery: (1) an integrated structured solution built on top of Ekta (Ekta-RD), and (2) an integrated unstructured solution directly built on top of physical layer broadcast (DSR-RD). Both versions are modeled after the Service Location Protocol framework.[24]

In DSR-RD, the resource discovery application is integrated with the DSR routing protocol. It essentially uses physical layer broadcast augmented with source routing to perform resource discovery as follows. Each node transmits RESOURCE REQUEST packets (similar to ROUTE REQUEST packets of DSR), and each node that does not own the resource requested rebroadcasts the RESOURCE REQUEST packet after encoding its IP address into the source route. If a node in the network owns that resource, it responds with a service reply packet (RESOURCE REPLY) that is unicast back to the requester similar to the ROUTE REPLY packet of DSR. RESOURCE REQUEST packets contain sequence numbers to ensure that the overhead incurred is at most N (network size) packets. The remaining overhead is the number of RESOURCE REPLY packet transmissions, which is determined by the degree of replication of the resource being requested.

In Ekta-RD, the resource discovery application is built on top of the Ekta DHT substrate. Ekta provides three DHT APIs, *route(Message, Key)*, *route(Message, IP Address,)* and *broadcast(Message, Broadcast Address)*, as well as an additional API *Proximity(IP Address)*. The *Proximity(IP Address)* interface returns the hop distance of the node specified by looking up the locally cached routes. If no such route is cached, Ekta returns null.

Ekta-RD simply relies on Ekta to route a RESOURCE REQUEST packet to the correct directory agent and receive a reply. The announcement of each available resource is inserted into Ekta via hashing the resource identifier into a *Key*, and invoking *route(Message, Key)* of Ekta with the *Message* containing the resource description. Similarly, when a resource is required, Ekta-RD hashes the resource identifier into a *key* and invokes *route(Message, key)* of Ekta. Because of the DHT abstraction, Ekta will route the announcement and queries for the same resource to the same node in Ekta—the directory agent for the mapped resource. The node replies to the requester with a list of nodes that own the resource in the network. The requester then finds the closest node out of this list using Ekta's *Proximity(IP Address)* interface and contacts the chosen node to use the resource. If Ekta cannot determine the proximity of any node in the list returned in the RESOURCE REPLY, the application randomly selects a node out of the list and contacts that node to use the resource. This will trigger a ROUTE REQUEST for that randomly chosen node by Ekta.

The experimental results[19] show that for the resource discovery application, the DHT-based approach Ekta-RD consistently outperforms the broadcast-based approach DSR-RD for a wide range of application parameters. Specifically, DSR-RD incurs comparable routing overhead to Ekta-RD for high inter-arrival time between resource queries. As the inter-arrival time decreases, Ekta-RD incurs up to an order of magnitude lower overhead compared to DSR-RD. Furthermore, these results hold true for a wide range of mobilities. These results suggest that efficient structured P2P substrates such as Ekta provide a viable and efficient approach to building distributed applications in mobile ad hoc networks.

47.6.3   Multicast

A wide variety of work has concentrated on developing overlay multicast protocols for the Internet,[5,11,26,52,70,72] due to the difficulties inherent in the router assisted IP multicast. Among these, many protocols specifically use structured P2P routing techniques for scalable multicasting.[10,11,52,70] Although no prior work has adapted structured P2P-based multicast protocols to MANETs, we discuss the issues associated with this approach using Scribe[11] as an example. Scribe is built on top of Pastry, which provides it with a generic object location and routing substrate. It leverages Pastry's reliability, self-organization, and locality properties to create and manage groups and to build efficient per-group multicast trees.

Ekta provides a structured object location and routing substrate for MANETs, similar to Pastry for the Internet, and thus can potentially be leveraged by Scribe to provide multicast services in MANETs. The reverse path tree built by Scribe/Ekta would be an overlay tree based on nodeIDs. Each node performs prefix-based routing to a root node (based on the hash of the group identifier). Consequently, many group members in a local region choose a nearby common node whose nodeID matches a longer prefix to the group identifier's hash than their own nodeID. That node, in turn, chooses another such node until the root node is reached and the tree is formed. Note that non-group members may also be chosen as part of the tree, depending on their nodeIDs and locations.

Although there has not been any study on P2P-based multicast in MANETs as described above, the performance benefit of this approach is potentially limited for the following reasons.

First, due to the constraints imposed by prefix-based routing in the namespace by Scribe/Ekta, the tree constructed for a group may not be as efficient as one constructed taking into account the physical connectivity among the nodes. MANET multicast protocols typically use some sort of broadcast mechanism to construct a multicast tree. For example, consider two non-leaf nodes (A and B) in a multicast tree that are within transmission range of each other. In Scribe/Ekta, leaf nodes can select either of these two nodes as parents, depending on their routing tables and nodeIDs. This can result in a multicast tree where both these nodes are internal nodes of the tree. However, in a broadcast-based approach to tree construction (e.g., Ref. 67), the nearby nodes would choose only one of these two nodes (say A) as a parent. This chosen node would then be an internal node of the tree. Note that in the broadcast case, one multi-hop route is needed to connect node A to its parent node. For Scribe/Ekta, both nodes A and B need to connect to their parent nodes via multi-hop routing, resulting in increased bandwidth consumption. Additionally, some tree building approaches for MANETs also take advantage of Wireless Multicast Advantage,[66] which further reduces their bandwidth costs by allowing multiple receivers to receive packets with one transmission.

Second, with node mobility, the overlay tree built by Scribe/Ekta must be continuously maintained and updated, which may cause increased overhead. Although low overhead prefix requests can be used to repair the overlay tree, Scribe/Ekta is likely to be less robust with increased mobility as compared to *mesh*-based MANET multicast protocols.[28,37] Mesh-based multicast protocols are better at dealing with link breaks due to mobility. Note that this trade-off is not specific to using a P2P overlay but is more generally about tree-based versus mesh-based schemes. Thus, to maintain a certain degree of robustness, Scribe/Ekta-based multicast is likely to pay a higher cost than previously proposed mesh-based MANET multicast protocols.

In general, most applications that involve all or a large subset of nodes in the MANET, such as broadcast and multicast, are likely to be more efficient if built directly on top of the physical broadcast mechanism compared to built on top of a structured P2P substrate. This is because an efficient structured P2P substrate will experience many flooding-based route discoveries in maintaining individual links in the overlay, while each such flooding can already reach all or a large number of nodes in the MANET.

47.7   Other Overlay-Based Approaches in MANETs

There have been several studies on building overlays in MANETs that involve a subset of nodes, that is, similar to the overlays in the Internet. So far, all of these studies have focused on constructing overlays to directly support multicast or unicast. That is, the overlays built in such studies are not meant to be general-purpose substrates for building distributed applications and services.

47.7.1   Multicast

Many multicast protocols have been proposed for MANETs[14,18,29,37,57] that maintain state at both group members and non-group members to support a multicast session. This potentially lowers the resilience of the protocols to node mobility due to the higher requirements of state maintenance. Subsequently, overlay multicast protocols for MANETs[23,69] have been proposed to improve robustness and reduce overhead at the cost of potentially sub-optimal multicast trees.

AMRoute[69] uses bi-directional unicast tunnels to organize the multicast group members into a virtual mesh. After mesh creation, one group member, designated the logical core, initiates the creation and maintenance of a shared data delivery multicast tree. To deal with node mobility, the tree is periodically rebuilt. AMRoute can be inefficient due to the use of a static virtual mesh in building a shared tree under network topology changes.

PAST-DM[23] was subsequently proposed to deal with these inefficiencies. PAST-DM continuously optimizes the quality of the multicast tree to reconcile the overlay and current physical topology. The idea is not use a static virtual mesh, but rather allow the virtual mesh to gradually adapt to changes in the underlying network topology. An adapted mesh can then result in better adapted multicast trees. The multicast trees in PAST-DM are locally constructed, source-based Steiner trees built using the Takahashi-Matsuyama heuristic.[63] PAST-DM also tries to eliminate redundant physical links to reduce the bandwidth cost of the multicast session.

A location-guided tree[13] is a multicast tree constructed using location information (e.g., available from GPS) in a MANET, and thus is also a form of overlay multicast. In this approach, a source-based Steiner tree is constructed using the Takahashi-Matsuyama heuristic[63] for performing multicast. Unlike PAST-DM, geometric distances are used as link costs.

NICE-MAN[6] is an overlay-based multicast protocol for MANETs based on the NICE[5] application layer multicast for the Internet. NICE is a multi-source application layer multicast that creates a hierarchy of fully meshed clusters, that is, the sibling nodes sharing the same parent node are fully connected. Because of the full-meshed clusters, NICE is much more resilient to node failures than a simple tree-based multicast protocol. To adapt NICE for MANETs, NICE-MAN makes three changes. First, a cluster leader continuously monitors the cluster members and transfers the cluster leader role to another cluster member if that cluster member becomes closer to the center of the cluster than itself. That is, the cluster leader is continuously reselected to maintain the parent-children locality of the overlay tree. Second, instead of RTT, the network distance metric is changed to use the number of hops in a multi-hop path provided by the underlying unicast routing protocol of MANETs (e.g., AODV). Third, the broadcast capability of the medium is exploited using the overlay NICE tree as a backbone and nodes within the transmission range of an overlay node do not explicitly join the NICE hierarchy.

47.7.2 Virtual Infrastructure

Many routing protocols developed, such as DSR and AODV, involve all nodes in route creation and maintenance while relying on broadcast relaying for route creation. Subsequently, many researchers have worked on routing solutions that enable protocols to use only a subset of the nodes and avoid broadcast relays. To this end, routing based on a connected dominating set (subset) was proposed, where the search space for a route is reduced to the nodes in the set. A set is defined as *dominating* if all the nodes in the system are either in the set or are neighbors (within the transmission range) of nodes in the set.

Constructing a connected dominating set effectively *overlays* a virtual infrastructure (core) in the MANET. Distributed and efficient computation of the connected dominating set has been studied extensively, and many routing protocols based on this approach have been proposed.[3,17,31,35,41,58–60,62,65,68] Virtual infrastructures have been proposed for handling unicast, multicast, as well as QoS traffic.

Virtual infrastructures have also been proposed to provide various services in MANETs, such as mobility management[40] and service discovery.[36]

47.8 Summary and Future Research

In this chapter we presented a survey of support for peer-to-peer overlay abstractions in mobile ad hoc networks fostered by the recent advent of peer-to-peer overlay networks and systems in the Internet. We classify the existing approaches for supporting P2P overlay abstractions in MANETs according to the type of P2P overlays (structured versus unstructured) being modeled and the layer at which the P2P overlay protocol operates. In particular, if the protocol is layered over an existing routing protocol for MANETs, we classify it as the *layered design*; and if the protocol is integrated with a MANET multi-hop routing protocol, we classify it as the *integrated design*. Our classification excludes numerous works that build applications by directly exploiting the inherent peer-to-peer nature of ad hoc networks, for example, gossip-based data dissemination, or construct overlays to directly support multicast or unicast, for example, virtual infrastructures. Such studies, many of which were proposed before the advent of the Internet P2P overlays, do not leverage the P2P overlay abstraction.

The subject of exploiting peer-to-peer overlays in MANETs is relatively new. Many interesting problems require further research, including:

- Given the high dynamics in MANETs due to node mobility, which of the unstructured or structured overlay abstractions is more efficient in supporting common distributed applications such as file sharing?

- How can one efficiently integrate a DHT for the Internet with MANET routing protocols other than DSR to support the DHT abstraction in MANETs? For example, the use of hop-by-hop routing as in AODV[50] and DSDV[49] requires that all nodes along a route to a destination maintain this route. This implies that an integration with Pastry would require these nodes to have a prefix match with the destination nodeID in order to contain the destination nodeID in their prefix-based routing structures. With source routing, however, the intermediate physical hops need not maintain a route to the destination.

- How can one efficiently integrate other DHTs such as CAN and Chord with MANET routing protocols? A Chord-based DHT integrated with DSR would be similar to Ekta in that each node would store in its successor list and finger table a list of source routes to the corresponding nodeIDs. However, because the routing table entries for Chord are required to refer to specific points in the namespace, proximity-aware selection of overlay hops would be less flexible than in Pastry. This same problem would exist in a CAN-based integrated DHT.

- Can DHTs be leveraged to support scalable unicast in MANETs? Current unicast protocols for MANETs can support up to a few hundred nodes, while DHTs in the Internet have been shown to scale to millions of nodes.

- Can incentive techniques developed for P2P overlays in the Internet for encouraging peering nodes to cooperate be applied to MANETs?

- Can distributed security, trust, and reputation techniques developed in P2P overlays in the Internet be applied to MANETs?

- Can routing protocols developed in MANETs be leveraged to improve the performance and functionalities of P2P overlays in the Internet?

Acknowledgment

This research was supported by NSF grant ANI-0338856.

References

1. The Gnutella protocol specification, 2000. http://dss.clip2.com/GnutellaProtocol04.pdf.
2. The Gnutella 0.6 protocol draft, 2002. http://rfc-gnutella.sourceforge.net/.
3. K.M. Alzoubi, P.-J. Wan, and O. Frieder. Message-optimal connected dominating sets in mobile ad hoc networks. In *Proc. of ACM MobiHoc,* June 2002.
4. D. Andersen, H. Balakrishnan, F. Kaashoek, and R. Morris. Resilient overlay networks. In *Proc. of ACM SOSP,* October 2001.
5. S. Banerjee, B. Bhattacharjee, and C. Kommareddy. Scalable application layer multicast. In *Proc. of ACM SIGCOMM,* August 2002.
6. S. Blodt. Efficient end system multicast for mobile ad hoc networks. In *Proc. of the 1st International Workshop on Mobile Peer-to-Peer Computing (MP2P'04),* March 2004.

7. L. Breslau et al. Advances in network simulation. *IEEE Computer,* 33(5):59–67, May 2000.

8. J. Broch et al. A performance comparison of multi-hop wireless ad hoc network routing protocols. In *Proc. of ACM MobiCom,* October 1998.

9. M. Castro, P. Druschel, Y.C. Hu, and A. Rowstron. Exploiting network proximity in peer-to-peer overlay networks. Technical Report MSR-TR-2002-82, 2002.

10. M. Castro, P. Druschel, A.-M. Kermarrec, A. Nandi, A. Rowstron, and A. Singh. Splitstream: high-bandwidth multicast in a cooperative environment. In *Proc. of ACM SOSP,* October 2003.

11. M. Castro, P. Druschel, A.-M. Kermarrec, and A. Rowstron. Scribe: a large-scale and decentralized application-level multicast infrastructure. *IEEE Journal on Selected Areas in Communication (JSAC),* 20(8), October 2002.

12. Y. Chawathe, S. Ratnasamy, L. Breslau, N. Lanham, and S. Shenker. Making Gnutella-like P2P systems scalable. In *Proc. of ACM SIGCOMM,* September 2003.

13. K. Chen and K. Nahrstedt. Effective location-guided tree construction algorithms for small group multicast in MANET. In *Proc. of IEEE INFOCOM,* June 2002.

14. C. Chiang, M. Gerla, and L. Zhang. Forwarding Group Multicast Protocol (FGMP) for multihop, mobile wireless networks. *Baltzer Cluster Computing,* 1(2):187–196, 1998.

15. I. Clarke, O. Sandberg, B. Wiley, and T.W. Hong. Freenet: a distributed anonymous information storage and retrieval system. In *Workshop on Design Issues in Anonymity and Unobservability,* July 2000. ICSI, Berkeley, CA.

16. F. Dabek, M.F. Kaashoek, D. Karger, R. Morris, and I. Stoica. Wide-area cooperative storage with CFS. In *Proc. of ACM SOSP,* October 2001.

17. B. Das and V. Bharghavan. Routing in ad-hoc networks using minimum connected dominating sets. In *Proc. of IEEE ICC,* Volume 1, pages 376–380, Montreal, Canada, June 8–12, 1997.

18. S.K. Das, B.S. Manoj, and C.S.R. Murthy. A Dynamic core based multicast routing protocol for ad hoc wireless networks. In *Proc. of ACM MobiHoc,* June 2002.

19. S.M. Pucha, H. Das, and Y.C. Hu. Ekta: An efficient DHT substrate for distributed applications in mobile ad hoc networks. In *Proc. of IEEE WMCSA,* December 2004.

20. S.R. Das, C.E. Perkins, and E.M. Royer. Performance comparison of two on-demand routing protocols for ad hoc networks. In *Proc. of IEEE INFOCOM,* March 2000.

21. S. Deering, C. Partridge, and D. Waitzman. Distance vector multicast routing protocol. Technical Report 1075, IETF, November 1988.

22. FIPS 180-1. Secure Hash Standard. Technical Report Publication 180-1, Federal Information Processing Standard (FIPS), NIST, U.S. Department of Commerce, Washington, D.C., April 1995.

23. C. Gui and P. Mohapatra. Efficient overlay multicast for mobile ad hoc networks. In *Proc. of IEEE WCNC,* March 2003.

24. E. Guttman. Service location protocol: automatic discovery of IP network services. *IEEE Internet Computing,* 3(4):71–80, 1999.

25. C. Hendrick. Routing Information Protocol. RFC 1058, IETF, June 1988.

26. Y. hua Chu, S.G. Rao, and H. Zhang. A case for end system multicast. In *Proc. of ACM Sigmetrics,* pages 1–12, June 2000.

27. J. Jetcheva, Y.C. Hu, D. Maltz, and D. Johnson. A Simple Protocol for Multicast and Broadcast in Mobile Ad Hoc Networks, IETF Draft, July 2001.

28. J.G. Jetcheva and D.B. Johnson. Adaptive demand-driven multicast routing in multi-hop wireless ad hoc networks. In *Proc. of ACM MobiHoc,* October 2001.

29. L. Ji and M.S. Corson. A lightweight adaptive multicast algorithm. In *Proc. of IEEE GLOBECOM,* November 1998.

30. D.B. Johnson and D.A. Maltz. *Dynamic Source Routing in Ad Hoc Wireless Networks.* Kluwer Academic, 1996.

31. J. Wu, M. Gao, and I. Stojmenovic. On calculating power-aware connected dominating sets for efficient routing in ad hoc wireless networks. In *Proc. of ICPP,* September 2001.

32. B. Karp and H. Kung. GPSR: greedy perimeter stateless routing for wireless networks. In *Proc. of ACM MobiCom*, August 2000.

33. KaZaA. Homepage http://www.kazaa.com.

34. A. Klemm, C. Lindemann, and O. Waldhorst. A special-purpose peer-to-peer file sharing system for mobile ad hoc networks. In *Proc. of IEEE VTC*, October 2003.

35. U.C. Kozat, G. Kondylis, B. Ryu, and M.K. Marina. Virtual dynamic backbone for mobile ad hoc networks. In *Proc. of IEEE ICC*, June 2001.

36. U.C. Kozat and L. Tassiulas. Network layer support for service discovery in mobile ad hoc networks. In *Proc. of IEEE INFOCOM*, April 2003.

37. S.-J. Lee, W. Su, and M. Gerla. On-demand multicast routing protocol in multihop wireless mobile networks. *Kluwer Mobile Networks and Applications*, 2000.

38. J. Li, J. Jannotti, D.S.J.D. Couto, D.R. Karger, and R. Morris. A scalable location service for geographic ad hoc routing. In *Proc. of ACM MobiCom*, August 2000.

39. Z. Li, B. Li, D. Xu, and X. Zhou. iFlow: middleware-assisted rendezvous-based information access for mobile ad hoc applications. In *Proc. of MobiSys*, 2003.

40. B. Liang and Z.J. Haas. Virtual backbone generation and maintenance in ad hoc network mobility management. In *Proc. of IEEE INFOCOM*, March 2000.

41. C.R. Lin and M. Gerla. Adaptive clustering for mobile wireless networks. *IEEE Journal of Selected Areas in Communications*, 15(7):1265–1275, 1997.

42. I.F. Matei Ripeanu. Mapping the Gnutella network macroscopic properties of large-scale peer-to-peer systems. In *Proc. of IPTPS*, 2002.

43. M. Mauve, J. Widmer, and H. Hartenstein. A survey on position-based routing in mobile ad hoc networks, *IEEE Network Magazine*, 15(6):30–39, November 2001.

44. J. Moy. Multicast extensions to OSPF. Technical Report 1584, IETF, March 1994.

45. J. Moy. OSPF version 2. RFC 1583, IETF, March 1994.

46. J.C. Navas and T. Imielinski. Geographic addressing and routing. In *Proc. of ACM MobiCom*, 1997.

47. L.B. Oliveira, I.G. Siqueira, and A.A. Loureiro. Evaluation of ad-hoc routing protocols under a peer-to-peer application. In *Proc. of IEEE WCNC*, March 2003.

48. M. Papadopouli and H. Schulzrinne. Effects of power conservation, wireless coverage and cooperation on data dissemination among mobile devices. In *Proc. of ACM MobiHoc*, October 2001.

49. C.E. Perkins and P. Bhagwat. Highly dynamic destination-sequenced distance-vector routing (DSDV) for mobile computers. In *Proc. of ACM SIGCOMM*, August 1994.

50. C.E. Perkins and E.M. Royer. Ad hoc on-demand distance vector routing. In *Proc. of the IEEE WMCSA*, February 1999.

51. S. Ratnasamy, P. Francis, M. Handley, R. Karp, and S. Schenker. A scalable content-addressable network. In *Proc. of ACM SIGCOMM*, August 2001.

52. S. Ratnasamy, M. Handley, R. Karp, and S. Shenker. Application-level multicast using content-addressable networks. In *Proc. of the Third International Workshop on Networked Group Communication*, November 2001.

53. S. Ratnasamy, B. Karp, L. Yin, F. Yu, D. Estrin, R. Govindan, and S. Shenker. GHT: a geographic hash table for data-centric storage in sensornets. In *Proc. of 1st ACM WSNA*, September 2002.

54. Y. Rekhter and T. Li. A Border Gateway Protocol 4 (BGP-4). RFC 1771, IETF, March 1995.

55. A. Rowstron and P. Druschel. PAST: a large-scale, persistent peer-to-peer storage utility. In *Proc. of ACM SOSP*, October 2001.

56. A. Rowstron and P. Druschel. Pastry: scalable, distributed object location and routing for large-scale peer-to-peer systems. In *Proc. of Middleware*, November 2001.

57. R. Royer and C. Perkins. Multicast using ad-hoc on demand distance vector routing. In *Proc. of ACM MobiCom*, August 1999.

58. P. Sinha, R. Sivakumar, and V. Bharghavan. Enhancing ad hoc routing with dynamic virtual infrastructures. In *Proc. of IEEE INFOCOM*, Anchorage, AK, April 2001.

59. R. Sivakumar, B. Das, and V. Bharghavan. Spine routing in ad hoc networks. *ACM/Baltzer Publications Cluster Computing Journal, Special Issue on Mobile Computing,* 1(2):237–248, 1998.

60. R. Sivakumar, P. Sinha, and V. Bharghavan. CEDAR: a core-extraction distributed ad hoc routing algorithm. *IEEE Journal on Selected Areas in Communications,* 17(8):1454–1465, August 1999.

61. I. Stoica, R. Morris, D. Karger, M.F. Kaashoek, and H. Balakrishnan. Chord: a scalable peer-to-peer lookup service for internet applications. In *Proc. of ACM SIGCOMM,* August 2001.

62. I. Stojmenovic, M. Seddigh, and J. Zunic. Dominating sets and neighbor elimination-based broadcasting algorithms in wireless networks. *IEEE Trans. Parallel Distrib. Syst.,* 13(1):14–25, 2002.

63. H. Takahashi and A. Matsuyama. An approximate solution for the Steiner problem in graphs. *Mathematica Japonica,* 24:573–577, 1980.

64. C. Tang, Z. Xu, and S. Dwarkadas. Peer-to-peer information retrieval using self-organizing semantic overlay networks. In *Proc. of ACM SIGCOMM,* August 2004.

65. P.-J. Wan, K.M. Alzoubi, and O. Frieder. Distributed construction of connected dominating set in wireless ad hoc networks. *Mob. Netw. Appl.,* 9(2):141–149, 2004.

66. J.E. Wieselthier, G.D. Nguyen, and A. Ephremides. On the construction of energy-efficient broadcast and multicast trees in wireless networks. In *INFOCOM 2000,* Tel-Aviv, Israel, March 26–30, 2000.

67. J.E. Wieselthier, G.D. Nguyen, and A. Ephremides. Energy-efficient broadcast and multicast trees in wireless networks. *Mob. Netw. Appl.,* 7(6):481–492, 2002.

68. J. Wu and H. Li. A dominating-set-based routing scheme in ad hoc wireless networks. *Wireless Networks in the Telecommunication Systems,* 3:63–84, 2001.

69. J. Xie, R.R. Talpade, A. Mcauley, and M. Liu. AMRoute: ad hoc multicast routing protocol. *Mob. Netw. Appl.,* 7(6):429–439, 2002.

70. R. Zhang and Y.C. Hu. Borg: a hybrid protocol for scalable application-level multicast in peer-to-peer systems. In *Proc. of NOSSDAV,* June 2003.

71. B.Y. Zhao, J.D. Kubiatowicz, and A.D. Joseph. Tapestry: an infrastructure for fault-resilient wide-area location and routing. Technical Report UCB//CSD-01-1141, U.C. Berkeley, April 2001.

72. S.Q. Zhuang, B.Y. Zhao, A.D. Joseph, R.H. Katz, and J. Kubiatowicz. Bayeux: an architecture for scalable and fault-tolerant wide-area data dissemination. In *Proc. of NOSSDAV,* June 2001.

Index

Index

T

T - #0293 - 101024 - C0 - 254/178/48 [50] - CB - 9780849328329 - Gloss Lamination

T - #0293 - 101024 - C0 - 254/178/48 [50] - CB - 9780849328329 - Gloss Lamination